PRACTICAL ASPECTS
OF
TRAPPED ION MASS
SPECTROMETRY

Volume IV

*Theory and
Instrumentation*

PRACTICAL ASPECTS OF TRAPPED ION MASS SPECTROMETRY

Volume IV

Theory and Instrumentation

Edited by

Raymond E. March
John F. J. Todd

CRC Press
Taylor & Francis Group
Boca Raton   London   New York

CRC Press is an imprint of the
Taylor & Francis Group, an **informa** business

CRC Press
Taylor & Francis Group
6000 Broken Sound Parkway NW, Suite 300
Boca Raton, FL 33487-2742

First issued in paperback 2017

© 2010 by Taylor and Francis Group, LLC
CRC Press is an imprint of Taylor & Francis Group, an Informa business

No claim to original U.S. Government works

ISBN 13: 978-1-138-11344-2 (pbk)
ISBN 13: 978-1-4200-8371-2 (hbk)

Library of Congress Cataloging-in-Publication Data

Practical aspects of ion trap mass spectrometry / edited by Raymond E. March, John F.J. Todd.
    p. cm. -- (Modern mass spectrometry)
    Includes bibliographical references and index.
    ISBN 0-8493-4452-2 (vol. 1)
    1. Mass spectrometry. I. March, Raymond E. II. Todd, John F.J. III. Series.

QD96.M3P715 1995
539.7'.028'7--dc20   95-14146

Visit the Taylor & Francis Web site at
http://www.taylorandfrancis.com

and the CRC Press Web site at
http://www.crcpress.com

To ion trappers, young and old, everywhere.

Contents

Preface...xi
About Volume V xvii
Editors...xxi
Contributors xxv

PART I Fundamentals

Chapter 1 An Appreciation and Historical Survey of Mass Spectrometry 3

Raymond E. March and John F. J. Todd

Chapter 2 Ion Traps for Miniature, Multiplexed, and Soft-Landing
Technologies.. 169

*Scott A. Smith, Christopher C. Mulligan, Qingyu Song,
Robert J. Noll, R. Graham Cooks, and Zheng Ouyang*

PART II New Ion Trapping Techniques

Chapter 3 Theory and Practice of the Orbitrap Mass Analyzer 251

Alexander Makarov

Chapter 4 Rectangular Waveform Driven Digital Ion Trap (DIT)
Mass Spectrometer: Theory and Applications.................................273

Francesco L. Brancia and Li Ding

Chapter 5 High-Field Asymmetric Waveform Ion Mobility
Spectrometry (FAIMS) 309

Randall W. Purves

Chapter 6 Ion Traps with Circular Geometries.. 373

Daniel E. Austin and Stephen A. Lammert

PART III   Fourier Transform Mass Spectrometry

Chapter 7   Ion Accumulation Approaches for Increasing Sensitivity
and Dynamic Range in the Analysis of Complex Samples..............401

   Mikhail E. Belov, Yehia M. Ibrahim, and Richard D. Smith

Chapter 8   Radio Frequency-Only-Mode Event and Trap Compensation
in Penning Fourier Transform Mass Spectrometry.........................433

   Adam M. Brustkern, Don L. Rempel, and Michael L. Gross

Chapter 9   A Fourier Transform Operating Mode Applied to a
Three-Dimensional Quadrupole Ion Trap.......................................469

   Y. Zerega, J. Andre, M. Carette, A. Janulyte, and C. Reynard

PART IV   Quadrupole Rod Sets

Chapter 10 Trapping and Processing Ions in Radio Frequency Ion Guides.......525

   *Bruce A. Thomson, Igor V. Chernushevich,
   and Alexandre V. Loboda*

Chapter 11 Linear Ion Trap Mass Spectrometry with Mass-Selective
Axial Ejection..545

   James W. Hager

Chapter 12 Axially Resonant Excitation Linear Ion Trap (AREX LIT) 573

   Yuichiro Hashimoto

PART V   3D-Quadrupole Ion Trap Mass Spectrometry

Chapter 13 An Examination of the Physics of the High-Capacity
Trap (HCT)...593

   *Desmond A. Kaplan, Ralf Hartmer, Andreas Brekenfeld,
   Jochen Franzen, and Michael Schubert*

Chapter 14 Electrically Induced Nonlinear Ion Traps.......................................619

   Gregory J. Wells and August A. Specht

Chapter 15 Fragmentation Techniques for Protein Ions Using Various
Types of Ion Trap 671

Jochen Franzen and Karl Peter Wanczek

Chapter 16 Unraveling the Structural Details of the Glycoproteome by
Ion Trap Mass Spectrometry 707

Vernon Reinhold, David J. Ashline, and Hailong Zhang

Chapter 17 Collisional Cooling in the Quadrupole Ion Trap
Mass Spectrometer (QITMS) 739

Philip M. Remes and Gary L. Glish

Chapter 18 'Pressure Tailoring' for Improved Ion Trap Performance 769

Dodge L. Baluya and Richard A. Yost

Chapter 19 A Quadrupole Ion Trap/Time-of-Flight Mass Spectrometer
Combined with a Vacuum Matrix-Assisted Laser
Desorption Ionization Source 793

*Dimitris Papanastasiou, Omar Belgacem, Helen Montgomery,
Mikhail Sudakov, and Emmanuel Raptakis*

PART VI Photochemistry of Trapped Ions

Chapter 20 Photodissociation in Ion Traps 827

Jennifer S. Brodbelt

Chapter 21 Photochemical Studies of Metal Dication Complexes
in an Ion Trap 847

Guohua Wu, Hamish Stewart, and Anthony J. Stace

Author Index 863

Subject Index 871

Preface

This monograph is Volume IV of a mini-series devoted to (i) practical aspects of applications of mass spectrometry for the study of gaseous ions confined in ion traps; and (ii) treatments of the theory of ion confinement in each ion-trapping device. Volumes I–III were published in 1995 under the title *Practical Aspects of Ion Trap Mass Spectrometry*. Volume III, *Chemical, Environmental and Biomedical Applications*, is a companion to Volumes I and II, subtitled *Fundamentals of Ion Trap Mass Spectrometry* and *Ion Trap Instrumentation*, respectively. Volumes I–III are concerned principally with the history, theory, and applications of the quadrupole ion trap and, to a lesser degree, of the quadrupole mass filter.

The practice of trapping gaseous ions and the applications thereof have expanded considerably during the past decade or so, as witnessed by the substantial growth in popularity of quadrupole ion traps and of Fourier transform ion cyclotron resonance (FT-ICR) mass spectrometers, instruments that hitherto were regarded as being rival rather than complementary technologies. In addition, we have seen the nascence of new methods for trapping ions, such as the Orbitrap™, the digital ion trap (DIT), the rectilinear ion trap (RIT), and the toroidal ion trap. Furthermore, during this period, there have been significant advances in the development and application of the quadrupole ion trap and of the quadrupole mass filter, both standalone and in concatenation with other mass spectrometric instruments, for example, with FT-ICR and with time-of-flight (TOF) mass spectrometers. New and/or modified existing methods for ion processing have been developed and applied; these methods include electron capture dissociation (ECD), electron transfer dissociation (ETD), charge inversion, proton transfer reaction (PTR), electron transfer (ET), and ion attachment (IA). Other recent advances involving the coupling of ion mobility spectrometry (IMS) with mass spectrometry have brought about the introduction of high-field asymmetric waveform ion mobility spectrometry (FAIMS) and traveling wave ion mobility mass spectrometry (TWIM-MS). Indeed, so many advances have occurred in the ion trapping field that we needed to consider a somewhat broader definition of ion trapping compared to that which has been employed hitherto; after several iterations, we arrived at the definition proposed in Section 1.1, *"an ion is 'trapped' when its residence time within a defined spatial region exceeds that had the motion of the ion not been impeded in some way."* Clearly, this definition includes those various forms of IMS mentioned above. Armed with this definition of 'trapped ions,' it seemed appropriate to the editors that a further volume in this mini-series could be undertaken, not limited to quadrupole devices but encompassing advances in all aspects of trapped ion mass spectrometry. When a commercial product has achieved a degree of market acceptance, which we believed was the case for the three volumes of *Practical Aspects of Ion Trap Mass Spectrometry*, one is reluctant to lose the connectivity within the mini-series upon embracing an expansion of the field in question. Fortunately, a minor word change to *Practical Aspects of Trapped Ion Mass Spectrometry* saved the day. With this small but significant change in title, the expanded field could be considered and included within the 'practical aspects of ion trapping' rubric. The collective response to our subsequent approaches to potential authors in the expanded ion-trapping field was nearly overwhelming, so much so that in fact two monographs, Volumes IV and V, have resulted from this endeavor.

Volume IV is entitled *Theory and Instrumentation*, and is composed of six parts: Fundamentals, New Ion Trapping Techniques; Fourier Transform Mass Spectrometry, Quadrupole Rod Sets, 3D-Quadrupole Ion Trap Mass Spectrometry, and Photochemistry of Trapped Ions. Volume V is entitled *Applications*, and features four parts: Ion Reactions, Ion Conformation and Structure, Ion Spectroscopy, and Practical Applications.

Within Volume IV, *Part I. Fundamentals* is composed of an introductory chapter and a discussion of ion traps for miniature, multiplexed and soft-landing mass spectrometers. Chapter 1 presents a brief history of our understanding of atoms and charged particles, together with an account of experiments with particle beams. We have attempted to recapture some of the early excitement that resulted from the application of magnetic and electrostatic fields in efforts to separate charged particle species. The development of the mass spectrograph and, later, the mass spectrometer is traced up to the middle of the twentieth century. We then follow three streams in the development of 'modern mass spectrometry' that occurred simultaneously following this initial period of activity. The three streams are (i) the development of analytical mass spectrometry; (ii) the development of gas-phase ion chemistry utilizing largely commercially available sector mass spectrometers; and (iii) the emergence and development of 'dynamic mass spectrometers,' in which the time-dependence of electric field strength, magnetic field strength, or ion movement is utilized for mass analysis. Developments in each of these three streams of mass spectrometry contributed to the emergence of the multiple techniques for the study of trapped ions. In addition, we have laid out a detailed exposition of the theory of quadrupole devices and a somewhat less detailed exposition of the FT-ICR instrument.

In Chapter 2, the drive to chemical analysis *in situ* has been directed to the miniaturization of Paul ion traps, cylindrical ion traps and, more recently, to rectilinear ions traps as lightweight tandem mass spectrometers for field use. Advances in capabilities for microfabrication have led to an interest in arrays of ion traps, where this interest was fueled by the need to compensate for the diminution in ion trapping capacity associated with miniaturization. Arrays of ions can be used to separate ion species and to collect the purified compound after soft-landing the corresponding ions onto appropriate surfaces. New experiments of sets of traps assembled in various geometrical arrangements, from simple parallel and serial arrangements for use in different types of mass spectrometry experiments are described.

Part II. New Ion Trapping Techniques presents discussions of theory and practice of the Orbitrap mass analyzer, the rectangular waveform-driven DIT mass spectrometer, the FAIMS, and ion traps with circular geometries. The Orbitrap mass analyzer, described in Chapter 3, represents only the first commercial implementation of a highly promising class of devices where ions are separated on the basis of their *m/z*-values over the course of multiple reflections or deflections in an electrostatic field. The term 'Orbitrap' was coined to define a mass analyzer based on orbital trapping in an electrostatic trap with harmonic oscillations and image current detection; the harmonicity of the oscillations of the ions ensures the superior performance of the Orbitrap analyzer with respect to dynamic range, mass accuracy and resolving power. With image current detection, the Orbitrap analyzer extends the family of Fourier transform mass analyzers, which until recently contained only one widely used analyzer, the FT-ICR mass spectrometer.

Chapter 4 describes the design and novel operation of a DIT. Ion motion in a Paul or quadrupole ion trap driven by a rectangular wave quadrupolar field was described in the early 1970s, but it was not until 2000 that the mass-selective instability method utilizing digital operating conditions was described. The concept of the DIT was first introduced with a theoretical study on ion motion in a quadrupolar field driven by a digital waveform, where the adjective 'digital' is derived from the digital circuitry used to produce the rectangular waveforms. The circuits of the DIT switch very rapidly between discrete DC high voltage levels in order to generate the trapping waveform voltage applied to the ring electrode.

In Chapter 5, high-field FAIMS, which is operated typically in the atmospheric pressure regime, is discussed. This method for ion separation exploits the electric field dependence of ion mobility through combinations of oscillating and variable direct current electric field gradients, which act perpendicularly to the direction of a flow of bath gas containing the analyte ions. The reader is introduced to the basic concepts of FAIMS, and is made familiar with the diversity and scope of this gas-phase ion separation technique. Following an account of the historical timeline for the development of FAIMS, the basic principles of IMS that are necessary for the introduction of FAIMS, including a discussion of ion properties under the influence of high electric fields, are reviewed. The roles of the bath gas and gas number density are examined together with the diversity of hardware used in FAIMS including variants of the electrode set and the waveform generator. A brief discussion of commercially available FAIMS devices precedes a presentation of a wide range of applications in which FAIMS has been used in conjunction with mass spectrometry.

A new type of ion trap, the toroidal ion trap, is discussed in Chapter 6. Despite the advances made in recent years in quadrupole ion traps, there remains a fundamental limitation to ion trap mass spectrometers (ITMS) in general, that of space charge. Ion traps can be viewed as 'ion flasks' and, as such, are capable of storing a limited number of ions before ion/ion repulsion imposes a ceiling on the number of ions that can be contained physically in the device. Through attempts to reduce the space-charge problem, new ion trap geometries, such as those of the linear and RITs, have been explored so as to increase ion populations. From a pencil-and-paper examination of a 3D ion trap, it was observed that when one rotated the device on an edge axis, a toroidal ion trap mass analyzer was obtained. Clearly, the device has a larger volume yet the main dimension, r_0, which governs the operation of the ion trap, has not changed. The device has a single ion storage region with no entrance or exit apertures, where field perturbations are inevitable, and the ions experience essentially the same field everywhere within the ion storage volume. The toroidal ion trap has a compact, spatially efficient geometry.

Part III. Fourier Transform Mass Spectrometry. The application in 1973 of a Fourier transform to the signal output of an ion cyclotron resonance cell led to the formation of the high-performance mass spectrometer known as the FT-ICR instrument. The three chapters of Part *III* are devoted to discussions of ion accumulation for increasing sensitivity in FT-ICR spectrometry, the RF-Only-Mode Event for Penning Traps in Fourier Transform Mass Spectrometry, and a Fourier transform operating mode applied to a three-dimensional quadrupole ion trap.

In Chapter 7 attempts in increase sensitivity by the use of external ion creation and accumulation of ions in a 2D ion trap prior to their admission to the FT-ICR are described.

Chapter 8 examines the application of a radio-frequency-only-mode event in an attempt to apply electrical compensation to an ion trap as a means to solve the two problems that stand in the way of higher performance; first, the trap must be operated at low pressure, making difficult the incorporation of high-pressure events to stabilize ions and to manipulate ion clouds and, second, the requirement to use a trapping electric field.

In Chapter 9 a new operating mode of an RF quadrupole ion trap for mass analysis is described wherein a signal generated by the ejected ions is subjected to Fourier transformation in order to create an 'image' of their motion prior to ejection. This new FT-operating mode requires a set of elementary experiments with the addition of the processing of the TOF histograms of the ejected ions recorded for each of them. The amplitudes and frequency of the potentials applied to the trap electrodes remain constant during confinement. Ion motion must be as pure as possible and have a large amplitude. An image signal of the motion of the simultaneously trapped ions versus confinement time is computed from these histograms, and the secular frequency spectrum is obtained by means of a Fourier transformation.

Part IV. Quadrupole Rod Sets, presents three aspects of the behavior of quadrupole rod sets, which are now ubiquitous in mass spectrometry laboratories. The trapping and processing of ions in RF ion guides of a tandem mass spectrometer is followed by two discussions of the linear ion trap (LIT): first, mass-selective axial ion ejection in the LIT and, second, axially resonant excitation in the linear ion trap (AREX LIT).

The quadrupole mass filter gained rapid acceptance both as a mass analyzer and as a low-pressure ion gauge; thereafter, its function as an RF-only collision cell for tandem mass spectrometry led to the discovery of collisional focusing of ions within the confines of the four parallel rods and the propensity of the mass analyzer for transmitting ions through the rod set with high efficiency. Today, multiple quadrupole (and octopole and hexapole) ion guides are employed in tandem mass spectrometers to bridge regions of high pressure between the source and mass-selective analyzers. The recognition of the ion-confining properties of the quadrupole rod set led to the creation of a new ion-trapping device, the LIT(or 2D ion trap), where DC voltages are applied to constrain ion motion in the axial direction. Chapter 10 discusses the facility with which ions can be trapped and processed in such radiofrequency ion guides. Here, processing involves the transfer of trapped ions either against the normal direction of gas and ion flow or with the flow in a tandem LIT mass spectrometer. The theme of LIT mass spectrometry is continued in Chapters 11 and 12. The former presents an examination of mass-selective axial ejection of ions, as opposed to mass-selective radial ejection of ions that is possible also, and a description of the fundamentals of trapping and moving ions from place to place, along with some methods and applications. In the latter chapter, Chapter 12, an 'axially-resonant excitation' linear ion trap, which has been termed the 'AREX LIT' is described. The AREX LIT has an axial harmonic potential that is produced by applying a DC voltage to vane-shaped electrodes inserted between the rod electrodes. The axial harmonic potential permits mass-selective operations such as ion isolation, activation, and ion ejection.

Part V. 3D-Quadrupole Ion Trap Mass Spectrometry is composed of seven chapters that illustrate and exemplify the extent of development of the 3D-QIT in recent years. Chapter 13, which is concerned with the physics behind the high ion storage capability that has been achieved for the commercially available High Capacity Trap (HCT), is of particular interest because it describes a technique for delaying the inevitability of the onset of space charge. The following chapter, Chapter 14, builds upon the relatively recent utilization of nonlinear fields for the rapid mass-selective ejection of ions from the ion trap with a discussion of electrically induced nonlinearities in ion traps. These first two chapters are concerned principally with the development and explanation of the modes of operation of new commercial ion trap instruments. We would like to acknowledge here the enormous contribution of manufacturers of mass spectrometric instruments both to the field of mass spectrometry and, in particular, to Volumes IV and V. A total of ten chapters have been contributed from six manufacturers. We appreciate keenly the difficulties encountered by authors in industrial settings to find sufficient time to devote to the preparation of manuscripts; we thank them and applaud their efforts.

At this juncture, one might ask, justifiably, just how effective are such ion trap instruments, and this question is answered in good measure by the following two chapters. First, in Chapter 15, there is a detailed review of fragmentation techniques for protein ions in different kinds of ion traps and, second, in Chapter 16, is a quite extraordinary account of the utilization of up to MS^7 in an ITMS for the unraveling of structural details of the glycoproteome. Chapters 15 and 16 are free of lists of performance statistics and have focused principally on the methods employed in the complex analysis of protein ions and glycoprotein ions. The complex roles of ion/neutral collisions, which are critical to the overall performance of the quadrupole ion trap, are examined in the following two chapters; first, in Chapter 17, an examination of collisional cooling in the 3D-QIT is presented and, second, in Chapter 18, methods are proposed for the tailoring of buffer gas pressure during various segments of a scan function in attempts to achieve improved 3D-QIT performance. Part V concludes with Chapter 19 that is concerned with the performance of a new hybrid instrument formed by the combination of a 3D-QIT with a TOF mass spectrometer and a vacuum matrix-assisted laser desorption ionization (vMALDI) source. This is the sole chapter wherein a vMALDI source is combined with a QIT that, in turn, is combined with a TOF mass spectrometer.

Part VI. Photochemistry of Trapped Ions introduces the general topic of photodissociation in ion traps, which is followed by a discussion of chemical and photochemical studies of metal dication complexes in a 3D-QIT. In Chapter 20, photodissociation is presented as an alternative ion activation method that entails exposing an ion population to photons until sufficient internal energy has been accumulated to cause dissociation. Ion activation requires the absorption of one or more ultraviolet or visible photons, or dozens of infrared photons, and the activation period may range from nanoseconds to hundreds of milliseconds. For compounds of interest of low absorptivity, such as drugs, peptides and proteins, oligosaccharides, and nucleic acids, a strategy is presented whereby the infrared absorptivities are enhanced by non-covalent addition of strong infrared chromophores. An alternative strategy is presented in Chapter 21 wherein metal dication complexes are formed for photochemical studies. Here, a pick-up technique is described that is capable of producing multiply-charged metal/ligand complexes from a wide variety of metals and ligands; examples of such complexes are $[Cu(Ar)_n]^{2+}$ for $n = 1-6$ and $[Al(CH_3CN)_4]^{3+}$. The pick-up technique employs a quadrupole mass filter coupled with a QIT. The ions thus formed are subjected to photodissociation using infrared and ultraviolet/visible radiation.

We wish to thank the many people who have assisted us in one way or another with myriad tasks that must be carried out in order to arrive at the publication of a monograph from a collection of manuscripts in a variety of formats and styles. First of all, to our contributors, without whom this monograph would not have appeared in print. We give thanks for their individual inspiration; we thank them for the fruits of their labors, and for their patient toleration of the idiosyncrasies of our editing, often involving repeated iterations between the two of us and the authors themselves. The 21 chapters that constitute Volume IV have originated from 55 authors and co-authors. For many of these co-authors this project has been a novel experience, thus we thank our lead authors for responding to our urging that they collaborate with young scientists in their laboratories. From where else will the monographs of tomorrow originate?

At CRC Press, we thank Fiona Macdonald, Hilary Rowe, Pat Roberson, Rachael Panthier, Lindsey Hofmeister, and Jennifer Derima; at Datapage, we thank Ramkumar Soundararajan, the Project Manager. Finally, we express our sincere appreciation for the tolerance of our respective spouses, Kathleen March and Mavis Todd, and for their patience, support, and sacrifices while this project, known informally as "PRATIMS," took over our lives.

Raymond E. March
John F. J. Todd

About Volume V

Practical Aspects of Trapped Ion Mass Spectrometry

Volume V: Applications of Ion Trapping Devices

Edited by Raymond E. March and John F. J. Todd

Found in Volume V

PART I Ion Reactions

Chapter 1 Ion/Ion Reactions in Electrodynamic Ion Traps 3

Jian Liu and Scott A. McLuckey

Chapter 2 Gas-Phase Hydrogen/Deuterium Exchange in Quadrupole-
Ion Traps..35

Joseph E. Chipuk and Jennifer S. Brodbelt

Chapter 3 Methods for Multi-Stage Ion Processing Involving Ion/Ion
Chemistry in a Quadrupole Linear Ion Trap......................................59

Graeme C. McAlister and Joshua J. Coon

PART II Ion Conformation and Structure

Chapter 4 Chemical Derivatization and Multistage Tandem Mass
Spectrometry for Protein Structural Characterization.......................83

Jennifer M. Froelich, Yali Lu, and Gavin E. Reid

Chapter 5 Fourier Transform Ion Cyclotron Resonance Mass Spectrometry
in the Analysis of Peptides and Proteins..121

Helen J. Cooper

Chapter 6   MS/MS Analysis of Peptide–Polyphenols Supramolecular
                Assemblies: Wine Astringency Approached by ESI-IT-MS 153

                Benoît Plet and Jean-Marie Schmitter

Chapter 7   Structure and Dynamics of Trapped Ions 169

                Joel H. Parks

Chapter 8   Applications of Traveling Wave Ion Mobility-Mass
                Spectrometry 205

                Konstantinos Thalassinos and James H. Scrivens

PART III   Ion Spectroscopy

Chapter 9   The Spectroscopy of Ions Stored in Trapping Mass
                Spectrometers 239

                Matthew W. Forbes, Francis O. Talbot, and Rebecca A. Jockusch

Chapter 10   Sympathetically Cooled Single Ion Mass Spectrometry 291

                Peter Frøhlich Staanum, Klaus Højbjerre, and Michael Drewsen

Chapter 11   Ion Trap: A Versatile Tool for the Atomic
                Clocks of the Future! 327

                Fernande Vedel

PART IV   Practical Applications

Chapter 12   Boundary-Activated Dissociations (BAD) in a Digital Ion
                Trap (DIT) 367

                *Francesco L. Brancia, Luca Raveane, Alberto Berton, and
                Pietro Traldi*

Chapter 13   The Study of Ion/Molecule Reactions at Ambient Pressure
                with Ion Mobility Spectrometry and Ion Mobility/Mass
                Spectrometry 387

                Gary A. Eiceman and John A. Stone

Chapter 14   The Role of Trapped Ion Mass Spectrometry for Imaging 417

Timothy J. Garrett and Richard A. Yost

Chapter 15   Technology Progress and Application in
GC/MS and GC/MS/MS 439

Mingda Wang and John E. George III

Chapter 16   Remote Monitoring of Volatile Organic Compounds in
Water by Membrane Inlet Mass Spectrometry 491

Romina Pozzi, Paola Bocchini, Francesca Pinelli, and Guido C. Galletti

Author Index 509

Subject Index 513

Editors

Raymond E. March, PhD, DSc, D(hc), FCIC, is presently Professor Emeritus of Chemistry at Trent University in Peterborough, Ontario, Canada. He obtained a BSc (Hons) in Chemistry from Leeds University in 1957; a PhD from the University of Toronto in 1961 (supervised by Professor John C. Polanyi, Nobelist 1986); a DSc from Leeds University in 2000; and an honorary doctorate (D(hc)) from l'Université de Provence in 2008. From 1954 to 1957, he was a Cadet Pilot in the Leeds University Air Squadron Royal Air Force Volunteer Reserve (RAFVR) and, from 1958 to 1963, a Flight Lieutenant in the Royal Canadian Air Force (Auxiliary) (RCAF). From 1960 to 1961, he held a Canadian Industries Limited Research Fellowship. From 1962 to 1963, he was a Post-Doctoral Fellow with Professor H.I. Schiff at McGill University, and a Research Associate from 1963 to 1965, during which time he lectured at McGill University and Loyola College. In 1965, he joined the faculty of Trent University where he has conducted independent research for some 44 years in gas-phase kinetics, optical spectroscopy, gaseous ion kinetics, analytical chemistry, nuclear magnetic resonance spectroscopy, and mass spectrometry. Kathleen and Ray have been married for 51 years; they have three daughters, Jacqueline, Roberta, and Sally with spouses Paul, Stuart, and Lauren, respectively, and nine grandchildren, Shawn, Jessica, Thomas, Daniel, Rebecca, Sara, James, Madeline, and Carson, in order of appearance.

Dr. March has published and/or co-authored over 170 scientific papers and some 75 conference presentations in the above areas of research with emphasis on mass spectrometry, both with sector instruments and quadrupole ion traps. Dr. March is a co-author with Dr. Richard J. Hughes and Dr. John F. J. Todd of *Quadrupole Storage Mass Spectrometry*, published in 1989. A second edition of *Quadrupole Storage Mass Spectrometry*, co-authored by Dr. March and Dr. John F. J. Todd was published in 2005. Dr. March and Dr. John F. J. Todd co-edited three volumes entitled *Practical Aspects of Ion Trap Mass Spectrometry*, published in 1995. Dr. March is a co-author with Oscar V. Bustillos and André Sassine of *A Espectrometria de Massas Quadupolar*, published in Portuguese in 2005. Professor March is a Fellow of the Chemical Institute of Canada and a member of the American, British, and Canadian Societies for Mass Spectrometry. In 2009, he received the Gerhard Herzberg Award of the Canadian Society for Analytical Sciences and Spectroscopy (CSASS).

In 1995, he received the Distinguished Faculty Research Award from Trent University, and the Canadian Mass Spectrometry Society presented him with the Recognition Award and, in 1997, with the Distinguished Contribution Award. Dr. March is a member of the Editorial Advisory Boards for *Rapid Communications in Mass Spectrometry*, the *Journal of Mass Spectrometry*, and the *International Journal of Mass Spectrometry*. In 1975, Dr. March was an Exchange Fellow (NRC-CNRS) at Orsay, France, with Professor Jean Durup; in 1983, an Exchange Fellow (NRC-Royal Society of London) in Swansea, Wales, with Professor J. H. Beynon; in 1989 and 1992, a Visiting Professor, Université de Provence, Marseille, France, with Professor Fernande Vedel; in 1993 and 1995, a CNRS Visiting Professor, Université Pierre et Marie Curie, Paris, France, with Professor Jean-Claude Tabet; and in 1999, a Visiting Professor, Université de Provence, Marseille, France, with Yves Zerega. In 1987, Dr. March was a Distinguished Lecturer at the Universities of Berne, Neuchatel, and Lausanne, in Switzerland.

Dr. March's research in the field of mass spectrometry and gas-phase ion chemistry involved the development and application of mass spectrometric instruments, particularly quadrupole ion trap mass spectrometers and hybrid mass spectrometers, for both fundamental studies and the formulation of analytical protocols for the determination of compounds of environmental interest. His current research interests are focused within Trent University's Water Quality Centre (www.trentu.ca/wqc/). As a founding member of the Water Quality Centre his principal research interest lies in the mass spectrometric and nuclear magnetic resonance spectroscopic investigation of natural compounds that, having been formed by plants, may enter waterways and/or the water table. His current research involves the study of flavonoids and flavonoid glycosides; such compounds are often found in those products that have become known as neutraceuticals. Electrospray ionization combined with tandem mass spectrometry permits the investigation of ion fragmentation at high mass resolution and the derivation of possible ion fragmentation mechanisms using ion structures; these studies are supported by theoretical calculations carried out in collaboration with Professor E. G. Lewars. An important aspect of this research is the development of appropriate analytical protocols for flavonoid glycosides in water and in plant extracts. Nuclear magnetic resonance (NMR) studies of flavonoids and metabolites, carried out in collaboration with Professor D. A. Ellis and Dr. D. C. Burns, have permitted a rationalization of chemical shifts with product ion mass spectra and the development of a predictive model for ^{13}C chemical shifts in flavonoids. At present, Dr. March is carrying out an investigation of volatile compounds formed by Ash trees in response to an attack by the Emerald Ash Borer.

These researches are supported by the Natural Sciences and Engineering Research Council of Canada (Discovery Grants Program), the Canada Foundation for Innovation, the Ontario Research and Development Challenge Fund, Ontario Ministry of Natural Resources, and Trent University. Dr. March has enjoyed long-term collaborations with Professor John Todd, with colleagues at l'Université de Provence and l'Université Pierre et Marie Curie (France), and with colleagues in Padova (Italy).

John F. J. Todd, BSc, PhD, CChem, FRSC, CEng, FInstMC, is currently Emeritus Professor of Mass Spectroscopy at the University of Kent, Canterbury, U.K. He obtained his Class I Honours BSc degree in Chemistry in 1959 from the University of Leeds, from whence he also gained his PhD degree and was awarded the J. B. Cohen Prize in 1963; he was a member of the radiation chemistry group led by the late Professor F.S. (later Lord) Dainton, FRS. From 1963 to 1965, he was a Fulbright Research Scholar and Post-Doctoral Research Fellow in Chemistry with the late Professor Richard Wolfgang at Yale University. In 1965, he was one of the first faculty members appointed to the then new University of Kent at Canterbury, U.K. John and Mavis Todd have been married for 46 years and have three sons: John (Andrew), Eric, and Richard, two daughters-in-law Dorota and Marie, and six grandchildren, Alice, Max, Maja, Luke, Daniel, and Lara.

Professor Todd's research interests, spanning some 44 years, have encompassed positive and negative ion mass spectral fragmentation studies, gas discharge chemistry, ion mobility spectroscopy, analytical chemistry, and ion trap mass spectrometry. His work on 3D quadrupole (Paul) ion traps commenced in 1968, when he first developed the "Quistor/Quadrupole" instrument for the characterization of the behavior of ions confined in radiofrequency electric fields and as a vehicle for the study of gas-phase ion chemistry. As a consultant to Finnigan MAT during the 1980s and 1990s, he was a member of the original team that developed the first commercial ion trap mass spectrometer. In another consultancy role, Professor Todd is involved currently with one of the most extended single mass spectrometric investigations ever undertaken: the use of an ion trap mass spectrometer for the isotope ratio measurement of cometary material as part of the "Rosetta" project (launched 2004, scheduled arrival at its target comet in 2014). Professor Todd has published and/or co-authored some 116 scientific papers and over 118 conference contributions, concentrating mainly on various aspects of mass spectrometry. With Professor Dennis Price, he co-edited four volumes of *Dynamic Mass Spectrometry* and he edited *Advances in Mass Spectrometry 1985* (which contained the proceedings of the 10th International Mass Spectrometry Conference, Swansea, at which he was also a plenary lecturer). In addition, he was an editor of the *International Journal of Mass Spectrometry and Ion Processes* from 1985 to 1998, has served on the Editorial Boards of *Organic Mass Spectrometry/Journal of Mass Spectrometry* and *Rapid Communications in Mass Spectrometry*, and is currently a member of the Board for the *European Journal of Mass Spectrometry*. With Dr. Raymond E. March, Dr. Todd co-edited three volumes entitled *Practical Aspects of Ion Trap Mass Spectrometry*, published by CRC Press in 1995. In addition, Dr. Todd was a co-author with Dr. Raymond E. March and Dr. Richard J. Hughes of *Quadrupole Storage Mass Spectrometry*, published in 1989; a second edition of *Quadrupole Storage Mass Spectrometry*, co-authored by Dr. Todd and Dr. Raymond E. March was published in 2005.

Professor Todd is a Chartered Chemist and a Chartered Engineer, and has served terms as Chairman and as Treasurer of the British Mass Spectrometry Society. In 1988, he was a Canadian Industries Limited Distinguished Visiting Lecturer at Trent University, Peterborough, Ontario. In 1997, Dr. Todd was awarded the Thomson Gold Medal by the International Mass Spectrometry Society for "outstanding contributions to mass spectrometry," and in 2006 he was awarded the Aston Medal by the British Mass Spectrometry Society, of which he is also a Life Member. In 2008, he was accorded Honorary Life Membership of the Royal Society of Chemistry. Outside the immediate confines of his academic work, Professor Todd was appointed as Master of Rutherford College, University of Kent (1975–1985), and as the first Chairman of the newly created Canterbury and Thanet Health Authority (UK National Health Service) between 1982 and 1986. During the period 1995–2006 he was the founding Chairman of the newly established Board of Governors of St Edmund's School Canterbury, and until August 2007 he was a Governor of Canterbury Christ Church University; he was admitted as an Honorary Fellow of Canterbury Christ Church University in 2008. From 1979 to 1989, Professor Todd was Chairman of the Mass Spectrometry Sub-Committee, Commission I.5 of the International Union of Pure and Applied Chemistry (IUPAC), and between 1995 and 2007 he was Chairman of the Management Advisory Panel for the EPSRC National Mass Spectrometry Service Centre, based at the University of Wales Swansea. He has enjoyed long-term collaborations with co-editor Professor Raymond March, with colleagues at Finnigan MAT in the United Kingdom and the United States, and with groups in Nice (France) and Padova and Torino (Italy).

Contributors

J. Andre
Université de Provence - CNRS
Laboratoire Chimie Provence - UMR
    6264
Equipe Instrumentation et Réactivité
    Atmosphérique
Centre de Saint Jérôme
Marseille, France

David J. Ashline
The Glycomics Center
University of New Hampshire
Durham, New Hampshire

Daniel E. Austin
Department of Chemistry and
    Biochemistry
Brigham Young University
Provo, Utah

Dodge L. Baluya
Department of Chemistry
University of Florida
Gainesville, Florida

Omar Belgacem
Kratos Analytical
Shimadzu Biotech
Manchester, United Kingdom

Mikhail E. Belov
Biological Sciences Division
Pacific Northwest National
    Laboratory
Richland, Washington

Francesco L. Brancia
Shimadzu Research Laboratory
    (Europe)
Manchester, United Kingdom

Present address:
Aarhus University
Faculty of Agricultural Sciences
Department of Food Science
Tjele, Denmark

Andreas Brekenfeld
Bruker Daltonik Gmbh
Fahrenheitstr, Bremen, Germany

Jennifer S. Brodbelt
Department of Chemistry and
    Biochemistry
University of Texas
Austin, Texas

Adam M. Brustkern
Department of Chemistry
Washington University
St. Louis, Missouri

M. Carette
Université de Provence - CNRS
Laboratoire Chimie Provence - UMR
    6264
Equipe Instrumentation et Réactivité
    Atmosphérique
Centre de Saint Jérôme
Marseille, France

Igor V. Chernushevich
MDS Analytical Technologies
Concord, Ontario, Canada

R. Graham Cooks
Department of Chemistry
Purdue University, West Lafayette,
    Indiana

and

Center for Analytical Instrumentation
    Development
Purdue University
West Lafayette, Indiana

Li Ding
Shimadzu Research Laboratory
    (Shanghai) Co., Ltd.
Shanghai, People's Republic of China

Jochen Franzen
Bruker Daltonik Gmbh
Fahrenheitstr, Bremen, Germany

Gary L. Glish
Department of Chemistry
University of North Carolina
Chapel Hill, North Carolina

Michael L. Gross
Department of Chemistry
Washington University
St. Louis, Missouri

James W. Hager
MDS Analytical Technologies
Concord, Ontario, Canada

Ralf Hartmer
Bruker Daltonik Gmbh
Fahrenheitstr, Bremen, Germany

Yuichiro Hashimoto
Central Research Laboratory
Hitachi, Ltd.
Tokyo, Japan

Yehia M. Ibrahim
Biological Sciences Division
Pacific Northwest National
    Laboratory
Richland, Washington

A. Janulyte
Université de Provence - CNRS
Laboratoire Chimie Provence - UMR
    6264
Equipe Instrumentation et Réactivité
    Atmosphérique
Centre de Saint Jérôme
Marseille, France

Desmond A. Kaplan
Bruker Daltonics, Inc.
Billerica, Massachusetts

Stephen A. Lammert
Torion Technologies, Inc.
American Fork, Utah

Alexandre V. Loboda
MDS Analytical Technologies
Concord, Ontario, Canada

Alexander Makarov
Thermo Fisher Scientific
Bremen, Germany

Raymond E. March
Department of Chemistry
Trent University
Peterborough, Ontario, Canada

Helen Montgomery
Kratos Analytical
Shimadzu Biotech
Manchester, United Kingdom

Christopher C. Mulligan
Department of Chemistry
Purdue University
West Lafayette, Indiana

and

Department of Chemistry
Illinois State University
Normal, Illinois

Robert J. Noll
Department of Chemistry

and

Center for Analytical Instrumentation
    Development
Purdue University
West Lafayette, Indiana

Zheng Ouyang
Center for Analytical Instrumentation
    Development

and

Weldon School of Biomedical
    Engineering
Purdue University
West Lafayette, Indiana

Dimitris Papanastasiou
Shimadzu Research Laboratories
Manchester, United Kingdom

Randall W. Purves
Merck Frosst Canada Ltd.
Kirkland, Quebec, Canada

Present address:
National Research Council
Saskatoon, Saskatchewan, Canada

Emmanuel Raptakis
Kratos Analytical
Shimadzu Biotech
Manchester, United Kingdom

Vernon Reinhold
The Glycomics Center
University of New Hampshire
Durham, New Hampshire

Philip M. Remes
Department of Chemistry
University of North Carolina
Chapel Hill, North Carolina

Don L. Rempel
Department of Chemistry
Washington University
St. Louis, Missouri

C. Reynard
Université de Provence - CNRS
Laboratoire Chimie Provence - UMR
    6264
Equipe Instrumentation et Réactivité
    Atmosphérique
Centre de Saint Jérôme
Marseille, France

Michael Schubert
Bruker Daltonik Gmbh
Fahrenheitstr, Bremen, Germany

Richard D. Smith
Biological Sciences Division
Pacific Northwest National Laboratory
Richland, Washington

Scott A. Smith
Department of Chemistry
Purdue University
West Lafayette, Indiana

Qingyu Song
Department of Chemistry
Purdue University
West Lafayette, Indiana

August A. Specht
Varian Inc.
Scientific Instruments
Walnut Creek, California

Anthony J. Stace
Department of Physical Chemistry
School of Chemistry
The University of Nottingham
Nottingham, United Kingdom

Hamish Stewart
Department of Physical Chemistry
School of Chemistry
The University of Nottingham
Nottingham, United Kingdom

Mikhail Sudakov
Shimadzu Research Laboratories
Manchester, United Kingdom

Bruce A. Thomson
MDS Analytical Technologies
Concord, Ontario, Canada

John F. J. Todd
School of Physical Sciences
University of Kent
Canterbury, Kent, United Kingdom

Karl Peter Wanczek
Institute of Inorganic and Physical
   Chemistry
University of Bremen
Bremen, Germany

Gregory J. Wells
Varian Inc.
Scientific Instruments
Walnut Creek, California

Guohua Wu
Department of Physical Chemistry
School of Chemistry
The University of Nottingham
Nottingham, United Kingdom

Richard A. Yost
Department of Chemistry
University of Florida
Gainesville, Florida

Y. Zerega
Université de Provence - CNRS
Laboratoire Chimie Provence - UMR
   6264
Equipe Instrumentation et Réactivité
   Atmosphérique
Centre de Saint Jérôme
Marseille, France

Hailong Zhang
The Glycomics Center
University of New Hampshire
Durham, New Hampshire

Part I

Fundamentals

1 An Appreciation and Historical Survey of Mass Spectrometry

Raymond E. March and John F. J. Todd

CONTENTS

1.1 Introduction 6
1.2 Early History of Mass Spectrometry...8
    1.2.1 Molecular Beam Instruments 8
    1.2.2 The Genesis of 'Mass Spectrometry' 9
        1.2.2.1 Cathode Rays 10
        1.2.2.2 Canal Rays 11
        1.2.2.3 Ion/Molecule Reactions 12
        1.2.2.4 The Proton 14
    1.2.3 Mass Spectrometers with Ion Focusing.............................15
        1.2.3.1 Electron Bombardment Source............................18
        1.2.3.2 Utilization of Double-Focusing Mass Spectrometers..........19
        1.2.3.3 Hydrocarbon Mass Spectrometric Analysis 19
1.3 Mass Spectrometry 'Comes of Age'..21
    1.3.1 Introduction 21
    1.3.2 The Development of Analytical Mass Spectrometry 21
        1.3.2.1 Some Early Instrumentation 21
        1.3.2.2 Metastable Ion Peaks 22
        1.3.2.3 Accurate Mass, High Resolution and Isotopic Ratios..........24
        1.3.2.4 Qualitative Mass Spectral Analysis and
            Fragmentation Mechanisms...27
        1.3.2.5 Collision-Induced Dissociation and the Beginnings
            of Tandem Mass Spectrometry 30
        1.3.2.6 Methods of Ionization..34
            1.3.2.6.1 Electron Impact Ionization...............................35
            1.3.2.6.2 Spark Source Ionization 35
            1.3.2.6.3 Surface (Thermal) Ionization............................37
            1.3.2.6.4 Field Ionization/Field Desorption 38
            1.3.2.6.5 Chemical Ionization 40
            1.3.2.6.6 Bombardment Ionization...................................43
                 1.3.2.6.6.1 Introduction 43

　　　　　　　　　　　1.3.2.6.6.2　Secondary Ion Mass
　　　　　　　　　　　　　　　　　Spectrometry 44
　　　　　　　　　　　1.3.2.6.6.3　Bombardment with Fast
　　　　　　　　　　　　　　　　　Neutral Atom Beams.......................45
　　　　　　　　　　　1.3.2.6.6.4　Californium-252 Plasma
　　　　　　　　　　　　　　　　　Desorption Mass Spectrometry...... 47
　　　　　　　　　　　1.3.2.6.6.5　Laser Ionization............................47
　　　　　　　　1.3.2.6.7　'Spray' Ionization 49
1.4　The Study of Gas-Phase Ion Chemistry with Sector
Mass Spectrometers..53
　　1.4.1　Ion/Molecule Reactions...53
　　1.4.2　Space-Charge Ion Trapping in a Conventional Ion Source 53
　　1.4.3　Metastable Ions...55
　　1.4.4　Field Ionization Kinetics...60
　　1.4.5　Acceleration Region Kinetics 60
　　1.4.6　Collision-Induced Dissociation 62
　　　　　1.4.6.1　Neutralization-Reionization Mass Spectrometry 62
　　　　　1.4.6.2　Translational Energy-Loss Spectroscopy 63
　　1.4.7　Ion Thermodynamic Properties...64
　　1.4.8　Summary 66
1.5　The Development of Dynamic Mass Spectrometers 67
　　1.5.1　Introduction and Definitions...67
　　　　　1.5.1.1　'Dynamic' and 'Static' Mass Spectrometers 67
　　　　　1.5.1.2　Classification of Types of Dynamic Mass Spectrometer.....68
　　1.5.2　Instruments Based Upon Ion Cyclotron Resonance 69
　　　　　1.5.2.1　Introduction..69
　　　　　1.5.2.2　The Cyclotron 70
　　　　　1.5.2.3　The Omegatron 73
　　　　　　　　　1.5.2.3.1　The Omegatron as a Small Residual Gas
　　　　　　　　　　　　　　Analyzer.. 74
　　　　　1.5.2.4　Ion Cyclotron Resonance Mass Spectrometry....................75
　　　　　　　　　1.5.2.4.1　Introduction 75
　　　　　　　　　1.5.2.4.2　The Drift-Cell ICR Analyzer.............................75
　　　　　　　　　1.5.2.4.3　Trapped-Cell ICR Spectrometers 77
　　　　　　　　　1.5.2.4.4　Fourier Transform Ion Cyclotron Resonance
　　　　　　　　　　　　　　Mass Spectrometry 78
　　1.5.3　Time-of-Flight Mass Spectrometers.......................................80
　　　　　1.5.3.1　Introduction..80
　　　　　1.5.3.2　Magnetic Time-of-Flight Mass Spectrometers...................81
　　　　　　　　　1.5.3.2.1　A Helical Path Device 81
　　　　　　　　　1.5.3.2.2　Instruments Utilizing the Cyclotron
　　　　　　　　　　　　　　Principle... 81
　　　　　1.5.3.3　Linear Time-of-Flight Mass Spectrometers 83
　　　　　　　　　1.5.3.3.1　Basic Principle of Operation................................83
　　　　　　　　　1.5.3.3.2　Early Instruments 84
　　　　　　　　　1.5.3.3.3　Time-Lag Focusing................................86

	1.5.3.4	The 'Spiratron' 89
	1.5.3.5	The 'Reflectron' 90
	1.5.3.6	The Orthogonal-Acceleration Time-of-Flight Mass Spectrometer 91
1.5.4	Radiofrequency Devices 92
	1.5.4.1	Introduction 92
	1.5.4.2	The Use of Radiofrequency Fields for Velocity Selection 93
	1.5.4.3	Radiofrequency Ion Acceleration Spectrometers 94
		1.5.4.3.1 The Beckman Spectrometer 94
		1.5.4.3.2 The Bennett Spectrometer 94
		1.5.4.3.3 The Boyd Spectrometer 94
		1.5.4.3.4 The Redhead Spectrometer 94
	1.5.4.4	Instruments Employing Radiofrequency Quadrupolar Potentials 95
		1.5.4.4.1 Introduction 95
		1.5.4.4.2 Introduction Historical Development of RF Quadrupole Devices 95
		1.5.4.4.3 The Quadrupole Mass Filter 96
		1.5.4.4.4 The Monopole Mass Spectrometer 99
		1.5.4.4.5 The Three-Dimensional Quadrupole Ion Trap (QIT) 100
		1.5.4.4.5.1 Mass Spectrometry with the Quadrupole Ion Trap 101
		1.5.4.4.5.2 Tandem Mass Spectrometry with the Quadrupole Ion Trap 106
		1.5.4.4.5.3 The Cylindrical Ion Trap 106
		1.5.4.4.6 The Digital Ion Trap 107
		1.5.4.4.7 The Linear Ion Trap 108
		1.5.4.4.7.1 The Thermo Scientific Linear Ion Trap 109
		1.5.4.4.7.2 The AB Sciex Linear Ion Trap 110
		1.5.4.4.7.3 The Hitachi Axially Resonant Excitation Linear Ion Trap 110
		1.5.4.4.8 The Toroidal and Halo Ion Traps 111
		1.5.4.4.8.1 Introduction 111
		1.5.4.4.8.2 The Toroidal Ion Trap 111
		1.5.4.4.8.3 The Halo Ion Trap 112
1.5.5	The Kingdon Trap 112
	1.5.5.1	Introduction 112
	1.5.5.2	The Static Kingdon Trap 113
		1.5.5.2.1 Kingdon's Experiments 113
		1.5.5.2.2 Work in the 1960s 116
		1.5.5.2.3 Applications in Atomic Physics and FT-ICR 117
	1.5.5.3	The Knight Electrostatic Ion Trap 118
	1.5.5.4	The Orbitrap Mass Spectrometer 119

    1.5.5.5 Dynamic Kingdon Traps..123
       1.5.5.5.1 Introduction 123
       1.5.5.5.2 The Emergence and Physical Principles of
           The Dynamic Kingdon Trap............................124
1.6 Summary and Conclusions..126
Acknowledgments...128
References..128
Appendix: Theory of Radiofrequency Quadrupole Devices................................148
A1.1 Theory of Quadrupole Devices..148
   A1.1.1 Introduction..148
   A1.1.2 Quadrupolar Devices 149
A2.1 The Quadrupole Mass Filter (QMF) 151
   A2.1.1 The QMF with Round Rods...152
   A2.1.2 The Structure of the QMF 152
   A2.1.3 The Quadrupolar Potential 153
   A2.1.4 The Mathieu Equation 155
A3.1 The Quadrupole Ion Trap (QIT)..156
   A3.1.1 The Structure of the QIT 156
   A3.1.2 The Electrode Surfaces...158
   A3.1.3 The Quadrupolar Potential 158
   A3.1.4 The Mathieu Equation 160
A4.1 An Alternative Approach to Quadrupole Ion Trap Theory 161
A5.1 Regions of Ion Trajectory Stability 162
   A5.1.1 Secular Frequencies 165
References..168

1.1 INTRODUCTION

 "How do I love thee? Let me count the ways.
 I love thee to the depth and breadth and height
 My soul can reach, …"

In this poem* by Elizabeth Barrett Browning (1806–1861), the author seeks to define the dimensions through which her soul may reach as she examines the ways in which

---

* How do I love thee? Let me count the ways.
 I love thee to the depth and breadth and height
 My soul can reach, when feeling out of sight
 For the ends of Being and ideal Grace.
 I love thee to the level of everyday's
 Most quiet need, by sun and candle-light.
 I love thee freely, as men strive for Right;
 I love thee purely, as they turn from Praise.
 I love thee with a passion put to use
 In my old griefs, and with my childhood's faith.
 I love thee with a love I seemed to lose
 With my lost saints, — I love thee with the breath,
 Smiles, tears, of all my life! — and, if God choose,
 I shall but love thee better after death.

one human being can love another. Examination of the dimensions by which an objective may be approached, or even reached, is a common practice though few accounts of such examinations are presented so eloquently as in this Browning poem. The specific examination recounted in this volume and the succeeding volume is that of the practical aspects of carrying out trapped ion mass spectrometry. We propose that Browning's poem is a fitting analogy to our task of introducing the ways in which a bewildering variety of ions can be trapped, initially for the satisfaction of human curiosity and, ultimately, for the benefit of mankind. In so doing, we offer the following definition of a trapped ion: *"an ion is 'trapped' when its residence time within a defined spatial region exceeds that had the motion of the ion not been impeded in some way"*. In essence, the trapping of ions strives to enhance the duration of observation of the behavior of gaseous ions so that we may "witness this army, of such mass and charge".* Ions in the gas phase are tractable and the means by which an ion may be trapped are those of magnetic fields, electrostatic fields, electrodynamic fields, and collisions with neutral atoms and/or molecules. A major factor in the trapping of gaseous ions is the possibility of creating a pseudopotential well in which ions can be confined when their kinetic energies are less than the potential energy expressed as the depth of the well. While the ions are confined, they remain in motion and such motion may be characterized by secular frequencies that offer opportunities for resonance excitation. Alternatively, ions may be cooled, as in laser cooling or, in the absence of a pseudopotential well, they may be suspended in a dense stream of gas while being subjected to potentials to effect ion separation according to mass/charge ratio or conformation.

Ions exhibit both chemical reactivity and physical properties such as absorption of radiation from the ultra-violet region through the visible and into the infra-red region. By virtue of the constant motion of ions confined within an electrostatic or electrodynamic field, the resonant absorption of energy in the radiofrequency (RF) region can be employed to expand the trajectories of trapped ions against the focusing effects of ion/neutral collisions. Despite the ability to detect single ions that is relatively infrequently exercised, one encounters normally an ensemble of many ions that will inevitably give rise to space-charge repulsion between ions in the ensemble; such repulsive forces either must be harnessed or avoided. The two major advantages of experiments involving gaseous ions are that (i) ions can be separated according to their mass/charge, kinetic energy/charge, and momentum/charge ratios; and (ii) they can be detected with high efficiency. The contribution of detectors that employ electron multiplication and photon multiplication to the present state of mass spectrometry is enormous.

In presenting here a survey of the methods, instruments, and techniques employed in the study (and utilization) of trapped ions, it is apparent that experimentation moves through developmental stages in its evolution punctuated by discontinuities upon the nascence of new approaches for the achievement of objectives. In the Sections that follow we trace the routes by which the subject of 'trapped ion mass spectrometry' has evolved.

---

* Hamlet, Act IV, Scene IV.

1.2   EARLY HISTORY OF MASS SPECTROMETRY

Mass spectrometry is concerned with the study of charged particles which, be they atomic ions or ionized molecules, are composed of atoms, one or many, in which the total number of electrons is not equal to the total number of nuclear and covalently bound protons. The path to this level of understanding follows the work of Lavoisier in the 1780s on the compositions of compounds, Proust's Law of Definite Proportions in 1799, and John Dalton's formulation of his atomic theory in 1804, which was verified later by Berzelius. Of these researchers, only Dalton is commemorated in mass spectrometry in that the unit of mass is the dalton. On the basis of the atomic theory, an atom is defined as the basic unit of an element that can enter into chemical combination. Although the word 'atom' is derived from the Greek word άtomos, which means 'indivisible' or 'uncuttable', we now know well that the atom can be split; yet because the electron, proton, and neutron are atomic components, it is difficult to imagine that the concept of an 'atom' will ever be replaced.

During the nineteenth century, studies of electrolytic cells were accompanied by the development of batteries thus making direct current (DC) electrical sources available for research, particularly in the investigation of discharge tubes containing a gas under reduced pressure. A simple linear discharge tube to which a DC potential is applied consists of an anode, or positive electrode, and a cathode or negative electrode. A discharge through a low-pressure gas could produce various colored emissions that could be bent hither and thither upon being approached by a magnet. In 1886, Goldstein discovered that 'Kanalstrahlen' or 'canal rays' were positively charged because they were directed toward and passed through perforations in a cathode. Perrin studied anode rays and, upon passing them through an opening in an anode and using an electroscope, determined that they were negatively charged [1]. The early work with discharge tubes, and particularly those experiments involving passage of charged particles through an electrode perforation, set the stage for an assembly line approach to the study of these particles. It is of interest to note that while Henry Ford's novel approach to car manufacture was the introduction of the assembly line rather than static construction of each car, in many instances mass spectrometry has moved from being an 'ion flowing' system, or assembly line approach, to that of a static investigation of trapped ions.

1.2.1   MOLECULAR BEAM INSTRUMENTS

In 1911, a note by Louis Dunoyer appeared in the *Comptes Rendus de l'Academie des Sciences* to the effect that he had produced a beam of particles from a thermal source [2]. In later papers, the author described the formation of "molecular rays" of sodium emitted from a heated source, containing a vapor at moderate pressure, and directed through a hole into a vacuum. The particles (molecules or atoms) proceeded in straight trajectories as the kinetic theory predicted. A molecular beam instrument employs necessarily an in-line, linear, or axial arrangement where molecules are directed linearly from a source through the instrument to a detector. Subsequently, such instruments were employed frequently for the study of atoms emanating from a heated oven.

One of the earliest such studies is that by Max Born who reported on his observations of the attenuation of an atomic silver beam in air for the determination of the free-path of atoms [3]. In 1921, Otto Stern and Walter Gerlach, colleagues of Max Born, reported on their famous study of an atomic silver beam that was directed through an inhomogeneous magnetic field [4]. The interaction between the unpaired electron in each silver atom and the magnetic field caused the atom to be deflected from its straight-line path. Since the spinning motion of the electron is completely random, half of the atoms were deflected one way and half the other way, and two spots of equal intensity were observed on the detecting screen. These experiments gave conclusive proof of electron spin, $s = \pm \frac{1}{2}$, the fourth and final quantum number for the description of electrons. Stern was awarded the Nobel Prize in 1943. An appendix to this story is an account of a conversation of Otto Stern with Dudley Herschbach to the effect that the original deposits of two deflected silver beams were difficult to see on the metal detection plate. Only after exposure to the smoke from Stern's inexpensive cigars, which had a lot of sulfur in them, did the silver react to form silver sulfide, which is jet black, so easily visible. It was like developing a photographic film [5].

I.I. Rabi, who had become interested in molecular beam work while visiting Otto Stern in Hamburg, introduced molecular beam spectroscopy to the United States. Rabi was much taken with the simplicity of such experiments, coupled with the power to influence an atom or molecule. Rabi was awarded the Nobel Prize for Physics in 1944 for his work on developing the resonance method to record the magnetic properties of nuclei. One of Rabi's star graduate students at Columbia University, Norman Ramsey, also became a leader in the field of molecular beam spectroscopy. Ramsey went on to be a professor at Harvard, and his pursuit of molecular beams led to the development of the method of successive oscillatory fields in 1949; subsequently he was awarded the 1989 Nobel Prize in Physics for inventing this technique and applying it to the development of atomic clocks.* Ramsey's co-awardees of the Nobel Prize were Wolfgang Paul and Hans Dehmelt.

1.2.2 THE GENESIS OF 'MASS SPECTROMETRY'

Applied science makes improvements but pure science makes revolutions. J.J. Thomson. This quotation was reported by P.M.S. Blackett, *Times Educational Supplement.*

Beynon and Morgan [6] have given an interesting historical account of the development of mass spectrometry. We have drawn extensively on this article,[†] and recommend it for graduate courses in mass spectrometry. Clearly, the rate of development of mass spectrometry is dependent upon developments in other fields, for example, in photography and in the improvement in vacuum pumps. In 1879, Scientific American published an article on one of Edison's experiments [7]; this experiment was one of several which led to his creation of the first successful electric light, which would happen just a few months later on, October 22, 1879. The ready acceptance of electric light created a demand for improved, particularly faster, vacuum pumps for the evacuation of these light bulbs and, when mechanical pumps replaced the mercury Töpler pump, mass spectrometry was a beneficiary.

---

* See Volume 5, Chapter 11: Ion Trap: A Versatile Tool for the Atomic Clocks of the Future, by Fernande Vedel.
[†] With permission from Elsevier.

1.2.2.1 Cathode Rays

The genesis of the mass spectrometer lies in the early experiments to make quantitative measurements of the velocity, mass, and charge of cathode rays and canal rays. The properties of cathode rays were studied thoroughly by J.J. Thomson. Although it is a necessary consequence of their carrying a negative charge that they should be deflected by an electric field, only after several unsuccessful attempts to deflect cathode rays in this manner was Thomson able to demonstrate such behavior [8,9].

The ratio of the charge-to-mass, e/m_e, and the velocities of the cathode rays were measured assuming that they consisted of a stream of electrified particles, and this was achieved before the turn of the nineteenth century by several independent researchers. In J.J. Thomson's method, illustrated in Figure 1.1, the beam of cathode rays issuing through a perforated anode was collimated further by passing it through a metal slit at anode potential. An electric field was applied between the two plates A and B, thus producing a force in the plane of Figure 1.1 in a direction at right angles to the direction of motion of the rays. A magnetic field in a direction perpendicular to the plane of Figure 1.1 was produced by an electromagnet with its poles at right angles to the plates A and B and this produced a force, also in the plane of Figure 1.1, but in the opposite direction to that of the electric field. The method used by Thomson [10] involved direct measurement of the total charge received by an electrometer and the amount of heat produced when the rays fell on a thermocouple. The total charge and the total kinetic energy thus measured were combined with the results of magnetic deflection of the beam to give the required value of the charge-to-mass ratio.

The value for the ratio of the charge-to-mass for the electron e/m_e was almost 2000 times larger that Stoney's [11] value for the proton, e/m_H, which meant that either the charge carried by the particles (or 'corpuscles' as Thomson called them) in the cathode rays is large, or that the mass m_e is small compared to that of the proton. Thomson believed the latter to be the case and he made the momentous announcement of charged particles smaller than atoms in 1897 at a Friday evening Discourse at the Royal Institution [10]. In 1891, Stoney had suggested the name 'electron' for this definite, elementary quantity of electricity, without reference to the mass associated with it or the sign of the charge [12]. For many years, this original significance of the word was retained but later the word was used to denote the corpuscles themselves that make up the cathode rays.

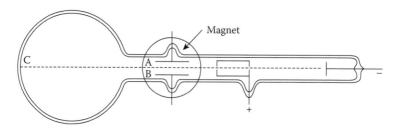

FIGURE 1.1 Thomson's apparatus for measuring the ratio of charge-to-mass for cathode rays. The deflection produced by a magnetic field on the collimated beam issuing from the discharge tube was balanced by that from an electric field. (Reproduced from Beynon, J.H. and Morgan, R.P., *Int. J. Mass Spectrom. Ion Phys.* 1978, *27*, 1–30. With permission from Elsevier.)

1.2.2.2 Canal Rays

The first thorough investigation of canal rays was made by Thomson [13], who called the rays 'positive rays' and who used the apparatus shown in Figure 1.2. The anode and cathode of his discharge tube are labeled A and B, respectively. A hole was bored through the cathode so as to accommodate a capillary, C, of fine bore. Such a capillary can be found today in some instruments that use electrospray ionization (ESI). The success of the whole experiment depended upon this bore being exceedingly small, because there was no focusing of the positive rays. The rays passed through region D, in which deflection by the coincident magnetic and electric fields took place, and fell on a screen E. Thomson used finely powdered willemite, a natural silicate of zinc, as a screen material to increase the brightness. With helium in his apparatus, Thomson was able to see three bands on his screen corresponding to three species of relative mass/charge ratio 4:2:1, which he identified as being due to the ions associated with helium, and with the molecule and atom of hydrogen, respectively. When air was introduced into the tube the most conspicuous bands on his screen were still those due to hydrogen, even at the lowest pressures that would permit a discharge; however, Thomson was able to see that the bands were, indeed, parts of parabolic curves, and other parabolas could be discerned also that he identified as being due to carbon, oxygen, neon, and mercury.

The big advance in the quality of spectra obtained came as the result of many improvements in the apparatus. A second instrument was constructed [14] which used a Gaede rotary mercury piston pump [15] as the primary means of evacuation, the newly invented method of Dewar [16] of adsorbing gases in coconut charcoal cooled in liquid air [17], and a photographic plate inside the vacuum chamber to detect the rays [18]. Thomson developed also an electrical method for detecting the ions, based on the use of the tilted electroscope developed by C.T.R. Wilson [19]. The ion currents were measured by observing, through a microscope, the movement of the end of the gold leaf and, in this way, currents as low as 10^{-13} A were measurable.

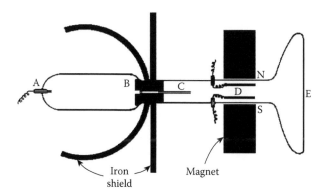

FIGURE 1.2 Thomson's apparatus for the study of positive rays. The collimated beam of positive rays issuing from the narrow tube, C, passes through coincident electric and magnetic fields at D. (Reproduced from Beynon, J.H. and Morgan, R.P., *Int. J. Mass Spectrom. Ion Phys.* 1978, *27*, 1–30. With permission from Elsevier.)

1.2.2.3 Ion/Molecule Reactions

The parabola instrument gave a variety of information not readily obtainable from other designs of mass spectrographs, one example of which concerns the then new field of ion/molecule reactions. Thus Thomson pointed out that ion/molecule reactions resulting in charge permutation, such as $M^+ \rightarrow M^\circ$ and $M^\circ \rightarrow M^-$, could be studied by this arrangement [20]. He built a special instrument, shown schematically in Figure 1.3, in which two electromagnets were placed between the perforated cathode of the discharge tube and the willemite screen. The first magnet deflected the beam in a vertical direction, the second in a horizontal direction, thus the spot A indicates an undeflected beam. The observation of a spot at position B that corresponded to maximum vertical and horizontal deflection shows that an ion has passed unchanged through both magnetic fields. Observation of a spot at position A indicates that the ion lost its charge in the field-free region (FFR) preceding the first magnet, the appearance of a spot at position C indicates that the ion lost its charge in the FFR between the magnets, and so on. The observation of spots corresponding to deflections in the opposite vertical and horizontal directions indicated the formation of negative ions in collisions with background gas.

Within the space of some 50 pages in his book [20] published in 1913, Thomson discussed all forms of charge permutation, and he estimated that the minimum energy a corpuscle must have to ionize a hydrogen atom is 11 eV. He considered the properties of negative ions, as well as ions carrying from two to eight positive charges, and pointed out that ion kinetic energy is shared between fragments resulting from dissociation in the ratio of their masses. Other topics included charge exchange of doubly charged ions, the emission of secondary electrons when ions strike a metal surface, how gases are desorbed from metals under bombardment by positive rays can be analyzed, and how the technique permitted the analysis of small amounts of gas arising from other

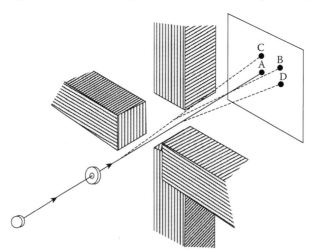

FIGURE 1.3 Thomson's apparatus in which a beam of positive rays passes successively between the poles of two perpendicular magnets and with which he was able to study ion/molecule reactions involving charge permutations. (Reproduced from Beynon, J.H. and Morgan, R.P., *Int. J. Mass Spectrom. Ion Phys.* 1978, *27*, 1–30. With permission from Elsevier.)

sources. He noted also the bright spots on some of his parabolas. We now know these to be due to fragmentation of 'metastable' or of collisionally activated polyatomic ions. Significantly, in this same book, Thomson described the most important of his discoveries at this period, namely the existence of the isotope of neon of mass-22.

Thomson's examination of neon in 1912 revealed an intense parabola corresponding to neon ions of mass-20 that was accompanied by a weaker parabola corresponding to ions of mass/charge ratio 22. He showed that this parabola was not due to ions carrying two charges [21], although he conceded that the parabola might possibly be due to a compound NeH_2. Thomson acquired a sample of pure neon and his assistant, F.W. Aston, attempted to separate it into its constituents by distillation and by diffusion through the stems of clay pipes; no change could be observed in the spectroscopic properties. It was concluded that the parabola corresponding to ions of mass/charge ratio 22 was indeed due to an isotope of neon. This was the first isotope of a stable element to be found, although unstable isotopes had been discovered in studies of radioactive elements. Thus, in 1910 Frederick Soddy [22] observed "The recognition that elements of different atomic weights may possess identical properties seems destined to have its most important application in the region [of the periodic table] of inactive elements where the absence of a second radioactive nature makes it impossible for chemical identity to be individually detected. Chemical homogeneity is no longer a guarantee that any supposed element is not a mixture of several different atomic weights or that any atomic weight is not merely a mean number." [22, p. 263]. Subsequently, in 1913, Soddy [23], coined the terms "isotopes" and "isotopic elements" [23, p. 263], from the Greek *isos topos* meaning the same place. Specifically, he stated "It now appears that the magnitude of this central positive charge, in terms of the unit of atomic charge, for example, that carried by the hydrogen atom, is probably the same as what is conveniently termed the "atomic number" [citation of five references in the original text]. The atomic number is the number of the place an element occupies in the periodic table...." [23, p. 271]. "It is the difference between positive and negative charges which gives the atomic number or the position of the element in the periodic table. Isotopes are those elements that have the same algebraic sum of the positive and negative charges in the nucleus but the arithmetic sum is different" [23, p. 272]. This was, of course, written without any knowledge of the existence of the neutron.

Lindemann and Aston [24] were the first scientists to investigate systematically from a theoretical standpoint the feasibility of different methods for isotope separation. The approaches considered were: Distillation ("to which chemical separation is closely allied thermodynamically"), Diffusion, Density, and Positive Rays. In respect of the last-mentioned method, the authors concluded "The chief difficulty is the excessive cold necessary in the receiving vessels which must be sufficient to fix the molecules even at the extremely low pressure of 10^{-4} mm in the camera." "No such difficulty occurs of course in the separation of the lead isotopes, but so far it has proved impossible to obtain lead positive rays quite apart from the difficulty of separating the parabolas when obtained." The overall conclusion was that "Centrifugal Separation appears promising, but would involve heavy outlay and elaborate preparations."

Joseph John Thomson received the Nobel Prize in Physics in 1906, while Nobel Prizes in Chemistry were awarded to Ernest Rutherford (1908), Frederick Soddy (1921), and to Francis William Aston (1922). Frederick Alexander Lindemann, later created Baron Cherwell and subsequently Viscount Cherwell, was a physicist by training. He was the first proponent of the mechanism of chemical reactions that involves a sequence of elementary steps, including the formation of an 'activated reaction intermediate' whose concentration exists in an approximately 'steady state' during the course of the reaction. He was the chief scientific advisor to Winston Churchill throughout the Second World War.

1.2.2.4 The Proton

Once the electron had been discovered, the search began for the electrically compensating positively charged species in discharges. As a background to this search were the findings of Rutherford, who had noticed that when alpha particles were shot into nitrogen gas, the signatures of hydrogen nuclei were shown by his scintillation detectors. Rutherford determined that hydrogen could have originated only from nitrogen, thus nitrogen must contain hydrogen nuclei. From the study of positive rays, the particle of lowest mass was found to be the hydrogen atom, thus it was concluded that the atom of positive electricity was the hydrogen positive ray, that is, the nucleus of the hydrogen atom. Rutherford suggested the name "proton"; the name comes from the Greek word πρστοσ meaning "first" to indicate that a proton is a primary substance [25].

In Thomson's descriptions of the parabolas that he observed, reference is made to bright spots or "beading" appearing in the parabolas. While no explicit explanation was forthcoming, it was observed that such beading occurred only in the parabolas of polyatomic ions. It is now understood that beading was a result of either metastable decomposition or collision-induced dissociation (CID) of polyatomic ions. In a 1920 paper [26], Aston described a brightening (or beading) of the parabolas of H^+ and H^- in the mass spectrum of H_2 at places where hydrogen ions with half the kinetic energy, corresponding to half the mass, would be expected. He invoked the process $H_2^+ + e^- \rightarrow H^+ + H^-$ as an explanation of this beading. In order to distinguish between H_2^+ and H^+, we must introduce the nomenclature for identifying ions. In mass spectrometry, the abbreviation m/z is used to denote the dimensionless quantity formed by dividing the mass number of an ion by its charge number. It has long been called the mass/charge ratio although m is not the ionic mass nor is z a multiple of the elementary (electronic) charge, e.* Thus, for example, for the ion $C_7H_7^{2+}$, m/z equals 45.5 [27]. As noted by Budzikiewicz and Grigsby [28], ironically, in the first (1922) [29] and second edition (1924) [30] of Aston's book the band at m/z 1 is not mentioned, and it cannot be seen in the hydrogen spectra depicted there. Nevertheless, the proton, H^+, is established firmly as an elementary particle.

Thomson showed remarkable foresight in his assessment of the long-term usefulness of the positive ray machine that he had developed. In the Preface to his 1913 book [20] he explained that one of the main reasons for writing this book was the hope that it might induce others, and especially chemists, to try this method of analysis. Later in his 'Recollections and Reflections' [31], he added "the spectrum enables us to determine the nature of gases inside the tube and thus provides a method of chemical

---

* The value of m/z is sometimes expressed in terms of the unit thomson (Th), where 1 Th = 1 u/e, where u is the atomic mass unit and e is the elementary (electronic) charge. [Cooks, R.G.; Rockwood, A.L. *Rapid Commun. Mass Spectrom.* 1991, 5, 93.]

analysis" and "The amount of gas required is very small, as the pressure of the gas is exceedingly low, generally less than one-hundredth part of a millimeter of mercury.... The rays are registered on the photograph within much less than a millionth of a second after their formation, so that when chemical combination or decomposition is going on in the gases in the tube, the method may disclose the existence of intermediate forms which have only a transient existence, as well as that of the final product, and may enable us to get a clearer insight into the process of chemical combination".

1.2.3 MASS SPECTROMETERS WITH ION FOCUSING

With the discovery of neon and the writing of his book, Thomson's great contributions to mass spectroscopy came to an end. The development of the subject then continued largely as a result of the work undertaken in Great Britain by F.W. Aston and in the United States by A.J. Dempster.

Aston and Dempster both performed their most important researches at the time when rapid advances in the techniques for obtaining high vacua were taking place, and these made it possible to improve instrument performance by avoiding scattering of the ion beam and unwanted ion/molecule reactions. The big advance came in 1915 with Gaede's invention of the mercury diffusion pump [32] by which gas was pumped by being swept away in a stream of mercury vapor. The design of this pump was improved upon by Langmuir [33,34] to yield the diffusion pump as we know it today. The cold trap continued in use until at least the 1970s, although the liquid air coolant was replaced for reasons of safety in the 1960s by liquid nitrogen.

Aston's first major contribution was to develop a new kind of instrument [30,35], which he named the 'mass spectrograph'; this instrument showed two great improvements over Thomson's parabola machine. First, Thomson's capillary, which produced a cylindrical ion beam, was replaced by a pair of slits that produced a beam in the form of a ribbon, and this arrangement permitted also a degree of differential pumping so as to reduce the number of ion/molecule collisions in the analyzer. Second, he discovered a method of velocity focusing of the ion beam [36] by passing it through successive electric and magnetic fields, as shown in Figure 1.4. With Aston's arrangement, all ions of a particular mass/charge ratio were brought to the same position on a photographic plate even though their velocities varied over a considerable range.

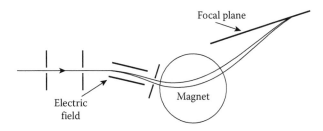

FIGURE 1.4 Aston's mass spectrograph, which could focus a collimated beam of ions even though the beam had a spread of kinetic energies, was the first use of 'velocity' focusing. (Reproduced from Beynon, J.H. and Morgan, R.P., *Int. J. Mass Spectrom. Ion Phys.* 1978, *27*, 1–30. With permission from Elsevier.)

FIGURE 1.5   A photograph of Aston's third mass spectrograph. (Reproduced with permission from the Cavendish Laboratory, Cambridge University.)

Figure 1.4 shows a departure from linear beam instruments, with the emergence of instruments constructed with sequential electric and magnetic fields, as opposed to coincident electric and magnetic fields in Thomson's machine, to obtain 'C'- or 'S'-shaped ion beams in attempts to obtain both velocity and direction-focusing.

When Aston introduced neon into his instrument he saw four lines corresponding to mass/charge ratios 10, 11, 20, and 22; the first pair corresponding to doubly charged ions, the second pair to singly charged ions. Thus the isotopic constitution of neon was settled beyond doubt. Since the atomic weight of chlorine showed a clear departure from pseudo-integral values, chlorine was the next element to be analyzed [37]. Its mass spectrum showed four singly charged ions at m/z 35, 36, 37, and 38; there was no sign of a line at m/z 35.46. The lines at m/z 35 and 37 were due to isotopes of chlorine and those at m/z 36 and 38 to their corresponding hydrogen acids.

By 1924, Aston had examined 50 of the 84 elements then known to exist. Three further instruments of the same design were built [38–40], the last of which achieved a mass resolution of 2000 and an accuracy of mass measurement of ca 2×10^5, an order of magnitude better than that of the first mass spectrograph. The most accurate measurements were made by the method of doublets, which is measuring the difference between the masses of two ions of nearly equal mass, such as D^+ and H_2^+. Aston's measurements were sufficiently accurate to determine the difference between the mass of a nucleus and the sum of the masses of the protons and neutrons of which it is composed. The energy equivalent of this difference is the *binding energy* of the nucleus. Thus mass measurements provide important information on the stability of nuclei. In Figure 1.5 is shown a photograph of Aston's third and final mass spectrograph; the overall length of the instrument, excluding the discharge bulb, is 105 cm. The instrument employed the same magnet as was used in Aston's second mass spectrograph, and W.G. Pye and Co., of Cambridge, carried out the necessary alterations to the pole pieces and construction of the rest of the apparatus [41]. In Figure 1.6 is shown a photograph of Aston working on the third mass spectrograph in his laboratory; the magnet obscures much of the mass spectrograph.

Aston's work was concerned primarily with measuring nuclidic masses, whereas Dempster's was concerned with measuring isotopic abundances. Aston's mass spectrograph permitted velocity focusing, whereas Dempster [42], using ion sources that produced but a small range of energies, developed an ion optical system that permitted direction-focusing. Dempster's mass spectrometer is shown in Figure 1.7; this is

FIGURE 1.6   Aston working on the third mass spectrograph. (Reproduced with permission from the Cavendish Laboratory, Cambridge University.)

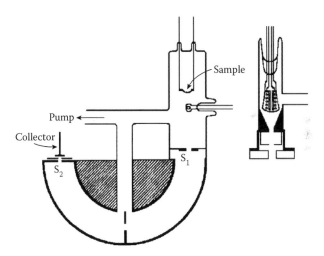

FIGURE 1.7   Dempster's mass spectrometer; this is the first example of a direction-focusing instrument. The direction of the magnet lies perpendicular to the plane of the diagram. (Reproduced from Beynon, J.H. and Morgan, R.P., *Int. J. Mass Spectrom. Ion Phys.* 1978, 27, 1–30. With permission from Elsevier.)

the first example of direction-focusing instrument. By accelerating ions of low kinetic energy through a common potential, a near monoenergetic beam of ions was obtained. Upon passing through the slit S_1, the ions entered a uniform magnetic field produced between two semi-circular iron plates and the ions were separated according to their momentum-to-charge ratios. The instrument made use of the property that beams of a given energy and mass diverging from a slit in a magnetic field cross again after deflection through 180°; that is to say, direction-focusing is achieved. Thus, for an ion of mass m and charge ze, where z is the charge number and e the electronic charge, the accelerating voltage V, magnetic field B, and radius of curvature of the ion beam R determine the mass-to-charge ratio according to the formula

$$\frac{m}{ze} = \frac{B^2 R^2}{2V} \tag{1.1}.$$

The beam diverging from slit S_1 was focused, by the action of the magnetic field, on slit S_2. A baffle, located halfway along the path, prevented reflected ions from reaching the exit slit. To obtain a mass spectrum, the magnetic field was kept constant and the accelerating voltage varied. Dempster's mass spectrometer was not suited for accurate mass measurements but it was notable for its simplicity compared to Aston's design and it became the model for many of the instruments developed later. By 1922, Dempster had already produced [43,44] accurate measurements of isotopic abundances for a large number of elements. The ion current was recorded initially using a quadrant electrometer hence the instrument was described as a mass spectrometer. In a later design [44], Dempster used a more refined electron bombardment source that is illustrated as an inset in Figure 1.7.

The sector magnetic field is still the basis for separation of ions of different mass/charge ratios in many mass spectrometers in use today. It was not until later, however, that it was realized that the focusing action of the 180° field used by Dempster was a particular case of focusing that occurs with any wedge-shaped (or sector) magnetic field [45], or that a radial electric field would give direction-focusing also. Most early instruments [46] were still based on 180° magnetic deflection; the exceptions were an instrument described by K.T. Bainbridge and E.B. Jordan (having somewhat appropriate initials!) [47] in which a 60° magnetic sector was preceded by a velocity selector, and a 60° instrument described by Nier [48].

1.2.3.1 Electron Bombardment Source

Steady improvement in the design of the electron bombardment source took place throughout the period described above. Dempster overcame a serious source of error in his earliest isotope abundance measurements [44], which arose from the fact that the change in accelerating voltage as the mass spectrum was scanned caused changes in the stray field at the source exit, where the ions were withdrawn from the source, and this affected the ion flow. This problem was overcome by the introduction of a screening gauze. This problem provides yet another example of the difficulties caused by the limited technology available, which precluded fixing the accelerating field and scanning the magnetic field. Gauze screening still has value in reducing unwanted stray fields.* The most noteworthy advances in source design are due to Bleakney [49], who introduced the idea of a magnetic field along the direction of the electron beam so as to collimate the electrons and thus to produce the ions more nearly on an equipotential surface in addition to increasing the electron path-length through the sample.

---

* See Volume 4, Chapter 8: Radiofrequency-only-Mode Event and Trap Compensation in Penning Fourier Transform Mass Spectrometry, by Adam Bruskern, Don L. Rempel, and Michael L. Gross.

1.2.3.2 Utilization of Double-Focusing Mass Spectrometers

Most of these instruments were used to seek new stable isotopes and to measure the relative abundances of the stable isotopes of various elements as they occur in nature. There was, at the same time, an increasing interest in the measurement of ionization energies and ionization efficiency [50] as a function of ionizing electron energy. Some instruments were built to give intense ion beams [51] and could produce on the order of 1 mg of a separated isotope per day. General improvements in the vacuum continued to be made so that, with baking the instrument to drive out adsorbed gas, a pressure of 10^{-6} Torr could be achieved. At such a pressure, the mean-free-path is *ca* 50 m and so problems associated with loss of performance due to scattering of the ion beam and to ion/molecule reactions were overcome almost completely.

The applications of mass spectroscopy to chemical analysis that had been pinpointed by Thomson were still neglected, although by the mid-1930s techniques for sample manipulation and for the production of high vacua were advanced sufficiently that much progress could have been made. The products formed by electron bombardment of some polyatomic molecules were discussed by Smyth [52] and mass spectra of propane and butane were obtained by Stewart and Olson [53].

Meanwhile, extremely important developments with regard to the understanding of focusing of ion beams were taking place. In 1929, Bartky and Dempster [54] published the theory of a double-focusing experimental arrangement in which the ion beam underwent both direction- and velocity-focusing by being passed through an electric sector immersed in a magnetic field perpendicular to the electric field. Later, instruments were constructed in which the ions passed successively through electric and magnetic fields in such a way that the combination of fields gave velocity-focusing whilst each individual field gave direction-focusing. Then in 1934, Herzog [55] and Mattauch and Herzog [56] published papers in which they described the general conditions for achieving double-focusing (direction- and velocity-focusing simultaneously) using a radial electric field and a homogeneous magnetic field with straight boundaries. Several versions of double-focusing instruments were built which gave very high mass resolution (>10,000) and led to a great increase in the accuracy with which nuclidic masses could be measured. The geometry of the 'C' type of double-focusing instrument is known as 'Nier-Johnson'. Here, the conventional arrangement was that of an electric sector followed by a magnetic sector but double focusing can be obtained with instruments of reverse geometry also (see Sections 1.3.2.4. and 1.3.2.5.). The geometry of the 'Z' type of double-focusing instrument is known as 'Mattauch-Herzog' in which a deflection of $\pi/(4\sqrt{2})$ radians in a radial electrostatic field is followed by a magnetic deflection of $\pi/2$ radians. It should be noted that the early instruments employed coincident electric and magnetic fields such that the mass-selected ion beam of interest was linear. Once the electric and magnetic fields had been separated in space, the ion beam of interest was no longer linear, and it was not until the emergence of triple-stage quadrupole instruments in the 1980s that we see a continuous linear ion beam being mass analyzed once more.

1.2.3.3 Hydrocarbon Mass Spectrometric Analysis

Until the early 1940s, little work had been reported on the fragmentation patterns of organic molecules, although molecules as large as cyclohexane [57] had been discussed and measurement of complex spectra was facilitated by the use of sensitive electrometer amplifiers. The pursuit of studies of hydrocarbons coincided with the recognition of an important industrial need, namely that of the analysis of the output of the catalytic crackers used in the petroleum industry. These studies led to rapid development of the capabilities of the mass spectrometer and to the manufacture of a large number of commercial instruments. The first account of the analysis of a hydrocarbon mixture by mass spectrometry was given by Hoover and Washburn [58,59]. Reference [58] might be considered to be something of a 'landmark publication' in terms of its content, and it is especially interesting in that the paper is written in a style directed at non-specialist scientists and engineers. Furthermore, it is the first report on the application of mass spectrometry to commercial work, as opposed to academic scientific research. Novel features include use of the dependence of the 'cracking pattern' for different hydrocarbons on the energy of the ionizing electrons as a means of resolving and analyzing quantitatively mixtures of these substances, the extremely low detection limits reported for trace hydrocarbons in soil samples, and the proposal that "A spectrometer with several exit slits, each slit so placed that it passes only the ions of a single mass, has many practical advantages." This last-mentioned arrangement, which includes the use of a separate amplifier circuit for each Faraday collector plate working independently, is the subject of a US Patent filed by Hoover [60] on 4 May, 1940. Interestingly, the affiliation of authorship by Hoover and Washburn [58] is given as President and Director of Research of United Geophysical Co., Pasadena, CA, respectively, yet within the same timeframe these individuals held the same respective positions of the Consolidated Engineering Corporation (CEC), Pasadena, CA. Both companies were, in fact, founded by Herbert Clark Hoover, Jr., a mining engineer and the eldest son of Herbert Clark Hoover, 31st President of the United States, who was, himself, a mining engineer by profession. The latter company (CEC) became the leading manufacturer of mass spectrometers in the United States for several decades. In 1943, it was shown that a mixture of C_5 and C_6 hydrocarbons containing nine components could be analyzed in just over 4 h, of which less than an hour was instrument time. This analysis time was *ca* 2% of that using the method of refractive index measurement.

For a graphic account of the early days of commercial mass spectrometry and its applications in the petroleum industry, where the war-time need for high quality aviation gasoline led the investment of large sums of money in developing the emerging technology, the reader is directed to the account first presented by Meyerson in 1984 [61]. The trials and tribulations experienced by these pioneers was summarized by E.B. Tucker, Group Leader in the Analytical Laboratory at the Standard Oil Company, in a statement that is still all too familiar to many practitioners of today: a mass spectrometer is a "machine that almost doesn't quite work"!

The many applications of mass spectrometry in so many scientific areas that followed the production of commercial mass spectrometers in the United States of America, England, and Germany still continue to expand. The details of this expansion, particularly in the area of ion trapping techniques, are described in this volume and in Volume 5. The works described in these two volumes have sprung from the firm foundation of mass spectrometry laid in the early part of the twentieth century.

1.3   MASS SPECTROMETRY 'COMES OF AGE'

1.3.1   INTRODUCTION

We turn now to events in the latter half of the twentieth century, and it is necessary to consider three concurrent streams of activity in order to appreciate the development of mass spectrometry and the advent of devices for trapping ions. The first stream follows the development of analytical applications of mass spectrometry, driven largely by the emergence of 'organic' and, to a lesser extent, by 'inorganic' mass spectrometry for applications outside the confines of the physics and chemical physics areas, while the second follows the detailed examination of the behavior of gaseous ions within, largely, mass spectrometers of more than one sector. This latter stream, which is discussed in Section 1.4, is classed broadly as the study of gas-phase ion chemistry, though it has had a marked effect on the development and understanding of new techniques for ionization, ion kinetics and decomposition, ion energetics, and structures. Within this time window, we have seen also the rapid development of bench-top computers and the advent of chemical computations. The third stream is an example of a discontinuity that can occur within the development of a field, in that new instrumental approaches to mass analysis relying on methods other than the use of magnetic sectors have been discovered; these new approaches are collected under the general title of 'dynamic mass spectrometry', and are considered in Section 1.5. Indeed, all the examples of the mass spectrometry-based ion trapping methods contained within these two companion volumes represent examples of different 'dynamic' devices.

1.3.2   THE DEVELOPMENT OF ANALYTICAL MASS SPECTROMETRY

1.3.2.1   Some Early Instrumentation

We have seen from Section 1.2.3.3 that the first widespread use of mass spectrometry for analytical purposes took place in the U.S. petroleum industry, but by the start of the 1950s the potential of the technique, both for analysis and for monitoring uranium hexafluoride enrichment, was being recognized in other countries. The birth of the mass spectrometry industry in the UK has been described in detail by the late Alan Quayle, who at the critical time was employed by Shell Research Ltd and was tasked with establishing mass spectrometry at the company's Thornton Research Centre [62]. Due to an embargo imposed at the time by the U.S. Government on the export of mass spectrometers and of high-speed recorders, it was necessary for the UK to be self-sufficient in terms of manufacturing the necessary instrumentation; as in the cases of CEC and Westinghouse in the United States and (later) Atlas Werke in Germany, it was a major electrical engineering concern that took up this challenge. Thus in 1944, with UK Government support, Metropolitan Vickers Electrical Company Limited (later to become Associated Electrical Industries Ltd (AEI), then Kratos Analytical and now Shimadzu) started to build four 60° magnetic sector 'MS1' mass spectrometers based upon the design of A.O. Nier [48,63]. Experience with the MS1 led to the development of the more advanced MS2 instrument, one of which was delivered to the Thornton laboratories in 1951. One particular improvement was the replacement of the glass-metal tube construction (which had to be broken open, re-sealed, and re-aligned periodically for maintenance) with an all-metal construction involving rubber 'O'-ring seals. The MS2 was a 90° magnetic sector instrument of 15-cm radius, operating with an accelerating voltage of 2 kV and fitted with a Faraday plate-type ion collector. A major difference between this instrument and the CEC mass spectrometers at the time was that with the MS2 a mass spectrum was scanned by varying the magnetic field at constant accelerating voltage, whereas with the former a mass scan was taken at constant magnetic field by varying the accelerating voltage. A summary of the construction details and performance of the MS2 mass spectrometer, and of its smaller companion the MS3, was presented by Blears at 'Pittcon' in March 1953 [64].

As with modern-day mass spectrometers, the users were seldom satisfied with what they had bought, and tended to embark on more or less ambitious schemes for adapting their instruments to more demanding applications. In the case of the petroleum industry this was to increase the upper mass limit of operation, combined with an increased scan speed: two aspirations that are often sought after today. In the case of the MS2 instrument, Shell and Metropolitan Vickers succeeded in extending the range to m/z 800, but because of the slow recording system the time taken to perform a full scan was between 1.5 and 2 hours! After numerous modifications, including the use of a photographic galvanometer recording unit, the scan time over the range m/z 25–700 was reduced to 15 minutes!

Having described briefly some of the instrumentation available at this time, we consider now how the mass spectral information generated by these early machines was being considered. Essentially there were two parallel tracks running: the first was the use of mass spectral data to provide both qualitative and quantitative analytical data about the identity and composition of industrial organic compounds, petroleum fractions, etc., the second was the measurement of isotopic abundances together with the continued refinement of mass spectrometer performance in order to obtain improved mass accuracy of isotope peaks and the resolution of closely separated doublets (see also Section 1.2.3.2). As we shall see, before long the second activity was to have an important influence on the first.

As the electron ionization (EI) mass spectra of more and more known (and hopefully pure) compounds were recorded libraries of mass spectral data were collected, such as the American Petroleum Institute Project 44 (API) tables [65]. This permitted the possibilities both of identifying unknown samples, and also the correlation of particular mass spectral features with molecular structure. The paper by Rock [66] represents a landmark publication in this area (see also Section 1.3.2.3.). However, without some more quantitative approaches being available, this was a tedious and potentially unreliable task. A summary of the methodology employed has been presented by Beynon [67].

1.3.2.2 Metastable Ion Peaks

The first major advance that facilitated mass spectral interpretation was the realization in 1945 by Hipple and Condon [68] (the latter author being of Franck–Condon Principle fame and the Director of the National Bureau of Standards at that time) that the presence of broad, low intensity peaks at non-integral values of m/z (m^*) could be correlated with the m/z-values of the precursor ion and of its fragmentation product ion. A comparison of the 'metastable' ion peak intensities observed with two alternative means of applying the accelerating potential (see Figure 1.8) confirmed this hypothesis, and it was shown [69] that, for singly charged m_1 ions 'decaying' to m_2 ions in the FFR between the ion source and the magnetic sector analyzer,

$$m^* = m_2^2 / m_1 \qquad (1.2).$$

Furthermore, by varying the pressure in the system they demonstrated that the linear dependence of the metastable peak height with pressure was consistent with the unimolecular process

$$m_1 = m_2 + m_n \qquad (1.3),$$

where m_n is the mass of a neutral fragment, not detected by the mass spectrometer. Thus, while the presence of metastable ion peaks in the mass spectrum was regarded

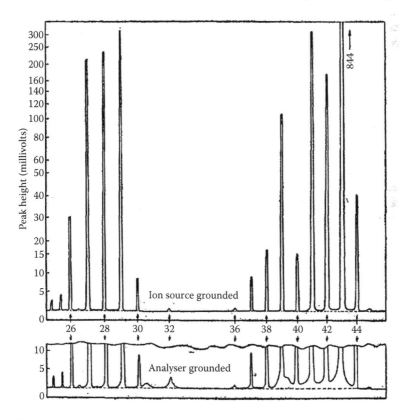

FIGURE 1.8 Portion of the mass spectrum of n-butane obtained on an automatic recorder employing a non-linear scale. The two curves show the effect on the spectrum of operation of the mass spectrometer with and without the ion source grounded. The disappearance of the diffuse peak at mass 32, for instance, is attributed to the presence of metastable ions in the mass spectrometer. The small residual peak at mass 32 is caused by O_2-impurity in the tube. The original charts have been retouched for reproduction. (Reproduced from Hipple, J.A. and Condon, E.U., *Phys. Rev.* 1945, *68*, 54–55. With permission from the American Physical Society.)

as a nuisance for quantitative measurements involving peak heights, their existence offered a very useful means of establishing the possible connectivity between different ion-pairs. One means of suppressing the appearance of metastable ions electronically, and thereby improving the resolution, developed by Craig is described in the article by Quayle [62]. Further consideration of the peak shapes, together with the quantitative information that can be gained there from, is presented in Section 1.4.

1.3.2.3   Accurate Mass, High Resolution and Isotopic Ratios

In 1954, John H. Beynon, working in the Dyestuffs Division of Imperial Chemical Industries Ltd, published a method that is still of the utmost importance to mass spectroscopists today [67]. This method is based upon the recognition that, because the atomic masses are not exactly integral multiples of a unit mass, different combinations of atoms of the same nominal mass will also have non-integral masses. Thus Beynon cited as an example two ions, $^{12}CH_2^+$ and $^{14}N^+$, the m/z-values of which are nominally 14, but actually had values of 14.0196 and 14.0070, respectively, [70]. In order to demonstrate the validity of this approach Beynon employed a single-focusing magnetic sector instrument having a resolution of 250. Working with a constant magnetic field, the ratio between the masses of two singly charged ions, m_1/m_2 is equal to the ratio of the accelerating potentials, V_2/V_1, and in order to determine an unknown mass one of the ions is chosen as a reference of known accurate mass; preferably the known standard should be as close in mass to the unknown as possible, yet still resolved from it. The principal inaccuracies were summarized as arising from (i) neglect of the kinetic energies possessed by the ions (see also below); (ii) variation of either the magnetic field or accelerating voltage during measurements; and (iii) difficulties in setting the mass spectrometer on the exact summits of the peaks. While factor (i) can be addressed by using a double-focusing instrument, the other two can be reduced by using more stable circuits and sharper-topped peaks. Nevertheless, using his single focusing instrument with an accelerating potential of 2000 V, Beynon obtained a mass accuracy of 1 part in 7000. It was suggested that accurate mass and fragmentation data combined with that obtained by infrared spectroscopy should aid the identification of small quantities of unknown organic compounds.

In the previous paragraph it was stated that in order to reduce the systematic uncertainty associated with the method of accurate mass measurement, it is necessary to employ a double-focusing instrument in order to compensate for any spread of kinetic energies of the ions in the beam. This spread of kinetic energies can arise both from the thermal motion of the sample atoms or molecule prior to ionization, and from inhomogeneities in the beam of ions as a result of the processes of extraction from the source and acceleration. Furthermore, in the case of product ions generated by fragmentation of ions formed from the original molecule, kinetic energy may be released during the dissociation step. In the case of the last-mentioned process, Nier and co-workers recognized that with a single focusing mass spectrometer, such as that used by Ney and Mann [71], those ions formed with additional kinetic energy would be focused at lower energies than are the 'molecular ions', and that a measurement of their relative positions in the mass spectrum will not be an accurate measure of the mass of the fragment [72]. Nier and Roberts concluded that "this limitation can be overcome only by using a double-focusing instrument, that is, one in which all the ions of the same mass leaving an ion source and having a range of range of energies as well as a range of angles will be refocused at the same point." The double-focusing instrument that Nier and Roberts designed [72], shown in Figure 1.9, had certain novel features, and differed from the mass spectrograph (that is, one with photographic recording), which was conventional at that time. The theoretical principles underlying the design of this instrument were discussed in detail by Johnson and Nier [73]. Thus, the ion beam from the source passed first through a 90° electrostatic sector to produce an energy spectrum with first-order velocity focusing in the focal plane containing the slit S_3 (Figure 1.9), and thence through a 60° magnetic sector that caused the ion beam to be directed with second-order angular focusing to a single point where the ion current could be measured electrically. A doublet of mass spectral peaks was examined by varying the potential applied to the

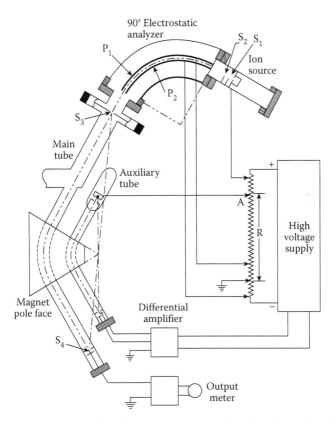

FIGURE 1.9 Schematic drawing of mass spectrometer tubes showing action of regulation circuit. Mean radius of cylindrical electrostatic analyzer 18.87 cm. Separation of plates P_1 and P_2, 1 cm. Radius of magnetic analyzer in main and auxiliary tubes, 15.24 and 7.62 cm, respectively. Asymmetrical construction reduces width of beam at S_4 owing to divergence of ion beam. S_1, S_2, S_3, and S_4 are 0.0025, 0.0033, 0.25, 0.0025 cm, respectively. S_1 and S_2 are separated by 3.71 cm. Main and auxiliary tube ion accelerating potentials are approximately 4000 and 1000 V, respectively. Hence, gas used in auxiliary tube has approximately the same molecular weight as the doublet to be studied. (Reproduced from Nier, A.O. and Roberts, T.R., *Phys. Rev.* 1951, *81*, 507–510. With permission from the American Physical Society.)

electrostatic analyzer (ESA) together with the accelerating potential applied to the ions, while maintaining the magnetic field constant. In order to minimize the effect of fluctuations in the magnetic field and in the ion deflecting voltage a second, auxiliary, mass spectrometer was mounted in the same magnetic field, as shown in Figure 1.9. Using a double collector arrangement in the auxiliary instrument, any minor shifts in the ion beam could be detected by means of a differential amplifier and the resulting signal fed to back to the control circuitry for the main mass spectrometer. This idea of transmitting two independent ion beams through the same magnetic field, albeit in the same analyzer tube, also formed the basis of the AEI MS30 dual beam double-focusing mass spectrometer developed by Brian Green and colleagues during 1968–1970 [74].

The instrument developed by Nier and Roberts has been described in some detail because it was this design that the engineers at Metropolitan-Vickers used as the basis for the double-focusing MS8 mass spectrometer that had been commissioned by Beynon in order to extend his work on the accurate mass measurement of organic ions. The MS8 was, in fact, a one–off model but is of importance because the experience gained with this instrument was incorporated into the later design of its larger successor, the 'MS9' (see below). A detailed description of the MS8 mass spectrometer was presented by Craig and Errock at what is generally regarded as being the first 'international mass spectrometry conference', held in London from 24th to 26th September 1958 [75].

The MS8 instrument comprised both 90° electric (radius 7.5 in.) and magnetic (radius 6.0 in.) sectors, with a maximum ion accelerating potential of 8 kV. Mass measurement could be made with a precision of about 1 part in 100,000, and the resolving power between adjacent peaks for equal height was (10% valley) was 10,000. Working with this mass spectrometer, Beynon produced the data that was discussed in yet another landmark publication [76]. In addition to re-affirming the use of accurate masses in order to determine the atomic composition of molecular ions, the author demonstrated how this method could be combined with a knowledge of the intensities of the peaks appearing at m/z, $(m + 1)/z$ and $(m + 2)/z$, as described earlier by Rock [66]. Thus, one can use the ratios of the intensities (P_m, P_{m+1}, and P_{m+2}, respectively) expressed as (P_{m+1}/P_m), (P_{m+2}/P_m), and (P_{m+1}/P_{m+2}) to distinguish between the various possible empirical formulae for the molecular ion by comparison with the corresponding calculated ratios computed from the known isotopic abundances. Using these two approaches, measurements of accurate mass and of isotope abundance ratios, Beynon was able to formulate a set of general rules as to how ions formed from different types of organic molecules break up when analysed in a mass spectrometer. Subsequently, Beynon and Williams produced a comprehensive listing of the accurate masses and isotopic abundance ratios of ions containing varying numbers of carbon, hydrogen, nitrogen, and oxygen atoms up to mass 500 u [77].

A further interesting topic included within this reference [76] is the pioneering combination of 'vapour-phase chromatography' (now known as gas chromatography (GC)) with mass spectrometry by trapping the individual constituents of a mixture as they emerged from the column and admitting them separately to the mass spectrometer. Given that the development of GC had only been reported by James and Martin in 1952 [78], three years before the submission date of Beynon's paper in *Mikrochimica Acta*, the implementation of this separation technique in combination with mass spectrometry surely represents a state-of-the-art approach to chemical analysis? A further, more detailed, account of the analytical applications of the MS8 instrument to the study of organic compounds was reported to the second international conference on mass spectrometry, held in Oxford in September 1961 [79].

Following from the evident success of the MS8 instrument, Robert Craig and colleagues at AEI (as the company was now called) went on to develop the MS9 double-focusing mass spectrometer [80]. This version also contained two 90° electrostatic and magnetic sectors, having radii of curvature equal to 15 in. and 12 in., respectively; these values are twice the dimensions employed in the MS8. The MS9 has been described by Quayle [62] as "one of the world's most successful high resolution mass spectrometers." In the early stages of development, as news about this new, larger, machine spread, more and more enthusiasts wished to view the first unit under construction. Apparently this influx of visitors must have alarmed one rather sceptical member of senior management, who was heard to remark "For goodness sake don't start telling people about it or they'll all want one!" [81]. A 1970 sales brochure for the product shows a guaranteed resolution of greater than 70,000 on a 10% valley definition (140,000 on a 5% contribution definition), and a mass measurement accuracy of better than 2 parts per million for a 2% difference between an unknown and a standard reference mass. At Kent University, UK, in 1968 a seismographic survey of the intended accommodation for the MS902 instrument had to be conducted while a nearby elevator (lift) was operated in order to ascertain that there were no low-frequency oscillations in the structure building before the performance specification could be quoted in the tender documents!

1.3.2.4 Qualitative Mass Spectral Analysis and Fragmentation Mechanisms

We have seen from the preceding discussion how the early applications of analytical mass spectrometry were directed mainly toward the characterization of hydrocarbon mixtures; in particular we have noted the difficulty of this task in the period before the advent of gas chromatography. One especially significant paper published toward the end of the 'pre-GC' era was that by S.M. Rock [66] of the CEC, Pasadena, CA. This publication was cited in the previous Section in connection with the discussion of the use of isotopic ratio measurements as an aid to compound identification. Rock noted that, at the time of his publication, whilst there had been numerous reports on quantitative determinations using mass spectrometry, there had been no systematic presentation of how mass spectrometry could be used qualitatively for compound identification. Thus he outlined some basic procedures that are still very much in use today: check the intervals between mass spectral peaks to identify (neutral) fragments and elements present in the compound; use peak-free regions of the mass spectrum to establish the absence of fragments; compare the peaks with those predicted from normal stable isotope ratios; recognize characteristic ions at 'half-mass' (arising from doubly-charged ions) and metastable ion transition peaks; be conscious that some, occasionally substantial, unpredicted ion peaks must arise from "rearrangement of the molecule"; and make use of mass spectral libraries. In a one-sentence paragraph that was also a forerunner of ideas to come, Rock noted that "Odd peaks are usually larger than even." We shall return briefly to the last point below in relation to the relative stability of even-electron (EE) and odd-electron (OE) ions.

Within the time-frame under consideration, and in parallel with the expanding work on mass spectral fragmentation processes, Rosenstock et al. [82] published the first of a series of papers on the extension of the absolute rate theory formulated by Eyring and co-workers [83] to the ionization and dissociation processes that occur within a mass spectrometer. [Anecdotally, this theory was developed while the research group was awaiting the delivery of its first mass spectrometer!]. The original underlying hypothesis for this theoretical approach is best summarized by this quotation from Rosenstock et al. [82]: "Following vertical ionization [that is, a Franck–Condon transition], the molecule-ion has a certain amount of excitation energy in its electronic and vibrational degrees of freedom, referred to the minimum of its lowest electronic state. Most of the time, the excited molecule-ion does not decompose immediately into ionized and neutral fragments but rather undergoes at least several vibrations. During these vibrations there is high probability of radiationless transitions among the many potential surfaces of the molecule-ion, resulting in a distribution of the excitation energy in completely random fashion. The molecule-ion decomposes only when the nuclei are in the proper configuration and a sufficient amount of vibrational energy has concentrated in the necessary degrees of freedom. The fragments in turn may have sufficient energy to decompose through a similar sequence of events. Rearrangement of the bonds, without dissociation, also may occur in the same fashion." This model, in some respects, represents the complete opposite of another widely made assumption, namely that the charge on an ion is 'localised' at the atom or bond from which the electron was removed during the ionization process, and that the site of the resulting net charge will direct any subsequent fragmentation. Debate concerning the relative merits of these two extreme approaches has continued unabated over the years. Interestingly, following a comprehensive and detailed consideration of the concepts of charge-localization and of radical localization as a means of predicting unimolecular fragmentation pathways of positive ions from organic compounds, Williams and Beynon [84] stated that their treatment was not at variance with a refined version of the quasi-equilibrium theory presented by Rosenstock and Krauss [85].

While Rock's account was confined mainly to the study of hydrocarbons, the application of mass spectrometry to the qualitative analysis of a much wider range of organic compounds was reported in detail by McLafferty, based on work at the Dow Chemical Company, Midland, MI, [86] where 10,000 samples per year were analysed in his laboratory. Echoing the frustration that Rock had expressed over the occurrence of rearrangement ions in mass spectra, McLafferty wrote "The main difficulty encountered in structure assignments from the ion fragments in the mass spectrum is that sometimes fragments occur that cannot arise from the simple cleavage of bonds in the molecule." Later, in the same section of the paper, he stated "Much of the analytical disadvantage incurred by these rearrangements would be removed if they could be predicted from the molecular structure of the compound. No such generalizations have been reported, though evidence has been collected for various mechanisms of rearrangements in hydrocarbons [references given]."

The issue of rearrangement processes occurring in the mass spectra of lactones had already been addressed by Friedman and Long [87], who, having invoked the, then recent, theoretical model for ionization and dissociation processes advanced by Rosenstock *et al.* [82] (see above) pointed out that "Such a theory does not suffice for prediction of what specific rearrangements will occur; for this one needs added assumptions. A general assumption whose utility is borne out by examination of large numbers of mass spectra is that there is a large probability of rearrangement only when one of the resulting fragments is of relatively high stability." The authors then went on to point out that in their study of the mass spectra of the selected lactones essentially all the observed rearrangements lead to ions having an even number of valence electrons. The question of whether ions having an even number of electrons were relatively more stable than those with an odd number had already been surmised by Delfosse and Bleakney in 1939 [88] in their interpretation of the 100 eV EI mass spectra of propane, propene (propylene), and allene, where they wrote "Another peculiar distribution of intensities is the small yield of the propylene and allene ions $C_3H_6^+$ and $C_3H_4^+$ coming from propane and $C_3H_4^+$ from propylene. Since these ions correspond to stable compounds, it might be supposed that they should be among the most abundant. No explanation of this behavior is known. Perhaps the ions with an even number of electrons are more stable than those with an odd number."

In 1959 McLafferty [89] returned to the issue of structure determination by mass spectrometry being hampered by the relatively unpredictable possibilities of molecular rearrangement. He proposed a classification of rearrangement processes as being either randomized, involving a higher energy pathway, or specific and occurring at lower internal energies. In relation to the latter, on the basis of a detailed consideration of five possible scenarios involving OE or EE ions, he showed that rearrangement product ions were formed in high abundance when the process occurred through formation of a sterically favored transition state of more stable products. This analysis, of what became known as the 'McLafferty Rearrangement' which was presented originally as shown in Scheme 1.1, has become one of the central tenets of mechanistic organic mass spectrometry. Indeed, unlike organic chemistry, to this day this process is the only named mass spectral reaction and, in July 2004, the *Journal of the American Society for Mass Spectrometry* dedicated a 'focus issue' in honor of this discovery.

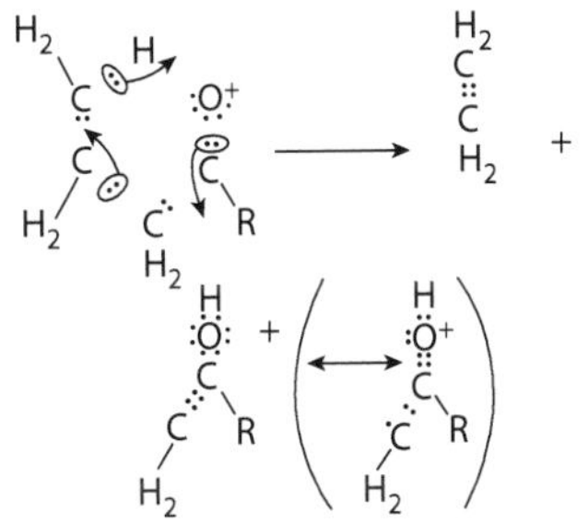

SCHEME 1.1 Original representation of a possible mechanism for what was named subsequently as the "McLafferty Rearrangement". (Reproduced from McLafferty, F.W., *Anal. Chem.* 1959, *31*, 82–87. With permission from the American Chemical Society.)

By the early 1960s, through the application of metastable ion analysis, accurate mass determination, and isotopic ratio analysis, together with a set of empirical ground rules for mass spectral interpretation (often confirmed through labeling studies with stable isotopes), the field was ripe for the publication of what, over the years, have become classic texts. Some examples of these books are those published by Beynon [90], Reed [91], Biemann [92], McLafferty [93,94] and Budzikiewicz, Djerassi and Williams [95–97], followed somewhat later by Beynon, Saunders and Williams [98] and a comprehensive text on instrumentation by Roboz [99].

Before leaving this aspect of the history of the evolution of 'organic mass spectrometry' in the 1950s and 1960s, it might be of some interest to consider briefly two examples of the rudimentary techniques and technology to which the pioneers resorted. The first of these concerns the problem of the adsorption of trace constituents during the mass spectral analysis of mixtures of hydrocarbons, as reported by Meyerson [100]. In this short paper the author explains that the composition of a sample admitted to the mass spectrometer could be changed as a result of the absorption of compounds present in a mixture, especially of the higher homologues, on the brass components of the inlet system, thereby causing inaccuracies in quantitative determinations. To overcome this problem, after each normal sample run water was admitted to the inlet system and the analysis repeated. Apparently water facilitated the desorption of the organic materials, yet on account of its low mass did not cause any interference with the mass spectral peaks of the desorbed analyte. For most mass spectroscopists, the prospect of deliberately admitting water to the instrument must appear to be suicidal!

The second example concerns what is probably the first report of the use of some form of injection system for the admission of vapor from solid organic materials to an EI source, the forerunner of the 'direct insertion probe'. This method was developed by Reed [101] for the analysis of some steroids and triterpenoids, and involved the arrangement shown in Figure 1.10. The device consisted of a re-entrant copper rod (C) mounted on the end of a copper block (B), which itself was screwed (with a gas-tight seal) on to a metal tube connected *via* a glass-metal seal (D) to a glass tube E mounted through a hole in the repeller plate (H) floating at 2000 V (for scans up to *m/z* 200). A sample container (F) equipped with a sinter (G) was held by means of a second screw thread on to the lower end of the rod. It was found that, in general, samples melting below 60°C were sufficiently volatile for the vapor to penetrate the sinter and pass into the electron beam (not shown), however for the less volatile analytes the copper block was heated by means of a gas micro-burner, the heat being conducted along the copper rod to the sample. For the particular compounds studied, it was observed that with a low-energy electron beam (9–15 eV) the ionized products of thermal decomposition during the ejection process could be detected, while at higher electron energies (30–70 eV) dissociations induced by electron impact were observed.

1.3.2.5 Collision-Induced Dissociation and the Beginnings of Tandem Mass Spectrometry

The influence of gas-phase collisions on the behavior of ions lies at the very heart of modern mass spectrometry, a statement that applies especially to studies involving trapped ions. Thus, as exemplified by many of the 37 chapters contributed to Volumes 4 and 5 of *Practical Aspects of Trapped Ion Mass Spectrometry*, we see the need to

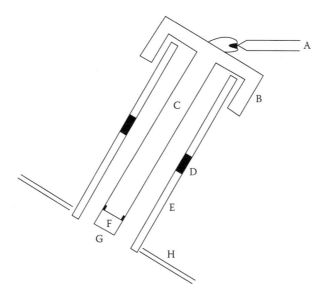

FIGURE 1.10　A, micro-burner. B, Copper block on the end of the copper rod forming a gas-tight seal. C, copper rod. D, Glass-metal seal. E, Glass inlet tube. F, Sample chamber. G, Terminal sinter. H, 2000 V, Repeller plate of the ion chamber. (Reproduced from Reed, R.I., *J. Chem. Soc.* 1958, 3432–3436. With the permission from the Royal Society of Chemical Society.)

(i) scrupulously avoid collisions within the timescale of the experiment, for example in the Fourier transform ion cyclotron resonance mass spectrometer (FT-ICR), the Orbitrap, atomic clocks, and single ion mass spectrometry; (ii) employ collisions to 'cool', to cause reactions, or to collisionally excite ions in 2D (two-dimensional) and 3D (three-dimensional) radiofrequency quadrupole traps; (iii) moderate the motion of ions in electric fields as in ion mobility spectrometry (IMS); and (iv) effect the longitudinal motion of ions in high-field asymmetric ion mobility spectrometry (FAIMS) through the use of a flowing gas stream.

The importance of collision processes is evident from a number of key text books and reviews that have appeared, including those by McLafferty [102], Busch *et al.* [103], Cooks [104], and McLuckey [105], from which the reader can develop an in-depth appreciation of the fundamentals and scope of the subject. The purpose of this Section is to present a brief summary of the most significant milestones as our understanding of the importance of collision phenomena has evolved.

We have seen in Section 1.2.2 how Thomson and Aston recognized that that the motion of ions within their early positive ray apparatus was influenced by collisions with the neutral sample species and with background gases, through the occurrence of what they termed 'beading' in their parabola mass spectra. A key paper by Aston in 1919 [26] provides a possible explanation of how the data observed could be attributed to the detection of ions arising from atoms and molecules of hydrogen. The aforementioned review by Cooks [104] provides a detailed commentary on this pioneering work. In 1925 Smyth [106], working also with hydrogen, showed that under ionizing conditions where only molecular $H_2^{+\bullet}$ ions were formed by electron impact, varying the pressure in the system led to the formation of H_3^+ and also to the appearance of H^+ through the collisional dissociation of H_2^+ and H_3^+ ions. In a study that is rather more recognizable as a forerunner of a modern CID experiment, Bainbridge and Jordan [107] investigated the dissociation of $CO^{+\bullet}$ to $C^{+\bullet}$ when an unspecified "scattering gas was introduced into the tube connecting the electric and magnetic deflection chambers" of the double-focusing mass spectrograph [47] that has already been mentioned in Section 1.2.3.

Moving through the 1940s we note that, as mentioned in Section 1.3.1.2, Hipple *et al.* [69], in their pioneering work on metastable ions, took care to ensure that the pressure-dependence of their ion intensity data was consistent with that of a unimolecular dissociation process. It was not until 1957 that Rosenstock and Melton [108] conducted a systematic study with a 60° magnetic sector single focusing instrument of the CID of ions formed from *n*-butane and from isobutane (1,1-dimethylpropane) following bombardment with 75-eV electrons and acceleration through a potential difference of 2100 V. Pressure-dependence measurements of the metastable ion intensities showed which products were formed by unimolecular decay of the precursor ion and which were formed as a result of collisional excitation. It was noted that there was a "striking parallel" between the yields of product ions resulting from the collisional dissociation processes and the results of electron-impact induced dissociations. A similar conclusion was reached later by Jennings [109] in respect of the CID of certain aromatic molecular ions, in a series of experiments that may be regarded as heralding the modern era of collisional activation studies [104,105].

Remaining in the 1950s, we note that by the end of this decade tandem mass spectrometers started to appear, an example of which was described by White *et al.* [110]. This group constructed a three-stage instrument comprising two consecutive magnetic focusing lenses of 20 in. radius followed by a 20 in. radius of curvature ESA. In a detailed report [111], White and co-workers described an extensive series of experiments in which a wide variety of mass-selected ions with kinetic energies in the *ca* 10 to *ca* 150 keV range were allowed to impinge upon a nickel film of thickness 500 Å or to interact with a gas (generally helium, oxygen, or argon) within a collision chamber; in each case, the target was positioned at the focal point between the two magnetic sectors. In this way various charge-exchange processes were investigated, as well as the collisional dissociation reactions of $CO^{+\bullet}$, NO^+, $N_2^{+\bullet}$, and of polyatomic hydrocarbon species.

Thus far, we have considered certain key investigations into the role of gas-phase collisions where each of the mass spectrometers was essentially a home-made, individual instrument, often designed for a specific purpose. The next major step forward in the evolution of collisional dissociation as a major tool in analytical mass spectrometer came in 1964, when Barber and Elliott [112] first described how the commercial MS9 double-focusing mass spectrometer could be operated in a modified way for the more detailed characterization and application of metastable ion decay processes. The essence of the 'Barber-Elliott scan' is that a product ion, m_2, formed from a precursor ion, m_1, as a result of metastable ion decay in the FFR between the ion source and ESA will have a lower kinetic energy than that of any undissociated m_1 ions and of those m_2 ions formed 'directly' within the ion source prior to acceleration. Consequently, if the double-focusing mass spectrometer is set up with an accelerating voltage V_1 and an ESA voltage E so that the 'direct' m_2 ions are transmitted through the magnetic field (B), the m_2 product ions formed by metastable ion decay, as above, will have insufficient kinetic energy to pass through the ESA and, therefore, will not be detected. However, if the accelerating voltage is increased while keeping E and B constant in order to compensate for this reduced kinetic energy, all the 'direct' ions will be defocused and cease to be transmitted while, ignoring any release of kinetic energy during the decomposition process, those m_2 ions formed *via* metastable ion decay will reach the detector when the accelerating voltage equals V_2, where

$$V_2/V_1 = m_1/m_2 \qquad (1.4).$$

Consequently, because the identity of m_2 is already known, the identity of the precursor ion, m_1, can be found unequivocally. This method offered, therefore, a distinct analytical advantage compared to attempting to deduce the connectivity between m_1 and m_2 from the value of m^* using Equation 1.2. In modern day parlance, the Barber-Elliott 'V-scan' allows one to perform a precursor ion scan in order to identify those ions which dissociate to form a specified product ion. A detailed example of the use of this approach was presented by Barber *et al.* [113].

In the ensuing years other types of so-called 'linked scan' were developed for investigating metastable ion decay processes occurring in different regions the ion flight path of the double-focusing mass spectrometer, and these have been assessed and discussed by Boyd and Beynon [114]. A later, and more comprehensive, account of 'linked scanning' methods was presented by Jennings and Mason [115], where it is shown that these various approaches not only allow the operator to record 'precursor ion' mass spectra, as explained above, but also 'product ion' mass spectra containing those ions derived from a common precursor, and 'constant neutral-loss' mass spectra in a manner analogous to the different scanning modes practised in contemporary tandem mass spectrometry instruments.

In our discussion so far we have considered only the simple case of metastable ion dissociation in which any gain in the kinetic energy of the product ions associated with the release of internal excitation energy of the precursor ion has been ignored. Furthermore, all the instruments utilized in the respective experimental studies have incorporated a magnetic sector that provides 'm/z-analysis'. In reality, the broad shapes 'metastable ions peaks' in the mass spectrum are evidence of the exoergic nature of unimolecular ion dissociation. In order to investigate these effects further Beynon *et al.* [116] devised the technique of 'ion kinetic energy spectroscopy' (IKES), in which they inserted an electron multiplier detector into the path of the ion beam as it was emerging from the ESA of a double-focusing mass spectrometer. Clearly, this arrangement interrupted the flow of ions into the magnetic sector, and in a typical experimental run the mass spectrometer was tuned up first using the normal double-focusing arrangement before the IKES multiplier was moved into position. With a fixed value of the accelerating potential, the voltage applied across the electrodes of the ESA was scanned down from a given value to zero so as to focus successively ion beams of various energies on the resolving slit and thence into the detector mounted immediately behind the slit. In this way, a kinetic energy spectrum of the transmitted ions was obtained, including those ions formed by metastable precursor decay in the FFR between the ion source and the ESA. Any kinetic energy released during the dissociation was immediately evident (and measurable) from the shapes of the peaks in the IKE spectrum. Further consideration of some examples of the physico-chemical information that can be gained from these experiments is considered in more detail in Section 1.4.

In discussing the method of IKES, Beynon *et al.* [116] recognized that there was a difficulty in identifying the *m/z*-values of the ions contained within an energy peak and, although they devised a means of overcoming this limitation in a later paper [117] from the same group the authors noted that "Ideally, using a double-focusing mass spectrometer specially designed for IKE studies, in which the magnetic sector precedes the electric sector, the same kind of measurement could be carried out by tuning the magnetic sector to one component of a doublet, then following the decomposition of this particular ion by scanning the electric sector." In this way the 'reversed geometry' double-focusing mass-analyzed ion kinetic energy spectrometer (MIKES), in which the magnetic sector precedes the ESA, was born [118,119]. Of the two groups pioneering the use of reversed geometry instruments, McLafferty and co-workers applied their instrument to the development of collisional activation as an analytical tool [120], whereas Beynon and co-workers concentrated their activities on the study of metastable ions [121].

Both the two reversed geometry instruments described in the preceding references were 'home-made': the first by the simple expedient of exchanging the positions of the ion source of a Hitachi Perkin-Elmer RMU-7 double-focusing mass spectrometer [118], and the second by reconfiguring the standard magnetic sector and ESA taken from a commercial AEI MS902 mass spectrometer [119]. Although two other commercial reversed geometry instruments (the Varian-MAT Models 311 and the CH5 D, and the Nuclide Corporation Model 12-90G(DF)) were already on the market at that time, it was in 1978 that the first commercial reversed geometry instrument designed specifically for MIKES work was described: this was the VG Micromass ZAB-2F mass spectrometer [122]. Although intended initially for MIKES studies, there has been widespread application of the ZAB instrument for structural and analytical studies utilizing CID, using the linked scanning techniques mentioned earlier [115], most of which apply both to reversed geometry and to normal, or 'forward geometry', double-focusing mass spectrometers.

1.3.2.6   Methods of Ionization

So far in our account of the historical development of analytical mass spectrometry we have concentrated on instrumentation and techniques associated with the study of organic molecules. However, in parallel with this work a strong and active research endeavor was continuing in what may collectively be called 'inorganic mass spectrometry'. Indeed, we should remember that the main stimulus for the early development of mass spectrometry was the need to characterize fully the abundances and accurate masses of all the isotopes of all the elements. Due to the wide variation in the different physical properties of the samples involved, it was necessary to explore, develop, and refine different methods by which these samples could be converted into gas-phase ions for subsequent mass analysis. General accounts of principal methods of ionization available by the end of the 1960s include those presented by Elliott [123] and by Roboz [99, Chapter 4].

1.3.2.6.1   Electron Impact Ionization

The original method by which Thomson and Aston created ions was by means of a low-pressure DC discharge [124]. Gaseous samples were admitted directly, while less volatile samples were suspended within a side arm of the glass envelope and vaporized through the heat of the "cathode ray stream"; alternatively, as Aston noted, Bainbridge and Mattauch used cavities drilled directly into the cathode in order to contain substances such as tellurium and compounds of the rare earths.

Ionization of the sample within a gas discharge system such as that described by Aston occurs *via* what is now called 'electron impact', although there was no attempt made to control the energies of the electrons causing the ionization. The issue of providing a much better controlled and reproducible EI source was addressed by Nier in two key papers. The first, published in 1940 [48], describes an EI source in which a fraction of the current of 90-eV electrons emitted by a tungsten ribbon 'hairpin' filament was collected on a 'trap plate'. The second (1947), and more detailed, account [125] shows a refined version of the earlier ion source, which now incorporated a magnetic field of 150 gauss, produced by a pair of 'Alnico' poles and a yoke, in order to send the electrons in a helical path so as to increase their path length in the source, The report contains also diagrams of the circuits employed. The electron emission current was regulated, and the electron accelerating potential could be varied between 7 and 75 V. "This feature is convenient in special problems if one is interested in studying appearance potentials or separating ions of the same mass but having different efficiency of ionization curves." To this day, EI sources are still based upon this original Nier design.

1.3.2.6.2   Spark Source Ionization

Although the first example of the mass spectrometric analysis of ions produced from solid samples by photon impact appears to be that published by Terenin and Popov [126], a major advance in the analysis of solids was the 'spark source', invented by Dempster [127,128]. The underlying principle is to strike a spark discharge between a pair of electrodes fabricated from the sample material. The abstract of the second of Dempster's two publications provides a concise summary of his work, in which three different types of device were investigated. "Three spectroscopic methods of exciting spectra have been examined as possible sources of positive rays. The vacuum vibrator, a mechanically interrupted low-potential spark, has been found to give positive ions although the source is not constant. The discharge of a large condenser charged to 60 kv between metal electrodes, has been found unsatisfactory as a source of positive ions. A high frequency alternating spark, coupled inductively to a primary oscillating spark circuit, is a strong convenient source. The ions may be accelerated to form a beam, and, with the Thomson parabola method of analysis, it has been found that multiply charged ions of the electrode elements are formed in all cases tried." (See Figure 1.11). Aston was clearly impressed by Dempster's work, and wrote ([124], p. 107) "When Dempster's new apparatus was built, four elements, palladium, iridium, platinum and gold, remained to be analysed. The peculiar chemical properties of these elements had defeated all attempts to obtain their mass spectra, though some indications of the complexity of iridium

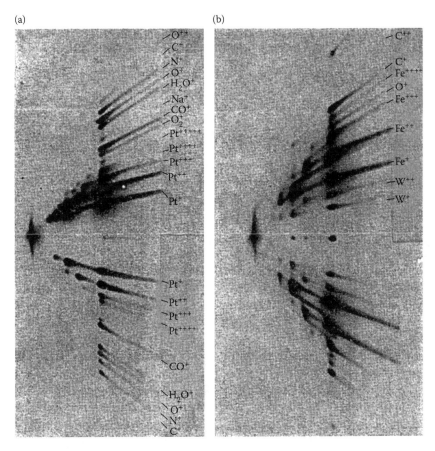

FIGURE 1.11 Ions from high frequency spark analyzed by electrostatic and magnetic deflections at right angles. (a) Platinum-platinum electrodes; (b) tungsten–steel electrodes. (Reproduced from Dempster, A.J., *Rev. Sci. Instrum.* 1936, *7*, 46–49. With permission from the American Institute of Physics.)

was [were] available from optical observations. With all four Dempster's spark method was brilliantly successful, ..." ([124], p. 107).

Although very rarely used now, if at all, the spark source proved to be a vital tool in advancing the application of mass spectrometry to metallic and other solid materials, and by the late 1950s commercially available instruments incorporating this method of ionization and employing Dempster's third method of maintaining a stable spark had appeared. A detailed description of the MS7 spark source mass spectrometer, which incorporated both photographic and electrical recording, together with the results obtained was presented by Craig *et al.* [129] at the first international mass spectrometry meeting held in London in 1958. In the following paper at the same conference, James and Williams [130] discussed the problem of analyzing non-conducting solids with the MS7 instrument, and described a method whereby electrodes fabricated from conducting and non-conducting materials were coupled together. Franzen and Schuy [131], in analyzing seven elements contained in an iron matrix, examined the relative merits of employing electrodes of different shapes. A general account of spark source mass spectrometry is to be found in Ahern [132], and a rather more recent review is that by Ramendik *et al.* [133]. Nowadays, the kinds of applications for which the spark source was invented have tended to be solved using inductively coupled plasma mass spectrometry (ICPMS), glow discharge mass spectrometry, or laser microprobe mass spectrometry (see for example, the accounts by Gray [134], Harrison [135], and Verbueken *et al.* [136], respectively).

1.3.2.6.3   Surface (Thermal) Ionization

Notwithstanding the later commercial success of the spark source method, in a review on trace element determination by the mass spectrometer in 1953, Inghram [137] published a somewhat critical assessment of the 'Dempster vacuum spark method', and also of the use of ion bombardment, reserving his praises for an isotope dilution technique in which known amounts of the sample and of an enriched isotope of the sample are dissolved and mass analysed using an unspecified ionization technique, which was presumably surface (or 'thermal') ionization in which the sample solution is deposited onto a filament, evaporated and the filament then heated. As we shall see, Mark Inghram was, in fact, a major force behind the development of two ionization techniques used for the analysis of solids and inorganic materials; he was also a member of the team that first established the age of the Earth. Judging by the concluding remarks of his review, Inghram displayed clearly a somewhat jaundiced, but nevertheless honest, view of mass spectrometry, and one that must have been shared by others over the years: "in these early stages of applying mass spectrometric methods to solids analysis the sensitivities and accuracies are directly proportional to the competence of the personnel assigned to them. The major problem with a mass spectrometer is that it always gives an answer. It is up to the laboratory worker to know what, if anything, that answer means."

The method of surface ionization, as mentioned in the previous paragraph, was in fact the means by which Dempster created ions in his first direction-focusing sector mass spectrometer (see Section 1.2.3. and Figure 1.7) [42]. It should be noted, however, that Dempster himself appears to have drawn his inspiration from the work of Richardson [138,139], who had made a detailed study of the emission of positive ions from hot metal surfaces. The major problem with the single filament design used originally by Dempster is that in many cases it is impossible to control simultaneously the respective temperature conditions for the evaporation and for the ionization processes. This is a particular problem in instances where the ionization energy of the sample exceeds that of the work function of the evaporating surface, needing a higher temperature to effect ionization, yet the substance is volatile and the evaporation rate is too high. In addition, the chemical nature of different samples at a single filament temperature may be different, making comparison between the mass spectral characteristics of different analytes difficult to interpret. The positioning of a single filament with respect to the source electrodes is shown in Figure 1.12a [140], and an improved 'boat' design offering improved performance is shown in Figure 1.12b.

As a solution to this problem, Inghram and Chupka [141] proposed a three-filament design which is still in use (See Figure 1.12c; the filaments are made normally from tungsten or rhenium. The idea is that the analyte is deposited on one of the side

(a)   (b)   (c)

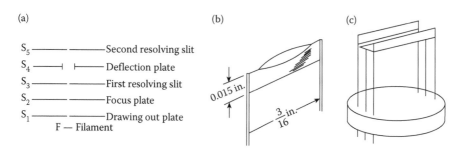

S_5 ———— ————Second resolving slit
S_4 ——— ⊢ ⊣ ——Deflection plate
S_3 ———— ————First resolving slit
S_2 ———— ————Focus plate
S_1 ———— ————Drawing out plate
          F — Filament

0.015 in.

$\dfrac{3}{16}$ in.

FIGURE 1.12   Diagram of (a) a typical surface ionization source assembly; (b) a 'boat'-type filament; (c) a normal triple filament arrangement. (Reproduced from Wilson, H.W. and Daly, N.R., *J. Sci. Instrum.* 1963, *40*, 273–285. With permission from the Institute of Physics.)

filaments, whose temperature is controlled to maintain the optimum conditions for evaporation, while the inner filament is run at a higher temperature in order to effect the maximum ionization efficiency of the sample that has been evaporated across; all the filaments are operated at the same potential. In an example quoted by Inghram and Chupka, gadolinium evaporates from a single filament of tungsten at 1250 K with an ionization efficiency of *ca* 10^{-9}, whereas with the center filament at 2500 K the ionization efficiency is 3×10^{-5}; at the same time, with the single, lower tempera- ture, filament most of the gadolinium appears as GdO$^+$, whereas from the multiple filament arrangement Gd$^+$ predominates. The second sample filament is partly to provide electrical symmetry about the normal ion path, but also offers the possibility for purifying the sample by successive evaporation from one filament to the other. A paper describing some early results on the commercial Metropolitan-Vickers MS5 thermal emission mass spectrometer incorporating the triple filament design was presented at the first international mass spectrometry meeting by Palmer [142], and an alternative design, comprising three parallel ribbons, was developed by Patterson and Wilson [143]; this new arrangement allowed a known standard and an unknown sample to be investigated at the same time.

1.3.2.6.4   *Field Ionization/Field Desorption*

Although based upon completely different physical principles, Field Ionization (FI) and Field Desorption (FD) both possess a characteristic similar to that of surface ion- ization, namely that the sample, depending upon the experiment, may be mounted on some form of filament; in this case, however, ionization occurs through the influence of a very intense electric field. The invention of FI is another ionization technique whose origin may be credited to Inghram and co-workers [144,145]. In this instance, it was field ion microscopy, developed by Muller and Bahadur [146,147] that acted as the inspiration. In field ion microscopy the sample surface, which is exposed to an 'imaging gas', such as helium or neon, is cooled to a low temperature and held at a high positive potential. Under the influence of the intense electric field, the adsorbed gas atoms become positively charged and are ejected in a direction essentially nor- mal to the surface toward a detection system. The spatial distribution of the collected ions represents a magnified image of the surface of the sample. Inghram and Gomer adapted this arrangement by allowing a fraction of these ejected ions to pass through

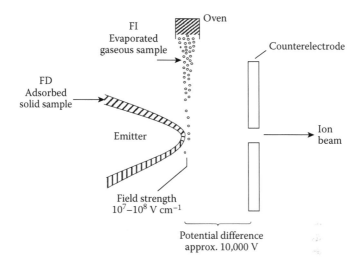

FIGURE 1.13 Schematic representation of the difference between field ionization (FI) and field desorption (FD). Note the different solid sample supplies in the two modes. (Reproduced from Schulten, H-R., *Int. J. Mass Spectrom. Ion Phys.* 1979, *32*, 97–283. With permission from Elsevier.)

a hole in the detector screen and into a mass spectrometer, fitted with an electron multiplier, in order to afford a means of their identification. Their initial results [144] yielded the intensity distributions of H^+ and of $H_2^{+\bullet}$ ions as a function of the electric field strength, from which it was deduced that FI occurred by electron tunneling. In their second paper [145], the authors examined a number of other diatomic gases, as well as some simple hydrocarbons, methanol and acetone (propanone). They noted that, in contrast to electron impact ionization, the molecular ion predominated in almost all cases and they commented that "The extreme simplicity of the spectra often permits the use of isotopic peaks for the direct determination of chemical formulas." A simple representation of the differences between the FD and FI techniques is shown in Figure 1.13.

From this small beginning, a whole new and lively area of mass spectrometry developed rapidly, especially under the leadership of Hans-Dieter Beckey [148] whose book has been a standard source of reference for nearly 40 years. From reading the literature on the subject, one senses rapidly that the preparation of efficient and reproducible emitters is almost as much an art as a science. A highly detailed review on both FI and FD was published by Schulten [149] (at the time a member of the Beckey group at Bonn), and a more comprehensive text is that by Prókai [150]. With the advent of newer methods of ionizing complex and non-volatile molecules, such as matrix-assisted laser desorption (MALDI) (Section 1.3.2.6.6.5) and ESI (see Section 1.3.2.6.7), the rather specialized and less user-friendly techniques of FI and FD have tended to fade away except for the analysis of certainly applications involving intractable material such as certain kinds of polymers (see, for example, the text by Ashcroft [151]).

1.3.2.6.5   Chemical Ionization

Since leaving the consideration of EI in the early part of this Section (see Section 1.3.2.6.1), our discussion has concentrated on the development of means for ionizing non-volatile, and often thermally unstable, materials; in some instances these have involved 'difficult' samples, such as metals and minerals. We have also examined a means of investigating inorganic salts in aqueous solutions. However, with regard to the mass spectral analysis of certain simple, 'conventional', organic materials, including hydrocarbons, there remained a problem with EI, namely that when using electrons with sufficient energy to produce a reasonable signal intensity, the extent of fragmentation resulted in the intensities of the undissociated molecular ions being too low to provide useful analytical information. This situation arises because the interaction of the sample molecule with energetic electrons leads to significant internal excitation of the former, in what has become known as a 'hard' ionization process. A solution to this problem resulted from an accidental discovery in 1965 by Munson and Field of what they termed 'chemical ionization' (CI) [152].

Working in the Esso Research and Engineering Company in Baytown, Texas (yet another petroleum company to have been the source of a major advance in mass spectrometry), Frank Field and Burnaby Munson were engaged upon an extensive research program into the study of ion/molecule reactions. Although, as we have seen in Section 1.2.2.3, the possibility that ions and molecules could react in the gas phase had been recognized by Thomson [20] the systematic study of gas-phase ion chemistry did not become a major area of activity until the 1950s and 1960s (see also Section 1.4). During this period an essentially 'second front', populated mainly by physical chemists and chemical physicists, opened up in mass spectrometry running in parallel with the studies in analytical mass spectrometry and the fragmentation mechanisms of an increasingly wide range of classes of organic compounds. It was the advent of chemical ionization that brought these two streams back together.

In the published version of a paper presented to the Fourth International Mass Spectrometry Conference, held in Berlin in 1967, Field [153] wrote: "The discovery of chemical ionization mass spectrometry was a development of a programme [program] of investigation of ion-molecule [ion/molecule] reactions which has been going in our laboratory for a number of years. In the course of attempting to determine the mass spectrum of methane at pressures of the order of 1 to 2 torr [Torr], we noticed that trace impurities had marked effects on the spectra obtained. This initially constituted an undesirable phenomenon, but the investigations undertaken to elucidate it led to the concept of chemical ionization."

In seeking an explanation for their results, Munson and Field drew upon the extensive literature on ion/molecule reactions that had already appeared at that time, but one publication in particular provided the key: this was the landmark report by Tal'roze and Lyubimova [154] in 1952 of the discovery of the new species CH_5^+. These authors showed that the ratio of intensities of the ion peaks at m/z 17 and m/z 16 followed a first-order dependence upon methane pressure which extrapolated to the natural abundance of $^{13}C/^{12}C$ at zero pressure, and that the ionization efficiency curve for the species at m/z 17 was, within experimental error, identical to that of methane

(m/z 16). These observations led the authors to conclude that the ion CH_5^+ was being formed by the reaction

$$CH_4^+ + CH_4 \rightarrow CH_5^+ + CH_3 \qquad (1.5),$$

a suggestion that was met initially with a great deal of skepticism by some organic chemists, who were reluctant to accept that a 'five-substituted' carbon species could exist [155]!

In their pressure-dependence studies of the ion/molecule reactions occurring in methane, Munson and Field showed [153] that at 1 Torr pressure, the ions in the mass spectrum occurred predominantly at 17 Th (48% total abundance) and 29 Th (41%), and attributed the formation of the latter species to the reaction [152]

$$CH_3^+ + CH_4 \rightarrow C_2H_5^+ + H_2 \qquad (1.6).$$

Subsequent reactions of these two product species with a trace substance, BH, present in the gas then occurred *via* proton- or hydride-transfer reactions, represented as

$$CH_5^+ + BH \rightarrow BH_2^+ + CH_4 \qquad (1.7),$$

$$C_2H_5^+ + BH \rightarrow BH_2^+ + C_2H_4 \qquad (1.8),$$

and

$$C_2H_5^+ + BH \rightarrow B^+ + C_2H_6 \qquad (1.9).$$

It was concluded that the relative facilities of these reactions will depend upon the relative basicities of the species involved. For example, if BH is a stronger base than is CH_4, then Reaction 1.7 will proceed readily, effecting the transfer of the proton from the reagent ion CH_5^+ to an analyte BH. Because a process such as Reaction (1.7) is not very exoergic, relatively little internal excitation energy will remain in the product ion BH_2^+, so that the extent of subsequent fragmentation will be much less than that which results normally from ionization *via* electron impact. For this reason, CI is known as a 'soft' ionization process. As a result of a comprehensive series of investigations, Field was able to draw a number of general conclusions about this new means of ionization [153].

   (i) The acid/base nature of the chemistry involved was recognized.
  (ii) In chemical ionization, the processes are not 'adiabatic', and are not governed by the Franck–Condon principle, as in EI or photoionization (PI).
 (iii) The CI product ions tend to be EE species, in contrast to EI, PI, and FI.
  (iv) In CI the amounts of energy involved tend to be quite low by mass spectrometric standards, although variation occurs depending upon the identity of the reactant material. The resulting mass spectra are markedly different

from those obtained by EI on the one hand and FI on the other hand. Often the mass spectra give evidence for different aspects of the structure of the molecule than do other modes of ionization.

For their investigations, the authors used a 'tight' ion source with high-capacity differential pumping of the source region; with a source pressure of 1 Torr, the pressure immediately outside the ionization chamber was about 2×10^{-3} Torr, and that in the analyzer about 2×10^{-5} Torr. It was noted further that the CI mass spectral peaks are quite conventional and acceptable, and that the operation did not cause a loss of performance or increase maintenance problems. Indeed, because the necessary technical modifications to the instrument were relatively modest and easy to implement, the new area of 'high pressure' chemical ionization mass spectrometry became popular rapidly, and instrument manufacturers soon incorporated the facility into their commercial offerings. Although the pioneering applications had involved volatile samples, higher molecular weight, less volatile compounds including some natural products such as alkaloids, sugars and steroids, organometallics, and thermally sensitive compounds could be analysed through the process of 'desorption' or 'direct' chemical ionization (DCI)[156], in which the sample is loaded into a suitable direct insertion probe and gently volatilized directly into the CI plasma [see 151, p. 82]. In a variant of the original means for effecting chemical ionization, Horning *et al.* [157] designed an ion source operating at atmospheric pressure using nitrogen as the reagent gas and a ^{63}Ni foil as the primary source of ionization. This has been termed 'atmospheric pressure chemical ionization', APCI, sometimes contracted to 'API'.

In his somewhat prophetic conclusion, Field [153] stated "Chemical ionization mass spectrometry is still very new, and the determination of the extent of its utility, both scientific and practical, will require much further work. However, on the basis of the results obtained so far we are quite optimistic about the prospects. Of interest and potential utility is the fact that different substances can be used as the reactant gas, and these different reactants will generate different mass spectra with a given additive substance. One can expect in this way to obtain more information about a substance that [than?] could be obtained from a single spectrum alone."

From the foregoing discussion, it is apparent that for chemical ionization to take place conditions must exist within the ion source for (i) a sufficient number of collisions to occur between the primary ions and the neutral reagent gas in order for there to be an appropriate concentration of reactant ions; and (ii) that the subsequent reaction by which the analyte molecules are ionized must proceed whilst the reactant ions so-formed are still present within the source. As we have seen, the discovery of the CI process arose because the experiments were being conducted at high source pressures. However, the underlying requirement for the processes to occur efficiently is that the residence times of ions involved are sufficiently long. Thus McIver *et al.* [158] showed that ions trapped at low pressures for a sufficiently long time (typically at 4.7×10^{-7} Torr for 3 s) in the cell of an ion cyclotron resonance (ICR) mass spectrometer could undergo chemical ionization reactions, and a similar observation was made by Todd and co-workers [159–162] in respect of the quadrupole ion trap (QIT operating typically at a reactant gas pressure of 10^{-5} Torr, with storage times of up to *ca* 3.5 ms).

We have already seen [153] that Field envisaged the potential utility of employing different reagent gases for chemical ionization. An excellent example of the diversity such experiments was provided by Hunt in a paper presented at the sixth International Mass Spectrometry Conference, held in Edinburgh in 1974 [163]. In this study, deuterium oxide, argon–water, ammonia, and nitric oxide were used as reagent gases. With D_2O in the CI source, because of the large number of collisions that occur between sample and neutral deuterium oxide, all the active hydrogen atoms in the organic molecule are exchanged. Thus one performs two CI analyses: one with water, methane, or isobutane to establish the molecular weight, and the other with deuterium oxide to count the number of active hydrogen atoms that undergo exchange.* In this way one can, for example, differentiate between primary, secondary, and tertiary amines. When a mixture of argon and water is used as the CI gas, the mass spectra exhibit all the features that are characteristic of both conventional EI and of methane CI. The H_3O^+ ion acts as a Brønsted acid and protonates most organics to give abundant $[M + H]^+$ ions, while reaction of the sample with the other reagent species occurs *via* charge-exchange, Penning ionization (through collisions with metastable argon atoms), or low-energy electron impact. It was noted that the mass spectra of perfluorokerosene (PFK) obtained with argon–water CI and with normal EI were identical, so that this compound could still be used as an internal mass standard in high resolution work.

By contrast with the foregoing results, the use of ammonia as the reagent gas caused almost no fragmentation, and was capable of selective ionization of highly basic compounds. EI of nitric oxide under CI conditions was found to generate NO^+, and this acted as an oxidizing agent, hydride and hydroxide extractor, and electrophile toward organic molecules. This reagent ion was found to be capable of differentiating between primary, secondary, and tertiary alcohols.

Many papers and reviews have been written about chemical ionization over the years, and an authoritative monograph is that by Harrison [164].

1.3.2.6.6   Bombardment Ionization

1.3.2.6.6.1   Introduction   From the preceding discussion we have seen that as attempts were made to expand the range of possible materials that could be examined by mass spectrometry, there was a growing need for implementation of ionization techniques that provided a stable source of ions from involatile materials that, in many cases, were thermally sensitive. We have already noted that FD offers a means of doing this; in this Section we highlight key aspects of the development of other ionization methods that rely on desorption processes, but in which the necessary energy is supplied through a 'bombardment' process.

The title 'bombardment ionization' embraces several different kinds of approach that have evolved and expanded over the past 50 years. In this discussion we shall explore the modest beginnings of these different techniques and highlight briefly the key advances, however in an account of this nature we can but 'scratch the surface' (no pun intended!) of all the refinements that have been made. Specifically, the methods that we shall mention include ion bombardment, that is, 'secondary ion

---

* See also Volume 5, Chapter 2: Gas-Phase Hydrogen/Deuterium Exchange in Quadrupole Ion Traps, by Joseph E. Chipuk and Jennifer S. Brodbelt.

mass spectrometry' (SIMS), californium-252 plasma desorption mass spectrometry (^{252}Cf PDMS), fast neutral atom bombardment (FAB), and photon bombardment, that is laser desorption and matrix-assisted laser desorption ionization (MALDI). Numerous reviews, contributed chapters and texts have appeared over the years, and while some of these might now appear dated the reader may wish to examine publications such as those by Morris [165], Lyon [166], and Benninghoven *et al.* [167] to gain a flavor of the kinds of applications and associated experimental issues that concerned investigators in the early to mid-1980s.

1.3.2.6.6.2 Secondary Ion Mass Spectrometry The name SIMS describes a technique in which a beam of high energy 'primary' ions, usually an atomic species such as Ar$^+$ accelerated to several kiloelectronvolts, impinges upon a solid material or surface film and causes 'secondary' ions that are characteristic of the target sample to be 'sputtered' there from; these secondary ions are subsequently mass-analysed. In addition to ions being desorbed, normally an abundance of secondary neutral species is ejected also from the surface, and these can be characterized by subsequent ionization (often by EI) followed by mass analysis. Secondary neutral mass spectrometry (SNMS) is an extremely valuable tool in the study of surface science, but because it has hitherto played no apparent rôle as a mass spectrometric technique in its own right, it will not be considered further. We shall see that within the SIMS technique, there are essentially two extreme operating conditions: 'static' SIMS, in which the primary ion current density is of the order of 10^{-9} A cm^{-2}, and 'dynamic' SIMS, where the primary ion current densities may be as high as several amperes per square centimetre. Static SIMS is used for gently sampling a monolayer of material in a virtually non-destructive manner, and dynamic SIMS causes erosion deep into the sample and is of value indepth-profiling.

An excellent history of SIMS is to be found in the 'Retrospective Lecture' given by Honig [168] who, apart from providing a chronological account of the early developments, provides a valuable table of what he calls "The Great Acronym Stew" that this subject has spawned. A variant of this lecture was presented also in 1985 [169]. As with many other aspects of modern mass spectrometry, the existence of 'secondary Canalstrahlen' was noted by J.J. Thomson [14], who, in 1910, wrote "I had occasion in the course of the work to investigate the secondary Canalstrahlen produced when the primary Canalstrahlen strike against a metal plate. I found that the secondary rays which were emitted in all directions were for the most part uncharged, but that a small fraction carried a positive charge." Although there were a few similar observations made by researchers in the 1920s and 1930s, the birth of SIMS might be considered to be the research described briefly by Herzog and Viehböck [170] in 1949. Interestingly, in this 'Letter to the Editor', which is completely devoid of references, the authors refer to 'canal-rays'. Key features in the design of their experimental system were (i) the electric field of the primary ion source was completely separated from the electric field in front of the target material, (ii) the sputtering took place in a high vacuum so that losses by collisions with the background gas species were avoided, and (iii) all the secondary ions were accelerated by the same voltage, consequently a magnetic field on its own was sufficient to produce a mass spectrogram. This design of sputtering ion source was subsequently refined by Liebl and Herzog [171].

Commencing at the RCA Laboratories in Princeton, NJ, in 1950, Richard Honig built up a major research activity in SIMS, working initially on the surface composition of oxide-coated cathodes. Writing in 1958 [172], he reported that, with inert gas ions in the 30–400 eV range impinging on silver, germanium, and germanium-silicon alloy surfaces, sputtered neutral particles characteristic of the sample were one hundred times more abundant than were the secondary ions, echoing the observation made by Thomson nearly 50 years before. With Honig's system, around 80% of the sputtered particles had energies less than 5 eV. The progress of Honig's work can be found in the numerous review articles that he has published [173–175].

The application of SIMS to the analysis of organic and biologically important compounds witnessed a major advance in 1969–1970 when Benninghoven [176] described the use of static SIMS (see above) in which very low current densities of 2.5 keV Ar^+ ion beams were allowed. Using samples prepared by dipping a silver target in solutions of the analyte (concentrations typically 10 µg µL^{-1}), Benninghoven and Sichtermann [177–178] were able to obtain strong mass spectral signals containing $[M + H]^+$ and $[M–H]^-$ ions up to *ca* 300 Th together with structurally significant fragment ions from some 40 biologically important compounds, such as amino acids, peptides, drugs, vitamins, etc. Their data indicated that they could achieve unambiguous identification of all the samples tested with detection limits in the 10^{-12} g range, without restrictions as to volatility and thermal stability of the analytes. A full account of the technology and applications of SIMS within the time period under consideration is to be found in Reference [167].

1.3.2.6.6.3   Bombardment with Fast Neutral Atom Beams   One of the potential difficulties with using ion beams as the bombarding primary species is that electrical charging of the target may occur, leading to instabilities in the beam of sputtered particles. A solution to this problem is to use beams of fast neutral atoms; these can be prepared by first forming the corresponding positive atomic ions from the source gas, accelerating them to high energy (several kiloelectronvolts), and then allowing the ion to undergo resonant charge exchange during collisions with other neutral atoms of the gas present within the 'ion gun'. Oliphant [179] appears to have pioneered the use of neutral beams to study the emission of electrons and other particles sputtered from a metal surface, but relatively little further work was done in this area until the 1960s. For example, Dillon *et al.* [180] explored the possibility of using neutral atom beams in their study of the emission of positive ions from polymers. However it was not until the work of Barber *et al.* [181] that the technique of 'fast atom bombardment' came to the attention of the mass spectrometry community. Using a beam of energetic argon atoms, typically in the range 2–8 keV, they were able to obtain mass spectra showing, for example, the $[M + H]^+$ ion (*m/z* 1319) from met-lys-bradykinin. For their experimental system they modified two commercial instruments: the AEI MS902 (forward geometry) and the VG ZAB-1F (reversed geometry) double-focusing mass spectrometers. With each of these machines they were able to perform various metastable ion scanning experiments. However, because these scans are relatively slow procedures, it was necessary to ensure stability of the ion beam over a prolonged period. Intriguingly, in Reference [181] there is only a brief, rather uninformative, reference as to how this was done: "Furthermore, if attention is paid to the solvent and support system in our sample handling, then we can obtain mass spectra which are stable for hours compared to the transience of FD spectra." In a second publication [182], submitted some five weeks after Reference [181], the authors were slightly more explicit in their discussion: "During our work we have noted, as others have reported [Reference to work of Benninghoven] that in many cases the spectra obtained are transient; rapid 'fading' ensues with a half-life for the pseudomolecular ion of the order of a few minutes. The origin of this fading is obscure, being possibly a mixture of surface damage by the primary atom beam, and surface contamination by the residual gases from the relatively poor vacuum obtainable in the ion source region in most conventional large mass spectrometers.

"We have, however, largely overcome this effect by judicious use of solvent and support systems, paying particular attention to the viscosity and volatility of the medium from which the sample is deposited on the stage. By this means we have preserved samples, with no depletion of 'parent' ion sensitivity over periods of hours." Note that there is no specific indication as to precisely which 'solvent and support systems' were being used. However, the well-read mass spectrometric sleuth might have been aware of the parallel publication in the *Biochemical Journal* [183] where it indicates that "The solution (1 μL) was deposited on the probe and diluted with glycerol (2 μL); the probe was then inserted *via* a vacuum lock into the modified source of a Vacuum Generator[s] ZAB 1F mass spectrometer, where it intercepted the fast-atom beam." It was not until the April 1982 issue of *Analytical Chemistry* [184] that a full account of the use of non-volatile matrices for supporting the FAB sample was given: it appears that the use of such a method was inspired by the adventitious observation that certain materials, such as pumping fluids, diffusion pumping oils, and siloxanes, frequently found as contaminants in organic samples, gave mass spectra that lasted for hours. Evidently the glycerol solvent matrix works by continually presenting a replenished surface layer of analyte molecule to the primary fast atom beam. Not all materials are soluble in glycerol, of course, and in such instances one may have to use mixed solvent matrices. One such example [185] is the addition of N,N-dimethylformamide (DMF) to the glycerol in the ratio *ca* 1:2 to assist in the dissolution of certain molybdenum and tungsten complexes.

As an alternative to fast atoms, in a comparative study Aberth *et al.* [186] showed that fast Cs^+ ions could be employed and that with a cesium ion beam accelerated to 6 keV the analyte ion sensitivity was at least a factor of three better than with the equivalent beam of fast xenon atoms, yet there was no electrical charging of the matrix. Nowadays commercial systems employ exclusively fast cesium ion beams, of energy up to 35 keV, and the title of 'FAB' has been replaced by 'LSIMS' – meaning 'liquid secondary ion mass spectrometry', reflecting both the use of a primary ion beam and a liquid target matrix.

Following these initial publications, there was an 'explosion' in the use of FAB as a means of extending mass spectrometry to the analysis of substances, both organic and inorganic, that could be studied previously only by FD, static SIMS, or by derivatization to produce a variant on the original analyte that was sufficiently volatile for analysis by, for example, chemical ionization. Furthermore, the technology required was relatively simple and inexpensive, meaning that existing instruments could be modified by means of 'retro-fits'. Subsequently, a method of continuous sample injection was developed [187] ('flowing FAB'), where the different samples could be supplied to the FAB probe *via* injection through a capillary tube without having to interrupt the analytical process. An example of the use of flowing FAB mass spectrometry to analyse the eluent from a directly coupled ion chromatography column for the analysis of inorganic ions, water soluble vitamins, and glucose derivatives was described by Al-Omair and Todd [188]. A collection of articles on the methodology and some practical applications of flowing FAB has appeared under the editorship of Caprioli [189].

In presenting this account of the emergence of fast atom bombardment as a major technological advance in widening the analytical scope of mass spectrometry, attention should be drawn to the work of Devienne and his group [190], who had been active in the same field since the mid-1960s, which they termed 'molecular beam solid analysis' (MBSA). This work appears to have concentrated mainly on inorganic materials, and did not utilize any form of solvent matrix for supporting the analyte.

1.3.2.6.6.4 Californium-252 Plasma Desorption Mass Spectrometry Another chance observation that led to the development of an alternative desorption ionization method for large, thermally sensitive, biomolecules occurred, surprisingly, in experiments concerned with nuclear chemistry [191]. Macfarlane and co-workers [192] were determining the masses of the fission fragments of short-lived isotopes deposited as monolayers on thin foil using a time-of-flight (TOF) technique. In addition to the anticipated results, they found that their mass spectra contained other peaks that they surmised were due to the desorption of molecular species from the surface caused by the heavily ionizing radiation associated with the recoiling fragments (for example ^{16}O from the decay of ^{20}Na). On checking this suggestion further by placing a ^{252}Cf spontaneous fission source (half-life 2.65 y), from which the heavy fission fragments are emitted with energies of several hundred megaelectronvolts, behind the foil they observed orders of magnitude increase in the molecular ion yield. Coating the thin film with arginine and cystine gave rise to both positive and negative ion mass spectra containing the respective molecular ions and various fragments [193]. A later publication [194] describes how ^{252}Cf-PDMS can be applied to the study of moderately large peptides, nucleotides, and natural products in what is essentially a non-destructive technique on account of the small amount of analyte used. The resolution of their TOF instrument was sufficient to provide elemental composition information up to 500 Th. Among numerous other publications, Macfarlane has written a valuable account [195] of the possible mechanisms that may operate in the various different types of particle-induced desorption process.

The use of 252-californium plasma desorption is clearly a highly specialized technique, and not many other groups have made use of this method of ion production. Compounds of biological interest, such as steroids and nucleotides, have been studied, for example, by Krueger and co-workers [197–198]. Demirev *et al.* [199] explored the effects of using different matrices on the internal energies of protonated insulin molecules formed by plasma desorption.

1.3.2.6.6.5 Laser Ionization There are many similarities between the history of the use of laser radiation for the ionization of large, involatile, thermally sensitive compounds and that of employing energetic beams of primary ions or neutral atoms:

in each case, the scope and versatilities of the methods only became evident and of real practical use as a result of an adventitious discovery of the need to support the analyte in an appropriate matrix, comprising a 'facilitating' substance. Another similarity is that, in each case, the interaction of the primary beam with the target material gives rise to abundances both of neutral species and of ions. In the case of laser radiation, the extent of ionization depends upon the photon energy and on the nature of the analyte, and some experiments employ laser 'desorption' to eject a plume of neutrals that is then 'post-ionized' either by a pulse of photons from a second laser, or by some other means such as EI.

As with SIMS, experiments in which laser irradiation of the sample is combined with mass analysis of the ejected material has become a standard method in surface studies, and systems for 'microprobe mass-analysis' are commercially available. Here the aim is to use microscopic observation to position the laser spot on a particular region of the target area and then to observe the ejected material, usually by means of TOF mass spectrometry. A description of such an instrument has been presented by Feigl et al. [200]. In an alternative type of investigation, the laser may be used to pyrolyze the sample and to analyse the resulting products by mass spectrometry.

Various reviews of the literature on laser ionization appeared in the 1980s, two of which are those by Conzemius and Capellen [201], which are especially comprehensive, and the more specialized survey by Verbueken et al. [202]. An early series of mass spectral studies on the analysis of alkali organic salts by mass spectrometry was carried out by Vastola and co-workers [203–205] using a small pulsed ruby laser, and Posthumus et al., [206] employing a TEA-CO_2 laser, examined a range of polar involatile bioorganic molecules including oligosaccharides, glycosides, nucleotides, amino acids, and oligopeptides. An interesting account of the applications up to ca 1980 of laser desorption from a mass spectrometry perspective is that contributed by Kistemaker et al. [207].

As noted earlier in this Section, it was an unintended event that completely transformed the effectiveness of laser desorption ionization; in this case an 'accident' led to the development of a technique so profound that its discoverer, Koichi Tanaka, was awarded a share of the Nobel Prize for Chemistry in 2002. This method is now known as 'matrix-assisted laser desorption ionization' (MALDI). Tanaka, an Electrical Engineer by profession, has presented an account of his work in his Nobel Lecture [208], in which he described how, at the time of the discovery, he was a member of a team of engineers and scientists at Shimazu Corporation, Kyoto, Japan working on the design and construction of a new type of TOF mass spectrometer. It was his remit to develop sample preparation and ionization technologies. Following work on desorption by 'rapid heating' that was current in the 1980s, and at the suggestion a colleague, Yoshikazu Yoshida, Tanaka mixed the sample material with ultrafine metal powder (UFMP) in the hope that powder would absorb the laser light and cause rapid heating of the sample. Whilst this use of a solid sample matrix containing UFMP did have a beneficial effect, for example in the analysis of polyethylene glycol (PEG), there was no evidence of the production of ions beyond ca 2000 Th. Within the same timeframe as this work, the FAB technique using a glycerol matrix had been discovered (see Section 1.3.2.6.6.3) and Tanaka tried using this matrix in his laser desorption ionization source without any enhancement of molecular-ion generation. In a bid to improve the performance of UFMP, and using what Tanaka recalls was vitamin B_{12} as a test sample, various solvents such as acetone (propanone) were employed to disperse the powder as a suspension, in the hope that energy transfer to the analyte would be increased. By mistake, on one occasion Tanaka used glycerol instead of acetone but, rather than waste the expensive UFMP, he placed the sample into the mass spectrometer and exposed the suspension to laser radiation in the vacuum in order to drive off the glycerol, monitoring the mass spectra at the same time. To his surprise, where unfragmented ions of vitamin B_{12} were expected to appear in the mass spectrum, peaks that seemed to be noise were observed. Further experimentation confirmed that these signals were indeed genuine mass spectral peaks, and optimization of the operating conditions resulted ultimately in mass spectra showing ion clusters formed from lysozyme (M = 14,306 Da) in excess of 100,000 Th [209]. Within the same time-frame as the work by Tanaka, Karas and co-workers [210] used the aforementioned microprobe instrument (the LAMMA 1000) to explore different organic compounds for use as matrices in what is now the MALDI technique, and one of these materials (nicotinic acid) was used subsequently by them in the analysis of large proteins [211,212].

1.3.2.6.7 'Spray' Ionization

We commenced this chapter with a brief appreciation of some of the early aspects of molecular beam research, and it was with a background of expertise in this field that John Fenn developed successfully the method of ESI, an achievement for which he also was a co-recipient of the 2002 Nobel Prize for Chemistry. Indeed, it was in an Honor Issue of the Journal of Physical Chemistry dedicated to Fenn for his work on molecular beams that he and Yamashita [213,214] published the first two papers describing this new electrospray technique.

An illustration of the electrospray ion source employed by Yamashita and Fenn is shown in Figure 1.14. The operational features were described as follows [213]. "Liquid sample was introduced through a stainless steel hypodermic needle, of internal diameter 0.1 mm, chamfered at the end to form a sharp-edged conical tip and maintained at 3–10 kV relative to ground. The distance of the needle to an end plate containing the nozzle was in the range from 20 to 30 mm. The surrounding

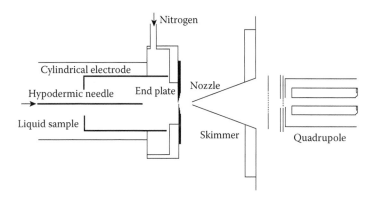

FIGURE 1.14 Schematic diagram of an electrospray ion source. See text for operational details. (Reproduced from Yamashita, M. and Fenn, J.B., *J. Phys. Chem.* 1984, *88*, 4451–4459. With permission from the American Chemical Society.)

region was a cylindrical electrode 30 mm in diameter and maintained at 500–600 V relative to ground. Dry nitrogen typically at 1050 Torr was passed through the region at a velocity of several cm s^{-1} in order to sweep the solvent vapor from the evaporating droplets." Essentially, as a fine jet of a solution of the analyte leaves the conical tip under the influence of a strong electric field gradient, evaporation occurs leading to the formation of charged droplets and, ultimately, to ions that are sampled *via* the nozzle and the skimmer into the mass spectrometer. Fenn's initial aim in designing this system was to undertake infrared spectroscopic studies in free jets of some complex and relatively non-volatile molecules in the vapur phase having masses sufficiently small for the ionic species to be analyzed with a quadrupole mass filter of mass/charge range up to 450 Th. The mass spectra that Yamashita and Fenn obtained exhibited the prolific formation of adduct and cluster ions from the range of organic compounds they studied (sprayed generally using a 1:1 methanol–water solvent), the relative amount of which depended upon the potential applied to the needle. These observations led them to realize that the system might have potential for use as an ion source, stating "We believe that the results we have obtained with EPSI [now 'ESI'] … open up an extremely promising approach to mass spectrometric analysis of samples containing organic molecules that are too large, complex, involatile, or fragile for ionization by conventional methods. We are eagerly pursing this approach."

These experiments were not, of course, performed in isolation from the extensive earlier investigations on what might be called 'spray ionization' techniques, or from the historical background relating to water jets and the electrical charging of water droplets (see below). Numerous reviews and summary accounts of the general area of spray ionization have appeared in the literature. The review by Vestal [215] describes the position immediately prior to the advent of ESI, whilst Fenn *et al.* [216,217] provided more general accounts of their work up to late 1990. A further review, by Hamdam and Curcuruto [218] and published in 1991, surveys the development of the ESI technique in relation to other research concerning droplets of various kinds, but the most recent and informative is that by Fenn [219] who, in his Nobel Lecture, describes at first hand both the stages by which his own career and scientific interests developed and the sequence of events that led to the invention of the ESI source.

As with many other aspects of mass spectrometry, some of the physical principles involved can be traced back over a hundred years. In this instance, it is the work of John William Strutt, the third Baron Rayleigh, that provides a defining feature. Lord Rayleigh (1842–1919) was Cavendish Professor of Physics at Cambridge and Head of the Department when J.J. Thomson began his work there; Lord Rayleigh received the Nobel Prize in Physics in 1904. Among Rayleigh's many interests was the electrical charging of water droplets [220]. Realising that if the size of the charged droplet is reduced then a limit is reached when the repulsive force between the electrical charges exceeds the 'cohesive tension' (that is, surface tension) so that instability results and "Under these circumstances the liquid is thrown out in fine jets, whose fineness, however, has a limit." Applying his model to a rain-drop of diameter 1 mm, Rayleigh stated that "The electromotive force of a Daniell cell is about .004 [no units given]; so that an electrification of about 5000 cells would cause the division of the drop in question". Subsequent experimental studies by Zeleny [221–223], conducted partly at Cambridge, UK, through the hospitality of J.J. Thomson, demonstrated the validity of Rayleigh's approach.

Some of Fenn's work prior to his devising the ESI source was stimulated by the work of Dole *et al.*, [224], who attempted to generate beams of 'macroions' formed by electrospraying dilute solutions of polystyrene molecules (solvent benzene:acetone 3:2) into a bath gas of nitrogen at atmospheric pressure. Dole's model for how the ESI process occurred was based on the Rayleigh argument that as the droplets became smaller and smaller with the continued evaporation of solvent the ultimate droplet would become a free gas-phase ion. The hypothesis is sometimes called the 'charge residue model'. Assuming that the macroions were singly charged and achieved their terminal velocity in the free jet, and using a retarding potential method, Dole and co-workers obtained apparent values for the masses of the polystyrene molecules in the range of *ca* 50 kDa. Fenn and colleagues attempted to replicate this work, but they felt that the retarding potential method was not a very satisfactory way of measuring *m/z*, hindered also by the fact that, at the accelerating voltages available to them, the large macroions did not achieve sufficiently high velocities to be detected by electron multipliers.

As noted above, Fenn and co-workers returned ultimately to employing an electrospray technique in conjunction with samples of much lower molecular weight. However in an intervening stage other approaches by various groups to using spray ionization were attempted. Evans and Hendricks [225] developed an electrohydrodynamic ionization (EHI) source, initially for liquid metals, having been stimulated by the use of this technology in research into space thruster propulsion systems. Their ionization system was enclosed in a vacuum chamber and consisted of a motor-driven hypodermic needle floating at 13 kV and directed toward a hole in an extractor electrode at −500 V; the ions formed were then admitted directly to a modified AEI MS7 (Mattauch-Herzog geometry) mass spectrograph (see Section 1.3.2.6.2.). In a later paper, Evans and co-workers [226] described the application of their EHI source to the analysis of organic compounds. Fenn [219] has pointed out a key difficulty with EHI in that because the sample solution is dispersed in a vacuum, unless the solvent has a very low vapor pressure, there is a 'freeze drying' effect; as a result within the evacuated environment the ions are generally solvated. Furthermore, the ions have a high kinetic energy, in a manner similar to those generated by FD (see Section 1.3.2.6.4), so that some form of double-focusing instrument is required. A review of this technique has been published by Cook [227].

Iribarne and Thomson [228] and Thomson and Iribarne [229], as part of their meteorological research, described an atmospheric pressure ion evaporation (APIE) technique in which nebulized water droplets were ionized at atmospheric pressure by means of an 'induction electrode' to which a potential of 3.5 kV was applied. One important outcome of this research has been in the development of an alternative to the charge residue model (see above) of ion formation from charged droplets. This is the 'charge evaporation model', in which it is argued that before ultimate droplets containing only one solute molecule are formed, the field at the surface of the droplet is sufficiently intense to eject a solute ion from the droplet surface into the ambient gas. Fenn has argued [213,219] that the 'desorption' model of ion evaporation provides the best explanation of the results that they have seen with their ESI source.

Before returning to complete our discussion of ESI, mention should be made also of the thermospray (TSP) technique developed by Vestal and co-workers [215,230]. This method was designed originally as a separation/ionization interface in order to transfer the eluent from a liquid chromatograph directly into a mass spectrometer. In the first embodiment of the system the idea was to effect rapid vaporization of the solvent under strictly controlled temperature collisions followed by ionization, by means of EI or CI, followed by orthogonal injection of the ions *via* a differentially pumped interface into the mass spectrometer. It was discovered subsequently [231] that efficient ionization occurred within the TSP source without the need for the direct ionization facility to be turned on.

The thermospray ionization method was developed into a highly successful commercial product, but its use has now been superseded largely by the electrospray technique; apart from other factors, one principal reason for this shift in popularity is the facility with which the ESI process leads to the formation of multiply charged ions. Since the mass spectrometer measures the mass/charge ratio of an ion, an increase in the charge means that more massive ions can be analyzed without having to increase physically the performance specifications of the instrument.

This multiple charging occurs through the addition of cations, such as H^+ or Na^+, to the neutral molecule. Thus it was found [232], for example, that the electrospray mass spectrum of gramicidin S using a methanol-water solvent mixture (1:1) had a much more intense peak at m/z 571, arising from the $[M + 2H]^{2+}$ ion, than that at m/z 1141 from the $[M + H]^+$ ion. The subject of multiple charging in the electrospray mass spectrum of poly(ethylene glycols), here involving the addition of up to 23 sodium ions, was the subject of a much more detailed investigation [233], and this led to a realization that useful information on the mass of the ionic species could be gained from the peak multiplicity [234]. Thus, in the case of ions formed by protonation, each peak within a group corresponding to the 'parent ion' peak represents the original molecule to which different numbers of protons (n_i) have been added. If it is assumed that the values of n_i differ by one between adjacent peaks, then from the measured values of m/z for each of the peaks it is a simple matter to calculate the values both of n_i and of M, the molecular weight of the original molecule. By repeating the calculation for successive pairs of peaks within the entire envelope it is possible to determine several values for M independently of one another, and then to calculate the mean value for the molecular weight of the analyte together with the corresponding standard deviation. Covey *et al.*, [235], using an 'ion-spray' source (which is a proprietary variant of the original electrospray design), have shown how this approach can be applied to the determination of the molecular weights of selected compounds of biological interest, achieving accuracies and precisions of $\pm$ 1 Da. Mann *et al.* [236], in a more extensive treatment, demonstrated that the sequences of peaks for multiply charged ions can be treated in two alternative ways: as a means of obtaining a series of independent measures of the molecular weight as described above and, *via* a "deconvolution algorithm", transforming the measured mass spectrum into the single peak that would be expected for the original molecule with a single (mass-less) charge. The latter approach can be useful in revealing if other sets of peaks corresponding to multiply-charged ions from different species are overlapping within the same mass spectral envelope. Whilst the 'adjacent peak' method for determining molecular weights works successfully with relatively low resolution instruments such as the quadrupole mass filter (QMF) and the QIT (see Section 1.5.4.4), under higher resolution it is possible to resolve the peak corresponding to a single charge state (n_i) into a multiplet of peaks corresponding to successively differing numbers of ^{13}C atoms in the original molecule. In this case, the multiplet peak spacing expressed in thomsons is equal to ($1/n_i$). From this 'measured' value of the charge state together the measured value of m/z of the unresolved multiplet the mass (M) of the analyte molecule can be determined and, as in the previously described method, several different independent values of M obtained by examining the peak-spacing between different component peaks of the multiplet. The resolution of modern mass filters and ion traps is generally sufficient to resolve multiplet peaks corresponding to $n_i = 2$ or 3, but for values of n_i much in excess of this range it is necessary to employ higher resolving power mass analyzers, for example double-focusing sector (see Sections 1.2.3.2 and 1.3.2.3), Orbitrap (Section 1.5.5.4) and FT-ICR (Section 1.5.2.4.4) instruments.

1.4 THE STUDY OF GAS-PHASE ION CHEMISTRY WITH SECTOR MASS SPECTROMETERS

We turn now to the question of the utility of sector mass spectrometers for investigation of thermochemical properties of ions and the kinetics of ion reactions, both unimolecular and bimolecular. While the development of multi-sector mass spectrometers continued apace and the analytical applications of these instruments continued to widen, were opportunities being provided for other types of investigation?

1.4.1 Ion/Molecule Reactions

The study of ion/molecule reactions, which had lain largely moribund since the early studies by Thomson, experienced a renaissance in the 1950s due, in part, to the recent findings of atmospheric explorations, interest in flames and explosions, and experimental opportunities brought about by combining high-pressure (*ca* 1 Torr) sources with conventional mass spectrometers. The flowing afterglow (FA) apparatus and, later, the selected ion flow tube (SIFT) apparatus were employed for such studies and the drift tube. Ion cyclotron resonance mass spectrometry (see Section 1.5.2.4.) was applied also to the study of ion/molecule reactions. Many exothermic ion/molecule reactions have been shown to proceed at rates in excess of the molecular collision rate due to the long-distance attraction between an ion and the ion-induced dipole in the neutral reactant. The high collision cross-section for ion/molecule reactions had been derived by Langevin, in 1905, using an ion/ion-induced dipole model [237]. Thermodynamically allowed ion/molecule reactions have high rate constants such that these reactions can dominate the chemistry in, for example, flames, planetary atmospheres, and explosions.

1.4.2 Space-Charge Ion Trapping in a Conventional Ion Source

In 1946, Heil showed that positive ions could be trapped in the negative space charge created by a magnetically confined electron beam within a conventional Nier-type ion source [238]. In the model for ion trapping proposed by Baker and Hasted [239], the space charge of an electron beam forms an electric potential well and positive ions may become entrapped within the beam. Ion trapping occurs when the trap depth, V_T,

$$V_T = Ka^2 \tag{1.10},$$

exceeds the kinetic energy of the nascent ion (*ca* 0.04 eV), where K is a constant that depends on the operating conditions of the source and a is the electron-beam radius. Cylindrical potential traps of depth >0.1 eV can be realized readily. In Figure 1.15 are shown [240] radial potential functions, which are similar to those calculated by Baker and Hasted [239]. The function labeled A is due to electrons only and defines a potential well, B is due to ions only and C is the sum of A and B. As all ions are formed within the electron beam, and thus within the region of the trapping potential, exchange among ions may occur such that a nascent ion may become trapped, while an 'aged' ion is liberated from the trapping region and leaves the source. Thus, a distribution of ion ages may be found within the ion beam leaving the source. Bourne and Danby have pointed out the usefulness of a modified age distribution for the study of unimolecular ion decomposition [241].

Harrison and co-workers made use of the ion-trapping properties of electron space charge in the source of a conventional magnetic deflection mass spectrometer for the study of ion/molecule reactions [242]. Absolute rate coefficients and relative rate coefficients were obtained for molecular ions of, for example, methane, ethylene, acetylene, methyl fluoride, and methyl chloride. The ions could be trapped for

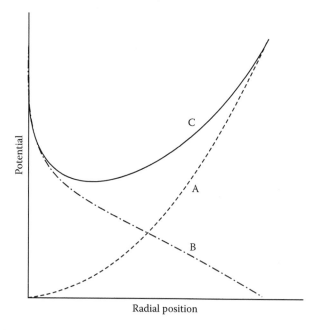

FIGURE 1.15 Typical radial potentials for an electric potential well formed by the space charge of an electron beam. A, Due to electrons only; B, due to positive ions only; C, the sum of A and B. (Reproduced from Morgan, T.G., March, R.E., Harris, F.M., and Beynon, J.H., *Int. J. Mass Spectrom. Ion Processes* 1984, *61*, 41–58. With permission from Elsevier.)

periods up to 2 ms, permitting the study of the reactions not only of primary ions but of products formed in the primary reactions. The results indicated that the reactant ions have kinetic energies in the range 0.3 to 0.5 eV.

1.4.3 METASTABLE IONS

In Section 1.3.2.2 we discussed the observation of broad, low intensity peaks at non-integral values of m/z (m^*) that could be correlated with the m/z-values of the precursor ion and of its fragmentation product ion. Such precursor ions were described as 'meta-stable' ions. The relatively large width of these ion peaks of low intensity is due to the release of kinetic energy upon fragmentation of the precursor ion. The fragmentation product ions are scattered isotropically, reminiscent of an exploding grenade, with the result that some product ions are forward scattered while some are backward scattered leading to a broadening of the observed product ion peak. The observation of metasta-ble ions facilitated mass spectral interpretation as was described in Section 1.3.2.2. Let us commence further examination of metastable ions by considering a typical BE sec-tor mass spectrometer (magnetic sector B and ESA sector E, ESA), such as is shown in Figure 1.16, and in which an ion encounters three non-field-free regions and three FFR between ion source and ion detector [121,243]. The three non-field-free regions are the accelerating region, the region within the magnetic, and the region within the electrostatic sector, while the FFRs lie between the end of the accelerating region and the entrance to the magnetic sector, the exit of the magnetic sector to the entrance to the end of the ESA, and the exit to the ESA to the exit slit that precedes the detector. An additional mass analyzer, for example, a quadrupole mass filter, is required in order to monitor ion decomposition in the third FFR. The original purpose of the 'reversed' or 'inverse Nier-Johnson' geometry for a tandem mass spectrometer was to use the double-focusing properties to obtain greatly enhanced mass resolution. However, they provide also a unique method for studying in detail the decay of metastable ions by using the magnet for mass (or momentum-to-charge) separation and then performing experiments on these mass-selected ions in the second FFR of the instrument.

The flight time of an ion is shown in Figure 1.17, which summarizes the pro-cesses that take place in a BE mass spectrometer. The lower part of Figure 1.17 shows

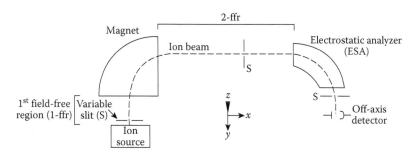

FIGURE 1.16 Schematic drawing of a mass spectrometer of reverse geometry wherein the magnetic sector (B) is followed by the electrostatic sector (E) to give a BE arrangement. (Reproduced from Holmes, J.L., Aubry, C., and Mayer, P.M., *Assigning Structures to Ions in Mass Spectrometry*, CRC Press, Boca Raton, FL, 2007. With permission from CRC Press.)

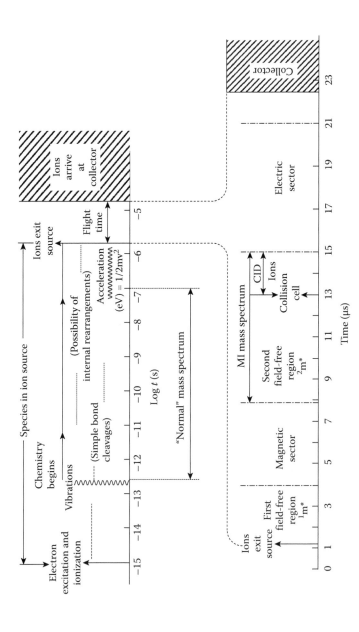

FIGURE 1.17 Time-scale for transit through a VG Micromass ZAB-2F reversed geometry mass spectrometer of an ion of *m/z* 100 accelerated by application of a potential of 8 kV. (Reproduced from Holmes, J.L., Aubry, C., and Mayer, P.M., *Assigning Structures to Ions in Mass Spectrometry*, CRC Press, Boca Raton, FL, 2007. With permission from CRC Press.)

the timescale for an ion of m/z 100 accelerated through 8 kV to travel through the VG Micromass ZAB-2F mass spectrometer (see also Section 1.3.2.5); after *ca* 1 μs the ion leaves the source, travels through the second FFR during the interval 8–15 μs, and reaches the detector after some 22–23 μs. The upper timescale shows the time-dependence of the processes of ionization, ion vibrational motion, simple bond cleavages, and ion rearrangements that occur in the source. It follows then, that for a unimolecular event to occur in the second FFR after, say, 10 μs, the decay constant of the excited species must be of the order of 10^{-5} s.

Immediately following the cessation of hostilities in WWII, Hipple and co-workers [68,69] discussed the dissociation of ions in a non-field-free region. While the first observation of metastable ions may be claimed by Thomson [20], it was Hipple and Condon who established clearly, in 1945, the existence of metastable ions in the mass spectrum of butane [68]. The following year, Hipple *et al.* [69] proposed that metastable hydrocarbon ions may lose hydrogen atoms during transit through the non-field-free acceleration region, giving rise to ions with unconventional momenta and kinetic energy. Decomposition of ions in a strong field such as an acceleration field permits measurement of short lifetimes that experience the strong field immediately after formation.

'Metastable' is the adjective describing those ions that dissociate unimolecularly during their flight through the mass spectrometer from the source to the detector. Ion dissociation occurs by virtue of the internal energy (that is, rotational, vibrational, and electronic) acquired by an ion during the ionization process. The subsequent rates of dissociation of the metastable ion are immutable in the absence of collisions, hence first-order dissociation can occur throughout the flight path that, normally, is fixed and is the total path length through a concatenation of non-field-free regions and FFRs. By definition, metastable ions cannot be detected directly because they cease to be metastable once they reach the detector. In one sense, metastable ions are trapped in the mass spectrometer and, although the product ions of unimolecular decay can be detected when they are formed in a FFR, the rates of decay cannot be influenced. The sole variable is the flight time along the path length; however, as the effective range of variation of the acceleration voltage in most sector instruments is of the order of 3–8 kV, the total flight can be varied only by a factor <2.

The observation of metastable ion decay in the second FFR of a *BE* mass spectrometer is made by transmitting a mass-selected ion species through the magnetic sector and monitoring the product ion while scanning the ESA; the result is a mass-selected ion kinetic energy spectrum, or MIKES spectrum (see also Section 1.3.2.5.). The ion signals (or peaks) arising from product ions of metastable decay have characteristic shapes, and an analysis of these can provide remarkable insight into many physico-chemical aspects of ion fragmentation chemistry. It is the plethora of information of ion fragmentation and the structures of ions that sustained prolonged research into metastable ions during the period 1960 to *ca* 1985, despite the relatively low ion signal intensities observed, *ca* 0.1% of the base peak intensity. Energy-resolved metastable ion peaks, that is, energy-resolved ion signals from the product ions of metastable decay, appear in two basic shapes: they can be either Gaussian or flat-topped or dish-topped, as shown in Figure 1.18, but they are always quasi-symmetric. Narrow Gaussian peaks are characterized by a low kinetic energy

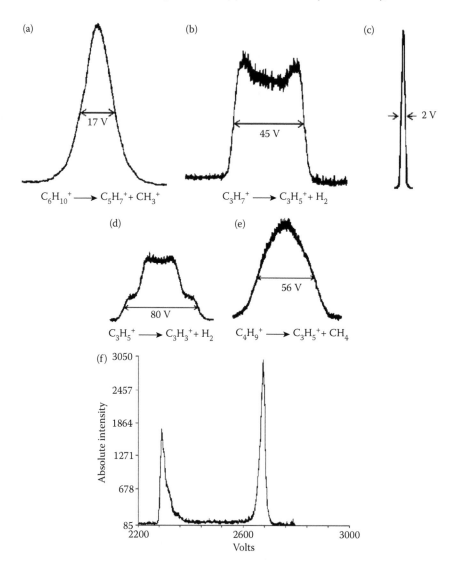

FIGURE 1.18 Various shapes of metastable ion peaks: (a) typical Gaussian peak; (b) dished metastable ion peak; (c) the signal for a beam of mass-selected non-fragmenting ions showing their much smaller translational energy spread; (d) composite peak showing two dished peaks for the reaction shown; (e) composite peak showing a mixed Gaussian and dished peak for the reaction shown; and (f) partial MIKE ($< E$) spectrum (that is, one observed by reducing the normal operating value of the electrostatic sector voltage, E) of triply charged benzo[a]pyrene, $C_{12}H_{20}^{3+}$, showing charge separation process, where $C_2H_2^+$ is the ion detected in this spectrum (note that the peak-width exceeds 400 V). (Figures 1.18(a)–(e) Reproduced from Holmes, J.L., Aubry, C., and Mayer, P.M., *Assigning Structures to Ions in Mass Spectrometry*, CRC Press, Boca Raton, FL, 2007. With permission from CRC Press; Figure 1.18(f) Reproduced from March, R.E. and Hughes, R.J., *Int. J. Mass Spectrom. Ion Processes* 1986, *68*, 167–182. With permission from Elsevier.)

release and near-zero reverse activation energy of the order of millielectronvolts. As the observed Gaussian peaks broaden and become flat-topped, the kinetic energy release increases along with the reverse activation energy. The observed peak broadening is due to the conversion of a fraction of the excess energy of the fragmenting metastable ion into the translational degrees of freedom of the charged and neutral products. Since the release of translational energy is isotropic, the symmetry of kinetic energy release peaks arises from the almost equal efficiencies for detecting product ions that have been either forward scattered or backward scattered in the unimolecular fragmentation process.

A dish-topped peak is usually wide and indicates the release of a large amount of translational kinetic energy. Product ions that are principally either forward or backward scattered have little radial kinetic energy, pass through the beam-defining slit, and are detected with relatively high signal intensity. However, product ions that have relatively high radial kinetic energy from the fragmentation process can move away from the ion beam axis to such an extent that they can no longer pass through the beam-defining slit; they will not reach the detector and so the detected ion signals show dishing in the central portion of the peak due to the ions lost. For metastable decay of doubly- or triply-charged ions, dishing is so severe that the metastable ion peak shows only the leading and trailing edges of the peak [244].

The Coulombic energy of repulsion, ϕ_r in joules, between two elementary positive charges is given by

$$\phi(r) = -\frac{e^2}{4\pi\varepsilon_0 r} \tag{1.11}$$

where e is the electronic charge, 1.602×10^{-19} C, ε_0 is the permittivity of free space, 8.854×10^{-12} J^{-1}C^{2}m^{-1} and r is the intercharge separation in meters. Thus, r may be calculated directly from the relation

$$r(\mathring{A}) = \frac{14.40}{T} \tag{1.12}$$

when T, the kinetic energy release, is expressed in electronvolts. For dissociation of a triply charged metastable ion, the numerical value in Equation 1.12 is 28.79, assuming that the intercharge distances are equal such that the charges form an equilateral triangle. For a planar triply-charged ion such as benzo[a]pyrene, $C_{12}H_{20}^{3+}$, which has an elliptical shape, there are four equilateral triangular arrangements of three charges on the boundary of the ion.

In the interpretation of intercharge separations from ion kinetic energy release data, an assumption is made that for the intercharge distances to be meaningful the positive charges are localized at the instant of ion fragmentation. The steepness of the sides of the peaks due to charge separation indicates that the distribution of ion kinetic energies is narrow, which suggests that the charges are localized. Furthermore, delocalized charges constrained within a given carbon skeletal arrangement should give rise to enhanced kinetic energy releases.

1.4.4 FIELD IONIZATION KINETICS

Metastable ion dissociation has been exploited in two methods for measuring decompositions in the nanosecond time-frame (1 ns to 1 μs) using electron impact ionization, and in the FI (see also Section 1.3.2.6.4) kinetics method for the picosecond time range. Tal'roze and Karachevtsev [245] initiated ionization by electron impact at, or near, a filament mounted axially in a cylinder. A potential was applied to the filament so as to produce a field of ca 10^4 V cm^{-1}. Decomposition of molecular ions during flight from the filament to the exit slit at the cylinder wall produced fragment ions deficient in kinetic energy. The potential at the point of decomposition, as defined by the kinetic energy deficiency, permitted calculation of the lifetime prior to decomposition. Ottinger [246] directed a collimated molecular beam along an equipotential surface within a similar electric field of ca 10^4 V cm^{-1}. Electrons generated externally to the field were directed into the molecular beam, brought about ionization of molecules in the molecular beam, and were then repelled by the field. Again, decomposition of molecular ions during acceleration toward the exit slit leading to a mass filter produced fragment ions with an energy deficiency dependent upon the age of the molecular ion at the point of decomposition. The brevity of lifetimes achievable by this type of approach depends upon the electric field strength and the magnitude of the ionization region. When the electric field is increased to 10^8 V cm^{-1} and the ionization region reduced to 1 nm, lifetimes approaching the period of a molecular vibration should be observed. The considerable activity in FI kinetics has been reviewed by Derrick [247] and Nibbering [248].

1.4.5 ACCELERATION REGION KINETICS

Koyanagi et $al.$ [249] presented a method for the investigation of the kinetics of metastable ion decomposition in a non-field-free region, which they described as acceleration region kinetics (ARK). The location of a specific unimolecular fragmentation event may be identified when this fragmentation occurs in a non-field-free region such as the accelerating field, and may be changed by variation of source residence time or acceleration voltage. They demonstrated the utility of a conventional Nier-type ion source, attached to a double-focusing mass spectrometer, for the investigation of events occurring in the $10^{-9} - 10^{-7}$ s in the acceleration region, without having to resort to any other type of specialized equipment or procedure.

Fragment ions formed throughout a non-field-free region exhibit a limited continuum of apparent mass/charge ratios as ion momenta are determined by the locations at which fragment ions are formed. The relative signal intensities of isotopic fragment ions are thus controlled not only by the usual isotopic abundances but also by the lifetimes of excited precursors.

Consider the reaction

$$m_1^{+*} \rightarrow m_2^+ + m_3 \qquad (1.13)$$

where dissociation of the excited, or metastable (*), primary ion occurs as the ion is being accelerated following emergence from the source. Fragmentation occurs when m_1^{+*} has fallen through a fraction b of the accelerating potential V_0. After fragmentation, the nascent product ion m_2^+ has a kinetic energy of m_2beV_0/m_1. Zero kinetic energy release is assumed; the case for fragmentation with release of kinetic energy is given elsewhere [250]. The product ion m_2^+ falls through the remaining fraction, $(1-b)$, of the accelerating potential. The apparent mass/charge ratio, m_b^*, at which m_2^+ is transmitted through the magnetic sector is given by

$$\frac{m_b^+}{e} = \frac{m_2^2}{em_1}\left[\left(m_2b + m_1(1-b)\right)/m_2\right] \tag{1.14},$$

as discussed by Cooks *et al.* [121].

The ARK method has been applied to the singly-charged states of N_2O and *n*-butylbenzene and to the doubly-charged state of N_2O. From the unimolecular decomposition of N_2O^+, half-lives of 190 ns and 90 ns were determined for the decay of excited states of N_2O^+ to produce NO^+ and N_2^+, respectively, as shown in Figure 1.19 [250]. Metastable decomposition of the molecular ion of *n*-butylbenzene exhibits competitive dissociation leading to the formation of $C_7H_7^+$ (*m/z* 91) and $C_7H_8^+$ (*m/z* 92). With the signal intensity of the former as a percentage of that of the latter, the relative abundances of the peaks at *m/z* 91 and 92 were observed as 183% directly from the source, 5.4% in the first FFR, and 1.5% in the 2nd FFR. When these competing fragmentation processes were observed in the accelerating region, the intensity ratios were found to be 48.2% at $b = 0.333$ and 30.1% at $b = 0.624$. Thus, with a standard ion source, it is possible to apply ARK to observe competing fragmentations at short, though uncertain, reaction times.

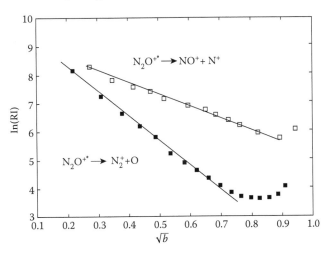

FIGURE 1.19 Natural logarithm of ion fragment signal intensity *versus* the square root of b, the fraction of the accelerating voltage experienced by N_2O^{+*} prior to dissociation. The upper trace corresponds to formation of fragment ion NO^+; the lower trace corresponds to formation of fragment ion N_2^+. (Reproduced from Koyanagi, G.K., McMahon, A.W., and March, R.E., *Rapid Commun. Mass Spectrom.* 1987, *1*, 132–135. With permission from J. Wiley & Sons.)

1.4.6   Collision-Induced Dissociation

The adventitious admission of a gas (or air) into a FFR of a double sector mass spectrometer can result in the dissociation of an ion in flight to form a product ion and a neutral moiety (see also Section 1.3.2.5). Once the possibility for changing the chemical nature of an ion in flight in this manner was recognized, gas collision cells were introduced into tandem mass spectrometers, particularly into the second FFR. Gas collision cells were some 20–50 mm in length with piping for the admission of gas and, usually, they were located close to a pumping outlet so as to maintain the required low pressure throughout the instrument. Collision-induced dissociation proved to be a valuable technique for assigning ion structures in the gas phase [243,251]. Thus was born the technique of tandem mass spectrometry, abbreviated as MS/MS; this general method involves at least two stages of mass analysis, either in conjunction with a dissociation process within a collision cell or a chemical process in which an ion undergoes a change of mass and/or charge [102,103,121] (see also Section 1.3.2.5). Thomson can claim not only the first observation of metastable ions, but the first observations of CID and charge permutations in ion/molecule reactions also [20].

1.4.6.1   Neutralization-Reionization Mass Spectrometry

Following the maxim that 'if one toy is good, two toys will be better', the 1980s saw the introduction of two gas collision cells in tandem. For example, shown in Figure 1.20 is a schematic diagram of a double-focusing (2F) hybrid BEQQ instrument where the BE combination is followed by a quadrupole gas collision cell and a quadrupole mass filter; the instrument is known as a ZAB-2FQ, where ZAB (see Sections 1.2.3 and 1.3.2.6.6.3) stands for zero alpha and beta focusing errors. The instrument is equipped with two collision cells and an intermediate deflector electrode in the 2nd FFR [252,253]. Gas

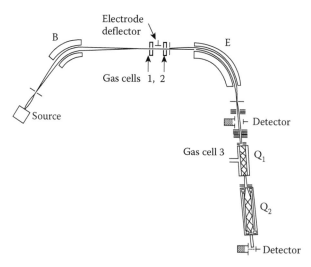

FIGURE 1.20   Schematic diagram of a BEQQ mass spectrometer, ZAB-2FQ, of reverse geometry. (Reproduced from Harrison, A.G., Mercer, R.S., Reiner, E.J., Young, A.B., Boyd, R.K., March, R.E., and Porter, C.J., *Int. J. Mass Spectrom. Ion Processes* 1986, *74*, 13–31. With permission from Elsevier.)

cell 2, the bias voltage on which is continuously variable up to ±5 kV with respect to ground, is the standard cell, 20 mm long by 1 mm wide as used for high-energy collisional studies and is located at the first focal point of the instrument. Gas cell 1 is 40 mm long and 2.5 mm wide and is not isolated electrically from ground; a distance of 56 mm separates the collision cells. Gas cell 1 and the ion beam deflector electrode, which can be floated at $ca \pm 3$ kV, are used in neutralization-reionization experiments; such experiments comprise the technique known as neutralization-reionization mass spectrometry (NRMS) Ion neutralization, for example of H^+ as in Equation 1.15 (see below), can occur in gas cell 1. All ions exiting gas cell 1 can be deflected electrically out of the beam such that only neutral moieties of high velocity enter gas cell 2. Reionization, for example of $H\cdot$ to H^- as in Equation 1.16, can occur in gas cell 2.

1.4.6.2 Translational Energy-Loss Spectroscopy

Professor Graham Cooks of Purdue University, and an author in this series (see the following chapter), has likened mass spectrometrists to the Pre-Raphaelites who had a predilection for painting on each other's canvases. In this manner, we have itinerant mass spectrometrists who visit laboratories, use the apparatus or instruments found there, and contribute to the research of the host laboratory. In some cases, the purpose of the visitors is to acquaint themselves with the instruments or to make an entrée to the field of mass spectrometry. Such was the situation of one of us (REM).

In 1972–1973, the focus of the laboratory of Professor Jean Durup, of the Université de Paris-Sud at Orsay, was translational energy-loss spectrometry (TELS) [256–259]. A schematic diagram of an apparatus for TELS is shown in Figure 1.21. The two analyzers in this apparatus are shown schematically so as to emphasize the molecular beam origin of the assembled instrument. Briefly, a beam of protons of known translational energy is extracted linearly from a Wien filter, directed through a collision cell containing a target gas at low pressure and, upon the exiting the collision cell, the kinetic energy/charge ratio is determined with a 127° ESA. A key feature of the instrument was the extensive angular collimation that was fitted. The objective of the early experiments by Durup and co-workers was the investigation of the ionization energies (IE) of neutral molecules of N_2, NO, and O_2 to form doubly-charged species; hence the early studies were described as 'double-charge transfer' experiments. The translational energy is measured of H^- ions that have gained two electrons in the processes

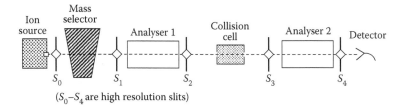

FIGURE 1.21 Schematic diagram of the generic design of a translational-energy loss spectrometer. S_0–S_4 are high-resolution resolving slits. Analyzer 1 can be a Wien filter and Analyzer 2 can be a 127° electrostatic analyzer. (Reproduced from Brenton, A.G., *Int. J. Mass Spectrom.* 2000, *200*, 403–422. With permission from Elsevier.)

$$H^+ + N_2 \rightarrow H^. + N_2^{+.} \tag{1.15},$$

$$H^. + N_2 \rightarrow H^- + N_2^{+} \tag{1.16},$$

and in the double-charge transfer reaction shown in Equation 1.17

$$H^+ + N_2 \rightarrow H^- + N_2^{2+} \tag{1.17}.$$

Clearly, the translational energy loss in Equations 1.15 and 1.16, where two nitrogen molecules were ionized to the singly-charged state, is less than that in Equations 1.17 where two electrons were removed from a single nitrogen molecule. The translational energy of the H^- ion (E_1) after two collisions as described by Equations 1.15 and 1.16, is given by

$$E_1 = E_1^0 - (2E_2 + 2Q_1 - EA(H^+) - (EA(H^.))) \tag{1.18},$$

where $E_1{}^0$ is the initial translational energy of the proton, E_2 is the energy converted into translational energy of each nitrogen molecule, Q_1 is the amount of kinetic energy converted to internal energy in each nitrogen molecule (that is, the ionization energy of nitrogen into the ground or excited singly charged state), EA (H^+) is the electron affinity of the proton, and EA (H^-) is the electron affinity of the hydrogen atom. The translational energy of the H^- ion (E_1) after a single collision as described by Equation 1.17 is given by

$$E_1 = E_1^0 - (E_3 + Q_2) + (EA(H^+) + (EA(H^.))) \tag{1.19},$$

where E_3 is the energy converted into translational energy of a single nitrogen molecule, and Q_2 is the amount of kinetic energy converted to internal energy in a single nitrogen molecule (that is, the ionization energy of nitrogen into the ground or excited doubly-charged state).

The translational-energy loss spectra of the product H^- ions formed revealed the vertical excitation energies of the doubly-charged target. This method has been shown to be robust for obtaining data on doubly-charged ions, and it compares favorably with data from Auger spectroscopy. The method is accurate to better than 0.5 eV. Table 1.1 shows a typical set of data obtained by double-charge transfer spectroscopy [258]. TELS provides information on state-to-state cross-sections; applicability of quantum selection rules in collisions; metastable fractions in an ion beam; and thermochemical data such as ionization, excitation, and electron attachment energies [260].

1.4.7 ION THERMODYNAMIC PROPERTIES

An essential part of a vigorous examination of gas-phase ion chemistry was the experimental determination of thermodynamic properties of neutrals and ions; for

TABLE 1.1

Comparison of Second Ionization Energies (eV) of N_2, NO, and O_2 Obtained by Theory, Translational Energy-Loss Spectroscopy (TELS), Electron Impact Ionization (EI), and Auger Electron Spectroscopy.

Target (M)	States of M^{2+}	Theory	TELS[a]	EI	Auger Electron Spectroscopy
O_2	$\tilde{X}\,^3\Sigma_g^-$				
	$\tilde{X}\,^1\Sigma_g^+$	35.5		36.3 ± 0.5	37.4
	$\tilde{B}\,^3\Pi_g$	42.15	43.0 ± 0.5		
	$\tilde{C}\,^3\Pi_u$	48.4	48.0 ± 1		
N_2	$\tilde{X}\,^1\Sigma_g^+$				
	$\tilde{X}\,^3\Pi_u$	42.6		42.7 ± 0.1	43.0
	$\tilde{a}\,^1\Sigma_g^+$	42.7	43.1 ± 0.5		
	$^3\Sigma_g^-$	43.8		43.8 ± 0.1	
	$\tilde{b}\,^1\Pi_u$	44.0	45.2 ± 0.5		
NO	$\tilde{X}\,^2\Pi$				
	$\tilde{X}\,^2\Sigma^+$	38.1	39.3 ± 0.5	39.8 ± 0.3	40.1
	$\tilde{B}\,^2\Sigma^+$	42.9	42.4 ± 1.0		
	$\tilde{D}\,^2\Pi$		47.2 ± 0.5		

[a] TELS data taken from Appell, J.; Durup, J.; Fehsenfeld, F.C.; Fournier, P. *J. Phys. B. At. Mol. Phys.* 1973, *6*, 197–205.

example, the determination of ionization energy, IE, proton affinities and hydride affinities, and enthalpies of clustering. The enthalpy of a chemical compound is defined as the enthalpy difference between the compound and its constituent elements; the enthalpy of formation of a gaseous cation must include the ionization energy, that is the energy required to remove an electron. There are three basic methods for determining gaseous ion energetics. First, the application of monoenergetic electron impact [261] or resolved PI [262] to determine the IE-value of a fragment ion from a stable molecule. Second, determination of the equilibrium constant, K_{eq}, for a proton-transfer reaction (PTR)

$$AH^+ + B = A + BH^+ \,(\Delta H) \qquad (1.20),$$

where K_{eq} is given by

$$K_{eq} = \frac{[BH^+]\,[A]}{[AH^+]\,[B]} \qquad (1.21).$$

This type of experiment can be carried out in a high-pressure mass spectrometer, flowing afterglow apparatus, or ICR mass spectrometer where the neutral reactants' ratio, [A]/[B], is held constant. The Gibbs free energy, ΔG^0, for the reaction can be determined as

$$\Delta G^0 = -RT \ln K_{eq} \qquad (1.22),$$

and, when the equilibrium constant is examined over a range of temperature, both ΔS^0 and ΔH^0 can be determined as

$$\Delta G^0 = \Delta H^0 - T\Delta S \qquad (1.23).$$

The third method utilizes direct computation of the enthalpy change for protonation [263–264]

$$A + H^+ = AH^+ \; (\Delta H) \qquad (1.24),$$

where the proton affinity of reactant A is defined as the negative of ΔH for Reaction (1.24). In this manner, once the proton affinities are known for reactants A and B, a ladder, or scale, of proton affinities spanning a range of over 100 kcal mol^{-1} (418 kJ mol^{-1}) can be obtained [265].

Once ΔH is known for Reaction (1.24), this value can be regarded as the first stage enthalpy change for the series of clustering reactions of the type

$$AH^+ + nA = A_{n+1}H^+ \qquad (1.25),$$

where A may represent H_2O for the water cluster series, or CH_3OH for the methanol cluster series, and so on. In similar manner, for example, electron affinities, chloride affinities, and fluoride affinities have been measured and re-measured in order to obtain self-consistency within each series.

1.4.8 SUMMARY

We have traced briefly the development of beam instruments for neutral particles and the development of analogous ion beam instruments, that is, mass spectrometers. We have shown that, during the period *ca* 1960–1985 wherein the rate of development of sector mass spectrometers slowed moderately, there was a period of intense activity in physical chemistry research; particularly in gas-phase ion chemistry, using the available mass spectrometers. Such activity has not ceased but the advent of quadrupole instruments (see Section 1.5.4), which have brought about a revolution in mass spectrometry, has offered new instruments for both mass analysis and the pursuit of gas-phase ion chemistry. Accounts of research activity into gas-phase ion chemistry concerning the three-dimensional quadrupole ion trap (3D-QIT) prior to and subsequent to the introduction of the commercial ion trap in 1983 have been given elsewhere [266–269]. During the 1990s, began the modern era of trapped ion mass spectrometry with the advent of new ion trapping methods and instruments, and it is with the fruits of ion trapping researches during this modern era that Volumes 4 and 5 of this series are concerned.

1.5   THE DEVELOPMENT OF DYNAMIC MASS SPECTROMETERS

1.5.1   Introduction and Definitions

With the exception of the occasional reference *en passant* to other types of mass spectrometer, the entire discussion on analyzers within this chapter has, so far, been focused upon the development and applications of magnetic sector instruments, used either alone in a single-focusing configuration, or in conjunction with electric fields as in the early parabola mass spectrometers and in the later double-focusing designs. However, during the period covered by this historical account of the evolution of mass spectrometry, many active lines of research into other types of mass spectrometer were being pursued, with the result that while some of these designs have now become extinct, rather like the mass spectrometric equivalent of the *Dodo**, others are the very lifeblood of the subject today. Indeed, as we shall see, all the mass-analysis devices based upon ion trapping that form the subject of Volumes 4 and 5 of *Practical Aspects of Trapped Ion Mass Spectrometry* fall into this latter category. In this Section we present an account of how some of these various types of instrument, known collectively as 'dynamic' mass spectrometers, have emerged to occupy their current dominant position. In this year of the sesquicentennial celebration of Darwin's birth, the instrumental equivalent of the aphorism "survival of the fittest" must surely be "survival of the fittest for purpose"? In the account that follows we have drawn extensively on the slim volume entitled *Dynamic Mass Spectrometers* by Erich Blauth [270], a book that contains a wealth of information on instrumentation, both current and forgotten.

1.5.1.1   'Dynamic' and 'Static' Mass Spectrometers

Blauth [270] has defined 'dynamic' mass spectrometers as "those systems in which the time-dependence of one or more parameters of the system, *e.g.*, electrical field strength, magnetic field strength, or ion movement, is fundamental to the mass analysis [process]". This definition contrasts with that of the 'static', classical continuous beam, sector mass spectrometers, where "the parameters of the apparatus remain constant in time, except for the slow change required for the purpose of recording a mass spectrum".

In his survey of dynamic mass spectrometers, Blauth introduced a systematic classification system based upon the various physical principles involved. Surprisingly, he listed some 47 different types of instrument. Several of these were simply individual one-off creations, others were developed to a sufficiently advanced stage that they became, albeit briefly in some instances, commercial products, while others have survived and, after many further refinements, are used commonly today. Interestingly, it is probably the modern quadrupole mass filter that matches most closely the form in which it was originally conceived. Having said this, it should be noted also that the Orbitrap™ analyzer, invented by Makarov[†] has its origins in the Kingdon trap [271], which does not appear to have been covered explicitly by Blauth, who, perhaps, did not regard it as being a mass spectrometer?

---

* *Raphus cucullatus*: an extinct species of bird that inhabited Mauritius, reportedly last observed in the second half of the 17th century (*The IUCN List of Threatened Species 2009.1*. International Union for Conservation of Nature and Natural Resources. (http://www.iucnredlist.org/details/143436/0).

[†] Volume 4, Chapter 3: Theory and Practice of the Orbitrap™ Mass Analyzer, by Alexander Makarov.

In the light of the discussion in the earlier Sections of this chapter, in which we have sung repeatedly the praises of sector mass analyzers, a reader new to this field might wonder why there was, in fact, any need to invent alternative means for effecting the mass analysis of ions. However, there are numerous types of applications for which the static mass spectrometers are not best-suited. Blauth [270] has summarized some of these as:

1. the precision determination of large masses (where the resolvable mass difference is independent of the mass);
2. the demand for small, light weight mass spectrometers in space and upper atmosphere research;
3. the demand for small, light weight instruments of relatively low resolving power for residual gas analysis (RGA) and leak detection applications;
4. the capability of providing very fast and comprehensive analysis of products of explosions and combustion;
5. isotope separation.

An early example of application (3) was the commercial development of the QMF for RGA in the manufacture of thin films under ultra-high vacuum conditions, as reported by Bob Finnigan [272].

1.5.1.2   Classification of Types of Dynamic Mass Spectrometer

The underlying principle of dynamic mass spectrometers is the use of time dispersion in the motion of the ions to separate the mass spectrum into its individual components. The scheme for classifying dynamic mass spectrometers devised by Blauth [270] is essentially two-dimensional: the nature of the motion exhibited by the ions during the analysis process, and the means by which time dispersion is achieved. The method of classification can be summarized as follows; the reader is referred to the original reference for a more detailed account and explanation of the different categories.

The three basic types of motion considered are:

1. Linear direct: in which the ions start from a point and travel in a straight line.
2. Linear periodic: the ions oscillate in straight lines about a point in space, about a straight line, perpendicular to a plane, or along a straight line about a point. In some instances there may be a superimposed component of motion of the ions that is perpendicular to the direction of oscillation.
3. Circular periodic: here the motion of the ions is influenced by the superposition of magnetic and/or electric fields that may themselves be varying with time.

There are four methods by which time dispersion is achieved:

A. Energy balance: if the ions fulfil certain 'resonance conditions' energy can be supplied continuously to or withdrawn from them during each of the types of motion listed above.
B. Time-of-flight: if ions of different masses but with the same energy or momentum start out at the same time from the same place their TOF over a

TABLE 1.2
Classification of Dynamic Mass Spectrometers

Section	Mass Spectrometer	Classification*
1.5.2.3	Omegatron	3A
1.5.2.4	Ion Cyclotron Resonance	3A
1.5.3.2	Magnetic Time-of-Flight	3B
1.5.3.3	Linear Time-of-Flight	1B
1.5.3.4	"Spiratron"	3B
1.5.4.3.1	Beckman	1A
1.5.4.3.2	Bennett	1A
1.5.4.3.3	Boyd	1A
1.5.4.3.4	Redhead	1A
1.5.4.4.3	Quadrupole Mass Filter	2C
1.5.4.4.4	Monopole	2C
1.5.4.4.5	Quadrupole Ion Trap (QIT)	2C
1.5.4.4.6	Digital Ion Trap	2C
1.5.4.4.7	Linear Ion Trap	2C
1.5.4.4.8	Toroidal and Halo Ion Traps	2C
1.5.5.4	Orbitrap	2C

*See Section 1.5.1.2 for details.

specified distance depends upon the mass/charge ratio of the ion for each of the three basic types of ion motion.

C. Path stability spectrometers: for each of the three types of motion, stable or unstable paths can be created using suitable combinations of electromagnetic fields.

D. Characteristic frequency generator spectroscopes: here, under certain conditions of focusing and separation, ions having one of the three basic types of motion may excite oscillations, in suitable devices, that are characteristic of the mass/charge ratio of the ion.

A table showing each of the 47 types of dynamic mass spectrometer on can be found on pages 168–170 of Reference [270].

In the survey that follows we shall restrict our discussion to a brief consideration only of those types of dynamic mass spectrometer that at some point in the past have been thought to be sufficiently viable for (limited) commercial manufacture, whilst offering a more detailed historical account of those instruments that are in current use today (see Table 1.2)

1.5.2 INSTRUMENTS BASED UPON ION CYCLOTRON RESONANCE

1.5.2.1 Introduction

As implied by its name, the underlying principle of operation of the ICR mass spectrometer is derived from that of the cyclotron accelerator, which is used for generating beams of high-energy ions. When discussing the instrumentation and applications of ICR mass spectrometry, the authors of most modern texts seem to commence their discussions immediately with an account of the most recent, and most successful, manifestation of this method of mass analysis, namely FT-ICR which was invented in 1974 (see Section 1.5.2.4.4), ignoring the previous 25 years during which the basic method evolved through various different stages. In this Section we shall explore briefly the earlier generations of ICR analyzer, instruments that were, in the main, seen as being rather specialized and expensive means of investigating the kinetics and mechanisms of gas-phase ion/molecule reactions (see Section 1.4.1).

1.5.2.2   The Cyclotron

The idea of confining protons in a magnetic field, in such a manner that they can be accelerated repeatedly in small incremental steps by means of resonant energy transfer from an oscillating electric field, was perfected originally in the early 1930s by E.O. Lawrence and his group working at what was then called the Radiation Laboratory at the University of California Berkeley. At that time, there was considerable international interest in accelerating light ions to energies in excess of 1 MeV in order to create new nuclides (isotopes) by bombardment of the nuclei of the stable elements. Recognizing that to achieve such high energies by means of a single-stage accelerator was technically impossible to achieve, the race was on to devise some method of accelerating the ions incrementally. The obvious solution was to build a multi-stage linear accelerator where the ions were given successive increases in energy as they passed between different sections of the machine to which high amplitude RF potentials were applied. A key requirement was the matching between the frequency of the applied voltage, the lengths of each of the accelerating sections and the continually increasing velocity of the accelerated particles. While it would be possible to achieve 1 MeV for heavy ions such as Hg^+, for protons, which would have considerably greater velocities, the required overall length of the accelerator was unrealistic. An alternative solution was to confine the ions radially by means of a magnetic field and to find a means of supplying successive increments of energy from a relatively low voltage source so that there was a 'multiplying factor'. A number of rival groups had considered this general approach theoretically, but Lawrence was the first person to achieve success with this method experimentally.

The principle of the method can be seen from Figure 1.22a, taken from the patent granted to Lawrence in 1934 [273]. The basic idea is to confine the ions within a thin cylindrical volume created by two 'D'-shaped electrodes, marked 1 and 2 in Figure 1.22a; the whole structure, in an evacuated enclosure, is immersed in a magnetic field acting perpendicular to the plane containing the electrodes. Figure 1.22b shows a cross-sectional view indicating the resultant magnetic and electric lines of force. If the same potential is applied to electrodes 1 and 2, an ion injected into the cylindrical volume from the region near to the central hole will follow, under the influence of the magnetic field, a circular path determined by the balance between the centrifugal force of an ion and the magnetic force upon it. For an ion of mass m and total charge (ze) moving along a circular path of radius r in a magnetic field B, this condition is represented by the expression

(a)

(b)

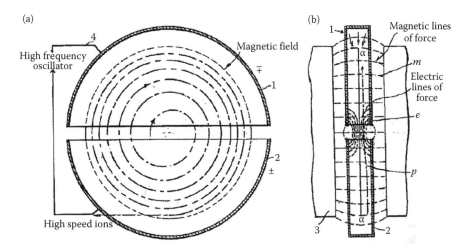

FIGURE 1.22 Diagrammatic elevations showing the electrode configuration together with the magnetic and electric lines of force in the cyclotron. (a) Plan view, in which the $\mp$ and $\pm$ symbols marked against electrodes 1 and 2, respectively, indicate the alternating potential gradient provided by the high frequency generator; (b) cross-sectional view. For details of the other labels on figures see the US Patent from which the diagrams have been reproduced. (Reproduced from Lawrence, E.O., *U.S. Patent* 1934, 1,948,384. With permission from the US Patent and Trademark Office.)

$$\frac{mv^2}{r} = \frac{B(ze)v}{c} \tag{1.26},$$

where v is the velocity of the ion and c is the speed of light.

If each of the electrodes 1 and 2 is now coupled to the opposite phases of an RF power supply (represented in Figure 1.22a by the $\mp$ and $\pm$ symbols marked against electrodes 1 and 2, respectively) having the appropriate frequency, when an ion crosses the gap between the two electrodes it will receive an additional amount of energy corresponding to the potential difference at the moment of crossing; this will result in the ion following an outwardly spiraling path.

A key feature of this arrangement is that the time t for traversal of a semi-circular path, given by

$$t = \frac{\pi r}{v} = \frac{\pi mc}{B(ze)} \tag{1.27},$$

is independent of the radius r of the path and of the velocity v of the ion. Consequently the beauty of the system is that as the ion moves along its spiral path its motion will always be in phase with the oscillating electric field provided that the magnetic field is adjusted correctly; ultimately, the ion reaches the boundary of the structure, where it strikes a suitably positioned target or detector.

For one complete circular orbit, the value of t must be equal to half the period of the oscillation frequency, f, of the accelerating voltage, so that we can calculate

$$f = \frac{B(ze)}{2\pi mc} = \frac{\omega_c}{2\pi} \qquad (1.28),$$

where ω_c is the so-called cyclotron frequency, expressed in radians s^{-1}. The equivalent wavelength, λ, of the RF potential, a quantity used by Lawrence in his papers, is

$$\lambda = \frac{2\pi mc^2}{B(ze)} \qquad (1.29).$$

In a presentation of the proposed method to a meeting of the US National Academy of Sciences in September 1930, Lawrence and his student Edlefsen stated [274] "For example, oscillations of 10,000 volts and 20 meters wave-length impressed upon plates of 10 cm radius in a magnetic field of 15,000 gauss will yield protons having about one million volt-electrons [electronvolts] of kinetic energy. Thus the 'amplification factor' would be [10^6/10,000 =] 100. The method is being developed in this laboratory, and preliminary experiments indicated that there are probably no serious difficulties in the way of obtaining high enough speeds to be useful for studies of atomic nuclei."

It appears [275] that Lawrence's first prototype machine was in fact four inches (10 cm) in diameter, and it is interesting to note here the parallel situations between what might be regarded as the births of the ICR mass spectrometer and that of the QMF, invented by Wolfgang Paul some 23 years later (see Section 1.5.4.). In order to test the theoretical design parameters for their respective high-energy ion accelerating machines, both types of instrument were created as small-scale working models that became eventually viable mass spectrometers in their own right, and each of the principal scientists behind each of these endeavors was ultimately a Nobel Laureate.

Although Edlefsen may have obtained some preliminary success with his prototype device, it was a fresh graduate student, M. Stanley Livingston, who was given the challenge of getting the "magnetic resonance accelerator" to work, and by April/May 1931, Lawrence and Livingston were able to report preliminary success with their new machine to the meeting of the American Physical Society in Washington, DC [276]. Evidently Lawrence was a person who believed in ensuring that all eventualities were covered and so, for good measure, at the same time as Livingston commenced his project he had another graduate student, David H. Sloan, construct a 1 MeV "linear resonance accelerator"! However, Lawrence's preferred project appears to have been the magnetic device, as evidenced by a letter written by him (at the age of 29) in February 1930 to his parents: "I have started an experimental research based on a very interesting and important idea. If the work should pan out the way I hope it will be by all odds the most important thing I will have done. The project has fascinating possibilities." (As quoted in Ref. 275.). Clearly the two students each achieved their respective objectives of constructing accelerators that worked as predicted, with full papers under the authorship of Sloan and Lawrence [277] and of Lawrence and Livingston [278] appearing in *Physical Review* for 1931 and 1932, respectively. Compared to the current situation, is it interesting to see two landmark publications appearing under only dual authorship.

Today, projects of this magnitude would be achieved as a result of a substantial team effort, as exemplified by the construction of the large hadron collider (LHC) at CERN, in Geneva, Switzerland! The name 'cyclotron' did not reach the general public until a later publication in 1935 [279], where, in a footnote, it is stated "The word "cyclotron", of obvious derivation, has come to be used as a sort of laboratory slang for the magnetic device."

1.5.2.3 The Omegatron

The invention of what is now the ICR mass spectrometer is generally credited to Hipple, Sommer and Thomas [280] who, working in 1949 at the National Bureau of Standards in Washington, DC, employed the cyclotron principle to determine certain fundamental physical constants. They designed an electrode arrangement that could be inserted into the magnetic field of a nuclear magnetic resonance (NMR) spectrometer in place of the normal NMR probe. A uniform RF electric field of variable frequency applied at right angles to a magnetic field accelerated ions of a given m/z at resonance until their spiral trajectories attained a radius of 1 cm, at which point they struck a collector and the ion current was measured. The trajectories of ions other than of the given m/z could not attain a radius of this magnitude unless the applied frequency was tuned to the appropriate value. As the frequency was varied a 'resonance peak' in the ion signal was detected, the width of which could be reduced by decreasing the amplitude of the RF voltage, thereby increasing the length the number of cycles before the ions are detected.

Hipple *et al.* were able to determine the ratio of the cyclotron frequency of the proton, ω_c, to its nuclear resonance precessional frequency, ω_n, (measured using the NMR probe in the same magnetic field, see above), and hence obtain a value for the faraday using an expression given in Ref. 280. The result they obtained agreed well with a value obtained electrolytically using an iodine voltameter.

In a second, more detailed account, the same group [281] improved upon these physical measurements using what, by now, they had called the 'omegatron' (see Figure 1.23, and stating that "One of the most important applications of the omegatron

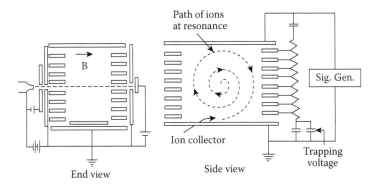

FIGURE 1.23 Simplified diagram of the omegatron showing the method of applying RF and DC voltages; the direction of the magnetic field, B, is shown by the arrow marked on the 'end view'. (Reproduced from Sommer, H., Thomas, H.A., and Hipple, J.A., *Phys. Rev.* 1951, *82*, 697–702. With permission from the American Physical Society.)

should be in the field of precise mass measurement." Using their system, Sommer *et al.* [281] were able to determine the mass separation of the $(D^+ - H_2^+)$ doublet as being equal to $(15.45 \pm 0.08) \times 10^{-4}$ u, in excellent agreement with that obtained by Roberts and Nier [282] equal to 15.49×10^{-4} u, with a somewhat smaller experimental error. It was predicted that the capability of the method could be improved so as to enable precision measurements to be made on higher mass ions.

It should be noted from the experimental accounts given in Refs. 280 and 281 that Hipple, Sommer and Thomas were evidently measuring the frequencies at which resonance occurred by observing the point at which the trajectories expanded such that the ions struck a collector. However, in a short Letter to the Editor published in 1950, Sommer and Thomas [283] described a method of combining the techniques of nuclear resonance absorption and the omegatron: "If these parallel plates [to which the RF voltage is applied] are made part of the capacitance of a parallel resonant circuit, the energy absorbed by the ions can be detected in exactly the same manner as nuclear resonance absorption." Having noted that even without optimizing the operating parameters the maximum sensitivity obtained by resonant energy absorption was considerably greater than that achieved by collecting the ions, they concluded "The sensitivity, simplicity, and speed of response of this [resonance absorption] method of detection should make it highly desirable for any application where the principle of the omegatron might be used, such as high resolution mass analyzers, analytical mass spectrometers, and mass spectrometer type leak detectors."

Perhaps the next stage in the evolution of the ICR method can be considered to be the work published by Wobschall *et al.* [284,285], who employed a modified ICR cell in order to determine the collision cross-sections at near thermal energies of N_2^+ and Ar^+ with their respective parent gases. Recognizing that the ion production region and the measurement regions are not separated in the omegatron, and that the device is operated under conditions of low pressure and high RF fields with the presence of trapping voltages, they designed a new structure in which ion/molecule interactions could be studied. Their measurement chamber consisted of a cylindrical ion source and drift region in a strong magnetic field within which the exciting RF electric field could be applied. They utilized also a method of measuring the ICR power absorption in order to monitor the effects of collisions. As the ions moved into resonance with the applied RF field, under the influence of a frequency sweep, and experienced continuous acceleration the energy extracted from the electric field was monitored and the effects of collisional broadening of the absorption peak observed. A key feature of this method of detection by energy absorption is that the ions are not destroyed in the process, provided that they do not reach the boundary of the device.

1.5.2.3.1   *The Omegatron as a Small Residual Gas Analyzer*

At the same time as these high performance applications of the omegatron continued to be extended, the instrument was proving also to be an ideal means of providing RGA for high vacuum systems, as described first by Alpert and Buritz [286] of Westinghouse Research Laboratories. They used the omegatron in conjunction with what is now the well-known Bayard-Alpert ionization gauge [287] to measure the partial pressures of residual gases in a highly evacuated system. Niemann and Kennedy [288] built an omegatron for rocket flight applications, and commercially this type of analyzer was still being manufactured by Leybold-Heraeus for use as an RGA until 1971: it offered an m/z-range of 1–250, with a resolution (FWHM) of 60 at 60 Th and a minimum detectable partial pressure of 10^{-5} times the total pressure within the range $10^{-7} - 10^{-5}$ Torr. More details on the performance of the omegatron as a gas analyser, including applications to space research, can be found in the review by Ball *et al.* [289].

1.5.2.4   Ion Cyclotron Resonance Mass Spectrometry

1.5.2.4.1   Introduction

In tracing the history of mass spectrometers employing the ICR principle, in addition to the aforementioned applications we note that the mid-1960s represents also the stage of development of the instrument when it can first be described as being a true mass spectrometer. As we shall see, there are in fact three landmark points along the path that leads to the current form of the instrument: these are the drift-cell ICR mass spectrometer, the trapped-ion cell analyzer, and the highly successful FT-ICR method of operation. In this account we shall concentrate our attention upon the earlier versions of the instrumentation: the reader is referred to Volume 4, Chapters 8 and 9, and to Volume 5, Chapter 5* in order to see examples of some current applications of FT-ICR. It should be noted that, unlike the modern FT-ICR mass spectrometers, which are designed as high-performance instruments for a wide range of analytical applications, those employing the drift-cell and the trapped-ion cell analyzers were aimed mainly at the rather specialized field of ion/molecule reaction studies.

Numerous reviews covering ICR have appeared over the years. As indicated earlier (Section 1.5.2.1), while more recent surveys tend to be confined to the advances that have occurred since FT-ICR was first discovered in 1974 (see Section 1.5.2.4.4), reviews that cover the earlier forms of the instrumentation include those by Baldeschwieler [290], Baldeschwieler and Woodgate [291], Futrell [292,293], and Wanczek [294,295].

1.5.2.4.2   The Drift-cell ICR Analyzer

This instrument was developed by P.M. Llewellyn of Varian Associates, Palo Alto, CA [296], and was first announced at the 13th Conference on Mass Spectrometry and Allied Topics held by committee ASTM-E-14 in St Louis, MA, in 1965. This announcement was noted by Rinehart and Kinstle [297] in their *Annual Review of Physical Chemistry* for 1968, in which they wrote: "One of the most interesting new types of mass spectrometer is the ICR mass spectrometer [citation of Ref. 296 above]. The ion source and entire spectrometer are placed in the magnetic field. Ionization is

---

* See Volume 4, Chapter 8: Radio Frequency-Only-Mode Event and Trap Compensation in Penning Fourier Transform Mass Spectrometry by Adam M. Brustkern, Don L. Rempel and Michael L. Gross; Volume 4, Chapter 9: A Fourier Transform Operating Mode Applied to a Three-Dimensional Quadrupole Ion Trap by Y. Zerega, J. Andre, M. Carette, A. Janulyte and C. Reynard; Volume 5, Chapter 5: Fourier Transform Ion Cyclotron Resonance Mass Spectrometry in the Analysis of Peptides and Proteins, by Helen J. Cooper.

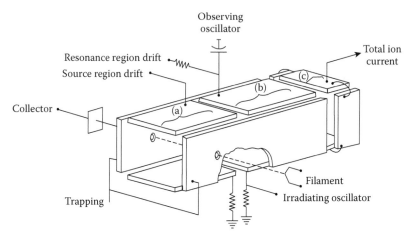

FIGURE 1.24   Cutaway view of the cyclotron resonance cell. The electron beam is co-linear with the magnetic field. (a) Ion souce; (b) ion analyzer; (c) ion collector. For further details see text. (Reproduced from Beauchamp, J.L., Anders, L.R., and Baldeschwieler, J.D., *J. Am. Chem. Soc.* 1967, *89*, 4569–4577. With permission from the American Chemical Society.)

effected in the source by an electron beam which is collinear with the magnetic field direction; then the ions produced are drifted gently in a direction perpendicular to both the magnetic and electric fields to the resonance region by a weak applied volt-age. Resonance of the ions then can be effected in much the same way as in an NMR or electron paramagnetic resonance (EPR) spectrometer, *i.e.*, by sweeping the magnetic field while applying irradiating oscillator and detection; double resonance techniques are also possible." Varian marketed their ICR instrument as the 'Syrotron', and it can, in some respects, be considered as being essentially a rectangular version of the cylindrical design developed first by Wobschall *et al.* [284] (see Section 1.5.2.3). A cutaway view of the cell designed by Baldeschwieler *et al.* [298] is shown in Figure 1.24. In this design the cell is divided equally into a source and resonance region together with a final, smaller, ion collection region; the overall dimensions are $2.54 \times 2.54 \times 12.7$ cm. An electron beam collinear with the magnetic field causes the continuous formation of ions in the source region, where they are confined laterally by potentials (of polarity appropriate to the sign of the ionic charge) applied to the trapping plates; the application of static voltages to the top plates of the cell causes the ions to drift into the resonance region. The top and bottom plates of the resonance region form part of the capacitance of a tank circuit of a limited oscillator, based on a design used in NMR instruments [299], and capable of high sensitivity in detecting small imped-ance changes. As with the experiments of Wobschall *et al.* described earlier, ions of a specified value of *m/z* are detected as energy absorption peaks when their cyclotron motion comes into resonance with the frequency of the observing oscillator.

The cell assembly was mounted between the pole caps of an electromagnet, and the overall circuitry of the instrument was built so as to permit three different types of modulation experiment. In the first method, the magnetic field was modulated by means of Helmholtz coils located on the pole caps; in the same manner as in NMR and EPR, changing the magnetic field causes the oscillating ions to move in and out of resonance. The second type of modulation involved varying the drift velocity (and hence transit time) of the ions, thereby causing a modulation of the number of ions in the cell and hence a change in the detected signal. The third type of modulation provided a highly specific so-called 'double resonance' technique. A square wave modulation was applied to the irradiator oscillation signal, so that when the frequency of the latter matched that of a precursor ion involved in an ion/molecule reaction, the detected signal of the product ion was modulated also. Using this connectivity, it was possible to establish the detailed mechanism of the reaction. Detailed accounts of this technique, together with examples, have been presented by Anders *et al.* [300], Beauchamp *et al.* [301], and Bowers *et al.* [302]. Beauchamp and Armstrong [303] described subsequently a resonant ion ejection technique which complemented the double resonance experiments, permitting the determination of ion/molecule reaction products as well as their partitioning with ion energy. A short 'progress report' by Goode *et al.* [304] published in 1971 contains references to various other modulation and multi-pulse techniques for studying ion/molecule reaction kinetics and mechanisms by means of ICR.

Notwithstanding the evident success of the three-region design of the ICR cell described above, a number of short-comings of this design were indicated by Clow and Futrell [305]. These issues included the undesirable effects of field-penetration between the different regions, resulting in distortion of the direction of net migration of the ions and the detected peak shapes, together with the perturbation of the supposed-thermal energies of the ions by the oscillator signal connected to the reaction/observation region. As a solution to this, Clow and Futrell [305] placed an additional 'reaction region' between the source and observation (analyzer) regions, to create a four-cell system. A further refinement to this design was the incorporation of an electron gun control grid in order to facilitate the creation of a pulse of ions in a defined time period within the overall duty cycle of a discrete experiment, rather than having ions being formed continuously.

1.5.2.4.3   Trapped-cell ICR Spectrometers

A major conceptual advance in ICR mass spectrometry occurred also in 1970, when McIver [306] published his design for a trapped-ion analyzer cell. In place of the standard three-region cell of his Varian Associates ICR-9 machine he mounted a single-cell electrode arrangement in which not only were the ions created by a pulse of electrons, as in the Futrell four-region cell, see Section 1.5.2.4.2, but an additional pair of 'end plates' was fitted in order to confine the ions in the longitudinal direction (see Figure 1.25). McIver [306] cited the differences between the single-cell and the previous three-region design as being (1) ions were now created and detected in the same region; (2) the ions were confined within the cell, instead of drifting out: in this way much longer observation times were possible (for example, 0.10 s as against 1 to 2 ms); and (3) the trapped ion cell relied entirely on pulsed modes of operation, as opposed to having a continuous drift of ions. A further advantage was that the space-charge effects of the electron beam were obviated by the pulsed mode of operation, so that there was no interference with the detection of ions later in the sequence. Notwithstanding these changes, the ions were still detected using the cyclotron resonance absorption method, and double-resonance experiments were possible with this new arrangement. In some

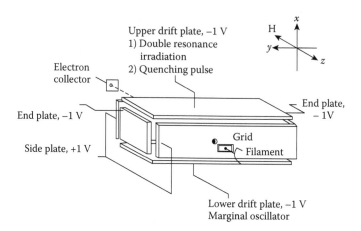

FIGURE 1.25   Schematic diagram of the trapped ion analyzer cell. The DC voltages applied to each of the cell plates are suitable for trapping positive ions. (Reproduced from McIver, R.T., *Rev. Sci. Instrum.* 1970, *41*, 555–558. With permission from the American Institute of Physics.)

respects the idea of a single-cell design looks both backwards to the original omega-tron arrangement of Hipple *et al.* [280] (but now with pulsed ion creation and a different means of detection), and forwards to the design of the cell utilized in the Fourier transform (FT) mode of operation (see Section 1.5.2.4.4.). A model for the motion of ions in the trapped ICR cell has been published by Sharp *et al.* [307], and McMahon and Beauchamp [308] have described how what was then the 'conventional' three-region design was converted to trapped-ion operation.

1.5.2.4.4   *Fourier Transform Ion Cyclotron Resonance Mass Spectrometry*

Although by the early 1970s ICR mass spectrometry was a well-established technique, especially in the realm of ion/molecule reaction chemistry, the different instruments offered by a number of manufacturers were relatively quite expensive. Furthermore, with an operating pressure in the analyser region of the order of 10^{-7} Torr, necessary in order not to degrade the resolution through collisional broadening of the resonance peaks, and the relatively slow mass spectral scan speed, these systems were not considered seriously as general-purpose analytical machines. However, in 1974 the position changed dramatically with the pivotal invention by Comisarow and Marshall [309–311] of the FT mode of operation.

The basic idea is that, rather than scan the magnetic field or the frequency of the exciting signal slowly and measure the power absorption as ions of successive m/z-values come into resonance, the excitation frequency is swept rapidly and the consequential coherent expansion of the different ion trajectories into larger orbits is detected by means of measuring the 'image current' in an external circuit.

Although different forms of ICR cell have been designed for use in the FT mode, conceptually it is easiest to envisage a cubic structure, having edge dimensions equal typ-ically to 2.54 cm, comprising three pairs of parallel plate electrodes (see Figure 1.26a). A gated electron beam running parallel to the magnetic lines of force is admitted through holes in the two 'trapping plates', to which appropriate potentials are applied,

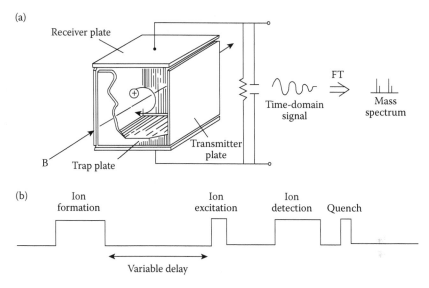

FIGURE 1.26 (a) Schematic representation of a cubic trapped ion cell commonly used in Fourier transform mass spectrometry. The direction of the magnetic field, B, is shown by the arrow. Coherent motion of the ions in the cell induces an image current in the receiver plates. The time domain signal is subjected to a Fourier transform algorithm to yield a mass spectrum. For further details, see text. (b) Simple pulse sequence used in elementary Fourier transform mass spectrometry experiments. For further details, see text. (Reproduced from Buchanan, M.V. (Ed.) *Fourier Transform Mass Spectrometry – Evolution, Innovation, and Applications.* ACS Symposium Series No. 359, American Chemical Society, Washington, DC 1987. With permission from the American Chemical Society.)

and thence to a collector to stabilize emission. Under the influence of the magnetic field the ions start to precess with their characteristic cyclotron frequency close to the center of the device. One of the remaining two pairs of electrodes (one of which is labeled 'transmitter plate' in Figure 1.26a) is coupled to the opposing polarities of a generator supplying the RF excitation potential, and the other pair of electrodes ('receiver plates') form part of the image current detection circuit. In the simplest mode of operation, represented by Figure 1.26b, following a variable delay period after ion formation, a short, high intensity, broadband RF signal (a 'chirp') is applied to the transmitter plates, after which the transient image current response is recorded and the ions then removed from the cell by 'quenching'. The frequency components of the image current correspond to the cyclotron frequencies of the different ions present in the cell, and deconvolution of the signal by means of Fourier transformation converts the data into a mass spectrum. The signal-to-noise ratio of the output may be improved by repeating the cycle several times and averaging the resultant data, and various forms of ion manipulation such as double-resonance experiments may be carried out during the time interval between ion formation and ion excitation (see Figure 1.26b). In contrast to the modern instrumentation now available, Alan Marshall recalls [312] that the Nicolet storage oscillograph used by him and Mel Comisarow in their earliest experiments had 1024 channels covering the frequency range available. The intensity of each channel was read by Mel and recorded by Alan; these intensities were to the base 10. Alan then calculated the intensity manually to the base 8 for entry to a Varian 621 computer from which the first Fourier transform mass spectrum of methane was obtained. Three excellent accounts in which further details of the technique, performance of the instrumentation, and some applications may be found are those by Marshall [313], by Buchanan [314], by Marshall and Verdun [315], and by Guan *et al.* [316].

Before leaving the subject of FT-ICR, we should mention two particular enhancements in the technology of the instrumentation that have assisted greatly in improving the suitability of the method for analytical applications: these are the 'dual cell' design [317,318], in which the ions are formed within a higher pressure source region before being transferred to the differentially pumped analyzer region operating at a pressure some thousand times lower, and the capability to transfer ions formed in an externally mounted source, such as one for ESI or MALDI, into the FT-ICR cell [319].

Current performance levels that can be achieved by FT-ICR mass spectrometry are outstanding. For example, Schaub *et al.* [320] have described recently a high performance instrument operating with a 14.5 T magnet in which a resolving power of 200,000 (FWHM) at m/z 400 is achieved at an acquisition rate of greater than one mass spectrum per second. The mass accuracy shows a four-fold improvement, with a two-fold higher resolving power, compared to that which can be obtained with similar 7 T systems at the same scan rate. In terms of analytical capability, this means that closely similar compounds, for example a group of cockroach neuropeptides, can be distinguished, and highly complex mixtures, such as petroleum fractions, resolved (*ca* 50,000 detected peaks in a mass spectrum), with elemental composition assignment.

1.5.3 TIME-OF-FLIGHT MASS SPECTROMETERS

1.5.3.1 Introduction

It is now some 160 years since Fizeau [321] devised a means for measuring the velocity of light using a toothed wheel of known speed of rotation to interrupt a beam that was reflected from a mirror over a known distance and detected as it passed through the same opening through which it had been transmitted [322]. Much the same approach is adopted in the TOF mass spectrometer, when the time taken for an ion of known kinetic energy or momentum to travel a specified distance is used to determine its mass/charge ratio. This simple concept of operation makes the TOF analyzer probably the easiest to understand of all the different types of mass spectrometer. However, as we shall see, there have been several crucial refinements in design that have been introduced over the years in order to achieve the high level of performance exhibited by modern commercially available TOF instruments.

In examining the evolutionary stages through which TOF mass measurement has passed, one is struck by the distinct parallels with the development of the ICR technique (see Section 1.5.2); indeed the reader might be surprised to learn that some of the very first TOF mass spectrometers utilized magnetic fields in a way that was fundamentally an adaptation of the cyclotron principle. This is not the only similarity however: for example, after early 'home-built' instruments were reported in the literature, both the TOF and the ICR methods reached stages of commercial availability, but generally for limited fields of application. Thus, the ICR mass spectrometer was seen as being primarily a tool for the study of ion/molecule reactions, whilst the fast acquisition of 'complete' mass spectra that is possible with a TOF machine made it ideal for monitoring the chemistry of fast-changing systems, such as combustion, explosions, shock-waves, etc., and, in a few laboratories, even effluents from GC columns. Then, just as the low volumes of sales for these specialist applications were threatening their continued commercial viability, other developments occurred that expanded the fields of application and levels of analytical performance. In the case of TOF technology, these were the need for analyzers of essentially infinite mass range that were compatible with pulsed ionization sources, such as laser desorption, plasma desorption, and MALDI, the invention of the 'reflectron' ion mirror for enhancing resolution, and the introduction of 'orthogonal-acceleration' (oa) for the sampling of ions from continuous beams, such as electrospray sources or in a hybrid configuration with a beam analyser, such as a quadrupole mass filter. In this Section we shall consider each of these stages in turn.

1.5.3.2   Magnetic Time-of-Flight Mass Spectrometers

1.5.3.2.1   A Helical Path Device

One of the earliest designs of mass spectrometer based upon the TOF principle is that described by O.G. Koppius [323] in a U.S. Patent that was filed in October 1946 and in which the instrument is described as "a mass-velocity electron discharge device." In particular, the inventor had the objective of producing an instrument that was inexpensive, simple to manufacture, compact and portable. Essentially, the idea was to inject a pulse of ions at an angle to the axis of an evacuated tube that was contained within a magnetic field generated by a solenoid. In this way the ions would assume a multi-turn helical path and then impinge upon a collector electrode. Consequently, the total path length of the ions would be many times longer than the linear distance between the source and the collector (see also Section 1.5.3.4 for a description of the 'spiratron', which uses an electrostatic field for a similar purpose). In the patent, typical design parameters are quoted as being a linear distance equal to 30 cm, an injection angle of 20° relative to the axis, and a magnetic field of 1000 gauss. However, no actual performance data for this system appear to have been published.

1.5.3.2.2   Instruments Utilizing the Cyclotron Principle

In 1948, S.A. Goudsmit [324], working at the Brookhaven National Laboratory, New York, noted that there appeared to have been no design of mass spectrometer published that made use of the constancy of the TOF in a magnetic field. As we have seen from Equation 1.28, the angular velocity (that is, the cyclotron frequency) of an ion in a magnetic field is independent of the velocity of the ion; it is also unaffected by the direction in which the ion is released. Thus the time for a complete revolution (the period) is directly proportional to the value of m/z and inversely proportional to the magnetic field strength, B. Furthermore, a divergent ion beam is re-focused to a point every complete turn. Goudsmit calculated that for an ion of ca 150 Th in a field of 100 gauss the flight time for a single revolution would be about 1000 μs, so that the time interval between successive values of m/z would be ca 6.7 μs, a figure that was well within the precision of the measured pulse width (0.1 μs). If the ions were collected after several revolutions, the time intervals would be proportionately longer, and the accuracy of determination greater.

In Goudsmit's design of analyzer [325], which he termed the 'chronotron', the ions are injected along a generally helical path such that there will be points of focus along the lines of magnetic flux passing through the source, called 'focal lines of flux'. Ions of different m/z, different velocities, or different directions of emission from the source, will all focus at points along the focal lines of flux, so that by positioning a detector at a point displaced from the source along the axis of the helical path by an integral number of helical turns it is possible to measure the flight times of the ions that have been emitted in short, accurately timed pulses. A more detailed account of the system, together with the experimental values of the accurate masses they determined for 14 nuclides, was published by Hays *et al.* [326]. The estimated precision of the data they reported lay in the range from ±0.0005 to ±0.0025 u.

Within the same timeframe as Goudsmit was developing his chronotron, L.G. Smith [327], also working at Brookhaven, devised a similar, but in some respects, a more sophisticated way of performing the same experiment. In designing his modified approach, Smith was concerned that simple electrostatic and magnetic focusing was impossible in the chronotron, causing ion losses through spreading along the axis of the magnetic field. In the smaller volume occupied by his instrument, inhomogeneities in the magnetic field were easier to minimize, and the spreading of the beam due to scattering was reduced.

The basic design of the experiment is shown in Figure 1.27. The magnetic field acts in a direction perpendicular to the cross-section through the cylindrical

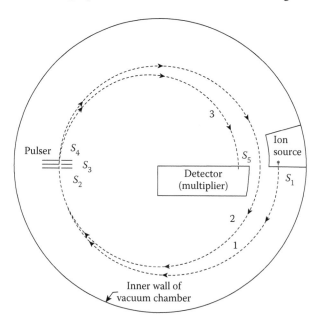

FIGURE 1.27 Schematic cross-section of the magnetic period mass spectrometer. For further details, see text. (Reproduced from Smith, L.G., *Rev. Sci. Instrum.* 1951, *22*, 115–116. With permission from the American Institute of Physics.)

vacuum chamber, and operates with a pulsed sequence as follows: (1) a positive voltage pulse is applied to the source to cause a bunch of ions to be ejected, and to follow a circular trajectory shown as '1'; (2) the ions are allowed to remain solely under the influence of the magnetic field as they move through 180° and reach the 'pulser'; (3) the outer electrodes S_2 and S_4 of the pulser are maintained at ground potential while a negative square wave voltage pulse is applied to the central electrode, S_3, causing those ions between S_3 and S_4 to be decelerated so that they follow a new circular trajectory '2' that misses both the ion source and the detector; and (4) after an adjustable time, during which the ions undergo several orbits, a second negative pulse is applied to S_3 to decelerate them further on to pathway '3', and thence into the detector.

The relative timings of the first and second negative pulses applied to S_3 are adjusted so as to maximize the detected signal, and the time-delay between the two pulses can be related to the m/z-values of the ions. The main function of this instrument was not for mass-scanning in analytical applications, but rather for determining the accurate masses and associated 'packing fractions' of selected nuclides. The patent filed by Smith [328] in 1951 describes, as an example, the determination of the mass of ^{32}S relative to the then accepted standard of $^{16}O = 16.0000$, where, by admitting sulfur dioxide as a sample, the relative pulse delay for maximizing the signals from the $^{32}S^+$ and $^{16}O_2^+$ ions gave a mass of ^{32}S equal to 31.9823 ± 0.001 u. This figure is close to the currently accepted value of 31.972072, relative to $^{12}C = 12.000000$ u)[329]. Due to the superficial resemblance of his machine to the synchrotron, Smith named his device the "synchrometer". Further results obtained with this system were reported by Smith [330] and by Smith and Damm [331], who presented a more extensive description of their instrument, together with a theoretical analysis of its operation. Interestingly, in their work these authors studied also molecular materials, and when plotting the ratio of the background current to peak current of $B_5H_9^+$ ions at different pressures observed a second-order pressure-dependence of the background intensity, indicating contributions arising from CID.

1.5.3.3 Linear Time-of-Flight Mass Spectrometers

1.5.3.3.1 Basic Principle of Operation

Before considering the early instruments in detail, it will be helpful for us to examine at its simplest level the underlying theory behind the linear TOF mass spectrometer. The principle of the method is to measure the times taken for ions of different m/z-values to travel in a straight line between the ion source and the detector: for a given charge, the lighter the ion the shorter the flight time.

Suppose an ion of mass m and charge (ze), starting from a defined point and with zero thermal energy, is accelerated by a potential V such that it travels a distance L with a uniform velocity v before reaching the detector. The flight time, t, will be given by

$$t = \frac{L}{v}$$

(1.30).

The value of v can be found from the expression for the kinetic energy of the ion, given by

$$\frac{1}{2}mv^2 = (ze)V \tag{1.31},$$

so that

$$v = \left(\frac{2(ze)V}{m}\right)^{\frac{1}{2}} \tag{1.32},$$

whence

$$t = L\left(\frac{m}{2(ze)V}\right)^{\frac{1}{2}} \tag{1.33}.$$

Thus, the flight time is directly proportional to the square root of the mass/charge ratio of the ion, and the time interval between the arrival of ions of successive mass/charge ratios is directly proportional to the flight distance.

1.5.3.3.2 Early Instruments

The first linear TOF mass spectrometers, namely instruments based upon designs not employing magnetic fields, were developed independently and essentially simultaneously by two groups: W.E. Stephens, working at the University of Pennsylvania, and A.E. Cameron and D.F. Eggers, Jr., at the Tennessee Eastman Corporation, Oak Ridge, TN. At the meeting of the American Physical Society held in Cambridge, MA, in May 1946, Stephens [332] reported that an instrument was under construction in which it was intended that microsecond-wide pulses of ions would be selected every millisecond and then allowed to travel down a "vacuum tube" whereupon "ions of different M/e have different velocities and consequently separate into groups spread out in space." On detection *via* a Faraday cage, it should be possible to display the resulting amplified signal upon an oscillograph, where "The indication would be continuous and visual and easily photographed." This method was described subsequently by Stephens in a very brief patent [333], filed in April 1947, where it was noted that the anticipated low cost of the device contrasted with that of instruments utilizing electric or magnetic fields, for which the initial cost "is seldom less than \$25,000." Furthermore, it was noted that among other advantages, as a result of the absence of magnets and the associated stabilizing equipment, the instrument was lighter and also that the "resolution does not depend upon the smallness and alignment of slits, etc."; additionally, "the ion spectrum is scanned at a rapid scan rate". The performance of the Stephens instrument was reported subsequently by Wolff and Stephens [334] in 1953. It should be noted that, in contrast both to their earlier model and to modern TOF instruments, in which all the ions being pulsed from the ion source are accelerated in such a way that they all have the same kinetic energy regardless of m/z-value, by arranging that the accelerating field was turned off before any of the ions had passed completely through the field the ions all acquired the same momentum: in this way, the mass scale displayed on the oscillograph was linear.

The principle of the constant momentum TOF mass spectrometer can be derived as follows. Consider an ion of mass m_i and charge (ze) accelerated through a potential V_i over a distance d_i, from Equation 1.32 the average velocity v_i is given by

$$v_i = \left(\frac{2(ze)V_i}{m_i} \right)^{\frac{1}{2}} \tag{1.34}.$$

The time t_{min} taken by the ion of lowest mass m_{min} to traverse the acceleration region of distance d at a maximum average velocity v_{max} is given by

$$t_{min} = \frac{d}{v_{max}} \tag{1.35}$$

and is a constant when the acceleration potential is arrested at time t_{min}, that is, when the ion of m_{min} has reached the end of the accelerating region. Thus, for all ions,

$$t_{min} = \frac{d_i}{v_i} = d_i \left(\frac{m_i}{2(ze)V_i} \right)^{\frac{1}{2}} \tag{1.36},$$

where d_i is the distance traveled by m_i through a potential difference V_i. The field of the accelerating region, E, is constant at a value of V_i/d_i so that

$$d_i = \frac{V_i}{E} \tag{1.37},$$

such that

$$t_{min} = \frac{V_i}{Ev_i} = \frac{V_i}{E} \left(\frac{m_i}{2(ze)V_i} \right)^{\frac{1}{2}} = \frac{1}{E} \left(\frac{m_i V_i}{2(ze)} \right)^{\frac{1}{2}} \tag{1.38}.$$

Therefore, from Equations 1.34 and 1.38

$$t_{min} = \frac{v_i}{E} \left(\frac{m_i}{2(ze)} \right)^{\frac{1}{2}} \times \left(\frac{m_i}{2(ze)} \right)^{\frac{1}{2}} = \frac{m_i v_i}{2E(ze)} \tag{1.39}.$$

Thus, all ions that have been subjected to an arrested acceleration potential have a common momentum given by

$$m_i v_i = 2E\, t_{min}(ze) \tag{1.40}$$

If the length of the accelerator, d, is insignificant compared to the length of the flight tube, L, then the flight time, t_i, of an ion with velocity v_i will be

$$t_i = \frac{L}{v_i} = \frac{Lm_i}{2t_{min}E(ze)} \tag{1.41}.$$

Equation 1.41 agrees with the expression given by Wolff and Stephens [334], except that their equation lacks the factor of 2 in the denominator.

Within the same time-frame as the above work, Cameron and Eggers [335] were developing their "Velocitron", which apparently was near to completion when Stephens first made his announcement. In reporting upon their instrument, the authors noted that mass spectrometers that used a magnetic field for ion separation did not lend themselves easily to instantaneous display of the entire mass spectrum, and that the observation of individual peaks was too slow when it was desirable to monitor a system in which the composition was changing as a function of time. It may be noted that the time interval between the arrival of ions of different m/z-ratio will be determined by the length of the (glass) analyzer tube, which in their case was 3.17 m, as well as by the value of the accelerating voltage (see Equation 1.33) and by the width of the voltage pulse applied to inject the ions into the flight tube. The authors recognized also that it was necessary for the ion source to produce ions that were nearly monoenergetic, otherwise there would be broadening of the "received pulse". Notwithstanding this appreciation by the authors of these instrumental requirements, however, the mass spectra contained in their report show extremely poor resolution.

The issue of allowing for the thermal velocities of the ions prior to acceleration was considered in detail by Katzenstein and Friedland [336], who developed a TOF analyzer for appearance energy measurements. They employed a one meter-long stainless steel analyzer tube, and an ion source fitted with a more sophisticated grid arrangement for accelerating the ions. They noted a five-fold improvement in resolution compared to the models described by Cameron and Eggers [335] and by Wolff and Stephens [334].

1.5.3.3.3   Time-Lag Focusing

We have already seen from the preceding discussion in this Section that the major limiting factors governing the resolution attainable with the linear TOF analyzer are the spatial dispersion of the ions ('space resolution') within the ion source and the thermal energies ('energy resolution') of the ions at the moment they are accelerated into the flight tube. Thus, if the ions are not all situated on a plane perpendicular to the direction of travel then they will not all experience the same accelerating potential and so will exhibit a spread in kinetic energies, even if they have the same value of m/z; this problem will be exacerbated further by any non-uniformities in the accelerating field. Similarly, because of the random thermal motion of the sample molecules prior to ionization *via* electron impact, a fraction of the ions will possess components of motion opposing the direction of acceleration, while an equal fraction will be moving forward in the direction of ion ejection at the moment the accelerating voltage pulse is applied. Consequently the former ions will have to be 'turned around' and will fall behind relative to the latter ions, which will have a favorable contribution from their initial velocity causing them to move ahead. These issues of spatial and energy focusing were addressed and essentially solved by Wiley [337] in a U.S. Patent filed in October 1951, and by Wiley and McLaren [338], from the Bendix Aviation Corporation, Detroit, MI, in a somewhat later publication. A more discursive account was presented also by Wiley [339].

The method developed by Wiley and McLaren is basically quite simple, and is known as 'time-lag focusing'. The key features are the use of a 'double-field' ion source, with two separate accelerating regions, and the timing sequence of the pulses applied to the various electrodes. A diagram detailing the overall system is shown in Figure 1.28; while various versions of this illustration appeared in the aforementioned publications by Wiley and McLaren, this one reproduced from Blauth [270] is based upon the description of the commercial system presented by Harrington [340], and is considered to be more informative.

A complete duty cycle for a single mass spectral event ('scan') involves a number of steps, as follows. (1) Admitting a pulse (width typically 0.25 μs) of electrons *via* the 'control grid' adjacent to the filament, such that ions are created within the source region, labeled 's'. During this time no other electric fields are applied in the source region, and the ions as they are formed will tend to remain trapped within the negative space charge arising from the electron beam (see also Section 1.4.2.); this arrangement provides a measure of spatial focusing along the line of the beam. (2) The electron beam is now gated off *via* the control grid, and an adjustable time-delay (the 'time-lag') allowed; during this time ions may diffuse away from the region

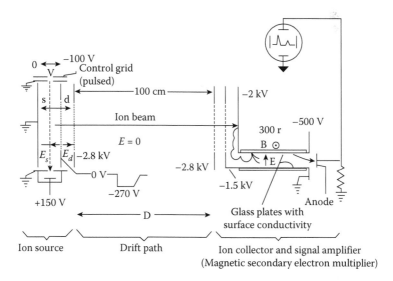

FIGURE 1.28 Schematic diagram of the Bendix time-of-flight mass spectrometer. Note that the direction of the magnetic field, B, employed in the secondary electron multiplier lies perpendicular to the plane of the diagram. For further details, see text. (Reproduced from Blauth, E.W., *Dynamic Mass Spectrometers*, Elsevier, Amsterdam, 1966.)

where they were trapped in the electron beam. (3) Following the delay, a negative-going pulse (here-270 V, typical width 2.5 μs) is applied to the first grid, as shown, and this generates a field E_s within the source region that directs the ions into the main acceleration region in which there is a static field E_d. The effect of E_s is such that any ions that lie to the left of the original axis of the electron beam receive a proportionately greater accelerating force than do the ions that lie to the right, and with the correct adjustment of the ratio of E_d/E_s it is possible to obtain the same flight times for ions of the same value of m/z but which started at different positions. From Figure 1.28 it will be noted that the final acceleration is produced by the second grid, which in this instance is at a potential of −2.8 kV: the ions then pass through the evacuated field-free flight tube (shown as $E = 0$) and impinge upon the detector (see also below). The precise geometry of this electrode arrangement means that the ions reach their maximum kinetic energy within about 5% of their flight time, as opposed to *ca* 50% of the time in the earlier published accounts. As a consequence, any contribution to the flight time arising from the initial thermal energy is very small, giving improved energy focusing over that exhibited by single-field sources. However, a further adjustable parameter available is the time-delay, τ, after the cessation of the electron pulse, so that by optimizing the settings of τ and of E_d/E_s it is possible to achieve the best compromise conditions between ideal energy and ideal spatial focussing.

Reference was made in the preceding discussion to the detector system shown in Figure 1.28. Due to the rapid changes in ion intensity associated with the arrival of successive ion pulses (typically 10,000 complete mass spectral recordings are obtained per second) it is essential to have an electron multiplier and associated electronics that have sufficient bandwidth. Wiley adapted a design published by Smith [341] in a manner described briefly by Harrington [340]. The multiplier itself comprises two parallel sheets of glass coated with a chemically inert resistive material. Potentials are applied along the surfaces and between the surfaces, as shown, and the assembly is mounted in a magnetic field of 300 G acting into the plane of the diagram. The arrival of positive ions on the plate floating at −2 kV causes the ejection of secondary electrons which, under the influence of the various electric fields and the magnetic field, are accelerated and directed in a series of semi-circular 'loops' along the electrode surfaces such that each impact causes the emission of further secondary electrons. Ultimately the resulting avalanche of electrons is intercepted by an anode connected to an oscillographic display. In later versions of this arrangement, several anodes were fitted so that with appropriately timed pulses applied to gating electrodes, the signals arising from one or more selected ion types could be monitored individually.

Using this instrument, Wiley indicated [339] that a typical resolution of 600 (FWHM) at 600 Th should be achievable, a considerable enhancement over the previous designs. Early examples of experimental studies carried out using Bendix TOF mass spectrometers are those by White *et al.* [342] and by Homer *et al.* [343].

Further, more detailed, investigations into the theory and applications of the Wiley–McLaren time-lag focusing technique were carried out by Sanzone and co-workers [344–346].

1.5.3.4 The 'Spiratron'

In Section 1.5.3.2. we have seen examples of TOF instruments in which the trajectories of the ions are confined to helical paths by means of the applications of magnetic fields. The 'Spiratron', first proposed by Bakker [347] has certain similar features, except that in this case the helical path is created by injecting the ions into a radial electric field between two concentric cylindrical electrodes between which is maintained a DC potential. If the centripetal force, mv^2/r of an ion of mass m and charge (ze) moving at radial distance r with a velocity v is balanced by the force zeE arising from the radial electric field E then the trajectory will form a circular orbit, and if the ion is injected with a small velocity component along the axis of the electrode structure it will describe a helical path around the inner cylinder. However, in addition to allowing one to 'compress' the equivalent of a long flight path into a relatively small volume, the electrostatic field has focusing properties that can help to remove the initial spread in kinetic energies of the ions. The basic physics of the focusing of charged particles in a cylindrical field of this kind was formulated originally in 1929 by Hughes and Rojansky [348], who compared theoretically the use of magnetic fields and/or electrostatic fields for the analysis of the velocities of electrons. A key distinction between the two cases is that when a beam of electrons of given velocity is projected into a uniform magnetic field acting in a direction perpendicular to the lines of magnetic flux, refocusing of the beam occurs after rotation through an angle of 180° (that is π radians), and integral multiples thereof, relative to the entrance slit, whereas when the same beam of electrons is confined within a radial electrostatic field, refocusing occurs after rotation through an angle of 127°17' ($\pi / \sqrt{2}$ radians) and multiples thereof. Furthermore, for maximum energy resolution the angle of deflection through the electric field must be $(2k - 1)$ $(\pi / \sqrt{2})$, where k is an integer. Bakker constructed such a device [347] in which the overall length of the cylindrical assembly was 26 cm such that the total angle of deflection was $17\pi / \sqrt{2}$ (equal to 6.01 revolutions) and the equivalent linear path was 142 cm. The mass/charge range of the instrument was 1–1000 Th at a repetition rate of 20,000 scans s^{-1}, with a maximum obtainable resolution for complete separation of adjacent masses at 630 Th.

In a subsequent paper, Bakker *et al.* [349] extended this approach, and showed how the design parameters of the spiratron could be specified so that 'time focusing' was attained, that is ions of the same value of m/z but with a spread of velocities that were injected at precisely the same time all arrived at the detector at the same time. Essentially, it was found that by selecting the correct angle of injection and arranging for the number of revolutions in the cylindrical electric field to equal an even multiple of $\pi / \sqrt{2}$, relative to the 'correct' ions, the faster ions will move ahead along the linear component of motion and will lag behind with respect to the angular component, and *vice versa* for the relatively slow ions. Consequently, time-focusing can be achieved by optimizing the relationship between the orbital and linear path lengths within the analyzer, taking into account any additional linear sections between the ion source and the analyzer, and between the analyzer and the detector. An instrument based upon this theory was constructed, but never fully characterized.

Recently, Satoh *et al.* [350], in a paper that contains numerous references to other multi-pass TOF analyzers, described an elaborate arrangement using four toroidal electrostatic sectors through which the ions pass in a corkscrew trajectory comprising a succession of 15 figure-of-eight shaped orbits. The flight path per orbit was 1.308 m, making the total path length *ca* 20 m. A mass resolution of 35,000 (FWHM) at m/z greater than 300 was obtained, with a mass accuracy of 1 ppm and an ion transmission of *ca* 100%. The ions emerging from the EI source were injected into the analyzer using the 'orthogonal-acceleration' technique (see Section 1.5.3.6. below).

1.5.3.5   The 'Reflectron'

We have seen already from the two preceding Sections that in order to achieve time-focusing, some form of compensation method has to be introduced into the flight path of the ions in order to ensure that within an ion 'packet' all the ions of the same m/z-value that leave the ion source at the same time arrive simultaneously at the detector, even though there is a distribution in the kinetic energies, and hence velocities, of the ions. The spiratron represents just one of numerous attempts to do this (see, for example, the review by Wollnik [351]). Without doubt, the most successful approach is that of the 'ion mirror', or 'reflectron'. Indeed, like the time-lag focusing arrangement, which is now over 50 years old, the reflectron represents one of the landmark contributions to TOF mass spectrometry and, in conjunction with MALDI (see Section 1.3.2.6.6.5), has enabled the technique to become of such major importance today.

The reflectron was invented by Mamyrin *et al.* [352], and is basically a very simple concept: by projecting the ion packet into a reflecting electric field, ions penetrate the field according to their kinetic energies, so that those ions traveling fastest penetrate the furthest and, therefore, possess longer path lengths. With appropriate potentials and other instrumental design parameters, for ions that set off from the same plane within the ion source, compensation for the kinetic energy spread occurs so that all the ions of the same value of m/z arrive at the detector together.

A schematic diagram of the arrangement is illustrated in Figure 1.29, and an account of the theory of operation of the reflectron is contained within the 2001

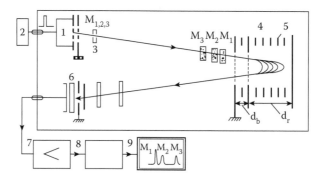

FIGURE 1.29   Time-of-flight mass spectrometer with a two-section reflector having plane electric fields. Components: 1–pulsed ion source; 2–power supply (laser); 3–focused ion packed near the ion source; 4–double sectional reflector; 5–electrodes for creation of plane electric fields; 6–detector; 7–wide band amplifier; 8–computer; 9–display; d_b–the ion deceleration section; d_r the ion reflection section. (Reproduced from Mamyrin, B.A., *Int. J. Mass Spectrom.* 2001, *206*, 251–266. With permission from Elsevier.)

review by Mamyrin [353]. A packet of ions represented as $M_{1,2,3}$ is accelerated out of the ion source and into the first section, d_b, of the reflector; the ions then pass into the reflection section, d_r. The extent to which ion trajectories experience the decelerating potential within the reflectron depends upon the kinetic energies of the ions. Furthermore, the more the decelerating potential is experienced on entry the greater will be the acceleration gained upon leaving the reflectron as the ions are directed to the detector. It will be noted that separation of the different mass/charge ratios, M_1, M_2, and M_3, is already occurring during the first linear section of the flight path, but after reflection the widths of the individual packets are reduced before they reach the detector. In some instruments, what appears as a back plate in the reflectron is replaced by an additional detector into which the ions can pass if the reflecting field is turned off. Any unimolecular dissociation of metastable ions occurring during the first linear section can be detected also by reflecting the remaining precursor ions and any product ions, and observing the impact of fast neutral species on the additional detector; correlation techniques enable connectivity to be established between the precursor ions and the various product species.

The introduction of reflectron technology has enabled resolving powers well in excess of 10^4 to be achieved, and the stability and quality of modern high-speed digital electronic circuits, together with high-capacity data storage, have permitted accurate mass determinations to be made with TOF mass spectrometers on a routine basis. This level of performance contrasts markedly with the figures of merit for the pioneering instrumentation. Cotter and Cornish [354] have described a tandem TOF system employing two stages of mass analysis, each with a reflectron, arranged in 'Z' configuration, with a higher pressure collision cell situated in between the analyzers along the oblique portion the 'Z'.

1.5.3.6 The Orthogonal-Acceleration Time-Of-Flight Mass Spectrometer

The final stage in our account of the evolution of TOF mass spectrometry concerns the introduction of the oa technique, invented by Guilhaus and Dawson [355,356], working at the University of New South Wales, Australia. We have seen already in the preceding Section how the reflectron enables high resolution to be achieved with a TOF instrument, provided that the ion packets are formed in a well-defined plane. The oa method offers a means of injecting a pulse of ions formed from a continuous ion beam, for example from an ESI source, into the TOF analyzer.

The basic layout of the system is shown schematically in Figure 1.30. Ions emerging from the source (along the x-direction) at the top left hand side of the diagram are focused toward the 'pusher' region, where (for positive ions) a positive-going push-out pulse causes the ions to be ejected orthogonally toward the drift region (y-direction). While these ions are being analysed, the push-out potential is returned to zero so that the pusher region can be re-filled with ions. Although the collimating lens immediately after the source broadens the energy spread in the x-direction, the ions have very little energy in the y-direction. In order to produce a well-defined pulse of ions, the rise-time of the push-out pulse should be short, but the amplitude quite low (typically 100 V). The potentials applied to grids 2 and 3 are optimized to meet the requirements of spatial focusing. Obviously this oa technique can be combined advantageously with a reflectron to reduce further the deleterious effects upon the resolution

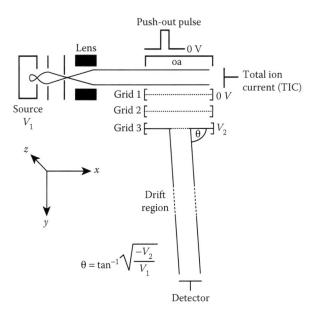

FIGURE 1.30 Schematic diagram of the orthogonal acceleration time-of-flight mass spectrometer. For details, see text. (Reproduced from Guilhaus, M., and Dawson, J.H., *Rapid Commun. Mass Spectrom.* 1989, *3*, 155–159. With permission from J. Wiley & Sons, Inc.)

of kinetic energy spread of the ions in the *y*-direction. Some further details of the oa TOF mass spectrometry, together with examples of its applications, have been given by Guilhaus and co-workers [357,358]. Recent reports by Blake *et al.* [359,360] have described the combination of an oa-TOF mass spectrometer with a PTR instrument for use in the analysis of trace volatile organic compounds.

1.5.4 RADIOFREQUENCY DEVICES

1.5.4.1 Introduction

Mass spectrometers that utilize radiofrequency quadrupole electric fields form one of the most versatile and diverse classes of analyzers in current use. Indeed, over two-thirds of the chapters in Volumes 4 and 5 of *Practical Aspects of Trapped Ion Mass Spectrometry* are concerned with a wide range of applications of these devices, especially of the 3D QIT and the more recent addition to the range of trapping instruments, the 2D linear ion trap (LIT). However, in concentrating on these means of trapping ions, there is a danger of forgetting the immense contribution that the QMF has made to the advancement of analytical mass spectrometry, especially in association with chromatographic separation techniques and in the serial combination of several analyzers, both quadrupole and non-quadrupole, to tandem mass spectrometry.

Due to the importance of RF quadrupole-based instruments, rather than include an abbreviated account of the theory in this chapter or provide various alternative derivations in the other individual chapters in the two volumes, we have elected to attach an Appendix that contains a detailed development of the theory, starting from the basic underlying physical principles. Consequently, in examining the historical evolution of these systems there are instances where we have referred the reader to the relevant equations and figures presented in the Appendix, rather than risk duplication of the material or find that the logical exposition of the theory is too dispersed.

A common feature of the different RF quadrupole systems under consideration is that they all rely upon *trajectory stability* for effecting mass analysis, ion selection, and ion trapping. As we shall see, thoughts about employing this physical principle to mass spectrometry began to emerge in the early 1950s, essentially from groups who were concerned with developing accelerator technology. At the same time, other instruments based upon the use of RF electric fields were also being developed; in the main these alternative types can be regarded as relying on *ion acceleration* or *phase stability*, operating basically on the same principle as the linear accelerator (see also Section 1.5.2.2). The key distinction between these acceleration devices and the transmission quadrupole analyzers, for example the QMF, is that in the former the radiofrequency potential is applied in the direction of travel of the ions in order to increase (or decrease) their velocity, whereas in the latter oscillating electric fields are applied orthogonally to the direction of travel so as to influence the transverse components of the motion of the ions.

In the main, these alternative forms of radiofrequency mass spectrometer were far less successful than the QMF, although it was not immediately evident at the time which ones would succeed and which would be consigned to the history books. Thus, while the performance of the quadrupole devices has continued to improve as a result of increased sophistication of the associated technology, leading to an increased range of applications, instruments such as the Redhead spectrometer and the Topatron (see Section 1.5.4.3.4), while being manufactured commercially for a period, have now ceased to exist. A common feature is, however, that like the omegatron discussed in Section 1.5.2.3, the main target markets for these early mass spectrometers were as low-specification residual gas analyzers and leak detectors for high vacuum systems. Notwithstanding the eventual demise of the 'acceleration spectrometers', before embarking upon the main subject of this Section, namely instruments based on the use of RF quadrupolar potentials, it is, perhaps, appropriate to consider briefly some these *Dodos*, in order to place the successful story of quadrupole development in context.

We shall begin, in fact, with what Blauth has described (see [270, p. 270]) as being the very first dynamic mass spectrometer to be developed, namely an instrument that combined a RF velocity selector with an electrostatic sector, the latter for analysis of the kinetic energies of the ions and hence providing mass analysis.

1.5.4.2 The Use of Radiofrequency Fields for Velocity Selection

In the RF velocity filter mass spectrometer described by Smythe [361–363], a continuous beam of positive ions emerging from the ion source was subjected to a number of alternating electric fields, generated between a series of pairs of parallel plates, applied at right angles to the beam in such a way that all but certain velocities were removed from the beam. The velocity-selected beam was then passed through an electrostatic sector so that detection of the transmission of ions of energy equal to $\frac{1}{2} mv^2$ would immediately give a value for m when v was known. There were certain defects in the earliest version [361], which resulted in some ions with undesired velocities being transmitted, and this problem was addressed in the later version of the system [362,363].

1.5.4.3 Radiofrequency Ion Acceleration Spectrometers

1.5.4.3.1 The Beckman Spectrometer

In some respects this instrument was similar to the velocity selection analyzer described in the previous Section (Section 1.1.5.4.2.), and was, in fact, a 'decelerating' spectrometer [364]. A beam of ions formed by EI was accelerated to 2500 V, collimated, and passed through a series of 24 RF 'gaps', spaced such that the preferred (or resonant) ions were decelerated by an amount equal to the root mean square (RMS) value of the RF voltage at each gap; all other particles were emitted from the RF analyzer with correspondingly higher energies. The resultant 'energy dispersed' beam was then separated by injection at an angle of 45° into an electrostatic deflecting system, the preferred (lowest energy) ion beam being collected in a Faraday cup. A linear mass scan could be achieved by varying the frequency of the alternating voltage applied to the gaps.

1.5.4.3.2 The Bennett Spectrometer

In the Bennett spectrometer [365,366], ions were passed through a series of three-grid sets, with DC voltages applied to the outer pair of grids of each set and an RF potential to the central grid. Ions were either accelerated or retarded, depending upon their velocity, and on the frequency and amplitude of the RF voltage at the instant at which the ions pass through the grids (phase angle). Those ions having the greatest kinetic energies were allowed to pass through a retarding grid and recorded as the amplitude or frequency of the RF supply was scanned. The performance could be enhanced by increasing the number of acceleration stages.

1.5.4.3.3 The Boyd Spectrometer

The Boyd mass spectrometer [367,368] works on a similar principle to that of the Bennett analyzer but comprises a stack of equally spaced electrodes with circular apertures (diameter *ca* 3 mm), which are connected alternately either to RF or to DC voltages. Again a retarding field is applied in order to detect only the most energetic ions, and either the amplitude or the frequency of the RF potential is scanned in order to record a mass spectrum.

1.5.4.3.4 The Redhead Spectrometer

The Redhead design of acceleration mass spectrometer [369,370] differed from the Bennett design in that the analyzer comprised a stack of some 20 grids connected alternately either to a DC potential supply or to one comprising a combined DC and RF voltage. Sinusoidal or rectangular wave potentials were applied, and again only those ions that had acquired sufficient kinetic energy were able to overcome the retarding field in front of the detector. This principle of operation was employed commercially by Leybold-Heraeus in their 'Topatron' (standing for *to*tal and *par*tial pressure measurement) residual gas analyzer. Further details of the operation of the

Topatron can be found on page 41 of Blauth's book [270] and page 106 of the survey by Ball *et al.* [289]. Typically a scan range of 2–150 Th was achieved, with a resolution of 40 (FWHM).

1.5.4.4   Instruments Employing Radiofrequency Quadrupolar Potentials

1.5.4.4.1   Introduction

In this part of our discussion we offer a brief account of the different variations in mass spectrometer instrumentation that have been based upon the concept of containing ions by the use radiofrequency quadrupolar potentials. As noted earlier, a detailed exposition of the basic physics behind this approach is presented in the Appendix to this chapter; at this stage we have attempted to trace the historical evolution of the technology in a more discursive and essentially non-mathematical style.

There is, perhaps, one point that should be clarified at this stage, namely the fact that the term *quadrupolar* refers to the situation in which the potential at a point within the device, and to which an ion may be subjected, depends upon the square of the distance of the point from the origin of reference; it does not arise because the QMF comprises an array of four rod-shaped electrodes! The electric field generated by a difference in potential between two points is equal to the first derivative of the potential with respect to displacement along the direction of the potential and, therefore, when the potential is quadrupolar as defined in the previous sentence, the strength of the electric field has a linear dependence upon distance. On occasions one reads about 'quadrupole field' devices as a short-hand description of the electric field generated by the applied quadrupolar potential. The term 'quadrupole field' is, of course, a misnomer in this context; indeed the main reason for ions behaving as they do within 'RF quadrupole' devices is that they are subject to one or more *linearly* dependent electric fields. As we shall see later in the discussion (Sections 1.5.4.4.5. and A4.1.), imperfections in the quadrupolar potential, deliberate or unintentional, can give rise to contributions from higher-order, non-linear, fields being superimposed upon the dominant linear fields, the results of which may enhance or diminish the performance of the device depending upon the circumstances (see, for example, Volume 4, Chapters 13 and 14)*.

1.5.4.4.2   Introduction Historical Development of RF Quadrupole Devices

The summary account that follows draws extensively on other surveys and reviews that have appeared in the literature over the years, and particular mention should be made of the texts by Dawson [371], March, Hughes, and Todd [266], and March and Todd [269]; readers are recommended to refer to these publications for further details and background information on the various topics covered.

The two-dimensional quadrupole mass filter, the three-dimensional radiofrequency quadrupole ion trap, and the LIT and are, in fact, the only three 'surviving'

---

* See Volume 4, Chapter 13: An Examination of the Physics of the High-Capacity Trap (HCT), by Desmond A. Kaplan, Ralf Hartmer, Andreas Brekenfeld, Jochen Franzen, and Michael Schubert; Volume 4, Chapter 14: Electrically-induced Nonlinear Ion Traps, by Gregory J. Wells and August A. Specht.

members of a family of devices that utilize path stability as a means of separating ions according to the ratio mass/charge-number *(m/z)*. Those that have 'disappeared' include the monopole mass spectrometer and a device comprising three pairs of parallel plate electrodes arranged in a cubic configuration. The original public disclosure of the operating principle of the quadrupole mass spectrometer (as the QMF was described initially) was the patent by Paul and Steinwedel [372], working at the University of Bonn and filed in 1953, and to Paul *et al.* [373], filed in 1958. Paul's discovery of the principle of confining ions by means of strong-focusing electric fields arose as he and the members of his group were designing a new high-energy accelerator that was to be constructed under the streets of Bonn. Once he had established the fundamental characteristics of quadrupole devices and built his accelerator, Paul's immediate interest in quadrupole mass spectrometry disappeared. Only when he was preparing his Nobel Lecture did Paul discover, to his amazement, what remarkable things analytical chemists were doing with 'his' invention [374]!

In fact, the same ideas were also put forward in 1953 by Post and Heinrich [375] for a "mass spectrograph using strong focusing principles", and by Good [376] for "a particle containment device". It would appear that Courant, Livingston* and Snyder [377] stimulated these proposals through the publication, in the previous year, of the theory of strong focusing of charged particle beams using alternating gradient quadrupole magnetic fields. Yet N. C. Christofilos [378], an electrical engineer working in Athens, Greece, had discovered the strong-focusing technique two years earlier. Despite his applications for patents, only one was granted [379]; a report on his work was submitted of to the University of California Radiation Laboratory but his work was largely overlooked at that time. Wolfgang Paul and his colleagues at the University of Bonn recognized the principle of using strong-focusing fields for mass analysis [380], and the first detailed account of the operation of a QIT appeared in the thesis of Berkling [381] in 1956.

In the account that follows we consider first two two-dimensional mass analyzers, the QMF and the monopole mass spectrometer, followed by a discussion of the three-dimensional quadrupole ion trap, the cylindrical ion trap (CIT) and the digital ion trap (DIT). Finally, we consider three variants of the relatively new 'linear ion trap'.

1.5.4.4.3   The Quadrupole Mass Filter

A mathematical treatment of the theory underlying the operation of the QMF is presented in Section A2.1 of the Appendix to this chapter; this account is intended to be more discursive in order to provide the reader with a general appreciation of the principles involved.

The QMF comprises a square array of four accurately machined and aligned conducting rods. This array is interposed between an ion source and a detector, usually an electron multiplier of some type, as shown in Figure 1.31. Opposite pairs of rods are coupled together and an electrostatic radiofrequency potential applied between

---

* M.S. Livingston, a co-author of Reference [377], is the same person who, with E.O. Lawrence, first developed the cyclotron (see Section 1.5.2.2). Livingston received the Enrico Fermi award in 1986, the same year in which he died.

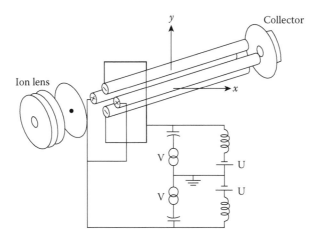

FIGURE 1.31   Schematic diagram of the electrode arrangement of the quadrupole mass filter. The symbols U and V denote, respectively, DC and RF potentials applied between opposite pairs of rod electrodes. For further details, see text. (Reproduced from March, R.E. and Todd, J.F.J., *Quadrupole Ion Trap Mass Spectrometry*, 2nd ed., John Wiley & Sons, Hoboken, NJ, 2005. With permission.)

the pairs. In addition, the opposite pairs of electrodes may be biased with equal and opposite DC potentials, applied positively in the *x*-direction and negatively in the *y*-direction. The intention is to create equal but opposite quadrupolar potential distributions in each of the *x*- and *y*-directions, taken with respect to the central axis (*z*-direction) as being the origin. Ideally the inner surfaces of the rods should be hyperbolic in cross-section in order to maintain the correct quadrupolar shape of the potential distribution. For ease of manufacture it has been found that, in practice, round rods may be employed provided the ratio of the diameters of the rods to that of the inscribed circle that is tangential to the inner surfaces of the electrodes is of the correct value.

Let us consider initially the case where the DC bias applied between the *x*- and- and *y*-pairs of electrodes is set to zero. When a positive ion is injected (at energies of a few electronvolts) from the source so that it moves through the electrode array (but not precisely along the *z*-axis itself), it will be subjected to electrical forces acting orthogonally to the direction of motion along the axis. The ideal quadrupolar potential distribution will result in these forces varying linearly with the displacements of the ion in the *x*- and *y*-directions, and the three components of the motion of the ion (that is, *x*, *y*, and *z*) can be considered independently. Thus, with the application of a sinusoidally varying RF potential between the *x*- and *y*-electrode pairs, when the potential on the *x*-electrodes is positive the force acting upon the ion along the *x*-direction will cause *focusing* toward the *z*-axis; at the same instant the potential on the *y*-electrodes has an equal and opposite negative value, causing the ion to be *defocused* away from the *z*-axis, toward one of the *y*-electrodes. As the RF potential changes phase, so will the directions of focusing and defocusing change. As noted by Paul [382], in a periodic inhomogeneous electric field, like that acting within a quadrupole device, "there is a small average force left, which is always in the direction of the lower field, in our case toward the center." Consequently, certain conditions exist that enable the ions to be transmitted through the array without hitting the electrodes. These conditions depend upon maintaining the correct balance between the *m/z*-values of the ions, the inscribed diameter of the electrode array, the frequency and amplitude of the RF potential applied between the electrode pairs, and the magnitude of any DC potential bias between the electrode pairs. Ions that satisfy these conditions are said to possess *stable* trajectories, while those that do not satisfy these conditions develop trajectories that are *unstable*, leading to losses of the ions, generally by collision with the electrodes. Mathematically, the behavior of ions when subjected to the forces acting within an RF quadrupole device is described by the Mathieu equation. This equation contains two parameters, symbolized normally as *a* and *q*, which define the stability criteria. While both parameters include the values of *m/z*, the inscribed radius (r_0), and the frequency of the oscillating potential, *a* is directly proportional to the magnitude of the DC bias potential and *q* is directly proportional to the amplitude of the RF potential applied between the electrode pairs. An important property of the Mathieu equation is that regions of *stability* and of *instability* exist in (*a*, *q*) space, known as 'stability diagrams', that define whether, in the quadrupole devices, ions will remained confined within the electric field or be lost. A more detailed mathematical account and examples of stability diagrams are given in the Appendix.

In our discussion so far we have considered the particular case where the DC potential applied between the pairs of electrodes is set to zero, that is *a* = 0. If we now apply a non-zero value of the DC potential bias between the pairs of rod electrodes, the symmetry of the combined DC and RF electric fields acting, respectively, in the *x*- and *y*-directions will no longer exist: there will always be a net focusing force acting in the *x*-direction and a net defocusing force acting in the *y*-direction that will be superimposed upon the small average force resulting from the periodic inhomogeneous field (see earlier). If, for a given value of the amplitude of the RF potential, fixed inscribed radius and oscillation frequency, and for a given value of *m/z*, we increase the magnitude of the DC potential progressively there will come a point where the ion trajectory is no longer stable. Changing the value of the amplitude of the RF potential and repeating the experiment results in trajectory instability occurring at different values of the DC potential. When ions of differing values of *m/z* are contained within the device, conditions may be selected in which some ions maintain stable trajectories, whereas other ions develop unstable trajectories and are lost. In order to operate the electrode array as a quadrupole mass filter, the values of the DC and the RF amplitudes are scanned linearly, keeping a constant ratio between, them so that at any one time only ions corresponding to one *m/z*-value are transmitted to the detector. As the scan continues a succession of ions with increasing values of *m/z* each satisfy briefly the stability criteria and are recorded, resulting in a mass spectrum. The 'filtering' action works in such a way that during the transmission of any one *m/z*-value, ions of lower values are lost in the *x*-direction, whereas ions of higher values are lost in the *y*-direction.

The resolution of the mass filter is determined by the ratio of the amplitudes of the DC and RF potentials, that is, by the quotient of *a/q*. When *a* = 0, corresponding to the condition we discussed earlier, and the RF amplitude only is scanned, an essentially unresolved ion beam extending over a wide range of *m/z*-values is transmitted, giving rise to a broad signal trace recorded on the detector system. Historically, this trace was called the 'total pressure curve', and was used as a measure of the pressure in a vacuum system when the QMF was being employed as a residual gas analyzer. A detailed illustration of this mass filtering process has been presented by Lawson and Todd [383]. Nowadays, in 'triple quadrupole' tandem mass spectrometers, the central quadrupole rod set ('Q2') is run in the RF-only mode, possibly with scanning of the RF amplitude in order to optimize the conditions, while is it acting as a collision cell so that the remaining precursor ions and any product ions are guided toward the third mass filter (Q3) for mass analysis.

It should be noted that, unlike most mass spectrometers we have considered thus far, the kinetic energy of the ions being analyzed is not a fundamental parameter responsible for the analysis process itself. For this reason, the 2D quadrupole device that we have been discussing is referred to generally as being a 'filter', rather than a 'spectrometer'. It should be noted though the kinetic energy of the ion beam is not a primary factor in determining the mass-analysis process, it is not entirely without influence. This is because, other things being equal, the resolution of the mass filter is determined by the number of RF cycles to which the ion is exposed. Thus a longer electrode array should result in improved resolution and, similarly, slower-moving ions should be better resolved. However this improvement in performance comes at a price: the greater the length of time taken for the mass-analysis process, the greater the risk of ion losses, resulting in lower ion transmission and hence reduced sensitivity. These ion losses may be the result of scattering collisions with background gases (although the pressure constraints for the operation of the QMF are not generally as stringent for other types of analyzer), imperfections in the electric fields, and a combination of adverse initial entry conditions relating to the RF phase as well as the angular spread and off-axis velocity components of the ions. A detailed appraisal of the factors affecting the performance of the QMF has been provided by Austin, Holme, and Leck [384]. A further complication is that electric fringing fields through which the ions must pass between the ion source and the electrode array of the mass filter can cause losses of ions that would otherwise have retained stable trajectories. Brubaker [385] provided a solution to this problem by mounting a small RF-only 'pre-filter' section on the main electrode array, which has the effect of delaying the exposure of the ions to the influence of the DC potentials applied to the latter during the mass spectral analysis processes. This approach is incorporated into many modern commercial mass filter instruments. The effects of the entrance field on the performance of the QMF have recently been re-examined in detail by Banner [386].

1.5.4.4.4   The Monopole Mass Spectrometer

Some 10 years after Paul and co-workers' original description of the QMF [372,373], von Zahn [387,388] described an instrument based on the operating conditions of the former, but with important differences. For the analysis of positive ions, effectively one negative (y) quadrant of the normal QMF rod set was formed by removing three of the rod electrodes and replacing them by a grounded rectangular 'V'-shaped electrode, with the open side facing toward the remaining rod, so that the 'V' was aligned with the original asymptotic planar zero-potential surfaces. With a unipolar source of RF potential applied to the remaining negatively biased rod, the same stability conditions apply as in the normal QMF, provided that the displacement condition $|x| < y$ is maintained. However, analysis of the ion motion reveals that after a certain interval the ion will be driven into the minus y-quadrant and, therefore, collide with the 'V', unless the nodal cross-over point of the trajectory coincides with the exit from the field into the detector. This behavior, therefore, places theoretical and practical limitations on the operation of the monopole when compared to the QMF. A key difference is that the kinetic energy of the injected ions is now a *fundamental* operating parameter, that ensures correct positioning of the nodal cross-over point, rather than being *incidental* as in the QMF. For this reason the monopole is referred to as a 'spectrometer', rather than being a 'filter'. Mass spectral scanning with the monopole is performed in exactly the same manner as with the QMF, that is by sweeping the amplitudes of the DC and RF potentials together with a constant ratio between them, although because of changes in the stability criteria that result from the new electrode arrangement the resolution is much less dependent upon the precise value of the ratio (that is, a/q). Furthermore, for a given operating frequency, the mass/charge range of the monopole is greater than that of the QMF, although because of the more stringent entry criteria the transmission efficiency is reduced by *ca* 50%. More detailed discussions on the monopole and its performance can be found in the accounts by Lawson and Todd [383] and by Herzog [389].

1.5.4.4.5   *The Three-dimensional Quadrupole Ion Trap (QIT)*

This device, which has been called also the QUISTOR [390], standing for *qu*adrupole *i*on *stor*age trap, was described as "another construction of the electrodes" in the original patent for the QMF [372]. Essentially it is a cylindrically symmetric three-electrode structure comprising a single sheet hyperboloid (the 'ring' electrode) mounted between the two components of a two-sheet hyperboloid (the 'end-cap' electrodes). Normally the end-caps are connected together electrically, and combined DC and RF potentials applied between the ring electrode and the two end-caps. Figure A3 shows a vertical section through a QIT in photographic and diagrammatic form. In practice, the end-caps are usually maintained at ground potential and a unipolar RF voltage source connected to the ring electrode, although in many experiments potentials may be applied between the two end-caps while still being referenced to ground. In an ideally formed electrode structure of infinite extent, connecting the voltage sources in this way creates a three-dimensional quadrupolar potential within the ion trap so that an ion is subjected to electrical forces acting independently in each of the orthogonal directions x, y, and z. Conventionally z is taken as the direction of the rotational axis of symmetry, and x and y form the radial plane; for ease of mathematical treatment this three-dimensional co-ordinate system is often reduced to a description in terms of cylindrical (r, z) co-ordinates.

Since ions within the trap are subject to the oscillating quadrupolar potentials in a manner analogous to those operating within the QMF, but acting in three dimensions rather than two, their motion is described by appropriate Mathieu equations and exactly the same kinds of stability/instability criteria apply. An ion having a trajectory that is stable both radially and axially will, in theory, remain trapped indefinitely, although in practice ion losses may occur through collisional scattering and space-charge effects. A further factor is that imperfections in the potential distribution can lead to the presence of higher-order fields (see also Sections 1.5.4.4.1 and A.4.1.), and these in turn may cause the phenomenon of 'non-linear resonances'. In effect, under certain conditions where an ion trajectory is predicted to be stable within a pure quadrupole potential, resonant energy transfer may occur such that ion trajectories become unstable rapidly. Several detailed accounts of the history, theory, instrumentation, and applications of the QIT have been published, including those by Todd *et al.* [391], March *et al.* [266], and March and Todd [269], as well as Volumes 1 to 3 of the present Series [268].

Operation of the ion trap in a manner analogous to the total pressure mode of the QMF enables the storage of ions over a wide range of *m/z*-values, providing the basis for a variety of physical and chemical experiments to be carried out. Many investigations into various aspects of ion/molecule reaction chemistry, the infrared multiphoton dissociation of ions (IRMPD), and the dynamics of trapped ions were carried out by the current authors in the 1970s and 1980s [392], using mainly a system in which the QUISTOR was used in association with a quadrupole mass filter. In addition, the atomic physics literature abounds with accounts of the use of the QIT for high resolution spectroscopic studies [393]. Several contemporary examples of the way in which the QIT has enabled some novel and sophisticated experiments are described within different contributions to Volumes 4 and 5 of this Series.

1.5.4.4.5.1 Mass Spectrometry with the Quadrupole Ion Trap   So far we have considered only the trapping of ions within the QIT; in terms of the operation of the QIT as a mass spectrometer, its historical evolution can be divided into three separate 'ages of the ion trap' [266]. These are 'mass-selective detection', 'mass-selective storage', and mass-selective ejection'. Of these three, only the third has proved to be a commercially viable approach: indeed, without it the QIT may have become yet another *Dodo!*

1.5.4.4.5.1.1   Mass-Selective Detection   The method of mass-selective detection was indicated in the first patent awarded to Paul and Steinwedel [372], and its implementation was described in the pioneering publications by Paul *et al.* [394] and by Fischer [395]. For this technique, in which the ions were formed continuously, an auxiliary oscillation circuit supplying a low-amplitude voltage of fixed frequency was connected between the end-cap electrodes, and the ions maintained in stable trajectories by means of a combined RF and DC potential applied between the ring electrode and the pair of the end-cap electrodes. For a fixed RF amplitude, the DC potential was scanned slowly in order to change the values of the *a* parameter until a point was reached when a frequency component of the motion of the first ion type to be detected came into resonance with that of the auxiliary circuit. The response within the auxiliary circuit to the resulting energy take-up was detected, and fed to the *y*-deflection of an oscilloscope whose *x*-deflection was driven by the scanned DC potential, and as successive ions came into resonance this led to the display of a mass spectrum. Although this method of detection was essentially non-destructive, in that the ion was not physically collected on an electrode, after detection the ions would be lost ultimately as the additional energy they had acquired would lead to trajectory instability. As we shall see below (Sections 1.5.4.4.5.1.3 and 1.5.4.4.5.2), in current ion trap instrumentation this method of resonantly exciting the ion trajectories may still be part of the overall experimental procedure, but ion losses are prevented by collisional cooling through the presence of buffer gas.*

In practise this method of utilizing the ion trap as a mass spectrometer was not particularly successful: the resolution and mass range were severely limited, and because the heavier ions were present during the detection of the lighter ions but not *vice versa* the conditions under which the ions were detected changed during a scan.

1.5.4.4.5.1.2   Mass-Selective Storage   This method of generating a mass spectrum with the QIT was developed by Dawson and Whetten [396–398] in 1968, and is notable because of two conceptually different changes in approach. Firstly, the ions were created as a discrete packet, using a gated electron beam rather as in the early TOF mass spectrometers (see Section 1.5.3.) and, secondly, the ions are detected by pulse-ejecting them from the trap through holes in one of the end-cap electrodes and into an external multiplier. As we shall see (Section 1.5.4.4.5.1.3), both these features are incorporated into modern ion trap mass spectrometers that use mass-selective ejection. The actual mass spectral scanning process resembles closely that employed for the QMF (see Section 1.5.4.4.3). Combined RF and DC potentials are applied to the ring electrode and their amplitudes scanned relatively slowly, maintaining a constant ratio between them. The value of this ratio is adjusted so that at any one time only ions of one m/z-value are retained within the trap. One end-cap electrode is grounded while the other, nearest to the detector, is connected to a pulse generator. The cycle of operation comprises creating ions (by admitting electrons) for a define period (typically 5 ms), storing them for *ca* 25 μs, and then applying a rectangular DC pulse to the 'exit' end-cap electrode in order to eject the ions into the electron multiplier detector. Because of the significant number of ions being ejected from the trap during the storage period on account of their trajectory instability, it is desirable to minimize the loading on the detector by use of a gating arrangement together with 'box-car' detection that is only operational when the intentionally ejected ions are expected.

Many useful developments and refinements to the use of the QIT as a mass spectrometer based upon the mass-selective storage method were carried out by Sheretov and his group [399–401]. Compared to the mass spectra produced by the mass-selective detection method, the quality of the data obtained by this new method was excellent; however, in terms of modern requirements the scan speed was still relatively slow. In carrying out their own experiments using mass-selective storage, Mather, Waldren, and Todd [402] found that the resolution and peak shape could be improved further by superimposing a small negative DC potential (*ca* 5 V) on the ring electrode, which probably counteracted space-charge effects. Although both Dawson and Whetten were working at the General Electric Company at the time of the invention of the mass-selective scanning mode, no commercial instruments appear to have been developed. However, although the possible mass spectral potential of the ion trap operated in this manner was never fully realized, this method of pulse-ejecting the ions from the trap did provide the stimulus for the creation of the

---

* See Volume 4, Chapter 17: Collisional Cooling in the Quadrupole Ion Trap Mass Spectrometer (QITMS), by Philip M. Remes and Gary L. Glish; Volume 4, Chapter 18: 'Pressure Tailoring' for Improved Ion Trap Performance, by Dodge L. Baluya and Richard A. Yost.

QUISTOR and the research conducted by the current authors utilizing this device as a storage source in conjunction with an external mass analyzer (see above).

1.5.4.4.5.1.3 Mass-Selective Ejection   As noted above (see Section 1.5.4.4.5.1), at this stage in the evolution of the QIT as a mass spectrometer all prospects of the device being turned into a commercially successful instrument capable of offering a level of performance exceeding that which was already available through the use of other technology appeared to have been exhausted. However, in the quest for a new type of cheap, small-scale, instrument that could be integrated into a gas chromatography/mass spectrometer (GC/MS) system, George Stafford and colleagues from the Finnigan Corporation became aware of a presentation by one of us (REM) at the Seattle meeting of the American Society for Mass Spectrometry in 1979. Armitage and March [403] had coupled their QUISTOR/quadrupole mass spectrometer system to the outflow of a gas chromatograph and were performing GC/MS analyses by repeatedly ionizing the eluent (using both EI and CI) in the ion trap and pulse-ejecting the ions into the repetitively scanning quadrupole mass filter. Stimulated by the prospect of using the ion trap in this way, but unimpressed by the need for an external means of mass analysis, Stafford, in association with Paul Kelley and David Stephens, proceeded to devise a new way of generating mass spectral scans with the ion trap, resulted in the filing of a landmark patent in December 1982 [404]. This new method was the mass-selective ejection scan [405].

Like all good inventions, the underlying concept of the mass-selective ejection scan is really very simple, and is best seen in terms of the stability diagram for the QIT shown in Figure 1.32. The reader is referred to the Appendix to this chapter for a detailed account how the stability diagram is derived. Essentially Figure 1.32 is a plot of the Mathieu stability parameters (see Sections 1.5.4.4.3. and A5.1.) *a versus q*; the subscripts 'z' denote that the values refer to stability in the axial direction of motion; the equivalent values for stability parameters in the radial direction may be found by dividing a_z and q_z by two, respectively. As with the QMF, for an ion of given *m/z*-value situated within an ion trap of defined geometry and subject to RF potential having a given frequency, the value of *a* is determined by the magnitude of any DC potential applied between the ring and the end-cap electrodes, and the value of *q* is determined similarly by the amplitude of the RF potential. For an ion to remain confined within the trap, its corresponding values of a_z and q_z must lie within the outer boundaries of the diagram. The lines running diagonally across the diagram and labeled β_r or β_z denote certain frequencies that are present within the radial and axial components of the trajectories of the ions, respectively. General reference to these frequency characteristics was made earlier in our brief account of the mass-selective detection method (Section 1.5.4.4.5.1.1).

When operating the QIT in the mass-selective ejection mode, the RF voltage source is connected to the ring electrode only, and in the simplest, and original, version of the method the two end-cap electrodes were grounded. Improved performance and increased versatility of the range of experiments available is achieved by coupling an auxiliary, low-amplitude, RF power supply between the end-cap electrodes. Through adjustment of the values of the applied RF and DC trapping voltages, the (a_z, q_z) co-ordinates for ions with a specified *m/z*-value can be positioned so as to coincide

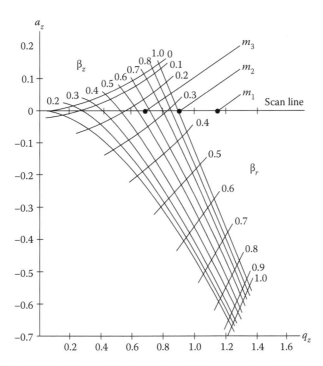

FIGURE 1.32   Stability diagram for the quadrupole ion trap plotted in (a_z, q_z) space. The points labeled m_1, m_2, and m_3 denote ions of successively greater values of m/z as the respective values of q_z move along the q_z-axis (indicated as the 'scan line') as a result of increasing the amplitude of the RF potential in order to effect mass-selective ejection. For further details, see text. (Reproduced from March, R.E. and Todd, J.F.J., *Quadrupole Ion Trap Mass Spectrometry*, 2nd ed., John Wiley & Sons, Inc., Hoboken, NJ, 2005. With permission from John Wiley & Sons, Inc.)

with a particular value of β_z so that, when the auxiliary power supply is tuned to the corresponding frequency, resonant excitation of the axial motion of the ion occurs.

In describing the sequence of events during the generation of a mass spectrum using the mass-selective ejection mode, it is easiest to envisage the behavior of three types of ions having m/z-values $m_1 < m_2 < m_3$ confined within the trap through application of an RF potential only, that is the values of a_z for each are equal to zero. As a result the corresponding values of q_z for each of the ions map on to what is labeled the 'scan line' in Figure 1.32. Initially, with a low-amplitude of the RF potential, all the three points for m_1, m_2, and m_3 will lie close to the origin, well within the envelope of the stability diagram. However, if the amplitude of the RF trapping potential is now increased linearly, the corresponding values of q_z for each of the ions will move along the scan line toward the boundary labeled $\beta_z = 1.0$; the ions of lowest value of m/z will reach the boundary first, whereupon their components of motion in the axial direction will suddenly become unstable so that the ions fly out of the trap and on to a suitably positioned detector. Continuing the scan of the RF potential causes the remaining ions to move toward the boundary and ultimately be ejected, resulting in the generation of a mass spectrum. Figure 1.32 shows the situation where the m_1

ions have already been ejected (mass-selectively) and detected, the m_2 ions are on the point of ejection and the m_3 ions are still confined within the trap. Once complete, the RF potential is returned to its starting value and the operational cycle re-commenced by either forming a fresh packet of ions within the trap (for example by EI), or by injection from an externally mounted ion source.

While the earliest mass spectra recorded by this method were tolerable, a number of subsequent technical and operational changes have enhanced the quality of the mass spectral performance enormously. The first of these, discovered somewhat fortuitously, was that the collisional cooling of the ions through addition of a light buffer gas, typically helium at a pressure of *ca* 1 m Torr, improves both the mass spectral resolution and sensitivity of the device. Chapters 17 and 18 of this Volume describe this particular aspect in more detail. Further refinements have included controlling accurately the number of ions present in the trap ('automatic gain control', AGC) in order to reduce adverse effects on performance caused by space-charge interactions, and the application of an auxiliary RF voltage at the appropriate frequency applied between the end-cap electrodes so as to sharpen the ion ejection process and, thereby, improve resolution ('axial modulation'). Extension of the mass spectral range of the ion trap has been made possible through the creation of instability zones within the normal stability diagram through resonant excitation of the axial motion of the ions at lower values of β_z. It has been found also that reducing the scan speed can improve the resolution substantially. Of particular significance have been the improvements gained by the use of contributions of higher-order fields to the normal trapping field: these have been effected through adjustments of the electrode geometry of the ion trap electrode arrangement, either by changes to the spacing of the electrodes or through deliberate changes to shapes of the electrodes. Further details of these aspects, including literature citations, are to be found in [269]. Chapter 13 of this Volume describes the role played by the judicious use of higher-order fields in enhancing mass spectral performance of the QIT, and Chapter 14 describes how such fields can be induced by the superposition of different types of oscillating potentials coupled to the ion trap.

1.5.4.4.5.1.4   A Fourier Transform Operating Mode Quadrupole Ion Trap   A fourth method of undertaking mass analysis of ions stored in the QIT is the subject of Chapter 9 of this Volume,* to which the reader is directed for a full and detailed account of this novel approach. A schematic representation of the apparatus and the timing sequence is shown in Figure 9.1. In essence, the overall process involves a series of 'elementary experiments' where, following ion injection and confinement for a range of different times under defined conditions, the bunch of trapped ions to be mass-analyzed is pulse-ejected, as in the ejection stage of the mass-selective storage experiment but without prior mass-selection (see Section 1.5.4.4.5.1.2), and the arrival time profile of the different species at a fast-response detector measured. The resulting TOF histogram represents an 'image signal' of the motions of the simultaneously trapped ions, Fourier transformation which gives the axial secular frequencies with different *m/z*-values at the moment, thus generating a mass spectrum.

---

* See Volume 4, Chapter 9: A Fourier Transform Operating Mode Applied to a Three-Dimensional Quadrupole Ion Trap, by Y. Zerega, J. Andre, M. Carette, A. Janulyte and C. Reynard.

1.5.4.4.5.2   Tandem Mass Spectrometry with the Quadrupole Ion Trap   In addition to the basic use of the QIT as a mass spectrometer, a major contributing factor to the success of the QIT as an analytical device is its facility for performing tandem mass spectrometry (mass spectrometry/mass spectrometry, MS/MS or MS^2) experiments, as described initially by Louris *et al.* [406]. As the ion trap works on a repetitive time-sequence of individual steps, it is possible to incorporate easily additional stages into the overall duty cycle. In the MS/MS mode of operation, ions are first created within or injected into the ion trap, allowed to 'cool' collisionally, and then subjected to a first stage of mass analysis, wherein the intended precursor ion is selectively retained while the unwanted ions are removed, usually by a combination of RF potential scanning and resonant excitation. The precursor ion is then itself resonantly excited ('tickled') with a low-amplitude auxiliary dipolar wave form of the appropriate frequency applied between the end-cap electrodes in such a way that instead of the ions being ejected from the trap they become energized by the main RF trapping potential and undergo repetitive collisional excitation with the buffer gas atoms. After an appropriate interval the resulting product ions are then mass-analyzed by scanning them out of the trap for detection. This 'tandem-in-time' experiment is extremely versatile, and additional mass-selection/collisional excitation steps can be built into the duty cycle so as to offer multiple stages of MS/MS (MS^n). A favorable factor is the high efficiency of CID compared to that normally obtainable with beam ('tandem-in-space') instruments. Chapter 16 of this volume* illustrates wonderfully just how sophisticated the application of the QIT to the tandem mass spectral investigation of biologically important compounds has become.

1.5.4.4.5.3   The Cylindrical Ion Trap   The CIT is essentially a geometrical approximation to the idealized QIT, with the normal hyperboloidal end-cap electrodes being replace by planar round discs and the ring electrode by a cylindrical tube. The earliest published accounts of such a device were the paper by Wuerker *et al.* [407] and the patent by Langmuir *et al.* [408], filed in August 1959. Although containment of massive charged particles within the device was demonstrated, there did not appear to be any direct mass spectrometric applications demonstrated. Interestingly, the patent contains also a description of an ion trap having a cubic geometry, comprising three opposing pairs of planar electrodes, each pair being coupled to the output from a three-phase RF power supply. Although the shape of the stability diagram for this device was determined experimentally [409], no other reports on ion traps based on this approach appear to have been reported.

Some 10 years after the studies by Langmuir and co-workers, Benilan and Audoin [410], and subsequently Bonner *et al.* [411] fabricated a cylindrical trap that was then characterized in some detail in a joint research program between the current authors' respective groups [412,413]. A number of other 'physical applications' of the CIT have been summarized in Chapter 6 of [269].

---

* See Volume 4, Chapter 16: Unraveling the Structural Details of the Glycoproteome by Ion Trap Mass Spectrometry, by Vernon Reinhold, David J. Ashline, and Hailong Zhang.

The return of the CIT to applications in the mass spectrometric context was inspired essentially by the discovery of the mass-selective ejection scan (see Section 1.5.4.4.5.1.3), with the group of Graham Cooks at Purdue University engaging in a major research program concerned with the miniaturization of ion traps, especially for monitoring applications involving the use of hand-held mass spectrometers. A full account of this work, including the development of the rectilinear ion trap (RIT), is presented in Chapter 2 of this Volume.*

1.5.4.4.6   The Digital Ion Trap

The DIT is the name given to a relatively recent addition to the range of ion trapping devices, in which the geometry of the electrode arrangement is essentially the same as for the 3D QIT but, rather than having the usual sinusoidally varying RF potential, the waveform of the trapping potential is rectangular, having been generated by means of digital electronics.

The concept of using rectangular waves for driving quadrupole devices is not new. For example, in a patent filed in 1971, Hiller [414] proposed that the potential applied to the electrodes of a quadrupole mass filter "may be formed by a repetitive sequence of segments each one of which is composed of either one or a number of linearly varying functions of time and/or one or a number of exponentially varying functions of time where the exponents of each portion are either real functions of time or complex functions of time but not purely imaginary functions of time." This patent covered rectangular and trapezoidal waveforms in addition to other functions. Richards *et al.* [415] derived stability criteria for a square waveform of fixed frequency and demonstrated a mass scan of a quadrupole mass filter with such a square waveform, and these authors [415] explored also the application of rectangular and trapezoidal waveforms to the QIT and described the corresponding stability diagrams. More recently, Sheretov *et al.* [416], using a switching circuit to generate a frequency scan of a pulsed waveform, operated a QIT in the mass-selective instability mode. However, overall, the use of rectangular waveforms for operating quadrupole devices had received only limited attention until the advent of the DIT. A full account of the theoretical and technical aspects of this instrument, together with examples of mass spectral data, is presented in Chapter 4 of this Volume, and a discussion of boundary-activated dissociation (BAD) experiments performed with the DIT is given in Chapter 12 of Volume 5.† The following brief description is included here for completeness, and reference will be made to some of the illustrations that form part of Chapter 4.

The principles upon which the DIT is based were first published in 2000 by Ding and Kumashiro [417], who described ion motion in a quadrupolar electric field that was driven by a rectangular wave voltage and suggested the mass-selective resonance method utilizing the ion secular frequency under digital operation conditions. The basis of this approach is shown schematically in Figure 4.1, in which the digital

---

* See Volume 4, Chapter 2: Ion Traps for Miniature, Multiplexed and Soft Landing Technologies, by Scott A. Smith, Chris C. Mulligan, Qingyu Song, Robert J. Noll, R. Graham Cooks and Zheng Ouyang.

† See Volume 4, Chapter 4: Rectangular Waveform Driven Digital Ion Trap (DIT) Mass Spectrometer: Theory and Applications, by Francesco Brancia and Li Ding; Volume 5, Chapter 12: Boundary-Activated Dissociations (BAD) in a Digital Ion Trap (DIT), by Francesco Brancia, Luca Raveane, Alberto Berton and Pietro Traldi.

signal generator supplies waveforms of the types shown in Figure 4.2. The circuits of the DIT switch very rapidly, with precisely controlled timing, between discrete DC high voltage levels in order to generate the trapping waveform voltage applied to the ring electrode. Referring to Figure 4.2a, the period of the asymmetric rectangular 'drive' waveform is T, where Td (with $d < 1$) is the duration of the positive voltage, V_1, and $T(1 - d)$ is the duration of the negative voltage, V_2; generally the values of the voltages are set so that $V_1 = -V_2 = V_0$. The timing parameter d determines whether or not there is the equivalent of a DC potential superimposed upon the amplitude of the drive potential: when $d = 0.5$, that is a duty cycle of 50%, the resulting symmetry of the positive- and negative-going portions means that the effective DC amplitude is zero, corresponding to a value of the stability parameter $a = 0$ (see Section 1.5.4.4.5.1.3.). Increasing the value of d or reducing it will introduce effective positive or negative DC levels, respectively. In the basic scanning mode, a mass spectrum can be generated with the DIT by ramping linearly the amplitude of V_0, thereby driving the trapped ions to the stability limit sequentially in order of increasing values of m/z. This process is exactly analogous to the basic mass-selective ejection scan in the 3D QIT. Alternatively, the frequency of the drive potential may be scanned in order to produce a mass spectrum.

Figure 4.2b shows the waveform of a supplementary dipolar excitation pulse that may be applied to the end-cap electrodes. Adjustable parameters are the delay of the midpoint of the DC pulse with respect to the rising edge of the rectangular waveform, given by t_d, and the width w_d of the DC pulse of amplitude V_d. The repetition rate of the supplementary DC pulse can be set so that it is a sub-multiple of that of the drive voltage, and in this way one can adjust the effective value of β at which ions of a specified value of m/z may be resonantly excited.

As noted above, reference should be made to Chapter 4 of this Volume for further details of this novel instrument, including a list of relevant literature citations.

1.5.4.4.7 *The Linear Ion Trap*

We have noted in Section 1.5.4.4.3. that when operated in the RF potential-only mode, that is, in the absence of any DC potential applied between the x- and y-pairs of rod electrodes, the QMF is capable of transmitting ions having a broad range of m/z-values. The LIT is essentially an extension of this technology, except that instead of allowing the ions to continue their axial motion toward a detector, they are retained within the electrode array through the application of appropriate axial trapping fields. Once ions are trapped, a range of experiments is possible, including MS/MS analyses and studies involving the interaction with oppositely charged ions, such as in electron transfer dissociation.* One of the advantages of the LIT over the conventional 3D QIT is the greater volume available in which to trap the ions, leading to increased capacity before the onset of space-charge limitations. However, depending upon the precise mode of operation, this gain in performance may not be so significant when set against the capabilities of the High-Capacity Trap (HCT),

---

* See Volume 5, Chapter 1: Ion/Ion Reactions in Electrodynamic Ion Traps, by Jian Liu and Scott A. McLuckey; Volume 5, Chapter 3: Methods for Multi-Stage Ion Processing Involving Ion/Ion Chemistry in a Quadrupole Linear Ion Trap, by Graeme C. McAllister and Joshua J. Coon.

described in Chapter 13 of this Volume. A further advantage of the LIT is the ease with which it may be coupled to a triple quadrupole instrument, or to a high resolution mass analyzer, such as the Orbitrap or an FT-ICR mass spectrometer. Other experimental arrangements that utilize the relatively open access afforded by the LIT include the facility to attach several ion sources. A detailed account the LIT is to be found in Chapter 5 of [269].

Returning to the specific operation of the LIT as a mass spectrometer, there are in fact three alternative types of instrument commercially available, one developed by Thermo Scientific, one by AB Sciex, and one by Hitachi.

1.5.4.4.7.1   The Thermo Scientific Linear Ion Trap   The first LIT to be disclosed publicly was the Thermo Scientific instrument, detailed by Bier and Syka [418] in a U.S. Patent filed in May 1994, and its performance was described in depth by Schwartz *et al.* in 2002 [419]. A schematic representation of the LIT itself is shown in Figure 1.33. In reality the device is mounted between an ion injection system, normally comprising an ESI source and a series of multipole ion guides so that ions can enter the trap along the *z*-axis, and one or two detectors are positioned laterally so as to intercept ions ejected along the *x*-direction through the slots shown in the 'central section'. A supply of helium buffer gas is connected to the system, and RF potentials are applied to all three pairs of *x*- and *y*-electrodes; in addition, for positive ions, the 'front' and 'rear' sections can be floated at an appropriate positive DC potential in order to confine the trapped ions axially. In order to scan a mass spectrum, the amplitude of the RF potential applied to the center section is increased linearly, and mass-selective ion ejection through the slots in the central *x*-pair of electrodes achieved by the application of an auxiliary dipolar excitation waveform between the two *x*-electrodes. This mode of operation is exactly analogous to that of the 3D QIT mass spectrometer, and the usual range of experiments

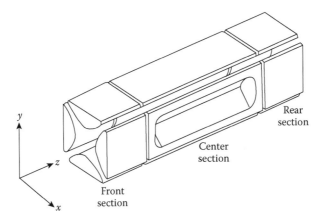

FIGURE 1.33   Angled view of three sections of the Thermo Scientific Linear Ion Trap. The detector faces the slot in the 'center section', and ions are injected along the *z*-axis through the 'front section'. For further details, see text. (Reproduced from March, R.E. and Todd, J.F.J., *Quadrupole Ion Trap Mass Spectrometry*, 2nd ed., John Wiley & Sons, Hoboken, NJ, 2005. With permission from John Wiley & Sons, Inc.)

involving MS^n can be performed also in the same manner. However, the additional flexibility arises from the ease with which a second ion source or an external, high performance, mass analyzer can be mounted along the z-axis at the 'free' end of the electrode array. Chapter 3 of Volume 5 of this series describes a variety of experiments performed with an instrument incorporating the Thermo Scientific LIT. Cooks and co-workers have developed the 'rectilinear ion trap' (RIT), which is the analogue of the CIT. Further information on the RIT is presented in Chapter 2 of this Volume.

1.5.4.4.7.2 The AB Sciex Linear Ion Trap   The AB Sciex LIT mass spectrometer, assembled as a complete instrument, is shown in Figure 11.2. This LIT was first disclosed publicly by Hager [420] in a U.S. Patent filed in June 1998; articles describing the system have been published by Hager [421] and by Londry and Hager [422].

At first sight one might think that the instrument shown in Figure 11.2 is simply an elaborate triple quadrupole mass spectrometer for tandem mass spectrometry studies, and in some respects it is. However, while the various rod sets can be used in the 'traditional triple quadrupole mode', those sections labeled Q2 and Q3 can also act, respectively, as high- and low-pressure linear ion traps by the appropriate biasing of the various interspace electrodes. The novel feature of this type of ion trap is that mass analysis is effected by 'mass-selective axial ejection' (MSAE). The idea is that ions can be induced to leave Q3 in the axial direction, toward the detector at the right hand side of the diagram, by ramping the amplitude of the drive RF voltage in the presence of a resonance excitation signal applied between one of the pair of rod electrodes. As a result of the combined fields, ions near to the exit end of Q3 having a large radial amplitude gain sufficient energy from the fringing fields (see also Section 1.5.4.4.3) to overcome a repulsive potential applied to the 'exit' lens and reach the detector. Details of this method, together with relevant citations, are given in Chapter 11 of this Volume, and an account of research carried out with an AB Sciex LIT instrument is given in Chapter 1 of Volume 5.

1.5.4.4.7.3 The Hitachi Axially Resonant Excitation Linear Ion Trap   The most recent addition to the range of linear ion traps is the 'axially-resonant excitation' (AREX) LIT developed by Hashimoto *et al.* [423] at Hitachi, Tokyo, Japan. Full details of the construction and operation of this instrument are given in Chapter 12 of this Volume.* This abbreviated account is included here in order to ensure that this survey of dynamic mass spectrometers is complete, and the reader should refer to Figures 12.1 and 12.2 as part of this discussion.

The AREX LIT, shown in Figure 12.1, comprises four round quadrupole rod electrodes together with four pairs of vane electrodes that are inserted between the rods, the whole assembly being mounted between a pair of 'Incap' and 'Endcap' electrodes. The two phases of the RF trapping voltage are applied to the rod pairs in the normal manner in order to confine the ions in the radial direction. The vane electrodes consist of a set of four 'front' vanes and another set of four 'rear' vanes,

---

* See Volume 4, Chapter 12: Axially-resonant Excitation Linear Ion Trap (AREX LIT), by Yuichiro Hashimoto.

the shapes of which were optimized by calculations of the electric potential; these are illustrated in the lower half of Figure 12.1. The four vane electrodes comprising each set are at the same potential, and the effect of this electrode arrangement is to create an essentially parabolic potential along the z-axis of the rod set (see Figure 12.2); as a result the ions develop secular oscillations along the z-axis, with each value of m/z having a characteristic oscillation frequency. In order to eject the ions, a supplemental AC voltage is applied between the front and rear vanes and the frequency of the AC voltage scanned so that when it matches the frequency of the secular motion corresponding to a given m/z-value these ions are excited resonantly and ejected out of the trap in an axial direction. A mass range from m/z 100 to m/z 1000 has been achieved at a scan speed of 5.7 Th ms^{-1}. A high ion-ejection efficiency is achieved by applying an appropriate potential between the incap and endcap electrodes.

Examples of mass spectra obtained with this new system, together with technical details of the instrument and its performance are given in Chapter 12.

1.5.4.4.8 The Toroidal and Halo Ion Traps

1.5.4.4.8.1 Introduction   Two other recent additions to the range of ion traps based upon the use of three-dimensional radiofrequency electric fields are those based upon a cylindrically circular geometry. These are the toroidal ion trap and the novel halo ion trap and are described in detail in Chapter 6 of this Volume.*

1.5.4.4.8.2 The Toroidal Ion Trap   The relationship between the toroidal ion trap [424] and the normal 3D QIT is illustrated in Figure 6.1. This figure shows that in the former device the space containing the ions is now a toroidal volume bounded by four continuous surfaces: two end-cap electrodes, a ring electrode, and a central 'spindle' electrode. In some respects this arrangement is similar in nature to a design described by Bier and Syka in the patent on linear ion traps [418], and indeed the toroidal trap might be considered to be a linear trap of infinite length in which the rod electrodes have been bent into a circle.

The underlying physical principle of achieving mass analysis with the toroidal trap is the same as the QIT, namely using the method of mass-selective ejection by scanning the amplitude of the RF potential applied to the ring and spindle electrodes, combined with a supplementary AC potential applied between the end-cap electrodes in order to effect resonant excitation. Ions formed in an external source may be injected into the trap through one or more apertures in one of the end-cap electrodes, and ion ejection occurs through slots in the other end-cap electrode, see Figure 6.2. The main virtue of the toroidal trap is that for the same value of the equivalent of the inscribed radius, r_0, in the QIT (see Section A3.1.1), and hence amplitude of the RF trapping voltage, the volume available for containing the ions is significantly greater while retaining the relatively compact nature of the structure.

In reality, it was found that the curvature of the electrode surfaces in the symmetrical electrode arrangement shown in Figure 6.1 introduced imperfections in the

---

* Volume 4, Chapter 6: Ion Traps with Circular Geometry, by Daniel E. Austin and Stephen A. Lammert.

trapping field that resulted in poor quality mass spectra. Modeling of the trapping field combined with ion trajectory simulation led to the creation of an asymmetric design of trap with which the mass spectral resolution was much enhanced. Further details of this instrument, and of a reduced-size toroidal ion trap, may be found in Chapter 6.

1.5.4.4.8.3   The Halo Ion Trap   The halo ion trap is a variant on the toroidal ion trap, in which the trapping fields are made using planar equipotential electrodes. Construction of the device makes use of microfabrication technology, and the electrodes comprise two ceramic plates whose inner-facing surfaces have been imprinted with closely spaced concentric metal rings and overlaid with a resistive layer of germanium. The appropriate toroidal trapping field is generated by a suitable potential function applied to the set of rings on each plate [425].

Schematic representations of the electrode arrangement and mass spectral operation are shown in Figure 6.15, and the reader is referred to Section 6.3.2 of Chapter 6 for further details. One intriguing possibility that results from the flexible nature of the electrode arrangement is the ability to 'construct' totally different forms of ion trap (see Section 6.3.3). For example, Figure 6.19 illustrates a 'double trap' in which a QIT is nested within a toroidal ion trap, raising the possibility of conducting different mass spectral operations, for example CID, in one trap followed by ion transfer and mass analysis in the other trap.

1.5.5   THE KINGDON TRAP

1.5.5.1   Introduction

One of the most exciting events in twenty first century mass spectrometry must surely be the introduction of the Orbitrap trap mass spectrometer, marketed by Thermo Scientific. This new type of analyzer was invented by Alexander Makarov, and first disclosed by him in a U.S Patent filed in 1996 [426]. The impact that the Orbitrap has had on the analytical capability of high-performance mass spectrometry, especially in the biomedical area, has been so great that Alexander was the recipient of the Distinguished Contribution Award of the American Society for Mass Spectrometry in 2008, and the Editors offer him our warmest congratulations on his being honoured in this way. Members of the ASMS are able to view Dr. Makarov's Award Lecture on the ASMS web site www.asms.org, and we are delighted especially that Dr. Makarov has contributed Chapter 3, describing the theory, operation and applications of the Orbitrap analyzer, to this current Volume of *Practical Aspects of Trapped Ion Mass Spectrometry*.*

Readers of Chapter 3, Volume 4, will note that, surprisingly, the recent literature upon which the novel concept of the Orbitrap is based is remarkably sparse. In fact, the underlying physical principle of operation, namely the oscillation of ions in a cylindrically symmetric DC electric field without the presence of a magnetic field or of RF voltages, is precisely that employed in what appears to be the earliest ion trapping

---

* See Volume 4, Chapter 3: Theory and Practice of the Orbitrap Mass Analyzer, by Alexander Makarov.

device, the Kingdon trap first described in 1923 [427]. Although atomic physicists have made intermittent use of the Kingdon trap for the capture and storage of ions, for example for spectroscopic studies [393], the potential of this kind of trap for mass spectrometric applications is now starting to be recognized, as witnessed by the discussion by Franzen and Wanczek in Section 15.4.3. of Chapter 15 of this Volume.*

In this final Section concerned with Dynamic Mass Spectrometers, we examine first the early pioneering work of Kingdon and mention briefly some subsequent applications of the Kingdon trap in atomic physics and in FT-ICR. Two of these experiments in particular paved the way for the invention of the Orbitrap, the underlying operational principles of which are then considered. Up to this point we shall have been concerned with the *static* Kingdon trap (SKT), that is one in which the trapping of ions is effected with the application of DC electric fields only. Recent research with the Orbitrap has made use of the appropriate addition of RF potentials to excite or to de-excite ions mass-selectively [428], and this topic links naturally with a further aspect of the Kingdon trap, namely the *dynamic* Kingdon trap (DKT) [429], in which ions are confined by means of superimposed RF and DC potentials—an approach that provides some fascinating contrasts with the behavior of radiofrequency quadrupolar devices.

1.5.5.2 The Static Kingdon Trap

When reading what appears to be the earliest report on a device that turns out ultimately to be the foundation stone for some particularly striking later development, no matter how long the intervening time interval may have lasted, one inevitably scrutinizes the publication for details of the background to the original research: the context in which it was carried out, references to prior work, and inputs from other workers mentioned within the text or in an acknowledgment. In reading the first description of what has become known as the 'Kingdon trap', or more specifically the SKT [427], it is the final short paragraph that indicates the environment within which Kingdon was working: "It is a pleasure to express my thanks to Dr Irving Langmuir for many valuable suggestions during the course of this work. The theory developed here is based on his suggestion that the effects observed were due to imprisoned positive ions."

This reference is to the famous scientist Irving Langmuir, who was appointed to the General Electric Company, Schenectady, NY, in 1912 and retired as Associate Director in 1950. Langmuir was born in 1881, and died in 1957; he was the inventor of the gas-filled incandescent light bulb [430], and was awarded the Nobel Prize for Chemistry in 1932 for his work on Surface Chemistry. So, as with almost every other major development in mass spectrometry, here we have yet another example of where the 'genealogical roots' of the invention can be traced back to the influence of a Nobel Laureate.

1.5.5.2.1 Kingdon's Experiments

One of the major activities at General Electric during the 'teens and twenties' of the twentieth century was the development of valve tubes for electronic circuits,

---

* See Volume 4, Chapter 15: Fragmentation Techniques for Protein Ions Using Various Types of Ion Traps, by J. Franzen and K. P. Wanczek.

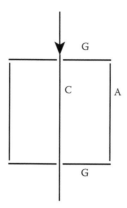

FIGURE 1.34   Basic design of the original Kingdon trap, in which A is a cylindrical cathode and is C a central anode in the form of a wire. The two electrodes marked G are 'guards' to which a potential is applied in order to 'imprison' ions within the trap. For further details, including the dimensions and materials employed in two traps of this design, see text. (Reproduced from Kingdon, K.H., *Phys. Rev.* 1923, *21*, 408–418. With permission from the American Physical Society.)

and the wider research program within which Kingdon was engaged was concerned with understanding the physics of these devices, and in particular discovering means of mitigating the space-charge effects in electron clouds. Prior to his publication describing his own research [427], Kingdon assisted A. W. Hull, also in the GE Research Laboratory, in his work on "The effect of a uniform magnetic field on the motion of electrons between coaxial cylinders" [431]. Indeed, it should be noted that the starting point for Kingdon's own investigation was the theoretical analysis published by Hull [431], but with the strength of the magnetic field applied to the coaxial cylinders set to zero.

Conceptually, the original Kingdon trap was a very simple device, as shown schematically in Figure 1.34. A filament, acting as the cathode C, was positioned along the axis of a cylindrical anode, A. Two guards ('end-cap electrodes' in modern parlance) to "imprison" the ions were mounted at GG. Two versions of the trap, "Tube K-26" and "Tube K-27" were used in the experiments reported in the paper. In K-26 the cathode was made from 0.01 cm (4 mil) tungsten wire and the cylindrical anode from thin molybdenum, 1.9 cm (0.75 in.) diameter and 3.6 cm (1.4 in.) long. The guards were made of six spokes of 0.025 cm (10 mil) tungsten wire evenly spaced around a nickel ring, 1.9 cm in diameter. The spokes did not reach the center of the ring, and a space 2 mm in diameter was left for the cathode to pass through. The whole unit was assembled on a glass frame and mounted as one unit in a glass envelope. Tube K-27 had the same sized filament and an anode cylinder 2 in. (5.1 cm) in diameter and 2 in. (5.1 cm) long. The guards in this case were molybdenum discs, 2.0 in. (5.1 cm) in diameter, pierced with 0.13 cm (52 mil) holes to admit the cathode filament. The gases studied were helium, hydrogen, neon, and mercury vapor, and the pressures were measured with an ion gauge (for low pressures) and a McLeod gauge for higher pressures.

Although not specifically mentioned in the papers by Hull [431] and by Kingdon [427], the potential within the cylindrical volume defined by the central wire and the outer electrodes has a logarithmic dependence upon the radial distance from the center of the trap, in a manner analogous to that between the cylindrical electrodes of the spiratron TOF mass spectrometer [347,349] (see Section 1.5.3.4.). Derivations of the appropriate equations can be found in the publications by Langmuir and Compton [432] and by Cady [433]. 'College – level' derivations have also been published by Tipler and Mosca [434], and by Pace [435] on the internet.

The overall purpose of Kingdon's investigation was to determine the effect that the presence of different gases at various pressures had on the voltage–current characteristics of the device. These characteristics are plots of the measured current of electrons that had been emitted from the cathode (filament) and collected on the anode as a function of the potential applied between the anode and the cathode. In the absence of gas, the electron current was limited by the electron space charge that developed within the cylindrical cavity, and the measurements were intended to show how the presence of positive ions led, hopefully, to the neutralization of the electron space charge. It will, of course, be realized that the positive ions themselves would be expected to simply be attracted toward the negative cathode and be discharged on the wire. However, based upon the work of Hull [431], Kingdon argued that under the conditions of the experiment the majority of ions acting under the influence of the electric field would fail to strike the filament; indeed, conservation of angular momentum requires that any ion that has a radial component of velocity and just misses the filament will continue to move around in a spiral trajectory until neutralized in some way. Kingdon suggested that, unless an ion strikes the filament on its first traversal of the annular space, it should remain orbiting around the filament up to 300 times.

Two types of experiment were carried out. In the first, the guards were both connected to the negative end of the filament so that any positive ions could discharge at the end of the cylinder. In the second series of experiments, the guards were connected to the anode cylinder so that the escape of ions through the end of the cylinder was prevented.

In summary, it was found that the collected anode current at a given potential applied between the electrodes was always greater when there were ions present and, from the two sets of results obtained, Kingdon engaged in a detailed discussion of the lifetimes of the positive ions, and the factors which influenced them. The factors considered were (1) discharge of the ions to the cathode; (2) leakage of the ions through the holes in the guard plates; (3) recombination with electrons; and (4) the effects of collisions with gas molecules. Overall, it was argued that the experimental evidence pointed to factor (4) being the chief cause limiting the lives of the positive ions.

The subsequent detailed analysis of factor (4) involved modeling the environment within the trap using the kinetic theory of gases, and a consideration of the earlier literature on the effects of energy loss by ions on colliding with gas molecules. In fact, the existing literature appears to have comprised only two publications! The first was a paper by Horton and Davies [436] on the ionizing power of positive ions in helium who, in explaining their results, stated that "This is no doubt due to the positive ions losing energy by collisions with helium atoms", providing an early rationale for the use of helium as a buffer gas in modern trapping devices! The second citation was of a paper by Saxton [437], which was concerned with the impact ionization by low-speed positive hydrogen ions in hydrogen. This consideration led to the following interesting paragraph concerning the effects of collisions in his device: "In the present experiments it seems likely that at the first collision the ion will lose a considerable part of its energy, and will not be able to move out as close to the anode as it did before. Its total velocity will be smaller, but its transverse velocity will probably be larger than before the collision. It is therefore unlikely that the ion will strike the filament immediately after the first collision. Succeeding collisions reduce the energy of the ion still further, and force it to follow paths which on the average get closer and closer to the filament. Finally after perhaps three or four collisions the total energy of the ion is so low that it is able to strike the cathode." What more precise allusion could there be to the collisional migration of ions?

In comparing the results with helium and with mercury vapor in the trap, Kingdon noted that there was a far greater neutralization of space charge with mercury. Even allowing for the lower ionization energy, greater mass, and lower velocity (at the same temperature) of mercury atoms, it was noted that the difference between the effects of the two gases was so marked it was thought that some new factor must be entering into play in the case of mercury. It was suggested that, at the low pressures of mercury studied (10^{-6}–10^{-7} Torr (quoted as "mm")), there was an effect arising from the pressure of residual gas (10^{-6} "mm") whose presence had been ignored: "The mercury ions were therefore not able to transverse their calculated mean free paths." Again, there is a hint of collisional cooling.

From the data obtained in a further series of experiments, in which the anode voltage was applied intermittently to the guards by means of a commutator (in other words, a pulsed experiment), the time taken for the anode current to reach half its value suggested that at a pressure of 4.2×10^{-7} 'mm', the half-life of a mercury ion was 7.8×10^{-3} s, compared to a value calculated by kinetic theory considerations of 1.4×10^{-3} s.

1.5.5.2.2   Work in the 1960s

As this story of the Kingdon trap progresses, one becomes aware of what appears to be a significant 40-year time-gap during which relatively little activity was reported in the literature. Interest was re-kindled in 1963 with the suggestion by Herb and co-workers [438,439] that a concentric cylinder arrangement would form the basis for an ion gauge, which they called the "Orbitron". At the same time that these short abstracts appeared, a more extensive account of the motion of ions under the influence of the logarithmic potential that exists within the cylindrical electrodes of the orbitron was published by Hooverman [440], who provided examples of plots of the orbits of ions within the device. None of these three communications makes any reference to the Kingdon trap. Other experiments with the Kingdon trap included its use by Brooks and Herschbach [441] as a detector in molecular beam experiments; this reference contains a number of citations to work in which the trap was used also as a positive ion detector and for making ionization energy measurements. As an extension of their earlier work, Herb and co-workers [442] described the features of an orbitron vacuum pump. In this application, the Kingdon trap comprised a titanium rod together with a titanium cylinder (diameter 10 cm), with the pumping effect being due mainly to the gettering action of the titanium that had been freshly sputtered by ion impact. Unfortunately the research program had to be terminated before the system was characterized fully.

In 1966, a variant on the Kingdon trap was described by McIlraith [443], in which two parallel wires were mounted parallel to the axis of a conducting cylinder, but the principle of trapping in the device does not seem to have progressed beyond the construction of mechanical/gravitational analogues of the forces acting in the device, and speculation about the holding of particles ranging from the size of dust to small planets in the gravitational field of double stars.

1.5.5.2.3   Applications in Atomic Physics and FT-ICR

The next most significant landmark in the progression from the original Kingdon trap to the Orbitrap was the brief account in 1981 by Knight [444] of a modified version of the design, suggested initially by M.H. Prior, which incorporated an axial harmonic potential. Further details of this device and its application to the storage of laser-produced metal ions are considered in the next Section (see Section 1.5.5.3.). An experimental system incorporating a "standard" Kingdon trap is that of Church and his group [445], working at the Texas A&M University. They have described the confinement of 2 keV ions produced by a Calutron ion source using an ion trap that was quite large compared to most of the previous examples we have discussed: the aluminium cylinder had a length of 10 cm and a diameter of 10 cm, and was capped by plane aluminium electrodes that were insulated both from the cylinder and the central wire. The cylinder had a number of holes through which the incident beam could pass, ejected ions detected, and photons emitted from the excited species observed. Key features of the operational duty cycle were a means of gating the incident beam on and off and a means of decelerating and defocusing the ion beam as it entered the trap. While ions were being admitted to the trap, the central wire was held at the same positive potential as the cylinder, after which the potential of the central wire was reduced rapidly to zero typically for 10–100 ms. During this time the ions were trapped, after which they were then released for detection on a microchannel plate (MCP) by returning the wire potential to that of the cylinder. A typical operating pressure was 4.0×10^{-7} Torr, and in some experiments a background pressure of argon at 1.0×10^{-6} Torr was introduced into the system. In the first test experiments H^+ ions were studied, and trapped ion lifetimes measured. Later work involved the study of highly-charged ions, such as Ar^{q+}, with ($2 \leq q \leq 11$). A summary of the work on the lifetimes of metastable levels of atomic ions confined in the Kingdon trap, as well as in the Paul and in the Penning traps, has been presented by Church [446].

The use of the Kingdon trap as the cell for an FT-ICR mass spectrometer has been reported by a number of workers, as summarized in the review by Wanczek [447]. One particular example is the research of Russell and co-workers [448], who designed what they called 'wire-ion guide' (WIG) cell. This cell took the form of the normal 'Kingdon shape' of a long cylinder with a central wire, except that the cylinder, instead of being a continuous surface, was divided into four curved plates, which acted as the transmitter and receive electrodes for the FT-ICR experiment (see

Section 1.5.2.4.4.). In this application, the WIG cell is not acting as a true SKT, on account of the presence of the magnetic field, nevertheless this comprehensive study included numerous SIMION equipotential plots [449,450] that are of direct interest to the design of the Orbitrap (see Section 1.5.5.4.) and of the DKT (see Section 1.5.5.5.) [448]. A resolution of $m/\Delta m = ca$ 300,000 was observed in the mass spectrum of C_{60}^{+}, and the presence of the wire to which a potential could be applied provided an additional parameter that could be optimized to improve performance.

The use of a standard Kingdon trap for the storage of singly, doubly, and triply charged positive argon ions for subsequent analysis by TOF mass spectrometry was reported by Sekioka *et al.* [451,452]. In this experiment the outer cylindrical electrode contained a hole through which the ions could be ejected for mass analysis through the application of a positive 'dump' pulse to the central wire.

1.5.5.3   The Knight Electrostatic Ion Trap

We have already noted in Section 1.5.5.2.1 that the potential distribution within the cylindrical volume between the central wire and outer electrode of the Kingdon trap is logarithmic. The novelty of the electrostatic trap described by Knight [444], to which reference has already been made in Section 1.5.5.2.3, is that an additional quadrupolar term has been added to give an cylindrically symmetric potential of the form

$$\phi = A(z^2 - \frac{r^2}{2} + B \ln r) \tag{1.42},$$

in which A and B are appropriately chosen constants.

A schematic diagram of the experimental system of the electrode configuration needed to achieve the potential distribution given by Equation 1.42, is shown in Figure 1.35. The aim of this design was to retain the cylindrical symmetry of the Kingdon trap, including the use of the central wire electrode, but to enable the ions to undergo harmonic oscillations in the axial direction. Thus we see (Figure 1.35) that the outer cylinder of the standard Kingdon trap was replaced by two equatorially separated, approximately conical, outer electrodes, which, in fact, were fabricated from 'loose mesh'; a radiofrequency potential could be applied between the two outer electrodes. The "midplane" radius of the device was 3.0 cm. A laser-produced plasma generated on a metal target made from the sample material, which was mounted directly on to one of the outer electrodes, resulted in positive ions entering the trap through the mesh, and after a defined storage time the trapped ions could be detected either by axial resonant ejection on to an electron multiplier, or by pulsing the wire positively and ejecting the ions equatorially on to a collector plate. Although no mass spectral data were reported, Knight stated that he was able to identify the mass/charge ratio "for the stored ions through the harmonic nature of the axial potential", given by Equation 1.42.

With the outer electrodes referenced to ground potential, it was found that a minimum trapping voltage applied to the wire, V_{wire}, equal to ca −50 V was necessary, and that the ion storage properties were roughly constant over the range −500 V ≤ V_{wire} ≤ −100 V. The trap was operated at a pressure of ca 3×10^{-8} Torr, and the species studied included Be^+, C^+, Al^+, Fe^+, and Pb^+. It was estimated that the initial

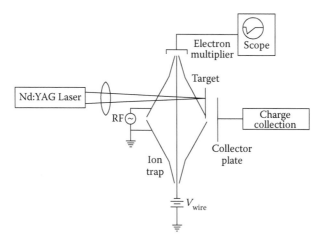

FIGURE 1.35 Schematic diagram of the electrostatic ion trap employed by Knight [444] For details, see text. (Reproduced from Knight, R.D., *Appl. Phys. Lett.* 1981, *38*, 221–223. With permission from the American Institute of Physics.)

number of ions stored was typically *ca* 2×10^8; non-exponential decays in the ion population over periods up to *ca* 145 ms were observed, with values of up to *ca* 500 ms for ions created by electron impact and stored at the same operation pressure.

Before concluding this account of the Knight electrostatic ion trap, it should be noted that, as pointed out in a recent review by Perry *et al.* [453], in modeling the electric field within his device Knight observed that the cylindrical wire approxima-tion for the inner electrode distorted the harmonic nature of the potential and caused the resonances in the ion motion to be much weaker than those observed normally in a quadrupole potential trap. Furthermore, in their account of FT-ICR experiments employing the WIG cell (see Section 1.5.5.2.3), Gillig *et al.* [448] showed a SIMION equipotential plot of an "ideal" cylindrical Kingdon trap in which the potentials applied to the end-plates had been adjusted to form the field lines described by Equation 1.42, and concluded that "the optimum equipotential lines are created by a superposition of quadrupolar and logarithmic potentials, creating a spindle-like trapping volume" in which there would be harmonic motion in the axial direction. Such an ideal trapping volume can be also created by modifying the shapes of the central and outer elec-trodes of the trap, as employed by Makarov in his creation of the Orbitrap [426,454].

1.5.5.4 The Orbitrap Mass Spectrometer

The aim of the discussion thus far within this Section on the Kingdon trap has been to provide an evolutionary setting within which to examine the Orbitrap mass spectrometer. In this present Section, the intention is not to duplicate the account presented by Dr. Makarov in Chapter 3 of this Volume, but rather to emphasize the fundamental characteristics of this method of mass analysis and to indicate some recent developments that may lead to future generations of instruments based upon Orbitrap technology. To avoid replication, the reader is referred to some of the figures in Chapter 3, and one of the equations is reproduced in the current discussion, but numbered as in Chapter 3. The authors are extremely grateful to Dr Makarov for his extremely helpful comments during the writing of this Section. An account of the Orbitrap has been given also by Hu *et al.* [455].

In the preceding Section (Section 1.5.5.3) we have already seen how Knight introduced the idea of utilizing combined logarithmic and quadrupolar potentials within a Kingdon-type trap of the form given by Equation 1.42 in order to define more precisely the harmonics of the axial motion of the trapped ions. The design of Orbitrap follows essentially the same approach, with the potential $U(r,z)$ within the cavity of the ion trap being given by [456]

$$U(r,z) = \frac{k}{2}\left(z^2 - \frac{r^2}{2}\right) + \frac{k}{2}\cdot(R_m)^2 \cdot \ln\left[\frac{r}{R_m}\right] + C \qquad (3.1),$$

where r, z are cylindrical coordinates ($z = 0$ being the plane of symmetry of the field), C is a constant, k is the field curvature, and R_m is the characteristic radius. As noted in Chapter 3, the 'characteristic radius', which is unique to this device, has also a physical meaning in that the radial force is directed toward the axis for $r < R_m$, and away from it for $r > R_m$, while at $r = R_m$ the radial force equals 0. We shall examine this parameter is slightly more detail presently. Within the literature on the Orbitrap, the form of Equation 3.1 has been termed both 'quadro-logarithmic' and 'hyperlogarithmic', with an apparently recent preference for the latter.

In order to generate the appropriate potential distribution within the volume of the trap, the shapes the electrode surfaces for the axially symmetric central and the outer electrodes, denoted by the subscripts 1 and 2, respectively, have to be fabricated according to the expression

$$z_{1,2}(r) = \left(\frac{r^2}{2} - \frac{(R_{1,2})^2}{2} + R_m^2 \ln\left[\frac{R_{1,2}}{r}\right]\right)^{\frac{1}{2}} \qquad (1.43),$$

where $R_{1,2}$ are the maximum radii of the corresponding electrodes, and the equatorial plane of symmetry through the structure is at $z = 0$ [454]. An important novel feature of the Orbitrap resulting from this geometry is that the central electrode is no longer a thin wire, as in all the examples of Kingdon-type traps considered thus far, but is a spindle that is wider in the center and narrower at the ends. Essentially this geometry accords with the SIMION modeling of an ideal trap undertaken by Gillig *et al.* [448] as discussed in the preceding Section 1.5.5.2.3. A diagram of the Orbitrap mass analyzer fabricated according to Equation 1.43 is shown in Figure 3.1. It should be noted that the outer electrode is split equatorially into two halves, labeled OE-1 and OE-2 in Figure 3.1, in a manner analogous to the design of the Knight trap [444], and this feature is essential for the detection of the ions by either of two methods (see below).

Returning briefly to the concept of the characteristic radius, R_m, it is instructive to examine more closely the precise nature of the separate dimensionally dependent terms involving r and z in Equation 3.1. In fact, the potential described by Equation 3.1 is derived from what Korsunskii and Bazakutska [456] described as being an electric field of the 'difference' or 'subtractive' type, in which the spatial components of the field are given by the equations of the form

$$E_r = E_1 / r - E_2 r \tag{1.44}$$

and

$$E_z = 2E_2 z \tag{1.45},$$

where E_1 and E_2 are constants. Accordingly, in Equation 3.1 we note that (1) there are no 'cross-terms' involving both r and z together (for example, rz, r^2z, etc.); (2) there is only one quadratic term in z, which in fact defines the axial electric field responsible for the axial harmonic motion and which is independent of the radial dimension, r; and (3) that there are two separate expressions for the radial component of the potential, one quadratic and one logarithmic; again, in the absence of cross-terms, there is no dependency upon the axial displacement. Since the value of the electric field at a point r is given by $(\partial U(r,z)/\partial r)_z$, and noting that the first differential of r^2 is $2r$ whereas that of $\ln(r)$ is $1/r$, we see that, in accordance with Equation 1.44, the total radial trapping field gradient is a combination of two opposing terms: one that increases linearly with r as the ion moves further away from the central electrode and one which decreases inversely with r. This effect can be seen in Figure 1.36, which shows the potential distribution corresponding to Equation 3.1 in both the axial (z) and radial (r) dimensions. Thus the axial potential is of a quadrupolar nature, whereas the radial potential extending from the central axis first increases, passes through a maximum, and then decreases toward zero. The value of r at which the potential maximizes is the characteristic radius, R_m, and when $r < R_m$ the potential is focusing toward the axis, whereas when $r > R_m$ it is defocusing. Clearly, if an ion were placed at rest within the cavity of the trap, provided $r < R_m$ it would simply be attracted toward the central electrode

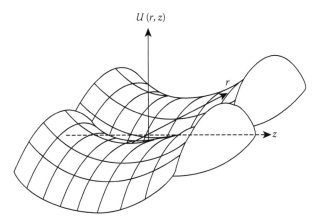

$U(r, z)$

FIGURE 1.36 Graphical representation of one form of the electrostatic potential ($U(r, z)$) distribution in the axial (z) and radial (r) directions of the Orbitrap. (Reproduced from Makarov, A.M., *U.S. Patent*, 5,886,346 1999. With permission from the US Patent and Trademark Office.)

and be discharged. In reality, as in a normal Kingdon trap, the centrifugal force of the rotational motion of the ions contributes a potential energy term that is additional to the electrostatic potential, creating an effective trapping potential within which the ion is confined as it rotates in a stable orbit around the central electrode.

The precise nature of the motion of ions within the Orbitrap is somewhat complex, and comprises three separate contributions: one resulting from the harmonic oscillations in the axial direction, and the other two from the combined fields in the radial direction. Reference to Equations 3.2 through 3.4 in Chapter 3 shows that whereas the two radial oscillation frequencies depend upon the initial radial displacement, R, the axial oscillation frequency depends inversely only upon the square root of the mass/charge (m/z) ratio of the ions. It is the high level of coherence of the axial oscillations, requiring crucial control over the conditions under which the ions are injected into the trapping cavity, that endows the Orbitrap with its ability to detect ions with very high resolution.

Conceptually there are, in fact, at least three methods for detecting ions in the Orbitrap that have been described. The first is to eject the ions mass-selectively on to an external detector mounted axially by superimposing an oscillating potential of swept amplitude or frequency on the static voltage on the central electrode [454], thus in effect creating a DKT, as discussed in the following Section (Section 1.5.5.5). The second is to apply an oscillating potential of swept frequency between the two halves of the outer electrodes in order to eject the ions resonantly on to an axially mounted detector [457], in a manner analogous to that employed by Knight [444]. The third is to detect the image current on the split outer electrodes resulting from the different axial oscillation frequencies of the ions of differing m/z-ratios and then Fourier transform the resulting transient signal in the same manner as employed in FT-ICR mass spectrometers. Different approaches have been adopted for exciting the axial oscillations of a packet of ions, including the application of a voltage pulse between the split outer electrodes and the off-equator tangential injection of ions ('excitation by injection'). The latter is generally preferred because it minimizes the perturbations of the electric field within the trap, although it is more demanding on the technology of ensuring the fast ejection of a large ion population from an external RF storage device [458,459]. Further details concerning ion injection, including use of the new 'C-trap', and detection can be found in Chapter 3.

Before considering some of the more recent experiments and further developments concerning the Orbitrap, it is, perhaps, worth mentioning that the idea of using image current detection arising from rotating pulses of ions combined with Fourier transformation of the signal was proposed in 1995 by Oksman [460] as the basis of a TOF mass spectrometer. In this approach pulses of ions were injected into the Kingdon-like electric field generated between two concentric cylinders closed with parallel plates at both end, in a manner somewhat analogous to that employed by Bakker in the spiratron [347]. The principle was to force ions of equal energy to circulate on paths of equal radius and, through image current detection, to measure their flight times per revolution as reflected by their frequency of oscillation. Due to the complexity of the radial motion, as opposed to the well-defined axial motion in the Orbitrap, it is difficult to see that the Oksman method would have made a viable mass spectrometer.

Apart from the wealth of experimental data that are now being obtained with instruments in which the Orbitrap has been employed to provide high resolution, accurate mass-analysis of ions that have been stored in a LIT, recent experiments by Cooks, Makarov and co-workers [457] have provided an interesting insight as to how the axial motion of ion within the Orbitrap may be controlled by the application of resonant alternating potentials to the split outer electrodes. Thus, if an alternating potential of the appropriate frequency is applied so that it is 180° out of phase with the axial oscillations of the ions, it is possible to de-excite, that is quench, the trajectories of the ions mass-selectively so that they collapse to the equatorial plane; this effect is detected by monitoring the reduction in strength of the detected image current signal, and results in the ions being brought to a "dark state". Continued application of the resonant oscillating voltage causes the ions to 're-phase' so that the detected signal returns to its original level before de-excitation commenced. Techniques of this kind hold promise for ion isolation and collisional activation, including surface-induced dissociation experiments, within the Orbitrap.

Other prospective developments for the future may be gained from examination of current patent applications, including the application of appropriate potentials to multi-electrode arrays [461] that mimic a continuous surface, thereby allowing corrections to be made for field imperfections or the introduction of 'virtual' electrodes having novel shapes, and the deliberate introduction of field distortions [462] that may improve certain operational features of the Orbitrap in the same way as the presence of contributions from non-linear fields may endow RF quadrupole ion traps with beneficial characteristics (see Section 1.5.4.4.5).

1.5.5.5 Dynamic Kingdon Traps

1.5.5.5.1 Introduction

Thus far our discussion has focused on the so-called *static* Kingdon traps (of which the Orbitrap is a particularly important example), in which trapping in the radial direction is determined by the balance between the centrifugal force arising from the rotational motion of the ions and the centripetal force acting toward the central electrode that results from the static (DC) potential difference applied between the central and the outer electrodes. In the DKT, radial stability is dependent upon the superposition of an oscillating RF potential on the DC trapping potential. As will be seen in the later discussion, the principle advantage of doing this is to enable radial stability of the ions without it being necessity for them to orbit around the central electrode. Reference has already been made [454] to the idea of resonantly ejecting ions from the Orbitrap by the application of an RF potential to the central (spindle) electrode, but this does not appear to have been explored in practice. However, because of the potential applicability of DKTs to mass analysis and because no previous discussion of these devices has, to our knowledge, appeared in the mass spectrometry literature, we conclude our survey of dynamic mass spectrometers by considering briefly the relevant physics of the DKT and in particular examining how the behavior of ions in the DKT contrasts with that in traps that utilize RF quadrupolar potentials, a process that will lead to a brief consideration of Chaos Theory! In presenting this material the authors have drawn extensively on the publications by Reinhold Blümel and his group at the Wesleyan University, Middletown, CT, and we wish to acknowledge the considerable assistance of Professor Blümel through his discussions with one of us (JFJT). In the account that follows, an attempt has been made explicitly to describe the principal aspects of the work in non-mathematical terms: for a more rigorous treatment of the physics behind some of the statements, the reader should consult the ensuing references. In this respect, it should be noted that the primary research foci in these studies were the crystallization of laser-cooled ions and the chaos theory of ion traps, not mass spectrometry!

1.5.5.5.2   The Emergence and Physical Principles of the Dynamic Kingdon Trap

Recognizing that the original description of the SKT occurred in 1923 [427], it is somewhat surprising to learn that the 'discovery' of the dynamic variant did not occur until over 70 years later when Reinhold Blümel set out to explore what he termed the 'dynamic Kingdon trap' (DKT), and to compare its characteristics with the radiofrequency quadrupole (Paul) trap [429]. Subsequently it transpired [463] that this report represented essentially a 'rediscovery' of the DKT, in that unpublished studies on the device had been undertaken by two students, R.E Bahr and E. Behre, working in the laboratory of Dr. Ernst Teloy at the University of Freiburg, Germany, in the period 1969–1972 [464,465].

As we have seen already (Section 1.5.5.2.1), a key feature of the SKT is that for ions to be held in stable trajectories they must possess an angular momentum around the central wire electrode; without this rotational motion the ions would simply collide with the wire and be discharged. In the DKT, an RF potential is superimposed upon the original DC voltage applied to the central wire, so that now ions within the trap are subjected both to a static electric field and a dynamic (RF) field. As with the SKT, in the following discussion of the DKT no account is taken of any motion of the ions in the axial direction.

On its own, the RF potential would drive the ions in the direction of the weaker field, that is away from the wire toward the cylindrical electrode, in accordance with the principle enunciated by Paul and noted earlier [382] (see Section 1.5.4), namely that in an inhomogenous electric field the trapped ion is a 'low-field seeker'. This effect occurs regardless of the sign of the charge on the ion. Thus we note immediately that there is a contrast between the DKT and the Paul trap: in the latter, the direction of the weaker field is always *toward* to the center of the device, that is the RF potential is *focusing*, whereas in the DKT the direction of the weaker field is *away* from the center so that the superimposed RF potential is *defocusing*.

However, if an attractive potential is applied to the wire, then there can be a balance between the focusing effect of the centrally directed DC field and the defocusing effect of the RF field. By a process somewhat analogous to that adopted in the treatment of the motion of ions within an RF quadrupole device, the radial motion of ions within the DKT was expressed in terms of what Blümel [429] called the "Kingdon equation". Consequently ions can now be stabilized in the Kingdon trap so that they move radially, and there is no need for them to possess angular momentum about the central wire electrode. This situation is analogous to the derivation of the equations of ion motion in the 3D quadrupole ion trap, where any contribution from angular velocity around the axis of symmetry (that is, the z-axis) is assumed implicitly to be zero (see Section A3.1.2).

While the non-linear dynamics of the Kingdon equation were examined by Linz [466], to develop the treatment of the DKT further, in their earlier papers Blümel and co-workers [429,467–470] treated the influence of the RF potential in terms of a pseudopotential approach, developed originally by Dehmelt [471], in a manner analogous to that used in RF quadrupole devices (see [269, in Chapter 3]). Thus ions were considered to move under the influence of an *effective* potential comprising the sum of the DC potential and the pseudopotential.

The result of this procedure was to deduce that, in the case of the DKT, there was a single trapping parameter, symbolized sometimes as η and sometimes as q, which is equal to the ratio of $V_{RF}/2U_{DC}$. This deduction contrasts with the quadrupolar situation in the Paul trap, where there are two control parameters, a and q (see Appendix). From the concept of the effective potential, one deduces that the ions will exhibit oscillations within a potential well together with a characteristic micromotion (caused by the effect of the RF potential).

In commenting on his conclusions, Blümel [467] states that there are many similarities between the Paul trap and the DKT, but also that there are marked differences, the main one being in the topology of the potential minima. "While the pseudopotential minimum in the Paul trap is a single point, the Kingdon trap in its dynamical form shows a minimum only in the radial direction, leaving ions to roam in the axial direction. Thus the locus of the potential minima is a cylinder and the main consequence of this difference is that ions trapped in the DKT have much more space to avoid each other, resulting in a cylindrical *sheet* of crystallized ions as the minimum energy configuration." Interestingly, Blümel suggests also [467] that the three-dimensional confinement of ions could be achieved by bending the central wire electrode into a ring and the cylindrical conductor into a torus, so that on the preceding premise the potential minima would be a torus also.

A consequence of this treatment of the physics of the DKT is that, on this basis, it is impossible to eliminate the micromotion so that, although the DKT no longer requires there to be an angular component of motion to ensure stability, there is, nevertheless, always a residual micromotion, the existence of which makes the DKT unattractive for spectroscopy experiments, but is of interest in the study of 'crystal formation' of trapped ions and chaos theory. In this latter connection, Blümel [469] noted that in the Paul trap the motion of a single ion can always be defined, and that chaotic behavior only arises when two or more ions are trapped; however in the DKT chaotic behavior can be exhibited even by a single trapped ion.

As a result of further computations on the Kingdon equation, but without recourse to the pseudopotential approximation, Garrick-Bethell and Blümel [472] discovered unexpected instabilities in the trapping of ions in the DKT. The existence of these was masked previously by the use of the pseudopotential approximation. They argued [472] that, when considered in terms of phase space plots of [radial velocity] *versus* [radial displacement], 'trapping islands' existed outside of which there was a 'sea of chaos'. For a given value of ($V_{RF}/2U_{DC}$), ions with trajectories that started inside the island stay trapped inside the island for ever, leading to the concept of a 'primary trapping island'. "Apart from a small set of trajectories launched at specific initial conditions, most trajectories that start outside the island, after a brief chaotic transient, either reach the outer cylindrical electrode or the axial wire. In both cases the particle discharges and is lost from the trap." However at certain values of $(V_{RF}/2U_{DC})$ these islands 'collapse' to a single point in phase space in which no combination of the radial velocity and radial displacement yields a stable trajectory. This observation was interpreted in terms of the collapse occurring when the 'winding frequency' of the phase-space plot (that is the rate, in terms of the frequency of the RF potential, at which the ions that appear at a starting point in phase-space of the trajectory return to the same point) enters into resonance with a sub-multiple of the drive frequency, for example $\Omega/3$. On the basis of the pseudopotential approach, this behavior is totally unexpected.

In extending this work further and as a result of a more extensive, generalized, consideration of radiofrequency traps, including the DKT, the Penning trap, and the Paul trap, Blümel and co-workers [473] concluded that "Trapping in the DKT is a non-linear effect which depends decisively on the existence of stable islands in phase space. In this respect the DKT is fundamentally different from the Paul trap or the Penning trap, in which, in principle, particles can be stored irrespective of their positions and momenta." "Compared with the Paul trap the instabilities of the DKTs are particularly devastating. In the Paul trap encountering an $N = 3$ or an $N = 4$ instability [analogous to the island collapse in the DKT] merely means the breakup of an ordered ion configuration; the ions themselves remain trapped. In the DKTs, however, encountering such an $N = 3$ or an $N = 4$ instability means the complete loss of the particle from the trap."

More recently Garrick-Bethell and Blümel [474] have extended their considerations of the trapping of ions with the DKT by exploring the stability criteria in terms of quantum mechanics.

In concluding this short account one is left, therefore, pondering as to whether the DKT might have some relevant application to mass spectrometry. In leaving this almost open question, one is prompted by a remark made by Peik and Fletcher [475]. In one of the few experimental studies on traps utilizing a combined electric field of the kind discussed above, these workers studied charged steel spheres, and also charged *lycopodium* spores, suspended in a field created between the hemispherical end of a copper wire and a curved copper disk, and concluded "The dynamic Kingdon trap and its numerous possible geometrical variants offer a large storage volume, a deep pseudopotential, and small mass selectivity, and can in this respect even be superior to traditional quadrupole traps."

1.6   SUMMARY AND CONCLUSIONS

When first contemplating this opening chapter to *Practical Aspects of Ion Trapping Mass Spectrometry, Volume 4: Theory and Instrumentation* and to the companion *Volume 5: Applications*, our initial thoughts turned naturally to preparing an outline to what would be essentially a short guide to ion trapping techniques, but placed in the wider context of contemporary practices in mass spectrometry and its applications. However, as our thoughts began to crystallize it became clear to us that, as in any scientific discipline, much of what is possible currently is rooted very much in the inspirations and endeavors of those who have gone before us.

We were, of course, not ignorant of the fact that 2009 marks the sesquicentenary of the birth of Charles Darwin, the creator of the theory of biological evolution. In many ways the development of new approaches and instrumentation in mass spectrometry follows also an evolutionary path. Furthermore, one is conscious that the time-frame for the publication of our current two Volumes coincides precisely with the 100th anniversary of the period during which the underlying physical basis of mass spectrometry itself was being established by J. J. Thomson, F.W. Aston, and others. It soon became evident to us, therefore, that we needed to widen our historical perspective and take this opportunity to present an "appreciation" of how the modern methods and applications in mass spectrometry, many of which are represented by the 36 contributed chapters in Volumes 4 and 5, have been made possible through the efforts of our predecessors.

In approaching this task, we have been inspired greatly by the excellent account by John H. Beynon and Roger P. Morgan [6], in which they traced in detail the development of magnetic sector mass spectrometers up to the early 1950s. We have, in fact, drawn extensively on this article in Section 1.2 of the current chapter. To us, one of the most engaging aspects of the account by Beynon and Morgan was the way in which they engendered a personal 'feel' for the various historical events by including contemporary verbatim quotations made by the original practitioners.

We have, therefore, elected to follow this pattern, and in so doing have adopted a form of 'genealogical' approach to the various topics that we have covered. This, in part, reflects the personal interest of one of us (JFJT) in Family History, but with the difference that here we are tracing the origins and emergence of ideas, rather than distant ancestors. At the same time, the use of direct quotations and of reported anecdotes enlivens the story, and makes it 'real'. In many instances this has meant visiting and re-visiting in depth very many of the early publications that we have cited, rather than just repeating references given in other historical accounts. An essential aspect of this has been the rigor with which we have striven to impose upon ourselves the task that we required of our contributing authors, namely the inclusion of the full bibliographic details of the publications that we have cited no matter how obscure they may be. In this task we have been assisted greatly by a number of individuals named in the Acknowledgments.

In writing this chapter we have, ourselves, learned a great deal. For example, we have 'discovered' the origin of the term 'cyclotron', and the ways in which some kinds of mass analyzer have survived (or been re-born) and are currently available, while others have become extinct, like the *Dodo*. In the final major Section (1.5.5) of the chapter we have included deliberately a detailed account of the Kingdon trap. Not only is this, arguably, the earliest form of ion trap (1923 [427]), but is conceptually the precursor of the latest commercial offering, the highly successful Orbitrap mass spectrometer.

Of some interest is the current research into the fundamentals of the 'dynamic Kingdon trap', in which an RF potential is superimposed upon the static electric field of the Kingdon trap. It will be interesting to see whether this fascinating device will, at some point, form the basis of some new type of mass spectrometer, as yet unborn.

In presenting this account, we hope that our readers, especially those just embarking upon their careers as professional scientists, will share with us our appreciation of the fascinations of mass spectrometry.

ACKNOWLEDGMENTS

The authors wish to express their sincere thanks to the following persons for their assistance in checking drafts of parts of the manuscripts and/or providing details of certain references and other information used in the preparation of the Chapter, or within Volumes 4 and 5 generally: Reinhold Blümel, Wesleyan University, Middletown, CT, USA; Michael A. Grayson, Archivist, American Society for Mass Spectrometry, Santa Fe, NM, USA; Jane Harrison, Royal Institution of Great Britain, London, UK; Alexander Makarov, Thermo Scientific, Bremen, Germany; Alan G. Marshall, National High Magnetic Field Laboratory, Florida State University, Tallahassee, FL, USA; Andrew R. Rickard, National Centre for Atmospheric Science, University of Leeds, UK; Natasha Serne, Royal Dublin Society, Dublin, Ireland; and David J. Wineland, National Institute for Standards and Technology, Boulder, CO, USA.

REFERENCES

1. Perrin, J. New experiments on the kathode rays, trans. *Nature* 1896, *53*, 298–299.
2. Dunoyer, L. Sur la theorie cinetique des gaz et la realization d'un rayonnement material d'origine thermique. *C. R. Acad. Sci. (Paris)* 1911, *152*, 592–595.
3. Born, M. Eine direkte Messung der freien Weglänge neutraler Atome. *Phys. Zeits.* 1920, *21*, 578–581. Reprinted in Ausgewählte Abhandlungen I, pp. 694–697.
4. Gerlach, W.; Stern, O. Das magnetische Moment des Silberatoms. *Z. Phys.* 1922, *9*, 353–355.
5. Friedrich, B.; Herschbach, D. Stern and Gerlach: How a bad cigar helped reorient atomic physics. *Phys. Today* 2003, *December*, 53–59.
6. Beynon, J.H.; Morgan, R.P. The development of mass spectrometry; an historical approach. *Int. J. Mass Spectrom. Ion Phys.* 1978, *27*, 1–30.
7. The Edison electric light, in *Scientific American*, March 22, 1879, New York.
8. Thomson, J.J. XL. Cathode rays. *Philosoph. Mag. V* 1897, *44*, 293–316.
9. Thomson, J.J. On the cathode rays. *Proc. Camb. Philosoph. Soc.* 1897, *9*, 243–244.
10. Thomson, J.J. Cathode rays. Presented at a Discourse of the Royal Institution of Great Britain, London, on 30 April, 1897. (Published in *Proc. Roy. Inst. Great Britain* 1896–1898, *15*, 419–432.)
11. Stoney, G.J. LII. On the physical units of nature. *Philosoph. Mag. V* 1881, *11*, 381–390.
12. Stoney, G.J. On the cause of double line and of equidistant satellites in the spectra of gases. Paper read at meetings on 26 March and 22 May, 1891. (Published in *Sci. Trans. Roy. Dublin Soc. Series II* 1888–1992, *4*, 563–608.).
13. Thomson, J.J. XLVII. On rays of positive electricity. *Philosoph. Mag. VI* 1907, *13*, 561–575.
14. Thomson, J.J. LXXXIII. Rays of positive electricity. *Philosoph. Mag. VI* 1910, *20*, 752–767.
15. Gaede, W. Demonstration einer rotierenden Quecksilberluftpumpe. *Phys. Z* 1905, *6*, 758–760.
16. Dewar, J. Charcoal vacua. *Nature.* 1875, *12*, 217–219. (Report on a paper by P.G. Tait and J. Dewar, presented by J. Dewar at a discourse of the Royal Society of Edinburgh, 12 July 1875).
17. Rayleigh Lord, *Life of Sir J.J. Thomson.* University Press, Cambridge, 1942, p. 170.
18. Thomson, J.J. XXVII. Rays of positive electricity. *Philosoph. Mag. VI* 1911, *21*, 225–249.

19. Pidduck, F.B. *A treatise on electricity*. 2nd ed. 1925, Cambridge: Cambridge University Press, p. 80.
20. Thomson, J.J. *Rays of Positive Electricity and Their Application to Chemical Analyses*, London: Longmans Green and Co., 1913. p. 56.
21. Thomson, J.J. Further applications of the method of positive rays. Presented at a Discourse of the Royal Institution of Great Britain, London, on 17 January, 1913. (Published in *Proc. Roy. Inst. Great Britain* 1911–1913, *20*, 591–600.)
22. Soddy, F. Radioactivity. In J.C. Cain (Ed.) *Ann. Rep. Progress Chem. (Chem. Soc.)* 1910, *7*, 256–286.
23. Soddy, F. Radioactivity. In J.C. Cain (Ed.) *Ann. Rep. Progress Chem. (Chem. Soc.)* 1913, *10*, 262–288.
24. Lindemann, F.A.; Aston, F.W. XLVIII. The possibility of separating isotopes. *Philosoph. Mag.* 1919, *37*, 523–534.
25. Rutherford, E. Meeting of the British Association for the Advancement of Science, Cardiff, 1920.
26. Aston, F.W. The distribution of intensity along the positive ray parabolas of atoms and molecules of hydrogen and its possible explanation. *Proc. Camb. Philosoph. Soc.* 1920, *19*, 317–323.
27. Todd, J.F.J. Recommendations for nomenclature and symbolism for mass spectroscopy. *Pure Appl. Chem.* 1991, *63*, 1541–1566.
28. Budzikiewicz, H.; Grigsby, R.D. Half protons or doubly charged protons? The history of metastable ions. *J. Am. Soc. Mass Spectrom.* 2004, *15*, 1261–1265.
29. Aston, F.W. *Isotopes*, London: Edward Arnold and Co., 1922.
30. Aston, F.W. *Isotopes*. 2nd ed., London: Edward Arnold and Co., 1924.
31. Thomson, J.J. *Recollections and Reflections*. Cambridge: Cambridge University Press 1937.
32. Gaede, W. Die Diffusion der Gase durch Quecksilberdampf bei niederen Drucken und die Diffusionsluftpumpe. *Ann. Physik.* 1915, *46*, 357–392.
33. Langmuir, I. The condensation pump: An improved version of high vacuum pump. *J. Franklin Inst.* 1916, *182*, 719–743.
34. Langmuir, I. A high vacuum mercury vapor pump of extreme speed. *Phys. Rev.* 1916, *8*, 48–51.
35. Aston, F.W. LXXIV. A positive ray spectrograph. *Philosoph. Mag. V* 1919, *38*, 707–714.
36. Aston, F.W.; Fowler, R.H. LIX. Some problems of the mass-spectrograph. *Philosoph. Mag. VI.* 1922, *43*, 514–528.
37. Aston, F.W. Isotopes and atomic weights. Presented at a Discourse of the Royal Institution of Great Britain, London, on 11 February, 1921. (Published in *Proc. Roy. Inst. Great Britain* 1920–1922, *23*, 299–310.)
38. Costa, J-L. Spectres de masse de quelques elements légers. *Ann. de Physique.* 1925, *4*, 425–456.
39. Aston, F.W. A new mass-spectrograph and the whole number rule. (Bakerian Lecture). *Proc. Roy. Soc. London, Ser. A.* 1927, *115*, 487–514.
40. Aston, F.W. A new mass-spectrograph and the whole number rule. *Proc. Roy. Soc. London, Ser. A* 1927, *115*, 487–514.
41. Aston, F.W. A second-order focusing mass spectrograph and isotopic weights by the doublet method. *Proc. Roy. Soc. A.* 1937, *163*, 391–404.
42. Dempster, A.J. A new method of positive ray analysis. *Phys. Rev.* 1918, *11*, 316–325.
43. Dempster, A.J. Positive ray analysis of lithium and magnesium. *Phys. Rev.* 1921, *18*, 415–422.
44. Dempster, A.J. Positive-ray analysis of potassium, calcium and zinc. *Phys. Rev.* 1922, *20*, 631–638.

45. Barber, N.F. Note of the shape of an electron beam bent in a magnetic field. *Proc. Leeds Philosoph. Lit. Soc. (Scientific Section)* 1933, *2*, 427–434.
46. Nier, A.O. A mass-spectrographic study of the isotopes of Hg, Xe, Kr, Be, I, As, and Cs. *Phys. Rev.* 1937, *52*, 933–937.
47. Bainbridge, K.T.; Jordan, E.B. Mass spectrum analysis 1. The mass spectrograph. 2. The existence of isobars of adjacent elements. *Phys. Rev.* 1936, *50*, 282–296.
48. Nier, A.O. A mass spectrometer for routine isotope abundance measurements. *Rev. Sci. Instrum.* 1940, *11*, 212–216.
49. Bleakney, W. The ionization potential of molecular hydrogen. *Phys. Rev.* 1932, *40*, 496–501.
50. Tate, J.T.; Smith, P.T. Ionization potentials and probabilities for the formation of multiply charged ions in the alkali vapors and in krypton and xenon. *Phys. Rev.* 1934, *46*, 773–776.
51. Smythe, W.R.; Rumsbaugh, L.H.; West, S.S. A high intensity mass-spectrometer. *Phys. Rev.* 1934, *45*, 724–727.
52. Smyth, H.D. Products and processes of ionization by low speed electrons. *Rev. Mod. Phys.* 1931, *3*, 347–391.
53. Stewart, H.R.; Olson, A.R. The decomposition of hydrocarbons in the positive ray tube. *J. Am. Chem. Soc.* 1931, *53*, 1236–1244.
54. Bartky, W.; Dempster, A.J. Paths of charged particles in electric and magnetic fields. *Phys. Rev.* 1929, *33*, 1019–1022.
55. Herzog, R.F.K. Ionen-und electronenoptische Zylinderlinsen und Prismen. I. *Z. Phys.* 1934, *89*, 447–473.
56. Mattauch, J.; Herzog, R.F.K. Über einen neuen Massenspektrographen. *Z. Phys.* 1934, *89*, 786–795.
57. Conrad, R. Decomposition of hexane, cyclohexane and benzene in the positive ray tube. *Trans. Faraday Soc.* 1934, *30*, 215–220.
58. Hoover, Jr., H.; Washburn, H.W. *Am. Inst. Mining Met. Engrs. Tech. Publ. No.* 1205 (1940). (Published in *Petroleum Technology*, May 1940, 100–106.)
59. Hoover, Jr., H.; Washburn, H.W. Analysis of hydrocarbon gas mixtures by mass spectrometry. *Calif. Oil World* 1941, *34*, 21–22.
60. Hoover, Jr., H. Mass spectrometer. *U.S. Patent,* 1944, 2,341,551.
61. Meyerson, S. Reminiscences of the early days of mass spectrometry in the petroleum industry. *Org. Mass Spectrom.* 1986, *21*, 197–208.
62. Quayle, A. Recollections of mass spectrometry of the fifties in a UK petroleum laboratory. *Org. Mass Spectrom.* 1987, *22*, 569–585.
63. Nier, A.O. A mass-spectrographic study of the isotopes of argon, potassium, rubidium, zinc and cadmium. *Phys. Rev.* 1936, *50*, 1041–1045.
64. Blears, J. Metropolitan-Vickers mass spectrometers. *Anal. Chem.* 1953, *25*, 522.
65. *American Petroleum Institute Project 44: Selected Mass Spectral Data.* Thermodynamics Research Center, Texas A & M University, TX, 1947, 1948, 1963, and 1966.
66. Rock, S.M. Qualitative analysis from mass spectra. *Anal. Chem.* 1951, *23*, 261–268.
67. Beynon, J.H. Qualitative analysis of organic compounds by mass spectrometry. *Nature.* 1954, *174*, 735–737.
68. Hipple, J.A.; Condon, E.U. Detection of metastable ions with the mass spectrometer. *Phys. Rev.* 1945, *68*, 54–55.
69. Hipple, J.A.; Fox, R.E.; Condon, E.U. Metastable ions formed by electron impact in hydrocarbon gases. *Phys. Rev.* 1946, *69*, 347–356.
70. Ogata, K.; Matsuda, H. Masses of light atoms. *Phys. Rev.* 1953, *89*, 27–32.
71. Ney, E.P.; Mann, A.K. Mass measurement with a single field mass spectrometer. *Phys. Rev.* 1946, *69*, 239.
72. Nier, A.O.; Roberts, T.R. The determination of atomic mass doublets by means of a mass spectrometer. *Phys. Rev.* 1951, *81*, 507–510.

73. Johnson, E.G.; Nier, A.O. Angular aberrations in sector shaped electromagnetic lenses for focusing beams of charged particles. *Phys. Rev.* 1953, *91*, 10–17.

74. Elliott, M. An appreciation of Brian Green. *Rapid Commun. Mass Spectrom.* 1996, *10*, 1563–1565.

75. Craig, R.D.; Errock, G.A. Design and performance of a double-focusing mass spectrometer for analytical work. In J.D. Waldron (Ed.) *Advances in Mass Spectrometry*, Vol. 1, pp. 66–85. London: Pergamon Press, 1958.

76. Beynon, J.H. The use of the mass spectrometer for the identification of organic compounds. *Mikrochim. Acta.* 1956, *44*, 437–453.

77. Beynon, J.H.; Williams, A.E. *Mass and Abundance Tables for Use in Mass Spectrometry*, Amsterdam: Elsevier, 1963.

78. James, A.T.; Martin, A.J.P. Gas-liquid chromatography: The separation and micro-estimation of volatile fatty acids from formic acid to dodecanoic acid. *Biochem. J.* 1952, *50*, 679–690.

79. Beynon, J.H. Experience in using a double focusing mass spectrometry for organic chemical analysis. In R.M. Elliott (Ed.) *Advances in Mass Spectrometry*, Vol. 2, pp. 216–229. Oxford: Pergamon Press, 1963.

80. Craig, R.D.; Green, B.N.; Waldron, J.D. Application of high resolution mass spectrometry in organic chemistry. *Chimia.* 1963, *17*, 33–42.

81. Related by R.D Craig in a personal conversation with one of us (JFJT).

82. Rosenstock, H.M.; Wallenstein, M.B.; Wahrhaftig, A.L.; Eyring, H. Absolute rate theory for isolated systems and the mass spectra of polyatomic molecules. *Proc. Nat. Acad. Sci.* 1952, *38*, 667–678.

83. Glasstone, S.; Laidler, K.J.; Eyring, H. *The Theory of Rate Processes.* New York: McGraw-Hill, 1941.

84. Williams, D.H.; Beynon, J.H. The concept and rôle of charge localization in mass spectrometry. *Org. Mass Spectrom.* 1976, *11*, 103–116.

85. Rosenstock, H.M.; Krauss, M. Quasi-equilibrium theory of mass spectra. In F.W. McLafferty (Ed.) *Mass Spectrometry of Organic Ions.* Chapter 1. New York: Academic Press, 1963. pp 1–64.

86. McLafferty, F.W. Mass spectrometric analysis – broad applicability to chemical research. *Anal. Chem.* 1956, *28*, 306–316.

87. Friedman, L.; Long, F.A. Mass spectra of six lactones. *J. Amer. Chem. Soc.* 1953, *75*, 2832–2836.

88. Delfosse, J.; Bleakney, W. Dissociation of propane, propylene and allene by electron impact. *Phys. Rev.* 1939, *56*, 256–260.

89. McLafferty, F.W. Mass spectrometric analysis – molecular rearrangements. *Anal. Chem.* 1959, *31*, 82–87.

90. Beynon, J.H. *Mass Spectrometry and its Applications to Organic Chemistry.* Amsterdam: Elsevier, 1960.

91. Reed, R.I. *Ion Production by Electron Impact.* 1962. London: Academic Press.

92. Biemann, K. *Mass Spectrometry.* McGraw-Hill, New York, 1962.

93. Mclafferty, F.W.; (Ed.) *Mass Spectrometry of Organic Ions.* New York: Academic Press, 1963.

94. McLafferty, F.W. *Mass Spectral Correlations.* Washington, DC: American Chemical Society, 1963.

95. Budzikiewicz, H.; Djerassi, C.; Williams, D.H. *Interpretation of Mass Spectra of Organic Compounds.* 1964, San Francisco, CA: Holden-Day.

96. Budzikiewicz, H.; Djerassi, C.; Williams, D.H. *Structure Elucidation of Natural Products by Mass Spectrometry Volume I: Alkaloids.* 1964, San Francisco, CA: Holden-Day.

97. Budzikiewicz, H.; Djerassi, C.; Williams, D.H. *Structure Elucidation of Natural Products by Mass Spectrometry Volume II: Steroids, Terpenoids, Sugars, and Miscellaneous Classes.* 1964, San Francisco, CA: Holden-Day.

98. Beynon, J.H.; Saunders, J.H.; Williams, A.E. *The Mass Spectra of Organic Molecules.* 1968. Amsterdam: Elsevier.

99. Roboz, J. *Introduction to Mass Spectrometry – Instrumentation and Techniques.* 1968, New York: Wiley Interscience.

100. Meyerson, S. Trace components by sorption and vaporization in mass spectrometry. *Anal. Chem.* 1956, *28*, 317–318.

101. Reed, R.I. Electron impact and molecular dissociation. Part I. Some steroids and triterpenoids. *J. Chem. Soc.* 1958, 3432–3436.

102. McLafferty, F.W. (Ed.) *Tandem Mass Spectrometry.* 1983, New York: Wiley.

103. Busch, K.L.; Glish, G.L; McLuckey, S.A. *Mass Spectrometry/Mass Spectrometry – Techniques and Applications in Tandem Mass Spectrometry.* 1988, New York: VCH Publishers.

104. Cooks, R.G. Collision-induced dissociation: Readings and commentary. *J. Mass Spectrom.* 1995, *30*, 1215–1221.

105. McLuckey, S.A. Principles of collisional activation in analytical mass spectrometry. *J. Am. Soc. Mass Spectrom.* 1992, *3*, 599–614.

106. Smyth, H.D. Primary and secondary products of ionization in hydrogen. *Phys. Rev.* 1925, *25*, 452–468.

107. Bainbridge, K.T.; Jordan, E.B. Mass-spectrographic measurement of the C^+ band resulting from the dissociation of CO^+. *Phys. Rev.* 1937, *51*, 595.

108. Rosenstock, H.M.; Melton, C.E. Metastable transitions and collision-induced dissociations in mass spectra. *J. Chem. Phys.* 1957, *26*, 314–322.

109. Jennings, K.R. Collision-induced decomposition of aromatic molecular ions. *Int. J. Mass Spectrom. Ion. Phys.* 1968, *1*, 227–235.

110. White, F.A.; Rourke, F.M.; Sheffield, J.C. A three stage mass Spectrometer. *Appl. Spectroscopy.* 1958, *12*, 46–52.

111. Rourke, F.M.; Sheffield, J.C.; Davis, W.D.; White, F.A. Charge permutation and dissociation of molecular ions by impact with neutral molecules. *J. Chem. Phys.* 1959, *31*, 193–199.

112. Barber, M; Elliott, R.M. Comparison of metastable spectra from single and double focusing mass spectrometers. *Proc. 12th Annual Conference on Mass Spectrometry and Allied Topics,* ASTM Committee E-14, Montreal, Canada, 1964, pp. 150–157.

113. Barber, M.; Wolstenholme, W.S.; Jennings, K.R. Metastable ions in a double focusing mass spectrometer. *Nature* 1967, *213*, 664–666.

114. Boyd, R.K.; Beynon, J.H. Scanning of sector mass spectrometers to observe the fragmentation of metastable ions. *Org. Mass Spectrom.* 1977, *12*, 163–165.

115. Jennings, K.R.; Mason, R.S. Tandem mass spectrometry utilizing linked scanning of double focusing instruments. In F.W. McLafferty (Ed.) *Tandem Mass Spectrometry.* Chapter 9, pp. 197–222, New York: Wiley, 1983.

116. Beynon, J.H.; Caprioli, R.M.; Baitinger, W.E.; Amy, J.W. The ion kinetic energy spectrum and the mass spectrum of argon. *Int. J. Mass Spectrom. Ion Phys.* 1969, *3*, 313–321.

117. Beynon, J.H.; Caprioli, R.M.; Ast, T. The effect of deuterium labelling on the width of a 'metastable peak'. *Org. Mass Spectrom.* 1971, *5*, 229–234.

118. Wachs, T.; Bente III, P.F.; McLafferty, F.W. Simple modification of a commercial mass spectrometer for metastable data collection. *Int. J. Mass Spectrom. Ion Phys.* 1972, *9*, 333–341.

119. Beynon, J.H.; Cooks, R.G.; Amy, J.W.; Baitinger, W.E.; Ridley, T.Y. Design and performance of a mass-analyzed ion kinetic energy (MIKE) spectrometer. *Anal. Chem.* 1973, *45*, 1023A–1031A.

120. McLafferty, F.W.; Bente III, P.F.; Kornfeld, R.; Tsai, S-C.; Howe, I. Collisional activation spectra of organic ions. *J. Am. Chem. Soc.* 1973, *95*, 2120–2129. (Reprinted in *Org. Mass Spectrom.* 1995, *30*, 797–806.)

121. Cooks, R.G.; Beynon, J.H.; Caprioli, R.M.; Lester, G.R. *Metastable Ions.* 1973, Amsterdam: Elsevier Scientific.

122. Morgan, R.P.; Beynon, J.H.; Bateman, R.H.; Green, B.N. The MM-ZAB-2F double-focussing mass spectrometer and MIKE spectrometer. *Int. J. Mass Spectrom. Ion Phys.* 1978, *28*, 171–191.

123. Elliott, R.M. Ion sources. In C.A. McDowell (Ed.) *Mass Spectrometry.* Chapter 4. New York: McGraw-Hill, 1963 pp. 69–103.

124. Aston, F.W. *Mass Spectra and Isotopes,* 2nd ed. London: Edward Arnold, 1942.

125. Nier, A.O. A mass spectrometer for isotope and gas analysis. *Rev. Sci. Instrum.* 1947, *18*, 398–411.

126. Terenin, A.N.; Popov, B. Photodissociation of salt molecules into ions. *Phys. Z. Sowjet-union* 1932, *2*, 299–318.

127. Dempster, A.J. New ion sources for mass spectroscopy. *Nature* 1935, *135*, 542.

128. Dempster, A.J. Ion sources for mass spectroscopy. *Rev. Sci. Instrum.* 1936, *7*, 46–49.

129. Craig, R.D.; Errock, G.A.; Waldron, J.D. Determination of impurities in solids by spark source mass spectrometry. In J.D. Waldron (Ed.) *Advances in Mass Spectrometry.* Vol. 1, pp. 137–156. London: Pergamon Press, 1959.

130. James, J.A.; Williams, J.L. The analysis of non-conducting solids by the mass spectrometer. In J.D. Waldron (Ed.) *Advances in Mass Spectrometry.* Vol. 1, pp. 157–161. Pergamon Press: London, 1959.

131. Franzen, J.; Schuy, K.D. The effect of electrode shape on analytical precision in spark source mass spectrometry. In E. Kendrick (Ed.) *Advances in Mass Spectrometry,* Vol. 4, pp. 449–456. London: Institute of Petroleum, 1968.

132. Ahern, A.J. (Ed.) *Mass Spectrometric Analysis of Solids.* 1966. Amsterdam: Elsevier.

133. Ramendik, G.; Verlinden, J.; Gijbels, R. Spark source mass spectrometry. In F. Adams, R. Gijbels, R. Van Grieken, (Eds.) *Inorganic Mass Spectrometry.* Chapter 2. New York: Wiley, 1988 pp. 17–84.

134. Gray, A.L. Inductively coupled plasma mass spectrometry. In F. Adams, R. Gijbels, R. Van Grieken, (Eds.) *Inorganic Mass Spectrometry.* Chapter 6. New York: Wiley, 1988.

135. Harrison, W.W. Glow discharge mass spectrometry. In F. Adams, R. Gijbels, R. Van Grieken, (Eds.) *Inorganic Mass Spectrometry.* Chapter 3. New York: Wiley, 1988 pp. 85–123.

136. Verbueken, A.H.; Bruynseels, F.J.; Van Grieken, R.; Adams, F. Laser microprobe mass spectrometry. In F. Adams, R. Gijbels, R. Van Grieken, (Eds.) *Inorganic Mass Spectrometry.* Chapter 5. New York: Wiley, 1988 pp. 173–256.

137. Inghram, M.G. Trace determination by the mass spectrometer. *J. Phys. Chem.* 1953, *57*, 809–814.

138. Richardson, O.W. The specific charge of the ions emitted by hot bodies. *Phil. Mag.* 1908, *16*, 740–767.

139. Richardson, O.W. *The Emission of Electricity from Hot Bodies.* London: Longmans Green, 1916, p. 196.

140. Wilson, H.W.; Daly, N.R. Mass spectrometry of solids. *J. Sci. Instrum.* 1963, *40*, 273–285.

141. Inghram, M.G.; Chupka, W.A. Surface ionization source using multiple filaments. *Rev. Sci. Instrum.* 1953, *24*, 518–520.

142. Palmer, G.H. High sensitivity solid source mass spectrometry. In J.D. Waldron (Ed.) *Advances in Mass Specrometry.* Vol. 1, pp. 89–102. London: Pergamon Press, 1959.

143. Patterson, H.; Wilson, H.W. New mass spectrometer solid source arrangement for the simultaneous analysis of two samples. *J. Sci. Instrum.* 1961, *39*, 84–85.

144. Inghram, M.G.; Gomer, R. Mass spectrometric analysis of ions from the field microscope. *J. Chem. Phys.* 1954, *22*, 1279–1280.

145. Gomer, R.; Inghram, M.G. Applications of field ionization to mass spectrometry. *J. Am. Chem. Soc.* 1955, *77*, 500.

146. Muller, E.W. Field desorption. *Phys Rev.* 1956, *102*, 618–624.

147. Muller, E.W.; Bahadur, K. Field ionization of gases at a metal surface and the resolution of a field ion microscope. *Phys. Rev.* 1956, *102*, 624–631.

148. Beckey, H.D. *Field Ionization Mass Spectrometry*. Oxford: Pergamon Press, 1971.

149. Schulten, H-R. Biochemical, medical, and environmental applications of field-ionization and field-desorption mass spectrometry. *Int. J. Mass Spectrom. Ion Phys.* 1979, *32*, 97–283.

150. Prókai, L. *Field Desorption Mass Spectrometry*. 1990. New York: Marcel Dekker.

151. Ashcroft, A.E. *Ionization Methods in Organic Mass Spectrometry*. Cambridge, UK: The Royal Society of Chemistry, 1997.

152. Munson, M.S.B.; Field, F.H. Chemical ionization mass spectrometry. I. General introduction. *J. Am. Chem. Soc.* 1966, *88*, 2621–2630.

153. Field, F.H. Chemical ionization mass spectrometry. In E. Kendrick (Ed.) *Advances in Mass Spectrometry*. Vol. 4, pp. 645–665. London: Institute of Petroleum, 1968.

154. Tal'roze, V.L.; Lyubimova, A.K. Secondary processes in the ion source of a mass spectrometer. *Dokl. Akad. Nauk SSSR* 1952, *86*, 909–912. (Translated version in Franklin, J.L. (Ed.) *Ion-Molecule Reactions, Part I. Kinetics and Dynamics*. pp. 39–42. Stroudsburg, PA: Dowden, Hutchinson & Ross, 1979).

155. As related by V.L. Tal'roze to JFJT.

156. Baldwin, M.A.; McLafferty, F.W. Direct chemical ionization of relatively involatile samples. Application to underivatized peptides. *Org. Mass Spectrom.* 1973, *7*, 1353–1356.

157. Horning, E.C.; Horning, M.G.; Carroll, D.I.; Dzidic, I.; Stillwell, R.N. New picogram detection system based on a mass spectrometer with an external ionization source at atmospheric pressure. *Anal. Chem.* 1973, *45*, 936–943.

158. McIver, Jr., R.T.; Ledford, Jr., E.B.; Miller, J.S. Proposed method for mass spectrometric analysis for ultra-low vapor pressure compounds. *Anal. Chem.* 1975, *47*, 692–697.

159. Bonner, R.F.; Lawson, G.; Todd, J.F.J. A low-pressure chemical ionization source: An application of a novel type of ion storage mass spectrometer. *J. Chem. Soc. Chem. Commun.* 1972, 1179–1180.

160. Bonner, R.F.; Todd, J.F.J.; Lawson, G.; March, R.E. Ion storage mass spectrometry: Applications in the study of ionic processes and chemical ionization reactions. In A.R. West (Ed.) *Advances in Mass Spectrometry*, Vol. 6, pp. 377–384. London: Institute of Petroleum, 1974.

161. Mather, R.E.; Lawson, G.; Todd, J.F.J.; Bakker, J.M.B. The quadrupole ion storage trap (QUISTOR) as a low pressure chemical ionization source for a magnetic sector mass spectrometer. *Int. J. Mass Spectrom. Ion Phys.* 1979, *39*, 1–37.

162. Boswell, S.M.; Mather, R.E.; Todd, J.F.J.; Chemical ionization in the ion trap: A comparative study. *Int. J. Mass Spectrom. Ion Phys.* 1990, *99*, 139–149.

163. Hunt, D.F. Reagent gases for chemical ionization mass spectrometry. In A.R. West (Ed.) *Advances in Mass Spectrometry*, Vol. 6, pp. 517–522. London: Institute of Petroleum, 1974.

164. Harrison, A.G. *Chemical Ionization Mass Spectrometry*, 2nd ed. Boca Raton, FL: CRC Press, 1992.

165. Morris, H.R. (Ed.) *Soft Ionization Biological Mass Spectrometry*. London: Heyden, 1981.

166. Lyon, P.A. (Ed.) *Desorption Mass Spectrometry – are SIMS and FAB the same?* ACS Symposium Series No. 291, Washington, DC: American Chemical Society, 1985.

167. Benninghoven, A.; Rüdenauer, F.G.; Werner, H.W. *Secondary Ion Mass Spectrometry – Basic Concepts, Instrumental Aspects, Applications and Trends*. New York: Wiley, 1987.

168. Honig, R.E. The development of secondary mass spectrometry (SIMS). *Retrospective Lectures, 32nd ASMS Conference on Mass Spectrometry and Allied Topics*. San Antonio, TX, 1984, pp. 19–46.

169. Honig, R.E. The growth of secondary ion mass spectrometry (SIMS): A personal view of its development. In A. Benninghoven, R.J. Colton, D.S. Simons, H.W. Werner (Eds.) *Secondary Ion Mass Spectrometry SIMS V*, Berlin: Springer-Verlag, 1986, pp. 2–15.

170. Herzog, R.F.K.; Viehböck, F.P. Ion source for mass spectrometry *Phys. Rev.* 1949, *76*, 855–856.
171. Liebl, J.; Herzog, R.F.K. Sputtering ion source for solids. *J. Appl. Phys.* 1963, *34*, 2893–2896.
172. Honig, R.E. Sputtering of surfaces by positive ion beams of low energy. *J. Appl. Phys.* 1958, *29*, 549–555.
173. Honig, R.E. The application of mass spectrometry to study of surfaces by sputtering. In J.D. Waldron (Ed.) *Advances in Mass Spectrometry,* Vol. 1, pp. 162–171. London: Pergamon Press, 1959.
174. Honig, R.E. Mass spectrometric studies of solid surfaces. In R.M. Elliott (Ed). *Advances in Mass Spectrometry,* Vol. 2, pp. 25–37. Pergamon Press: Oxford, 1963.
175. Honig, R.E. Analysis of solids by mass spectrometry. In W.L. Mead, (Ed.) *Advances in Mass Spectrometry,* Vol. 3, pp. 101–129. London: Institute of Petroleum, 1966.
176. Benninghoven, A. Die Analyse monomolekularer Festkörperoberflächenschichten mit Hilfe der Sekundärionenemission, *Z. Phys.* 1970, *230*, 403–417.
177. Benninghoven, A.; Sichtermann, W. Secondary ion mass spectrometry: A new analytical technique for biologically important compounds. *Org. Mass Spectrom.* 1977, *12*, 595–597.
178. Benninghoven, A.; Sichtermann, W. Detection, identification and structural investigation of biologically important compounds by secondary ion mass spectrometry. *Anal. Chem.* 1978, *50*, 1180–1184.
179. Oliphant, M.L.E. The action of metastable atoms of helium on a metal surface. *Proc. Roy. Soc. London, Ser. A* 1929, *124*, 228–242.
180. Dillon, A.F.; Lehrle, R.S.; Robb, J.C.; Thomas, D.W. Positive ion emission from polymers under ion impact. In E. Kendrick, (Ed.) *Advances in Mass Spectrometry.* Vol. 4, pp. 477–490. London: Institute of Petroleum, 1968.
181. Barber, M.; Bordoli, R.S.; Sedgwick, R.D.; Tyler, A.N. Fast atom bombardment of solids (F.A.B.): A new ion source for mass spectrometry. *J. Chem. Soc. Chem. Commun.* 1981, 325–327.
182. Barber, M.; Bordoli, R.S.; Sedgwick, R.D.; Tyler, A.N. Fast atom bombardment of solids as an ion source in mass spectrometry. *Nature.* 1981, *293*, 270–275.
183. Barber, M.; Bordoli, R.S.; Garner, G.V.; Gordon, D.B.; Sedgwick, R.D.; Tetler, L.W.; Tyler, A.N. Fast-atom-bombardment mass spectra of enkephalins. *Biochem. J.* 1981, *197*, 401–404.
184. Barber, M.; Bordoli, R.S.; Elliott, G.J.; Sedgwick, R.D.; Tyler, A.N. Fast atom bombardment mass spectrometry. *Anal. Chem.* 1982, *54*, 645A–657A.
185. Tolun, E.; Proctor, C.J.; Todd, J.F.J.; Walshe, J.M.A.; Connor, J.A. Mixed solvent matrices for fast atom bombardment mass spectrometry of (4,4'-disubstituted-2,2'-bipyridine) tetracarbonyl-metal complexes of molybdenum and tungsten. *Org. Mass Spectrom.* 1984, *19*, 294.
186. Aberth, W.; Straub, K.M. Burlingame, A.L. Secondary ion mass spectrometry with cesium ion primary beam and liquid target matrix for analysis of bioorganic compounds. *Anal. Chem.* 1982, *54*, 2029–2034.
187. Caprioli, R.M.; Fan, T.; Cottrell, J.S. Continuous-flow sample probe for fast atom bombardment mass spectrometry. *Anal. Chem.* 1986, *58*, 2949–2954.
188. Al-Omair, A.S.; Todd, J.F.J. Preliminary studies on ion chromatography/fast-atom bombardment mass spectrometry for the characterization of various ionic compounds. *Rapid Commun. Mass Spectom.* 1989, *3*, 405–409.
189. Caprioli, R.M., (Ed.) *Continuous-Flow Fast Atom Bombardment Mass Spectrometry.* Chichester, UK: Wiley, 1990.
190. Devienne, F.M.; Roustan, J-C. 'Fast atom bombardment' – a rediscovered method for mass spectrometry. *Org. Mass Spectrom.* 1982, *17*, 173–181.

191. Macfarlane, R.D. ^{252}Californium plasma desorption mass spectrometry (^{252}Cf-PDMS). In H.R. Morris (Ed.) *Soft Ionization Biological Mass Spectrometry.* pp. 110–119. London: Heyden, 1981.

192. Macfarlane, R.D.; Torgerson, D.F.; Fares, Y.; Hassell, A. "On-line" beta recoil spectrometry (maggie). *Nucl. Instrum. Meth.* 1974, *116*, 381–388.

193. Torgerson, D.F.; Skowronski, R.P.; Macfarlane, R.D. New approach to the mass spectroscopy of non-volatile compounds. *Biochem. Biophys. Res. Commun.* 1974, *60*, 616–621.

194. Macfarlane, R.D.; Torgerson, D.F. Californium-252 plasma desorption mass spectroscopy. *Science.* 1976, *191*, 920–925.

195. Macfarlane, R.D. Particle-induced desorption mass spectrometry of large involatile biomolecules: surface chemistry in the high-energy short-time domain. *Acc. Chem Res.* 1982, *15*, 268–275.

196. Becker, N.; Fürstenau, N.; Krueger, F.R.; Weiß, G.; Wien, K. Ionization of non-volatile organic compounds by fast heavy ions and their separation by mass spectrometry. *Nucl. Instrum. Meth.* 1976, *139*, 195–201.

197. Becker, O.; Fürstenau, N.; Knippelberg, W.; Krueger, F.R. Mass spectra of substances ionized by fission fragment induced desorption. *Org. Mass Spectrom.* 1977, *12*, 461–464.

198. Krueger, F.R.; Wien, K. Time-of-flight spectrometry of non-volatile organic compounds by fast heavy-ion-induced volatilization and ionization. In N.R. Daly (Ed.) *Advances in Mass Spectrometry.* Vol. 7B, pp. 1429–1432. London: Heyden, 1978.

199. Demirev, P.; Fenselau, C.; Cotter, R.J. High mass fragmentation. Matrix effects in plasma desorption mass spectrometry. *Int. J. Mass Spectrom. Ion Processes.* 1987, *78*, 251–258.

200. Feigl, P.; Schueler, B.; Hillenkamp, F. LAMMA 1000, a new instrument for bulk microprobe analysis by pulsed laser irradiation. *Int. J. Mass Spectrom. Ion Phys.* 1983, *47*, 15–18.

201. Conzemius, R.J.; Capellen, J.M. A review of the applications to solids of the laser ion source in mass spectrometry. *Int. J. Mass Spectrom. Ion Phys.* 1980, *34*, 197–271.

202. Verbueken, A.H.; Bruynseels, F.J.; Van Grieken, R.E. Laser microprobe analysis: a review of applications in the life sciences. *Biomed. Mass Spectrom.* 1985, *12*, 438–463.

203. Vastola, F.J.; Pirone, A.J. Ionization of organic solids by laser irradiation. In E. Kendrick (Ed.) *Advances in Mass Spectrometry.* Vol. 4, pp. 107–111. London: Institute of Petroleum, 1968.

204. Vastola, F.J.; Mumma, R.O.; Pirone, A.J. Analysis of organic salts by laser ionization. *Org. Mass Spectrom.* 1970, *3*, 101–104.

205. Mumma, R.O.; Vastola, F.J. Analysis of organic salts by laser ionization mass spectrometry. Sulfonates, sulfates and thiosulfates. *Org. Mass Spectrom.* 1972, *6*, 1373–1376.

206. Posthumus, M.A.; Kistemaker, P.G.; Meuzelaar, H.L.C.; Ten Noever de Brauw, M.C. Laser-desorption-mass spectrometry of polar nonvolatile bio-organic molecules. *Anal. Chem.* 1978, *50*, 985–991.

207. Kistemaker, P.G.; van der Peyl, G.J.Q.; Haverkamp, J. Laser desorption mass spectrometry. In H.R. Morris (Ed.) *Soft Ionization Biological Mass Spectrometry.* pp. 120–136. London: Heyden, 1981.

208. Tanaka, K. The origin of macromolecule ionization by laser irradiation (Nobel Lecture). *Angew. Chem. Int. Ed.* 2003, *42*, 3861–3870.

209. Tanaka, K.; Waki, H.; Ido, Y.; Akita, S.; Yoshida, Y. Protein and polymer analyses up to *m/z* 100,000 by laser ionization time-of-flight mass spectrometry. *Rapid. Commun. Mass Spectrom.* 1988, *2*, 151–153.

210. Karas, M.; Bachmann, D.; Bahr, U.; Hillenkamp, F. Matrix-assisted ultraviolet laser desorption of non-volatile compounds. *Int. J. Mass Spectrom. Ion Processes.* 1987, *78*, 53–68.

211. Karas, M.; Hillenkamp, F. Laser desorption ionization of proteins with molecular masses exceeding 10,000 daltons. *Anal. Chem.* 1988, *60*, 2299–2301.

212. Karas, M.; Ingedoh, A.; Bahr, U.; Hillenkamp, F. Ultraviolet-laser desorption/ionization mass spectrometry of femtomolar amounts of large proteins. *Biomed. Env. Mass Spectrom.* 1989, *18*, 841–843.

213. Yamashita, M.; Fenn, J.B. Electrosray ion source. Another variation on the free-jet theme. *J. Phys. Chem.* 1984, *88*, 4451–4459.

214. Yamashita, M.; Fenn, J.B. Negative ion production with the electrospray ion source. *J. Phys. Chem.* 1984, *88*, 4671–4675.

215. Vestal, M.I. Ionization techniques for non-volatile materials. *Mass Spectrom. Rev.* 1983, *2*, 447–480.

216. Fenn, J.B.; Mann, M.; Meng, C.K.; Wong, S.F.; Whitehouse, C.M. Electrospray ionization for mass spectrometry of large biomolelcules. *Science.* 1989, *246*, 64–71.

217. Fenn, J.B.; Mann, M.; Mong, C-K.; Wong, S-F.; Whitehouse, C.M. Electrospray ionization – principles and practice. *Mass Spectrom. Rev.* 1990, *9*, 37–70.

218. Hamdam, M.; Curcuruto, O. Development of electrospray ionisation technique. *Int. J. Mass Specrom. Ion Processes.* 1991, *108*, 93–113.

219. Fenn, J.B. Electrospray wings for molecular elephants (Nobel Lecture). *Angew. Chem. Int. Ed.* 2003, *42*, 3871–3894.

220. Lord Rayleigh, On the equilibrium of liquid conducting masses charged with electricity. *Philosoph. Mag.* 1882, *14*, 184–186.

221. Zeleny, J. The electrical discharge from liquid points, and a hydrostatic method of measuring the electric intensity at their surfaces. *Phys. Rev.* 1914, *2*, 69–91.

222. Zeleny, J. On the conditions of instability of electrified drops, with applications to the electrical discharge from liquid points. *Proc. Camb. Philosoph. Soc.* 1915, *18*, 71–83.

223. Zeleny, J. Instability of electrified liquid surfaces. *Phys. Rev.* 1917, *10*, 1–6.

224. Dole, M.; Mack, L.L.; Hines, R.L.; Mobley, R.C.; Ferguson, L.D.; Alice, M.B. Molecular beams of macroions. *J. Chem. Phys.* 1968, *49*, 2240–2249.

225. Evans Jr., C.A.; Hendricks, C.D. An electrohydrodynamic ion source of the mass spectrometry of liquids. *Rev. Sci. Instrum.* 1972, *43*, 1527–1530.

226. Simons, D.S.; Colby, B.N.; Evans, Jr., C.A. Electrohydrodynamic ionization mass spectrometry–the ionization of liquid glycerol and non-volatile organic solids. *Int. J. Mass Spectrom. Ion Phys.* 1974, *15*, 291–302.

227. Cook, K.D. Electrohydrodynamic mass spectrometry. *Mass Spectrom. Rev.* 1986, *5*, 467–518.

228. Iribarne, J.V.; Thomson, B.A. On the evaporation of small ions from charged droplets. *J. Chem. Phys.* 1976, *64*, 2287–2294.

229. Thomson, B.A.; Iribarne, J.V. Field induced ion evaporation from liquid surfaces at atmospheric pressure. *J. Chem. Phys.* 1979, *71*, 4451–4463.

230. Blakley, C.R.; McAdams, M.J.; Vestal, M.L. Crossed-beam liquid chromatograph–mass spectrometer combination. *J. Chromatogr.* 1978, *158*, 261–276.

231. Blakley, C.R.; Carmody, J.J.; Vestal, M.L. A new soft ionization technique for mass spectrometry of complex molecules. *J. Am. Chem. Soc.* 1980, *102*, 5931–5933.

232. Whitehouse, C.M.; Dreyer, R.N.; Yamashita, M.; Fenn, J.B. Electrospray interface for liquid chromatographs and mass spectrometers. *Anal. Chem.* 1985, *57*, 675–679.

233. Wong, S.F.; Meng, C.K.; Fenn, J.B. Multiple charging in electrospray ionization of poly(ethylene glycols). *J. Phys. Chem.* 1988, *92*, 546–550.

234. Mann, M.; Meng, C.K.; Fenn, J.B. Parent mass information from sequences of peaks of multiply charged ions. *Proc. 36th ASMS Conference of Mass Spectrometry and Allied Topics*, San Francisco, CA, 1988, pp. 1207–1208.

235. Covey, T.R.; Bonner, R.F.; Shushan, B.I.; Henion, J. The determination of protein, oligonucleotide and peptide molecular weights by ion-spray mass spectrometry. *Rapid Commun. Mass Spectrom.* 1988, *7*, 249–256.

236. Mann, M.; Meng, C.K.; Fenn, J.B. Interpreting mass spectra of multiply charged ions. *Anal. Chem.* 1989, *61*, 1702–1708.

237. Langevin, P. *Ann. Chim. Phys. Ser. 8*, 1905, *8*, 245–264. Translated in E.W. McDaniel, *Collision Phenomena in Ionized Gases*, Appendix II, New York: Wiley, 1964, pp. 701–726.

238. Heil, O. *Electron. Ind.* 1946, *2*, 81.

239. Baker, F.A.; Hasted, J.B. Electron collision studies with trapped positive ions. *Philosoph. Trans. Roy. Soc. London.* 1966, *261*, 33–65.

240. Morgan, T.G.; March, R.E.; Harris, F.M.; Beynon, J.H. Photodissociation of positive ions trapped within the space charge of a magnetically confined electron beam. *Int. J. Mass Spectrom. Ion Processes* 1984, *61*, 41–58.

241. Bourne, A.J.; Danby, C.J. *J. Sci. Instrum. Ser. 2*, 1968, *1*, 155.

242. Herod, A.S.; Harrison, A.G. Bimolecular reactions of ions trapped in an electric space charge. *Int. J. Mass Spectrom. Ion Phys.* 1970, *4*, 415–431.

243. Holmes, J.L.; Aubry, C.; Mayer, P.M. *Assigning Structures to Ions in Mass Spectrometry.* Boca Raton, FL: CRC Press, 2007.

244. March, R.E.; Hughes, R.J. Charge separation of metastable triply charged ions derived from Benzo[a]pyrene. *Int. J. Mass Spectrom. Ion Processes* 1986, *68*, 167–182.

245. Tal'rose, V.I.; Karachevtsev, G.V. Ion-Molecule Reactions. In W.L. Mead, (Ed.). *Advances in Mass Spectrometry.* Vol. 3, pp. 211–233. London: Institute of Petroleum, 1966.

246. Ottinger, C. Messung der Zerfallszeiten von Molecülionen *Z. Naturforsch. Teil A.* 1967, *22*, 20–40.

247. Derrick, P.J. Ion lifetimes. In A. Maccoll (Ed.) *International Reviews of Science, Mass Spectrometry, Physical Chemistry, Ser. II.* pp. 1–46. London: Butterworths, 1975.

248. Nibbering, N.M.M. Mechanistic studies by field ionization kinetics. *Mass Spectrom. Rev.* 1984, *3*, 445–447.

249. Koyanagi, G.K.; McMahon, A.W.; March, R.E.; Harrison, A.G. N_2O: a case study for acceleration region kinetics in a double-focusing mass spectrometer with a conventional Nier-type ion source. *Int. J. Mass Spectrom. Ion Processes.* 1989, *87*, 249–274.

250. Koyanagi, G.K.; McMahon, A.W.; March, R.E. Unimolecular dissociation of ions in the acceleration region of a mass spectrometer. *Rapid Commun. Mass Spectrom.* 1987, *1*, 132–135.

251. Holmes, J.L. Assigning structures to ions in the gas phase. *Org. Mass Spectrom.* 1985, *20*, 169–183.

252. Harris, F.M.; Beynon, J.H. Photodissociation in beams: Organic ions. In M.T. Bowers (Ed.) *Gas Phase Ion Chemistry.* Vol. 3, Chapter 19, pp. 100–128. Orlando, FL: Academic Press, 1984.

253. Harrison, A.G.; Mercer, R.S.; Reiner, E.J.; Young, A.B.; Boyd, R.K.; March, R.E.; Porter, C.J. A hybrid BEQQ mass spectrometer for studies in gaseous ion chemistry. *Int. J. Mass Spectrom. Ion Processes.* 1986, *74*, 13–31.

254. Terlouw, J.K.; Kieskamp, W.M.; Holmes, J.L.; Mommers, A.A.; Burgers, P.C. The neutralisation and re-ionisation of mass-selected positive ions by inert gas atoms. *Int. J. Mass Spectrom. Ion Processes* 1985, *64*, 245–250.

255. Cooks, R.G. (Ed.) *Collision Spectroscopy.* New York: Plenum, 1978.

256. Brenton, A.G. Translational energy spectroscopy: a personal perspective of its development. *Int. J. Mass Spectrom.* 2000, *200*, 403–422.

257. Lee, A.R.; Wilkins, A.C.R.; Enos, C.S.; Brenton, A.G. Translational energy spectroscopy of single-electron capture by C^{2+} ions in He, Ne and Ar. *Int. J. Mass Spectrom. Ion Processes* 1994, *134*, 213–220.

258. Appell, J.; Durup, J.; Fehsenfeld, F.C.; Fournier, P. Double charge transfer spectroscopy of diatomic molecules. *J. Phys. B. At. Mol. Phys.* 1973, *6*, 197–208.

259. Fournier, P.; March, R.E.; Benoît, C.; Govers, T.R.; Appell, J.; Fehsenfeld, F.C.; Durup, J. Double charge transfer $H^+ + Xe \rightarrow H^- + Xe^{++}$ using a monoenergetic H^+ beam. *Proc. 8th Int. Conf. Phys. Electr. At. Coll.,* Oct. 1973, Belgrade. *Electronic and Atomic Collisions* 1973, *1*, 753–758.

260. Fournier, P.; Benoît, C.; Durup, J.; March, R.E. Spectroscopie de double transfert de charge: séparation des termes de multiplet de Xe^+ et Xe^{++} produits par collison $H^+ + Xe$. *C. R. Acad. Sc. Paris B.* 1974, *278*, 1039–1041.

261. Lossing, F.P. Heats of formation of some isomeric $[C_nH_{2n+1}O]^+$ ions. Substitutional effects on ion stability. *J. Am. Chem. Soc.* 1977, *99*, 7526–7530.

262. Rosenstock, H.M.; Buff, R.; Ferreira, M.A.A.; Lias, S.G.; Parr, A.C.; Stockbauer, R.L.; Holmes, J.L. Fragmentation mechanism and energetics of some alkyl halide ions. *J. Am. Chem. Soc.* 1982, *104*, 2337–2345.

263. Ruscic, B.; Berkowitz, J.; Curtis, L.A.; Pople, J.A. The ethyl radical: photoionization and theoretical studies. *J. Chem. Phys.* 1989, *91*, 114–121.

264. Pople, J.A.; Curtiss, L.A. The energy of N_2H_2 and related compounds. *J. Chem. Phys.* 1991, *95*, 4385–4388.

265. Szulejko, J.E.; McMahon, T.B. Progress toward an absolute gas-phase proton affinity scale. *J. Am. Chem. Soc.* 1993, *115*, 7839–7848.

266. March, R.E.; Hughes, R.J.; Todd, J.F.J. *Quadrupole Storage Mass Spectrometry.* New York: Wiley Interscience, 1989.

267. March, R.E. An introduction to quadrupole ion trap mass spectrometry. *J. Mass Spectrom.* 1997, *32*, 351–369.

268. March, R.E.; Todd, J.F.J. (Eds.) *Practical Aspects of Ion Trap Mass Spectrometry:* Vol. 1, *Fundamentals*; Vol. 2, *Instrumentation*; Vol. 3, *Chemical, Biomedical, and Environmental Applications.* Boca Raton, FL: CRC Press, 1995.

269. March, R.E.; Todd, J.F.J. *Quadrupole Ion Trap Mass Spectrometry.* 2nd ed. Hoboken, NJ: John Wiley & Sons, 2005.

270. Blauth, E.W. *Dynamic Mass Spectrometers.* Amsterdam: Elsevier, 1966.

271. Kingdon, K.H. A method for the neutralization of electron space charge by positive ionization at very low gas pressures. *Phys. Rev.* 1923, *21*, 408–418.

272. Finnigan, R.E. Early developments of the quadrupole mass spectrometer. *Retrospective Lectures, 32nd ASMS Conference on Mass Spectrometry and Allied Topics,* San Antonio, TX, 1984, pp. 117–118.

273. Lawrence, E.O. Method and apparatus for the acceleration of ions. *U.S. Patent* 1934, 1,948,384.

274. Lawrence, E.O.; Edlefsen, N.E. On the production of high speed protons. *Science* 1939, *72*, 376–377.

275. Heilbron, J.L.; Seidel, R.W. *Lawrence and His Laboratory: A History of the Lawrence Berkeley Laboratory, Volume I.* Berkeley, CA: University of California Press, 1989, http://ark.cdlib.org/ark:/13030/ft5s200764/ (Accessed May 28, 2009).

276. Lawrence, E.O.; Livingston, M.S. A method for producing high speed hydrogen ions without the use of high voltages. *Phys. Rev.* 1931, *27*, 1707.

277. Sloan, D.H.; Lawrence, E.O. The production of heavy high speed ions without the use of high voltages. *Phys. Rev.* 1931, *38*, 2021–2032.

278. Lawrence, E.O.; Livingston, M.S. The production of high speed light ions without the use of high voltages. *Phys. Rev.* 1932, *40*, 19–35.

279. Lawrence, E.O.; McMillan, E.; Thornton, R.L. The transmutation functions for some cases of deuteron-induced radioactivity. *Phys. Rev.* 1935, *48*, 493–499.

280. Hipple, J.A.; Sommer, H.; Thomas, H.A. A precise method of determining the faraday by magnetic resonance. *Phys. Rev.* 1949, *76*, 1877–1878.

281. Sommer, H.; Thomas, H.A.; Hipple, J.A. The measurement of the *e/M* by cyclotron resonance. *Phys. Rev.* 1951, *82*, 697–702.

282. Roberts, T.R.; Nier, A.O. Atomic mass determination with a mass spectrometer. *Phys. Rev.* 1950, *77*, 746.

283. Sommer, H.; Thomas, H.A. Detection of magnetic resonance by ion resonance absorption. *Phys. Rev.* 1950, *78*, 806.

284. Wobschall, D.; Graham, Jr., J.R.; Malone, D.P. Ion cyclotron resonance and the determination of collision cross sections. *Phys. Rev.* 1963, *131*, 1565–1571.

285. Wobschall, D. Ion cyclotron resonance spectrometer. *Rev. Sci. Instrum.* 1965, *36*, 466–475.

286. Alpert, D.; Buritrz, R.S. Ultra-high vacuum. II. Limiting factors on the attainment of very low pressures. *J. Appl. Phys.* 1954, *25*, 202–209.

287. Bayard, R.T.; Alpert, J. Extension of the low pressure range of the ion gauge. *Rev. Sci. Instrum.* 1950, *21*, 571–572.

288. Niemann, H.B.; Kennedy, B.C. Omegatron mass spectrometer for partial pressure measurements in upper atmosphere. *Rev. Sci. Instrum.* 1966, *37*, 722–728.

289. Ball, G.W.; Lawson, G.; Todd, J.F.J. The application of dynamic mass spectrometers to problems in gas analysis. In D. Price (Ed.) *Dynamic Mass Spectrometry*. Vol. 3, pp. 99–182. London: Heyden, 1972.

290. Baldeschwieler, J.D. Ion cyclotron resonance spectroscopy. *Science* 1968, *159*, 263–273.

291. Baldeschwieler, J.D.; Woodgate, S.S. Ion cyclotron resonance spectroscopy. *Acc. Chem. Res.* 1971, *4*, 114–120.

292. Futrell, J.H. Ion cyclotron resonance mass spectroscopy. In D. Price (Ed.) *Dynamic Mass Spectrometry*, Vol. 2, pp. 97–133. Heyden & Son, London, 1971.

293. Futrell, J.H. Ion cyclotron resonance. In J.H. Futrell (Ed.) *Gaseous Ion Chemistry and Mass Spectrometry*. pp. 127–138. New York: Wiley, 1986.

294. Wanczek, K-P. Ion cyclotron resonance spectrometry: A review of instrumentation and theory. In D. Price and J.F.J. Todd (Eds.) *Dynamic Mass Spectrometry*, Vol. 6, pp. 14–32. London: Heyden, 1981.

295. Wanczek, K.P. Ion cyclotron resonance spectrometry – a review. *Int. J. Mass Spectrom. Ion Phys.* 1984, *60*, 11–60.

296. Llewellyn, P.M. Ion cyclotron resonance mass spectrometer. *Proc. 13th Annual Conference on Mass Spectrometry and Allied Topics,* ASTM Committee E-14, St Louis, MA, pp. 313–318, 1965,

297. Rinehart Jr., K.L.; Kinstle, T.H. Mass spectrometry. In H. Eyring, C.J. Christensen, H.S. Johnston (Eds.) *Annual Review of Physical Chemistry,* Vol. 19, pp. 301–342. Palo Alto, CA: Annual Reviews, Inc., 1968.

298. Baldeschwieler, J.D.; Benz, H.; Llewellyn, P.M. Ion-molecule reactions in an ion cyclotron resonance mass spectrometer. In E. Kendrick (Ed.) *Advances in Mass Spectrometry*. Vol. 4, pp. 113–120. London: Institute of Petroleum, 1968.

299. Robinson, F.N.H. Nuclear resonance absorption circuit. *J. Sci. Instrum.* 1959, *36*, 481–487.

300. Anders, L.R.; Beauchamp, J.L.; Baldeschwieler, J.D. Ion cyclotron double resonance. *J. Chem. Phys.* 1966, *45*, 1062–1063.

301. Beauchamp, J.L.; Anders, L.R.; Baldeschwieler, J.D. The study of ion-molecule reactions in chloroethylene by ion cyclotron resonance spectroscopy. *J. Am. Chem. Soc.* 1967, *89*, 4569–4577.

302. Bowers, M.T.; Elleman, D.D.; Beauchamp, J.L. Ion cyclotron resonance of olefins. I. A study of the ion-molecule reactions of electron-impacted ethylene. *J. Phys. Chem.* 1968, *72*, 3599–3612.

303. Beauchamp, J.L.; Armstrong, J.T. An ion ejection technique for the study of ion-molecule reactions with ion cyclotron resonance spectroscopy. *Rev. Sci. Instrum.* 1969, *40*, 123–128.

304. Goode, G.C.; O'Malley, R.M.; Ferrer-Correia, A.J.; Jennings, K.R. Ion cyclotron resonance mass spectrometry. In A. Quayle (Ed.) *Advances in Mass Spectrometry,* Vol. 5, pp. 195–210. London: Institute of Petroleum, 1971.

305. Clow, R.P.; Futrell, J.H. Ion-cyclotron resonance study of the kinetic energy dependence of ion-molecule reaction rates. *Int. J. Mass. Spectrom. Ion Phys.* 1970, *4*, 165–179.

306. McIver, Jr., R.T. A trapped ion cell for ion cyclotron resonance spectroscopy. *Rev. Sci. Instrum.* 1970, *41*, 555–558.

307. Sharp, T.E.; Eyler, J.R.; Li, E. Trapped-ion motion in ion cyclotron resonance spectroscopy. *Int. J. Mass Spectrom. Ion Phys.* 1972, *9*, 421–439.

308. McMahon, T.B.; Beauchamp, J.L. A versatile trapped ion cell for ion cyclotron resonance spectroscopy. *Rev. Sci. Instrum.* 1972, *43*, 509–512.

309. Comisarow, M.B.; Marshall, A.G. Fourier transform ion cyclotron resonance spectroscopy. *Chem. Phys. Lett.* 1974, *25*, 282–283.

310. Comisarow, M.B.; Marshall, A.G. Selective-phase ion cyclotron resonance spectroscopy. *Can. J. Chem.* 1974, *52*, 1997–1999.

311. Comisarow, M.B.; Marshall, A.G. Selective-phase ion cyclotron resonance spectroscopy. *Can. J. Chem.* 1974, *52*, 1997–1999.

312. As related by Alan Marshall to REM.

313. Marshall, A.G. Fourier transform ion cyclotron resonance mass spectrometry. *Acc. Chem. Res.* 1985, *18*, 316–322.

314. Buchanan, M.V. (Ed.) *Fourier Transform Mass Spectrometry – Evolution, Innovation, and Applications.* ACS Symposium Series No. 359, Washington, DC: American Chemical Society, 1987.

315. Marshall, A.G.; Verdun, F.R. *Fourier Transforms in NMR, Optical, and Mass Spectrometry.* Amsterdam: Elsevier, 1990.

316. Guan, S.; Kim, H.S.; Marshall, A.G.; Wahl, M.C.; Wood, T.D.; Xiang, X. Shrink-wrapping an ion cloud for high-performance Fourier transform ion cyclotron resonance mass spectrometry. *Chem. Rev.* 1994, *94*, 2161–2182.

317. Cody, R.B.; Kinsinger, J.A.; Ghaderi, S.; Amster, I.J.; McLafferty, F.W.; Brown, C.E. Developments in analytical Fourier-transform mass spectrometry. *Anal. Chim. Acta* 1985, *178*, 43–66.

318. Cody, R.B. Method for external calibration of ion cyclotron resonance mass spectrometers. *U.S. Patent* 1990, 4,933,547.

319. Hunt, D.F.; Shabanowitz, J.; Mc Iver, Jr., R.T.; Hunter, R.L.; Syka, J.E.P. Ionization and mass analysis of nonvolatile compounds by particle bombardment tandem-quadrupole Fourier transform mass spectrometry. *Anal. Chem.* 1985, *57*, 765–768.

320. Schaub, T.M.; Hendrickson, C.L.; Horning, S.; Quinn, J.P.; Senko, M.W.; Marshall, A.G. High performance mass spectrometry: FT-ICR at 14.5 tesla. *Anal.Chem.* 2008, *80*, 3985–3990.

321. Fizeau, H.L. Sur une expérience relative à la vitesse de propagation de la lumière. *Comp. Rend. Acad. Sci. (Paris)* 1849, *29*, 90–92.

322. See, for example, http://scienceworld.wolfram.com/physics/FizeauWheel.html

323. Koppius, O.G. Mass spectrometer. *U.S. Patent,* 1952, 2,582,216.

324. Goudsmit, S.A. A time-of-flight mass spectrometer. *Phys. Rev.* 1948, *74*, 622–623.

325. Goudsmit, S.A. Magnetic time-of-flight mass spectrometer. *U.S. Patent,* 1955, 2,298,905.

326. Hays, E.E.; Richards, P.I.; Goudsmit, S.A. Mass measurements with a magnetic time-of-flight mass spectrometer. *Phys. Rev.* 1951, *84*, 824–829.

327. Smith, L.G. A new magnetic period mass spectrometer. *Rev. Sci. Instrum.* 1951, *22*, 115–116.
328. Smith, L.G. Magnetic-period mass spectrometer. *U.S. Patent,* 1955, 2,709,750.
329. http://www.sisweb.com/referenc/source/exactmas.htm
330. Smith, L.G. Recent developments with the mass synchrometer. *Phys. Rev.* 1952, *85*, 767.
331. Smith, L.G.; Damm, C.C. Mass synchrometer. *Rev. Sci. Instrum.* 1956, *27*, 638–649.
332. Stephens, W.E. A pulsed mass spectrometer with time dispersion. *Phys. Rev.* 1946, *69*, 691.
333. Stephens, W.E. Mass spectrometer. *U.S. Patent* 1952, 2,612,607.
334. Wolff, M.M.; Stephens, W.E. A pulsed mass spectrometer with time dispersion. *Rev. Sci. Instrum.* 1953, *24*, 616–617.
335. Cameron, E.A.; Eggers, Jr., D.F. An ion "velocitron". *Rev. Sci. Instrum.* 1948, *19*, 605–607.
336. Katzenstein, H.S.; Friedland, S.S. New time-of-flight mass spectrometer. *Rev. Sci. Instrum.* 1955, *26*, 324–327.
337. Wiley, W.C. Mass spectrometer. *U.S. Patent* 1954, 2,685,035.
338. Wiley, W.C.; McLaren, I.H. Time-of-flight mass spectrometer with improved resolution. *Rev. Sci. Instrum.* 1955, *26*, 1150–1157.
339. Wiley, W.C. Bendix time-of-flight mass spectrometer. *Science* 1956, *124*, 817–820.
340. Harrington, D.B. The time-of-flight mass spectrometer. In J.D. Waldron (Ed.) *Advances in Mass Spectrometry,* Vol. 1, pp. 219–265. London: Pergamon Press, 1959.
341. Smith, L.G. Magnetic electron multipliers for detection of positive ions. *Rev. Sci. Instrum.* 1951, *22*, 166–170.
342. White, D.; Sommer, P.N.; Walsh, P.N.; Goldstein, H.W. The application of the time-of-flight mass spectrometer to the study of inorganic materials at elevated temperatures. In R.M. Elliott (Ed.) *Advances in Mass Spectrometry,* Vol. 2, pp. 110–127. Oxford: Pergamon Press, 1963.
343. Homer, J.B.; Lehrle, R.S.; Robb, J.C.; Takakhasi, M.; Thomas, D.W. Application of a time-of-flight mass spectrometer to the examination of ion-molecule interactions. In R.M. Elliott (Ed.) *Advances in Mass Spectrometry,* Vol. 2, pp. 503–521. Oxford: Pergamon Press, 1963.
344. Sanzone, G. Energy resolution of the conventional time-of-flight mass spectrometer. *Rev. Sci. Instrum.* 1970, *41*, 741–742.
345. Marabale, N.L.; Sanzone, G. High-resolution time-of-flight mass spectrometry. Theory of the impulse-focused time-of-flight mass spectrometer. *Int. J. Mass Spectrom. Ion Phys.* 1974, *13*, 185–194.
346. Browder, J.A.; Miller, R.L.; Thomas, W.A.; Sanzone, G. High-resolution TOF mass spectrometry. II. Experimental confirmation of impulse-field focusing theory. *Int. J. Mass Spectrom. Ion Phys.* 1981, *37*, 99–108.
347. Bakker, J.M.B. The spiratron. In A. Quayle (Ed.) *Advances in Mass Spectrometry,* Vol. 5, pp. 278–282. London: Institute of Petroleum, 1971.
348. Hughes, A.; Rojansky, V. On the analysis of electronic velocities by electrostatic means. *Phys. Rev.* 1929, *34*, 284–290.
349. Bakker, J.M.B.; Freer, D.A.; Todd, London J.F.J. Preliminary studies on a new time-focusing time-of-flight mass spectrometer. In D. Price, J.F.J. Todd (Eds.) *Dynamic Mass Spectrometry,* Vol. 6, pp. 91–110. Heyden & Son, London, 1981.
350. Satoh, T.; Tsuno, H.; Iwanaga, M.; Kammei, Y. The design and characteristic features of a new time-of-flight mass spectrometer with a spiral ion trajectory. *J. Am. Soc. Mass Spectrom.* 2005, *16*, 1969–1975.
351. Wollnik, H. Time-of-flight mass analyzers. *Mass Spectrom. Rev.* 1993, *12*, 89–114.

352. Mamyrin, B.A.; Karataev, V.I.; Shmikk, D.V.; Zagulin, V.A. The mass-reflectron, a new nonmagnetic time-of-flight mass spectrometer with high resolution. *Sov. Phys. JETP* 1973, *37*, 45–48.

353. Mamyrin, B.A. Time-of-flight mass spectrometry (concepts, achievements, and prospects). *Int. J. Mass Spectrom.* 2001, *206*, 251–266.

354. Cotter, R.J.; Cornish, T.J. Tandem time-of-flight mass spectrometer. *U.S. Patent*, 1993, 5,202,563.

355. Guilhaus, M.; Dawson, J.H. Mass spectrometer. *U.S. Patent,* 1992, 5,117,107.

356. Guilhaus, M.; Dawson, J.H. Orthogonal-accleration time-of-flight mass spectrometer. *Rapid Commun. Mass Spectrom.* 1989, *3*, 155–159.

357. Guilhaus, M. Principles and instrumentation in time-of-flight mass spectrometry – physical instrumentation and concepts. *J. Mass Spectrom.* 1995, *30*, 1519–1532.

358. Guilhaus, M.; Mylinski, V.; Selby, D. Perfect Timing: time-of-flight mass spectrometry. *Rapid Commun. Mass Spectrom.* 1997, *11*, 951–962.

359. Blake, R.S.; Whyte, C.; Hughes, C.O.; Ellis, A.M.; Monks, P.S. Demonstration of proton-transfer reaction time-of-flight mass spectrometry for real-time analysis of trace volatile organic compounds. *Anal. Chem.* 2004, *76*, 3841–3845.

360. Blake, R.S.; Monks, P.S.; Ellis, A.M. Proton-transfer reaction mass spectrometry. *Chem. Rev.* 2009, *109*, 861–896.

361. Smythe, W.R. A velocity filter for ions and electrons. *Phys. Rev.* 1926, *28*, 1275–1286.

362. Smythe, W.R.; Mattauch, J. A new mass spectrometer. *Phys. Rev.* 1932, *40*, 429–433.

363. Smythe, W.R. On the isotopic ratio in oxygen. *Phys. Rev.* 1934, *45*, 299–303.

364. Donner, W. An RF linear decelerator mass spectrometer. *Applied Spectrosc.* 1954, *8*, 157–162.

365. Bennett, W.H. Radio frequency mass spectrometer. *Phys. Rev.* 1948, *74*, 1222.

366. Bennett, W.H. Radio frequency mass spectrometer. *J. Appl. Phys.* 1950, *21*, 143–149.

367. Boyd, R.L.F. A mass-spectrometer probe method for the study of gas discharges. *Nature* 1950, *165*, 142–144.

368. Boyd, R.L.F.; Morris, D. A radio-frequency probe of the mass-spectrometric analysis of ion concentrations. *Proc. Phys. Soc.* 1955, *A 68*, 1–10.

369. Redhead, P.A. A linear radio-frequency mass spectrometer. *Can. J. Phys.* 1952, *30*, 1–13.

370. Redhead, P.A.; Crowell, C.R. Analysis of the linear R.F. mass spectrometer. *J. Appl. Phys.* 1953, *24*, 331–337.

371. Dawson, P.H. (Ed.) *Quadrupole Mass Spectrometry and Its Applications*. Amsterdaml: Elsevier, 1976. Reprinted as an "American Vacuum Society Classic" by the American Institute of Physics under ISBN 1563964554.

372. Paul, W.; Steinwedel, H. Verfahren zur Trennung bzw. zum getrennten Nachweis von Ionen verschiedener spezifischer Ladung. *German Patent,* 1956, 944,900; Apparatus for separating charged particles of different specific charges. *U.S. Patent,* 1960, 2,939,952.

373. Paul, W.; Reinhard, H-P.; Frohlich, H. Method of separating ions of different specific charges. *U.S. Patent,* 1960, 2,950, 389.

374. Paul, W. Personal communication to JFJT.

375. Post, R.F.; Heinrich, L. *University of California Radiation Laboratory Report* (S. Shewchuck) UCRL 2209, Berkeley, CA, 1953.

376. Good, M.L. *University of California Radiation Laboratory Report* UCRL-4146, Berkeley, CA, 1953.

377. Courant, E.D.; Livingston, M.S.; Snyder, H.S. The strong-focusing synchrotron: a new high energy accelerator. *Phys. Rev.* 1952, *88*, 1190–1196.

378. Dawson, P.H.; Whetten, N.R. Quadrupoles, monopoles and ion traps. *Res./Dev.* 1969, *19*(2), 46–50.

379. Christofilos, N. Focusing system for ions and electrons. *U.S. Patent*, 1956, 2,736,799.

380. Paul, W.; Steinwedel, H. Ein Neues Massenspektrometer ohne Magnetfeld. *Z. Naturforsch.* 1953, *8a*, 448–450.

381. Berkling aus Leipzig, K. Der Entwurf eines Partialdruckmessers. *Thesis*, Physikalisches Institut der Universität, Bonn, West Germany, 1956.

382. Paul, W. Electromagnetic traps for charged and neutral particles (Nobel Lecture). *Rev. Mod. Phys.* 1990, *62*, 531–540.

383. Lawson, G.; Todd, J.F.J. Radiofrequency quadrupole mass spectrometers. *Chem. Brit.* 1972, *8*, 373–380.

384. Austin, W.E.; Holme, A.E.; Leck, J.H. The mass filter: Design and performance. In P.H. Dawson (Ed.) *Quadrupole Mass Spectrometry and its Applications*. Chapter VI, pp. 121–152. Amsterdam: Elsevier, 1976.

385. Brubaker, W.M. An improved quadrupole mass analyzer. In E. Kendrick (Ed.). *Advances in Mass Spectrometry*, Vol. 4, pp. 293–299. London: Institute of Petroleum, 1968.

386. Banner, A.E. The effects of entrance fields on the performance of quadrupole mass spectrometers. *Eur. J. Mass Spectrom.* 2009, *15*, 445–458.

387. von Zahn, U. Monopole spectrometer, a new electric field mass spectrometer. *Rev. Sci. Instrum.* 1963, *34*, 1–4.

388. von Zahn, U. Method and apparatus for separating ions of different specific charges. *U.S. Patent*, 1965, 3,197, 633.

389. Herzog, R.F. The monopole: Design and performance. In P.H. Dawson (Ed.) *Quadrupole Mass Spectrometry and its Applications*, Chapter VII, pp. 153–180. Amsterdam: Elsevier, 1976.

390. Lawson, G.; Todd, J.F.J. *Mass Spectroscopy Group Meeting*. 1971. Bristol, UK, Abstract No. 44.

391. Todd, J.F.J.; Lawson, G.; Bonner, R.F. Quadrupole ion traps. In P.H. Dawson (Ed.) *Quadrupole Mass Spectrometry and its Applications*, Chapter VIII, pp. 180–224. Amsterdam: Elsevier, 1976.

392. Todd, J.F.J.; March, R.E. A retrospective review of the development and application of the quadrupole ion trap prior to the appearance of commercial instruments. *Int. J. Mass Spectrom.* 1999, *190,191*, 9–35.

393. Church, D.A. Collision measurements and excited-level lifetime measurements of ions stored in Paul, Penning and Kingdon ion traps. *Phys. Rep.* 1993, *228*, 253–358.

394. Paul, W.; Osberghaus, O.; Fischer, E. *Forschungsberichte des Wirtschaft und Verkehrministeriums Nordrhein Wesifalen*, No. 415, Westdeutscher Verlag, Köln and Opladen, 1958.

395. Fischer, E. Die dreidimensionale Stabilisirung von Ladunssträern in einem Vierpoldfeld, *Z. Phys.* 1959, *156*, 1–26.

396. Dawson, P.H.; Whetten, N.R. Three-dimensional mass spectrometer and gauge. *U.S. Patent*, 1970, 3,527,939.

397. Dawson, P.H.; Whetten, N.R. Ion storage in three-dimensional, rotationally symmetric, quadrupole fields. I. Theoretical treatment. *J. Vac. Sci. Technol.* 1968, *5*, 1–10.

398. Dawson, P.H.; Whetten, N.R. Ion storage in three-dimensional, rotationally symmetric, quadrupole fields. II. A sensitive mass spectrometer. *J. Vac. Sci. Technol.* 1968, *5*, 11–18.

399. Sheretov, E.P.; Zenkin, V.A. Shape of the mass peak in a three-dimensional quadrupole mass spectrometer. *Sov. Phys. Tech. Phys.* 1972, *17*, 160–162.

400. Sheretov, E.P.; Zenkin, V.A.; Boligatov, O.I. Three-dimensional quadrupole mass spectrometer with ion storage. *Gen. Exp. Tech.* 1971, *14*, 195–197.

401. Sheretov, E.P.; Zenkin, V.A.; Samodurov, V.F.; Veselkin, N.D. Three-dimensional quadrupole mass spectrometer with sweep of the mass spectrum by variation of the frequency of the supply signal. *Gen. Expt. Tech.* 1973, *16*, 194–196.

402. Mather, R.E.; Waldren, R.M.; Todd, J.F.J. The characterization of a quadrupole ion storage mass spectrometer. In D. Price, J.F.J. Todd (Eds.) *Dynamic Mass Spectrometry.* Vol. 5, pp. 71–85. London: Heyden, 1978.

403. Armitage, M.A.; Fulford, J.E.; Hughes, R.J.; March, R.E. Quadrupole ion store characterization and application. *Proc. 27th ASMS Conference on Mass Spectrometry and Allied Topics,* Seattle, WA, 1979, 449–450.

404. Stafford, G.C.; Kelley, P.E.; Stephens, D.R. Method of mass analyzing a sample by use of a quadrupole ion trap. *U.S. Patent,* 1985, 4,540,884.

405. Stafford ,Jr., G.C.; Kelley, P.E.; Syka, J.E.P.; Reynolds, W.E.; Todd, J.F.J. Recent improvements in and analytical applications of advanced ion trap technology. *Int. J. Mass Spectrom. Ion Processes* 1984, *60*, 85–98.

406. Louris, J.N.; Cooks, R.G.; Syka, J.E.P.; Kelley, P.E.; Stafford, Jr, G.C.; Todd, J.F.J. Instrumentation, applications and energy deposition in quadrupole ion trap MS/MS spectrometry. *Anal. Chem.* 1987, *59*, 1677–1685.

407. Wuerker, R.F.; Shelton, H.; Langmuir, R.V. Electrodynamic containment of charged particles. *J. Appl. Phys.* 1959, *30*, 342–349.

408. Langmuir, D.B.; Langmuir, R.V.; Shelton, H.; Wuerker, R.F. Containment device. *U.S. Patent,* 1962, 3,065,640.

409. Wuerker, R.F.; Goldenberg, H.M.; Langmuir, R.V. Electrodynamic containment of charged particles by three-phase voltages. *J. Appl. Phys.* 1959, *30*, 441–442.

410. Benilan, M-N.; Audoin, C. Confinement d'ions par un champ électrique de radiofréquence dans une cage cylindrique. *Int. J. Mass Spectrom. Ion Phys.* 1973, *11*, 421–432.

411. Bonner, R.F.; Fulford, J.E.; March, R.E.; Hamilton, G.F. The cylindrical ion trap. Part I. General introduction. *Int. J. Mass Spectrom. Ion Phys.* 1977, *24*, 255–269.

412. Mather, R.E.; Waldren, R.M.; Todd, J.F.J.; March, R.E. Some operational characteristics of a quadrupole ion storage mass spectrometer having cylindrical geometry. *Int. J. Mass Spectrom. Ion Phys.* 1980, *33*, 201–230.

413. Fulford, J.E.; March, R.E.; Mather, R.E.; Todd, J.F.J.; Waldren, R.M. The cylindrical ion trap: a theoretical and experimental study. *Can. J. Spectrosc.* 1980, *25*, 85–97.

414. Hiller, J. Means for effecting improvements to mass spectrometers and mass filters. *U.K. Patent,* 1971, 1,346,393.

415. Richards, J.A.; Huey, R.M.; Hiller, J. Waveform parameter tolerances for the quadrupole mass filter with rectangular excitation. *Int. J. Mass Spectrom. Ion Processes* 1974, *15*, 417–428.

416. Sheretov, E.P.; Rozhkov, O.W.; Kiyushin, D.W.; Malutin, A.E. Mass selective instability mode without a light buffer gas. *Int. J. Mass Spectrom.* 1999, *190/191*, 103–111.

417. Ding, L.; Kumashiro, S. Ion motion in the rectangular wave quadrupole field and digital operation mode of a quadrupole ion trap mass spectrometer. *Rapid Commun. Mass Spectrom.* 2006, *20*, 3–8.

418. Bier, M.E.; Syka, J.E.P. Ion trap mass spectrometer system and method. *U.S. Patent,* 1995, 5,420, 425.

419. Schwartz, J.C.; Senko, M.W.; Syka, J.E.P. A two-dimensional quadrupole ion trap mass spectrometer. *J. Am. Soc. Mass Spectrom.* 2002, *13*, 659–669.

420. Hager, J.W. Axial ejection in a multipole mass spectrometer. *U.S. Patent,* 2001, 6,177, 668.

421. Hager, J.W. A new type of linear ion trap mass spectrometer. *Rapid Commun. Mass Spectrom.* 2002, *16*, 512–526.

422. Londry, F.A.; Hager, J.W. Mass-selective axial ion ejection from the linear quadrupole ion trap. *J. Am. Soc. Mass Spectrom.* 2003, *14*, 1130–1147.

423. Hashimoto, Y.; Hasegawa, H.; Baba, T.; Waki, I. Mass selective ejection by axial resonant excitation from a linear ion trap. *J. Am. Soc. Mass Spectrom.* 2006, *17*, 685–690.

424. Lammert, S.A.; Plass, W.R.; Thompson, C.V.; Wise, M.B. Design, optimization and initial performance of a toroidal rf ion trap mass spectrometer. *Int. J. Mass Spectrom.* 2001, *212*, 25–40.

425. Austin, D.E.; Wang, M.; Tolley, S.E.; Maas, J.D.; Hawkins, A.R.; Rockwood, A.L.; Tolley, H.D.; Lee, E.D.; Lee, M.L. Halo ion trap mass spectrometer. *Anal. Chem.* 2007, *79*, 2927–2932.

426. Makarov, A.A. Mass spectrometer. *U.S. Patent,* 1999, 5,886, 346.

427. Kingdon, K.H. A method for the neutralization of electron space charge by positive ionization at very low pressures. *Phys. Rev.* 1923, *21*, 408–418.

428. Hu, Q.; Makarov, A.A.; Cooks, R.G.; Noll, R.J. Resonant ac dipolar excitation for ion motion control in the Orbitrap mass analyzer. *J. Phys. Chem. A.* 2006, *110*, 2682–2689.

429. Blümel, R. Dynamic Kingdon trap. *Phys. Rev. A.* 1995, *51*, R30–R33.

430. Langmuir, I. Incandescent electric lamp. *U.S. Patent,* 1916, 1,180,159 (Filed April 19, 1913).

431. Hull, A.W. The effect of a uniform magnetic field on the motion of electrons between coaxial cylinders. *Phys. Rev.* 1921, *18*, 31–57.

432. Langmuir, I.; Compton, K.T. Electrical discharges in gases. Part II. Fundamental phenomena in electrical discharges. *Rev. Mod. Phys.* 1931, *3*, 191–257.

433. Cady, W.G. The potential distribution between parallel plates and concentric cylinders due to any arbitrary distribution of space charge. *J. Appl. Phys.* 1935, *6*, 10–13.

434. Tipler, P.A.; Mosca, G. *Physics for Scientists and Engineers.* 5th ed. (Extended), New York, W.H. Freeman and Company, pp. 752–755, 2004.

435. http://www.davidpace.com/physics/em-topics/capacitance-cylinders.htm

436. Horton, F.; Davies, A.C. An investigation of the ionising power of the positive ions from a glowing tantalum filament in helium. *Proc. Roy. Soc. A* 1919, *95*, 333–353.

437. Saxton, A.J. LXX. Impact ionization by low-speed positive H-ions in hydrogen. *Phil. Mag. Ser. 6* 1922, *44*, 809–823.

438. Herb, R.G.; Pauly, T.; Fisher, K.J. Electrostatic electron containment. *Bull. A. Phys. Soc. Ser. 2* 1963, *8*, 336.

439. Mourad, W.G.; Pauly, T.; Herb, R.G. Orbitron-ionization gauge. *Bull. A. Phys. Soc. Ser. 2* 1963, *8*, 336–337.

440. Hooverman, R.H. Charged particle orbits in a logarithmic potential. *J. Appl. Phys.* 1963, *34*, 3505–3508.

441. Brooks, P.R.; Herschbach, D.R. Kingdon cage as a molecular beam detector. *Rev. Sci. Instrum.* 1964, *35*, 1528–1533.

442. Douglas, R.A.; Zabritski, J.; Herb, R.G. An orbitron vacuum pump. *Rev. Sci. Instrum.* 1965, *36*, 1–5.

443. McIlraith, A.H. A charged particle oscillator. *Nature* 1966, *212*, 1422–1424.

444. Knight, R.D. Storage of ions from laser-produced plasmas. *Appl. Phys. Lett.* 1981, *38*, 221–223.

445. Yang, L.; Church, D.A. Confinement of injected ions in a Kingdon trap. *Nucl. Inst. Meth. Phys. Res.* 1991, *B56/57*, 1185–1187.

446. Church, D.A. Lifetimes of metastable levels of atomic ions. *Phys. Scr.* 1995, *T59*, 216–220.

447. Wanczek, K.P. ICR spectrometry – a review of new developments in theory, instrumentation, and applications. I. 1983–1986. *Int. J. Mass Spectrom. Ion Processes* 1989, *95*, 1–38.

448. Gillig, K.J.; Bluhm, B.K.; Russell, D.H. Ion motion in a Fourier transform ion cyclotron resonance wire ion guide cell. *Int. J. Mass Spectrom. Ion Processes* 1996, *157/158*, 129–147.

449. http://www.simion.com (accessed 9 July 2009).

450. Appelhans, A.D.; Dahl, D.A. Measurement of external ion injection and trapping efficiency in the ion trap mass spectrometer and comparison with a predictive model. *Int. J. Mass Spectrom.* 2002, *216*, 269–284.

451. Sekioka, T.; Terasama, M.; Awaya, Y. Ion storage in Kingdon trap. *Radiat. Eff. Defects Solids* 1991, *117*, 253–258.

452. Sekioka, T.; Terasama, M.; Awaya, Y. Ion storage in Kingdon trap. Spring-8 Annual Report, 1994, 220–221. Spring-8, Kamigori-cho, Ako-gun, Hyogo-ken 678-12, Japan. Available at: http://www.spring8.or.jp/pdf/en/ann_rep/94/p220-221.pdf

453. Perry, R.H.; Cooks, R.G.; Noll, R.J. Orbitrap mass spectrometry: instrumentation, ion motion and applications. *Mass Spectrom. Rev.* 2008, *27*, 661–699.

454. Makarov, A.A. Electrostatic axially harmonic orbital trapping: a high-performance technique of mass analysis. *Anal. Chem.* 2000, *72*, 1156–1162.

455. Hu, Q.; Noll, R.J.; Li, H.; Makarov, A.; Cooks, R.G. The orbitrap: a new mass spectrometer. *J. Mass Spectrom.* 2005, *40*, 430–443.

456. Korsunskii, M.I.; Basakutsa, V.A. A study of the ion-optical properties for a sector-shaped electrostatic field of the difference type. *Soviet Physics-Tech. Phys.* 1958, *3*, 1396–1409.

457. Hu, Q.; Makarov, A.A.; Cooks, R.G.; Noll, R.J. Resonant ac dipolar excitation for ion motion control in the Orbitrap mass analyzer. *J. Phys. Chem. A.* 2006, *110*, 2682–2689.

458. Hardman, M.; Makarov, A.A. Interfacing the Orbitrap mass analyzer to an electrospray ion source. *Anal. Chem.* 2003, *75*, 1699–1705.

459. Makarov, A.A.; Hardman, M.E.; Schwartz, J.C.; Senko, M.W. Mass spectrometry method and apparatus. *U.S. Patent,* 2007, 7,265,344.

460. Oksman, P. A Fourier transform time-of-flight mass spectrometer. A SIMION calculation approach. *Int. J. Mass Spectrom. Ion Processes* 1995, *141*, 67–76.

461. Makarov, A.A. Multi-electrode ion trap. *U.S Patent Application* 2008/0203293.

462. Makarov, A.; Denisov, E.V.; Jung, G.; Balschun, W.; Horning, S.R. Electrostatic trap. *U.S Patent Application* 2008/0315080.

463. Blümel, R.; Bonneville, E.; Carmichael, A. Chaos and bifurcations in ion traps of cylindrical and spherical design. *Phys. Rev. E* 1998, *57*, 1511–1518.

464. Bahr, R.E. Untersuchung der Eigenschaften eines Hochfrequenz-Ionen-Speichers. *Diplomarbeit*, Physikalisches Institut der Universität Freiburg, 1969 (unpublished).

465. Behre, E. Bau und Erprobung einer massenselektiven Speicherionenquelle. *Zulassungsarbeit*, Physikalisches Institut der Universität Freiburg, 1972 (unpublished).

466. Linz, S.J. Nonlinear dynamics of the Kingdon trap. *Phys. Rev. A* 1995, *52*, 4282–4284.

467. Blümel, R. The dynamic Kingdon trap: a novel design for the storage and crystallization of laser-cooled ions. *Appl. Phys. B* 1995, *60*, 119–122.

468. Blümel, R. An introduction to chaos in dynamic ion traps. *Phys. Scr.* 1995, *T59*, 126–130.

469. Blümel, R. Nonlinear dynamics of trapped ions. *Phys. Scr.* 1995, *T59*, 369–379.

470. Blümel, R.; Bonneville, E.; Carmichael, A. Chaos and bifurcations in ion traps of cylindrical and spherical design. *Phys. Rev. E* 1998, *57*, 1511–1518.

471. Dehmelt, H.G. Radiofrequency spectroscopy of stored ions I: Storage. In D.R.Bates (Ed.) *Advances in Atomic and Molecular Physics, Volume 3*. New York: Academic Press, 1967, pp. 53–72.

472. Garrick-Bethell, I.; Blümel, R. Unexpected instabilities in the dynamic Kingdon trap. *Phys. Rev. A* 2003, *68*, 031404, (4 pages).

473. Garrick-Bethell, I.; Clausen, Th.; Blümel, R. Universal instabilities in radio-frequency traps. *Phys. Rev. E* 2004, *69*, 056222, (15 pages).

474. Garrick-Bethell, I.; Blümel, R. Quantum mechanics of the dynamic Kingdon trap. *Phys. Rev. A* 2006, *73*, 023411, (5 pages).

475. Peik, E.; Fletcher, J. Electrodynamic trapping of charged particles in a monopole field. *J. Appl. Phys.* 1997, *82*, 5283–5286.

APPENDIX: THEORY OF RADIOFREQUENCY QUADRUPOLE DEVICES

A1.1 THEORY OF QUADRUPOLE DEVICES

This Appendix, Theory of Radiofrequency Quadrupole Devices, is presented here because of the several accounts in this Volume, and in Volume 5, that have arisen from the use of radiofrequency quadrupole ion trapping devices. In addition, other ion trapping devices such as the digital ion trap* and the toroidal ion trap† make use of the concept of an ion trajectory stability region in the discussion of the theory and practice of ion trapping device operation. This presentation of the theoretical treatment or derivation flows through three stages. First, the general expression for the potential is modified by the Laplace condition. Second, it is recognized that the structure of each quadrupolar device is constrained by the requirement that the hyperbolic electrodes share common asymptotes. Third, the derivation is based on a demonstration of the equivalence of the force acting on an ion in a quadrupole field and the force derived from the Mathieu equation; this equivalence permits the application of the solutions of Mathieu's equation to the confinement of gaseous ions.

A1.1.1 Introduction

Quadrupole devices are dynamic instruments in which ion trajectories are influenced by a set of time-dependent or dynamic forces. Ions in fields that arise from the application of a quadrupolar potential experience strong focusing in which the restoring force, which drives the ions back toward the center of the device, increases linearly with displacement from the origin.

The motion of ions in such fields is described mathematically by the solutions to a second-order linear differential equation [1]. These differential equations have been treated in an introductory fashion [2–4] and in detail [5–10]. The relevant mathematics has also been examined [11,12]. From Mathieu's investigation of the mathematics of vibrating stretched skins, he was able to describe solutions in terms of regions of stability and instability. These solutions can be applied here, and lead to the concepts of stability and instability to describe the trajectories of ions confined in quadrupole devices: for stable solutions, the displacement of the ion periodically passes through zero, whereas for unstable solutions the displacement increases without limit to infinity.

The path that we shall follow is to equate an expression for a force (mass × acceleration) in Mathieu's equation with an expression that gives the force on an ion in an electric field that results from the application of a quadrupolar potential. This comparison, that is laid out below in simple mathematical terms, allows us to express the magnitudes and frequencies of the potentials applied to the electrodes of quadrupolar

---

* See Volume 4, Chapter 4: Rectangular Waveform Driven Digital Ion Trap (DIT) Mass Spectrometer: Theory and Applications, by Francesco Brancia and Li Ding.
† See Volume 4, Chapter 6: Ion Traps with Circular Geometries, by Daniel E. Austin and Stephen A. Lammert.

devices, the sizes of these quadrupolar devices, and the mass/charge ratio of ions confined therein in terms of Mathieu's trapping parameters, a_u and q_u, where u represents the co-ordinate axes x, y, r, and z, as appropriate to the geometry being considered. On this basis, the concept of stability regions in (a_u,q_u) space is adopted in order to discuss the confinement, and limits thereto, of gaseous ions in quadrupole devices.

Let us consider an expression for the electric potential within a quadrupolar device, that is, *between* the electrodes. When this expression is subjected to the Laplace condition, the electric field in each direction can be derived, in turn, for each of the two-dimensional quadrupole mass filter (QMF) and the three-dimensional quadrupole ion trap (QIT). For each of these devices, an expression for the force within a two-dimensional field and within a three-dimensional field, respectively, is derived and is compared with the corresponding expression from the Mathieu equation. In this manner, the stability parameters for each of the QMF and the QIT are derived. By examination of stability criteria, we proceed to a discussion of regions of stability and instability in which we develop stability diagrams. Particular attention is devoted to the quadrupole ion trap stability diagram that lies closest to the origin in stability parameter, (a_u,q_u), space. Finally, a cursory treatment of the general expression for the potential is presented so as to introduce the concept of contributions of fields of higher order than that resulting from a quadrupolar potential, that is, from hexapolar and octopolar potentials.

In a logical mathematical approach to the theory of quadrupolar devices, one should begin with an examination of the field in one dimension and proceed to examine two- and three-dimensional devices.

A1.1.2   QUADRUPOLAR DEVICES

The theory presented here is based on the behavior of a single ion in an infinite, ideal, field resulting from the application of a quadupolar potential, in the total absence of any background gas. The term 'quadrupolar' refers to the dependence of the potential at a point within such a device upon the square of the distance from the origin of reference. In a quadrupolar device described with reference to rectangular coordinates, the potential ($\phi_{x,y,z}$ at any given point within the device can be expressed in its most general form as

$$\phi_{x,y,z} = A(\lambda x^2 + \sigma y^2 + \gamma z^2) + C \tag{A1},$$

where A is a term independent of x, y, and z that includes the electric potential applied *between* the electrodes of opposing polarity (either a radiofrequency (RF) potential alone or in combination with a direct current (DC) potential), C is a 'fixed' potential applied effectively to all the electrodes so as to 'float' the device, and λ, σ, and γ are weighting constants for the x, y, and z co-ordinates, respectively. It can be seen from Equation A1 that in each co-ordinate direction the potential increases quadratically with x, y, and z, respectively, and that there are no 'cross terms' of the type xy, etc. This property of Equation A1 has important implications for the treatment of the motion of ions within the field, in that we can consider the components of motion in the x-, y-, and z-directions to be independent of each other.

In an electric field, it is essential that the Laplace condition

$$\nabla^2\phi_{x,y,z} = 0 \tag{A2}$$

be satisfied; that is, the second differential of the potential at a point be equal to zero, where

$$\nabla^2 = \frac{\partial^2}{\partial x^2} + \frac{\partial^2}{\partial y^2} + \frac{\partial^2}{\partial z^2} \tag{A3}.$$

Once the quadrupole potential given in Equation A1 is substituted into the Laplace equation (A2) we obtain

$$\nabla^2\phi = \frac{\partial^2\phi}{\partial x^2} + \frac{\partial^2\phi}{\partial y^2} + \frac{\partial^2\phi}{\partial z^2} = 0 \tag{A4}.$$

The partial derivatives of the field are found as

$$\frac{\partial\phi}{\partial x} = \frac{\partial}{\partial x}\left(A\lambda x^2\right) = 2A\lambda x \tag{A5}$$

and

$$\frac{\partial^2\phi}{\partial x^2} = 2\lambda A \tag{A6}.$$

Likewise

$$\frac{\partial^2\phi}{\partial y^2} = 2\sigma A \quad \text{and} \quad \frac{\partial^2\phi}{\partial z^2} = 2\gamma A \tag{A7}.$$

Substitution of Equations A6 and A7 into Equation A4 yields

$$\nabla^2\phi = A\left(2\lambda + 2\sigma + 2\gamma\right) = 0 \tag{A8}.$$

Clearly, A is nonzero, therefore we obtain

$$\lambda + \sigma + \gamma = 0 \tag{A9}.$$

While an infinite number of combinations of λ, σ, and γ exist which satisfy Equation A9, the simplest that have been chosen in practice are, for the two-dimensional QMF and the linear ion trap (LIT)

$$\lambda = -\sigma = 1; \quad \gamma = 0 \tag{A10},$$

and for the cylindrically symmetric QIT

$$\lambda = \sigma = 1; \; \gamma = -2 \tag{A11}.$$

A2.1   THE QUADRUPOLE MASS FILTER (QMF)

Substituting the values in Equation A10 into Equation A1 gives

$$\phi_{x,y} = A(x^2 - y^2) + C \tag{A12}.$$

In order to establish a quadrupolar potential of the form described by Equation A12 we have to consider a configuration comprising two pairs of electrodes having hyperbolic cross-sections formed according to equations of the general type

$$\frac{x^2}{x_0^2} - \frac{y^2}{a^2} = 1 \tag{A13}$$

for the x-pair of rod electrodes, and

$$\frac{x^2}{b^2} - \frac{y^2}{y_0^2} = -1 \tag{A14}$$

for the y-pair of rod electrodes corresponding, respectively, to the conditions $x = \pm x_0$ when $y = 0$ and $y = \pm y_0$ when $x = 0$. Equations A13 and A14 describe two complementary rectangular hyperbolae and, in order to establish a quadrupolar potential, it is a condition that the hyperbolae share common asymptotes. Since Equations A13 and A14 are rectangular hyperbolae the asymptotes will, therefore, have slopes of $\pm 45°$, so that $a = \pm x$ and $b = \pm y$.

In practice, all the commercial mass filters described to date and using hyperbolic electrodes have been constructed symmetrically according to the condition

$$x_0 = y_0 = r_0 \tag{A15},$$

where r_0 is the radius of the inscribed circle tangential to the inner surface of the electrodes. From Equation A15 the equations for the electrode surfaces become

$$x^2 - y^2 = r_0^2 \tag{A16}$$

for the x-pair of electrodes, and

$$x^2 - y^2 = -r_0^2 \tag{A17}$$

for the y-pair of electrodes.

A2.1.1 THE QMF WITH ROUND RODS

It should be noted that many modern instruments use arrays of round rather than hyperbolic rods to reduce costs and to simplify construction. When circular rods of radius r are used, a good approximation to a quadrupole field can be obtained when the radius of each rod is made equal to $1.148 \times$ the desired r_0 value [13]. Recently, this assertion has been questioned by Gibson and Taylor [14], and it has been claimed that it is not possible to give a single figure for r/r_0 because the results are influenced to a small extent by the form of the ion beam entering the QMF. Gibson and Taylor found that a value in the range $r = 1.12 \times r_0$ to $r = 1.13 \times r_0$ produces the best performance.

A2.1.2 THE STRUCTURE OF THE QMF

A QMF comprised of circular rods is shown in Figure A1. Each pair of opposite rods is connected electrically, such that a two-dimensional quadrupole field is established in the x-y plane. The ions enter and travel in the z-direction. While traveling in the z-direction, the ions oscillate in the x-y plane also due to the potentials applied to the rods. This oscillation, which is described as the secular frequency, is a property of the mass/charge ratio of a given ion species. Therefore, all ions of a specific mass/charge ratio will react equally to the electric potentials imposed by the quadrupole assembly. Under appropriate electrical conditions, ions of a single mass/charge ratio will have a stable trajectory for the entire length of the quadrupole. A QMF can be operated so as to transmit either all ions or a specific range of mass-to-charge ratios and to focus them at the exit aperture. The ions transmitted impinge subsequently onto the detector.

An electric potential, $\phi_{y\text{-}y}$, is applied to the vertical rod pair in Figure A1 and $\phi_{x\text{-}x}$ is applied to the horizontal rod pair ring electrode, such that the pairs of rods are

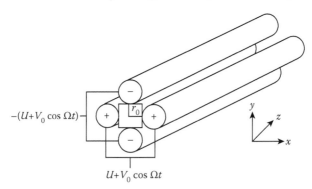

$$-(U+V_0 \cos \Omega t)$$

$$r_0$$

$$U+V_0 \cos \Omega t$$

FIGURE A1 Quadrupole mass filter. The ions enter and travel in the z-direction, while oscillating in the x-y plane. The oscillation is controlled by the DC (U) and RF (V) potentials applied to each pair of rods. Only those ions with stable trajectories at the selected U and V values will travel the length of the quadrupole mass filter and be detected. (Reproduced from March, R.E. and Todd, J.F.J., *Quadrupole Ion Trap Mass Spectrometry*, 2nd ed., John Wiley & Sons, Inc., Hoboken, NJ, 2005. With permission from John Wiley & Sons, Inc.)

out-of-phase with each other. Two points should be noted here: first, each rod (or electrode) in a rod array is either in-phase or out-of-phase with the remaining three rods and, second, the potential along the z-axis of the rod array is zero.

A2.1.3 THE QUADRUPOLAR POTENTIAL

As noted earlier, the actual quadrupolar potential to which an ion is subjected, ϕ_0, is given by the *difference* between the potentials applied to the x-pair and the y-pair of electrodes. Thus

$$\phi_0 = \phi_{x-x} - \phi_{y-y} \tag{A18}.$$

Considering now, say, the x-pair of electrodes, since the potential must be the same across the whole of the electrode surface, we can write that when $y = 0$, $x_0^2 = r_0^2$, so that substituting in Equation A2 we have

$$\phi_{x-x} = A\left(r_0^2\right) + C \tag{A19}.$$

Likewise we have

$$\phi_{y-y} = A\left(-r_0^2\right) + C \tag{A20},$$

so that from Equations A18, A19, and A20 we find

$$\phi_0 = 2Ar_0^2 \tag{A21},$$

$$\therefore A = \frac{\phi_0}{2r_0^2} \tag{A22}.$$

Hence Equation A12 becomes

$$\phi_{x,y} = \frac{\phi_0}{2r_0^2}\left(x^2 - y^2\right) + C \tag{A23},$$

whence we note that at the origin ($x = 0$, $y = 0$) we have

$$\phi_{0,0} = C \tag{A24}.$$

If the electrode structure is floated at zero (ground) potential then $C = 0$, so that Equation A23 becomes

$$\phi_{x,y} = \frac{\phi_0}{2r_0^2}\left(x^2 - y^2\right) \tag{A25},$$

which is the expression for the potential often found in standard texts.

We proceed now to examine the motion of an ion when subjected to the potential given by Equation A25. If we consider first the component of motion in the x-direction, then putting $y = 0$ in Equation A25 gives

$$\phi_{x,0} = \frac{\phi_0 x^2}{2r_0^2} \tag{A26},$$

so that the electric field at the point $(x,0)$ is

$$\left(\frac{d\phi}{dx}\right)_y = \frac{\phi_0 x}{r_0^2} \tag{A27}.$$

As a result, the force acting on an ion, $\mathbf{F}_x$, at a point $(x,0)$ is given by

$$\mathbf{F}_x = -e\left(\frac{d\phi}{dx}\right)_y = -e\frac{\phi_0 x}{r_0^2} \tag{A28},$$

where the negative sign indicates that the force acts in the opposite direction to increasing x. Since force = mass × acceleration, from Equation A28 we can write

$$m\left(\frac{d^2x}{dt^2}\right) = -e\frac{\phi_0 x}{r_0^2} \tag{A29}.$$

Let us now consider a real system in which

$$\phi_0 = 2(U + V\cos\Omega t) \tag{A30},$$

where V is the zero-to-peak amplitude of a RF potential oscillating with angular frequency Ω (expressed in radian s^{-1}) and $+U$ is a DC voltage applied to the x-pair of electrodes while a DC voltage of $-U$ volts is applied to the y-pair of electrodes.

Thus from Equations A29 and A30

$$m\left(\frac{d^2x}{dt^2}\right) = -2e\frac{(U + V\cos\Omega t)x}{r_0^2} \tag{A31},$$

which may be expanded to

$$\left(\frac{d^2x}{dt^2}\right) = -\left[\frac{2eU}{mr_0^2} + \frac{2eV\cos\Omega t}{mr_0^2}\right]x \tag{A32}.$$

A2.1.4 THE MATHIEU EQUATION

The canonical or commonly accepted form of the Mathieu equation is

$$\frac{d^2u}{d\xi^2} + (a_u - 2q_u \cos 2\xi)u = 0 \tag{A33},$$

where u is a displacement, ξ is a dimensionless parameter equal to $\Omega t/2$ such that Ω must be a frequency as t is time, and a_u and q_u are additional dimensionless stability parameters which, in the present context of quadrupole devices, are in fact 'trapping' parameters. It can be shown by substituting $\xi = \Omega t/2$ and using operator notation to find

$$\frac{d}{dt} = \frac{d\xi}{dt}\frac{d}{d\xi} = \frac{\Omega}{2}\frac{d}{d\xi} \tag{A34},$$

so that

$$\frac{d^2}{dt^2} = \frac{d\xi}{dt}\frac{d}{d\xi}\left\{\frac{d}{dt}\right\} = \frac{\Omega^2}{4}\frac{d^2}{d\xi^2} \tag{A35},$$

we can write

$$\frac{d^2u}{dt^2} = \frac{\Omega^2}{4}\frac{d^2u}{d\xi^2} \tag{A36}.$$

Substitution of Equation A36 into Equation A33, substituting Ωt for 2ξ and rearranging yields

$$\frac{d^2u}{dt^2} = -\left[\frac{\Omega^2}{4}a_u - 2\times\frac{\Omega^2}{4}q_u\cos\Omega t\right]u \tag{A37}.$$

We can now compare directly the terms on the right hand sides of Equations A32 and A37, recalling that u represents the displacement x, to obtain

$$-\left[\frac{2eU}{mr_0^2} + \frac{2eV\cos\Omega t}{mr_0^2}\right]x = -\left[\frac{\Omega^2}{4}a_x - 2\times\frac{\Omega^2}{4}q_x\cos\Omega t\right]x \tag{A38},$$

whence one deduces the relationships

$$a_x = \frac{8eU}{mr_0^2\Omega^2} \tag{A39},$$

and

$$q_x = \frac{-4eV}{mr_0^2\Omega^2} \tag{A40}.$$

When this derivation is repeated to obtain the force on an ion in the y-direction in a QMF, one finds that $a_x = -a_y$ and $q_x = -q_y$; this relationship is obtained because $\lambda = -\sigma = 1$. The a_u, q_u trapping parameters are particularly interesting because they are functions of the magnitude of either the DC voltage or the RF voltage applied to a quadrupolar device, respectively, the RF frequency (in radian s^{-1}), the mass/charge ratio of a given ion species, and the size (r_0) of the device; that is, the trapping parameters are functions of the instrumental parameters that govern the various operations of a quadrupole device. Other parameters of interest, β_u and ω_u, can be derived from the a_u, q_u trapping parameters. Both β_u and ω_u describe the nature of the ion trajectories: β_u is a complex function of a_u and q_u, and ω_u is the so-called secular frequency of the ion motion in the u-direction.

A3.1   THE QUADRUPOLE ION TRAP (QIT)

The quadrupole ion trap may function both as an ion store, in which gaseous ions can be confined for a period of time, and as a mass spectrometer of large mass range, variable mass resolution, and high sensitivity. As a storage device, the quadrupole ion trap confines gaseous ions, either positively or negatively charged and, on occasion, ions of each polarity. The confining capacity of the quadrupole ion trap arises from the formation of a trapping potential well when appropriate potentials are applied to the electrodes of the ion trap.

A3.1.1   THE STRUCTURE OF THE QIT

The quadrupole ion trap mass spectrometer consists essentially of three shaped electrodes that are shown in open array in Figure A2. Two of the three electrodes are virtually identical and, while having hyperboloidal geometry, resemble small inverted saucers; these saucers are the so-called end-cap electrodes and each has one or more holes in the center for transmission of electrons and ions. One end-cap electrode contains the 'entrance' aperture through which electrons and/or ions can be gated periodically while the other is the 'exit' electrode through which ions pass to a detector. The third 'ring' electrode has an internal hyperboloidal surface; in some early designs of ion trap system, a beam of electrons was gated through a hole in this electrode rather than an end-cap electrode. The ring electrode is positioned symmetrically between two end-cap electrodes as shown in Figure A3. Figure A3a shows a photograph of an ion trap cut in half along the axis of cylindrical symmetry, while Figure A3b is a cross-section of an ideal ion trap showing the asymptotes and the dimensions r_0 and z_0, where r_0 is the radius of the ring electrode in the central horizontal plane and $2z_0$ is the separation of the two end-cap electrodes measured along the axis of the ion trap.

FIGURE A2 The three electrodes of the quadrupole ion trap shown in open array. (Reproduced from March, R.E. and Todd, J.F.J., *Quadrupole Ion Trap Mass Spectrometry*, 2nd ed., John Wiley & Sons, Inc., Hoboken, NJ, 2005. With permission from John Wiley & Sons, Inc.)

(a)   (b)

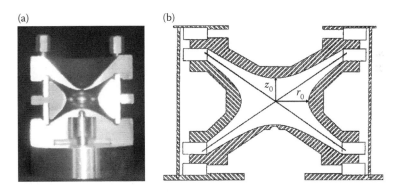

FIGURE A3 Quadrupole ion trap: (a), photograph of an ion trap cut in half along the axis of cylindrical symmetry; (b), a schematic diagram of the three-dimensional ideal ion trap showing the asymptotes and the dimensions r_0 and z_0. (Reproduced from March, R.E. and Todd, J.F.J., *Quadrupole Ion Trap Mass Spectrometry*, 2nd ed., John Wiley & Sons, Inc., Hoboken, NJ, 2005. With permission from John Wiley & Sons, Inc.)

The electrodes in Figure A3 are truncated for practical purposes, but in theory they extend to infinity and meet the asymptotes. The asymptotes arise from the hyperboloidal geometries of the three electrodes. The geometries of the electrodes are defined so as to produce an ideal quadrupole potential distribution that, in turn, will produce the necessary trapping field for the confinement of ions.

A3.1.2 THE ELECTRODE SURFACES

We have seen that for the cylindrically symmetric QIT the values of λ, σ, and γ given in Equation A11 satisfy the Laplace condition (Equation A2) when it is applied to Equation A1. Thus we have

$$\phi_{x,y,z} = A\left(x^2 + y^2 - 2z^2\right) + C \tag{A41}.$$

To proceed, we must convert Equation A41 into cylindrical polar co-ordinates employing the standard transformations $x = r\cos\theta$, $y = r\sin\theta$, $z = z$. Thus Equation A41 becomes

$$\phi_{r,z} = A\left(r^2 - 2z^2\right) + C \tag{A42}$$

It should be noted that, in making this transformation, there is an implicit assumption that the ion possesses zero angular velocity around the z-axis.

We can write the equations for the electrode surfaces as

$$\frac{r^2}{r_0^2} - \frac{2z^2}{r_0^2} = 1 \quad \text{(ring electrode)} \tag{A43},$$

and

$$\frac{r^2}{2z_0^2} - \frac{z^2}{z_0^2} = -1 \quad \text{(end-cap electrode)} \tag{A44}.$$

The gradients of the asymptotes, m, are given by

$$m = \pm\frac{1}{\sqrt{2}} \tag{A45}.$$

This relationship corresponds to the asymptotes having an angle of 35.264° with respect to the radial plane of the ion trap.

A3.1.3 THE QUADRUPOLAR POTENTIAL

As in the case of the QMF, one proceeds to evaluate the constants A and C in the general expression for the potential within the ion trap, Equation A42. By analogy with Equation A18 we define a quadrupolar potential ϕ_0 in terms of the difference between the potentials applied to the ring and the pair of end-cap electrodes

$$\phi_o = \phi_{\text{ring}} - \phi_{\text{end-caps}} \tag{A46}.$$

Recalling that $r = \pm r_0$ when $z = 0$ and $z = \pm z_0$ when $r = 0$, substitution into Equation A42 gives

$$\phi_{ring} = A(r_0^2) + C \qquad \text{(A47),}$$

and

$$\phi_{end\text{-}caps} = A(-2z_0^2) + C \qquad \text{(A48),}$$

so that from Equation A46

$$\phi_0 = A(r_0^2 + 2z_0^2) \qquad \text{(A49),}$$

whence

$$A = \frac{\phi_0}{(r_0^2 + 2z_0^2)} \qquad \text{(A50),}$$

and therefore from Equation A42

$$\phi_{r,z} = \frac{\phi_0(r^2 - 2z^2)}{(r_0^2 + 2z_0^2)} + C \qquad \text{(A51).}$$

As with the QMF, the constant C is evaluated by taking account of the potentials actually connected to the electrodes of opposing polarity. In the quadrupole ion trap the end-cap electrodes are normally held at ground (that is, zero) potential while a unipolar RF potential plus any DC voltage are applied to the ring electrode only. Thus Equation A51 becomes

$$\phi_{0,z_0} = \phi_{end\text{-}caps} = \frac{\phi_0(0 - 2z_0^2)}{(r_0^2 + 2z_0^2)} + C = 0$$

$$\therefore C = \frac{2\phi_0 z_0^2}{(r_0^2 + 2z_0^2)} \qquad \text{(A52),}$$

whence

$$\phi_{r,z} = \frac{\phi_0(r^2 - 2z^2)}{(r_0^2 + 2z_0^2)} + \frac{2\phi_0 z_0^2}{(r_0^2 + 2z_0^2)} \qquad \text{(A53).}$$

From Equation A53 we see that the potential at the center of the ion trap $(0,0)$ is no longer zero, but is at a potential equal to a fraction $(= 2z_0^2/(r_0^2 + 2z_0^2))$ of that applied to the ring electrode.

Having set the pair of end-cap electrodes at ground potential, we must now define the value of ϕ_0 in terms of the real-system potentials applied to the ring electrode. Thus

$$\phi_0 = (U + V\cos\Omega t) \tag{A54},$$

where the quantities are defined in the same manner as those in Equation A30. Substituting from Equation A54 into Equation A53 yields

$$\phi_{r,z} = \frac{(U + V\cos\Omega t)(r^2 - 2z^2)}{(r_0^2 + 2z_0^2)} + \frac{2(U + V\cos\Omega t)z_0^2}{(r_0^2 + 2z_0^2)} \tag{A55}.$$

As with the QMF, the components of the ion motion in the radial (r) and the axial (z) directions may be considered independently, so for the axial direction we can write, by analogy with Equations A28 and A29,

$$\mathbf{F}_z = -e\left(\frac{d\phi}{dz}\right)_r = e\frac{4\phi_0 z}{(r_0^2 + 2z_0^2)}$$

$$= m\left(\frac{d^2z}{dt^2}\right) \tag{A56}.$$

Hence from Equation A54 we have

$$m\left(\frac{d^2z}{dt^2}\right) = \frac{4e(U + V\cos\Omega t)z}{(r_0^2 + 2z_0^2)} \tag{A57},$$

which may be expanded to give

$$\left(\frac{d^2z}{dt^2}\right) = \left[\frac{4eU}{m(r_0^2 + 2z_0^2)} + \frac{4eV\cos\Omega t}{m(r_0^2 + 2z_0^2)}\right]z \tag{A58}.$$

A3.1.4   THE MATHIEU EQUATION

As with the mass filter, we recognize the similarity of Equation A58 with the Mathieu equation (A33)

$$\frac{d^2u}{d\xi^2} + (a_u - 2q_u\cos 2\xi)u = 0 \tag{A33},$$

so that using the transformations given in Equations A34, A35, and A36 and replacing u by z we can write

$$\left[\frac{4eU}{m(r_0^2 + 2z_0^2)} + \frac{4eV\cos\Omega t}{m(r_0^2 + 2z_0^2)}\right]z = -\left[\frac{\Omega^2}{4}a_z - 2\times\frac{\Omega^2}{4}q_z\cos\Omega t\right]z \tag{A59},$$

whence one deduces the relationships

$$a_z = -\frac{16eU}{m(r_0^2 + 2z_0^2)\Omega^2}$$ (A60)

and

$$q_z = \frac{8eV}{m(r_0^2 + 2z_0^2)\Omega^2}$$ (A61).

When this derivation is repeated for the radial component of motion at a fixed value of z one finds that

$$a_z = -2a_r \quad and \quad q_z = -2q_r$$ (A62),

again arising from the values of λ, σ, and γ inserted into the general equation A1 when it is applied to the quadrupole ion trap.

So far in this formulation of the motion occurring within the ion trap we have made no assumption concerning the relationship between the dimensions r_0 and z_0. Historically we see that ever since the early descriptions of the ion trap [15,16] the relationship

$$r_0^2 = 2z_0^2$$ (A63)

has been selected as a requirement for forming the ideal quadrupolar potential distribution. Furthermore, we note that with the identity given in Equation A63 the asymptotes with the gradients $\pm 1/\sqrt{2}$ (Equation A45) will pass through the co-ordinates $\pm r_0$, $\pm z_0$. Knight has shown that in practical ion trap systems with truncated electrodes, under the conditions of Equation A63 the asymptotes bisect the gaps between the ring electrode and the end-cap electrodes at high values of r and z, thereby minimizing the contributions to the potential of higher-order terms [17]; the personal communication is reproduced elsewhere [10]. It should be noted that inserting the relationship given in Equation A63 into Equation A53 shows that under the conditions where the end-cap electrodes are held at earth potential, the potential at the center of the ion trap is equal to half that applied to the ring electrode.

A4.1 AN ALTERNATIVE APPROACH TO QUADRUPOLE ION TRAP THEORY

This approach has the advantage of introducing components of the trapping potential of order higher than quadrupolar. A solution of Laplace's equation in spherical polar co-ordinates (ρ, θ, φ) for a system with axial symmetry (such as is the case for the quadrupole ion trap) is obtained from the theory of differential equations and has the general form

$$\phi(\rho,\theta,\varphi) = \phi_o \sum_{n=0}^{\infty} A_n \frac{\rho^n}{r_0^2} P_n(\cos\theta)$$ (A64),

where A_n are arbitrary coefficients and $P_n(\cos \theta)$ denotes a Legendre polynomial. When $\rho^n P_n(\cos \theta)$ is expressed in cylindrical polar co-ordinates, Equation A65 is obtained as

$$\phi_{r,z} = \phi_0 \left[A_2 \frac{(r^2 - 2z^2)}{2r_0^2} + A_3 \frac{(3r^2 z - 2z^3)}{2r_0^3} + A_4 \frac{(3r^4 - 24r^2 z^2 + 8z^4)}{8r_0^2} \right.$$
$$\left. + A_5 \frac{(15r^4 z - 40r^2 z^3 + 8z^5)}{8r_0^5} + A_6 \frac{(5r^6 - 90r^4 z^2 + 120r^2 z^4 - 16z^6)}{16r_0^6} \right]$$

(A65).

The values of $n = 0$, 1, 2, 3, 4, 5, and 6 correspond to the monopole, dipole, quadrupole, hexapole, octopole, decapole, and docecapole components, respectively, of the applied potential, ϕ. Higher-order components such as hexapole and octopole can play important roles in the operation of modern ion trap mass spectrometers.

A5.1   REGIONS OF ION TRAJECTORY STABILITY

Quadrupole ion trap operation is concerned with the criteria that govern the stability (and instability) of the trajectory of an ion within the field, that is, the experimental conditions that determine whether an ion is stored within the device or is ejected from the device and either lost or detected externally.

The solutions to Mathieu's equation are of two types: (i) periodic but unstable; and (ii) periodic and stable. Solutions of type (i) are called Mathieu functions of integral order and form the boundaries of unstable regions on the stability diagram. The boundaries, that are referred to as 'characteristic curves' or 'characteristic values', correspond to those values of the new trapping parameter, β_z, that are integers, that is, 0,1,2,3, ... ; β_z is a complex function of a_z and q_z but can be approximated as given in Equation A66 for $q_r<0.2$ and $q_z<0.4$. The boundaries represent, in practical terms, the point at which the trajectory of an ion becomes unbounded.

$$\beta_u \approx \sqrt{\left(a_u + \frac{q_u^2}{2} \right)}$$

(A66).

It should be noted that the approximation for β_u given in Equation A66 is known as the 'Dehmelt approximation'. A precise value of β_u is obtained from a continued fraction expression in terms of a_u and q_u.

Solutions of type (ii) determine the motion of ions in an ion trap. The stability regions corresponding to stable solutions of the Mathieu equation in the z-direction are shaded and labeled z-stable in Figure A4a. The stability regions corresponding to stable solutions of the Mathieu equation in the r-direction are shaded and labeled r-stable in Figure A4b; it can be seen that they are doubled in magnitude along the ordinate and inverted, that is, multiplied by -2. It is seen from Equation A62 that $a_z = -2a_r$ and $q_z = -2q_r$, that is, the stability parameters for the r- and z-directions differ by a factor of -2.

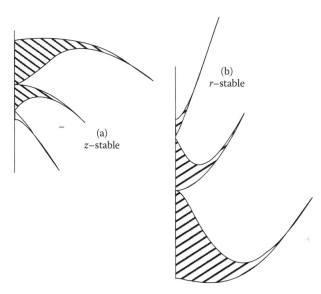

FIGURE A4 Graphical representation of three Mathieu stability regions: (a) z-stable; (b) r-stable. The r-stable region is obtained as $-2 \times$ the z-stable region. (Reproduced from March, R.E. and Todd, J.F.J., *Quadrupole Ion Trap Mass Spectrometry*, 2nd ed., John Wiley & Sons, Inc., Hoboken, NJ, 2005. With permission from John Wiley & Sons, Inc.)

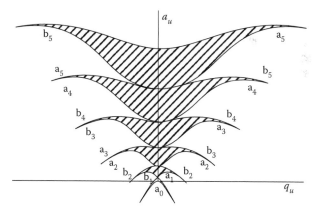

FIGURE A5 Mathieu stability diagram in one dimension of (a_u, q_u) space. The characteristic curves a_0, b_1, a_1, b_2, ... divide the plane into regions of stability and instability. The even-order curves are symmetric about the a_u-axis, but the odd-order curves are not. The diagram itself, however, appears to be symmetric about the a_u-axis. (Reproduced from March, R.E. and Todd, J.F.J., *Quadrupole Ion Trap Mass Spectrometry*, 2nd ed., John Wiley & Sons, Inc., Hoboken, NJ, 2005. With permission from John Wiley & Sons, Inc.)

Presented in Figure A5 is the Mathieu stability diagram [1] in one dimension of $(a_u,\ q_u)$ space showing the regions delineated by characteristic numbers of a cosine-type function (a_m) of order m and a sine-type function (b_m) of order m. This diagram is labeled in the terminology used by McLachan [11,12]. The boundaries of even-order are symmetric about the a_u-axis, but the boundaries of odd-order are

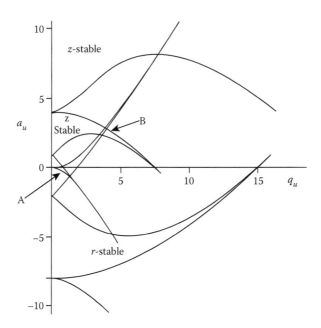

FIGURE A6   The Mathieu stability diagram in (a_z, q_z) space for the quadrupole ion trap in both the r- and z-directions. Regions of simultaneous overlap are labeled A and B. While the axes are labeled a_u and q_u, the diagrammatic representation shown here shows the ordinate and abscissa scales in units of a_z and q_z, respectively. (Reproduced from March, R.E. and Todd, J.F.J., *Quadrupole Ion Trap Mass Spectrometry*, 2nd ed., John Wiley & Sons, Inc., Hoboken, NJ, 2005. With permission from John Wiley & Sons, Inc.)

not; however, the diagram appears to be symmetric about the a_u-axis. A diagram that represents ion trajectory stability (and instability) regions in both the r- and z-directions can be constructed by overlapping parts a and b of Figure A4 as shown in Figure A6. Here, two stability regions labeled A and B are identified; region A lies on the q_u axis and region B is displaced away from the q_u axis. When one envisages the overlap of Figure A5 with the inverse of Figure A5, it is clear that there are many regions of stability though most of them are not accessible at this time [5].

Ions can be stored in the ion trap provided that their trajectories are stable in the r- and z-directions simultaneously; such trajectory stability is obtained in the region closest to the origin, that is, region A in Figure A6. Regions A and B are referred to as stability regions; region A is of the greatest importance at this time and is shown in greater detail in Figure A7. The co-ordinates of the stability region in Figure A7 are the Mathieu parameters a_z and q_z. Here, we plot a_z vs q_z rather than using the general parameters a_u vs q_u in order to avoid confusion. In Figure A7, the $\beta_z = 1$ stability boundary intersects with the q_z axis at $q_z = 0.908$; this working point is that of the ion of lowest mass/charge ratio, that is, low-mass cut-off, that can be stored in the ion trap for given values of r_0, z_0, V, and ω.

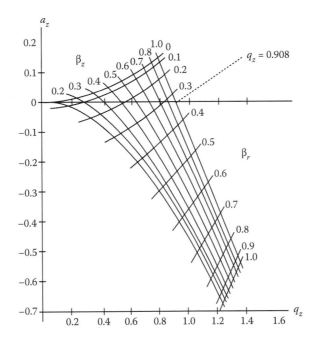

FIGURE A7 Stability diagram in (a_z, q_z) space for the region of simultaneous stability A (Figure A6) in both the r- and z-directions near the origin for the three-dimensional quadrupole ion trap; the iso-β_r and iso-β_z lines are shown in the diagram. The q_z-axis intersects the $\beta_z = 1$ boundary at $q_z = 0.908$, that corresponds to q_{max} in the mass-selective instability mode. Conventionally, it is the stability diagram in (a_z, q_z) space that is presented. (Reproduced from March, R.E. and Todd, J.F.J., *Quadrupole Ion Trap Mass Spectrometry*, 2nd ed., John Wiley & Sons, Inc., Hoboken, NJ, 2005. With permission from John Wiley & Sons, Inc.)

A5.1.1 SECULAR FREQUENCIES

A three-dimensional representation of an ion trajectory in the ion trap, as shown in Figure A8, has the general appearance of a Lissajous curve or figure-of-eight composed of two fundamental frequency components, $\omega_{r,0}$ and $\omega_{z,0}$ of the secular motion [18]. The description 'fundamental' implies that there exist other higher-order (n) frequencies and the entire family of frequencies is thus described by $\omega_{r,n}$ and $\omega_{z,n}$. These secular frequencies are given by

$$\omega_{u,n} = (n + \tfrac{1}{2}\beta_u)\Omega, \ 0 \le n \le \infty \tag{A67}.$$

While the fundamental axial secular frequency, $\omega_{z,0}$, is usually given in units of Hertz in the literature and referred to simply as ω_z, it should be expressed in radians s^{-1}. At this time, the higher-order frequencies are of little practical significance.

The resemblance of the simulated ion trajectory shown in Figure A8 to a roller coaster ride is due to the motion of an ion on the potential surface shown in Figure A9. The oscillatory motion of the ion results from the undulations of the potential surface

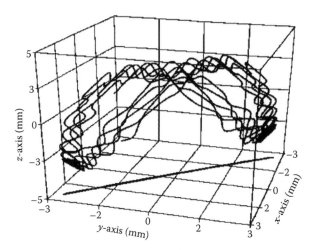

FIGURE A8 Trajectory of a trapped ion of *m/z* 105. The initial position was selected randomly from a population with an initial Gaussian distribution (FWHM of 1 mm); $q_z = 0.3$; zero initial velocity. The projection onto the *x*-*y* plane illustrates planar motion in three-dimensional space. The trajectory develops a shape that resembles a flattened boomerang. (Reproduced from Nappi, M., Weil, C., Cleven, C.D., Horn, L.A., Wollnik, H., and Cooks, R.G., *Int. J. Mass Spectrom. Ion Processes,* 1997, *161*, 77–85. With permission from Elsevier.)

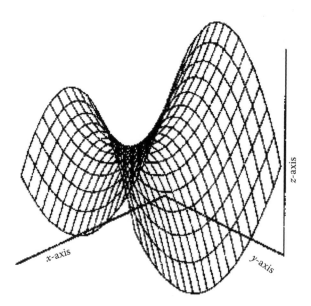

FIGURE A9 Pure quadrupole potential surface for a quadrupole ion trap. Note the four poles of the surface and the similarity of the field shape to the trajectory in Figure A8. (Reproduced from March, R.E. and Todd, J.F.J., *Quadrupole Ion Trap Mass Spectrometry*, 2nd ed., John Wiley & Sons, Inc., Hoboken, NJ, 2005. With permission from John Wiley & Sons, Inc.)

that can be envisaged as rotation of the potential surface. The simulation of the ion trajectory was carried out using the Ion Trajectory SIMulation program (ITSIM) [19], while the potential surface was generated [20] from Equation A65 by calculating $\phi_{r,\varphi,z}$ for $A_2^{\circ} = 1$ and all of the other coefficients equal to zero for increment steps of 1 mm in both radial and axial directions.

Some of the secular frequencies defined by Equation A67 are illustrated in Figure A10. In this figure are shown the results of a power spectral Fourier analysis of the trajectory of an ion of m/z 100 calculated using ITSIM [19]. The essential trapping parameters were $r_0 = 10$ mm, $z_0 = 7.071$ mm, $\Omega/2\pi = 1.1$ MHz, $q_z = 0.40$, and the background pressure was zero. The ion's fundamental axial secular frequency, $\omega_{z,0}$, is of interest because it is the axial motion of an ion that is excited during axial modulation (see Equation A67); this frequency, $\beta_z\Omega/2$, was observed at 160.91142 kHz and is shown at the LHS of Figure A10a. Also shown here are two sets of complementary frequencies corresponding to $\Omega \pm \beta_z\Omega/2$ and $2\Omega \pm \beta_z\Omega/2$. In Figure A10b where the

(a)

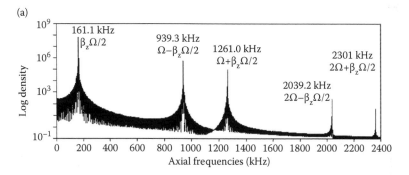

(b)

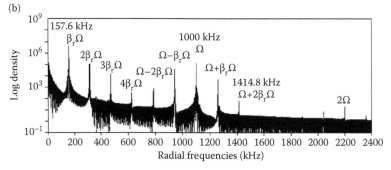

FIGURE A10 Power-spectral Fourier analysis of the trajectory of an ion of m/z 100 calculated using ITSIM. The essential trapping parameters were $r_0 = 10$ mm, $z_0 = 7.071$ mm, $\Omega/2\pi = 1.1$ MHz, $q_z = 0.40$, and the background pressure was zero. The data were collected at intervals of 100 ns for 1 ms: (a) axial frequencies; (b) radial frequencies. The magnitude of the frequency band is plotted on a logarithmic scale and shows the intensity of each harmonic lower by several orders of magnitude from the fundamental secular frequency. Note that the radial position data were obtained from $r = \sqrt{(x^2+y^2)}$, and not from either x or y; thus the radial fundamental secular frequency is observed at $\beta_r\Omega$ rather than $\beta_r\Omega/2$. (Reproduced from March, R.E. and Todd, J.F.J., *Quadrupole Ion Trap Mass Spectrometry*, 2nd ed., John Wiley & Sons, Inc., Hoboken, NJ, 2005. With permission from John Wiley & Sons, Inc.)

radial frequencies are shown, the central feature is the RF drive frequency Ω; also shown are two sets of complementary frequencies corresponding to $\Omega \pm \beta_r \Omega$ and $2\Omega \pm 2\beta_r \Omega$, the harmonic frequencies $\beta_r \Omega$, $2\beta_r \Omega$, and $3\beta_r \Omega$, and the set of frequencies corresponding to 2Ω and $2\Omega \pm \beta_r \Omega$.

REFERENCES

1. Mathieu, E. Mémoire sur le mouvement vibratoire d'une membrane de forme elliptique. *J. Math. Pures Appl. (J. Liouville)* 1868, *13*, 137–203.
2. Dawson, P.H.; Whetten, N.R. Ion storage in three-dimensional, rotationally symmetric, quadrupole fields. I. Theoretical treatment. *J. Vac. Sci. Technol.* 1968, *5*, 1–10.
3. Dawson, P.H.; Whetten, N.R. Ion storage in three-dimensional, rotationally symmetric, quadrupole fields. II. A sensitive mass spectrometer. *J. Vac. Sci. Technol.* 1968, *5*, 11–18.
4. Campana, J.E. Elementary theory of the quadrupole mass filter. *Int. J. Mass Spectrom. Ion Phys.* 1980, *33*, 101–117.
5. Dawson, P.H. (Ed.). *Quadrupole Mass Spectrometry and Its Applications.* Elsevier, Amsterdam, 1976. Reprinted as an "American Vacuum Society Classic" by the American Institute of Physics under ISBN 1563964554.
6. March, R.E.; Hughes, R.J.; Todd, J.F.J. *Quadrupole Storage Mass Spectrometry.* Wiley Interscience, New York, 1989.
7. Lawson, G.; Todd, J.F.J.; Bonner, R.F. Theoretical and experimental studies with the quadrupole ion storage trap ('QUISTOR'). In D. Price and J.F.J. Todd (Eds.) *Dynamic Mass Spectrometry.* Vol. 4, pp. 39–81. Heyden, London, 1976.
8. March, R.E. An introduction to quadrupole ion trap mass spectrometry. *J. Mass Spectrom.* 1997, *32*, 351–369.
9. March, R.E.; Todd, J.F.J. (Eds.). *Practical Aspects of Ion Trap Mass Spectrometry.* Vols. I–III, Modern Mass Spectrometry Series. CRC Press, Boca Raton, FL, 1995.
10. March, R.E.; Todd, J.F.J. *Quadrupole Ion Trap Mass Spectrometry.* 2nd ed., John Wiley & Sons, Hoboken, NJ, 2005.
11. McLachan, N.W. *Theory and Applications of Mathieu Functions.* Clarendon Press, Oxford, 1947.
12. Campbell, R. *Théorie Générale de l'Equation de Mathieu.* Masson (*General Theory of the Mathjen Equation*), Paris, 1955.
13. Denison, D.R. Operating parameters of a quadrupole in a grounded cylindrical housing. *J. Vac. Sci. Technol.* 1971, *8*, 266–269.
14. Gibson, J.R.; Taylor, S. Numerical investigation of the effect of electrode size on the behaviour of quadrupole mass filters. *Rapid Commun. Mass Spectrom.* 2001, *15*, 1960–1964.
15. Paul, W. ; Osberghaus, O.; Fischer E. Ein Ionenkäfig (An Ion Cage). *Forschungsberichte Wirtschafts- und Verkehrministeriums, Nordrhein-Westfalen, Nr. 415 (Research Reports, Department of Commerce and Transport, North Rhine-Westphalia, No. 415), Westdeutscher Verlag*, Köln und Opladen, 1958, pp. 6–7.
16. Wuerker, R.F.; Shelton, H.; Langmuir, R.V. Electrodynamic containment of charged particles. *J. Appl. Phys.* 1959, *30*, 342–349.
17. Knight, R.D. Personal communication, 2003.
18. Nappi, M.; Weil, C.; Cleven, C.D.; Horn, L.A.; Wollnik, H.; Cooks, R.G. Visual representations of simulated three-dimensional ion trajectories in an ion trap mass spectrometer. *Int. J. Mass Spectrom. Ion Processes* 1997, *161*, 77–85.
19. Reiser, H-P.; Kaiser, Jr. R.E.; Cooks, R.G. A versatile method of simulation of the operation of ion trap mass spectrometers. *Int. J. Mass Spectrom. Ion Processes* 1992, *121*, 49–63.
20. Splendore, M. constructed the general potential surface shown in Figure A9; personal communication, 2003.

2 Ion Traps for Miniature, Multiplexed, and Soft-Landing Technologies

Scott A. Smith, Christopher C. Mulligan,
Qingyu Song, Robert J. Noll,
R. Graham Cooks, and Zheng Ouyang

CONTENTS

2.1 Overview, Geometries, and Scaling 170
    2.1.1 Overview.. 170
    2.1.2 New Geometries and Configurations... 171
        2.1.2.1 Cylindrical Ion Traps 173
        2.1.2.2 Rectilinear Ion Traps 175
        2.1.2.3 Toroidal Ion Traps... 177
    2.1.3 Scaling Considerations 178
        2.1.3.1 Fabrication Tolerances 178
        2.1.3.2 Capacitance... 180
        2.1.3.3 Power... 180
        2.1.3.4 Pseudopotential Well Depth... 181
        2.1.3.5 Trapping Efficiency, Trapping Capacity, and
                Extraction Efficiency 182
        2.1.3.6 Pressure... 182
        2.1.3.7 Mass Range... 182
        2.1.3.8 Resolution 183
2.2 Simulations of Ion Trap Performance... 183
    2.2.1 Approach.. 184
    2.2.2 Ion Trajectory Simulation (ITSIM) 184
    2.2.3 Example of Geometry Optimization: Miniature Cylindrical
        Ion Traps 186
    2.2.4 Chemical Mass Shifts... 187
    2.2.5 Rectilinear Ion Traps 190
    2.2.6 The Orbitrap 190
2.3 Fabrication of Miniature Ion Traps by Modern Methods 191
    2.3.1 Semiconductor Processes 192
    2.3.2 Printed Circuit Board (PCB) Processes...200
    2.3.3 Laser-Machined Ceramics...202

2.3.4   Low-Temperature Co-fired Ceramic (LTCC) 203
2.3.5   Stereolithography Apparatus (SLA) 205
2.4   Multiplexed Ion Trap Arrays 207
2.4.1   Cylindrical Ion Trap Arrays 209
2.4.1.1   Parallel Arrays of CITs of the Same Size 210
2.4.1.2   Parallel Arrays of CITs of Different Size 216
2.4.1.3   Serial Arrays of CITs 219
2.4.2   Rectilinear Ion Trap Arrays 221
2.4.2.1   Parallel Arrays of RITs 221
2.4.2.2   Serial RITs and Ion Soft-Landing 227
2.5   Conclusions and Future Outlook 230
References 230
Appendix 237
References (Appendix) 246

2.1   OVERVIEW, GEOMETRIES, AND SCALING

2.1.1   OVERVIEW

During the past decade, the development of ion trap instruments has continued on a trajectory toward high performance in chemical analysis [1–3]. This objective has been pursued with emphases on high ion trapping capacity, high-mass range, increased mass resolution and high-speed operation. Such efforts are embodied in an increasing number of more complex instruments employing multiple ion traps or combinations of ion traps with other types of mass analyzers [4–6]. In parallel with these trends, there has been a rapid increase in interest in chemical analysis *in situ*; as a generally applicable method with high sensitivity and specificity, mass spectrometry has been a natural candidate to meet this challenge. The obvious solution has been to develop miniature, portable mass spectrometers, a development that was reviewed [7] in 2000 and which has accelerated strongly since then.

Ion traps were selected quickly, along with time-of-flight systems, [8] as the prime candidates for miniaturization. Two factors favor the use of ion traps: (i) an inherent capability for performing tandem mass spectrometry, MS/MS, in a single device; and (ii) their pressure tolerant operation. Shrinking a 3D Paul trap in each linear dimension by factors of up to four while maintaining the geometry [9] has been demonstrated to have the obvious advantage of allowing the use of RF voltages of lower amplitude, according to the quadratic relationship between the size and RF amplitude in the Mathieu equation. The difficulty involved in manufacturing accurately ion traps with hyperbolic electrodes, first recognized in 1960 by Paul and Steinwedel, [10] led to the development of simplified ion trap geometries, which are the preferred geometries for miniaturization [11] (see Figure 2.1). Simulation tools were developed and used to optimize the geometries of these simplified devices in order to achieve performance with non-ideal fields in which the non-idealities mutually compensate for each other and the net result was unit mass resolution over a mass/charge ratio range of interest [12]. These objectives were met in the case of cylindrical ion traps (CITs) [13] where the simulation-based optimization routines described below were first implemented.

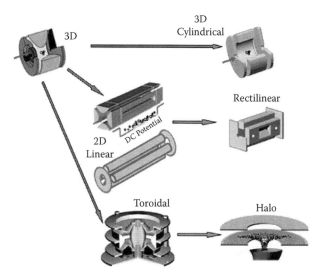

FIGURE 2.1 Geometric evolution of ion traps for use in miniature and multiplexed mass spectrometers showing development from 3D (hyperbolic) to 2D (linear), and 3D (toroidal) systems, and simplifications in form. (Reprinted from Ouyang, Z. and Cooks, R.G., *Ann. Rev. Anal. Chem.* 2009, 2, 197–214. With permission from Annual Reviews of Analytical Chemistry.)

Meanwhile, the demonstration of the advantages in ion trapping efficiency and total ion trapping capacity associated with linear (2D) ion traps (LITs) [5,6] encouraged the incorporation of this even more favorable geometry into miniature ion trap mass analyzers again in a simplified form, in this case that of the rectilinear ion trap (RIT). As previously, optimization of the geometry provided a simple electrode structure, associated with a quadrupole field in which the effects of the non-idealities are cancelled [14]. A similar development of a simplified structure [15] occurred in the form of the halo trap in response to the advantages of the toroidal geometry ion trap [15].

Advances in capabilities for microfabrication led naturally to an interest in arrays of ion traps. This interest was fueled by the need to compensate for the diminution in ion trapping capacity associated with miniaturization. In addition, new experiments are made possible when sets of traps are assembled in various geometrical arrangements, from simple parallel and serial arrangements of use in different types of mass spectrometry experiments to complex arrangements aimed at allowing advanced computations and even data storage. An early micro ion trap array [16] used individual traps as small as 1 μm, with the expectation of ion trap operation at extremely low RF amplitudes (several volts), high ion storage capacity, and low-cost large-scale production of the mass analyzer. The potential also exists for applications in quantum computing and ion clocks.

2.1.2 New Geometries and Configurations

The optimization of the ion traps of new geometry is a process that combines theoretical modeling, numerical simulation, and experimental characterization. With the availability of modern mechanical design tools, electric field solvers, and continually evolving simulation programs such as Ion Trajectory Simulation

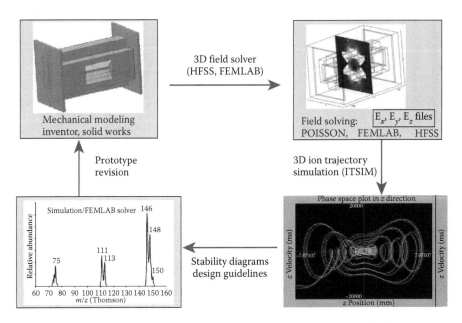

FIGURE 2.2   Schematic representation of the ion trap mass analyzer optimization procedure, involving cycles of field calculations, simulations, and experimental verification.

(ITSIM), procedures involving multiple optimization cycles as shown in Figure 2.2 have been developed gradually and used successfully for the optimization of ion traps with simplified geometries [12,13,17]. A summary of some of the traps developed at Purdue since 2000 is presented in the Appendix.

Multiplexing of ion traps in parallel arrays meets the needs of several types of application and is based on several different concepts (see Figure 2.3). An array of ion traps of the same size (Figure 2.3b) enhances overall trapping capacity, which otherwise would be compromised due to reductions in trap volume and RF amplitude [16]. In one experiment using a parallel trap array, simple single-channel ion detection is used to detect all the ions ejected from the array. Multiple-channel ion detection, on the other hand, occurs when signals from individual traps are measured separately; that is, an array of detectors is used such that data resulting from distinct channels are distinguishable and essentially non-interfering. It often makes sense that multi-channel analysis using the latter operation be performed simultaneously rather than sequentially in order to maximize the information obtained per unit time. For example, an array of ion traps all of the same size, when coupled with an array of ionization sources, atmospheric pressure interfaces, and ion detectors, can be used to perform high-throughput analysis [18–21]. Simultaneous mass analysis of multiple samples, or alternatively of the same sample subjected to the same or different ionization and mass analysis modes, can be applied using this type of multiplexed platform. An array of ion traps of different sizes (Figure 2.3c) can also be used in parallel channel experiments; such an array has been used with a single ion source and ion beam to achieve approximate mass analysis without requiring RF scanning [22].

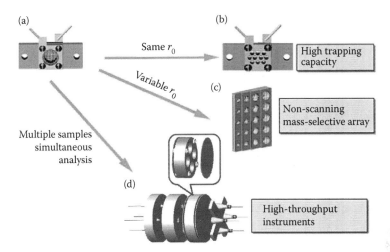

FIGURE 2.3 Parallel ion trap arrays targeted at different types of analytical performance. Starting (a) with a single ion trap one can elaborate into, (b) an array of identical traps operated as a single unit to increase ion trapping capacity, or (c) an array of traps of different radii to allow non-scanning selection of ions of different masses, or (d) an array of identical traps operated with different samples to achieve higher throughput.

The need for serial multiplexing of ion traps was originally less obvious, because unlike ion beam-type mass analyzers including the quadrupole mass filter, multiple-stage ion isolation, excitation, and mass analysis can be conducted in a single ion trap, a widely appreciated and unique feature of ion traps. However, the coupling of two or more ion traps in series (Figure 2.3d) allows different environments (such as differences in operating pressure) to be set appropriately for ion manipulation with high efficiency while allowing the possibility of improving the duty cycle of the ion trap instrument by a 'one accumulates while the other mass analyzes' fashion [23]. The use of 2D linear ion traps allows much higher ion transfer efficiency when compared to 3D traps, a trait which has led to the exploration of serially multiplexed 2D linear traps [24]. In comparison to the original Paul trap and the CIT, a unique feature of the RIT is the fact that the entrance and ejection directions are orthogonal. This arrangement eliminates line-of-sight concerns in ion optics experiments and allows new possibilities (mostly unexplored) for ion manipulation, transport, separation, and collection.

2.1.2.1 Cylindrical Ion Traps

Before CITs were used as mass analyzers, they had been introduced as ion storage devices, in both non-mass-selective modes [25,26] and mass-selective modes [27,28]. Wells *et al.* designed a CIT for mass analysis with a radial half-distance (r_0) of 10 mm, the so-called 'full size' CIT, and optimized the geometry to give an effectively quadrupolar field by stretching the axial dimension z_0 [29]. By changing the contributions of the higher-order field components, such as octopole and dodecapole fields as shown in Figure 2.4, the mass resolution of CITs could be improved significantly by empirically and mutually compensating for the effects of these higher-order fields. This process involves iterative simulations followed by experiments, as detailed in

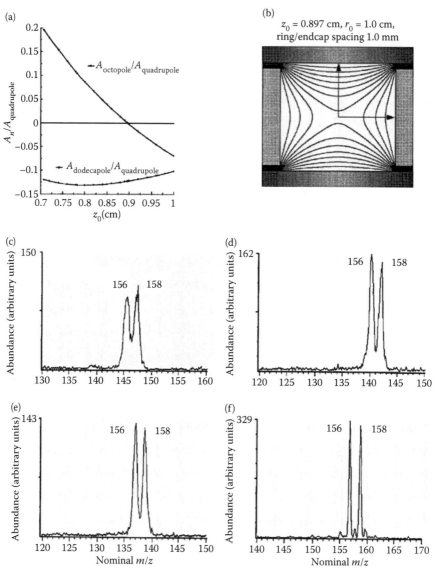

FIGURE 2.4 Cylindrical ion trap: (a) electrical field components; (b) internal electrical field; and (c)–(f) optimization of resolution by stretching z_0. For all mass spectra (c)–(f), $P_{bromobenzene} = 1 \times 10^{-6}$ Torr, $P_{He} = 8 \times 10^{-4}$ Torr, and ionization times were 500 μs. For (c) and (d), z_0 was held constant at 0.783 cm, while the ring/end-cap electrode spacing was varied from 3.3 to 1.0 mm. No axial modulation was used. For (e) and (f), the ring/end-cap electrode gap was held constant at 1.0 mm. (e) $z_0 = 0.783$ cm, axial modulation at 470 kHz, 2 V. (d) $z_0 = 0.897$ cm, axial modulation at 450 kHz, 2.5 V. (Reproduced from Wells, J.M., Badman, E.R., and Cooks, R.G., *Anal. Chem.* 1998, *70*, 438–444. With permission from the American Chemical Society.)

Section 2.2. This practice has been applied for the development of CITs at half size ($r_0 = 5.0$ mm) and quarter size ($r_0 = 2.5$ mm). Empirical correlations were established between the higher-order field contributions and the ion trap performance and applied in the course of optimization procedures (Figure 2.4) for CITs [12] and, subsequently, for RITs [14]. Starting with the characterization of the first full-size CIT as a mass analyzer, unit mass resolution and a mass/charge range up to at least m/z 500 has been the design objective for a family of simplified geometry ion traps and their arrays. With these developments, the CIT rapidly became the basis for mass analyzer miniaturization due to the ease of fabrication in single unit or array formats. For example, a half-size CIT using mesh end-cap electrodes was used in a serial array and found to allow efficient ion trapping and transfer of ions between two CITs [30]. A quarter-size CIT was characterized for chemical analysis applications using membrane sample introduction; the device was capable of near-unit resolution and a limit of detection at 500 ppb for toluene in water [13]. Using conventional precision machining, CITs as small as $r_0 = 0.5$ mm have been made from stainless steel; one such CIT provided unit resolution at low mass when operated at an RF frequency of 5.8 MHz [31]. CITs with internal radii in the sub-millimeter to 1 μm range have been fabricated subsequently using a variety of methods [16,32,33].

2.1.2.2 Rectilinear Ion Traps

The RIT is a simplified version of the linear ion trap [14,34]. It is comprised of two pairs of flat electrodes (the x- and y-pairs) to which RF potentials are applied and one pair of DC-only end-cap electrodes (see Figure 2.5). In this, the usual six-electrode configuration, ions are trapped by the pseudopotential well created by the RF electric field in the x–y plane and by the DC trapping potential in the z-direction. In an alternative configuration [35], the end-cap electrodes are omitted and a single-phase RF potential suffices to create a z-directional time-varying barrier to ion loss. During an RF scan performed to achieve mass analysis by the mass-selective instability experiment of Stafford *et al.* [4], ions are ejected through slits in the x-electrodes and detected by one or two electron multiplier detectors. The mass resolution of the RIT has been improved by a 25%-stretch in the inter-electrode distance in the x-direction; this allows the achievement of unit resolution when using the mass-selective instability scan and resonance ejection at 1.0 MHz RF. As is the case for other linear ion traps, the RIT has enhanced trapping capacity when compared with 3D traps having comparable critical dimensions. In the further evolution of RITs, miniature RITs with $x_0 = 1.7$ mm have been fabricated using stereolithography apparatus (SLA) which employs laser-induced polymerization to develop the electrode assembly followed by a metal sputtering to coat the surfaces [36]. Arrays of RITs in linear and circular configurations have also been built using the SLA processes [37].

The RIT has been used as the basis for a number of instruments in various configurations that were built for different analytical purposes. The optimized RIT geometry has a length of 40 mm (the full z-dimension) and radial half-distances (denoted u_0, u is the dimension) of $x_0 = 5$ mm and $y_0 = 4$ mm. An RIT instrument with a mass/charge range up to m/z 2000 and equipped with electrospray ionization

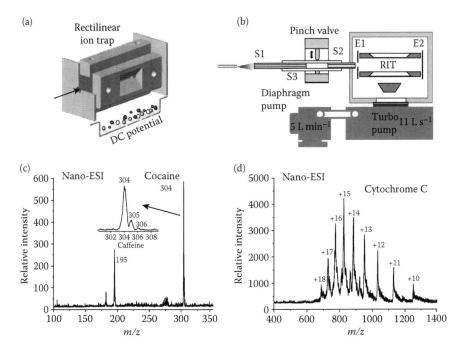

FIGURE 2.5 (a) Configuration of the rectilinear ion trap; (b) discontinuous atmospheric pressure interface (DAPI) coupled to a miniature mass spectrometer equipped with an RIT; (c) mass spectrum recorded for cocaine in water using a Mini 10 instrument; and (d) mass spectrum of cytochrome c using a Mini 11 instrument.

(ESI) was built to analyze biological molecules including intact proteins. The performance of this RIT at high pressure was also explored and protein mass spectra were acquired at pressures as high as 48 mTorr with air as the buffer gas; a resolution (FWHM) of 2 Th was obtained with this device at high pressure for small organics [38]. A soft-landing instrument (see Section 2.4.2.2) was constructed using an RIT to separate a mixture of ESI-generated species according to their m/z-values and to collect the purified compound after soft-landing of the corresponding ions onto appropriate surfaces [23]. In addition to using the RIT in the conventional mode to mass-select trapped ions, mass analysis of ion beams has also been achieved by applying notched waveforms to the RIT to isolate selected ions using mass-selective resonance ejection at the frequencies corresponding to all those ions that were not to be retained in the beam. Multiple ions of different m/z-values can be selected simultaneously by this waveform method allowing, for example, ions derived from the same peptide or protein but having different charge states to be selected at the same time. Note that RF/DC isolation of ions falling in a continuous range of masses has been implemented also using an RIT-based on the original mass-selective stability mode of mass filter operation [39]. Multiplexed instruments with four separate analyzers have been developed also using RITs as the mass analyzers [20,21].

RITs with $x_0 = 5$ mm and $y_0 = 4$ mm have been used in the battery-powered hand-held miniature mass spectrometers built in these laboratories and known as the 'Mini 10' and 'Mini 11.' The complete systems weigh 10 kg and 5.0 kg, respectively [40,41]. Unit resolution and an m/z-range up to 800 Th has been achieved using compact RF circuits. In order to couple the small vacuum systems of these mass spectrometers with atmospheric pressure ionization sources, one can use a capillary of suitably small diameter and continuously introduce sample. However, such an approach is highly ineffective in that a trade-off exists between sensitivity (determined by capillary dimensions) and analytical performance (determined by pressure). For this reason, an alternative has been developed in which a capillary of much larger dimensions may be used while still avoiding overloading of the pumping system by discontinuous introduction of externally generated ions (see Figure 2.5). After a short period of ion (and gas) introduction, a pinch valve seals the capillary inlet (Figure 2.5b) and the system is allowed to pump down to base pressure (ca 1 second). The gas is removed but the ions are trapped in the RIT during this period and they are then subjected to mass analysis. In spite of a low duty cycle, this device, termed a discontinuous atmospheric pressure interface (DAPI), [42] allows a much improved ion transfer efficiency. In the open phase of the DAPI cycle the pressure exceeds 0.1 Torr, but mass analysis is performed after the pressure has dropped to a few mTorr. Hence, miniature mass spectrometers with low pumping capacity are now more amenable to coupling with atmospheric pressure ionization, such as nano-ESI, ESI, atmospheric pressure chemical ionization (APCI), as well as ambient sampling and ionization methods [43] including desorption electrospray ionization (DESI) [42,44] and the low temperature plasma (LTP) probe [45]. Figure 2.5c and d show the analysis of cocaine and protein in water using nano-ESI in conjunction with the Mini 11 hand-held mass spectrometer [41].

2.1.2.3   Toroidal Ion Traps*

The toroidal geometry [1] (see Figure 2.6) is an alternative to the linear geometry for improving the trapping capacity of the quadrupole ion trap mass analyzer. It is a three-dimensional device but, in terms of the dimensions over which the trapping electric field operates it can be considered to act as a 2D trap, in a manner similar to the LIT and RIT. The performance of a full-scale toroidal trap was optimized by Lammert et al. in 2001 [46]. The radial field had to be optimized (compare radial stretching in the RIT and axial stretching—equivalent to radial compression—in the CIT) to produce an asymmetric radial field which compensates for other field distortions and optimizes the net field as indicated by simulations. The resulting system provides unit resolution and high trapping capacity. More recently, a new toroidal trap, miniaturized by a factor of five ($r_0 = 2$ mm) but of the same geometry, was characterized using the mass-selective instability scan [47]; subsequently, the trap was used in a portable GC/MS system which featured mass analysis via scanning of the resonance ejection AC frequency [48]. The halo trap, comprised of concentric arrays of flat ring electrodes integrated into two opposing laser-cut ceramic disks, uses differentially applied potentials on the electrodes in order to effect the geometry of

---

* See Volume 4, Chapter 6: Ion Traps with Circular Geometries, by Daniel E. Austin and Stephen A. Lammert.

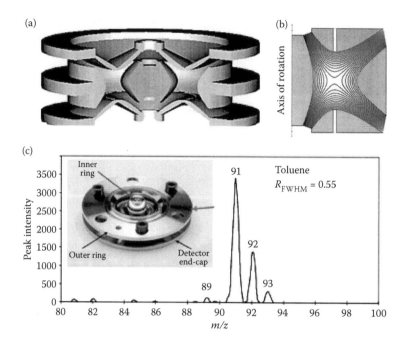

FIGURE 2.6 Toroidal ion trap: (a) configuration; (b) electric field showing optimized inner and outer radii in ratio of 0.55; and (c) mass spectrum recorded using a miniature toroidal ion trap ($r_0 = 2$ mm). ((a) and (b) Reproduced from Lammert, S.A., Plass, W.R., Thompson, C.V., and Wise, M.B., *Int. J. Mass Spectro.*, 212, 25–40, 2001. With permission from American Chemical Society; (c) Reproduced from Contreras, J.A., Murray, J.A., Tolley, S.E., Oliphant, J.L., Tolley, H.D., Lammert, S.A., Lee, E.D., Later, D.W., and Lee, M.L., *J. Am. Soc. Mass Spectrom.* 2008, *19*, 1425–1434. ASMS, with permission from Elsevier.)

a toroidal trap; the mass analyzer has been characterized using frequency-scanning and resonance ejection [15].

2.1.3 Scaling Considerations

With the recent increased interest in miniaturized ion traps for the purposes of portability, multiplexing, and power reduction, it is relevant to review the effects of the scaling when the dimensions are reduced. There are many properties of an ion trap that may change as its dimensions are reduced. Some parameters affected by a decrease in the critical trapping dimensions are examined below.

2.1.3.1 Fabrication Tolerances

As the dimensions of ion traps are reduced, non-conventional methods of trap fabrication need to be used in order to meet requirements of absolute tolerance. These methods of microfabrication, well established in the integrated circuit (IC) and micro-electromechanical systems (MEMS) communities, include photolithography and micromachining in a variety of materials. These techniques make small feature sizes available, though often at the cost of accuracy. Hence, the relative magnitudes

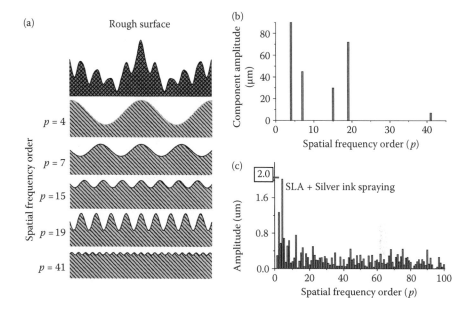

FIGURE 2.7 (a) Model rough surface and five of its spatial frequency components and (b) its spatial frequency spectrum. (c) Spatial frequency spectrum of an actual surface fabricated by stereolithography apparatus (SLA) and coated by spraying with silver ink. (Reproduced from Xu, W., Chappell, W.J., Cooks, R.G., and Ouyang, Z., *J. Mass Spectrom.* 2008, *44*, 3, 353–360. With permission from Wiley.)

of defect features are greater on these scales and the performance of miniature traps could suffer as a result.

Recently, the concept of spatial roughness frequency has been introduced to allow a quantitative characterization of electrode surface roughness and its impact on the electric field inside an electrical assembly such as an ion trap [49]. As shown in Figure 2.7, a rough surface can be expressed using components of simple sinusoidal waves using the equation

$$g_r(x) = \sum_{p=0}^{\infty} R_p \cos\left(\frac{p\pi}{a}x\right)$$ (2.1),

where $g_r(x)$ is the surface roughness function along the surface x-direction, $\omega_p = (p\pi/a)$ is defined as the p-th order 'Spatial Frequency' component for a surface length of a, and R_p is the amplitude of this component. Using the spatial frequency concept, methods have been developed to characterize the internal electric field of ion traps of simple geometries, such as cylindrical or RITs. It was found that roughness components of relatively low order (<10) result in severe effects on of the electric fields, including the relative magnitudes of quadrupole, octopole, and dodecapole fields (Figure 2.8a). In addition to the broadening of peaks, mass shifts due to these field variations occur also. For an array of ion traps fabricated with a given electrode surface roughness, peak broadening due to the random mass shifts associated with

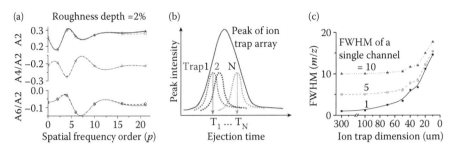

FIGURE 2.8   (a) Roughness spatial frequency effects on quadrupole electric field coefficient 'A2' (given as A_2 in Equation (2.6)), calculated for ion trap with fixed roughness depth (2%): theoretical calculation (solid curves) compared with numerical simulation (dashed curves). The 'A4/A2' and 'A6/A2' curves indicate the relative magnitudes of the octopole and dodecapole electric field coefficients with respect to A2. (b) Schematic view of the array effect, which is the relationship between single ion trap resolution and ion trap array resolution. (c) Calculated resolution of ion trap arrays with different single channel resolution and size (roughness representative of electrodes with gold sputtering metallization; ion m/z 1000 used in calculation). (Reproduced from Xu, W., Chappell, W.J., Cooks, R.G., and Ouyang, Z., *J. Mass Spectrom.* 2008, *44*, 3, 353–360. With permission from Wiley.)

each individual ion trap could occur (Figure 2.8b). When a large number of traps are fabricated on a small scale to make up an array, this effect can be overwhelming and become the dominant factor that determines the mass resolution of the ion trap array. As shown in Figure 2.8c, for an array of ion traps fabricated with a given surface roughness, the overall mass analysis resolution is determined by the mass resolution of each individual ion trap only when the trap size is relatively large. When each individual channel of the trap array is significantly miniaturized, the random variations caused by the surface roughness may become the dominant factor and the mass resolution of the array can be much poorer than that of each individual trap.

2.1.3.2   Capacitance

Several groups have reported high parasitic capacitance in microfabricated traps [16,50]. When the capacitance is too high, impedance mismatching will result and the ultimate voltage delivered to the traps can be lower than desired. In order to minimize capacitance, it is critical to avoid placing conductors which are not essential to the trapping field in proximity to the electrodes.

2.1.3.3   Power

Power is an important aspect of miniature traps, particularly for those destined for use as field devices. A general formula for power dissipated (P_D) in the resonant circuit of an RF ion trap is [51]

$$P_D = \frac{kq_z^2 M^2 r_N^4 f^6 C^2 R_s}{A_2^2}$$

(2.2),

where P_D is defined as the dissipated power in the resonant circuit, k is a constant, q_z is the value of the Mathieu parameter corresponding to axial motion, M is the upper mass to be analyzed, r_N is the normalization radius used in the multipole expansion calculation of the electric potential (see Equation (2.6), f is the RF frequency, C is the capacitance, R_s is the equivalent series resistance, and A_2 is the coefficient of the quadrupole field in the multipole expansion. Because pumps are the most power-intensive components of most mass spectrometers, efforts are being made to reduce pumping loads and to eliminate turbo pumps. However, the RF circuit can also draw a significant portion of the total power; this demand may be remedied, in part, by scale reduction, because the power has a fourth-order dependence upon scale. Likewise, it may be seen that power also scales to the sixth-order with RF frequency. For low-power RF amplification, a helical resonator having a high Q-factor has a lower power consumption than has a digitally amplified power supply. It is typical for the RF amplifier to be located immediately adjacent to the vacuum manifold in which the trap resides in order to minimize resistance and capacitance. Furthermore, the cabling used for the RF power transmission often has its shielding removed, and efforts are made to keep the wires as far away as possible from conductors in order to avoid parasitic capacitance. To keep the power dissipation to a minimum, trap geometries which are 'open' between the electrodes (that is, with little or no spacer material) may consume less power than in instances where the spacer material can absorb RF radiation. An array offers energy efficiency by reducing the amount of power required to operate it; that is, electrodes can be shared between adjacent analyzers [52,53]. Furthermore, it is important that an ion trap array be of smaller dimension than the wavelength of the RF drive potential, otherwise, power loss across the array will result [16].

2.1.3.4 Pseudopotential Well Depth

The pseudopotential well depth determines the magnitude of the effective potential experienced by a trapped ion. The formulae for pseudopotential well depth in the radial ($\bar{D}_r$) and axial ($\bar{D}_z$) dimensions for a 3D Paul trap are [54]

$$\bar{D}_r = \frac{zeV_{RF}^2}{4mr_0^2\Omega^2} = \frac{1}{8}q_rV_{RF} \tag{2.3}$$

and

$$\bar{D}_z = \frac{zeV_{RF}^2}{4mz_0^2\Omega^2} = \frac{1}{8}q_zV_{RF} \tag{2.4},$$

where z is the number of electronic charges e on the ion, V_{RF} is the applied RF 'drive' potential (zero-to-peak), m is the ion mass, r_0 is the inscribed radius of the ring electrode, z_0 is the axial half-distance between the end-cap electrodes, ω is the RF angular frequency, q_r is value of the Mathieu parameter corresponding to radial motion, and q_z is that corresponding to axial motion. One consequence of a shallow well depth is that resolution may be poorer due to ion trajectories more closely approaching rough electrode surfaces. Also, the charge density of a trap having low pseudopotential well depth is diminished, further limiting the number of ions that may be contained.

2.1.3.5   Trapping Efficiency, Trapping Capacity, and Extraction Efficiency

Often, as trap dimensions are reduced, the value of the amplitude, V_{RF}, used and the resulting pseudopotential well depth are also reduced. The practical consequence is that ions with a few electronvolts of kinetic energy entering a reduced-size trap (*e.g.*, a CIT) will have an increased likelihood of not being contained in stable trajectories. For commercially available instrumentation, it is known that the trapping efficiency of externally generated ions captured by 2D traps (>95%) is superior to that of 3D traps (<5%) [5,6]. In addition, the trapping capacities of 2D traps are larger than those of 3D traps (*ca* 40 × in the case of commercial versions) while still having comparable extraction efficiencies (and hence higher sensitivities). Note that as r_0 decreases, V_{RF} will likely also be decreased to provide a selected mass range. With a lower value of V_{RF} comes a shallower pseudopotential well depth, and hence fewer ions will be contained in the trap, that is, sensitivity is lost. For example, Blain *et al.* calculated that a CIT with $r_0 = 1$ μm, $q_z = 0.3$, and $\omega = 1$ GHz will have a pseudopotential well depth of $\bar{D}_z = 0.11$ V, a value which is likely to be problematic for successful trapping of externally generated ions [16]. One partial solution to this problem is to use 2-D traps, which are known to have better trapping efficiency during ion injection [6]. Another sensitivity-impacting consequence of reduced-scale traps is the poor sampling of the incident ion beam due to the dimensions of the trap end-cap aperture being smaller than the width of the incident beam.

2.1.3.6   Pressure

Higher-pressure analyses could allow lower-power operation (in terms of pumping demands) and hence are advantageous in some experiments. As the pressure increases, the collision cross-sections of ions will play a greater role; for example, acceleration of ions through a high field at elevated pressures could cause collision-induced dissociation (CID). Furthermore, a high abundance of polar molecules (for example, ambient water vapor or solvent) at high pressures could cause chemical effects that might result in mass shifts. One advantage to high-pressure operation is the potential to open further the inlet between the atmospheric interface with the vacuum manifold to allow a greater rate of ion transport into the vacuum system and, hence, to improve ion introduction efficiency. Alternatively, the number of pumping stages could be minimized or eliminated to reduce power consumption and to decrease the occurrence of inter-stage ion beam expansion.

2.1.3.7   Mass Range

It can be demonstrated by re-arrangement of the formula for the Mathieu q parameter that, with all the other parameters remaining constant, the m/z-range may be extended by reducing the size of an ion trap [55]. For the mass-selective instability scan, the maximum value of m/z of ions that may be extracted from a trap, $(m/z)_{max}$, can be calculated with the following equation, which is shown here for the case of 3D Paul traps:

$$\left(\frac{m}{z}\right)_{max} = \frac{8 \cdot V_{max}}{q_{eject} \ r_0^2 \ \Omega^2}$$

(2.5),

where V_{max} is the amplitude of the RF drive potential (zero-to-peak) necessary to eject ions of maximum m/z, and q_{eject} is the resonance ejection point on the Mathieu stability diagram.

2.1.3.8 Resolution

According to Goeringer et al., ion trap resolution is a function of pressure, charge, instrument stability, signal-to-noise ratio (S/N) selected, and residence time [56]. Fischer demonstrated mathematically that the resolution of a given m/z-value is ultimately proportional to the number of resonance cycles experienced prior to ejection [57]. This reasoning has been applied to conventional ion traps operated in the mass-selective instability mode using resonance ejection [4]. It has been found that the number of secular oscillations is maximized by minimizing the scan rate, by using low auxiliary AC potentials, and by using high auxiliary excitation frequencies (corresponding to a high q near but not at the stability limit of $q_z = 0.908$) [55,58]. Using these principles, Schwartz et al. achieved a resolution $(R) > 30,000$ $(m/\Delta m$, FWHM) at m/z 502 with a scan rate of 27.8 Th s^{-1} [58]. On a similar basis, higher resolution values have been achieved by very slow scans of isotopic envelopes, including $R > 10^5$ for the protonated peptide Substance P $(m/z$ 1349), $R > 10^6$ for a CsI salt cluster $(m/z$ 3510), and even $R > 10^7$ for the calibrant PFTBA $(m/z$ 614) [59]. In the context of miniature traps, resolution always depends to some extent on the stability of the electronic circuitry (voltage and frequency resolution), and this dependence may be heightened when a very small range of a low-amplitude voltage is scanned. Furthermore, the precision with which miniature analyzers are machined is poor in comparison to commercial conventional-scale traps; this issue is relevant particularly when the proximity with which ions pass near electrodes increases with reduced scale of the device, and perturbations in the electric field due to defects in the electrode become more likely. For miniature trap arrays, different traps could have different effective potentials (where effective potential is the product of pseudopotential well depth and the relative displacement of ion trajectory from the trap center) due to differences in fabrication errors; hence, ejection of ions of a particular m/z-value from multiple traps could lead to a distribution of ejection times and peak broadening of the summed signal.

2.2 SIMULATIONS OF ION TRAP PERFORMANCE

As mass spectrometry becomes even more widely used, two major thrusts in instrumentation development increasingly will depend on simulations. On the one hand, high performance instruments which attempt to break performance barriers in mass range, resolution, or accuracy, will benefit from such simulations. On the other hand, miniaturization of mass spectrometers raises many issues, both well known (geometric effects) and newer (surface imperfections), which must be considered in optimizing new devices for fieldable chemical analysis. The development strategy for these small mass spectrometers—our focus here—is to provide adequate analytical performance with instruments of reduced size, weight and power consumption. Such a goal benefits greatly from computer simulations of ion motion including full simulations of mass spectra.

Numerical simulations have been pursued to study the effects of various factors, including the effects of buffer gas collisions, space charge, non-linear fields, and the influence of resonance excitation. Computer simulations have been demonstrated as an effective approach to explore ion trajectories in a variety of mass analyzers, including quadrupole ion trap, sector, time-of-flight and Fourier transform ion cyclotron resonance instruments. Numerical simulations facilitate performance improvement in existing devices and the discovery of new operating modes. They also allow performance characteristics of as-yet-unbuilt instruments to be assessed and thus provide guidance for improved designs. They have been particularly effective with quadrupole ion traps where the motion of multiple particles can be simulated in programs like ITSIM, providing a realistic means of calculating mass spectra. (The calculations do not include fragmentation kinetics but this could in principle be accomplished by adding standard RRKM kinetics and internal energy deposition information.)

2.2.1 Approach

Computer simulations of ion motion in newly proposed ion trap instrumentation allow reasonably rapid exploration of the effects of changes in electrode geometry, as opposed to the conventional process of designing, constructing, installing, and then properly testing new electrode geometries (usually custom made from stainless steel). The overall process (compare Figure 2.2) employs simulations to best advantage and entails a first feedback loop, where performance *in the simulation* is optimized with respect to the particular application. Once the exact proposed electrode geometry is thus chosen, fabrication and testing can proceed. The results of this testing constitute the second feedback loop, the cycle being completed with an improved design and further computer simulation. Thus, it is imperative that simulation software should strive for accuracy and precision in its ultimate numerical results.

2.2.2 Ion Trajectory Simulation (ITSIM)

The ITSIM program, has been developed in our laboratory and that of Wolfgang Plass over nearly two decades [17,60]; the timeline of ITSIM development is summarized in Table 2.1. The ITSIM software can be used to facilitate instrumentation development by using simulations to elucidate the complicated ion behavior under various conditions. It has been applied successfully to study various phenomena, such as chemical mass shifts, and the performance characteristics of quadrupole ion traps of hyperbolic (QIT) and cylindrical (CIT) geometry, for example resolution and mass accuracy. It has been used also to develop a new toroidal geometry ion trap and to optimize the cylindrical ion trap geometry. As discussed later in this chapter, the orbitrap geometry has been implemented in ITSIM, which has been applied to simulations of ion motion inside the device. The latest versions (5.0 and 6.0) of ITSIM feature code written in C as a 32-bit Windows program. The capabilities of the publically available ITSIM 5.0, as well as the research versions 5.1 and 6.0, are summarized in Table 2.2. In 1999, March and coworkers compared the performance of ITSIM Version 4.1 [60] to that of ISIS [61], and of SIMION 6.0 [62] (Version 8.0

TABLE 2.1
Development Timeline of ITSIM

Version	Description	Years	Representative References
1.0–2.0	MS-DOS version	1990–1993	116
2.01–2.5	Various MS-DOS versions	1993–1996	117,118
3.1	16-Bit Windows version	1997	60
3.32, 4.0	32-Bit Windows version	1997–1998	13
4.1	Intermediate reprogramed Windows 95/98/NT version	1998–1999	119
5.0	Completely reprogramed and significantly expanded Windows 95/98/NT version	1999–2000	69,12
5.1	Various improvements, addition of orbitrap geometry	2002–2003	14
6.0	Importing of electrode geometries directly from solvers; most analyzer types configurable from menus	2004–present	17, 77

TABLE 2.2
Features of ITSIM 5.0

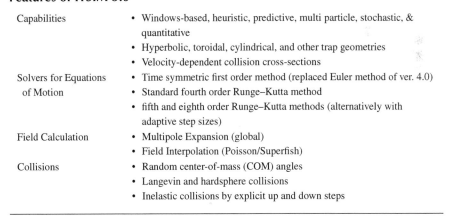

Capabilities	• Windows-based, heuristic, predictive, multi particle, stochastic, & quantitative
	• Hyperbolic, toroidal, cylindrical, and other trap geometries
	• Velocity-dependent collision cross-sections
Solvers for Equations of Motion	• Time symmetric first order method (replaced Euler method of ver. 4.0)
	• Standard fourth order Runge–Kutta method
	• fifth and eighth order Runge–Kutta methods (alternatively with adaptive step sizes)
Field Calculation	• Multipole Expansion (global)
	• Field Interpolation (Poisson/Superfish)
Collisions	• Random center-of-mass (COM) angles
	• Langevin and hardsphere collisions
	• Inelastic collisions by explicit up and down steps

is the most current), and found all three programs to be generally comparable [63], although only ITSIM allows stochastic collision multi-particle simulations.

Exact electrode geometries of arbitrary ion optical assemblies, mass analyzers, and ion traps can be modeled and ion motion simulated in the corresponding electric fields. Electric fields may be obtained by one of three methods: (i) by utilizing the corresponding analytical electrical potential; (ii) by modeling the electrode geometry in an external Laplace solver, such as SIMION 7 or 8 [64,65], COMSOL 3.0a (COMSOL, Inc., US), or the Poisson/Superfish package [66] (Los Alamos National Laboratory, Los Alamos, NM), and importing an array of empirical electric potential points into ITSIM; or (iii) by describing the electrical potential using a multipole expansion [67]

$$\Phi(\rho,\vartheta) = \sum_{l=0}^{\infty} A_l \left(\frac{\rho}{r_N} \right)^l P_l(\cos\vartheta), \tag{2.6}$$

where $\Phi(\rho,\vartheta)$ is the potential as a function of spherical coordinates, l represents the multipole order and takes on integral values from 0 to infinity (for example, 1 = dipole, 2 = quadrupole, 3 = hexapole, 4 = octapole, 5 = decapole, 6 = dodecapole, etc.), $P_l(\cos\vartheta)$ are Legendre polynomials of order l, A_l are the expansion coefficients, and r_N is a normalization radius. Note that in practice often only the even-order polynomial terms are of importance owing to the symmetry of the problem and the series is usually terminated after $l = 6$. The last is perhaps the most common method.

Computation using ITSIM allows the motion of a large number of ions inside any ion optical instrument to be followed under user-defined conditions. Initial ion velocities and spatial coordinates may be specified individually, randomly, or as distributions (for example, Gaussian). Thus, initial conditions may be specified that match practically any scenario for ion creation, for example, *internal* electron ionization (EI), or *transfer* into the device, for example, of ions generated by ESI. Ion trajectories are then calculated numerically by propagating Newton's equations of motion for each ion using Runge–Kutta methods [68]. The accurate treatment of collisions is essential in any mass spectrometer simulation program. Ion traps are particularly complicated since the ions undergo rapid changes in velocity during each cycle, meaning that combinations of hard sphere and Langevin cross-sections must be used to describe even elastic scattering. This complication has led in ITSIM to a treatment of velocity-dependent elastic scattering cross-sections which combines Langevin and hard sphere predictions in a simple 'hybrid' fashion. Additionally, ion/ion (that is, space charge) interactions can be taken into account also and must be used to explain some phenomena, such as mass shifts and peak coalescence [67,69].

2.2.3 EXAMPLE OF GEOMETRY OPTIMIZATION: MINIATURE CYLINDRICAL ION TRAPS

An example of the application of ITSIM is presented here. Wu *et al.* used ITSIM to predict mass spectral performance resulting from miniature CITs of several different geometries [12]. Cylindrical ion traps, as already noted, are three-dimensional (in terms of field) QITs of simplified geometry, and thus are easily machined and hence amenable to miniaturization. The mass resolution of a miniature mass analyzer is determined in large part by the successful optimization of its simplified geometry; adequate mass resolution is probably the decisive factor in determining the success of miniaturization. In all work of this type, experimental tests were carried out in parallel with simulations. The results from both simulations and experiments were correlated to facilitate the design of improved CITs.

It is well known that end-cap electrode holes and truncation of the electrodes of a Paul trap add higher-order terms in Equation (2.6) that, in some cases, can cause undesirable effects, such as the delay of ion ejection [67] and the deterioration of mass resolution. Among all the odd- and even-multipole expansion coefficients, the quadrupolar contribution (A_2) and the next two even higher-order components, octapolar (A_4) and dodecapolar (A_6), are of particular importance. The other higher-order terms are normally very small in weight and their effects are usually (and safely) ignored. Terms having odd values of l are zero, as the trap is symmetric with respect to reflection in the $z = 0$ plane; the monopolar term in the potential may be ignored.

Previous studies on the potential distribution inside the CIT show that it is primarily quadrupolar in the central region [70–72]. However, the non-hyperbolic shape of the electrodes and the presence of holes in the end-cap electrodes introduce a large negative dodecapolar component [29]. To compensate for the weakened strength of the field, the geometry of the CIT is usually adjusted to introduce a positive octopolar component to compensate partially for the negative higher-order field component. In our experience, the best resolution in a CIT under boundary or nonlinear resonance ejection conditions is obtained usually when the sum of the strengths for the positive octopolar and the negative dodecapolar components, relative to the quadrupolar strength, is about –10% [13,29].

Six parameters are needed to describe the CIT geometry: half-length of the ring electrode (z_b), inner radius of the ring electrode (r_0), distance between the ring electrode and the end-cap (d_s), center-to-end-cap distance (z_0), thickness of the ring electrode (r_b), end-cap hole radius (r_H), and the thickness of the end-cap (d_E). The first three parameters were found to have the most effect on the CIT potential. In addition, both the entrance and exit end-cap holes are often proportionally much larger for a miniature CIT in order to accommodate the incident ion beam, and hence must be included also in this group of critical parameters.

To compare the mass resolution of the experimental mass spectra obtained with CITs with that of the mass spectra obtained *via* simulations, the extent of separation between the isotopic peaks at m/z 146 and 148 in the EI mass spectrum of 1,3-dichlorobenzene was used, as measured by the percentage of the valley between the two peaks. The ratio of the height of the valley to the average height of the isotope peaks was compared for five different CIT geometries. The RF potential applied to the ring electrode was 1.1 MHz, with $r_0 = 5.0$ mm and z_0 ranging between 5.0 and 5.5 mm. The resolution with both boundary and resonant ion ejection was studied. The most successful geometry obtained in this study reaffirmed the '–10% rule' previously found empirically, with $A_2 = 0.736$, $A_4 = 0.055$ (giving, expressed as a percentage, $A_4/A_2 = 7.47\%$) and $A_6 = -0.131$ ($A_6/A_2 = -17.80\%$); the sum of the last two percentages being –10.33%.

2.2.4 CHEMICAL MASS SHIFTS

Considerable work, entailing both simulations and experiments, has been performed to understand the phenomenon of chemical mass shifts. The proposed mechanism involves a delay in ion ejection during mass analysis, typically of the order of some hundreds of microseconds, relative to the ideal case, caused by field penetration into the end-cap apertures [13]. This weakening of the electric field is shown in Figure 2.9. The resulting ejection delay can be removed or reduced by adding higher-order field components to the trapping field to offset the effect of the field penetration. Such

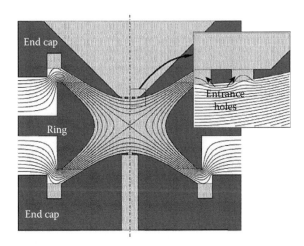

FIGURE 2.9 Equipotential plot of the electric potential distribution in the quadrupole ion trap (3D Paul trap) showing the weakening of the field at the entrance and, by implication, exit end-cap holes. The penetration of the potential into the holes causes this weakening of the field (*viz.* potential gradient). This weakened field causes an ejection delay for all ions. (Reproduced from Plass, W.R., Li, H.Y., and Cooks, R.G., *Int. J. Mass Spectrom.* 2003, *228*, 237–267. With permission from Elsevier.)

higher-order components to the electric potential are added by increasing the end-cap spacing of the trap, hence explaining the observation that chemical mass shifts decrease as the ion trap end-cap electrode spacing is increased. The ejection delay is modified also by collisions with the neutral bath gas, usually helium, which is used in ion traps to improve trapping efficiency through cooling and, by reducing the dimensions of the initial ion cloud and ion velocity, to enhance mass resolution. It was found through simulations and experiments that the probability of a collision that modifies the ejection delay is compound-dependent, hence a difference in collision probability between the ions used to calibrate the mass scale and an analyte will lead to an apparent shift of the recorded *m/z* ratio of the analyte. Note that, with respect to reference ions for which the delay has some particular value, other ions can display either positive or negative mass shifts.

Both the elastic collision cross-section, which is dominated by the size of the ions, and the propensity of ions to fragment have been implicated as important factors in determining the probability of a delay-modifying collision. Shown in Figure 2.10 is a comparison of simulation and experimental data illustrating the effects of the ejection delay, as a function of helium buffer gas pressure, for the PFTBA fragment ion at *m/z* 131. At first sight, the considerable and rich fine structure of the peak at very low pressure is somewhat surprising. Note that this fine structure is well-reproduced by the simulations. Figure 2.11 shows the result for two different chemical entities, both at *m/z* 134. In this case, the *n*-butylbenzene ion, which fragments relatively readily, shows an ejection delay that is shorter than that of $^{134}Xe^+$, a monatomic ion. This difference in ejection delays is rationalized by the fact that fragmentation takes place during ejection (a lengthy process, given the field weakening); the sudden change in ion mass from the parent mass to the fragment causes the fragment ion to

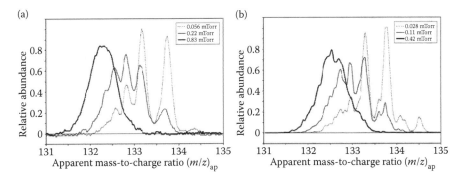

FIGURE 2.10 Boundary ejection signal for the m/z 131 fragment ion of PFTBA recorded using the Finnigan ITMS ion trap for different helium buffer gas pressures. (a) Experiment with $z_0 = 7.07$ mm. (b) Simulation with $z_0 = 6.8$ mm. A constant 0.3 mm offset in the value of z_0 and a decrease by a factor of 2 in pressure was needed to obtain good agreement with experimental results. The very rich peak structure, for a single m/z ion population, is unexpected. Increasing pressure washes out the fine structure, in both the simulation and the experiment. (Reproduced from Plass, W.R., Li, H.Y., and Cooks, R.G., *Int. J. Mass Spectrom.* 2003, *228*, 237–267. With permission from Elsevier.)

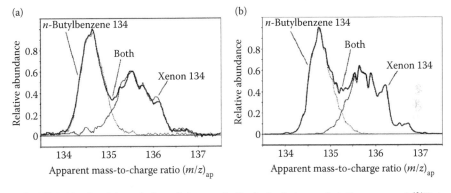

FIGURE 2.11 Partial resolution of the nominally isobaric ions n-butylbenzene and $^{134}Xe^+$, both at m/z 134, due to the operation of chemical shifts in the course of boundary ejection from the Finnigan ITMS ion trap. (a) Experiment with $z_0 = 7.07$ mm and He pressure $= 0.83$ mTorr. (b) Simulation with $z_0 = 6.8$ mm and a helium pressure of 0.42 mTorr. The thick line shows the signal when both samples were admitted to the trap, the thin lines show the signal obtained when the individual compounds were admitted to the trap, or simulated, respectively; both types of ions are ejected together in the valley between the peaks. (Reproduced from Plass, W.R., Li, H.Y., and Cooks, R.G., *Int. J. Mass Spectrom.* 2003, *228*, 237–267. With permission from Elsevier.)

be ejected rapidly, thus experiencing a shorter overall ejection delay than for $^{134}Xe^+$, which can undergo no such fragmentation. The working point (q_{eject}-value) used for resonance ejection and the direction of the RF mass analysis scan also affect the results. Chemical mass shifts are present when resonance ejection is used, unless q_{eject} is chosen to correspond to a nonlinear resonance point, where the shifts are effectively removed. The shifts are removed also by performing the mass analysis scan in the reverse direction, that is, from high mass to low mass [73,74]. Subsequent work using simulations and experiment have shown that chemical mass shifts may have some analytical utility, based on the characteristic values seen for particular classes of compounds. One example is represented by the nitroaromatic compounds, which show very large apparent mass shifts, typically 1–3 Th [75].

2.2.5   RECTILINEAR ION TRAPS

During the development of the CIT, changes in the $z_o - r_o$ ratio were found to be an effective means to vary the mass resolution due to the effect that the ratio has on the higher-order field contributions. This method was adopted when optimizing the cross-sectional geometry of the RIT. While holding x_o constant at 5.0 mm, three distinct RIT devices were constructed with slits for ion ejection in the x-electrodes and y_o = 5.0 mm, 4.2 mm, and 4.0 mm, the last two geometries resulting in devices stretched in the x-direction, analogous to stretching the CITs and QITs in the z-direction. The relative strengths of the octapolar and dodecapolar fields were calculated using the Poisson software (LANL, Los Alamos). The RIT with y_o = 4.0 mm yielded the best resolution. This device's multipole expansion coefficients were 0.654, 0.052, and –1.113, for the quadrupolar, octapolar, and dodecapolar components, respectively. As with the CITs, the sum of the octapolar and dodecapolar coefficients is approximately –10% (7.9% – 17.2% = –9.3%) and the octapolar coefficient is positive, having started as –0.036 in the case $x_o = y_o$ = 5.0 mm. The results of this theoretical/calculational approach agree well with those of the same approach applied previously to CITs. A four-electrode RIT (that is, without the z-end lenses providing axial trapping), was shown also to operate with good performance. Simulations and calculations of potentials revealed that considerable fringing fields in the axial direction provided adequate axial trapping, such that the need for z-end lenses was obviated [35]. Computations using ITSIM have been applied also to the miniature versions of the RIT, providing the optimal cross-sectional geometry in various sizes [14]. RITs have been fabricated using various materials, including metal-coated polymers made using SLA prototyping methods, [36,76] see also Section 2.3.

2.2.6   THE ORBITRAP*

Simulation of the AC dipolar excitation of ion axial motion in the Orbitrap™ has been performed using ITSIM 6.0 [77]. The Orbitrap inner and outer electrodes were generated in AutoCAD, a 3D drawing program. The electrode geometry was imported into the 3D field solver COMSOL; the field array was then imported into ITSIM 6.0. The empirical electric field so imported was checked against that obtained by taking the derivative of the analytic potential distribution expression [78] and they were found to agree to within 0.5%. Calculations simulating experiments where an AC signal was applied to the split outer electrodes at the axial resonant frequency of a selected ion showed axial excitation, resulting in eventual ion ejection when the AC was in phase with ion axial motion (that is, 0° phase relative to ion axial motion). De-excitation of the ion axial motion until the ions were at rest at the equator ($z = 0$) was observed

---

* See Volume 4, Chapter 3: Theory and Practice of the Orbitrap Mass Analyzer, by Alexander Makarov.

when the applied AC was out of phase with the ion axial motion, with re-excitation of the axial motion occurring when the dipolar AC was continued beyond this point in time. Both de-excitation and re-excitation could be achieved mass-selectively, and each depended upon the amplitude and duration (number of cycles) of the applied AC signal. The effects of changing the AC amplitude, frequency, phase relative to ion motion, and bandwidth of the applied waveform have all been simulated. In each case the simulation results were compared directly with the experimental data and good agreement was observed. Such ion motion control experiments and their simulation provide the possibility to improve Orbitrap performance, and to develop tandem mass spectrometry (MS/MS) capabilities inside the Orbitrap.

2.3 FABRICATION OF MINIATURE ION TRAPS BY MODERN METHODS

Conventionally fabricated 3D QITs comprise precision-machined stainless steel electrodes, typically with a critical dimension of $r_0 \approx 1$ cm. Variants of such 'full scale' QITs include the following geometries: *3D traps* (hyperbolic [10,79,80] and simplified [25,29]), *2D traps* (hyperbolic [1,5,6] and simplified [14]), and *toroidal traps* (hyperbolic [46] and simplified [15]). The ceramic insulating mounts and spacers for these traps are likewise fabricated by precision machining processes. More recently, traps have also been constructed using new fabrication methods in attempts to take advantage of the benefits of reduced scale, of multiplexing, or of particular fabrication processes (for example, integrated circuitry). As new designs for miniaturized ion trap geometries require smaller feature sizes or more complex arrayed geometries, they become much more difficult, or even impossible, to create using conventional precision machining. Hence, microfabrication technologies, including those related to integrated circuit (IC) and micromachining (sub-millimeter features) methods, are being utilized increasingly.

The new methods of ion trap fabrication offer advantages over conventional methods in terms of achievable geometry, minimal feature sizes, production time, and material cost. Conventional machining, being a '2.5D' technique (that is, 3D without the capacity for creating overhanging elements), is limited in the extent to which it can be used to create complex geometries; for example, some geometries that might be of interest may require building layer-upon-layer, or the use of 'sacrificial' internal element-characteristics available only with certain methods of microfabrication. Additionally, IC and MEMS technologies are suitable for batch fabrication of ion traps, which could help greatly to reduce production costs and times, and improve reproducibility. Another noteworthy feature of some microfabrication processes is their amenability to monolithic fabrication, in which the entire assembly (including electrodes and insulating materials) comprises a single contiguous object. The consequences of scaling laws at reduced ion trap dimensions introduce potential new issues as well as topics for research; for example, electrode surface roughness that is negligible in a large trap might have significant consequences in a small device. The following account describes some unconventional ion trap fabrication methods, including the steps involved in production, considerations of advantages and constraints, and examples of ion traps produced by these methods.

2.3.1 Semiconductor Processes

Lithography may be defined as the transfer of an image or pattern from one object to another [81]. Essentially all microfabricated traps use photolithographic processes, which comprise the patterning of a substrate using photons as the agents of pattern transfer. Typically, a substrate (for example, an oxidized silicon wafer) is first covered with a photo-responsive thin film (the 'photoresist') and then covered with a stencil (the 'mask') consisting of patterned transparent and opaque regions. A light source is allowed to shine through the mask to expose selectively the underlying photoresist. Next, either the exposed or unexposed regions of the photoresist are dissolved in solution and rinsed away. The areas of exposed substrate may then undergo chemical attack by wet or dry etching to form channels. A variety of thin or thick film deposition steps may follow in addition to further etching steps. This process of selective addition and/or removal of material results in a patterned and often multiply layered object composed of conductive and non-conductive materials, which may be interspersed with lateral and/or vertical air gaps, see Figure 2.12. With photolithography, commercially achievable feature sizes are on the order of several tens of nanometers.

A notable delineation between micromachining techniques is that of surface micromachining *versus* bulk micromachining [81,82]. In *surface micromachining*, features are built up from iterative deposition and etching of multiple layers of thin films (1–100 μm) of various materials. The lowest substrate layer serves only as a structural function. Often, a sacrificial layer within the stack of films is etched selectively, allowing the creation of functional gaps. While surface micromachining allows very detailed layering, the deposition processes can be very lengthy, for example, addition of polysilicon by Low-Pressure Chemical Vapor Deposition (LPCVD), has typical rates of 1 μm h^{-1}. *Bulk micromachining*, on the other hand, includes processes whereby the substrate layer is made functional *via* deep etching, for example, reactive ion etching. Often, photolithographic patterning and metallization occurs after machining. Since bulk micromachining is not limited by long surface deposition times, devices can easily have a greater vertical dimension than a surface-machined structure. A negative aspect of bulk micromachined devices is the lack of easy integration with integrated circuitry attainable by surface micromachining. Often, a combination of both methodologies is used to manufacture devices to maximize the benefits of each process.

One application of semiconductor processing methodologies applied to ion trap fabrication is to be found in the arrays of micro-CITs reported by Cruz *et al.* [32]. These authors used a combination of MEMS and IC technology to fabricate a massively parallel array of CITs with $r_0 = 1$ μm (10^6 traps in 0.25 cm^2) as well as one with $r_0 = 5$ μm (20,736 traps), see Figure 2.13. The intention was to develop a chip-based platform capable of performing comparably to a conventionally scaled trap yet with lower power and operation at much higher pressures. Fabrication began by bulk machining a silicon wafer to provide an exit aperture for the ions as they move from the ion traps to a set of Faraday cup detectors. The trap electrodes were constructed using surface micromachining, where the back end-cap electrodes were made from the layer nearest the substrate, the middle layer(s) contained the ring electrodes, and

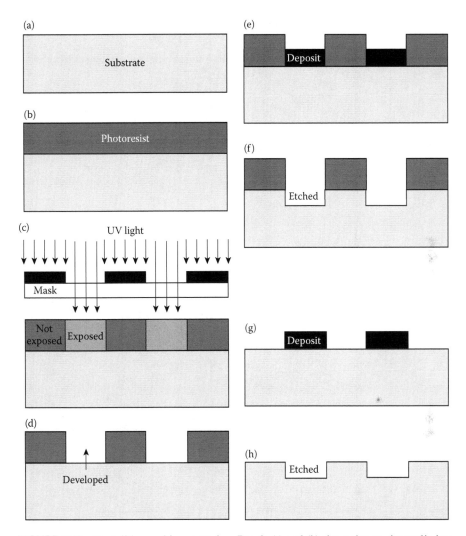

FIGURE 2.12 Photolithographic processing. Panels (a) and (b) show photoresist applied to substrate, (c) selective exposure of photoresist, (d) development of photoresist. After a pattern is made, material may be either deposited followed by photoresist removal, (e) and (g), or etched followed by photoresist removal, (f) and (h). (Reproduced from Judy, J.W., *Smart Mater. Struct.* 2001, *10*, 1115–1134. With permission from IOP Publishing Ltd.)

the top-most layer contained the front end-cap electrodes. To keep power consumption to a minimum, tungsten was chosen to be the conductive material, on account of its electrical and thermal properties (resistivity, thermal conductivity, melting point, and skin depth—a measure of the distance that the AC current density penetrates into a conductive surface). To fabricate the electrodes, the authors first deposited a layer of SiO_2 by plasma-enhanced chemical vapor deposition (PECVD), followed by photolithographic patterning and etching by fluoromethane plasma to remove unwanted material. Next, a 25 nm adhesion layer of TiN was sputtered, and tungsten

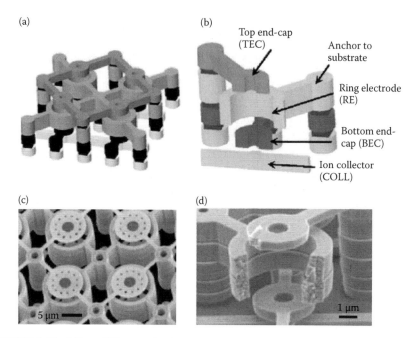

FIGURE 2.13 Micro-CIT array fabricated in silicon and tungsten as a nine-layer structure with overhanging elements using both bulk and surface micromachining, and incorporating steps of CVD, sputtering , photolithography, and chemical-mechanical polishing. (a) Unit cell of the array, (b) trap electrode description, (c) top-down view of the array, (d) cross-sectional view of a single micro-CIT element. (Reproduced from Cruz, D., Chang, J.P., Fico, M., Guymon, A.J., Austin, D.E., and Blain, M.G., *Rev. Sci. Instr.* 2007, *78*, 015107. With permission from the American Institute of Physics.)

was deposited by chemical vapor deposition (CVD). Lastly, the surface underwent chemical–mechanical polishing (CMP), which used a polyurethane pad and an oxidized/acidified colloidal alumina slurry to polish or plane the surface, leaving a flat surface with the metal remaining only in the pre-etched trenches. This process of etching, overfilling, and polishing or planing, known as 'damascene processing', was carried out for all layers of the array. During fabrication, structural supports ('anchor points') consisting of stacked dielectric (800 nm of SiN on 630 nm of SiO_2) were integrated into the array and run down to the silicon substrate. After a layer of aluminum (8000 Å) had been sputtered and plasma etched to form contact pads, the chip was immersed in hydrofluoric acid solution (49%) to remove the sacrificial SiO_2 separating the tungsten electrodes. As determined by scanning electron microscopy (SEM), the final structure had an end-cap-to-ring electrode spacing of 0.5 μm vertically and 4.5 μm horizontally. Layer-by-layer misalignment tolerance defined by the photolithography tool was 0.15 μm (3σ). In addition, the ring electrode walls were seen to be slightly non-vertical; simulations comparing 1 μm sloped and vertical traps suggested this manufacturing error would cause 24% fewer ions to exit the trap through the bottom toward the detector. The installed electrical characteristics for the 5 μm trap array were as follows: resistance, $R = 1.77$ ω, impedance, $L = 5386$ pH, and capacitance, $C = 49.6$ pF. However, the device suffered from impedance mis-matching with the signal transmission line, hence as much as 66% of the power supplied was reflected. Despite the limitation of applied voltage, the traps were evaluated using toluene ionized by external EI. With an RF drive potential of 4 V_{0-p} and a frequency of 200 MHz, the authors trapped ions for 30 µs and then shut off the applied RF for 1 ms to allow detection of the ions at the collector elements integrated into the chip beneath the CITs. The small signal seen was attributed to >3000 ions, which indicated that about 15% of the traps contained ions. The main problem with this device was that the trapping pseudopotential well-depth was only a factor of three or so greater than the ion thermal kinetic energy. Larger volume traps utilizing correspondingly greater values of V_{RF} would be expected to work much better, but the IC fabrication methods are not amenable to structures in which r_0 is greater than about 20 microns.

Van Amerom *et al.* have employed a different method of trap fabrication in silicon to construct micro-CIT arrays, see Figure 2.14 [33]. The 5×5 micro-CIT array featured much larger critical half-distances of $r_0 = z_0 = 360$ µm. To fabricate a complete array, two identical structures were built, each consisting of a substrate, an end-cap electrode layer, and half of a ring electrode layer. A 350 µm thick n-type [100] polished silicon wafer was coated with Si3N$_4$ (3 µm thickness by LPCVD) and then SiO$_2$ (5 µm thickness) insulating layers; the SiO$_2$ was 'thermally grown' by exposing a silicon wafer to high temperatures in an oxygenated environment. Next, the wafer was patterned photolithograpically and etched using deep reactive

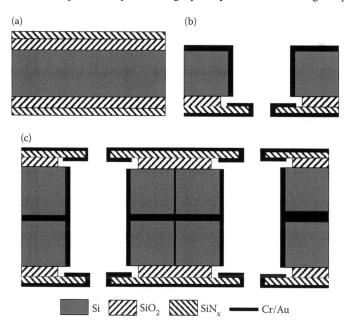

FIGURE 2.14 Process flow for micro-CIT array fabrication. (a) Silicon wafer coated with Si$_3$N$_4$ and SiO$_2$ layers, (b) one-half of a trap after etching and metalization, (c) two completed traps after bonding. (Reproduced from van Amerom, F.H.W., Chaudhary, A., Cardenas, M., Bumgarner, J., and Short, R.T., *Chem. Eng. Comm.* 2008, *195*, 98–114. With permission from Taylor & Francis.)

ion etching (DRIE). The process of DRIE uses a high-density/low-energy plasma, for example, generated by an inductively coupled plasma source, to achieve a high etch rate (several micrometers/minute), with reported aspect ratios between 10:1 and 50:1, depending upon the conditions used [81]. As is the case here, DRIE etches selectively one material over another (for example, Si versus SiO_2 selectivity of 100:1). After the etching was completed, the half-structures were masked photolithographically and selected areas were metalized with a layer of Cr/Au using a 'lift-off' methodology, that is during metallization, a resist existed only on the areas where metal was not wanted; the resist and any metal deposited onto it were later removed by means of chemical dissolution. Care was taken to minimize metallization of the open areas of the substrate surrounding the end-cap electrode holes, resulting in the final version of the trap having a capacitance of only 45 pF. After the half-structures had been completed, the two halves were aligned and bonded together using 'flip chip' bonding. This is a common IC industry technique where one chip is 'flipped' over and aligned onto the other one, and the process is completed by the making of mechanical and electrical connections; in this instance the thin gold film on each half of the assembly was bonded together at sufficient heat and pressure. Lastly, the assembled trap array was bonded to a printed circuit board (PCB) using silver epoxy and the wiring connections were made. Unfortunately, the trap failed during testing and much of it was destroyed. The authors suggest this may have been due to sharp corners on the ring electrodes, which lead to an electrical breakdown; subsequent versions will have these corners etched off.

Pau *et al.* have made micro-scale CIT arrays composed of phosphorous-doped polysilicon electrodes with SiO_2 insulators [50]. Doped polysilicon is an attractive material for micromachined electrodes because it can be engineered to have resistivities of $\leq 10^{-4}$ ω cm, and it can be patterned with good precision (>95% uniformity, aspect ratios >20:1 (aspect ratios are the ratio of length in one dimension to the deviation in an orthogonal dimension), and side wall tolerances <10 nm (the maximum deviation in the dimension normal to the surface)). By occasionally inserting phosphorus atoms (five valence electrons) into a crystal lattice of silicon (four valence electrons), one electron associated with each phosphorus atom is only very weakly bound and hence available for conduction. Fabrication of the arrays on silicon wafers involved six iterations of material deposition by CVD followed by DRIE, see Figure 2.15. After the layers were deposited, annealing (heating to remove mechanical stresses) was carried out on the polysilicon at 1050°C for 10 h. The authors demonstrated the detection of approximately 10 xenon ions (generated by EI) using a device having 256 channels, each having critical dimensions of 20 μm for both r_0 and z_0, see Figure 2.16. For the experiment, an RF potential of 45 V_{0-p} driven at 100 MHz was applied to the ring electrode while the end-cap electrodes were grounded; ion ejection required ramping V_{RF} from 45 to 90 V_{0-p} over 200 μs while a high voltage gate mesh between the array and an electron multiplier was switched to pass mode.

Stick and coworkers have fabricated a monolithic array of micro-scale 2D ion traps by the photolithographic processing of a semiconductor substrate (gallium–arsenide), see Figure 2.17 [83,84]. The intention of fabricating this array was to lay the groundwork for a quantum computer chip. The first step in fabrication was the deposition, by molecular beam epitaxy (MBE), of four alternating layers of aluminum

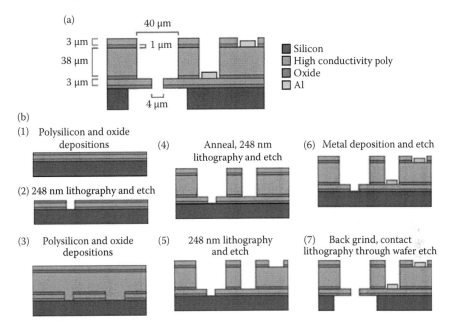

FIGURE 2.15 Micro-CIT array fabricated in doped polysilicon with SiO_2 insulators using CVD and DRIE. (a) Finished trap and dimensions, (b) Fabrication process (shown in steps (1)–(7)). (Reproduced from Pau, S., Pai, C.S., Low, Y.L., Moxom, J., Reilly, P.T.A., Whitten, W.B., and Ramsey, J.M., *Phys. Rev. Lett.* 2006, *96*, 120801. With permission from The American Physical Society.)

gallium–arsenide insulator (AlGaAs: 4 μm thickness, hereafter termed layers 1 and 3) and gallium–arsenide conductor (GaAs: 2.3 μm thickness, hereafter termed layers 2 and 4) on a GaAs substrate. Next, the back-side of the GaAs substrate was etched isotropically (this creates non-vertical walls) through to the first MBE-deposited layer (AlGaAs) to allow eventual optical access to the trapping region for laser-based experiments. Next, the application of DRIE through layers three and four allowed access to layer two for electrical connections. Sets of silver/germanium bond pads were deposited on layers two and four to provide electrical connections to the electrodes (seven on each level and on each side of the trap). Further DRIE was employed to define the adjacent electrodes (130 μm wide with 25 μm gaps between adjacent electrodes), and to etch all the way through the chip to the GaAs substrate level (60 μm gaps between opposing electrodes). A hydrofluoric acid etch of the AlGaAs was used to undercut the GaAs layers by 15 μm, providing an air gap between top and bottom electrode pairs; removal of this insulating material from the vicinity of the trapping region prevents distortion of the trapping field through charge buildup. The monolithic fabrication used here is ideal for batch fabrication and avoids manual handling during assembly and alignment. Finally, the chip was mounted in a ceramic chip carrier and the wiring to the electrodes was completed with 25 μm gold wires. All the RF electrodes were connected to a single wire, while the DC-only electrodes were grounded through 1000 pF capacitors (the measured RF pickup was <1% of potential applied to the RF electrodes). The trap RF potential of 8.0 V_{0-p} was driven at 100 MHz,

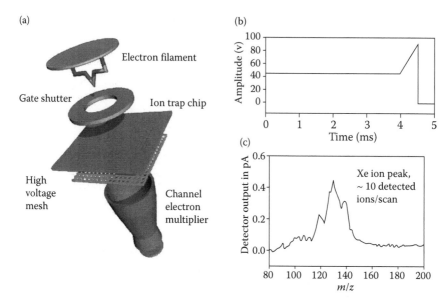

FIGURE 2.16   Micro-CIT array, timing diagram and data. (a) Array assembly with ion source, gate, and detector. (b) RF timing diagram for trapping and scan-out periods. (c) Mass spectrum of Xe⁺. (Adapted from Pau, S., Pai, C.S., Low, Y.L., Moxom, J., Reilly, P.T.A., Whitten, W.B., and Ramsey, J.M., *Phys. Rev. Lett.* 2006, *96*, 120801. With permission from The American Physical Society.)

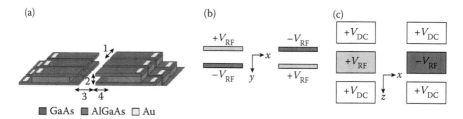

FIGURE 2.17   Monolithic ion trap array, (a) Microscale 2D semiconductor trap array. (a-1) electrode width = 130 μm; (a-2) 4 μm-thick AlGaAs inter-electrode spacing layer; (a-3) tip-to-tip separation between cantilever electrodes = 60 μm; (a-4) 23 μm-thick GaAs cantilevered electrodes with 15 μm cantilever. (b) Configuration of V_{RF} applied to the trapping electrodes in the x-y plane. (c) Configuration of V_{DC} and V_{RF} applied to the trapping electrodes in the x-z plane. ((a) Reproduced from Stick, D., Hensinger, W.K., Olmschenk, S., Madsen, M.J., Schwab, K., and Monroe, C., *Nat. Phys.* 2006, *2*, 36–39. With permission from Nature Publishing Group; (b) and (c) Reproduced from Madsen, M.J., Hensinger, W.K., Stick, D., Rabchuk, J.A., and Monroe, C., *Appl. Phys. B.* 2004, *78*, 639–651. With permission from Springer-Verlag.)

with DC potentials of +1.00 V on the end-cap electrodes and –0.33 V on the trapping electrodes. The authors noted that though the cantilevered electrodes should have mechanical resonance in the 1–10 MHz range, the RF frequency applied caused no significant mechanical resonance. The trap performance was tested by confining, cooling, heating, and the 'shuttling' of single ¹¹¹Cd⁺ ions (shuttling is a quantum

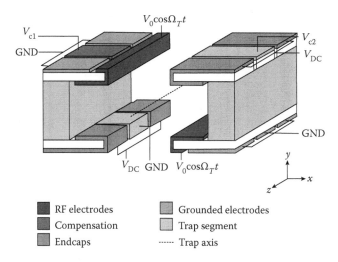

FIGURE 2.18 Unit cell of linear trap array proposed for quantum computing. For structural details, see text. (Reproduced from Brownnutt, M., Wilpers, G., Gill, P., Thompson, R.C., and Sinclair, A.G., *New J. Phys.* 2006, *8*, 18. With permission from IOP Publishing Ltd.)

processing term for the transport of ions between adjacent traps). Using fluorescence to observe the trajectory of a single Doppler-cooled $^{111}Cd^+$ ion in one trap, the authors were able to determine the Mathieu stability parameter to be $q_r = 0.62$. Simulations suggest that the trap has a potential well-depth of *ca* 0.08 eV, a value so close to room temperature ($3/2kT = 0.025$ eV) that continuous laser cooling was required to prevent ion heating and escape. The electrodes described were tested for electrical properties, and unfavorable resistance and capacitance values determined along the length of a single electrode (20 ohms and 34 pF, respectively) indicated high power loss, which should be addressable in later iterations through the choice of different materials and/or dimensions.

A design for a chip-based monolithic 2D trap has been proposed by Brownnutt *et al.* using silica-on-silicon processes and photolithographic patterning, Figure 2.18 [85]. In this device, the authors' intention was to develop a platform for a quantum processor. The design consists of a silicon wafer (40–500 µm thick) coated with SiO_2 on both sides (15 µm thick, thermally grown). Using DRIE, a channel is etched through the wafer and the silicon layer within the cavity is undercut by wet etching to leave the SiO_2 as cantilevers, which extend part way into the cavity. The electrodes are constructed by evaporating gold (100 nm thick) onto the photolithographically patterned SiO_2 cantilevers. Two RF electrodes run the entire length of the trap and are located both on opposite sides of the trap and opposite faces of the chip. In the corresponding inverse locations are two sets of three adjacent electrodes (six in total), where the four outermost electrodes (along the z-axis) are the trap DC electrodes and the two center electrodes are meant to provide compensation potentials for precision control of unwanted ion micromotion. Line-of-sight to the trapping cavity is not blocked from either the top or bottom of the chip to allow easy access for optical manipulation and measurement. Achievable critical dimensions of the trap are $20 \leq$

$r_0 \leq 350$ μm and $80 \leq z \leq 3030$ μm. The authors have considered RF power absorption by the materials comprising the chip and have calculated also the mechanical resonance frequency range of the cantilever electrodes; neither will limit significantly the operability of the trap.

2.3.2 Printed Circuit Board (PCB) Processes

Huber *et al.* fabricated a multi-segment LIT comprising four PCBs, see Figure 2.19 [86]. The PCBs were aligned in holders at their ends such that one edge of each board faced a central axis, with the boards spread at 90° intervals. The PCB substrate was standard copper-coated polyimide (18 μm copper thickness). Fifteen isolated DC electrodes were defined by etching 120 μm-wide grooves into the copper of two opposing boards; these electrodes were spread along three trap sections: (1) wide-acceptance loading zone ($r_0 = 2$ mm); (2) taper zone (gradient r_0); (3) experimental zone ($r_0 = 1$ mm). The other two boards were unmodified and provided the single-phase driving RF supply. Such PCB fabrication of ion traps is simple, fast, and cheap, and the traps are ultra-high vacuum-compatible. The authors note that this fabrication methodology is limited to $r_0 > 50$ μm. The open geometry between the trapping electrodes is ideal for spectroscopic studies. The electrode segments in the experimental section were closely spaced (120 μm) and had short lengths (500 μm) because the authors were interested in developing an array for ion shuttling. Because unintended oscillatory ion micromotion was critically undesirable, two stainless steel compensation electrodes, in parallel and opposite each other across the trap axis, were installed to provide compensation potentials to offset stray charge buildup on the etched insulator or geometric imperfections. The trap was demonstrated through operation with an RF supply

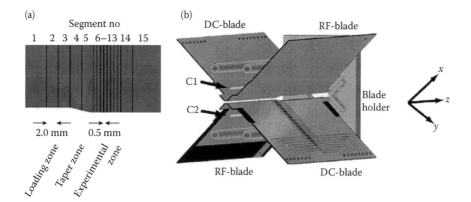

FIGURE 2.19 Depiction of PCB trap. (a) Finished DC blade, (b) Assembled trap (front blade holder not shown). Note compensation rods C1 and C2 used to control ion micromotion. For further details, see text. (Reproduced from Huber, G., Deuschle, T., Schnitzler, W., Reichle, R., Singer, K., and Schmidt-Kaler, F., *New J. Phys.*, 10, 15, 2008. With permission from IOP Publishing Ltd and Deutsche Physikalische Gesellschaft.)

running at 74.2 MHz and 204 V_{0-p} to allow manipulation of Doppler-cooled ion crystals of $^{40}Ca^+$.

Pau *et al.* have constructed a QIT with planar geometry, see Figure 2.20 [87]. The intention of the authors was to test a simple trap amenable to implementation using ICs and MEMS. To do this, the researchers patterned and etched concentric rings on one or both sides of a double-sided PCB where both faces of the insulating substrate were metalized. One trap design contained three concentric ring electrodes on one face of a PCB; RF power was applied to the middle ring only while the other two were grounded. A second design comprised a single ring electrode on one face of the PCB, with the other face acting as a ground plane. The device was demonstrated by experiments that included: (i) trapping of diamond microcrystals at 760 Torr; and (ii) mass analysis of both xenon and dimethyl methylphosphonate (DMMP) fragment ions at *ca* 10^{-4} Torr (see Figure 2.20d and e). For the microparticle trapping experiment, sufficient trapping forces were provided with an RF power source of 500 V_{rms} running at a frequency of 60 Hz, together with an added DC potential of 0–100 V to offset gravity; the particles were observed, by means of illumination by a helium–neon laser, to be trapped stably for one week before the experiment was terminated. Testing the trap as a mass analyzer in vacuum required

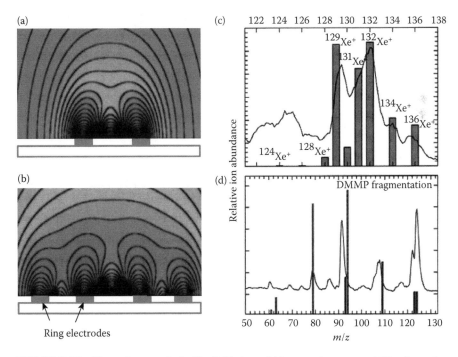

FIGURE 2.20 Planar ion trap. Left side: field plots of (a) a one-ring trap, and (b) a three-ring trap. Right side: mass spectra recorded using a planar toroidal ion trap, (c) EI-MS spectrum of xenon with calculated isotopic abundance overlaid as a histogram, and (d) EI-MS spectrum of DMMP fragments with calculated isotopic abundances overlaid as a histogram. (Reproduced from Pau, S., Whitten, W.B., and Ramsey, J.M., *Anal. Chem.* 2007, *79*, 6857–6861. With permission from the American Chemical Society.)

an RF drive frequency of 2.8 MHz. Xenon ions and DMMP fragment ions generated by external EI were guided to the trap with appropriate ion optical lensing,
and mass analysis was effected by mass-selective instability scanning (for example,
detection of the DMMP-derived ions required ramping the RF amplitude from 88
to 350 V_{0-p} in 8 ms), and detection was provided by an electron multiplier. A typical
peak width of *ca* 1.5 Th (FWHM) was observed for DMMP (*m/z* 124). The xenon
isotopic distribution agreed with the expected relative abundances, though some
extraneous peaks appeared to be due to off-gassing of the PCB materials due as the
board was heated by RF absorption. A significant advantage that planar traps offer
over three-dimensional traps is that a simple two-dimensional electrode pattern is
required, which means that the potential for errors in machining a third dimension
does not exist.

2.3.3 LASER-MACHINED CERAMICS

Austin *et al.* have fabricated a simplified geometry toroidal QIT, dubbed the 'halo
trap' [15]. The halo trap consists of two parallel planar ceramic discs (6.35 mm thick,
4 mm apart), each with 15 integrated concentric gold ring electrodes spaced 0.5 mm
apart and having radii between 5 and 12 mm, see Figure 2.21. Circular channels
were laser-cut into the ceramic discs to define the ring electrode perimeters. Holes
were laser-cut along the ring electrode channel perimeters to provide vias (holes

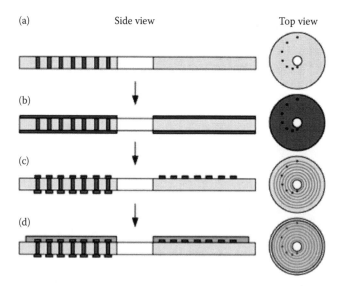

FIGURE 2.21 Fabrication of a halo trap ring electrode disc. (a) Vias are cut and filled with
a Au/W alloy; (b) layers of chromium and gold are deposited to fill ring electrode channels,
(c) ring electrodes are photolithographically isolated; (d) semi-conductive germanium coating
is applied. (Reproduced from Austin, D.E., Wang, M., Tolley, S.E., Maas, J.D., Hawkins, A.R.,
Rockwood, A.L., Tolley, H.D., Lee, E.D., and Lee, M.L., *Anal. Chem.* 2007, *79*, 2927–2932.
With permission from the American Chemical Society.)

made to allow electrical connections to be made between two faces of a substrate) to each ring electrode. Also, laser-cut holes were made for alignment rods, and to provide a hole in the center of a disc for ion ejection to the detector. The vias were filled with a gold/tungsten alloy and the discs were coated on each side with evaporated chromium followed by gold (in order to fill the ring electrode channels). Next, the discs were polished to 1 μm surface roughness (Figure 2.21a). After polishing, a photoresist was spun and patterned to define the electrodes as 100 μm wide and 500 μm apart (Figure 2.21b). Development and etching isolated each electrode while providing a lead to the non-trapping faces of the discs for connections with the control electronics. After removing the photoresist (Figure 2.21c), a 500 nm-thick layer of germanium was deposited on the trapping side of the plates (Figure 2.21d) to minimize charge buildup on the alumina and to provide a smoothing of the equipotential surfaces. As with previous toroidal traps, [46] the doughnut-shaped trapping field is spread over a circumference which, when compared to conventional 3D traps, allows greater trapping capacity, reduced space-charge effects, and enhances dynamic range. The open structure of the trap allows for internal EI; on the other hand, the presence of an RF potential barrier could make injection of externally generated ions difficult. The RF signal was applied to each electrode quasi-independently through a capacitor circuit, which allowed arbitrary electronic shaping of the trapping field by DC voltage adjustments; that is, the equipotential surfaces need not conform to the electrode geometry. In order to effect ion ejection into the detector, the trapped ion population was ejected resonantly by sweeping the ion secular frequencies (50–600 kHz). The dipolar excitation was carried out by superimposing an AC waveform (hundreds of mV_{0-p}) onto the RF potential applied to the first and last electrodes. The trap was found to produce a resolution of *ca* 77 ($m/\Delta m$, FWHM) for the m/z 49 fragment of dichloromethane with an RF drive frequency of 1.9 MHz at $V_{RF} = 250$ V_{0-p}, Figure 2.22c and d.

2.3.4   LOW-TEMPERATURE CO-FIRED CERAMIC (LTCC)

Chaudhary *et al.* have fabricated a CIT using a gold-plated ceramic ring electrode and stainless steel end-cap electrodes, see Figure 2.23 [88]. The ceramic used was in the form of low-temperature co-fired ceramic (LTCC) tape, a material used commonly for packaging in the semiconductor industry. The LTCC tape comprises glass-encapsulated ceramic particles suspended in an organic matrix, making it a pliable material that is easily cut or molded. After shaping, heating LTCC at several hundred degrees Celsius followed by cooling causes it to harden into its final form: the glass coating of the ceramic particles becomes free-flowing and ultimately hardens upon cooling as a continuous structure. The interest in using LTCC here was due to its ease of manufacture and amenability to batch fabrication (and economies of scale) for miniature ion traps and arrays. In order to form the ring electrode, 52 rings punched from commercial LTCC tape were stacked and compressed into a stainless steel mold. This procedure was followed by lamination through hydraulic compression at 8000 psi and 85°C, leaving a single multi-layer piece in the form of a ring electrode. The outer edge of the electrode was tapped subsequently to attach the electrical connections during final assembly. The piece was then fired at 850°C

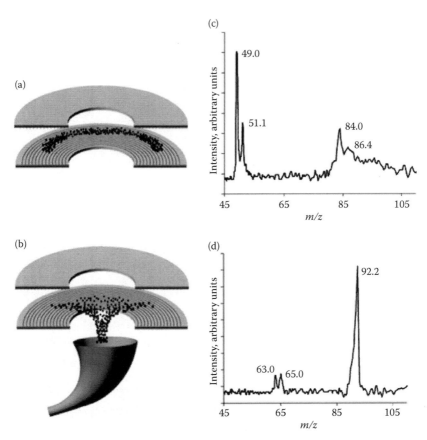

FIGURE 2.22 Depiction of the halo trap and its use for mass analysis: (a) trapped ion population within the structure; (b) radial ejection to a central detector; (c) mass spectrum of dichloromethane (FWHM 0.64 Th at 49.0 Th); and (d) mass spectrum of toluene (FWHM 1.3 Th at 92.2 Th). (Reproduced from Austin, D.E., Wang, M., Tolley, S.E., Maas, J.D., Hawkins, A.R., Rockwood, A.L., Tolley, H.D., Lee, E.D., and Lee, M.L., *Anal. Chem.* 2007, *79*, 2927–2932. With permission from the American Chemical Society.)

for final curing, following which the ring electrode underwent 'electroless plating', an autocatalytic process in which metal is deposited from solution onto an object by chemical reduction. The plating (2.5 μm total thickness) consisted of a phosphorous-doped nickel adhesion layer followed by a layer of gold for the final electrode surface. Photoresist was then spun onto the metalized structure and computer-directed UV light exposed those areas to be de-metalized (that is, those areas destined to contact the end-cap electrodes). After removal of the exposed photoresist, the object was placed in *aqua regia* to remove exposed metal. Subsequent immersion in acetone and methanol removed the remaining photoresist. Final assembly included wiring the connections and clamping the 100 μm thick end-cap electrodes onto the LTCC piece. The designed CIT radial and z-axis half-distances were 1.375 and 1.580 mm, respectively; however, the final device had shrunk during curing by factors

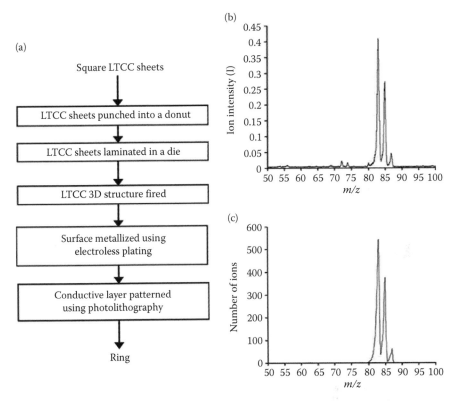

FIGURE 2.23 Low-temperature co-fixed ceramic (LTCC) ion trap. (a) Fabrication process for an LTCC CIT ring electrode; (b) experimental; and (c) simulated mass spectra for $CHCl_3$. (Reproduced from Chaudhary, A., van Amerom, F.H.W., Short, R.T., and Bhansali, S., *Int. J. Mass Spectrom.* 2006, *251*, 32–39. With permission from Elsevier.)

of $10 \pm 0.5\%$ and $9.5 \pm 0.5\%$ in the radial and z-axis dimensions, respectively. The authors noted that their processes could be improved to provide tolerances of 0.1% or better (achievable in industry). The performance of the trap was tested in the mass-selective instability mode [4] with an RF drive frequency of 3.926 MHz and a trapping voltage, V_{RF}, of 114 V_{0-p}; a typical peak width for chloroform fragments and the fragment ion peaks of PFTBA (from m/z 69 to 219) was *ca* 1.8 Th (FWHM), giving ($R = 122$ for 219 Th ($m/\Delta m$) (See Figure 2.23b and c).

2.3.5 STEREOLITHOGRAPHY APPARATUS (SLA)

Yu *et al.* and Fico *et al.* have used SLA rapid prototyping methodologies to manufacture a variety of full and reduced-scale RITs and RIT arrays, see Figure 2.24a [39,86]. Stereolithography is a branch of lithography capable of creating a 3D physical object from a 3D virtual object designed by CAD. Using SLA fabrication can provide complex monolithic 3D structures by computer-automated point-to-point and layer-by-layer UV laser polymerization of a pool of liquid monomer. As each

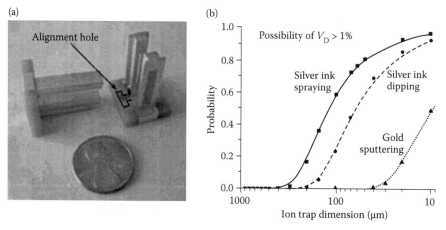

FIGURE 2.24 Stereolithographic apparatus (SLA) ion traps: (a) Disassembled two-piece SLA RIT. (Reproduced from Fico, M., Yu, M., Ouyang, Z., Cooks, R.G., and Chappell, W.J., *Anal. Chem.* 2007, *79*, 8076–8082. With permission from the American Chemical Society.); (b) Calculated probability of dimensional variation (V_D) exceeding a given threshold (>1%) for various SLA metallization processes and trap sizes. (Reproduced from Xu, W., Chappell, W.J., Cooks, R.G., and Ouyang, Z., *J. Mass Spectrom.* 2008, *44*, 3, 353–360. With permission from Wiley.)

layer is completed, a stage just under the surface is lowered incrementally, for example, in 50 μm-increments. During the photopolymerization process, each succeeding layer bonds with the preceding one. After UV curing, the SLA object is cured thermally for several hours. Advantages of prototyping with SLA include laser-defined precision (50–75 μm lateral resolution) and monolithic fabrication. Also, SLA allows the fabrication of complex array geometries, which cannot be machined conventionally. The RITs constructed using this method each had 5 mm × 4 mm 'stretched' geometries (*x* by *y*) and critical dimensions (x_0) of 5.0, 2.5, or 1.7 mm. A custom polymer, Nanoform 15120 (DSM Somos, New Castle, DE), was prepared with a high thermal stability to avoid potential deformation due to RF heating of the insulating material. The most difficult part of the process was metallization; unless performed carefully, the metal-on-polymer processing was prone to peeling or bubbling. Various methods of metallization were used, including gold sputtering, silver ink spraying followed by copper electroplating, and 'electroless' copper deposition followed by copper electroplating. A comparison of the empirically tested relative roughness (or dimensional variation, V_D) of several methods of SLA metallization demonstrates that the smoothest surfaces are achieved through sputtering [49] (Figure 2.24b). In order to prevent inter-electrode contact, either a mask or a sacrificial layer was incorporated prior to the metallization steps. The trap performances have been evaluated qualitatively using ESI-generated protonated arginine examined by mass-selective instability scanning with resonant ion ejection at $q_x = 0.83$ and with an RF drive frequency of 1.2 MHz, see Figure 2.25. Since the value of V_{RF} was reduced with physical scale, the smaller traps demonstrated poorer resolution due to shallower pseudopotential well-depths.

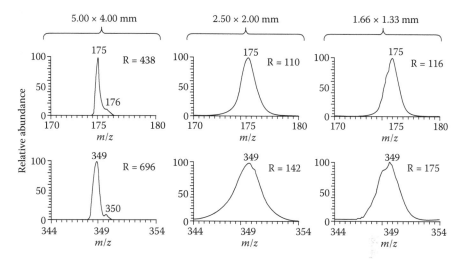

FIGURE 2.25 Comparison of the mass spectral quality obtained with the SLA ion traps for singly protonated (top, 175 Th) and doubly protonated ESI-generated arginine ions (bottom, 349 Th) as the '$x \times y$'-radial half-distances of the RIT vary. Resolution values are displayed next to the peaks and represent $m/\Delta m$ for FWHM. Data were obtained with helium buffer at a pressure of 7.5×10^{-5} Torr. (Reproduced from Fico, M., Yu, M., Ouyang, Z., Cooks, R.G., and Chappell, W.J., *Anal. Chem.* 2007, *79*, 8076–8082. With permission from the American Chemical Society.)

2.4 MULTIPLEXED ION TRAP ARRAYS

Multiplexed mass spectrometers, in which each instrument comprises more than one ionization source, mass analyzer, and/or detection system, have advantages over traditional mass spectrometers that analyze a single sample *via* a single ionization method each time the instrument is scanned. Depending upon the configuration, multiplexing can lead to gains in sensitivity, selectivity, throughput, and versatility of analysis. For applications that require high-throughput analysis of a large number of samples, such as combinatorial library screening [90], proteomics [91], and metabolomics [92], multiplexed analytical instruments lead to significant savings in time for sample batches; for this reason, it is advantageous to multiplex as much of the total experiment as possible.

To illustrate the concept of mass spectrometer multiplexing, a depiction of the seven possible multiplex configurations of the three main parts of the mass spectrometer-the ion source (S), the mass analyzer (A), and the detector (D) – is shown in Figure 2.26. This schematic depiction shows the instruments that could result conceptually from multiplexing one to all of the elements of the mass spectrometer. For simplification, only two of the multiplexed elements for one-element multiplexed systems are shown explicitly (that is, $S_n AD$, $SA_n D$, and SAD_n). The configurations end with the fully multiplexed mass spectrometer $(SAD)_n$, such that n ion source/analyzer/ion detector channels operate simultaneously. This configuration of n parallel channels processing n independent ion populations differs from using n separate instruments in terms of the space, hardware, and power savings associated with the use of a common vacuum system and common control electronics.

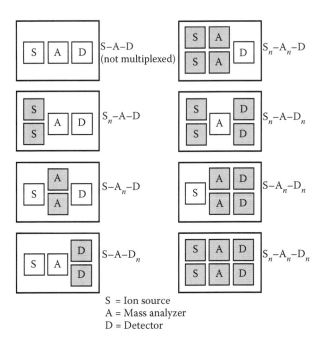

S = Ion source
A = Mass analyzer
D = Detector

FIGURE 2.26   Possible configurations of multiplexed mass spectrometers, where S, A, and D represent the ion source, mass analyzer, and detector, respectively.

Optimum versatility requires independent operation of all n mass analyzers but, in practice, the use of a single RF power supply and identical mass analyzer operation should be suitable for some applications [21].

One concern for multiplexed mass spectrometers in which ions travel in generally parallel paths between the ionization source and mass analyzer is cross-talk between adjacent channels of mass analysis. That is, the presence of signals due to compounds that are introduced into one channel appearing in another channel. There are two possible sources of cross-talk; in *neutral cross-talk*, neutral molecules pass from one channel of mass analysis (channel A) to another channel (channel B) resulting in ion/molecule reactions involving the neutrals from channel A and ions in channel B. In *ionic cross-talk*, ions cross channels either before or after mass analysis. These processes can occur in different regions of a multiplexed mass spectrometer from the ion sources to the detectors or anywhere in-between. Ions (or neutrals) crossing from one channel of mass analysis into another can either be detected in that channel, or they can undergo reactions (ion/ion or ion/molecule), leading to mis-characterization of the sample originally intended for the analytical channel. Ionic cross-talk can be caused by poor ion optics and can even occur in the detection system when ions ejected from one ion trap reach the detector of a different channel.

For the purposes of this chapter, we will focus on the multiplexing of ion trap mass analyzers into parallel or serial arrays. Figure 2.27 shows a schematic representation of parallel and serial combinations of pre-MS operations (e.g., sample introduction and/or ionization), MS operations (mass analysis), and post-MS operations

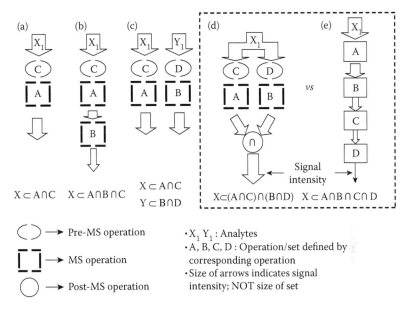

FIGURE 2.27   Depiction of parallel/serial combinations of MS operations, where ⊂ represents a possible operation and ∩ is the logic expression for 'and.' (a) Single MS operation preceded by single pre-MS operation. (b) Serial combination of two MS operations, coupled to single pre-MS operation. (c) Parallel operations on two different analytes; each MS operation is coupled serially to its own pre-MS operation. (d) Parallel operations on a single analyte, with post-detection signal processing. Each parallel channel consists of a pre-MS operation coupled serially to an MS operation. (e) Complete serial coupling of four (unspecified) operations. Note the small signal for (e) compared with parallel analysis, (d). (Reproduced from Tabert, A.M., Misharin, A.S., and Cooks, R.G., *Analyst*, 2004, *129*, 323–330. With permission from The Royal Society of Chemistry.)

(e.g., post-detection signal processing) [93]. Figure 2.27d illustrates the concept of parallel operations being performed on a single sample with summed post-detection signal processing. In this example, each parallel channel consists of a pre-MS operation, coupled serially to a single MS operation. Complete serial coupling of four (unspecified) operations is shown schematically in Figure 2.27e. Note the lower signal intensity for (Figure 2.27e) compared with parallel analysis, (Figure 2.27d), because, in addition to a serial configuration sampling only a single ion beam, serial transport of ions between traps is subject to the product of transfer, trapping, and ejection efficiencies of each trap [93].

2.4.1   CYLINDRICAL ION TRAP ARRAYS

This section discusses the parallel and serial coupling of CITs to enhance the sensitivity or selectivity of analysis, or to add versatility to the experiments that can be done. The simplified geometry and ease of machining of the CIT make it a good candidate for MS arrays; micromachined arrays of micro-scale CITs have been constructed (as detailed in Section 2.3).

2.4.1.1 Parallel Arrays of CITs of the Same Size

The conceptual end-point of a parallel array MS experiment used to examine a single sample would be for each parallel channel of a multiplexed instrument to employ a unique combination of sample introduction, ionization, mass selection, and ion detection steps in series, producing high selectivity for different analytes of interest in the sample. Such an instrument would be simpler than a set of independent instruments because many components (vacuum system, some electronics, and some of the signal readout system) would be shared in the single instrument. The signals from each channel could be processed in an integrated fashion, not independently. In this way, detection of particular analytes could require appropriate signals in all the parallel channels simultaneously. This single sample experiment is to be contrasted with the ideal version of a multiple sample experiment in which identical operations in parallel channels are performed simultaneously on different samples. In the discussion that follows both types of experiments, single sample and multiple sample, are presented, together with hybrid versions.

Tabert and co-workers developed a fully multiplexed CIT array mass spectrometer with four parallel channels each consisting of an ion source, mass analyzer, and ion detection system [18]. A multi-element external chemical ionization/electron ionization (CI/EI) source was coupled to a parallel array of CITs each of equal size (internal radius 2.5 mm), and the signal was recorded using an array of four miniature (2-mm inner diameter) electron multipliers. This parallel configuration allowed high-throughput experiments to be performed in which multiple samples were subjected to external EI, the separate ion populations being injected simultaneously into a CIT array and the mass spectra recorded using a four-channel data acquisition program written in LabVIEW Version 6.0.2i for this purpose. Ion and neutral cross-talk between adjacent channels of mass analysis was investigated. Each parallel channel of this instrument showed reasonable performance, providing a mass range of approximately m/z 50–500, a resolution of 1000 at m/z 300 (FWHM), and MS/MS ability. Figure 2.28 shows the results of simultaneous data acquisition of all four channels during EI of 1,3-dichlorobenzene (identical concentrations leaked into each channel's ionization region), displaying similar results for each channel. To demonstrate a high-throughput experiment in which differing analytes are used, 1,3-dichlorobenzene was introduced into channel **3**, bromobenzene was introduced into channel **2**, and acetophenone was introduced into channel **1**, with the results seen in Figure 2.29. For this experiment, channel **4** was left idle. Some neutral cross-talk is evident, but the proof of principle was established. An m/z 120 peak from acetophenone was observed in the 1,3-dichlorobenzene spectrum. This small amount of cross-talk is believed to be the result of neutral acetophenone molecules passing into the adjacent ionization volume. The molecular radical cation (m/z 120) of acetophenone (ionization energy (IE) 8.9–9.6 eV) is generated due to a charge exchange reaction with 1,3-dichlorobenzene (IE 9.10–9.28 eV). Resonance (zero enthalpy change) maximizes the charge exchange rate [94]. The absence of the fragment ion m/z 105 (electron affinity (EA) 9.6–10.5 eV) supports this interpretation. The values for IE and EA were obtained from the NIST Chemistry WebBook [95]. Differential pumping of the source and analyzer regions of the instrument would help to prevent undesired ion/

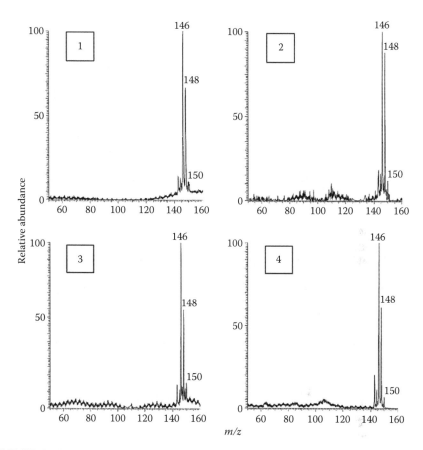

FIGURE 2.28   Normalized mass spectra obtained in each of the four channels for EI-MS analysis of 1,3-dichlorobenzene using a fully multiplexed CIT array mass spectrometer (for details, see text). (Reproduced from Tabert, A.M., Griep-Raming, J., Guymon, A.J., and Cooks, R.G. *Anal. Chem.* 2003, *75*, 5656–5664. With permission from the American Chemical Society.)

molecule reactions during trapping. Neutral cross-talk can only be avoided entirely by complete physical separation of the four channels. In further experiments using an RIT array and a new vacuum manifold design to eliminate all line-of-sight trajectories between channels, the cross talk was reduced to less than 5% of ion signal [20].

To increase flexibility in the type of ionization methods that can be utilized, a multiplexed, four-channel CIT array mass spectrometer capable of ambient ionization was constructed [43]. This instrument, equipped with ESI sources, comprised four complete parallel channels of mass analysis, two of which were operated during the experiments described here. All components of the instrument, with the exception of the ESI sources, were housed in a common vacuum manifold and operated using common control electronics. A schematic representation of the major electrical and vacuum components and their interconnections is shown in Figure 2.30. A two-stage, differentially pumped vacuum system transferred the ions created in the

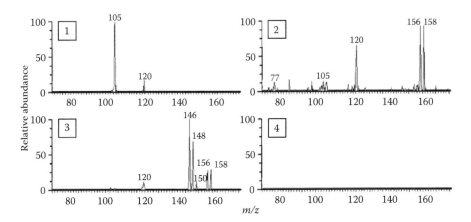

FIGURE 2.29 High-throughput experiment in which three channels of the parallel four-channel CIT array were used to analyze different chemical vapors. Acetophenone was introduced into channel **1**, bromobenzene into channel **2**, and 1,3-dichlorobenzene into channel **3**, with channel **4** left idle. Cross-talk contamination can be seen between the channels. (Reproduced from Tabert, A.M., Griep-Raming, J., Guymon, A.J., and Cooks, R.G. *Anal. Chem.* 2003, 75, 5656–5664. With permission from the American Chemical Society.)

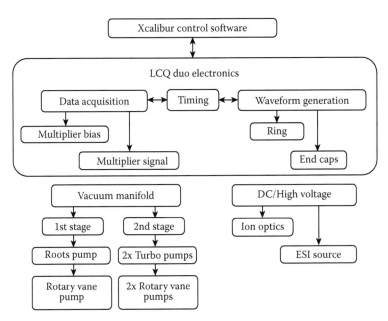

FIGURE 2.30 Schematic of the major electrical and vacuum components of the CIT array mass spectrometer capable of ambient sampling, depicting their interconnections. (Reproduced from Misharin, A.S., Laughlin, B.C., Vilkov, A., Takáts, Z., Ouyang, Z., and Cooks, R.G., *Anal. Chem.* 2005, 77, 459–470. With permission from the American Chemical Society.)

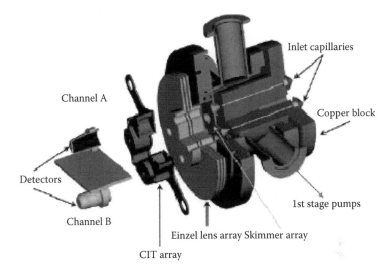

Channel A

Inlet capillaries

Copper block

Detectors

1st stage pumps

Channel B

Einzel lens array Skimmer array

CIT array

FIGURE 2.31 Schematic of the ion introduction system, ion optics, array of CIT analyzers, and detection system of the ambient sampling CIT array mass spectrometer. (Reproduced from Misharin, A.S., Laughlin, B.C., Vilkov, A., Takáts, Z., Ouyang, Z., and Cooks, R.G., *Anal. Chem.* 2005, *77*, 459–470. With permission from the American Chemical Society.)

ESI sources into the CIT array. The atmospheric interface contained stainless steel sampling capillaries that served to sample ions from the electrospray plume while limiting the gas flow into the vacuum system. The capillaries, 254 μm inner diameter by 80 mm in length, were held in position by a copper block, shown in Figure 2.31, in a coaxial arrangement with the apertures of an array of skimmers serving to limit gas transfer into the second vacuum stage. The CIT array utilized with this instrument was constructed of stainless steel. The four identical CITs were spaced symmetrically about a 28-mm-diameter circle as pictured in Figure 2.32. Each CIT had an inscribed radius (r_0) of 5.00 mm and a center-to-end-cap distance (z_0) of 4.93 mm. The apertures in the end-cap electrodes for ion entrance and egress were 2.0 mm in diameter. Furthermore, this array was specially machined to remove most of the unnecessary metal from both the ring and the end-cap electrodes, in an effort to reduce the total trap capacitance and, thus, the capacitive load on the RF drive electronics.

To investigate the performance of the multiplexed instrument in the case of small molecules, solutions of the amino acids arginine (MW = 174.20 u) and glutamine (MW = 146.14 u) were prepared and electrosprayed from the multi-channel source into the instrument operated in both single-channel and multi-channel modes. Figure 2.33a shows a mass spectrum recorded during a single-channel experiment in which a 10 mM arginine solution in methanol/water (1:1 v/v) containing 1% acetic acid was electrosprayed into channel A of the instrument. Correspondingly, Figure 2.33b shows the mass spectrum obtained during a single-channel experiment in which a 10 mM solution of glutamine in methanol/water (1:1 v/v) containing 1% acetic acid was electrosprayed into channel B. Both mass spectra are dominated by peaks corresponding to the protonated amino acids (nominally, 175 Th for arginine and 147 Th for

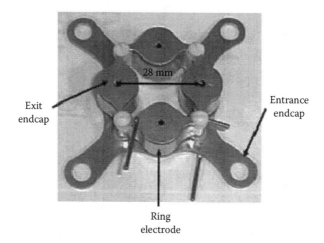

Exit
endcap

Entrance
endcap

Ring
electrode

FIGURE 2.32   Parallel array of improved CIT mass analyzers. The inner area of the array has been removed to reduce total capacitance of the device. (Reproduced from Misharin, A.S., Laughlin, B.C., Vilkov, A., Takáts, Z., Ouyang, Z., and Cooks, R.G., *Anal. Chem.* 2005, *77*, 459–470. With permission from the American Chemical Society.)

glutamine). In addition, Figure 2.33b shows the formation of the proton-bound dimer of glutamine at 293 Th. Panels c and d of Figure 2.33 show mass spectra obtained from channels A and B of the instrument, respectively, during the corresponding two-channel experiment. Long ion injection times (2 s), high concentrations of analytes, and high electron multiplier voltages were chosen to characterize the extent of cross-talk between channels of the instrument; that is, it was desirable to maximize the ion signal intensity in each channel and also to maximize the *S/N* to enhance cross-talk signals if present. Figures 2.33c and d both show ionic cross-talk between the channels, although to a relatively small degree (*ca.* 1% relative abundance).

To elucidate the sources of cross-talk present in both channels of mass analysis during the two-channel experiment, a single-channel experiment, where arginine was sprayed into channel A of the instrument and detected in both channels while nothing was sprayed into channel B, was performed. The observation of the same type of behavior in the level of cross-talk *versus* injection time for both the one and two-channel experiments over a wide range of injection times (100–5000 ms) suggests that the cross-talk is not due to neutral cross-talk followed by ion/molecule reactions, that is, proton transfer from protonated glutamine (PA = 224.1 kcal mol^{-1}) [96] to arginine (PA = 251.20 kcal mol^{-1}), see Figure 2.34 [96]. The thermochemical data suggest that the presence of the [glutamine +H]$^+$ ion signal in channel A of the instrument, in which arginine was analyzed during a two-channel experiment, could be caused only by ionic cross-talk in which ions from channel B are detected in channel A; this is because proton transfer from [arginine +H]$^+$ to glutamine is endothermic by 113 kJ mol^{-1}. The leveling-off of the curve in Figure 2.34 suggests that the traps reach their ion capacity limit at long injection times. The space-charge effects due to the large number of ions in the trap can influence the spread of ion trajectories during ejection and, hence, the amount of ionic cross-talk seen in the detection system, that is, the

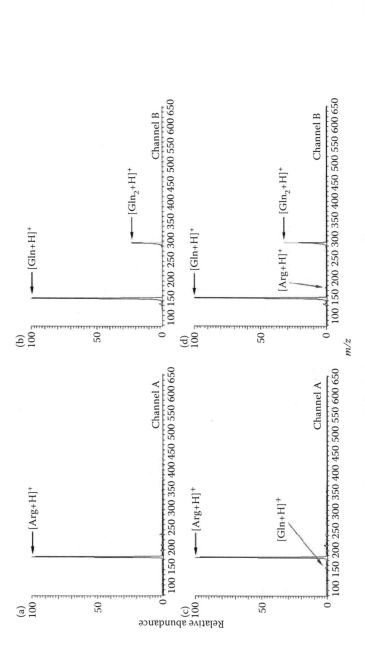

FIGURE 2.33 Performance of parallel array of CIT mass analyzers. (a) Single-channel experiment, 10 mM arginine (protonated molecule at m/z 175) electrosprayed into channel A; (b) single-channel experiment, 10 mM glutamine (protonated molecule at m/z 147 and protonated dimer at m/z 293) electrosprayed into channel B; (c) and (d) two-channel experiment in which 10 mM arginine and 10 mM glutamine are electrosprayed into channels A and B, respectively. The inner area of the array has been removed to reduce total capacitance of the device. Cross-talk is quantified as less than 1%. (Reproduced from Misharin, A.S., Laughlin, B.C., Vilkov, A., Takáts, Z., Ouyang, Z., and Cooks, R.G., *Anal. Chem.* 2005, 77, 459–470. With permission from the American Chemical Society.)

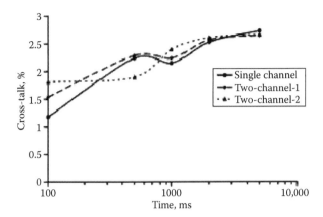

FIGURE 2.34 Dependence of the level of cross-talk *vs* injection time during one and two-channel experiments with a parallel CIT array mass spectrometer. (Reproduced from Misharin, A.S., Laughlin, B.C., Vilkov, A., Takáts, Z., Ouyang, Z., and Cooks, R.G., *Anal. Chem.* 2005, *77*, 459–470. With permission from the American Chemical Society.)

portion of ions that will reach a detector in another channel. For this instrument, the sources of cross-talk were identified and minimized, but this phenomenon is inherent to the overall design of the instrument, which provides opportunities for ion trajectories to cross into other channels in the ion optics and the ion detection system. Though zero cross-talk is desired, small amounts of cross-talk can be corrected numerically through post-acquisition data processing, as the information necessary to make the correction is available in the channels from which the cross-talk originated.

In summary, the two instruments described above show how parallel ion trap arrays can add both sensitivity and selectivity to the overall experiment. Multiple, independently operated MS channels allow a higher throughput, with either the same or different samples being run concurrently. With four separate channels, one could envision a single analyte experiment where mass spectra are recorded in both EI and CI mode, product ion mass spectra from an MS/MS experiment is gathered for structural elucidation, and an ion/molecule reaction is performed, all simultaneously in a single sampling run. Furthermore, coupling a multi-channel instrument with the flexibility of an atmospheric inlet allows ambient sampling of externally generated ions (97), which could prove useful to any field that requires mass spectral data collection of samples in their native environment.

2.4.1.2 Parallel Arrays of CITs of Different Size

Parallel arrays can be constructed also to allow novel mass spectrometric experiments. To demonstrate this, a multiplexed mass spectrometer in which each array element is a CIT of different size was constructed so that when the array is operated with a fixed RF potential, ions of different masses (or mass ranges) are stored in each trap [98]. By choosing the dimensions of each CIT element in the array, a multiple ion monitoring experiment can be performed, allowing mass-selected detection. Ion storage used either RF/DC (apex) isolation or the stored waveform inverse Fourier transform (SWIFT) method [99] for ion isolation. Of note is the fact that the array reduces the complexity of the electronics needed to operate the ion trap, which should make it suitable for use in a miniature mass spectrometer system. This CIT array was designed so that all ions could be ejected simultaneously using a fast DC pulse [100] applied to the entrance end-cap electrodes of all array elements. The ejected ions are detected using a position-sensitive detector that integrates the ion signal from each individual trap and yields the total ion intensity *versus* the CIT from which the signal was generated. Then, depending on the type of array used ('selected ion' or full mass spectrum), data analysis methods such as pattern recognition, partial least squares, or neural networks can be used to recognize compound classes or individual compounds.

To test the concept of mass analysis by varying trap dimensions, a two-CIT array was built with r_0 values of 5.00 and 4.00 mm and z_0 values of 4.61 and 3.71 mm, respectively. These sizes were chosen based on the criterion that a common isolation window of 10 Th or less be used for the isolation of diagnostic ions for each species of interest. *n*-Nonane and toluene were chosen as test compounds for evaluation of the array, and Figure 2.35 shows EI mass spectra for each compound: *n*-nonane in the 5.00 mm radius CIT (Figure 2.35a), as well as in the 4.00 mm radius CIT (Figure 2.35b), and toluene in the 4.00-mm-radius CIT (Figure 2.35c). All data were recorded with resonant ion ejection at 630 kHz, 6.0 V_{p-p}. Each trap was examined separately through application of –12 V DC to the exit end-cap electrode of the CIT not

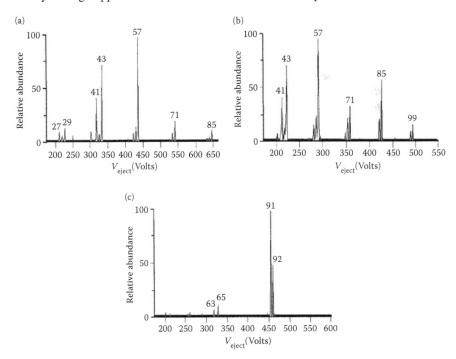

FIGURE 2.35 Electron impact mass spectrum of *n*-nonane recorded using an array consisting of two CIT elements of different size but each of optimized geometry: (a) Data from a single 5.0-mm-radius CIT; (b) data from a single 4.0-mm-radius CIT; and (c) mass spectrum of toluene from a 4.0-mm-radius CIT. (Reproduced from Badman, E.R. and Cooks, R.G., *Anal. Chem.* 2000, *72*, 5079–5086. With permission from the American Chemical Society.)

being examined. This arrangement prevented ions from leaving this trap and allowed independent accurate mass assignment and visualization of each trap's contents. The abscissa for each mass spectrum is reported in terms of the applied RF ramp voltage so that comparison between the ejection times of the ions from both traps can be made. In the range of 425–475 V_{RF}, it can be seen that the diagnostic ions from both CITs are ejected. This scale shows also the relative shift in ejection time as a result of the variation in the dimensions of the CITs. The isolated mass windows for each of the traps (Figure 2.36) show m/z 57 ions from n-nonane (in the 5.00 mm radius CIT) and m/z 91 and 92 ions from toluene (in the 4.00-mm-radius CIT). This figure demonstrates that variation in the trap dimensions can be used as a mass-selection device under conditions where the same RF and DC voltages are applied to all traps being analyzed. From Figures 2.35 and 2.36, it is apparent that although the ions of interest derived from the chosen model compounds are isolated, the actual isolation window ranges are larger than that necessary to isolate only ions of the two mass-to-charge ratios of interest. Of course, a small adjustment in geometry (for example, reducing the value of r_0 of the 4.00-mm CIT slightly) would bring the ions closer together in (a_z, q_z) space so that an even tighter isolation window could be used. This arrangement would provide a simple way of developing an automated compound-specific identification system. Simply by adding more CIT elements of appropriate size more diagnostic ions could be trapped providing, therefore, greater specificity. Hence, a multiple ion monitoring CIT array could be created that would be appropriate for the analysis of a specific compound or class of compounds. Such a system could be useful in industry,

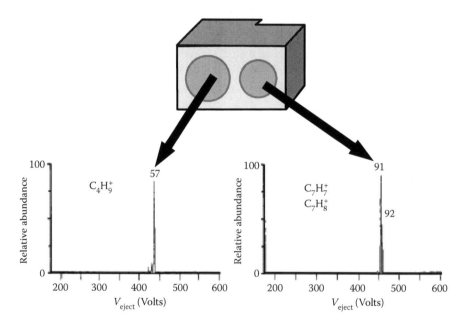

FIGURE 2.36 Mass spectra from CITs of different size under identical RF/DC isolation conditions. The 5.0 mm radius CIT traps m/z 57 from n-nonane, while the 4.0 mm radius CIT traps m/z 91 and 92 from toluene. (Reproduced from Badman, E.R. and Cooks, R.G., *Anal. Chem.* 2000, *72*, 5079–5086. With permission from the American Chemical Society.)

for example, to monitor for polycyclic aromatic hydrocarbons (PAHs) in factory environments in order to ensure compliance with permissible exposure limits.

2.4.1.3 Serial Arrays of CITs

Linear CIT arrays are of interest as devices in which the individual traps are used for ion accumulation, mass selection, ion reaction, and product ion analysis, as in a triple quadrupole mass spectrometer. Spatial separation of the ionization region from those used for mass analysis and ion/molecule reactions should have analogous advantages to those seen when internal ionization is replaced by external ionization for CI experiments on an ion trap instrument [101]. A serial device could also provide the basis for implementation of such alternative tandem mass spectrometric scan modes as precursor ion scans [102], that are used to screen for compounds that have the same sub-structure, an experiment which now can be carried out in ion traps only by reconstruction of the data taken using product ion scans [103]. Additional advantages of serial arrangements of trapping devices have been noted in the introductory discussion in Section 2.1.2.

A serial array consisting of two identical miniature CITs (radius = 5.0 mm) was built and operated with a single RF source and independent control of the DC potentials applied to each of the four end-cap electrodes [30]. Stored ions were transferred from one CIT to the other using DC pulses in an effort to determine the transfer efficiency, the accompanying CID, and the temporal distribution of ion transference. During EI, the electrons were driven by the electric field between the end-cap electrodes and the filament into both ion traps. By applying a DC pulse to the end-cap electrode of one of the CITs, all the ions trapped in it could be ejected before the RF amplitude applied to this trap was scanned in order to detect the ions in the other CIT. To transfer the ions ejected from the first CIT into the second CIT and trap them there, a DC pulse was applied to the exit end-cap electrode of the second CIT to prevent the ions from simply passing through it. A schematic of this serial arrangement of CITs can be seen in Figure 2.37.

In ion transfer experiments, attempts were made to transfer ions trapped in CIT **1** to CIT **2** where they were again trapped prior to detection. This transfer was achieved by using pulsed DC potentials applied to the gate mesh electrodes. To compare ion transfer under different conditions, including different DC pulse amplitudes and widths, an experiment was performed, the results of which formed a standard for comparison. In this experiment, a separate DC pulse was applied on the end-cap electrode **c** in order to empty CIT **2** before the RF scan was performed. The ions in CIT **1** were ejected during the subsequent RF scan and some of them passed through the end-cap electrodes **b**, **c**, and **d** to reach the electron multiplier detector (Figure 2.38a). The intensities of the peaks are proportional to the total number of the ions in CIT **1** and are approximately constant at fixed ionization time, low-mass cut-off (a measure of the RF voltage amplitude), sample, and helium pressure. The transfer of ions from CIT **1** to CIT **2** was then achieved by applying simultaneously DC pulses **a** and **d**, after ion ejection from CIT **2** using pulse **c**. The ions trapped in CIT **2** by this procedure were ejected subsequently in the course of an RF scan and detected using the electron multiplier (Figure 2.38b). The peak intensities in the mass spectrum shown in Figure 2.38b are proportional to the abundance of the ions trapped in CIT **2**. A comparison between the peak intensities in Figure 2.38b and a

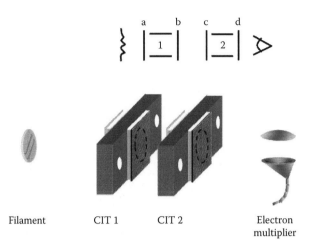

FIGURE 2.37   Schematic of a serial CIT array (for details, see text). (Reproduced from Ouyang, Z., Badman, E.R., and Cooks, R.G., *Rapid Commun. Mass Spectrom.* 1999, *13*, 2444–2449. With permission from Wiley.)

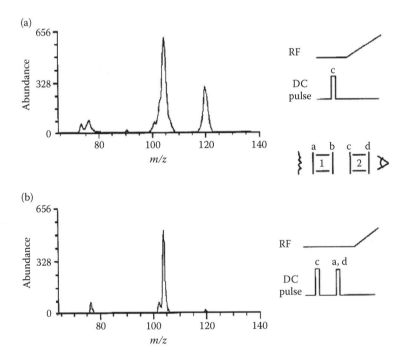

FIGURE 2.38   Ion transfer within the serial CIT array: (a) Mass spectrum of ions trapped in CIT **1**; (b) mass spectrum of ions transferred to CIT **2**; DC pulse **a**, V = 187.0 V, width = 2.84 μs; DC pulse **d**, V = 140 V, width = 2.62 μs. (Reproduced from Ouyang, Z., Badman, E.R., and Cooks, R.G., *Rapid Commun. Mass Spectrom.* 1999, *13*, 2444–2449. With permission from Wiley.)

gives an approximate value of the relative transfer efficiency for the ions of each m/z-value. The total efficiency is approximately 50% under these specific conditions.

Overall, this serial instrument demonstrated that the transfer efficiency is highly dependent upon the voltage of the ejection DC pulse, and on the voltage and width of the repulsive DC pulse. By optimizing these variables, novel experiments can be accomplished.

2.4.2 RECTILINEAR ION TRAP ARRAYS

The advantages of RITs in ion capacity, resolution, and efficiency of trapping externally generated ions are such that they are now the elements of choice in multiplexed arrays in this laboratory. In this section, the details of parallel and serial coupling of RITs are discussed as a means of illustrating the importance of this research area.

2.4.2.1 Parallel Arrays of RITs

Tabert et al. expanded on their work with EI/CI ionization in CIT arrays [20] to utilize RITs so as to increase mass spectral resolution and ion capacity. Of note, this instrument employed a single vacuum chamber that was designed to minimize the cross-talk between its four parallel ion source/mass analyzer/detector channels, as seen in Figure 2.39. Each channel operated using a common set of control electronics, including a single RF amplifier and transformer coil. High-throughput analysis can be achieved by operating all four channels simultaneously, with single and multiple stages of mass analysis. Figure 2.40 shows data obtained from the simultaneous

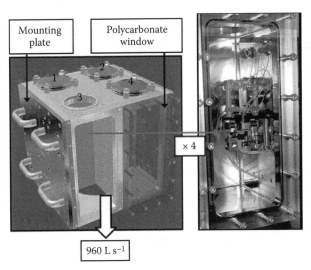

FIGURE 2.39 Four-channel, parallel RIT mass spectrometer. Mechanical drawing of stainless steel vacuum manifold designed to minimize cross-talk between the four parallel channels of analysis, labeled 1–4. An elongated cross was welded in place to divide the chamber into four quadrants while still permitting a single turbomolecular drag pump to be used. One source/analyzer/detector assembly is pictured as viewed through the polycarbonate window of one channel. (Reproduced from Tabert, A.M., Goodwin, M.P., Duncan, J.S., Fico, C.D., and Cooks, R.G., *Anal. Chem.* 2006, *78*, 4830–4838. With permission from the American Chemical Society.)

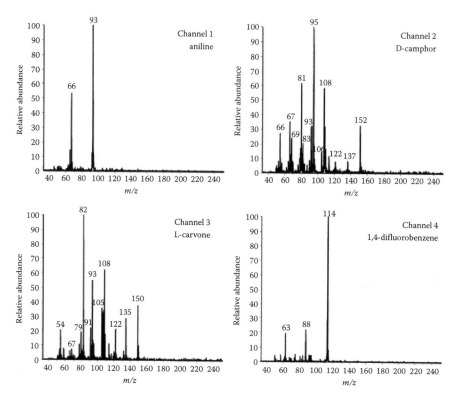

FIGURE 2.40 Mass spectra recorded using a four channel, parallel RIT mass spectrometer during simultaneous analysis of aniline, D-camphor, L-carvone, and 1,4-difluorobenzene in channels **1**, **2**, **3**, and **4**, respectively. (Reproduced from Tabert, A.M., Goodwin, M.P., Duncan, J.S., Fico, C.D., and Cooks, R.G., *Anal. Chem.* 2006, *78*, 4830–4838. With permission from the American Chemical Society.)

mass analysis of aniline (MW = 93.13 u), D-camphor (MW = 152.23 u), L-carvone (MW = 150.22 u), and 1,4-difluorobenzene (MW = 114.09 u) in channels **1**, **2**, **3**, and **4**, respectively. The *m/z* 114 peak of 1,4-difluorobenzene (introduced into channel **4**) is present as a cross-talk peak in the spectra of D-camphor in channel **2** and L-carvone in channel **3**. The ratio of the peak area of *m/z* 114 to the total peak area in channel **2** is 1.8%, while that in channel **3** was calculated to be 1.3%. As expected, there is a correlation between volatility of each species and its probability of cross-talk, as high volatility species are more likely to pass in the gas phase into the ionization volume of other channels. While the problem of cross-talk was not remedied totally in this instrumental design, it was reduced significantly in comparison with the earlier CIT effort [19] discussed in Section 2.4.1.1.

Fico *et al.* have constructed arrays of miniature RITs using SLA [104]. The fabrication process, used previously for individual traps, [36,89] allows rapid prototyping of complex arbitrary geometries which are not readily machinable with conventional methods. A 12-channel array was built consisting of RITs spaced about a circle with the *z*-axis of each trap oriented orthogonally to the plane of the circle. Tangential orientation

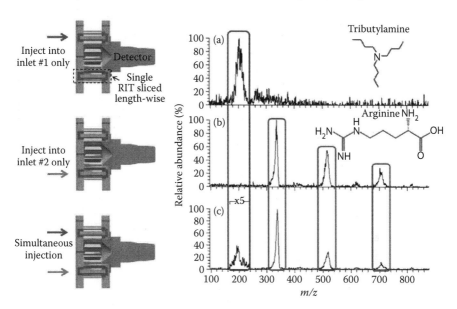

FIGURE 2.41 High-throughput experiment using two channels of a 12-channel RIT array fabricated by SLA: (a) Analysis of protonated tributylamine from a single trap; (b) Analysis of protonated arginine clusters from a single trap; (c) simultaneous analysis of protonated tributylamine and protonated arginine clusters introduced through different inlets and analyzed by two distinct traps. (Reproduced from Fico, M., Maas, J.D., Smith, S.A., Costa, A.B., Chappell, W.J., and Cooks, R.G., *Analyst,* 2009, *134*, 1338–1347. With permission from the Royal Society of Chemistry.)

of the x-electrodes with respect to the array z-axis allows ion ejection into a central cavity, wherein an electron multiplier is positioned for ion detection (Figure 2.41). Hence, an array of this design is both space-saving and minimizes the number of detectors needed for such a multi-channel array. The 'stretched' RITs had geometric half-distances of $x_0 = 1.7$ mm, $y_0 = 1.3$ mm, and $z_0 = 8.3$ mm, making them one-third-scale of the original RIT [14]. To simplify the assembly and electrode positioning, only two parts were fabricated, each containing both sections needed for either the x- or y-electrodes, respectively. The inner and outer x-electrodes were isolated from each other to allow application of a dipolar excitation waveform for mass-selective instability scanning. Metallization of the polymer to define the electrodes occurred either through gold sputtering, silver spray-paint followed by gold electroplating, or copper electroless deposition followed by gold electroplating. The end-cap electrodes consisted of either copper plates or PCBs, which were isolated electrically and fixed by screws to the array assembly. The device resided in a single-stage vacuum with air buffer at pressures of 15–20 mTorr. Operating parameters included an RF drive frequency of 760 kHz at a scan rate of 12,500 Th s^{-1}. An m/z range of 2000 Th was achievable using an RF maximum amplitude of only 158 V$_{0-p}$. Full-scan and tandem mass spectrometry (MS/MS) data were recorded for a variety of analyte ions, including the peptide bombesin seen in Figure 2.42. Peak widths of several Th were typical (FWHM) with S/N better than 100; such performance values are expected to improve with operation at higher frequencies,

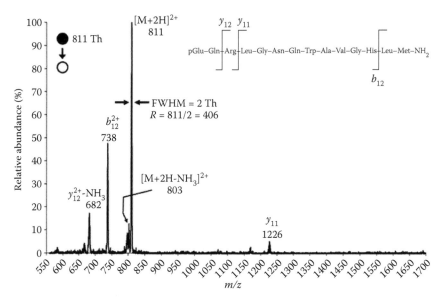

FIGURE 2.42 MS/MS analysis of the doubly-charged tetradecapeptide bombesin performed in a single miniature SLA fabricated RIT ($x_0 = 1.7$ mm). (Reproduced from Fico, M., Maas, J.D., Smith, S.A., Costa, A.B., Chappell, W.J., and Cooks, R.G., *Analyst,* 2009, *134*, 1338–1347. With permission from the Royal Society of Chemistry.)

higher RF voltages, and lower manifold pressures. Operation of the device for mass analysis of species generated by one and two ESI sources and various configurations of the atmospheric interface (including multiple inlets) has been demonstrated (see Figure 2.41). In addition to exhibiting the capability of arbitrary complex geometries accessible through SLA rapid prototyping, such traps of modestly reduced scale allow low-power operation and have the potential for high-throughput mass analysis.

A four-channel multiplexed mass spectrometer with RIT analyzers was designed and built to incorporate an atmospheric pressure interface [105], similar to the work with CIT arrays by Misharin *et al.* described previously [19]. All the traps and ion optics were housed in a single, four-chamber vacuum manifold, with each chamber sharing a common pumping system. The mass spectrometer was equipped with four atmospheric pressure ionization sources, with each source providing a different ion type (positive ESI, negative ESI, positive APCI, and negative APCI). Simultaneous data acquisition capabilities were demonstrated with this four-channel configuration.

During the characterization of this instrument, it was observed that due to variation in geometries of the ion traps as a result of machining and assembling tolerances, ions of the same *m/z* ratio can appear at different apparent *m/z*-values in different traps (due to differences in effective ejection potentials, indicating discrepancies in electrode finish and/or trap assembly alignment). In addition to peak mis-assignment, this issue is also detrimental to the performance of isolation and activation steps. To solve this problem, a supplementary DC voltage was applied to the individual traps and scanned along with the main RF scan. With careful calibration of the supplementary DC potential, the secular frequency of ions can be changed in such a way that the

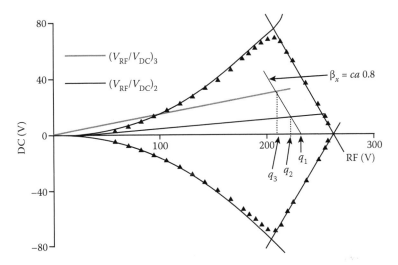

FIGURE 2.43 Procedure used for channel-to-channel mass correction in a high performance four channel array RIT mass spectrometer showing supplementary DC offset scans used in conjunction with RF scan (for details, see text). (Adapted by R.E. March from Kothari, S., Song, Q., Xia, Y., Fico, M., Taylor, D., Amy, J.W., Stafford, G., and Cooks, R.G., *Anal. Chem.* 2009, *81*, 1570–1579. With permission from the American Chemical Society.)

working point for ions of any given m/z is located on the same iso-β_x line in all the traps, resulting in accurate mass assignment. The ability to add this supplementary DC potential to an individual trap was achieved by making a coil with four filars, and applying an independently controlled DC potential (V_{DC}) to each filar. The coil allowed independent supplemental DC voltages to be applied to each trap such that the analyzers each had their own V_{DC}/V_{RF} operating lines. With such independent control, a common iso-β_x line could be established so that ions having the same value of m/z in all traps are excited at or near the same frequency. An example of this configuration is depicted in Figure 2.43 for three groups of ions of the same m/z ratio, one group in each of three ion traps, that appear at three different apparent m/z-values: q_1 in trap #1, q_2 in trap #2, and q_3 in trap #3. The line for $\beta_x = ca$ 0.8 passes through q_1. In order to eject ions at q_2 corresponding to at $\beta_x = ca$ 0.8, a DC ramp is applied at $(V_{RF}/V_{DC})_2$. In order to eject ions at q_3 corresponding to $\beta_x = ca$ 0.8, a higher ramped DC potential is applied at $(V_{RF}/V_{DC})_3$. Note that $(V_{RF}/V_{DC})_3 <(V_{RF}/V_{DC})_2$. In this manner, with the V_{RF} alone applied to trap #1, $(V_{RF}/V_{DC})_2$ applied to trap #2, and $(V_{RF}/V_{DC})_3$ applied to trap #3, all of the ions in each trap should be ejected at points along the iso-β_x-line equal to ca 0.8. After application of these ramped DC potentials with the newly designed coil, the mass inaccuracy among the four traps was reduced to <0.5 u.

Four channels of simultaneous mass analysis with different ionization modes offer the user the ability to generate and to analyze more data on a single sample in less time. A number of experiments were performed to demonstrate these abilities. One of these experiments is the mapping of lipids from an *Escherichia coli* extract. Figure 2.44 shows data (analog) acquired simultaneously from each of the four channels for the analysis of lipid extract from *E. coli* using four distinct ionization

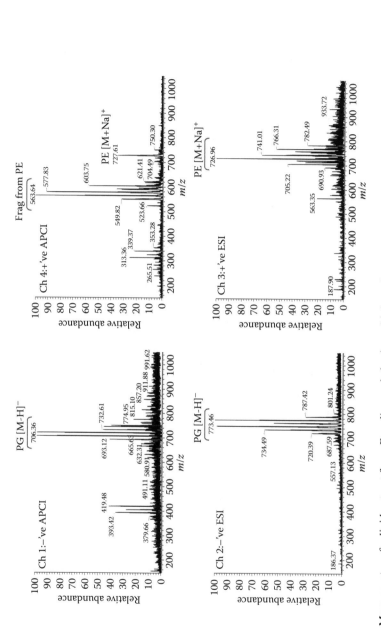

FIGURE 2.44 Mass spectra of a lipid extract from *E. coli* recorded using a high performance multiplexed four-channel RIT array instrument. PE: phosphoethanolamine; PG: phosphatidylglycerol. Lipid solution: 1 mg/mL of total lipid extract from *E. coli* (bought from Avanti) in 50:50 CH₃OH/ CHCl₃. For ESI, the analyte solution is pumped at 3 μL min⁻¹ and for APCI at 8 μL min⁻¹. (Reproduced from Kothari, S., Song, Q., Xia, Y., Fico, M., Taylor, D., Amy, J.W., Stafford, G., and Cooks, R.G., *Anal. Chem.* 2009, *81*, 1570–1579. With permission from the American Chemical Society.)

modes, that is, +/– APCI and +/– ESI. It is evident from the data in Figure 2.44 that complementary information can be obtained from different ionization modes. Therefore, by combining all the information from the four channels, comprehensive lipid mapping can be achieved in a high-throughput fashion.

2.4.2.2   Serial RITs and Ion Soft-Landing

Recently, Song *et al.* have coupled two RITs into a serial array, see Figure 2.45 [23]. The purpose of this work was to enable single or multi-species isolation from an ionic mixture to be injected and passed as a continuous beam. The first RIT, Q1, was operated in either a filter-only mode or used to trap, filter, and analyze. The second RIT, Q2, was used either to analyze the processed product passing from Q1 or to act as an RF-only ion guide for selected ions to be soft-landed at a surface just beyond Q2. By applying a SWIFT waveform to Q1, resonant excitation resulting in ejection occurs for all ions except those which are of interest for further processing; the ions retained will have experienced no excitation as the waveform exhibited zero amplitude for their secular frequencies. A multiple ionic species selection mode, termed 'summed product ion scan', was implemented and represents a variant form of product ion scan. Such multi-species selection allows several experiments to be carried out, including selection of ions from the same precursor compound which differ in charge state (e.g., $[M+nH]^{n+}$; $n \geq 1$), oligomeric form (e.g., $[nM+H]^{+}$; $n \geq 1$), or ionizing agent (e.g., $[M+H]^{+}$, $[M+Na]^{+}$, $[M+K]^{+}$, etc.).

The primary purpose for developing the continuous waveform isolation methodology was for ion soft-landing (SL) applications [106–110]. For a typical SL experiment,

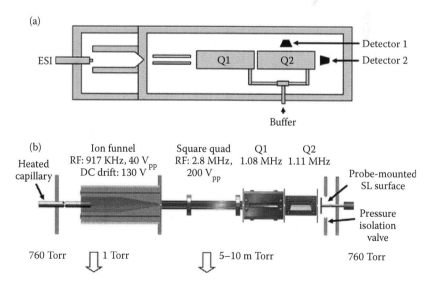

FIGURE 2.45   Serial tandem RIT instrument used for mass selection during continuous ion transmission: (a) Schematic showing the two RITs (Q1 and Q2); (b) detailed depiction of the ion optics and analyzers, including ion funnel, and high transfer efficiency ion guide. (Reproduced from Song, Q., Smith, S.A., Gao, L., Volný, M., Ouyang, Z., and Cooks, R.G., *Anal. Chem.* 2009, *81*, 1833–1840. With permission from the American Chemical Society.)

samples are first ionized in the gas phase followed by m/z-selection and deposited subsequently as the intact purified species onto a surface at hyperthermal energies. An SL experiment is characterized by rapid separations (milliseconds timescale), high resolving capabilities, and the non-destructive collection of the purified species of interest at a surface for subsequent recovery or analysis. Ion soft-landing has potential use for applications in fields as diverse as protein chip production, [108,111] enantiomer amplification [112], catalyst surface fabrication [39], thin film organic electronics [109], and covalent surface modification for sensor preparation [113] or bio-compatible surface passivation for medical implants [114]. Compared to traditional chromatographic or electrophoretic separation methods, mass spectrometry has superior resolving power (particularly useful for combinatorial chemistry which examines large batteries of structurally similar compounds), rapid isolation speed, and extremely high isolation efficiency. The underlying reason for SL technology not being used widely as yet is the small final yield (typically *ca* 1% or less), which is a direct result of the overall inefficiency inherent in the ionization, ion transportation, mass analysis, and ion collection processes.

The RIT has been adopted as a mass analyzer in soft-landing instruments, where its resolving power meets the requirements of many ion SL experiments. The high-order fields introduced in the RIT by the simplified geometry allow the use of higher working pressures up to 50 mTorr (which may be of importance to preserve conformations of biologically active species in the ionic state) [38]; in addition, operating at elevated pressures helps to reduce the length of the ion optical path and hence increase the ion current flux. However, despite the above advantages associated with the use of ion traps, there are several disadvantages. Previous ion trap SL instrumentation suffered low duty cycle (the extent to which the ion beam is used) due to a methodology which involved sequential trapping, mass-selection, and DC pulse-out of the purified species to a surface [108,111]. In order to provide a higher ion flux into the trap, lower injection times were used, but this resulted in lower duty cycles as the mass selection steps could not be reduced proportionately. Hence, a continuous waveform isolation methodology was developed which provided high duty cycle (100%) while also allowing high-pressure operations. Furthermore, the method was designed to enable an arbitrary number of m/z-values to be selected and collected in a simultaneous fashion; this is in contrast with traditional ion isolation methods used by quadrupole SL instruments, such as SWIFT and mass-selective stability scanning (that is, RF/DC scanning), where ions of one m/z-value can be selected. A good example of the utility of multi-m/z selection is for proteins ionized by ESI; for SL methods limited to selecting only one species at a time, there is a reduction of the total potential SL yield when multiple charge states exist for large species ionized by ESI in that only one charge state can be selected at a given time.

To implement the continuous waveform RIT method, a system comprising two serial RITs ($x_0 = 5$ mm; $y_0 = 4$ mm; $z_{total} = 40$ mm) was constructed. Each analyzer had an independent RF power supply with a frequency of *ca* 1 MHz. A manifold pressure of 3–5 mTorr with helium buffer gas was used. The multi-species isolation experiment was demonstrated on a mixture of drugs, see Figure 2.46. Although it was found that the ion populations entering Q1 experienced an insufficient number

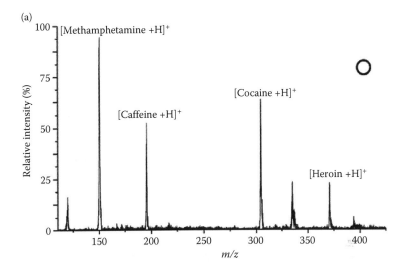

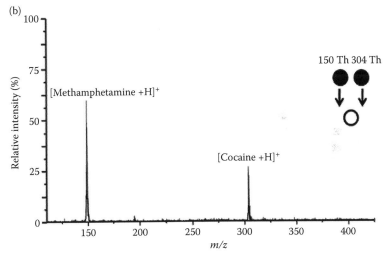

FIGURE 2.46 Mass selection during continuous ion transmission in a serial tandem RIT instrument capable of atmospheric pressure ionization. As indicated by the symbol, two ions, protonated methamphetamine and protonated cocaine are selected from a drug mixture: (a) drug mixture prior to selection; (b) post-selection mass analysis. (Reproduced from Song, Q., Smith, S.A., Gao, L., Volný, M., Ouyang, Z., and Cooks, R.G., *Anal. Chem.* 2009, *81*, 1833–1840. With permission from the American Chemical Society.)

of secular cycles for complete isolation, this situation could be remedied with higher frequencies or longer residence times (achievable by lengthening the device or the addition of a Brubaker lens). Transfer efficiencies could be improved also by implementing phase-matching between the two ion traps. Furthermore, transfer efficiencies between Q1 and Q2 could be improved if Q1 had symmetric dimensions in x and y; the existing '5 × 4' dimensions for Q1 were meant to facilitate analytical performance during radial ejection but, unfortunately, they introduce a non-zero effective potential along the trap axis equivalent to *ca* 30% of the applied RF potential, as calculated using SIMION 8.0 (Scientific Instrument Services, Ringoes, NJ).

2.5 CONCLUSIONS AND FUTURE OUTLOOK

The ideal miniature ion trap has the following attributes: (i) low power consumption; (ii) high pressure operation; (iii) sufficient trapping capacity and pseudopotential well-depth to provide the desired performance; (iv) compact size; (v) easy integration with other system components; and (vi) values of V_{RF} and electrode spacing which will not initiate electrical discharge at any pressure. Though 3D traps are attractive for their ability to eject ions axially, 2D geometries should also be considered, as they have been demonstrated to have higher trapping efficiencies, trapping capacities, and extraction efficiencies when compared to commercial 3D traps; in fact, 2D traps have been reported to be six to eight times more sensitive than conventional 3D traps [5,6]. Moreover, as the trapping field volume in 2D traps is nominally independent of the *z*-dimension, an arbitrarily long micro-scale 2D trap can be envisioned; such a trap in an array would reduce dead space otherwise occupied by electrodes for adjacent traps. This reduction in dead space is achieved further through the use of shared-electrode trap array geometries. Simplified IC-defined fabrication, for example as used in the halo trap [15], could afford the advantages of easy ion injection/ejection and could lessen the imprecision of potential surfaces due to errors in microfabrication, an issue with great significance for traps of small scale. Should micro ion traps prove to be practical at elevated pressures, miniature pumps capable of low-Torr operation could be used [115]. To prevent RF breakdown at high pressures, particularly in the regime at which Paschen curves 'bottom out', frequency-scanning methods of mass analysis could be used which do not involve high voltage RF power; that is, scanning the auxiliary frequency at constant V_{RF}.

The future of micro ion traps and trap arrays includes, perhaps, applications to *in vivo* medical diagnostics, hand-held or remote battery-powered environmental monitoring, and quantum computing. An all-in-one integrated system design is likely to have many or all system components and circuitry on a single chip. If predictions based on scaling laws hold true, these devices will perform analyses at near-ambient pressures, eliminating the need for high-power vacuum pumps.

REFERENCES

1. Bier, M.E.; Syka, J.E.P. Ion trap mass spectrometer system and method. *U.S. Patent* 1995, *5*, 420,425.
2. March, R.E. Quadrupole ion trap mass spectrometry: Theory, simulation, recent developments and applications. *Rapid Commun. Mass Spectrom.* 1998, *12*, 1543–1554.
3. Douglas, D.J.; Frank, A.J.; Mao, D.M. Linear ion traps in mass spectrometry. *Mass Spectrom. Rev.* 2005, *24*, 1–29.
4. Stafford, G.C.; Kelley, P.E.; Syka, J.E.P.; Reynolds, W.E.; Todd, J.F.J. Recent improvements in and analytical applications of advanced ion trap technology. *Int. J. Mass Spectrom. Ion Processes* 1984, *60*, 85–98.
5. Hager, J.W. A new linear ion trap mass spectrometer. *Rapid Commun. Mass Spectrom.* 2002, *16*, 512–526.

6. Schwartz, J.C.; Senko, M.W.; Syka, J.E.P. A two-dimensional quadrupole ion trap mass spectrometer. *J. Am. Soc. Mass Spectrom.* 2002, *13*, 659–669.
7. Badman, E.R.; Cooks, R.G. Special feature: Perspective-miniature mass analyzers. *J. Mass Spectrom.* 2000, *35*, 659–671.
8. Cotter, R.J. Time-of-Flight mass spectrometry-basic principles and current state. In *Time-of-Flight Mass Spectrometry.* American Chemical Society: Washington, DC, 1994; Vol. 549, pp. 16–48.
9. Kaiser, R.E.; Cooks, R.G.; Moss, J.; Hemberger, P.H. Mass range extension in a quadrupole ion-trap mass spectrometer. *Rapid Commun. Mass Spectrom.* 1989, *3*, 50–53.
10. Paul, W.; Steinwedel, H. Apparatus for separating charged particles of different specific charges. *U.S. Patent* 1960, 2,939,952.
11. Ouyang, Z.; Cooks, R.G. Miniature mass spectrometers. *Annu. Rev. Anal. Chem.* 2009, *2*, 187–214.
12. Wu, G.X.; Cooks, R.G.; Ouyang, Z. Geometry optimization for the cylindrical ion trap: Field calculations, simulations and experiments. *Int. J. Mass Spectrom.* 2005, *241*, 119–132.
13. Badman, E.R.; Johnson, R.C.; Plass, W.R.; Cooks, R.G. A miniature cylindrical quadrupole ion trap: Simulation and experiment. *Anal. Chem.* 1998, *70*, 4896–4901.
14. Ouyang, Z.; Wu, G.X.; Song, Y.S.; Li, H.Y.; Plass, W.R.; Cooks, R.G. Rectilinear ion trap: Concepts, calculations, and analytical performance of a new mass analyzer. *Anal. Chem.* 2004, *76*, 4595–4605.
15. Austin, D.E.; Wang, M.; Tolley, S.E.; Maas, J.D.; Hawkins, A.R.; Rockwood, A.L.; Tolley, H.D.; Lee, E.D.; Lee, M.L. Halo ion trap mass spectrometer. *Anal. Chem.* 2007, *79*, 2927–2932.
16. Blain, M.G.; Riter, L.S.; Cruz, D.; Austin, D.E.; Wu, G.X.; Plass, W.R.; Cooks, R.G. Towards the hand-held mass spectrometer: Design considerations, simulation, and fabrication of micrometer-scaled cylindrical ion traps. *Int. J. Mass Spectrom.* 2004, *236*, 91–104.
17. Wu, G.X.; Cooks, R.G.; Ouyang, Z.; Yu, M.; Chappell, W.J.; Plass, W.R. Ion trajectory simulation for electrode configurations with arbitrary geometries. *J. Am. Soc. Mass Spectrom.* 2006, *17*, 1216–1228.
18. Tabert, A.M.; Griep-Raming, J.; Guymon, A.J.; Cooks, R.G. High-throughput miniature cylindrical ion trap array mass spectrometer. *Anal. Chem.* 2003, *75*, 5656–5664.
19. Misharin, A.S.; Laughlin, B.C.; Vilkov, A.; Takáts, Z.; Ouyang, Z.; Cooks, R.G. High-throughput mass spectrometer using atmospheric pressure ionization and a cylindrical ion trap array. *Anal. Chem.* 2005, *77*, 459–470.
20. Tabert, A.M.; Goodwin, M.P.; Duncan, J.S.; Fico, C.D.; Cooks, R.G. Multiplexed rectilinear ion trap mass spectrometer for high-throughput analysis. *Anal. Chem.* 2006, *78*, 4830–4838.
21. Kothari, S.; Taylor, D.; Amy, W.J.; Stafford, G.; Cooks, R.G. Development of a multiplexed high throughput mass spectrometer. *Proc. 55th ASMS Conference on Mass Spectrometry and Allied Topics*, Indianapolis, IN, June 3–7, 2007.
22. Zhang, C.; Chen, H.W.; Guymon, A.J.; Wu, G.X.; Cooks, R.G. Instrumentation and methods for ion and reaction monitoring using a non-scanning rectilinear ion trap. *Int. J. Mass Spectrom.* 2006, *255*, 1–10.
23. Song, Q.; Smith, S.A.; Gao, L.; Volný, M.; Ouyang, Z.; Cooks, R.G. Mass-selection of ion beams using waveform isolation in radiofrequency quadrupoles. *Anal. Chem.* 2009, *81*, 1833–1840.
24. Schwartz, J.C.; Syka, J.E.P.; Quarmby, S. A new dual cell linear ion trap configuration for improved quadrupole ion trap performance. *Proc. 56th ASMS Conference on Mass Spectrometry and Allied Topics*, Denver, CO, June 1–5, 2008.

25. Langmuir, D.B.; Langmuir, R.V.; Shelton, H.; Wuerker, R.F. Containment device. *U.S. Patent* 1962, 3,065,640.

26. Benilan, M-N.; Audoin, C. Confinement d'ions par un champ electrique de radio-frequence dans une cage cylindrique. *Int. J. Mass Spectrom. Ion Phys.* 1973, *11*, 421–432.

27. Fulford, J.E.; March, R.E.; Mather, R.E.; Todd, J.F.J.; Waldren, R.M. The cylindrical ion trap—a theoretical and experimental study. *Can. J. Spectros.* 1980, *25*, 4, 85–97.

28. March, R.E.; Todd, J.F.J. *Quadrupole Ion Trap Mass Spectrometry*. 2nd ed.; John Wiley & Sons: Hoboken, NJ, 2005.

29. Wells, J.M.; Badman, E.R.; Cooks, R.G. A quadrupole ion trap with cylindrical geometry operated in the mass selective instability mode. *Anal. Chem.* 1998, *70*, 438–444.

30. Ouyang, Z.; Badman, E.R.; Cooks, R.G. Characterization of a serial array of miniature cylindrical ion trap mass analyzers. *Rapid Commun. Mass Spectrom.* 1999, *13*, 2444–2449.

31. Kornienko, O.; Reilly, P.T.A.; Whitten, W.B.; Ramsey, J.M. Micro ion trap mass spectrometry. *Rapid Comm. Mass Spectrom.* 1999, *13*, 50–53.

32. Cruz, D.; Chang, J.P.; Fico, M.; Guymon, A.J.; Austin, D.E.; Blain, M.G. Design, microfabrication, and analysis of micrometer-sized cylindrical ion trap arrays. *Rev. Sci. Instr.* 2007, *78*, 015107.

33. Van Amerom, F.H.W.; Chaudhary, A.; Cardenas, M.; Bumgarner, J.; Short, R.T. Microfabrication of cylindrical ion trap mass spectrometer arrays for handheld chemical analyzers. *Chem. Eng. Comm.* 2008, *195*, 2, 98–114.

34. Song, Q.Y.; Kothari, S.; Senko, M.A.; Schwartz, J.C.; Amy, W.J.; Stafford, G.; Cooks, R.G.; Ouyang, Z. Rectilinear ion trap mass spectrometer with atmospheric pressure interface and electrospray ionization source. *Anal. Chem.* 2006, *78*, 718–725.

35. Song, Y.S.; Wu, G.X.; Song, Q.Y.; Cooks, R.G.; Ouyang, Z.; Plass, W.R. Novel linear ion trap mass analyzer composed of four planar electrodes. *J. Am. Soc. Mass Spectrom.* 2006, *17*, 631–639.

36. Fico, M.; Yu, M.; Ouyang, Z.; Cooks, R.G.; Chappell, W.J. Miniaturization and geometry optimization of a polymer-based rectilinear ion trap. *Anal. Chem.* 2007, *79*, 8076–8082.

37. Smith, S.A.; Fico, M.; Maas, J.D.; Chappell, W.J.; Cooks, R.G. Circular array of miniature rectilinear ion traps. *Proc. 56th ASMS Conference on Mass Spectrometry and Allied Topics*, Denver, CO, June 1–5, 2008.

38. Song, Q.Y.; Xu, W.; Smith, S.A.; Gao, L.; Chappell, W.J.; Cooks, R.G.; Ouyang, Z. Ion trap mass analysis at high pressure: an experimental characterization. *J. Mass Spectrom.* 2009, published online: DOI 10.1002/jms.1684.

39. Peng, W-P.; Goodwin, M.; Nie, Z.; Volný, M.; Ouyang, Z.; Cooks, R.G. Ion soft landing using a rectilinear ion trap mass spectrometer. *Anal. Chem.* 2008, *80*, 6640–6649.

40. Gao, L.; Song, Q.; Patterson, G.E.; Cooks, R.G.; Ouyang, Z. Handheld rectilinear ion trap mass spectrometer. *Anal. Chem.* 2006, *78*, 5994–6002.

41. Gao, L.; Sugiarto, A.; Harper, J.D.; Cooks, R.G.; Ouyang, Z. Design and characterization of a multisource hand-held tandem mass spectrometer. *Anal. Chem.* 2008, *80*, 7198–7205.

42. Gao, L.; Cooks, R.G.; Ouyang, Z. Breaking the pumping speed barrier in mass spectrometry: Discontinuous atmospheric pressure interface. *Anal. Chem.* 2008, *80*, 4026–4032.

43. Cooks, R.G.; Ouyang, Z.; Takáts, Z.; Wiseman, J.M. Ambient mass spectrometry. *Science* 2006, *311*, 5767, 1566–1570.

44. Takats, Z.; Wiseman, J.M.; Gologan, B.; Cooks, R.G. Mass spectrometry sampling under ambient conditions with desorption electrospray ionization. *Science* 2004, *306*, 5695, 471–473.

45. Harper, J.D.; Charipar, N.A.; Mulligan, C.C.; Zhang, X.; Cooks, R.G.; Ouyang, Z. Low-temperature plasma probe for ambient desorption ionization. *Anal. Chem.* 2008, *80*, 9097–9104.

46. Lammert, S.A.; Plass, W.R.; Thompson, C.V.; Wise, M.B. Design, optimization and initial performance of a toroidal rf ion trap mass spectrometer. *Int. J. Mass Spectrom.* 2001, *212*, 25–40.

47. Lammert, S.A.; Rockwood, A.A.; Wang, M.; Lee, M.L.; Lee, E.D.; Tolley, S.E.; Oliphant, J.R.; Jones, J.L.; Waite, R.W. Miniature toroidal radio frequency ion trap mass analyzer. *J. Am. Soc. Mass Spectrom.* 2006, *7*, 916–922.

48. Contreras, J.A.; Murray, J.A.; Tolley, S.E.; Oliphant, J.L.; Tolley, H.D.; Lammert, S.A.; Lee, E.D.; Later, D.W.; Lee, M.L. Hand-portable gas chromatograph-toroidal ion trap mass spectrometer (GC-TMS) for detection of hazardous compounds. *J. Am. Soc. Mass Spectrom.* 2008, *19*, 1425–1434.

49. Xu, W.; Chappell, W.J.; Cooks, R.G.; Ouyang, Z. Characterization of electrode surface roughness and its impact on ion trap mass analysis. *J. Mass Spectrom.* 2008, *44*, 353–360.

50. Pau, S.; Pai, C.S.; Low, Y.L.; Moxom, J.; Reilly, P.T.A.; Whitten, W.B.; Ramsey, J.M. Microfabricated quadrupole ion trap for mass spectrometer applications. *Phys. Rev. Lett.* 2006, *96*, 120801.

51. Laughlin, B.C. Development of High-Throughput and Miniaturized Mass Spectrometers Using Miniature Cylindrical ion Trap Analyzers. Ph.D. Thesis, Purdue University, West Lafayette, 2005.

52. Paul, W.; Reinhard, H.P.; von Zahn, U. Das elektrische Massenfilter als Massenspektrometer und Isotopentrenner. *Z. Phys.* 1958, *152*, 143–182.

53. Boumsellek, S.; Ferran, R.J. Trade-offs in miniature quadrupole designs. *J. Am. Soc. Mass Spectrom.* 2001, *12*, 633–640.

54. Major, F.G.; Dehmelt, H.G. Exchange-collision technique for RF spectroscopy of stored ions. *Phys. Rev.* 1968, *170*, 91–107.

55. Kaiser, R.E.; Cooks, R.G.; Stafford, G.C.; Syka, J.E.P.; Hemberger, P.H. Operation of a quadrupole ion trap mass-spectrometer to achieve high mass charge ratios. *Int. J. Mass Spectrom. Ion Processes* 1991, *106*, 79–115.

56. Goeringer, D.E.; Whitten, W.B.; Ramsey, J.M.; Mcluckey, S.A.; Glish, G.L. Theory of high-resolution mass-spectrometry achieved via resonance ejection in the quadrupole ion trap. *Anal. Chem.* 1992, *64*, 1434–1439.

57. Fischer, E. Die dreidimensionale Stabilisierung von Ladungstragern in einem Vierpolfeld. *Z. Phys.* 1959, *156*, 1–26.

58. Schwartz, J.C.; Syka, J.E.P.; Jardine, I. High-resolution on a quadrupole ion trap massspectrometer. *J. Am. Soc. Mass Spectrom.* 1991, *2*, 198–204.

59. Williams, J.D.; Cox, K.A.; Schwartz, J.C.; Cooks, R.G. High mass, high resolution mass spectrometry. In *Practical Aspects of Ion Trap Mass Spectrometry,* Vol. 2, Chapter 1 R.E. March, and J.F.J. Todd, Eds. CRC Press: Boca Raton, FL, 1995, pp. 3–48.

60. Bui, H.A.; Cooks, R.G. Windows version of the ion trap simulation program ITSIM: A powerful heuristic and predictive tool in ion trap mass spectrometry. *J. Mass Spectrom.* 1998, *33*, 297–304.

61. Londry, F.A.; Alfred, R.L.; March, R.E. Computer simulation of single ion trajectories in Paul type ion traps. *J. Am. Soc. Mass Spectrom.* 1993, *4*, 687–705.

62. Dahl, D.A. SIMION 3D Version 6.0. *Proc. 43rd ASMS Conference on Mass Spectrometry and Allied Topics* Atlanta, GA, May 21–26, 1995, 717.

63. Forbes, M.W.; Sharifi, M.; Croley, T.; Lausevic, Z.; March, R.E. Special feature tutorial article: Simulation of ion trajectories in a quadrupole ion trap: A comparison of three simulation programs. *J. Mass Spectrom.* 1999, *34*, 1219–1239.

64. Dahl, D.A.; McJunkin, T.R.; Scott, J.R. Comparison of ion trajectories in vacuum and viscous environments using SIMION: Insights for instrument design. *Int. J. Mass Spectrom.* 2007, *266*, 156–165.

65. Manura, D. SIMION 8.0. www.simion.com (Accessed December 4, 2009).

66. Billen, J.H.; Young, L.M. POISSON/SUPERFISH on PC compatibles. *Proceedings of the 1993 Particle Accelerator Conference*, Washington, DC, 1993, 790.

67. Plass, W.R.; Li, H.Y.; Cooks, R.G. Theory, simulation and measurement of chemical mass shifts in RF quadrupole ion traps. *Int. J. Mass Spectrom.* 2003, *228*, 237–267.

68. Press, W.H.; Teukolsky, S.A.; Vetterling, W.T.; Flannery, B.P. *Numerical Recipes: The Art of Scientific Computing*. 3rd ed.; Cambridge University Press, New York, NY, 2007.

69. Jungmann, K.; Hoffnagle, J.; DeVoe, R.G.; Brewer, R.G. Collective oscillations of stored ions. *Phys. Rev. A* 1987, *36*, 3451–3454.

70. Bonner, R.F.; Fulford, J.E.; March, R.E.; Hamilton, G.F. Cylindrical ion trap 1. General introduction. *Int. J. Mass Spectrom. Ion Processes* 1977, *24*, 255–269.

71. Lagadec, H.; Meis, C.; Jardino, M. Effective potential of an RF cylindrical trap. *Int. J. Mass Spectrom. Ion Processes* 1988, *85*, 287–299.

72. Lee, W.W.; Oh, C.H.; Kim, P.S.; Yang, M.; Song, K.S. Characteristics of cylindrical ion trap. *Int. J. Mass Spectrom.* 2003, *230*, 25–31.

73. Wells, J.M.; Plass, W.A.; Cooks, R.G. Control of chemical mass shifts in the quadrupole ion trap through selection of resonance ejection working point and rf scan direction. *Anal. Chem.* 2000, *72*, 2677–2683.

74. Li, H.Y.; Plass, W.R.; Patterson, G.E.; Cooks, R.G. Chemical mass shifts in resonance ejection experiments in the quadrupole ion trap. *J. Mass Spectrom.* 2002, *37*, 1051–1058.

75. Li, H.Y.; Peng, Y.N.; Plass, W.R.; Cooks, R.G. Chemical mass shifts in quadrupole ion traps as analytical characteristics of nitro-aromatic compounds. *Int. J. Mass Spectrom.* 2003, *222*, 481–491.

76. Ouyang, Z.; Gao, L.; Fico, M.; Chappell, W.J.; Noll, R.J.; Cooks, R.G. Quadrupole ion traps and trap arrays: Geometry, material, scale, performance. *Eur. J. Mass Spectrom.* 2007, *13*, 13–18.

77. Wu, G.X.; Noll, R.J.; Plass, W.R.; Hu, Q.Z.; Perry, R.H.; Cooks, R.G. Ion trajectory simulations of axial ac dipolar excitation in the Orbitrap. *Int. J. Mass Spectrom.* 2006, *254*, 53–62.

78. Makarov, A. Electrostatic axially harmonic orbital trapping: A high-performance technique of mass analysis. *Anal. Chem.* 2000, *72*, 1156–1162.

79. Paul, W.; Steinwedel, H. A new mass spectrometer without a magnetic field. *Z. Naturforsch., A: Phys. Sci.* 1953, *8*, 448–450.

80. Paul, W. Electromagnetic traps for charged and neutral particles. *Rev. Mod. Phys.* 1990, *62*, 531–540.

81. Madou, M.J. *Fundamentals of Microfabrication*. 2nd ed.; CRC Press: Boca Raton, FL, 2002.

82. Judy, J.W. Microelectromechanical systems (MEMS): Fabrication, design and applications. *Smart. Mater. Struct.* 2001, *10*, 1115–1134.

83. Madsen, M.J.; Hensinger, W.K.; Stick, D.; Rabchuk, J.A.; Monroe, C. Planar ion trap geometry for microfabrication. *Appl. Phys. B* 2004, *78*, 639–651.

84. Stick, D.; Hensinger, W.K.; Olmschenk, S.; Madsen, M.J.; Schwab, K.; Monroe, C. Ion trap in a semiconductor chip. *Nat. Phys.* 2006, *2*, 36–39.

85. Brownnutt, M.; Wilpers, G.; Gill, P.; Thompson, R.C.; Sinclair, A.G. Monolithic microfabricated ion trap chip design for scaleable quantum processors. *New J. Phys.* 2006, *8*, Article No. 232 (18 pp).

86. Huber, G.; Deuschle, T.; Schnitzler, W.; Reichle, R.; Singer, K.; Schmidt-Kaler, F. Transport of ions in a segmented linear Paul trap in printed-circuit-board technology. *New J. Phys.* 2008, *10*, Article No. 013004 (15 pp).

87. Pau, S.; Whitten, W.B.; Ramsey, J.M. Planar geometry for trapping and separating ions and charged particles. *Anal. Chem.* 2007, *79*, 6857–6861.

88. Chaudhary, A.; van Amerom, F.H.W.; Short, R.T.; Bhansali, S. Fabrication and testing of a miniature cylindrical ion trap mass spectrometer constructed from low temperature co-fired ceramics. *Int. J. Mass Spectrom.* 2006, *251*, 32–39.

89. Yu, M.; Fico, M.; Kothari, S.; Ouyang, Z.; Chappell, W.J. Polymer-based ion trap chemical sensor. *IEEE Sens. J.* 2006, *6*, 1429–1434.

90. Enjalbal, C.; Martinez, J.; Aubagnac, J.L. Mass spectrometry in combinatorial chemistry. *Mass Spectrom. Rev.* 2000, *19*, 139–161.

91. Aebersold, R.; Goodlett, D.R. Mass spectrometry in proteomics. *Chem. Rev.* 2001, *101*, 269–295.

92. Yanagida, M. Functional proteomics; current achievements. *J. Chrom. B.* 2002, *771*, 89–106.

93. Tabert, A.M.; Misharin, A.S.; Cooks, R.G. Performance of a multiplexed chemical ionization miniature cylindrical ion trap array mass spectrometer. *Analyst* 2004, *129*, 323–330.

94. Futrell, J.H. *Gaseous Ion Chemistry and Mass Spectrometry.* Wiley: New York, 1986.

95. *NIST Chemistry WebBook, NIST Standard Reference Database Number 69.* National Institute of Standards and Technology, Gaithersburg MD. June 2005. (http://webbook. nist.gov). Eds: Linstrom, P.J. and Mallard, W.G.

96. Hunter, E.P.L.; Lias, S.G. Evaluated gas phase basicities and proton affinities of molecules: An update. *J. Phys. Chem. Ref. Data* 1998, *27*, 413–656.

97. Venter, A.; Nefliu, M.; Cooks, R.G. Ambient desorption ionization mass spectrometry. *TrAC, Trends Anal. Chem.* 2008, *27*, 284–290.

98. Badman, E.R.; Cooks, R.G. Cylindrical ion trap array with mass selection by variation in trap dimensions. *Anal. Chem.* 2000, *72*, 5079–5086.

99. Guan, S.; Marshall, A.G. Stored waveform inverse Fourier transform (SWIFT) ion excitation in trapped-ion mass spectrometry: Theory and applications. *Int. J. Mass Spectrom. Ion Processes* 1996, *157/158*, 5–37.

100. Lammert, S.A.; Cooks, R.G. Surface-induced dissociation of molecular-ions in a quadrupole ion trap mass-spectrometer. *J. Am. Soc. Mass Spectrom.* 1991, *2*, 487–491.

101. Brodbelt, J.S.; Louris, J.N.; Cooks, R.G. Chemical ionization in an ion trap mass-spectrometer. *Anal. Chem.* 1987, *59*, 1278–1285.

102. Schwartz, J.C.; Wade, A.P.; Enke, C.G.; Cooks, R.G. Systematic delineation of scan modes in multidimensional mass-spectrometry. *Anal. Chem.* 1990, *62*, 1809–1818.

103. Johnson, J.V.; Pedder, R.E.; Yost, R.A. MS-MS parent scans on a quadrupole ion trap mass-spectrometer by simultaneous resonant excitation of multiple ions. *Int. J. Mass Spectrom. Ion Processes* 1991, *106*, 197–212.

104. Fico, M.; Maas, J.D.; Smith, S.A.; Costa, A.B.; Chappell, W.J.; Cooks, R.G. Circular arrays of polymer-based miniature rectilinear ion traps. *Analyst 2009*, *134*, 1338–1347.

105. Kothari, S.; Song, Q.; Xia, Y.; Fico, M.; Taylor, D.; Amy, J.W.; Stafford, G.; Cooks, R.G. Multiplexed four-channel rectilinear ion trap mass spectrometer. *Anal. Chem.* 2009, *81*, 1570–1579.

106. Franchetti, V.; Solka, B.H.; Baitinger, W.E.; Amy, J.W.; Cooks, R.G. Soft landing of ions as a means of surface modification. *Int. J. Mass Spectrom. Ion Processes* 1977, *23*, 29–35.

107. Grill, V.; Shen, J.; Evans, C.; Cooks, R.G. Collisions of ions with surfaces at chemically relevant energies: Instrumentation and phenomena. *Rev. Sci. Instr.* 2001, *72*, 3149–3179.

108. Ouyang, Z.; Takats, Z.; Blake, T.A.; Gologan, B.; Guymon, A.J.; Wiseman, J.M.; Oliver, J.C.; Davisson, V.J.; Cooks, R.G. Preparing protein microarrays by soft-landing of mass-selected ions. *Science* 2003, *301*, 1351–1354.

109. Rader, H.J.; Rouhanipour, A.; Talarico, A.M.; Palermo, V.; Samori, P.; Mullen, K. Processing of giant graphene molecules by soft-landing mass spectrometry. *Nat. Mater.* 2006, *5*, 276–280.

110. Laskin, J.; Wang, P.; Hadjar, O. Soft-landing of peptide ions onto self-assembled mono-layer surfaces: An overview. *Phys. Chem. Chem. Phys.* 2008, *10*, 1079–1090.

111. Blake, T.A.; Ouyang, Z.; Wiseman, J.M.; Takáts, Z.; Guymon, A.J.; Kothari, S.; Cooks, R.G. Preparative linear ion trap mass spectrometer for separation and collection of puri-fied proteins and peptides in arrays using ion soft landing. *Anal. Chem.* 2004, *76*, 21, 6293–6305.

112. Nanita, S.C.; Takats, Z.; Cooks, R.G. Chiral enrichment of serine via formation, dis-sociation, and soft-landing of octameric cluster ions. *J. Am. Soc. Mass Spectrom.* 2004, *15*, 1360–1365.

113. Wang, P.; Hadjar, O.; Gassman, P.L.; Laskin, J. Reactive landing of peptide ions on self-assembled monolayer surfaces: An alternative approach for covalent immobilization of peptides on surfaces. *Phys. Chem. Chem. Phys.* 2008, *10*, 1512–1522.

114. Volný, M.; Elam, W.T.; Ratner, B.D.; Turuček, F. Enhanced in-vitro blood compatibil-ity of 316L stainless steel surfaces by reactive landing of hyaluronan ions. *J. Biomed. Mater. Res. B* 2007, *80B*, 505–510.

115. Whitten, W.B.; Reilly, P.T.A.; Ramsey, J.M. High-pressure ion trap mass spectrometry. *Rapid Commun. Mass Spectrom.* 2004, *18*, 1749–1752.

116. Reiser, H.P.; Julian, R.K.; Cooks, R.G. A versatile method of simulation of the operation of ion trap mass spectrometers. *Int. J. Mass Spectrom. Ion Processes* 1992, *121*, 49–63.

117. Julian, R.K.; Nappi, M.; Weil, C.; Cooks, R.G. Multiparticle simulation of ion motion in the ion-trap mass-spectrometer-resonant and direct-current pulse excitation. *J. Am. Soc. Mass Spectrom.* 1995, *6*, 57–70.

118. Nappi, M.; Weil, C.; Cleven, C.D.; Horn, L.A.; Wollnik, H.; Cooks, R.G. Visual repre-sentations of simulated three-dimensional ion trajectories in an ion trap mass spectrom-eter. *Int. J. Mass Spectrom. Ion Processes* 1997, *161*, 77–85.

119. Wells, J.M.; Plass, W.R.; Patterson, G.E.; Zheng, O.Y.; Badman, E.R.; Cooks, R.G. Chemical mass shifts in ion trap mass spectrometry: Experiments and simulations. *Anal. Chem.* 1999, *71*, 3405–3415.

APPENDIX

Instrument/ Analyzer	Inlet	Ionization method[a]	Analytes/Samples	LOD[b]	R[c]	m/z range	Comments	Figure	Ref
Mini 5 and Mini 7 CIT, $r_0 = 2.5$ mm	Leak	EI	PFTBA[d], 1,2-dichlorobenzene		Unit	48–140; 100–264			[1]
Mini 5 and Mini 7 CIT, $r_0 = 2.5$ mm	Leak	EI	methyl salicylate (wintergreen)	1 pg	152	48–140; 100–264	background-corrected signal for 30 s		[2]
	Leak	EI	p-nitrotoluene, acetophenone, methyl salicylate, DMMP[e]				MS, MS[2], MS[3]		
	MIMS	EI	10 ppb methyl salicylate in air				trap and release MIMS		
	MIMS	EI	100 ppm aq. Toluene		91				
Mini 7 CIT, $r_0 = 2.5$ mm	MIMS	EI	3-phenyl-2-propenal (cinnamon)	<38 ppm			sample for 5 s (trap & release MIMS)		[3]
	MIMS		methyl salicylate	<45 ppm			temporal profile 4 min		
Mini 7 CIT, $r_0 = 2.5$ mm	Leak	EI	various ion/molecule reactions		2 Th	185	SWIFT[f] prep/ isolation of reagent ions, MS[2] and MS[3] characterization of products		[4]

(Continued)

APPENDIX (Continued)

Instrument/ Analyzer	Inlet	Ionization method.[a]	Analytes/Samples	LOD[b]	R[c]	m/z range	Comments	Figure	Ref
Mini 7 CIT, $r_0 = 2.5$ mm	FIMS (SPME)	EI	methyl salicylate in air	25 ppb			3 min sampling time, 200 mL min⁻¹ flow rate; LDR = 50 to 800 ppb		[5]
			naphthalene in air	25 ppb			3 min sampling time, 200 mL min⁻¹ flow rate		
			headspace over aq. toluene solutions	1 ppb			5 min sampling time LDR[g] = 1 to 200 ppb		
			saturated vapors of acetophenone (522 ppm), toluene (37 ppth), methyl salicylate (45 ppm), α-terpinene (unknown), 3,4-dichlorotoluene (414 ppm), DMMP (1266 ppm), p-nitrotoluene (216 ppm), naphthalene (111 ppm), nitrobenzene (322 ppm)				5 s sampling times		
			α-terpinene (aq.)	100 ppb			LOD for MIMS estimated as 10 ppb		
			toluene from aq. BTX[i] mixtures				500–4000 ppb calibration		

Device	Ionization	Analyte	Value	Unit / Range	MS²	Notes	Ref
Mini 8 CIT, $r_0 = 2.5$ mm	DAPI, APCI	methyl salicylate, nitrobenzene	1.24 ppb, 630 pptr	Unit 45–450, 372		3-stage differential pumping; inlet to atmosphere is 254 μm id. × 5 cm long capillary; 1st stage is 700 mTorr, ions exit capillary and impinge on skimmer, 500 μm orifice, ions focused by tube lens; square quad guides ions through 2nd stage and into 1 mm aperture into CIT	[6]
	ESI	arginine, 100 uM					
		(−)-ephedrine, m/z 166, 10 μM, 3 μL min⁻¹	LOD 1 μM est. retrospectively				
		lomefloxacin, m/z 352, 10 μM, 3 μL min⁻¹	LOD 1 μM est. retrospectively				
Mini 8 CIT, $r_0 = 2.5$ mm	DPAPI, APCI	Arsine, 9%, RHᵇ, m/z 91	6.7 ppb				[7]
		Arsine, 50% RH, m/z 143	6.8 ppb				
		DMMP, 9% RH, m/z 125	2.2 ppb				
		DMMP, 83% RH, m/z 125	26 ppb				
		benzene, M⁺, m/z 78	12 ppb				
		toluene, [M−H]⁺, m/z 91	2.3 ppb				
		pyridine, [M + H]⁺, m/z 80	27 ppb				
		vinyl acetate, [M + H]⁺, m/z 87	15 ppb				
		OSHAᵈ D air + 0.1% Windex	2.2 ppb		2.2 ppb using MS²		
		OSHA D air + 0.1% WD-40	7.1 ppb		2.2 ppb using MS²		
		OSHA D air + 0.1% diesel fuel	7.1 ppb		2.2 ppb using MS²		
Mini 8 CIT, $r_0 = 2.5$ mm	DPAPI, DESI	DEETᵏ, (N,N-diethyl-m-toluamide)	< 10 ng				[8]

alachlor (2-chloro-N-(2,6-diethylphenyl)-N-(methoxymethyl)acetamide), atrazine (2-chloro-4-(ethylamino)-6-(isopropylamino)-1,3,5-triazine); Excedrin tablet: acetaminophen, aspirin, and caffeine

(Continued)

APPENDIX (Continued)

Instrument/Analyzer	Inlet	Ionization method.[a]	Analytes/Samples	LOD[b]	R[c]	m/z range	Comments	Figure	Ref
RIT, $x_0 = 5$ mm	MIMS	EI, GDEI	headspace toluene vapor		91	69–502			[9]
	MIMS	EI	PFTBA						
			methyl salicylate from candy mint	< 45 ppm			MS^3 demonstrated		
			naphthalene from mothball	< 112 ppm			1.3 s sampling time, 175 s clear time		
			headspace vapor from liquid mixture: DMMP 330 ppm, n-butylbenzene 350 ppm, methyl salicylate 10 ppm, 1,3-dichlorobenzene 710 ppm				MS^2 demonstrated for each analyte		
			aq. naphthalene	50 ppb					
RIT, $x_0 = 5$ mm		GDEI	PFTBA, headspace vapors 690 ppm			414	LDR up to 1 ppm; highest m/z observed, most likely due to non-optimized electron kinetic energy, preventing observation of higher m/z ions		[10]

Mini 10 RIT, x_0 = 5 mm	CAPI	ESI	dibutylamine, 100 nM, [M + H]+, m/z 130; tributylamine, 100 nM, [M + H]+, m/z 186	3 Th wide	127 μm × 10 cm tube, crude lensing from capillary and gate bias; 13 atm mL/min throughput; operating pressure 15 mTorr; MS² spectrum of tributyl amine	[11]
			protonated methamphetamine (m/z 150), cocaine (304), and heroin (370)	'ppb'	LDR = 3 orders of magnitude	
			bradykinin (1 μg/mL), synthetic peptide KGAILKGAILR (1 ug/mL), and mixture		+2 and +3 charge states observed	
Mini 10 RIT, x_0 = 5 mm	DESI	EI	cocaine, 50 ng on Teflon or paper		LDR: 61	[12]
	sorbent tube		acrolein, m/z 57	30 ppb	LDR: 200	
			acrylonitrile, m/z 53	15 ppb	10	
			cyanogen chloride, m/z 61	20 ppb	10	
			ethylene oxide, m/z 45	200 ppb	100	
			formaldehyde, m/z 31	3 ppm	14	
			hydrogen cyanide, m/z 28	180 ppb	10	
			ethyl parathion, m/z 97	1 ppb	67	
			phosgene, m/z 63	12 ppb	10	
			sulfur dioxide, m/z 48	1.2 ppm	61	
			acrolein, m/z 57	30 ppb		

(Continued)

APPENDIX (Continued)

Instrument/Analyzer	Inlet	Ionization method.[a]	Analytes/Samples	LOD[b]	R[c]	m/z range	Comments	Figure	Ref
Mini 10 RIT, $x_0 = 5$ mm	DAPI	nano	5 ppm caffeine (m/z 195), cocaine		Unit	> 304			[13]
			methamphetamine, cocaine, heroin	50 ppb	Unit	> 370			
			methamphetamine, cocaine, heroin	500 ppb	Unit	~400	MS²		
		ESI	Lysine	500 ppb	Unit	> 293			
		APCI	DMMP	3 ppb					
		DESI	Teflon 25 ng		> 304				
		DESI	basic violet 3, basic blue 26		Unit	470			
Mini 10 RIT, $x_0 = 5$ mm	Tenax (sorbent)	EI	benzene in BTX mixtures in vapor phase	0.6 ppb	Unit		10 s sampling time		[14]
			toluene				0.2 ppb		
	MIMS		benzene vapor				0.2 ppm to 60 ppm LDR		
			acetone vapor				0.1 ppm LOD, 1-400 ppm LDR		

Mini 11 RIT, x₀ = 5 mm	MIMS	EI	PFTBA vapor		614		1041 kHz		[15]
	DAPI	nano	Ultramark	100 ppm	1422	180 @ 1422	695 kHz		
			cytochrome c, + 10 to + 18 charge states	200 ppm	1250				
			myoglobin, + 13 to + 22 charge states	200 ppm	1300				
	MIMS	EI	aq. naphthalene	150 ppb			LDR 200 ppb – 5 ppm		
	DAPI	ESI	Cocaine	80 ppb			LDR 200 ppb – 5 ppm		
	DESI		Ink				not quantitative		
			pharmaceutical tablets- Claritin, aspirin, Excedrin (aspirin, acetaminophen, caffeine)				not quantitative		
Mini 10 RIT, x₀ = 5 mm	DAPI	nano	cocaine			Unit		2.5	
Polymer RITs x₀ = 5, 2.5, 1.7 mm	CAPI	ESI	protonated arginine monomer and dimer		> 1200	696	(SLA[1] fabricated) full, half, and third-sized RITs	2.24	[16]
Ion Trap Arrays									
CIT array, r₀ = 2.5 mm	Leak	EI, CI	1,3 dichlorobenzene in 4 channels					2.28	[17]
CIT array, r₀ = 2.5 mm	Leak	EI, CI	acetophenone, bromobenzene, 1,3-dichlorobenzene					2.29	[17]
CIT array, r₀ = 2.5 mm	DPAPI	ESI	10 mM arginine (m/z 175), 10 mM glutamine (m/z 147, dimer at m/z 293)			Unit	1% cross-talk 2 stages of diff. pumping	2.33	[18]

(Continued)

APPENDIX (Continued)

Instrument/ Analyzer	Inlet	Ionization method.[a]	Analytes/Samples	LOD[b]	R[c]	m/z range	Comments	Figure	Ref
CIT array, diff sized traps	Leak	EI	n-nonane, toluene				5 mm gets m/z 57 from nonane (alkane); 4 mm traps m/z 91,92 for toluene (aromatic)	2.35, 2.36	[19]
Serial CIT array, $R_0 = 5$ mm	Leak	EI	acetophenone, ca 5 × 10⁻⁷ Torr					2.37	[20]
Parallel RIT array, $x_0 = 5$ mm	Leak	EI	aniline (m/z 93); D-camphor (m/z 152); L-carvone (m/z 150); 1,4-difluorobenzene (m/z 114)		Unit			2.39, 2.40	[21]
Polymer RIT array, $X_0 = 1.7$ mm	CAPI	ESI	bombesin + 2 charge state (m/z 811); tributylamine; protonated monomer, dimer, and trimer of arginine		2 Th	> 1200	circular array of 1/3 scale RITs; fabricated by SLA; MS, MS²	2.41	[22]

'multiplexed' 4-channel RIT, $x_0 = 5$ mm	CAPI	ESI, APCI	lipid extract from *E. coli*		2.44	*
Serially coupled RITs, $X_0 = 5$ mm	CAPI	ESI	methamphetamine, cocaine, caffeine, heroin	for ion soft-landing	2.45, 2.46	[23]

Notes: Inlet: DPAPI = differentially pumped atmospheric pressure inlet; CAPI = continuous atmospheric pressure inlet; DAPI = discontinuous atmospheric pressure inlet; Leak = low conductivity leak valve; MIMS = membrane inlet mass spectrometry; FIMS = fiber inlet mass spectrometry (*e.g.*, direct insertion of solid phase microextraction [SPME] fiber).

a Ionization Method: EI = electron impact; CI = chemical ionization; ESI = electrospray ionization; APCI = atmospheric pressure chemical ionization; nano = nanospray; DESI = desorption electrospray ionization; GDEI = glow discharge electron impact.

b Limit of detection.

c Resolution ($m/\triangle m$).

d Perfluorotributylamine.

e Dimethyl methylphosphonate.

f Stored waveform inverse fourier transform.

g Linear dynamic range.

h Relative humidity.

i BTX = benzene, toluene, and xylene mixture.

j Occupational Safety & Health Administration.

k N,N-Diethyl-meta-toluamide.

l Stereolithography apparatus.

REFERENCES (APPENDIX)

1. Patterson, G.; Guymon, A.; Riter, L.S.; Everly, M.; Griep-Raming, J.; Laughlin, B.C.; Zheng, O.Y.; Cooks, R.G. Miniature cylindrical ion trap mass spectrometer. *Anal. Chem.* 2002, *74*, 6145–6153.

2. Riter, L.S.; Peng, Y.A.; Noll, R.J.; Patterson, G.E.; Aggerholm, T.; Cooks, R.G. Analytical performance of a miniature cylindrical ion trap mass spectrometer. *Anal. Chem.* 2002, *74*, 6154–6162.

3. Riter, L.S.; Laughlin, B.C.; Nikolaev, E.; Cooks, R.G. Direct analysis of volatile organic compounds in human breath using a miniaturized cylindrical ion trap mass spectrometer with a membrane inlet. *Rapid Commun. Mass Spectrom.* 2002, *16*, 2370–2373.

4. Riter, L.S.; Meurer, E.C.; Handberg, E.S.; Laughlin, B.C.; Chen, H.; Patterson, G.E.; Eberlin, M.N.; Cooks, R.G. Ion/molecule reactions performed in a miniature cylindrical ion trap mass spectrometer. *Analyst* 2003, *128*, 1112–1118.

5. Riter, L.S.; Meurer, E.C.; Cotte-Rodriguez, I.; Eberlin, M.N.; Cooks, R.G. Solid phase micro-extraction in a miniature ion trap mass spectrometer. *Analyst* 2003, *128*, 1119–1122.

6. Laughlin, B.C.; Mulligan, C.C.; Cooks, R.G. Atmospheric pressure ionization in a miniature mass spectrometer. *Anal. Chem.* 2005, *77*, 2928–2939.

7. Mulligan, C.C.; Justes, D.R.; Noll, R.J.; Sanders, N.L.; Laughlin, B.C.; Cooks, R.G. Direct monitoring of toxic compounds in air using a portable mass spectrometer. *Analyst* 2006, *131*, 556–567.

8. Mulligan, C.C.; Talaty, N.; Cooks, R.G. Desorption electrospray ionization with a portable mass spectrometer: In situ analysis of ambient surfaces. *Chem. Commun.* 2006, *16*, 1709–1711.

9. Gao, L.; Song, Q.; Patterson, G.E.; Cooks, R.G.; Ouyang, Z. Handheld rectilinear ion trap mass spectrometer. *Anal. Chem.* 2006, *78*, 5994–6002.

10. Gao, L.; Song, Q.Y.; Noll, R.J.; Duncan, J.; Cooks, R.G.; Zheng, O.Y. Glow discharge electron impact ionization source for miniature mass spectrometers. *J. Mass Spectrom.* 2007, *42*, 675–680.

11. Keil, A.; Talaty, N.; Janfelt, C.; Noll, R.J.; Gao, L.; Ouyang, Z.; Cooks, R.G. Ambient mass spectrometry with a handheld mass spectrometer at high pressure. *Anal. Chem.* 2007, *79*, 7734–7739.

12. Keil, A.; Hernandez-Soto, H.; Noll, R.J.; Fico, M.; Gao, L.; Ouyang, Z.; Cooks, R.G. Monitoring of toxic compounds in air using a handheld rectilinear ion trap mass spectrometer. *Anal. Chem.* 2008, *80*, 734–741.

13. Gao, L.; Cooks, R.G.; Ouyang, Z. Breaking the pumping speed barrier in mass spectrometry: Discontinuous atmospheric pressure interface. *Anal. Chem.* 2008, *80*, 4026–4032.

14. Sokol, E.; Edwards, K.E.; Qian, K.; Cooks, R.G. Rapid hydrocarbon analysis using a miniature rectilinear ion trap mass spectrometer. *Analyst* 2008, *133*, 1064–1071.

15. Gao, L.; Sugiarto, A.; Harper, J.D.; Cooks, R.G.; Ouyang, Z. Design and characterization of a multisource hand-held tandem mass spectrometer. *Anal. Chem.* 2008, *80*, 7198–7205.

16. Fico, M.; Yu, M.; Ouyang, Z.; Cooks, R.G.; Chappell, W.J. Miniaturization and geometry optimization of a polymer-based rectilinear ion trap. *Anal. Chem.* 2007, *79*, 8076–8082.

17. Tabert, A.M.; Griep-Raming, J.; Guymon, A.J.; Cooks, R.G. High-throughput miniature cylindrical ion trap array mass spectrometer. *Anal. Chem.* 2003, *75*, 5656–5664.

18. Misharin, A.S.; Laughlin, B.C.; Vilkov, A.; Takáts, Z.; Ouyang, Z.; Cooks, R.G. High-throughput mass spectrometer using atmospheric pressure ionization and a cylindrical ion trap array. *Anal. Chem.* 2005, *77*, 459–470.

19. Badman, E.R.; Cooks, R.G. Cylindrical ion trap array with mass selection by variation in trap dimensions. *Anal. Chem.* 2000, *72*, 5079–5086.

20. Ouyang, Z.; Badman, E.R.; Cooks, R.G. Characterization of a serial array of miniature cylindrical ion trap mass analyzers. *Rapid Mass Spectrom.* 1999, *13*, 2444–2449.

21. Tabert, A.M.; Goodwin, M.P.; Duncan, J.S.; Fico, C.D.; Cooks, R.G. Multiplexed rectilinear ion trap mass spectrometer for high-throughput analysis. *Anal. Chem.* 2006, *78*, 4830–4838.

22. Fico, M.; Maas, J.D.; Smith, S.A.; Costa, A.B.; Chappell, W.J.; Cooks, R.G. Circular arrays of polymer-based miniature rectilinear ion traps. *Analyst*, 2009, *134*, 1338–1347.

23. Song, Q.Y.; Smith, S.A.; Gao, L.; Volný, M.; Ouyang, Z.; Cooks, R.G. Mass-selection of ion beams using waveform isolation in radiofrequency quadrupoles. *Anal. Chem.* 2008, *81*, 1833–1840.

Part II

New Ion Trapping Techniques

3 Theory and Practice of the Orbitrap Mass Analyzer

Alexander Makarov

CONTENTS

3.1 Introduction: Orbitrap Analyzer as a Special Case of
Electrostatic Ion Traps 252
    3.1.1 What is an Electrostatic Ion Trap? 252
    3.1.2 Taxonomy of Electrostatic Traps (ESTs) 252
    3.1.3 Introduction to the Orbitrap Mass Analyzer..................................253
3.2 Operation of the Orbitrap Mass Analyzer 254
    3.2.1 Motion of Trapped Ions 254
    3.2.2 Ion Detection 256
    3.2.3 Formation of Coherent Ion Packets 256
    3.2.4 Basic Analytical Parameters of the Orbitrap Mass Analyzer.........257
3.3 Common Properties of the Orbitrap and other ESTs...................................257
    3.3.1 An Overview...257
    3.3.2 Decay of Ion Packets in an EST 257
    3.3.3 Space-Charge Effects in an EST 258
    3.3.4 MS/MS Inside an EST 259
3.4 Orbitrap Mass Analyzer with External RF Storage 259
    3.4.1 'Slow' *versus* 'Fast' Ion Injection 259
    3.4.2 'Fast' Injection: Axial *versus* Radial Ejection from an
          RF-Storage Device..261
3.5 The Orbitrap Analyzer as an Accurate Mass Detector in the LTQ
Orbitrap Hybrid Mass Spectrometer and its Extensions 263
    3.5.1 Instrument Configuration and Operation 263
    3.5.2 Additional Analytical Capabilities Provided by the C-Trap...........264
    3.5.3 Applications of the LTQ Orbitrap Instrument................................264
    3.5.4 Extensions of the LTQ Orbitrap 267
3.6 The Orbitrap Mass Analyzer as an Accurate Mass Detector in a
Stand-Alone Mass Spectrometer 266
3.7 Further Developments of the Orbitrap Mass Analyzer 267
3.8 Conclusion 268
Acknowledgments...268
References...269

3.1 INTRODUCTION: ORBITRAP ANALYZER AS A SPECIAL CASE OF ELECTROSTATIC ION TRAPS

3.1.1 WHAT IS AN ELECTROSTATIC ION TRAP?

Being in many ways an unusual newcomer, the Orbitrap™ mass analyzer [1] nevertheless represents only the first commercial implementation of a highly promising class of devices where ions are separated on the basis of their m/z-values over the course of multiple reflections or deflections in an electrostatic field.

A very large number of oscillations require inherent stability of ion motion. In electrostatic fields, this requirement comes seemingly into contradiction with Earnshaw's theorem [2]: "a collection of point charges cannot be maintained in a stable stationary equilibrium configuration solely by the electrostatic interaction of the charges." This contradiction is, in reality, circumvented by making the equilibrium non-stationary, that is, by providing the ions with substantial velocities. Typically, ions are accelerated by potentials of a few hundreds to few thousands of volts. As a consequence of this requirement, it becomes important to avoid significant reduction of ion energy over the duration of trapping. Therefore, collisions with background gas should be minimized always in such devices in order to provide mean free paths that are much greater typically than the overall path length.

The term 'electrostatic traps' (ESTs) is proposed to denote such substantially electrostatic ion-optical systems where the ions experience multiple changes of direction (reflections or deflections) along at least one spatial dimension. When ions change direction multiple times along *all* dimensions over the duration of m/z-separation, such traps can be called 'closed electrostatic traps', otherwise they are referred to as 'open electrostatic traps'. The latter utilize only one pass (or one reflection) of ions along one of the spatial dimensions and, therefore, allow typically a much smaller number of direction changes along the dimension of m/z-dispersion. On the other hand, such traps reduce overlapping of the trajectories of ions having different values of m/z.

3.1.2 TAXONOMY OF ELECTROSTATIC TRAPS (ESTs)

Three major types of mass-analyzing ESTs have been described in the literature [1,3–23]:

(a) Reflection ESTs, where the axis of m/z-separation is straight;
(b) Deflection ESTs, where the axis of m/z-separation is curved;
(c) Orbital ESTs, where the axis of m/z-separation is straight but ions move along a curved axis in a perpendicular direction.

Ions may be mass-analyzed in such traps using very different principles, such as:

- Image current detection with subsequent Fourier or Hadamard transform to yield the frequency spectrum and hence m/z [24];
- Mass-selective ejection, wherein ions are excited by additional pulsed or radio frequency (RF) voltages and are detected directly using a secondary electron multiplier;

TABLE 3.1
Taxonomy of Electrostatic Traps (ESTs)

Type of EST	Principle of Detection	Examples of Closed Type	Examples of Open Type
Reflection	Image current	Melzner [3], Benner [4], May [5]	Could be done using approach of reference [4]
	Resonant excitation	Behrisch [6; (Figure 6),7]	
	TOF with secondary electron detection	Gridless mirrors by Wollnik [8],Dahan [9]	Multi-mirror by Wollnik [8; (Figure 1)], parallel-mirror by Su [10], Nazarenko [11], Verentchikov [12]
Deflection	Image current	Moller [13], Oksman [14]	
	Resonant excitation	Moller [13], Yamaguchi [15]	
	TOF with secondary electron detection	Moller [13], Yamaguchi [15], MULTUM [16], and MULTUM-2 [17]	Sakurai [18], Satoh [19]
Orbital	Image current	Orbitrap [1]	
	Resonant excitation	Knight [20]	
	TOF with secondary electron detection	As used for detection of products by Yang [21]	Spiratron (though no reflection) [22], ion mirror by Alimpiev [23]

- Time-of-flight (TOF) separation of pulsed ion packets with the direct detection of multiple values of m/z on a secondary electron multiplier;
- Potentially, TOF separation of a modulated ion beam with direct detection and subsequent Hadamard de-modulation [25] to yield ion TOFs and hence the values of m/z (not implemented yet).

Examples of such traps are presented in Table 3.1. At the time of this publication, only the Orbitrap mass analyzer has made its way into mainstream mass spectrometry, consequently this chapter focuses on the properties and applications of this instrument.

3.1.3 Introduction to the Orbitrap Mass Analyzer

The term "Orbitrap" was first coined to define a mass analyzer based on orbital trapping in an EST with harmonic oscillations and image current detection. Therefore, the Orbitrap analyzer extends the family of Fourier transform mass analyzers, which until recently contained only one widely used analyzer: the Fourier transform ion cyclotron resonance (FT-ICR) mass spectrometer.

The harmonic nature of the oscillations, which enables the Orbitrap mass analyzer to provide, for a given size, the highest frequency, and the highest quality of focusing within this class of ESTs makes image current detection a logical choice for it. Furthermore, the harmonicity of the oscillations of the ions ensures the superior performance of the Orbitrap analyzer with respect to the dynamic range, mass accuracy, and resolving power.

The roots of this analyzer can be traced back to 1923, when the principle of orbital trapping was proposed by Kingdon [26]. Over the next half a century it was used for spectroscopy experiments but it proved to be really difficult to turn the method into anything resembling a mass spectrometer. It was the quadro-logarithmic field used for orbital trapping of laser-produced ions which enabled Knight to perform crude mass analysis by applying axial resonant excitation to the trapped ions [20]. At the same time, this attempt demonstrated that the quest for a new mass analyzer would require very many improvements in all the key areas, most notably a more accurate definition of the quadro-logarithmic field, ion injection from an external ion source, and an improvement in ion detection.

These issues have been addressed successfully by Makarov and co-workers [1,27]. Since then, a number of very significant technological advances have been implemented, especially the development of pulsed injection from an external ion-storage device [27–29]. The advent of pulsed injection has allowed the Orbitrap analyzer to enter mainstream mass spectrometry as a part of a hybrid instrument [29]. Both proof-of-principle and product development have been carried out entirely by the same scientific team, just within the time-span of a few years.

3.2 OPERATION OF THE ORBITRAP MASS ANALYZER

3.2.1 MOTION OF TRAPPED IONS

The Orbitrap mass analyzer consists of an outer barrel-like electrode and a central spindle-like electrode along the axis. These electrodes are shaped in such a way that the quadro-logarithmic potential distribution, $U(r, z)$, is formed [30,31]

$$U(r,z) = \frac{k}{2}\left(z^2 - \frac{r^2}{2}\right) + \frac{k}{2}(R_m)^2 \cdot \ln\left[\frac{r}{R_m}\right] + C \tag{3.1},$$

where r, z are cylindrical coordinates ($z = 0$ being the plane of the symmetry of the field), C is a constant, k is the field curvature, and R_m is the characteristic radius. The last-mentioned has also a physical meaning: the radial force is directed toward the axis for $r < R_m$, and away from it for $r > R_m$, while at $r = R_m$ it equals 0.

When ions start their motion at the correct energy and radius, stable trajectories are formed which combine three cyclic motions (see Figure 3.1):

- Rotational motion with frequency of rotation ω_φ;
- Radial motion with frequency ω_r;
- Axial oscillations along the central electrode with frequency ω.

In practice, it is preferable to have the resulting spiral motion as close to circular as possible, because this reduces the influence of field imperfections. For a radius R of this spiral,

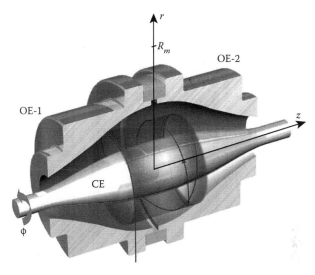

FIGURE 3.1 Diagram of the Orbitrap mass analyzer showing a stable spiral trajectory of an ion between the central electrode (CE) and split outer electrodes (OE-1 and OE-2). The value of R_m is indicated on the radial axis, r.

$$\omega_\varphi = \omega\sqrt{\frac{\left(\dfrac{R_m}{R}\right)^2 - 1}{2}} \qquad (3.2),$$

$$\omega_r = \omega\sqrt{\left(\frac{R_m}{R}\right)^2 - 2} \qquad (3.3),$$

and

$$\omega = \sqrt{\frac{e}{(m/z)} \cdot k} \qquad (3.4),$$

where e is the elementary charge (1.602×10^{-19} C).

Even in this simplest form rotational and radial frequencies show dependence upon the initial radius R, while the axial frequency is completely independent of all initial velocities and coordinates of the ions. Therefore, only the axial frequency can be used for determinations of mass-to-charge ratios, m/z.

Another peculiar consequence of the spiral motion is the increased requirement on radial field strength: rotation around the electrode will be stable only when Equation 3.3 is positive, which requires $R < R_m/\sqrt{2}$, that is ions should reside deep inside the potential well in order not to fly over its ridge.

3.2.2   Ion Detection

The axial oscillation frequencies can be detected directly by measuring the image current on the outer Orbitrap electrodes, as shown in Figure 3.1. A broadband detection is followed by a fast Fourier transform (FFT) to convert the recorded time-domain signal into a mass-to-charge ratio spectrum [24]. The image current is amplified and processed exactly in the same way as for FT-ICR, resulting in a similar sensitivity and signal-to-noise ratios. There is a minor but important distinction, however: the square-root dependence originating from the electrostatic nature of the field causes a much slower drop in resolving power observed for ions of increased m/z-value. As a result, the Orbitrap analyzer may outperform FT-ICR in this respect for ions above a particular m/z (typically, above 1000–2000 Th).

There could be an alternative way of detecting ions which follows the original proposal of Knight [20]: to excite the ions axially using a voltage at a resonant frequency and to scan the mass range by sweeping this frequency. As shown in reference [1], the main advantage of this approach over conventional traps is the ability to eject any ions, including those of very high m/z-value, using only very low RF voltages. However, such a trap would not be able to carry out MS^n experiments (see Section 3.4 below), which severely limits its appeal in comparison to a Paul trap. In comparison to the image current detection, this method provides significantly lower resolving power for the same scan time and wide mass range; it is also expected to suffer from higher susceptibility to space charge. Therefore, the image current detection remains the major mode of operation for the Orbitrap mass spectrometers.

3.2.3   Formation of Coherent Ion Packets

The most important pre-requisite for image current detection is the relatively small size of the ion cloud: its axial size in the analyzer must be significantly smaller than the amplitude of oscillations we wish to detect. The requisite cloud size can be achieved in one of two ways:

- Excitation of ions from the equatorial plane by applying excitation to the split outer electrodes of the trap. This approach is traditional in FT-ICR analyzers and has been applied also for resonant excitation in the Orbitrap [32,33].
- Excitation by off-axis injection of pulsed ion packets (excitation by injection). This approach minimizes perturbations of the quadro-logarithmic field, but requires a very fast ejection of a large ion population from an ion source or from an external RF-storage device.

The second approach was used first with a pulsed laser source [1], which provides naturally the required parameters of ion packets. When the second approach was attempted using continuous ion sources, the challenge was shifted to designing an appropriate external storage device with matching characteristics.

3.2.4   BASIC ANALYTICAL PARAMETERS OF THE ORBITRAP MASS ANALYZER

In common with some other mass analyzers (for example, the quadrupole mass filter), analytical parameters of the Orbitrap are, to a large extent determined, by the present status of manufacturing technology and electronics. The current level of precision machining enables the resolving power of the Orbitrap mass analyzer to reach several hundred thousand. Internal and thermal noise of electronic components of the preamplifier impacts upon the sensitivity of image current detection, making the limit of detection of the order of five to 10 elementary charges for one-second acquisition. The mass error using external mass calibration is a few ppm, and remains stable over a 24-hour period in a typical laboratory environment. The accuracy is limited principally by the noise of electronic components of the high voltage power supply. An acquisition speed up to 10 spectra s^{-1} reflects the present speed at which we are able to apply a high voltage to the central electrode. While the maximum resolving power of the Orbitrap analyzer remains inferior to that of FT-ICR mass analyzers, the combination of analytical parameters appears to be sufficiently attractive, given the absence of a superconducting magnet and of lower m/z ion losses suffered during transfer from an external ion-storage device.

Other analytical parameters (such as dynamic range) are much more dependent on the system integration of the final instrument and, therefore, are discussed later.

3.3   COMMON PROPERTIES OF THE ORBITRAP AND OTHER ESTs

3.3.1   AN OVERVIEW

It should be kept in mind that many of Orbitrap properties and associated technical solutions are common to all ESTs, for example:

- The necessity to form short ion packets along the direction of m/z dispersion;
- The use of time-dependent fields to introduce ions into closed traps;
- The link between accuracy of m/z measurement and stability of voltages on electrodes;
- The limitations on resolving power imposed by accuracy of electrode shape and location;
- As already mentioned, the necessity to provide sufficiently long mean-free-path;
- Space-charge effects, etc.

3.3.2   DECAY OF ION PACKETS IN AN EST

As mentioned earlier, under ideal conditions the ions could remain in an EST indefinitely. Unfortunately, collisions with residual gas species cause the ions to scatter and limit the time that a signal can be detected. The signal is reduced for two main reasons. (1) The loss of momentum increases the size of the ion packet, thus increasing aberrations and accelerating loss of coherence caused by field imperfections. (2) Collisions could lead to prompt or metastable ion fragmentation. Both processes are random in time and, therefore, produce a non-coherent cloud of ions that cannot be detected by image current detection, even when ions are still stable within the trap.

The time between collisions is inversely proportional to the residual pressure inside the EST and to the cross-section of an ion. For a pressure of 10^{-10} mbar, the interval ranges from several seconds for small molecules to <1 s for small proteins. By improving the vacuum in the analyzer below this level, isotopic resolution of proteins up to several tens kilodaltons has been demonstrated [34].

When collisions are not a limiting factor, the resolving power of ESTs is limited by their field aberrations for given spreads of initial parameters of an ion packet. In the case of the Orbitrap analyzer, these aberrations appear only due to miniscule imperfections of the electrode manufacturing and to the presence of the slot for ion injection.

3.3.3   SPACE-CHARGE EFFECTS IN AN EST

Space-charge effects are the Achilles' heel of any trap. In ESTs, these effects appear typically at higher charge densities than in RF ion traps, and are especially pronounced in traps with a strong dependence of the period of oscillation on ion energy. These effects include:

- Mass shifts, that is dependence of the period of oscillation on the total charge in the trap and the intensity of individual peaks [35]. Unlike in other types of traps, these mass shifts are typically independent of m/z;
- Coalescence, that is the mutual locking of closely separated m/z-packets at a certain phase difference. For image current detection, coalescence could even make these packets undetectable [36];
- Diffusion, that is increase of packet size by space–charge interactions and hence progressive growth of aberrations [37];
- Synchronization (self-bunching), that is decrease of packet size when a certain space-charge density is exceeded in a nonlinear field [37]. This counter-intuitive effect results in an undesirable difference of packet widths for m/z-values of different intensities (for example, isotopes).

Similar effects have been observed in other traps, for example in FT-ICR mass analyzers.

In the Orbitrap, all space-charge effects are reduced greatly by the shielding action of the central electrode that screens ions on one side of the ion ring from influencing the ions on the other side. The closer the trajectories move toward the central electrode, the lower is the influence of space charge. Any remaining mass shifts could be eliminated by an appropriate calibration procedure that is dependent linearly on the total ion charge inside the trap.

To avoid diffusion and self-bunching, a slight, controlled, distortion of the ideal electrode shape is introduced to provide the appropriate dependence of the oscillation period on ion energy [38].

3.3.4 MS/MS INSIDE AN EST

The ions trapped in ESTs have kinetic energies typically in the kiloelectron-volt range, therefore when ions suffer the inevitable high-energy collisions with background gas fragmentation of the ion occurs; the rate at which collisions occur can be regulated by controlling the gas pressure. Pulsed lasers offer another way to induce fragmentation of the ions. Unfortunately, when an ion decays in an EST its product ions will all have the same velocity. As their kinetic energy is proportional to their individual m/z-values, the trajectories become very distorted and different relative to the trajectories of the precursor ions.

In the case of the Orbitrap, low-mass product ions will fall onto the central electrode, while lower-charge state product ions will hit the outer electrodes. The ratio χ of the maximum to minimum values of m/z of product ions can be estimated approximately as $\chi \approx (R/R_1)^2$, where R is the radius of the trajectory of the precursor ions and R_1 is the maximum radius of the central electrode. In practice, $\chi = 3...4$.

This property limits seriously the utility of all ESTs for MS^n, bearing in mind especially other disadvantages, such as the absence of collisional cooling, increased cycle time, inferior resolving power, and sensitivity due to larger size of product ion packets, cost, and complexity of such an apparatus. This is why in all practical applications any EST (including the Orbitrap analyzer) is likely to be used only as an accurate mass detector, rather than an MS/MS device in its own right.*

3.4 ORBITRAP MASS ANALYZER WITH EXTERNAL RF STORAGE

3.4.1 'SLOW' VERSUS 'FAST' ION INJECTION

Continuous or quasi-continuous ion sources can be interfaced to the Orbitrap mass analyzer using only an additional means of external storage, which is provided normally by RF-only gas-filled multipoles, preferably a linear ion trap. In principle, once ions are trapped in the external storage device, any further manipulations of the ions, including isolation, fragmentation, MS^n, are possible. The ability to eject subsequently the stored ions into the Orbitrap analyzer required extensive development of the instrumentation. Two major approaches were considered:

(1) 'Slow injection', where ions are transferred from an external circular trajectory into an orbit in the hyper-logarithmic field [32]. While the ions are being injected inside a cylindrical gap on the outside of the outer electrodes, the voltages applied to all the concentric electrodes are ramped so that the ions spiral through a slot between detection electrodes (see Figure 3.2). As a result, a flat but wide ring of ions is formed on the equator of the Orbitrap; the ring of ions is then excited by applying voltages to the split outer electrodes. The injection process can take up to few hundred microseconds, hence the use of the term 'slow injection'. Due

---

* See also Volume 4, Chapter 2: Ion Traps for Miniature, Multiplexed and Soft Landing Technologies, by Scott A. Smith, Christopher C. Mulligan, Qingyu Song, Robert J. Noll, R. Graham Cooks and Zheng Ouyang.

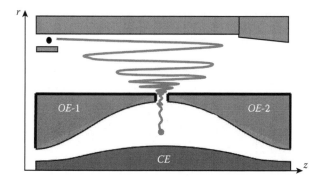

FIGURE 3.2 Schematics of ion-squeezing within the framework of 'slow injection'. The evolution of the ion's radius of rotation over many hundreds of microseconds is presented by the curved line superimposed over a cross-section of the appropriate hypothetical electrode structure. This approach has never been tested in practice.

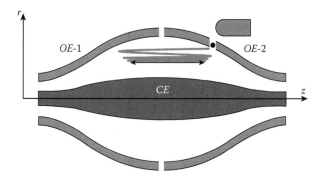

FIGURE 3.3 Schematics of ion-squeezing within the framework of 'fast injection'. The evolution of ion's radius of rotation over a few tens of microseconds is presented by the curved line superimposed over a cross-section of the Orbitrap analyzer. The final oscillatory movement of the ion following stabilization of the central electrode voltage is shown by a double-sided arrow. This approach is used currently in all practical Orbitrap analyzers.

to its complexity and possible ion losses, to date it has not been tested experimentally.

(2) 'Fast injection', where ions are injected into the Orbitrap analyzer at an offset from its equator in very short packets [1,27–29]. This mode of injection provides naturally "excitation by injection" (see Figure 3.3). Due to the strong dependence of the rotational frequencies on the ion energies, angles, and initial positions, each ion packet soon spreads over the angular coordinate and forms a thin rotating ring. Ultimately, the second approach has

proven to be more practical and robust, so that all practical Orbitrap mass analyzers use this approach.

In both cases, the radius of ion packet rotation is 'squeezed' down by increasing the electric field in the analyzer. This increase is created by ramping the voltage on either the central electrode during 'fast' injection or all electrodes during 'slow' injection. Following the excitation of the ion motion, all the voltages are stabilized so that no frequency drift can take place during image current detection.

It should be noted that prior to detection, ion packets form thin rings rotating around the central electrode and bouncing along it. The most important consequences of the ring structure are (i) many more ions can be present in the Orbitrap mass analyzer before the space-charge effects start influencing the mass resolution and accuracy of the measurement, and (ii) the radial or rotational frequencies never appear in the spectrum [1].

3.4.2 'FAST' INJECTION: AXIAL *VERSUS* RADIAL EJECTION FROM AN RF-STORAGE DEVICE

The high kinetic energy and short packet-width required by 'fast' injection comes into conflict with frequent gas collisions and low ion energies typical for any gas-filled RF-storage device. This conflict can be resolved by minimizing the path length until the ions are ejected from the RF storage field, either by concentrating the ions near to the exit from the storage region [27,28], or by the extracting ions through apertures in the RF electrodes [29]. Pulsed voltages are applied, therefore, to the end-electrodes [27,28] or across the RF electrodes [29] so that the ions find themselves in a strong extraction field. Additional lenses are used for the final spatial focusing of the ion beam into the entrance of the Orbitrap analyzer, as well as to facilitate differential pumping in order to achieve the very high vacuum necessary for effective mass measurement. Ions of individual mass-to-charge ratios arrive at the entrance of the Orbitrap analyzer as a tight packet with dimensions of a few millimeters, which is considerably smaller than the amplitude of their axial oscillations.

Extensive experiments with both schemes of ejection have shown that the axial ejection of ions from an RF-only quadrupole is affected severely by space-charge effects, which depend upon the number of ions, the intensity at each m/z-value, and other parameters. The resulting severe discriminations during ion injection into the Orbitrap proved to be the decisive factor in favor of the radial ejection.

The method of radial ejection is best implemented in a curved RF-only quadrupole (C-trap, see Figure 3.4). Ions are ejected along lines converging to the pole of curvature which coincides with the Orbitrap entrance. As the ions enter the Orbitrap, they are picked up and squeezed by its electric field. As the result, ions remain concentrated (within <1 mm^3) only for a very short time, so that space-charge effects do not have time to develop. Thus the C-trap allows one to interface the Orbitrap analyzer to any type of ion stream. It is important to note that this approach is applicable to almost all types of EST.

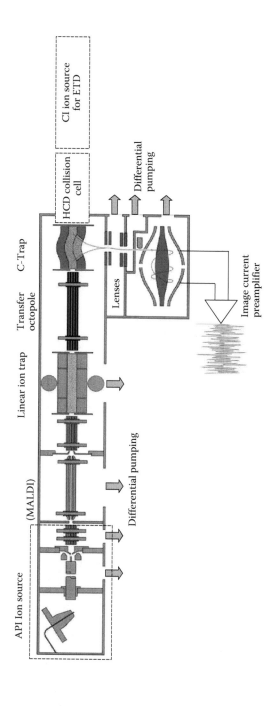

FIGURE 3.4 Schematic layout of the LTQ-Orbitrap mass spectrometer, with later extensions indicated by dashed lines: low-pressure MALDI source instead of the atmospheric-pressure ionization (API) source, HCD collision cell behind the C-trap and chemical ionization (CI) ion source for ETD behind the HCD cell.

3.5 THE ORBITRAP ANALYZER AS AN ACCURATE MASS DETECTOR IN THE LTQ ORBITRAP HYBRID MASS SPECTROMETER AND ITS EXTENSIONS

3.5.1 INSTRUMENT CONFIGURATION AND OPERATION

The entry of the Orbitrap analyzer into mainstream mass spectrometry took place in 2005 with the introduction of a hybrid instrument where it was combined, *via* the C-trap, with a linear ion trap (Figure 3.4). The linear ion trap with radial ion ejection was chosen because of its very high sensitivity, superb control of the ion population, short cycle time, and MS^n capability [39]. A detailed description of the instrument performance is given in reference [9].

Depending upon the analysis requirements, the two analyzers can be used independently or in concert, and provide quite similar MS^n spectra, the only major difference being the resolution and mass accuracy of the peaks observed. A true parallel operation is achieved by allowing for a short pre-view of the ions being measured in the Orbitrap analyzer; this pre-view then permits definition of the precursor ions for the linear ion trap to fragment (see Figure 3.5). One Orbitrap mass spectrum acquired

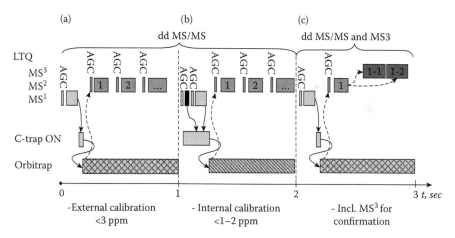

FIGURE 3.5 Schematics of parallel operation of the LTQ™ and Orbitrap mass spectrometers for the following typical data-dependent (dd) experiments. (a) Precursor high-resolution scan using the Orbitrap analyzer, with several MS/MS experiments initiated in the LTQ in parallel using a short initial portion of the Orbitrap scan. No internal calibrant is used. This experiment is typical for proteomics applications; (b) The same as (a), but with an additional stage of isolation of an internal calibrant (black rectangle) and mixing it with analyte in the C-trap prior to injection into the Orbitrap analyzer. This experiment allows maximum mass accuracy (typically sub-ppm); (c) The same as (a), but followed by more elaborate use of the LTQ for several dd MS^3 scans depending upon the results of the MS^2 scan. This experiment is especially useful for reliable confirmation, for example, in doping or metabolite studies. Automatic gain control (AGC) denotes a pre-scan using the LTQ, solid arrows indicate physical transfer of ions from LTQ to the C-trap or from the C-trap to the Orbitrap analyzer. Dashed arrows represent data-dependent decisions. A typical time-scale in seconds is shown on the time-line.

at a nominal resolving power of 60,000 (FWHM at $m/z = 400$) together with three linear ion trap product ion mass spectra are obtained within one second [29].

3.5.2 Additional Analytical Capabilities Provided by the C-Trap

The C-trap enables several intriguing modes of operation.

- The C-trap supports multiple fills. An injection of a fixed number of ions of a known reference compound can be followed by injection of analyte ions. Both sets of ions are then injected simultaneously into the Orbitrap. This procedure allows for a robust internal calibration of each mass spectrum, with r.m.s. errors below 1 ppm [40].
- Multiple injections of ions fragmented or selected under different conditions can be stored together and acquired in a single Orbitrap mass spectrum.
- Ions can be fragmented by injecting them into the C-trap at higher energies to yield fragmentation patterns similar to those in triple-stage quadrupole mass spectrometers [41].
- The C-trap represents a 'T-piece' which allows one to interface it to additional devices, such as collision cells, ion/molecule reaction cells, electron transfer dissociation (ETD), etc.

The last-mentioned quality has been utilized extensively for later extension of the LTQ Orbitrap instrument.

3.5.3 Applications of the LTQ Orbitrap Instrument

The key attributes of the Orbitrap analyzer, namely ruggedness, high mass accuracy, sensitivity, dynamic range, and excellent resolving power, make it very powerful for use in both proteomics and in the analysis of small molecules, especially when dealing with very complex mixtures. The list of publications describing applications of the LTQ Orbitrap mass spectrometer is growing quickly. A few examples are listed below.

In proteomics, major applications include:

- High throughput bottom-up techniques, with [42–44] or without [45] isotopic labeling. For the former, the SILAC (Stable Isotope Labeling using Amino Acids in Cell Culture) [42,43] approach is especially notable, as it requires all capabilities of the instrument to be fully utilized;
- Post-translational modifications, [46] especially phosphorylation;
- Middle-down analysis of small proteins [47];
- *De novo* sequencing of peptides [41,48];
- Analysis of intact proteins such as antibodies [49].

With respect to small molecule analyses, applications include:

- Biotransformation profiling *via* MS^n coupled with accurate mass measurement, all in a single experiment [50,51];

- Metabolic networks [52];
- Drugs of abuse [53–55].

In some areas of research the use of a high resolution/accurate mass analyzer with MS^n capabilities promises to cause a major shift in experimental approaches and strategies. For instance, the study of lipid mixtures, traditionally using precursor ion and neutral loss ion scanning on triple-quadrupole mass spectrometers, is now able to adopt a 'profiling' approach relying on high resolution and accurate mass [56,57]. This method has the potential to revolutionize the study of lipids (lipidomics), making it a high-throughput global approach akin to biomarker discovery in the areas of proteomics and metabolomics.

3.5.4 EXTENSIONS OF THE LTQ ORBITRAP

Since the introduction of the original Orbitrap instrument, further developments have taken place toward expanding its analytical capabilities.

First, the addition of a collision cell behind the C-trap has allowed the use of collisions at energies higher than those accessible in the linear trap, hence the use of the term *higher-energy collisional dissociation* (HCD) [41]. Ions are allowed to pass through the C-trap, enter the acceleration gap and then fragment in an RF-only multipole collision cell in a way similar to fragmentation in triple-quadrupole instruments. Unlike the latter, the product ions are then trapped and cooled inside the multipole. The ions are then returned back to the C-trap, from which they are injected into the Orbitrap mass analyzer in the usual manner. This novel procedure enables collection of almost all of the product ions without low mass cut-off and, therefore, makes possible the analysis of immonium ions, quantitation with iTRAQ™ or TMT™ tags, *de-novo* sequencing, etc.

A second important extension includes the addition of ETD capabilities, and even more extensive use of the C-trap as a T-piece device [58]. For ETD, reagent ions (such as the fluoroanthene anion) are produced by a chemical ionization (CI) ion source behind the HCD collision cell, pass through the HCD cell and the C-trap into the linear ion trap, where ion/ion reactions take place. The ETD product ions are then analyzed by the LTQ or transferred to the C-trap for further Orbitrap analysis. ETD is a powerful fragmentation method that allows one to analyze a much greater variety of post-translational modifications than is possible with collision-induced dissociation (CID) or HCD, for example multiple phosphorylation, glycosylation, sulfation, etc.*

Due to high resolving power and mass accuracy of the Orbitrap analyzer, this analysis can be carried out not only for peptides but also for small and medium-size

---

* See also Volume 4, Chapter 15: Fragmentation Techniques for Protein Ions Using Various Types of Ion Trap, by J. Franzen and K. P. Wanczek; Volume 4, Chapter 16: Unraveling the Structural Details of the Glycoproteome by Ion Trap Mass Spectrometry, by Vernon Reinhold, David J. Ashline, and Hailong Zhang; Volume 5, Chapter 1: Ion/Ion Reactions in Electrodynamic Ion Traps, by Jian Liu and Scott A. McLuckey; Volume 5, Chapter 3: Methods for Multi-Stage Ion Processing Involving Ion/Ion Chemistry in a Quadrupole Linear Ion Trap, by Graeme C. McAlister and Joshua J. Coon.

proteins. The resulting instrument configuration permits the combination of CID with HCD and ETD using a data-dependant decision tree.

The addition of a matrix-assisted laser desorption ionization (MALDI) source operating at reduced pressure is another important extension of the capabilities of the LTQ Orbitrap instrument. Trapping in a gas-filled RF-only multipole removes limitations on laser power and the production of greater ion populations with a smaller number of shots [59]. This procedure permits a very high dynamic range of many thousands in just a single Orbitrap scan. At the same time, high ion transmission to the Orbitrap affords also high sensitivity. Important applications include peptide mass fingerprinting, imaging, and liquid chromatography-matrix-assisted laser desorption ionization (LC-MALDI) [59].

3.6 THE ORBITRAP MASS ANALYZER AS AN ACCURATE MASS DETECTOR IN A STAND-ALONE MASS SPECTROMETER

Following the introduction of the LTQ Orbitrap, a non-hybrid mass spectrometer, the *Exactive*™, has been developed wherein a stand-alone Orbitrap mass analyzer is combined with an atmospheric-pressure ionization source (API) [60].

Figure 3.6 shows the schematic layout of the instrument. Samples are introduced into the API source and the ions formed are transferred from the source through four stages of differential pumping using RF-only multipoles into a curved RF-only trapping quadrupole (the C-trap). In the C-trap the ions are accumulated and their energy dampened using a bath gas (nitrogen). Ions are then injected through three further stages of differential pumping using a curved lens system into the Orbitrap analyzer,

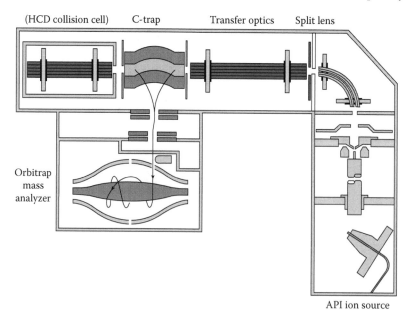

FIGURE 3.6 Layout of the *Exactive*™ mass spectrometer (including an optional HCD collision cell).

where mass spectra are acquired *via* image current detection. The vacuum inside the Orbitrap mass analyzer is maintained in the 10^{-10} mbar range.

A requirement of any ion trapping device is the ability to control the ion population within the trap. The AGC feature in combination with the precise determination of the ion injection time allows the instrument to be used for accurate quantitative analyses. Automatic control of the number of ions in the Orbitrap is achieved by measuring the total ion charge using a pre-scan and then using this quantity to calculate the ion injection time for the subsequent analytical scan. For very high scan rates, the previous analytical scan is used as a pre-scan in order to optimize the scan cycle time without compromising AGC. Ion gating is performed using a fast split lens setup that ensures the precise determination of the ion injection time.

The key performance features of the instrument are:

- Mass resolving power of up to 100,000 (FWHM at *m/z* 200);
- Scan speed of up to 10 spectra s^{-1} at nominal resolving power 10,000;
- Mass range *m/z* 50–4000;
- High in-scan dynamic range of signal intensity (four orders of magnitude)
- Mass accuracies of better than 2 ppm;
- Fast polarity switching (full cycle of one positive and one negative scan within one second).

In addition, the instrument introduces a feature of broad-band fragmentation without mass selection ("All Ions MS/MS") which can be implemented using an optional HCD collision cell after the C-trap.

The instrument appears to be very well suited for discovery work, screening applications [61], quantitative analyses [62], and elemental composition determinations.

3.7 FURTHER DEVELOPMENTS OF THE ORBITRAP MASS ANALYZER

All trapping mass analyzers are known to benefit from increasing field strength, major advantages being higher dynamic range, repetition rate, resolving power, and tolerance to space charge. Therefore, a new design of the Orbitrap mass analyzer has been developed, capable of providing maximum field strength at a given voltage. This design features a 50% thicker central electrode used with almost unchanged outer electrodes (see Figure 3.7 [63]). Similarly to high-field FT-ICR instruments, the new design has been named the High-Field (HF) Orbitrap.

As expected, the increased field strength inside the new Orbitrap mass analyzer has enabled a significant increase in resolving power for a given acquisition time: by about 50% at the same voltage, and by 80% with an increased voltage on the central electrode. It has been discovered that relative shifts of *m/z*-value in the Orbitrap due to space charge are substantially *m/z*-independent, proportional to total ion signal, and compare very favorably even with the highest-field FT-ICR instruments. The reduction of space-charge effects is a direct result of a smaller gap and the resulting enhanced ion shielding in the HF Orbitrap compared to the standard Orbitrap or FT-ICR analyzer.

(a)   (b)

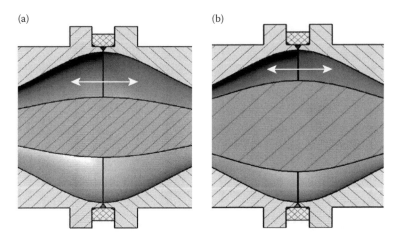

FIGURE 3.7 Comparison of a standard (a) and a high-field Orbitrap™ analyzer (b). The double-sided arrows indicate the average radius of ion trajectory and amplitude of ion oscillations.

By careful balancing of construction tolerances, a maximum resolving power in excess of >600,000 has been achieved for m/z 195 and >350,000— for m/z 524. It is still far below record values obtained in FT-ICR mass spectrometry, but it is more than adequate even for the most demanding complex mixtures, such as petroleum.

3.8   CONCLUSION

It is clear that the Orbitrap mass analyzer has become a unique and powerful addition to the arsenal of mass spectrometric techniques for probing biological systems and increasing the selectivity of, and confidence in, routine analyses.

Orbitrap analyzer technology is expected to continue evolving toward higher acquisition speed, higher resolving power and mass accuracy, and higher sensitivity. This evolution will undoubtedly give rise to exciting new applications of this analyzer. This progress is expected to be accompanied by an increasing use of other forms of ESTs in mass spectrometry.

ACKNOWLEDGMENTS

The author expresses his deep gratitude to all those who ensured the success of the Orbitrap development: S. Davis, A. Hoffmann, R. Lawther, N. Demetriades, J. Hughes of HD Technologies (Manchester, UK); M. Hardman, and B. McKnight of Thermo Masslab Ltd (Manchester, UK); E. Hemenway, M. Antonczak, M. Senko, J.E.P. Syka, J. Schwartz, V. Zabrouskov, T. Second, I. Jardine, T. Ziberna, L. Taylor, J. Hurst, and C. Katz of ThermoFisher Scientific (San Jose, CA); E. Denisov, A. Kholomeev, W. Balschun, O. Lange, K. Strupat, S. Horning, R. Pesch, J. Srega, G. Jung, W. Huels, F. Czemper, O. Hengelbrock, A. Wieghaus, J. Griep-Raming,

E. Schroeder, U. Froehlich, D. Nolting, R. Malek, T. Moehring, H. Muenster, M. Kellmann, M. Zeller, S. Strube, and M. Scigelova of ThermoFisher Scientific (Bremen, Germany). The author is also very grateful to his wife, Anna Makarova, for her unfaltering support of the Orbitrap endeavor from its very first days.

REFERENCES

1. Makarov, A. Electrostatic axially harmonic orbital trapping: a high-performance technique of mass analysis. *Anal. Chem.* 2000, *72*, 1156–1162.
2. Earnshaw, S. On the nature of the molecular forces which regulate the constitution of the luminiferous ether. *Trans. Camb. Phil. Soc.* 1842, *7*, 97–112.
3. Melzner, F. Pulsed time of flight mass spectrometers. *U.S. Patent* 1965, 3,226,543.
4. Benner, W.H. A gated electrostatic ion trap to repetitiously measure the charge and *m/z* of large electrospray ions. *Anal.Chem.* 1997, *69*, 4152–4168.
5. May, M.A.; Marshall, A.G.; Wollnik, H. Spectral analysis based on bipolar time-domain sampling: a multiplex method for time-of-flight mass spectrometry. *Anal.Chem.* 1992, *64*, 1601–1605.
6. Behrisch, R.; Blauth, E.W.; Melzner, F.; Meyer, E.H. Mass spectrometer. *U.S. Patent* 1965, 3,174,034.
7. Makarov, A.A.; Giannakopulos, A. Method of operating a multi-reflection ion trap. *Patent* 2008, WO2008059246.
8. Wollnik, H.; Przewloka, A. Time-of-flight mass spectrometers with multiply reflected ion trajectories. *Int. J. Mass Spectrom.* 1990, *96*, 267–274.
9. Dahan, M.; Fishman, R.; Heber, O.;Rappaport, M.; Altstein, N.; Zajfman, D.; van der Zande, W.J. A new type of electrostatic ion trap for storage of fast ion beams. *Rev. Sci. Instrum.* 1998, *69*, 76–83.
10. Su, C-S. Multiple reflection type time-of-flight mass spectrometer with two sets of parallel-plate electrostatic fields. *Int. J. Mass Spectrom.* 1989, *88*, 21–28.
11. Nazarenko, L.M.; Sekunova, L.M.; Yakushev, E.M. Time-of-flight mass spectrometer with multiple reflections. *Russian Patent*, SU 1992, 1,725,289.
12. Verentchikov, A.N.; Yavor, M.I.; Hasin, Y.I.; Gavrik, M.A. Multi-reflection planar time-of-flight mass analyzer, Parts I and II. *Tech. Phys.* 2005, *50*, 73—86.
13. Moller, S.P. ELISA, an electrostatic storage ring for atomic physics. *Nucl. Instrum. Methods in Phys. Research Section A*, 1997, *394*, 281–286.
14. Oksman, P. A Fourier transform time-of-flight mass spectrometer. A SIMION approach. *Int. J. Mass Spectrom.* 1995, *141*, 67–76.
15. Yamaguchi, S.; Ishihara, M.; Toyoda, M.; Okumura, D. Time-of-flight mass spectrometer. *US Patent* 2006, 7,148,473.
16. Matsuo, T.; Toyoda, M.; Sakurai, T.; Ishihara, M. Ion optics for multi-turn time-of-flight mass spectrometers with variable mass resolution. *J. Mass Spectrom.* 1997, *32*, 1179–1185.
17. Okumura, D.; Toyoda, M.; Ishihara, M.; Katakuse, I. A compact sector-type multi-turn time-of-flight mass spectrometer MULTUM II. *Nucl. Instrum. Methods in Phys. Research Section A* 2004, *519*, 331–337.
18. Sakurai, T.; Ito, H.; Matsuo, T. MS/MS and MS/MS/MS analyses in a multisector time-of-flight mass spectrometer. *Anal. Chem.* 1994, *66*, 2313–2317.
19. Satoh, T.; Tsuno, H.; Iwanaga, M.; Kammei, Y. The design and characteristic features of a new time-of-flight mass spectrometer with a spiral ion trajectory. *J. Am. Soc. Mass Spectrom.* 2005, *16*, 1969–1975.
20. Knight, R.D. Storage of ions from laser-produced plasmas. *Appl. Phys. Lett.* 1981, *38*, 221–222.

21. Yang, L.; Church, D.A.; Tu, S.; Jin, J. Measured lifetimes of selected metastable levels of Ar^q+ ions (q = 2,3,9 and 10) stored in an electrostatic ion trap. *Phys. Rev. A* 1994, *50*, 177–186.

22. Bakker, J.M.B. The Spiratron. In: *Advances in Mass Spectrometry.* Vol. 5, ed. A. Quayle, London: Institute of Petroleum, 1971, 278–280.

23. Alimpiev, S.S.; Makarov, A.A.; Mlynski, V.V.; Nikiforov, S.M.; Sartakov, B.G. Time-of-flight mass spectrometer. *Russian Patent,* SU 1991, 1,716,922.

24. Marshall, A.G.; Verdun, F.R. *Fourier Transforms in NMR, Optical, and Mass Spectrometry: A User's Handbook.* Elsevier: Amsterdam, 1990.

25. Brock, A.; Rodriguez, N.; Zare, R.N. Hadamard transform time-of-flight mass spectrometry. *Anal. Chem.* 1998, *70*, 3735–3741.

26. Kingdon, K.H. A method for the neutralization of electron space charge by positive ionization at very low gas pressures. *Phys. Rev.* 1923, *21*, 408–418.

27. Hardman, M.; Makarov, A.A. Interfacing the Orbitrap mass analyzer to an electrospray ion source. *Anal. Chem.* 2003, *75*, 1699–1705.

28. Hu, Q.; Noll, R.J.; Li, H.; Makarov, A.; Hardman, M.; Cooks, R.G. The Orbitrap: a new mass spectrometer. *J. Mass Spectrom.* 2005, *40*, 430–443.

29. Makarov, A.; Denisov, E.; Kholomeev, A.; Balschun, W.; Lange, O.; Horning, S.; Strupat, K. Performance evaluation of a hybrid linear ion trap/orbitrap mass spectrometer. *Anal. Chem.* 2006, *78*, 2113–2120.

30. Korsunskii, M.I.; Bazakutsa, V.A. A study of the ion-optical properties for a sector-shaped electrostatic field of the difference type. *Soviet Physics-Tech. Phys.* 1958, *3*, 1396–1409.

31. Gall, L.N.; Golikov, Y.K.; Aleksandrov, M.L.; Pechalina, Y.E.; Holin, N.A. Time-of-flight mass spectrometer. *USSR Inventor's Certificate* 1986, 1,247,973.

32. Makarov, A.A. Mass spectrometer. *US Patent* 1999, 5,886,346.

33. Hu, Q.; Li, H.; Makarov, A.; Cooks, R.G.; Noll, R.J. Resonant AC dipolar excitation for ion motion control in the Orbitrap mass analyzer. *J. Phys. Chem. A* 2006, *110*, 2682–2689.

34. Denisov, E.; Strupat, K.; Makarov, A.A.; Zabrouskov, V. Pushing intact protein detection limits of the Orbitrap mass analyzer. *Proc. 55th ASMS Conference on Mass Spectrometry and Allied Topics,* Indianapolis, June 3–7, 2007.

35. Masselon, C.; Tolmachev, A.V.; Anderson, G.A.; Harkewicz, R.; Smith, R.D. Mass measurement errors caused by "local" frequency perturbations in FTICR mass spectrometry. *J. Am. Soc. Mass Spectrom.* 2002, *13*, 99–106.

36. Mitchell, D.W.; Smith, R.D. Cyclotron motion of two Coulombically interacting ion clouds with implications to Fourier-transform ion cyclotron resonance mass spectrometry. *Phys. Rev. E* 1995, *52*, 4366–4386.

37. Pedersen, H.B.; Strasser, D.; Amarant, B.; Heber, O.; Rappaport, M.L.; Zajfman, D. Diffusion and synchronization in an ion-trap resonator. *Phys. Rev. A* 1992, *65*, 042704.

38. Makarov, A.A.; Denisov, E.V.; Jung, G.; Balschun, W.; Horning, S. Improvements in an electrostatic trap. *Patent* 2006, WO2006129109.

39. Schwartz, J.C.; Senko, M.W.; Syka, J.E.P. A two-dimensional quadrupole ion trap mass spectrometer. *J. Am. Soc. Mass Spectrom.* 2002, *13*, 659–669.

40. Olsen, J.V.; de Godoy, L.M.; Li, G.; Macek, B.; Mortensen, P.; Pesch, R.; Makarov, A.A., Lange, O., Horning, S., Mann, M. Parts per million mass accuracy on an orbitrap mass spectrometer *via* lock mass injection into a C-trap. *Mol. Cell. Proteomics* 2005, *4*, 2010–2021.

41. Olsen, J.V.; Macek, B.; Lange, O.; Makarov, A.; Horning, S.; Mann, M. Higher-energy C-trap dissociation for peptide modification analysis. *Nature Methods* 2007, *4*, 709–712.

42. Graumann, J.; Hubner, N.C.; Kim, J.B.; Ko, K.; Moser, M.; Kumar, C.; Cox, J.; Scholer, H.; Mann, M. Stable isotope labeling by amino acids in cell culture (SILAC) and proteome quantitation of mouse embryonic stem cells to a depth of 5,111 proteins. *Mol. Cell. Proteomics* 2008, *7*, 672–683.

43. Hanke, S.; Besir, H.; Oesterhelt, D.; Mann, M. Absolute SILAC for accurate quantitation of proteins in complex mixtures down to the attomole Level. *J. Proteome Research* 2008, *7*, 1118–1130.

44. Bantscheff, M.; Boesche, M.; Eberhard, D.; Matthieson, T.; Sweetman, G.; Kuster, B. Robust and sensitive iTRAQ quantification on an LTQ-Orbitrap mass spectrometer. *Mol. Cell. Proteomics* 2008, *7*, 1702–1713.

45. Lu, B.W.; Motoyama, A.; Ruse, C.; Venable, J.; Yates, J.R.; III. Improving protein identification sensitivity by combining MS and MS/MS information for shotgun proteomics using LTQ-Orbitrap high mass accuracy data. *Anal. Chem.* 2008, *80*, 2018–2025.

46. Cantin, G.T.; Yi, W.; Lu, B.W.; Park, S.K.; Xu, T.; Lee, J.D.; Yates, J.R. III. Combining protein-pased IMAC, peptide-based IMAC, and MudPIT for efficient phosphoproteomic analysis. *J. Proteome Research* 2008, *7*, 1346–1351.

47. Waanders, L.F.; Hanke, S.; Mann, M. Top-down quantitation and characterization of SILAC-labeled proteins. *J. Am. Soc. Mass Spectrom.* 2007, *18*, 2058–2064.

48. DiMaggio, P.A.; Floudas, C.A.; Lu, B.W.; Yates, J.R. III. A hybrid method for peptide identification using integer linear optimization, local database search, and quadrupole time-of-flight or Orbitrap tandem mass spectrometry. *J. Proteome Research* 2008, *7*, 1584–1593.

49. Bondarenko, P.V.; Zabrouskov, V.; Makarov, A.; Zhang, Z. LC/MS top-down analysis and intact mass analysis of recombinant immunoglobulin gamma antibodies on Orbitrap. *Proc. 56th ASMS Conference on Mass Spectrometry and Allied Topics*, Denver, June 1–5, 2008.

50. Peterman, S.M.; Duczak, N.; Kalgutkar, A.S.; Lame, M.E.; Soglia, J.R. Application of a linear ion trap/Orbitrap mass spectrometer in metabolite characterization studies: examination of the human liver microsomal metabolism of the non-tricyclic anti-depressant nefazodone using data-dependent accurate mass measurements. *J. Am. Soc. Mass Spectrom.* 2006, *17*, 363–375.

51. Lim, H-K.; Chen, J.; Sensenhauser, C.; Cook, K.; Subrahmanyam, V. Metabolite identification by data-dependent accurate mass spectrometric analysis at resolving power of 60000 in external calibration mode using an LTQ/Orbitrap. *Rapid Comm. Mass Spectrom.* 2007, *21*, 1821–1832.

52. Breitling, R.; Pitt, A.R.; Barrett, M.P. Precision mapping of the metabolome. *Trends Biotechnol.* 2006, *24*, 543–548.

53. Thevis, M.; Kamber, M.; Schaenzer, W. Screening for metabolically stable arylpropionamide derived selective androgen receptor modulators for doping control purposes. *Rapid Commun. Mass Spectrom.* 2006, *20*, 870–876.

54. Thevis, M.; Krug, O.; Schaenzer, W. Mass spectrometric characterization of efaproxiral (RSR13) and its implementation into doping controls using liquid chromatography–atmospheric pressure ionization-tandem mass spectrometry. *J. Mass Spectrom.* 2006, *41*, 332–338.

55. Thevis, M.; Thomas, A.; Schanzer, W. Mass spectrometric determination of insulins and their degradation products in sports drug testing. *Mass Spectrom. Rev.* 2008, *27*, 35–50.

56. Ejsing, C.; Moehring, T.; Bahr, U.; Duchoslav, E. Collision-induced dissociation pathways of yeast sphingolipids and their molecular profiling in total lipid extracts: a study by quadrupole TOF and linear ion trap-orbitrap mass spectrometry. *J. Mass Spectrom.* 2006, *41*, 372–389.

57. Schwudke, D.; Hannich, J.T.; Surendranath, V.; Grimard, V.; Moehring, T.; Burton, L.; Kurzchalia, T.; Shevchenko, A. Top-down lipidomic screens by multivariate analysis of high-resolution survey mass spectra. *Anal. Chem.* 2007, *79*, 4083–4093.

58. McAlister, G.; Berggren, W.; Griep-Raming, J.; Horning, S.; Makarov, A.; Phanstiel, D.; Stafford, G.; Swaney, D.; Syka, J.; Zabrouskov, V.; Coon, J. A proteomics grade electron transfer dissociation-enabled hybrid linear ion trap-Orbitrap mass spectrometer. *J. Proteome Research* 2008, *7*, 3127–3136.

59. Strupat, K.; Bui, H.; Kovtoun, V.; Viner, R.; Prieto Conaway, M.C.; Izgarian, N.; Lange, O.; Stafford, G.; Horning, S.; Phillips, J.J.; Moehring, T. MALDI produced ions inspected with a linear ion trap-Orbitrap mass analyzer. *Proc. 56th ASMS Conference on Mass Spectrometry and Allied Topics*, Denver, June 1–5, 2008.

60. Wieghaus, A.; Makarov, A.A.; Froehlich, U.; Kellmann, M.; Denisov, E.; Lange, O. Development and applications of a new benchtop Orbitrap mass spectrometer. *Proc. 56th ASMS Conference on Mass Spectrometry and Allied Topics*, Denver, June 1–5, 2008.

61. Kellmann, M.; Taylor, L.; Ghosh, D.; Wieghaus, A.; Muenster, H. High resolution and high mass accuracy: a perfect team for food and feed analysis in complex matrices. *Proc. 56th ASMS Conference on Mass Spectrometry and Allied Topics*, Denver, June 1–5, 2008.

62. Bateman, K.; Muenster, H.; Kellmann, M.; Tiller, P.; Papp, R.; Taylor, L.C. Full scan data acquisition for rapid quantitative and qualitative analysis using a bench-top non-hybrid ESI-Orbitrap mass spectrometer. *Proc. 56th ASMS Conference on Mass Spectrometry and Allied Topics*, Denver, June 1–5, 2008.

63. Denisov, E.; Lange, O.; Makarov, A.; Balschun, W.; Griep-Raming, J. Performance of a high-field Orbitrap mass analyzer. *Proc. 56th ASMS Conference on Mass Spectrometry and Allied Topics*, Denver, June 1–5, 2008.

4 Rectangular Waveform Driven Digital Ion Trap (DIT) Mass Spectrometer: Theory and Applications

Francesco L. Brancia and Li Ding

CONTENTS

4.1 Introduction 273
4.2 Ion Motion in a Rectangular Wave Quadrupole Field 276
4.3 Secular Motion and the Pseudopotential Well 279
4.4 Digital Asymmetric Waveform 284
4.5 Digital Frequency Scan and Extended Mass Range for
     High Mass Detection 285
4.6 Non-Stretched Geometry and Field-Adjusting Electrode 286
4.7 Mass Resolution 290
4.8 Forward and Reverse Scanning 291
4.9 Scan Direction and Space Charge 295
4.10 Precursor Ion Isolation 298
4.11 Tandem Mass Spectrometry in Conjunction with Electron Capture
     Dissociation (ECD) 299
4.12 Digital Linear Ion Trap 302
4.13 Conclusion 305
References 306

4.1 INTRODUCTION

The use of radiofrequency (RF) quadrupole fields for mass spectral analsysis was invented by Wolfgang Paul and others in the 1950s. The method can be implemented as a linear, two-dimensional (2D) electrode array, in the quadrupole and the monopole, or as a three-dimensional (3D) 'ion trap,' comprising a ring electrode and two end-cap electrodes. The monopole has never been developed as an analytical instrument, and while the quadrupole mass spectrometer grew rapidly in terms of performance and range of applications, it was only with the introduction of the mass-selective instability scan mode [1] three decades later that the ion trap became a powerful and inexpensive alternative to other mass analyzers, exploiting its high mass resolution and high sensitivity for the analysis and detection of high mass compounds. A periodically-alternating quadrupole field is generated by application of RF and direct current (DC) voltages to appropriate electrodes. Under the influence of the quadrupole field, ions with different mass-to-charge ratios undergo stable or unstable oscillations, achieving either confinement within the trap or ejection. Since the development of the axial resonant modulation technique, the performance of the 3D quadrupole ion trap has been improved dramatically, transforming this mass spectrometer, originally introduced as a simple gas chromatography (GC) detector, into a much more versatile analytical tool. Ion traps combine ion storage with mass analysis, exhibiting an acceptable mass (m/z) range, applicable even for the analysis of high mass biomolecules, mass resolution up to 1×10^5, and with high sensitivity [2]. In addition, ion isolation followed by multi-stage tandem mass spectrometric analysis (MS^n) [3] via collisional-induced dissociation (CID), ion/ion electron transfer dissociation (ETD) [4], or electron capture dissociation (ECD) [5] can provide additional structural information on the analyte of interest.

In an RF-only ion trap, a mass spectrum is obtained by sweeping the RF voltage amplitude. The common feature among different types of quadrupole ion traps is the high frequency oscillator employed to generate the RF voltage. In the circuitry, the output waveform and frequency must be kept constant due to the limitation imposed by the resonant network. Under these conditions a mass spectrum can be generated only by changing the RF voltage (see also below). To understand better the effect on the mass range, it is easier to refer to the 'stability parameter,' q_z, used to express the Mathieu equation, which can be written as

$$m/e = \frac{4V}{q_z r_0^2 \Omega^2} \qquad (4.1),$$

where r_0 is the inscribed radius of the ring electrode, the closest separation of the end-cap electrodes, $2z_0$, is given by $r_0^2 = 2z_0^2$, and Ω is the drive frequency. The value of q_z lies within the range 0–0.908 for ions to possess stable trajectories and remain confined within the trap, and from Equation 4.1 it can be seen that there is a direct relationship between mass-to-charge ratio (m/e) and the zero-to-peak amplitude of the RF voltage (V). The upper limit of this range is the point of instability, at which the ions of a given m/e-value are ejected as the amplitude of the RF voltage (V) is increased. As a consequence, in amplitude-scanning devices detection of high mass-to-charge ratio ions requires high voltages (tens of kilovolts) which can cause discharge problems in the ion optics and electronics of the mass spectrometer. Resonant ejection at a lower q_z-value or at a lower drive frequency can provide a partial solution to the problem. Scanning the value of Ω of the sinusoidal trapping waveform at fixed amplitude is an alternative approach for extending the mass range in a quadrupole ion trap; this can be achieved by combining a waveform generator with a power amplifier [6]. The main problems associated with this approach are, however, a excessive power consumption and waveform distortions at high voltage, which affect the robustness of the frequency scan.

The preceding discussion starts from the assumption that sinusoidal waveforms are used to provide both the driving potential to the ring electrode and the resonant excitation to the two end-cap electrodes for ion isolation, ion excitation, and axial modulation. However, in principle, any type of waveform may be utilized to drive a quadrupole ion trap. In the late 1990s, Sheretov and collaborators developed hyperboloid mass analyzers for space exploration under the "VEGA" and "Mars-96" programs, in which ions were ejected *via* the mass-selective instability scan mode using pulsed or rectangular RF waveforms [7]. Ion motion in an ion trap driven by a rectangular wave quadrupolar field was described in the early 1970s [8,9], but since the stability parameters used were different from the conventional Mathieu parameters a and q, a direct comparison with the data obtained employing a sinusoidal waveform was difficult to establish. Hence, the work received limited attention from the scientific community. In 2000, Ding and Kumashiro described ion motion in a quadrupolar electric field that was driven by a rectangular wave voltage and suggested the mass-selective resonance method with the ion secular frequency under digital operation conditions [10].

The concept of the digital ion trap (DIT), described here, was first introduced with a theoretical study on ion motion in a quadrupolar field driven by a digital waveform. For the reader's understanding it is crucial to define precisely what a 'digital waveform' is. The adjective 'digital' takes its inspiration from the digital circuitry used to produce the rectangular waveforms. The circuits of the DIT switch very rapidly between discrete DC high voltage levels in order to generate the trapping waveform voltage applied to the ring electrode. The timing of the switches can be controlled precisely by specially designed digital circuitry (see Figure 4.1). When this idea was first conceived, an extensive simulation study was carried out in order to determine the *in silico* performance of the digital quadrupole ion trap [11]. Starting from what was demonstrated at a theoretical level, the first prototype was built in accordance with the output of this simulation work [12].

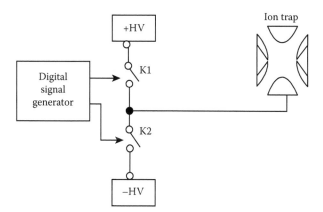

FIGURE 4.1   Schematic diagram of the DIT circuitry; K1 and K2 represent the two switches.

4.2   ION MOTION IN A RECTANGULAR WAVE QUADRUPOLE FIELD

Similar to a conventional 3D quadrupole ion trap, ion motion in a DIT can be studied using the ion stability conditions. In general, ion motion in a pure quadrupole field with a digital waveform time-dependence, such as displayed in Figure 4.2a, can be derived analytically, and different parameters can be chosen so that the stability conditions can be displayed graphically. The independent parameters of stability proposed by Sheretov and collaborators, where pulse intensities a_1 and a_2 are used for each of the positive and negative excursions, were not adopted widely since the conventional Mathieu parameters a and q have been preferred to describe the motion of trapped ions. In order to compare the performance of the DIT with that of a conventional RF-driven ion trap, it is useful to express the stability conditions in terms of similar parameters (a, q).

In this derivation, we select an 'unstretched' 3D quadrupole ion trap with the inscribed radius of the ring electrode $r_0^2 = 2z_0^2$, where z_0 is the axial distance from center of an ion trap to the inner surface of an end-cap electrode. For an ion of mass m and charge e moving in a trap driven by a periodic rectangular waveform (as seen in Figure 4.2), the stability parameters can be expressed as

$$a_z = \frac{8eU}{m\Omega^2 r_0^2} \tag{4.2a}$$

$$q_z = -\frac{4eV}{m\Omega^2 r_0^2} \tag{4.2b},$$

where $\Omega = 2\pi/T$ is the frequency, in which T is the period of the rectangular wave. While these expressions look exactly the same as those for the sinusoidal wave voltage, we must clarify what the DC and AC components, U and V, represent for this

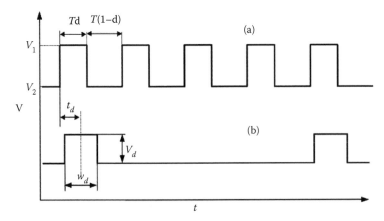

FIGURE 4.2   Rectangular wave voltages applied to the digital ion trap. (a) The drive voltage on the ring electrode; (b) The excitation voltage applied to the end-cap electrodes.

waveform. While the definitions of U and V as the DC voltage and AC amplitude (zero-to-peak), respectively, on the ring electrode are straightforward for a sinusoidal voltage, the situation can become complicated for other waveforms. For the DIT, we define usually U as the DC component of the waveform having duty cycle d,

$$U = dV_1 + (1-d)V_2 \qquad (4.3a)$$

and V to be the average AC amplitude over the waveform period, i.e., $|V_1 - U|d + |V_2 - U|(1-d)$, or

$$V = 2(V_1 - V_2)(1-d)d \qquad (4.3b)$$

(see Figure 4.2).

In the case of a square wave $d = 0.5$ (i.e., 50% duty cycle rectangular wave), U is the mean value between the positive and negative voltage levels and V corresponds to half of the voltage difference between the positive and negative voltage levels; when $V_1 = -V_2$ (symmetric digital levels) Equation 4.3 shows that for the square wave $U = 0$ for $d = 0.5$.

The quadrupolar potential in the DIT can be described as

$$\phi(r,z) = \frac{V_{rec}}{2r_0^2}(r^2 - 2z^2) + V_{rec}/2 \qquad (4.4a)$$

where V_{rec} is the rectangular wave voltage applied to the ring electrode. The motion in the z-direction can be expressed as

$$\frac{d^2z}{dt^2} = \frac{2eV_{rec}}{r_0^2 m}z \qquad (4.4b)$$

where during the high voltage level

$$V_{rec} = V_1 = U + \frac{V}{2d} \qquad (4.5a),$$

and during the low voltage level

$$V_{rec} = V_2 = U - \frac{V}{2(1-d)} \qquad (4.5b).$$

The solution to Equation 4.4b is therefore

$$z = Cch\left(\sqrt{\frac{2eV_{rec}}{r_0^2 m}}t\right) + Dsh\left(\sqrt{\frac{2eV_{rec}}{r_0^2 m}}t\right) \qquad (4.6).$$

Here C and D are constants determined by the initial conditions used. Since V_{rec} takes two constant values in two segments of one periodical cycle, the above solution can be expressed in two forms: as a hyperbolic function when $V_{rec} > 0$ and as a sinusoidal function when $V_{rec} < 0$. The ion motion is in fact defined repeatedly by these two mathematical functions. To study the stability of motion, the Mathieu equation is not longer applicable and the matrix transform method must be used [13]. Introducing the parameters

$$\xi = \Omega T / 2 \tag{4.7}$$

$$K_1 = \sqrt{\frac{q_z}{d} - a_z} \tag{4.8a}$$

and

$$K_2 = \sqrt{\frac{q_z}{d-1} - a_z} \tag{4.8b}$$

and assuming initial conditions

$$z\Big|_{\xi=0} = z_0; \quad \frac{dz}{d\xi}\Big|_{\xi=0} = \dot{z}_0 \tag{4.9}$$

the solution of Equation 4.6 becomes

$$z = z_0 ch(K_i\xi) + \dot{z}_0 sh(k_i\xi)/k_i \tag{4.10}$$

in which $K_i = K_1$ for the high voltage level V_1 and $K_i = K_2$ for the lower voltage level V_2.

From this solution we can obtain the transposed matrices in phase space, Φ_1 and Φ_2, for the high level excursion Td and the low level excursion $T(1-d)$, respectively. Taking one period T as a step indexed by n, the position and velocity of ion after one periodic cycle can be expressed as

$$\begin{bmatrix} z_{n+1} \\ \dot{z}_{n+1} \end{bmatrix} = \Phi \begin{bmatrix} z_n \\ \dot{z}_n \end{bmatrix} \tag{4.11}$$

where $\Phi = \Phi_1 \Phi_2 = \begin{bmatrix} \phi_{11} & \phi_{12} \\ \phi_{21} & \phi_{22} \end{bmatrix}$ can be obtained from Equation 4.10 as

$$\phi_{11} = ch(K_1\pi d)ch[K_2\pi(1-d)] + sh(K_1\pi d)sh[K_2\pi(1-d)]K_2/K_1 \tag{4.12a}$$

$$\phi_{12} = ch(K_1\pi d)sh[K_2\pi(1-d)]/K_2 + sh(K_1\pi d)ch[K_2\pi(1-d)]/K_1 \tag{4.12b}$$

$$\phi_{21} = sh(K_1\pi d)ch\left[K_2\pi(1-d)\right]K_1 + ch(K_1\pi d)sh\left[K_2\pi(1-d)\right]K_2 \qquad (4.12c)$$

$$\phi_{22} = sh(K_1\pi d)sh[K_2\pi(1-d)]K_1/K_2 + ch(K_1\pi d)ch[K_2\pi(1-d)] \qquad (4.12d).$$

Under the continuous rectangular wave electric field, the complete and necessary conditions to stabilize the ion motion are met when the modulus of the eigenvalues of the above transposed matrices are either less than or equal to unity.

The eigenfunction of the matrix can be written as

$$\lambda^2 - \left\{2ch(K_1\pi d)ch\left[K_2\pi(1-d)\right] + \frac{K_1^2 + K_2^2}{K_1K_2}sh(K_1\pi d)sh\left[K_2\pi(1-d)\right]\right\}\lambda + 1 = 0 \quad (4.13).$$

For this equation to have roots $|\lambda| \le 0$, the value of Δ for the quadratic equation must be zero or negative, $i.e.$,

$$-1 \le ch(K_1\pi d)ch\left[K_2\pi(1-d)\right] + \frac{K_1^2 + K_2^2}{K_1K_2}sh(K_1\pi d)sh\left[K_2\pi(1-d)\right] \le 1 \qquad (4.14),$$

which defines the boundaries of the stability region. By substitution of Equations. 4.8a and b into Equation 4.14, we can obtain the boundary function of the stable region in terms of a_z and q_z coordinates, with d as an additional parameter. The stability diagram of ion motion is plotted in the z-direction, as shown in Figure 4.3.

Under the same rectangular voltage, the ion motion function in the *radial* (*r*) direction is similar to that in the axial (*z*) direction, although the potential compared to that in Equation 4.3 is halved and has the opposite polarity. Using the approach adopted for conventional ion traps, the stable region of radial motion can be plotted as a function of a_z and q_z, so that the combined stability regions in both the z- and r-directions are obtained. A portion of the combined stability diagram is displayed in Figure 4.4. In the case of a symmetric square wave, where $d = 0.5$, the intersection of the boundary of the first stability region with the q_z-axis gives a q_z-value equal to 0.712, which is smaller than that obtained using the sinusoidal waveform.

4.3   SECULAR MOTION AND THE PSEUDOPOTENTIAL WELL

The secular frequencies that characterize ion oscillations in a rectangular wave quadrupole field, as displayed in Figure 4.5, are quite similar to those observed in a sinusoidal wave quadrupole field. Different (a_z, q_z) points on the stability diagram generate different oscillation patterns. In this section we now study the frequency of this motion in the case of the rectangular drive waveform.

From Equation 4.12, it can be proved further that

$$\phi_{11}\phi_{12} - \phi_{12}\phi_{21} = 1 \qquad (4.15).$$

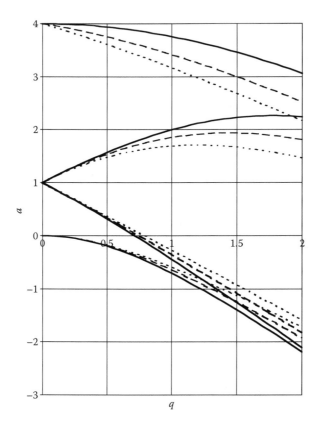

FIGURE 4.3 The stability diagram for ion motion in the z-direction in the rectangular wave-form quadrupole ion trap. The solid line, the dashed line, and dotted line are for a duty cycle of 0.5. (Reproduced from Ding, L., Sudakov, M., and Kumashiro, S., *Int. J. Mass Spectrom.* 2002, *221*, 117–138. With permission from Elsevier.)

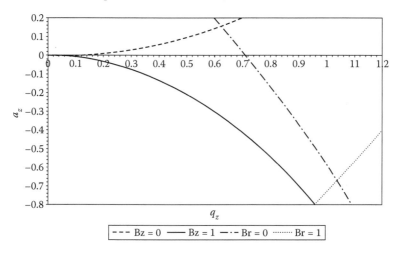

FIGURE 4.4 First region of the stability diagram for the DIT.

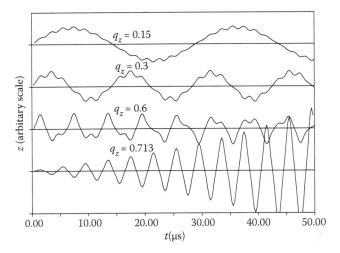

FIGURE 4.5 Trajectories of ions showing the different secular frequencies at different working points on the (a, q) stability diagram.

By using Equation 4.11 for the two sequential rectangular wave cycles we obtain

$$z_{n+1} + z_{n-1} = (\phi_{11} + \phi_{22})z_n \tag{4.16}.$$

This equation is satisfied at the end of every rectangular driven wave cycle (where $\Omega T_n = 2n\pi$) for a stable ion trajectory by the following solutions

$$z = A_l \cos(2l \pm \beta_z)\frac{\Omega T}{2} + B_l \sin(2l \pm \beta_z)\frac{\Omega T}{2} \tag{4.17},$$

where $l = 0,1,2$, and β_z should satisfy $\cos\beta_z\pi = \dfrac{\phi_{11} + \phi_{22}}{2} < 1.$

Therefore, the ion motion is the composite of harmonic oscillations at the frequencies of

$$\omega_{z,l} = \frac{(2l \pm \beta_z)\pi}{T} = (l \pm \frac{1}{2}\beta_z)\Omega \tag{4.18}.$$

It should be noted that the solution of Equation 4.17 has the same form as for the solution of the Mathieu equation, and the ion motion has frequency components similar to those obtained using the sinusoidal driven waveform. For $l = 0$, the fundamental frequency component is also called the 'secular motion frequency.' Here, we have unified the two concepts of secular motion, and the secular frequency can be expressed as

$$\omega_z = \frac{1}{2}\beta_z\Omega$$

$$\beta_z = \frac{1}{\pi}\arccos\frac{\phi_{11}+\phi_{22}}{2} \tag{4.19}.$$

With Equations 4.8 and 4.12, we can replace the ϕ_{ij} with a, q parameters. In the case of a DC-free (symmetric) square wave, where $d = 0.5$ and $V_1 = -V_2$,

$$\beta_z = \frac{1}{\pi}\arccos\left[\cos\left(\pi\sqrt{q_z/2}\right)\cosh\left(\pi\sqrt{q_z/2}\right)\right] \tag{4.20}.$$

The secular frequency of an ion in the square wave quadrupole field can be calculated directly using the above analytical expression. This direct calculation differs from the case of sinusoidal wave, in which an iterative process needs to be used in order to achieve an acceptable approximation.

The ion oscillation at this frequency is caused by the effect of the rectangular wave electric field. An ion's frequency of motion is a function of both the mass-to-charge ratio and the repetition rate of the driving rectangular wave form. Each β_z-value corresponds to a frequency of secular motion, and the conditions $\beta_z = 0$ and 1 correspond to the stability boundaries. For $a_z = 0$ and $q_z = 0.712$, Equation 4.20 returns a value $\beta_z = 1$, which corresponds to the boundary of the stability diagram in the z-direction. At the same time, when $a_z = 0$ and $q_z < 0.3$, Equation 4.19 may be simplified further by approximation as

$$\omega_s \approx 0.454 q_z\Omega \tag{4.21}.$$

Figure 4.6 shows the secular frequencies obtained between Equations 4.20 and 4.21, as shown by the dashed line, and the ion oscillation frequencies obtained by numerical ray tracing of the ion motion generated by the digital waveform is displayed by means of crosses (+).

Normally the depth, $\overline{D}$, of the "effective potential well" (also termed the 'pseudo-potential well') is used to estimate the maximum kinetic energy of ions that remain trapped. For the motion of an ion in the z-direction, the depth can be expressed as follows

$$\overline{D}_z = \frac{m\omega_s^2}{2e}z_0^2 \tag{4.22}.$$

Here, z_o is the maximum displacement of an ion from the ion trap center in the z-direction. For a sinusoidal RF trap, the approximation $\beta_z \approx q_z/\sqrt{2}$ is valid for $q_z < 0.4$, and combining Equations 4.22 and 4.2b provides the expression

$$\overline{D}_z = \frac{q_z V}{8} = 0.125 q_z V[eV] \tag{4.23}.$$

The depth of the pseudopotential well for a square wave digital waveform can be estimated for small values of q_z ($q_z < 0.3$) as

$$D_z \approx 0.206 q_z V[eV] \tag{4.24}.$$

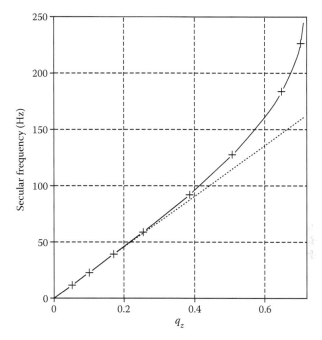

FIGURE 4.6 Secular frequency predicted by Equations 4.20 and 4.21, as shown by the dashed line, together with the ion oscillation frequency values obtained by a numerical ray tracing of the ion motion generated by the digital waveform displayed by means of crosses (+).

Equation 4.24 indicates that the pseudopotential well of the DIT is deeper compared to that obtained in a sinusoidal RF trap for the same given values of q_z and V. This result can be attributed to a larger averaged value of the periodical trapping voltage for the same amplitude V. At the same time, the maximum q_z-value for stable ion motion with a square waveform potential is correspondingly reduced: $q_z = 0.712$, compared with $q_z = 0.908$ for a sinusoidal RF waveform.

As in the sinusoidal waveform-driven quadrupole field, the secular motion can be altered using various methods of excitation, *i.e.*, dipolar excitation or quadrupolar excitation, in order to achieve different kinds of performance for mass analysis.

During a resonant ion ejection scan in the DIT, it is preferable to apply a digital waveform to the end-cap electrodes for exciting the secular motion of ions. The digital waveform is a rectangular pulse signal (as displayed in Figure 4.2b), with a repetition rate that is a fraction of the driving digital waveform frequency. Figure 4.7 shows the frequency components of the digital excitation signal having a frequency equal to one quarter of that of the drive voltage. In such a case, not only is the fundamental frequency component, ω_0, of the excitation waveform used to excite the ion secular motion, but additional higher harmonic components can also match other higher-order components (say, $\Omega(1 - \beta_z/2)$) of the ion motion of the same value of q_z. The final aim for generating the digital excitation signal is to minimize and ultimately avoid that one of these components induce resonant excitation in ions having a different q_z-value. For an excitation pulse rate corresponding to a small β_z-value, the higher

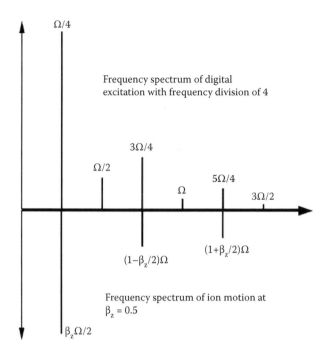

FIGURE 4.7 Frequency components of a digital excitation signal with the frequency $\omega_{ex} = \Omega/4$. The frequency components corresponding to ion motion are scaled relative to that of the drive rectangular wavefrequency, Ω.

harmonic components may be effective in exciting other trapped ions within the first stability region. In order to avoid this, the digital waveform can be tailored to cancel out the second and/or third harmonic components. Hence a pure square wave is just an example of a waveform without the second harmonic component.

4.4 DIGITAL ASYMMETRIC WAVEFORM

With the duty cycle fixed at 50%, the stability parameter a_z is determined only by the DC offset of the rectangular waveform (floating of the rectangular waveform relative to the zero potential) [14]. From a practical point of view, it is convenient to fix the high voltage level V_1 and the low voltage level V_2 to be $V_1 = -V_2 = V_0$.

In the DIT prototype, two values of V_0 equal to $+/-500$ V and to $+/-1$ kV are used typically, depending on the required mass range and the achievable switching voltage at which the fast metal–oxide–semiconductor field-effect transistor (MOSFET) switches can work. The MOSFET is a device used to amplify or switch electronic signals. It is by far the most common field-effect transistor in digital circuits. As noted previously, when the duty cycle is 50% and $U = 0$, the ion motion can be described using only the q_z-axis. However, with a duty cycle diverging from 50%, a DC quadrupole component is introduced. Therefore in the DIT, the position, or working point, of an ion on the stability diagram can be determined by varying two parameters: the waveform period, T, and the duty cycle d, when V_0 and the mass-to-charge ratio are known. When d is fixed, the ratio a_z/q_z, corresponding to the slope of the scan line is also fixed, and

$$\frac{a_z}{q_z} = -\frac{2U}{V} = \frac{1-2d}{4d(1-d)} \tag{4.25}$$

At a constant value of d, the locus of working points lies on a straight scan line passing through the origin of the (a_z, q_z) stability diagram. Hence the slope of the straight line, corresponding to the U/V ratio, is a function of d. By varying T, ions moving along a scan line in the (a_z, q_z) diagram have a stable trajectory at each point that can be determined. Compared to the sinusoidal wave RF ion trap, in which superimposition of a DC potential on the main trapping field requires the use of an additional DC power supply, in the DIT the DC component can be generated easily by using an asymmetric digital waveform, where the duty cycle $d \neq 0.5$. This operation can be achieved simply by varying the parameters entered into the control software of the mass spectrometer.*

4.5 DIGITAL FREQUENCY SCAN AND EXTENDED MASS RANGE FOR HIGH MASS DETECTION

The ability to extend the mass scan range is a direct consequence of the digital frequency scan. The switches for generating the high voltage digital trapping waveform are constructed with power MOSFETs. The upper and lower DC levels can be set between ±250 V and ±1000 V, with typical values of ±250 V, ±500 V, and ±1000 V being used. Control of the switch-timing is achieved by means of a specially designed digital signal generator (DSG), based on 'direct digital synthesizing technology.' The design of the DSG provides precise control of the waveform period at a resolution of 50 ps, with a maximum inter-scan jitter of ±250 ps. The DSG provides also the trigger signal for the pulsed dipole excitation and ejection waveforms, and allows eight-bit control of the pulse width. The phase delay relative to the trapping waveform is controlled digitally also by the DSG. In a typical mass scan, the frequency of the dipole excitation is obtained digitally by division of the frequency of the trapping waveform. For example, division rates $n = 3$ and $n = 4$ correspond to resonance excitation points at fixed β_z-values of 2/3 and 1/2, respectively. As the mass scan is obtained by scanning the period of the digital waveform at fixed amplitude, rather than by scanning the amplitude at fixed frequency, high voltage breakdown does not occur when the scan is extended to high mass at constant q_z. Therefore, the DIT is capable of achieving high mass analysis with a low trapping voltage and a fixed value of q_z. To demonstrate this capability, matrix-assisted laser desorption ionization (MALDI) ions were generated at atmospheric pressure and introduced into the ion trap. A trapping voltage of $V_0 = \pm 1$ kV was set for these experiments. Figure 4.8 shows the MALDI mass spectrum of 5 pmol of horse heart myoglobin in which the singly protonated species [M + H] + and the doubly protonated species [M + 2H]$^{+2}$ are observed (note

---

* See Volume 5, Chapter 12: Boundary-Activated Dissociations (BAD) in a Digital Ion Trap (DIT), by Francesco Brancia, Luca Raveane, Alberto Berton and Pietro Traldi.

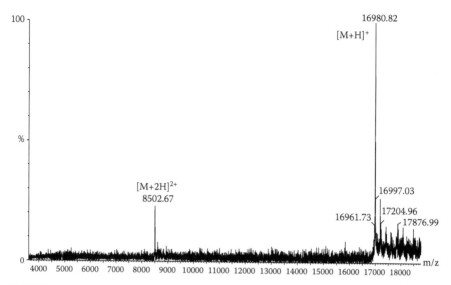

FIGURE 4.8 MALDI mass spectrum of singly- and doubly-protonated horse heart myoglobin molecules. A total of 5 pmol was loaded on to the target. The sample was prepared in saturated sinapinic acid. Both the singly and doubly charged ions show tailing on the high mass side of the ion signals due to the formation of water and matrix clusters. When the source temperature was increased the scattered background ion intensities decreased, and the singly charged species was detected with higher abundance. This mass spectrum was acquired with the source temperature of 300°C.

that the *m/z*-values are not calibrated). The peak corresponding to the singly protonated species has a width, expressed as full-width at half-maximum (FWHM), of 7.8 Th. At the scan speed employed (40,000 Th s^{-1}), the individual isotope peaks can not be resolved, but the measured peak width is equal to that defined by the envelope of the naturally occurring isotopic distribution, indicating a mass spectral resolution of >2100. To produce this mass spectrum, the period of the digital waveform was varied from 4.3 to 9.8 ns, with fixed step of 0.5 ns and five waves per step. Resonant ion excitation is set for $\beta_z = 1/2$, resulting in a scanned mass-to-charge ratio range between 3600 and 18,700 Th in a single scan of 388 ms duration. Recent results generated using a high vacuum MALDI DIT prototype incorporating ion optics similar to those employed on the Shimadzu quadrupole ion trap (QIT) [15]* show that the DIT can detect readily protein ions up to 66 kDa [16], as shown in Figure 4.9.

4.6 NON-STRETCHED GEOMETRY AND FIELD-ADJUSTING ELECTRODE

The geometry of the electrodes has a profound impact on the quadrupole field generated and *ergo*, on the mass resolution and mass measurement accuracy obtained in

---

* See Volume 4, Chapter 19: A Quadrupole Ion Trap/Time-of-Flight Mass Spectrometer Combined with a Vacuum Matrix-Assisted Laser Desorption Ionization Source, by Dimitris Papanastasiou, Omar Belgacem, Helen Montgomery, Mikhail Sudakov and Emmanuel Raptakis.

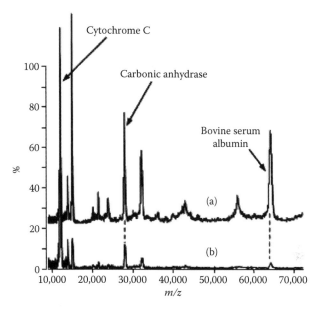

FIGURE 4.9   MALDI mass spectra of protein mixtures acquired on a DIT mass spectrometer incorporating a high vacuum ion source with conversion dynode voltage fixed during detection (a) and changed during detection (b) (Reproduced from Tanaka, K., Sekiya, S., Jinno, M., Hazama, M., Kodera, K., and Iwamoto, S., *Proc. 56th ASMS Conference on Mass Spectrometry and Allied Topics*, Denver, CO, 2008. With permission.)

the ion trap. According to the theory, a pure quadrupole field is generated when the inner surface of the ion trap is hyperboloidal with the end-cap spacing $2z_0 = \sqrt{2}r_0$. In reality, the two apertures through which ions enter and are ejected from the ion trap provide the primary cause of distortions of the quadrupole field [17]. To circumvent this problem in commercial three-dimensional ion traps, the theoretical geometry is modified by increasing the distance between the two end-cap electrodes by 10.7% or by employing specially shaped electrodes. These distortions from the pure quadrupolar geometry result in modification of the trapping field, in which inhomogeneities due to the apertures in the end-cap electrodes are compensated partially by superimposing higher-order multipole fields on the main trapping quadrupole field.

When the decision was taken to build the DIT, different geometries for the electrodes were tested using simulations. In the case of rectangular waveforms applied to an ion trap with a 10% stretched geometry, simulation indicated that the secular frequency of an ion increases continually as a function of ion oscillation amplitude. This frequency increase is due to higher-order multipole fields caused by the electrode geometry and such positive high-order fields do indeed increase the ejection speed and offer better mass resolution with respect to the unstretched ion trap. Figure 4.10 shows the amplitude evolution during a resonance ejection scan using a DIT with 10% axially stretched geometry. Three distinct simulations are performed using the same ion and identical initial conditions.

With the stretched ion trap, however, a relatively large dipolar excitation intensity, V_{dip}, is needed (50 V dipole pulse in the simulation), otherwise the ion trajectory

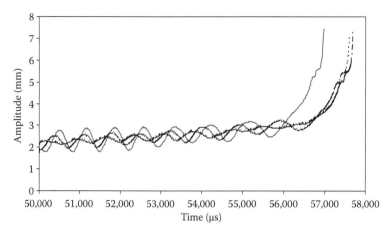

FIGURE 4.10 Amplitude evolutions during a resonant ion ejection scan with a 10% stretched digital ion trap. The dipolar excitation is applied at $q_z = 0.5$, with a pulse width of 0.2 T; $m = 3500$ Da, $z = 2$ for all ions. The helium buffer gas pressure was 1×10^{-3} mbar at 300 K, with an amplitude of dipolar excitation $V_d = 50$ V. (Reproduced from Ding, L., Sudakov, M., and Kumashiro, S., *Int. J. Mass Spectrom.* 2002, *221*, 117–138. With permission from Elsevier.)

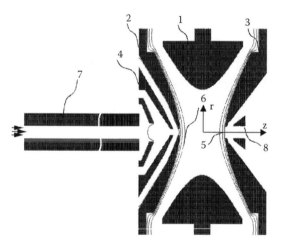

FIGURE 4.11 A modified digital ion trap structure, including 1 the ring electrode held at ±500 V, 2 and 3 the end-cap electrodes, and 4 the field-adjusting electrode that is supplied with an adjustable DC voltage. Also shown are: 5 the mesh used to cover the ejection end-cap aperture, 6 the equipotential lines of the electric field generated in the ion trap, 7 an octopole lens for ion introduction, and 8 the extraction electrode.

expansion may terminate as the secular frequency of the ion shifts away from the excitation frequency at increased amplitudes of ion motion. However, stretching the ion trap in the axial direction is not the only way to reduce the effect of the end-cap holes. Instead, a different geometry comprising a non-stretched geometry with an external field-adjusting electrode (FAE) was decided upon. Figure 4.11 shows the arrangement of the electrodes used in the DIT. In order to correct the defect in the

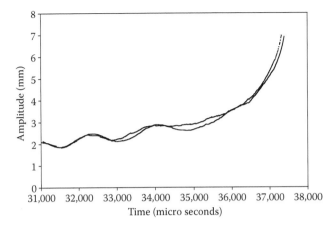

FIGURE 4.12   Amplitude evolutions during a resonance ejection scan using an ion trap with a field-adjusting electrode. Rectangular drive voltage ±1000 V, dipolar excitation at $q_z = 0.5$ with width 0.6 T. $m = 3500$ Da, $z = 2$, the helium buffer gas pressure was 1×10^{-3} mbar at 300 K (a) Forward mass scan with $V_d = 5$ V, $V_{fa} = 1.5$ kV. The two traces represent two identical simulations performed using the same initial conditions. (Reproduced from Ding, L., Sudakov, M., and Kumashiro, S., *Int. J. Mass Spectrom.* 2002, *221*, 117–138. With permission from Elsevier.)

quadrupole electric field in the proximity of the end-cap hole, an adjustable DC voltage is applied to the FAE electrode so that the secular frequency shift can be adjusted adequately prior to ejection. The exit end-cap hole is covered by a thin mesh to screen the negative field caused normally by the extraction electrode positioned next to the detector. In the DIT prototype, the diameters of the ion injection and ejection apertures are 1.5 and 2.4 mm, respectively.

Evolution of the amplitudes of ion oscillation during ion resonant ejection has been studied by simulations performed using the new geometry. Figure 4.12 illustrates the case when a field adjusting voltage, V_{fa}, of 1.5 kV is applied to the FAE. The other parameters are identical to those employed in the simulation displayed in Figure 4.10. In the new geometry, the expansion of ion oscillation is steadier than that obtained in a stretched ion trap. The improvement can be explained as follows: for resonance at an ejection point corresponding to a relatively large value of β_z ($\beta_z = 0.5$), ions approach the aperture only at the negative phase of the trapping field. When a 1.5 kV DC voltage is applied to the FAE in order to correct the field (Figure 4.13a and b, electrode labeled 1), the equipotential lines are not distorted near the apertures. In contrast, the field would be distorted, as shown in Figure 4.13b, if a negative voltage, typically used for the introduction of positive ions, is applied to the FAE near the end-cap electrode. Under these conditions, ions oscillate up to the maximum excursion permitted within the corrected field region, however, during the positive phase of the rectangular waveform, when the field faults are not corrected, ions are outside the corrected field. The combined effect in both phases is that ions approaching ejection are subjected to a quadrupole field with a controlled amount of higher-order components. It is important to point out that, in principle, the FAE could be implemented also in amplitude-scanning devices (as in the commercial

(a)

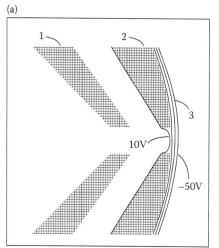

(b)

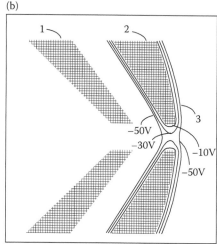

FIGURE 4.13   Field-adjusting electrode (FAE) (1) end-cap electrode (2) and equipotential lines (3) (corresponding to 30 V, −30 V, −100 V, −200 V) showing the electric field near the end-cap hole when the ring electrode is charged at −1 kV and the field adjusting electrode is charged at (a) + 1.5kV and at (b) −400V, respectively.

instruments), with the DC value applied to the external electrode varying as a function of the RF voltage amplitude applied on the ring electrode. Obviously, in practice such a procedure is difficult to achieve because perfect synchrony between the DC and RF voltages is required.

4.7   MASS RESOLUTION

Good mass resolution with the non-stretched DIT geometry can be achieved only with an appropriate voltage applied to the FAE. Figure 4.14 shows a plot of the FWHM resolution obtained experimentally and *via* simulations for singly charged MALDI ions of Glu-fibrinopeptide B for different voltages set on the FAE. All other parameters remained unchanged and are: scan speed 977 Th s^{-1}, dipolar excitation equal to ±1 V applied for 39% of the digital waveform period, and a buffer gas pressure of *ca* 1×10^{-3} mbar. Under these conditions, a mass resolution of *ca* 8000 was achieved for a field-adjusting voltage of 1132 kV. The mass resolution trend line predicted by simulation using different field-adjusting voltages is in good agreement with that determined experimentally.

Figure 4.15 depicts the mass spectra of doubly-charged bradykinin ions acquired with different field-adjusting potentials applied to the FAE. As can be seen, variation in the DC potentials induce a profound effect upon the mass resolution. As expected variation of the DC potential from the optimum value applied to the FAE results in perturbation of the quadrupole field producing lower mass resolution. Different scan speeds require a different field-adjusting voltage V_{fa} for maximum resolution, but precise control of the field-adjusting voltage is not crucial as the resolution and mass shift are not critically sensitive to V_{fa}. Mass spectra of Glu-fibrinopeptide B

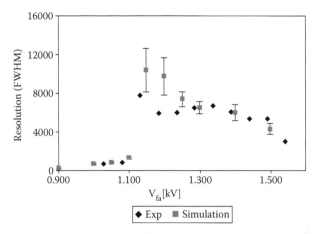

FIGURE 4.14   A plot of resolution vs. field-adjusting voltage. The experimental and simulated resolution (FWHM) data are given for the [M + H]$^+$ ion of Glu-fibrinopeptide B at m/z 1570. The conditions are: scan speed 977 Th s^{-1}, dipolar excitation ±1 V applied for 39% of the digital waveform period, and helium buffer gas pressure ca 1×10^{-3} mbar. (Reproduced from Ding, L., Sudakov, M., Brancia, F.L., Giles, R., and Kumashiro, S., *J. Mass Spectrom.* 2004, *39*, 471–484. With permission from Wiley.)

collected with optimized V_{fa} at scan speeds of 977, 200, and 39 Th s^{-1} are shown in Figure 4.16. A maximum mass resolution of 19,000 was achieved at a scan speed of 39 Th s^{-1}, equivalent to a FWHM peak width of 0.08 Th. In conclusion, for the majority of applications the mass resolution achieved by our current DIT prototype, which employed a lower maximum trapping voltage, is comparable to that of the conventional RF trap.

4.8   FORWARD AND REVERSE SCANNING

The combination of forward and reverse scanning has been exploited to mitigate the chemical mass shift in mass spectra [18]. With the ion trap of stretched geometry, when ions are scanned resonantly out of the trap in the reverse direction (from high mass to low mass), the mass resolution obtained is not comparable to that observed in mass spectra acquired with a forward scan (low mass to high mass) [19]. Deviations in the spacing and shapes of the ion trap electrodes from the theoretical geometry introduce a positive octopole component which improves ejection speed in the forward scan but causes ejection delay in the reverse scan.

In the DIT, because of the electronic control of the fringing field generated, it is possible to switch readily to the reverse scan mode, in which mass spectra are generated by scanning the frequency of the digital waveform from a low to a high value. Figure 4.17 illustrates the scan functions used in both scan regimes in order to obtain mass spectra. On the x-axis the time of each operation is displayed, while the y-axis corresponds to the period of the trapping voltage. Since the period is inversely proportional to the frequency of the digital waveform, there is a direct relationship between mass-to-charge and period so that y-axis can be considered equivalent to $(m/z)^{1/2}$.

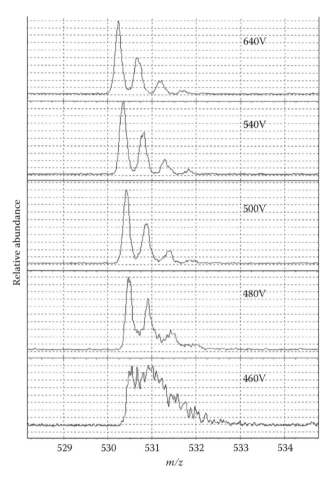

FIGURE 4.15   Electrospray mass spectra of the $[M + 2H]^{2+}$ ion rom bradykinin acquired using different field-adjusting potentials.

We have shown above (Figure 4.15) that the DIT forward scan mass resolution improves dramatically when V_{fa} is increased above a certain value. Under these conditions, the resulting field has a net positive higher-order component. When an ion approaches resonance ejection under conditions where a higher-order field exists, it will suffer a shift in secular frequency as the amplitude of oscillation increases. Therefore, the differences between the forward and reverse scan can be understood as follows. Consider a fixed-frequency resonance excitation signal applied to the end-cap electrodes with V_{fa} set to give a positive higher-order field. Under these conditions, trapped ions with a secular frequency that is lower than the excitation frequency come rapidly into resonance and are ejected. Thus it can be envisaged that there exists a sharp cut-off for ions approaching the resonance point from the lower secular frequency (high mass) side. Whereas trapped ions at higher proximal secular frequencies (lower mass) repeatedly fall out of resonance each time the amplitude

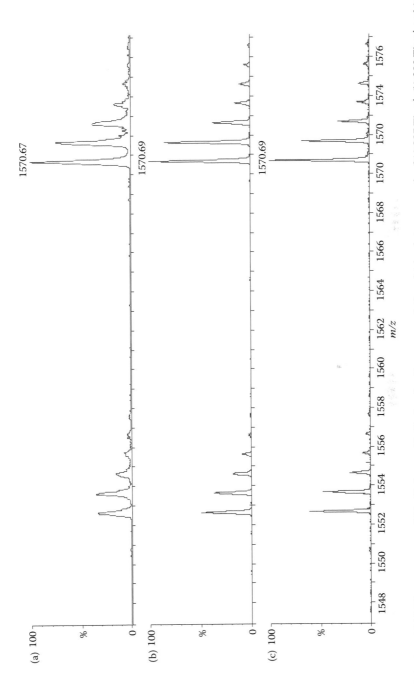

FIGURE 4.16 MALDI mass spectra of Glu-fibrinopeptide B, showing the mass resolution obtained at scan speeds of (a) 997 Th s⁻¹, (b) 200 Th s⁻¹ and (c) 39 Th s⁻¹. The field-adjusting voltage was optimized separately for each scan speed. The spectral resolution is 8000, 12,000, and 19,000 (FWHM), respectively. (Reproduced from Ding, L., Sudakov, M., Brancia, F.L., Giles, R., and Kumashiro, S., *J. Mass Spectrom.* 2004, *39*, 471–484. With permission from Wiley.)

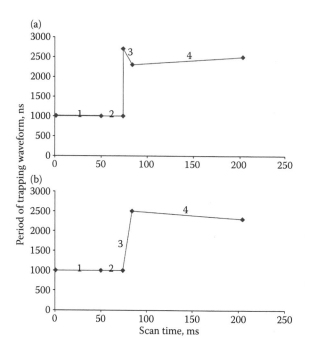

FIGURE 4.17 Scan functions used for obtaining mass spectra by varying the period of the trapping voltage. In the forward scan (a), an introduction time of between 1 and 100 ms (1) and a cooling time using helium (2) is used before ramping the frequency up to the maximum value corresponding to the smallest ion within the mass-to-charge range (3). Subsequently, ions are ejected from the trap by scanning the period of the waveform from low to high values (4). In the reverse scan (b), ion introduction (1) and ion cooling (2) are performed accordingly using the same period. After adjusting the voltage applied to the external electrode (3), the period is then scanned from high to low values over the same mass-to-charge range (4).

of their oscillatory motion increases. Consequently, ions, which fall in and out of resonance, take an unpredictable time to be ejected (due to random factors, including collisions with the buffer gas). As a consequence of ions falling in and out of resonance, there is a blurred boundary for ion ejection at the high frequency (low mass) side of the applied excitation frequency. Therefore, high-mass resolution is achieved only in the forward scan (when trapped ions are brought into resonance approaching from the lower secular frequency side), and poor mass resolution is observed in the reverse scan (when the ions are brought into resonance approaching from the higher secular frequency side).

However, if a lower value of V_{fa} is applied to produce a net negative higher-order field component, trapped ions approaching resonance are subjected to a secular frequency shift toward lower values as their amplitude of secular oscillation increases. Now the scenario described above is reversed: a sharp 'cut' is generated on the high secular frequency (low mass) side, and a blurred boundary for ion ejection occurs on the low secular frequency (high mass) side. During a resonant ion ejection mass scan, poor mass resolution is observed for the forward scan and good mass resolution

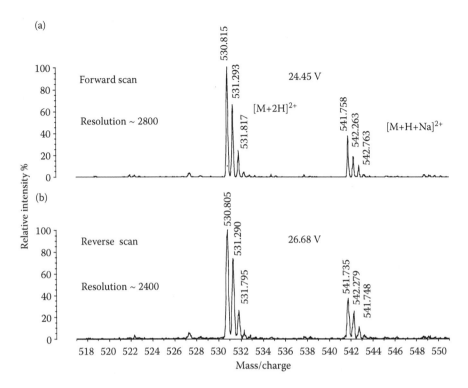

FIGURE 4.18   Electrospray mass spectra of 5 pmol μL^{-1} bradykinin acquired in (a) forward and (b) reverse scan using an introduction time of 10 ms. During resonant ion ejection at $\beta_z = 0.5$, a scan rate of 782 Th s^{-1} was used. The mass spectra shown were the sum of 30 scans each. The ion abundances, expressed in volts and shown on the mass spectra represent the peak heights of the most abundant ions in the isotopic envelopes of the doubly protonated bradykinin $[M + 2H]^{2+}$ molecule. (Reproduced from Brancia, F.L., Giles, R., and Ding, L., *J. Mass Spectrom.* 2004, *39*, 702–704. With permission from Wiley.)

is obtained for the reverse scan. Figure 4.18 shows the electrospray mass spectra of bradykinin acquired in the forward (a) and reverse scan (b) modes, accumulating ions in the trap for a period of 10 ms [20]. The ion signals observed in the mass spectra correspond to the doubly-protonated bradykinin species and its sodium-cationized analog. The mass resolution of the $[M + 2H]^{2+}$ ion obtained with a reverse scan (2400; FWHM) is comparable to the value for a forward scan (2800; FWHM). The ion abundances expressed in volts over each mass spectrum are measured as the peak heights of the most abundant ions (^{12}C-only ion within the isotopic envelope of the doubly protonated bradykinin molecule, $[M + 2H]^{+2}$).

4.9   SCAN DIRECTION AND SPACE CHARGE

Among all parameters responsible for errors in correct mass assignment, space-charge effects (ion–ion interactions among species of identical polarity) play a crucial role in affecting the position on the *m/z*-axis at what an ion is observed. During resonant ejection, each ion of a specific m/z-value is subjected to charge interactions due to the charge density generated by (i) the total number present in the trap, (ii) the total number of ions at that specific value of m/z and, by (iii) the abundances and m/z differences of neighboring ions [21,22]. When the ion trap approaches the space-charge limit, an effective DC component is produced in the quadrupole electric field, which will affect the secular frequency of the ion of interest. The concomitant shift toward a lower value of frequency will result in a higher observed mass-to-charge value for the ion under investigation [13]. The scan direction can change the effect of space charge on mass measurement accuracy. The correlation between the number of trapped ions and the mass measurement accuracy of the monoisotopic $^{12}C_{50}$-containing peak of doubly-protonated bradykinin, $[M + 2H]^{2+}$, under both scan directions has been studied by adjusting the number of ions introduced in the ion trap.

When the total number of ions is increased, no evident effects are observed on the mass resolution of the ion signals acquired in the reverse scan (2400; FWHM). As illustrated in Figure 4.19b, the monoisotopic ion containing $^{12}C_{50}$ is still resolved

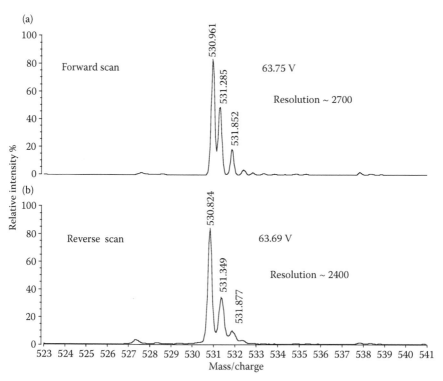

FIGURE 4.19 Electrospray mass spectra of 5 pmol μL⁻¹ bradykinin acquired in the (a) forward and (b) reverse scan modes when the ion trap is operated at higher charge density (as represented by the peak heights of the most abundant ions in the isotopic envelopes of the doubly protonated bradykinin [M + 2H]²⁺ molecule, *ca* 64 V). Due to the space-charge effect, the m/z-spacing between $^{12}C_{50}$- and $^{12}C_{49}{}^{13}C$-containing ions is reduced to 0.33 Th in the forward scan, (a). (Reproduced from Brancia, F.L., Giles, R., and Ding, L., *J. Mass Spectrom.* 2004, *39*, 702–704. With permission from Wiley.)

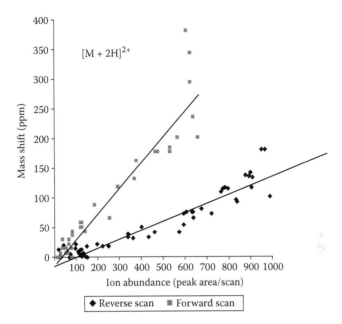

FIGURE 4.20   Effect of ion abundance on the mass displacement of the $[M + 2H]^{2+}$ ion peak (m/z 530.785) of bradykinin in the forward and reverse scanning modes. The ion abundance, expressed in peak area per scan, was varied by manipulating the introduction time. The mass shift is expressed in ppm.

clearly from the adjacent ion incorporating one ^{13}C isotope when the ion abundance corresponds to an observed base peak intensity of 63.69 V [20]. However, a reduction in the m/z-spacing between the $^{12}C_{50}$- and the $^{12}C_{49}{}^{13}C$-containing ions is observed in the forward scan in the presence of an excess of charge (see Figure 4.19a). Correlation between the ion intensity and mass accuracy has generated the graph depicted in Figure 4.20. It is important to note that, regardless of the scan direction, during resonant ejection at $\beta_z = 0.5$, when ions approach the space-charge limit the $^{12}C_{50}$-containing ions still experience the DC potential induced by themselves, the adjacent ions of the isotopic envelope ($^{12}C_{49}{}^{13}C$ and $^{12}C_{49}{}^{13}C_2$) and the remainder of the ions present in the ion trap. This space charge shifts the secular frequency of the bradykinin $[M + 2H]^{2+}$ ion to a lower value, inducing delayed or advanced ejection depending upon the direction of scan. In the forward scan, a more severe shift is observed compared to that in the reverse scan. Mass displacement reaches a maximum of 360 ppm. In the reverse scan, deterioration of mass accuracy is less severe and a mass shift corresponding to the maximum value of 180 ppm is obtained at higher peak intensity. The different mass displacements observed under the two scan modes are attributable to the order in which ions are ejected from the ion trap. During the forward scan, the Coulombic interactions between all isotopic ions existing in the ion cloud influence the secular frequency of the monoisotopic ion until this species is scanned out of the ion trap. In the reverse scan, because the $^{12}C_{50}$-containing ion is the last ion species to be ejected, the space-charge effect deriving from the presence of all adjacent ions is minimized and limited fluctuations in mass shift are observed for the monoisotopic ion population. As a consequence of this minimization of space charge, in the reverse scan the ion of interest reaches the space-charge limit at higher value of relative ion abundance (1000) with respect to that obtained in the forward scan (657).

4.10 PRECURSOR ION ISOLATION

In trapping devices, precursor ion isolation is achieved by ejecting all ions with mass-to-charge ratios different from that of the precursor ion. Conventionally, precursor ion isolation is performed by applying to the end-cap electrodes one or more broadband dipolar excitation waveforms, such as stored waveform inverse Fourier transform (SWIFT) [23] and filtered noise field (FNF) [15,24]. However, when these approaches are used, a maximum isolation resolution of ca 1300 can be achieved [15] (expressed as a ratio in which the denominator is the baseline width of the isolation window). In the DIT, a combination of using both forward and reverse scan regimes in sequence provides very high mass resolution for precursor ion isolation. This 'scan clipping' procedure is obtained by setting different voltages on the FAE for forward and reverse scans. As a consequence of the different voltages used, a sharp isolation edge is obtainable in both scan directions. The efficacy of such an approach has been demonstrated isolating the $^{13}C_1$-containing ion within the isotopic envelope of the $[M + 3H]^{3+}$ insulin-β-chain. In this experiment, the trapping rectangular wave voltage, V_{rec}, was held at ±500 V, and the mass spectra obtained at each stage of the ion isolation are shown in Figure 4.21. Initially a low resolution SWIFT waveform is used to isolate the triply charged isotopic envelope with an estimated mass window of ca 30 Th (Figure 4.21a). Then the β_z-value of the precursor ion is increased above the $\beta_z = 0.5$ ejection point and V_{fa} is set to 280 V in order to eject the ions of higher mass-to-charge ratio. A digital dipolar excitation signal is then applied at $\beta_z = 0.5$ and the digital waveform period, T, is scanned down to eject the unwanted ions (Figure 4.21b and c). The remaining ions are then cooled again (for 15 ms), and the working point of the precursor ion is moved to a value of β_z slightly below the $\beta_z = 0.5$ ejection point. The voltage V_{fa} is now changed to 580 V and the digital dipolar waveform re-applied at $\beta_z = 0.5$; T is scanned up to remove the unwanted monoisotopic ion, as shown in Figure 4.21d. Note the procedures depicted in Figure 4.21c and d follow independently from that of Figure 4.21b. The result demonstrates that a scan-clipping procedure using dynamic fringing field modification with the FAE, can provide precursor ion isolation with a mass resolution exceeding 3500. In the above procedure, the monoisotopic ion is isolated with an efficiency approaching 100% and isolation of the first ^{13}C isotope-containing ion is achieved with an efficiency of 50%.

In the DIT, a resolving DC can be generated easily using a digital asymmetric waveform. This approach can be used as an alternative method for precursor ion isolation termed 'digital asymmetric waveform isolation' (DAWI). As shown above, variation of the duty cycle allows introduction of a DC component in the quadrupole field. Hence the a/q scan line moves toward the apex of the stability diagram so that the region of stability decreases. When the scan line approaches the apex, ions within a small mass range are retained in the trap allowing precursor ion isolation. The main advantage of DAWI is that the isolation time is around 1 ms. However, the resolution of precursor ion isolation is not as high as that obtained using forward/reverse scan.

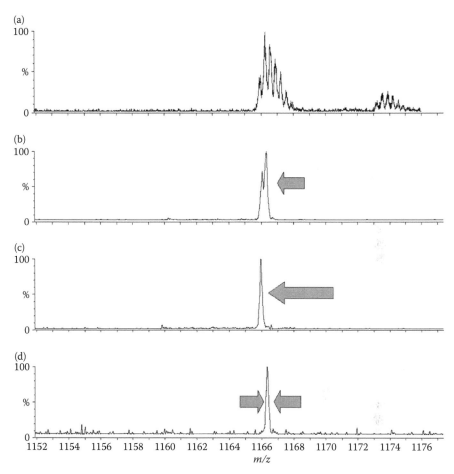

FIGURE 4.21 High resolution precursor ion isolation using a combination of forward and reverse scans of the insulin-β-chain $(M + 3H)^{3+}$ ion, molecular formula of $C_{157}H_{232}N_{40}O_{47}S_2$. (a) Mass spectral analysis following application of a low resolution SWIFT waveform; (b) The reverse scan 'clip' is used to isolate the two $^{12}C_{157}$- and $^{12}C_{156}{}^{13}C_1$-containing ions; (c) The reverse scan 'clip' is used to isolate just the $^{12}C_{157}$-containing ion peak; (d) Reverse and forward scan 'clippings' are used in sequence to create a 0.3 Th isolation window containing only the $^{12}C_{156}{}^{13}C_1$ isotopic ion peak.

4.11   TANDEM MASS SPECTROMETRY IN CONJUNCTION WITH ELECTRON CAPTURE DISSOCIATION (ECD)

Tandem mass spectrometry *via* collisions either with an inert gas [25], or with a surface [26], or through IR multiphoton excitation [27] has become the method of choice for peptide sequencing in the post-genomic era. The common feature of these techniques is the increase of the internal energy of the analyte that results in a higher level of charge mobility. Transfer of the proton on the peptide bond induces charge-directed cleavage of the peptide backbone. Under low energy multiple collisions conditions mostly b- and y-type fragment ions together with fragment ions resulting from neutral losses [28,29]. A fragment ion in which the charge is retained on the N-terminus is described as b-ion whereas one in which the proton is borne by the C-terminus is defined as y-ion. Since activation caused by thermal processes favors the lowest energy fragmentation pathway, loss of labile conjugated groups (for example, phosphate) is observed readily also [30,31]. Unlike thermal processes, ECD appears to be a non-ergodic technique in which fragmentation occurs before the internal excitation energy distributes itself over all vibrational degrees of freedom [32]. Energy randomization shows its utility in providing a more extensive sequence coverage. To date the majority of data published have been generated on Fourier transform ion cyclotron resonance (FT-ICR) mass spectrometers. [25]*

Due to the presence of a radio-frequency voltage with an amplitude of hundreds of volts, ECD is difficult to implement in quadrupole ion traps for the following reason. When electrons are injected into a quadrupole ion trap, either they are accelerated above the energy threshold for electron capture (10 eV), or they are repelled so that interactions between trapped ions and electrons are precluded. However, in sinusoidal waveform-driven ion traps two different approaches for generating ECD can be implemented. In the first approach, peptide ions are isolated and irradiated with electrons in a linear ion trap, to which an RF potential and a magnetic field are applied simultaneously. Subsequently mass analysis of the product ion population is performed in a time-of-flight (TOF) analyzer [33],[†] In the second approach, ECD is carried out inside a three-dimensional quadrupole ion trap to which a magnetic field generated by permanent magnets is applied. Electrons are injected at the beginning of the positive-going portion of the RF potential waveform, allowing interactions with the trapped ions within the time window of the RF half-period [34].

Due to the type of waveforms, the DIT offers an opportunity for generating more efficient ECD in an electrodynamic ion trap. Since discrete voltage levels are employed, in specific time frames (of a few microseconds), the electric field is constant so that electrons can be injected into the ion trap; following deceleration the electrons reach the precursor ion cloud situated in the center of the ion trap. To design the electron gun, electron/ion optical simulations (SIMION 7) [35] have been used to optimize the most effective geometry. Figure 4.22 displays the simulated trajectories of electrons injected into the DIT obtained using different voltage settings with the design depicted in the figure (see caption for details). The configuration of the electron gun takes into account the design of the DIT, in which the conversion dynode and electron multiplier are positioned next to the exit hole in order to collect the ejected ions. Therefore, the electron source is placed further behind the ion detector in the attempt to reduce the expected random noise detected in the mass spectral signal. The simulations assumed that the electrons are accelerated in the vicinity of the emitter from a surface of radius of 0.3 mm. The initial kinetic energy is expected to be around 0.5 eV with angular distribution + /−50 degrees from normal, and the

---

* See Volume 5, Chapter 5: Fourier transform ion cyclotron resonance mass spectrometry in the analysis of peptides and proteins, by Helen J. Cooper.

† See Volume 4, Chapter 12: Axially-resonant Excitation Linear Ion Trap (AREX LIT), by Yuichiro Hashimoto.

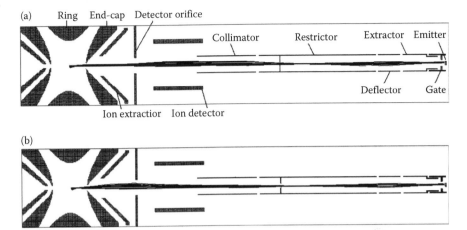

FIGURE 4.22 Electron injection trajectories into the digital ion trap performed using SIMION 7.0. The voltage settings are as follows: electron emitter −250 V, gate −225 V, extractor 100 V, deflector 60 V, restrictor 1300 V, ion detector 0 V, cap electrode 0 V, and DIT ring electrode −500 V. The voltage applied to the ion extractor cone and detector orifice is set at 1300 V for simulation (a) and −200 V for case (b). (Reproduced from Ding, L. and Brancia, F.L., *Anal. Chem.* 2006, *78*, 1995–2000. With permission from the American Chemical Society.)

electrons are focused subsequently into the collimator through the restrictor lens. Large variation in the voltage applied to the deflector has an effect on the trajectory, up to the point that no electrons are injected into the ion trap. When the trapping voltage applied to the ring electrode enters its negative excursion (−500 V), the potential in the center of the ion trap is about −250 V. Electrons accelerated from the emitter with a potential of −250 V are slowed down until they stop. When the electrons change direction, their kinetic energy is approximately 1 eV, which corresponds to the electron energy used for ECD experiments. Variation of the potentials applied to the ion extraction cone and ion detector has been shown to affect the trajectory of electrons. When a high positive voltage (+1300 V) is applied (see Figure 4.22a) the electron beam is focused more narrowly. Although such focusing facilitates electron injection through the end-cap hole, unwanted ionization of the residual gas can occur, generating ions that can produce background noise in the ion trap. To avoid this inconvenience, a negative voltage setting applied to the extractor is preferable. The simulation indicates that limited differences in the trajectory of the electron beam are observed when a potential of −200 V, as shown in Figure 4.22b, is used instead of a potential of 1300 V.

The ECD spectrum of the doubly protonated substance P, a peptide whose fragmentation pattern is well known, is dominated by a series of c-type ions from which it is possible to identify a five-amino acid sequence tag (Figure 4.23). An ECD interaction period of 416 ms was used and the mass spectrum displayed was accumulated for 250 scans. However, accumulation after 20 scans produces a mass spectrum in which fragment ions can be detected with sufficiently good signal-to-noise ratio. Figure 4.24 shows a comparison of the intensities of each of the c_5, c_6, and c_7 fragment ions obtained with 20 scans (upper) and 250 scans (lower). The analysis

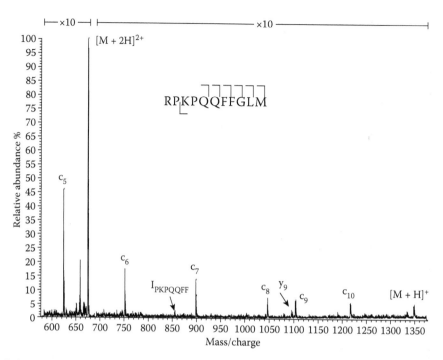

FIGURE 4.23 ECD mass spectrum of $[M + 2H]^{2+}$ ions generated by electrospray ionization of substance P acquired after the accumulation of 250 scans. An ECD interaction period of 416 ms was used in the experiment (Reproduced from Ding, L. and Brancia, F.L., *Anal. Chem.* 2006, *78*, 1995–2000. With permission from the American Chemical Society.)

time necessary to generate this ECD mass spectrum accumulated over 20 scans (20 s) is compatible to the time scale used in online LC (liquid chromatography) experiments.

4.12 DIGITAL LINEAR ION TRAP

The linear ion trap possesses many advantages with respect to its three-dimensional counterparts, and such ion traps are becoming increasingly popular. The linear ion trap can also be driven digitally, using two pairs of switches that generate two opposite phases of digital waveform for the *x*-poles and *y*-poles, respectively.

Figure 4.25 shows a schematic diagram of the switching circuits used to provide the digital waveform to the *x*- and *y*-electrodes of a linear ion trap in order to create the radial confinement electric field. Confinement along the axis of the linear ion trap can be achieved using two end-electrodes to provide the axial potential barrier.

For the rectangular wave voltage supplied as shown in Figure 4.25, the voltages on the *x*-poles and *y*-poles can be described as

$$V_x = V_{rec} = \begin{Bmatrix} V_0, & nT < t < nT + Td \\ -V_0, & nT + Td < t < (n+1)T \end{Bmatrix}$$

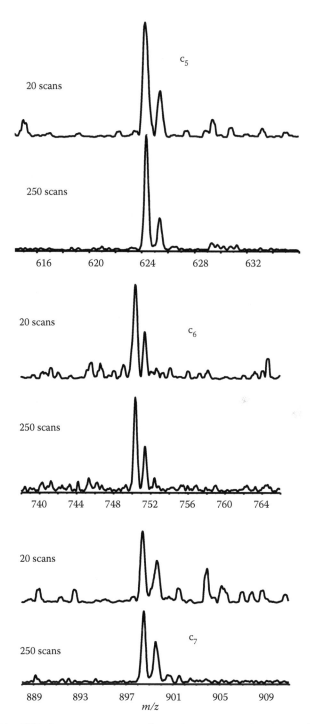

FIGURE 4.24 ECD fragments corresponding to ions c_5, c_6, and c_7 from substance P produced after 20 and 250 scans.

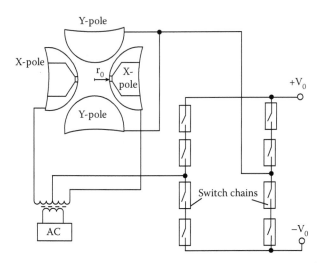

FIGURE 4.25   Schematic diagram of the linear ion trap driven digitally using two pairs of switching circuits working between one high voltage level V_0 and one low voltage level $-V_0$.

$$V_y = -V_{rec} \tag{4.26}.$$

The ion motion in the x- and y-directions can be described by

$$\frac{d^2x}{dt^2} = -\frac{2eV_{rec}}{r_0^2 m} x \tag{4.27a}$$

and

$$\frac{d^2y}{dt^2} = \frac{2eV_{rec}}{r_0^2 m} y \tag{4.27b}.$$

Equation 4.27a is written in the same form as Equation 4.4b for ion motion in a 3D quadrupole ion trap along the symmetric z-axis. If we keep the definitions of U and V as employed previously in Equation 4.2, and define

$$a_y = -a_x = -\frac{8eU}{m\Omega^2 r_0^2} \tag{4.28a}$$

$$-q_y = q_x = -\frac{4eV}{m\Omega^2 r_0^2} \tag{4.28b},$$

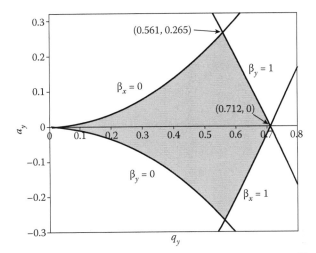

FIGURE 4.26 The first region of the stability diagram for a standard linear ion trap (pure 2D quadrupole field) driven by two opposing digital waveforms generated by switching between one high voltage level and one low voltage level, without using an additional DC component.

the stability diagram for the digital linear ion trap, as displayed in Figure 4.26, can be obtained in the same way as for 3D ion trap. The diagram is plotted for the case of equal plus and minus voltage levels, in which the resolving DC level is created only using an asymmetric duty cycle waveform between the x- and y-poles. When the duty cycle corresponds to $d = 0.5$, the working points of all ions sit on the q_z-axis, and $q_z = 0.712$ is the low mass cut-off point corresponding to the stability boundaries in both the x- and y-directions. The apex of the first stability region is located at ($q_y = 0.561$, $a_y = 0.265$) and to reach this point the duty cycle of the digital waveform needs to be 0.612.

4.13 CONCLUSION

As shown, the DIT possesses unique features with respect to the other three-dimensional ion traps. The utilization of rectangular waveforms expands the capabilities of ion trap devices toward new applications. The possibility to scan the frequency instead of the amplitude demonstrates that ions of higher mass-to-charge ratios can be analyzed with only 1 kV quadrupole trapping voltage, providing in principle a novel tool for analysis of protein samples in the megadalton region [16]. A non-stretched ion trap geometry in combination with the field adjusting electrode to compensate for distortion of the quadrupole field provides good resolution in both the forward and reverse scan modes. In tandem mass spectrometry, high resolution precursor ion selection, the capability to analyze higher mass-to-charge ratio fragments combined with the possibility to perform ECD in the trap opens up new prospects for the topdown analysis of proteins in quadrupole ion traps.

REFERENCES

1. Stafford, G.C., Jr.; Kelley, P.E.; Syka, J.E.P.; Reynolds, W.E., Todd, J.F.J. Recent improvements in and analytical application of advanced ion trap technology. *Int. J. Mass Spectrom. Ion Processes* 1984, *60*, 85–98.

2. Williams, J.D.; Cox, K.A.; Cooks, R.G.; Kaiser, R.E.; Schwartz, J.C. High mass-resolution using a quadrupole ion trap mass spectrometer. *Rapid Commun. Mass Spectrom.* 1991, *5*, 327–329.

3. Louris, J.N.; Brodbelt-Lustig, J.S.; Cooks, R.G.; Glish, G.L.; Van Berkel, G.J.; McLuckey, S.A. Ion isolation and sequential stages of mass spectrometry in a quadrupole ion trap mass spectrometer. *Int. J. Mass Spectrom. Ion Processes* 1990, *96*, 117–137.

4. Syka, J.E.; Coon, J.J.; Schroeder, M.J.; Shabanowitz, J.; Hunt, D.F. Peptide and protein sequence analysis by electron transfer dissociation mass spectrometry. *Proc. Natl. Acad. Sci. USA* 2004, *101*, 9528–9533.

5. Ding, L.; Brancia, F.L. Electron capture dissociation in a digital ion trap mass spectrometer. *Anal. Chem.* 2006, *78*, 1995–2000.

6. Schlunegger, U.P.; Stoeckli, M.; Caprioli, R.M. Frequency scan for the analysis of high mass ions generated by matrix-assisted laser desorption/ionization in a Paul trap. *Rapid Commun. Mass Spectrom.* 1999, *13*, 1792–1796.

7. Sheretov, E.P.; Gurov, V.S.; Safonov, M.P.; Philippov, I.W. Hyperboloid mass spectrometers for space exploration. *Int. J. Mass Spectrom.* 1999, *189*, 9–17.

8. Richards, J.A.; Huey, R.M.; Hiller, J. A new operating mode for the quadrupole mass filter. *Int. J. Mass Spectrom. Ion Phys.* 1973, *12*, 317–339.

9. Sheretov, E.P.; Terentiev, V.I. The fundamentals of the theory of quadrupole mass spectrometry with pulse drive. *J. Technical Physics* 1972, *42*, 953–956.

10. Ding, L.; Kumashiro, S. Ion motion in the rectangular wave quadrupole field and digital operation mode of a quadrupole ion trap mass spectrometer. *Rapid Commun. Mass Spectrom.* 2006, *20*, 3–8.

11. Ding, L.; Sudakov, M.; Kumashiro, S. A simulation study of the digital ion trap mass spectrometer. *Int. J. Mass Spectrom.* 2002, *221*, 117–138.

12. Ding, L.; Sudakov, M.; Brancia, F.L.; Giles, R.; Kumashiro, S. A digital ion trap mass spectrometer coupled with atmospheric pressure ion sources. *J. Mass Spectrom.* 2004, *39*, 471–484.

13. Konenkov, N.V.; Sudakov, M.; Douglas, D.J. Matrix methods for the calculation of stability diagrams in quadrupole mass spectrometry. *J. Am. Soc. Mass Spectrom.* 2002, *13*, 597–613.

14. Berton, A.; Traldi, P.; Ding, L.; Brancia, F.L. Mapping the stability diagram of a digital ion trap (DIT) mass spectrometer by varying the duty cycle of the trapping rectangular waveform. *J. Am. Soc. Mass Spectrom.* 2008, *19*, 620–625.

15. Martin, R.L.; Brancia, F.L. Analysis of high mass peptides using a novel matrix-assisted laser desorption/ionisation quadrupole ion trap time-of-flight mass spectrometer. *Rapid Commun. Mass Spectrom.* 2003, *17*, 1358–1365.

16. Tanaka, K.; Sekiya, S.; Jinno, M.; Hazama, M.; Kodera, K.; Iwamoto, S. Analysis of macromolecular ions using a MALDI DIT MS mass spectrometer. *Proc. 56th ASMS Conference on Mass Spectrometry and Allied Topics*, Denver, CO, June 1–5, 2008.

17. Syka, J.E.P. Commercialization of the quadrupole ion trap. In: *Practical Aspects of Ion Trap Mass Spectrometry*. Volume 1, Chapter 3, eds. R.E. March and J.F.F. Todd, Boca Raton, FL: CRC Press, 1995, 169–205.

18. Wells, J.M.; Plass, W.R.; Cooks, R.G. Control of chemical mass shifts in the quadrupole ion trap through selection of resonance ejection working point and RF scan direction. *Anal. Chem.* 2000, *72*, 2677–2683.

19. Williams, J.D.; Cox, K.A.; Cooks, R.G.; McLuckey, S.A.; Hart, K.J.; Goeringer, D.E. Resonant ejection ion trap mass spectrometry and non linear field contributions: the effect of scan direction on mass resolution. *Anal. Chem.* 1994, *66*, 725–729.

20. Brancia, F.L.; Giles, R.; Ding, L. Effect of reverse scan on mass measurement accuracy in an ion trap mass spectrometer. *J. Mass Spectrom.* 2004, *39*, 702–704.

21. Cleven, C.D.; Cox, K.A.; Cooks, R.G.; Bier, M.E. Mass shifts due to ion/ion interactions in a quadrupole ion-trap mass spectrometer. *Rapid Commun. Mass Spectrom.* 1994, *8*, 451–454.

22. Cox, K.A.; Cleven, C.D.; Cooks, R.G. Mass shifts and local space charge effects observed in the quadrupole ion trap at higher resolution. *Int. J. Mass Spectrom.* 1995, *144*, 47–65.

23. Guan, S.; Marshall, A.G. Stored waveform inverse Fourier transform (SWIFT) ion excitation in trapped-ion mass spectrometry: theory and applications. *Int. J. Mass Spectrom. Ion Processes* 1996, *157–158*, 5–37.

24. Hoekman, D.J.; Kelley, P.E. Method for generating filtered noise signal and broadband signal having reduced dynamic range for use in mass spectrometry. *US Patent* 1997, 5,703,358.

25. McLuckey, S.A. Principles of collisional activation in analytical mass spectrometry. *J. Am. Soc. Mass Spectrom.* 1992, *3*, 599–614.

26. Cooks, R.G.; Ast, T.; Mabud, A. Collisions of polyatomic ions with surfaces. *Int. J. Mass Spectrom. Ion Processes* 1990, *100*, 209–265.

27. Little, D.P.; Speir, J.P.; Senko, M.W.; O'Connor, P.B.; McLafferty, F.W. Infrared multiphoton dissociation of large multiply charged ions for biomolecule sequencing. *Anal. Chem.* 1994, *66*, 2809–2815.

28. Dongre, A.R.; Jones, J.L.; Somogyi, A.; Wysocki, V.H. Influence of peptide composition, gas-phase basicity, and chemical modification on fragmentation efficiency: evidence for the mobile proton model. *J. Am. Chem. Soc.* 1996, *118*, 8365–8374.

29. Tsaprailis, G.; Nair, H.; Somogyi, A.; Wysocki, V.H.; Zhong, W.Q.; Futrell, J.H.; Summerfield; S.G.; Gaskell, S.J. Influence of secondary structure on the fragmentation of protonated peptides. *J. Am. Chem. Soc.* 1999, *121*, 5142–5154.

30. Annan, R.S.; Carr, S.A. Phosphopeptide analysis by matrix-assisted laser desorption time-of-flight mass spectrometry. *Anal. Chem.* 1996, *68*, 3413–3421.

31. Zubarev, R.A. Reactions of polypeptide ions with electrons in the gas phase. *Mass Spectrom. Rev.* 2003, *22*, 57–77.

32. Zubarev, R.A.; Kelleher, N.L.; McLafferty, F.W. Electron capture dissociation of multiply charged protein cations. A non-ergodic process. *J. Am. Chem. Soc.* 1998, *120*, 3265–3266.

33. Baba, T.; Hashimoto, Y.; Hasegawa, H.; Hirabayashi, A.; Waki, I. Electron capture dissociation in a radio frequency ion trap. *Anal. Chem.* 2004, *76*, 4263–4266.

34. Silivra, O.A.; Kjeldsen, F.; Ivonin, I.A.; Zubarev, R.A., of polypeptides in a three-dimensional quadrupole ion trap: implementation and first results. *J. Am. Soc. Mass Spectrom.* 2005, *16*, 22–27.

35. Dahl, D.A. SIMION for the personal computer in reflection. *Int. J. Mass Spectrom.* 2000, *200*, 3–25.

5 High-Field Asymmetric Waveform Ion Mobility Spectrometry (FAIMS)

Randall W. Purves

CONTENTS

5.1 Introduction 310
    5.1.1 Ions at Atmospheric Pressure 310
        5.1.1.1 Introduction to Ion Mobility 311
        5.1.1.2 Ion Mobility at High Electric Fields 313
    5.1.2 Origins of FAIMS 315
5.2 FAIMS: Underlying Principles 316
    5.2.1 Ion Separation in FAIMS and Basic Terminology 316
    5.2.2 Ion Focusing in FAIMS 321
    5.2.3 Ion Trapping in FAIMS 327
    5.2.4 Modeling Ion Behavior and Ion Losses 328
5.3 Role of Bath Gas and Variables Affecting Gas Number Density
within FAIMS 330
    5.3.1 Role of the Bath Gas 330
        5.3.1.1 Bulk Gases 331
        5.3.1.2 Trace Gases 333
    5.3.2 Gas Number Density (N) 334
        5.3.2.1 The Importance of E/N 334
        5.3.2.2 Role of Temperature in Ion Separation 335
5.4 FAIMS Hardware 337
    5.4.1 Electrodes 337
        5.4.1.1 Planar Electrodes 338
        5.4.1.2 Curved Electrodes 340
        5.4.1.3 Tandem FAIMS Systems 342
        5.4.1.4 Other FAIMS Electrode Designs 343
    5.4.2 Asymmetric Waveform 344
        5.4.2.1 Types of Waveforms 344
        5.4.2.2 Considerations and Performance 344
    5.4.3 Commercial FAIMS Systems for Use with
Mass Spectrometry 347
        5.4.3.1 Mine Safety Appliances Company (MSA) 347
        5.4.3.2 Ionalytics/Thermo 348

         5.4.3.3   Sionex Corporation 350
         5.4.3.4   Owlstone Nanotech 350
5.5   FAIMS Applications in Mass Spectrometry 351
      5.5.1   Large Molecules 351
              5.5.1.1   Gas-Phase Protein Structures 352
              5.5.1.2   Peptides 355
      5.5.2   Small Molecules 358
              5.5.2.1   Inorganic 358
              5.5.2.2   Environmental 360
              5.5.2.3   Bioanalytical 362
      5.5.3   Other Applications 364
5.6   Summary and Outlook 364
References 366

5.1   INTRODUCTION

In the Prefaces to Volumes 4 and 5 of *Practical Aspects of Trapped Ion Mass Spectrometry* the Editors have explained that, in defining the scope of these publications, it is considered that "an ion is 'trapped' when its residence time within a defined spatial region exceeds that had the motion of the ion not been impeded in some way." High-field asymmetric waveform ion mobility spectrometry (FAIMS), which is operated typically in the atmospheric pressure regime, falls clearly within this definition. This method for separating different types of ions exploits the electric field dependence of ion mobility through combinations of oscillating and variable DC electric field gradients that act perpendicularly to the direction of a flow of bath gas containing the analyte ions.

The goal of this chapter is to introduce the reader to the basic concepts of FAIMS, and to familiarize the reader with the diversity and scope of this gas-phase ion separation technique. Section 5.1 will review briefly the basic principles of ion mobility spectrometry (IMS) that are necessary for introducing FAIMS, including a discussion of ion properties under the influence of high electric fields. In addition, a historical timeline for the development of FAIMS will be given. Section 5.2 will introduce a simplified FAIMS system that will serve to illustrate the fundamental concepts and terminology. The discussion in this chapter will progress through various examples of increasing complexity. Section 5.3 will expand on the role of two critical variables in FAIMS separations: the bath gas and the gas number density. Section 5.4 will examine more closely the diversity of hardware used in FAIMS, exploring different variants of the electrode set and the waveform generator, ending with a brief discussion of some of the commercially available FAIMS devices. Section 5.5 will present a wide range of different applications in which FAIMS has been used in conjunction with mass spectrometry. The chapter will conclude (Section 5.6) with the author's view on the future direction of FAIMS.

5.1.1   IONS AT ATMOSPHERIC PRESSURE

In this volume, several different concepts have been described that illustrate how to confine ions under high vacuum conditions, for example, at 10^{-8}–10^{-10} Torr, which is the pressure range for maximum system performance in Fourier transform ion cyclotron resonance mass spectrometry (FT-ICR MS). In this chapter, the focus will be on manipulations of ions at pressures that are orders of magnitude greater. Ion mobility spectrometry experiments have been carried out at low milliTorr pressures (see, for example, Viggiano et al. [1]) but are done typically at pressures ranging from about ca 1 Torr to atmospheric pressure. As a consequence, the requirements for ion focusing at these pressures will be quite different from requirements in vacuum, as collisions dominate at atmospheric pressure and momentum arguments are no longer relevant. The basic principles of IMS presented in this chapter will serve as a necessary foundation for the presentation of FAIMS, since FAIMS can be categorized as a specialized type of IMS.

5.1.1.1 Introduction to Ion Mobility

More thorough descriptions of ion mobility can be found in the literature (for example, Mason and McDaniel [2] or Eiceman et al. [3]).* For the purposes of presenting FAIMS, the basic concepts of ion mobility will be considered in this Section. A conventional drift-tube IMS experiment (also referred to as DT-IMS), as described by Borsdorf and Eiceman [4] is illustrated in Figure 5.1, which is reproduced from O'Donnell et al. [5]. The sample is introduced into the ionization region (Figure 5.1a). Ionization is achieved using sources amenable to high pressures, such as corona discharge, electrospray, and radioactive foils (Figure 5.1b). The drift tube in this example consists of equally spaced plates with equal voltage drops across consecutive plates. By performing the experiment in this manner, the electric field within the drift tube will remain constant, a necessary condition for a DT-IMS experiment. A Faraday plate was used as the 'detector' in this example, but for coupling with mass spectrometry this plate will contain a pin hole in order to allow ions to pass into the mass spectrometer (for example, a time-of-flight (TOF) or other mass analyzer suitable for analyzing ion packets). A key component in the design is the shutter grid, which plays a critical role in determining the peak shape/width in an ion mobility spectrum. By applying a voltage pulse to the shutter grid (also referred to as a gating plate) ions will enter the drift tube (Figure 5.1c), and their velocities in the drift tube are dictated by the fundamental ion mobility equation [2].

$$v = KE \tag{5.1}$$

where v represents the velocity of an ion in a bath gas (cm s^{-1}), E represents the electric field (V cm^{-1}), and K is the ion mobility constant. Note that this form of the equation is a simplification, and is better expressed as $v = KE/N$, where N is the gas number density that will be described later, see Section 5.3. The value of the reduced mobility, K_0, can be calculated from the drift time, t (s), measured under defined experimental conditions, namely the length of the drift tube, d (cm), the gas

---

* See Volume 5, Chapter 13: The Study of Ion/Molecule Reactions at Ambient Pressure with Ion Mobility Spectrometry and Ion Mobility/Mass Spectrometry, by Gary A. Eiceman and John A. Stone.

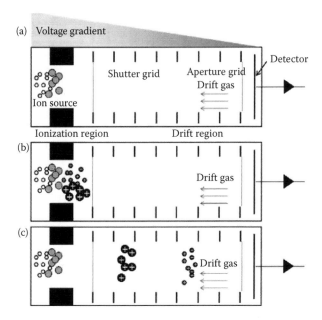

FIGURE 5.1 A conventional drift-tube ion mobility spectrometry (DT-IMS) experiment. A voltage gradient (constant electric field) is applied as is illustrated by the triangle. (a) Two different compounds are introduced into the ionization region; (b) ionization occurs; and then (c) two groups of ions enter the drift region when the shutter grid opens and become separated as they flow against the drift gas to reach the detector, whereas neutral samples do not enter the drift region. (Reproduced from O'Donnell, R.M., Sun, X., and Harrington, P.B., *Trends Anal. Chem.* 2008, *27*, 44–53. With permission from Elsevier Limited.)

pressure, p (Torr), and the temperature, T (K). Assuming no changes in ion identity occur during passage through the drift cell (for example, solvation, decomposition, etc.), K_0, is given by [2]

$$K_0 = \left(\frac{d}{tE}\right)\left(\frac{p}{760}\right)\left(\frac{273}{T}\right) \quad (\text{cm}^2\ \text{V}^{-1}\text{s}^{-1}) \qquad (5.2).$$

A common application of DT-IMS is in the determination of average collision cross-sections, Ω_D, for investigations of gas-phase ion conformations, especially for proteins and peptides.* The expression used for this application is [3]

$$K = \left(\frac{3q}{N}\right)\left(\frac{2\pi}{\mu kT}\right)^{\frac{1}{2}}\left(\frac{(1+\alpha)}{\Omega_D}\right) \qquad (5.3),$$

---

* See Volume 5, Chapter 8: Applications of Traveling Wave Ion Mobility Mass Spectrometry, by Konstantinos Thalassinos and James H. Scrivens.

where q is the charge of the ion, N is the neutral gas density (to be discussed later in Section 5.3.2), k is Boltzmann's constant, and μ, the reduced mass of the ion (m) in bath gas (M), is given by

$$\mu = \left(\frac{mM}{m+M} \right) \qquad (5.4).$$

Note that α, which is a correction factor, is less than 0.02 if $m > M$ [4].

As was mentioned earlier, the width of the gating time of the shutter grid plays a critical role in determining the resolution, peak width, and sensitivity of the device. Typically, longer gating times will improve sensitivity, but at the cost of resolution. Longer drift tubes can be used also to increase resolution, but at the expense of sensitivity as losses due to diffusion can be significant.

5.1.1.2   Ion Mobility at High Electric Fields

It would appear that one way to reduce diffusion would be by increasing the electric field, thereby reducing the time that the ions spend in the drift tube. However, an important observation to note here is that the ion mobility is not a constant, but is a function of the electric field, expressed as

$$K(E/N) = K(0)[1 + f(E/N)] \qquad (5.5).$$

Here $K(0)$ is the mobility at the low-field limit (that is, $K(0) = K$ as E/N approaches zero, and N is the gas number density. The role of N will be examined more thoroughly in Section 5.3.2, however, for the purposes of this discussion, because N can be expressed in cm^{-3}, the quotient E/N is expressed in units of V cm^2 or townsend (Td), where 1 Td $= 1 \times 10^{-17}$ V cm^2. Note that for simplicity, the use of the term 'electric field' in the text will imply the quantity (electric field)/(gas number density).

This dependence of mobility on electric field can be expressed also using an infinite series of even powers of E/N, as [2,6,7]

$$K(E/N) = K(0)[1 + a(E/N)^2 + b(E/N)^4 + c(E/N)^6 + d(E/N)^8 + ...] \qquad (5.6),$$

where the factors a, b, c, d, ... are functions of the ion–gas interaction [2].

Thus, in order to relate the drift time to ion properties such as cross-section (as described by Equation 5.3 above), the value of $f(E/N)$ in Equation 5.5 (or the terms $a(E/N)^2$, $b(E/N)^4$, ... in Equation 5.6) must be <<1 such that the expressions within the square brackets in Equations 5.5 and 5.6 are *ca* 1 at the value of E/N employed in the experiment.

The dependence of mobility as a function of E/N has been investigated for several ions (using reduced pressures), and tables of data can be found in some of the pioneering work dating back to the 1970s (for example, Ellis *et al.* [8]). Although the mobility is ion-dependent, for typical IMS/MS experiments reported in the literature, it has been observed that $K(E/N)$ varies by <1% for systems operated under 20 Td when $p \approx 1$–5 Torr [7]. Above this value the deviations can be significant, with the first term (*i.e.*, $a (E/N)^2$ in Equaton 5.6) commonly dominating up to 100 Td, but with other terms become more important and even dominant at higher values of E/N (e.g., >120 Td) [9].

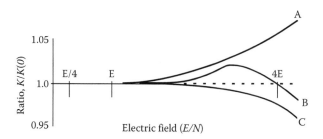

FIGURE 5.2 Hypothetical dependence of the mobility of three different ions, A, B, and C, on E/N. The ratio $K/K(0)$ represents the ratio of the mobility of an ion at a given E/N divided by the mobility of the same ion as the value of E/N approaches zero (*i.e.*, $K(0)$). The values indicated as E/4, E, and 4E are used for illustrative purposes (see text). (Adapted from Purves, R.W., Guevremont, R., Day, S., Pipich, C.W., and Matyjaszczyk, M.S., *Rev. Sci. Instrum.* 1998, *69*, 4094–4105. With permission from the American Institute of Physics.)

Figure 5.2 shows a plot of the ion mobility as a function of electric field strength that has been adapted from the literature [10]. The scale on the y-axis is expressed as $K/K(0)$, which represents the ratio of the mobility of an ion at a given E/N (x-axis) divided by the mobility of the same ion at E/N approaching zero. Note that the expression K_h/K, where the subscript 'h' implies ion mobility at 'high electric fields,' has also been used in the literature to describe this ratio. The plot shows the mobility of three different hypothetical ions; the figure has been adapted to serve as an illustration for different examples presented throughout this chapter. The assignment of these ions as 'A,' 'B,' and 'C' is consistent with terminology that was used in the literature for describing these different ion types [10,11]. In particular, type A ions experience increases in mobility with increases in electric field strength, type C ions experience decreases in mobility with increases in electric field strength, and type B ions initially experience increases before experiencing decreases in mobility as E/N is increased. This terminology will be retained for the discussion herein, although this model is an oversimplification as, for example, it is believed that all type A ions would eventually experience decreases in mobility if high enough values of E/N could be applied [2]. The change in mobility of an ion as a function of E/N is dependent upon complex interactions between the ion and the bath gas, as is described extensively by Mason and McDaniel [2]; a detailed description of these interactions is beyond the scope of this present chapter. Although predicting these changes *a priori* currently remains elusive, it has been observed experimentally that type A and B ions are typically low molecular weight species (for example, perchlorate, haloacetic acids), whereas type C ions are typically larger species such as peptides and proteins. Although this observation applies generally, there have been some noticeable exceptions. As will be discussed in Section 5.5.1.1, recent work [12] has shown that some larger proteins can also experience increases in their mobilities as a function of electric field.

Both increases (type A behavior) and decreases in mobility (type C behavior) can be explained by examining the complex interactions between an ion and the bath gas [2]. In some cases, increases in mobility arise due to the existence of ion-bath gas clusters (present at low electric fields) that are dissociated at higher field strengths.

That is, because the ion is 'smaller' at high fields, it exhibits an apparent increase in mobility. This clustering is typically more apparent with small ions as larger ions do not show as significant a deviation. Not only does a larger ion have the ability to dissipate better energy resulting from heating by the bath gas, but the changes in the size of the ion arising from clustering and declustering will not have such a significant effect upon the overall 'size' of the ion.

5.1.2   ORIGINS OF FAIMS

The invention and early development of FAIMS occurred in Russia. The concept of measuring differences in the mobilities of ions at high electric fields relative to low electric fields was introduced by Gorshkov, who received an Inventors Certificate in 1982 [13]. The first refereed publication in the English literature was in 1993 by Buryakov *et al.* [14] who described "a new method of ion separation based on the non-linear dependence of the mobility coefficient on the electric field density." The authors demonstrated the technique for detecting trace amounts of amines in atmospheric air.

An important development in the evolution of the FAIMS technology was the introduction of a cylindrical FAIMS device by Buryakov [15]. As will be described in Section 5.2.2, the non-homogeneous fields between the cylinders allow for increased sensitivity due to ion focusing.

In the early 1990s, the technology was brought to North America and worked on initially by the Mine Safety Appliances Company (MSA, Pittsburg, PA), who trademarked the name Field Ion Spectrometer® (FIS®). The company MSA manufactured a device that was based on the cylindrical design and used electrometric detection [16]. Several FIS beta instruments were distributed to a selected number of researchers, with the primary focus being trace gas analysis, especially detection of explosives (see for example Ref. [17]). Although the MSA effort was eventually halted in the early 2000s, the progress made by MSA was nonetheless a critical step in the evolution of the FAIMS technology.

One of the MSA FIS instruments was sent to researchers at the National Research Council of Canada (NRC, Ottawa, ON, Canada). In collaboration with MSA, a prototype was manufactured that modified the FIS device such that it could be coupled to a mass spectrometer ('FAIMS-MS'). The characterization of this prototype (using a corona discharge source) coupled to a FAIMS-MS system was reported in 1998 [10], and the subsequent use of this device with electrospray ionization mass spectrometry was reported in 1999 [11]. The complementary nature of FAIMS and MS enabled the researchers at NRC to advance FAIMS instrumentation and applications using prototypes having non-homogeneous electric fields between the electrodes. At the same time, a group of researchers at Charles Stark Draper Laboratory (Cambridge, MA) were developing micromachined planar FAIMS devices having homogeneous fields between the electrodes. In 2001, the company Sionex Corporation (Cambridge, MA) was founded based on technology licensed from Charles Stark Draper Laboratory, while researchers from NRC founded Ionalytics Corporation (Ottawa, ON, Canada) based on technology licensed from NRC. In Section 5.4, these different FAIMS devices will be described further, and a short section on commercially available instruments will be presented. In addition, it is important to note that simultaneously with the development that has been occurring in North America, groups within Russia (for example, Buryakov) have remained actively involved with FAIMS.

Ionalytics continued to develop FAIMS for mass spectrometry, in part through collaborations with academic, government, and industrial researchers. Many of these collaborators, such as R.D. Smith and A.A. Shvartsburg at Pacific Northwest National Laboratories (Richland, WA), have since made key contributions to furthering the understanding and development of FAIMS.

A pivotal development at Ionalytics was the creation of a temperature-controlled electrode that was necessary for using FAIMS with heated ionization sources, which attracted the interest of the pharmaceutical industry. Shortly after this development, Ionalytics was acquired by Thermo Electron in 2005 (now Thermo Fisher, San Jose, CA), who introduced FAIMS on the 'TSQ Quantum' triple quadrupole mass spectrometer in 2006. Although, Ionalytics had produced a limited number of instruments, the introduction of FAIMS (with integrated software) on Thermo instrumentation has made FAIMS available to an increased number of users of this technology, especially in the field of bioanalysis.

As with many new instrumental developments, several names have been used to describe the same technique or variations of the same technique. In addition to FAIMS, the most common name appearing in the literature that has been used to reference the technique is differential mobility spectrometry or DMS. Other common names by which FAIMS has also been called include transverse field compensation IMS, field ion spectrometry, and radio frequency-IMS. As noted previously, the name FIS was trademarked by MSA and was ambiguous because of confusion with field ionization techniques [18]. The name 'RF-IMS' [19] also leads to confusion because the RF waveforms used in, for example, triple quadrupole mass spectrometry, could not be used for FAIMS (see Section 5.2). In any case, regardless of the name, FAIMS is a gas-phase ion separation technique that appears poised to make a significant impact in the field of mass spectrometry.

5.2 FAIMS: UNDERLYING PRINCIPLES

5.2.1 ION SEPARATION IN FAIMS AND BASIC TERMINOLOGY

Figure 5.3a shows a simplified FAIMS system consisting of three regions: ionization, ion separation, and ion detection. These three regions have been separated by the dashed lines in the figure. The ionization and ion detection regions are straightforward in this set-up. The FAIMS device can be coupled readily to atmospheric pressure ionization sources including, but not limited to, radioactive foils [14], corona discharge [10], and electrospray ionization [11]. The detection region has primarily involved coupling FAIMS with electrometric [13,14,16] and mass spectrometric detection [10,14]; the use of the latter being illustrated in the figure and also the primary focus of this chapter.

The ion separation region (also called the *analyzer region*) is the heart of the FAIMS device. As will be shown later in this chapter, this region can consist of dramatically different geometries, all of which have the primary purpose of being used for ion separation. In Figure 5.3a, the separation region consists of two parallel, flat plates, the distance between them being referred to as the *gap width*. The upper

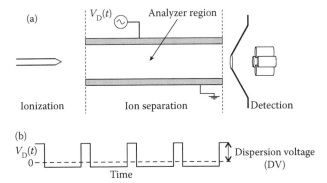

FIGURE 5.3   (a) A schematic of the basic components of a FAIMS system, namely ionization, ion separation, and ion detection. The system shown here has a corona discharge ionization source, flat, parallel electrodes, and detection is achieved using a quadrupole mass filter. (b) An example of an asymmetric waveform, $V_D(t)$, used to exploit the difference in ion mobility as a function of E/N. The rectangular asymmetric waveform is applied to the upper FAIMS electrode and the maximum amplitude of the waveform is referred to as the DV, or dispersion voltage.

plate has a waveform, $V_D(t)$, applied to it, while the bottom plate is maintained at ground potential. The waveform has two requirements: it must be asymmetric (that is, $V_{max} \neq |V_{min}|$) and the voltage-time product for one complete cycle, of period τ, of the waveform is zero. That is [13,14]

$$\int_0^\tau Vt = 0 \qquad (5.7).$$

An example of such a waveform, given in Figure 5.3b, is rectangular in nature. The high-voltage portion of this waveform lasts only one quarter as long as the low voltage portion, however, the magnitude of the high voltage portion is four times as great. Since $V_{max} \neq |V_{min}|$ and the product of voltage $\times$ time during the high field portion ($4V \times \frac{1}{4} t = V \times t$) equals that during the low voltage portion ($V \times t$), this waveform satisfies the conditions set out above. Note that the maximum absolute value of the waveform, referred to as the *Dispersion Voltage* (DV), is indicated in Figure 5.3b. At present, electronic constraints limit the output on the asymmetric waveform; for example, maximum DV values of up to *ca* 8 kV [20] have been reported. To generate electric fields that can cause significant changes in ion mobility, and thus achieve ion separation, requires typically at least 40 Td, or *ca* 10 kV cm^{-1} at atmospheric pressure and room temperature. Consequently, to achieve these fields, the plates must be spaced closely together (for example, 2 mm has been commonly used). The properties of the asymmetric waveform and its influence on ion separation are described in more detail in Section 5.4.2.

Ion separation in FAIMS has been described by Buryakov *et al.* [14] and Purves *et al.* [10]. For illustrative purposes, consider three hypothetical (positive) ions, A, B, and C; the dependence of the mobility of each ion on electric field is described by

Figure 5.2 (see Section 5.1.2). Consider also the situation in which a gas is used to transport the ions through the FAIMS ion separation region (from left to right in Figure 5.3a). The asymmetric waveform (Figure 5.3b) is applied to the upper plate whilst the lower plate is set to ground potential; the plates are 2 mm apart. The DV-value has been selected to produce an electric field having magnitude E during the high-field portion of the waveform, thus during the low voltage portion (having opposite polarity), the electric field will be −E/4. As the ions are carried between the plates from left to right by the bath gas, they will oscillate under the influence of the waveform as is illustrated in Figure 5.4a. The distance that the ion moves in the y-plane during each portion of the waveform is dependent on the ion's mobility in the gas medium (which remains constant in this example), the waveform function (a rectangular wave having DV = E), and the waveform frequency. For example, if $K(0)$ for ion B is 2.0 cm^2 V^{-1} s^{-1}, E = 20 Td (*ca* 5 kV cm^{-1} at room temperature and

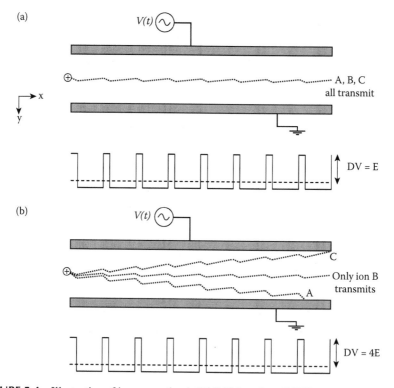

FIGURE 5.4 Illustration of ion separation in FAIMS based on $K/K(0)$ vs. E/N plots for ions A, B, and C shown in Figure 5.2 (for simplicity, all three ions have to same $K(0)$ value); (a) application of an asymmetric waveform with dispersion voltage E that is insufficient to cause a change in the mobility of ions A, B, or C. Consequently, all three ions are transmitted successfully through the device. (b) When the dispersion voltage is increased to 4E, only ion B has the appropriate $K/K(0)$ required for successful ion transmission. Since the ion mobility of ion A increases at 4E, it is lost to the lower electrode; similarly, as the mobility of ion C decreases at 4E, it is lost to the upper electrode (see text for details).

atmospheric pressure), and the waveform has a frequency of 1 MHz ($t = 0.2$ μs during the high voltage portion and 0.8 μs during the low voltage portion), the distance traveled during the high voltage portion can be calculated as follows. From Equation 5.1, the ion velocity, $v = KE = (2.0$ cm^2 V^{-1} s^{-1}) × (5000 V cm^{-1}) = 10,000 cm s^{-1}, and thus the distance traveled, $d_t = v \times t = 0.002$ cm, or 0.02 mm. During the low voltage portion of the waveform, the time will increase by a factor of four, but the magnitude of the velocity of the ion (now in the opposite direction) will decrease by the same factor of four, thereby giving the same distance (i.e., 0.02 mm) traveled in the opposite direction. Therefore, because there is *no* change in the mobility of ion B (or, in other words, the ratio of the high-field to low-field mobility is exactly one), or in the mobility of any of the three ions in this example at the values of E/N employed (see Figure 5.2), in the absence of ion losses (for example, space-charge repulsion, diffusion) all three ions will be transmitted successfully through the device. Note that a waveform with a frequency of 10 kHz would not transmit ion B under the same conditions, because the lower frequency would give $d_t = 2$ mm, the same distance as the spacing between the electrodes in this example.

It is important to emphasize that unlike DT-IMS, the ions in this illustration of FAIMS are not transported through the device by an electric field, but instead are carried through the ion separation region (from left to right) by a flowing bath gas (for example, nitrogen) typically called the *carrier gas*. The electric fields act perpendicularly to the ion flow and cause the ions to oscillate back and forth between the plates as the gas carries them through the device.

Consider now the same example, except that the magnitude of the waveform has been increased by a factor of 4. Now during the high-field portion of the waveform, the ions will experience an electric field of 4E (80 Td in the numerical example) and during the low voltage portion, the field will be −E. During the high-field portion of the waveform, the mobility of ion A will increase (see Figure 5.2) and as a consequence ion A travels further than during the low voltage portion of the waveform. As a result, over many cycles of the waveform, ion A will begin to move toward the lower plate, eventually being lost to the lower electrode (see Figure 5.4b). The larger the difference in ion mobility from the low-field value, the faster the ion will collide with the wall (the difference in the low-to-high field mobility has been exaggerated in the figure to illustrate the point). A similar argument can be made for ion C; that is ion C will exhibit a decrease in mobility at the higher electric field (4E) and will have a net movement toward the opposite plate as ion A (that is, the upper plate), again being discharged at the electrode surface (Figure 5.4b). For ion B, the behavior is more complex because although the mobility of ion B is in a non-constant region of the plot in Figure 5.2, at the electric fields applied in this example (4E and −E), the mobility of ion B does not change (i.e., $K(E/N) = K(0)$). Hence, ion B will be transmitted successfully through the device, as is illustrated in Figure 5.4b.

Consider now the motion of ion A in the above example, but with a DC offset voltage, called the *Compensation Voltage*, or CV, superimposed upon the waveform. To compensate for the increase in mobility of ion A at the higher value of the electric field, the CV-value must be negative as this will pull the ion away from the lower plate. By a similar argument, the CV must be positive for ion C (in order to push

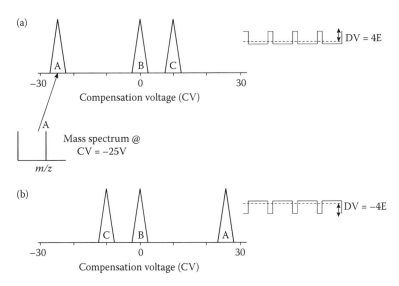

FIGURE 5.5 CV spectra for hypothetical ions A, B, and C (based on *K/K(0)* vs *E/N* plots shown in Figure 5.2) using flat plate electrodes and (a) a waveform with a positive DV (= 4E); and (b) a waveform with a negative DV (= −4E). Note that by setting the CV to a given value, ions can be continuously transmitted through the device and into a mass spectrometer for detection. For example, a mass spectrum can be acquired by setting the CV to −25 V (see inset in (a)). At this CV-value ion A, occurring at *m/z*-value 'A,' is observed.

the ion away from the upper plate), and CV = 0 for ion B (since there is no net drift, because the high-field and low-field mobilities are equal).

Now consider a continuous source of ions having properties defined by either A, B, and/or C from Figure 5.2. By scanning the value of CV, one can generate a *CV spectrum*, such as the one shown in Figure 5.5a for this hypothetical example; note that the idealized triangular peak shape ignores the effects of diffusion and space-charge repulsion. When generating the CV spectrum, the scan speed must take into account the *residence time*, which is the time that the ions spend within the FAIMS analyzer region; this is, typically, on the millisecond timescale (for example, the Thermo FAIMS system described in Section 5.4.3.2 has a residence time of *ca* 100 ms). Scans that are performed too quickly may distort the appearance of the CV spectra if changes in CV occur before the ions can successfully pass through the device. Also note that if the asymmetric waveform is inverted (as is illustrated in the figure), the net result is that the CV would be inverted also and thus the CV spectrum would look like the one shown in Figure 5.5b.

Due to the continuous nature of FAIMS, it can be compared with a quadrupole mass filter, whereas DT-IMS is better compared with a TOF mass spectrometer, in which the supply of ions is pulsed in each case. That is, the FAIMS analyzer can be scanned (CV spectrum) to detect various ions present in a mixture or can be set to transmit ions continuously at a particular CV. The continuous stream of ions exiting FAIMS can then be allowed to pass into a mass spectrometer to determine their identity through the generation of a 'mass spectrum.' Recall that only ions with the correct ratio of high-to-low field mobility will be transmitted successfully. Thus, for example, when the CV is set to −25 V and a mass spectrometer is used to detect the ions transmitted through FAIMS, as is also illustrated in Figure 5.5a, a mass spectrum containing only the m/z-value of ion A is observed in this hypothetical example. In contrast to this arrangement, in DT-IMS, the ions are pulsed and arrive at the detector as a function of time.

Figure 5.5 serves also to illustrate the concept of *resolution* and *resolving power* using FAIMS. Although the term 'resolving power' has been used to describe FAIMS spectra as the ratio of peak value of the CV to the full-width at half-maximum (FWHM) of the ion peak [21], it is apparent that this interpretation must be used with caution as it can lead to meaningless results. For example, ion B in Figure 5.5 would give a resolving power of zero. On the other hand, the resolution between two peaks can be treated in a similar fashion to that which is used typically in chromatography. For example, the resolution, R_s, between peaks (A and B) in a CV spectrum was calculated as [22]

$$R_s = 2(CV_B - CV_A) / (PW_A + PW_B)$$
$$(5.8),$$

where CV_A and CV_B represent the CV-values of maximum intensity for peaks A and B, respectively, and PW_A and PW_B represent the peak widths at 10% height for peaks A and B, respectively.

For the flat plate device described above, it is apparent that when the net ion drift is small, and there are not enough cycles of the waveform to discharge the ion on the electrode surface, the ion will be transmitted successfully. Thus, in the absence of diffusion and space-charge repulsion, an infinitely long device will be required to give an infinitely narrow peak. Returning to our previous example, the width of the peaks in Figure 5.5 can be narrowed by either increasing the length of the device or reducing the gas flow rate. Of course, because of ion-loss mechanisms, such as diffusion and space-charge repulsion, increases in resolution are accompanied by decreases in the signal intensity.

5.2.2   ION FOCUSING IN FAIMS

Consider now the same continuous source generating ions having properties defined by either A, B, or C, but with the flat, parallel plates replaced by concentric cylinders, again with spacing, d, as is illustrated in a cross-sectional view in Figure 5.6a. When the same waveform having the same magnitude, which was applied to obtain the result shown in Figure 5.5a, is applied to the inner cylinder, this time the resulting CV spectrum for ions A, B, and C, shown in Figure 5.6b, is much different as the peaks corresponding to both ion B and ion C are absent. Furthermore, the peak shape for ion A differs considerably, having a flat top, again in the absence of diffusion and space-charge repulsion, and an increased width compared to that shown in Figure 5.5a. In addition, when the polarity of the waveform is inverted, the CV spectra are again very different as the polarity of the CV-values for the peaks do not simply invert. As shown in Figure 5.6c, this time the peaks corresponding to ions

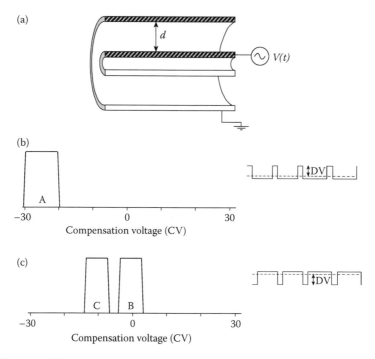

FIGURE 5.6 CV spectra for hypothetical ions A, B, and C (based on $K/K(0)$ vs. E/N plots shown in Figure 5.2) using a cylindrical FAIMS electrode shown in cross section in (a). (b) CV spectrum using a waveform with a positive DV; and (c) CV spectrum using a waveform with a negative DV. The appearance of the spectra is vastly different from those obtained for the flat plate electrodes (see text for details).

B and C are present and ion A is not detected. Again, the peaks in Figure 5.6c are wider and have flat tops; note that in practice, because of diffusion and space-charge effects, these peaks will generally appear to be Gaussian, as will become apparent in Figure 5.7.

Guevremont and Purves [23] reported the behavior of the CV spectra based on the polarity of the waveform applied to the cylindrical electrodes described above, along with two other observations that were inconsistent with literature results obtained for flat plates. These two observations are readily apparent in Figure 5.7, which shows the effect of the DV-value on the peak width and signal intensity for CV scans of a given mixture of oxoanions (XO_3^-) [24]. The m/z-values for four oxoanions were monitored using FAIMS-MS as the CV-value was scanned from 0 to 40 V to produce *ion selected* CV ('IS-CV') spectra (termed 'ion selected' because only four ion species were monitored specifically). The figure shows very large increases in sensitivity with increases in the DV, in contrast to the case with flat plates where the peak intensity can only remain the same or become lower as the *effective gap width* decreases. Note that the 'effective' gap width is the space between the electrodes minus the 'net' distance that the ion travels radially (or orthogonally) with respect to the gas flow direction during the distance that the ion travels in one direction perpendicular to the gas flow during the application of one cycle of the waveform. In addition, significant increases

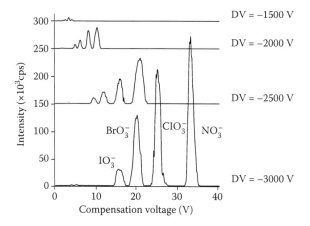

FIGURE 5.7 Effect of dispersion voltage (DV) on CV spectra. Ion-selected compensation voltage (IS-CV) spectra (m/z 62, 83, 127, and 175) acquired in negative ion mode using a 62 μM mixture of nitrate, chlorate, bromate, and iodate in a 9:1 methanol/water (v/v) solution containing 0.2 mM ammonium acetate (see text for details). (a) DV = −1500 V; (b) DV = −2000 V; (c) DV = −2500 V; and (d) DV = −3000 V. (Reproduced from Barnett, D.A., Guevremont, R., and Purves, R.W., *Appl. Spectrosc.* 1999, 53, 1367–1374. With permission from Society for Applied Spectroscopy.)

in the peak widths at higher CV-values, which should be approximately the same with flat plate electrodes, are observed. These observations all supported the authors' conclusions that atmospheric pressure ion focusing occurs within the FAIMS analyzer region when using cylindrical electrodes [23].

A description of the atmospheric pressure ion focusing effect was reported by Guevremont and Purves [23] and also by Krylov [25]. The reason for the existence of the ion focusing in the cylindrical device, but not in the flat plate arrangement, is a direct result of an electric field gradient, that is a non-homogeneous field, that exists between the two cylinders in contrast to the constant electric field that is present between two flat parallel plates. Specifically, the electric field between two flat parallel plates is simply $E = V_a/d$, where V_a is the applied voltage and d is the distance between the plates, at all points between the plates; in contrast, for two concentric cylinders the electric field at a radial location, r, is given by [23]

$$E(r) = -V_a / [r \ln(a/b)] \tag{5.9},$$

where a is the radius of the inner electrode wall, and b is the radius at the outer electrode wall. For the electrodes used to generate the data in Figure 5.7, $a = 7$ mm and $b = 9$ mm.

The concept of ion focusing was first illustrated experimentally using data that were acquired for water cluster ions of the form $[H_2O]_nH^+$, generated with a corona discharge source in nitrogen [23]. The authors first acquired CV spectra at different DV values, ranging from 500 to 3300 V in 100 V increments, to determine the optimal value of CV for each selected value of DV. Figure 5.8 shows a subset of these plots, that is, eight CV spectra in which DV is increased in 100 V increments from

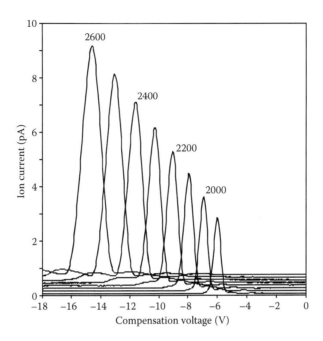

FIGURE 5.8 CV spectra acquired using electrometer-based detection for water cluster ions. The spectra were acquired using a corona discharge ionization source in purified nitrogen and with 100 V DV increments from DV 1900 to DV 2600 V. (Reproduced from Guevremont, R. and Purves, R.W., *Rev. Sci. Instrum.* 1999, *70*, 1370–1383. With permission from American Institute of Physics.)

1900 to 2600 V, with the 'optimal' CV-values being taken as the CV corresponding to maximal ion intensity for each of these curves. These optimal conditions represent a 'balanced' situation because, at these specific CV/DV values, the ion will experience no net motion toward either electrode [23]. From these optimum values, a plot of compensation field (CF) *versus* dispersion field (DF) was derived, as shown in Figure 5.9a, where CF = CV/*d* and DF = DV/*d*. For the water cluster ions, only when the fields that are applied to the FAIMS device correspond to points that fall on the line in Figure 5.9a, will they experience no net drift toward either electrode, that is, there is a balanced condition. If the applied fields correspond to a location on the plot to the left of the line, these ions will experience a net drift toward the inner electrode and, conversely, if the fields correspond to a location to the right of the line these ions will experience a net drift toward the outer electrode.

Consider the application of the optimum applied DV and CV-values corresponding to the point labeled 'X' in Figure 5.9a, that is, DV = 2500 V and CV = –12.8 V. Figure 5.9b shows the 'actual' strength of the electric field experienced by an ion at a series of radial locations between the two cylinders from very near the inner electrode (*ca* 7 mm) to very near the outer electrode (*ca* 9 mm); these values were calculated using Equation 5.9. When the plots from the actual conditions corresponding to point X on Figure 5.9a are superimposed on the balanced conditions (see Figure 5.9c), the concept of ion focusing becomes apparent. For example, consider when the ion is near

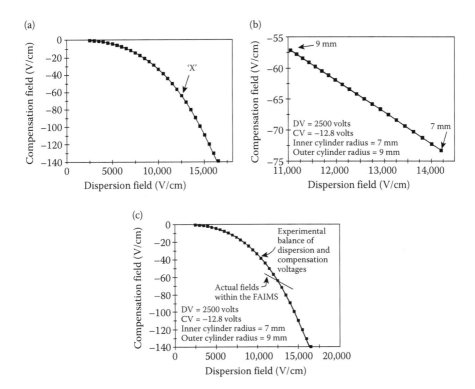

FIGURE 5.9 Illustration of ion focusing in a cylindrical FAIMS device. (a) Experimentally determined conditions of optimum transmission ('balance points') between CF and DF for water cluster ions. The values were obtained from the ion current peak maxima for the water cluster ions from CV spectra collected at 100 V increments from DV = 500 to DV = 3300 V (a subset of these data is shown in Figure 5.8). (b) The calculated electric fields within the FAIMS analyzer region from very near the inner electrode (*ca* 7 mm) to very near the outer electrode (*ca* 9 mm) when DV = 2500 V and CV = –12.8 V. (c) Superimposition of the plots showing conditions of 'balance' (from (a)) with the 'actual' (calculated) conditions within the FAIMS analyzer region (from (b)). (Reproduced from Guevremont, R. and Purves, R.W., *Rev. Sci. Instrum.* 1999, *70*, 1370–1383. With permission from American Institute of Physics.)

the wall of the outer electrode (that is, r is almost equal to 9 mm), the ion will have a net movement (recall that superimposed on this net movement is the oscillation back and forth) toward the inner electrode as discussed earlier. The strength of the focusing field at any radial location can be estimated by the difference of the actual and balanced fields within the FAIMS device. Thus, near the outer electrode, from Figure 5.9c the net field acting on the ion can be estimated to be about –10 V cm^{-1}, since near the outer electrode, in which the DF is *ca* 11 kV cm^{-1}, the corresponding 'balance CF' is *ca* 47 V cm^{-1} and the 'actual CF' is *ca* –57 V cm^{-1}). This net field will result in a net movement of the ion away from the outer electrode. Similarly, near the inner electrode there is a net field of about 15 V cm^{-1} that will move the ion away from the inner electrode. Thus, at any point between the electrodes, the ions will experience a focusing field that pushes them toward a balanced radial location. Note that for the flat plates, because the actual field is homogeneous, it would be represented by a single point on the plot of CF *versus* DF. Consequently for flat plates, there is NO ion focusing! Since the ion focusing will counteract the effects of diffusion and space-charge repulsion, this means that the optimum signal intensities for cylindrical FAIMS electrodes are higher than they are for planar electrodes. However, because of this ion focusing, the peak width for an infinitely long cylindrical FAIMS device is no longer infinitely small. Using Figure 5.9 again as an example, keeping the value of DV constant at 2500 V, as the CV-value is changed, the actual fields can be recalculated, resulting in the 'actual' plot being simply shifted upward or downward in Figure 5.9c, as the DV value remains the same. As long as the balance and actual lines intersect, there will be a balance condition at a radial location between the two electrodes and the ions will be transmitted. For the data in Figure 5.9c, this CV-range can be estimated roughly by drawing vertical lines from the ends of the line for the actual fields (which represents locations near the inner and outer electrode surfaces), and taking the intersection points on the balance line. Thus, for the inner electrode this corresponds to CF $\approx$ −90 V cm^{-1}, and for the outer electrode CF $\approx$ −45 V cm^{-1}. These values correspond to CV-values of *ca* −18 V and −9 V, which gives a first approximation of the peak width (*ca* 9 V). Note that the actual peak width shown in Figure 5.8 (*ca* 3V wide at the baseline) is significantly narrower then the 9 V estimate at DV = −2500 V as several effects have been ignored. These effects include the loss of ions near the electrode surfaces due to the finite distances that ions travel during application of the waveform, ion extraction efficiency, electrode imperfections, etc. In any case, it is important to note that a flat plate design will offer better resolution, but this will come at the expense of signal intensity.

A closer examination of Equation 5.9 reveals that as the radius of the inner electrode gets increasingly smaller, the effect on the electric field becomes increasingly significant. Thus, it could be expected that the peak widths could be changed by changing the electrode diameters, but keeping a constant gap between them. Cylindrical prototypes having inner/outer electrode radii values of 4/6 mm, 8/10 mm, and 12/14 mm have been investigated [26], and narrowing of the peak shapes obtained with electrodes having wider diameters was observed experimentally for the bromochloroacetate anion. Thus, in general, when using very large electrode radii, one would expect to see peak widths approaching those of a flat plate design, with a lower sensitivity as the focusing becomes weaker. With very small electrode radii, very large peak widths would result, accompanied by higher sensitivity as a result of the stronger focusing conditions.

Returning to Figure 5.9, in order to explain the dependence of the behavior of the FAIMS device on the polarity of the waveform, consider changing the polarity of the waveform to DV = −2500 V and CV = 12.8 V whilst continuing to inject $[H_2O]_nH^+$ ions; the result will be simply a change of sign on both axes in all the traces in the figure. This time, it can be shown that when the ion is to the left of the line for the 'balance' conditions, it will be moved toward the *outer* electrode. Thus, a water cluster ion near the outer electrode will experience a field of about 10 V cm^{-1}, which means that it will be pushed toward the outer electrode and lost. Hence, in an analogous fashion, all $[H_2O]_nH^+$ ions not at the balance point will be repelled. Due to the dependence of the CV spectrum on the polarity of the waveform for devices having non-homogeneous fields, there are two distinct modes of operation, Mode 1 and Mode 2; for positive ions these have been called P1 and P2, and for negative ions, N1 and N2 [10,23]. In short, P1 and N2 waveforms are of the type shown in Figure 5.6b and P2 and N1 waveforms of the type shown in Figure 5.6c.

5.2.3 ION TRAPPING IN FAIMS

The capability of the cylindrical FAIMS configuration to focus ions along a radial surface between the electrodes gave rise to the possibility of extending further this concept to three-dimensions and achieving ion trapping at atmospheric pressure. Although several unique trapping devices had been reported at reduced pressures (as described in this book, for example), trapping ions at atmospheric pressure had remained relatively unexplored. To demonstrate ion trapping at atmospheric pressure, a FAIMS device was designed in which the inner electrode had a 1 mm radius and a hemispherical terminus, and the outer electrode was a cylinder with a 3 mm radius, as is shown schematically in Figure 5.10. The inner electrode was given a very small

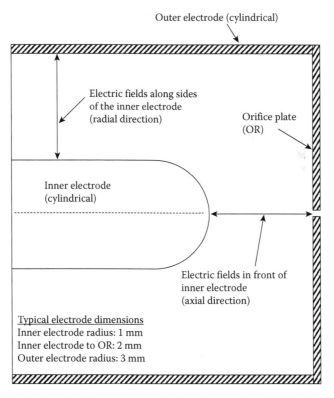

FIGURE 5.10 Schematic of a cross-sectional view of a trapping FAIMS device. The spacing between the tip of the inner electrode and the orifice plate is 2 mm. The figure illustrates the axial direction between the inner electrode and orifice plate along which the virtual 'net' electric field in Figure 5.11(c) was calculated. (Reproduced from Guevremont, R., Ding, L., Ells, B., Barnett, D.A., and Purves, R.W., *J. Am. Soc. Mass Spectrom.* 2001, *12*, 1320–1330. With permission from Elsevier Science Inc.)

radius in order to maximize the virtual 'net' fields in the region near the hemispherical terminus. This device was operated in a similar manner to the cylindrical FAIMS device in that the focusing fields were perpendicular to the gas flow. However, when the gas transported the ions to the hemispherical terminus, the ions were either allowed to pass through the orifice of the mass spectrometer, or the voltage on the orifice plate could be ramped positive (for positive ions) to selectively 'trap' the ions. Ion trapping on this system was demonstrated initially using a background ion generated by corona discharge in positive mode (m/z 380) [27], and the half-life of the exponential decay of the ion in the trapping FAIMS was found to be *ca* 5 ms.

In a subsequent publication, a tandem FAIMS arrangement shown in Figures 5.11a and b (see also Section 5.4.1.2) was used to characterize a trapping FAIMS device for studies with electrospray-generated gramicidin S ions [28]. The first FAIMS unit (labeled sFAIMS) in the tandem combination had wider electrode radii and hence better ion separation efficiency than the trapping FAIMS (labeled tFAIMS). The function of the first FAIMS was to lower the number of background ions captured in the trap and, thereby, reduce space-charge effects [28]. The trapping portion of the FAIMS retained the same dimensions that were used in the first study. Using computational methods, the virtual net fields for the $[M + 2H]^{2+}$ ion of gramicidin S were calculated for the region between the hemispherical tip and the ion inlet to the mass spectrometer. Simulations of the experimental conditions that were used for trapping are shown in Figure 5.11c for different gas flows (with 600 mL min^{-1} being an estimate of actual experimental conditions). In this figure, the tip of the inner electrode is at an axial distance of 0.0 mm and the orifice plate is at 2.0 mm. Positive virtual net fields will move the $[M + 2H]^{2+}$ gramicidin S ions toward the OR plate, whereas negative virtual net fields will move the $[M + 2H]^{2+}$ gramicidin S ions toward the inner electrode. Using the conditions defined in the figure, at gas flows of *ca* 600 mL min^{-1}, ions at an axial distance of greater than about 1.3 mm will move toward the orifice plate, and will not be trapped, but ions at an axial distance of less than 1.3 mm will move toward the inner electrode. In the latter example, because of the small radius of the inner electrode, there is a steep increase in the virtual field near the inner electrode (*ca* 0.2 mm), thus ions can remain trapped between axial distances of *ca* 0.2 to 1.3 mm. Note that by changing to a negative OR value, the virtual net field becomes positive at all axial distances in the figure, and in this way ions can be extracted. The ion-trapping kinetics for the $[M + 2H]^{2+}$ ion of gramicidin S showed a much longer half-life than for the m/z 380 ion in the first study, and was estimated to be *ca* 2 s. Although other FAIMS trapping devices can be readily envisioned, accounts of their development have yet to be reported.

5.2.4   Modeling Ion Behavior and Ion Losses

Thus far, ion losses have not been discussed, but these will affect significantly the performance of the FAIMS, for example, sensitivity, peak shape, etc. The three main ion-loss processes within a FAIMS device that need to be considered are diffusion, space-charge effects, and chemical alteration. Examples of chemical alteration include changes in clustering, fragmentation, or changes in conformation. In essence, when an ion is stable initially at a given value of CV within the FAIMS device and then

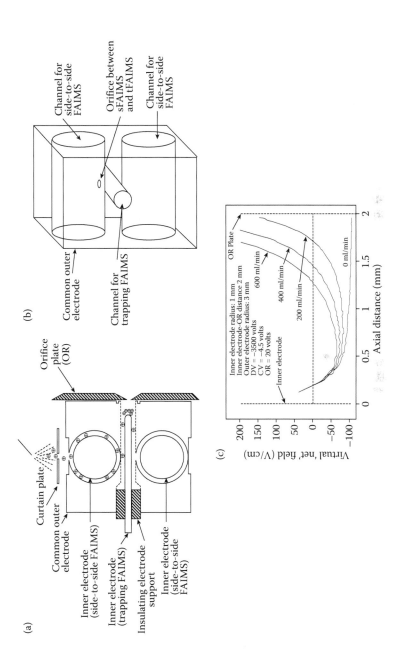

FIGURE 5.11 (a) Schematic of a side-to-side FAIMS and trapping FAIMS operating in series; and (b) schematic of the common outer electrode. (c) Effect of the gas flow on trapping in the trapping FAIMS device. As referenced in Figure 5.10, virtual 'net' fields for the [M + 2H]²⁺ ion of gramicidin S are calculated along the axis between the inner electrode and the orifice plate with gas flows from left to right (see text for details). (Reproduced from Guevremont, R., Ding, L., Ells, B., Barnett, D.A., and Purves, R.W., *J. Am. Soc. Mass Spectrom.* 2001, *12*, 1320–1330. With permission from Elsevier Science Inc.)

undergoes chemical alteration such that the 'modified' ion is no longer stable at this given CV-value, then the ion will be lost to the walls of the device. However, a modified ion could be detected if the change occurred 'instantaneously,' that is the change occurred before the original ion could discharge at the electrode walls when the CV-value that would transmit the modified ion is applied. Note that although chemical alteration is an important consideration, it is not modeled readily and, therefore, has not been discussed in detail in the literature.

Although modeling becomes more complex at high electric fields because, for example, diffusion will depend on the electric field strength, both diffusion and space-charge effects can be considered for planar or cylindrical FAIMS devices. Mathematical models [29–34], and numerical simulations using computational methods [26,35–38] that incorporate these ion losses have been reported to help understand, and ultimately to predict, FAIMS performance. A detailed comparison of these models and their assumptions is beyond the scope of this chapter, however, two important observations are that: (i) the saturation current within the FAIMS increases as the degree of ion focusing increases, and (ii) diffusion and Coulomb repulsion always act to expand an ion packet. Whilst the latter effects act so as to decrease the resolving power of mass spectrometry techniques in general, they will improve the resolution in FAIMS devices by eliminating preferentially ions at the edges of CV spectral peaks that pass near the electrode surfaces [37].

5.3   ROLE OF BATH GAS AND VARIABLES AFFECTING GAS NUMBER DENSITY WITHIN FAIMS

In the previous section, the basics of ion separation, focusing, and trapping within a FAIMS device were introduced along with the corresponding terminology. This Section explores the role of the bath gas in FAIMS in more detail. In particular, the influence of the composition of the gas, which includes the effect of trace gases (either present by impurity or added deliberately), and the importance of the number gas density will be described. Note that the bath gas within FAIMS is typically called the *carrier gas* as it transports ions through the FAIMS analyzer region. This is not the same as the *curtain gas* (gas introduced into the desolvation region), which consists of two components; the *desolvation gas* that flows countercurrent to the ions entering the desolvation region, and the *carrier gas* as defined above.

5.3.1   ROLE OF THE BATH GAS

As was mentioned in the Introduction (Section 5.1), collisions dominate ion mobility experiments near or at atmospheric pressure. Consequently, it is not surprising that the presence of the gas will play a critical role in ion separation. The first part of this Section will discuss the role of 'bulk' gases, which for the purposes of this discussion will be those that constitute greater than 1% of the total bath gas composition. The second part of the Section describes some interesting results that show the significant impact that trace amounts of gases (for example at parts per million (ppm) levels) may have on the CV spectra.

5.3.1.1 Bulk Gases

Early FAIMS publications involved the use of air [14] or nitrogen [10]. The use of different pure gases, including N_2, SF_6, O_2, CO_2, and N_2O, to transport ions through FAIMS was explored by Barnett *et al.* [39]. This work, which also examined the theory of the mobility of ions at high values of E/N in different gases, concluded that [39] "In general, oxygen caused the largest field-induced deviations in mobility" and that "The use of oxygen as a carrier gas may be expected to offer analytical advantages given the direct correlation between K_h/K [which represents the ratio of high-to-low field mobility] and CV."

An important development was the implementation of gas mixtures, which resulted in significant improvements for FAIMS analyses. Although a study involving a ternary gas mixture of $N_2/CO_2/He$ has been reported [40], binary gas mixtures by far have been used most commonly, and will be the focus of this discussion. The two binary gas mixtures that have shown the largest benefits include both N_2/He mixtures and N_2/CO_2 mixtures. The addition of helium to the nitrogen bath gas was first shown to give a substantial improvement in the analysis of morphine and codeine [41]. By changing from nitrogen to 60% helium in nitrogen (note that higher helium compositions resulted in electrical discharge within the FAIMS analyzer region), the signal intensity of the analytes increased by *ca* 50 times! [41] The use of N_2/He mixtures generally will permit an increase in the magnitude of the CV-values (and hence usually ion focusing strength, and thus sensitivities), such mixtures are used commonly in commercial instrumentation.

Binary N_2/CO_2 mixtures were used to separate *ortho-*, *meta-*, and *para*-phthalic acids [42]. In this work, all three acids could be separated in a carrier gas of 5% CO_2 in nitrogen, whereas in the pure gases, only one isomer could be separated from the other two, which remained overlapped in the CV spectrum [42]. An N_2/CO_2 mixture was also used for improving detection of trihalogenated acetic acids with FAIMS [43]. As is shown in Figure 5.12, employing a mixture containing 3% CO_2 in nitrogen improved dramatically the detection of trihalogenated acetic acids. Using nitrogen alone (Figure 5.12a), the trihalogenated species were detected as their decarboxylated anions and no signal was observed from bromodichloroacetic acid (BDCAA). With the gas mixture (Figure 5.12b), the signals increased markedly and BDCAA is readily detected in the form of BDCA⁻, the bromodichloroacetate anion, as is shown in the figure.

Particularly in the examples for the N_2/CO_2 mixtures, the behavior of the species was not intuitive because it did not agree with theoretical predictions that can best be described using Blanc's Law [2]

$$\frac{1}{K_{mix}} = \sum_j \frac{X_j}{K_j}$$

(5.10).

Here, the inverse of the mobility of an ion in a mixture of gases, K_{mix}, is proportional to the sum of the abundance (mole fraction), X_j, of each gas divided by the mobility of the ion in each gas, K_j.

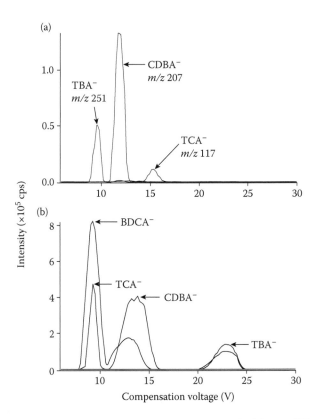

FIGURE 5.12 Effect of a binary gas mixture (3% carbon dioxide) on CV spectra for a solution containing four trihalogenated species, namely 1 ppm of each bromodichloroacetic acid (BDCAA), chlorodibromoacetic acid (CDBAA), trichloroacetic acid (TCAA), and tribromoacetic acid (TBAA). (a) IS-CV spectra using pure nitrogen as the carrier gas and (b) IS-CV spectra using a mixture of 3% carbon dioxide in nitrogen. In (a) all species were detected as their decarboxylated anions (no signal was observed for BDCAA) whereas in (b) all species were detected as their haloacetate anions. (Reproduced from Ells, B., Barnett, D.A., Purves, R.W., and Guevremont, R., *Anal. Chem.* 2000, 72, 4555–4559. With permission from American Chemical Society.)

Shvartsburg *et al.* [44] have proposed a 'universal' model to describe FAIMS separations in gas mixtures. The basis of this model was the recognition that Blanc''s Law is rigorous only at values of E/N approaching zero, and that departures from Blanc's law (which have long been known) are of similar magnitudes to typical deviations between high- and low-field mobilities (which are only a few percent). Therefore, deviations from Blanc's law will have minimal effect on IMS separations but may have huge effects on FAIMS performance. By modifying Blanc's law for high values of E/N using momentum transfer theory, the authors derived a model that is 'universally' applicable. The authors tested this model with literature results, and then used the model to predict optimal gas combinations. They concluded that [44] "the optimum FAIMS buffer would be (i) a binary mixture of gases with a huge disparity of molecular masses and collision cross-sections, and (ii) strongly resist electrical breakdown." Consequently, the authors suggested the combination of He/SF$_6$ to improve resolution and peak capacity of FAIMS. To date, the use of this gas mixture for FAIMS applications has yet to be tested fully.

5.3.1.2   Trace Gases

In addition to gases present in bulk quantities, the addition of a gas modifier at trace levels (*e.g.*, ppm) can have a profound effect on the FAIMS analysis. The first use of a modifier at low concentrations in the gas to dramatically affect ion separation was disclosed in US Patent 7,026,612 by Guevremont *et al.* [45], which was published in August 2003. In this patent, multiply-charged states of bovine ubiquitin were examined while varying the amount of 2-chlorobutane (to produce mixtures ranging from 0 to 1000 ppm) that was added to the nitrogen carrier gas. For the +5 charge state, while only two peaks were observed in nitrogen alone, up to five peaks were observed when 1000 ppm of 2-chlorobutane were added to the carrier gas. This result implies that at least five different forms of the +5 charge state are present in the gas-phase. Many of these forms overlap in nitrogen alone, however, with the addition of the trace gas, these five forms are separated.

In a subsequent publication by Eiceman and coworkers [46], organic vapors were added into the drift gas to improve the separation capabilities of nitro-organic explosives. In the typical purified air bath gas, the nitro-organic explosives being studied were transmitted over a narrow CV-range of four volts, from −1 to 3 V. However, the authors found that by adding 1000 ppm of methylene chloride to the drift gas that the range could be increased dramatically to 18 volts (from 3 to 21 V). More recently, Levin *et al.* [47] reported the use of gas modifiers (dopants) in the quantitation of a peptide from a prepared sample containing seven peptides. The authors claimed that the gas dopants reduced the size of peptide aggregates formed during nanoelectrospray, thereby enabling enhanced peptide ion separation in FAIMS.

Also disclosed in the aforementioned patent by Guevremont *et al.* [45] was the critical role that the solvent could play in the FAIMS analysis. Significant amounts of water had been shown previously to have catastrophic results on FAIMS spectra [10]. However, by using a gas purification cartridge containing molecular sieves, that had been compromised by water (the water concentration introduced into the FAIMS analyzer was estimated to be at sub-ppm levels), signal intensities of amphetamine, methamphetamine, and their methylenedioxy derivatives could be improved dramatically compared with spectra that were acquired with new cartridges; however, the improvements were not predictable [45]. The use of water in this way illustrates the need to control strictly the quality of the bath gas entering the FAIMS, and can explain readily why, at high flow rates, 'CV shifting' has been observed [48]. Indeed, the present author has numerous experiences in these types of solvent effects from his own attempts to improve signal intensity by moving the needle closer to the FAIMS inlet (especially at high liquid flow rates), thereby exceeding the capabilities of the desolvation region and introducing solvent vapors into the gas stream. In addition to observing the onset of CV shifting for many compounds, the author has observed also that some compounds (for example, reserpine) are much less sensitive to the needle position (and hence desolvation) than are other compounds, such as acetaminophen.

5.3.2 Gas Number Density (N)

As was mentioned in the Introduction (Section 5.1), the mobility of an ion is proportional to the electric field strength divided by the gas number density (N). The gas number density, corresponding to n moles of gas contained within a volume V, can be calculated from

$$N = (n/V) \times N_A \qquad (5.11),$$

where N_A is Avogadro's constant (6.022×10^{23} mol^{-1}). The quotient n/V can be calculated from the ideal gas law

$$(n/V) = p/RT \qquad (5.12),$$

where R, the gas constant, is 0.08206 L atm^{-1} mol^{-1} K^{-1}, and p (atm) and T (K) are measurable quantities. Thus, the units of N are expressed as inverse volume (for example, cm^{-3}) and, as noted in Section 5.1.1.2, E/N is expressed in units of V cm^2, or Td (recalling that 1 Td = 10^{-17} V cm^2).

5.3.2.1 The Importance of E/N

Up to now, the examples discussed have assumed that the FAIMS experiments were carried out at atmospheric pressure and room temperature. However, from Equations 5.11 and 5.12, it is clear that higher values of E/N can be achieved by reducing the pressure and/or by increasing the temperature of the system. In fact, lowering the pressure was the approach that was used initially by Viehland and others to investigate the behavior of ions at higher E/N values [8,49]. Of course the pressure must be sufficient for collisions to dominate, as described earlier. These experiments were carried out typically at 1–2 Torr.

The effect of temperature on FAIMS analyses has also been explored recently [22,50,51]. Heated ionization sources are required particularly for the analysis of small molecules by mass spectrometry, where higher liquid flow rates (e.g., 0.5 mL min^{-1}, or greater) are used typically, and thermal heating is necessary in order to achieve sufficient desolvation. These sources heat the FAIMS electrodes indirectly and, without controlling the electrode temperature, the number gas density within the FAIMS analyzer region, and hence E/N, will change, thereby causing shifts in the CV-values of transmission of the analytes of interest. Using a FAIMS system mounted on an MDS Sciex API 3000 triple quadrupole mass spectrometer, Barnett and coworkers showed that, through indirect heating by the ion source, the FAIMS device required approximately two hours in order to reach equilibrium temperature [51].

A temperature-controlled system, which uses room temperature and/or heated compressed air to maintain a constant temperature of the electrodes between 40 and 120°C, was reported recently by Barnett et al. [22] This arrangement requires approximately five minutes to reach the equilibrium temperature, and remains stable to within ±1°C. A key feature of the device is the ability to effect separate temperature control of the outer and inner electrodes. The advantages of maintaining the electrodes at independent temperatures are described in Section 5.3.2.2, below.

5.3.2.2   Role of Temperature in Ion Separation

Recall Figure 5.8 that showed CV/DV plots for the water cluster ions. These CV spectra will be valid only at one given value of N, as defined by the atmospheric pressure and room temperature that were used for the experiment. Due to the dependence of ion mobility on N (described above), it follows that CV/DV plots acquired at different temperatures will not have the same N-value and, therefore, would not be expected to overlap on the same graph. This effect is illustrated in Figure 5.13a that shows five CV/DV plots that were generated at five different temperatures for the cesium ion [22]. However, by incorporating this change in the parameter N into the calculation of CF and DF, Barnett *et al.* [22] demonstrated for the cesium ion that changes in the CV/DV plots observed as a function of temperature were readily accounted for through changes in number gas density, as is shown in Figure 5.13b. For some ion species, changes in number gas density were not sufficient to explain completely changes in the optimum CV/DV values as the CF/DF traces at different temperatures did not overlap, in contrast to the behavior observed for cesium ions in Figure 5.13. For example, non-overlapping CF/DF plots acquired at two different temperatures for iodide ions, were observed. This deviation was explained by changes in the types of clusters formed between the ion and the bath gas at the two different temperatures [22]. Also, for gramicidin S, discrepancies in CF/DF plots at different temperatures were observed and these were attributed to changes in ion–neutral interactions at the different temperatures [22].

The ability to control individually the inner and outer electrode temperatures has important consequences for FAIMS experiments. These implications were illustrated

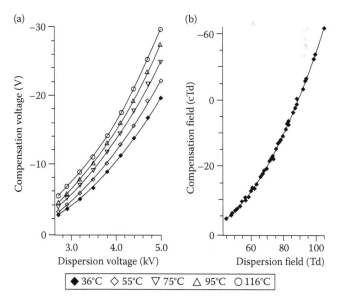

FIGURE 5.13   (a) ESI-FAIMS-MS data acquired for the cesium cation in nitrogen using electrode temperatures of 36, 55, 75, 95, and 116°C. (b) Resulting conversion of CV/DV data pairs to field strengths (CF/DF) in units of centitownsends. (Reproduced from Barnett, D.A., Belford, M., Dunyach, J.J., and Purves, R.W., *J. Am. Soc. Mass Spectrom.* 2007, *18*, 1653–1663. With permission from Elsevier Inc.)

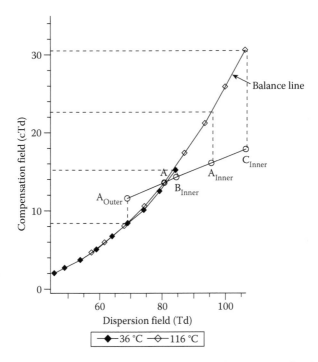

FIGURE 5.14 CF/DF plots derived for ESI-FAIMS-MS data collected for the [M-H]⁻ ion of taurocholic acid (*m/z* 514) in a gas mixture of 50/50 helium/nitrogen with the temperature of both electrodes set at 36°C and then at 116°C. Changing the inner electrode temperature will change the electric fields near the inner electrode wall as is illustrated by the points A_{Inner}, B_{Inner}, and C_{Inner} which are the result of employing inner electrode temperatures of 76, 36, and 116°C, respectively (B_{Outer} remains fixed at 76°C). Using the temperature to change the electric fields within the FAIMS will effect parameters such as peak width and signal intensity as is described in the text. (Reproduced from Barnett, D.A., Belford, M., Dunyach, J.J., and Purves, R.W., *J. Am. Soc. Mass Spectrom.* 2007, *18*, 1653–1663. With permission from Elsevier Inc.)

by Barnett *et al.* using deprotonated taurocholic acid [22]. Figure 5.14 shows CV/DV data for the [M-H]⁻ anion of taurocholic acid, examined in a gas mixture of 1:1 helium:nitrogen (a typical medium for analysis) acquired with both electrodes set to 36°C (closed diamonds) or 116°C (open diamonds), and which were transformed subsequently to CF/DF data. Because the plots acquired at 36 and 116°C overlap, as shown in the figure, it is implicit that for any temperature within this range the graph can be used to calculate the optimum CV for a given DV. For example, consider the mid-temperature (that is, 76°C) between these two temperatures, taking point A on the plot in Figure 5.14 corresponds to DV = 4235 V and CV = 7.15V). For these conditions, the fields within the analyzer region can be calculated at any radial point, as was illustrated previously in Figure 5.9b; the points A_{outer} and A_{inner} have been added to represent the fields very near the outer and inner electrodes, respectively. When the CV-value is changed, the straight line connecting A_{outer} and A_{inner} will effectively shift upward or downward, and the values where A_{outer} and A_{inner} intersect the 'balance line' can be used as a rough approximation for the peak width. From the CF-values, the peak can be calculated to extend from $CV = ca$ 5.2 to $CV = ca$ 10.0, giving a baseline width equal to ca 4.8 V. To illustrate the effect of changing the temperature of the inner electrode (the outer electrode remains at 76°C), consider the minimum and maximum temperature values, 36°C (B_{inner}) or 116°C (C_{inner}) that could be applied in this experiment. The magnitude of the fields near the inner electrode will change: DF = 84.5 Td and CF = 14.3 cTd for B_{inner}, and DF = 106.4 Td and CF = 17.9 cTd for C_{inner}; points corresponding to these values have been added to Figure 5.14. Note that the fields near the outer electrode will remain the same. With the new conditions, the same method can be used to obtain approximate peak widths. The CV-value for B_{inner} is ca 7.7 V and that for C_{inner} is ca 12.1 V, thus, the peak width for the temperatures $T_{inner} = 36$°C and $T_{outer} = 76$°C is ca 2.5 V, whereas the peak width for $T_{inner} = 116$°C and $T_{outer} = 76$°C is ca 6.9 V. Also note that, in addition to the peak width, the apparent optimum CV-value will shift also, as this value is approximately the midpoint between the minimum and maximum CV-values. Experimentally, an example of the effect of electrode temperature on resolution (calculated using Equation 5.8) is shown in Figure 5.15 by means of an example of the CV spectra for taurocholic acid (dashed line) and methotrexate (solid line), both of which exhibit type 'C' behavior (see Section 5.1.1.2). In this example, the resolution is seen to improve as the temperature on the inner electrode is lowered relative to that of the outer electrode. Note, however, that this improvement in resolution is occurring to the detriment of signal intensity.

In general, predicting the behavior of an ion based on changes in the electrode temperatures can be carried out as long as the shape of the balance line is known, because this shape will determine the variation of peak width as a function of temperature. For example, for an ion exhibiting type C behavior, improvements in resolution will be achieved by lowering the inner electrode temperature relative to the outer electrode temperature; however, for a type B ion the behavior will be more complex, as it is a function of the value of E/N employed in the experiment.

This ability to control resolution through independent temperature control of the inner and outer electrodes has important implications also for flat plate electrodes. With different values of N near the inner and outer electrodes, the E/N value will no longer be constant across the analytical gap, thereby enabling, in theory, ion focusing to occur in a flat plate design!

5.4   FAIMS HARDWARE

The operating principles presented in the previous Sections have described the requirements for FAIMS operation. As these requirements can be met readily using several alternative approaches, many different types of systems have been reported with undoubtedly many more to come. This Section will describe the hardware of some of the more commonly used devices, including those found in commercial FAIMS instruments.

5.4.1   ELECTRODES

Endless different, diverse electrode geometries can be envisioned, and already a wide variety of designs have been reported, especially in patents and patent applications.

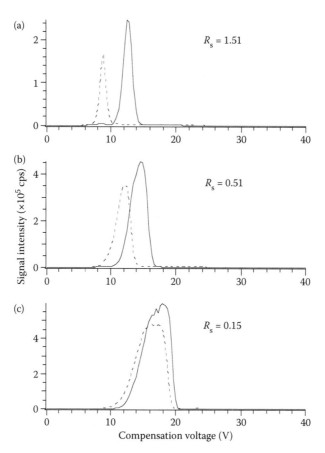

FIGURE 5.15 Resolution of the [M–H]⁻ ions of taurocholic acid (dashed line) and methotrexate (solid line) using differential electrode temperatures of (T_{Inner}/T_{Outer}): (a) 36/76°C; (b) 76/76°C; (c) 116/76°C. The resolution values were calculated using Equation 5.8. (Reproduced from Barnett, D.A., Belford, M., Dunyach, J.J., and Purves, R.W., *J. Am. Soc. Mass Spectrom.* 2007, *18*, 1653–1663. With permission from Elsevier Inc.)

Clearly this Section cannot discuss all of these possible fabrications, but will explore some of the more commonly used electrodes found in the published literature, and will examine also two designs from the patent literature that may play important roles in future FAIMS work.

5.4.1.1 Planar Electrodes

The planar electrode design ('flat plate FAIMS') [13,14] will have a constant electric field in the ion separation region, in the absence of temperature control, which means that ion focusing will not occur. Although this design will suffer from very significant ion losses due to diffusion and space-charge repulsion, it nonetheless offers the capability for higher resolution analyses when compared with systems having electrodes that generate non-homogeneous fields.

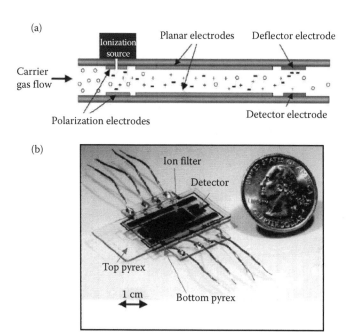

FIGURE 5.16 (a) Schematic showing a cross-section of the differential mobility spectrometry (DMS) drift tube; and (b) photograph of a micromachined chip. (Reproduced from Miller, R.A., Nazarov, E.G., Eiceman, G.A., and King, A.T., *Sens. Actuators, A.* 2001, *91*, 307–318. With permission from Elsevier Science B.V.)

Miller and coworkers have adapted the planar design for use with microelectromechanical systems ('MEMS') technology to fabricate a planar MEMS FAIMS device [19]. A schematic showing a cross-sectional view is given in Figure 5.16a and a photograph of the device shown in Figure 5.16b gives an indication of the relative scale. The metal electrodes (indicated as 'planar electrodes' in the figure) were formed through a coating process of two Pyrex wafers, resulting in a miniature FAIMS device having a total volume of 0.6 cm^3. This design is the basis of the commercial Sionex DMS 'FAIMS system' described in Section 5.4.3.3).

Optimization of the resolution for flat plate designs was investigated by Shvartsburg *et al.*, who compared the resolution obtainable for two different commercial systems, one based on flat plates (DMS, Sionex) and one based on a cylindrical design (Selectra, Ionalytics) [52]. Experimentally, improvements in resolution were not observed for the DMS instrument over that manufactured by Selectra. The authors noted that the resolution for a flat plate design is dependent strongly on ion residence time within the analyzer region, and suggested that the residence time within the DMS device was too short (*ca* 2–5 ms) to make proper comparisons with the residence time within the Selectra design (*ca* 0.2 s). Consequently, the authors designed a flat plate system with a gap width of 2 mm and a length of 50 mm with the goal of increasing the ion residence time. With this new device, the ion residence time was increased to *ca* 0.1 s and with the longer residence times, the authors observed that the resolution could be increased by up to a factor of four compared to that obtained with cylindrical FAIMS devices [52].

5.4.1.2 Curved Electrodes

The devices described within this Section are presented in the order in which they have appeared in the literature. The primary advantage of FAIMS devices with non-homogeneous electric fields is that ion focusing can occur. Especially in cases where the ion focusing is strong, these devices will result in much higher signal intensities compared with flat plate devices.

The electrodes shown in Figure 5.17a were part of the MSA design, which was the first commercially produced FAIMS system. In this device, ions are carried by the carrier gas flow through the analyzer region, which is located between the two concentric cylinders, that is, the inner and outer electrodes, where ion separation occurs. Ions that are transmitted successfully through the length of the device (that is, those with the appropriate ratios of high-to-low field mobility) can then pass out through one of four openings in the outer electrode where they are detected using an electrometer (not shown in the figure; the electrometer extends all the way around the outer electrode). Although this device was sensitive in trace gas analysis, a limitation arose from the difficulty in identifying peaks unambiguously because of unpredictable changes in ion mobility, clustering behavior, and/or the chemical structure of the ion [10].

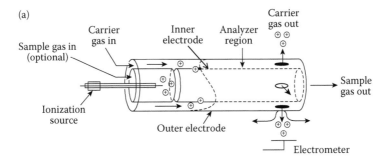

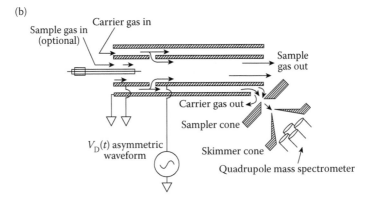

FIGURE 5.17 (a) Three-dimensional schematic of the cylindrical FAIMS electrodes, having electrometric detection, that were offered by MSA Company. (b) Cross-sectional view of the MSA cylindrical electrodes adapted for use with mass spectrometric detection. (Adapted from Purves, R.W., Guevremont, R., Day, S., Pipich, C.W., and Matyjaszczyk, M.S., *Rev. Sci. Instrum.* 1998, *69*, 4094–4105. With permission from the American Institute of Physics.)

In an attempt to overcome this limitation, the MSA electrode set was modified for detection using a quadrupole mass filter, and this design is shown schematically in Figure 5.17b. This modification involved replacing the electrometer with a 'sampler orifice' cone placed at the end at a 45° angle relative to the axis of the FAIMS cylinders [10]. Since this device is coupled to a mass spectrometer, ions can be sampled only from one exit in the analyzer region unlike the electrometer arrangement, which contained four openings leading to the detector. Nonetheless, this device was used for FAIMS-MS characterization [10,11], and was employed also in many of the early FAIMS-MS applications that will be discussed in Section 5.5.

Figure 5.18a shows a cross-sectional view of a 'dome' electrode, which consists of an inner electrode that is cylindrical with the exception of having a hemispherical terminus. The performance of the device was reported to be superior to that of the FAIMS-MS design shown in Figure 5.17b for two main reasons. First, the source was located externally to the device, thereby making the device less prone to solvent contamination and, therefore, applicable for use with higher sample flow rates. Second, the ions could be sampled more efficiently by the analyzer, because all stable ions are brought to the central axis of the inner electrode and are focused toward the opening in the MS orifice plate [53]. For this FAIMS-MS design, the gas intake into the mass analyzer and the 'effective' volume of the FAIMS electrodes (that is the volume of the space within which the ions travel, but excluding the 'dead' space to the left of the ion inlet in the figure) will dictate the residence time. For work on the MDS Sciex API 300 mass spectrometer (flow *ca* 0.5 mL min^{-1}), a typical residence time was of the order of 250 ms. Prototypes with different cylindrical radii have since been used [26] and, as was mentioned in Section 5.2.2, these different radii have profound effects on both the sensitivity and the peak shape.

Figure 5.18b shows a schematic diagram of a different type of cylindrical electrode, which has been referred to as the 'cube' and also as 'side-to-side.' This design also has two concentric cylinders, however in this device, as indicated by the arrows in the figure, the ions move around the surface of the inner electrode, as opposed to along the length of the inner cylindrical electrode. Thus, unlike the device in Figure 5.18a, there is no gas flow or other mechanism for preventing ions from moving toward the ends of the cylinders. Despite this apparent limitation, the sensitivity with this electrode was found to be similar to that observed for 'dome' devices having the same electrode radii (unpublished results). The driving force for the creation of this design was that mechanical coupling to high liquid flow rate ionization sources could be accomplished more readily because the ion inlet and ion outlet were parallel to each other, whereas with the dome design the ion inlet was perpendicular to the ion outlet. Thus conventional ion sources, such as those on the MDS Sciex API instrumentation, could be readily adapted to this design.

The need to control the temperature (described in Section 5.3.2) was the major driving force in the evolution of the design of the cube. As the use of resistive elements was not feasible in the cube because of the intense RF field from the applied asymmetric waveform [22], the electrode set was modified further. In particular, the new electrodes had channels incorporated in them that could be used to carry gas for cooling or heating the electrodes. These electrodes represent the basis of the commercial FAIMS electrode set offered by Thermo and are described in Section 5.4.3.2.

5.4.1.3   Tandem FAIMS Systems

Strictly speaking, the dome electrode shown above in Figure 5.18a is a tandem FAIMS arrangement because there are two distinct ion separation regions: the cylindrical region along the length of the electrode set, and the spherical region near the tip of the device. Since the equations describing the electric fields for the spherical geometry differ from those for the cylindrical geometry, the optimum values for transmission will be different also and, therefore, the net CV spectrum will be an

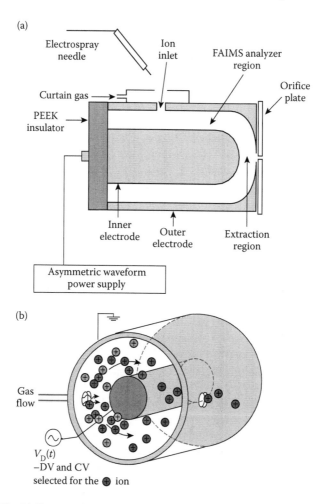

FIGURE 5.18   (a) Cross-sectional view of a 'dome' FAIMS device having an inner electrode terminating in a hemisphere. Ions enter through the ion inlet, are transported along the length of the inner cylinder by the carrier gas and, using the appropriate CV, are focused toward the opening in the MS orifice plate. (Adapted from Guevremont, R., and Purves, R., *J. Am. Soc. Mass Spectrom.* 2005, *16*, 349–362. With permission from Elsevier Inc.) (b) A 'cube' or 'side-to-side' FAIMS device. Ions enter through the ion inlet in the outer electrode, are transported around both sides of the inner cylinder by the carrier gas as is indicated by the arrows and, using the appropriate CV, exit *via* the opening in the outer electrode.

intersection of the CV 'windows' of these two regions. Virtual net potential fields were shown to be modified by longitudinally displacing the tip, thereby leading to enhanced resolution by truncating one side of the peak [35]. This type of device was employed in the early investigations of gas-phase protein conformations (see Section 5.5.1.1), and the longitudinal displacement of the tip was used to improve resolution in an effort to distinguish between conformers.

A tandem system that combined two different types of FAIMS analyzers was described in Section 5.2.3, see Figure 5.11a and b. The first FAIMS analyzer consisted of larger radii electrodes that offered improved resolution and the capability to remove background ions, whereas the second FAIMS analyzer consisted of smaller radii electrodes having stronger focusing fields suitable for trapping ions. Other tandem FAIMS systems can be envisioned combining either different geometries tailored for a given analysis or multiple units of similar geometry. Examples of the later situation involve changing critical parameters in each electrode set, such as gas composition, temperature combination, etc. In short, there are endless possibilities for tandem FAIMS work and, as the technique matures, it is expected that more tandem devices will be reported in the literature.

5.4.1.4 Other FAIMS Electrode Designs

Several interesting designs of FAIMS electrodes have been reported in patent publications, and undoubtedly there will be more to come. Although some of these designs hold promise and intrigue, the present discussion will be restricted to two versions of FAIMS electrodes that have appeared in the patent literature but have not been reported yet in refereed publications. These two geometries include segmented FAIMS electrodes [54,55] and FAIMS electrodes that can be operated at reduced pressures [56,57].

A segmented FAIMS device to trap ions was disclosed first in a patent by Guevremont and Purves [54]. A segmented FAIMS device can be used also to create a field-driven FAIMS analyzer [55]. In a conventional flow-FAIMS, that is, where gas is used to carry ions through the analyzer region, all of the ions will have the same residence time, whereas with the field-driven FAIMS, electric fields are used to transport the ions through the FAIMS and, as a consequence, the residence time is inversely proportional to the mobility. The performance of a field-driven analyzer has been modeled by Shvartsburg and Smith and the authors found that, in the analyses of complex mixtures having a broad range of mobilities, such as proteolytic digests, a field-driven FAIMS analyzer could offer some advantages [58].

A reduced pressure FAIMS device could offer benefits also. As will be described in the following Section, waveform generators are restricted in their output voltages, especially for rectangular waveforms. A reduced pressure FAIMS (lower N) offers the advantage of employing a lower waveform amplitude (reduced by the same factor as for N) to achieve the same E/N values (or potentially even larger values) as those obtained at atmospheric pressure. Although the idea of a reduced pressure FAIMS had been reported previously [56], the concept of a 'Quadrupole FAIMS apparatus' disclosed recently by Belford et al. [57] is particularly intriguing because it includes also another important feature, namely the 'on-off FAIMS.' That is, the mass spectrometer would be used in its conventional configuration, and then switched to use with FAIMS through software control only, that is, without having to modify the hardware. The current Thermo FAIMS/MS system requires a manual change of the ionization source by the user, because simply setting the voltages to zero with the FAIMS analyzer present results in a loss in sensitivity of approximately an order of magnitude compared with conventional operation of the mass spectrometer.

5.4.2   Asymmetric Waveform

5.4.2.1   Types of Waveforms

The asymmetric waveform is a critical component of the FAIMS separation technique, and waveform characteristics such as peak amplitude (DV), frequency (ω), and even the profile, $V_D(t)$, will affect the analytical response [38]. Although one can envision an infinite number of waveform types, each having very different characteristics that satisfy the requirements set out in Section 5.2.1, the present discussion will be restricted to three main types of waveforms that have been described in the literature and have been used experimentally. These waveform types, shown at different frequency ratios in Figure 5.19, include bisinusoidal (Figure 5.19a), which consists of a sinusoidal wave and its phase-shifted harmonic [59], rectangular (Figure 5.19b) [13], and a clipped displaced sinusoidal waveform (Figure 5.19c) [60,61]. For simplicity in illustrating the fundamental concepts of FAIMS, a rectangular asymmetric waveform has been used. However, in practice, generating this waveform at the desired voltages and frequencies has not been achieved as yet because of electronic power consumption constraints [6]. Thus, the vast majority of published FAIMS work has utilized a type of the bisinusoidal waveform (generally the sum of a sinusoidal waveform and its first harmonic). The general form of the bisinusoidal waveform can be expressed as [6,21]

$$V_D(t) = \left[f \sin \omega t + \sin(2\omega t - \pi / 2) \right] V_{max} / (f + 1) \qquad (5.13),$$

where V_{max} is the maximum voltage output of the waveform (that is, DV). More specifically in these publications, f, the high-to-low field ratio is 2 thereby simplifying the equation to

$$V_D(t) = \left[2 \sin \omega t + \sin(2\omega t - \pi / 2) \right] V_{max} / 3 \qquad (5.14).$$

5.4.2.2   Considerations and Performance

Shvartsburg and coworkers have developed and refined a 'first-principles computational' model for simulating operation of the FAIMS analyzer [9,21,37]. The model incorporates a variety of geometries and involves injecting an ensemble of ions that evolves under the influence of external electric fields (resulting from the application of an asymmetric waveform), diffusion, and Coulomb repulsion [37]. Based on this model, the authors concluded that "optimization of the asymmetric waveform has shown an essentially equal merit of bisinusoidal and clipped sinusoidal waveforms, but a substantial advantage of a rectangular profile over both in terms of resolution

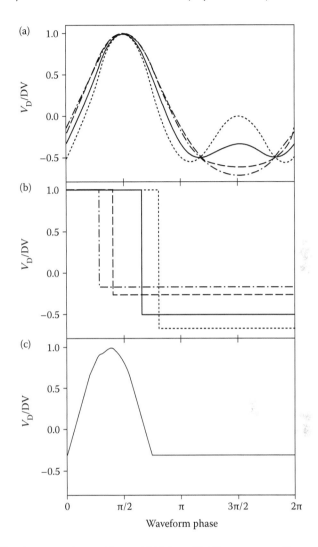

FIGURE 5.19 Asymmetric waveforms: (a) bisinusoidal from superpositions of a sinusoidal and scaled second harmonic; (b) rectangular; and (c) clipped sinusoidal. In (a) and (b), the frequency ratios, f (Equaton 5.13) are: 2 (solid line), 4 (dash line), and 6 (dash dotted line). The value of f for the dotted line is: (a) 1.0 and (b) 1.5. In (c), $f = 1.7$. (Reproduced from Shvartsburg, A.A., Tang, K., and Smith, R.D., *J. Am. Soc. Mass Spectrom.* 2005, *16*, 2–12. With permission from Elsevier Inc.)

and/or sensitivity. In particular, maximum FAIMS resolution could be improved by over a third" [21]. A major reason for the superior performance of the rectangular waveform is that, other factors remaining constant, this waveform maximizes the net displacement of an ion because E is fixed, whereas other waveforms (for example, bisinusoidal) are composed of a range of values of E and hence, exhibit a lower degree of asymmetry [9]. Maximization of the net displacement will result in stronger ion focusing, which generally improves ion transmission.

The voltages required for FAIMS analysis are limited by the onset of electrical breakdown within the FAIMS analyzer region. The value at which breakdown occurs can be determined from Paschen's Law, which is dependent on the gas type, the gap size, and the pressure. The breakdown in air, at atmospheric pressure, using a 2 mm gap will occur at *ca* 30 kV cm^{-1}. As was mentioned earlier, due to electronic constraints, maximum voltages of rectangular asymmetric waveforms (*e.g.*, DV *ca* 1 kV [62]) are currently much lower than the maximum voltages of bisinusoidal waveforms (*e.g.*, DV *ca* 5 kV for FAIMS on the Thermo TSQ Quantum instrument or DV *ca* 8 kV (*ca* 12 kV$_{p-p}$) reported by Ridgeway [20]). As a result, for many of the typical FAIMS analyzers described in Section 5.4.1 (that is with a gap width of 2 mm) the resulting fields with a rectangular waveform (*ca* 5 kV cm^{-1}) would also be much less than with a bisinusoidal waveform, which would lead to dramatically decreased ion separation/focusing capabilities. For example, Figure 5.7 illustrates the large difference in performance at DV = −1500 V and DV = −3000 V for a 2 mm gap. However, recalling that *E/N* is the important parameter, as technology advances and smaller gap sizes [19] and/or lower pressures [57,63] are used more routinely, rectangular waveforms should play an increasingly important role as larger values of *E/N* would be readily achievable. Such a system has been employed commercially, and is discussed in Section 5.4.3.4.

Another important consideration is the frequency of the waveform, as was discussed earlier. Since the analytical gap between the electrodes (gap width) must be kept relatively small (typically 2 mm or less) to generate fields sufficient for FAIMS separation to occur, the waveform frequency (ω) must be large enough such that the ion moves a distance less than the gap while oscillating during each portion of the waveform. An extreme example of the effect of frequency was given in Section 5.2.1, which showed that for a 10 kHz waveform generating a field of *ca* 20 Td that the movement during one polarity of one cycle of the waveform would be about 2.0 mm (for an ion having a *K(0)* of 2.0 cm^2 V^{-1} s^{-1}). Since this movement is about the same distance as the gap width, the result would be that the ion would collide with one of the electrode walls and would not be detected. Conversely for the same conditions using a 1 MHz frequency, the ion movement would only be *ca* 0.02 mm during each polarity of the waveform, enabling detection at the appropriate CV-value. In other words, as ω is lowered the ion movement during each polarity of the waveform becomes greater, which means that the gap width is being effectively narrowed as well because the distance traveled by the ion is proportional to 1/ω. Thus, a reduction in waveform frequency will, in effect, generally improve resolution, but any improvement will come at the expense of sensitivity.

The value of the high-to-low field ratio, *f*, in Equation 5.13 is an optimizable parameter that also has been examined by Shvartsburg and coworkers. In their initial model, in which higher-order terms in Equation 5.6 were ignored [21], the optimum value for *f* was determined to be 2, the value currently used in the Thermo FAIMS systems, and the optimum value for *f* was found to be independent of the waveform type. However, when the model was refined [9] to include higher-order terms in Equation 5.6, which are especially relevant for small gaps and reduced pressures that can generate very large values of *E/N*, the optimum value of *f* was found to vary depending on the ion type (that is, A, B, or C) and the value of the applied field. In general, for type C ions, the optimum value of f is ca 2, whereas for type B ions, or type A ions that convert to type B ions at high values of E/N, the optimum value of f is ca 1.35 [9].

Shvartsburg and coworkers have also extended their model to look at waveform imperfections, for example noise and ripple [38]. The authors found that waveform perturbations are significant, even at low levels; they attributed this finding to the FAIMS process being based on a small difference between two large quantities and, therefore, highly sensitive to minor variations of either one. In addition, the authors found that electronic noise at random frequencies (white noise) or at fixed frequency (ripple) will decrease sensitivity, but can be used to improve resolution because, in essence, the noise acts so as to narrow the range of conditions that allow ion transmission. In fact, the ripple voltage was a feature that was available on the MSA design that could be used for this purpose (see Section 5.4.3.1).

Recently, Shvartsburg and coworkers [7] reported on the feasibility of using higher-order differential ion mobility separations ('HODIMS'). This approach uses time-dependent electric fields comprising more than two intensity levels that could effect distinct higher-order separations based on higher-order terms of expression for ion mobility (see Equation 5.6). Currently IMS is considered 1st order and FAIMS is considered 2nd order; that is $K(E/N)$ is dominated by $K(0)$ for IMS, in Equation 5.6, and $K(E/N)$ is dominated by 'a' for FAIMS (for 2nd order). In the same way that FAIMS, IMS, and MS are orthogonal to each other, the advantage of HODIMS is expected to arise because these higher-order separations (for example, 3rd order would be dominated by 'b' in Equation 5.6) also would be largely orthogonal to each other, as well as to FAIMS, IMS, and MS. In addition, hardware similar to that employed in existing FAIMS systems could be used, so that switching between separation orders could be carried out within the same analyzer [7]. However, thus far, this concept has not been demonstrated experimentally.

5.4.3   COMMERCIAL **FAIMS** SYSTEMS FOR USE WITH MASS SPECTROMETRY

This Section gives a brief overview of some of the commercial FAIMS systems that currently are, or have been available, focusing on those for use with mass spectrometry. Within this section, a wide variety of devices illustrate the diversity of the FAIMS technique. A key step in the evolution of FAIMS was the availability of the 'beta' FAIMS unit from MSA, and our overview will begin with examination of this system. Systems made available by Ionalytics and Sionex are described also. With the acquisition of Ionalytics by Thermo and the greater availability of FAIMS devices, and hardware in general, new systems and more companies have been appearing, and an interesting FAIMS device from a new company, Owlstone Nanotech, is discussed briefly.

5.4.3.1   Mine Safety Appliances Company (MSA)

The MSA (Pittsburg, PA, USA) design consisted of a bisinusoidal waveform (maximum DV = +/– 3300 V) and an electrode set with concentric cylinders, this design is shown in Figure 5.17a. The target application of this device was trace gas analysis, especially detection of explosives. The MSA effort was halted in the early 2000s.

5.4.3.2   Ionalytics/Thermo

Ionalytics (Ottawa, ON, Canada) was the first company to offer FAIMS devices for use with mass spectrometry, and was acquired by Thermo Electron (now Thermo Fisher, San Jose, CA, USA) in August 2005. Initially, Ionalytics made two different electrode geometries available: a 'dome' design, based on Figure 5.18a, was used for low flow rate work mainly on Waters Q-TOF mass spectrometers (Milford, MA, USA), and a 'cube' design, based on Figure 5.18b, for work mainly with small molecules on various triple quadrupole instruments, including those manufactured by Applied Biosystems (Foster City, CA, USA), Thermo, and Waters.

Figure 5.20a shows a schematic representation of the dome design coupled to the Waters Q-TOF micro for use with a nanoESI source. The dome electrode itself is shown in Figure 5.20b and an exploded view of the electrode set is shown in Figure 5.20c. Since the inlet and outlet on the FAIMS are orthogonal, the source (which sprayed off-axis to the cone without FAIMS present) sprayed directly at the opening in the curtain plate and, therefore, could be used only with low liquid flow rates. This FAIMS set-up was used primarily for protein and proteomic analysis. Since the acquisition of Ionalytics by Thermo, this system is no longer available.

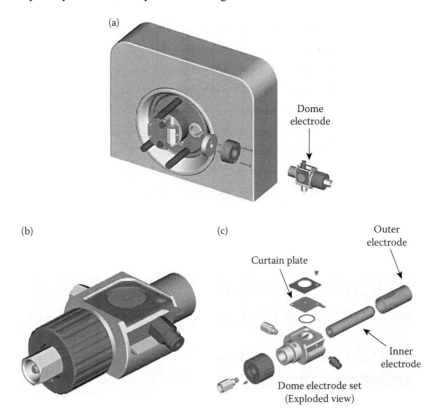

FIGURE 5.20   (a) Schematic illustrating the coupling of a 'dome' FAIMS to a Waters Q-Tof micro for use with nanoESI; (b) expanded view of the 'dome' electrode set; and (c) exploded view of the 'dome' electrode set showing the individual components.

The original 'cube' electrode set (Figure 5.18b) did not have temperature control and thus could not be used with high-liquid flow rates, used typically for small molecule analysis, because of fluctuations caused in the analyte CV-values by indirect heating of the electrodes by the heated ionization source [51]. The development of the temperature-controlled electrode was a key step in the evolution of the Ionalytics system, and the subsequent acquisition of the company by Thermo Electron. Thermo now offers FAIMS devices on a variety of their mass spectrometers including the TSQ Quantum, Vantage, and LTQ systems.

A picture of the FAIMS analyzer for use with the TSQ Quantum mass spectrometer is shown in Figure 5.21a, and a schematic representation of the temperature-controlled electrode set is shown in Figure 5.21b. The analytical gap width is 2.5 mm, and a bisinusoidal waveform with a frequency of 750 kHz and a DV of ±5 kV is used. For these instruments, the FAIMS system is being used largely as an additional component in the toolkit of the mass spectrometrist. That is, FAIMS is being used on these instruments to solve problems that arise from issues such as elevated background noise, endogeneous interferences, etc. These types of issues are common

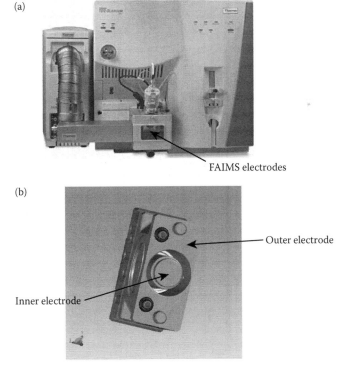

(a)

FAIMS electrodes

(b)

Outer electrode

Inner electrode

FIGURE 5.21   (a) Photograph of the Thermo Fisher FAIMS system offered on a TSQ Quantum. (Reproduced from Thermo Scientific-FAIMS Operator's Manual, page 1. With permission from Thermo Fisher Scientific Inc.). (b) Three-dimensional schematic of a temperature-controlled FAIMS device with one of the end-caps removed to illustrate better the inside of the device. (Reproduced from Barnett, D.A., Belford, M., Dunyach, J.J., and Purves, R.W., *J. Am. Soc. Mass Spectrom.* 2007, *18*, 1653–1663. With permission from Elsevier Inc.)

where matrices are complex, for example in biological samples such as plasma and urine; these applications will be further described in Section 5.5.

5.4.3.3 Sionex Corporation

Unlike the MSA or Ionalytics/Thermo systems, the Sionex (Bedford, MA, USA) version is based on the flat plate design. Initially the Corporation products were geared toward trace detection of chemical vapors, particularly for use with homeland security, using FAIMS with electrometric detection. These products have evolved and more selective products, for example the 'DMS/IMS²' instrument (which combines differential mobility spectrometry, DMS or FAIMS, with IMS, giving two-dimensional data), are now also available. The electrodes for these products are based on the design shown in Figure 5.16, which was fabricated through a micromachining process. The analytical gap width is 0.5 mm and a clipped sinusoidal waveform having a frequency of 1.18 MHz is used. This waveform can generate a DV of 1500 V, giving a maximum field of ca 30 kV cm⁻¹.

More recently, Sionex has entered the mass spectrometry market by coupling their electrodes to a Waters Micromass ZQ (shown schematically in Figure 5.22a), and also to a JEOL (Tokyo, Japan) time-of-flight mass spectrometer (a photograph of this arrrangement is shown in Figure 5.22b). The main use of the mass spectrometry-based product is to increase sensitivity for biomarker and peptide detection by prefiltering ions in the gas-phase.

5.4.3.4 Owlstone Nanotech

Of late, additional companies have been using FAIMS in their platforms, including an interesting variant developed by Owlstone Nanotech (Cambridge, UK), founded in 2004, with electrometric detection. The microchip-sized spectrometer, shown in Figure 5.23a is fabricated using processes similar to those used in the semiconductor industry. The device has a small gap size (tens of micrometers), an expanded view of which is shown in Figure 5.23b. This small gap size enables the generation of

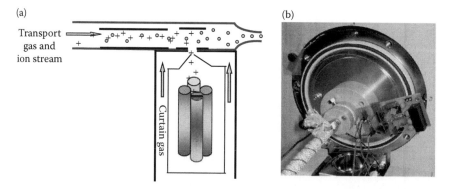

FIGURE 5.22   (a) Schematic of a right-angle configuration of the Sionex MicroDMx™ (planar electrodes) coupled to a single quadrupole Waters Micromass ZQ; (b) Photograph of the Sionex MicroDMx pre-filter mounted on a Jeol time-of-flight mass spectrometer. (Reproduced with permission from Sionex Corporation.)

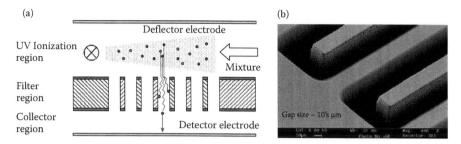

FIGURE 5.23   (a) Schematic of the microchip FAIMS spectrometer offered in the Owlstone Nanotech 'Lonestar.' Ions are transported through the ion separation region by an electric field instead of by a gas flow. (b) A photograph of a small region of the device showing the presence of multiple ion channels with each channel having a small gap size (tens of micrometers). (Reproduced with permission from Owlstone Nanotech Inc.)

high electric fields even with the low voltages produced by rectangular asymmetric waveforms (amplitude *ca* 250 V and frequency 28 MHz). In fact, the reduced gap size results in a significant increase in the value of *E/N* at which the onset of breakdown of the electric field occurs, characterized by the Paschen curve, meaning that the device can attain *E/N*-values not accessible to conventional devices with 2 mm gaps. The device also has multiple ion channels to maintain high sensitivity and, in addition, ions are transported through the ion separation region by an electric field instead of by a gas flow (that is, 'field-driven'), which eliminates the need for mechanical pumps.

Although Owlstone has presented recently a proof-of-principle system coupling their FAIMS system to a Thermo LXQ mass spectrometer [64], the primary focus of the Company has been, thus far, on products not coupled to a mass spectrometer. The 'Lonestar,' which incorporates their FAIMS design, is being used in security applications, including the detection of chemical warfare agents, toxic industrial chemicals, and trace explosive vapors. The Lonestar is being used also to monitor a variety of different chemical vapors in applications ranging from on-line process monitoring to laboratory-based research and development.

5.5 FAIMS APPLICATIONS IN MASS SPECTROMETRY

The continuous nature of FAIMS allows it to be coupled readily between atmospheric pressure ionization sources and mass spectrometry, serving as a gas-phase ion filter. Due to this simplicity and the complementary mechanism of FAIMS, combinations with different mass spectrometric applications have been diverse. As a consequence, this Section is intended only to highlight key milestones and publications, and will not serve as a complete review of all work involving FAIMS devices.

5.5.1 LARGE MOLECULES

For the purpose of this discussion, the term 'large molecules' refers to peptides and proteins. Typically, protein and peptide ions exhibit type C behavior, and are detected in the P2 and N2 modes. As was alluded to earlier, recent work [12] has shown that some larger proteins can also initially experience increases in their high-field to low-field mobility ratios with increasing electric field strength before experiencing decreases in this ratio at yet higher fields (type B behavior). This phenomenon is believed to be due to reversible 'locking' (that is, alignment) of the electric dipoles of the protein ions by the FAIMS electric field. That is, proteins with large dipole moments will experience a reversible locking (alignment) of their electric dipoles at high-electric fields leading to these observed increases in mobility.

5.5.1.1 Gas-Phase Protein Structures

The study of structures of gas-phase protein ions can provide information on the roles of intermediates and pathways in protein folding. This observation is supported by Ashcroft and coworkers who used FAIMS to study β_2-microglobulin conformers in the gas-phase and concluded that "ESI-FAIMS-MS offers significant opportunities for the study of the conformational properties of proteins and thus may present valuable insights into the roles that different conformers play in diseases related to protein folding" [65].

The study of gas-phase protein structures began almost immediately with the coupling of an electrospray ionization source to FAIMS, as changes in the charge-state distribution of cytochrome c were observed at different CV-values. This observation suggested that FAIMS was able to separate different gas-phase forms of the same ions (*i.e.*, conformers) [11]. Subsequent work with bovine ubiquitin [66] revealed multiple conformations for the same charge state, including at least three for the +8 charge state. These conformers were shown to be sensitive to solution conditions, such as pH and solvent composition [66]. Accounts of the early work with proteins (namely ubiquitin) resulted in two further publications [67,68]. With advancements in the technology, higher voltages were achievable with the asymmetric waveform, which allowed higher electric fields to be achieved [67] and enabled the detection of multiple elongated conformers of bovine ubiquitin, including four each for both the +11 and +12 charge states. In the other work [68], FAIMS was coupled with energy-loss techniques pioneered by Covey and Douglas [69] to enable collision cross-sections to be measured for the different conformers. This coupling of FAIMS with energy-loss techniques was the first illustration of combining FAIMS with other techniques that probe conformational information, and illustrated the use of multidimensional separations. Figure 5.24 shows a CV spectrum for the +8 charge state of bovine ubiquitin with collision cross-section measurements superimposed. Note that without FAIMS, only an average value would be obtainable for the cross section for the +8 charge state as the mass spectrometer (measuring mass/charge ratio) cannot differentiate the different conformations of this charge state.

Since the continuous nature of FAIMS enables it to be coupled readily with other techniques, several multidimensional approaches with FAIMS have been reported, many of which are described below. For example, H-D exchange was carried out within the FAIMS device in order to probe conformers of equine cytochrome c [70]. This approach illustrated that the complementary nature of the two techniques enabled detection of a greater number of conformers than would have been detected by either FAIMS or H-D exchange alone. The use of H/D exchange within FAIMS followed by energy-loss measurements was employed also to probe the peptide bradykinin [71]. Figure 5.25 illustrates that because of the FAIMS separation (Figure 5.25a), the

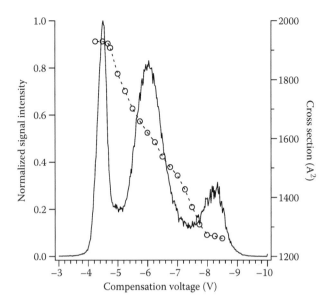

FIGURE 5.24 An IS-CV spectrum obtained for the +8 charge state of bovine ubiquitin upon which is superimposed the collision cross-section measurements (indicated by the circles) obtained using the energy-loss method. The change in cross-section across the peak centered at CV *ca*−6V indicates the presence of more than one conformer within this peak. (Reproduced from Purves, R.W., Barnett, D.A., Ells, B., and Guevremont, R., *J. Am. Soc. Mass Spectrom.* 2000, *11*, 738–745. With permission from Elsevier Science Inc.)

presence of four different species of doubly charged bradykinin can be derived from the H-D exchange mass spectra (Figure 5.25b through f). In particular, one conformer is observed in the mass spectrum acquired at CV = −6.9 V (Figure 5.25b) in which the most intense peak is *m/z* 533.5. Two other conformers become apparent from examination of the mass spectra acquired at increasingly less negative CV-values (Figure 5.25 c through e). These conformers have *m/z* distributions centered at 532.0 and 535.0 with maximal intensities at CV *ca* −6.1, and −6.5V (not shown), respectively. All of these three aforementioned conformers are different from the deuterium incorporation level of the species at CV = −5.4 V (Figure 5.25f) [71]. Note that in the absence of FAIMS, only a broad H-D mass spectrum was observed. Robinson and Williams performed H/D experiments on gas-phase conformers of bovine ubiquitin in an FT/ICR MS after using the FAIMS analyzer to give an upfront separation [72]. The results from these two techniques were not strongly correlated, rather they showed that the two methods were highly complementary. Williams and coworkers looked also at the role of electron capture dissociation (ECD) on the conformation of ubiquitin [73]. They found that the ECD spectra of conformer-selected ions of the same charge states differed both in electron capture efficiency and in fragment ion intensities.

Shvartsburg *et al.* used FAIMS followed by IMS separations in conjunction with mass spectrometry to investigate conformers of bovine ubiquitin and cytochrome c [74]. More conformers were distinguished using the FAIMS/IMS combination, as is illustrated for bovine ubiquitin in Table 5.1. There was, however, a substantial

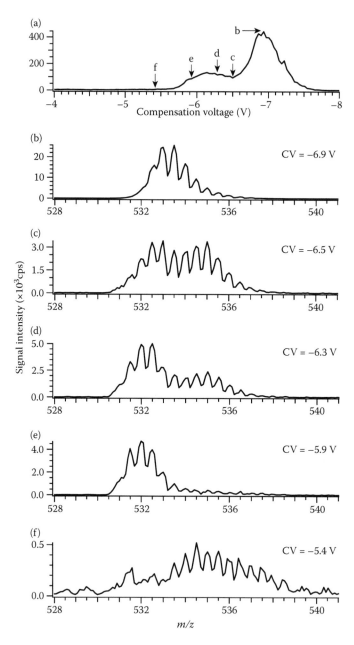

FIGURE 5.25 The use of FAIMS, mass spectrometry, and H/D exchange combined to investigate conformers of doubly-charged bradykinin; (a) CV spectrum of 2 μM bradykinin with D$_2$O present in the nitrogen curtain gas. Mass spectra acquired at (b) CV = −6.9 V; (c) CV = −6.5 V; (d) CV = −6.3 V; (e) CV = −5.9 V; and (f) CV = −5.4 V (as indicated by the labels marked on the trace in (a)). (Reproduced from Purves, R.W., Barnett, D.A., Ells, B., and Guevremont, R., *Rapid Commun. Mass Spectrom.* 2001, *15*, 1453–1456. With permission from John Wiley and Sons, Ltd.)

TABLE 5.1

Number of Ubiquitin Ion Conformers Distinguished by Ion Mobility Methods (Positive Ion Mode)

	Charge State (z)									
	6	7	8	9	10	11	12	13	14	Σall z
IMS	3–4	3	2–3	2	2	1	1	1	1	16–18
FAIMS	2–3	3	2	2	1	1	1	1	1	14–15
FAIMS/IMS	7	6	6	6	5	3	3	3	1	40

Source: Shvartsburg, A.A., Li, F., Tang, K., and Smith, R.D., *Anal. Chem.* 2006, *78*, 3304–3315. With permission.

correlation between FAIMS and IMS separations. In an elegant approach, this correlation was used to elucidate individual conformer structures using 'tandem IMS' experiments. In particular, a specific species separated by FAIMS was unfolded by controlled heating in the ion funnel placed between the FAIMS and the IMS cell and the resulting structure was probed by IMS.*

The relation of gas-phase ion structure to that in the solution phase has been a topic of debate. This debate is clouded further by RF heating caused by the asymmetric waveform of FAIMS. The extent of heating (ΔT) can be calculated [2,68,75], using the two-temperature kinetic theory of gaseous ion transport [75], from

$$\Delta T(E) = T_{ion}(E) - T = M[K(E)E]^2 / (3k) \qquad (5.15),$$

where T_{ion} is the ion temperature. The important thing to note from Equation 5.15 is that E (and thus ΔT) will change as a function of time because of the periodic nature of the waveform (see, for example, Figure 5.19). Initial work [68] suggested that a mean ΔT (that is, the average ΔT during the cycle of the waveform) could be used to estimate ion heating. Shvartzburg *et al.* [75] used a FAIMS-IMS-MS system to compare IMS spectra, with and without FAIMS present. They observed that changes due to heating were more consistent with a maximum ΔT (*i.e.*, ΔT_{max}) that can be calculated at the peak voltage of the asymmetric waveform. The observation that ΔT_{max} was more consistent with ion heating was supported also by subsequent work using a temperature-controlled FAIMS device [76], in which CV spectra were collected as a function of waveform amplitude and gas temperature, T.

5.5.1.2 Peptides

The potential of FAIMS for the analysis of tryptic peptides was examined initially using a tryptic digest of pig hemoglobin and a triple quadrupole mass spectrometer [53]. This work illustrated the ability of the FAIMS to reduce dramatically chemical background noise, and to enable the detection of low level peptides, which were

---

* For more details on the ion funnel, see Volume 4, Chapter 7: Ion Accumulation Approaches for Increasing Sensitivity and Dynamic Range in the Analysis of Complex Samples, by Mikhail E. Belov, Yehia M. Ibrahim and Richard D. Smith.

obscured by the background using ESI-MS. The role of FAIMS in peptide analysis was investigated further by Barnett *et al.* using a total of 282 peptides obtained from the digests of 13 different proteins [77]. As is shown in Figure 5.26, this work illustrated that the majority of the doubly and triply charged peptide ions where separated (in terms of CV) from the singly charged ions. The authors suggested that, by monitoring two or three CV-values, the majority of the doubly- and triply-charged ions could be transmitted through the FAIMS device. Although some of the peptides in a mixture

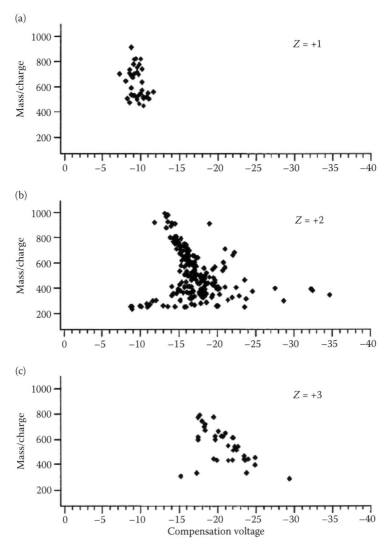

FIGURE 5.26 Summary plots of *m/z* as a function of CV for (a) 38 singly; (b) 208 doubly; and (c) 36 triply-charged peptide ions from 13 protein digests using a 1:1 helium/nitrogen carrier gas mixture and DV = 4000 V. (Reproduced from Barnett, D.A., Ells, B., Guevremont, R., and Purves, R.W., *J. Am. Soc. Mass Spectrom.* 2002, *13*, 1282–1291. With permission from Elsevier Science Inc.)

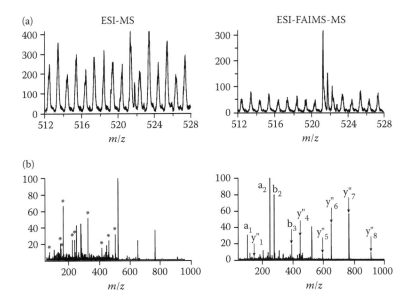

FIGURE 5.27   (a) Comparison of mass spectra of a 1.7 fmol μL⁻¹ tryptic digest of pig hemo-globin collected on the QqTOF instrument with resolution (FWHM) of approximately 6000 using ESI-MS and ESI-FAIMS-MS (same MS acquisition time). (b) Comparison of product ion mass spectra (MS/MS) for the doubly-charged tryptic ion of MFLGFPTTK (*m/z* 521.3) col-lected with identical operating conditions on the QqTOF instrument using either conventional ESI or ESI-FAIMS. The peaks labeled by asterisks indicate fragment ions resulting from CID of 'isobaric' background species. (Reproduced from Barnett, D.A., Ding, L., Ells, B., Purves, R.W., and Guevremont, R., *Rapid Commun. Mass Spectrom.* 2002, *16*, 676–680. With permission from John Wiley & Sons Ltd.)

would not be transmitted at these values, a liquid chromatography (LC)-FAIMS-MS analysis would detect peptides of lower abundance that may be lost in the noise of a conventional LC-MS analysis, thereby improving sequence coverage [77]. The subsequent coupling of FAIMS to a QqTOFMS instrument [78] illustrated how the FAIMS could be used to simplify the MS/MS spectra and, consequently, to aid peptide identification. Figure 5.27 illustrates the simplification in the mass spectral data for a multiply-charged tryptic peptide ion. Although the resolution of the TOF analyzer (*ca* 6000 FWHM) enables visual discrimination of a multiply-charged species at *m/z* 521.8 in the ESI-MS spectrum in Figure 5.27a, an improvement in the signal-to-background (S/B) ratio of approximately six-fold for this ion using ESI-FAIMS-MS was achieved [78]. However, not only are the intense background ions present in the ESI-MS data removed using FAIMS (Figure 5.27a), but the product ion mass spectrum is simplified also, as is shown in Figure 5.27b, as it lacks many of the background-related fragment ions observed with conventional ESI-TOFMS that would otherwise complicate peptide identification. In this figure, the asterisks in the ESI-MS/MS spectrum indicate frag-ment ions that do not result from cleavage of the peptide ion but instead indicate ions resulting from CID of 'isobaric' background species. These ions, that would compli-cate peptide identification, are absent in the ESI-FAIMS-MS spectrum.

Investigations with nanoLC-MS and FAIMS have been carried out by Thibault and coworkers [79] for complex tryptic digests using both global and targeted proteomics approaches. By implementing a function whereby the acquisition cycle consisted of sequential steps through three different CV voltages (synchronized with collection of the data), an increase of 20% in the number of peptides detected compared with conventional nanoelectrospray alone was observed [79]. Excellent reproducibility was reported, as replicate LC-FAIMS-MS runs had more than 90% of the peptide ions showing less than 30% variation in ion intensity [79]. The use of FAIMS for proteomics has been shown to be particularly well suited for targeted applications, in which the CV can be fixed and low abundance peptides can be filtered from the background noise [79,80]. This approach was used by Klaassen and Romer, who reported a Good Laboratory Practice (GLP)-validated quantitation method for a peptide in a biological matrix using FAIMS [81]. Recently, MacCoss and coworkers evaluated nanoLC-FAIMS-MS on an ion trap mass spectrometer [82] and found that increases in peak capacity (> eight-fold) and dynamic range (> five-fold) could readily be achieved compared to the non-FAIMS analysis, that is nanoLC-MS.

The use of FAIMS coupled with IMS and mass spectrometry for the analysis of tryptic peptides has also been investigated by Tang *et al.* [83]. The authors reported a high orthogonality between FAIMS and IMS and, as a result, an effective increase in the analysis of complex mixtures of tryptic peptides.

5.5.2   SMALL MOLECULES

The term 'small molecule' has been applied arbitrarily in this discussion to molecules with molecular weights of less than 1000 Da. These include small biological molecules, as well as environmental/inorganic species. Unlike the peptides and proteins, the behavior of these ions is much less predictable, with applications being carried out using both polarities of the waveform, that is in both Mode 1 and Mode 2.

5.5.2.1   Inorganic

Reduction of background noise using Mode 1 for inorganic ions has been, in some cases, truly remarkable. The reason for this observation is that much of the chemical background (for ESI) consists of solvent clusters and impurities that are transmitted through the FAIMS in Mode 2, whereas only very specific background ions (for example, acetate) are transmitted through FAIMS in Mode 1. Consequently, several inorganic species are commonly transmitted in regions in which virtually no other ions pass through the analyzer. An example of background noise reduction in Mode 1 can be found in work by Barnett *et al.* that examined nitrate, chlorate, bromate, and iodate ions by ESI-FAIMS-MS [83]. Figure 5.28 shows mass spectra obtained using the same solution containing nitrate, chlorate, bromate, and iodate ions in 0.2 mM ammonium acetate. Without FAIMS (Figure 5.28a), the chemical background is intense and this is illustrated by the expanded view in the inset. However with ESI-FAIMS-MS (Figure 5.28b) at CV = 20 V (N1 mode) the background has essentially been eliminated and low abundance isotopes can be identified readily, as is shown in the inset. Note that the reader may recall that this same mixture of oxoanions was

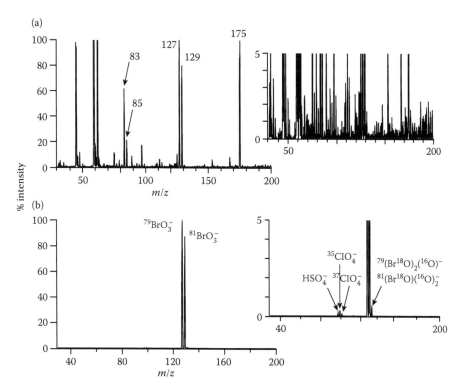

FIGURE 5.28　(a) Mass spectrum of a 62 μM mixture of nitrate, chlorate, bromate, and iodate obtained using conventional ESI-MS on an API-300 triple-quadupole mass spectrometer. (b) Mass spectrum of the same solution using ESI-FAIMS-MS at DV = −3000, CV = 20.0 V. The inset in both (a) and (b) show a 20 × expanded view of the baseline. (Reproduced from Barnett, D.A., Guevremont, R., and Purves, R.W., *Appl. Spectrosc.* 1999, *53*, 1367–1374. With permission from Society for Applied Spectroscopy.)

used to illustrate ion focusing in Figure 5.7. From the trace acquired at DV = −3000 V in Figure 5.7, the optimum CV-values for the each of the oxoanions can be determined readily, and mass spectra acquired at each of these optimal CV-values.

McSheehy and Mester reported the use of FAIMS for trace element speciation in biological tissues [84]. Despite an extensive clean-up/separation protocol, several minor arsenic-containing compounds could not be detected using electrospray mass spectrometry. However, with the additional separation in the gas-phase, many low-level compounds, including arsenic-containing compounds, were detected and structural determinations using ESI-FAIMS-MS/MS were carried out [84].

An interesting application involves the enrichment of isotopes which was first illustrated for $^{35}Cl^-$ and $^{37}Cl^-$ [85]. Typically, the separation of isotopes improves as the percentage change in molecular weight of the two species is increased. Isotope separation may appear to be a complicating issue when using isotopically labeled internal standards, which are employed commonly in the analysis of biomolecules. This would be especially the case for applications in which only one value of CV is used, and CV-switching is not desirable. Recall the rate at which CV-switching can be

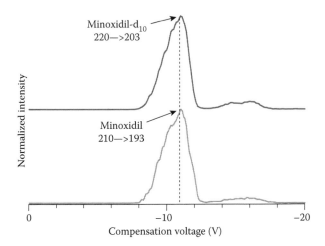

FIGURE 5.29 CV spectra acquired in positive ion mode for d_{10}-minoxidil $[M + H]^+ = 220.3$ Th and minoxidil $[M + H]^+ = 210.3$ Th using a temperature-controlled electrode set with both electrodes set to 80°C.

carried out (that is, stepping the CV to different values to monitor multiple analytes) depends upon the residence times of the ions in the FAIMS analyzer, which is typically on the order of *ca* 100 ms. That is, after a CV-value has been stepped to monitor an analyte, the mass spectrometer must remain idle for 100 ms (the residence time) before this analyte will be transmitted. Ultimately this idle time will reduce the duty cycle of the mass spectrometer. However, since the percentage change in molecular weight for deuterated internal standards is usually small, the difference in transmission is typically not an issue. Consider the extreme (in terms of percent difference in *m/z*) example shown in Figure 5.29 for minoxidil (*m/z* 210) and d_{10}-minoxidil (*m/z* 220.3). These CV spectra were acquired using a temperature-controlled electrode set with both electrodes set to 80°C, and show virtually no difference between the optimal CV-value for the analyte and for the deuterated internal standard. Note, however, that the peaks in Figure 5.29 are 'relatively broad,' and when different temperatures are applied on the inner and outer electrodes to create narrow peak widths for optimizing ion separation (recall Section 5.3.2.2), it is prudent to check the transmission of both species since optimal transmission may not be obtained for both compounds at the same CV-value.

5.5.2.2 Environmental

Applications of ESI-FAIMS-MS involving environmentally relevant compounds have used both Modes 1 and 2, as well as a variety of carrier gases. Early applications with ESI-FAIMS-MS included work with chlorinated and brominated byproducts of water disinfection [86] and the detection of perchlorate in water [87]. Both of these applications were improved upon significantly in subsequent work. For the analysis of perchlorate, improved hardware enabled the isobaric interference caused by the sulfate isotope (that is, $H^{34}SO_4^-$) to be eliminated completely from the analysis, as is illustrated by the CV spectrum in Figure 5.30 [88]. For the work with haloacetic acids, as was

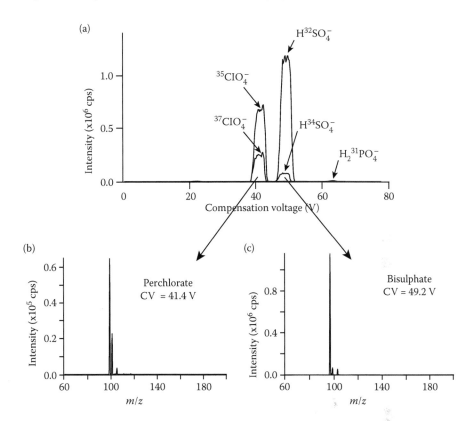

FIGURE 5.30 (a) IS-CV spectra (m/z 97, 99, 101) acquired using negative ion electrospray of a solution of perchlorate. Mass spectra were acquired at CV-values of (b) 41.4 and (c) 49.2 V. (Reproduced from Ells, B., Barnett, D.A., Purves, R.W., and Guevremont, R., *J. Environ. Monit.* 2000, 2, 393–397. With permission from The Royal Society of Chemistry.)

discussed in Section 5.3.1.1, through the addition of a small amount of carbon dioxide to the FAIMS carrier gas, detection of bromodichoroacetic acid (BDCAA) as BDCA⁻ was made possible and sensitivities for other trihalogenated species were increased significantly [43]. Comparisons of detection limits using ESI-FAIMS-MS with existing gas chromatography (GC) methods were made [89], and it was found that "the detection limits with this method (using FAIMS) were much lower than [with] previously existing GC and GC/MS methods, and quantitation results compared favorably with other existing methods." In addition, the analysis was simplified as the FAIMS method did not require sample preparation or chromatographic separation.

The use of FAIMS has been applied also to the detection of microcystins [90], and Mester and coworkers have used FAIMS for several environmental analyses, including the analysis of the mycotoxin zearalenone [91]. Sultan and Gabryelski, using ESI-FAIMS-TOFMS, performed a detailed analysis of contaminants in drinking water [92]. Not only were a large number of components detected at trace levels without sample clean up, but many previously unknown disinfection byproducts were discovered also, including glycolic acid.

5.5.2.3   Bioanalytical

Although LC-MS/MS is a very sensitive and selective technique, FAIMS has found a niche in the analysis of biomolecules, particularly in the pharmaceutical industry, because of its orthogonality to both LC and MS. In general, bioanalytical challenges in which FAIMS has been used to improve the analysis can be separated into three types. These categories include: reduction/removal of interferences; detection of ions that do not produce good product ion mass spectra; and the need to increase throughput.

Applications for removal of interferences such as isobaric ions, endogenous interferences, and elevated background have been the most prevalent due to the complex nature of biological fluid samples such as plasma and urine. The utility of using FAIMS to separate isobaric ions was first illustrated through the separation of leucine and isoleucine [93]. Early applications involving the analysis of human urine showed a dramatic reduction in chemical background and a subsequent improvement in detection limits for morphine and codeine [41] as well as amphetamine and various derivatives [94] for ESI-FAIMS-MS compared with ESI-MS.

Despite the specificity of MS/MS techniques, the analysis of some compounds, especially in complex matrices, has proven challenging and FAIMS has been shown to improve detection limits in some of these applications. Although much work has been presented at conferences, merely a handful of papers have appeared in the refereed literature, and the focus here will be on this published material. Kapron and coworkers reported the use of FAIMS to quantitate a compound in the presence of its N-oxide metabolite [95]. The compound and metabolite coeluted by LC and without FAIMS, the metabolite would undergo in-source fragmentation and contribute to the compound's signal intensity, thereby leading to erroneous quantitative results. With FAIMS, the compound and metabolite transmitted at different CV-values consequently, by selecting the appropriate CV-value of the compound, the metabolite would not enter the mass spectrometer and could not interfere with the quantitation. The use of FAIMS was incorporated into the analysis of prostanoids in order to reduce chemical background noise, separate the isobaric ions PGE_2 and PGD_2, and separate dynamically interchanging TXB2 anomers [96].

Mahan and King used LC-APCI-FAIMS-MS to measure taxol in plasma [97]. The measurement of taxol levels in plasma is important in the development of anti-cancer drugs, however, because taxol does not produce good product ion mass spectra, the specificity of MS/MS cannot be used and the analysis is plagued by a large chemical background. With FAIMS, the lower quantitation limit was improved by over an order of magnitude [97].

Mehl et al. evaluated the use of FAIMS in the quantitative analysis of acetaminophen in CYP 1A2 marker assays [98]. Since FAIMS was able to remove the background from the sample matrix and, in addition, remove phenacetin so that it could not interfere with the quantitation of acetaminophen, the authors were able to use flow injection analysis instead of LC and, thereby, reduced their run times from five minutes to less than one minute [98].

Hatsis et al. investigated the use of nanoESI with FAIMS to improve sample throughput [99]. Figure 5.31a shows the separation of morphine from its glucuronide metabolite at different inner and outer electrode temperatures. Using 60°C on each

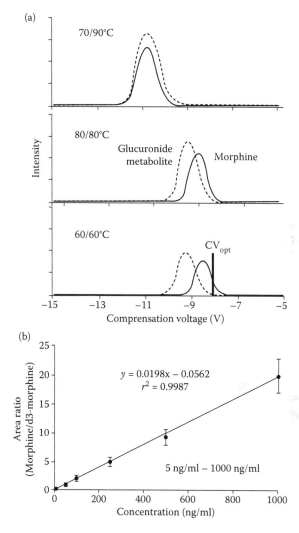

FIGURE 5.31   (a) Compensation voltage scans acquired to determine the effect of inner and outer electrode temperature for morphine and its glucuronide metabolite (infused at 100 ng mL^{-1}). (b) Standard curve for morphine in extracted mouse plasma using nanoESI-FAIMS showing linearity over three orders of magnitude. (Reproduced from Hatsis, P., Brockman, A.H., and Wu, J.T., *Rapid Commun. Mass Spectrom.* 2007, *21*, 2295–2300. With permission from John Wiley & Sons Ltd.)

electrode, morphine can be transmitted selectively through the device, reducing drastically the effect of the glucuronide; note that temperatures lower than 60°C affected the results negatively. The authors also looked at the analytical figures of merit, and found the response to be linear over the range investigated (Figure 5.31b) [99].

In situations such as the example for morphine and its glucuronide metabolite, optimizing the FAIMS parameters can sometimes be challenging. Typically, the parameters are investigated one at a time, but recently Champarnaud and coworkers reported a 'design of experiment' approach (DoE) in which multiple factors were investigated simultaneously to carry out method development more efficiently [100].

5.5.3 OTHER APPLICATIONS

Although the first part of this Section dealt with the main application areas of FAIMS, with the wide diversity of the utility of this technique many other niche applications can be imagined and some of these are touched upon briefly below.

The coupling of FAIMS and FTICR-MS has been used by Williams and coworkers to investigate poly(ethylene glycol) (PEG) [101,102]. The FAIMS stage was found to enhance the detection capability, resulting in improved detection limits and the ability to detect low abundance PEG molecular cations that had not been detected previously.

Capillary electrophoresis electrospray mass spectrometry (CE-ES-MS) is an inherently noisy technique, but used commonly to analyze complex bacterial lipopolysaccharides (LPS). Li *et al.* reported the use of a CE-FAIMS-MS system to investigate LPS and reported a significant reduction in the mass spectral noise, leading to improved detection limits [103].

Cluster ions of leucine enkepthalin produced by ESI were also investigated using FAIMS [104]. The authors demonstrated that analysis by means of FAIMS could be used to separate four complex ions of leucine enkephalin, all having the same value of m/z (that is, $[2M + H]^+$, $[4M + 2H]^{2+}$, $[6M + 3H]^{3+}$, $[8M + 4H]^{4+}$).

Mie and coworkers investigated the use of FAIMS for the separation of enantiomers [105,106]. By adding an appropriate chiral reference compound to the analyte solution, upon ionization, the enantiomers formed diastereomeric complexes. The authors showed that terbutaline enantiomers were complexed with metal ions and an amino acid to form diastereomeric complexes that were separated with FAIMS, as is shown in Figure 5.32.

5.6 SUMMARY AND OUTLOOK

Although interfacing mass spectrometers with atmospheric pressure ionization sources has been revolutionary, until recently the physical space between these two entities had remained largely underdeveloped. However, with the implementation of FAIMS as an interface in this region, gas-phase ions can be 'processed' at atmospheric pressure before their introduction into the mass spectrometer. The complementary natures of the FAIMS ion separation mechanism and mass spectrometry, combined with the additional capability of an atmospheric pressure ion focusing mechanism enables FAIMS to maintain similar signal intensities compared to those that would be observed without a FAIMS device being present. These inherent properties have given rise to the potential for this technique to be a very powerful tool for use with mass spectrometry.

As illustrated in Section 5.5, FAIMS has been used already in a very broad range of applications, from the separation of chlorine isotopes to improved detection of environmental contaminants. Separations of mixtures of isobaric ions and investigations

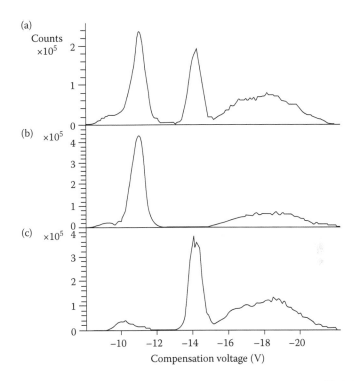

FIGURE 5.32 Separation of (+)/(−)-terbutaline as trimeric cluster ions [Cu(L-Trp)$_2$(terb)-H]$^+$. IS-CV spectra from (a) a mixture of (+)- and (−)-terbutaline; (b) (+)-terbutaline individually; and (c) (−)-terbutaline individually. (Reproduced from Mie., A., Ray, A., Axelsson, B.O., Jornten-Karlsson, M., and Reimann, C.T.,, *Anal. Chem.* 2008, *80*, 4133–4140. With permission from the American Chemical Society.)

of gas-phase conformations of proteins have been illustrated also. The capability of FAIMS to reduce background noise in LC-MS and LC-MS/MS analyses has proven to be an important capability for users whose applications focus on the analysis of small molecules.

Despite the utility of the applications presented herein, FAIMS is still a relatively new technique, and as the understanding of FAIMS continues to move forward, the applications of the technique will evolve also. Already a wide variety of intriguing designs have been proposed, with some of these having been commercialized. Multichannel devices with gap sizes of the order of tens of microns should play a significant role when coupled to mass spectrometry, as the electronic requirements are reduced and higher values of *E/N* can be achieved. Historically, mass spectrometry interfaces have been 'pin-hole-based,' on account of the need to introduce ions from atmospheric pressure into lower pressure regions and, for the most part, this approach has been used to interface ionization sources with FAIMS. However, because FAIMS operates at atmospheric pressure, the potential exists to redesign this interface and increase dramatically the number of ions that can be sampled from the ionization source. In this way, the capability of FAIMS to sample ions should be improved significantly, further enhancing the sensitivity of the analysis.

Finally, despite its utility as an interface between the ionization source and the mass spectrometer, FAIMS can be employed also either at reduced pressures within the mass spectrometer or, potentially, with other non-mass spectrometry means of detection. It is evident, therefore, that with the wide applicability of the technique FAIMS appears to be poised to make a significant impact.

REFERENCES

1. Viggiano, A.A.; Morris, R.A.; Mason, E.A. Mobilities and interaction potentials for the O^+-He and O^--He systems. *J. Chem. Phys.* 1993, *98*, 6483–6487.
2. McDaniel, E.W.; Mason, E.A. *Transport Properties of Ions in Gases.* Wiley, New York, 1988.
3. Eiceman, G.A.; Karpas, Z. *Ion Mobility Spectrometry.* CRC Press, Boca Raton, 2005.
4. Borsdorf, H.; Eiceman, G.A. Ion mobility spectrometry: principles and applications. *Appl. Spectrosc. Rev.* 2006, *41*, 323–375.
5. O'Donnell, R.M.; Sun, X.; Harrington, P.B. Pharmaceutical applications of ion mobility spectrometry. *Trends Anal. Chem.* 2008, *27*, 44–53.
6. Guevremont, R.; Barnett, D.A.; Purves, R.W.; Viehland, L.A. Calculation of ion mobilities from electrospray ionization high-field asymmetric waveform ion mobility spectrometry mass spectrometry. *J. Chem. Phys.* 2001, *114*, 10270–10277.
7. Shvartsburg, A.A.; Mashkevich, S.V.; Smith, R.D. Feasibility of higher-order differential ion mobility separations using new asymmetric waveforms. *J. Phys. Chem. A* 2006, *110*, 2663–2673.
8. Ellis, H.W.; Pai, R.Y.; McDaniel, E.W.; Mason, E.A.; Viehland, L.A. Transport properties of gaseous ions over a wide energy-range. *At. Data Nucl. Data Tables* 1976, *17*, 177–210.
9. Shvartsburg, A.A.; Smith, R.D. Optimum waveforms for differential ion mobility spectrometry (FAIMS). *J. Am. Soc. Mass Spectrom.* 2008, *19*, 1286–1295.
10. Purves, R.W.; Guevremont, R.; Day, S.; Pipich, C.W.; Matyjaszczyk, M.S. Mass spectrometric characterization of a high-field asymmetric waveform ion mobility spectrometer. *Rev. Sci. Instrum.* 1998, *69*, 4094–4105.
11. Purves, R.W.; Guevremont, R. Electrospray ionization high-field asymmetric waveform ion mobility spectrometry-mass spectrometry. *Anal. Chem.* 1999, *71*, 2346–2357.
12. Shvartsburg, A.A.; Bryskiewicz, T.; Purves, R.W.; Tang, K.; Guevremont, R.; Smith, R.D. Field asymmetric waveform ion mobility spectrometry studies of proteins: dipole alignment in ion mobility spectrometry? *J. Phys. Chem. B* 2006, *110*, 21966–21980.
13. Gorshkov, M.P. Method of analysis of impurities in gases. *U.S.S.R. Inventor's Certificate* 1982, 966,583.
14. Buryakov, I.A.; Krylov, E.V.; Nazarov, E.G.; Rasulev, U. Kh. A new method of separation of multi-atomic ions by mobility at atmospheric pressure using a high-frequency amplitude asymmetric strong electric field. *Int. J. Mass Spectrom. Ion Processes* 1993, *128*, 143–148.
15. Buryakov, I.A.; Krylov, E.V.; Soldatov, V.P. Method of analyzing traces of substances in gases. *U.S.S.R. Inventor's Certificate* 1989, 1,485,808.
16. Carnahan, B.L.; Tarassov, A.S. Ion mobility spectrometer. *U.S. Patent* 1995, 5,420,424.
17. Riegner, D.E.; Harden, C.S.; Carnahan, B.; Day, S. Qualitative evaluation of field ion spectrometry for chemical warfare agent detection. *Proc. 45th ASMS Conference of Mass Spectrometry and Allied Topics,* Palm Springs, CA June 1–5, 1997.
18. Beckey, H.D. *Principles of Field Desorption and Field Ionization Mass Spectrometry,* Permagon, Oxford, 1977.
19. Miller, R.A.; Nazarov, E.G.; Eiceman, G.A.; King, A.T. A MEMS radio-frequency ion mobility spectrometer for chemical vapor detection. *Sens. Actuators, A* 2001, *91*, 307–318.

20. Ridgeway, M.; Remes, P.; McKinney, C.; Glish, G.L. Radio frequency power supply for the production of high amplitude asymmetric waveforms. *Proc. 56th ASMS Conference of Mass Spectrometry and Allied Topics,* Denver, CO June 1–5, 2008.

21. Shvartsburg, A.A.; Tang, K.; Smith, R.D. Optimization of the design and operation of FAIMS analyzers. *J. Am. Soc. Mass Spectrom.* 2005, *16*, 2–12.

22. Barnett, D.A.; Belford, M.; Dunyach, J.J.; Purves, R.W. Characterization of a temperature-controlled FAIMS system. *J. Am. Soc Mass Spectrom.* 2007, *18*, 1653–1663.

23. Guevremont, R.; Purves, R.W. Atmospheric pressure ion focusing in a high-field asymmetric waveform ion mobility spectrometer. *Rev. Sci. Instrum.* 1999, *70*, 1370–1383.

24. Barnett, D.A.; Guevremont, R.; Purves, R.W. Determination of parts-per-trillion levels of chlorate, bromate, and iodate by electrospray ionization/high-field asymmetric waveform ion mobility spectrometry/mass spectrometry. *Appl. Spectrosc.* 1999, *53*, 1367–1374.

25. Krylov, E.V. A method of reducing diffusion losses in a drift spectrometer. *Tech. Phys.* 1999, *44*, 113–116.

26. Guevremont, R.; Purves, R. Comparison of experimental and calculated peak shapes for three cylindrical geometry FAIMS prototypes of differing electrode diameters. *J. Am. Soc. Mass Spectrom.* 2005, *16*, 349–362.

27. Guevremont, R.; Purves, R.W.; Barnett, D.A.; Ding, L. Ion trapping at atmospheric pressure (760 Torr) and room temperature with a high-field asymmetric waveform ion mobility spectrometer. *Int. J. Mass Spectrom.* 1999, *193*, 45–56.

28. Guevremont, R.; Ding, L.; Ells, B.; Barnett, D.A.; Purves, R.W. Atmospheric pressure ion trapping in a tandem FAIMS-FAIMS coupled to a TOFMS: studies with electrospray generated gramicidin S ions. *J. Am. Soc. Mass Spectrom.* 2001, *12*, 1320–1330.

29. Krylov, E.V.; Nazarov, E.G.; Miller, R.A. Differential mobility spectrometer: model of operation. *Int. J. Mass Spectrom.* 2007, *266*, 76–85.

30. Krylov, E.V. Comparison of the planar and coaxial field asymmetrical waveform ion mobility spectrometer (FAIMS). *Int. J. Mass Spectrom.* 2003, *225*, 39–51.

31. Buryakov, I.A. Mathematical analysis of ion motion in a gas subjected to an alternating-sign periodic asymmetric-waveform electric field. *Tech. Phys.* 2006, *51*, 1121–1126.

32. Elistratov, A.A.; Sherbakov, L.A.; Nikovaev, E.N. Analysis of non-linear ion drift in spectrometers of ion mobility increment with cylindrical drift chamber. *Eur. J. Mass Spectrom.* 2006, *12*, 153–160.

33. Elistratov, A.A.; Sherbakov, L.A. Space charge effect in spectrometers of ion mobility increment with planar drift chamber. *Eur. J. Mass Spectrom.* 2007, *13*, 115–123.

34. Elistratov, A.A.; Sherbakov, L.A. Space charge effect in spectrometers of ion mobility increment with a cylindrical drift chamber. *Eur. J. Mass Spectrom.* 2007, *13*, 259–272.

35. Guevremont, R.; Thekkadath, G.; Hilton, C.K. Compensation voltage (CV) peak shapes using a domed FAIMS with the inner electrode translated to various longitudinal positions. *J. Am. Soc. Mass Spectrom.* 2005, *16*, 948–956.

36. Pervukhin, V.V.; Sheven, D.G. Suppression of the effect of charge cloud in an ion mobility increment spectrometer to improve its sensitivity. *Tech. Phys.* 2008, *78*, 110–116.

37. Shvartsburg, A.A.; Tang, K.; Smith, R.D. Modeling the resolution and sensitivity of FAIMS analyses. *J. Am. Soc. Mass Spectrom.* 2004, *15*, 1487–1498.

38. Shvartsburg, A.A.; Tang, K.; Smith, R.D. FAIMS operation for realistic gas flow profile and asymmetric waveforms including electronic noise and ripple. *J. Am. Soc. Mass Spectrom.* 2005, *16*, 1447–1455.

39. Barnett, D. A.; Ells, B.; Guevremont, R.; Purves, R.W.; Viehland, L.A. Evaluation of carrier gases for use in high-field asymmetric waveform ion mobility spectrometry. *J. Am. Soc. Mass Spectrom.* 2000, *11*, 1125–1133.

368 Practical Aspects of Trapped Ion Mass Spectrometry, Volume IV

40. Cui, M.; Ding, L.; Mester, Z. Separation of cisplatin and its hydrolysis products using electrospray ionization high-field asymmetric waveform ion mobility spectrometry coupled with ion trap mass spectrometry. *Anal. Chem.* 2003, *75*, 5847–5853.

41. McCooeye, M.A.; Ells, B.; Barnett, D.A.; Purves, R.W.; Guevremont, R. Quantitation of morphine and codeine in human urine using high field asymmetric waveform ion mobility spectrometry (FAIMS) with mass spectrometric detection. *J. Anal. Toxicol.* 2001, *25*, 81–87.

42. Barnett, D.A.; Purves, R.W.; Ells, B.; Guevremont, R. Separation of o-, m-, and p-phthalic acids by high-field asymmetric waveform ion mobility spectrometry (FAIMS) using mixed carrier gases. *J. Mass Spectrom.* 2000, *35*, 976–980.

43. Ells, B.; Barnett, D.A.; Purves, R.W.; Guevremont, R. Detection of nine chlorinated and brominated haloacetic acids at part-per-trillion levels using ESI-FAIMS-MS. *Anal. Chem.* 2000, *72*, 4555–4559.

44. Shvartsburg, A.A.; Tang, K.; Smith, R.D. Understanding and designing field asymmetric waveform ion mobility spectrometry separations in gas mixtures. *Anal. Chem.* 2004, *76*, 7366–7374.

45. Guevremont, R.; Purves, R.; Barnett, D.; Ells, B. FAIMS apparatus and method using carrier gases that contain a trace amount of a dopant species. *U.S. Patent* 2006, 7,026,612.

46. Eiceman, G.A.; Krylov, E.V.; Krylova, N.S.; Nazarov, E.G.; Miller, R.A. Separation of ions from explosives in differential mobility spectrometry by vapor-modified drift gas. *Anal. Chem.* 2004, *76*, 4937–4944.

47. Levin, D.S.; Miller, R.A.; Nazarov, E.G.; Vouros, P. Rapid separation and quantitative analysis of peptides using a new nanoelectrospray-differential mobility spectrometer-mass spectrometer system. *Anal. Chem.* 2006, *78*, 5443–5452.

48. Kolakowski, B.M.; McCooeye, M.A.; Mester, Z. Compensation voltage shifting in high-field asymmetric waveform ion mobility spectrometry-mass spectrometry. *Rapid Commun. Mass Spectrom.* 2006, *20*, 3319–3329.

49. Viehland, L.A.; Mason, E.A. Transport properties of gaseous ions over a wide energy-range. 4. *At. Data Nucl. Data Tables* 1995, *60*, 37–95.

50. Purves, R.W.; Barnett, D.A.; Kolakowski, B.M.; Kapron, J.T.; McRae, G. Evaluation of a temperature controlled FAIMS system for enhancing bioanalytical LC-MS/MS. *Proc. 53rd ASMS Conference on Mass Spectrometry and Allied Topics,* San Antonio, TX, June 5–9, 2005.

51. Barnett, D.A.; Guevremont, R.; Purves, R.W. Design and characterization of a temperature controlled FAIMS system for small molecule analysis. *Proc. 53rd ASMS Conference on Mass Spectrometry and Allied Topics,* San Antonio, TX, June 5–9, 2005.

52. Shvartsburg, A.A.; Li, F.; Tang, K.; Smith, R.D. High-resolution field asymmetric waveform ion mobility spectrometry using new planar geometry analyzers. *Anal. Chem.* 2006, *78*, 3706–3714.

53. Guevremont, R.; Barnett, D.A.; Purves, R.W.; Vandermey, J. Analysis of a tryptic digest of pig hemoglobin using ESI-FAIMS-MS. *Anal. Chem.* 2000, *72*, 4577–4584.

54. Guevremont, R.; Purves, R. Apparatus and method for atmospheric pressure-3-dimensional ion trapping. *U.S. Patent* 2003, 6,621,077.

55. Miller, R.A.; Zahn, M. Longitudinal field driven field asymmetric ion mobility filter and detection system. *U.S. Patent* 2003, 6,512,224.

56. Stott, W.R.; Javahery, G. Mass spectrometer, including coupling of an atmospheric pressure ion source to a low pressure mass analyzer. *PCT Patent* 2000, WO 00,063,949.

57. Belford, M.W.; Dunyach, J.J.; Purves, R.W. Quadrupole FAIMS apparatus. *PCT Patent* 2008, WO 08,067,331.

58. Shvartsburg, A.A.; Smith, R.D. Scaling of the resolving power and sensitivity for planar FAIMS and mobility-based discrimination in flow- and field-driven analyzers. *J. Am. Soc. Mass Spectrom.* 2007, *18*, 1672–1681.

59. Kouvnetsov, V. High voltage waveform generator. *U.S. Patent* 1998, 5,801,379.
60. Buryakov, I.A. Qualitative analysis of trace constituents by ion mobility increment spectrometer. *Talanta*, 2003, *61*, 369–375.
61. Nazarov, E.G.; Miller, R.A.; Eiceman, G.A.; Stone, J.A. Miniature differential mobility spectrometry using atmospheric pressure photoionization. *Anal. Chem.* 2006, *78*, 4553–4563.
62. Papanastasiou, D.; Wollnik, H.; Rico, G.; Tadjimukhamedov, F.; Mueller, W.; Eiceman, G.A. Differential mobility separation of ions using a rectangular asymmetric waveform. *J. Phys. Chem. A* 2008, *112*, 3638–3645.
63. Nazarov, E.G.; Coy, S.L.; Krylov, E.V.; Miller, R.A.; Eiceman, G.A. Pressure effects in differential mobility spectrometry. *Anal. Chem.* 2006, *78*, 7697–7706.
64. Rush, M.; Thompson, A.; Holden, M.; Toutoungi, D. Development of a rapid, high-field field-asymmetric ion mobility pre-filter for improved LC-MS analysis of complex samples. *59th Pittsburgh Conference*, New Orleans, LA, March 1–7, 2008.
65. Borysik, A.J.H.; Read, P.; Little, D.R.; Bateman, R.H.; Radford, S.E.; Ashcroft, A.E. Separation of β_2-microglobulin conformers by high-field asymmetric waveform ion mobility spectrometry (FAIMS) coupled to electrospray ionization mass spectrometry. *Rapid Commun. Mass Spectrom.* 2004, *18*, 2229–2234.
66. Purves, R.W.; Barnett, D.A.; Guevremont, R. Separation of protein conformers using electrospray-high field asymmetric waveform ion mobility spectrometry-mass spectrometry. *Int. J. Mass Spectrom.* 2000, *197*, 163–177.
67. Purves, R.W.; Barnett, D.A.; Ells, B.; Guevremont, R. Investigation of bovine ubiquitin conformers separated by high-field asymmetric waveform ion mobility spectrometry: cross section measurements using energy-loss experiments with a triple quadrupole mass spectrometer. *J. Am. Soc. Mass Spectrom.* 2000, *11*, 738–745.
68. Purves, R.W.; Barnett, D.A.; Ells, B.; Guevremont, R. Elongated conformers of charge states +11 to +15 of bovine ubiquitin studied using ESI-FAIMS-MS. *J. Am. Soc. Mass Spectrom.* 2001, *12*, 894–901.
69. Covey, T.; Douglas, D.J. Collision cross sections for protein ions. *J. Am. Soc. Mass Spectrom.* 1993, *4*, 616–623.
70. Purves, R.W.; Ells, B.; Barnett, D.A.; Guevremont, R. Combining H-D exchange and ESI-FAIMS-MS for detecting gas-phase conformers of equine cytochrome c. *Can. J. Chem.* 2005, *83*, 1961–1968.
71. Purves, R.W.; Barnett, D.A.; Ells, B.; Guevremont, R. Gas-phase conformers of the $[M + 2H]^{2+}$ ion of bradykinin investigated by combining high-field asymmetric waveform ion mobility spectrometry, hydrogen/deuterium exchange, and energy-loss measurements. *Rapid Commun. Mass Spectrom.* 2001, *15*, 1453–1456.
72. Robinson, E.W.; Williams, E.R. Multidimensional separations of ubiquitin conformers in the gas phase: relating ion cross sections to H/D exchange measurements. *J. Am. Soc. Mass Spectrom.* 2005, *16*, 1427–1437.
73. Robinson, E.W.; Leib, R.D.; Williams, E.R. The role of conformation on electron capture dissociation of ubiquitin. *J. Am. Soc. Mass Spectrom.* 2006, *17*, 1469–1479.
74. Shvartsburg, A.A.; Li, F.; Tang, K.; Smith, R.D. Characterizing the structures and folding of free proteins using 2-D gas-phase separations: observation of multiple unfolded conformers. *Anal. Chem.* 2006, *78*, 3304–3315.
75. Shvartsburg, A.A.; Li, F.; Tang, K.; Smith, R.D. Distortion of ion structures by field asymmetric waveform ion mobility spectrometry. *Anal. Chem.* 2007, *79*, 1523–1528.
76. Robinson, E.W.; Shvartsburg, A.A.; Tang, K.; Smith, R.D. Control of ion distortion in field asymmetric ion mobility spectrometry via variation of dispersion field and gas temperature. *Anal. Chem.* 2008, *80*, 7508–7515.
77. Barnett, D.A.; Ells, B.; Guevremont, R.; Purves, R.W. Application of ESI-FAIMS-MS to the analysis of tryptic peptides. *J. Am. Soc. Mass Spectrom.* 2002, *13*, 1282–1291.

370   Practical Aspects of Trapped Ion Mass Spectrometry, Volume IV

78. Barnett, D.A.; Ding, L.; Ells, B.; Purves, R.W.; Guevremont, R. Tandem mass spectra of tryptic peptides at signal-to-background ratios approaching unity using electrospray ionization high-field asymmetric waveform ion mobility spectrometry/hybrid quadrupole time-of-flight mass spectrometry. *Rapid Commun. Mass Spectrom.* 2002, *16*, 676–680.

79. Venne, K.; Bonneil, E.; Eng, K.; Thibault, P. Improvement in peptide detection for proteomics analyses using nanoLC-MS and high-field asymmetry waveform ion mobility mass spectrometry. *Anal. Chem.* 2005, *77*, 2176–2186.

80. Xia, Y.Q.; Wu, S.T.; Jemal, M. LC-FAIMS-MS/MS for quantification of a peptide in plasma and evaluation of FAIMS global selectivity from plasma components. *Anal. Chem.* 2008, *80*, 7137–7143.

81. Klaassen, T.; Romer, A. FAIMS: GLP-validated quantitation method for a peptide in biological matrix. *Proc. 56th ASMS Conference on Mass Spectrometry and Allied Topics,* Denver, CO, June 1–5, 2008.

82. Canterbury, J.D.; Yi, X.; Hoopman, M.R.; MacCoss, M.J. Assessing the dynamic range and peak capacity of nanoflow LC-FAIMS-MS on an ion trap mass spectrometer for proteomics. *Anal. Chem.* 2008, *80*, 6888–6897.

83. Tang, K.; Li, F.; Shvartsburg, A.A.; Strittmatter, E.F.; Smith, R.D. Two-dimensional gas-phase separations coupled to mass spectrometry for analysis of complex mixtures. *Anal. Chem.* 2005, *77*, 6381–6388.

84. McSheehy, S.; Mester, Z. Arsenic speciation in marine certified reference materials Part 1. Identification of water-soluble arsenic species using multidimensional liquid chromatography combined with inductively coupled plasma, electrospray and electrospray high-field asymmetric waveform ion mobility spectrometry with mass spectrometric detection. *J. Anal. At. Spectrom.* 2004, *19*, 373–380.

85. Barnett, D.A.; Purves, R.W.; Guevremont, R. Isotope separation using high-field asymmetric waveform ion mobility spectrometry. *Nucl. Instrum. Methods Phys. Res., Sect. A* 2000, *450*, 179–185.

86. Ells, B.; Barnett, D.A.; Froese, K.; Purves, R.W.; Hrudey, S.; Guevremont, R. Detection of chlorinated and brominated byproducts of drinking water disinfection using electrospray ionization – High-field asymmetric waveform ion mobility spectrometry – mass spectrometry. *Anal. Chem.* 1999, *71*, 4747–4752.

87. Handy, R.; Barnett, D.A.; Purves, R.W.; Horlick, G.; Guevremont, R. Determination of nanomolar levels of perchlorate in water by ESI-FAIMS-MS. *J. Anal. At. Spectrom.* 2000, *15*, 907–911.

88. Ells, B.; Barnett, D.A.; Purves, R.W.; Guevremont, R. Trace level determination of perchlorate in water matrices and human urine using ESI-FAIMS-MS. *J. Environ. Monit.* 2000, *2*, 393–397.

89. Gabryelski, W.; Wu, F.; Froese, K.L. Comparison of high-field asymmetric waveform ion mobility spectrometry with GC methods in analysis of haloacetic acids in drinking water. *Anal. Chem.* 2003, *75*, 2478–2486.

90. Ells, B.; Froese, K.; Hrudey, S.E.; Purves, R.W.; Guevremont, R.; Barnett, D.A. Detection of microcystins using electrospray ionization high-field asymmetric waveform ion mobility mass spectrometry/mass spectrometry. *Rapid Commun. Mass Spectrom.* 2000, *14*, 1538–1542.

91. McCooeye, M.A.; Kolakowski, B.; Boison, J.; Mester, Z. Evaluation of high-field asymmetric waveform ion mobility spectrometry mass spectrometry for the analysis of the mycotoxin zearalenone. *Anal. Chim. Acta.* 2008, *627*, 112–116.

92. Sultan, J.; Gabryelski, W. Structural identification of highly polar nontarget contaminants in drinking water by ESI-FAIMS-Q-TOF-MS. *Anal. Chem.* 2006, *78*, 2905–2917.

93. Barnett, D.A.; Ells, B.; Guevremont, R.; Purves, R.W. Separation of leucine and isoleucine by electrospray ionization-high field asymmetric waveform ion mobility spectrometry – mass spectrometry. *J. Am. Soc. Mass Spectrom.* 1999, *10*, 1279–1284.

94. McCooeye, M.A.; Mester, Z.; Ells, B.; Barnett, D.A.; Purves, R.W.; Guevremont, R. Quantitation of amphetamine, methamphetamine, and their methylenedioxy derivatives in urine by solid-phase micro extraction coupled with electrospray ionization – high-field asymmetric waveform ion mobility spectrometry – mass spectrometry. *Anal. Chem.* 2002, *74*, 3071–3075.

95. Kapron, J.T.; Jemal, M.; Duncan, G.; Kolakowski, B.; Purves, R. Removal of metabolite interference during liquid chromatography/tandem mass spectrometry using high-field asymmetric waveform ion mobility spectrometry. *Rapid Commun. Mass Spectrom.* 2005, *19*, 1979–1983.

96. Kapron, J.; Wu, J.; Mauriala, T.; Clark, P.; Purves, R.W.; Bateman, K.P. Simultaneous analysis of prostanoids using liquid chromatography/high-field asymmetric waveform ion mobility spectrometry/tandem mass spectrometry. *Rapid Commun. Mass Spectrom.* 2006, *20*, 1504–1510.

97. Mahan, E.A.; King, R. The use of LC-APCI-FAIMS-MS for the quantitative analysis of taxol in mouse plasma. *Proc. 53rd ASMS Conference on Mass Spectrometry and Allied Topics,* San Antonio, TX, June 5–9, 2005.

98. Mehl, J.T.; Crathern, S.J.; King, R.C. Rapid quantitative analysis of acetaminophen in CYP 1A2 marker assays by FIA-FAIMS-MS. *Proc 53rd ASMS Conference on Mass Spectrometry and Allied Topics,* San Antonio, TX, June 5–9, 2005.

99. Hatsis, P.; Brockman, A.H.; Wu, J.T. Evaluation of high-field asymmetric waveform ion mobility spectrometry coupled to nanoelectrospray ionization for bioanalysis in drug discovery. *Rapid Commun. Mass Spectrom.* 2007, *21*, 2295–2300.

100. Champarnaud, E.; Laures, A.M.F.; Borman, P.J.; Chatfield, M.J.; Kapron, J.T.; Harrison, M.; Wolff, J.C. Trace level impurity method development with high-field asymmetric waveform ion mobility spectrometry; systematic study of factors affecting the performance. *Rapid Commun. Mass Spectrom.* 2009, *23*, 181–193.

101. Robinson, E.W.; Garcia, D.E.; Leib, R.D.; Williams, E.R. Enhanced mixture analysis of poly(ethylene glycol) using high-field asymmetric waveform ion mobility spectrometry combined with fourier transform ion cyclotron resonance mass spectrometry. *Anal. Chem.* 2006, *78*, 2190–2198.

102. Robinson, E.W.; Sellon, R.E.; Williams, E.R. Peak deconvolution in high-field asymmetric waveform ion mobility spectrometry (FAIMS) to characterize macromolecular conformations. *Int. J. Mass Spectrom.* 2007, *259*, 87–95.

103. Li, J.; Purves, R.W.; Richards, J.C. Coupling capillary electrophoresis and high-field asymmetric waveform ion mobility spectrometry mass spectrometry for the analysis of complex lipopolysaccharides. *Anal. Chem.* 2004, *76*, 4676–4683.

104. Guevremont, R.; Purves, R.W. High field asymmetric waveform ion mobility spectrometry-mass spectrometry: an investigation of Leucine enkephalin ions produced by electrospray ionization. *J. Am. Soc. Mass Spectrom.* 1999, *10*, 492–501.

105. Mie, A.; Jornten-Karlsson, M.; Axelsson, B.O.; Ray, A.; Reimann, C.T. Enantiomer separation of amino acids by complexation with chiral reference compounds and high-field asymmetric waveform ion mobility spectrometry: preliminary results and possible limitations. *Anal. Chem.* 2007, *79*, 2850–2858.

106. Mie, A.; Ray, A.; Axelsson, B.O.; Jornten-Karlsson, M.; Reimann, C.T. Terbutaline enantiomer separation and quantification by complexation and field asymmetric ion mobility spectrometry-tandem mass spectrometry. *Anal. Chem.* 2008, *80*, 4133–4140.

6 Ion Traps with Circular Geometries

Daniel E. Austin and Stephen A. Lammert

CONTENTS

6.1 Introduction 373
    6.1.1 The Problem of Space Charge and Ion Storage Capacity 374
6.2 Toroidal RF Ion Trap 376
    6.2.1 Full-Sized Toroidal Ion Trap: Conception and Theory 376
    6.2.2 Full-Sized Toroidal Ion Trap: Development and Performance 380
    6.2.3 Reduced-Size Toroidal RF Ion Trap 385
6.3 The Halo Ion Trap and Related Devices 388
    6.3.1 Non-Equipotential Electrodes 389
    6.3.2 The Halo Ion Trap 391
    6.3.3 Planar Quadrupole and Coaxial Ion Traps 393
6.4 Conclusions and Outlook 396
References 396

6.1 INTRODUCTION

By the mid-1990s ion trap mass spectrometers had become widely accepted in the mass spectrometry market due, in large part, to their inherent high sensitivity, capability of MS^n experiments, simplicity of construction, and lower instrument costs. By this era, many of the early limitations of ion traps had been overcome. Mass range had been extended well beyond the *m/z* 650 limit of the first Finnigan ITD systems, and demonstrations of *m/z*-range well beyond 100,000 Th had been shown. Concurrently, mass resolution had been improved and commercial instruments delivered routinely mass resolutions in excess of 100,000 (albeit only over short mass ranges). Other types of mass analyzers had higher performance in terms of mass accuracy and quantitation, but ion traps had improved dramatically in these areas.

Despite these advances, there remained a fundamental limitation to ion trap mass spectrometers in general, that of space charge. Ion traps can be viewed as 'ion flasks' and, as such, are capable of storing a limited number of ions before ion/ion repulsion eventually forces a ceiling on the number of ions that can be contained physically in the device. In addition, ions within a trap modify the electric field experienced by other ions, thereby degrading performance as a mass analyzer.

6.1.1　The Problem of Space Charge and Ion Storage Capacity

Dehmelt [1] developed a detailed model for treating space charge in ion traps based on his pseudopotential-well model. The ion population in a trap reaches a maximum when the Coulombic repulsion between ions is balanced exactly by the trapping potential. Following the derivation given by March and Todd [2], the maximum ion density in a trap, N_{max}, is given by

$$N_{max} = \frac{3}{64\pi}\frac{m\Omega^2}{e^2}q_z^2 \tag{6.1},$$

where m is the mass of the ion, e is the electronic charge, Ω is the frequency of the RF drive voltage, and q_z is the stability parameter for ion motion in the axial (z) direction as described by the Mathieu equation.

This 'maximum storage limit' is far in excess of other, more analytically significant limits to ion density such as the ion isolation space-charge limit and the mass analysis storage limits.* Ghost peaks, mass shifts, and peak broadening may all occur when charge densities in traps are too high [3]. For high-resolution analysis, the maximum permissible ion density is the lowest of all applications [4]. In addition to space-charge effects, high ion or sample concentrations also cause problems through the occurrence of ion/molecule reactions in traps. Although methods such as automatic gain control (AGC) have been developed to address this limitation [5], ultimately the small trapping volume and limited number of ions analyzed result in reduced sensitivity.

One obvious method to increase the ion storage capacity of a 3D ion trap would be to simply increase the size of the device. Practical limitations derived from the fundamental ion trap equations, however, control the maximum size of the ion trap. Ion motion in traps is described by the Mathieu equation [2], which has been developed fully in a previous chapter. For a given device and operation parameters, the solutions to the Mathieu equation predict whether an ion will have a stable and bounded trajectory in a particular trapping device. These solutions can be summarized and plotted in terms of the two transformation parameters, a_u and q_u (where the subscript 'u' denotes either the radial (r) dimension, or the axial (z) dimension) that relate the operating parameters and ion characteristics to stability in a given dimension. Shown below are the 'a' and 'q' transformation expressions for a and q for the axial (z) dimension of a traditional 3D ion trap

$$a_z = -\frac{16eU}{m(r_0^2 + 2z_0^2)\Omega^2} \tag{6.2},$$

$$q_z = \frac{8eV}{m(r_0^2 + 2z_0^2)\Omega^2} \tag{6.3}.$$

---

* See Volume 4, Chapter 13: An Examination of the Physics of the High-Capacity Trap (HCT), by Desmond A. Kaplan, Ralf Hartmer, Andreas, Brekenfeld, Jochen Franzen, and Michael Schubert.

In the above equations, m is the ion mass (kg molecule^{-1}), r_0 (m) is the inner radius of the ring electrode and $2z_0$ (m) is the separation of the inner surfaces of the end-cap electrodes in the axial dimension of the device, $+U$ is a DC voltage applied to the ring electrode in combination with V, which is the zero-to-peak amplitude of the RF potential (kg m^2 s^{-2} C^{-1}), Ω is the applied RF frequency (radians s^{-1}), and e (C) is the ion charge. As can be seen in the two equations, the stability parameter a is directly proportional to the applied DC voltage (U, volts) while the stability parameter q is directly proportional to the applied RF voltage (V, volts, zero-to-peak). The set of stable values for both a and q is shown graphically in the familiar stability diagram given in Chapter 1 of this volume.*

As can be seen from the squared relationship between r_0 and V in Equation 6.3, in order to hold q_z constant even a small increase in the radius must be offset with much larger increases in the amplitude of the RF operating voltage for a constant value of Ω. Current commercially available ion trap mass spectrometers with r_0 of the order of 1 cm operate with RF voltages as high as 15 kV$_{p-p}$. Given the magnitude of these voltages, there is little headroom left as higher RF voltages will approach the break-down voltage for insulators and other components (e.g., electrical feedthroughs) in the vacuum system. Limited by the Mathieu equation, the paradox is how to increase the ion storage capacity of the ion trap without increasing the size of the trap.

Space charge also limits efforts to develop miniature versions of ion traps for portable mass spectrometers [6,7]. Although the voltage-size relationship of Equation 6.3 works in favor of smaller, portable instruments, two other issues arise. Firstly, smaller ion traps have smaller trapping volumes and, therefore, can contain fewer ions. For instance, going from a trap with $r_0 = 1$ cm to one having $r_0 = 1$ mm results in a 1000-fold decrease in trapping volume and, for a given ion density, a similar decrease in the number of ions. Secondly, it is generally desirable to operate smaller ion traps at lower voltages, resulting in a lower tolerable charge density within the trap. Thus the ion capacity in small traps is limited by reductions in both the ion density and the trapping volume. In the extreme case of micron-sized cylindrical ion traps operated at low voltages (8 V), simulations have shown a space-charge limit of a single ion [8].

Several methods have been explored to increase ion populations in ion trap mass spectrometry. Newer trap geometries, such as the linear and rectilinear ion traps, extend the trapping volume in one dimension [9–11], thereby increasing the number of ions that can be stored and analyzed. The space-charge problem has been addressed also by using arrays of smaller ion traps [12–15]. Arrays of traps recover the loss of sensitivity produced from small ion populations in miniature traps. One limitation to arrayed devices is that physical variations among the devices within an array will result in reduced mass resolution unless each trap has its own detector. Good performance from an array of traps requires a high degree of uniformity among individual traps.†

---

* Volume 4, Chapter 1: An Appreciation and Historical survey of mass spectrometry (Appendix), by Raymond E. March and John F.J. Todd.

† See Volume 4, Chapter 2: Ion Traps for Miniature, Multiplexed and Soft Landing Technologies, by Scott A. Smith, Christopher C. Mulligan, Qingyu Song, Robert J. Noll, R. Graham Cooks and Zheng Ouyang.

6.2   TOROIDAL RF ION TRAP

6.2.1   Full-Sized Toroidal Ion Trap: Conception and Theory

At the 47th Pittsburgh Conference (Chicago, IL, 1996) Ferran Scientific of San Diego, CA, displayed an arrayed quadrupole residual gas mass analyzer [16]. In contrast to a conventional 2D quadrupole mass filter with four rods, the Ferran version contained 16 rods arranged in a 4×4 array, creating nine parallel mass analysis paths as opposed to one. The radial dimensions of this device were considerably smaller than a typical laboratory-scale quadrupole and, as such, the operating RF voltage was much lower. One of the authors (Lammert) noted with some curiosity this new form of an old technology and, upon returning to Oak Ridge National Laboratory, discussed it with his colleague, Marc Wise. Remembering that a simplified 'description' of a 3D ion trap was similar to what would be obtained if a 2D quadrupole cross-section were rotated about its central axis, the two of them embarked on a paper exercise of rotating various forms of this device in space. The results were sometimes whimsical, and often un-buildable or impractical, but one version stood out. It was observed that if, rather than rotating the device on a central axis, one rotated the device on an edge axis, the result was a series of stacked or concentric circular volumes. When this same logic was applied to the conventional 3D ion trap cross-section, the toroidal ion trap mass analyzer (Figure 6.1) was born. It was obviously immediately that the device had a larger volume, yet the main dimension which governed the operation of the ion trap (r_0) had not changed! Another advantage of this arrangement was that there was only a single ion storage region as opposed to an array of trap regions, each of which must be matched exactly and coupled equivalently to ionization and detection optics. This single storage region had no entrance or exit apertures (as in linear ion traps), where field perturbations are inevitable. The ion should experience essentially the same field everywhere within the ion storage volume. Finally, the toroid is a compact, spatially efficient geometry that can couple well with other ion optic elements, most notably detectors.

The toroidal ion trap, based on an off-central-axis rotation of a 3D quadrupole ion trap cross-section, is similar in nature to a device described by Bier and Syka in their patent [17] on linear ion traps. In the development of their linear ion trap concept, Bier and Syka introduced curvature as a means to converge a wide, spatially dispersed ejected ion population to a point in space where physically more compact detectors could be employed. One of the conceptual embodiments in their patent shows this curvature reaching its logical endpoint where both ends of a 2D ion trap are brought together in a circular geometry. Wise and Lammert did not become aware of this conceptual device until after assembly of the first version of a toroid mass spectrometer had begun. If one considered an infinitely long linear quadrupole curved back upon itself in a circle, the disruption to the essentially quadrupolar field would be negligible. As the annular radius of curvature (R) becomes finite, the non-linear distortions to the quadrupolar field become more and more pronounced. Without correction these deleterious non-linear fields are dominant, and both trapping and mass analysis performance will suffer in any device of a practical size (that is $R/r_0 < ca$ 10).

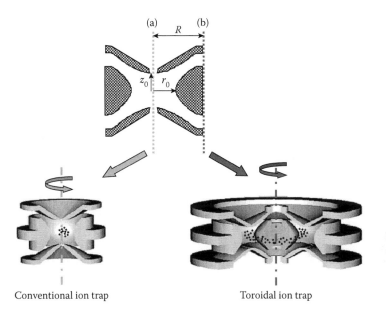

FIGURE 6.1 Rotation of a conventional 3D ion trap electrode cross-section generates the electrode geometries as shown. Rotation about axis (a) gives the conventional 3D ion trap device with two end-cap electrodes and a ring electrode. Rotation about axis (b) generates a toroidal trapping field, but with the same cross-sectional dimensions as the conventional 3D ion trap. The toroidal device has two end-cap electrodes together with two ring electrodes (outer and inner) that form the trapping field. The radial dimension (r_0) and axial dimension (z_0) define the distance from the ring and end-cap electrodes, respectively, to the minimum of the potential field. Further annotations show the definition of R, the radius of curvature from the central axis for the toroidal geometry. (Adapted from Lammert, S.A., Plass, W.R., Thompson, C.V., and Wise, M.B., *Int. J. Mass Spectrom.* 2001, *212*, 25–40. With permission from Elsevier.)

Curved quadrupolar devices were not new at this stage of development. As far back as 1969, Church [18] demonstrated his 'racetrack' version of an ion storage ring, which employed a curved quadrupole rod set with a 2-inch radius of curvature. This device, however, was used solely for ion storage and not mass analysis. In the mid-1980s, Finnigan-MAT introduced the TSQ-70 triple quadrupole mass spectrometer, in which the center quadrupole rod set employed for collision-induced dissociation was curved slightly [19] in order to prevent line-of-sight access to the detector by neutrals from desorption ion sources. In this case, the RF-only quadrupole, Q2, was a transmission device that neither stored nor mass-analyzed ions. The first curved mass analysis quadrupoles [20] were introduced by Steiner at Bear Instruments (now part of Varian Inc., Walnut Creek, CA). The original triple quadrupole instrument employed all three quadrupole units in a continuous arc of *ca* 270° from source to detector. Later versions of this approach still appear in current Varian Inc. triple quadrupole products, although the mass analysis quadrupoles (Q1 and Q3) are now linear in geometry. Again, as in the TSQ-70, the curved quadrupole devices are used in the traditional transmission mode.

Dawson and others have examined extensively the theory of quadrupolar devices, especially as it relates to the two most common versions of these devices, the 2D linear quadrupole mass filter and the 3D quadrupole ion trap [21]. Since the toroidal geometry is new, a brief summary for these devices follows.

The electric potential for any ion trap with cylindrical symmetry can be expressed as a multipole expansion, which in spherical coordinates (ρ, θ, ϕ) has the form [22]

$$\Phi(\rho,\theta,\phi,t) = \Phi_0(t) \sum_{l=0}^{\infty} A_l \left(\frac{\rho}{r_N}\right)^l P_l(\cos\theta) \qquad (6.4),$$

where Φ_0 is the electric potential applied to the ring electrode (end-cap electrodes grounded), r_N is a normalization radius (related to the trap dimensions), A_l is the expansion coefficient of order l, and $P_l(\cos\theta)$ is the Legendre polynomial of order l. The monopolar ($l = 1$) and all odd higher-order terms for l can be ignored for ion traps with reflection symmetry with respect to the central ($z = 0$) plane. When $l = 2$, and all other even higher-order terms are set to zero, a quadrupolar field is produced, and the electric fields in the r- and z-dimensions are linear. The potential of a quadrupolar field expressed in cylindrical coordinates (r, ϕ, z) is

$$\Phi(r,\phi,z,t) = \Phi_0(t)\frac{A_2}{r_N^2}\left(z^2 - \frac{1}{2}r^2\right) \qquad (6.5).$$

The presence of even, higher-order terms ($l = 4,6,8...$) gives rise to the addition of non-linear fields.

For the discussion of toroidal ion traps it is convenient to use cylindrical coordinates (r, ϕ, z) with some additional parameters (see Figure 6.2). The dimensional constant, R, is the major radius of the toroid, and represents the distance from the z-axis to the potential minimum in the toroidal trapping volume. The variable r represents the distance between any point and the z-axis. Note that this gives r a different meaning than in conventional ion traps, in which r is the displacement from the trapping center. At the center of the trapping volume, $r = R$. The value of r_N, analogous to r_0 in conventional ion traps, is the distance from the center of the trapping volume to the nearest electrode surface in the r-direction. The radial component of distance between any point and the center of the trapping volume, designated s, is given by $s = r - R$. The displacement of any point from the potential minimum is thus given by the values of z and s. Note that s is positive for points farther out from the z-axis than R, and negative for points that lie within the distance R.

For a quadratic potential well whose minimum falls on a circle with radius R, the form of the potential is given by

$$\Phi(r,\phi,z) = \lambda(r - R)^2 + \mu z^2 \qquad (6.6),$$

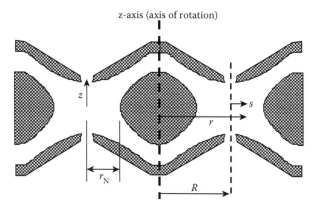

FIGURE 6.2 Coordinate system used in discussion of toroidal ion trap. The dimension R is the distance from the rotational axis to the minimum of the trapping potential. The variable s designates the radial distance from the minimum of the trapping potential. Note that s can be either positive or negative.

where λ and μ are arbitrary constants. The Laplace equation expressed in cylindrical coordinates,

$$\nabla^2\Phi_{r,\phi,z} = \frac{1}{r}\frac{\partial\Phi}{\partial r}\left(r\frac{\partial\Phi}{\partial r}\right) + \frac{1}{r^2}\frac{\partial^2\Phi}{\partial\theta^2} + \frac{\partial^2\Phi}{\partial z^2} = 0 \tag{6.7},$$

can be applied to Equation 6.6 to give

$$\nabla^2\Phi = 2\lambda(2-\frac{R}{r}) + 2\mu = 2\lambda\left(2-\frac{R}{R+S}\right) + 2\mu = 0 \tag{6.8}.$$

The implication of Equation 6.8 is that the potential distribution can be generated by electric fields for two cases only, namely

$$R = 0 \text{ (quadrupole ion trap)} \tag{6.9},$$

and

$$R \to \infty \text{ (linear quadrupole)} \tag{6.10}.$$

From this consideration it can be seen that a toroidal ion trap with purely quadrupolar (linear) fields is not possible. However, when s is small compared to R (that is, the ion excursion is small compared with the major toroidal radius), then the electric field in the trapping region is approximately linear. In this case, Equation 6.8 reduces to

$$\lambda + \mu = 0 \tag{6.11},$$

and Equation (6.5) can be rewritten as

$$\Phi(s,z,t) = \Phi_0(t)\frac{A_2}{r_N^2}(z^2 - s^2) \tag{6.12},$$

where Φ_0 is the potential on the ring and the central (spindle) electrode, and A_2 is a dimensionless coefficient. The respective fields in cylindrical coordinates are then

$$E_s = -2\Phi_0(t)\frac{A_2}{r_N^2}s \tag{6.13}$$

and

$$E_z = 2\Phi_0(t)\frac{A_2}{r_N^2}z \tag{6.14}.$$

Similar to linear quadrupoles, these fields are of the same absolute value and differ only in sign. In this respect, the toroidal ion trap is more like a 2D quadrupole than a 3D ion trap.

6.2.2 FULL-SIZED TOROIDAL ION TRAP: DEVELOPMENT AND PERFORMANCE

Once funds had been secured for the project from the United States Department of Energy, fabrication of a device began in earnest, and by time of the American Society for Mass Spectrometry Conference in 1998, a first generation toroidal ion trap mass spectrometer had been introduced. This device (now referred to as the symmetrical toroidal RF ion trap) was based on a toroidal configuration using a traditional 3D ion trap cross-section with an r_0-value r_N of 1 cm and a curvature of radius (R) of 3 cm (Figure 6.3). A commercial Finnigan-MAT ITMS mass spectrometer system was equipped with a modified vacuum chamber, and the original electronics, data acquisition and software package were used with only minor modifications. As the value of r_0 for the new device was the same as that for the commercial version, the only necessary modifications to the RF power supply were a change in operating frequency and re-tapping of the primary coil in the RF transformer because the capacitance of the toroidal ion trap mass analyzer was larger than that of the commercial 3D ion trap. The conventional RF amplitude scan with axial modulation resonance ejection was employed. The commercial ion source was modified to fit over one portion of the annular slit in the top of the toroidal mass analyzer. A custom-built multi-channel plate detector was fabricated, so that it had a diameter sufficient to sample the entire 6+ cm diameter array of the exit slits.

The initial performance (sensitivity, mass resolution, peak shapes, and robustness) for this device was poor. As can be seen in the mass spectrum for n-butylbenzene (Figure 6.4), the peaks at m/z 91 and m/z 92 are unresolved, and the shape and intensity of the mass peaks were strongly dependent on the ejection q_z value. At some values of q_z symmetrical peaks were obtained, while at other values the mass peaks were split. At some ejection q_z values, the mass spectrum disappeared altogether.

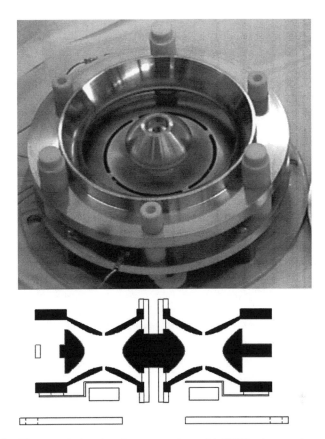

FIGURE 6.3 Photograph showing the symmetric toroidal RF ion trap analyzer with the top end-cap electrode removed. Below the photograph is a cross-section schematic of the device showing the shape of the two ring electrodes and the two end-cap electrodes. (Reproduced from Lammert, S.A., Plass, W.R., Thompson, C.V., and Wise, M.B., *Int. J. Mass Spectrom.* 2001, *212*, 25–40. With permission from Elsevier.)

After some time, it became clear that the trapping field imperfections, introduced by rotating the 3D ion trap quadrupole cross-section on a spatially different axis, were degrading the performance of the device. Therefore these fields, presumed to be even, higher-order non-linear fields, would have to be addressed.

At the time, deriving an analytical solution to the required shape for the four separate toroid ion trap electrodes in order to create an essentially quadrupolar field seemed to be a daunting task. What was known, however, was that in order to have an essentially quadrupolar field, the potential inside various devices could be analyzed using various field simulation programs and, in order for the trapping region to be quadrupolar, the fields had to be linear. In the 1980s, Finnigan-MAT encountered a similar situation with the early ion traps [23]. Mass shifts were observed for ions of the same nominal mass, but different chemical composition. Even-order non-linear fields were introduced into the ion trap by the presence of the exit holes. These non-linear fields were, in fact, sub-linear in that the field potential fell off negatively

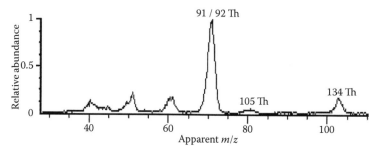

FIGURE 6.4 Mass spectrum of *n*-butylbenzene; note the unresolved peaks at *m/z* 91/92. The mass scale is uncalibrated due to differences in the RF frequency between the commercial ion trap electronics and the modified electronics for the toroidal version. (Reproduced from Lammert, S.A., Plass, W.R., Thompson, C.V., and Wise, M.B., *Int. J. Mass Spectrom.*, 2001, *212*, 25–40. With permission from Elsevier.)

from a linear trend with increasing distance from the trap center. Finnigan solved this problem by stretching the axial separation of the end-cap electrodes by approximately 11% until the trapping field was slightly super-linear. Bruker addressed the same issue by changing the angle of the asymptotes of the hyperbolic surfaces that comprise the trap electrodes [22]. With the help of Wolfgang Plass (at that time, a graduate student in the Cooks' group at Purdue University), an effort was undertaken to 'correct' the fields in the toroidal ion trap. The variables that were available were the separation of the end-cap electrodes ($2z_0$), the asymptote angle, and the entrance/exit slit widths. The original toroidal ion trap (just like the early 3D ion traps) had a sub-linear trapping field. Changes were made to a simulated toroidal cross-section in the form of increased end-cap electrode separation, as well as increasing the asymptote angle for the center ring electrode and decreasing the asymptote angle for the outer ring electrode. For each of these candidate geometries, equipotential lines were derived using the POISSON software (Los Alamos National Laboratory, Los Alamos, NM), and the resulting fields were evaluated and plotted as a function of both radial distance (*r*) and axial distance (*z*). This process was repeated iteratively until the axial fields were linear across the entire trapping region. Then, a slight super-linear non-linear field was added to mimic the Finnigan ion trap systems. In this exercise, only the axial fields were optimized because the quality of these fields is the more important to the mass ejection process during the analytical scan. In this early stage of the optimization exercise, no attempt was made to correct or to optimize the radial fields. That effort remains to be undertaken. As discussed above, ultimately, a fully corrected toroidal ion trap field would look, in large part, like that of a linear quadrupole (rather than a quadrupole ion trap), with a small axial non-linear component added. The closer one gets to that goal, the more similar the axial and radial frequencies of ion motion will become. This similarity of axial and radial frequencies poses the potential problem that as ions are excited resonantly in the axial dimension for the mass ejection portion of an analytical scan, the small, but present residual non-linear fields will couple some portion of that motion into the radial dimension and thus degrade the ejection profile. This effect was observed during the optimization process using a modified version of the ion trap trajectory

(a) (b)

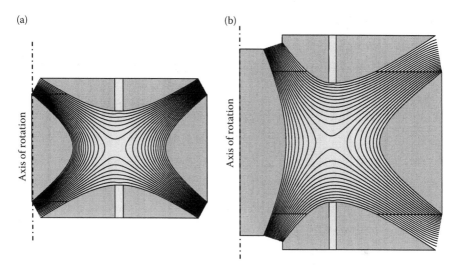

FIGURE 6.5 Comparison of the trapping fields for the (a) symmetric toroidal RF ion trap; (b) the field-corrected asymmetric toroidal RF ion trap. Note that the inner ring electrode on the left in (b) has a much steeper hyperbolic asymptote than the outer ring electrode on the right in (b). Each half of the end-cap electrodes' asymptotes corresponds to the accompanying ring electrode. (Reproduced from Lammert, S.A., Plass, W.R., Thompson, C.V., and Wise, M.B., *Int. J. Mass Spectrom.*, 2001, *212*, 25–40. With permission from Elsevier.)

simulation program ITSIM (Purdue University, West Lafayette, IN). Adding a small DC quadrupolar component (altering the Mathieu a-parameter [2]) will move the operation line in the stability diagram away from the $a = 0$ axis and will increase the difference between the axial and radial secular frequencies; in essence, placing the radial frequency off-resonance. An improvement in mass resolution under these conditions was predicted by the simulations and later observed experimentally.

Based on the results of the field optimization exercise, a new set of toroidal RF ion trap electrodes was fabricated. This corrected analyzer (dubbed the 'asymmetrical' toroidal RF ion trap owing to the different curvature between the inner and outer ring electrode surfaces) is shown schematically in Figure 6.5b as compared to the original symmetrical version shown in Figure 6.5a. A photograph of the new design is shown in Figure 6.6. Results obtained using this new geometry confirmed the simulation predictions and demonstrated unit mass resolution for the m/z 91/92 pair (Figure 6.7). In addition, the strong dependence of peak shape on the mass ejection q_z-value was eliminated. The first paper on toroidal ion traps [24] discussed in detail the design, performance, theory, and optimization of both the symmetric and asymmetric versions.

In commercial ion trap instruments, trapped ions are localized at the center of the trap to a small volume roughly 1.0 mm in radius [25], corresponding to a volume of *ca* 0.5 mm³. Assuming that ions in the toroidal ion trap occupy a toroidal volume with width of 1.0 mm and a major radius (R) of 3 cm, the corresponding trapping volume is 148 mm³. Accordingly, this full-size toroidal ion trap, with $r_0 = 1$ cm, has a significantly larger trapping volume than a standard 3D quadrupole ion trap with $r_0 = 1$ cm.

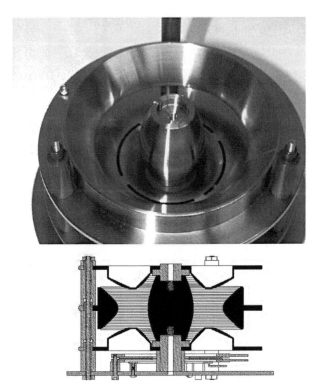

FIGURE 6.6 Photograph showing the asymmetric toroidal RF ion trap analyzer with the top end-cap electrode removed. Below the photograph is a cross-section schematic of the device showing the shape of the two ring electrodes and the two end-cap electrodes. (Reproduced from Lammert, S.A., Plass, W.R., Thompson, C.V., and Wise, M.B., *Int. J. Mass Spectrom.* 2001, *212*, 25–40. With permission from Elsevier.)

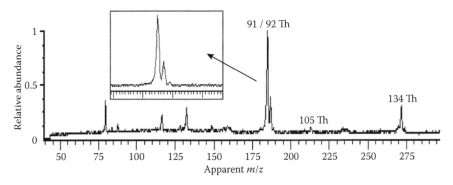

FIGURE 6.7 Mass spectrum of *n*-butylbenzene. The inset shows the improved mass resolution for the *m/z* 91/92 pair. As in Figure 6.3, the mass scale is uncalibrated. (Adapted from Lammert, S.A., Plass, W.R., Thompson, C.V., and Wise, M.B., *Int. J. Mass Spectrom.* 2001, *212*, 25–40. With permission from Elsevier.)

6.2.3 Reduced-Size Toroidal RF Ion Trap

One of the more compelling features of the toroidal RF ion trap is the possibility to trade off the increased trapping volume for a smaller analyzer assembly. As was discussed above, the squared relationship between the RF drive voltage (V) and the radius of the mass analyzer (r_0) as shown in Equation 6.3 implies that a small change in radius will reduce the necessary RF drive voltage. Commercial ion trap RF power supplies are typically large, high-power electronic assemblies and usually require a sizable coil as part of the tuned RF tank circuit. In addition, the reduced analyzer dimensions yield a shorter ion path length that should reduce the already modest vacuum requirements of ion traps even further, permitting the use of even smaller turbo-molecular pumps.

A 1/5-scale version of the asymmetric RF ion trap (Figure 6.8) has been fabricated which has an ion trapping volume comparable to commercial, 1 cm radius 3D ion traps. The miniature toroidal RF ion trap analyzer [26] (Figure 6.9) is utilized in a portable gas chromatograph/mass spectrometer (GC/MS) ion trap system commercialized by Torion Technologies, Inc. (American Fork, UT). The Guardion-7™ is a completely self-contained portable GC/MS system [27] weighing 13 kg and comprising a volume of approximately 30,500 cm³ (47 cm × 36 cm × 18 cm). The size and weight specifications include on-board computer data acquisition and processing as well as the batteries and all utilities (helium gas supply) sufficient for four hours of operation (approximately 50 samples). The 0.2-cm-value of r_0 of this miniature RF toroidal ion trap permits RF voltages well below 1.5 kV$_{p-p}$; a level that allows the use of direct-drive RF circuits in place of the larger transformer versions. The use of direct-drive RF circuits together with heating zone design efficiencies allow the system to operate on less than 80 W. A photograph of the complete system is shown

FIGURE 6.8 Photograph of the miniature, 1/5th-scale toroidal RF ion trap with the top endcap electrode removed. A U.S. quarter is shown for scale. (Reproduced from Lammert, S.A., Rockwood, A.A., Wang, M., Lee, M.L., Lee, E.D., Tolley, S.E., Oliphant, J.R., Jones, J.L., and Waite, R.W., *J. Am. Soc. Mass Spectrom.* 2006, *17*, 916–922. With permission from Elsevier.)

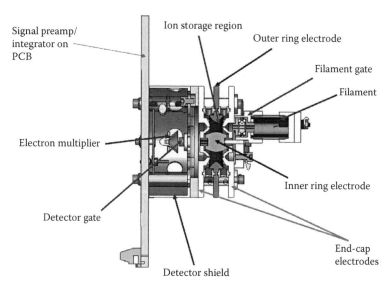

FIGURE 6.9   Schematic of the major components of the miniature toroidal RF ion trap mass analyzer.

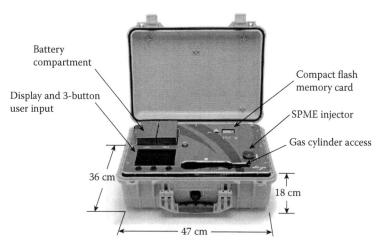

FIGURE 6.10   The Guardion-7, highlighting the major visible system components. The entire system, including batteries and utilities weighs 13 kg, consumes < 80 W of power, and occupies a volume of only 47 cm × 36 cm × 18 cm.

in Figure 6.10, and the primary operational figures of merit for the ion trap are summarized in Table 6.1.

The analytical performance of this miniature toroidal RF ion trap has been demonstrated to be comparable to commercial ion trap mass analyzers. Better than unit mass resolution is obtained, as is shown for the toluene molecular ion

TABLE 6.1
Asmmetric Toroidal RF Ion Trap Analyzer Operating Parameters

Analyzer Radius (r_0)	2.0 mm
RF Voltage	600 V_{0-p}
RF Frequency	4 MHz
Mass/Charge Range	45–500 Th
Analytical Scan Type	Frequency Scan (1800 kHz–110 kHz)
Scan segment time	60 ms
Peak Width, FWHM	0.25 Th at m/z 92

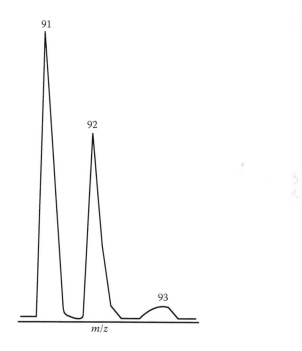

FIGURE 6.11 The molecular ion region from a mass spectrum of toluene showing better than unit mass resolution.

region spectrum in Figure 6.11. Figure 6.12 reveals good agreement between the acquired full scan mass spectrum for the chemical warfare agent VX (a dangerous nerve gas) and the corresponding reference NIST (National Institute of Standards and Technology) library spectrum. Finally, a system detection limit of 250 pg (*S/N* >3) has been demonstrated by solid phase micro extraction (SPME) direct injection after depositing a known volume of a quantitative standard solution of dodecane directly onto the SPME fiber.

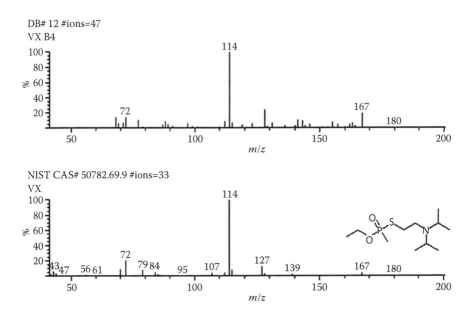

FIGURE 6.12   Comparison of the (top) mass spectrum for the chemical warfare agent VX acquired from the Guardion-7 compared to the (bottom) NIST reference mass spectrum.

Several challenges accompany the miniaturization of mass analyzers, including ion traps. Errors in the hyperbolic surfaces due to machining tolerances (typically of the order of ± 0.0005 inches) are more important as the analyzer size is reduced. In order to maintain an exact scaling of the original toroidal analyzer, entrance and exit slits with widths of the order of 0.010 inches must be cut into the end-cap electrodes using an electrical discharge machining (EDM) technique. These dimensions are near the limits of the technology, and produce surfaces with considerably more roughness than conventional machining. Since the ion trap analyzer is used as a part of a GC/MS instrument, the reduced dimensions of the toroid analyzer give rise to gas conductance and pumping considerations that are not as important in the larger version. Insulating spacers in the analyzer must not only have high tolerances, but must allow gas conductance out of the trap cavity. Finally, even surface roughness tolerances (machining marks, etc.) take on added significance at this scale. While these obstacles have been overcome to a large part in the 1/5th scale toroidal trap, further reductions in analyzer dimensions using this approach will become more problematic.

6.3   THE HALO ION TRAP AND RELATED DEVICES

As mass analyzer miniaturization efforts push the limits of conventional machining, many groups are turning to microfabrication methods to produce smaller ion traps. The cylindrical ion trap geometry is the simplest to produce, and arrays of microfabricated cylindrical ion traps have been reported by Pau [28], Blain [15], and van Amerom [29]. Microfabrication methods have also been used to produce time-of-flight mass analyzers, such as those produced by Muller's group [30] and by Verbeck [31]. Microfabrication alleviates some of the difficulties of miniaturization using conventional machining, particularly issues with fabrication tolerances and surface roughness. However, as noted above, space charge becomes an even greater obstacle to instrument performance for small ion traps.

A variation on the toroidal ion trap, called the halo ion trap [32], combines microfabrication technology with the large ion trapping capacity of the toroidal geometry. The halo ion trap utilizes also a new approach to ion trap fabrication in which electric fields are made using planar non-equipotential surfaces. The halo trap uses two planar ceramic substrates, the facing surfaces of which have been imprinted with closely spaced concentric metal rings, and overlaid with a thin resistive layer of germanium. A toroidal quadrupolar trapping field is produced by applying a specific potential function to the set of rings on each plate. In the resulting trap, the electric fields are independent of the substrate geometry, and are defined entirely by the choice of potentials on each metal ring and the resulting potential function on the resistive layer. With this approach, it is possible to produce various trapping geometries and both to optimize and to modify electric fields in ways not possible with conventional traps made using shaped metal electrodes.

6.3.1 NON-EQUIPOTENTIAL ELECTRODES

Electric fields in traditional ion traps are established using metal electrodes that define equipotential (isopotential) boundary conditions in the Laplace equation. The trapping fields made from metal electrodes are constrained by the electrode geometry. Different trap geometries (for example, the hyperbolic Paul trap, cylindrical, toroidal, rectilinear, etc.) are made using electrodes of corresponding shape and spatial arrangement. Curved electrode surfaces must be carefully designed, machined, and assembled for successful trap operation. However, even with accurately designed electrodes, the necessary physical truncation of the electrodes results in imperfect fields; furthermore, it is not readily possible to vary individual field components independently (*e.g.*, quadrupole, octopole, dodecapole). The difficulties of accurate electrode machining and alignment become even more limiting as the size of the analyzer is reduced, most notably for field-portable and miniature mass spectrometers.

Electric fields for ion trapping and mass analysis can be established also by non-equipotential surfaces. In the absence of space charge, an electric field must obey the Laplace equation,

$$\nabla^2 \Phi = 0 \tag{6.15}$$

Surfaces with defined potentials act as boundary conditions, and, with the Laplace equation, define the specific shape and magnitude of the electric field everywhere in space. Boundary conditions need not be equipotential surfaces, such as metal electrodes, but can be surfaces on which the potential varies over one or more spatial dimensions. Consider the example shown in Figure 6.13. At the left, hyperbolic metal electrodes produce the familiar quadrupolar potential functions of the Paul trap. On the right, an arbitrary closed surface, in this case a cylinder, is defined within the

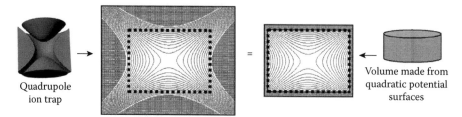

FIGURE 6.13 Quadratic potential functions on two planes and a cylinder produce a quadrupolar potential distribution identical to that in a trap made using hyperbolic metal electrodes. (Reproduced from Austin, D.E., Peng, Y., Hansen, B.J., Miller, I.W., Rockwood, A., Hawkins, A.R., and Tolley, S.E., *J. Am. Soc. Mass Spectrom.*, 2008, *19*, 1435–1441. With permission of Elsevier.)

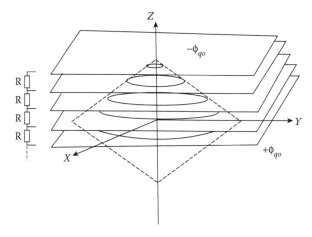

FIGURE 6.14 Realization of the electrode structure by densely spaced equidistant metallic sheets with circular holes to form the inner surface of the cone. The electrodes are connected by a network of resistors each with resistivity R. (Reproduced from Wang, Y. and Wanzcek, K.P., *J. Chem. Phys.*, 1993, *98*, 2647–2652. With permission of the American Institute of Physics.)

quadrupole ion trap. The potential at every point on this surface is chosen to be equal to the potential in a quadrupolar potential distribution, such that

$$\Phi = A(r^2 - 2z^2) \tag{6.16}.$$

The resulting potential distribution within such a volume will be quadrupolar at every point. The geometry of this surface can be chosen based on convenience or practical considerations. Whereas electrode truncation produces higher-order multipoles in the trap at the left, the potentials in the idealized surface at the right can be perfectly quadrupolar, or can be made to match the fields of the truncated trap, depending on the potential function applied to the closed surface.

An ion trap made of non-equipotential surfaces was first proposed by Wang and Wanczek [33]. In their device, a Paul-type ion trap could be made using two conical arrangements of holes in stacked metal plates (Figure 6.14). Identical resistors between adjacent plates provide the linear potential function that, for conical surfaces, produces a quadrupolar field within the trap. Although the potential function near the metal plates is jagged due to the discrete nature of the plates, the field quickly averages out to a smooth function toward the middle of the trap. Wang and Wanczek extended further this idea to a similar trap made using two thick sheets of resistive material. Conical holes drilled into two such sheets could be assembled into a quadrupole ion trap. Since such a device would be completely enclosed, they proposed detecting ions using Fourier transformation of the image current signal arising from the ions' secular motion. Due to the principle of superposition, it is possible in theory to produce simultaneously exact quadrupolar and dipolar fields within this trap. In a later paper, Wang *et al.* used the same principle to design and test an ion guide using four planar resistive sheets [34]. Each of the four planar resistive sheets had a linear potential gradient along the surface, and the square arrangement of these sheets resulted in a quadrupole potential distribution within the device.

Using non-equipotential boundary conditions to establish electric fields releases the fundamental constraint between field geometry and electrode geometry. Electrodes with simple planar geometry can produce quadrupolar potentials if the potential function on the planar electrode is quadratic. Similarly, planar electrodes can produce toroidal or other trapping fields with appropriate choice of potential function along the electrode surfaces. Closed, idealized surfaces can be replaced with a more open electrode arrangement as long as the resulting edge effects are far from the trapping and analyzing regions.

6.3.2 The Halo Ion Trap

In the halo ion trap [32], toroidal electric fields are made using an arrangement of two planar electrodes (Figure 6.15). Each electrode is imprinted with 15 gold rings, 0.1 mm wide and ranging in radius (in 0.5-mm increments) from 5 to 12 mm. A capacitive voltage divider is used to control the amplitude of the RF voltage applied to each ring electrode. These amplitudes are chosen so as to produce fields resembling those of the toroidal ion trap. A thin layer of a semiconductor (germanium), deposited on the plates, acts to prevent charge buildup and to smooth out the potential applied by the ring electrodes. Around the exterior of the trapping region, between the plates, is a grounded metal spacer, 4 mm wide, to define further the electric field. The fields in the trap are thus the result of the potential distribution on the germanium layer and the spacer. Holes in the spacer are used to introduce an ionizing electron beam into the trap, and for sample introduction and pressure measurement. As with the toroidal ion trap, ions are trapped and cooled into a ring. The performance of the halo ion trap was demonstrated using a resonant ejection scan in which ions were ejected radially toward the trap center and out axially into an electron multiplier detector. (Figure 6.15c–d). The RF amplitude used in these experiments was 650 V_{p-p}. A representative mass spectrum of dichloromethane taken using the halo ion trap is shown in Figure 6.16.

Electric fields and ion behavior in the halo ion trap have been simulated using SIMION 7.0 (Scientific Instrument Services, Ringoes, NJ) ion trajectory software. Resistive materials can be modeled by creating electrodes with the same potential

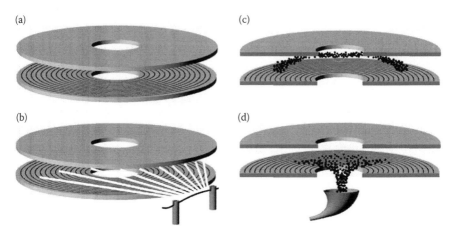

FIGURE 6.15 (a) The Halo IT mass analyzer consists of two parallel ceramic plates; the inside surfaces of each are patterned with metal ring electrodes and overlaid with germanium; (b) *in situ* electron ionization; (c) cut-away showing ion trapping between plates; (d) ion ejection into an electron multiplier. (Detector position and size are not to scale) (Reproduced from Austin, D.E., Wang, M., Tolley, S.E., Maas, J.D., Hawkins, A.R., Rockwood, A.L., Tolley, H.D., Lee, E.D., and Lee, M.L., *Anal. Chem.* 2007, *79*, 2927–2932. With permission of the American Chemical Society.)

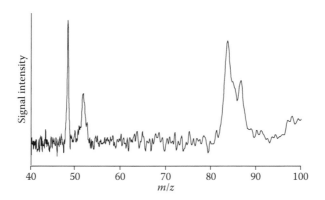

FIGURE 6.16 Mass spectrum of dichloromethane recorded using the halo ion trap.

functions as the resistive surfaces. For simulation of the halo ion trap, the potentials on the germanium layer between rings were determined using the potentials in the spaces between concentric conducting cylinders. The potential of each ring is an independent variable, so the potential distribution of the entire trap is a linear combination of these variables. Although optimized fields in the halo ion trap could be determined by finding an analytical solution for these variables, this approach has not yet been attempted. Rather, ring potentials have been chosen based on visually matching the fields between the halo and toroidal ion traps in SIMION. Determination of A_n terms [22], as is done typically with quadrupole ion traps, is not strictly possible in a toroidal coordinate system, although they can be approximated.

The relationship between plate spacing and the higher-order multipole components (*e.g.*, octopole) in the halo trap is quite different from that in conventional ion traps. In ion traps made using shaped metal electrodes, higher-order multipoles can be controlled to some extent by changing the spacing between electrodes. For example, stretching or 'shimming' the end-cap electrodes in quadrupole ion traps increases the positive octopole contribution, compensating for electrode truncation and the ion entrance/exit holes, thereby improving ion ejection and mass resolution [35]. In contrast, with the halo ion trap, the plate spacing has little effect on higher-order multipoles [36]. Changing the plate spacing leaves the field quadrupolar, modifying only the edge effects, with their corresponding higher-order multipoles. In planar electrode traps, higher-order multipoles can be added by modifying the potential function on the resistive material. Since higher-order multipoles can be modified electronically, and are not constrained by physical geometry, new opportunities exist for producing and experimenting with custom fields.

Simulations indicated that ions in the halo trap were cooled collisionally to a toroid with major radius (R) of 8.5 mm, similar to the 6.0 mm major radius of ions in the reduced-size toroidal ion trap. The resulting ion storage volumes are comparable.

Simulations have identified also a difficulty with ion ejection from the halo ion trap. Ejecting ions through the central hole in one of the trap plates requires bending the ion trajectories through a right angle. However, the fields needed to bend ion trajectories themselves disrupt ion motion during resonant ejection by breaking the plane of symmetry between the plates. In addition, due to the variation in ion kinetic energy during ejection, many ions do not reach the detector. A new halo ion trap design in which ions are ejected axially through slits, as is done in the toroidal ion trap, is under development, and should address both the ejection issues and other edge effects.

Several aspects of the halo ion trap design may be advantageous for some types of analysis. The toroidal storage volume results in a large ion storage capacity. The open space between the two plates provides convenient access for ion injection, optical or electron beam access, or faster pumping. Electrode alignment is simplified with only two pieces to position. The use of polished ceramic plates and lithographically deposited surface features results in a highly planar electrode surface, avoiding problems associated with surface roughness or machining. Finally, electric fields can be modified or adjusted electronically, without adjustment of plate position or design. Two disadvantages to the electric fields are the large edge effects from the electrode plate edges and the field roughness near the electrode surface resulting from discrete rings. It remains to be seen whether these effects can be overcome with improved potential functions or improved design.

6.3.3 PLANAR QUADRUPOLE AND COAXIAL ION TRAPS

Due to the independence between field geometry and electrode geometry using non-equipotential electrodes, other types of ion traps can be made using this method. For example, Peng *et al.* reported [37] on the development of a quadrupole (Paul) ion trap (Figure 6.17) made using two plates with the same design as the halo trap plates, but with a smaller central hole. Trap operation has been demonstrated using a

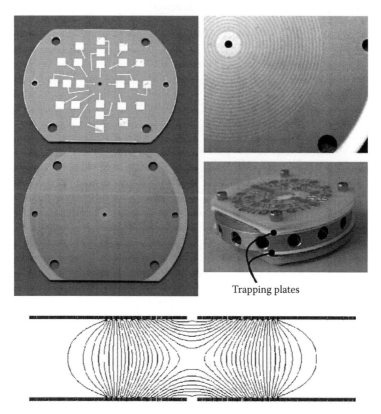

FIGURE 6.17 Quadrupole ion trap made using planar electrodes. At left are shown top and back sides. Electrode rings are under the germanium and, therefore, cannot be seen easily, but can be seen in enlargement at upper right. Lower right shows trap assembly without voltage divider or connectors. Below is shown the electric field within the trap. (Reproduced from Austin, D.E., Peng, Y., Hansen, B.J., Miller, I.W., Rockwood, A., Hawkins, A.R., and Tolley, S.E., *J. Am. Soc. Mass Spectrom.* 2008, *19*, 1435–1441. With permission from Elsevier.)

simple mass-selective instability scan based on ramping the RF amplitude, and also using dipolar resonant ejection. Ions are ejected straight through the central holes, without the ion trajectory-bending requirement of the halo ion trap. A representative mass spectrum of trichloroethylene is shown in Figure 6.18. The development of this 'planar quadrupole' trap was motivated by the desire for a planar electrode trap in which the ion behavior was well understood. Although this trap does not have the larger storage capacity of the toroidal trap, the complications of toroidal fields are avoided. Many of the advantages of the halo trap are retained, including higher precision in electrode fabrication and alignment, open trap access, and the ability to explicitly add or to subtract individual higher-order multipoles. Of particular interest is the ability to produce a nearly perfect dipole field between the plates, which may be of use in resonant ejection or other experiments.

Tolley *et al.* reported [38] the development of a coaxial ion trap in which both a halo and Paul trap are produced simultaneously between two electrode plates. The

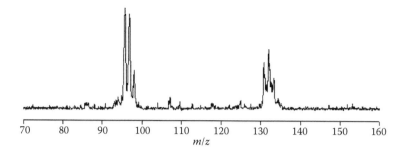

FIGURE 6.18 Mass spectrum (uncalibrated) of trichloroethylene made using the planar quadrupole ion trap.

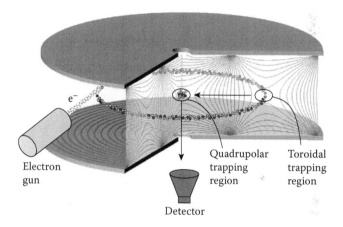

FIGURE 6.19 The coaxial ion trap consists of a toroidal trapping volume surrounding a quadrupole trapping volume. Both trapping volumes are created simultaneously using a single pair of plates. Ions can be transferred between the two trapping regions.

quadrupole ion trap is at the center of the halo ion trap. This 'double trap', shown in Figure 6.19, was made using the same two plates as the planar quadrupole trap. Ion trapping and mass analysis in each trap, as well as transfer of ions from the halo to the quadrupole trap, have been demonstrated. A pseudo-tandem mass analysis, in which ions are mass-selectively ejected from the halo trap to the quadrupole trap, stored in the quadrupole trap, and then ejected from the quadrupole trap to a detector, has also been demonstrated (see Figure 6.20), although a dissociation step between the mass analysis steps has not yet been attempted. This coaxial arrangement of two traps would facilitate tandem-in-space experiments, such as neutral-loss scans, that are difficult using conventional ion traps.

The coaxial trap possesses the advantages of the halo trap and the planar quadrupole trap, as well as offering new opportunities arising from the combination. For instance, ions can be trapped and stored in the toroidal region, which has a larger ion capacity, then directed to the quadrupole trap for mass analysis. Simultaneous storage and analysis would enable a high duty cycle.

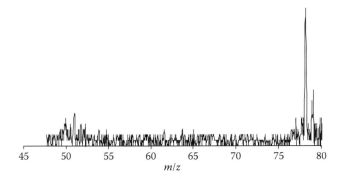

FIGURE 6.20 Mass spectrum of benzene recorded using the coaxial ion trap. Ions were created and trapped in the toroidal trapping region, ejected to and trapped in the central quadrupolar region, then ejected.

6.4 CONCLUSIONS AND OUTLOOK

Ion traps with circular geometries have been developed primarily to address the limited ion storage capacity and space-charge-limited performance of the quadrupole ion trap, especially in the context of ion trap miniaturization. Common to all ion traps with circular geometry are terms in the electric field resulting from the curvature of the trapping region about an axis outside the trapping volume. Opportunities for future research include understanding the effect of these curved fields on ion motion, and optimization of fields and electrode structure for maximum analytical performance. Ion traps made using planar non-equipotential electrodes, such as the halo ion trap, provide opportunities to explore new combinations of electric fields and trap geometries.

REFERENCES

1. Dehmelt, H.G. Radiofrequency Spectroscopy of Stored Ions I: Storage, in *Advances in Atomic and Molecular Physics, Vol 3*. Bates, D.R., Ed., Academic Press, New York, 1967, pp. 53–72.
2. March, R.E.; Todd, J.F.J. *Quadrupole Ion Trap Mass Spectrometry, 2nd edn.*, Wiley, Hoboken, NJ, 2005.
3. Kocher, F.; Favre, A.; Gonnet, F.; Tabet, J-C. Study of ghost peaks resulting from space charge and non-linear fields in an ion trap mass spectrometer. *J. Mass Spectrom.* 1998, *33*, 921–935.
4. Cox, K.A.; Cleven, C.D.; Cooks, R.G. Mass shifts and local space charge effects observed in the quadrupole ion trap at higher resolution. *Int. J. Mass Spectrom.* 1995, *144*, 47–65.
5. Schwartz, J.C.; Zhou, X.G.; Bier, M.E. Method and apparatus of increasing dynamic range and sensitivity of a mass spectrometer. *U.S. Patent*, 1996, 5,572,022.
6. Patterson, G.E.; Guymon, A.J.; Riter, L.S.; Everly, M.; Griep-Raming, J.; Laughlin, B.C.; Ouyang, Z.; Cooks, R.G. Miniature cylindrical ion trap mass spectrometer. *Anal. Chem.* 2002, *74*, 6145–6153.
7. Yang, M.; Kim, T-Y.; Kim, H-G.; Hwang, H-C.; Yi, S-K.; Kim, D-H. Development of a palm portable mass spectrometer. *J. Am. Soc. Mass Spectrom.* 2008, *19*, 1442–1448.

8. Austin, D.E.; Cruz, D.; Blain, M.G. Simulations of ion trapping in a micrometer-sized cylindrical ion trap. *J. Am. Soc. Mass Spectrom.* 2006, *17*, 430–441.

9. Schwartz, J.C.; Senko, M.W.; Syka, J.E.P. A two-dimensional quadrupole ion trap mass spectrometer. *J. Am. Soc. Mass Spectrom.* 2002, *13*, 659–669.

10. Hager, J.W. A new linear ion trap mass spectrometer. *Rapid Comm. Mass Spectrom.* 2002, *16*, 512–526.

11. Ouyang, Z.; Wu, G.X.; Song, Y.S.; Li, H.Y.; Plass, W.R.; Cooks, R.G. Rectilinear ion trap: concepts, calculations, and analytical performance of a new mass analyzer. *Anal. Chem.* 2004, *76*, 4595–4605.

12. Badman, E.R.; Cooks, R.G. Cylindrical ion trap array with mass selection by variation in trap dimensions. *Anal. Chem.* 2000, *72*, 5079–5086.

13. Tabert, A.M.; Griep-Raming, J.; Guymon, A.J.; Cooks, R.G. High-throughput miniature cylindrical ion trap array mass spectrometer. *Anal. Chem.* 2003, *75*, 5656–5664.

14. Tabert, A.M.; Goodwin, M.P.; Duncan, J.S.; Fico, C.D.; Cooks, R.G. Multiplexed rectilinear ion trap mass spectrometer for high-throughput analysis. *Anal. Chem.* 2006, *78*, 4830–4838.

15. Blain, M.G.; Riter, L.S.; Cruz, D.; Austin, D.E.; Wu, G.; Plass, W.R.; Cooks, R.G. Towards the hand-held mass spectrometer: design considerations, simulation, and fabrication of micrometer-scaled cylindrical ion traps. *Int. J. Mass Spectrom.* 2004, *236*, 91–104.

16. Ferran, R.J.; Boumsellek, S. High-pressure effects in miniature arrays of quadrupole analyzers for residual gas analysis from 10(-9) to 10(-2) torr. *J. Vac. Sci. Technol. A.* 1996, *14*, 1258–1265.

17. Bier, M.E.; Syka, J.E.P. Ion trap mass spectrometer system and method. *U.S. Patent* 1995, 5,420,425.

18. Church, D.A. Storage-ring ion trap derived from the linear quadrupole radio-frequency mass filter. *J. Appl. Phys.* 1969, *40*, 3127–3134.

19. Syka, J.E.P.; Schoen, A.E. Characteristics of linear and nonlinear rf-only quadrupole collision cells. *Int. J. Mass Spectrom. Ion Processes* 1990, *96*, 97–109.

20. Steiner, U. Ion filter and mass spectrometer using accurate hyperbolic quadrupoles. *U.S. Patent* 1996, 5,559,327.

21. Dawson, P.H. Ed., *Quadrupole Mass Spectrometry and Its Applications*, Elsevier, Amsterdam, 1976; reprinted by the American Institute of Physics Press, Woodbury, 1995.

22. Franzen, J.; Gabling, R-H.; Schubert, M.; Wang, Y. Non-Linear Ion Traps. in *Practical Aspects of Ion Trap Mass Spectrometry, Vol 1.* March, R.E.; Todd, J.F.J., Eds., CRC Press, Boca Raton, FL,1995, pp 49–167.

23. Syka, J.E.P. Commericalization of the Quadrupole Ion Trap. in *Practical Aspects of Ion Trap Mass Spectrometry, Vol 1.* March, R.E.;Todd, J.F.J., Eds., CRC Press, Boca Raton, FL, 1995, pp 169–205.

24. Lammert, S.A.; Plass, W.R.; Thompson, C.V.; Wise, M.B. Design, optimization and initial performance of a toroidal rf ion trap mass spectrometer. *Int. J. Mass Spectrom.* 2001, *212*, 25–40.

25. Hemberger, P.H.; Nogar, N.S.; Williams, J.D.; Cooks, R.G. Laser photodissociation probe for ion tomography studies in a quadrupole ion-trap mass-spectrometer. *Chem. Phys. Lett.* 1992, *191*, 405–410.

26. Lammert, S.A.; Rockwood, A.A.; Wang, M.; Lee, M.L.; Lee, E.D.; Tolley, S.E.; Oliphant, J.R.; Jones, J.L.; Waite, R.W. Miniature toroidal radio frequency ion trap mass analyzer. *J. Am. Soc. Mass Spectrom.* 2006, *17*, 916–922.

27. Lammert, S.A.; Contreras, J.A.; Murray, J.A.; Tolley, H.D.; Tolley, S.E.; Lee, E.D.; Lee, M.L.; Later, D.W. Totally self-contained gas chromatograph-toroidal mass spectrometer for field application. Presented at the Pittsburgh Conference on Analytical Chemistry and Applied Spectroscopy, New Orleans, LA, 3 March 2008.

28. Pau, S.; Pai, C.S.; Low, Y.L.; Moxom, J.; Reilly, P.T.A.; Whitten, W.B.; Ramsey, J.M. Microfabricated quadrupole ion trap for mass spectrometer applications. *Phys. Rev. Lett.* 2006, *96* (12), 120801-1–120801-4.

29. van Amerom, F.H.W.; Chaudhary, A.; Cardenas, M.; Bumgarner, J.; Short, R.T. Microfabrication of cylindrical ion trap mass spectrometer arrays for handheld chemical analyzers. *Chem. Eng. Comm.* 2008, *195* (2), 98–114.

30. Hauschild, J.P.; Wapelhorst, E.; Muller, J. Mass spectra measured by a fully integrated MEMS mass spectrometer. *Int. J. Mass Spectrom.* 2007, *264*, 53–60.

31. Verbeck, G.F. MEMS assembled ion optical devices: current technology and a look at advantages and disadvantages. Presented at the 6th Harsh Environment Mass Spectrometry Workshop, Cocoa Beach, Fl, 17–20 September 2007.

32. Austin, D.E.; Wang, M.; Tolley, S.E.; Maas, J.D.; Hawkins, A.R.; Rockwood, A.L.; Tolley, H.D.; Lee, E.D.; Lee, M.L. Halo ion trap mass spectrometer. *Anal. Chem.* 2007, *79*, 2927–2932.

33. Wang, Y.; Wanczek, K.P. Generation of an exact three-dimensional quadrupole electric field and superposition of a homogeneous electric field within a common closed boundary with application to mass spectrometry. *J. Chem Phys.* 1993, *98* (4), 2647–2652.

34. Wang, Y.; Wanczek, K.P.; Jiang, X.Q.; Hua, Z.Y. Exact two-dimensional quadrupole field and superposition of a homogeneous field. *Rev. Sci. Instrum.* 1993, *64*, 2585–2590.

35. Louris, J.; Schwartz, J.C.; Stafford, G.; Syka, J.E.P.; Taylor, D. *Proc. 40th ASMS Conf. Mass Spectrometry and Allied Topics,* Washington, DC, 1992, p. 1003.

36. Austin, D.E.; Peng, Y.; Hansen, B.J.; Miller, I.W.; Rockwood, A.; Hawkins, A.R.; Tolley, S.E. Novel ion traps using planar resistive electrodes: implications for miniaturized mass analyzers. *J. Am. Soc. Mass Spectrom.* 2008, *19*, 1435–1441.

37. Peng, Y.; Miller, I.W.; Zhang, Z.; Hansen, B.J.; Wang, M.; Tolley, S.E.; Lee, M.L.; Hawkins, A.R.; Austin, D.E. Planar resistive electrode ion traps, *Proc. 56th ASMS Conf. Mass Spectrometry and Allied Topics,* Denver, CO, 2008, MPA 008.

38. Tolley, S.E.; Hawkins, A.R.; Austin, D.E.; Hansen, B.J.; Lee, E.D.; Lee, M.L. Coaxial Ion trap using concentric planar electrode arrays. Presented at the Pittsburgh Conference on Analytical Chemistry and Applied Spectroscopy, New Orleans, LA, 3 March 2008, Abstract 1930-6.

Part III

Fourier Transform Mass Spectrometry

7 Ion Accumulation Approaches for Increasing Sensitivity and Dynamic Range in the Analysis of Complex Samples

Mikhail E. Belov, Yehia M. Ibrahim, and Richard D. Smith

CONTENTS

7.1 Introduction 401
7.2 External Accumulation of Ions for FT-ICR Detection 402
    7.2.1 ICR Trap 402
    7.2.2 Accumulation of Ions in an External Ion Trap 408
7.3 Ion Trapping with Time-of-Flight Mass Spectrometry 414
7.4 Ion Trapping with Ion Mobility-Time-of-Flight Mass Spectrometry 419
7.5 Conclusion 425
References 425

7.1 INTRODUCTION

The aim of this chapter is to provide the reader with an insight into methods for increasing sensitivity and dynamic range in the analysis of complex samples by utilizing data-directed mass spectrometry in conjunction with selective ion accumulation techniques. Three specific areas are considered: (i) the external accumulation of ions for detection by Fourier transform ion cyclotron resonance (FT-ICR) mass spectrometry (Section 7.2); (ii) the use of ion trapping prior to time-of-flight (TOF) mass analysis, including the use of the ion funnel trap (IFT), (Section 7.3); and (iii) the combination of ion mobility spectrometry (IMS) with TOF mass spectrometry (Section 7.4). From the range of applications discussed, it is evident that these different approaches are making significant strides toward the goal of achieving higher analysis throughput, accompanied by improved sensitivity and mass accuracy.

7.2   EXTERNAL ACCUMULATION OF IONS FOR FT-ICR DETECTION

The need for higher mass accuracy and precision in analysis of, for example, bio-logical compounds is now commonly addressed by storing charged particles in an ion trap using a superposition of the constant static magnetic and spatially inho-mogeneous electric fields, often referred to as Penning trap. The Penning trap has been successfully employed in FT-ICR Mass Spectrometry instrumentation. Since its inception in 1973 [1], FT-ICR has been the subject of multiple reviews [2–9], several journal issues [10, 11], and books [12, 13] that give a full-range technical introduction to both single ion and more populated ion clouds' (a group of ions whose motion is influenced by the other ions in the cloud) behavior in combined magnetic and electric fields, subsequent signal processing, and technique applications. The principles and design of geometric and electrical configurations of ICR traps have been reviewed also [14]. The reader is referred to these publications for more infor-mation. In this chapter, we give a brief overview of the ion cloud behavior in the superimposed magnetic and electric fields, with an emphasis on ICR trap design, and highlight FT-ICR performance improvements due to accumulation of ions in another ion trap external to FT-ICR.

7.2.1   ICR TRAP*

An ion cloud of a given m/z-value trapped in a combined trap experiences four basic types of motion that include cyclotron, magnetron, axial oscillations, and rotation around its central axis [15]. Figure 7.1 shows the schematic trajectory of an ensem-ble of such ions exhibiting both cyclotron and magnetron motions. The interaction between the spacecharge of an ion cloud and its image charge in the trap walls causes a slow drift around the trap's central axis, in addition to the magnetron drift caused by the trapping fields [16–18]. In conventional non-neutral plasma experiments, this image-induced drift is dominant and the motion it causes is called diocotron motion [15]. As a result, the detected cyclotron frequency, ω_{ICR}, is a superposition of the fast and slow oscillation frequencies in the trap:

$$\omega_{ICR} = \Omega - \omega_M - \omega_D - \delta_{sc} \tag{7.1},$$

$$\omega_M = \frac{\Omega}{2} - \frac{\Omega}{2}\sqrt{1 - \frac{2\omega_z^2}{\Omega^2}} \approx \frac{V_t}{2|B|d^2} \tag{7.2},$$

$$\omega_z = \sqrt{\frac{azeV_t}{md^2}} \tag{7.3},$$

---

* See also Volume 4, Chapter 8: Radio Frequency-Only-Mode Event and Trap Compensation in Penning Fourier Transform Mass Spectrometry, by Adam M. Brustkern, Don L. Rempel, and Michael L. Gross.

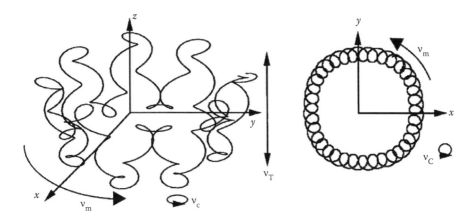

FIGURE 7.1 Ion motion in a two-inch cubic Penning trap in a perfectly homogeneous magnetic field of 3T for an ion of m/z 2300, for 10 V trapping voltage. The three natural motional frequencies and amplitudes have relative magnitudes given by: $\omega_+ = 4.25\ \omega_z$, $\omega_z = 8.5\ \omega_-$, $\rho_- = 4z_{max} = 8\rho_+$ [18]. Note that the field produced by a cubic trap is not perfectly quadrupolar as manifested by the shape of the magnetron orbit (not a perfect circle) in the xy plane. The magnetic field points in the negative z-direction (Reproduced from Marshall, A.G., Hendrickson, C. L., and Jackson, G. S., *Mass Spectrom. Rev.* 1998, *17*, 1–35. With permission from John Wiley and Sons. Inc.)

$$\omega_D \approx \left(\frac{\rho_c^2}{r_w^2}\right)\omega_R \qquad (7.4).$$

Here ω_{ICR}, ω, ω_M, and ω_D are the detected, unperturbed, magnetron, and diocotron frequencies, respectively; ω_z is the frequency of axial oscillation, δ_{sc} is the space-charge term [19], a is the geometry factor, V_t is the trapping voltage, B is the uniform magnetic field, d is the characteristic length of the trap, m/z is the mass-to-charge ratio of an ion, e is the electronic charge, ρ_c and r_w are the ion cloud and ion trap wall radii [20], respectively, and ω_R is the ion cloud rotation frequency due to $E \times B$ drift. Equations 7.1–7.4 show that the detected cyclotron frequency depends on the axial oscillation frequency, the number of ions in the trap, and the ion cloud interaction with its image charge. Because the detected cyclotron frequency of an ion ensemble, ω_{ICR}, is practically always lower than the unperturbed cyclotron frequency, ω, due to the magnetron motion and space-charge repulsion (see Equation 7.1), the former is often referred to as the reduced cyclotron frequency. Ions of low m/z-value also experience relativistic shifts in the measured cyclotron frequency [21]; the effect is typically ignored in experiments with ions of higher m/z-value.

Following the ion trapping event, signal acquisition is performed in two steps that are referred to as *ion excitation* and *detection*. Ion cloud excitation is accomplished with an RF waveform at a frequency resonant with the reduced cyclotron frequency of the ion ensemble, and of amplitude sufficient to cause an increase in the ion cloud cyclotron radius up to about half the radius of the ICR trap over the excitation period.

It is important to note that if resonant energy is added too slowly, ion–ion collisions will dissipate it into random thermal motion, resulting in lower quality signals. One explanation for this observation is that upon excitation the axial space-charge field no longer cancels the trapping field, so that the excess axial energy must be converted to the thermal motion of individual ions in their respective clouds, causing minority species in the initial ion cloud to lose their coherence [22]. Signal detection is obtained by recording the image charge induced on detection segments of the ion trap. In theory, an ion post-excitation radius, r, is independent of the m/z-value [4,23]:

$$r = \frac{V_{p-p}T_{excite}}{2dB_0} \tag{7.5},$$

where V_{p-p} is the peak-to-peak voltage, T_{excite} is the excitation period, d is the distance between the excitation electrodes, and B_0 is the magnetic field. However, in practice, because of the non-uniform spatial distribution of the excitation field within an ICR trap, ions have some narrow radial distribution that will tend to be broadened by space-charge effects.

Challenges for ICR trap design thus encompass reduction of the radial component of the electric trapping field in the detection region, linearization of the excitation field, and reduction of the anharmonic terms of the constant axial and alternating azimuthal electric fields, particularly at greater ion displacements from the trap center. A trade-off between geometric configurations that produce ideal harmonic trapping potentials, optimally 'open' access for externally generated ions, and uniform RF excitation and detection electric fields was a limiting factor in the earlier ICR trap designs. The ICR trap (or cell) has evolved continuously from the simple six-electrode cubic cell [24] to a wide array of cell geometries including orthorhombic [25,26], cylindrical [27,28], hyperbolic [29,30], and multiple-electrode geometries [31–34]. In addition, the open-ended trapped-ion cell [31] has been applied to external ion injection [35] because of the inherent advantages of the collinear electrode geometry. Design improvements were aimed at addressing the limitations of non-ideal potential distributions within the ICR cell, and at decoupling of the ion motions in the combined trap.

The radial electric field contributes to a number of undesirable effects, including spectral sidebands [32], cyclotron frequency shifts [26], reduced mass accuracy [33], and diminished upper m/z limit [36]. To reduce the magnetron motion, a 'screened' electrostatic cubic trap, in which grounded screens inside the end-cap electrodes produce an approximate 'particle-in-a-box' potential, was developed [35]. The screened trap was found to have 100-fold smaller shift in the detected cyclotron frequency than that of a similar cubic trap, with a cyclotron frequency almost independent of cyclotron radius. Screening of the trapping potentials was then explored in the open trapped-ion cell geometry with compensation electrodes [36]. The function of the compensation electrodes in reducing the radial electric field is analogous to the grounded transmissive screen inserted between the trap electrodes and the trapping volume in the closed screened cell, whereby the interior of the cell is shielded from the trapping field. The supplementary voltage applied to the inner set of compensation electrodes reduces the radial electric field by nearly two orders of magnitude and

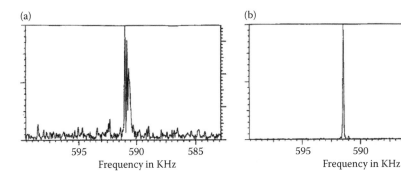

FIGURE 7.2 Effect of increasing dynamically the particle-in-a-box character of the potential well to reduce space-charge-induced peak splitting and frequency shift for improved dynamic range. Frequency spectra of benzene: (a) obtained by applying + 6 V to both the trap and the compensation electrodes; (b) obtained by applying + 6 V to the trap electrodes and –2.2 V to the compensation electrodes. The improvement in signal-to-noise ratio exceeds an order of magnitude. (Reproduced from Vartanian, V.H., Hadjarab, F., and Laude, D. A., *Int. J. Mass Spectrom. Ion Processes.* 1995, *151*, 175–187. With permission from Elsevier).

increases the potential well-depth for greater ion capacity. Figure 7.2 shows a comparison of FT-ICR signals obtained with open and compensated cylindrical cells. An order of magnitude improvement in signal-to-noise ratio (*S/N*) was obtained by screening the detection region from the trapping electrodes.

Linearization of the excitation field in the open cylindrical cell was achieved by the capacitive coupling of the excitation and trapping segments [37]. In the closed cylindrical cell, the excitation field has been linearized successfully by depositing thin conductive electrodes on ceramic trapping plates so that electrode shapes reflect the theoretical potential distribution of the infinitely long cylindrical cell, the design being referred to as the 'infinity' cell [38]. The net result of the linearized excitation field is the proportional increase in the absolute mass abundances observed over the entire range of excitation levels. In addition, rapid truncation of the signals at an excitation potential corresponding to the cell radius corresponds to the expected mass-independent radial ejection of all ions.

Despite substantial differences in ICR cell designs, the constant trapping and alternating azimuthal electric fields of various geometries remain harmonic near the cell center (that is, at small radial and axial displacements from the cell center) [5]. When excited to higher cyclotron radii, ions experience electric fields that deviate from the ideal harmonic axial distribution, resulting in dependence of the axial oscillation frequency on the ion's axial position, which causes the loss of coherence by an ion cloud. It is noteworthy that the ion's axial oscillation frequency is independent of its axial position only in the ideal harmonic well. Anharmonicities of the trapping DC field cause two major undesirable effects. First, the mode frequencies depend on the excitation amplitude. Second, all the motions are coupled. In practice, deviations of the axial field distribution from the ideal harmonic potential lead to the shifts in the measured cyclotron frequency and often contribute to the accelerated loss of coherence by an ion ensemble. As a result, ions positioned at the axial periphery of an ion cloud 'evaporate' from the coherent ensemble, creating comet-like structures that were observed with supercomputer modeling [39]. Concurrently, Mitchell and Smith have developed an analytical model of Coulombic interaction of two ion clouds that predicts both positive and negative frequency shifts, phase-locking and phase-modulation [40]. An increase in the total number of trapped ions results in further elongation of an ion cloud along the trap axis and extension of the ion cloud into the trap regions with more anharmonic fields, thus exacerbating further the observed frequency shifts. These deleterious effects of the non-ideal quadrupolar DC potential are well-recognized and various approaches to address cell anharmonicity have been explored. Although a Penning trap with hyperboloidal electrodes generates a near-perfect three-dimensional axial quadrupolar trapping potential, there are several reasons why hyperbolic traps are not used widely in FT-ICR experiments. First, the curved surfaces of the ring and end-cap electrodes reduce the available trap volume thus exacerbating space-charge effects that are detrimental for high accuracy measurements. Second, azimuthal RF excitation field is spatially quite inhomogeneous leading to severe distortion of ion relative abundances and axial ejection. Therefore, non-hyperbolic geometries suit better broadband FT-ICR experiments, and the challenge is to solve 'the inversion problem' by finding boundary conditions defined by the ICR trap geometry that would yield the harmonic electric fields. In the pioneering work by Gabrielse et al. [41], anharmonicities of an open-endcap Penning trap were considered analytically. The electric potential V near the center of a Penning trap can be expanded in Legendre polynomials due to azimuthal symmetry of the cylindrical design

$$V = \frac{1}{2}V_0 \sum_{k_{even}=0}^{\infty} C_k \left(\frac{r}{d}\right)^k P_k(\cos\theta) \tag{7.6},$$

where V_0 is the trapping potential, and the characteristic distance, d, given by

$$d^2 = \frac{1}{2}(z_0^2 + \frac{1}{2}\rho_0^2) \tag{7.7},$$

is chosen to be similar to the hyperbolic trap treatment, and only even C_k coefficients are non-zero when reflection symmetry across the $z = 0$ plane is assumed. The parameters z_0 and ρ_0 denote the distance along z-axis from the trap center to the outer edge of a compensation electrode and the trap radius, respectively. In a perfect quadrupole potential, $C_2 = 1$ and all other terms equal zero. When $C_4 \neq 0$, the axial oscillator is anharmonic, with the oscillation frequency depending on the amplitude. The shift in frequency is

$$\frac{\Delta\omega_z}{\omega_z} = \frac{3}{2}\left(\frac{C_4}{C_2}\right)\frac{E_z}{zeV_oC_2} \tag{7.8},$$

where E_z is the kinetic energy in the particle's axial oscillation.

Using a set of compensation electrodes, the lowest anharmonic terms (C_4 and C_6) were drastically reduced, thereby enabling high precision measurements. In the other detailed account [22] of the trapping field anharmonicity, solutions of cylindrically and azimuthally symmetric Laplace's equation for the trapping DC fields were considered. Equation 7.2 was expanded using a Taylor series, and the lowest non-linear term (C_4) was set to zero by choosing the optimum aspect ratio of the ICR trap (that is, the ratio of the cell length z_0 to the radius r_0). As a result, the trap's performance was found to be essentially independent of the trapping potential, making possible precise measurements of ions formed from small molecules in a 30 V well. The compensated cylindrical ICR cell design has been revisited recently and examined in liquid chromatography/mass spectrometry (LC/MS) experiments with a mixture of proteins digested with trypsin [42]. Following capillary LC separation, peptide ions were generated with an electrospray ionization (ESI) source and then detected with an FT-ICR mass spectrometer. A mass accuracy histogram was obtained by matching the retention times and m/z-values of the detected LC/MS features against the entries in the database which, in turn, were generated in a number of tandem, MS/MS, experiments. Figure 7.3 shows the mass accuracy histograms obtained with the open, infinity, and compensated-open cylindrical ICR cells. This initial evaluation has shown a better mass precision for the compensated cylindrical cell as compared to the capacitively coupled open and infinity ICR cell configurations.

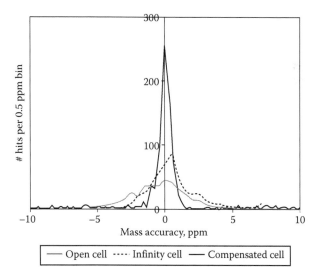

FIGURE 7.3 Mass accuracy histograms obtained for LC-MS separation of a standard mixture of proteins digested with trypsin, number of matched elution features per mass residual bin of 0.5 ppm. Open cell, compensation is not applied, gray curve: 3.6 ppm FWHM, 7.6 ppm baseline, 356 hits. Infinity cell, dotted curve: 2.2 ppm FWHM, 5.8 ppm baseline, 446 hits. Compensated cell, black curve: 0.9 ppm FWHM, 2.1 ppm baseline, 465 hits (bin reduced to 0.2 ppm, with vertical scale adjusted × 2.5 to compensate for the different bin size). (Reproduced from Tolmachev, A.V., Robinson, E.W., Wu, S., Kang, H., Lourette, N.M., Paša-Tolić, L., and Smith, R.D., *J. Amer. Soc. Mass Spectrom.* 2008, *19*, 586–597. With permission from Elsevier.)

We should note that reducing anharmonicities of the three-dimensional axial trapping potential and linearizing the one-dimensional azimuthal dipolar electric potential (for ICR excitation/detection) are critically important for attaining precise mass measurements at high dynamic range in FT-ICR mass spectrometry. The 'orthogonalized' ICR trap (that is a trap where the frequency of axial oscillation is independent of changes in trapping potential) is less susceptible to higher levels of space-charge, making it highly attractive for complex bioanalytical applications.

7.2.2 ACCUMULATION OF IONS IN AN EXTERNAL ION TRAP

In a number of applications, including gas and liquid chromatography (GC and LC), ions are generated outside the ICR cell. Kofel et al. [43] and McIver et al. [44] have developed two approaches for introducing externally generated ions into an ICR cell. The former employed a system of electrostatic lenses, while the latter made use of an RF-only quadrupole ion guide. Though efficiency of ion transfer to the ICR cell was reportedly high (no numbers were provided), only a small portion of the ions generated in the ion source could be trapped in the ICR cell. The inefficiency of trapping ions from the continuous beam in the ICR cell at a pressure of 10^{-9} Torr was circumvented initially by the dual-cell design where the source cell could operate at substantially higher pressure than that of the analyzer cell [45]. With this dual-cell instrument, ions could be trapped in both the source and analyzer cells. However, because both of these cells needed to be located inside the magnet bore, access to the source cell was difficult.

Concurrently, a gated-trapping technique was developed by Smalley and co-workers in experiments with metal clusters produced by laser vaporization of a metal target in a pulsed nozzle [46]. This technique requires reduction of the electric potential of the entrance end-cap electrode to ground in order to admit ions to the cell, followed by rapid re-establishment of the optimum trapping potential prior to data acquisition. A variation of the gated-trapping technique was reported later by Caravatti [47], who proposed to transfer the kinetic energy of the motion parallel to the magnetic field into directions perpendicular to the magnetic field, that is, the 'side-kick' approach. Although side-kick deflection increased the ion injection period to 200 ms [48], undesirable side effects, including increased radial diffusion [49] and enhancement of harmonic signals due to larger magnetron radius [50], were reported. Accumulated trapping is the third technique for capturing externally generated ions in the ICR cell [51]. Ion trapping efficiency was enhanced by pulsing an inert gas into the ICR cell resulting in increased ion/neutral interactions and accumulation of ions [52]. It is significant that ions produced by a continuous source, such as ESI, could be accumulated directly in the ICR cell with the efficiency of accumulated ion trapping approaching 1%. Incorporation of an electrodynamic ion funnel into an FT-ICR instrument was shown to increase dramatically ion signals generated with FT-ICR employing accumulated trapping [53], and, under optimized conditions, yielded an ultra-high sensitivity in the analysis of small proteins [54]. Figure 7.4 shows portions of the mass spectra obtained with 130 zmol* of cytochrome c and with myoglobin

---

* 1.0 zmol (zeptomol) is 10^{-21} mol.

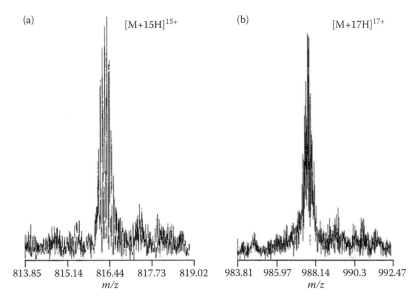

FIGURE 7.4 Portions of mass spectra showing [M + 15H]$^{15+}$ ions from horse cytochrome c (a) and [M + 17H]$^{17+}$ ions from horse myoglobin (b). Sample concentrations were 0.4 nM, and the total amount consumed for each protein was 135 zmol (*ca* 80,000 molecules). (Reproduced from Belov, M.E., Gorshkov, M.V., Udseth, H.R., Anderson, G.A., and Smith, R.D., *Anal. Chem.*, 2000, 72, 2271–2279. With permission from the American Chemical Society.)

electrosprayed at concentrations of 0.4 nM [54]. The drawback of this approach is the need for highly efficient pumping, as trapping is accomplished at a pressure of *ca* 10^{-5} Torr, while optimal data acquisition requires the ICR cell pressure to be in the range of 10^{-9}–10^{-10} Torr. Though cryopanels were shown to provide a pumping speed of 10^5 L s^{-1} [55], the need for regular panel 'regeneration' makes this approach impractical for high throughput and chromatographic applications.

A significant increase in duty cycle has been demonstrated using ion accumulation in an external octopole trap followed by gated trapping [56]. Other improvements derived from external accumulation were reported to be enhanced *S/N* and mass-resolving power [56]. The benefits of external ion accumulation were enhanced further by accumulating selectively the species of interest in the external ion trap prior to FT-ICR detection [57]. The usefulness of ion pre-selection with FT-ICR was recognized initially by McIver [58] who proposed using quadrupole mass filtering (RF/DC) of ions generated by an external source. A technique complementary to quadrupole mass filtering is based on resonant dipolar (or quadrupolar) excitation in an RF-only quadrupole. Quadrupolar excitation is accomplished by applying two 180° out-of-phase excitation waveforms to the pairs of opposite quadrupole rods in the manner similar to that by which the main RF field is generated [59]. In the adiabatic approximation, ion motion in an RF-only quadrupole can be represented by superposition of rapid oscillations at the main RF frequency and a smooth harmonic drift motion in a pseudopotential (or 'effective potential') well [60]. To eject resonantly ions of a specific *m/z*-value, dipolar excitation is applied at a frequency matching the period of the ion's oscillation in the pseudopotential well. Given that the time-averaged pseudopotential is harmonic, the oscillatory secular motion of the ion in the pseudopotential well is mass-dependent, with the frequency of motion, ω, [ω$_r$] governed by

$$\omega_r \approx \frac{q_r \Omega}{\sqrt{8}}$$ (7.9).

The dimensionless Mathieu parameter q_r is given by

$$q_r = \frac{4zeV_{rf}}{m\Omega^2 r_0^2}$$ (7.10),

where V_{rf} is the zero-to-peak RF voltage amplitude, *m/z* is the ion's mass-to-charge ratio, *e* is the electronic charge, ω is the main RF frequency, and r_0 is the inscribed quadrupole radius.

In a particular trap geometry (for example, using quadrupolar excitation), an ion's motion can be driven parametrically at a frequency $\omega_{param} = 2 \times \omega$ [61]. Parametric excitation [62] offers the advantages of reducing the number of possible resonances and achieving higher resolution for an ion species of given *m/z*-value. Welling *et al.* [61], Campbell *et al.* [63], Collings and Douglas [59], Cha *et al.* [64], and Belov *et al.* [65] have shown that, by introducing a supplementary RF-field, individually mass-selected species stored in a linear quadrupole ion trap can be ejected efficiently using either resonant or parametric excitation. When the frequency of the auxiliary RF field is equal to the secular frequency (that is, resonant excitation) or to twice the secular frequency (that is, parametric excitation) of a particular ion species, the auxiliary RF field causes the selected species to oscillate with increased amplitude. For lower ion populations in the linear quadrupole ion trap (that is, in the absence of significant space-charge effects), the increase in the amplitude of the auxiliary RF field results in effective ion ejection from the trap. When a superposition of excitation sine waveforms $\omega_1, \omega_2...\omega_n$ is applied to the pair of rods of the selection quadrupole, the ions with corresponding $(m/z)_1, (m/z)_2,. (m/z)_n$ will be ejected *concurrently* from the quadrupole without having any influence upon the transmission of other ions, provided the mass resolution is sufficient. This technique of concurrent ejection of multiple ion species is in contrast with the conventional quadrupole RF/ DC filtering or ion packet isolation in a 3D quadrupolar trap, where a relatively narrow *m/z*-range is targeted for transmission or further activation.

By ejecting higher-abundance species and accumulating lower-abundance ions for a longer accumulation time in a linear ion trap (LIT), the effective dynamic range of the ICR cell was expanded by up to two orders of magnitude [57]. Importantly, ion transfer from the external accumulation trap to the ICR cell is most efficient when the temporal profile of an ion packet incorporating all ion species matches the period of ion axial oscillation in the ICR cell. To expedite ion ejection from an LIT, an axial DC electric field between the end-cap electrodes of the trap was generated either by segmenting the quadrupole rods [66] or by introduction of thin wires between the octopole rods [67]. Electrodynamic ion funnel-assisted [54] selective external accumulation of low-abundance proteins in an RF-only quadrupole followed by FT-ICR detection yielded a limit-of-detection of *ca* 20 zmol (12,000 molecules, *S/N* ratio of 3:1) [66]. Using ion accumulation in a short octopole, three peptides in a mixture totaling 500 attomoles (amol) each in water (10 μL, 50 amol μL^{-1}) were separated and detected, demonstrating detection from a mixture at low endogenous biological concentration. The highest sensitivity was attained with arg(8)-vasotocin, in which a total of 300 amol was detected in artificial cerebrospinal fluid (1 μL, 300 amol μL^{-1}) and a total of 100 amol in water (1 μL, 100 amol μL^{-1}) [68].

To expand further the dynamic range of FT-ICR equipped with an LIT, *data-dependent* ion ejection of high-abundance ion species in the course of capillary LC separation was developed [69]. This approach, also referred to as dynamic range expansion applied to mass spectrometry ('DREAMS'), is based on alternation between a survey pre-scan, with a short accumulation time, and a scan where species in high abundance are removed prior to accumulation in the LIT, while the remaining ions that were in low abundance are accumulated for extended periods. Concurrent data-dependent ejection of multiple high-abundance ion species during the DREAMS scans is accomplished with RF-only excitation, so that those species of interest that are in low abundance can be accumulated for 10–100-fold longer periods, resulting in a significant increase in the number of components identified (for example, peptides). In the initial application of this approach [70], a sample derived from a tryptic digest of proteins from mouse B16 cells cultured in both natural isotopic abundance and ^{15}N-labeled media was studied. Figure 7.5 shows a comparison of the total ion chromatograms acquired in LC-ESI-FT-ICR experiments with the conventional and DREAMS-enhanced ion trapping approaches. Peptide pairs of low abundance not detected with the conventional technique were readily detected and identified using DREAMS. The use of DREAMS enabled assignment of approximately 80% more peptide pairs, thus providing quantitative information for approximately 18,000 peptide pairs in a single analysis [70]. Data-dependent ion accumulation in the LIT has been used efficiently in 'top-down' proteomic studies also [71]. Proteins introduced by a nanospray robot were selected using RF/DC quadrupole filtering, dissociated either in an external octopole trap or ICR cell, and then matched against a database sequence with characterization of N-terminal post-translational modifications [72].

The advantages of external ion accumulation are negated substantially when this process is accompanied by a significant bias across the *m/z*-range. Two major mechanisms of ion discrimination that apply when accumulating ion populations in an RF-only quadrupole were reported [73]. These mechanisms are: (1) radial separation due to *m/z*-dependent balance between the force arising from the pseudopotential and space-charge repulsion (that is, radial stratification [74]), resulting in discrimination against higher *m/z*-value ions; and (2) the space-charge-induced instability of lower *m/z*-value ions. In addition, fragmentation has been observed when a greater number of ions was trapped in an RF-only hexapole [75–77] or quadrupole. The latter, however, could be controlled [78] as a function of the ratio of the radial (that is, the pseudopotential) and the axial potential wells, and used to an advantage in tandem mass spectrometry experiments. In addition, the accumulation of an excessive number of ions in an external trap and transfer of a condensed ion packet to an

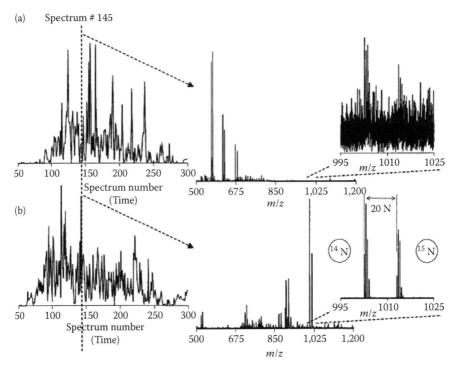

FIGURE 7.5   Capillary LC-ESI-FT-ICR results showing total ion chromatogram (TIC) and portions of the mass spectra acquired during the normal spectrum acquisition process and the alternating DREAMS spectrum acquisition process obtained using a 50-cm long column and 120 min reverse-phase gradient. (a) TIC reconstructed from the FT-ICR spectra acquired during RPLC separation of the mixture of identical aliquots of a natural isotopic abundance and ^{15}N-labeled version of a mouse B16 cells and representative spectrum obtained using broadband mode acquisition and a 100 ms accumulation time. (b) Corresponding TIC and representative spectrum obtained using RF-only selective acquisition and 300 ms accumulation time. Resonant frequencies for RF-only dipolar excitation were identified during a broadband ion acquisition (a) and up to five species having relative abundance of 10% were then data-dependently ejected during a selective acquisition that followed immediately after (b). Total amount of sample injected is 5 micrograms. (Reproduced from Paša-Tolic, L., Harkewicz, R., Anderson, G.A., Tolic, N., Shen, Y.F., Zhao, R., Thrall, B., Masselon, C., and Smith, R.D., *J. Amer. Soc. Mass Spectrom.* 2002, *13*, 954–963. With permission from Elsevier.)

ICR cell has a major impact upon the accuracy of mass measurement. Space-charge-induced cyclotron frequency shifts (see Equation 7.1) become pronounced with a dynamic ion source, such as the capillary LC/ESI of a complex biochemical sample. To mitigate the effect of space-charge, automatic gain control (AGC), along with independent introduction and control of the calibrant ions, has been implemented [79]. The AGC approach was developed earlier to control ion populations in 3D ion traps [80]. In LC-ESI experiments with a tryptic digest of bovine serum albumin (BSA), AGC-assisted external ion accumulation with internal calibration provided a 10-fold increase in the number of identified BSA tryptic peptides compared to that obtained with a fixed ion accumulation time and external calibration methods. External ion accumulation with FT-ICR has been advanced further with the development of a hybrid instrument, incorporating a linear quadrupole ion trap coupled to an FT-ICR analyzer [81]. This configuration provides rapid and automated MS and MS/MS analyses, similar to the 'data-dependent scanning' found on standard 3-D Paul traps, but with substantially improved dynamic range, mass measurement accuracy, mass resolution, and detection limit. The recently released, commercial version of this instrument (LTQ-FT) operates in the LC/MS mode (1 s per scan) with a mass resolution of 100,000 and is equipped with AGC to provide mass measurement accuracy of 2–3 ppm without the use of an internal standard. In recent experiments with the LTQ-FT instrument, the dynamic range and sensitivity were improved further by incorporating DREAMS, augmented with AGC capability. Data-dependent ejection of the more abundant species was performed in an RF-only quadrupole preceding the linear quadrupolar ion trap. Figure 7.6 shows the results of experiments with human blood plasma samples using an LC-ESI-LTQ-FT instrument that was operated in both the conventional and DREAMS modes [82]. Peptide identification was based on the AMT tag approach [83], where the detected LC/MS features were matched against the human plasma database using retention time and mass accuracy information. False discovery rates were calculated by matching LC/MS features against the database shifted by 11 Da [84]. The data indicate that a 15–20% greater number of

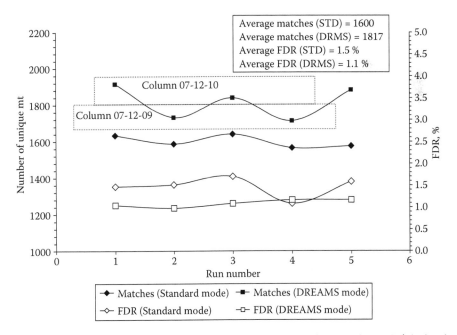

FIGURE 7.6 The number of identified unique tryptic peptides from a 0.5 mg mL^{-1} depleted human blood plasma sample as a function of the LC-FT-ICR run number. Data were obtained with an LTQ-FT-ICR (ThermoFisher, San Jose, CA) instrument operating in the conventional AGC (♦) and DREAMS-enhanced (■) AGC modes. A two-column capillary 5000 psi LC system was employed. (Reproduced from Belov, M.E., Robinson, E.W., Liyu, A.V., Prior, D.C., and Smith, R.D., *J. Proteome Res.* 2009, manuscript in preparation. With permission.)

uniquely identified peptides was obtained in the DREAMS mode, at a comparable or lower false discovery rate.

As a final remark in this section, we would emphasize that external ion accumulation has expanded greatly the capabilities of FT-ICR mass spectrometry in the analysis of complex biochemical mixtures. A plethora of analytical characteristics, including sensitivity, mass accuracy, mass resolution, dynamic range, and throughput has been improved dramatically. In addition, hybrid design enables concurrent handling of two independent ion packets, so that analyte sequence information (for example, product ion mass spectra) can be obtained while acquiring high resolution MS data. Developments of advanced methods that would allow more efficient use of the existing hardware are underway, and this work is expected to have a great impact on further applications of FT-ICR mass spectrometry in biochemical research.

7.3   ION TRAPPING WITH TIME-OF-FLIGHT MASS SPECTROMETRY

Hybrid quadrupole TOF mass spectrometers have been used increasingly in a number of applications because of their speed of data acquisition, high mass accuracy, resolving power, and high sensitivity [85]. A TOF mass analyzer separates ions according to their mass-to-charge ratio, m/z, in a field-free drift tube ($t \propto \sqrt{m/z}$, where t is the TOF). TOF mass spectrometry is inherently a pulsed technique so that pulsed-ion sources, such as matrix-assisted laser desorption ionization (MALDI), represent a natural fit to the analyzer. However, coupling a continuous ion source such as ESI to TOFMS has been challenging. Chopping or modulating the continuous ion beam results in inefficient ion utilization. The challenge has been addressed by sampling ions from the continuous ion beam into TOF orthogonally to the ions' initial trajectory, the design being referred to as orthogonal acceleration TOF (oa-TOF) MS [86]. Optimization of the resolving power of a TOFMS requires a well-defined start time of the ion packet. Ideally, all ions of different m/z-value should have zero initial velocity spread immediately prior to their flight. As orthogonally introduced ions have initial velocity distributions parallel to the TOF axis, this velocity spread brings about a degradation in the resolving power [86]. The initial velocity distribution along the TOF axis leads to the so-called 'turnaround time' which is constituted in the temporal spread of the ion packet at the TOF detector [86,87]. Since the initial velocity spread cannot be compensated by an ion mirror [88], other factors need to be brought into consideration. The turnaround time can be reduced either by increasing the extraction field, though this method has a practical limit, or by reducing the spread of the initial velocity distribution along the TOF axis. The initial velocity spread was minimized by using an RF-only ion guide that operated at an elevated pressure [89]. In a quadrupole rod set at a pressure of few milliTorr, ions are cooled and focused collisionally near to the quadrupole axis, so that the initial velocity spread is reduced [89,90]. Collisional cooling can be achieved also using other RF ion guides, such as hexapoles, octopoles, or a stacked-ring assembly. Indeed, practically all commercial ESI-oa-TOF instruments employ an RF ion guide upstream of the TOF extraction region in order to achieve high mass-resolving power.

Because TOFMS utilizes a release-and-wait technique, ions are lost between extraction pulses. Thus, the accumulation of ions in an ion trap during TOF analysis ensures an increased duty cycle as compared to the continuous beam instrument [91,92]. However, as ions of different m/z-values have different flight times from the trap to the TOF extraction region, the high duty cycle is achieved usually for a narrow range of m/z-values [93]. A set of delay times is required to cover a broader m/z-range [92]. An excellent review on LITs with a section on TOF mass analyzers has been published recently, and the reader is referred to this publication for more information [93]. The discussion below will focus on the new developments in ion trap/TOF technology since 2005.*

Incorporation of an ion trap with a TOFMS has led not only to an increase in the instrument duty cycle and mass-resolving power, but has enabled MS/MS experiments also. Because quadrupole ion traps are capable of MS^n analysis, while oa-TOFMS instruments exhibit a mass-resolving power in excess of 10,000 and a mass measurement accuracy of less than 5 ppm, a hybrid ion trap/oa-TOFMS represents an attractive platform for bioanalytical applications [86]. Early implementation encompassed a single LIT interfaced to an oa-TOFMS [94]. Precursor ion isolation, fragmentation (through dipole excitation using a supplementary RF waveform) and cooling were performed in the LIT (at a pressure of 7 mTorr), whereas ion detection was accomplished with the oa-TOFMS. In an effort to achieve high excitation resolution and maintain optimum ion transmission to an oa-TOFMS, Hashimoto *et al.* introduced a collisional RF-only quadrupole between a Paul trap and the oa-TOFMS [95]. Following MS^n analysis in the Paul trap, the product ions were then focused collisionally in the RF-only quadrupole that resulted in improvement in oa-TOFMS sensitivity and mass-resolving power.

In a modified design shown in Figure 7.7, Hashimoto *et al.* [96] replaced the Paul trap with an LIT and also trapped ions in the collisional RF-only quadrupole.† In this dual ion trap/oa-TOFMS instrument, ions were accumulated initially in an RF-only octopole, released and then trapped in a subsequent first LIT for precursor ion isolation and fragmentation. The product ions were focused collisionally in the second LIT for a short period comparable with the spectrum acquisition time of the oa-TOFMS, and then released into the oa-TOFMS for mass analysis. The optimum pressure ranges were found to be different for MS^n, and for effective collisional cooling. The isolation resolution for the fragmentation of reserpine ions was enhanced by a factor of three as the helium pressure was decreased from 20 to 0.2 mTorr. On the other hand, a pressure of *ca* 60 mTorr in the collisional LIT was found to be optimal in terms of the sensitivity and resolving power ($m/\Delta m = 10,000$) of the oa-TOFMS. The ion packets released from the second LIT were synchronized and time-delayed relative to the TOF extraction pulses, and the increase in sensitivity was attributed to the improvement in the duty cycle (2.5–4 times for a 200–700 m/z-range), as well as to the enhanced trapping efficiency of the LIT compared to that of the Paul trap [95]. The dual ion trap/oa-TOFMS capability was demonstrated in MS^3 experiments with 20 amol of reserpine. As mentioned above, time-delayed release of ions from a trap into the TOF extraction region results in improvements in duty cycle for ions in

---

* See also Volume 4, Chapter 19: A Quadrupole Ion Trap/Time-of-Flight Mass Spectrometer Combined with a Vacuum Matrix-Assisted Laser Desorption Ionization. Source, by Dimitris Papanastasiou, Omar Belgacem, Helen Montgomery, Mikhail Sudakov, and Emmanuel Raptakis.

† See also Volume 4, Chapter 12: Axially-resonant Excitation Linear Ion Trap (AREX LIT), by Yuichiro Hashimoto.

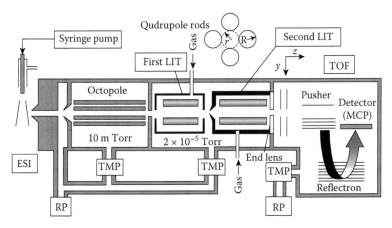

FIGURE 7.7 Schematic of the dual-LIT/oa-TOF. (Reproduced from Hashimoto, Y., Hasegawa, H., and Waki, I., *Rapid Commun. Mass Spectrom.* 2005, *19*, 1485–1491. With permission from John Wiley and Sons Inc.)

a limited *m/z*-range, implying the need to use a set of delays to cover a broad range of *m/z*-values [92]. It is worth noting that the charge density of the ion cloud, generated by ion accumulation in an ion trap prior to oa-TOFMS analysis, exceeds that of the continuous beam and can pose a challenge for the TOF acquisition system that is manifested as a signal saturation effect. Because of the inherently lower dynamic range, time-to-digital converters (TDCs) are more prone to this saturation effect than are analog-to-digital converters (ADC). In order to minimize the signal saturation effect during the recording of signals from higher ion density packets, relatively narrow ion packets were transformed into a quasi-continuous beam using a collisional RF-only quadrupole prior to TOFMS [95]. Such an approach, however, reduces the gain in duty cycle obtained by synchronizing the trap release with the TOF extraction pulses. To maximize the TDC dynamic range, Hashimoto *et al.* [96] trapped and released ions from the collisional quadrupole at a repetition rate equal to that of the TOFMS. During the trapping and release of the ions from the collisional quadrupole, the first LIT (in the dual trap/oa-TOFMS system) was performing the subsequent MS^n analysis, thus maximizing the duty cycle of the experiment.

Electron capture dissociation (ECD) [97] in an RF-only LIT has been developed recently, and coupled to the oa-TOFMS. Electron capture induces backbone cleavage of multiply protonated peptides and proteins by rupturing amine bonds, with the reaction cross-section at a maximum when induced by low kinetic energy electrons [98,99]. This approach is useful particularly for detecting peptide post-translational modification sites, the information being lost usually in collision-induced dissociation (CID) experiments [99]. Thus far, ECD had been limited mainly to FT-ICR in a way that constrained its utility for the broad bioanalytical community.* Baba *et al.* [100,101] and Satake *et al.* [102] have reported recently on the implementation of ECD

---

* See also Volume 4, Chapter 15: Fragmentation Techniques for Protein Ions Using Various Types of Ion Trap, by J. Franzen and K. P. Wanczek; Volume 5, Chapter 5: Fourier transform ion cyclotron resonance mass spectrometry in the analysis of peptides and proteins, by Helen J. Cooper.

in an RF-only LIT. Ions were confined radially by the force arising from the pseudo-potential, while axial confinement was accomplished by the DC potentials applied to the end-cap electrodes. Low kinetic energy electrons (< 1 eV) were confined to the trap axis by a magnetic field (150 mT) created by a cylindrical permanent magnet that enclosed the LIT. Accumulation times ranging from 7 to 10 ms were found to be optimum for the maximum ECD product yield. Precursor ions from a nanoESI source were selected first in a LIT (the CID trap) and then either were transmitted directly to an oa-TOFMS or were deflected orthogonally into the ECD trap using a quadrupole beam bender. Figure 7.8 shows the instrument, which can perform both CID (in the CID trap) or ECD (in the ECD trap), with the dissociation products being focused collisionally for high resolution detection with the oa-TOFMS [103].

Ibrahim *et al.* have developed an IFT coupled to an oa-TOFMS [104]. The IFT design is based on a stacked-ring assembly that incorporates a set of thin ring electrodes separated by dielectric spacers of the same thickness. A 180° phase-shifted RF field is applied to the adjacent electrodes to confine the ions radially [105]. The charge capacity of an RF field-focusing device is based on the pseudopotential, which is created by the oscillating electric field [106]. The pseudopotential in the IFT increases exponentially near the electrode surface, and has a 'flat-bottom' distribution around the axis of the device as compared to the lower order multipoles (for example, the parabolic profile obtained with an RF-only quadrupole). In addition, the pseudopotential in the IFT is governed by $V^*_{funnel} \approx 1/\delta^2$, where δ is the spacing between the adjacent electrodes of the funnel (typically much less than the funnel elements' inscribed radius) [107]. The pseudopotential in, for example, a quadrupole is $V^*_{quad} \approx 1/r_0^2$, where r_0 is the inscribed

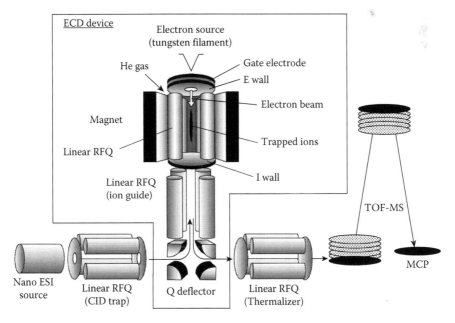

FIGURE 7.8 ECD-TOF mass spectrometer. (Reproduced from Satake, H., Hasegawa, H., Hirabayashi, A., Hashimoto, Y., and Baba, T., *Anal. Chem.* 2007, *79*, 8755–8761. With permission from the American Chemical Society.)

radius of the quadrupole. Therefore, when the regions of the ion funnel and quadrupole having the same inscribed radius are compared, the pseudopotential for the ion funnel is considerably greater than that of the quadrupole, provided that the same RF potential is applied to both the funnel and the quadrupole. The efficiency of trapping in RF devices depends upon the residual gas number density. Since the operational pressure (*ca* 1 Torr) of the IFT is 3–4 orders of magnitude higher than pressures used typically in a quadrupolar LIT, the IFT trapping efficiency is significantly greater than that of commercial quadrupolar ion traps. In order for the pseudopotential (and charge capacity) of the quadrupolar LIT ($r_0 = 4.1$ mm) to match that of the IFT ($\delta = 0.5$ mm), the RF potential applied to the quadrupole needs to be 64-fold greater than that applied to the IFT, limited practically by the onset of corona discharge in the intermediate pressure region. Shown in Figure 7.9, the IFT device encompasses an electrodynamic ion funnel [107,108] and the IFT. The ion funnel compresses radially a diffuse ion beam at the funnel exit. Inside the IFT, ions are confined radially by the applied RF field, while an axial barrier is generated by the potentials applied to terminal grids. Once ejected from the IFT by means of a 100 μs-wide pulse, an ion cloud is focused radially in the funnel-like converging section, introduced into a quadrupole interface and then into oa-TOFMS equipped with an ADC-based data acquisition system [104]. Ion current measurements indicated an IFT charge capacity of *ca* 10^7 charges, with the trapping efficiency ranging from 20 to 80%. Mass spectral measurements showed an increase in the signals obtained from model peptides, as well as a reduction in the baseline noise as compared to the continuous (no ion trapping) mode. The enhancement, by

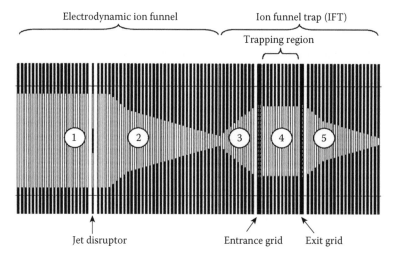

FIGURE 7.9 Schematic diagram of the ion funnel trap (IFT) configuration. The numbers refer to the different sections of the funnel and IFT. Sections (1) and (2) constitute the electrodynamic ion funnel where diffuse ion beam is focused. The focused ion beam is guided through section (3) into the trapping section (4). Ion packets exiting the trapping section (4) are focused in section (5) and then released into the subsequent stages of the oa-TOF instrument. (Reproduced from Ibrahim, Y., Belov, M.E., Tolmachev, A.V., Prior, D.C., and Smith, R.D., *Anal. Chem.* 2007, *79*, 7845–7852. With permission from the American Chemical Society.)

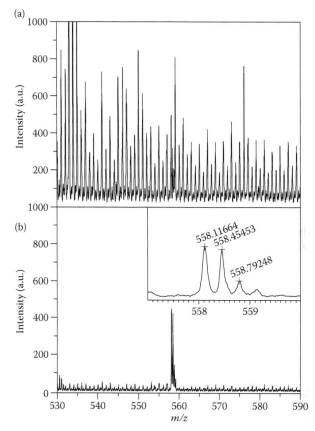

FIGURE 7.10 Improvement in the signal-to-noise ratio (*S/N*) of $[M + 3H]^{3+}$ for ESI of a 0.1 nM neurotensin solution: (a) Continuous regime; (b) Trapping mode. Accumulation time = 100 ms, extraction time = 70 μs. In both the continuous and trapping modes the sample infusion rate was 60 nL min^{-1} and the TOF acquisition time was 3 s. (Reproduced from Ibrahim, Y., Belov, M.E., Tolmachev, A.V., Prior, D.C., and Smith, R.D., *Anal. Chem.* 2007, *79*, 7845–7852. With permission from the American Chemical Society.)

a factor of up to 50, in the *S/N* was found to be more pronounced for low concentration samples, as shown in Figure 7.10. Ibrahim *et al.* [109] have implemented also an AGC capability with the IFT, and have evaluated the performance of the instrument in capillary LC experiments with tryptic digests of a *Shewanella oneidensis* proteome. As a result of sensitivity improvement and the related increase in mass measurement accuracy, the number of unique peptides identified in the AGC-IFT mode was *ca* fivefold greater than that obtained in the continuous regime.

7.4 ION TRAPPING WITH ION MOBILITY-TIME-OF-FLIGHT MASS SPECTROMETRY

As with the TOF mass spectrometer, IMS is an inherently pulsed analytical technique, in which a weak electric field is used to separate rapidly ions traversing a drift tube filled with a homogenous neutral gas [110]. Each cycle of operation in the IMS is initiated with injection of a discrete ion packet through an ion gate into the drift tube. Spatial separation of gas-phase ions is achieved through the combination of a drag force exerted on an ion by the gaseous medium and the strength of the electric field.* Under the condition of negligible spacecharge, the temporal full-width at half-maximum (FWHM), Δt, of the IMS signal is governed by [111]

$$\Delta t = \sqrt{t_{\text{gate}}^2 + \left(\frac{t_{\text{drift}}}{R_{\text{d}}}\right)^2} \tag{7.11}$$

and

$$R_{\text{d}} = \sqrt{\frac{LEze}{16k_b T \ln 2}} \tag{7.12},$$

where R_{d} is the thermal diffusion-limited maximum resolution, t_{gate} is the width of the IMS gate, t_{drift} is the drift time of the ion, L is the drift tube length, E is the electric field strength, z is the ion's charge number, e is the electronic charge k_b is Boltzmann's constant, and T is the absolute temperature.

Typically, an ion packet pulse width of < 1 ms and a total drift time of 10–100 ms are reported [111]. Because an IMS signal is obtained at a gate pulse width of 100–200 µs, the implied duty cycle of less than 1% restricts severely the instrument sensitivity and throughput and requires prolonged signal averaging to obtain useful information. Coupling an ion mobility spectrometer to a mass spectrometer represents another analytical challenge because of the pressure differential between an IMS drift tube (1–760 Torr) and a mass analyzer (10^{-6}–10^{-7} Torr). This pressure gradient imposes the need for a small aperture at the drift tube terminus that limits drastically ion transmission through an IMS–MS interface. The pulsing of ion packets into the drift tube may be conducted by chopping the continuous ion beam either mechanically or electrically (for example using a Bradbury-Nielsen type gate [112]). After admission of an ion packet into the drift tube, the continuous beam is diverted until the IMS separation is completed. Clearly, this pulsing mechanism results in severe ion losses. In order to improve the duty cycle without affecting the ion mobility resolving power, ion traps have been utilized to accumulate and to store the ions between injections of the ion packets. In addition to storing ions between gate releases, these traps can focus tightly the ion packets, both spatially and temporally, and so improve ion transmission through the entrance aperture of the drift tube. Hoaglund *et al.* [113,114] and Creaser *et al.* [115] coupled a Paul trap to IMS and demonstrated the benefits of the trap in improving the instrument duty cycle. The trap was pressurized to 10^{-4}–10^{-5} Torr by the helium gas escaping from the drift tube

---

* See also Volume 4, Chapter 5: High-Field Asymmetric Waveform Ion Mobility Spectrometry (FAIMS), by Randall W. Purves; Volume 5, Chapter 8: Applications of Traveling Wave Ion Mobility Mass Spectrometry, by Konstantinos Thalassinos and James H. Scrivens; Volume 5, Chapter 13: The Study of Ion/Molecule Reactions at Ambient Pressure with Ion Mobility Spectrometry and Ion Mobility/ Mass Spectrometry, by Gary A. Eiceman and John A. Stone.

entrance aperture when the drift tube pressure was 1–2 Torr. The signal intensity for all ions released from the trap reached 60% of that observed for a continuous beam of ions over a similar time period. In addition to improving the ion mobility duty cycle, the ion trap was used as an ion source to form those ions that are difficult to generate other than through ion/molecule reactions [115]. The study of non-covalent complexes of ethylmethylamine and *n*-propylamine with 18-crown-6 represents an example of these studies [115]. The duty cycle of the IMS was increased further with an octopole LIT. Because ions could be trapped along the octopole axis instead of in a small volume within the Paul trap, the octopole LIT offered higher charge capacity. Myung *et al.* [116] have estimated that improvements in signal levels of *ca* 50–200 times that obtained using the Paul trap, thus enabling a 10-fold lower detection limit. Apart from multipoles, RF-energized stacked-ring electrode arrangements have been used also to trap ions emitted from a continuous source prior to IMS separation [117–120]. As mentioned above, these traps employ a 180° phase-shifted RF field to confine the ions radially, whilst axial confinement is provided by the DC potential applied to the exit electrode. At higher pressures, the ion motion is relaxed by multiple collisions with the background gas, and the ions move along the axial DC field lines through the stacked-ring assembly that is aligned coaxially with the drift tube. Wyttenbach *et al.* [117] and Pringle *et al.* [118] have employed stacked-ring ion guides to accumulate and to pulse ion packets into an ion mobility spectrometer. The pressure in these traps was in the range of 10^{-2}–10^{-3} Torr (which is lower than that within the IMS drift tube) and an RF frequency and amplitude were 1–2.7 MHz and of 50–400 V (peak-to-peak), respectively. In an effort to minimize ion transfer losses into and from the IMS drift tube, Tang *et al.* [119] reported on a design in which the drift tube was capped between two electrodynamic ion funnels. This arrangement allowed removal of the entrance and exit apertures used conventionally in the drift tubes. The ion funnel that preceded the drift tube had an 'hour-glass' design and acted as an ion trap, while the second ion funnel at the drift tube terminus collimated the diffusively dispersed ion clouds into a TOFMS. Eliminating the drift tube entrance aperture resulted in an increase in the ion trap pressure to 4 Torr, which was challenging experimentally in terms of the previous designs that had no axial DC gradient. The hour-glass IFT design is essentially a stacked-ring trap of conical shape coupled to an ion funnel. The hour-glass IFT was separated from the drift tube by only a single grid, whose potential was pulsed to release trapped ion packets into the IMS. As reported by Tang *et al.* [119], an axial DC gradient of 24 V cm^{-1} was used in the IFT, which is higher than the value of 0.6 V cm^{-1} reported by Wyttenbach *et al.* [117]. Although the hour-glass IFT design enabled ion trapping and release at a pressure of 4 Torr, the trapping efficiency was low. To increase further the ion trapping and extraction efficiencies at a pressure of 4 Torr, Clowers *et al.* [120] developed an electrodynamic IFT that enabled the incorporation of additional features found to be critically important for the efficient IFT operation at high pressures. As shown in Figure 7.11, an additional 10 ring electrodes of constant inner diameter were incorporated into the diverging section of the hour-glass design, which was followed by a converging section. This modification resulted in several improvements. First, the ions were trapped in the cylindrical rather than in the conical section of the original hour-glass design, which reduced ion losses on the trap electrodes close to

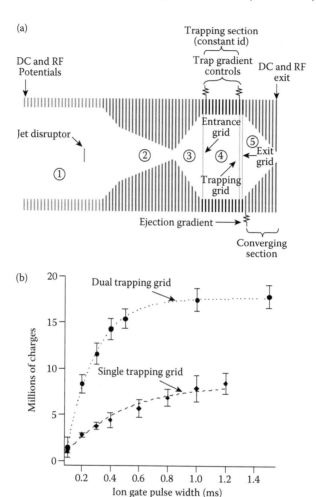

FIGURE 7.11 (a) Annotated schematic of the ion funnel trap consisting of five distinct regions that: (1) accept the rapidly expanding free jet of ions; (2) focus ions entering the chamber that houses the next three regions that (3) guide ions into (4) the ion trapping region; and (5) the final converging region and funnel exit. Regions (3)–(5) are separated by a series of DC-only grids used to trap and to eject the ions accumulated in region (4). (b) A plot illustrating the efficiency of ejection of the ion packet from the IFT as a function of the gate pulse-width. (Reproduced from Clowers, B.H., Ibrahim, Y.M., Prior, D.C., Danielson, W.F., Belov, M.E., and Smith, R.D., *Anal. Chem.* 2008, *80*, 612–623. With permission from the American Chemical Society.)

the grid. Second, the axial DC field in the IFT was decoupled from that applied to the ion funnel, ensuring independent control over the axial DC gradient in the IFT. While a DC gradient of 25 V cm^{-1} was applied to the ion funnel, the IFT axial DC gradient was kept at 1 V cm^{-1} increasing significantly the trapping efficiency [104]. A combination of the low axial DC gradient and the cylindrical trap geometry led to the trapping efficiency approaching unity for short accumulation times, increasing the IFT charge capacity to *ca* 10^7 elementary charges. Third, a dual exit grid configuration was implemented that facilitated ion accumulation in close proximity to the trap exit. This arrangement was found to increase the number of ions injected into the IMS over a short gate open event (*ca* 200 μs) by a factor of approximately three as compared to the single grid configuration (Figure 7.11). The IFT charge capacity could be reached in a few milliseconds during the analysis of higher concentration samples. In experiments with a 200 nM tryptic digest of BSA protein, the maximum sequence coverage (*ca* 75%) was achieved at accumulation times of 10–20 ms. Although this is a highly efficient IFT, it was found that such a trap could be filled with ions to its capacity in a small fraction of the IMS separation timescale, limiting the instrument duty cycle to 1–5%. To increase further the sensitivity of the IMS-TOF instrument, Belov *et al.* [121] combined the benefits of ion accumulation with multiple releases of ion packets into the IMS drift tube using an elaborate multiplexing scheme. Multiplexing implies introduction of multiple ion packets into the IMS drift tube on the timescale of a single separation. The timing for ion packet release into the drift tube is determined by a modified binary pseudo-random sequence. The premise for sensitivity improvement in the multiplexing experiment is rather simple: when the spectral line intensities are detected simultaneously in N measurements characterized by the uncorrelated noise, the theoretical increase in S/N over a single measurement is then expected to be $\approx \sqrt{N}$. The challenges in signal reconstruction are associated with diffusion and space-charge-induced broadening of ion packets in the IMS drift tube, and the enormous over-sampling due to signal acquisition with a TOF digitizer. A combination of ion trapping with the multiplexing scheme was shown to enhance the ion utilization efficiency beyond the 50% level that is predicted theoretically for modulation of the continuous beam using a Simplex matrix. The resultant raw mass spectrum in the multiplexed IMS-TOFMS experiments represents the sum of the intensities of all the ion species (often coming from different ion packets) that strike the TOF detector at a given time. The encoded raw IMS-TOFMS mass spectra were reconstructed successfully using an inverse transform algorithm. In the initial implementation of the multiplexing algorithm, the accumulation periods throughout the encoding sequence were varied, which required knowledge of a weighed function for signal reconstruction. To make the reconstruction routine more robust in the analysis of complex samples, the accumulation periods throughout the sequence were held constant and set to be equal to the duration of the shortest interval between two adjacent gate-open events (for example, 2 ms for 5-bit encoding sequence) [122]. In experiments with a mixture of several peptides, enhancement factors of 7–13 in the S/N ratio as a result of multiplexing were reported. In addition, both multiplexed and signal-averaging modes were used to obtain 70% sequence coverage of tryptically digested BSA. It was found that the same sequence coverage could be achieved 10 times faster using multiplexed IMS-TOFMS, as compared to the signal-averaged data from the IMS-TOFMS operating with a similar accumulation time of 2 ms. The key components of the multiplexed IMS-TOFMS technology are: (1) efficient ion accumulation in the IFT; (2) temporal separation between the release of adjacent ion packets that takes into account the ion packet broadening due to thermal diffusion and space-charge repulsion; and (3) constant and short gate-open events throughout the encoding sequence to ensure high IMS resolution. In LC experiments with highly complex proteolytic digests (for example, depleted human blood plasma), accumulation periods in the IFT need to be adjusted to accommodate the alternating ion production rates. As discussed in the previous section, the AGC capability with an ion trap fulfilled this purpose. Belov *et al.* [123] have reported recently on a dynamically multiplexed IMS-TOFMS approach that accounts for source function variation in a manner similar to the AGC feature in quadrupolar ion traps. In these new experiments, the bit number of the encoding sequence, corresponding to the ion accumulation intervals between adjacent gate-open events, was varied to account for the changes in the ion flux entering the IFT. Figure 7.12 shows the dynamically encoded and reconstructed signals for one of the reverse-phase fractions of the depleted human blood plasma sample in the course of a 15-min

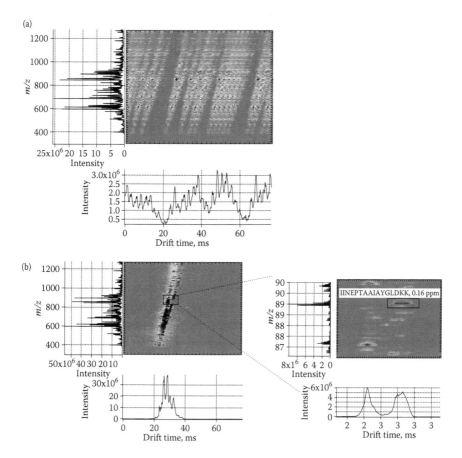

FIGURE 7.12 IMS-TOFMS signals detected and reconstructed in dynamic multiplexed IMS-TOFMS experiments with a 0.5 mg mL^{-1} depleted human blood plasma sample. (a) 5-bit encoded IMS-TOFMS signal from human blood plasma fraction 14 acquired in technical replicate #4; (b) reconstructed IMS-TOFMS signal from the 2D map in A. The inset in (b) shows one of the identified peptides from the depleted human blood plasma (sequence: IINEPTAAAIAYGLDKK) at a mass measurement accuracy of 0.16 ppm. (Reproduced from Belov, M.E., Clowers, B.H., Prior, D.C., Danielson III, W.F., Liyu, A.V., Petritis, B.O., and Smith, R.D., *Anal. Chem.* 2008, *80*, 5873–5883. With permission.)

experiment with 25 fractions. Based on signal intensity, the encoding sequence was adjusted to avoid overfilling the IFT and to provide maximum duty cycle per IMS separation. The use of a calibration function ensured that signals of high intensity were acquired with higher-bit encoding sequences that yield shorter accumulation times, while signals of low intensity were recorded with a low-bit sequences corresponding to longer accumulation times. As a result, a 15 minute-long dynamically multiplexed IMS-TOFMS experiment yielded the number of unique peptide identifications comparable to that obtained in two-hour long LC-FT-ICR run at an estimated false discovery rate of 4%. This multiplexed IMS-TOFMS instrument represents an ultra-high sensitivity high-throughput platform that could be employed potentially in applications such as candidate biomarker discovery experiments.

7.5 CONCLUSION

The rapidly evolving approaches for handling ion populations in diverse ion traps have shown that a combination of different techniques and the use of versatile hybrid instrumentation are presently important drivers for improving the quality of measurements obtainable with a range of mass spectrometry platforms. With the increasing need for higher analysis throughput, sensitivity, and mass accuracy, the complementary separation approaches characterized by higher speed and reproducibility are likely to provide an additional impetus for improving further ion trapping and related technologies (for example, selective trapping such as in DREAMs) for a wide range of important applications.

REFERENCES

1. Comisarow, M.B.; Marshall, A.G. Fourier transform ion cyclotron resonance mass spectroscopy. *Chem. Phys. Lett.* 1974, *25*, 282–283.
2. Wilkins, C.L.; Chowdhury, A.K.; Nuwaysir, L.M.; Coates, M.L. Fourier-transform mass-spectrometry-current status. *Mass Spectrom. Rev.* 1989, *8*, 67–92.
3. Dienes, T.; Pastor, S.J.; Schurch, S.; Scott, J.R.; Yao, J.; Cui, S.L.; Wilkins, C.L. Fourier transform mass spectrometry-advancing years (1992 mid 1996). *Mass Spectrom. Rev.* 1996, *15*, 163–211.
4. Marshall, A.G.; Hendrickson, C.L.; Jackson, G.S. Fourier transform ion cyclotron resonance mass spectrometry: a primer. *Mass Spectrom. Rev.* 1998, *17*, 1–35.
5. Marshall, A.G. Milestones in Fourier transform ion cyclotron resonance mass spectrometry technique development. *Int. J. Mass Spectrom.* 2000, *200*, 331–356.
6. Marshall, A.G.; Hendrickson, C.L. Fourier transform ion cyclotron resonance detection: principles and experimental configurations. *Int. J. Mass Spectrom.* 2002, *215*, 59–75.
7. Zhang, L.K.; Rempel, D.; Pramanik, B.N.; Gross, M.L. Accurate mass measurements by Fourier transform mass spectrometry. *Mass Spectrom. Rev.* 2005, *24*, 286–309.
8. Holliman, C.L.; Rempel, D.L.; Gross, M.L. Mass spectrometry in biomolecular sciences. *NATO ASI SERIES, Series C: Math. Phys. Sci.* 1996, *475*, 147–175.
9. Amster, I.J. Fourier transform mass spectrometry. *J. Mass Spectrom.* 1996, *31*, 1325–1337.
10. Wilkins, C.L. Special issue: Fourier transform mass spectrometry. *Trends Anal. Chem.* 1994, *13*, 223–251.
11. Marshall, A.G. Special issue: Fourier transform ion cyclotron resonance mass spectrometry. *Int. J. Mass Spectrom. Ion Processes.* 1996, *157/158*, 1–410.

12. Buchanan, M.V. *Fourier Transform Mass Spectrometry: Evolution, Innovation, and Applications (ACS Symposium Series).* Oxford University Press: New York, 1987; Vol. 359, p. 1–205.

13. Marshall, A.G.; Verdun, F.R. *Fourier Transforms in NMR, Optical and Mass Spectrometry: A User's Handbook.* Elsevier: Amsterdam, 1990.

14. Guan, S.; Marshall, A.G. Ion traps for Fourier transform ion cyclotron resonance mass spectrometry: principles and design of geometric and electric configurations. *Int. J. Mass Spectrom. Ion Processes* 1995, *146/147*, 261–296.

15. Peurrung, A.J.; Kouzes, R.T.; Barlow, S.E. The non-neutral plasma: an introduction to physics with relevance to cyclotron resonance mass spectrometry. *Int. J. Mass Spectrom. Ion Processes* 1996, *157/158*, 39–83.

16. White, W.D.; Malmberg, J.H.; Driscoll, C.F. Resistive-wall destabilization of diocotron waves. *Phys. Rev. Lett.* 1982, *49*, 1822–1826.

17. Jeffries, J.B.; Barlow, S.E.; Dunn, G.H. Theory of space-charge shift of ion cyclotron resonance frequencies. *Int. J. Mass Spectrom. Ion Processes* 1983, *54*, 169–187.

18. Schweikhard, L.; Ziegler, B.; Bopp, H.; Lutzenkirchen, K. The trapping condition and a new instability of the ion motion in the ion cyclotron resonance trap. *Int. J. Mass Spectrom. Ion Processes* 1995, *141*, 77–90.

19. Beachamp, J.L.; Armstrong, J.T. An ion ejection technique for the study of ion-molecule reactions with ion cyclotron resonance spectroscopy. *Rev. Sci. Instrum.* 1969, *40*, 123–128.

20. Jackson, J.D. *Classical Electrodynamics.* 2nd ed.; John Wiley and Sons: New York, 1975.

21. Gorshkov, M.V.; Nikolaev, E.N. Optimal cyclotron radius for high resolution FT-ICR spectrometry. *Int. J. Mass Spectrom. Ion Processes* 1993, *125*, 1–8.

22. Barlow, S.E.; Tinkle, M.D. "Linearizing" an ion cyclotron resonance cell. *Rev. Sci. Instrum.* 2002, *73*, 4185–4200.

23. Schweikhard, L.; Marshall, A.G. Excitation modes for Fourier transform-ion cyclotron resonance mass spectrometry. *J. Am. Soc. Mass Spectrom.* 1993, *4*, 433–452.

24. Comisarow, M.B. Cubic trapped-ion cell for ion-cyclotron resonance. *Int. J. Mass Spectrom. Ion Phys.* 1981, *37*, 251–257.

25. McIver, R.T. A trapped ion analyzer cell for ion cyclotron resonance spectroscopy. *Rev. Sci. Instrum.* 1970, *41*, 555–558.

26. Hunter, R.L.; Sherman, M.G.; McIver, R.T. An elongated trapped-ion cell for ion cyclotron resonance mass spectrometry with a superconducting magnet. *Int. J. Mass Spectrom. Ion Phys.* 1983, *50*, 259–274.

27. Gabrielse, G.; Mackintosh, F.C. Cylindrical Penning traps with orthogonalized anharmonicity compensation. *Int. J. Mass Spectrom. Ion Processes* 1984, *57*, 1–17.

28. Kofel, P.; Allemann, M.; Kellerhals, H.P.; Wanczek, K.P. Coupling of axial and radial motions in ICR cells during excitation. *Int. J. Mass Spectrom. Ion Processes* 1986, *74*, 1–12.

29. Brown, L.S.; Gabrielse, G. Geonium theory: physics of a single electron or ion in a Penning trap. *Rev. Mod. Phys.* 1986, *58*, 233–311.

30. Schweikhard, L.; Lindinger, M.; Kluge, H-J. Parametric-mode-excitation/dipole-mode-detection Fourier-transform–ion-cyclotron-resonance spectrometry. *Rev. Sci. Instrum.* 1990, *61*, 1055–1058.

31. Beu, S.C.; Laude, D.A. Open trapped ion cell geometries for Fourier transform ion cyclotron resonance mass spectrometry. *Int. J. Mass Spectrom. Ion Processes* 1992, *112*, 215–230.

32. Allemann, M.; Kellerhals, H.P.; Wanczek, K-P. Sidebands in the ICR spectrum and their application for exact mass determination. *Chem. Phys. Lett.* 1981, *84*, 547–551.

33. Alber, G.M.; Marshall, A.G.; Hill, N.C.; Schweikhard, L.; Ricca, T.L. Ultra-high resolution Fourier transform ion cyclotron resonance mass spectrometer. *Rev. Sci. Instrum.* 1993, *64*, 1845–1852.

34. Grosshans, P.B.; Marshall, A.G. Theory of ion cyclotron resonance mass spectrometry: resonant excitation and radial ejection in orthorhombic and cylindrical ion traps. *Int. J. Mass Spectrom. Ion Processes* 1990, *100*, 347–379.

35. Wang, M.D.; Marshall, A.G. Elimination of z-ejection in Fourier transform ion cyclotron resonance mass spectrometry by radio-frequency electric-field shimming. *Anal. Chem.* 1990, *62*, 515–520.

36. Vartanian, V.H.; Hadjarab, F.; Laude, D.A. Open cell analog of the screened trapped-ion cell using compensation electrodes for Fourier transform ion cyclotron resonance mass spectrometry. *Int. J. Mass Spectrom. Ion Processes* 1995, *151*, 175–187.

37. Beu, S.C.; Laude, D.A. Elimination of axial ejection during excitation with a capacitively coupled open trapped-ion cell for Fourier-transform ion-cyclotron resonance mass-spectrometry. *Anal. Chem.* 1992, *64*, 177–180.

38. Caravatti, P.; Allemann, M. The 'infinity cell': a new trapped-ion cell with radiofrequency covered trapping electrodes for Fourier-transform ion cyclotron resonance mass spectrometry. *Org. Mass Spectrom.* 1991, *26*, 514–518.

39. Nikolaev, E.N.; Miluchihin, N.V.; Inoue, M. Evolution of an ion cloud in a Fourier transform ion cyclotron resonance mass spectrometer during signal detection: its influence on spectral line shape and position. *Int. J. Mass Spectrom. Ion Processes* 1995, *148*, 145–157.

40. Mitchell, D.W.; Smith, R.D. Cyclotron motion of 2 coulombically interacting ion clouds with implications to Fourier-transform ion-cyclotron resonance mass-spectrometry. *Phys. Rev. E* 1995, *52*, 4366–4386.

41. Gabrielse, G.; Haarsma, L.; Rolston, S.L. Open-endcap Penning traps for high precision experiments. *Int. J. Mass Spectrom. Ion Processes* 1989, *88*, 319–332.

42. Tolmachev, A.V.; Robinson, E.W.; Wu, S.; Kang, H.; Lourette, N.M.; Paša-Tolić, L.; Smith, R.D. Trapped-ion cell with improved dc potential harmonicity for FTICR MS. *J. Am. Soc. Mass Spectrom.* 2008, *19*, 586–597.

43. Kofel, P.; Allemann, M.; Kellerhals, H.P.; Wanczek, K.P. External generation of ions in ICR spectrometry. *Int. J. Mass Spectrom. Ion Processes* 1985, *65*, 97–103.

44. McIver, R.T.; Hunter, R.L.; Bowers, M.T. Coupling a quadrupole mass spectrometer and Fourier transform mass spectrometer. *Int. J. Mass Spectrom. Ion Processes* 1985, *64*, 67–77.

45. Littlejohn, D.P.; Ghaderi, S. Mass spectrometer and method. *U.S. Patent* 1986, 4,581,533.

46. Alford, J.M.; Williams, P.E.; Trevor, D.J.; Smalley, R.E. Metal cluster ion cyclotron resonance. Combining metal cluster beam technology with FTICR. *Int. J. Mass Spectrom. Ion Processes* 1986, *72*, 33–51.

47. Caravatti, P. Method and apparatus for the accumulation of ions in a trap of an ion cyclotron resonance spectrometer, by transferring the kinetic energy of the motion parallel to the magnetic field into directions perpendicular to the magnetic field. *U.S. Patent* 1990, 4,924,089.

48. Stacey, C.C.; Kruppa, G.H.; Watson, C.H.; Wronka, J.; Laukien, F.H.; Banks, J.F.; Whitehouse, C.M.; Voyksner, R.D. Reverse-phase liquid chromatography electrospray ionization Fourier transform mass spectrometry in the analysis of peptides. *Rapid Commun. Mass Spectrom.* 1994, *8*, 513–516.

49. Grosshans, P.B.; Chen, R.; Limbach, P.A.; Marshall, A.G. Linear excitation and detection in Fourier transform ion cyclotron resonance mass spectrometry. *Int. J. Mass Spectrom. Ion Processes* 1994, *139*, 169–189.

50. Francl, T.J.; Fukuda, E.K.; McIver, R.T. A diffusion-model for non-reactive ion loss in pulsed ion cyclotron resonance experiments. *Int. J. Mass Spectrom. Ion Phys.* 1983, *50*, 151–167.

51. Beu, S.C.; Laude, D.A. Ion trapping and manipulation in a tandem time-of-flight-Fourier transform mass spectrometer. *Int. J. Mass Spectrom. Ion Processes* 1991, *104*, 109–127.

52. Beu, S.C.; Senko, M.W.; Quinn, J.P.; McLafferty, F.W. Improved Fourier-transform ion-cyclotron-resonance mass spectrometry of large biomolecules. *J. Am. Soc. Mass Spectrom.* 1993, *4*, 190–192.

53. Belov, M.E.; Gorshkov, M.V.; Udseth, H.R.; Anderson, G.A.; Tolmachev, A.V.; Prior, D.C.; Harkewicz, R.; Smith, R.D. Initial implementation of an electrodynamic ion funnel with Fourier transform ion cyclotron resonance mass spectrometry. *J. Am. Soc. Mass Spectrom.* 2000, *11*, 19–23.

54. Belov, M.E.; Gorshkov, M.V.; Udseth, H.R.; Anderson, G.A.; Smith, R.D. Zeptomole-sensitivity electrospray ionization-Fourier transform ion cyclotron resonance mass spectrometry of proteins. *Anal. Chem.* 2000, *72*, 2271–2279.

55. Winger, B.E.; Hofstadler, S.A.; Bruce, J.E.; Udseth, H.R.; Smith, R.D. High-resolution accurate mass measurements of biomolecules using a new electrospray ionization ion cyclotron resonance mass spectrometer. *J. Am. Soc. Mass Spectrom.* 1993, *4*, 566–577.

56. Senko, M.W.; Hendrickson, C.L.; Emmett, M.R.; Shi, S.D.H.; Marshall, A.G. External accumulation of ions for enhanced electrospray ionization Fourier transform ion cyclotron resonance mass spectrometry. *J. Am. Soc. Mass Spectrom.* 1997, *8*, 970–976.

57. Belov, M.E.; Nikolaev, E.N.; Anderson, G.A.; Auberry, K.J.; Harkewicz, R.; Smith, R.D. Electrospray ionization-Fourier transform ion cyclotron mass spectrometry using ion preselection and external accumulation for ultrahigh sensitivity. *J. Am. Soc. Mass Spectrom.* 2001, *12*, 38–48.

58. McIver, R.T. Apparatus and method for injection of ions into an ion cyclotron resonance cell *U.S. Patent* 1985, 4,535,235.

59. Collings, B.A.; Douglas, D.J. Observation of higher order quadrupole excitation frequencies in a linear ion trap. *J. Am. Soc. Mass Spectrom.* 2000, *11*, 1016–1022.

60. Dehmelt, H.G. Radiofrequency spectroscopy of stored ions I: storage. *Adv. At. Mol. Phys.* 1967, *3*, 53–72.

61. Welling, M.; Schuessler, H.A.; Thompson, R.I.; Walther, H. Ion/molecule reactions, mass spectrometry and optical spectroscopy in a linear ion trap. *Int. J. Mass Spectrom. Ion Processes* 1998, *172*, 95–114.

62. Landau, L.D.; Lofshitz, E.M. *Mechanics*. Pergamon Press: Oxford, 1976; Vol. 1.

63. Campbell, J.M.; Collings, B.A.; Douglas, D.J. A new linear ion trap time-of-flight system with tandem mass spectrometry capabilities. *Rapid Commun. Mass Spectrom.* 1998, *12*, 1463–1474.

64. Cha, B.C.; Blades, M.; Douglas, D.J. An interface with a linear quadrupole ion guide for an electrospray-ion trap mass spectrometer system. *Anal. Chem.* 2000, *72*, 5647–5654.

65. Belov, M.E.; Anderson, G.A.; Smith, R.D. Higher-resolution data-dependent selective external ion accumulation for capillary LC-FTICR. *Int. J. Mass Spectrom.* 2002, *218*, 265–279.

66. Belov, M.E.; Nikolaev, E.N.; Anderson, G.A.; Udseth, H.R.; Conrads, T.P.; Veenstra, T.D.; Masselon, C.D.; Gorshkov, M.V.; Smith, R.D. Design and performance of an ESI interface for selective external ion accumulation coupled to a Fourier transform ion cyclotron mass spectrometer. *Anal. Chem.* 2001, *73*, 253–261.

67. Wilcox, B.E.; Hendrickson, C.L.; Marshall, A.G. Improved ion extraction from a linear octopole ion trap: SIMION analysis and experimental demonstration. *J. Am. Soc. Mass Spectrom.* 2002, *13*, 1304–1312.

68. Quenzer, T.L.; Emmett, M.R.; Hendrickson, C.L.; Kelly, P.H.; Marshall, A.G. High sensitivity Fourier transform ion cyclotron resonance mass spectrometry for biological analysis with nano-LC and microelectrospray ionization. *Anal. Chem.* 2001, *73*, 1721–1725.

69. Belov, M.E.; Anderson, G.A.; Angell, N.H.; Shen, Y.; Paša-Tolić, L.; Udseth, H.R.; Smith, R.D. Dynamic range expansion applied to mass spectrometry based on data-dependent selective ion ejection in capillary liquid chromatography Fourier transform ion cyclotron resonance for enhanced proteome characterization. *Anal. Chem.* 2001, *73*, 5052–5060.

70. Paša-Tolić, L.; Harkewicz, R.; Anderson, G.A.; Tolic, N.; Shen, Y.F.; Zhao, R.; Thrall, B.; Masselon, C.; Smith, R.D. Increased proteome coverage for quantitative peptide abundance measurements based upon high performance separations and DREAMS FTICR mass spectrometry. *J. Am. Soc. Mass Spectrom.* 2002, *13*, 954–963.

71. Patrie, S.M.; Charlebois, J.P.; Whipple, D.; Kelleher, N.L.; Hendrickson, C.L.; Quinn, J.P.; Marshall, A.G.; Mukhopadhyay, B. Construction of a hybrid quadrupole/ Fourier transform ion cyclotron resonance mass spectrometer for versatile MS/MS above 10 kDa. *J. Am. Soc. Mass Spectrom.* 2004, *15*, 1099–1108.

72. Patrie, S.M.; Robinson, D.E.; Meng, F.Y.; Du, Y.; Kelleher, N.L. Strategies for automating top-down protein analysis with Q-FTICR MS. *Int. J. Mass Spectrom.* 2004, *234*, 175–184.

73. Belov, M.E.; Nikolaev, E.N.; Harkewicz, R.; Masselon, C.D.; Alving, K.; Smith, R.D. Ion discrimination during ion accumulation in a quadrupole interface external to a Fourier transform ion cyclotron resonance mass spectrometer. *Int. J. Mass Spectrom.* 2001, *208*, 205–225.

74. Tolmachev, A.V.; Udseth, H.R.; Smith, R.D. Radial stratification of ions as a function of mass to charge ratio in collisional cooling radio frequency multipoles used as ion guides or ion traps. *Rapid Commun. Mass Spectrom.* 2000, *14*, 1907–1913.

75. Sannes-Lowery, K.; Griffey, R.H.; Kruppa, G.H.; Speir, J.P.; Hofstadler, S.A. Multipole storage assisted dissociation, a novel in-source dissociation technique for electrospray ionization generated ions. *Rapid Commun. Mass Spectrom.* 1998, *12*, 1957–1961.

76. Sannes-Lowery, K.A.; Hofstadler, S.A. Characterization of multipole storage assisted dissociation: implications for electrospray ionization mass spectrometry characterization of biomolecules. *J. Am. Soc. Mass Spectrom.* 2000, *11*, 1–9.

77. Håkansson, K.; Axelsson, J.; Palmblad, M.; Håkansson, P. Mechanistic studies of multipole storage assisted dissociation. *J. Am. Soc. Mass Spectrom.* 2000, *11*, 210–217.

78. Belov, M.E.; Gorshkov, M.V.; Udseth, H.R.; Smith, R.D. Controlled ion fragmentation in a 2-D quadrupole ion trap for external ion accumulation in ESI FTICR mass spectrometry. *J. Am. Soc. Mass Spectrom.* 2001, *12*, 1312–1319.

79. Belov, M.E.; Zhang, R.; Strittmatter, E.F.; Prior, D.C.; Tang, K.; Smith, R.D. Automated gain control and internal calibration with external ion accumulation capillary liquid chromatography-electrospray ionization-Fourier transform ion cyclotron resonance. *Anal. Chem.* 2003, *75*, 4195–4205.

80. Schwartz, J.C.; Zhou, X.; Bier, M.E. Method and apparatus of increasing dynamic range and sensitivity of a mass spectrometer. *U.S. Patent* 1996, 5,572,022.

81. Syka, J.E.P.; Marto, J.A.; Bai, D.L.; Horning, S.; Senko, M.W.; Schwartz, J.C.; Ueberheide, B.; Garcia, B.; Busby, S.; Muratore, T.; Shabanowitz, J.; Hunt, D.F. Novel linear quadrupole ion trap/FT mass spectrometer: performance characterization and use in the comparative analysis of histone H3 post-translational modifications. *J. Proteome Res.* 2004, *3*, 621–626.

82. Belov, M.E.; Robinson, E.W.; Liyu, A.V.; Prior, D.C.; Smith, R.D. Enhanced dynamic range characterization of human plasma proteome with hybrid Fourier transform ion cyclotron resonance mass spectrometry. *J. Proteome Res.* 2009, manuscript in preparation.

83. Smith, R.D.; Anderson, G.A.; Lipton, M.S.; Paša-Tolić, L.; Shen, Y.F.; Conrads, T.P.; Veenstra, T.D.; Udseth, H.R. An accurate mass tag strategy for quantitative and high-throughput proteome measurements. *Proteomics* 2002, *2*, 513–523.

84. Petyuk, V.A.; Qian, W.J.C.M.; Wang, H.X.; Livesay, E.A.; Monroe, M.E.; Adkins, J.N.; Jaitly, N.; Anderson, D.J.; Camp, D.G.; Smith, D.J.; Smith, R.D. Spatial mapping of protein abundances in the mouse brain by voxelation integrated with high-throughput liquid chromatography-mass spectrometry. *Genome Res.* 2007, *17*, 328–336.

85. Steen, H.; Kuster, B.; Mann, M. Quadrupole time-of-flight versus triple-quadrupole mass spectrometry for the determination of phosphopeptides by precursor ion scanning. *J. Mass Spectrom.* 2001, *36*, 782–790.

86. Dodonov, A.F.; Kozlovski, V.I.; Soulimenkov, I.V.; Raznikov, V.V.; Loboda, A.V.; Zhen, Z.; Horwath, T.; Wollnik, H. High-resolution electrospray ionization orthogonal-injection time-of-flight mass spectrometer. *Eur. J. Mass Spectrom.* 2000, *6*, 481–490.

87. Wiley, W.C.; McLaren, I.H. Time-of-flight mass spectrometer with improved resolution. *Rev. Sci. Instrum.* 1955, *26*, 1150–1157.

88. Mamyrin, B.A.; Karataev, V.I.; Schmikk, D.V.; Zagulin, V.A. The mass-reflectron, a new non-magnetic time-of-flight mass spectrometer with high resolution. *Sov. Phys. JETP.* 1973, *37*, 45–48.

89. Douglas, D.J.; French, J.B. Collisional focusing effects in radio frequency quadrupoles. *J. Am. Soc. Mass Spectrom.* 1992, *3*, 398–408.

90. Krutchinsky, A.N.; Chernushevich, I.V.; Spicer, V.L.; Ens, W.; Standing, K.G. Collisional damping interface for an electrospray ionization time-of-flight mass spectrometer. *J. Am. Soc. Mass Spectrom.* 1998, *9*, 569–579.

91. Dresch, T.; Gulcicek, E.E.; Whitehouse, C. Ion storage time-of-flight mass spectrometer. *U.S. Patent* 2000, 6,020,586.

92. Chernushevich, I.V. Duty cycle improvement for a quadrupole-time-of-flight mass spectrometer and its use for precursor ion scans. *Eur. J. Mass Spectrom.* 2000, *6*, 471–479.

93. Douglas, D.J.; Frank, A.J.; Mao, D. Linear ion traps in mass spectrometry. *Mass Spectrom. Rev.* 2005, *24*, 1–29.

94. Collings, B.A.; Campbell, J.M.; Mao, D.; Douglas, D.J. A combined linear ion trap time-of-flight system with improved performance and MS^n capabilities. *Rapid Commun. Mass Spectrom.* 2001, *15*, 1777–1795.

95. Hashimoto, Y.; Waki, I.; Yoshinari, K.; Shishika, T.; Terui, Y. Orthogonal trap time-of-flight mass spectrometer using a collisional damping chamber. *Rapid Commun. Mass Spectrom.* 2005, *19*, 221–226.

96. Hashimoto, Y.; Hasegawa, H.; Waki, I. Dual linear ion trap/orthogonal acceleration time-of-flight mass spectrometer with improved precursor ion selectivity. *Rapid Commun. Mass Spectrom.* 2005, *19*, 1485–1491.

97. Zubarev, R.A.; Kelleher, N.L.; McLafferty, F.W. Electron capture dissociation of multiply charged protein cations. A nonergodic process. *J. Am. Chem. Soc.* 1998, *120*, 3265–3266.

98. Zubarev, R.A.; Horn, D.M.; Fridriksson, E.K.; Kelleher, N.L.; Kruger, N.A.; Lewis, M.A.; Carpenter, B.K.; McLafferty, F.W. Electron capture dissociation for structural characterization of multiply charged protein cations. *Anal. Chem.* 2000, *72*, 563–573.

99. Zubarev, R.A. Reactions of polypeptide ions with electrons in the gas phase. *Mass Spectrom. Rev.* 2003, *22*, 57–77.

100. Baba, T.; Hashimoto, Y. Mass spectrometer. *U.S. Patent* 2007, 7,166,835.

101. Baba, T.; Hashimoto, Y.; Hasegawa, H.; Hirabayashi, A.; Waki, I. Electron capture dissociation in a radio frequency ion trap. *Anal. Chem.* 2004, *76*, 4263–4266.

102. Satake, H.; Hasegawa, H.; Hirabayashi, A.; Hashimoto, Y.; Baba, T. Fast multiple electron capture dissociation in a linear radio frequency quadrupole ion trap. *Anal. Chem.* 2007, *79*, 8755–8761.

103. Deguchi, K.; Ito, H.; Baba, T.; Hirabayashi, A.; Nakagawa, H.; Fumoto, M.; Hinou, H.; Nishimura, S-I. Structural analysis of O-glycopeptides employing negative- and positive-ion multi-stage mass spectra obtained by collision-induced and electron-capture dissociations in linear ion trap time-of-flight mass spectrometry. *Rapid Commun. Mass Spectrom.* 2007, *21*, 691–698.

104. Ibrahim, Y.; Belov, M.E.; Tolmachev, A.V.; Prior, D.C.; Smith, R.D. Ion funnel trap interface for orthogonal time-of-flight mass spectrometry. *Anal. Chem.* 2007, *79*, 7845–7852.

105. Gerlich, D. Inhomogeneous RF fields: a versatile tool for the study of processes with slow ions. In *State-Selected and State-to-State Ion-Molecule Reaction Dynamics. Part 1. Experiment*, Ng, C-Y.; Baer, M., Eds. Wiley: New York, 1992; Vol. 82, pp. 1–176.

106. Tolmachev, A.V.; Kim, T.; Udseth, H.R.; Smith, R.D.; Bailey, T.H.; Futrell, J.H. Simulation-based optimization of the electrodynamic ion funnel for high sensitivity electrospray ionization mass spectrometry. *Int. J. Mass Spectrom.* 2000, *203*, 31–47.

107. Shaffer, S.A.; Tang, K.; Anderson, G.A.; Prior, D.C.; Udseth, H.R.; Smith, R.D. A novel ion funnel for focusing ions at elevated pressure using electrospray ionization mass spectrometry. *Rapid Commun. Mass Spectrom.* 1997, *11*, 1813–1817.

108. Shaffer, S.A.; Tolmachev, A.; Prior, D.C.; Anderson, G.A.; Udseth, H.R.; Smith, R.D. Characterization of a new electrodynamic ion funnel interface for electrospray ionization mass spectrometry. *Anal. Chem.* 1999, *71*, 2957–2964.

109. Ibrahim, Y.M.; Belov, M.E.; Liyu, A.V.; Smith, R.D. Automated gain control ion funnel trap for orthogonal time-of-flight mass spectrometry. *Anal. Chem.* 2008, *80*, 5367–5376.

110. Eiceman, G.A.; Karpas, Z. *Ion Mobility Spectrometry.* 2nd ed.; CRC Press/Taylor & Frances: Boca Raton, FL, 2005.

111. Revercomb, H.E.; Mason, E.A. Theory of plasma chromatography/gaseous electrophoresis – a review. *Anal. Chem.* 1975, *47*, 970–983.

112. Bradbury, N.E.; Nielsen, R.A. Absolute values of the electron mobility in hydrogen. *Phys. Rev.* 1936, *49*, 388–393.

113. Hoaglund, C.S.; Valentine, S.J.; Clemmer, D.E. An ion trap interface for ESI-ion mobility experiments. *Anal. Chem.* 1997, *69*, 4156–4161.

114. Hoaglund-Hyzer, C.S.; Clemmer, D.E. Ion trap/ion mobility/quadrupole/time-of-flight mass spectrometry for peptide mixture analysis. *Anal. Chem.* 2001, *73*, 177–184.

115. Creaser, C.S.; Benyezzar, M.; Griffiths, J.R.; Stygall, J.W. A tandem ion trap/ion mobility spectrometer. *Anal. Chem.* 2000, *72*, 2724–2729.

116. Myung, S.; Lee, Y.J.; Moon, M.H.; Taraszka, J.; Sowell, R.; Koeniger, S.; Hilderbrand, A.E.; Valentine, S.J.; Cherbas, L.; Cherbas, P.; Kaufmann, T.C.; Miller, D.F.; Mechref, Y.; Novotny, M.V.; Ewing, M.A.; Sporleder, C.R.; Clemmer, D.E. Development of high-sensitivity ion trap ion mobility spectrometry time-of-flight techniques: a high-throughput Nano-LC-IMS-TOF separation of peptides arising from a drosophila protein extract. *Anal. Chem.* 2003, *75*, 5137–5145.

117. Wyttenbach, T.; Kemper, P.R.; Bowers, M.T. Design of a new electrospray ion mobility mass spectrometer. *Int. J. Mass Spectrom.* 2001, *212*, 13–23.

118. Pringle, S.D.; Giles, K.; Wildgoose, J.L.; Williams, J.P.; Slade, S.E.; Thalassinos, K.; Bateman, R.H.; Bowers, M.T.; Scrivens, J.H. An investigation of the mobility separation of some peptide and protein ions using a new hybrid quadrupole/travelling wave IMS/oa-ToF instrument. *Int. J. Mass Spectrom.* 2007, *261*, 1–12.

119. Tang, K.; Shvartsburg, A.A.; Lee, H-N.; Prior, D.C.; Buschbach, M.A.; Li, F.; Tolmachev, A.; Anderson, G.A.; Smith, R.D. High-sensitivity ion mobility spectrometry/mass spectrometry using electrodynamic ion funnel interfaces. *Anal. Chem.* 2005, *77*, 3330–3339.

120. Clowers, B.H.; Ibrahim, Y.M.; Prior, D.C.; Danielson, W.F.; Belov, M.E.; Smith, R.D. Enhanced ion utilization efficiency using an electrodynamic ion funnel trap as an injection mechanism for ion mobility spectrometry. *Anal. Chem.* 2008, *80*, 612–623.

121. Belov, M.E.; Buschbach, M.A.; Prior, D.C.; Tang, K.; Smith, R.D. Multiplexed ion mobility spectrometry-orthogonal time-of-flight mass spectrometry. *Anal. Chem.* 2007, *79*, 2451–2462.

122. Clowers, B.H.; Belov, M.E.; Prior, D.C.; Danielson, W.F.; Ibrahim, Y.; Smith, R.D. Pseudorandom sequence modifications for ion mobility orthogonal time-of-flight mass spectrometry. *Anal. Chem.* 2008, *80*, 2464–2473.
123. Belov, M.E.; Clowers, B.H.; Prior, D.C.; Danielson III, W.F.; Liyu, A.V.; Petritis, B.O.; Smith, R.D. Dynamically multiplexed ion mobility time of flight mass spectrometry. *Anal. Chem.* 2008, *80*, 5873–5883.

8 Radio Frequency-Only-Mode Event and Trap Compensation in Penning Fourier Transform Mass Spectrometry

Adam M. Brustkern, Don L. Rempel, and Michael L. Gross

CONTENTS

8.1 Introduction 434
8.2 Radio Frequency-Only-Mode Event 437
    8.2.1 Motivation 437
    8.2.2 Design and Implementation 437
        8.2.2.1 Stability Diagram of a Paul Trap in a B Field 438
    8.2.3 Application to Ion Chemistry 443
    8.2.4 Application to Ion Cloud Manipulation 445
        8.2.4.1 Collisional Focusing of Radially Dispersed
                Ion Clouds 445
        8.2.4.2 RF-Only-Mode Event with Electrical Trap
                Compensation 446
    8.2.5 Potential Analytical Applications of the RF-Only-
        Mode Event 447
        8.2.5.1 In-Field MALDI for Minimizing Mass
                Discrimination in Synthetic Polymer Analysis 447
        8.2.5.2 The Role of the RF-Only-Mode Event In MS/MS 448
8.3 Trap Compensation 448
    8.3.1 Introduction 448
    8.3.2 Particle-in-a-Box 449
        8.3.2.1 Field Shape and Design Goal 449
        8.3.2.2 Methods of Implementation 449

             8.3.2.2.1   An Elongated Trapped-Ion Trap for Ion
                                Cyclotron Resonance Mass Spectrometry 449
             8.3.2.2.2   A 'Screened' Electrostatic Ion Trap 452
             8.3.2.2.3   Field-Corrected Ion Trap 453
             8.3.2.2.4   Improvement of the Electric Field in the
                                Cylindrical Ion Trap 454
             8.3.2.2.5   Optimization of a Fixed-Volume, Open
                                Cylindrical Trap and an Open Cylindrical
                                Trap Analog of the Closed Screened Ion
                                Trap by using Compensation Electrodes 453
     8.3.3   Quadrupolar Well 457
             8.3.3.1   Field Shape and Design Goal 457
             8.3.3.2   Methods of Implementation 457
                       8.3.3.2.1   High Precision Mass Measurements using
                                Compensated Hyperbolic Penning Traps 455
                   8.3.3.2.2   Matrix-Shimmed Ion Cyclotron Resonance
                                Ion Trap 458
                   8.3.3.2.3   Ion Trap with Improved DC Potential
                                Harmonicity 458
                   8.3.3.2.4   An Electrically Compensated Trap
                                Designed to Eighth Order 461
8.4   Conclusions 466
References 466

8.1   INTRODUCTION

Fourier transform ion cyclotron resonance mass spectrometry (FT-ICR MS, or FTMS), first described by Comisarow and Marshall [1], is based on the circular motion of a charged particle in a uniform magnetic field. This motion is referred to as cyclotron motion and its frequency, ω, is given by the cyclotron equation

$$\omega = \frac{zeB}{m} \tag{8.1}$$

where z is the charge number of the ion charge, e is the electronic charge, m is the mass, and B is the magnetic field induction. As can be seen from this equation, the frequency of motion is strongly dependent upon the particle's mass-to-charge ratio (m/z), making FT-ICR MS analytically useful. In a modern FT-ICR MS instrument, built around a 7 tesla (T) superconducting magnet, for example, ion cyclotron frequencies are typically in a range from a several kilohertz (high m/z) to a few megahertz (low m/z). For example, an ion of m/z 1000 in a 7 T magnetic induction has a cyclotron frequency of approximately 100 kHz. If this ion has a cyclotron orbit radius of 1 cm, then it will travel a distance greater than six kilometers in a single second. Given that the ions are traveling such long distances during the time of detection, it is important that the background pressure in FT-ICR MS instruments be very low.

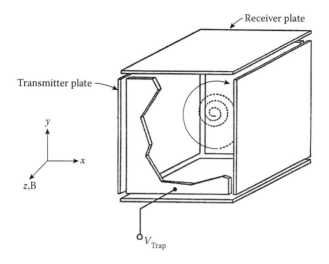

FIGURE 8.1 Schematic of a cubic trapped ion analyzer. (Adapted from Ledford, Jr., E.B., Ghaderi, S., White, R.L., Spencer, R.B., Kulkarni, P.S., Wilkins, C.L., and Gross, M.L., *Anal. Chem.* 1980, *52*, 463–468. With permission from the American Chemical Society.)

The magnetic field is successful in confining ion motion in the plane perpendicular to the field lines of the magnet, but motion along the field lines is relatively unaffected. To confine ions in the ion cyclotron resonance (ICR) trap for extended periods of time, it is necessary to create an electrostatic trapping well along the z-axis. This task is accomplished by applying a small voltage of the same polarity as the ions in the trap to two parallel plates that are positioned perpendicular to the magnetic field lines and form, in the axial direction, the outer boundaries of the ICR trap.

The simplest form of an ICR trap is the cubic trap. It is made up of three pairs of parallel plates that are arranged such that each plate forms the face of a cube, see Figure 8.1 [2]. Of the two pairs of plates that are parallel to the magnetic field, one pair is called the 'transmitter' (or 'excite') plates and the other the 'receiver' (or 'detect') plates. The two plates that are perpendicular to the magnetic field lines are the trapping plates, and they provide an electrostatic well for confining ion motion in the z-direction. Initially, ions have small orbital radii and must be excited to larger radii before detection by the application to the excite plates of an RF voltage that has the same frequency as the cyclotron motion of the ions. As the ions absorb energy, their cyclotron orbits increase. A signal is detected in the form of an image current induced on the detection plates by the motion of the ions. This signal is amplified by the circuit connected to the two detect plates and digitized. The idealized signal from a single ion is a sine wave. The signal from many ions of different m/z values is the sum of many sine waves of different frequencies. The signal is transformed from the time domain to the frequency domain by using a Fourier transform. The frequencies are then converted to values of m/z using an appropriate mass calibration equation.

The motion of the ions in the ICR trap is not governed solely by the presence of a strong magnetic field; in fact, cyclotron motion due to the magnetic field is only one of three modes of oscillation. Recall that to store ions in the ICR trap requires an electrostatic well that is oriented along the magnetic field or z-axis. This well is produced by applying to the trapping plates a DC potential (V_{TRAP}) of non-zero magnitude and the same sign as the ions of interest. An ion, once formed or injected into the trap, oscillates back and forth along the z-axis in this electrostatic well, executing simple harmonic motion. The electric field produced in the trap is neither perfectly linear nor oriented only in the z-direction. The radial component of the trapping electric field, E_r, supplies an outwardly directed force that opposes the inward force of the magnet leading to a third motion, called the magnetron motion, that results from the interaction of the magnetic and electric forces acting on the ion. Magnetron motion occurs in the same plane as the cyclotron motion, but is typically of much lower frequency. It can be described as the precession of the ion's guiding center about the trap's z-axis by moving along an equipotential line in the interior of the trap. The quadratic solution in Equation 8.2 describes the frequency of ion motion in the plane perpendicular to the magnetic field axis. As for all quadratic equations, there are two solutions. One is ω_+ the observed cyclotron frequency, and the other is ω_-, the magnetron mode frequency. The radial position of the ion is defined as r.

$$\omega_\pm = \frac{-zeB \pm \sqrt{z^2 e^2 B^2 - \dfrac{4mzeE_r}{r}}}{2m} \qquad (8.2).$$

In the absence of an electric field ($E_r = 0$), the observed cyclotron frequency, ω_+, reduces to the unperturbed cyclotron frequency ω in Equation 8.1, and the magnetron frequency, ω_-, is zero.

As an analytical instrument, FT-ICR MS developed slowly, owing to inadequate computers in the 1970s–1990s, lack of high induction, actively shielded magnets, and inadequate designs of external sources. These problems have been overcome during the early 2000s. Contemporary FT-ICR MS instruments now utilize many different ionization techniques (for example, electron ionization (EI), matrix-assisted laser desorption/ionization (MALDI), electrospray ionization (ESI), and nanoelectrospray ionization (nESI)), making it a very versatile tool. The technique of FT-ICR MS offers the highest mass-resolving powers and mass accuracies of any mass spectrometric technique. For a brief review of high precision mass spectrometry see Vedel and Werth [3].

Despite the success of the FT-ICR instrument as a high performance mass spectrometer, there are at least two problems that stand in the way of higher performance. The first is that the trap must be operated at low pressure, making difficult the incorporation of high-pressure events to stabilize ions and to manipulate ion clouds. The second problem is caused by the requirement to use a trapping electric field. As a result, ion cyclotron frequencies depend on the cyclotron, magnetron, and z-mode amplitudes of the trapped ions. For example, an ensemble of ions with large z-mode amplitude (coursing the full length of the trap) will have a different cyclotron frequency from that of an ensemble residing largely in the center of the trap. In this chapter, we describe both a radio-frequency-only-mode event and efforts to compensate electrically the trap as means to solve these problems.

8.2 RADIO FREQUENCY-ONLY-MODE EVENT

8.2.1 MOTIVATION

In the early days of FT-ICR MS, this technique served as a powerful tool for the study of gas-phase ion/molecule reactions. With the introduction of the double resonance experiment to FT-ICR MS, complex reaction pathways could be investigated with more certainty than with sector instruments and in much less time than with ICR experiments employing the drift cell [4]. The requirement of ultra high vacuum, 10^{-8}–10^{-10} Torr, for maximum system performance in FT-ICR MS is, however, a disadvantage in studies of ion chemistry. At the low background pressures required in FT-ICR MS, collisions with background neutrals are minimized, a consequence of which is that labile products formed during ion/molecule reactions are not always detected. In the absence of neutral collisions to remove internal energy, these metastable reaction products dissociate prior to detection. The existence of such products are known because they are seen easily in the instruments either incorporating high pressure (*ca* 0.1 Torr) ion sources on magnet sector or quadrupole analyzers or utilizing Paul traps, which operate at pressures in the range of 10^{-3}–10^{-4} Torr. Simply increasing the background pressure in the FT-ICR trap to see these metastable products is not a practical solution because ion/neutral collisions in the Penning trap lead to rapid radial ion loss [5] and consequent signal loss due to more rapid damping of cyclotron motion [6].

One possible solution is to switch temporarily the passive Penning trap to an active Paul trap, which can tolerate higher operating pressure and, thereby, stabilize transient reaction products. We termed this 'Paul mode' on an FT-ICR instrument the RF-only-mode event [7]. If this method is to be successful, one would expect to: (i) stabilize transient reaction products *via* collisional cooling, (ii) reduce ion loss due to magnetron mode expansion resulting from ion/neutral collisions, and (iii) achieve collisional focusing [8] of ions produced by various methods prior to excitation/detection, affording improved performance.

8.2.2 DESIGN AND IMPLEMENTATION

The early experiments involving the RF-only-mode event were carried out on an instrument built around a 1.2 T electromagnet. The instrument utilized a cubic trap of side 2.54 cm for ionization, storage, reaction, excitation, and detection. The RF-only-mode event was implemented by modulating the 1 V trapping well with a 1000 V_{0-p}, 1.1 MHz sine wave during the reaction delay interval of the sequence. The waveform was turned on and off exponentially with a tunable time constant that had a range of 30–3000 μs. During the reaction interval with the RF-only-mode event enabled, the trap can be pressurized momentarily with helium to approximately 10^{-3}–10^{-2} Torr by using a leak valve. The design of the transition from the normal operating mode to the RF-only-mode gives minimal ion loss (*e.g.*, < 15% [7]). Later the RF-only-mode event was implemented on an instrument utilizing a 3 T superconducting magnet. This implementation presented a modest challenge for this early design as the magnet was not shielded, and the components of the RF-only-mode event had to operate in the relatively strong fringe induction of the unshielded superconducting magnet.

8.2.2.1   Stability Diagram of a Paul Trap in a B Field

The use of the RF-only-mode event offers the possibility to focus collisionally in the radial direction the ions with mass-to-charge ratios over a wide range. This possibility is illustrated here by the calculation of RF parameters based on a 'stability' diagram for the radial ion motion. The m/z-range being focused collisionally is so large that the RF-voltage generator can be set up once (perhaps during installation) and, thereafter, no future intervention is required. There are other means of centering radially the ions; one approach is quadrupolar axialization. This method requires that the excitation waveform be tailored to the requirements of each experiment, unlike the RF-only-mode event.

The use of a stability diagram is preferred because it offers the economy of calculating the properties of the solutions of a normalized differential equation a minimum number of times. Thereafter, for the ion-motion analysis problem, the properties of solutions for a differential equation that describes a physical situation are understood by translating physical parameters for the physical differential equation to the equivalent parameters for the normalized differential equation. For the design problem, the physical parameters are unknown *a priori*, and the normalized differential equation is used as a surrogate for a physical situation that is to be solved. The requirements of the design are first worked out with the aid of the stability diagram of the normalized differential equation, after which the parameters for the normalized differential equation are translated into the parameters for the physical differential equation. This strategy is well established for periodic electrodynamic devices for which the normalized differential equation is the Mathieu equation and its accompanying Mathieu stability diagram.

The primary task for the RF-only-mode event is to overcome the effects of the outward radial components of the trapping electric field so that ions diffuse collisionally to the trap center. For this reason, we consider the radial motion of the ion. In addition to collisional damping (at reduced collision frequency for momentum transfer [6], ξ) of the motion of an ion [9] of charge e (commonly called q in the ICR literature) and mass m, the x- and y-components of the radial ion motion are coupled by the magnetic induction B. The RF waveform with amplitude V and angular frequency Ω is superimposed on the DC trapping voltage U, so that the resulting equations of motion are

$$\ddot{x} - (e/m)(U + V\cos(\Omega t))(2G_T)x = (e/m)\dot{y}B - \xi\dot{x} \tag{8.3}$$

and

$$\ddot{y} - (e/m)(U + V\cos(\Omega t))(2G_T)y = -(e/m)\dot{x}B - \xi\dot{y} \tag{8.4},$$

where the quantity G_T is a geometry constant that relates the applied trapping voltage to the quadrupolar component of the DC trapping electric potential in the trap. For example, if the trap is a 'hyperbolic' formed by a one sheeted-hyperboloid for the ring and a two-sheeted hyperboloid for the end-cap electrodes, then G_T is $1/(r_0^2 + 2z_0^2)$ (as reported by Jeffries *et al.* in their Table I [10]). With just the collisional damping or only the coupling caused by the magnetic induction, the treatment of the radial motion is fairly straightforward. With both damping and coupling, the treatment is slightly more complicated.

The differential equations for the x- and y-components of the radial motion are transformed into the form of the Mathieu equation. This transformation establishes the expressions for the 'stability' parameters. This exercise also puts the treatment into a form that results in a more familiar two-dimensional stability diagram with a parabolic operating line in the zero-pressure limit. The transformation is done in two stages. It first normalizes the time scale so that the modulation (RF) frequency becomes 2, and the x- and y-components are combined as real and imaginary parts of a new complex 'z' coordinate

$$z = x + iy \qquad (8.5)$$

$$t = (2/\Omega)\tau \qquad (8.6).$$

The second stage makes use of the collapsing Larmor [11] rotation, Equation 8.7, to transform another new complex 'Z' coordinate to the z-coordinate (*cf* [12])

$$z = \exp(-(iw + \xi_n)\tau/2)Z \qquad (8.7),$$

$$\xi_n \doteq (2/\Omega)\xi \qquad (8.8),$$

$$w \doteq (2/\Omega)(eB/m) \qquad (8.9).$$

In Equation 8.6, the rate of rotation is one half of the normalized unperturbed cyclotron frequency, w, and the rate constant for the collapse is one half of the normalized reduced collision frequency, ξ_n. The resulting Mathieu equation is shown in Equation 8.10

$$\ddot{Z} + (a_r - (iw + \xi_n)^2/4 - 2Q\cos(2\tau))Z = 0 \qquad (8.10).$$

The appearance of the Mathieu equation has been adjusted through the use of a number of equations. They make use of the critical mass, m_c, which is the highest dynamically stable mass ion that can be stored in the trap operated in the Penning mode; it is given in Equation. 8.11

$$m_c = eB^2/(8G_T U) \qquad (8.11).$$

The normalized unperturbed cyclotron frequency, w_{nc}, of the critical mass is shown in Equation 8.12

$$w_{nc} = (2/\Omega)(eB/m_c) \qquad (8.12).$$

The constant part of the coefficient for Z in the Mathieu equation reduces to a_r (see Equation 8.13) when the pressure and the magnetic induction are zero;

$$a_r = -w_{nc}w/4 \tag{8.13}$$

For a hyperbolic trap, a_r becomes $-8eU/(\Omega^2 m(r_0^2 + 2z_0^2))$. Compare this with Equation 2.74 on page 57 of March and Todd [13]. The constant terms of the coefficient for Z are grouped together as the stability parameter 'A,' which is complex. The real part of A is shown in Equation 8.14, and the imaginary part in Equation 8.15.

$$\text{Re}(A) = (w/2)^2(1 - w_{nc}/w) - (\xi_n/2)^2 \tag{8.14}$$

$$\text{Im}(A) = -(\xi_n/2)w \tag{8.15}$$

A factor of the time-varying (modulation) term of the coefficient for Z is the stability parameter 'Q,' which is real, and evaluates as shown in Equation 8.16:

$$Q = (w_{nc}w/8)(V/U) \tag{8.16}$$

For a hyperbolic trap, Q becomes $4eV/(\Omega^2 m(r_0^2 + 2z_0^2))$. Compare this with Equation 2.75 on page 57 of March and Todd [13].

The representation of a property of the solutions requires three dimensions to accommodate the variations that may occur in a given experiment. The dimensions correspond to the real part of A, the imaginary part of A, and Q. For each possible mass-to-charge ratio, there is a point with three coordinates that are the values of $\text{Re}(A)$, $\text{Im}(A)$, and Q, that are needed to parameterize the Mathieu equation: the coordinates are computed by using the value of w resulting from the mass-to-charge ratio in Equations 8.14 through 8.16, respectively. The operating line that results from the locus of these points becomes a curved 'path' in this three-dimensional (3D) space. If there are no collisions, the operating line is a parabola that fits into the $\text{Re}(A)$–Q plane [7]. If, in addition, the magnetic induction is zero, the operating line is the well-known straight line for the quadrupole ion storage trap [13].

The solution property of interest in the RF-only-mode event is the 'speed' with which the trajectories are damped to the center of the trap. The speed at an accessible point of the 3D space was evaluated as the most positive damping frequency (real part of the characteristic exponent) of the two independent solutions for the Mathieu equation. Numerical integration of the matrix form of the Mathieu equation for one period of the modulation gives the discrete transition matrix. The elements of the discrete transition matrix define the characteristic equation, whose roots are the characteristic multipliers from which the characteristic exponents are obtained [14]. Because the transformation of Equation 8.7 between the z- and Z-coordinates is collapsing, the characteristic exponents calculated in Z-coordinates have to be interpreted in terms of z-coordinates.

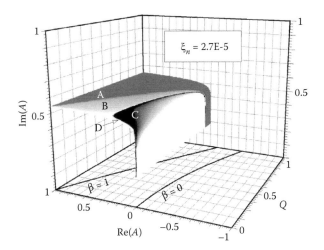

FIGURE 8.2 A section through the volume of accessible points for a constant normalized reduced collision frequency. The lines on the zero Im(A) plane are for β equal to 0 and β equal to 1 and mark off the region of stable radial trajectories in the zero pressure limit. A description for the shades of gray for Figure 8.3 applies here and can be found in the text.

The requirements for diagrammatic presentation are more than those for just answering the question of whether or not the solutions are stable. The rates for which the solutions converge to zero is important for determining whether the collisional focusing will be effective on the time scales of a reasonable experiment. In Figure 8.2, each point is represented in a shade of gray on the surface made up of the locus of accessible points for which the normalized reduced collision frequency is 2.7×10^{-5}. Starting at the corner of Re(A) equal to one and Q equal to one, this shade of gray represents rapidly divergent solutions and marks off a triangular area A that corresponds to unstable solutions resulting from the parametric resonance that touches the (A) axis at one. Adjacent to this triangle is an area B of a lighter shade that represents solutions that are still divergent, but the rate of divergence decreases as one moves across at constant Q to lower Re(A) on the surface. This lighter shade lies in an area that would have stable solutions in a normal stability diagram. The lighter shade is terminated at its lower right edge by the arc of the dark shade of area C; the arc rises quadradically from the graph origin as Q increases and then turns to level out before dropping slowly at first and then dropping faster as Q increases beyond 0.4, as seen in Figure 8.3, where the surface is viewed from above. At the top edge of the arc, the solutions either diverge or converge at very slow rates. As one moves down at a constant Q from the top edge of the arc, the rate at which the solution converges to zero increases and the shade of gray turns lighter. For small Re(A) the normalized rate approaches 0.5×10^{-5}. For value of Ω equal to $2\pi \times (1.1 \text{ MHz})$, this corresponds to a time constant of 58 ms according to Equation 8.6.

As noted in the previous paragraph, the surface, when Figure 8.2 is viewed from above, yields Figure 8.3, which is typical of surfaces obtained for other normalized reduced collision frequencies. The upper edge of the arc marks the transition from a pressure-focusing regime below the arc to one that is not pressure-focusing above it. The

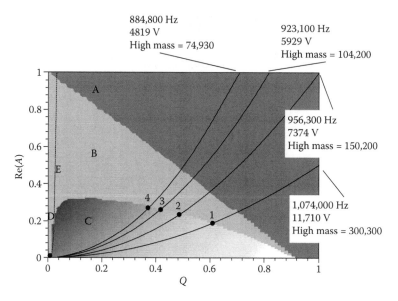

FIGURE 8.3 The projection of the surface shown in Figure 8.2 onto the zero Im(A) plane. The projection of candidate operating lines is also shown. The resulting RF frequency, amplitude and high mass-to-charge cut-off ratio are shown for each operating line when a low mass cut-off of *m/z* 100 is specified. The dotted line next to the Re(A) axis is the locus of points where the pseudopotential well balances the potential hill caused by the DC trapping potential in the zero pressure limit. A description for the shades of gray can be found in the text.

arc upper edge, Figure 8.3, remains essentially constant while the normalized reduced collision frequency is varied to change the locus of accessed points in the 3D space, except for operation under the very highest pressures. Because of this, a radio-frequency generator operating at a single frequency can be used for almost all experiments.

The area at low *Q* to the left of where the arc rises quadratically in Figure 8.3 represents points in the lighter shade of area D that are still divergent. For these points, the pseudopotential well formed by the RF potential is too shallow to overcome the potential hill, resulting from the electric DC trapping potential that is always present, and is used to confine the ions in the magnetic induction direction when the trap is operated in the Penning mode. As *Q* is increased at constant Re(A), the pseudopotential well becomes deeper and at some point balances the DC potential hill. In the zero-pressure limit, the higher of the two radial secular frequencies (in the 'z'-coordinates, which for this discussion is a composite of *x* and *y* coordinates) becomes the normalized cyclotron frequency, and a transition to a pressure-focusing regime becomes possible at this balance point. The locus of the balance points based on the pseudopotential well approximation is a parabola displayed as the dotted line E in Figure 8.3 and is near to the left hand edge of the arc.

The calculation of fixed RF frequency and amplitude begins with the understanding that the resulting operating line is a parabola when projected onto the zero-Im(A) plane and that the projection intersects the parabola of balance points close to the origin of the zero-Im(A) plane. The ion's mass at this intersection point approximates to the mass-to-charge ratio of the ion that can be focused collisionally (high mass cut-off). If we select a value of K that is greater than or equal to one so that the ion's mass at the intersection point is given by m_c/K, then the operating line that passes through this intersection point is governed by the Equation 8.17, where the second approximation is valid when w_{nc} is much less than Q

$$Re(A) \approx KQ^2/2 - \sqrt{K}Qw_{nc}/(2\sqrt{2}) \approx KQ^2/2 \qquad (8.17).$$

The operating line progresses up and to the right through the collisional focusing region as the ion's mass decreases, and intersects eventually with the arc at the low mass cut-off, m_l, where the coordinates are $Re(A_l)$ and Q_l. After the designer chooses the value for the low mass cut-off, the radio-frequency and amplitude are determined from the equations

$$\Omega = (eB/m_l)(1/\sqrt{Re(A_l)})\sqrt{1 - m_l/m_c} \qquad (8.18)$$

$$V = (Q_l/Re(A_l))(B^2/(4G_T))(e/m_l)(1 - m_l/m_c) \qquad (8.19).$$

The expression for the design value of Ω, Equation 8.18, is a rearrangement of Equation 8.14 with the use of Equations 8.9 and 8.12, assuming the effect of ξ_n is small. The expression for the design value of V, Equation 8.19, is a rearrangement of Equation 8.16 with the use of Equations 8.12, 8.11, 8.9, and 8.18. Operating lines are shown in Figure 8.3 for values of K that are 1 (point 1: $A_l = 0.184$, $Q_l = 0.607$), 2 (point 2: $A_l = 0.232$, $Q_l = 0.482$), 3 (point 3: $A_l = 0.259$, $Q_l = 0.416$), and 4 (point 4: $A_l = 0.271$, $Q_l = 0.368$) as one's eye moves from right to left. The numerical results shown in Figure 8.3 are for a cubic trap in a 3.0-T magnetic induction with a trapping voltage of 1 V and G_T is given by α/a^2 where α is 1.3869 (as reported by Jeffries *et al.* in their Table I [10]) and a, the trap size, is equal to 4.76 cm.

8.2.3 APPLICATION TO ION CHEMISTRY

An investigation of the reaction of a neutral 1,3-butadiene molecule and the vinyl methyl ether radical cation is a good example of how the RF-only-mode event can stabilize metastable reaction products [7]. The adduct of 1,3-butadiene and the methyl vinyl ether radical cation has an m/z equal to 112 and is observed easily in a conventional chemical ionization source where pressures are high, but is not seen in conventional FT-ICR MS. The only evidence for the existence of this adduct in standard FT-ICR experiments is the presence of its fragment ions at m/z 80 and 97. By inserting the RF-only-mode event into the experimental sequence, the detection of the adduct is possible [7]. Given that the reaction can now be carried out in the presence of helium bath gas, the adduct is stabilized by collisional cooling, affording the mass spectrum shown in Figure 8.4.

Another example of the success of the RF-only-mode event in stabilizing metastable reaction products is the observation of the β-distonic ion at m/z 107 formed from the reaction of pyridine radical cation and ethene [15], shown in Figure 8.5.

(a)

(b)

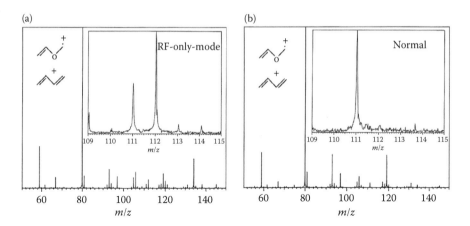

FIGURE 8.4 Mass spectra resulting from the reaction between neutral 1,3-butadiene and the vinyl methyl ether radical cation obtained by using (a) a sequence with a helium pulse and RF power modulation, and (b) a sequence with neither helium nor RF modulation. The overall reaction delay was 3.6 s in both cases. An expanded view of the area around m/z 112 is shown in the insets. (Reproduced from Rempel, D.L. and Gross, M.L., *J. Am. Soc. Mass Spectrom.* 1992, *3*, 590–594. With permission from Elsevier.)

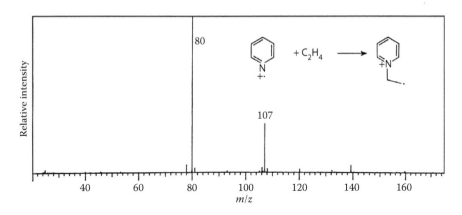

FIGURE 8.5 Mass spectrum illustrating the formation of the β-distonic ion of m/z 107 in the FTMS cell *via* collisional stabilization with the RF-only-mode event. The base peak at m/z 80 is attributed to protonated pyridine. The adduct is formed but not detected without the event. (Reproduced from Yu, S.J., Holliman, C.L., Rempel, D.L., and Gross, M.L., *J. Am. Chem. Soc.* 1993, *115*, 9676–9682. With permission from the American Chemical Society.)

The base peak in the spectrum, observed at m/z 80, is attributed to the presence of protonated pyridine. Without employment of the RF-only-mode event in the FT-ICR MS sequence, only indirect evidence of the formation of the β-distonic ion can be obtained. The RF-only-mode event was also used in the study of isomerization of 4-vinylcyclohexene radical cation [16] and the reaction of neutral ethene with each of propene and cyclopropene radical cations [17].

8.2.4   Application to Ion Cloud Manipulation

8.2.4.1   Collisional Focusing of Radially Dispersed Ion Clouds

The RF-only-mode event can also be used to focus collisionally an ion cloud to the center of the trap prior to excitation and detection. Such focusing was first demonstrated in FT-ICR MS with benzene radical cations that were dispersed intentionally by excitation of the magnetron mode. Using the RF-only-mode, we were able to focus collisionally the ions back to the trap center [18]. The radial extent of an ion cloud was probed by utilizing an ICR trap constructed with segmented trapping electrodes, as illustrated in Figure 8.6. The probing was accomplished by moving the ion cloud away from the center of the trap by using a second magnetron excitation, referred to as the 'probe' excitation, and applying a pulse of –2 V to the outer segment of the trapping electrode to quench ions with magnetron radii greater than the disk trapping segment. The amplitude of the detected signal after the quench, represented by the complex area, gave a measure of the number of ions that remained in the trap. The detected signal observed as a function of the amplitude of a magnetron excitation 'probe' gave a measure of the preprobe ion-cloud magnetron-mode amplitude distribution. Presumably, a radially diffused cloud would have a signal that grows slowly in intensity and then fades as a function of the magnetron excitation, whereas a tightly focused cloud would start at a high intensity, remain relatively constant and, finally, have a sharp decline when the probe excitation increases the magnetron radius beyond the radius of the disk. Implementation

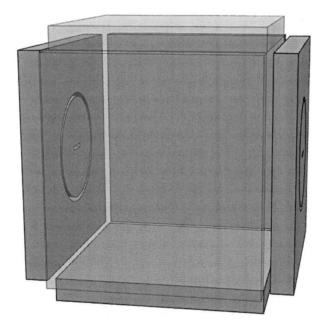

FIGURE 8.6   Diagram of cubic trap where the trapping electrodes have been divided into an inner disc and an outer segment. (Reproduced from Gooden, J.K., Rempel,D.L., and Gross, M.L., *J. Am Soc. Mass Spectrom.* 2004, *15*, 1109–1115. With permission from Elsevier.)

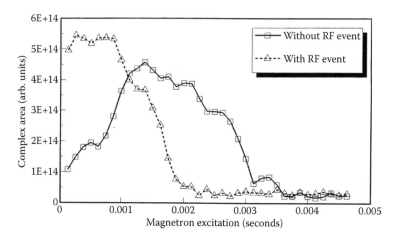

FIGURE 8.7   Plot of the complex area (initial amplitude of time-domain signal) *vs.* the magnetron excitation probe duration. Signal intensity curve for an ion cloud formed by EI of benzene and expanded to a magnetron radius of approximately 50% of the cell size; (⊟) unfocused ion cloud whose signal increases and decreases with time owing to the translation of the ion cloud to a radius greater than and less than that of the disk electrode; (△) ion cloud focused by a RF event whose signal starts level and drops off as the cloud is translated to a radius greater than that of the disk electrode. (Reproduced from Jacoby, C.B., Holliman, C.L., Rempel, D.L., and Gross, M.L., *J. Am. Soc. Mass Spectrom.* 1993, *4*, 186–189. With permission from Elsevier.)

of the RF-only-mode event was successful in reducing significantly the radial cloud size as evidenced by Figure 8.7. The RF-only-mode event produces a tightly focused ion cloud prior to excitation/detection, regardless of the initial state of the cloud. This observation allows the user to optimize the instrumental parameters by using a well-defined ion cloud produced from EI of an analyte (*e.g.*, benzene) without readjusting the method when using ionization methods that produce a more diffuse cloud (such as in-field MALDI).

8.2.4.2   RF-Only-Mode Event with Electrical Trap Compensation

Implementation of electrical trap compensation, which is discussed in Section 8.3, and the RF-only-mode offers significant advantages in FT-ICR MS [18]. Compensation increases the quadrupolar nature of the trapping well while the RF-only-mode pressure focuses the ions to the center of the trap. The outcome is illustrated by Figure 8.8. When a cloud of benzene ions produced by EI is expanded to approximately 50% of the trap size, no signal is observed in the absence of the RF-only-mode event, even with trap compensation (Figure 8.8a and c). A small peak can be detected when the RF-only-mode event is used to focus collisionally the ions toward the center of the trap (Figure 8.8b). The greatest improvement comes when the RF-only-mode event is used in conjunction with electrical trap compensation (Figure 8.8d). Under this set of conditions, the ion cloud is focused collisionally to the center of the trap, where detection is improved using electrical trap compensation. The idea of trap compensation will be explored more fully in Section 8.3.

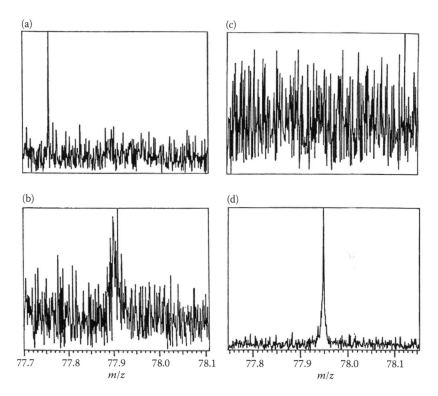

FIGURE 8.8 Mass spectra of a 1.25-kHz region of the mass range for the molecular ion of benzene for an ion cloud translated to approximately 50% of the cell size under different conditions of compensation and RF events: (a) neither compensation nor RF event; (b) only an RF event, with no compensation; (c) only compensation applied, with no RF event; (d) both compensation and RF event applied. (Reproduced from Jacoby, C.B., Holliman, C.L., Rempel, D.L., and Gross, M.L., *J. Am. Soc. Mass Spectrom.* 1993, *4*, 186–189. With permission from Elsevier.)

8.2.5 POTENTIAL ANALYTICAL APPLICATIONS OF THE **RF-ONLY-MODE** EVENT

Most contemporary FT-ICR instruments employ ion sources and RF-accumulation traps as interfaces to the ion sources that are both external to the FT-ICR trap. The external trap provides many of the cloud-focusing benefits that are also achievable with the RF-only-mode event. As such, the need for the RF-only-mode has decreased. Nevertheless, there are a few applications for which the RF-only-mode event may still be uniquely suited.

8.2.5.1 In-Field MALDI for Minimizing Mass Discrimination in Synthetic Polymer Analysis

To determine accurately the molecular weight distribution of a polymer, mass discrimination must be minimized [19]. One source of mass discrimination is the difference in flight times for the transfer of ions from the ionization source to the ICR trap. This form of mass discrimination can be minimized by using in-field MALDI.

Although this ionization technique may reduce greatly mass discrimination arising during the transfer process, it produces ions having a wide range of kinetic energies and initial directions; as a result, the MALDI-produced ions are difficult to catch in the trap. Stopping the ions in the trap with the use of collisions with a low-pressure buffer gas in the Penning mode alone is inefficient. Stopping the ions by having them do work in the axial (z) and radial directions against a retarding electric field, which is removed quickly when the ions are arrested, appears to be effective [20]. Once the ions are confined in the trap, the RF-only-mode event would be used to focus collisionally the widely dispersed ions back to the center of the trap for improved detection performance.*

8.2.5.2   The Role of the RF-Only-Mode Event in MS/MS

The RF-only-mode event can be used to transfer the well-known advantage of efficient collision-induced dissociation (CID) in Paul traps to FT-ICR MS. In the FT-ICR trap during the RF-only-mode event, excitation of the cyclotron mode (instead of the z-mode as is done in a Paul trap) may be used for ion activation. The use of high-pressure gas for CID is possible in the RF-only mode, as in the Paul trap. After fragmentation, the RF-only-mode event focuses collisionally the product ions to the center of the trap so that they can be detected with high resolving power, signal-to-noise, and mass measurement accuracy.

8.3   TRAP COMPENSATION

8.3.1   INTRODUCTION

Although FT-ICR mass spectrometry offers ultra high mass-resolving power and the ability to obtain routinely accurate mass measurements on the part-per-million scale [21], improvements are still needed especially for solving problems involving macromolecules. The route to improvements is to understand the heart of the FT-ICR MS instrument, namely the ICR trap, where ions are stored, excited, and detected. Within the trap there are the options of allowing the ions either to react with neutral species, or to undergo fragmentation induced by one of several different means, for example, Sustained Off-Resonance Irradiation-Collision-Induced Dissociation (SORI-CID), Infrared Multi-Photon Dissociation (IRMPD), and Electron Capture Dissociation (ECD).

In spite of its impressive performance, there exists a second problem that impacts resolving power, mass measurement accuracy, and dynamic range. This problem arises as a natural consequence of the electrostatic trapping well produced from finite electrodes, causing a frequency dependence on the mode amplitudes of an ion in the trap. As a reminder, the mode amplitudes refer to the three motions of ions in the trap: cyclotron, magnetron, and z-mode. These mode amplitudes describe the ever-changing position of an ion in the trap. The dependence of cyclotron frequency

---

* See also Volume 4, Chapter 9: A Fourier Transform Operating Mode Applied to a Three-Dimensional Quadrupole Ion Trap, by Y. Zerega, J. Andre, M. Carette, A. Janulyte, and C. Reynard; Volume 4, Chapter 19: A Quadrupole Ion Trap/Time-of-Flight Mass Spectrometer Combined with a Vacuum Matrix-Assisted Laser Desorption Ionization Source, by Dimitris Papanastasiou, Omar Belgacem, Helen Montgomery, Mikhail Sudakov, and Emmanuel Raptakis.

on mode amplitude arises from an outwardly directed radial electric field (E_r) that varies nonlinearly with position. To minimize this effect and to improve performance, several strategies have been employed, as evidenced by efforts to design new traps, some of which are depicted in Figure 8.9 (from a review on the subject of trap design published by Guan and Marshall [22] in 1995). The different approaches can be categorized in terms of their trapping electric field shape, design goal, and method of implementation. The two most commonly sought-after electric-field shapes correspond to the 'particle-in-a-box' (PIB) and the 3D quadrupolar potential wells. Regardless of the strategy employed, the overall goal is the same, namely to eliminate frequency shift as a function of mode amplitudes.

8.3.2 Particle-in-a-Box

8.3.2.1 Field Shape and Design Goal

As stated earlier, the observed cyclotron frequency is perturbed by the radial component of the trapping electric field. This perturbation would not present a huge problem if the effect of the radial electric field were constant over the volume of the trap. Because this is not the case, the cyclotron frequency of an ion is dependent not only upon its m/z-value but also upon its mode amplitudes (that is, ion position) in the trap. One way to address this effect is to make the radial component of the trapping electric field zero. If this were to be accomplished, Gauss' Law requires that the potential as a function of distance along the magnetic field axis (z) be constant. This situation, in an unmodified form, is not adequate for trapping ions along the z-axis. An ideal solution would be to have an abrupt potential barrier to reflect instantaneously the ions in the z-direction back to the trap interior without any perturbation of the cyclotron motion of the ion. This reflection potential is visualized as being zero at all points except at the edge of the trap, where it rises rapidly to a voltage sufficiently large to confine the ions in the manner of a 'particle-in-a-box.' Obviously this cannot be realized perfectly in practice, so any implementation is approximate.

A successful design of a PIB minimizes the radial electric field, thereby reducing the frequency-shift as a function of the trapping electrode voltage and ion position. One way to evaluate such a design is to calculate the radial electric-field strength inside the trap. Other ways to assess the success of a design are to measure the cyclotron frequency shift as a function of applied trap voltage or as a function of cyclotron orbit size. The PIB can be approximated by electrical or mechanical means; some specific approaches are discussed briefly below.

8.3.2.2 Methods of Implementation

8.3.2.2.1 An Elongated Trapped-Ion Trap for Ion Cyclotron Resonance Mass Spectrometry

An elongated trap is the first approach; it was introduced by Hunter *et al.* [23] and is shown schematically in Figure 8.9o and in more detail in Figure 8.10. Calculations of the equipotentials within the trap were done by using the exact analytical solution determined by Sharp *et al.* [24] The equipotential lines at a contour interval of 0.1 V

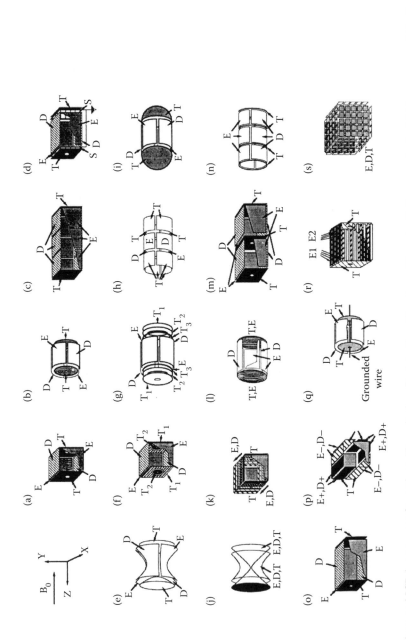

FIGURE 8.9 Assorted ICR ion trap configurations (E, excitation; D, detection; T, end-cap ('trapping'); S screen electrode) The purpose of the figure is to illustrate the variety of approaches to trap compensation. Some of the traps shown here are specifically addressed in the text. (Reproduced from Guan, S. and Marshall, A.G., *Int. J. Mass Spectrom. Ion Processes*, 1995, *146/147*, 261–296. With permission from Elsevier.)

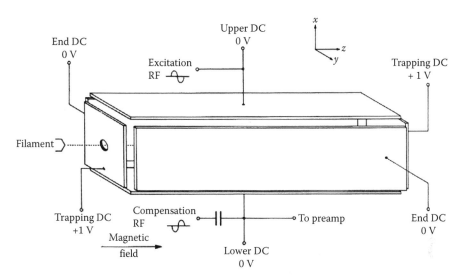

FIGURE 8.10 Schematic diagram of an elongated trapped-ion cell. (Reproduced from Hunter, R.L., Sherman, M.G., and McIver, Jr., R.T., *Int. J. Mass Spectrom. Ion Phys.* 1983, *50*, 259–274. With permission from Elsevier.)

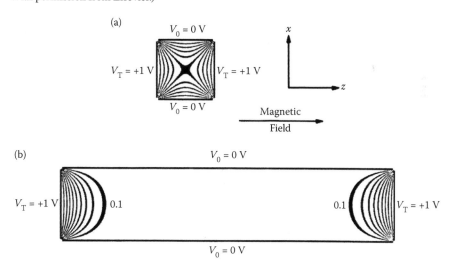

FIGURE 8.11 Equipotentials in trapped-ion cells in the (x, z) plane at $y = 0$: (a) conventional trapped-ion cell; (b) elongated trapped-ion cell. (Reproduced from Hunter, R.L., Sherman, M.G., and McIver, Jr., R.T., *Int. J. Mass Spectrom. Ion Phys.* 1983, *50*, 259–274. With permission from Elsevier.)

for the elongated trap and for the conventional ion trap are shown in Figure 8.11. As evidenced by the calculated equipotential lines in the elongated trap, the trapping potential does not penetrate very far into its interior. The potential calculated at the center is 77 μV, compared to 0.5 V for the conventional trapped-ion trap. According to Hunter *et al.*, [22] the elongated trap should be capable of storing more ions than

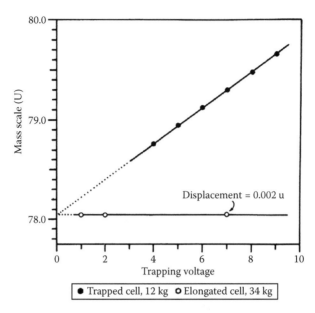

FIGURE 8.12   Plot of mass shift and, therefore, frequency shift, as a function of applied trapping voltage in a conventional trapped-ion cell and the elongated trapped-ion cell. The magnetic field strengths utilized were 12 and 34 kilogauss, respectively. (Reproduced from Hunter, R.L., Sherman, M.G., and McIver, Jr., R.T., *Int. J. Mass Spectrom. Ion Phys.* 1983, *50*, 259–274. With permission from Elsevier.)

a traditional design owing to this large region of low potential. Furthermore, given that the trapping potentials do not penetrate very far into the trap, there should be a reduced dependence of the cyclotron frequency of the ions on the trapping voltage and the *z*-mode amplitude. For these two reasons, the elongated trap should offer increased dynamic range and improved mass measurement accuracy.

Experiments carried out using benzene ionized by EI shows a linear increase in signal intensity as a function of the number of ions in the elongated trap up to *ca* 2×10^7 ions. Above this number of ions, space-charge effects become important. This number of ions represents by a factor of nearly 50 in improvement when compared to the capacity reported for the conventional trap [25].

Figure 8.12 compares the mass (cyclotron frequency) shift of the benzene molecular ion as a function of the applied trapping voltage in both the elongated and conventional trap: it is clear that the frequency shift is reduced greatly in the elongated trap. The *m/z* shift per volt applied to the trapping plates is *ca* 0.22 in the conventional cubic trap, compared to 0.002 in the elongated trap. This decrease is an order of magnitude greater than would be expected to arise from an increase in magnetic field induction from 1.2 to 3.4 T, the respective inductions used in evaluating the conventional and the elongated trap.

8.3.2.2.2   A 'Screened' Electrostatic Ion Trap

The screened trap introduced by Wang and Marshall [26] (see Figure 8.9d) seeks to approximate a PIB by placing a grounded screen 0.61 cm in front of the each of the trapping plate electrodes. In so doing, they effectively shield the interior of the trap from the potential on the trapping electrodes. SIMION 6.0 [27] calculations show that the electric potential in the interior of the trap is near zero, except for areas very close to the screen mesh. The potential inside the screened trap at a given position was calculated to be 15 to 20 times smaller than in an unscreened trap of the same dimensions.

Experimentally, the screened trap was effective in reducing the frequency shift as a function of trapping voltage by nearly two orders of magnitude when compared to an unscreened trap of the same dimensions. Perhaps a more important measure of its effectiveness is the frequency shift as a function of cyclotron-orbit size because the trapping voltage remains constant during the detection period of an experiment, whereas the ion-cloud distribution varies within any one experiment. The cyclotron orbit size was varied by changing the RF excitation amplitude, allowing a measurement of the frequency shift as a function of cyclotron orbit size. The frequency shift associated with cyclotron orbit size in the 5.08 cm screened trap is reduced by a factor of 10–20 compared to a 2.54 cm cubic trap. This finding is significant; one expects higher mass accuracy to be achieved because an ion's frequency is less dependent on its mode amplitudes (position). Another advantage of the screened trap is increased selectivity in precursor-ion selection for MS/MS; increased selectivity arises because an ion's frequency is nearly constant as it is excited to higher and higher cyclotron radii.

8.3.2.2.3 Field-Corrected Ion Trap

The field-corrected ICR trap introduced by Hanson *et al.* [28] was designed to produce uniform electric fields, both DC and RF, using a series of guard rings placed between the excitation plates and in front of the trapping plates (see Figure 8.13). The

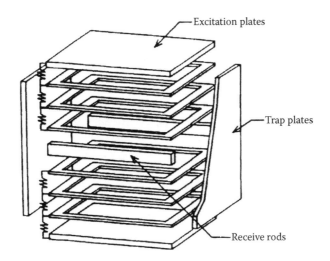

FIGURE 8.13 Schematic diagram of the field-corrected ion cell. (Reproduced from Hanson, C.D., Castro, M.E., Kerley, E.L., and Russell, D.H., *Anal. Chem.* 1990, *62*, 520–526. With permission from the American Chemical Society.)

shimming plates serve two purposes: to create a uniform excitation field to eliminate axial ejection during excitation, and to shield the trapping potential from the interior of the trap during detection, eliminating frequency shift as a function of mode amplitude.

Experimental data show that the shimming plates are successful at reducing the cyclotron frequency shift that is associated with the trapping well voltage. Image current detection can also be accomplished on a pair of small 'receive rods' without a noticeable effect on detection sensitivity. Narrow-band mass spectra acquired at high trapping voltage (10 V) in a normal cubic trap reveal peak splitting (Figure 8.14a) due to the inhomogeneity of the trapping electric field. Given that the shimming electrodes are capable of shielding the interior of the trap from the trapping electric field, peak splitting was eliminated in the field-corrected trap (Figure 8.14b).

8.3.2.2.4   Improvement of the Electric Field in the Cylindrical Ion Trap

The trap design by Naito *et al.* [29], illustrated in Figure 8.15, seeks to reduce the size of the radial electric field while maximizing the restoring force along the *z*-axis. This dual aim is accomplished by segmenting the trapping plate electrodes of a cylindrical ion trap into three pairs of circles with apertures of different radii to which independent voltages can be applied. Naito *et al.* demonstrated a reduced dependence of frequency on the trapping voltage and a decrease in frequency shift as a function of cyclotron orbit size. The reduction in frequency shift as a function of cyclotron orbit size resulted in a 10% increase in mass-resolving power. Given that the restoring force along the *z*-axis is also maximized, ions can be excited to larger cyclotron radii before axial ejection occurs, which translated into an increase of 65% in the signal-to-noise ratio (*S/N*).

8.3.2.2.5   Optimization of a Fixed-Volume, Open Cylindrical Trap and an Open Cylindrical Trap Analog of the Closed Screened Ion Trap by using Compensation Electrodes

As seen above, the effect of the trapping electric field in the elongated trap can be diminished by increasing the size of the trap in the *z*-direction [23]. In a 'closed' trap, where there are end-cap trapping electrodes, the aspect ratio is defined as the distance between the trapping electrodes divided by the diameter, and can only be varied by changing the physical size of the trap. In an 'open' trap, where the trapping well is provided by coaxial ring electrodes and the ends are open, the aspect ratio is defined as the distance between the trapping well maxima divided by the trap diameter, and can be changed by varying the width of the trapping electrodes while keeping the overall size of the trap constant. Vartanian and Laude [30] investigated open cylindrical traps with different aspect ratios in an effort to minimize the effect of the radial electric field. Using a trap with overall dimensions of 75 mm in length and 50 mm in diameter, they calculated that traps could be produced with aspect ratios that varied from 0.84 to 1.60. Perhaps surprisingly, the open cylindrical trap can achieve an aspect ratio that is larger than the ratio of its physical dimensions; this is because the potential maxima of the trapping well can actually exist beyond the physical dimensions of the trap.

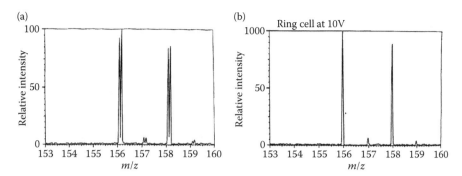

FIGURE 8.14 Narrow-band mass spectra of the bromobenzene molecular ion achieved at a trapping voltage of 10 V for (a) a cubic and (b) field-corrected ion cells. (Reproduced from Hanson, C.D., Castro, M.E., Kerley, E.L., and Russell, D.H., *Anal. Chem.* 1990, *62*, 520–526. With permission from the American Chemical Society.)

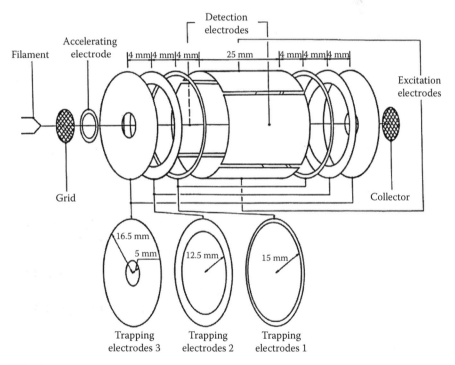

FIGURE 8.15 Schematic of the improved ion trap. (Reproduced from Naito, Y., Fujiwara, M., and Inoue, M., *Int. J. Mass Spectrom. Ion Processes.* 1992, *120*, 179–192. With permission from Elsevier.)

Vartanian and Laude built three open cylindrical traps with the same overall dimensions, but with different aspect ratios (Figure 8.16). The compressed trap has an aspect ratio of 0.84, the standard trap 1.20, and the elongated trap 1.60. These three designs were evaluated on the basis of the reduction of the radial electric field and the maximization of the axial well. It was determined by SIMION [31] calculations

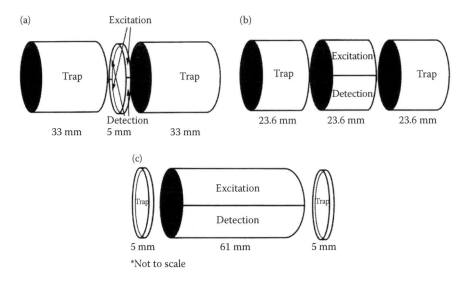

FIGURE 8.16 Three open cells with overall cell dimensions of 75 mm length and 50 mm diameter constructed for evaluation: (a) compressed cell with aspect ratio of 0.84; (b) standard cell with aspect ratio of 1.20; (c) elongated cell with aspect ratio of 1.60. (Reproduced from Vartanian, V.H. and Laude, D.A., *Int. J. Mass Spectrom. Ion Processes.* 1995, *141*, 189–200. With permission from Elsevier.)

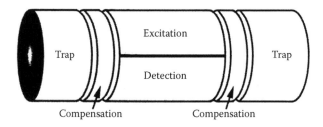

FIGURE 8.17 Open compensated trapped-ion cell of 75 mm length and 25 mm diameter showing the trap electrodes and compensation electrodes. (Reproduced from Vartanian, V.H., Hadjarab, F., and Laude, D.A., *Int. J. Mass Spectrom. Ion Processes.* 1995, *151*, 175–187. With permission from Elsevier.)

that the elongated trap, with an aspect ratio of 1.60, provided the best combination of these two factors.

Vartanian *et al.* [32] attempted to reduce the radial electric field in a cylindrical Penning trap by incorporating a single pair of compensation electrodes between the trapping rings and excite/detect electrodes (Figure 8.17). The potential produced in this trap is described as being very similar to that produced in a screened trap, but it does not suffer from some of the drawbacks of the screened trap (for example, ion neutralization caused by striking the wire mesh and contamination of the wire mesh with materials from the ion source). The compensation electrodes are effective in reducing the radial electric field at the center of the trap, not by the termination of the trapping field, but by the superposition of a field equal in magnitude but opposite in sign to that of the trapping field at the trap center. Calculations reveal that the optimum ratio of compensation voltage to trapping voltage is –0.33, and the experimentally determined optimum ratio is in close agreement at –0.36. The variation in frequency shift as a function of the applied trapping voltage is reduced by over two orders of magnitude in the compensated trap when compared to the uncompensated open cylindrical trap.

8.3.3 QUADRUPOLAR WELL

8.3.3.1 Field Shape and Design Goal

Another approach to eliminating the frequency shift as a function of mode amplitudes is to make the radial component of the trapping electric field proportional to the radius. In a rotationally symmetric system, a simple force-balance description produces an expression for the cyclotron frequency that depends on the ratio of the radial electric field, E_r, to the ion's cyclotron radius, r. To achieve a constant value of E_r/r the radial electric potential must be quadratic. If this is true, Gauss' Law requires that the electric potential along the z-axis be quadratic also, with a coefficient of opposite sign. As a consequence, the desired potential is described as a 3D quadrupolar well. Trap designs approximating a 3D quadrupolar well can be evaluated by evaluating E_r/r by calculation. In a perfect 3D quadrupolar well, the cyclotron frequency as a function of the different mode amplitudes is constant.

A quadrupole potential well can be produced by using accurately shaped-electrodes, such as those of hyperbolic traps [33,34], but the resulting instrument suffers from inefficient use of the magnet bore (that is, these traps have a small usable volume relative to the overall physical size of the trap) and a relatively inaccessible trap interior owing to the large electrodes. The quadrupolar potential well can be approximated also by using simpler electrode shapes and by optimizing the aspect ratio or by segmenting the electrode [35]. Examples of compensated traps that approximate a 3D quadrupolar well are discussed below.

8.3.3.2 Methods of Implementation

8.3.3.2.1 High Precision Mass Measurements using Compensated Hyperbolic Penning Traps

An ideal 3D quadrupolar potential can be realized in a Penning trap when the electrode surfaces are machined such that they follow exactly the desired equipotential lines. In this case, the electrodes surfaces would form hyperboloids of revolution (see Figure 8.9e). To form a functional trap for FT-ICR MS, one approach is to split the ring electrodes into four quadrants for excitation and detection. It is also necessary to have small holes in the end-cap electrodes for introduction of the electron beam. For these reasons, the field inside a hyperbolic FT-ICR trap is not perfectly quadrupolar. The potential in the trap also deviates from an ideal 3D quadrupolar potential because it is difficult to machine and to align the electrodes. To compensate for these aberrations, Van Dyck *et al.* incorporated guard electrodes into a modified hyperbolic Penning trap [33,34]. Utilizing this modified Penning trap, they observed increased mass-resolving power and improved mass measurement accuracy.

8.3.3.2.2 Matrix-Shimmed Ion Cyclotron Resonance Ion Trap

The matrix-shimmed trap constructed by Jackson *et al.* [36] takes electrode segmentation to the extreme. The matrix-shimmed trap (Figure 8.9s) is essentially a cubic trap in which each of the six standard electrodes is divided into a 5×5 grid of smaller electrodes. This complex segmentation allows, in principle, for the simultaneous optimization of the excitation, detection, and trapping fields. Theoretical calculations, the results of which are shown in Figure 8.18, reveal a closer achievement of ideal potentials when using the matrix-shimmed trap compared to using the traditional cubic trap.

The increased capacitance of the matrix-shimmed trap is a significant disadvantage given that the number of ions required for a detectable signal is directly proportional to the capacitance. The matrix-shimmed trap requires nearly three orders of magnitude more ions in the trap to detect a signal than does the normal cubic trap. Despite these limitations, the matrix-shimmed trap does eliminate harmonics (Figure 8.19), the frequency shift associated with cyclotron orbit size, and peak splitting due to inhomogeneities in the trapping electric field (Figure 8.20). The mass-resolving power achievable with the matrix-shimmed trap suffers owing to Coulombic repulsions among the increased number of ions required for signal detection.

8.3.3.2.3 Ion Trap with Improved DC Potential Harmonicity

Tolmachev *et al.* [37] designed and constructed a compensated open cylindrical ion trap that incorporates two pairs of compensation rings between the trapping rings and the excite/detect region at the center of the trap. This arrangement produced a potential that approximated more closely the 3D quadrupole well than that found in the standard open cylindrical trap. Various designs were evaluated theoretically on the basis of numerical calculations that aimed to minimize the variation in E_r/r. The chosen design struck a balance between improved performance and simplicity. The two compensation rings in the new design have the same width, but the compensation voltage supplied to each electrode is controlled independently. Figure 8.21 shows a comparison of the potentials calculated for the new trap operated in a conventional and a compensated mode; it also shows the magnitude of E_r/r as a function of z for the two traps. The calculations were done at several different cyclotron orbit sizes. There is no obvious improvement with the compensated trap in terms of the potentials but, in terms of E_r/r, the compensated mode produces less variation over a wider range of ion displacements in z and in r, when compared to the conventional trap. With this new trap design, Tolmachev *et al.* realized increased mass-resolving power, elimination of peak splitting, and increased mass accuracy through the reduction of the variation in E_r/r.

The transient acquired in the conventional trap reveals a beat pattern that is indicative of a mixture of species with slightly different frequencies (see Figure 8.22a and c for transients from the trap operated in the conventional manner and in the compensated mode, respectively). The mass spectrum that results from the FT of

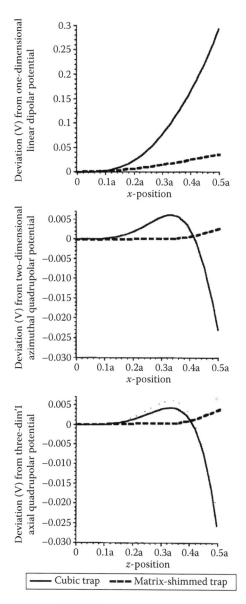

FIGURE 8.18 Deviation from dipolar linearity (excitation, *top*), two-dimensional quadrupo-larity (axialization, *middle*), and three-dimensional quadrupolarity (trapping, *bottom*) of electric potentials in cubic (———) and matrix-shimmed (– – –) traps. The deviations from dipolar linear-ity and two-dimensional azimuthal quadrupolarity were calculated along the line ($-a/2 \leq x \leq a/2$, $y = 0, z = 0$). Deviation from three-dimensional axial quadrupolarity was calculated along the line ($x = 0, y = 0, -c/2 \leq z \leq c/2$). Each potential was generated by applying +1 V or –1 V to appropriate pairs of opposed plates. Deviations from linearity and quadrupolarity are defined as the com-bined magnitude of all-but-linear or all-but-quadratic coefficients in the power series expansion of the potential in question. (Reproduced from Jackson, G.S., White, F.M., Guan, S., and Marshall, A.G., *J. Am. Soc. Mass Spectrom.* 1999, *10*, 759–769. With permission from Elsevier.)

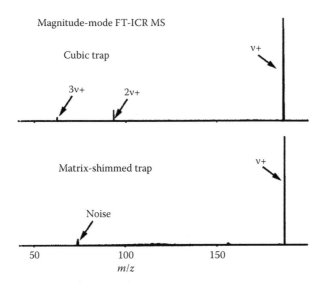

FIGURE 8.19   Experimental FT-ICR mass spectra for hexafluorobenzene molecular cations, $C_6F_6^+$, m/z 186, in cubic (*top*) and matrix-shimmed (*bottom*) Penning traps. The elimination of harmonics in the matrix-shimmed trap for ions at the same ICR radius (*ca* 70% of the trap radius) as in a cubic trap of the same size attests to the highly linear excitation/detection potentials in the matrix-shimmed trap. (Reproduced from Jackson, G.S., White, F.M., Guan, S., and Marshall, A.G., *J. Am. Soc. Mass Spectrom.* 1999, *10*, 759–769. With permission from Elsevier.)

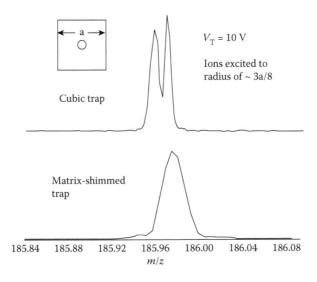

FIGURE 8.20   Peak splitting ($C_6F_6^+$ ions) due to electrostatic field inhomogeneity in a cubic trap (*top*) which is not present in a matrix-shimmed trap of the same size (*bottom*). (Reproduced from Jackson, G.S., White, F.M., Guan, S., and Marshall, A.G., *J. Am. Soc. Mass Spectrom.* 1999, *10*, 759–769. With permission from Elsevier.)

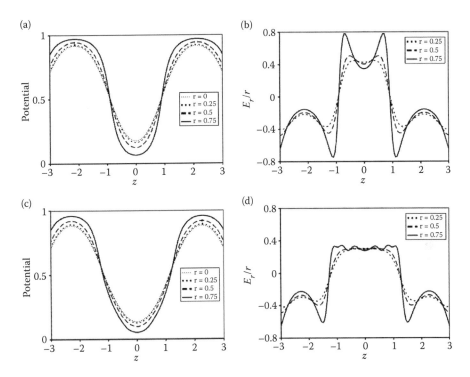

FIGURE 8.21 Plots of the calculated potentials for the two cell voltage configurations with units of length equal to the cell radius r_0: (a) potential distribution for the open cell configuration; (b) radial electric field divided by radius, calculated for the open cell; (c) potential distribution for the compensated cell configuration; and (d) radial electric field divided by radius, calculated for the compensated cell. (Reproduced from Tolmachev, A.V., Robinson, E.W., Wu, S., Kang, H., Lourette, N.M., Pasa-Tolic, L., and Smith, R.D., *J. Am. Soc. Mass Spectrom.* 2008, *19*, 586–597. With permission from Elsevier.)

the transient obtained with the conventional trap exhibits characteristic peak splitting attributable to the variation in frequency caused by inhomogeneity in the radial electric field (Figure 8.22b). The transient acquired in the compensated mode, however, shows a smooth decay, and the corresponding mass spectrum contains a well-defined single peak (Figure 8.22d). The mass-resolving power in the compensated mode approaches the theoretical limit of 400,000 at *m/z* 530 in a 12 T magnetic induction with a 1.3 s acquisition time.

8.3.3.2.4 An Electrically Compensated Trap Designed to Eighth Order

Another strategy, which we developed and evaluated [38], is to design a trap from first principles. To do this, we wrote an explicit expression of the cyclotron frequency *versus* all the mode amplitudes of a single ion. The design degrees of freedom are defined by four gap positions that isolate three compensating electrode pairs (see Figure 8.23), and three compensation voltages that were applied to the isolated electrode pairs. The trapping-electric potential for an ion with a given position amplitude in a closed cylindrical trap can be expressed by forcing the *z*-derivatives

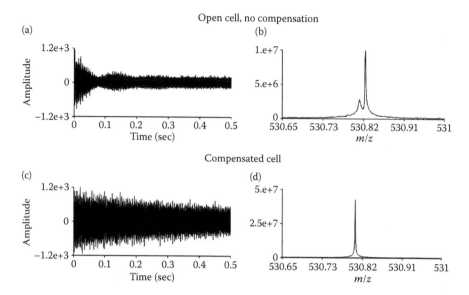

FIGURE 8.22 Ion transients and peak shapes obtained using an excitation power of 3 dB, which corresponds to total ion current and mass accuracy optimum for the compensated cell: (a) sample ion transient for the open cell configuration; (b) sample peak from Fourier-transformed data for the open cell configuration; (c) sample ion transient for the compensated cell configuration; (d) sample peak from Fourier-transformed data for the compensated cell configuration. (Reproduced from Tolmachev, A.V., Robinson, E.W., Wu, S., Kang, H., Lourette, N.M., Pasa-Tolic, L., and Smith, R.D., *J. Am. Soc. Mass Spectrom.* 2008, *19*, 586–597, With permission from Elsevier.)

of the spherical harmonic expansion to eighth order of the potential to match the z-derivatives of the modified Bessel function series solution for the Laplace equation with Dirichlet boundary conditions. To get from the electric potential, expressed at a point, to a cyclotron frequency for a single ion, expressed as a function of a set of mode amplitudes, the electric potential in the Hamiltonian was expressed for a single ion moving in a uniform magnetic induction. The Hamiltonian was simplified, converted to a function of action-angle coordinates, to remove the angle coordinates by using a canonical transform computed by the Deprit perturbation series to third order. The derivatives with respect to the mode actions of the cyclotron frequency (itself the derivative of the simplified Hamiltonian with respect to the cyclotron mode action) are functions of the z-mode, magnetron-mode, and the cyclotron-mode amplitudes. The cyclotron frequency as a function of various z-mode and cyclotron-mode amplitudes gives a frequency surface. We then changed systematically the design parameters of the trap, four gap positions and three voltages, until the three cyclotron frequency derivatives were minimized, indicating that the frequency surface had become at least 'locally flat' with an approximate first-order, triple-frequency focus. The minimization makes use of the root sum of squares (RSS) of the cyclotron frequency derivatives evaluated at 0.4, 0.0, and 0.6 for the cyclotron-mode, magnetron-mode, and z-mode amplitudes, respectively, all normalized to the trap inside radius.

(a)

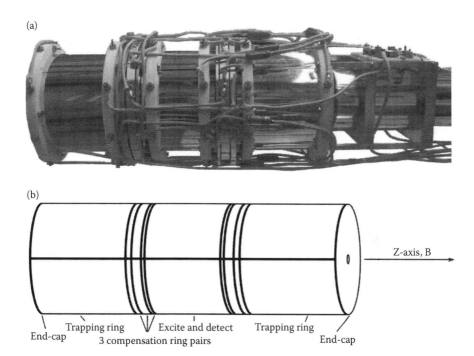

(b)

FIGURE 8.23 (a) Photograph of compensated trap mounted at the end of the quadrupole ion guide shown on the right hand side; (b) diagram of compensated trap. (Reproduced from Bruskern, A.M., Rempel, D.L., and Gross, M.L., *J. Am. Soc. Mass Spectrom.* 2008, *19*, 1281–1285. With permission from Elsevier.)

The design process produced many solutions; for each, the fourth-, sixth-, and eighth-order spherical harmonic contributions to the electric potential as originally expressed were effectively removed, giving solutions that are compensated to eighth-order in terms of the potential. From the many solutions, the design with gap positions greater or equal to 0.7 as normalized to the trap inside radius, minimum compensation voltages, and the smallest possible electric field along the z-axis near the trapping plate apertures was selected, see Figure 8.23.

The theory has not yet been able to predict accurately the best set of compensation voltages as determined by experiment. This small discrepancy between theory and experiment may arise because the manufacturing process cannot be executed in full accord with theory (for example, imperfections occur in the electrode thicknesses and spacing). Furthermore, the theory does not account for the work functions of the metal surfaces within the trap. Nevertheless, starting from the theoretical voltages, tuning enabled us to find a set of compensation ring voltages that produced a nearly flat frequency surface for a range of cyclotron and z-mode amplitudes. The trap can be operated in the compensated mode by applying the appropriate voltages to the compensation electrodes, or in the uncompensated mode by applying 0 V to the compensation electrodes. In this way, back-to-back comparisons can be made easily.

The MALDI mass spectra of vasopressin (*m/z* 1084) in Figure 8.24 illustrate the difference in performance with the compensated and uncompensated modes in the

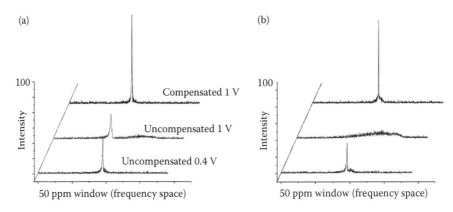

FIGURE 8.24 Signals from (a) hot ion clouds, formed by collisional cooling followed by re-excitation, and (b) collisionally cooled ion clouds. The three traces in each of (a) and (b) correspond to operation in the compensated mode at 1.0 V (top) and the uncompensated mode at 1.0 V (middle) and 0.4 V (bottom), respectively. Note that the mass spectrum acquired in the uncompensated mode from a cool cloud at 1.0 V has *ca* seven times the number of ions and its intensity scale is 20% of the other mass spectra shown here. (Reproduced from Brustkern, A.M. Rempel, D.L., and Gross, M.L., *J. Am. Soc. Mass Spectrom.* 2008, *19*, 1281–1285. With permission from Elsevier.)

narrowband acquisition. One set of mass spectra is from an ion cloud that was cooled collisionally with a pulse of nitrogen gas, and its z-mode amplitude then re-excited by pulsing the trapping plates; we refer to the outcome of this experiment as generating a 'hot' cloud (Figure 8.24a). The other set of mass spectra is of a collisionally cooled cloud (that is, one without z-mode re-excitation), designated appropriately as the 'cool' cloud (Figure 8.24b). A collisionally cooled cloud without re-excitation has an average z-position closer to the center of the trap than one excited by pulsing the end-cap electrodes. All of the mass spectra are of approximately the same number of ions, except the one acquired from a cool cloud in the uncompensated mode, where the trap was operated at 1.0 V trapping potential. This mass spectrum required seven times more ions and a vertical scale magnification of a factor of five to see easily the spectral features of interest.

At either trapping voltage and with either a hot or cool ion cloud, the compensated mode produces clearly improved mass-resolving power and spectra with better signal-to-noise ratio. The mass-resolving power of the uncompensated mode with inner rings (trapping rings) at 0.4 V approaches that of the compensated mode at a nominal well depth of 1.0 V, but because the well depth is reduced by more than half, the capacity of the uncompensated trap is reduced also.

The maximum achievable mass-resolving power in the compensated mode is at least three times that of the trap operated in the uncompensated mode, when both are run with nominal well depths of 1.0 V. Given that the nonlinearities in the electric field are reduced by the compensation voltages, ions of the same m/z-value within a cloud that has different axial and radial mode amplitudes oscillate with nearly the same cyclotron frequency. Because the distribution of frequencies for a given m/z within the cloud becomes narrower, the mass-resolving power increases. We observed mass-resolving powers (FWHH) as high as 1.7×10^7 for vasopressin at m/z 1084.5 in a 7.0-tesla magnetic induction in the compensated mode; this is the theoretical limit under these conditions when no spectral apodization (that is, smoothing of the ends of the time domain signal to remove the abrupt start and stop) is employed.

Trap compensation should lower the detection limit in FT-ICR MS. The ability of an ion cloud to phase-lock depends on both the frequency spread ($\triangle\omega$) across the cloud and the ion density within it [39]. Given that the compensation voltages reduce the nonlinearities in the electric field, thereby making $\triangle\omega$ small, the number of ions required for phase-locking in the compensated mode decreases. This observation means that a smaller number of ions is required to produce a well-defined, detectable signal, affording a higher signal-to-noise ratio for an equivalent number of ions; this outcome is illustrated by the mass spectra in Figure 8.24.

Fourier transform-ICR MS has difficulty detecting high m/z ions. As m/z increases, the magnetic force acting on the ions decreases, leaving the ions more vulnerable to nonlinearities in the electric field. Clouds composed of high m/z-value ions are axially and radially diffuse compared to clouds of low-m/z ions [40]. Therefore, ions with higher values of m/z occupy a larger volume of the trap than do low-m/z ions and, consequently, they experience a greater extent of both electric and magnetic-induction inhomogeneity.

Addressing problems at high m/z, we determined that the MALDI mass spectrum of cytochrome c in the compensated mode shows nearly three times higher mass-resolving power (Figure 8.25a and b, respectively) in the compensated mode compared to that achievable in the uncompensated mode. The difference in m/z assignment between the two spectra occurred because the instrument was not calibrated for operation in the compensated mode. Given the above considerations, it had been expected that the performance enhancement afforded by trap compensation would be greater for ions with higher m/z-values. This performance enhancement of greater than an order of magnitude increase in resolving power was not realized in these first experiments, indicating that the uncompensated mode is performing artificially well owing to phase locking, that tuning of the compensation voltages should be improved further, that magnetic induction inhomogeneities or other factors are operative, or that possibly a combination of all these factors may be operative.

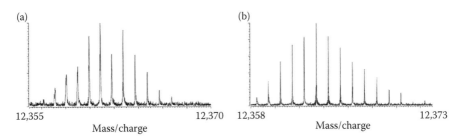

FIGURE 8.25 MALDI mass spectra of cytochrome c in the (a) uncompensated mode, and (b) compensated mode at 0.4 V. (Reproduced from Brustkern, A.M., Rempel, D.L., and Gross, M.L., *J. Am. Soc. Mass Spectrom.* 2008, *19*, 1281–1285. With permission from Elsevier.)

8.4   CONCLUSIONS

Although Fourier transform ion cyclotron mass spectrometry has become a powerful analytical technique, at least two problems are encountered as one seeks to develop a more versatile and accurate instrument. The first problem is that the trap must be operated at low pressure. This requirement limits its range of application in ion chemistry when intermediates or products require collisional stabilization to be observed and subsequently studied. Furthermore, tools for the precise tailoring of the ion-cloud distribution are limited without the use of collisions. One means of addressing these is the RF-only-mode event, effecting a marriage between the Penning and the Paul trap.

The second problem is that there is an ion cyclotron frequency shift as a function of ion mode amplitudes (position) in the trap. These shifts compromise mass-resolving power, signal-to-noise ratios, dynamic range, and mass measurement accuracy. One way of addressing this problem is through trap design, specifically of traps with compensated trapping fields. As evidenced by the outcomes of a number of efforts, trap compensation offers significant performance enhancements. As our understanding of the true behavior of the ions in the trap grows, trap designs will become increasingly better, and instrumental performance will improve even more.

REFERENCES

1. Comisarow, M.B.; Marshall, A.G. Fourier transform ion cyclotron resonance spectroscopy. *Chem. Phys. Lett.* 1974, *25*, 282–283.
2. Ledford, Jr., E.B.; Ghaderi, S.; White, R.L.; Spencer, R.B.; Kulkarni, P.S.; Wilkins, C.L.; Gross, M.L. Exact mass measurement by Fourier transform mass spectrometry. *Anal. Chem.* 1980, *52*, 463–468.
3. Vedel, F.; Werth, G. 1995. High precision mass spectrometry in the Penning traps. In *Practical Aspects of Ion Trap Mass Spectrometry Volume II Ion Trap Instrumentation.* Eds. R.E. March and J.F.J. Todd, pp. 237–262. Boca Raton: CRC Press.
4. Comisarow, M.B.; Grassi, V.; Parisod, G. Fourier transform ion cyclotron double resonance. *Chem. Phys. Lett.* 1978, *57*, 413–416.
5. Francl, T.J.; Fukuda, E.K.; McIver, Jr., R.T. A diffusion model for nonreactive ion loss in pulsed ion cyclotron resonance experiments. *Int. J. Mass Spectrom. Ion Phys.* 1983, *50*, 151–167.
6. Comisarow, M.B. Signals, noise, sensitivity and resolution in ion cyclotron resonance spectroscopy. *Lect. Notes Chem.* 1982, 31(Ion Cyclotron Reson. Spectrom. 2), 484–513.
7. Rempel, D.L.; Gross, M.L. High pressure trapping in Fourier transform mass spectrometry: a radiofrequency-only-mode event. *J. Am. Soc. Mass Spectrom.* 1992, *3*, 590–594.
8. Bonner, R.F.; March, R.E.; Durup, J. Effect of charge exchange reactions on the motion of ions in three-dimensional quadrupole electric fields. *Int. J. Mass Spectrom. Ion Phys.* 1976, *22*, 17–34.
9. Beauchamp, J.L. Theory of collision-broadened ion cyclotron resonance spectra. *J. Chem. Phys.* 1967, *46*, 1231–1243.
10. Jeffries, J.B.; Barlow, S.E.; Dunn, G.H. Theory of space-charge shift of ion cyclotron resonance frequencies. *Int. J. Mass Spectrom. Ion Processes* 1983, *54*, 169–187.
11. Goldstein, H. 1980. *Classical Mechanics.* Addison-Wesley Publishing Company, Reading, MA, 1980, pp 234–235.

12. Fischer, E. Die dreidimensionale Stabilisierung von Ladungsträgern in einem Vierpolfeld. *Z. Phys. A.* 1959, *156*, 1–26.

13. March, R.E.; Todd, J.F.J. *Quadrupole Ion Trap Mass Spectrometry,* 2nd edition. Hoboken, NJ: John Wiley and Sons, Inc., 2005.

14. Richards, J.A. *Analysis of Periodically Time-Varying Systems.* Berlin: Springer-Verlag, 1983.

15. Yu, S.J.; Holliman, C.L.; Rempel, D.L.; Gross, M.L. The β2-distonic ion from the reaction of pyridine radical cation and ethene: a demonstration of high-pressure trapping in Fourier transform mass spectrometry. *J. Am. Chem. Soc.* 1993, *115*, 9676–9682.

16. Vollmer, D.; Rempel, D.L.; Gross, M.L.; Williams, F. Isomerization of 4-vinyl-cyclohexene radical cation: a tandem mass spectrometry study. *J. Am. Chem. Soc.* 1995, *117*, 1669–1670.

17. Vollmer, D.L.; Rempel, D.L.; Gross, M.L. The r.f.-only-mode event for ion-chemistry studies in Fourier transform mass spectrometry: the reactions of propene and cyclopropane radical cations with neutral ethene. *Int. J. Mass Spectrom. Ion Processes* 1996, *157/158*, 189–198.

18. Jacoby, C.B.; Holliman, C.L.; Rempel, D.L.; Gross, M.L. Ion cloud manipulation using the radiofrequency-only-mode as an improvement for high mass detection in Fourier-transform mass spectrometry. *J. Am. Soc. Mass Spectrom.* 1993, *4*, 186–189.

19. Castoro, J.A.; Koster, C.; Wilkins, C.L. Investigation of a "screened" electrostatic ion trap for analysis of high mass molecules by Fourier transform mass spectrometry. *Anal. Chem.* 1993, *65*, 784–788.

20. Gooden, J.K.; Rempel, D.L.; Gross, M.L. Evaluation of different combinations of gated trapping, RF-only mode and trap compensation for in-field MALDI Fourier transform mass spectrometry. *J. Am. Soc. Mass Spectrom.* 2004, *15*, 1109–1115.

21. Easterling, M.L.; Mize, T.H.; Amster, I.J. Routine part-per-million mass accuracy for high-mass ions: space-charge effects in MALDI FT-ICR. *Anal. Chem.* 1999, *71*, 624–632.

22. Guan, S.; Marshall, A.G. Ion traps for Fourier transform ion cyclotron resonance mass spectrometry: principles and design of geometric and electric configurations. *Int. J. Mass Spectrom. Ion Processes* 1995, *146/147*, 261–296.

23. Hunter, R.L.; Sherman, M.G.; McIver, Jr., R.T. An elongated trapped-ion cell for ion cyclotron resonance mass spectrometry with a superconducting magnet. *Int. J. Mass Spectrom. Ion Phys.* 1983, *50*, 259–274.

24. Sharp, T.E.; Eyler, J.R.; Li, E. Trapped-ion motion in ion cyclotron resonance spectroscopy. *Int. J. Mass Spectrom. Ion Phys.* 1972, *9*, 421–439.

25. McIver, Jr., R.T.; Hunter, R.L.; Ledford, Jr., E.B.; Locke, M.J.; Francl, T.J. A capacitance bridge circuit for broadband detection of ion cyclotron resonance signals. *Int. J. Mass Spectrom. Ion Phys.* 1981, *39*, 65–84.

26. Wang, M.; Marshall, A.G. A "screened" electrostatic ion trap for enhanced mass resolution, mass accuracy, reproducibility, and upper mass limit in Fourier-transform ion cyclotron resonance mass spectrometry. *Anal. Chem.* 1989, *61*, 1288–1293.

27. Dahl, D.A.; Delmore, J.E. "Simion 3D Version 6.0". 1988, Idaho National Engineering Laboratory, P.O. Box 2726, Idaho Falls, ID 83403.

28. Hanson, C.D.; Castro, M.E.; Kerley, E.L.; Russell, D.H. Field-corrected ion cell for ion cyclotron resonance. *Anal. Chem.* 1990, *62*, 520–526.

29. Naito, Y.; Fujiwara, M.; Inoue, M. Improvement of the electric field in the cylindrical trapped-ion cell. *Int. J. Mass Spectrom. Ion Processes* 1992, *120*, 179–192.

30. Vartanian, V.H.; Laude, D.A. Optimization of a fixed-volume open geometry trapped ion cell for Fourier transform ion cyclotron mass spectrometry. *Int. J. Mass Spectrom. Ion Processes* 1995, *141*, 189–200.

31. Dahl, D.A.; Delmore, J.E. "Simion PC/PS2 Version 4.0". 1988, Idaho National Engineering Laboratory, P.O. Box 2726, Idaho Falls, ID 83403.

32. Vartanian, V.H.; Hadjarab, F.; Laude, D.A. Open cell analog of the screened trapped-ion cell using compensation electrodes for Fourier transform ion cyclotron resonance mass spectrometry. *Int. J. Mass Spectrom. Ion Processes* 1995, *151*, 175–187.

33. Van Dyck, Jr., R.S.; Schwinberg, P.B. Preliminary proton/electron mass ratio using a compensated quadring Penning trap. *Phys. Rev. Lett.* 1981, *47*, 395–398.

34. Van Dyck, Jr., R.S.; Wineland, D.J.; Ekstrom, P.A.; Dehmelt, H.G. High mass resolution with a new variable anharmonicity Penning trap. *Appl. Phys. Lett.* 1976, *28*, 446–448.

35. Gabrielse, G.; MacKintosh, F.C. Cylindrical Penning traps with orthogonalized anharmonicity compensation. *Int. J. Mass Spectrom. Ion Processes* 1984, *57*, 1–17.

36. Jackson, G.S.; White, F.M.; Guan, S.; Marshall, A.G. Matrix-shimmed ion cyclotron resonance ion trap simultaneously optimized for excitation, detection, quadrupolar axialization, and trapping. *J. Am. Soc. Mass Spectrom.* 1999, *10*, 759–769.

37. Tolmachev, A.V.; Robinson, E.W.; Wu, S.; Kang, H.; Lourette, N.M.; Pasa-Tolic, L.; Smith, R.D. Trapped-ion cell with improved DC potential harmonicity for FT-ICR MS. *J. Am. Soc. Mass Spectrom.* 2008, *19*, 586–597.

38. Brustkern, A.M.; Rempel, D.L.; Gross, M.L. An electrically compensated trap designed to eighth order for FT-ICR mass spectrometry. *J. Am. Soc. Mass Spectrom.* 2008, *19*, 1281–1285.

39. Mitchell, D.W. Realistic simulation of the ion cyclotron resonance mass spectrometer using a distributed three-dimensional particle-in-cell code. *J. Am. Soc. Mass Spectrom.* 1999, *10*, 136–152.

40. Wood, T.D.; Schweikhard, L.; Marshall, A.G. Mass-to-charge ratio upper limits for matrix-assisted laser desorption Fourier transform ion cyclotron resonance mass spectrometry. *Anal. Chem.* 1992, *64*, 1461–1469.

9 A Fourier Transform Operating Mode Applied to a Three-Dimensional Quadrupole Ion Trap

Y. Zerega, J. Andre, M. Carette,
A. Janulyte, and C. Reynard

CONTENTS

9.1 Introduction 471
9.2 Experimental Arrangement 472
9.3 Operating Mode Sequence 473
    9.3.1 Stages of an Elementary Experiment and Applied Potentials.......... 473
        9.3.1.1 Ion Creation.. 474
        9.3.1.2 Ion Injection 474
        9.3.1.3 Ion Confinement.. 474
        9.3.1.4 Ion Ejection 475
        9.3.1.5 TOF Recording 475
    9.3.2 Succession of the Elementary Experiments 475
    9.3.3 Succession Modes... 475
9.4 Axial Secular Frequency Measurement and Signal Processing................... 475
    9.4.1 Historical TOF Recording and Processing....................................... 476
    9.4.2 TOF Expression for One Ion... 477
    9.4.3 Image Signal of the Confined Ions 477
    9.4.4 Signal Sampling.. 478
    9.4.5 Shape of the Frequency Line and Discrete Fourier Transform 479
    9.4.6 Numerical Processing of TOF Histograms 479
9.5 Simulation.. 480
    9.5.1 General Remarks 480
    9.5.2 Initial Confinement Conditions without the Creation
        and Injection Stages.. 480
    9.5.3 Creation and Injection Stages 480
    9.5.4 Confinement Stage.. 482
    9.5.5 Ejection Stage and Detection.. 482

9.6   Metrological Parameters 482
       9.6.1   Signal and Noise..482
                9.6.1.1   Signal-to-Noise Ratio 482
                9.6.1.2   Noise Induced by the Shape of the Spectral Line.........482
                9.6.1.3   Amplitude of the Image Signal....................................483
                9.6.1.4   Noise Induced by the Operating Mode........................486
                9.6.1.5   Peak Amplitude Fluctuation..486
                9.6.1.6   Experimental Results..486
       9.6.2   Resolution...486
                9.6.2.1   Frequency Resolution 486
                9.6.2.2   Mass Resolution..488
                9.6.2.3   Increasing Resolution 490
                9.6.2.4   Optimal Mass Resolution 490
9.7   Perturbations of Ion Motion..491
       9.7.1   Collisions..491
       9.7.2   Space Charge..491
       9.7.3   Electrical Perturbations 491
       9.7.4   Mechanical Perturbations...492
9.8   Control, Supply and Measurement Devices..493
       9.8.1   Requirements of the Operating Mode 493
       9.8.2   Digital Time-Interval Generator 494
       9.8.3   Wave-Train Generator 494
       9.8.4   Measurement Device: TOF Digitization 495
9.9   Radial Secular Frequency Measurement 495
       9.9.1   Expression of the Radial Position of the Trap Exit 495
       9.9.2   Simulation Results..495
       9.9.3   Experimental Results..496
       9.9.4   Applications..498
9.10   Dual Quadrupole Ion Trap Mass Spectrometer 498
9.11   Ion Injection 500
       9.11.1   Introduction 500
       9.11.2   Equation Setting...501
       9.11.3   Results and Discussion 502
                9.11.3.1   Deceleration Potential Configurations........................502
                9.11.3.2   Ion Distribution at the End of Injection......................503
                9.11.3.3   Influences of the m/z Ratio 505
                9.11.3.4   Comparison of Simulation and Experimental
                           Results...505
                9.11.3.5   Conclusion 507
9.12   Ion Motion Nonlinearities Induced By Mechanical Imperfections...........507
       9.12.1   Introduction 507
       9.12.2   Theoretical Developments...507
       9.12.3   Results 509
                9.12.3.1   Peak Amplitude in the Vicinity of a Nonlinear
                           Resonance Line..509

9.12.3.2   Broadening of the Frequency...510
9.12.3.3   Measured Frequencies of an Ion Cloud Confined
            on a Nonlinear Resonance Line...............................510
9.13   Motion Modulations Induced By Electric Perturbation.............512
        9.13.1   Introduction 512
        9.13.2   Results 513
                9.13.2.1   Axial Ion-Motion Spectrum 513
                9.13.2.2   Spectrum Detected Using the FT Operating Mode......513
                9.13.2.3   Resolution and Signal-to-Noise Ratio 514
9.14   Conclusion...516
Acknowledgments...517
References...518

9.1   INTRODUCTION

Whereas Fourier Transform (FT) techniques are common in Ion Cyclotron Resonance (ICR) mass spectrometry [1], and more recently in the Orbitrap™ instrument,* only a few works refer to non-destructive FT techniques applied to an RF quadrupole ion trap. In most of the studies, ion image currents induced between the two end-cap electrodes are employed as the detection principle. At the beginning of confinement, a fast DC pulse or a dipolar AC frequency matching the frequencies of ion motion moves the ions out of the trap center and makes the motion coherent, as in FT-ICR instruments. Syka and Fies gave an account of the experimental use of this detection principle for mass analysis [2,3]. The trapping (or axial) motion has been detected in a Penning trap [4]. Two different schemes of excitation and detection were proposed and applied between the two end-caps of the trap. Wang and Wanczek proposed a new electrode configuration for an ion trap with cone-shaped boundaries [5]. An homogeneous field was then superimposed along the axial direction to the confinement field within the same region. As a result the excitation of the transient axial ion motion, which renders the motion coherent may attain a greater accuracy of mass selection. Consequently, enhanced resolution may be achieved with the image current detection method. Lammert et al. measured the frequencies of the radial and axial secular ion motion experimentally [6]. For this purpose, a fast DC pulse displaced a kinetically cooled ion cloud from the center of the trap. The ion cloud displacement was observed during confinement by means of a laser probe. Subsequently, Cooks et al. examined by simulation the operating methods of an ion trap involving various ion cloud excitations with non-destructive FT detection [7,8]. Other work followed concerning the influence of DC pulses on the coherence of motion and axial kinetic energy of the ions during confinement (DC pulse tomography) [9,10].

An ion trap was first used by our research team as a tool to study the lifetime of metastable ions [11,12]. The number of confined ions was measured as a function

---

* See Volume 4, Chapter 3: Theory and Practice of the Orbitrap Mass Analyzer, by Alexander Makarov.

of the confinement time by a DC switched-potential, total-ejection, method using a fast electron multiplier as the detector. As the detection process was destructive, this method employed a succession of elementary experiments (creation, confinement, ejection, and detection stages) with increasing confinement times. Periodic fluctuations were observed on the curves giving the number of ions *versus* confinement time. These fluctuations correlated with the positions and velocities of the ion cloud at the end of confinement because the FT of the number of confined ions as a function of time gave frequency peaks matching the axial secular frequency of motion of the ions. Following these observations, a new FT operating mode of an RF quadrupole ion trap for mass analysis was proposed to the scientific community [13].

As in the ion-lifetime studies, this new FT operating mode requires a set of elementary experiments with the addition of the processing of the Time-Of-Flight (TOF) histograms of the ejected ions recorded for each of them. The amplitudes and frequency of the potentials applied to the trap electrodes remain constant during confinement. Ion motion must be as pure as possible and have a large amplitude. An image signal of the motion of the simultaneously trapped ions *versus* confinement time is computed from these histograms, and the secular frequency spectrum is obtained by means of a Fourier transformation.

9.2   EXPERIMENTAL ARRANGEMENT

The experimental system comprises an external ion source and a 3D quadrupole ion trap for mass analysis. After confinement, the ions are detected by means of a fast electron multiplier. In order to ensure a high quality vacuum inside the trap, the source and the trap are mounted in two separate vacuum chambers (Figure 9.1). Each chamber is evacuated by means of a dry primary pump and a 250-L s^{-1} turbo-molecular pump. The nominal pressure inside the analysis chamber is *ca* 10^{-8} Torr.

The creation cell is a cylindrical grid 1 cm in diameter and 1 cm in height, with its axial position along the Oz axis. The repeller is a flat metal disk, having a central aperture 1 mm in diameter that closes the bottom of the creation cell. Its potential is $V_{rep} = 50$ V. Another flat metal disk, also having a central aperture of the same diameter, closes the top of the creation cell, both having the same potential, denoted as V_{cel}. Ions are transferred through the aperture in the top disk toward the trap. The electron gun is located 1 cm from the grid cylinder. Originally this cell was built to create negative ions by Rydberg electron attachment in a crossed-beam experiment. In this application atoms were maintained in their Rydberg states by reducing the number of collisions with background neutral species [14,15]. Here the cell is used to create positive ions by electron impact (EI) ionization. Other source types can be coupled to the analysis cell.

A molecular beam enters the creation cell *via* the aperture in the repeller disk along the Oz axis. During the ion creation stage, electrons enter the cell by the switching of the DC potentials applied to the source electrodes. When the injection phase occurs, the resulting positive ions move toward the trap and then enter through an aperture (diameter 1 mm) in the lower end-cap. This is the only aperture

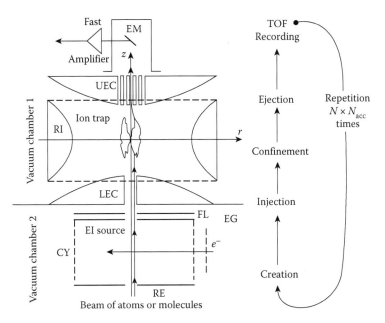

FIGURE 9.1 Schematic diagram of the experimental system: the EI source and the ion trap for mass analysis are in separate vacuum chambers. EM: Electron Multiplier; UEC: Upper End-Cap; RI: Ring; LEC: Lower End-Cap; CY: Cylindrical grid; RE: Repeller; FL: Focusing Lens and EG: Electron Gun. The stages of an elementary experiment are shown at the right-hand side.

through which the ions may pass from the creation cell to the analysis vacuum chamber.

The trap is quadrupolar, having a ring electrode with an inscribed radius r_0. The separation of the inner surfaces of the two end-cap electrodes is $2z_0$, with $z_0 = r_0/\sqrt{2} = 1$ cm (that is, 'unstretched' geometry). The surfaces of all the electrodes are truncated at $2r_0$. The trap has an axis of symmetry Oz and a radial plane xOy, with the radial direction $r = \sqrt{x^2 + y^2}$. Ions are ejected out of the trap by way of the upper end-cap electrode toward the electron multiplier; this electrode has a set of 'flues', each with a diameter of 0.5 mm, covering a circle 1 cm in diameter.

9.3   OPERATING MODE SEQUENCE

The amplitudes U_0 (DC) and V_0 (RF) and the angular frequency Ω of the confinement voltage remain constant during the confinement stage. A succession of elementary experiments is performed with increasing confinement times.

9.3.1   Stages of an Elementary Experiment and Applied Potentials

An elementary experiment comprises the following stages: creation, injection, confinement, ejection, and TOF recording (see Figure 9.1).

9.3.1.1 Ion Creation

The ion creation stage duration is denoted as t_{cre}, and during this time the potential V_{cel} (Section 9.2) is about 20 V. As the potential of the electron gun is about –35 V, the electrons move into the cell and ions are created by electron impact. In the center of the cylinder, the electron mean energy, V_{ene}, is about 70 eV; when $V_{cel} = 150$ V, the electrons do not enter the cell.

9.3.1.2 Ion Injection

The duration of the ion injection stage is denoted as t_{inj}, and the creation and injection stages are set so that $t_{cre} \leq t_{inj}$. The ions are moved from the creation cell toward the trap by means of switched DC voltages applied to the source and trap electrodes. In the source, with the potentials $V_{rep} = 50$ V and $V_{cel} = 20$ V (Section 9.2), positive ions are accelerated toward the top of the source cylinder and escape through the upper aperture. Between the creation cell and the trap they are also accelerated. Generally, during the injection stage the potential applied to the lower end-cap is $V_2 = -10$ V. Depending on V_1 and V_3, the DC potentials applied, respectively, to the upper end-cap and the ring electrodes, the ions are slowed down when they enter the trap. The ions are then stopped at the position z_s and then go back toward the lower end-cap.

9.3.1.3 Ion Confinement

The duration of confinement is expressed by, t_i with i being the index of the elementary experiment and of the confinement duration. During confinement, a combined DC and RF potential $U_0 + V_0 \cos(\Omega t + \varphi_0)$ (with Ω the angular frequency and φ_0 the initial phase of the RF potential) is applied to the ring electrode, the two end-caps being earthed.

The potential inside the trap is thus expressed by [16]

$$\phi(x,y,z,t) = \frac{U_0 + V_0 \cos(\Omega t + \varphi_0)}{2} \left[1 - \left(\frac{2z^2 - (x^2 + y^2)}{2z_0^2} \right) \right] \tag{9.1}.$$

The equation for ion motion along the Oz axis is

$$\frac{d^2 z(t)}{dt^2} + \frac{\Omega^2}{4} \left[a_z - 2q_z \cos(\Omega t + \varphi_0) \right] z(t) = 0 \tag{9.2},$$

with

$$a_z = -2a_r = -\frac{4zeU_0}{m\Omega^2 z_0^2} \quad \text{and} \quad q_z = -2q_r = \frac{2zeV_0}{m\Omega^2 z_0^2} \tag{9.3},$$

where m/z is the mass-to-charge ratio of the ions and e is the electronic charge. Ion motion is said to be 'stable' when the components of the trajectories are 'bound' in both the axial direction and the radial plane. The operating point of the trap is generally located inside the principal stability diagram which is defined in the plane (a_z, q_z) by the curves: $\beta_z = 0$, $\beta_z = 1$, $\beta_r = 0$, and $\beta_r = 1$. The values of β_z and β_r depend upon the parameters a_z, q_z and a_r, q_r, respectively.

The ion is confined when its motion is stable and when its trajectory amplitude does not exceed the limits fixed by the trap electrodes. The maximum amplitude of the trajectory is given by the initial conditions of the ion and the confinement parameters. Under typical operating conditions, the values of β_z, that is, the values of the operating points of the simultaneously confined ions lie in the range 0.5–0.75 when $\Omega/(2\pi) = 250\text{MHz}$.

9.3.1.4   Ion Ejection

After confinement, the ions are ejected from the trap by means of switched DC potentials. Generally, the potential V_e is applied to the lower end-cap and the potential $-V_e$ to the upper end-cap, the ring electrode being earthed. Other DC potential configurations are possible and will be investigated in future work. The ejection duration is denoted as t_{eje}.

9.3.1.5   TOF Recording

The fast electron multiplier gives an electric signal representative of the number of ion impacts upon the detector as a function of the flight time; this signal is thus a TOF histogram. The output is amplified and digitized by a high-speed transient recorder card. The quantity $t'_{i'}(= i'T'_e)$ is the sampled TOF variable, where i' is the TOF index and T'_e is the sampling period, generally equal to 4 ns. The number of TOF samples is denoted as N'.

9.3.2   SUCCESSION OF THE ELEMENTARY EXPERIMENTS

A sequence of N elementary experiments is performed with increasing confinement time t_i from $i = 1$ to N; usually, N ranges between 256 and 2000. The initial conditions of the ion clouds upon creation are taken to be the same statistically. This set of N elementary experiments can be repeated N_{acc} times in order to increase the signal-to-noise ratio. In this case, the N_{acc} TOF histograms are summed for each confinement time t_i (see Figure 9.2). The function $H(t_i,t'_{i'})$ is the resulting TOF histogram for the elementary experiment i and also for the confinement time, t_i.

9.3.3   SUCCESSION MODES

In the existing experimental device, the elementary experiments have a constant duration, denoted as t_{rep}, which must be greater than $(t_{inj} + NT_e + t_{eje})$. For $N = 1024$, the nominal duration of a total experiment is about 4.32 s. In order to reduce the total time, a new sequential device is under construction. Elementary experiments will be then concatenated so as to eliminate dead time after each ejection. In this case, the duration of the elementary experiment t_{rep} is variable, and the total duration is expressed by $\{N(t_{inj} + t_{eje}) + N/2(N + 1)T_e\}$. With $N = 1024$, the total duration is about 2.22 s.

9.4   AXIAL SECULAR FREQUENCY MEASUREMENT AND SIGNAL PROCESSING

The axial secular frequencies are obtained from this set of N TOF histograms. During confinement, the TOF of the ions evolves according to their axial secular frequencies (see Figure 9.2).

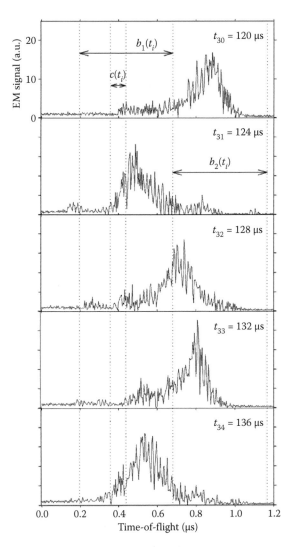

FIGURE 9.2 Experimental TOF histograms of SF_5^+ ions recorded for successive confinement times from $t_{30} = 120$ to $t_{34} = 136$ µs. The experimental conditions are: $z_0 = 1$ cm, $U_0 = 0.5$ V, $V_0 = 112.5$ V, $\Omega/(2\pi) = 250$ kHz, $\varphi_0 = \pi$ rad, $T_e = 4$ µs and $T'_e = 4$ ns.

9.4.1 Historical TOF Recording and Processing

Historically, the secular axial frequency was measured originally by TOF discrimination [13], in which a fast amplifier discriminator and a counter device are coupled to the electron multiplier. Only the first TOF values contained within a narrow range are added for each confinement time, and the FT of the resulting signal $c(t_i)$ gives the axial secular frequency spectrum.

Subsequently, another way of processing the TOF was employed in order to take into account all the flight time information [17]. Here, the electric signal from the electron multiplier is amplified and then digitized. The entire variation range is divided into two parts; in each part, the curve area is computed, which gives two temporal signals, $b_1(t_i)$ and $b_2(t_i)$. The maximum dynamics of the resulting temporal signal are obtained by adding the autocorrelation functions of $b_1(t_i)$ and $b_2(t_i)$, as the autocorrelation function removes the phase difference between the two signals. The FT of this sum then gives the axial secular frequency spectrum. More than two discriminations can be performed but they induce high-order frequencies in the measured spectrum.

The last treatment, described later, reduces greatly high-order frequency detection; it is also well adapted for processing experimental TOF histograms without reshaping the selected range for TOF recording. The previously mentioned treatments require that the TOF range be limited to the extreme variation range of the TOF histograms.

9.4.2 TOF EXPRESSION FOR ONE ION

First, the potential inside the trap is assumed to be dipolar during ion ejection. The ion TOF is the time which elapses from the instant t_i at the end of the confinement to the instant the ion reaches the upper end-cap. The time which an ion takes to reach the detector is negligible, as the electrical field between the upper end-cap and the detector is ten times higher than that in the trap. The flight time for an ion to reach $z = z_0$ is expressed by [18]

$$f(t_i) = \frac{1}{A}\left[-v_z(t_i) + \sqrt{v_z^2(t_i) - 2A(z(t_i) - z_0)}\right]$$ (9.4),

in which the ejection acceleration in the trap for an ion with charge-number z is given by $A = zeV_e/mz_0$. The ion TOF depends on $z(t_i)$ and $v_z(t_i)$, which are the position and the velocity at the end of confinement, respectively. Hence, the ion TOF $f(t_i)$ is a periodical function, the fundamental frequency of which is the secular axial frequency of the ion.

9.4.3 IMAGE SIGNAL OF THE CONFINED IONS

The ion impact on the first dynode of the electron multiplier gives an output pulse represented by the function $g(t' - f_i(t_i))$. The function $g(t')$ is centered on zero and describes the shape of the pulse. t' is the TOF variable. The function $f_i(t_i)$ expresses the TOF of the detected ion indexed by l at the confinement time t_i.

The TOF histogram, detected for a cloud composed of N_i^{conf} ions, is expressed by

$$H(t_i, t') = \sum_{l=1}^{N_i^{conf}} g(t' - f_i(t_i))$$ (9.5).

The image signal of the confined ions expressed as a function of confinement time is calculated by

$$A(t_i) = \int_0^{t'_{max}} t' H(t_i, t') dt'$$ (9.6),

where t'_{max} is the maximum value of the recorded TOF.

The signal $A(t_i)$, for a rectangular shape of the pulse generated by the electron multiplier and for only one detected ion, is expressed as

$$A(t_i) = h\, a f(t_i) + \frac{ha^2}{2}$$ (9.7),

where h is the height and a the width of the pulse. When two ions are detected, $A(t_i)$, is given by

$$A(t_i) = h\, a f_1(t_i) + h\, a f_2(t_i) + h\, a^2$$ (9.8).

In general for N_i^{conf} simultaneously confined ions, when the mean value is subtracted, the image signal, $b(t_i)$ is expressed by

$$b(t_i) = \beta \sum_{l=1}^{N_i^{conf}} f_l(t_i)$$ (9.9),

where β is the area of the pulse.

The periodic TOF function $f_1(t_i)$ contains the secular axial frequency information of each confined ion. For different m/z-values, the image signal is the superimposition of the secular axial frequencies of the confined species without any coupling peak between the secular axial frequencies of two different species.

The amplitude of the image signal varies according to the number of trajectories (or ions) and to the initial phase and amplitude distributions of the functions $f_i(t_i)$. For example, ions with a given mass-to-charge ratio confined within the same trajectory phase and amplitude lead to the same TOF function. Hence the image signal is expressed by $b(t_i) = \beta N_i^{conf} f(t_i)$. The influences of the phase and amplitude dispersions of the ion motion on the image signal amplitude will be discussed later in Section 9.6, with the axial trajectory instead of the TOF function in the expression for the image signal.

9.4.4 SIGNAL SAMPLING

The image signal of the ion trajectory is observed during the time $T_m = NT_e$, where T_e is the sampling period. Usually $T_e = T_\Omega$, where T_Ω is the period of the RF confinement voltage. However, the sampling period can be expressed generally by $T_e = jT_\Omega$. When j

is an integer, the signal is under-sampled and when $j = 1/2, 1/3, 1/4, \ldots$, the signal is sub-sampled [19]. Consequently, the confinement time is expressed by $t_i = ijT_\Omega$. In practical experiments, the minimum sampling period is $T_e = T_\Omega$, whereas for simulation purposes all cases are possible.

When the temporal signal is sampled, the Shannon or Nyquist criterion [19] limits frequency observation to the interval $[-1/(2T_e)$ to $1/(2T_e)]$. When the temporal signal is sampled at T_Ω, the maximum value of the frequency window is equal to $\Omega/2$. The secular axial frequency is expressed by $\omega_z = \beta_z\Omega/2$, with β_z ranging between 0 and 1. When $\beta_z = 1$, $\omega_z = \Omega/2$; consequently, the secular axial frequencies of the whole ensemble of the confinable ions are measured as the true values. Frequencies higher than $1/(2T_e)$ fold over in this interval at false frequency values. When $T_e = T_\Omega$, all of the coupling peaks between the secular axial motion and the confinement field Ω fold over at the frequency ω_z. One confined species in a pure quadrupole ion trap thus gives one axial secular frequency peak at ω_z.

9.4.5 SHAPE OF THE FREQUENCY LINE AND DISCRETE FOURIER TRANSFORM

Fourier transformation of the temporal image signal $b(t_i)$ gives the frequency spectrum of the simultaneously confined ions. The image signal is observed during the finite time T_m, hence the shape of each frequency line is expressed by $|\mathrm{sinc}(\pi^v T_m)|$ for the magnitude spectrum. The frequency line has a principal peak surrounded by auxiliary 'wiggles'. The amplitude of the first wiggle is equal to about 22% (as a percentage of the amplitude of the principal peak). In order to reduce the wiggles the temporal signal is multiplied by an apodization function before applying the Fourier transformation.

The Discrete Fourier Transform (DFT) algorithm computes the discrete frequency spectrum with a frequency step $\upsilon_m = 1/T_m$. This frequency step is not small enough to obtain the exact shape of the frequency line as the base width of the principal peak is $2/T_m$. In order to reduce the computed frequency step, zeroes are added to the temporal signal ('zero padding' or 'zero filling'). The peak amplitude and frequency are thus measured with greater accuracy.

9.4.6 NUMERICAL PROCESSING OF TOF HISTOGRAMS

A set of TOF histograms $H(t_i, t'_{i'})$ is recorded at confinement times t_i for $i = 1$ to N. The numerical processing of this set involves the following steps.

The number of confined ions *versus* confinement time is computed from

$$N_i^{\mathrm{conf}}(t_i) = \sum_{i'=1}^{N'} H(t_i, t'_{i'}) \tag{9.10}.$$

The image signal is expressed by

$$A(t_i) = \sum_{i'=1}^{N'} t'_{i'} \; H(t_i, t'_{i'}) \tag{9.11},$$

and $b(t_i)$ is computed by subtracting the mean value

$$b(t_i) = A(t_i) - \frac{1}{N}\sum_{i=1}^{N} A(t_i)$$ (9.12).

This temporal signal is multiplied by the apodization function $w(t_i)$ to give

$$b'(t_i) = w(t_i)b(t_i)$$ (9.13),

and Fourier transformation of the temporal signal $b'(t_i)$ then gives the secular axial frequency spectrum of the simultaneously confined ions.

9.5 SIMULATION

9.5.1 General Remarks

The framework of the computational code reproduces the stages of an elementary experiment, together with the succession of the N elementary experiments (Figure 9.3). Each stage processes an ion cloud, and the trajectory equations are expressed in the Cartesian coordinate system. Two frameworks are possible: (1) the code takes into account all the stages of the elementary experiment, that is, including the creation and injection stages; and (2) the code starts with the confinement stage, the initial confinement conditions (positions and velocities) in the trap being drawn from six Gaussian distributions. It is possible also to follow the positions and the velocities of the same ion cloud as a function of the confinement time from t_1 to t_N. This possibility is not described here. NI LabVIEW graphical development software (National Instruments Corporation, 11500 N Mopac Expwy, Austin, TX 78759-3504, U.S.A.) has been used to write the simulation code.

9.5.2 Initial Confinement Conditions without
the Creation and Injection Stages

As noted above, when the creation and injection stages are not simulated, the initial positions and velocities of the ions are drawn from six Gaussian distributions. The parameters are the mean values and standard deviations. For example, for the Oz axis, $\overline{z}(t_0)$ is the mean axial position, $\sigma[z(t_0)]$ the standard deviation of the axial position, $\overline{v}_z(t_0)$ the mean axial velocity, and $\sigma[v_z(t_0)]$ the standard deviation of the axial velocity. The number of ions is drawn from a Poisson distribution the parameter of which is the initial mean number of ions $\overline{N}_i$.

9.5.3 Creation and Injection Stages

When the creation and injection stages are taken into account, only the axial positions and velocities are computed during the injection process, because the neutral and ionic beams are concentrated along the Oz-axis. The initial axial position

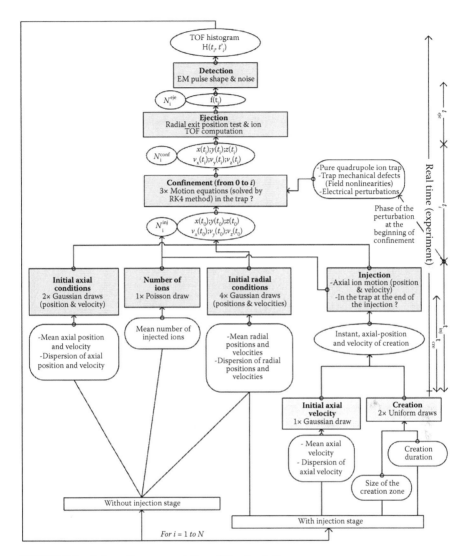

FIGURE 9.3   Schematic representation of the simulation process.

at creation, denoted as z_c, is drawn from a uniform distribution the boundaries of which are located in the creation zone. The N_i^{cre} instants of creation are either calculated with a constant temporal rate or drawn from a uniform distribution in the interval from 0 to t_{cre}. The number of ions created per second and per centimetre, denoted as D, is chosen for each simulation. The total number of created ions is calculated from: $N_i^{cre} = D \times t_{cre} \times 2dz_c$, where $2dz_c$ is the size of the creation zone. The initial axial velocity is the velocity of the neutral particles of the beam entering the source. It is drawn from a Gaussian distribution. The mean value of the velocity distribution, which is the mean value of the molecular beam, is calculated from $\bar{v}_c = \sqrt{2(\gamma/(\gamma-1))(kT/m)}$, where $T = 330$ K is the gas temperature; k is the Boltzmann constant; and γ is the ratio of the specific heats: for a xenon beam, a monatomic gas, $\gamma = 1.67$ [20].

From an elementary model, the analytical expressions of some important injection parameters, such as the stop position in the trap, and the injection trajectories are given in Section 9.11. Thus, the axial positions and velocities at the end of the injection process are calculated directly, and the number of injected ions, denoted as N_i^{inj}, is deduced from the ion locations. The radial injection trajectories are not computed; the radial positions and velocities at the beginning of confinement are drawn from four Gaussian distributions.

9.5.4 CONFINEMENT STAGE

During the confinement stage, ion trajectories are described by three Mathieu equations of the form given by Equation 9.2 for the axial direction in the case of an ideal quadrupole ion trap. The equations are solved numerically by a Runge–Kutta fourth-order (RK4) numerical method, with 400 points per period of the confinement voltage. During confinement, a test is performed in order to check that the ion position does not exceed the boundaries imposed by the trap electrodes.

9.5.5 EJECTION STAGE AND DETECTION

The ion TOF is calculated by Equation 9.4 for a dipolar ejection potential. The ion is ejected when its radial exit position at $z = z_0$ is located inside the circular ejection aperture defined by its center location and its diameter. The TOF histogram is constructed by taking into account the shape of the electron multiplier pulse. The same code is used to process ion TOF histograms (see Section 9.4.6) from both the experimental and the simulation results.

9.6 METROLOGICAL PARAMETERS

9.6.1 SIGNAL AND NOISE

9.6.1.1 Signal-to-Noise Ratio

Generally the signal-to-noise ratio, $R_{s/n}$, is given by

$$R_{s/n} = \frac{Y_s}{Y_n} \tag{9.14}$$

where Y_s is the amplitude of the signal peak and Y_n is the amplitude of the noise. Background noise is made up of a set of low-level peaks. Thus noise amplitude is defined by the mean value of the local maxima. The amplitude of a signal peak must be greater than four times the noise amplitude.

9.6.1.2 Noise Induced by the Shape of the Spectral Line

In FT analysis, the first source of background noise is induced by the shape of the spectral line. The noise level increases with the number of lines. However it can be diminished greatly by employing an apodization function. The Hamming function reduces the amplitude of the principal wiggle from 22 to 0.75%, and the exact Blackman function to 0.04%. In addition this apodization function induces a broadening of the main peak.

9.6.1.3 Amplitude of the Image Signal

The amplitude of the frequency peak is proportional to the number of confined ions. It depends also on the coherence of the motion of the ion cloud because the detected signal following confinement is a superimposition of all of the ion trajectories with different phases. In a preliminary step, which does not take into account the succession of the elementary experiments, 10 axial ion trajectories are simulated as a function of the confinement time by solving the equations for their axial motion by an RK4 method. The initial conditions of the ions are drawn from two Gaussian distributions with parameters $\bar{z}(t_0)$, $\sigma[z(t_0)]$ for the position and $\bar{v}_z(t_0)$, $\sigma[v_z(t_0)]$ for the velocity. The image signal $b(t_i)$ is calculated by adding up the 10 trajectories (see Figure 9.4). In order to obtain perfectly coherent motion (the same phase for all trajectories), $\bar{v}_z(t_0) = 0$ m s^{-1} and $\sigma[v_z(t_0)] = 0$ m s^{-1} (plot (a)). With these simulation conditions, the maximum amplitude of each trajectory does not exceed the initial axial position $z(t_0)$. The maximum amplitude of the image signal denoted as b_m is about 3 cm. The same value of the maximum amplitude is obtained for a non-coherent motion of the ion cloud with an initial velocity dispersion $\sigma[v_z(t_0)] = 600$ m s^{-1} (plot(b)). It is seen that another initial velocity dispersion, $\sigma[v_z(t_0)] = 900$ m s^{-1}, does not change this value (plot (c)).

An elementary analytical model involving ion trajectories under conditions of adiabatic confinement was developed and agrees with these simulation results [21]. With such a detection system (superimposition of the ion trajectories), the image signal of a single species can be expressed by

$$b(t_i) = \left[\sum_{l=1}^{N_i \text{ conf}} z_l(t_0) \right] \cos(\omega_z t_i) + \left[\sum_{l=1}^{N_i \text{ conf}} \frac{v_{z,l}(t_0)}{\omega_z} \right] \sin(\omega_z t_i) \qquad (9.15),$$

where l is the ion or trajectory index.

When the initial axial velocities $v_{z,l}(t_0)$ are drawn from a symmetric zero-centered distribution, the amplitude of the detected signal depends only on the sum of the initial axial conditions of each ion trajectory. With a set of initial positions drawn from a Gaussian distribution and initial velocities drawn from a zero-centered Gaussian distribution, the maximum amplitude of the image signal is then expressed by

$$b_m \approx N_i^{\text{conf}} \, |\bar{z}(t_0)| \qquad (9.16).$$

From this expression for the maximum image signal amplitude, it is concluded that it is not necessary to have coherent motion of the ion cloud in order to increase dynamic detection, provided that the initial axial positions have the same sign and

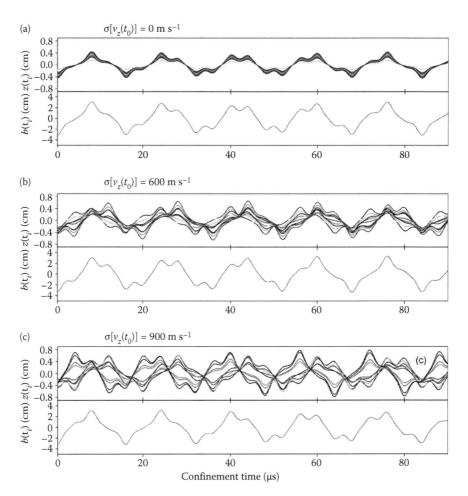

FIGURE 9.4 Simulated trajectories of 10 confined ions (upper plots) and the corresponding image signals (the sum of the 10 trajectories, lower plots) for (a): a coherent ion cloud motion with $\sigma[v_z(t_0)] = 0$ m s^{-1}; (b): a non-coherent ion cloud motion with $\sigma[v_z(t_0)] = 600$ m s^{-1}; and (c): a non-coherent ion cloud motion with $\sigma[v_z(t_0)] = 900$ m s^{-1}. The other conditions are: $\bar{z}(t_0) = -0.3$ cm, $\sigma[z(t_0)] = 0.1$ cm, $\bar{v}_z(t_0) = 0$ m s^{-1}, $m/z = 127$, $z_0 = 1$ cm, $U_0 = 0$ V, $V_0 = 100$ V, $\Omega/(2\pi) = 250$ kHz and $\varphi_0 = \pi$ rad.

that the distribution of the initial axial velocities is both symmetric and centered on zero. The ion injection protocol must locate the injected ion cloud in an off-center position (see the proposed injection protocol below in Section 9.11).

9.6.1.4 Noise Induced by the Operating Mode

The latter results do not take into account the protocol of the operating mode. Another source of noise is related to the construction of the periodical image signal as a function of confinement time, involving N elementary experiments. In the operating mode, for each elementary experiment a new ion cloud is trapped during the period t_i because all the ions are removed from the trap after confinement for the measurement stage. Hence, for each confinement time, the amplitude of the periodical image signal is computed from a different number of ions (that is, a different number of trajectories), and from a set of ion trajectories having different phases and amplitudes.

The dispersion of the initial conditions induces dispersion of the trajectory phase and amplitude. The operating mode is simulated under the usual operating conditions, with a constant number of confined ions and with Gaussian dispersions of the initial confinement conditions in positions and velocities (Figure 9.5). The signal-to-noise ratio increases in proportion to N and to the number of confined ions.

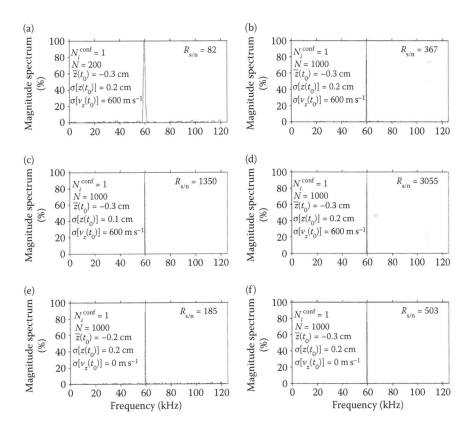

FIGURE 9.5 Simulated spectra calculated to estimate the signal-to-noise-ratio resulting from a succession of elementary experiments according to the number of ions, the number of elementary experiments and the initial confinement conditions. (a): $\bar{N}_i = 1$ ion, $N = 200$, $\bar{z}(t_0) = -0.3$ cm, $\sigma[z(t_0)] = 0.2$ cm; (b): $\bar{N}_i = 1$ ion, $N = 1000$, $\bar{z}(t_0) = -0.3$ cm, $\sigma[z(t_0)] = 0.2$ cm; (c): $\bar{N}_i = 1$ ion, $N = 1000$, $\bar{z}(t_0) = -0.3$ cm, $\sigma[z(t_0)] = 0.2$ cm; (d): $\bar{N}_i = 10$ ions, $N = 1000$, $\bar{z}(t_0) = -0.3$ cm, $\sigma[z(t_0)] = 0.2$ cm; (e): $\bar{N}_i = 1$ ion, $N = 1000$, $\bar{z}(t_0) = -0.2$ cm, $\sigma[z(t_0)] = 0.2$ cm; (f): $\bar{N}_i = 1$ ion, $N = 1000$, $\bar{z}(t_0) = -0.3$ cm, $\sigma[z(t_0)] = 0.2$ cm; $\sigma[v_z(t_0)] = 0$ m s⁻¹. The other parameters are: m/z 127, $z_0 = 1$ cm, $U_0 = 0$ V, $V_0 = 100$ V, $\Omega/(2\pi) = 250$ kHz, $\varphi_0 = \pi$ rad, $\bar{x}(t_0) = \bar{y}(t_0) = 0$ cm, $\sigma[v(t_0)] = 0.2$ cm, $\bar{v}_x(t_0) = \bar{v}_y(t_0) = \bar{v}_z(t_0) = 0$ m s⁻¹, $\sigma[v_x(t_0)] = \sigma[v_y(t_0)] = 300$ m s⁻¹ and $\sigma[v_z(t_0)] = 600$ m s⁻¹.

This ratio, $R_{s/n}$, is approximately equal to 82 for $N = 200$ (Figure 9.5a) and to 367 for $N = 1000$ (Figure 9.5b) for only one confined ion. With 10 ions with the same m/z ratio, $R_{s/n}$ is about 3055 (Figure 9.5d). The noise level diminishes as the initial axial position dispersion decreases: in Figure 9.5b with $\sigma[v_z(t_0)] = 0.2$ cm, $R_{s/n} = 367$, and in Figure 9.5c with $\sigma[v_z(t_0)] = 0.1$ cm, $R_{s/n} = 1350$. Between Figures 9.5b and e the mean initial axial position diminishes from -0.3 to -0.2 cm. As a consequence, $R_{s/n}$ diminishes as the maximum amplitude of the image signal decreases (Equation 9.15). When the results of Figures 9.5b and f are compared, it is seen that the background noise induced by the operating mode diminishes when the ion cloud motion is coherent as the initial axial velocity dispersion is reduced from 600 to 0 m s^{-1}.

These simulation results involve a small number of confined ions (in most cases only one ion is confined), which may not be ejected toward the electron multiplier to be detected. Thus, for example, between Figures 9.5b and d the increase of $R_{s/n}$ is not exactly 10-fold, as might be expected. However these results demonstrate the extreme sensitivity of the operating mode.

9.6.1.5   Peak Amplitude Fluctuation

The peak amplitude fluctuation induced by the fluctuation of the number of ions involved in each elementary experiment has been expressed [22]

$$\sigma[Y_s] = Y_s \frac{4}{\sqrt{N \, \bar{N}_i}} \tag{9.17}.$$

With $N = 1000$ and $\bar{N}_i = 100$ ions, the peak amplitude fluctuation is about 1%.

9.6.1.6   Experimental Results

Other noise sources are present in the experiment. An experimental result with $N_{acc} = 1$ shows the sensitivity of the processing method of the TOF histograms (Figures 9.6 and 9.7). The TOF histograms are recorded from about a hundred ejected ions. The number of ions can be estimated as ion pulses that can be distinguished in TOF histograms. In these experimental conditions, SF$_5^+$ ions were trapped with $\omega_z \approx 73.05$ kHz, and SF$_4^+$ ions with $\omega_z \approx 93.27$ kHz. The noise level is ca 5.3%. The SF$_5^+$ ion peak has $R_{s/n} = 19$, whereas the SF$_4^+$ peak does not appear because it is below the threshold for detection.

9.6.2   Resolution

9.6.2.1   Frequency Resolution

The spectral analysis is based upon the Fourier transformation of a periodic temporal signal, and the resolution increases with the duration of the signal observation, T_m. The Full Width at Half Maximum (FWHM) of the peak, δ_ω, is expressed by

$$\delta_\omega = \frac{\alpha 2\pi}{T_m} \, (rad) \tag{9.18},$$

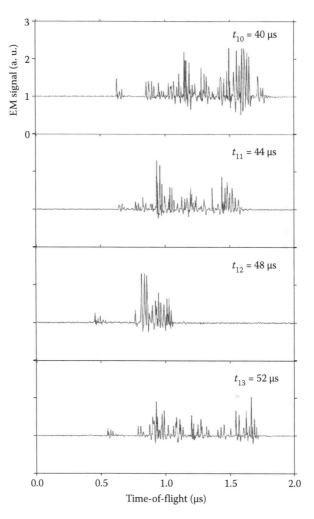

FIGURE 9.6 Experimental EI TOF histograms for positive ions formed from SF_6 for four successive confinement times from $t_{10} = 40$ to $t_{13} = 52$ μs. The experimental conditions are: $z_0 = 1$ cm, $U_0 = -2.8$ V, $V_0 = 100$ V, $\Omega/(2\pi) = 250$ kHz, $\varphi_0 = \pi$ rad, $N_{acc} = 1$, $N = 2000$, $T_e = 4$ μs, $N' = 500$ and $T'_e = 4$ ns.

where the quantity α depends on the shape of the time-domain function of finite duration. For a rectangular time-domain signal (representative of the observation window of finite duration), $\alpha \approx 1.2$ for the magnitude spectrum and 0.9 for the power spectrum. When the temporal signal is multiplied by the exact Blackman apodization function before applying the Fourier transform, $\alpha \approx 2.2$ for the amplitude spectrum and 1.6 for the power spectrum. With the Hamming function, $\alpha \approx 1.8$ for the amplitude spectrum and 1.3 for the power spectrum. The apodization function reduces the auxiliary wiggles but enlarges the main frequency peak. When

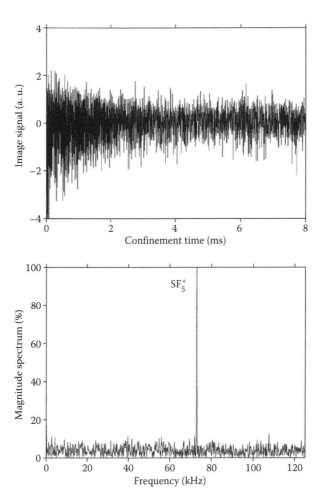

FIGURE 9.7   EI Image signal of the positive ions formed from SF_6 computed from the TOF histograms in Figure 9.6, together with the corresponding EI mass spectrum.

the separation criterion for two peaks is the FWHM, the frequency resolution is expressed by [23]

$$R_\omega = \frac{\omega_z}{\delta_\omega} = \frac{\beta_z Nj}{2\alpha} \tag{9.19}.$$

9.6.2.2   Mass Resolution

When $U_0 = 0$ the mass resolution can be calculated by [23]

$$R_m = \frac{\lambda(\beta_z)}{\beta_z} R_\omega = \frac{\lambda(\beta_z)Nj}{2\alpha} \tag{9.20},$$

where $\lambda(\beta_z)$ is equal to $q_z \delta\beta_z/\delta q_z$. The resolution increases with the operating point, and as confinement is performed with U_0, V_0, and Ω held constant, each ionic species is confined with its own operating point. Thus mass resolution is not constant and the frequency-to-mass conversion is not a linear operation. Another calculation of the mass FWHM is given by

$$\delta_{m/z} \approx \delta_\omega \frac{\Delta_{m/z}}{\Delta_{\omega_z}} \tag{9.21},$$

where $\Delta_{\omega z}$ is the difference between the axial-secular frequencies of two ions with a difference in m/z-values equal to $\Delta_{m/z}$.

From the preceding equation, Table 9.1 gives some values of the mass resolution for ions at three operating points and for two pairs of m/z ratios. The mass resolution is calculated with $N = 1000$. The mass resolution does not depend appreciably on m/z when each species is confined at the same operating point. However, for fixed confinement conditions (for $V_0 = 100$ V) the mass resolution decreases with the mass-to-charge ratio: $R_m = 494$ for m/z 100, and 43 for m/z 400. These working conditions of the trap can give high resolutions provided that the masses of the simultaneously confined singly charged ions are in a range lower than 100 Th.

TABLE 9.1

Peak Width (FWHM), $\delta_{m/z}$, and Resolution, R_m, for Two Pairs of m/z-values Computed from Equation 9.21 for an Amplitude Mass Spectrum for Three Operating Points. The Axial Secular Frequency ω_z is Measured From Spectra Computed with the Following Conditions: One Ion Confined in a Pure Quadrupole Ion Trap, a Hamming Apodization Function, $\Omega/(2\pi) = 250$ kHz, $z_0 = 1$ cm, $U_0 = 0$ V, $N = 1000$, $T_e = T_\Omega = 4\mu s$. δ_ω is Computed from Equation 9.18 and is Equal to 0.45 kHz.

m/z	V_0 (V)	β_z	ω_z (kHz)	$\Delta_{\omega z}$ (Hz)	$\delta_{m/z}$	R_m
100	80	0.49	60.8	0.75	0.6	165.8
101		0.48	60.1			
100	110	0.79	99.3	2.22	0.2	494.3
101		0.78	97.1			
400	320	0.49	60.8	0.19	2.39	167.6
401		0.49	60.6			
400	441	0.8	99.8	0.59	0.76	524.5
401		0.79	99.2			
400	110	0.15	19.19	0.05	9.23	43.32
401		0.15	19.14			

9.6.2.3　Increasing Resolution

In order to reduce the FWHM, N (Section 9.3.2) and j (Section 9.4.4) can be increased. When the confined ions are mass selected, the detected frequencies belong to a corresponding frequency range. The temporal image signal can be under-sampled without any overlapping of the frequency ranges of the different sampling orders [23]. The integer values of j can be chosen to be greater than 1. In this case, the criterion that limits the resolution is given by $(\beta_{z,2}-\beta_{z,1})j \leq 1$, where $\beta_{z,2}$ is the operating point corresponding to that of the confined ion having the lower m/z-value. This criterion is less restrictive for increasing the resolution than is the Shannon criterion, which can be written as $\beta_{z,2}j \leq 1$ (see Section 9.4.6.). However, with under-sampling, peaks are not located at the true frequency values as the frequency observation window is limited to $1/(2jT_\Omega)$. True frequency values υ_{true} can be deduced from $\upsilon_{\text{true}} = \pm \upsilon_{\text{app}} + C/(jT_\Omega)$, where υ_{app} is the apparent frequency of the peak, $\pm$ means that either a positive or a negative-frequency peak folds over, and C is the sampling order of the fold-over.

9.6.2.4　Optimal Mass Resolution

The under-sampling case is illustrated in an experimental result involving only singly charged isotopic Xe$^+$ ions (Figure 9.8) [24]. The operating conditions are: $N = 2000$ and $j = 4$, which gives the sampling period $T_e = 16$ µs and a frequency observation window in the range $0–1/(2jT_\Omega) = 31.25$ kHz. The positive frequency spectrum of the (-1) sampling order folds over within this frequency range. The frequency peak of m/z 136 appears at about 4 kHz, while its true frequency value is: $4–(-1) \times 62.5 = 66.5$ kHz. The peak FWHM is about 50 Hz, or 0.032 Th, giving a resolution of about 4000 for m/z 130, which is the optimum resolution that was achieved with the prototype device.

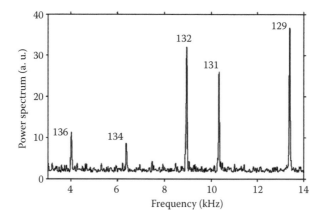

FIGURE 9.8 Optimum resolution obtained on experimental power mass spectrum of isotopic Xe$^+$ ions. The experimental conditions are: $U_0 = 0.5$ V, $V_0 = 118$ V, $\Omega/(2\pi) = 250$ kHz, $z_0 = 1$ cm, $\varphi_0 = \pi$ rad, $N = 2000$ and $T_e = 16$ µs. (Reproduced from Carette, M. and Zerega, Y., *J. Mass Spectrom.* 2006, *41*, 71–76. With permission from Wiley Inter Science.)

9.7 PERTURBATIONS OF ION MOTION

The signal $b(t_i)$ is the image of the ion trajectories. Hence during confinement, the ion trajectory must be as pure as possible. One species must lead to one spectral peak having a defined shape and an FWHM depending on the duration of the temporal observation window and the apodization function employed.

9.7.1 COLLISIONS

When collisions occur between confined ions and neutrals, ion losses, ion-motion phase-changes, and ion-motion amplitude-relaxation can be observed according to the mass difference between the ions and the neutrals [25].

The amplitude of the image signal decreases with the dynamics of the ion motion, for example when collisional cooling occurs with neutral particles that are lighter than the ions, or when the number of confined ions diminishes with collisions involving particles heavier than the ions. This amplitude decrease results in peak shape modification and peak-base enlargement. Thus when a rapid exponential decay of the ion population occurs, the peak shape in the amplitude spectrum becomes Lorentzian.

Under typical operating conditions at a pressure $p \approx 10^{-7}$ Torr inside the trap, ion collisions lead to an enlargement of the FWHM of about 10 Hz [23]. Phase variations in the ion motion during confinement induce background noise in the spectrum. In order to ensure a high quality vacuum in the analysis cell, the ion source and the trap for analysis are located in two different vacuum chambers in our experimental system.

9.7.2 SPACE CHARGE

With a sufficiently large number of confined ions, the space charge phenomenon occurs. There exists a Coulombic repulsion force between the ions which increases during periodical ion-cloud contractions. This shifts the frequency of motion [26]. In our experiments, the usual operating conditions lead to a small number of confined ions, so avoiding this space charge phenomenon.

9.7.3 ELECTRICAL PERTURBATIONS

Theoretically the potential $(U_0 + V_0 \cos \Omega t)$ is applied to the ring electrode of the trap. Since axial secular frequency of the ion motion depends on U_0, V_0, and Ω, perturbations of these three parameters impair frequency measurement.

Fast low-level random variations of the amplitudes or frequency of the potential during confinement are not integrated by the ion motion over a micro period, T_Ω. Low-level drifts of the amplitudes or frequency of the potential during the confinement time induce a broadening of the frequency peak. For example, with the operating conditions $U_0 = 0$ V, $V_0 = 100$ V, $\Omega/(2\pi) = 250$ kHz, $z_0 = 1$ cm for an ion of m/z 130, the secular axial frequency, ω_z, is *ca* 57.9 kHz. When U_0 increases by 50 mV, ω_z increases by 100 Hz. With such a variation in the magnitude of the potential it would not be possible to achieve an FWHM of 50 Hz, as described in the previously mentioned experimental result [24].

The source of the RF confinement voltage may have harmonic distortion. Higher frequency harmonics shift the value of the frequency of motion [27]. For example, with the operating conditions $U_0 = 0$ V, $V_0 = 100$ V, $\Omega/(2\pi) = 250$ kHz, $z_0 = 1$ cm for an ion having m/z 127, the axial secular frequency is equal to 59.65 kHz for an ideal AC voltage. When only the first harmonic (500 kHz) is added, $\omega_z/(2\pi) = 59.66$ kHz for a total harmonic distortion (ratio of the harmonic to the fundamental frequency) of 0.1%, and 59.74 kHz for 1%. Generally the harmonic distortion is constant, hence the frequency shift can be calculated and taken into account during mass calibration.

In the experimental device, the electrical perturbations have different origins. Switched power supplies can cause a ripple voltage at the switching frequency and at the harmonic frequencies thereof. The main power supply and its harmonics are a source of low-frequency noise at 50/100 Hz. Consequently, a mono-frequency perturbation applied to the trap electrodes in a quadrupolar configuration was studied in order to show its influence on ion motion (see Section 9.13) [28].

9.7.4 MECHANICAL PERTURBATIONS

von Busch and Paul first discovered nonlinear resonances in RF quadrupole devices [29,30]. Subsequently theoretical and experimental studies were carried out by Dawson and Whetten for an ion trap [31] and for a mass filter [32]. More recently Wang and Franzen have contributed to the understanding of nonlinearities by the establishment of mathematical formalisms and by simulation works [33–37].

Unexpected nonlinearities are induced by imperfections in the construction of the trap, for example apertures in the electrodes, electrode truncation, and electrode alignment [38,39]. The expression of the electric field generated by a nonlinear trap is not a linear function of the radial or axial position. For a long time nonlinearities were considered as defects, as they change the ion-motion frequency and can lead to ion losses [40,41]. However, the ions are influenced very weakly by nonlinearities when they are cooled in the center of the trap by means of a buffer gas.

The operating modes of the trap that involve voltage amplitude-scanning and resonant dipolar ejection of ions require the presence of nonlinearities in order to increase mass resolution, because the ion ejection velocity is increased [42]. In addition to these desired beneficial effects, nonlinearities can entail harmful effects, such as premature ion ejection during the mass scan and changes in ion secular frequency [43]. In collision-induced dissociation (CID) experiments, the precursor and product ions can be in off-center positions in the trap when dissociation occurs, in which case the nonlinearities in the electric field, in addition to the abrupt changes in values of the trapping parameter, can influence their motion. As a result, losses of ions are observed when the operating point in the stability diagram corresponds to a 'black hole' [44,45], or more generally to zones of poor confinement [46].

Nonlinearities are brought about by a modification of the position [47,48] and the shape [49,50] of the electrodes, or by the superimposition of an asymmetrical field onto the confinement field [51]. In the FT operating mode described here, nonlinearities impair the ion motion as trajectories with large amplitude are required in order to obtain good dynamics of the detected signal. A simulation study and some experimental results describing the influences of the nonlinearities, induced mainly by electrode truncation, are given in Section 9.12 [24,52].

9.8 CONTROL, SUPPLY AND MEASUREMENT DEVICES

9.8.1 REQUIREMENTS OF THE OPERATING MODE

In conventional 3D quadrupole ion trap devices, the amplitude and the frequency of the confinement voltages remain constant during ion creation, and then the amplitude of the RF voltage applied to the ring electrode is ramped in order to eject the ions sequentially for mass analysis. Slower RF scan rates are used to improve the resolution [53,54].

The FT-operating mode requires switched RF and DC potentials of constant amplitude and frequency applied to the electrodes in order to inject, confine, and eject the ions in and out of the trap in a set of repetitive elementary experiments (see Figure 9.9). Therefore, this operation requires particular types of control, measurement, and supply devices. Many switches of the potentials occur during the succession of elementary experiments. The number of switches per electrode signal is calculated being $4 \times N \times N_{acc}$. To achieve the best accuracy in defining the time-interval duration and phase-control of the RF confinement voltage at the beginning

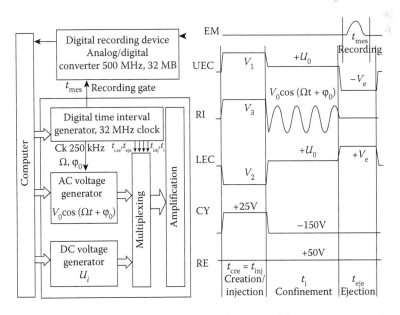

FIGURE 9.9 A typical wave train applied to the electrodes of the trap and the creation cell during an elementary experiment using electron ionization. The control, supply, and measurement units are shown on the left-hand side.

and at the end of confinement, the control and supply devices use central quartz clock technology. The device must provide adequate switching times: each electrode supply must have a slew rate of 500 to 1000 V μs^{-1}. The amplitudes of the potentials, time-interval durations and phases are controlled by computer. The control and interface program is written with NI LabVIEW graphical development software.

9.8.2 Digital Time-Interval Generator

The digital time-interval generator comprises the main clock at a frequency of 32 MHz and a set of digital frequency dividers to generate auxiliary clock signals between 4 MHz and 250 kHz. A set of down-counters generates the different time intervals of the stages of the elementary experiment: t_{cre}, t_{inj}, t_i, t_{eje}, and t_{mes}. The confinement duration is controlled by a clock signal at 250 kHz, and the others by a clock signal at 4 MHz. An up-counter and a comparator are required to control the increase and the number of confinement time intervals. A down-counter performs the count of the number of repetitions of a set of elementary experiments, N_{acc}.

A digital phase-control device adjusts the phase of a 250 kHz clock signal. As this signal will be used to generate the RF confinement voltage, this phase represents the phase of the RF confinement voltage at the beginning of the confinement.

9.8.3 Wave-Train Generator

The wave-train generator is composed of: (1) a low-level RF voltage generator; (2) a low-level DC voltage generator; (3) a multiplexing device; and (4) an amplification device (Figure 9.9). First, the low-level amplitudes of the RF and DC voltages are generated by RF and DC voltage generators. The 250 kHz clock signal from the digital time-interval generator enters the RF voltage generator. A square wave contains high-frequency harmonics, so the clock signal is filtered in order to retain only the fundamental frequency, and then amplified to the maximum amplitude of 5 V. The amplitude of the low-level RF confinement voltage, denoted as $(V_0\cos(\Omega t + \varphi_0))$, is adjusted in the range 0–5 V by a digital-to-analog converter used as an attenuator. The DC voltage generator comprises a set of digital-to-analog converters. Each converter uses the same voltage reference, and generates a stable low-level DC voltage in accordance with the DC potential that will be applied to a given electrode during a given time interval. Then, for each electrode, a low-level voltage wave-train is composed in accordance with the duration, the type of voltage (RF or DC), and the voltage amplitude required for each stage of the elementary experiment.

Lastly, each wave-train is amplified and applied to an electrode. Amplification is performed by a high-voltage high-slew-rate linear power operational amplifier, either the PA85 or the PA94 (Apex Precision Power, Cirrus Logic, Inc., 2901 Via Fortuna, Austin, TX 78746, U.S.A.); the maximum output voltages are 200 V and 400 V, respectively. The bandwidth is 20 MHz, and the slew rate is about 1000 V μs^{-1}. Currently, the output voltages of the wave-train generator are in the range ±350 V. The built-in operational amplifier is the PA94, and with such a maximum value of the confinement

RF voltage, ionic species of m/z 400 are trapped on the operating point $\beta_z \approx 0.5$. The resolution increases with the operating point. In order to confine a higher mass and to increase resolution, a new potential configuration will be applied to the trap electrodes during confinement: the ring will be supplied by the potential $(U_0 + V_0\cos(\Omega t + \varphi_0))$ and the two end-cap electrodes by $-(U_0 + V_0\cos(\Omega t + \varphi_0))$. Hence the same amplifier module can be employed. However, technical difficulties appear in generating the two voltages of opposite sign, in particular in obtaining the RF voltages that are out of phase by π radians.

9.8.4 MEASUREMENT DEVICE: TOF DIGITIZATION

Each TOF histogram is digitized and recorded by a high-speed transient card connected to the computer. The in-board analog-to-digital converter has a resolution of 8 bits, and a maximum sampling frequency of 500 MHz. With a memory size of 32 MB, the card can store $N' \times N \times N_{acc} \leq 2^{32}$ TOF histograms. Typically, with a sampling frequency of 250 MHz and a TOF variation in the range of 2 µs, $N' = 500$. Hence with $N = 1024$, 64 accumulations are possible.

9.9 RADIAL SECULAR FREQUENCY MEASUREMENT

When the ions are moved out of the trap a radial discrimination of the number of detected ions can occur, depending on the amplitude of their radial trajectories and the size of the ejection aperture.

9.9.1 EXPRESSION OF THE RADIAL POSITION OF THE TRAP EXIT

The potential inside the trap is assumed to be dipolar during ejection. The ion leaves the trap at $z = z_0$ for the radial position given by

$$x_e(t_i) = f(t_i)v_x(t_i) + x(t_i) \tag{9.22}$$

where $f(t_i)$ is the TOF function of the ion; $x(t_i)$ and $v_x(t_i)$ are the position and the velocity of the ion at the end of the confinement time t_i, respectively [18].

A mechanical discrimination of the number of detected ions can be performed by reducing the radial size of the aperture in the upper end-cap so as to make it smaller than the maximum radial size of the ion cloud during the confinement.

9.9.2 SIMULATION RESULTS

A simulation of the operating mode taking into account the TOF and the radial discrimination is shown in Figure 9.10. In the upper spectrum the aperture is centered, and has a diameter of 0.2 cm. During one radial period, the ion cloud has two symmetrical extensions and contractions around the Oz axis. As the aperture is centered and smaller than the size of the ion cloud, the number of detected ions is modulated at twice the radial secular frequency. In Equation 9.22, $f(t_i)$, a periodic function with

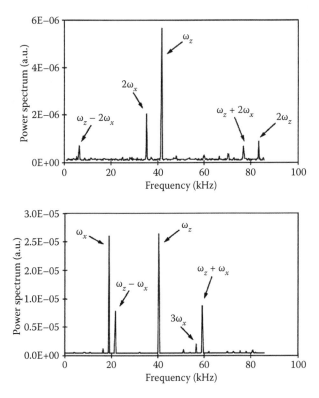

FIGURE 9.10 Simulated power mass spectra obtained for one confined ionic species (m/z 146) and for different hole configurations in the upper end-cap electrode: a small centered hole of 0.2 cm in diameter (upper); and for an off-center hole of 1 cm in diameter shifted 0.5 cm in the radial plane (lower). The common conditions are: $z_0 = 1$ cm, $U_0 = 0.5$ V, $V_0 = 54$ V, $\Omega/(2\pi) = 172$ kHz, $\varphi_0 = \pi$ rad, $\bar{N}_i = 10$ ions, and $N = 152$. The initial conditions of the ion cloud at the beginning of confinement are: $\bar{x}(t_0) = \bar{y}(t_0) = 0$ cm, $\sigma[x(t_0)] = 0.5$ cm, $\sigma[y(t_0)] = 0$ cm, $\bar{z}(t_0) = 0.35$ cm, $\sigma[z(t_0)] = 0.05$ cm, $\bar{v}_x(t_0) = \bar{v}_y(t_0) = 0$ m s^{-1}, $\sigma[v_x(t_0)] = \sigma[v_y(t_0)] = 100$ m s^{-1}, $\bar{v}_z(t_0) = 100$ m s^{-1}, and $\sigma[v_z(t_0)] = 100$ m s^{-1}. (Reproduced from Carette, M., Perrier, P., Zerega, Y., Brincourt, G., Payan, J.C., and Andre, J., *Int. J. Mass Spectrom. Ion Processes*, 1997, *171*, 253–261. With permission from Elsevier.)

a frequency of ω_z, is multiplied by $v_x(t_i)$, a periodic function with a frequency of ω_x. In consequence, peaks arising from coupling between the components of motion in the radial and axial directions are detected. In the lower spectrum the ejection aperture is off-center: it has a diameter of 1 cm and is shifted 0.5 cm in the radial plane. In this case the ω_x frequency is detected instead of $2\omega_x$. To observe only radial frequency peaks, the TOF discrimination must be suppressed.

9.9.3 EXPERIMENTAL RESULTS

Recent experimental works concerning the injection mode have given an interesting view of ion cloud behavior in the radial plane (Figure 9.11). Depending on the

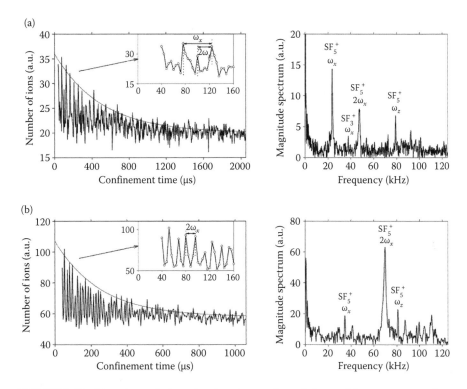

FIGURE 9.11 Experimental data showing the number of detected ions *versus* confinement time (left-hand plots) and the corresponding mass spectra. The ions were created by electron ionization of SF_6 molecules. (a) The electron multiplier has a detection area that is shifted slightly off-center; the experimental conditions were $z_0 = 1$ cm, $U_0 = -4.4$ V, $V_0 = 113.2$ V, $\Omega/(2\pi) = 250$ kHz, $\varphi_0 = \pi$ rad, $N = 512$, and $T_e = 4$ μs; (b) the experimental conditions were $U_0 = -0.2$ V, $V_0 = 125$ V, $\Omega/(2\pi) = 250$ kHz, $z_0 = 1$ cm, $\varphi_0 = \pi$ rad, $N = 256$, and $T_e = 4$ μs.

slowing-down of ions during injection, the ion cloud can have a large extension in the radial plane during confinement. The maximum radial amplitude of some trajectories can prevent transit through the upper end-cap ejection aperture, depending on the confinement time. In the experimental measurements, the minimum number of detected ions remains constant during confinement while the maximum has an exponential decay. Of these, the most off-center ions are lost during confinement. Any increase of the minimum number of detected ions during confinement would indicate a cooling process through collisions of the ions with lighter particles, whereas a decrease would show an ion loss by collision with heavier particles.

The injection conditions employed lead to large radial trajectories. Between 100 and 200 μs, the fluctuation in the number of detected ions is still about 30% of the total number for both cases. The maximum values of the two curves are fitted by a function $y_0 + Ae^{-t/\tau}$. The decay constant, τ, is equal to *ca* 500 and 250 μs for Figure 9.11a and b, respectively.

In Figure 9.11a, the main frequency peak is located at the radial secular frequency, which reveals non-symmetrical detection in the radial plane. This experiment enabled us to observe that the first dynode of the electron multiplier used in this experiment was misplaced upon assembly by the manufacturer. A new electron multiplier was mounted wherein the main frequency peak is located, Figure 9.11b, at twice the radial secular frequency.

9.9.4 APPLICATIONS

Ion loss and ion cooling may be found from the measurement of the fluctuation of the detected number of ions during confinement. Experimentally, difficulties can be encountered in measuring the value of the potentials applied to the trap electrodes, and a slight potential difference can be observed between the two end-cap electrodes. In conclusion, a greater accuracy in mass determination from the secular frequency of ions can be achieved with knowledge of both the radial and axial frequencies.

9.10 DUAL QUADRUPOLE ION TRAP MASS SPECTROMETER

Todd and co-workers were the first to suggest separating the creation and mass-analysis functions of a mass spectrometer. They used an ion trap as a reaction cell (QUISTOR mode) and a quadrupole mass filter to perform mass analysis [55–57].

As the FT operating mode requires external ion creation, it was proposed to employ another ion trap as a creation and reaction cell [58]. As well as these functions, the versatility of the ion trap operating modes offers other interesting preparation functions, such as ionic mass-selective enrichment, the main function required for the ion source in trace analysis.

An examination was carried out concerning an experimental device composed of one ion trap (T1) for ion creation and preparation, a transfer cell, and a second ion trap (T2) using the FT operating mode for mass analysis (see Figure 9.12). In this preliminary study, the role of trap T1 was limited to ion creation by electron ionization and ion cooling toward the center of the trap by collisions with helium gas in order to transfer the maximum number of ions toward trap T2.

This functionality of trap T1 is shown in an experimental result involving Xe^+ ions and helium buffer gas (Figure 9.13). The ions are created during a period of 1 ms in trap T1. They are confined immediately and subjected to cooling collisions, before being accelerated in the transfer cell and passing through trap T2 without being confined. The efficiency of the cooling process increases with confinement time. The ions are created in the whole of the trap domain. During confinement, more ions can be ejected through the upper aperture of trap T1, measuring 1 mm in diameter. The maximum number of ejected ions is reached for a confinement duration of about 5 ms. For long confinement times, ion loss processes predominate. High-resolution frequency spectra of isotopic Xe^+ ions have been obtained with this experimental device, and as a consequence, they have revealed for the first time peak splitting, arising from nonlinearity of the confinement field.

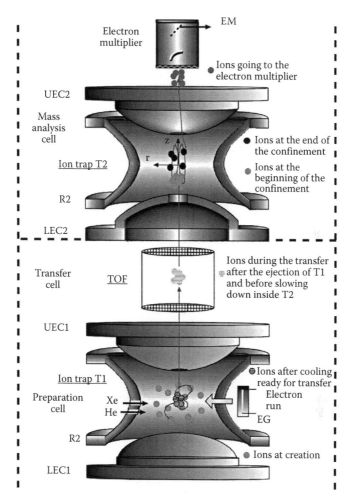

FIGURE 9.12 The dual ion trap mass spectrometer. (Reproduced from Zerega, Y., Perrier, P., Carette, M., Brincourt, G., and Nguema, T., Andre, J., *Int. J. Mass Spectrom.* 1999, *190/191*, 59–68. With permission from Elsevier.)

However, this study was restricted to the level of a feasibility test because several technological improvements were required before further progress could be made. The transfer cell was a basic flight tube without any focusing lenses at the exit of trap T1 and at the entrance of trap T2. At that time, the voltages for the trap electrodes were supplied from a different electronic module, including a resonant circuit and a power-switching transistor. The cut-off gave an exponential decrease of the RF confinement voltage over about two periods.

The present electronic circuit can switch voltages with a slew rate greater than 500 V μs⁻¹. Therefore, the use of a first trap, in place of a classical ion source, as a creation and mass-selective enrichment cell, continues to be of interest as a component for a mass spectrometer for trace analysis in order to increase the detection level.

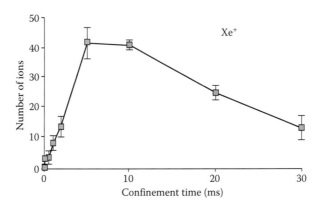

FIGURE 9.13   Plot showing the number of Xe⁺ ions detected *versus* confinement (or cooling) time in trap T1 for 1 ms of ion creation by electron ionization. (Reproduced from Zerega, Y., Perrier, P., Carette, M., Brincourt, G., Nguema, T., and Andre, J., *Int. J. Mass Spectrom.* 1999, *190/191*, 59–68. With permission from Elsevier.)

9.11   ION INJECTION

9.11.1   INTRODUCTION

Depending upon the nature of the ion source and the operating mode of the mass spectrometer, ion creation can be performed either internally or externally [16]. The first applications of ion traps were for determining the pressure and the proportions of gas constituents in high vacuum systems. Dawson and Whetten described the first use of an external source for an ion trap, using a conventional ion source such as a plasma discharge, an ionization chamber and a surface ionization electrode [59]. In 1983, the first commercially available ion trap instrument, the Ion Trap Detector (ITD), was developed by the Finnigan Corporation, using internal ion creation by means of electron ionization. The trap possessed both the ion creation and mass-analysis functions. Later, Chemical Ionization (CI) capability was added [60,61].

However, *in situ* ionization methods limit the performance of the instrument. The control of the EI collision parameters is difficult owing to the presence of the RF electric field. Moreover, modern ionization methods require external ion injection modes. Ion creation can be either continuous, for example for Atmospheric Pressure Glow Discharge Ionization (APGDI) and Electrospray Ionization (ESI), or pulsed, for example for Matrix-Assisted Laser Desorption/Ionization (MALDI) according to the nature of the neutral analyte and the most appropriate method of ionization [62]. When ions are injected into a trap, they cannot be confined because of their initial kinetic energy, even if they possess stable trajectories [63]. With the ITD, the confinement of externally created ions is assumed to be due to collisions with the helium buffer gas [64]. External axial injection of ions from a MALDI source has been improved by increasing the confinement RF voltage during injection [65]. Various scan functions of the RF confinement voltage, incorporating large zero-potential time-intervals, can be used in order to inject the ion during this instant without turning off the confinement voltage [66,67]. Another possible approach is to switch off the confinement potential during injection. Many publications describe this injection mode [68–72], which uses phase space dynamics [73–75].

The injection device must have easy-to-use potential and time controls. Moreover, ion positions at the beginning of confinement must have the same sign in order to attain the optimum dynamics of the detected image signal. For this purpose, the ions are accelerated before they enter the trap and then slowed down inside the trap. For example, the potential inside the trap is sufficient to stop all the injected ions before they reach the center of the trap. The proposed injection protocol leads to a steady ion flow into the trap, and an elementary analytical model has been developed for the axial trajectory. Hence, the axial position and velocity distribution of the injected ions and of those that will be trapped can be expressed according to the potentials applied to the electrodes and the injection duration [76].

Simulation tools such as ISIS (Integrated System for Ion Simulations) [25], ITSIM (Ion Trap Simulation program) [77], and SIMION (industry-standard charged particle optics simulation software) [78] can also solve ion trajectories in magnetic and/ or electric fields [79]. For electric fields, local values of the potential are computed by solving the Laplace equation. With our model, the analytical expressions of the ion parameters, such as the stop position and time, are calculated according to the initial conditions of the original neutral species prior to ionization and the injection parameters (the location of the electrodes and the applied potentials). Hence the influences of the injection parameters and of the initial conditions can be estimated more quickly than with the published simulation tools.

9.11.2 Equation Setting

Two zones are investigated: (i) from the source to the lower end-cap, where the ions are accelerated; and (ii) inside the trap, where the ions are first slowed down and stopped and then accelerated whereupon they go back. For each zone, equations are given for one ion and for the initial time $t = 0$.

As neutral and ionic beams are concentrated along the Oz axis during injection, it is assumed that the ion experiences a uniform dipolar electric field between the creation cell and the lower end-cap. The axial motion has only a positive constant acceleration given by

$$\gamma_z = \frac{ze}{m} \frac{(V_2 - V_c)}{(z_0 + z_c)} \qquad (9.23),$$

where $V_c = ((z_c - z_{cen} - dz)/2dz)(V_{cel} - V_{rep}) + V_{cel}$ is the potential at z_c, the position of the ion at the instant of creation. During injection, the potentials applied to the trap electrodes are denoted as V_1, V_2, and V_3 for the upper end-cap, the lower end-cap, and the ring, respectively.

The time required to reach the lower end-cap is expressed by

$$t_1 = \frac{-v_c + \sqrt{v_c^2 - 2\gamma_z(z_c + z_0)}}{\gamma_z} \quad \text{with} \quad v_c^2 > 2\gamma_z(z_c + z_0) \qquad (9.24),$$

and the ion enters the trap at the velocity

$$v_1 = \sqrt{v_c^2 - 2\gamma_z(z_c + z_0)} \tag{9.25}.$$

During injection into the trap, the ions are concentrated along the Oz axis. The multipole expansion in the Legendre polynomials of the electric potential is restricted to the second-order. The boundary conditions are calculated with the potential applied to the three trap electrodes, so that the potential inside the trap can be expressed approximately by

$$V(x,y,z) \approx \frac{V_1 + V_2 + 2V_3}{4} + \frac{V_1 - V_2}{2z_0}z$$

$$+ \frac{V_1 + V_2 - 2V_3}{8z_0^2}(2z^2 - (x^2 + y^2)) \tag{9.26}.$$

For small displacements in the radial plane, the axial motion equation is

$$\frac{d^2z(t)}{dt^2} + Bz(t) = -A \tag{9.27},$$

with

$$B = \frac{ze}{m} \frac{V_1 + V_2 - 2V_3}{2z_0^2} \quad \text{and} \quad A = \frac{ze}{m} \frac{V_1 - V_2}{2z_0} \tag{9.28}.$$

The dependencies of the ion position and velocity functions upon time are deduced from this equation. The expressions of the stop position and instant can be then calculated, as well as the position and velocity of the ion in the trap at the end of the injection process.

The repulsive field encountered by the ions as soon as they enter the trap can increase, decrease or be constant with the increasing of the axial direction according to the potentials applied to the electrodes. The acceleration must be negative for all t and $z(t)$ ranging between $-z_0$ and z_0. Hence first potential conditions must be $V_1 > V_3 > V_2$. For $V_1 + V_2 > 2V_3$, leading to $B > 0$, the deceleration of the ion increases in proportion to the displacement along the axial direction, whereas it decreases for $V_1 + V_2 < 2V_3$, leading to $B < 0$, and is constant for $V_1 - 2V_3$.

9.11.3 Results and Discussion

9.11.3.1 Deceleration Potential Configurations

In order to increase the number of ions that will be confined, the influences of the previously mentioned potential configurations have been compared for the same stop position in the trap (Figure 9.14). In this simulation result, the ions are created in the center of the source ($z_c = z_{cen}$) with an initial velocity given by $v_c = 3698\sqrt{1/m}$, where m is expressed in u, and with a constant time separating consecutive creations

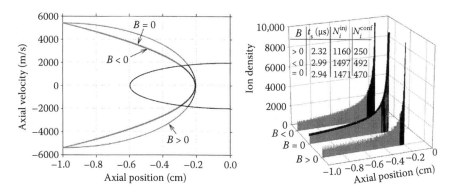

FIGURE 9.14 Phase space ion trajectory plots and ion density *vs* axial position for three configurations of the deceleration potentials: for $B > 0$, $V_1 = 95$ V and $V_3 = 0$ V; for $B = 0$, $V_1 = 38$ V and $V_3 = 19$ V; for $B < 0$, $V_1 = 35$ V and $V_3 = 20$ V and for $z \approx -2$ mm. The black parts of the curves represent the ions that will be confined. The other parameters are: m/z 100, $D = 5 \times 10^8$ ions s^{-1} cm^{-1}, $t_{cre} = t_{inj} = 100$ μs, $z_c = z_{cen} = -2.5$ cm, $dz_c = 0$ cm, $v_c = 370$ m s^{-1}, $V_{rep} = 20$ V, $V_{cel} = 10$ V, $V_2 = 0$ V, $z_0 = 1$ cm, $U_0 = 0$ V, $V_0 = 50$ V, $\Omega/(2\pi) = 250$ kHz and $\varphi_0 = \pi$ rad. (Reproduced from Janulyte, A., Zerega, Y., Carette, M., Reynard, C., and Andre, J., *Rapid Commun. Mass Spectrom.* 2008, 22, 2479–2480. With permission from John Wiley and Sons, Ltd.)

$\tau_c = 4$ ns. The rate of ion creation is chosen to ensure that around the stop position in the trap there is a sufficient distance between the ions to make Coulombic interactions negligible. Under these conditions, injected ions have the same trajectory in phase space. Hence ion distribution in the trap is located within this phase space trajectory at the end of injection. The shape of the injected ion trajectory influences the number of injected ions and the number that will be confined. The optimum number is obtained for the potential configurations leading to $B < 0$ and $B = 0$, as the resulting trajectory path is the shortest in phase space. In like manner, the influences of other experimental parameters, such as the confinement parameters and the potentials applied to the creation cell, have been examined.

9.11.3.2 Ion Distribution at the End of Injection

The distribution of the ions in the trap at the end of injection is determined by the distribution of the initial creation conditions and the transfer protocol used for injection. For example, phase space trajectories are shown for various injection parameters and creation conditions (see Figure 9.15). The initial axial positions in the source and the times of creation are drawn here from two uniform distributions, and the initial ion velocity is drawn from a Gaussian distribution. The ion cloud at the end of injection defines the two extreme ion trajectories in phase space. During injection, all the ion trajectories in phase space are located between these two extreme trajectories. The variation of the initial velocity of the molecule in the range of 0–200 m s^{-1} does not modify the ion trajectory during injection. Indeed, for this velocity range, the initial kinetic energy is negligible compared to electric potential energy. The dispersion of ions with zero velocity increases with the dispersion of the initial conditions

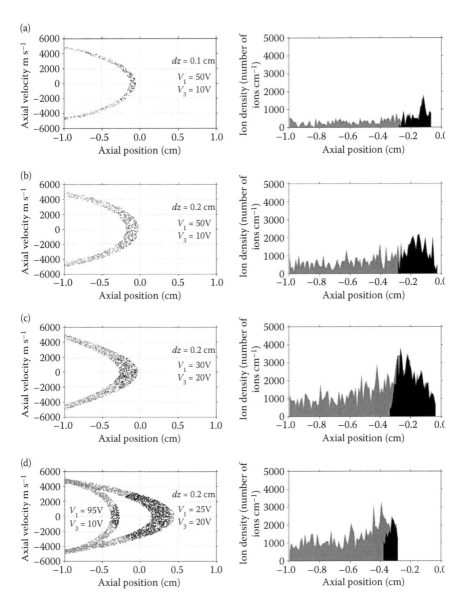

FIGURE 9.15 Phase space ion trajectory and ion density plots *vs* axial position obtained for different ion injection parameters and creation conditions. (a) $dz = 0.1$ cm, $V_1 = 50$ V, $V_3 = 10$ V; (b) $dz = 0.2$ cm, $V_1 = 50$ V, $V_3 = 10$ V; (c) $dz = 0.2$ cm, $V_1 = 30$ V, $V_3 = 20$ V; (d) $dz = 0.2$ cm, $V_1 = 95$ V and 25 V, $V_3 = 10$ V. The other parameters are: m/z 130, $D = 5 \times 10^8$ ions s^{-1} cm^{-1}, $t_{cre} = t_{inj} = 100$ µs, $z_c = z_{cen} = -2.5$ cm, $v_c = 324$ m s^{-1}, $V_{rep} = 20$ V, $V_{cel} = 10$ V, $V_2 = 0$ V, $z_0 = 1$ cm, $U_0 = 0$ V, $V_0 = 108.4$ V, $\Omega/(2\pi) = 250$ kHz, and $\varphi_0 = 0$ rad (Reproduced from Janulyte, A., Zerega, Y., Carette, M., Reynard, C., and Andre, J., *Rapid Commun. Mass Spectrom.* 2008, 22, 2479–2480. With permission from John Wiley and Sons, Ltd.)

of ions in the source (see cases Figure 9.15a and b). The mean stop-value does not change. Ion dispersion in the trap is linked also to the shape of the injection trajectory according to the potentials V_1 and V_3. For the same mean value of the stop position, reduced dispersions are obtained when V_3 tends to zero (see cases Figure 9.15b and c). The same behavior is observed when ions are stopped near the lower end-cap of the trap (see case Figure 9.15d). With the proposed injection protocol, the positions at the beginning of confinement of all the ions that will be confined can have either negative or positive values.

9.11.3.3   Influences of the *m/z* Ratio

The stop position in the trap does not depend on the *m/z*-value (for details of the calculation see reference [76]), hence the ion density increases with *m/z*-value (see Figure 9.16). Two configurations of the deceleration potentials are compared. With the lowest value of V_0, the number of confined ions is almost independent of *m/z*, but the number of confined ions is very small.

Choosing low values for the confinement voltage leads to low values of mass resolution. Other simulation results (not presented here) show that over smaller *m/z*-ranges of *ca* 100 Th, the number of confined ions can be independent of *m/z* for higher values of V_0.

9.11.3.4   Comparison of Simulation and Experimental Results

The validation of the injection model consists of a comparison of the simulation and experimental results concerning the evolution of the number of confined ions as a function of the deceleration potentials V_1 and V_3 (see Figure 9.17). The experimental results involve trapped SF_5^+ ions, and the values of the experimental parameters are

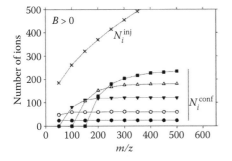
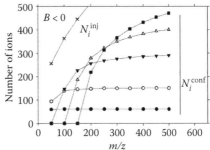

FIGURE 9.16   Plots showing the number of injected ions (×) *vs m/z*, together with the number of confined ions versus *m/z*: (•) for $V_0 = 20$ V; (○) for $V_0 = 50$ V; (▼) for $V_0 = 100$ V; (Δ) for $V_0 = 150$ V and (■) for $V_0 = 200$ V. For *B*>0, $V_1 = 74.4$ V and $V_3 = 0$ V; for *B*<0, $V_1 = 25$ V and $V_3 = 20$ V. Both configurations give $z_s = -0.1$ cm. The other parameters are: $D = 10^8$ ions s^{-1} cm^{-1}, $t_{cre} = t_{inj} = 100$ µs, $z_c = z_{cen} = -2.5$ cm, $v_c = 3698 \sqrt{1/mu}$ m s^{-1}, $V_{rep} = 20$ V, $V_{cel} = 10$ V, $V_2 = 0$ V, $z_0 = 1$ cm, $U_0 = 0$ V, $\Omega/(2\pi) = 250$ kHz and $\varphi_0 = \pi$ rad. (Reproduced from Janulyte, A., Zerega, Y., Carette, M., Reynard, C., and Andre, J., *Rapid Commun. Mass Spectrom.* 2008, *22*, 2479–2480. With permission from John Wiley and Sons, Ltd.)

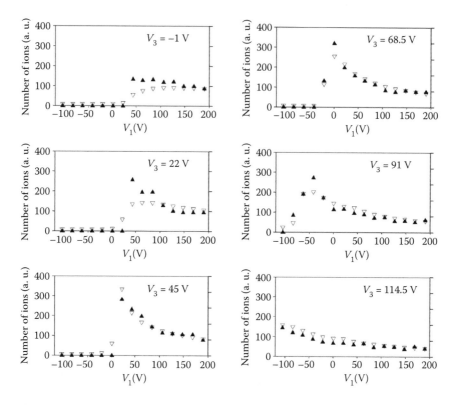

FIGURE 9.17 Plots showing the numbers of confined SF_5^+ ions *versus* the deceleration potentials V_1 and V_3: ($\triangledown$) experiment and ($\blacktriangle$) simulation. The common conditions are: $t_{cre} = t_{inj} = 100$ µs, $V_{ene} = 70$ eV, $V_{rep} = 60$ V, $V_{cel} = 25$ V, $V_2 = -10$ V, $z_0 = 1$ cm, $U_0 = -1.9$ V, $V_0 = 105$ V, $\Omega/(2\pi) = 250$ kHz, $\varphi_0 = \pi$ rad, $T_e = T_\Omega = 4$ µs and $N = 256$. The other simulation conditions are: m/z 127, $z_c = z_{cen} = -2.125$ cm, $dz_c = 0.125$ cm, $D = 8 \times 10^8$ ions s^{-1} cm^{-1}, $v_c = 3698 \sqrt{1/m}$ m s^{-1} and $T = 330$ K. (Reproduced from Janulyte, A., Zerega, Y., Carette, M., Reynard, C., and Andre, J., *Rapid Commun. Mass Spectrom.* 2008, 22, 2479–2480. With permission from John Wiley and Sons, Ltd.)

inserted into the simulation model. For both sets of results, the total number of ions is proportional to the sum of the areas of the TOF histograms recorded for the different durations of confinement.

For the highest value of V_3, the initial positions of the ions remain located between the lower end-cap and the center of the trap, whatever the value of V_1; the number of confined ions is almost constant. For lower values of V_3, ions can be stopped in the whole of the trap by varying V_1. The maximum number of ions is reached when the ion cloud is near to the center of the trap. With $V_3 < 70$ V, the lowest values of V_1 are insufficient to stop the ions before they reach the upper end-cap electrode.

Various creation conditions, such as size and location of the creation zone in the creation cell, have been simulated, and discrepancies exist between the experimental and simulation results when the removable ions in the simulation are distributed throughout the whole volume of the cell. The simulation curves and experimental curves fit one another when the removable ions in the simulation are limited to the upper quarter of the volume of the source (Figure 9.17).

9.11.3.5   Conclusion

The injection protocol is easy to establish. By simulation, the potential values are adjusted so as to have the maximum number of ions that will be confined with a distribution in position located typically in front of the trap center. Under typical experimental conditions the positions are distributed over a few millimeters. When $\varphi_0 = 0$, $\pi/2$, π, or $3\pi/2$, the velocity distribution is zero-centered. The adjustment of the injection duration is not critical, although this duration must be sufficient to have a steady flow of injected ions in the trap. With this injection mode, the initial-condition distributions of position and velocity lead to the maximum dynamic of the image signal (see Section 9.6.1.3).

9.12   ION MOTION NONLINEARITIES INDUCED BY MECHANICAL IMPERFECTIONS

9.12.1   Introduction

Mechanical defects, such as electrode truncation, electrode displacement, and the apertures in the two end-caps, lead to nonlinearity of the confinement field, resulting in the second-order differential equations of motion possessing polynomial coupling terms between the different perpendicular directions. Resonance phenomena result from coupling between the axial and radial directions (for the terms of the motion equations in $z^l \times x^m \times y^n$), and between a direction of the space and the excitation field (for the terms of the motion equations in z^l, x^m and y^n). Resonances are manifest as lines in the (a_z, q_z) plane by lines defined by the general equation $n_x \beta_x + n_y \beta_y + n_z \beta_z = k$, where n_x, n_y, n_z, and k are integers [34].

For the FT operating mode, the main harmful effect of nonlinearities is that the secular frequency depends on the maximum amplitude of the ion motion, as confinement involves a cloud of ions having large and different trajectory amplitudes. In consequence, the broadening of the maximum amplitudes of the motion induces a broadening of the detected secular frequencies. This broadening occurs whatever the value of the operating point β_z of the ion on the stability diagram. Another harmful effect is seen when the operating point is located near a nonlinear resonance line: the amplitude of the coupling peaks increases, and ion loss may occur by trajectory expansion.

In this work, nonlinearities induced by truncation of the electrodes have been studied mainly by simulation, and the effects appear in the experimental results obtained with a prototype ion trap having electrodes truncated at $2r_0$.

9.12.2   Theoretical Developments

The potential inside the ion trap is the solution of the Laplace equation, where the temporal variations of the electrical field are neglected. Wang and Franzen have given a development of the equation for a potential having nonlinearities, retaining the axial and radial plane symmetries which exist in the case of a pure quadrupolar trap [35]. Under this assumption, the odd terms of the development are zero. The potential can be then expressed by

$$\phi(x,y,z,t) = \phi_0(t)\left[A_0 + \frac{A_2}{2}\left(\frac{2z^2 - (x^2 + y^2)}{r_0^2} \right) \right.$$

$$\left. + \sum_{p=2}^{+\infty} A_{2p}\left(\frac{\rho}{r_0} \right)^{2p} P_{2p}^0(\cos\theta) \right]$$ (9.29).

where $\rho = \sqrt{x^2 + y^2 + z^2}$, $z = \rho \cos \theta$, $\phi_0(t) = U_0 + V_0 \cos(\Omega t + \varphi_0)$ is the potential applied to the ring electrode located at r_0 in the radial plane at $z = 0$, $P_{2p}^0(\cos\theta)$ is a Legendre polynomial of zero order and A_{2p} is a coefficient which depends on the boundary conditions.

In the pure quadrupole ion trap, $A_2 = 1$ and the other terms are zero. Franzen *et al.* have given the values of A_{2p}, restricted to $2p = 10$, for some types of nonlinearities which will be used in this study [80], and have presented a very interesting set of figures locating the resonance lines in the (β_r, β_z) plane [81].

The equations of ion motion are taken from the fundamental equation of dynamics. For example, the axial motion equation used in simulation is expressed by

$$\frac{d^2z}{d\tau^2} = -(a_z - 2q_z \cos 2\tau)\frac{r_0^2}{2}$$

$$\left[\frac{A_2}{2r_0^2}(4z) + \frac{A_4}{8r_0^4}(32z^3 - 48zr^2) \right.$$

$$+ \frac{A_6}{16r_0^6}(96z^5 - 480z^3r^2 + 180zr^4)$$

$$+ \frac{A_8}{128r_0^8}(87512z^7 - 54114z^5r^2 + 13440z^3r^4 - 2240zr^6)$$

$$+ \frac{A_{10}}{3840r_0^{10}}(38400z^9 - 691200z^7r^2 + 1814400z^5r^4$$

$$\left. -1008000z^3r^6 + 94500zr^8) \right]$$ (9.30),

with $\Omega t = 2\tau$ and $r = \sqrt{x^2 + y^2}$. The three equations are solved by an RK4 numerical method with 2000 points per period of the confinement voltage as higher frequencies are involved by the high order of the positions [82].

The secular axial frequency can be calculated for the case of the adiabatic approximation and without ion displacement in the radial plane. From the secular motion equation given by Makarov [42] and by employing a perturbation method [83], an approximate solution of the equation of secular axial motion can be written in the form

$$Z_s(t) \approx z(t_0)\cos\omega_z' t + \frac{z^3(t_0)}{4r_0^2}\frac{A_4}{A_2}\cos 3\omega_z' t + ...$$ (9.31),

where the secular frequency is expressed by [52]

$$\omega_z \approx \frac{\Omega |q_z|}{2\sqrt{2}}\left(A_2 + 3\frac{A_4}{r_0^2}z^2(t_0) + \ldots\right) \tag{9.32}$$

9.12.3 Results

9.12.3.1 Peak Amplitude in the Vicinity of a Nonlinear Resonance Line

When the motion frequency of trapped ions is measured with a fixed trap operating point, nonlinear resonance lines impair mass analysis. For example when, the operating point of the ion m/z 130 is located below the resonance line $\beta_z = 0.5$, the lower values of m/z are then subjected to the resonance effect. Ion trajectories were computed as a function of confinement time for different m/z ratios for the same operating point. The peak amplitudes of the secular frequency at ω_z and of the main coupling peak at $3\omega_z$ were evaluated from each spectrum of these trajectories, and the resulting evolution of the peak amplitude as a function of m/z is shown in Figure 9.18 for a trap having electrodes truncated at $3r_0$.

On resonance, the amplitude of the peak ω_z is not significant. The two peaks at ω_z and $3\omega_z$ overlap, and the amplitude of the coupling peak cannot be measured. An

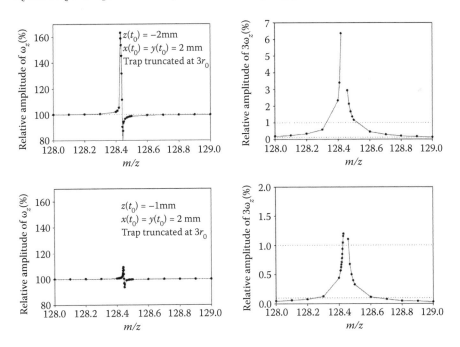

FIGURE 9.18 Curves showing the amplitude evolution of the peaks ω_z and $3\omega_z$ versus m/z for an ion trap having electrode surfaces truncated at $3r_0$ in the vicinity of the nonlinear resonance line $\beta_z = 0.5$ and for $z(t_0) = -1$ mm and -2 mm. The other conditions are: $z_0 = 1$ cm, $U_0 = 0$ V, $V_0 = 105$ V, $\Omega/(2\pi) = 250$ kHz, $T_e = 4$ μs, $N = 20,000$ and $x(t_0) = y(t_0) = 2$ mm. (Reproduced from Janulyte, A., Zerega, Y., and Carette, M., *Int. J. Mass Spectrom.* 2007, *263*, 243–259. With permission from Elsevier.)

increase of the coupling peak entails a decrease of the signal-to-noise ratio. When the minimum detection threshold is set at 1%, the resonance line impairs mass analysis over the m/z-range 128.32–128.52 Th for $z(t_0) = 2$ mm, and 128.425–128.462 Th for $z(t_0) = 1$ mm. With a minimum detection threshold set at 0.1%, the m/z-range is greater than 1 Th for $z(t_0) = -2$ mm and about 0.4 Th for $z(t_0) = -1$ mm.

The trap operating point can be chosen in order to avoid the nonlinear resonance line effect over a specified mass range. For example, under the conditions used in these simulation studies, the two nearest strong nonlinear resonance lines are $\beta_z = 0.5$ and $\beta_z + \beta_r = 1$. These two lines are located at frequencies of 62.500 kHz and 87.859 kHz, respectively. When the confinement potentials are $U_0 = 0$ V and $V_0 = 106$ V, ions whose m/z ratios range between ca 130 and 103, respectively, possess a secular frequency of the motion ranging between these two frequency values.

9.12.3.2   Broadening of the Frequency

The FT operating mode involves ions with large and different trajectory amplitudes. In the following result (Figure 9.19), the simulation takes this operating mode into account: the initial ion positions and velocities are drawn from six Gaussian distributions at the beginning of each confinement time. Figure 9.19 shows the amplitude spectrum of the image signal for a mean initial number of confined ions equal to 50 according to different observation times given by N. With an ion trap truncated at $2r_0$, modification of the peak shape and then peak splitting are observed for values of N greater than $N_{lim} = 2500$ as the peak FWHM decreases with N.

With an ion trap truncated at $3r_0$, $N_{lim} = 11,500$, the resolution limit is then more than four times greater than with an ion trap truncated at $2r_0$. At $\beta_z = 0.5$, with the operating conditions of Figure 9.19, the mass resolution limit is about 2321 for m/z 130 and a trap truncated at $3r_0$.

The mass resolution can be improved by increasing the amplitude of the RF voltage. At $\beta_z = 0.74$ with $V_0 = 138$ V, other sets of simulated spectra (not shown here) give $N_{lim} = 7500$ and a mass resolution of about 3250 for a trap truncated at $3r_0$. With the increase of β_z, N_{lim} decreases as the frequency broadening due to nonlinearity increases. However, the frequency difference between two confined species increases, which leads to an overall increase of mass resolution.

In conclusion, this simulation study has assumed large and extended initial positions given by a Gaussian distribution with parameters $\bar{z}(t_0) = -4$ mm and $\sigma[z(t_0)] = 2$ mm. Hence values of the mass resolution are obtained which are lower than in the corresponding experiment (Figure 9.8), where the value of about 4000 is attained with a trap truncated at $2r_0$. Other simulations need to be carried out with more realistic initial conditions, as given by the injection model (see Section 9.11).

9.12.3.3   Measured Frequencies of an Ion Cloud Confined
on a Nonlinear Resonance Line

The harmful effect of a nonlinear resonance line was demonstrated with an earlier ion trap which had a poor electrode alignment and mechanical design. The upper end-cap electrode was pierced with an aperture 1 cm in diameter in order to ensure

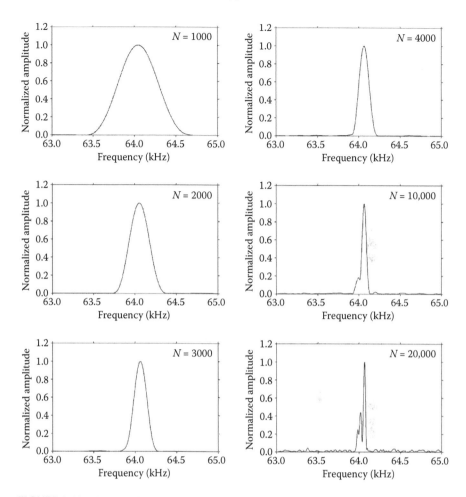

FIGURE 9.19 Curves showing the evolution of the amplitude spectrum of the axial trajectory of an ion cloud in a trap truncated at $2r_0$ for certain values of N. Simulation conditions are: m/z 130, $z_0 = 1$ cm, $U_0 = 0$ V, $V_0 = 108.4$ V, $\Omega/(2\pi) = 250$ kHz, $T_e = T_\Omega = 4$ μs, $\bar{N}_i = 50$ ions, $\bar{x}(t_0) = \bar{y}(t_0) = 0$ mm, $\sigma[x(t_0)] = \sigma[y(t_0)] = 2$ mm, $\bar{v}_x(t_0) = \bar{v}_y(t_0) = 0$ m s^{-1}, $\sigma[v_x(t_0)] = \sigma[v_y(t_0)] = 300$ m s^{-1}, $\bar{z}(t_0) = -4$ mm, $\sigma[z(t_0)] = 2$ mm, $\bar{v}_z(t_0) = 0$ m s^{-1} and $\sigma[v_z(t_0)] = 600$ m s^{-1}. (Reproduced from Janulyte, A., Zerega, Y., and Carette, M., *Int. J. Mass Spectrom.* 2007, *263*, 243–259. With permission from Elsevier.)

a large ion collection on the multiplier, and the shape of the grid covering the hole was not perfectly hyperboloidal. Two experimental operating conditions led to one operating point below and one on the nonlinear resonance line $\beta_z = 0.5$ for a confined cloud of SF$_5^+$ ions (see Figure 9.20). The resonance effects cause peak splitting, even at low-mass resolution.

Since the improvement of ion trap design, the resolution has been increased, however, the ion trap will soon be replaced by a new one truncated at $3r_0$ in order to reach higher resolution.

(a)
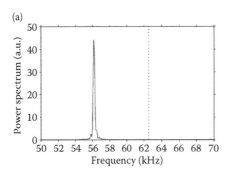

(b)
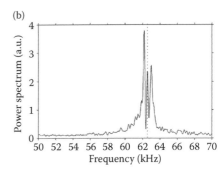

FIGURE 9.20 Experimental mass spectra of SF_5^+ ions confined with the following conditions: $z_0 = 1$ cm, $\Omega/(2\pi) = 250$ kHz, $T_e = 4$ μs and $N = 1024$. For spectrum (a) $U_0 \approx -0.7$ V and $V_0 \approx 98.1$ V; for spectrum (b) $U_0 \approx 0.25$ V, $V_0 \approx 104.7$ V (Reproduced from Carette, M., and Zerega, Y., *J. Mass Spectrom.* 2006, *41*, 71–76. With permission from Wiley InterScience.)

9.13 MOTION MODULATIONS INDUCED BY ELECTRIC PERTURBATION

9.13.1 INTRODUCTION

The influence of a mono-frequency perturbation superimposed on the confinement voltage in a quadrupolar configuration has been examined. The potential $\phi_0(t) = U_0 + V_0 \cos\Omega t + W_0 \cos(\omega t + \phi_0)$ was applied to the ion trap ring electrode, with the two end-caps being earthed. The perturbation voltage was $W_0 \cos(\omega t + \phi_0)$, with $W_0 \ll V_0$.

The potential inside the trap is expressed by Equation 9.1 with $\phi_0(t)$ defined as above. The ion motion equation along the Oz axis is then

$$\frac{d^2z}{dt^2} + \frac{\Omega^2}{4}\left[a_z - 2q_z \cos(\Omega t) - 2r_z \cos(\omega t + \phi_0)\right]z = 0 \tag{9.33},$$

with

$$a_z = -\frac{4zeU_0}{m\Omega^2 z_0^2}, \quad q_z = \frac{2zeV_0}{m\Omega^2 z_0^2} \quad \text{and} \quad r_z = \frac{2zeW_0}{m\Omega^2 z_0^2} \tag{9.34}.$$

The same equation was used to study the resonance excitation of confined ions by an RF 'tickle' potential, where various frequency values of the auxiliary potential, inducing resonant ion excitation, were determined [84]. Subsequently, Sudakov *et al.* gave a detailed theory of the resonant quadrupolar excitation of ions, leading to the expression of the frequencies of the axial ion motion [85]

$$\pm\omega_z + k\Omega \pm n\omega \quad k = 0,1,2,\ldots \quad n = 1,2,\ldots \tag{9.35}.$$

The additional excitation induces a further frequency modulation of the ion motion at ω.

9.13.2 RESULTS

9.13.2.1 Axial Ion-Motion Spectrum

The harmful effects of an electric perturbation on the ion-motion spectrum have been studied by simulation for which a single frequency modulation of the ion motion is assumed. The amplitude of the coupling peaks depends on the modulation index, γ, and is proportional to $\gamma \propto (W_0/V_0)(\Omega/\omega)$ with this assumption [28]. The axial trajectory given by Equation 9.33 is simulated by an RK4 numerical method, and the spectrum is then calculated using a Fourier transformation (Figure 9.21). Lower frequency perturbations induce higher-amplitude coupling peaks in the motion spectrum, and impair the frequency and amplitude measurements of the peak.

9.13.2.2 Spectrum Detected Using the FT Operating Mode

Mass analysis in the FT operating mode requires a succession of elementary experiments. The phase of the perturbation at the beginning of each confinement time can vary, and these phase variations induce a supplementary phase-modulation. The ion motion is then modulated at the frequency of the phase variation related to frequencies Ω and ω, and to t_{rep}, the duration of an elementary experiment, which can be either constant or variable. Hence, the simulation code (described previously in Section 9.5) must take into account the real time of the experiment and the succession mode of the elementary experiments in order to calculate the phase-value at the beginning of each confinement time.

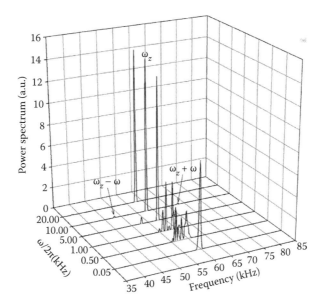

FIGURE 9.21 Simulation results showing the evolution of the power spectrum of the secular axial-motion of one ion *versus* $\omega/(2\pi)$, the frequency of the supplementary quadrupolar excitation. The conditions are: m/z 130, $z_0 = 1$ cm, $U_0 = 0$ V, $V_0 = 100$ V, $W_0 \approx 1$ V, $\Omega/(2\pi) = 250$ kHz, $T_e = T_\Omega = 4$ μs, $N = 1000$, $z(t_i) = -5$ mm and $v_z(t_i) = 0$ m s^{-1}. (Reproduced from Zerega, Y., Bouzid, S., Janulyte, A., Hallegatte, R., and Carette, M., *Meas. Sci. Technol.* 2005, *16*, 1201–1211. With permission from IOP Publishing Ltd.)

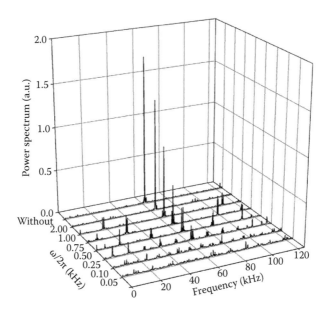

FIGURE 9.22 Experimental results showing the evolution of the power spectrum of SF_5^+ ions *versus* $\omega/(2\pi)$, the frequency of the supplementary quadrupolar excitation. 'Without' on the graph means without a supplementary quadrupolar excitation. The repetition time of the elementary experiments is constant $t_{rep} = 7.808$ ms. The experimental conditions are: m/z 127, $z_0 = 1$ cm, $U_0 \approx 0$ V, $V_0 \approx 106$ V, $W_0 \approx 1$ V, $\Omega/(2\pi) = 250$ kHz, $T_e = T_\Omega = 4$ μs and $N = 1000$. (Reproduced from Zerega, Y., Bouzid, S., Janulyte, A., Hallegatte, R., and Carette, M., *Meas. Sci. Technol.* 2005, *16*, 1201–1211. With permission from IOP Publishing Ltd.)

A comparison between an experimental result and a simulation result is shown with a constant duration of the elementary experiments ($t_{rep} = 7.808$ ms), see Figures 9.22 and 9.23, respectively. Both results show similar behavior of the spectra *versus* the frequency of the quadrupolar perturbation. The additional phase-modulation causes high-frequency peaks. In the spectrum, a fold-over of these peaks appears at lower frequency values due to sampling at 250 kHz. The signal-to-noise ratio increases with increasing values of ω.

As the actual sequential device runs with constant t_{rep}, only a simulation result gives the spectral behavior as a function of ω for a succession of elementary experiments without dead time (Figure 9.24). The frequencies of the phase-modulation are spread over a large range; consequently background noise appears instead of high-frequency peaks, and the signal-to-noise ratio falls off with decrease in ω.

9.13.2.3   Resolution and Signal-to-Noise Ratio

In our experimental work, the major perturbation source is the main power supply at 50 Hz, which can exert deleterious effects on frequency spectra. In order to avoid the phase-modulation induced by the succession of elementary experiments, the start of each elementary experiment can be phase-locked on to the main power supply with t_{rep} a constant multiple of 20 ms.

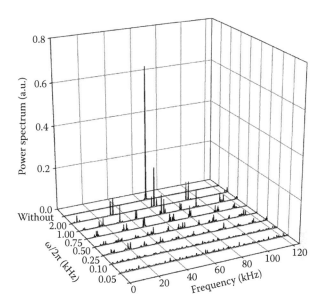

FIGURE 9.23 Simulation results showing the evolution of the power spectrum of ions *versus* the frequency of the supplementary quadrupolar excitation. 'Without' on the graph means without a supplementary quadrupolar excitation. The repetition time of the elementary experiments is constant $t_{rep} = 7.808$ ms. The conditions are: m/z 127, $z_0 = 1$ cm, $U_0 = 0$ V, $V_0 = 106$ V, $W_0 = 1$ V, $\Omega/(2\pi) = 250$ kHz, $T_e = T_\Omega = 4$ μs, $N = 1000$, $\bar{N}_i = 10$ ions, $\bar{x}(t_i) = \bar{y}(t_i) = 0$ mm, $\sigma[x(t_i)] = \sigma[y(t_i)] = 2$ mm, $\bar{v}_x(t_i) = \bar{v}_y(t_i) = 0$ m s⁻¹, $\sigma[v_x(t_i)] = \sigma[v_y(t_i)] = 300$ m s⁻¹, $\bar{z}(t_i) = -5$ mm, $\sigma[z(t_i)] = 2$ mm, $\bar{v}_z(t_i) = 0$ m s⁻¹, and $\sigma[v_z(t_i)] = 600$ m s⁻¹. (Reproduced from Zerega, Y., Bouzid, S., Janulyte, A., Hallegatte, R., and Carette, M., *Meas. Sci. Technol.* 2005, *16*, 1201–1211. With permission from IOP Publishing Ltd.)

The three simulation and experiment results presented previously were obtained with an additional potential with an amplitude of $W_0 = 1$ V and an RF confinement voltage with an amplitude of $V_0 = 106$ V. Here the limits of the resolution and the signal-to-noise ratio are estimated from simulated spectra with a perturbation at 50 Hz and lower amplitude values, $W_0 = 10$ and 50 mV. Two succession modes (see Section 9.3.3) were chosen with a constant duration of the elementary experiments, $t_{rep} = 80$ ms, and without any time between the elementary experiments in order to avoid high-amplitude peaks induced by the phase-modulation. The two major peaks induced by the frequency modulation are located on either side of the secular frequency peak at $\omega_z \pm \omega$. They can be separated if the resolution, which increases with N, is sufficient.

Other simulations (not presented here) were carried out with m/z 127, $U_0 = 0$ V and $V_0 = 106$ V, leading to $\beta_z = 0.51$. With $t_{rep} = 80$ ms and $W_0 = 50$ mV, the upper limit of the mass resolution was 560 with $N_{lim} = 1500$. When $W_0 = 10$ mV, the limit of the mass resolution was 1300 with $N_{lim} = 3500$ and $R_{s/n} = 7400$. Without any dead time between the elementary experiments, a broadening of the peak at ω_z occurs for lower values of N, induced by the additional phase-modulation at very low frequencies. Consequently, the resolution is worse than with $t_{rep} = 80$ ms. By contrast, with

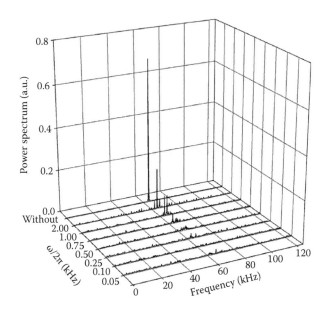

FIGURE 9.24 Simulation results showing the evolution of the power spectrum of ions *versus* the frequency of the supplementary quadrupolar excitation. The duration of the elementary experiments is variable according to confinement time. The conditions are: m/z 127, $z_0 = 1$ cm, $U_0 = 0$ V, $V_0 = 106$ V, $W_0 = 1$ V, $\Omega/(2\pi) = 250$ kHz, $T_e = T_\Omega = 4$ μs, $N = 1000$, $\bar{N}_i = 10$ ions, $\bar{x}(t_i) = \bar{y}(t_i) = 0$ mm, $\sigma[x_i(t_i)] = \sigma[y_i(t_i)] = 2$ mm, $\bar{v}_x(t_i) = \bar{v}_y(t_i) = 0$ m s^{-1}, $\sigma[v_x(t_i)] = \sigma[v_y(t_i)] = 300$ m s^{-1}, $\bar{z}(t_i) = -5$ mm, $\sigma[z(t_i)] = 2$ mm, $\bar{v}_z(t_i) = 0$ m s^{-1}, and $\sigma[v_z(t_i)] = 600$ m s^{-1}. (Reproduced from Zerega, Y., Bouzid, S., Janulyte, A., Hallegatte, R., and Carette, M., *Meas. Sci. Technol.* 2005, *16*, 1201–1211. With permission from IOP Publishing Ltd.)

$W_0 = 10$ mV a greater upper limit is attained for the mass resolution, which is 5600 with $R_{s/n} = 3700$. The resolution can be increased further by choosing higher values of V_0 for both succession modes.

Finally, in practice, a discrepancy can exist during confinement between the voltage applied to the two end-cap electrodes and the zero voltage reference as the two end-cap electrodes are not connected directly to this reference. Hence, the power supply may induce a low-amplitude voltage at 50 Hz between the two end-caps and the ring electrode, which leads to an additional quadrupolar excitation. It is important to take into account the harmful influences of such a perturbation, even at low level. Filters must be built into the electronic circuit in order to limit the maximum amplitude of this discrepancy to a few tens of millivolts.

9.14 CONCLUSION

The FT operating mode by described here provides a new means of detecting the axial and radial secular frequencies of simultaneously confined ions in a quadrupole trap. The ion motion must be as pure (or harmonic) as possible, with a large amplitude. Consequently, phenomena such as collisions, space charge, mechanical defects, and electrical perturbations must be reduced during ion confinement.

Recent work reveals that out-of-phase ion trajectories can lead to optimum dynamics of the detected signal, provided that the initial positions have the same sign and that the initial velocity distribution is symmetric and zero-centered. It is thus not necessary to employ a transient excitation at the beginning of confinement in order to increase the dynamics of the motion, if an adapted injection protocol is used. The ion injection protocol proposed locates the ions at the beginning of confinement in a narrow range of given off-center values by adjusting the deceleration potential during ion injection.

With the FT operating mode, the ion trap is a high performance tool. Our research prototype, which uses this operating mode, achieves high resolution and high sensitivity over m/z-ranges of a few tens of Th. For example, isotopic Xe^+ ions are mass-analyzed with an FWHM of about 0.03 Th. This performance is very promising for its future use as both a laboratory and as an on-line mass spectrometer.

The main research topic of the team concerns the direct analysis on-line of traces of persistent organic pollutants such as polychlorinated dibenzo-p-dioxins and polychlorinated dibenzofurans, known under the generic term 'dioxins'. Most of the pollution is due to local atmospheric emissions from industrial processes involving combustion, such as waste incineration plants. The aim of this project is to propose new methods and an on-line analyzer using new selective zeolitic materials in the sampling device. These zeolites may act as a selective sponge to increase the amount of selected molecules for analysis, as toxicity depends on the isomeric form (position of the chlorine atoms) of the dioxins. A thermal desorption device will be used in order to extract the trapped dioxin molecules, and it will be coupled to the FT mass analyzer in the on-line direct analyzer of dioxins.

ACKNOWLEDGMENTS

The main results concerning the mass spectrometer performance were obtained with the technical and financial support of the SERES Company (Aix-en-Provence, France) and the financial aid of the French National Centre for Scientific Research (CNRS) and the French Ministerial Delegation for Armament (DGA) (ACI 2002, Nouvelles Méthodologies Analytiques et Capteurs (NMAC), project: "analyseur de traces"). Currently, an extension of the work about the mass spectrometer is supported by the French Agency for Environment and Energy Management (ADEME) through a financial grant (convention ADEME/University of Provence #0674C0015) which also supports the on-line direct dioxin-trace analyser project.

In addition this project is supported through a PhD grant by the ADEME (2007–2010) and a financial grant of the French National Research Agency within the scope of the Environmental Technologies Programme and Sustainable Development (ANR PRECODD 2007, project: "Méthodes et moyens de mesure en temps réel de dioxines/furannes à l'émission de sources fixes", 2008–2012).

The French Competitiveness Clusters on Risk Management (Pôle de compétitivité Gestion des risques, vulnérabilité des territoires) have given the "METERDIOX" project their seal of approval.

We wish to express our deep gratitude to Ray March for the helpful discussions about 'trapped ions', and for his scientific support. We are extremely grateful to him for his frequent visits to our laboratory in Marseille as well, as for the warm welcome with which some members of our research team were received in Peterborough, Canada.

We greatly appreciated receiving John Todd in our laboratory in Marseille in March 2008 for the graduation ceremony at the Université de Provence, at which an Honorary Doctorate was conferred upon his friend and colleague Ray March.

REFERENCES

1. Marshall, A.G. Milestones in Fourier transform ion cyclotron resonance mass spectrometry technique development. *Int. J. Mass Spectrom.* 2000, *200*, 331–356.
2. Syka, J.E.P.; Fies, W.J. A Fourier transform quadrupole ion trap mass spectrometer. *Proc. 35th ASMS Conference on Mass Spectrometry and Allied Topics*, Denver, Colorado, May 24–29, 1987, pp. 767–768.
3. Syka, J.E.P.; Fies, W.J. Fourier transform quadrupole mass spectrometer and method. *U.S. Patent* 1988, 4,755,670.
4. Schweikhard, L.; Blundschling, M.; Jertz, R.; Kluge, H.J. Fourier transform mass spectrometry without ion cyclotron resonance: direct observation of the trapping frequency of trapped ions. *Int. J. Mass Spectrom. Ion Processes* 1989, *89*, R7-R12.
5. Wang, Y.; Wanczek, K.P. Generation of an exact three-dimensional quadrupole electric field and superposition of a homogeneous electric field within a common closed boundary with application to mass spectrometry. *J. Chem. Phys.* 1993, *98*, 2647–2652.
6. Lammert, S.A.; Cleven, C.D.; Cooks, R.G. Determination of ion frequencies in a quadrupole ion trap by using a fast direct-current pulse as pump and a laser probe. *J. Am. Soc. Mass Spectrom.* 1994, *5*, 29–36.
7. Cooks, R.G.; Cleven, C.D.; Horn, L.A.; Nappi, M.; Well, C.; Soni, M.H.; Julian, R.K. Non-destructive detection of ions in a quadrupole ion trap using a d.c. pulse to force coherent ion motion: a simulation study. *Int. J. Mass Spectrom. Ion Processes* 1995, *146–147*, 147–163.
8. Frankevich, V.E.; Soni, M.H.; Nappi, M.; Santini, R.E.; Amy, J.W.; Cooks, R.G. Non-destructive ion trap mass spectrometer and method. *U.S. Patent* 1997, 5,625,186.
9. Weil, C.; Wells, J.M.; Wollnik, H.; Cooks, R.G. Axial ion motion within the quadrupole ion trap elucidated by dc pulse tomography. *Int. J. Mass Spectrom.* 2000, *194*, 225–234.
10. Plass, W.R. Theory of dipolar dc excitation and dc tomography in the rf quadrupole ion trap. *Int. J. Mass Spectrom.* 2000, *202*, 175–197.
11. Vedel, M.; Andre, J.; Brincourt, G.; Zerega, Y.; Werth, G.; Schermann, J.P. Study of SF_6^- ion lifetime in a RF quadrupole trap. *Appl. Phys. B: Lasers Opt.* 1984, *34*, 229–235.
12. Brincourt, G.; Rajab, P.S.; Catella, R.; Zerega, Y.; Andre, J. Collision of SF_6 molecules with Xe(nf) Rydberg atoms in a quadrupole trap. Dependence of the SF_6^- ion lifetime on n. *Chem. Phys. Lett.* 1989, *156*, 573–577.
13. Brincourt, G.; Catella, R.; Zerega, Y.; André, J. Time-of-flight detection of ions ejected from a radiofrequency quadrupole trap: experimental determination of their f_z secular frequency. *Chem. Phys. Lett.* 1990, *174*, 626–630.
14. Carette, M.; Zerega, Y.; March, R.E.; Perrier, P.; Andre, J. Rydberg electron capture mass spectrometry of 1,2,3,4 tetrachlorodibenzo-*p*-dioxin. *Eur. J. Mass Spectrom.* 2000, *6*, 405–408.
15. Zerega, Y.; Carette, M.; Perrier, P.; Andre, J. Rydberg electron capture mass spectrometry of organic pollutants. *Organohalogen Compd.* 2002, *55*, 151–154.

16. March, R.E.; Hughes, R.J.; Todd, J.F.J. *Quadrupole Storage Mass Spectrometry*. John Wiley & Sons, New York, 1989.

17. Zerega, Y.; Andre, J.; Brincourt, G.; Catella, R. A new operating mode of a quadrupole ion trap in mass spectrometry: Part 2. Multichannel recording and treatment of the ion times of flight. *Int. J. Mass Spectrom. Ion Processes* 1994, *132*, 67–72.

18. Carette, M.; Perrier, P.; Zerega, Y.; Brincourt, G.; Payan, J.C.; Andre, J. Probing radial and axial secular frequencies in a quadrupole ion trap. *Int. J. Mass Spectrom. Ion Processes* 1997, *171*, 253–261.

19. Marshall, A.G.; Verdun, F.R. *Fourier Transforms in NMR, Optical, and Mass Spectrometry: A User's Handbook*. Elsevier, Amsterdam, 1990.

20. Sanna, G.; Tomassetti, G. The supersonic free jet. In *Introduction to Molecular Beams Gas Dynamics*. ed. T.K. Wei, Chapter 8, pp. 251–259. Imperial College Press, London, 2005.

21. Janulyte, A. *Etude physico-électrique et optimisation métrologique d'un spectromètre de masse pour la détection temps réel de traces de dioxines*. PhD thesis, Université de Provence, 2007.

22. Zerega, Y.; Andre, J.; Brincourt, G.; Catella, R. A new operating mode of a quadrupole ion trap in mass spectrometry: Part 1. Signal visibility. *Int. J. Mass Spectrom. Ion Processes* 1994, *132*, 57–65.

23. Zerega, Y.; Andre, J.; Brincourt, G.; Catella, R. New operating mode of a quadrupole ion trap in mass spectrometry Part 3. Mass resolution. *Int. J. Mass Spectrom. Ion Processes* 1994, *135*, 155–164.

24. Carette, M.; Zerega, Y. Non-linearity influences on the axial secular-motion spectrum of ions confined in a Fourier transform ion trap mass spectrometer. Experimental studies. *J. Mass Spectrom.* 2006, *41*, 71–76.

25. Londry, F.A.; Alfred, R.L.; March, R.E. Computer simulation of single-ion trajectories in Paul-type ion traps. *J. Am. Soc. Mass Spectrom.* 1993, *4*, 687–705.

26. Vedel, F.; Andre, J.; Vedel, M. Invariance temporelle et propriétés statistiques énergétiques et spatiales d'ions confinés dans une trappe quadrupolaire R.F. III – Fréquences séculaires du mouvement en présence de charge d'espace. *J. Phys.* 1981, *42*, 1611–1622.

27. Konenkov, N.V.; Sudakov, M.; Douglas, D.J. Matrix methods for the calculation of stability diagrams in quadrupole mass spectrometry. *J. Am. Soc. Mass Spectrom.* 2002, *13*, 597–613.

28. Zerega, Y.; Bouzid, S.; Janulyte, A.; Hallegatte, R.; Carette, M. Involvement of confinement voltage perturbations in the axial-motion spectrum of an ion confined in a quadrupole ion trap. *Meas. Sci. Technol.* 2005, *16*, 1201–1211.

29. von Busch, F.; Paul, W. Isotopentrennung mit dem elektrischen Massenfilter. *Z. Phys. A: Hadrons Nucl.* 1961, *164*, 581–587.

30. von Busch, F.; Paul, W. Uber nichtlineare Resonanzen im elektrischen Massenfilter als Folge von Feldfehlern. *Z. Phys. A: Hadrons Nucl.* 1961, *164*, 588–594.

31. Dawson, P.H.; Whetten, N.R. Non-linear resonances in quadrupole mass spectrometers due to imperfect fields I. The quadrupole ion trap. *Int. J. Mass Spectrom. Ion Phys.* 1969, *2*, 45–59.

32. Dawson, P.H.; Whetten, N.R. Non-linear resonances in quadrupole mass spectrometers due to imperfect fields II. The quadrupole mass filter and the monopole mass spectrometer. *Int. J. Mass Spectrom. Ion Phys.* 1969, *3*, 1–12.

33. Wang, Y.; Franzen, J. The non-linear resonance QUISTOR Part 1. Potential distribution in hyperboloidal QUISTORs. *Int. J. Mass Spectrom. Ion Processes* 1992, *112*, 167–178.

34. Wang, Y.; Franzen, J.; Wanczek, K.P. The non-linear resonance ion trap. Part 2. A general theoretical analysis. *Int. J. Mass Spectrom. Ion Processes* 1993, *124*, 125–144.

35. Wang, Y.; Franzen, J. The non-linear ion trap. Part 3. Multipole components in three types of practical ion trap. *Int. J. Mass Spectrom. Ion Processes* 1994, *132*, 155–172.

36. Franzen, J. The non-linear ion trap. Part 4. Mass selective instability scan with multipole superposition. *Int. J. Mass Spectrom. Ion Processes* 1993, *125*, 165–170.

37. Franzen, J. The non-linear ion trap. Part 5. Nature of non-linear resonances and resonant ion ejection. *Int. J. Mass Spectrom. Ion Processes* 1994, *130*, 15–40.

38. Eades, D.M.; Johnson, J.V.; Yost, R.A. Nonlinear resonance effects during ion storage in a quadrupole ion trap. *J. Am. Soc. Mass Spectrom.* 1993, *4*, 917–929.

39. Kaiser, R.E.; Cooks, R.G.; Stafford, G.C.; Syka, J.E.P.; Hemberger, P.H. Operation of a quadrupole ion trap mass spectrometer to achieve high mass/charge ratios. *Int. J. Mass Spectrom. Ion Processes* 1991, *106*, 79–115.

40. Alheit, R.; Hennig, C.; Morgenstern, R.; Vedel, F.; Werth, G. Observation of instabilities in a Paul trap with higher-order anharmonicities. *Appl. Phys. B: Lasers Opt.* 1995, *61*, 277–283.

41. Chu, X.Z.; Holzki, M.; Alheit, R.; Werth, G. Observation of high-order motional resonances of an ion cloud in a Paul trap. *Int. J. Mass Spectrom. Ion Processes.* 1998, *173*, 107–112.

42. Makarov, A.A. Resonance ejection from the Paul trap: a theoretical treatment incorporating a weak octapole field. *Anal. Chem.* 1996, *68*, 4257–4263.

43. Mo, W.; Langford, M.L.; Todd, J.F.J. Investigation of 'ghost' peaks caused by nonlinear fields in the ion trap mass spectrometer. *Rapid Commun. Mass Spectrom.* 1995, *9*, 107–113.

44. Guidugli, F.; Traldi, P. A phenomenological description of a black hole for collisionally induced decomposition products in ion-trap mass spectrometry. *Rapid Commun. Mass Spectrom.* 1991, *5*, 343–348.

45. Morand, K.L.; Lammert, S.A.; Cooks, R.G. Letter to the editor: concerning "black holes" in ion-trap mass spectrometry. *Rapid Commun. Mass Spectrom.* 1991, *5*, 491.

46. Eades, D.M.; Yost, R.A. Black canyons for ions stored in an ion-trap mass spectrometer. *Rapid Commun. Mass Spectrom.* 1992, *6*, 573–578.

47. Louris, J.N.; Schwartz, J.C.; Stafford, G.C.; Syka, J.E.P.; Taylor, D.M. The Paul ion trap mass selective instability scan: trap geometry and resolution. *Proc. 40th Conference on Mass Spectrometry and Allied Topics,* Washington, DC, May 31–June 5, 1992, pp. 1003–1004.

48. March, R.E.; Londry, F.A.; Alfred, R.L. Some thoughts on the 'stretched' ion trap. *Org. Mass Spectrom.* 1992, *27*, 1151–1152.

49. Walz, J.; Siemers, I.; Schubert, M.; Neuhauser, W.; Blatt, R. Motional stability of a nonlinear parametric oscillator: ion storage in the RF octupole trap. *Europhys. Lett.* 1993, *21*, 183–188.

50. Franzen, J. Simulation study of an ion cage with superimposed multipole fields. *Int. J. Mass Spectrom. Ion Processes* 1991, *106*, 63–78.

51. Splendore, M.; Marquette, E.; Oppenheimer, J.; Huston, C.; Wells, G. A new ion ejection method employing an asymmetric trapping field to improve the mass scanning performance of an electrodynamic ion trap. *Int. J. Mass Spectrom.* 1999, *190/191*, 129–143.

52. Janulyte, A.; Zerega, Y.; Carette, M. Harmful influences of confinement field non-linearities in mass identifying for a Fourier transform quadrupole ion trap mass spectrometer: simulation studies. *Int. J. Mass Spectrom.* 2007, *263*, 243–259.

53. Schwartz, J.C.; Syka, J.E.P.; Jardine, I. High resolution on a quadrupole ion trap mass spectrometer. *J. Am. Soc. Mass Spectrom.* 1991, *2*, 198–204.

54. Williams, J.D.; Cox, K.A.; Cooks, R.G.; Kaiser, R.E.; Schwartz, J.C.; Gaskell, S.J. High mass-resolution using a quadrupole ion-trap mass spectrometer. *Rapid Commun. Mass Spectrom.* 1991, *5*, 327–329.

55. Bonner, R.F.; Lawson, G.; Todd, J.F.J. Ion/molecule reaction studies with a quadrupole ion storage trap. *Int. J. Mass Spectrom. Ion Phys.* 1972/1973, *10*, 197–203.

56. Bonner, R.F.; Todd, J.F.J.; Lawson, G.; March, R.E. Ion storage mass spectrometry. Applications in the study of ionic processes and chemical ionization reactions. *Adv. Mass Spectrom.* 1974, *6*, 377–384.

57. Boswell, S.M.; Mather, R.E.; Todd, J.F.J. Chemical ionisation in the ion trap: a comparative study. *Int. J. Mass Spectrom. Ion Processes* 1990, *99*, 139–149.

58. Zerega, Y.; Perrier, P.; Carette, M.; Brincourt, G.; Nguema, T.; Andre, J. A dual quadrupole ion trap mass spectrometer. *Int. J. Mass Spectrom.* 1999, *190/191*, 59–68.

59. Dawson, P.H.; Whetten, N.R. Three-dimensional quadrupole mass spectrometer and gauge. *U.S. Patent* 1970, 3,527,939.

60. Stafford, G.C.; Kelley, P.E.; Stephens, D.R. Method of mass analyzing a sample by use of a quadrupole ion trap. *U.S. Patent* 1985, 4,540,884.

61. Louris, J.N.; Syka, J.E.P.; Kelley, P.E. Method of operating quadrupole ion trap chemical ionization mass spectrometry. *U.S. Patent* 1985, 4,686,367.

62. Kofel, P. Injection of mass selected ions into radiofrequency ion trap. In *Practical Aspects of Ion Trap Mass Spectrometry*, eds. R.E. March, J.F.J. Todd, Volume 2, Chapter 2, pp. 51–87. CRC Press, Boca Raton, FL, 1995.

63. Todd, J.F.J. Introduction to practical aspects of mass spectrometry. In *Practical Aspects of Ion Trap Mass Spectrometry,* eds. R.E. March, J.F.J. Todd, Volume 1, Chapter 1, pp. 18–19. CRC Press, Boca Raton, FL, 1995.

64. Louris, J.N.; Amy, J.W.; Ridley, T.Y.; Cooks, R.G. Injection of ions into a quadrupole ion trap mass spectrometer. *Int. J. Mass Spectrom. Ion Processes* 1989, *88*, 97–111.

65. Doroshenko, V.M.; Cotter, R.J. Injection of externally generated ions into an increasing trapping field of a quadrupole ion trap mass spectrometer. *J. Mass Spectrom.* 1997, *32*, 602–615.

66. Sadat Kiai, S.M.; Andre, J.; Zerega, Y.; Brincourt, G.; Catella, R. Study of a quadrupole ion trap supplied with a periodic impulsional potential. *Int. J. Mass Spectrom. Ion Processes* 1991, *107*, 191–203.

67. Sadat Kiai, S.M.; Zerega, Y.; Brincourt, G.; Catella, R.; Andre, J. Experimental study of a r.f. quadrupole ion trap supplied with a periodic impulsional potential. *Int. J. Mass Spectrom. Ion Processes* 1991, *108*, 65–73.

68. Kishore, M.N.; Ghosh, P.K. Trapping of ion injected from an external source into a three-dimensional r.f. quadrupole field. *Int. J. Mass Spectrom. Ion Phys.* 1979, *29*, 345–350.

69. O, C-S.; Schuessler, H.A. Confinement of pulse-injected external ions in a radiofrequency quadruple ion trap. *Int. J. Mass Spectrom. Ion Phys.* 1981, *40*, 53–66.

70. O, C-S.; Schuessler, H.A. Confinement of ions injected into a radiofrequency quadrupole ion trap: pulsed ion beams of different energies. *Int. J. Mass Spectrom. Ion Phys.* 1981, *40*, 67–75.

71. O, C-S.; Schuessler, H.A. Confinement of ions injected into a radiofrequency quadrupole ion trap: energy-selective storage of pulse injected ions. *Int. J. Mass Spectrom. Ion Phys.* 1981, *40*, 77–86.

72. O, C-S.; Schuessler, H.A. Trapping of pulse injected ions in a radio-frequency quadrupole trap. *Rev. Phys. Appl.* 1982, *17*, 83–88.

73. Todd, J.F.J.; Freer, D.A.; Waldren, R.M. The quadrupole ion store (QUISTOR). Part XII. The trapping of ions injected from an external source: a description in terms of phase-space 'dynamics. *Int. J. Mass Spectrom. Ion Phys.* 1980, *36*, 371–386.

74. Moore, R.B.; Lunney, M.D. The Paul trap as a collection device. In *Practical Aspects of Ion Trap Mass Spectrometry*, eds. R.E. March, J.F.J. Todd, Volume 2, Chapter 8, pp. 263–302. CRC Press, Boca Raton, FL, 1995.

75. Moore, R.B.; Ghalambor Dezfuli, A.M.; Varfalvy, P.; Zhao, H. Production, transfer and injection of charged particles in traps and storage rings. *Phys. Scr.* 1995, *T59*, 93–105.

76. Janulyte, A.; Zerega, Y.; Carette, M.; Reynard, C.; Andre, J. Model for ion injection into a quadrupole ion trap to assess the distribution of initial conditions of confinement. *Rapid Commun. Mass Spectrom.* 2008, *22*, 2479–2480.

77. ITSIM: Ion Trap Simulation program. http://aston.chem.purdue.edu/itsim. (Accessed November 2009.

78. SIMION: Industry standard charged particle optics simulation software. http://simion. com/. (Accessed Nov 2009)

79. Forbes, M.W.; Sharifi, M.; Croley, T.; Lausevic, Z.; March, R.E. Simulation of ion trajectories in a quadrupole ion trap: a comparison of three simulation programs. *J. Mass Spectrom.* 1999, *34*, 1219–1239.

80. Franzen, J.; Gabling, R.H.; Schubert, M.; Wang, Y. Nonlinear ion traps. In *Practical Aspects of Ion Trap Mass Spectrometry.* eds. R.E. March, J.F.J. Todd, Volume 1, Chapter 3, pp. 76–77. CRC Press, Boca Raton, FL, 1995.

81. Franzen, J.; Gabling, R.H.; Schubert, M.; Wang, Y. Nonlinear ion traps. In *Practical Aspects of Ion Trap Mass Spectrometry.* eds. R.E. March, J.F.J. Todd, Volume 1, Chapter 3, pp. 67–70. CRC Press, Boca Raton, FL, 1995.

82. Zerega, Y. *Méthodes Analytiques pour la Spectrométrie de Masse de Polluants Atmosphériques* (Analytical methods in mass Spectrometry of atmospheric polluants), HDR (post doctoral degree), Université de Provence, 2004.

83. Landau, L.D.; Lifshitz, E.M. *Mechanics,* 3rd Edn, Pergamon press, Oxford,1976.

84. Alfred, R.L.; Londry, F.A.; March, R.E. Resonance excitation of ions stored in a quadrupole ion trap. Part IV. Theory of quadrupolar excitation. *Int. J. Mass Spectrom. Ion Processes* 1993, *125*, 171–185.

85. Sudakov, M.; Konenkov, N.; Douglas, D.J.; Glebova, T. Excitation frequencies of ions confined in a quadrupole field with quadrupole excitation. *J. Am. Soc. Mass Spectrom.* 2000, *11*, 10–18.

Part IV

Quadrupole Rod Sets

10 Trapping and Processing Ions in Radio Frequency Ion Guides

Bruce A. Thomson, Igor V. Chernushevich, and Alexandre V. Loboda

CONTENTS

10.1 Introduction 525
10.2 Trapping and Moving Ions in a High-Pressure RF Quadrupole 528
10.3 Reversing Ion Flow for MS^n 531
10.4 Ion/Ion Reactions with Bi-Directional Ion Flow 534
10.5 Ion/Molecule Reactions of Trapped Ions.. 536
10.6 Branched Ion Guides for Multiple Sources 539
10.7 Conclusion 542
References.. 542

10.1 INTRODUCTION

Radio frequency (RF) ion guides have been used for many years for transporting and guiding ions along an ion optical path. Their use has become pervasive in commercial instruments since the enormous growth in atmospheric pressure ionization mass spectrometry, driven by both the popularity of electrospray ionization for biological analysis and by the widespread use of liquid chromatography/tandem mass spectrometry (LC/MS/MS) systems for the analysis of drugs and metabolites in the field of drug development. More recently, RF ion guides have begun to be adapted for use as ion traps in a number of applications: for temporary storage of ions while processing a previous batch at another location in the ion path, and to contain ions while they are processed by charge state separation techniques [1], by ion/photon reactions [2], by ion/molecule reactions (see Section 10.5), by ion/ion reactions [3], or even by ion/electron reactions [4]. In addition, ion guides now find use as local traps to store ions prior to mobility separation, and as containment devices for ion mobility drift tubes. Under software control, RF ion guides can be switched from transport mode to trapping mode and back to transport with a reversal of ion flow direction. Techniques of trapping ions inside the ion guide have expanded from the use of direct current (DC) barriers at the ends, to include the application of RF barriers at the end [5], axial fields combined with RF or DC barriers [6], and gas flow combined with DC or RF fields [7]. More techniques and applications will no doubt be developed in the future as the need for more and better methods of ion processing increases.

The initial use of RF ion guides was limited to low pressure applications. The excellent review by Gerlich [8] identifies many configurations for the use of inhomogeneous RF fields in a variety of geometries, including multipoles and ring guides, mainly for trapping and transporting ions without increasing their internal energy. Under the majority of practical conditions confinement of ions with oscillating fields can be described by the pseudopotential model [8]. In this model, the ion motion is represented by fast oscillations at the RF frequency and slow motion governed by a pseudopotential. The expression for the pseudopotential follows a mathematical treatment originally introduced by P.L. Kapitza [9]. One of the most important properties of the pseudopotential is that, unlike a static electrical potential, the pseudopotential is not governed by Earnshaw's theorem, which states that point charges cannot be maintained in a stable configuration by electrostatic fields alone. In contrast, RF fields can be configured to create pseudopotentials that confine ions in one, two, or three dimensions. Commonly, RF ion guides constrain ion motion in two dimensions while allowing ions to move freely along the axis. The first use of RF ion guides in a mass spectrometer system was as a containment device between two quadrupole mass filters, initially as a reaction region for photodissociation reactions [10] and, slightly later, as the first collision cell in a triple-quadrupole mass spectrometer system [11]. The early collision cells were used at moderately low pressure, where ions experienced only a few collisions at most, and where the RF field contained the fragment ions and transported them to the next analyzer [12]. The first use of an RF ion guide to transport ions from a high-pressure ion source into the analyzer was in the Trace Atmospheric Gas Analyzer (TAGA) 6000 triple-stage quadrupole mass spectrometer, where a quadrupole ion guide coupled the atmospheric pressure ion source to the quadrupole mass filter in a single cryogenic pumping stage [13]. The use of a single RF quadrupole to transport ions from the atmospheric pressure orifice into the quadrupole filter was maintained in the API III LC/MS/MS system through the 1980s and 1990s.

A significant step in the evolution of RF ions guides came in the early 1990's when Douglas and French introduced the use of high-pressure quadrupoles as ion guides [14]. They showed that focusing and cooling of the ion beam could be achieved in a RF ion guide through hundreds of collisions with the background gas particles, allowing efficient transport from a high-pressure region through a small aperture into a lower-pressure region. This discovery enabled the use of efficient differentially pumped vacuum systems with modest vacuum pumps for atmospheric pressure mass spectrometer systems. This discovery also led directly to the use of higher pressure in quadrupole collision cells in order to achieve the same benefits of collisional focusing (for efficient transport) and collisional cooling (for reduction of energy spread) [15]. Higher-order multipoles (hexapoles and octopoles) have also become common as ion guides and collision cells. The pseudopotential well model, which expresses the principle of ion confinement due to inhomogeneous RF fields, can be applied to multipoles under a wide range of operating conditions. This model shows that while higher-order multipoles provide stronger ion confinement (i.e., higher walls) than do quadrupoles, they also produce wider ion beams at the exit due to the flatter profiles of pseudopotential well. For a $2n$-pole multipole, the pseudopotential well potential can be expressed as

$$\Psi = \frac{n^2 e V^2}{4 m \Omega^2 r_0^2} \left(\frac{r}{r_0} \right)^{2n-2}$$ (10.1),

where $2n$ is the number of poles, e is the electron charge, V is the RF voltage (zero-to-peak) applied to one pair of rods, m is the ion mass, Ω is the RF angular frequency, and r_0 is the inscribed radius of the multipole [8]. Figure 10.1 compares the well profiles for quadrupole, hexapole, and octopole ion guides of the same interior radius with the parameters shown. The narrower beam that a quadrupole provides can be important in transmitting ions efficiently through a small aperture, or in confining ions to a smaller radial region for better interaction with a laser beam. On the other hand, higher-order multipoles provide stronger containment to prevent loss of ions with high radial energies as well as stable confinement conditions for a wider mass range at a given RF voltage.

Another form of RF ion guide is a ring guide, composed of a series of aperture plates with opposite RF phases applied to adjacent plates. The ring guide has been developed as a two-dimensional (2D) ion trap and as a transport device [8] with a DC axial field as well as with pulsed DC fields [16] to propel the ions along the axis. With optimally spaced plates, the ring guide provides a trapping potential with the flattest profile, but also the strongest confinement for a given RF voltage. A tapered configuration, known as an ion funnel [17], combines the strong ion containment of the ring guide with the focusing properties of a high-pressure quadrupole, providing a narrow ion beam at the small-aperture exit and a wide mass range for stable ion transmission; however, an ion funnel is mechanically more complex than is a quadrupole.

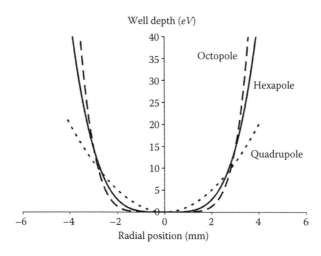

FIGURE 10.1 Pseudopotential well profile for an octopole, hexapole, and quadrupole with r_0 of 2.05 mm at a frequency of 2.5 MHz and RF voltage (zero-to-peak) of 200 V.

High-pressure ion guides have been embraced throughout the mass spectrometry community, and they are now used as collision cells, ion guides, ion traps, ion reaction cells, and mobility cells. Trapping, moving ions from one location to another, and processing ions has become straightforward. In this chapter, we describe the fundamentals of trapping and moving ions from place to place, along with some methods and applications.

10.2   TRAPPING AND MOVING IONS IN A HIGH-PRESSURE RF QUADRUPOLE

The electric fields experienced by ions in a 2D quadrupolar RF ion guide can be separated conveniently into the radial components (perpendicular to the axis) and the axial components (along the axis). In a pure 2D RF field in a multipole, there are no axial fields. Axial fields do exist at the entrance and exit (in the fringing field regions), and they can be introduced additionally in the interior by a variety of means to be described shortly.

In the early days of the use of RF ion guides, collisions were either used for fragmentation, or else avoided if possible, because they were viewed as potential sources of ion loss through scattering. In general, the ability of an ion guide to confine ions can be characterized by the depth of the pseudopotential well, given for a quadrupole by

$$\overline{D}_r = q_M V / 4 \qquad (10.2),$$

where q_M is the Mathieu parameter [18] and V is the RF voltage amplitude (zero-to-peak). This expression can be derived directly from Equation 10.1 for $r = r_0$ realizing that the Mathieu stability parameter $q_M = 4eV/m\Omega^2 r_0^2$.

An ion that enters with a radial energy of greater than $\overline{D}_r$, or acquires such a high radial energy by scattering inside the quadrupole, is likely to be lost from the ion guide. In addition, collisions in the interior of the ion guide cause ions to lose axial energy. In the early low-pressure collision cells used in triple-quadrupole systems, product ions of mass m formed by single or a few collision events from a precursor of mass M would exit from the cell with mass-dependent energies $E \approx E_0(m/M)^2$, where E_0 is the initial precursor energy. This condition required the rod offset of Q3 to be scanned with m/z to maintain constant ion energy through Q3 [19].

The discovery that many collisions in an ion guide would lead to an ion beam that was both focused toward the axis and cooled to near thermal energy was a breakthrough. Collisional cooling made it possible to avoid the limitations imposed by Liouville's theorem, which is used to describe the evolution of a large number of collision-free particles moving under the influence of gravitational and/or electromagnetic fields. According to this theorem, the phase-space volume of a collection of such particles is constant along their trajectory, therefore the focusing capabilities of electrostatic lenses are limited. In addition to allowing ion guides to be used to transport ions very efficiently through high-pressure vacuum stages, collisional focusing allowed the use of small apertures between stages and, thus, more modest vacuum pumps. The mechanism of collisional focusing can be viewed as collapse

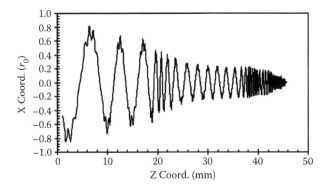

FIGURE 10.2 Calculated trajectory in the *x*-direction of an ion of *m/z* 609 in an RF-only quadrupole at a *q*-value of 0.3 and a pressure of 5 mTorr of nitrogen. The initial entrance energy was 10 eV. The ion experienced 95 collisions in the first 38 mm of this trajectory.

of the beam toward the center of the ion guide due to a combination of the inwardly directed pseudopotential force from the RF field, and the frictional effect of the multiple ion/gas collisions. Figure 10.2 shows an example of the calculated ion trajectory in a high-pressure ion guide, illustrating both the loss of axial energy (cooling) and the collapse of the radial ion motion (cooling with focusing). The reduced beam diameter shows clearly how collisional focusing helps improve the ion transmission through a conductance-limiting aperture into the next vacuum stage.

The loss of axial energy provides a thermalized ion beam, but slows transport through the ion guide, and may actually result in ions being trapped in the guide (or at least losing their net axial velocity so that diffusion dominates their axial motion). Such axial-energy loss was a problem in the first high-pressure collision cells, leading to cross-talk when switching rapidly between mass channels. Ions were able to exit from the ion guide only because trapping produced space charge, which established sufficient 'push' to move the ions out. Despite space-charge propulsion, the residence time became too long for rapid experiments.

To solve this problem, a collision cell with an axial electric field was developed, and introduced commercially as the LINAC® collision cell. Several configurations were developed and shown to be effective, including the use of auxiliary electrodes, tapered rods, segmented rods and resistively coated rods. However, the first commercial version used tilted rods with a small quadrupolar voltage applied between the rod pairs [20]. Field strengths of *ca* 5 V m⁻¹ were shown to be sufficient to move ions through a 20-cm rod set in a few milliseconds. While the axial field was used to minimize cross-talk in the collision cell, and to allow fast precursor and neutral-loss scans to be performed, a segmented quadrupole collision cell was shown to be useful also as a mobility or drift cell [21]. The group at York University (Ontario, Canada) subsequently adopted this mobility/collision cell for use in measuring collision cross-sections and separating protein conformers [22]. The combination of an RF ion guide with an ion mobility drift cell has the advantage of confining radially the ions for efficient transport and, therefore, providing high sensitivity when combined with a mass spectrometer (by overcoming the usual diffusion losses in ion mobility tubes); the disadvantage of this combination is that it provides only medium to low resolution because of the low pressure (10 mTorr–*ca* 4 Torr) and low electric field strength. A configuration that used an opposing flow of gas, with an axial field, was developed later by Loboda, providing much better mobility resolution for separation of matrix-assisted laser desorption ionization (MALDI)-generated ions [7].

The tilted-rod LINAC collision cell has the advantage of simplicity of construction, but requires the use of a small quadrupolar DC voltage to provide the axial field. This additional DC voltage can limit the mass range of ions that are transmitted simultaneously. While not a significant problem in a triple-quadrupole system, where only a limited mass range needs to be transmitted at one time, it becomes a limitation in quadrupole/time-of-flight and quadrupole/ion trap systems where a wide mass range must be transmitted. For that reason, the LINAC collision cell in the quadrupole/time-of-flight (QqTOF) system is of a different design, consisting of auxiliary electrodes located between the rods and tilted with respect to the axis [23]. This configuration provides the ability to change easily the strength and direction of the axial field without narrowing the mass range. It provides also the ability to trap ions and then to extract them rapidly by applying an axial field.

Trapping is managed efficiently by providing a DC barrier to an aperture lens at the end of the ion guide; in general, a barrier of a few volts is sufficient to trap ions once they are cooled collisionally. In addition, the entrance lens is usually made repulsive relative to the axial potential of the ion guide or collision cell, so that there is a barrier at each end. Ions enter the ion guide at an energy selected so as either to fragment them (20–100 eV) or to ensure their transmission through the ion guide without fragmentation (*ca* 10 eV). They are cooled during their first pass through the ion guide, which is normally at a pressure of 5–10 mTorr, reaching an axial potential that corresponds closely with the potential of the ion guide. The barriers at each end keep the ions from escaping. As ions accumulate in the ion guide, space charge will cause the ion beam to expand in diameter and, in the limit, ions will be lost either radially or axially, depending on the relative size of the axial and radial energy barriers. An estimate of the ion population in an ion guide can be obtained for a typical example of a well depth of 10 V. Considering that the electric field due to a cylinder of charge of diameter 2 mm and linear charge density λ C m^{-1} is $\lambda/2\pi\varepsilon_0 r$, the radial space-charge field is equal to the pseudopotential field at a charge density of *ca* 2.4×10^9 charges m^{-1}, or a total of 4.8×10^8 charges in a 200 mm-long quadrupole. This is a crude estimate because fringing field and RF heating effects are neglected. The charge capacity of RF ion guides was studied in considerable detail by Tolmachev *et al.* [24].

An RF ion guide combined with ion trapping has been used to improve the duty cycle of orthogonal time-of-flight (TOF) systems. Ions are trapped temporarily in the ion guide (or collision cell, in the case of a QqTOF system) by raising the voltage on the exit barrier, and then briefly lowering the voltage for a few microseconds to gate ions toward the TOF [25,26]. The orthogonal acceleration pulse is timed to accelerate the ions into the TOF as they reach the center of the pulsing region. Since the velocity of the mono-energetic ions after the ion guide is mass-dependent, the timing required is mass-dependent also. Correct selection of the width of the gating pulse and the delay time can provide 100% duty cycle into the TOF for a limited mass range. An axial field is required in the collision cell in order to move ions toward the exit and to prevent mixing and long delay times that could result in cross-talk.

So far we have shown how high-pressure RF ion guides provide collisional focusing and improved transmission, how axial electric fields help to keep ions moving and minimize cross-talk, and how momentary ion trapping works to improve the transmission efficiency in orthogonal TOF. In the next sections, we show some advanced methods and applications that are provided by combining these tools in the ion path of a tandem mass spectrometer.

10.3 REVERSING ION FLOW FOR MS^n

All beam-type mass spectrometers operate conventionally in a mode where ion flow is in a single direction, from the entrance (ion source) toward the exit (detector). The use of axial electric fields combined with ion trapping, however, provides the possibility of reversing the direction of ion motion, of moving ions back and forth in order to process them in different regions of the ion path, and of using the quadrupole mass filter multiple times [27].

For example, consider the following method of providing MS^3 in a standard QqTOF instrument (see Figure 10.3). Ions are transmitted from the ion source through Q1 into q2 where they are fragmented as they enter the collision cell (as usual). However, instead of allowing the product ions to flow through into the TOF analyzer, the ions are trapped and cooled in q2 by raising the potential of the exit lens, thus accumulating ions for a selected period of a few tens of milliseconds. After stopping the accumulation of ions in q2 (by turning off the flow of ions from the ion source), the product ions are moved back into q0 by reversing the direction of the axial field in q2, operating Q1 in an RF-only mode, and setting the potentials between the quadrupoles so that ions are attracted back into q0. By setting the q0 potential attractive by only a few volts relative to q2, the ions can be moved efficiently without fragmenting them as they enter q0. The ions are trapped in q0 (which operates at a pressure of *ca* 7 mTorr) by making the skimmer voltage and IQ1 voltage slightly repulsive with respect to the q0 voltage, so that an axial potential well is formed in q0. Once the ions have been transferred to q0, the ion flow can be reversed again by raising the q0 voltage above that of IQ1 and q2. The ions are then forced to move back toward q2 and, as they pass through Q1, Q1 can be used mass selectively to isolate a product ion that is fragmented subsequently in q2. The products of the initially selected product ion either can be transmitted into the oTOF mass analyzer (for MS^3) or they can be trapped again in q2 and put through another cycle to provide MS^4.

The speed and efficiency of the process is aided by using axial fields in both q0 and q2 in order to move ions in the desired direction. Axial fields of the order of 2.5–5 V m^{-1} are readily provided by using auxiliary electrodes between the rods that are tilted, and applying a voltage of 50–150 V to the electrodes (relative to the DC offset on the ion guide). Transit times through a 20 cm-long ion guide at a pressure of 7 mTorr are of the order of 10 ms with an axial field of 5 V m^{-1} for an ion of mobility 1 cm^2 V^{-1}s^{-1}, typical of small ions. Therefore total cycle times of <100 ms are feasible, and have been demonstrated experimentally.

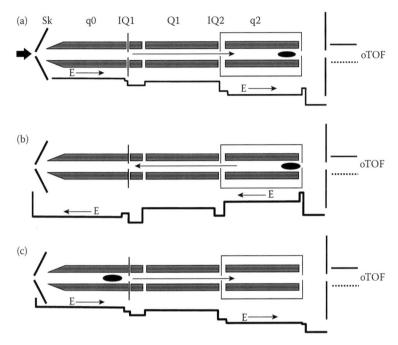

FIGURE 10.3   (a) Triple-quadrupole ion path as ions are introduced through Q1, filtering *m/z* 609 and fragmenting ions in q2. The product ions are trapped in q2 and confined near the exit by the axial field in q2. (b) Ions are moved back into q0 by reversing the direction of axial field in q2, and adjusting the ion path voltages as shown. (c) Ions are moved back through Q1 toward q2 selecting a product ion in Q1 and fragmenting it in q2 to provide MS³. Relative ion path voltages are indicated under each figure. The direction of the axial electric field in q0 and q2 during each step is indicated in each figure. SK is the skimmer in front of q0, IQ1 is the lens between q0 and Q1, and IQ2 is the lens between Q1 and q2.

Figure 10.4 shows the MS/MS spectrum of the *m/z* 786 doubly-charged precursor ion of [Glu¹]-fibrinopeptide B and two examples of MS³ of two product ions of *m/z* 684 and 1285. The product ions were selected for MS³ by trapping them in q2 and moving them back into q0 as described above. Quadrupole Q1 was operated at unit mass resolution for each experiment, both for the precursor fragmentation step (MS/MS) and for CID of the product ions (MS³). The efficiency of the process can be measured by summing the intensities of the second-generation product ions and comparing this to the intensity of the first-generation product ion in the MS/MS experiment. For example, the *m/z* 684 ions had an intensity (peak area) of 210 counts per second (cps) in an MS/MS experiment. The sum of the product ions from *m/z* 684 in the MS³ experiment is 60 cps, demonstrating an efficiency of 29%. Typical efficiencies, calculated in this way, range from 25 to 35%, depending upon the collision energy and the *m/z*-values of the product and precursor ions. Some of the losses are due to the additional step of resolving with Q1, and some are due to duty cycle losses in the orthogonal time-of-flight (oTOF) analyzer.

The efficiency of the transfer step alone can be measured by cycling the ions back and forth without fragmenting them. For example, when ions are trapped in

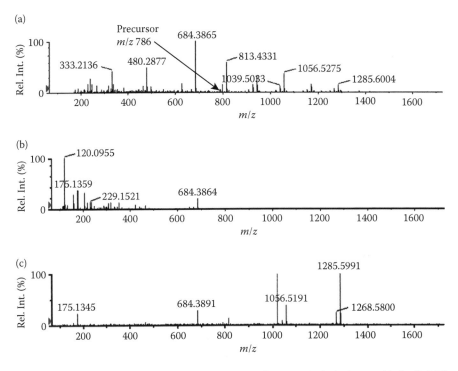

FIGURE 10.4 (a) MS/MS spectrum of m/z 786 (MH$_2^{2+}$) from [Glu^1]-fibrinopeptide B. (b) MS^3 of m/z 684 from [Glu^1]-fibrinopeptide B. (c) MS^3 of m/z 1285 from [Glu^1]-fibrinopeptide B.

q2, then moved back to q0 (with Q1 in the RF-only mode), then moved back to q2 and trapped again (without fragmentation), this can be considered one round trip. In this case, Q1 is resolving when the ions are moved from q0 to q2, but not when moved in the other direction. Figure 10.5 shows the efficiency of moving ions of m/z 609 through n cycles of this sort (measuring the ion intensity in the TOF at the end of each n cycles), indicating an efficiency of about 40% per round trip (($\bullet$) in Figure 10.5). When the same experiment is repeated with Q1 resolving when ions are moved from q2 into q0 (and RF-only when they move in the opposite direction) the efficiency is much lower—about 20% (($\blacktriangle$) in Figure 10.5). This difference in efficiencies can be explained by the fact that when the ions move into Q1 from q0, there is a short RF-only section in front of Q1 that reduces fringing field losses when Q1 is resolving. When ions move from q2 toward q0, there is no such RF-only lens to prevent losses into Q1. When the experiment is performed with Q1 resolving as ions move through it in both directions (($\blacksquare$) in Figure 10.5), the efficiency is very low (mostly due to the loss as ions move into Q1 from the q2 side). Figure 10.5 also shows ($\blacklozenge$) that if Q1 is not resolving when ions move in both directions, the efficiency is approximately 85% per round trip, remarkably high considering that ions move through two apertures (IQ1 at 1.4 mm diameter and IQ2 at 2.4 mm diameter).

The isolation step can be performed alternatively in Q1 by trapping ions in Q1 rather than moving them through Q1 into q0. Ions can be trapped efficiently in Q1 in

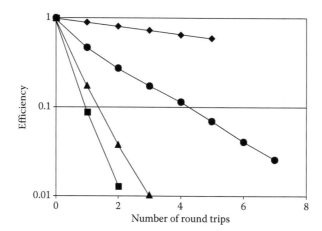

FIGURE 10.5 Efficiency of the transfer process between q2 and q0, counting one round trip as the process of moving ions that are trapped in q2 back into q0 and then back into q2: (◆) Q1 RF-only mode for travel in both directions; (●) Q1 at unit resolution for travel from q0 to q2; (▲) Q1 at unit resolution for travel from q2 to q0; (■) Q1 at unit resolution for travel in both directions.

an RF-only mode by raising the potential of the lens at the opposite end, as long as they experience a few collisions within Q1, sufficient to lose enough energy that they do not exit upon reflection. With q2 pressurized to about 7 mTorr, the small distance between Q1 and q2 ensures that enough collisions occur to provide efficient trapping when ions are moved at low energy into Q1. Once the product ions are trapped in Q1, the product ion of interest can be isolated using a variety of well-known techniques (developed for analytical 3D and 2D ion traps that are used for MS/MS). For example, radial excitation can be used to remove unwanted ions, leaving only the ion of interest. In the experiment described here an ion is isolated in Q1 by setting the Q1 m/z-value to the m/z of interest (with Q1 still in RF-only mode) and then briefly turning on the resolving DC for a period of 1.5 ms (the same resolving DC that is applied when Q1 is used in a conventional transmission mode as an ion filter). This procedure leaves only the ion of interest in Q1. Once the product ion is isolated in Q1, it can be accelerated back into q2 for fragmentation. Isolation in Q1 allows the cycle to be performed faster than when ions are moved back into q0; however, moving ions between q2 and q0 can, in principle, allow isolation and beam-type CID in both directions when the optics at both ends of Q1 are optimized, allowing even more rapid MS^n.

10.4 ION/ION REACTIONS WITH BI-DIRECTIONAL ION FLOW

Moving ions in both directions between linear ion traps allows other types of ion processing (beyond simply MS^n) to be performed. Ion/ion reactions have been shown to be a powerful tool in simplifying the interpretation of complex multiply charged mass spectra, and in providing structural information through specific fragmentation pathways. Yu *et al.* [28]. have shown that the bi-directional ion transfer technique described in the previous section can be used to enable and to control a variety of

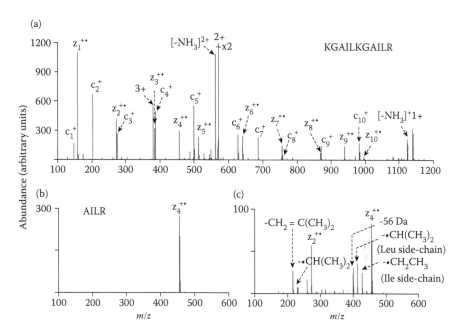

FIGURE 10.6 (a) ETD mass spectrum of the triply charged positive ion of peptide KGAILKGAILR; (b) selection (isolation) of the $z_4^{+\bullet}$ ion in Q1; (c) CID of the $z_4^{+\bullet}$ ion by accelerating the ion from Q1 into q2 at an energy of 17 eV. (Reproduced from Yu, X., Thomson, B. A., and McLuckey, S. A., *Anal. Chem.* 2007, *79*, 8199–8206. With permission from the American Chemical Society.)

useful ion/ion reactions in a QqTOF ion path. Ions can be trapped in q2 (with or without fragmentation), moved back either into or through Q1 for isolation of a particular product ion, and then the selected ion moved back into q2 for reaction with ions of the opposite polarity. The reagent ions can be introduced either through the entrance aperture or from a separate ion source that is coupled directly to q2. By using q2 as the reaction cell, and Q1 to select ions of interest, ions of the desired m/z can be reacted, providing better control and less ambiguity than when the entire mixture of ions from the source are reacted. Q1 can be used to select either an analyte ion or a reagent ion or both. In addition, q2 can be used to fragment the reaction products after Q1 has provided the isolation step.

An elegant example of this utility is shown in Figure 10.6 [28], where MS^3 is performed on the $z_4^{+\bullet}$ ion that is formed by electron-transfer dissociation (ETD), in order to provide more specific information on the peptide sequence. The triply charged positive ion of peptide KGAILKGAILR was selected with Q1 and trapped in q2. A second ion source was then used to generate azobenzene radical anions, which were selected by Q1 and introduced into q2 to react with the peptide ions. The ETD process formed a complete series of c and $z^{\bullet}$-type sequence ions. However, there is an ambiguity with respect to the leucine and isoleucine isomers, which is resolved by performing CID on the $z_4^{+\bullet}$ ion. After the ETD products were formed in q2, they were moved back into Q1, and the $z_4^{+\bullet}$ ion isolated by pulsing the resolving DC. The $z_4^{+\bullet}$ ion was then fragmented by accelerating it into q2, and the product ions were recorded by the TOF. The side-chain losses of 43 Da, specific to leucine, and 29 Da, specific to isoleucine, as shown in Figure 10.6c, confirm that $z_4^{+\bullet}$ contains a mixture of the two species. The loss of 43 Da from the $z_4^{+\bullet}$ ion and the absence of a loss of 29 Da from this ion then indicate the presence of leucine only, and inspire further confidence in the primary sequence identification. The ability to use q0 and q2 as ion traps and Q1 as a mass filter in either a beam mode or an LIT mode, provides significant flexibility that allows the ion isolation steps and the ion processing steps to be optimized independently.

10.5   ION/MOLECULE REACTIONS OF TRAPPED IONS

The rapidly increasing sensitivity of mass spectrometers does not solve automatically the problem of finding low abundance components in the forest of chemical noise. Various separation techniques used up-front of the MS (LC, ion mobility*, High-Field Asymmetric Waveform Ion Mobility Spectrometry (FAIMS†)) reduce chemical noise but, in complex mixtures concentrations at 10^{-8} M, the problem of low abundance component detection may remain. Small peaks that are hardly detectable in single-stage MS may be large enough to provide valuable information by MS/MS. One of the ways to clean up the complex mass spectra is to let multiply charged ions react with ions of the opposite charge, thus shifting their peaks to a higher m/z-range that is less congested with chemical noise. While this approach developed by McLuckey's group [29] is quite efficient, it requires an extra ion source and the ability to store simultaneously ions of opposite polarity in the same multipole.

We have developed a different concept of using ion/molecule charge transfer reactions to distinguish analyte ions from chemical noise ions. For example, a reagent gas can be selected that reduces the charge state of multiply charged ions and thus moves their peaks to higher m/z-range while leaving the charge state of chemical noise ions unaffected. This type of ion/molecule reaction is less efficient than its ion/ion equivalent, but this can be compensated for by the ability to add a reagent gas into RF-only quadrupoles q0 or q2 to a pressure of up to 10 mTorr.

One of the preferred ways to clean up mass spectra is to divide the m/z-range of interest into several 'windows' 100–300 Th wide, select them sequentially with Q1 in a very low resolution mode, perform proton transfer reactions in the reagent-filled q2 and analyze the resultant mixture of product ions with the second analyzer, which can be, for example, a quadrupole, a linear trap or an orthogonal TOF. Depending on the reagents, either reaction in q2 may be performed in a beam mode (on-the-fly with a time scale of the order of 10 ms) or ions may be trapped temporarily in q2 to allow more time for reactions to proceed.

Figures 10.7 and 10.8 demonstrate the effect of adding a reagent with high proton affinity (PA), in this case ammonia, to the collision cell. Panel A shows the initial precursor selection window (m/z from 450 to 550) before reactions, while panel B shows

---

* See Volume 5, Chapter 13: The Study of Ion/Molecule Reactions at Ambient Pressure with Ion Mobility Spectrometry and Ion Mobility/Mass Spectrometry, by Gary A. Eiceman and John A. Stone.

† See Volume 4, Chapter 5: High-Field Asymmetric Waveform Ion Mobility Spectrometry, by Randall W. Purves.

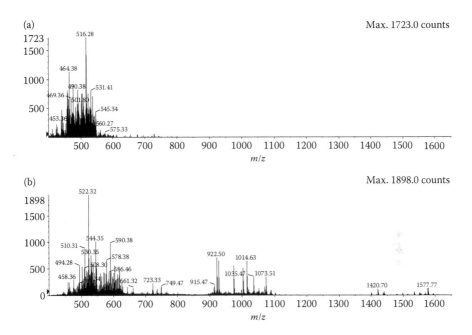

FIGURE 10.7 Mass spectra of BSA digest, 5×10⁻⁹ M concentration. A window was selected in Q1 with m/z from 450 to 550 Th; collision energy low (10 eV). (a) no reagents added; (b) ammonia added.

a mass spectrum of the same ions after charge reduction through proton transfer for a time period of a few seconds. The sample is a bovine serum albumin (BSA) digest diluted to 5×10^{-9} M concentration, in which the peaks of interest are already smaller than those of chemical noise. Three clearly separated groups of peaks are visible in Figure 10.7B: (i) 1+ ions remaining from the m/z-range selected initially with Q1; (ii) ions converted from 2+ into 1+ whereby their m/z-values were doubled; and (iii) ions converted from 3+ to 1+ whereby their m/z-values were tripled. Figure 10.8 shows expanded views of parts of the same two mass spectra. Panel A of Figure 10.8 shows an expanded view of the m/z 450–550 range from panel A of Figure 10.7. Panel B of Figure 10.8 is an expanded view of the m/z 900–1100 range from panel B of Figure 10.7 showing those ions that had undergone a 2+ to 1+ transition. Panel C of Figure 10.8 is an expanded view of the m/z 1350–1650 range from panel B of Figure 10.7 showing those ions that had undergone a 3+ to 1+ transition. Through appropriate scaling of the axis singly charged ions in panels B and C can be tracked back easily to their origin in A, where they were originally doubly or triply charged, respectively. As is clear from panel A, peaks of the 2+ and 3+ ions are barely visible in the forest of chemical noise. On the other hand, all of the major peaks in B and most of the major peaks in C originate from the tryptic peptides of BSA.

While a broad range of reagents can be used for proton transfer reactions, the essential criteria are that they should be reasonably volatile and have relatively high PA (PA >190 kcal mol⁻¹) in order to be able to abstract protons from multiply charged ions. However, a PA that is too high may be undesirable because it may lead to complete

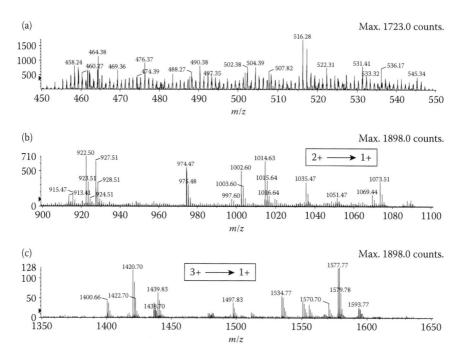

FIGURE 10.8   Segments of mass spectra shown in Figure 10.7. (a) no reagents added; (b) and (c) ammonia added. Panels (b) and (c), which are from the same mass spectrum, are aligned with (a) in such a way so that the original 2+ and 3+ ions in (a) are aligned vertically with their singly charged de-protonated 'relatives' in panels (b) and (c), respectively.

charge-stripping of analyte molecules. Some reagents with high PA (e.g., amines) form adducts with analyte ions, which makes them less useful for this method.

To date, the behavior of about 20 compounds as proton abstractors has been examined, and the most promising results were obtained with ammonia, acetone, and ethyl acetate. However, the reaction time scale is on the order of one second, whereas it is desirable to have proton transfer performed on a drive-through basis (that is, a time-scale of the order of a few milliseconds). More work is required to choose better reagents.

There is some loss of duty cycle involved in this method because several steps can be required to cover the whole m/z-range of interest. However, selection of a restricted mass range with Q1, followed by reaction in q2 realizes the significant advantage that the newly formed charge-reduced ions are moved to an empty and virtually clean area of the mass spectrum, thus improving significantly the signal:noise ratio and the ability to identify the presence of minor peptide components of a complex mixture.

A generalized method of utilizing charge transfer reactions to improve detection limits can, therefore, be formulated as follows:

- select ions within a desired m/z-range;
- add a reagent that reduces the charge state of ions of certain types (this can apply either to analyte ions or chemical noise ions);

- record a mass spectrum of the ions formed in Step 2;
- process the data taking into account that some ions have changed their *m/z*-value due to charge state reduction.

Once the peptide ions of interest are targeted in this survey scan, MS/MS or other ion processing techniques can be performed on the precursor ions in order to aid identification.

10.6 BRANCHED ION GUIDES FOR MULTIPLE SOURCES

Existing mass spectrometers usually analyze ions from one ion source at a time, with the exception of 3D ion traps [30] or when the second source is a built-in electron impact ionization source. Changing the ion source (e.g., from electrospray ionization (ESI) to MALDI) requires manual interaction and involves usually venting of at least part of the vacuum chamber, resulting in a significant pump-down time before the machine is back in operation.

It turns out, however, that RF multipole ion guides are flexible enough to be joined together in 'T', 'Y', or 'X' configurations, multiplexing them either at the input or output side. These arrangements allow either ion beams to be switched from one path into another, or beams from different sources to be selected as desired, or beams merged together. Several examples of such configurations of multipoles with multiple inputs and/or outputs are shown in Figure 10.9, with more found in the patent literature [31,32]. The plus and minus signs in Figure 10.9 designate the opposite phases of RF voltage applied to each of L, T, or X-shaped electrodes. Arrows indicate the directions of ion traffic, which can be controlled by the electric fields and/or gas flows at the entrances and exit(s) of each configuration. At the same time, ions are confined radially by the pseudopotential created by the RF field.

Configuration 1A can handle up to three ion sources whereas 1B, which is a 3D version of 1A, is able to couple upto five sources. Alternatively, some of the inputs can become outputs when the direction of the axial electric field is reversed. The application of an appropriate electric field to unused entrances and exits will prevent ions from flowing out. Additional flat electrodes can be used in 1A to prevent ion losses through the large opening in the middle of the 'X.' Configuration 2A can handle up to three sources but it differs from 1A in that its geometry allows reduction to two sources (as shown in configuration 2B) without the need to block an unused entrance.

As a proof of principle, we have developed and studied experimentally an ion guide similar to configuration 1A, with the exception that one of the side branches was blocked so as to allow two inputs and a single output (Figure 10.10). All experiments were carried out on a modified QStar® Elite quadrupole-TOF mass spectrometer. The vacuum chamber was modified to accommodate the new 'branched' q0 ion guide and second ion source. Either input side can accept a nanospray, Ionspray™ (a trademark of Applied Biosystems/MDS Analytical Technologies) or high-pressure MALDI source. In those cases when two atmospheric pressure sources are used simultaneously, the pressure increase in q0 from its normal value of 6 mTorr up to about 12 mTorr, had no noticeable effect on the instrument's performance. Flat

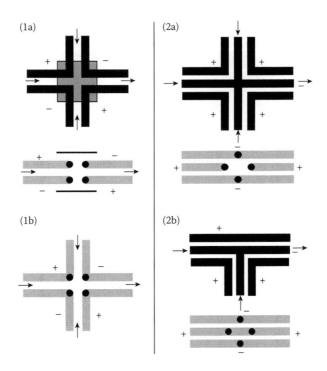

FIGURE 10.9 Examples of possible configurations of quadrupoles with multiplexed input/output.

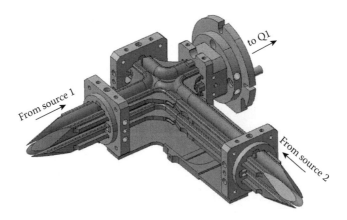

FIGURE 10.10 Dual source q0 interface.

electrodes (only one shown in Figure 10.10), to prevent ions from escaping vertically, were floated at +0.5 V with respect to q0 rod offset.

The first experiments with a dual interface were performed with two identical Ionspray sources at both input sides. Although there is an obvious asymmetry in the source positions, the performances of each of the two sources were found to be identical when the same calibrant solution was sprayed in each source under

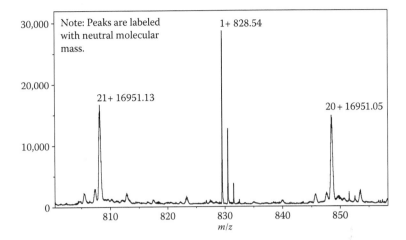

FIGURE 10.11 A mass spectrum recorded with two Ionspray sources working simultaneously, one of which was spraying a solution of myoglobin, while the other was loaded with a calibrant (peptide ALILTLVS). Note that for consistency mass peaks are labeled as the neutral molecular mass to indicate the accurate mass measurement of myoglobin.

identical conditions. Note that these sources can work either separately from each other, or simultaneously, dependent on the voltages applied between q0 and the skimmers (in the case of the ESI interface). One of the sources can be used for calibration purposes, as illustrated by Figure 10.11, where the TOF mass spectrum of myoglobin produced by one source was recalibrated with the help of a calibrant ion, m/z 828.54 from the peptide ALILTLVS, sprayed from another source. By the time ions enter the TOF, the ions originating from different sources become indistinguishable, and ions from any type of source can be used to recalibrate mass spectra obtained with any other type of source (ESI, high-pressure MALDI, high-pressure secondary ion mass spectrometry (SIMS), etc.).

Figure 10.12 illustrates the situation when ions from two different sources are mixed in q0, resulting in a single mass spectrum: source 1 was an Ionspray, while source 2 (on the side) was an oMALDI (a trademark of Applied Biosystems/MDS Analytical Technologies). Again, the sources can work either independently or simultaneously. It is worth mentioning, however, that nitrogen pressure of *ca* 6 mTorr necessary for collisional cooling and focusing in the oMALDI source [33] is provided automatically by the electrospray interface. No axial electric fields are created specifically to move ions from inputs to the exit inside this dual source configuration. Ion travel from both sources to the analyzer is driven by diffusion, gas flow and/or an electric field created by either field penetration from outside of the quadrupoles or by space charge. From this point of view, the position of the MALDI source on the side and electrospray source 'on-axis' is somewhat beneficial, because the gas flow from the electrospray interface assists in turning MALDI-generated ions toward Q1. In the absence of axial fields, ions may make many 'wrong turns' before they reach the only available exit. Several timing experiments have been performed, where ions were gated into q0 on either input side, and transit-time distributions were measured

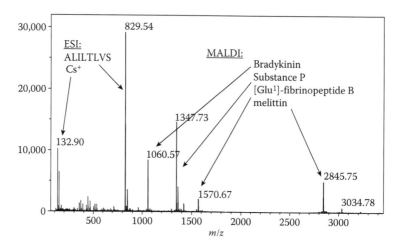

FIGURE 10.12 A mass spectrum obtained with simultaneous operation of Ionspray (as source 1) and oMALDI (as source 2). A calibration solution made of CsI (3×10^{-5}M) and peptide ALILTLVS (3×10^{-6}M) was sprayed in source 1; a mixture of four peptides was deposited on the MALDI plate of source 2: bradykinin, substance P, [Glu¹]-fibrinopeptide B and melittin.

for several ions. The results of those experiments showed that the average time ions spend traveling through q0 ranges from *ca* 5 ms for Cs⁺ to *ca* 30 ms for multiply-charged myoglobin ions, although the distribution for the latter tails up to >100 ms. In principle, the transit time can be improved either by cutting the rods into several segments and applying a voltage gradient, or by installing additional electrodes as described in the previous sections.

10.7 CONCLUSION

This chapter has described techniques for trapping and moving ions in RF-only mul-tipoles, as well as several applications where ions are trapped, processed, and moved within the vacuum system and ion path of a multipole mass spectrometer system. The new capabilities bring added dimensions to the analysis of ions by mass spec-trometry, allowing a variety of reactions and separations to be performed with great speed, either in sequence or in series with mass analysis. New combinations of mass analysis and ion processing will be developed in the future, and they will provide even more specific and sensitive methods of sample analysis in integrated systems. The ability to couple these orthogonal analytical methods together in a harmonized and application-directed workflow will lead to even wider application of mass spec-trometry in analytical chemistry and biology.

REFERENCES

1. Chernushevich, I.V.; Fell, L.M.; Bloomfield, N.; Metalnikov, P.S.; Loboda, A.V. Charge state separation for protein applications using a quadrupole time-of-flight mass spec-trometer. *Rapid Commun. Mass Spectrom.* 2003, *17*, 1416–1424.

2. Raspopov, S.A.; El-Faramawy, A.; Thomson, B.A.; Siu, K.W.M. Infrared multiphoton dissociation in quadrupole time of flight mass spectrometry: top-down characterization of proteins. *Anal. Chem.* 2006, *78*, 4572–4577.

3. Liang, X.; Han, H.; Xia, Y.; McLuckey, S.A. A pulsed triple ionization source for sequential ion/ion reactions in an electrodynamic ion trap. *J. Am. Soc. Mass Spectrom.* 2007, *18*, 369–376.

4. Baba, T.; Satake, H.; Manri, N.; Hirabayashi, A.; Hasegawa, H.; Hashimoto, Y. High speed electron capture dissociation in a radio frequency ion trap. *Proc. 55th ASMS Conference on Mass Spectrometry and Allied Topics*, Indianapolis, IN, 2007.

5. Syka, J.E.P.; Coon, J.J.; Schroeder, M.J.; Shabanowitz, J.; Hunt, D.F. Peptide and protein sequence analysis by electron transfer dissociation mass spectrometry. *Proc. Nat. Acad. Sci.* 2004, *101*, 9528–9533.

6. Thomson, B.A. Method of storing and reacting ions in a mass spectrometer. *U.S. Patent Application* 20080014656.

7. Loboda, A.V. Novel ion mobility setup combined with collision cell and time-of-flight mass spectrometer. *J. Am. Soc. Mass Spectrom.* 2006, *17*, 691–699.

8. Gerlich, D. Inhomogeneous RF fields: a versatile tool for the study of processes with slow ions, in *"State-Selected and State-to-State Ion-Molecule Reaction Dynamics"*, ed. Ng, C-Y. and Baer, M. John Wiley & Sons, NY, 1992.

9. Kapitsa, P.L. Dynamic stability of a pendulum in oscillations at the suspension point. *Zh. Eksp. Teor. Fiz.* 1951, *21*, 588–597. Kapitza, P.L. Collected papers of P.L. Kapitza, edited by D. TerHaar (Pergamon, NY, 1965) Vol 2, pp. 714–726.

10. McGilvery, D.C.; Morrison, J.D. Photodissociation of CHI and DHI. *J. Chem. Phys.* 1977, *67*, 368–369.

11. Yost, R.A.; Enke, C.G. Selected fragmentation with a tandem quadrupole mass spectrometer. *J. Am. Chem. Soc.* 1978, *100*, 2274–2275.

12. Yost, R.A.; Enke, C.G. Triple quadrupole mass spectrometry for direct mixture analysis. *Anal. Chem.* 1979, *51*, 1251A–1259A.

13. Thomson, B.A. Radio frequency quadrupole ion guides in modern mass spectrometry. *Can. J. Chem.,* 1998, *76*, 499–505.

14. Douglas, D.J.; French, J.B. Collisional focusing effects in radio frequency quadrupoles. *J. Am. Soc. Mass Spectrom.* 1992, *3*, 398–408.

15. Thomson, B.A.; Douglas, D.J.; Corr, J.J.; Hager, J.W.; Jolliffe, C.L. Improved collisionally activated dissociation efficiency and mass resolution on a triple quadrupole mass spectrometer system. *Anal. Chem.* 1995, *67*, 1696–1704.

16. Giles, K.; Pringle, S.D.; Worthington, K.R.; Little, D.; Wildgoose, J.L.; Bateman, R.H. Applications of a traveling wave-based radio-frequency-only stacked ring ion guide. *Rapid Commun. Mass Spectrom.* 2004, *18*, 2401–2414.

17. Shaffer, S.A.; Tang, K.; Anderson, G.A.; Prior, D.C.; Udseth, H.R.; Smith, R.D. A novel ion funnel for focusing ions at elevated pressure using electrospray ionization mass spectrometry. *Rapid Commun. Mass Spectrom.* 1997, *11*, 1813–1817.

18. Feser, K.; Kogler, W. The quadrupole mass filter for GC/MS applications. *J. Chromatog. Sci.* 1979, *17*, 57–63.

19. Shushan, B.; Douglas, D.J.; Davidson, W.R.; Nacson, S. The role of kinetic energy in triple quadrupole collision induced dissociation (CID) experiments. *Int. J. Mass Spectrom. Ion Phys.* 1983, *46*, 71–74.

20. Thomson, B.A.; Jolliffe, C.L. Spectrometer with axial field. *U.S. Patent* 1998, 5,847,386.

21. Javahery, G.; Thomson, B.A. A segmented radiofrequency-only quadrupole collision cell for measurements of ion collision cross section on a triple quadrupole mass spectrometer. *J. Am. Soc. Mass Spectrom.* 1997, *8*, 697–702.

22. Guo, Y.; Wang, J.; Javahery, G.; Thomson, B.A.; Siu, K.W.M. An ion mobility spectrometer with radial collisional focusing. *Anal. Chem.* 2005, *77*, 266–275.

23. Loboda, A.V.; Krutchinsky, A.; Loboda, O.; McNabb, J.; Spicer, V.; Ens, W.; Standing, K.G. Novel Linac II electrode geometry for creating an axial field in a multipole ion guide. *Eur. J. Mass Spectrom.* 2000, *6*, 531–536.

24. Tolmachev, A.V.; Udseth, H.R.; Smith, R.D. Charge capacity limitations of radio frequency ion guides in their use for improved ion accumulation and trapping in mass spectrometer. *Anal.Chem.* 2000, *72*, 970–978.

25. Whitehouse, C.M.; Guilcicek, E.; Andrien, B.; Banks, F.; Mancini, R. *Proc. 46th ASMS Conference on Mass Spectrometry and Allied Topics,* Orlando, FL, 1998, 39.

26. Chernushevich, I.V. Duty cycle improvement for a quadrupole time-of-flight mass spectrometer and its use for precursor ion scans. *Eur. J. Mass Spectrom.* 2000, *6*, 471–479.

27. Thomson, B.A.; Chernushevich, I.V. MS^nth on QqTOF tandem mass spectrometer. *Proc. 55th ASMS Conference on Mass Spectrometry and Allied Topics,* Indianapolis, IN, 2007.

28. Yu, X.; Thomson, B.A.; McLuckey, S.A. Bidirectional ion transfer between quadrupole arrays: MS^n ion/ion reaction experiments on a quadrupole/time-of-flight tandem mass spectrometer. *Anal. Chem.* 2007, *79*, 8199–8206.

29. He, M.; McLuckey, S.A. Charge permutation reactions in tandem mass spectrometry. *J. Mass Spectrom.* 2004, *39*, 1231–1259.

30. Badman, E.R.; Chrisman, P.A.; McLuckey, S.A. A quadrupole ion trap mass spectrometer with three independent ion sources for the study of gas-phase ion/ion reactions. *Anal. Chem.* 2002, *74*, 6237–6243.

31. Chernushevich, I.V.; Loboda, A.V.; Thomson, B.A.; Krutchinski, A.N. Mass spectrometer multiple device interface for parallel configuration of multiple devices. *U.S. Patent Application* 20070057178.

32. Kovtoun, V.V. Branched radio frequency multipole. *PCT Patent Application* WO2007/103489.

33. Krutchinsky, A.N.; Loboda, A.V.; Spicer, V.L.; Dworschak, R.; Ens, W.; Standing, K.G. Orthogonal injection of matrix-assisted laser desorption/ionization ions into a time-of-flight spectrometer through a collisional damping interface. *Rapid Commun. Mass Spectrom.* 1998, *12*, 508–518.

11 Linear Ion Trap Mass Spectrometry with Mass-Selective Axial Ejection

James W. Hager

CONTENTS

11.1 Introduction 545
11.2 Historical Development 546
11.3 Mass-Selective Axial Ion Ejection 549
11.4 Implementation within a QTRAP Instrument 555
11.5 In-Trap Ion Processing 559
    11.5.1 In-Trap Fragmentation in a Low Pressure LIT 559
    11.5.2 Multiply-Charged Ion Separation 560
    11.5.3 Time-Delayed Fragmentation 562
11.6 Charge Density Control 565
11.7 Applications 567
11.8 Conclusions 569
Acknowledgments 569
References 569

11.1 INTRODUCTION

Linear ion traps (LITs) have proliferated in mass spectrometry in the past decade, first as storage devices and then as mass spectrometers [1]. The major advantages of linear ion traps include high trapping efficiencies and large charge containment capacities, as well as the relative ease of hybridization with other ion processing units. This chapter provides a description of linear ion traps with mass-selective axial ion ejection as practiced in QTRAP® instruments.

Radiofrequency (RF) quadrupole linear ion traps differ from the more conventional three-dimensional (3-D) RF ion traps in several important aspects. Radial confinement is accomplished by the electrodynamic RF fields, while axial confinement is accomplished normally using electrostatic DC fields. The fact that the RF potentials of linear ion traps are applied generally in the radial dimension only has a profound effect on trapping efficiencies. Ions in conventional spherical ion traps are contained by three-dimensional quadrupole RF fields in all directions. Such 3-D trapping fields and the relatively small internal dimensions of such ion traps have limited trapping efficiencies for externally generated ions to a few percent. Externally generated ions must overcome the high voltage, rapidly oscillating RF fields to gain entry into the ion trap. Entry is optimum only at a few phases of the RF field, leading to the poor trapping efficiencies. Linear ion traps, on the other hand, have no on-axis RF field. That, coupled with the greater length over which to dissipate axial energy, means trapping efficiencies can be as high as 100% [1].

Another of the more obvious advantages of linear ion traps for mass spectrometry is their larger trapping capacity compared with conventional 3-D ion traps [1] due to the larger internal volume of the former. This geometric advantage differs among the commercially available linear ion traps. The radial dimensions of these devices are all similar, but the lengths differ. The length of the linear ion trap in the LTQ™ (Thermo Scientific) is about 37 mm [2], while that for the two different QTRAP devices is 127 mm for the 3200QTRAP and 203 mm for the 4000QTRAP.

Once the ions are trapped and cooled, mass-selective ion ejection can be performed either radially or axially. Mass-selective ion ejection also differs between the two types of commercially available linear ion traps. The LTQ ion trap employs radial ion ejection accomplished in a manner analogous to that used in conventional 3-D ion traps [2]. Ions are excited by a resonance field to the point where they gain sufficient amplitude along the excitation dimension to exit the linear ion trap through slots machined in one or two of the hyperbolic electrodes.*

Mass-selective axial ion ejection in the context of the QTRAP, on the other hand, is analogous to transmission RF-only mass spectrometers [3,4]. Ions are excited by a weak radially applied resonance field to the point where they gain some radial amplitude, but not enough to encounter the electrodes of the ion trap [5]. In contrast to the hyperbolic electrodes of the LTQ, the QTRAP instruments employ electrodes with a circular cross-sections. The ejection process occurs in the exit fringing-field due to coupling of the radial and axial degrees-of-freedom. This coupling is strongly dependent on the radial amplitude of the trapped ion. Since the ions are excited radially in a mass-selective manner, the ion ejection process itself is mass selective [5].

This chapter will be limited to descriptions of linear ion traps with mass-selective axial ejection using fringing-fields, including details of the mass-selective axial ejection process, and the implementation of such an ion trap within the ion path of a triple quadrupole mass spectrometer.

11.2   HISTORICAL DEVELOPMENT

Mass-selective axial ion ejection from a linear ion trap was first reported in 1998 [6]. The performance reported there was relatively poor, being limited to slow scan rates and significant spectral perturbations due to space charge. These first experiments were carried out using a much different approach than that employed currently. Then, the resonance excitation signal was applied directly to the end lens rather than to the linear ion trap electrodes [6]. In the region of the end lens, this mode of operation is equivalent to applying an auxiliary resonance excitation signal to all four electrodes

---

* See also Volume 4, Chapter 12: Axially-resonant Excitation Linear Ion Trap (AREX LIT), by Yuichiro Hashimoto; Volume 4, Chapter 14: Electrically-induced Nonlinear Ion Traps, by Gregory J. Wells and August A. Specht.

(a)   (b)

End lens

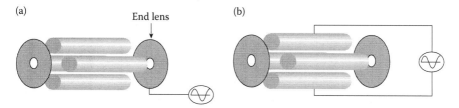

FIGURE 11.1   Two arrangements for effecting mass-selective axial ejection from a linear ion trap mass spectrometer. The auxiliary AC can be applied to (a) the end lens, or (b) the quadrupole electrodes in either a dipolar or quadrupolar (not shown) manner.

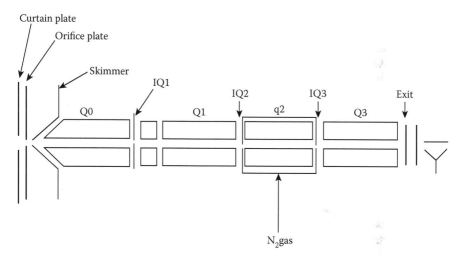

FIGURE 11.2   Schematic portrayal of the experimental apparatus based on the ion path of a triple quadrupole mass spectrometer. The linear ion trap mass spectrometer was created using either q2 or Q3. Q0 is the entry RF-only quadrupole ion guide. IQ1, IQ2, and IQ3 are interquadrupole apertures. (Reproduced from Hager, J.W., *Rapid Commun. Mass Spectrom.* 2002, *16*, 512–526. With permission from John Wiley & Sons Limited.)

in a quadrupolar fashion. Ions can be excited resonantly and gain radial amplitude only in this 'extraction region.' This mode of operation is contrasted to later work [5,7,8], and the current commercial implementation, in which resonance excitation is accomplished *via* auxiliary excitation signals applied to the linear ion trap electrodes themselves. These two approaches are illustrated in Figure 11.1.

The next report [7] appeared in 2002 and demonstrated much improved performance characteristics, using either the quadrupole collision cell (q2) or the final quadrupole (Q3) of a triple quadrupole ion path as the linear ion trap, as is illustrated in Figure 11.2. The basic scan function for either the q2 or Q3 ion trap was the same. Ions were admitted into the linear ion trap and thermalized *via* collisions with background nitrogen gas molecules. Axial ejection was accomplished by incrementing the RF drive voltage amplitude in the presence of an auxiliary quadrupolar resonance excitation signal for resonance excitation. The resonantly excited ions gain some radial amplitude, interact with the exit fringing-field and are ejected.

The magnitude of the auxiliary excitation voltage is lower than that typically used for resonance ejection in 3-D ion traps, because the desired outcome is radial excitation rather than radial ejection.

Both the q2 and Q3 linear ion trap mass spectrometers were pressurized with nitrogen, rather than helium. The pressure in the q2 linear ion trap was typically on the order of 4×10^{-4} Torr, while that of the Q3 linear ion trap was approximately an order of magnitude lower, about 3×10^{-5} Torr. These values are considerably lower than the milli-Torr values used in most other electrodynamic ion traps. Despite the order of magnitude difference in ion trap pressures, both the q2 and Q3 linear ion traps displayed the expected good trapping efficiencies of approximately 95% for the q2 trap and about 45% for the Q3 trap [7].

Ejection efficiencies for both linear ion traps were also characterized [7]. The mass-selective axial ejection efficiency of the q2 linear ion trap was somewhat less than that of the Q3 ion trap, probably due to the smaller end lens aperture size (3 mm *vs.* 8 mm). The ejection efficiency also was found to depend on the *q*-value used for resonance excitation as can be seen in Figure 11.3. The low ejection efficiencies observed at low *q*-values are due to the *q*-dependent axial forces, as discussed in Section 11.3. The optimum *q*-value was found to be approximately 0.85–0.87 for both linear ion traps [7].

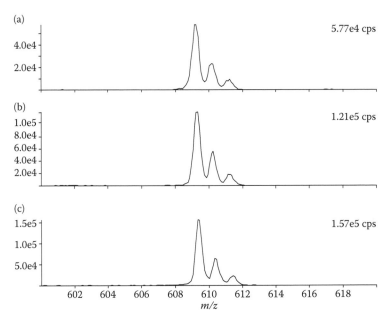

FIGURE 11.3 Examples of peak shapes and intensities for the electrospray-generated protonated reserpine ion at *m/z* 609 obtained using the q2 linear ion trap mass spectrometer with quadrupolar resonance excitation at ejection *q*-values of about, (a) 0.45, (b) 0.63, and (c) 0.84. The ion trap was scanned at 1000 Th s^{-1}. Sensitivity increases with increasing ejection *q*-value. (Reproduced from Hager, J.W., *Rapid Commun. Mass Spectrom.* 2002, *16*, 512–526. With permission from John Wiley & Sons Limited.)

There was found [7] to be a dependence of peak widths on mass scan rates, as is the case for most ion traps. The lower pressure of the Q3 linear ion trap was found to have a considerably more advantageous trade-off in this respect. The q2 linear ion trap generated peak widths of <0.7 Th full-width at half-maximum (FWHM) at scan rates of 1000 Th s^{-1}. In comparison, the Q3 linear ion trap displayed somewhat narrower peaks at a scan rate of 5200 Th s^{-1}, which was as fast as the instrument electronics allowed [7].

Charge capacities of the two linear ion traps were measured also at the point at which space-charge effects became observable on the mass spectral peaks [7]. The q2 linear ion trap showed space charge-induced peak broadening and mass assignment changes with approximately 10,000 trapped ions, although the storage capacity was orders of magnitude greater. The charge capacity of the Q3 linear ion trap seemed to be greater before the onset of space-charge effects were observed [7]. It was speculated that the reduced collision frequency in the lower pressure Q3 ion trap resulted in a more diffuse ion cloud, and thus a lower charge density for the same number of trapped ions [7].

Interestingly, the method of resonance excitation has a profound effect on the charge density at which the effects of space charge were observed. The early work published in 2002 described an instrument with quadrupolar (parametric) resonance excitation [7]. Later, it was found that dipolar excitation yielded much better performance with respect to space charge. Figure 11.4 shows a comparison of quadrupolar *versus* dipolar resonance excitation under the same charge density conditions. It is obvious that the spectral quality observed with dipolar excitation is superior. It is speculated that the difference is due to the fundamentally different ion trajectories that result from resonant dipolar excitation compared with resonant quadrupolar excitation. Ions that have been excited resonantly using a dipolar field display an initial linear increase in radial amplitude with time [1]. Quadrupolar excitation results in an exponential increase in radial amplitude [1]. It is suspected that ions subjected to quadrupolar resonance excitation spend more time near the linear ion trap axis than when dipolar resonance excitation is used. Thus, ions undergoing quadrupolar excitation are exposed to the space-charge field of cotrapped ions for a longer period of time.

One of the important findings of this early work was that a linear ion trap with mass-selective axial ejection could be integrated easily into the ion path of a triple quadrupole mass spectrometer. This instrumental arrangement will be discussed below.

11.3   MASS-SELECTIVE AXIAL ION EJECTION

RF fringing-fields are crucial to mass-selective axial ejection as practised on QTRAP instruments. Fringing-fields occur at the ends of multipole rod arrays due to the diminishing multipole potential [5]. These fringing-fields have been identified as being responsible for reducing the performance of transmission RF/DC quadrupole mass spectrometers [9] because of the reduction in the quadrupolar field, and the coupling of radial and axial fields. However, it is the exit fringing-field that enables axial ion ejection as practised in QTRAP instruments. The nature of the

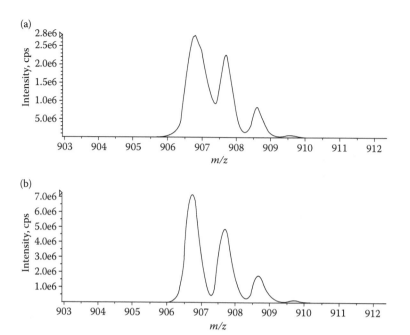

FIGURE 11.4 Comparison of the effect of space charge using the electrospray-generated *m/z* 906 polypropylene glycol ammonium adduct ion for (a) quadrupolar and (b) dipolar resonance excitation using the Q3 linear ion trap mass spectrometer. The linear ion trap was filled for 20 ms and scanned at 1000 Th s⁻¹ at an ejection *q*-value of 0.84 in both instances. The peak broadening and unusual isotope ratios in the upper spectrum are due to space-charge effects. The isotope ratios in lower spectrum are distorted slightly because of saturation of the electron multiplier, which has a maximum count rate of 7×10^6 cps.

axial forces leading to ion ejection has been investigated using a combination of analytic theory and computer simulations [5]. The results of that investigation will be summarized here.

　　To achieve mass-selective axial ejection, ions are first trapped and cooled *via* collisions with residual background gas molecules. Next, the amplitude of the drive RF voltage is ramped in the presence of a resonance excitation signal. Excited ions gain radial amplitude until they are ejected axially or neutralized on one of the trap electrodes. Away from the ends of the linear ion trap, the two-dimensional quadrupole potential can be written as [5]

$$\Phi_{2D} = \varphi_0 \frac{x^2 - y^2}{r_0^2} \tag{11.1},$$

where $2r_0$ is the inscribed diameter of the rod array and φ_0 is the electric potential, measured with respect to ground, applied with opposite polarity to each of the two pole pairs.

The reduction of the two-dimensional quadrupole potential in the fringing regions near the ends of the rods can be described by some function $f(z)$ so that the potential in the fringing regions can, to a first approximation, be written as [5]

$$\Phi_{FF} = \Phi_{2D} f(z) \tag{11.2},$$

and the axial component of the electric field due to the reduction of the two-dimensional quadrupole field is [5]

$$E_{z,\text{quad}} = -\Phi_{2D} \frac{\partial f(z)}{\partial z} \tag{11.3}.$$

The function $f(z)$ must be a decreasing function of z in the fringing-field. The value of E_z averaged over one RF cycle can be obtained for the case of no quadrupole DC component to be [5]

$$\left\langle E_{z,\text{quad}} \right\rangle_{RF} = \left| \frac{\partial f(z)}{\partial z} \right| \frac{m\Omega^2}{8e} q^2 \left(x^2 + y^2 \right) \tag{11.4},$$

where Ω is the angular frequency of the RF drive, q is the stability parameter, and $(x^2 + y^2)$ is the square of the radial coordinate.

Equation 11.4 shows the nature of the axial forces felt by a trapped ion near the ends of the linear ion trap. There is a net positive axial electric field directed out of the linear trap because of the fringing-field. To a first approximation, this field is proportional to the square of the radial displacement, the square of the stability coordinate q, and the mass-to-charge ratio (m/e) of the ion. This result is independent of the form of $f(z)$. The mass-selectivity of the ion ejection process is inherent in the fact that the collisionally cooled ions are given radial amplitude by means of resonance excitation, which is in itself a mass-selective process.

In addition to the axial field directed out of the linear ion trap due to the reduction of the quadrupolar potential, there is also an axial field directed in the opposite direction from the end lens trapping potential. These two axial fields can be summed to obtain the total axial field experienced by an ion, averaged over one RF cycle $<E_z>_{RF}$. Reference [5] shows that when the summed results are plotted as a function of axial position the resultant graph gives a representation of regions within the linear ion trap from which axial ejection will occur. The 'cone of reflection' is a surface on which the net axial force acting on an ion, over one RF cycle, is zero. Ions inside (to the left) the cone of reflection experience a negative axial force away from the end lens, while those outside (to the right) experience a net positive force toward the end lens. By appropriate choice of the end lens potential, ions with sufficient radial amplitude are able to penetrate this surface and be ejected axially, while all others are reflected [5]. Thus, the cone of reflection can be thought of as the boundary between the regions of ion reflection and ejection [5].

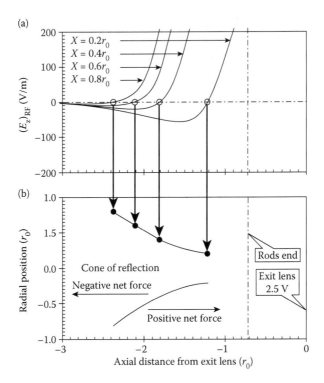

FIGURE 11.5 An example of mapping the cone of reflection. Frame (a) shows the total axial field felt by an ion of m/z 609 at (a, q) stability coordinates of (0, 0.45) averaged over 1 RF cycle. The locations of the zero-crossings, *i.e.*, coordinates where the net axial field is zero, in (a) are mapped onto a radial *vs.* axial position map shown in (b). (Reproduced from Londry, F.A. and Hager, J.W., *J. Am. Soc. Mass Spectrom.* 2003, *14*, 1130–1147. With permission from Elsevier.)

An example of a calculated cone of reflection is displayed in Figure 11.5. The zero-crossing axial coordinates for the total axial field felt by a test ion at a variety of radial excursions are mapped onto a radial *versus* axial position plot for a q-value of 0.45. The cone of reflection illustrated in Figure 11.5b demonstrates that as the radial amplitude of the trapped ion increases the positive extraction force moves deeper into the linear ion trap.

The cones of reflection can provide useful insights into the mass-selective axial ejection processes. For example, according to Equation 11.4 the strength of $< E_{z, quad} >_{RF}$ increases with the square of q. This relationship is illustrated in Figure 11.6 in which the cones of reflection for three different ejection q-values are shown. Ions at low q-values must penetrate furthest into the fringing-field region against the repulsive force of the end-lens potential and require very high radial amplitude to penetrate the cone and to be ejected. The cones of reflection for increasing ejection q-values extend increasingly farther into the linear ion trap. The greater penetration of the cone of reflection mapped out at higher q-values indicates an increasingly large volume from

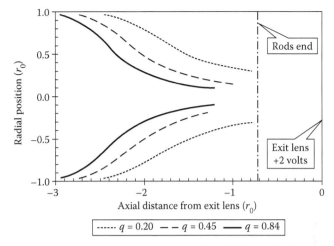

FIGURE 11.6 Cones of reflection for m/z 609 at three different ejection q-values. These cones were obtained analytically using the technique illustrated in Figure 11.5. (Reproduced from Londry, F.A., and Hager, J.W., *J. Am. Soc. Mass Spectrom.* 2003, *14*, 1130–1147. With permission from Elsevier.)

which axial ion ejection can occur. A larger volume for axial ion ejection suggests greater axial ejection efficiency corresponding to higher sensitivity [5].

Experimentally, the sensitivity and resolution of mass-selective axial ejection improve with increasing q, reaching broad maxima up to $q = 0.904$. Increasing resolution with increasing ejection q-value is most likely to be due to the associated increasing secular frequency dispersion with increasing q-value [10]. The increasing sensitivity for mass-selective axial ejection with increasing ejection q-value is demonstrated in Figure 11.7. These experimental results show a plot of the integrated ion signal for the mass-selected axial ejection of the protonated reserpine molecule at m/z 609 measured as a function of ejection q-value with an end-lens potential of 2 V. The fact that the intensity increases as a function of ejection q-value agrees qualitatively with the changes in the calculated cones of reflection illustrated in Figure 11.6.

It is possible to alter the cone of reflection by changing the fringing-fields. One of the simplest techniques for modifying the fringing-fields is by changing the RF voltage balance between the two pole pairs [4]. The addition of properly phased RF voltage of the same frequency to the end lens gives rise to an equivalent fringing-field modification [4]. When there is no reference to ground, such as deep within a linear ion trap, the balance condition between RF pole pairs is irrelevant. However, in the fringing-fields the balance condition becomes important and has a profound impact on mass-selective axial ejection [11,12]. Consider the case of three different RF balance conditions: (1) x-pole with + 15% more RF and y-pole with 15% less RF voltage, (2) x-pole with 15% less RF and y-pole with 15% more RF voltage, and (3) the RF-balanced condition in which x-pole and y-pole have the same RF voltage levels. Dipolar resonance excitation occurs between the x-pole rods. The mapped cones of reflection under these conditions are remarkably different as is

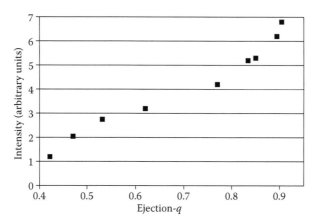

FIGURE 11.7 The integrated ion signal for mass-selective axial ejection of m/z 609 measured experimentally as a function of ejection q-value for an exit lens potential of 2 V. (Reproduced from Londry, F.A. and Hager, J.W., *J. Am. Soc. Mass Spectrom.* 2003, *14*, 1130–1147. With permission from Elsevier.)

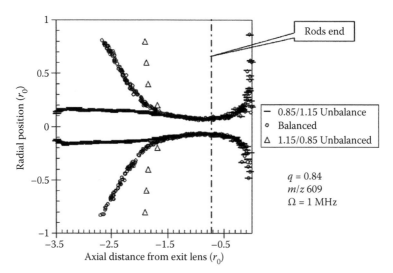

FIGURE 11.8 Cones of reflection for m/z 609 at an ejection q-value of 0.84 and an exit lens potential of 2.5 V under three different RF balance conditions. The heavy black trace corresponds to a 0.85/1.15 unbalanced configuration for the x-/y-pole pairs. Auxiliary excitation occurs along the x-pole pair. The (o) points correspond to the RF balanced configuration. The open triangle (Δ) points correspond to an x-/y-pole pair unbalanced configuration of 1.15/0.85.

shown in Figure. 11.8. Very poor performance would be expected for an RF unbalance condition in which the RF voltage along the x-pole pair is 1.15 times greater than the balanced configuration (y-pole pair would have 0.85 times the RF voltage of the balanced configuration). This variation in performance is seen by the fact that the dotted trace of Figure 11.8 does not extend into the linear ion trap as much as that of the RF balanced configuration (o). On the other hand, when the balance condition is reversed (the x-pole pair has the lower (0.85-times) balanced RF voltage and the y-pole pair has 1.15 times the balanced RF voltage value), (–), very high penetration of the cone of reflection into the linear ion trap is seen. This behavior is suggestive of much higher ion extraction efficiency than for the RF voltage balanced configuration. Additionally, the cone of reflection under the x-pole low condition becomes much more cylindrical in shape, which suggests that the ions are ejected over a much narrower range of radial amplitudes than for the configuration with balanced RF voltage [11,12].

11.4   IMPLEMENTATION WITHIN A QTRAP INSTRUMENT

Mass-selective axial ejection from a linear ion trap was introduced as a commercial product in 2002 as the QTRAP instrument. Implementation was as part of the ion path of a triple quadrupole mass analyzer in which the final quadrupole (Q3) could be operated as either a standard transmission quadrupole mass filter or as a linear ion trap mass spectrometer. There are several interesting capabilities of this Q-q-Q/LIT configuration that make it more than just another hybrid instrument. The ion path design allows the instrument to be used as either a fully functional triple quadrupole tandem mass spectrometer or as a hybrid quadrupole-LIT instrument with the ability to interconvert between the two configurations in less than 1 ms. This distinction is significant because quadrupole ion traps had not been capable previously of performing several important triple quadrupole operational modes, such as multiple reaction monitoring (MRM) as well as precursor ion and constant neutral loss scanning [13].

Figure 11.2 presents an illustration of a Q-q-Q/LIT QTRAP-type ion path; there is obviously a strong resemblance to a triple quadrupole ion path. The presence of Q1 and a collision cell (q2) upstream of the Q3 linear ion trap has meant that ion processing, such as precursor ion selection and fragmentation, can be conducted prior to filling the linear ion trap and final mass analysis [7,8].The ability to separate spatially these various ion processing steps affords the opportunity to optimize each individually, as well as to increase instrument functionality.

This spatial separation approach leads to some important differences between the Q-q-Q/LIT instruments and many other ion traps. In stand-alone ion traps, precursor ion isolation generally occurs after filling the ion trap with ions over a wide m/z range. The greater part of the original population of ions is discarded, and thus much of the finite ion trap capacity is unused. In the case of the Q-q-Q/LIT approach, however, the presence of upstream ion processing allows more efficient use of the linear ion trap capacity. For product ion scans using the hybrid configuration, Q1 selects the precursor ion in transmission mode, product ions are generated *via* energetic collisions with neutrals in q2 collision cell, and the product ions and residual precursor ions are transmitted into the linear ion trap [8].

In addition to the advantage of using all of the trap capacity for the species of interest, there are several other advantages of this approach. The use of the RF/DC quadrupole, Q1, for precursor ion isolation eliminates the requirement for a cooling step prior to ion isolation, since precursor ion selection occurs *via* transmission toward the linear ion trap [8]. The use of Q1 for precursor ion selection also eliminates the issue of inadvertent ion activation during isolation that has been shown to occur with in-trap isolation of fragile precursor ions [8].

The fact that the fragmentation step is decoupled from the linear ion trap offers the ability to obtain triple quadrupole-like fragmentation patterns and the ability to go below the traditional stand-alone ion trap 'low mass cut-off' [8]. The triple quadrupole-like fragmentation patterns are a direct result of the fact that the fragments are generated from Q1-to-q2 ion acceleration upstream of the Q3 linear ion trap. Multiple collision-induced dissociation processes within the collision cells of triple quadrupoles often give rise to product ions originating from higher energy dissociation processes than is the case for in-trap resonance excitation approaches [14]. Often the presence of these higher energy product ions is helpful for structural elucidation and library searching.

The low mass cut-off which occurs with in-trap product ion generation is a direct result of the fact that, in general, both the lowest m/z product ion and the precursor ion must be stable simultaneously within the ion trap. This requirement usually means conducting the resonance excitation collisional activation step at an RF voltage amplitude high enough to impart sufficient internal energy into the precursor ion for fragmentation, but also low enough to allow trapping of low m/z product ions. Invariably, there comes a point at which there is insufficient energy in the trapping field to generate product ions efficiently, and a low mass cut-off is observed. This condition corresponds generally to an RF voltage amplitude which is about 20–30% of that experienced by the precursor ion.

The Q-q-Q/LIT arrangement effectively eliminates the low mass cut-off, because of the spatial separation of the fragmentation step from the trapping step. Thus, the constraint found in stand-alone ion traps, that the precursor ion and the product ion of lowest m/z be stable simultaneously in the ion trap, is removed. To trap low m/z product ions, the stability parameters of the linear ion trap are adjusted simply to allow trapping of the targeted ions. There can be, however, a duty cycle penalty as can be demonstrated by reference to Figure 11.9, which displays the product ion mass spectra for taurocholic acid obtained with a conventional 3-D ion trap and a Q-q-Q/LIT instrument [8].

The top trace in Figure 11.9 shows the limitations of in-trap fragmentation of deprotonated taurocholic acid at m/z 514. This is a collisionally robust precursor ion which generates primarily low m/z product ions. In this case, there is insufficient coverage of the product ion mass spectrum to provide much structural information. The product ion mass spectrum obtained with the Q-q-Q/LIT instrument however, gives virtually complete coverage of the major product ions, including those at low m/z, as is seen in the lower trace of Figure 11.9. This composite product ion mass spectrum is obtained by using a portion of the total cycle time to trap and to scan out the low m/z fragments and a second portion to trap and to scan out the higher m/z fragments and residual precursor ions. The reason for the two segments is that under the conditions in which the linear ion trap is accepting the lowest m/z product ions, the higher m/z ions are not trapped efficiently because of the relatively low q-value of these higher m/z ions. This situation is discussed below.

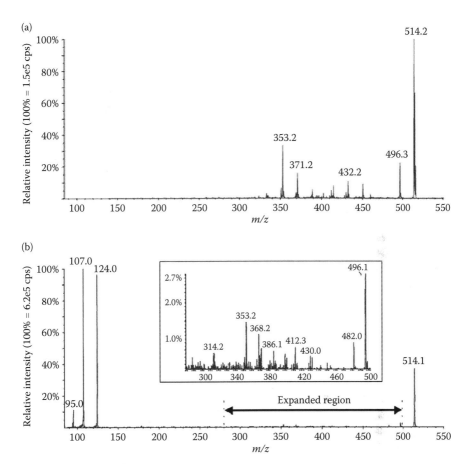

FIGURE 11.9 A comparison of the product ion mass spectra of taurocholic acid obtained with (a) a conventional three-dimensional ion trap and (b) a Q-q-Q/LIT operated in hybrid linear ion trap mode. The lower mass spectrum was recorded in a two-step manner as described in the text. (Reproduced from Hager, J.W. and LeBlanc, J.C.Y., *Rapid Commun. Mass Spectrom.* 2003, *17*, 1056–1064. With permission from John Wiley & Sons Limited.)

The duty cycle penalty with such an approach is an extra trap fill time (typically <50 ms), extra cool time (typically <50 ms), and extra overhead time (typically <10 ms). There is no additional scan time because the *m/z* range of the efficiently trapped ions only is scanned in each segment. The benefit of this operational mode is the ability to observe very low *m/z* product ions with high efficiency.

The fact that the trapping efficiency can be poor at very low *q*-values is due to the relatively low pseudopotential radial well-depths at these *q*-values. The pseudopotential is approximated by the expression [1]

$$\bar{D}_{x,y} \cong \frac{qV_{RF}}{4} \tag{11.6}.$$

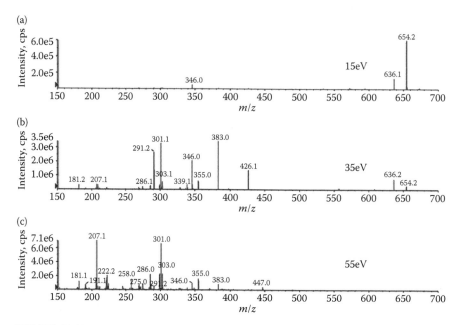

FIGURE 11.10 Product ion mass spectra for the electrospray-generated protonated bromocriptine ion at m/z 654 obtained using a Q-q-Q/LIT instrument in hybrid linear ion trap mode with Q1-to-q2 collision energy of (a) 15 eV, (b) 35 eV, and (c) 55 eV.

Even though the ions are admitted into the linear ion trap axially, collisions can lead to interconversion of the axial and radial energies of the ions. Trapping efficiency decreases when the radial pseudopotential well-depth becomes less than the axial ion kinetic energy. Since the point at which the trapping efficiency drops is predictable, the QTRAP instrument software makes the decision, when to segment the scan so that the resulting spectra reflect accurately the true relative ion abundances.

The spatial separation of the fragmentation and ion trapping steps inherent in the Q-q-Q/LIT design also means that the linear ion trap can be filled multiple times prior to the mass scan step. An example of this capability is filling the linear ion trap with products from different a variety of Q1-to-q2 collision energies [15]. As an example, consider the product ion mass spectra of protonated bromocriptine at m/z 654 using three different collision energies shown in Figure 11.10. The lowest collision energy mass spectrum shows insufficient product ion generation to be particularly useful, while the highest collision energy displayed shows predominantly low mass fragments. The 35 eV product ion mass spectrum is the most useful, but the most useful collision energy can be hard to predict *a priori*.

Figure 11.11 shows the product ion mass spectrum that results when the Q3 linear ion trap is filled sequentially with the product ions generated at the same three collision energies prior to the mass scan. The fill time for the linear ion trap has been split into three distinct periods. The first fill period corresponds to filling with the products of 15 eV Q1-to-q2 collisions. The next period fills the linear ion trap with the products of 35 eV Q1-to-q2 collisions. The final fill period introduces products

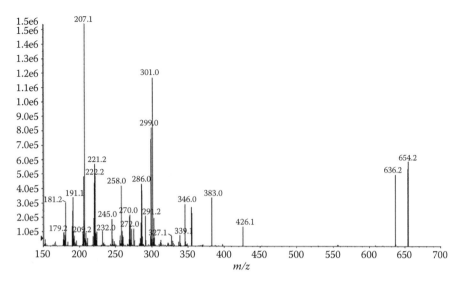

FIGURE 11.11 The product ion mass spectrum for the electrospray-generated protonated bromocriptine ion at m/z 654 obtained using a Q-q-Q/LIT instrument in hybrid linear ion trap mode with the linear ion trap fill, filled for 1/3 of the total fill time at each of the Q1-to-q2 collision energies of 15 eV, 35 eV, and 55 eV. Details in text.

from 55 eV Q1-to-q2 collisions. Next, all of the ions in the linear ion trap are cooled. Finally, the contents are scanned out yielding a composite product ion mass spectrum obtained in a single cycle. There is very high information content in the spectrum in Figure 11.11 since the three different collision energies were chosen to span the range of low-energy, mid-energy, and higher-energy fragmentation processes. This approach helps to reduce problems associated with inappropriate choices of collision energy during the analysis of unknowns as has been demonstrated for the screening of unknown reactive metabolites [16].

11.5 IN-TRAP ION PROCESSING

11.5.1 IN-TRAP FRAGMENTATION IN A LOW PRESSURE LIT

One of the simplest forms of in-trap ion processing is resonance excitation leading to fragmentation, which often provides critical information for structural elucidation, such as when performing MS^n where $n > 2$. Very efficient in-trap fragmentation within the Q3 LIT of a QTRAP instrument has been demonstrated [17–19] despite the fact that the pressure within the Q3 LIT is about two orders of magnitude lower than that of a conventional RF ion trap.

Dipolar resonance excitation of a trapped ion in a perfect quadrupole field leads to monotonic growth of the radial coordinate and eventual loss of the excited ion onto an electrode [17]. One would not expect the gas density at 0.03 mTorr to be sufficient to affect ion loss significantly because of the large mean free path at such a reduced pressure. However, the rod arrays used in the Q3 linear ion trap do not produce a purely quadrupole field because they do not have hyperbolic cross-sections. An accurate description of their round-rod geometry must include contributions from higher-order multipole fields [17,18]. These higher-order fields lead to changes in ion secular frequency as a function of radial displacement, as well as coupling between the x- and y-coordinates. Thus, an ion initially on-resonance will go out-of-resonance with the driving field as the degree of radial excitation is increased, lose radial amplitude to the point where it is back on-resonance with the exciting field at which point it will regain radial amplitude, etc. This oscillating behavior leads to a picture in which the radial amplitude of the resonantly excited ion oscillates in time enabling many more collisions than might be expected [17].

As implemented in the QTRAP instrument, in-trap fragmentation can be used either to generate MS^2 product ion mass spectra or for the second fragment generation step in an MS^3 approach. In practice, MS^3 spectra are generated in the following hybrid manner. The first-generation precursor ion is again selected *via* the Q1 resolving quadrupole, fragmented by acceleration into the pressurized collision cell, and the product and residual precursor ions are trapped in the Q3 linear ion trap. After an appropriate cooling period, the second-generation precursor ion is isolated at the tip of the (a, q) stability diagram by making use of the resolving DC connected to Q3; MS^3 fragments then are generated *via* dipolar resonance excitation at a q-value of about 0.24. Since the final fragmentation step is conducted using resonance excitation within the linear ion trap, there is a low mass cut-off corresponding to about 28% of the m/z of the precursor ion. As with conventional quadrupole ion traps, the use of higher resonance excitation q-values results in greater energy deposition with a concomitant increase in the low mass cut-off [19].

11.5.2 Multiply-Charged Ion Separation

A completely different type of in-trap ion processing is used to discriminate ions with higher charge states from those with lower charge states. Electrospray ionization has a propensity for creation of ions with a variety of charge states. In many proteomic analyses, it is the multiply-charged peptide and protein ions that are of interest. However, these multiply-charged ions can be obscured by singly-charged chemical background species, often completely. An in-trap ion manipulation technique has been developed in order to enhance the ratio of multiply- to singly-charged species within a linear ion trap [13,20,21]. Computer modelling has shown that there are subtle differences between the behavior of trapped singly- and multiply-charged ions at the same m/z-value in both the radial and axial dimensions [22]. The radial dependencies are somewhat similar to those proposed by Tolmachev *et al.* [23] for radial stratification of trapped ion populations in multipole ion guides, although they did not consider explicitly ions of similar m/z but different charge states.

The computer simulations show [22] that the radial excursions of singly-charged ions are greater than those of ions of higher charge, but of similar m/z. When the trapped ion cloud expands significantly, the singly-charged ions can be lost preferentially on the rods. Ions with greater radial extent are also more prone to coupling of their radial and axial degrees of freedom as discussed in Section 11.3. Larger radial excursions together with radial amplitude-dependent axial forces directed out of the linear ion trap will tend, therefore, to deplete preferentially the trapped ion population of lower charge-state species. In addition, simulations have shown [22] that discrimination between isobaric singly- and multiply-charged trapped ions can occur when all of the species are cooled to the similar kinetic energies. Under these conditions, singly-charged ions are observed to approach the end lenses more closely than the multiply charged ions and are thus more likely to be lost once the exit barrier is lowered beyond some threshold.

In practical terms, therefore, a population of ions from the ion source is admitted into the linear ion trap and cooled to the point where the constituent ions have similar kinetic energies, after which the voltage applied to one of the axial DC potential barriers is reduced to a pre-determined level. An example of the degree to which this technique can discriminate against low charge states is presented in Figure 11.12. The sample is a mixture of polypropylene glycols and contains a mixture of singly-, doubly, and triply-charged ions. The top trace shows a standard survey scan using the linear ion trap with no separation of charge state. The bottom trace shows the same mixture except after the charge state separation step has been implemented. The

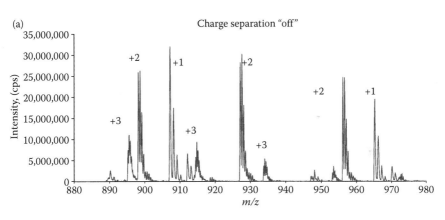

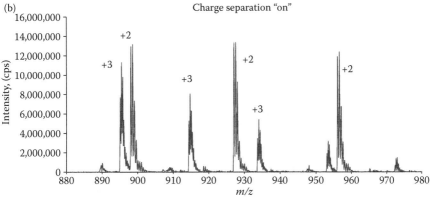

FIGURE 11.12 The Q-q-Q/LIT single MS spectra of a sample consisting of a mixture of polypropylene glycols with (a) no charge-separation step and (b) with a charge-separation step included in the scan function.

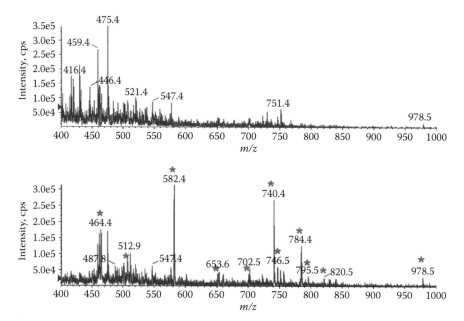

FIGURE 11.13 The Q-q-Q/LIT single MS spectra of a sample consisting of 20 fmol beta-case in digested with trypsin. The top trace shows the mass spectrum with no charge-separation step. The bottom trace illustrates the enhancement of multiply charged peptide ions with a charge-separation step included in the scan function. All peaks marked with asterisks are due to the multiply charged tryptic peptide ions.

relatively high signal intensities of the singly-charged ion species have been reduced by a factor of more than 100. The intensities of the doubly-charged ion species have been reduced by about a factor of two, while the triply-charged ion signals are virtually unchanged.

The utility of the separation technique for multiply-charged ions is shown in Figure 11.13 for a beta-casein trypsin digest. The top trace is the standard linear ion trap survey scan which displays a very complicated mass spectrum dominated by chemical noise due to singly charged ions. The bottom spectrum shows the degree of simplification that can be achieved using the separation technique for multiply charged ions. All of the peaks marked with an asterisk are multiply-charged tryptic peptide ions that would have been exceedingly difficult to identify in the standard survey scan.

11.5.3 Time-Delayed Fragmentation

In order to observe product ions with a triple quadrupole mass spectrometer, the ions must be formed during the approximately several hundred microseconds they are within the collision cell [24]. This time constraint limits the observable dissociation rates for product ion formation to $> ca$ 10^4 s^{-1}. Product ions that are formed with lower rates are typically not observed. Dissociation rates can be enhanced by increasing the internal energy of the precursor ion *via* the use of more energetic collision conditions, but at the cost of generating product ions with considerable internal energy. These excited product ions can dissociate also leading to the appearance of second- and third-generation product ions, thus adding a degree of complexity in structural elucidation studies such that MS^n experiments may be required.

The low pressure environment of the Q3 linear ion trap and the long observation time window make it possible to use time-resolution to distinguish between product ions generated from highly excited precursor ions and those generated from lower energy precursor ions [25].

Consider the internal energy distribution of a population of precursor ions shortly after collisional activation similar to that shown in Figure 11.14. Precursor ions with internal energies in the high-energy tail have sufficient energy to fragment most promptly. Those with less internal energy fragment on slower time scales, and so on. After a short period of waiting, the internal energy distribution will have changed due to the fact that some precursor ions will have fragmented and others will have been collisionally cooled. One can imagine a progression like that shown in Figure 11.14. Finally, a point will be reached in which the only remaining activated precursor ions are those that possess barely sufficient critical energy for dissociation.

Time-delayed fragmentation (TDF) makes use of the fact that the internal energy distribution of an activated trapped ion population evolves toward lower energies with time [25]. Thus it is possible to select different 'time slices' of precursor ions with different internal energies, and often different product ions. The process involves the acceleration of a short pulse of mass-selected ions from q2, which now operates as a linear storage ion trap, into the Q3 linear ion trap. The accelerated ions are activated collisionally and fragmentation occurs. Certain portions of the product ion mass spectrum are time-delayed by not allowing the capture of a specified *m/z* range through appropriate choices of the stability coordinates of the Q3 linear ion trap for the specified time-delay. After the delay time, the stability characteristics of the Q3 linear ion trap are changed to allow capture of the entire *m/z* product ion range. The result is often a simpler product ion mass spectrum with a much-reduced contribution of secondary (and greater) fragmentation products [25].

Figure 11.15 displays an example of the utility of the time-delayed fragmentation approach for the simplification of the low *m/z* portion of the product ion scan of a doubly-protonated tryptic peptide. The top mass spectrum in Figure 11.15 shows the low *m/z* portion of the hybrid linear ion trap product ion scan obtained using the collision cell for fragmentation. This mass spectrum shows a mixture of y- and b-sequence ions as well as internal fragment ions. The complexity in this mass spectral region is evident. The bottom trace is the time-delayed product ion mass spectrum with a delay time of 10 ms between precursor ion activation and product ion capture. This mass spectrum is simplified considerably compared to that in the upper trace. Virtually all of the peaks are assignable to y-sequence ions, which is typical of the behavior of time-delayed fragmentation of doubly-protonated tryptic peptides.

Figure 11.16 shows the high *m/z* portion of the triple quadrupole-like and time-delayed fragmentation product ion mass spectrum for the same peptide. In contrast to the behavior at low *m/z*, a time delay of 10 ms results in the observation of primarily b-type sequence ions.

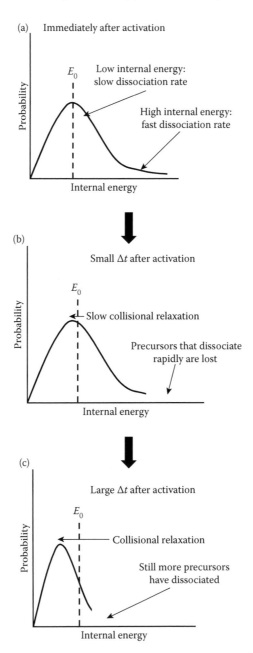

FIGURE 11.14 A schematic representation of the time evolution of an internal energy distribution of a collisionally activated precursor ion. E_0 is the critical energy threshold for generation of fragmentation products. (Reproduced from Hager, J.W., *Rapid Commun. Mass Spectrom.* 2003, *17*, 1389–1398. With permission from John Wiley & Sons Limited.)

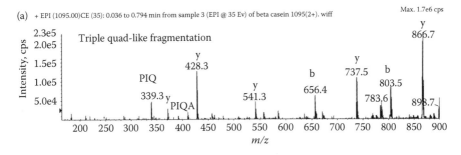

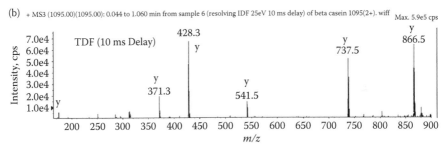

FIGURE 11.15   A comparison of two modes of product ion generation using the Q-q-Q/
LIT instrument in hybrid linear ion trap. The low *m/z* portion of the product ion mass spectra
for the beta casein doubly-protonated tryptic peptide DMPIQAFLLYQEPVLGPVR at *m/z*
1095 are presented. (a) The conventional triple quadrupole-like (Q1-to-q2) product ion mass
spectrum, and (b) the time-delayed product ion mass spectrum obtained with a delay time
of 10 ms.

Taken together, the data from these mass spectra suggest that the formation of
high *m/z* b-type and a low *m/z* y-type complementary ion pairs have the lowest critical energies for this particular doubly-protonated tryptic peptide. Larger degrees
of internal excitation imparted using axial acceleration through the collision cell
produce triple quadrupole-like fragmentation with a preponderance of high *m/z*
y-sequence ions and a mixture of b-, y-, and internal fragmentation products. The
time-delayed fragmentation approach allows the identification of low energy fragmentation pathways. Using a short or zero length delay time will shift the observation window to products from higher energy fragmentation pathways. The ability
to distinguish the two limiting fragmentation behaviors can be helpful in *de novo*
sequencing.

11.6   CHARGE DENSITY CONTROL

Space charge is the presence of excess ion density within an ion trap and, as
such, modifies the interaction of trapped ions with externally applied fields. This
modification can lead to degradation of the ejected ion mass spectra. From a
purely geometrical perspective, linear ion traps should be able to contain more
ions before the space-charge ion density limit is reached than does a conventional
ion trap. Even though linear ion traps can contain more ions than conventional

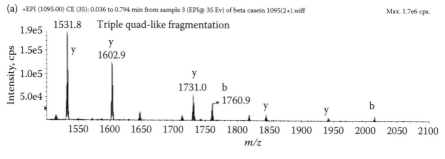

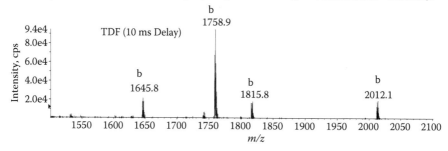

FIGURE 11.16 A comparison of two modes of product ion generation using the Q-q-Q/LIT instrument in a hybrid linear ion trap. The high *m/z* portions of the product ion mass spectra for the beta casein doubly protonated tryptic peptide DMPIQAFLLYQEPVLGPVR at *m/z* 1095 are presented. (a) The conventional triple quadrupole-like (Q1-to-q2) production mass spectrum, and (b) the time-delayed product ion mass spectrum obtained with a delay time of 10 ms.

ion traps, space-charge conditions can occur. There are also second level complications as Figure 11.4 illustrates. The method of ion excitation can also be an important factor determining whether the effects of space charge are observed, as described earlier.

The general approach for minimizing the effects of space charge on trapped ion mass spectra is to control the number of ions trapped. One of the first methods that was developed is to measure the number of trapped ions using a very fast 'pre-scan' and then to determine what the appropriate ion trap fill time would be for the analytical scan [26]. An alternative approach used in QTRAP instruments is to measure the ion flux in a transmission-only triple quadrupole experiment [27]. This way either MRM or quadrupole scanning can be used to estimate the number of ions that will be trapped in a specified period of time. The user selects an upper limit of either the total number of ions to be trapped or the linear ion trap fill-time. Both the fast pre-scan and the measurement of transmitted ion current are effective techniques for space-charge minimization. However, both have the drawback of increasing the effective cycle time by the time required to conduct the space-charge mitigation steps. Often this duty cycle penalty is worth being able to measure a space charge-free mass spectrum.

11.7   APPLICATIONS

The strength of the Q-q-Q/LIT instruments is the combination of fully functional triple quadrupole operational modes and a hybrid linear ion trap instrument within the same platform. Conversion from triple quadrupole operation to hybrid linear ion trap operation requires <1 ms, because only fast rise-time electronics are involved. This versatility provides the opportunity of using one or more of the triple quadrupole operational modes as the first step of an information-dependent experiment.

Triple quadrupole mass spectrometers are inherently insensitive while scanning, since the path stability requirements mean that a single m/z ion, or m/z ion pair, is stable throughout the length of the ion path at any one time. These requirements mean that most of the ions from the ion source are not used, or are wasted. Triple quadrupoles, do however, have two very selective scan modes that can be very useful when dealing with complex sample matrices: precursor ion and constant neutral loss scans. The selectivity of these scan modes makes them useful for investigations of bio-transformations in metabolism and proteomics [13,28]. However, their use requires some knowledge of the sample in order to select correctly the fragment ion or neutral loss of interest.

When interested in targeted precursor/product ion pairs (MRM) and not an entire MS/MS spectrum, triple quadrupoles can be very sensitive. With targeted analysis, it is desirable to discard the majority of the ions formed within the ion source in order to enhance the instrument duty cycle. This method significantly improves sensitivity, but at the expense of not observing unexpected analytes. The MRM operational mode is purely for targeted analysis. Many MRM transitions can be postulated for bio-transformations, but in order to be detected, the transition must be sought explicitly [28–30].

Figure 11.17 shows how the various Q-q-Q/LIT operational modes can be used in series for analyte detection, identification, and confirmation in an information-dependent manner. The initial scan mode, or survey scan, is obtained to determine whether there are potential candidates of interest for possible further investigation. The data from the survey scan are examined to determine whether additional dependent scans are warranted using predetermined selection criteria. Such selection criteria include peak intensity, charge state, and whether or not the detected ion signal falls within a pre-selected m/z-range. One of the strengths of the QTRAP approach is that the survey scans can be taken from one of several triple quadrupole operational modes, such as precursor ion or constant neutral loss scans or even MRM mode, or from the different linear ion trap scan modes described above.

The most general scan mode combination of survey and dependent scans involves using the linear ion trap single MS scan to trigger a hybrid product ion scan. This approach is very similar to that used with conventional ion trap instruments in which the survey scan provides high sensitivity, but also limited selectivity [28].

Selectivity can be increased by using one of the triple quadrupole scan modes discussed above as one of the survey scans. Phosphorylated peptides can be identified selectively using the negative polarity precursor ion scan detecting the (m/z –79)

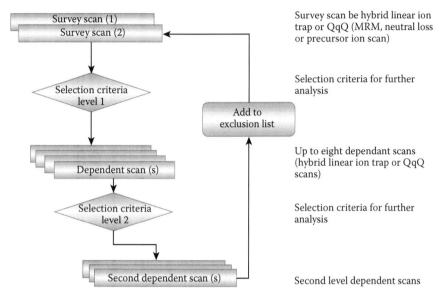

Survey scan (1) / Survey scan (2)	Survey scan be hybrid linear ion trap or QqQ (MRM, neutral loss or precursor ion scan)
Selection criteria level 1	Selection criteria for further analysis
Add to exclusion list	
Dependent scan (s)	Up to eight dependant scans (hybrid linear ion trap or QqQ scans)
Selection criteria level 2	Selection criteria for further analysis
Second dependent scan (s)	Second level dependent scans

FIGURE 11.17 An example of the way in which the various triple quadrupole and hybrid linear ion trap scan modes of the Q-q-Q/LIT instrument can be interleaved to obtain maximum information. The QTRAP instrument can be converted between triple quadrupole mode and the linear ion trap operational modes in <1 ms, allowing great flexibility for the nature of the survey and dependent scans. Very high sensitivity can be achieved when the linear ion trap MS scan is used as a survey scan. Often the most intense peaks in the survey scan are then chosen to undergo fragmentation in a product ion scan in Q-q-LIT mode. The triple quadrupole precursor ion and constant neutral loss scans are very useful survey scans when searching for biotransformation products. These scans are often followed by Q-q-LIT product ion and/or MS³ scans for detailed structural analysis. Although not a true 'scan' mode, triple quadrupole MRM operation is particularly powerful as the initial 'scan' in a targeted work flow. A response above a particular MRM threshold count rate would then trigger the product ion scan in Q-q-LIT mode for confirmation.

product ion [13,32]. Peptides containing a phosphotyrosine can be identified using a characteristic fragment ion at $(m/z + 216)$ [32]. Glutathione-trapped reactive metabolites can be detected using the precursor ion scan of the anion at m/z 272 [33]. When the instrument detects a specific bio-transformation product, a confirmatory hybrid product ion scan would be triggered.

The MRM mode of operation is very powerful, primarily due to the associated high sensitivity and fast measurement time (*ca* 10 ms). Then, detection of the analyte by MRM drives the acquisition of the hybrid linear ion trap product ion scan to identity definitively the analyte. This very sensitive and powerful approach has been exploited for determination of predicted metabolites [28,29], contaminants [30], as well as peptides and proteins [34,35]. This targeted, or hypothesis-driven [34], strategy makes use of information already known about the sample in order to maximize throughput and sensitivity.

11.8   CONCLUSIONS

The advent of linear ion trap mass spectrometers has led to resurgence in the use of RF ion traps in general. A major part of the interest in linear ion traps is the relative ease with which they can be incorporated into hybrid instruments with associated additional functionality. Mass-selective axial ejection as described here has allowed incorporation of the linear trap in a triple quadrupole ion path. The result is an instrument with the same triple quadrupole characteristics and performance levels as well as a hybrid quadrupole linear ion trap instrument with unique capabilities. The novelty of the Q-q-Q/LIT design lies with the ability to incorporate significant amounts of ion processing prior to injecting the ions into the linear ion trap. Spatial separation of each step allows a high degree of independent optimization not possible with a true tandem-in-time approach.

ACKNOWLEDGMENTS

I would like to thank my colleagues Yves LeBlanc, Bruce Collings, John Vandermey, and Frank Londry for their very important contributions in the development of the technology used in the QTRAP instruments.

REFERENCES

1. Douglas, D.J.; Frank, A.J.; Mao, D. Linear ion traps in mass spectrometry. *Mass Spectrom. Rev.* 2005, *24*, 1–29.
2. Schwartz, J.C.; Senko, M.W.; Syka, J.E.P. A two-dimensional quadrupole ion trap. *J. Am. Soc. Mass Spectrom.* 2002, *13*, 659–669.
3. Brinkmann, U. A modified quadrupole mass filter for separation of ions of higher masses with high transmission. *Int. J. Mass Spectrom. Ion Phys.* 1972, *9*, 161–166.
4. Hager, J.W. Performance optimization of fringing field modifications of a 24-mm long RF-only quadrupole mass spectrometer. *Rapid Commun. Mass Spectrom.* 1999, *13*, 740–748.
5. Londry, F.A.; Hager, J.W. Mass-selective axial ion ejection from a linear quadrupole ion trap. *J. Am. Soc. Mass Spectrom.* 2003, *14*, 1130–1147.
6. Hager, J. Mass spectrometry using a linear RF quadrupole ion trap with axial ion ejection. *Proc. 46th ASMS Conference on Mass Spectrometry and Allied Topics,* Orlando, FL, May 31–June 4, 1998, p. 490.
7. Hager, J.W. A new linear ion trap mass spectrometer. *Rapid Commun. Mass Spectrom.* 2002, *16*, 512–526.
8. Hager, J.W.; LeBlanc, J.C.Y. Product ion scanning using a Q-q-Q$_{linear ion trap}$ (QTRAP™) mass spectrometer. *Rapid Commun. Mass Spectrom.* 2003, *17*, 1056–1064.
9. Dawson, P.H. *Quadrupole Mass Spectrometry and its Applications.* Chapter V. Woodbury, NY: AIP Press, 1995.
10. Goeringer, D.E.; Whitten, W.B.; Ramsay, J.M.; McLuckey, S.A.; Glish, G.L. Theory of high-resolution mass spectrometry achieved via resonance ejection in the quadrupole ion trap. *Anal. Chem.* 1992, *64*, 1434–1439.
11. Londry, F.A. unpublished work.
12. Hager, J.W.; Londry, F.A. System and method for modifying the fringing fields of a radio frequency multipole. *U.S. Patent* 2006, 7,019,290.

13. LeBlanc, J.C.Y.; Hager, J.W.; Illisiu, A.M.P.; Hunter, C.; Zhong, F.; Chu, I. Unique scanning capabilities of a new hybrid ion trap mass spectrometer (QTRAP®) used for high sensitivity proteomics applications. *Proteomics* 2003, *3*, 859–869.

14. Busch, K.L.; Glish, G.L.; McLuckey, S.A. *Mass Spectrometry/Mass Spectrometry: Techniques and Applications of Tandem Mass Spectrometry.* New York: VCH, 1988.

15. Bloomfield, N.; LeBlanc, J.C.Y. Broad ion fragmentation coverage in mass spectrometry by varying the collision energy. *U.S. Patent* 2007, 7,199,361.

16. Wen, B.; Li, M.; Nelson, S.D.; Zhu, M. High-throughput screening and characterization of reactive metabolites using polarity switching of hybrid triple quadrupole linear ion trap mass spectrometry. *Anal. Chem.* 2008, *80*, 1788–1799.

17. Collings, B.A.; Stott, W.R.; Londry, F.A. Resonant excitation in a low-pressure linear ion trap. *J. Am. Soc. Mass Spectrom.* 2003, *14*, 622–634.

18. Douglas, D.J.; Glebova, T.A; Konenkov, N.V.; Sudakov, M.Y. Spatial harmonics of the field in a quadrupole mass filter with circular electrodes. *Tech. Phys.* 1999, *44*, 1215–1219.

19. Collings, B.A. Fragmentation of ions in a low-pressure linear ion trap. *J. Am. Soc. Mass Spectrom.* 2007, *18*, 1459–1466.

20. Hager, J.W. Method of mass spectrometry, to enhance separation of ions with different charges. *U.S. Patent* 2006, 7,041,967.

21. Chernushevich, I.V.; Fell, L.M.; Bloomfield, N.; Metalnikov, P.S.; Loboda, A.V. Charge state separation for protein applications using a quadrupole time-of-flight mass spectrometer. *Rapid Commun. Mass Spectrom.* 2003, *17*, 1416–1424.

22. Londry, F.A.; Hager, J.W. unpublished work.

23. Tolmachev, A.V.; Udseth, H.R.; Smith, R.D. Radial stratification of ions as a function of mass to charge ratio in collisional cooling radio frequency multipoles used as ion guides or ion traps. *Rapid Commun. Mass Spectrom.* 2000, *14*, 1907–1913.

24. Mauk, M.R.; Mauk, A.G.; Chen, Y.L.; Douglas, D.J. Tandem mass spectrometry of protein-protein complexes: cytochrome c-cytochrom-b5. *J. Am. Soc. Mass Spectrom.* 2002, *13*, 59–71.

25. Hager, J.W. Product ion spectral simplification using time-delayed fragment ion capture with tandem linear ion traps. *Rapid Commun. Mass Spectrom.* 2003, *17*, 1389–1398.

26. Weber-Grabau, M.; Bradshaw, S.C.; Syka, J.E.P. Method of increasing the dynamic range and sensitivity of a quadrupole ion trap mass spectrometer operating in the chemical ionization mode. *U.S. Patent* 1988, 4,771,172.

27. Hager, J.W. Method for reducing space charge in a linear ion trap mass spectrometer. *U.S. Patent* 2003, 6,627,876.

28. Hopfgertner, G.; Varesio, E.; Tschappat, V.; Grivet, C.; Bourgogne, E.; Leuthold, L.A. Triple quadrupole linear ion trap mass spectrometer for the analysis of small molecules and macromolecules. *Rapid Commun. Mass Spectrom.* 2004, *39*, 845–855.

29. Gao, H.; Materne, O.L.; Howe, D.L.; Brummel, C.L. Method for rapid metabolite profiling of drug candidates in fresh hepatocytes using liquid chromatography coupled with a hybrid quadrupole linear ion trap. *Rapid Commun. Mass Spectrom.* 2007, *21*, 3683–3693.

30. Bueno, M.J.M.; Aguera, A.; Gomez, M.J.; Hernando, M.D.; Garcia-Reyes, J.F.; Fernandez-Alba, A.R. Application of liquid chromatography/quadrupole-linear ion trap mass spectrometry and time-of-flight mass spectrometry to the determination of pharmaceuticals and related contaminants in wastewater. *Anal. Chem.* 2007, *79*, 9372–9384.

31. Li, A.C.; Gohdes, M.; Shou, W.Z. "N-in-one" strategy for metabolite identification using a liquid chromatograph/hybrid triple quadrupole linear ion trap instrument using multiple dependent product ion scans triggered with full mass spectra. *Rapid Commun. Mass Spectrom.* 2007, *21*, 1421–1430.

32. Witze, E.S.; Old, W.M.; Resing, K.A.; Ahn, N.G. Mapping protein post-translational modifications with mass spectrometry. *Nature Methods* 4, *10*, 798–806.

33. Wen, B.; Ma, L.; Nelson, S.D.; Zhu, M. High-throughput screening and characterization of reactive metabolites using polarity switching of hybrid triple quadrupole linear ion trap mass spectrometry. *Anal. Chem.* 2008, *80*, 1788–1799.

34. Domon, B.; Aebersold, R. Mass spectrometry and protein analysis. *Science* 2006, *312*, 212–217.

35. Anderson, L.; Hunter, C.L. Quantitative mass spectrometric MRM assays for major plasma proteins. *Mol. Cell. Proteomics* 2006, *5*, 573–588.

12 Axially Resonant Excitation Linear Ion Trap (AREX LIT)

Yuichiro Hashimoto

CONTENTS

12.1 Introduction 573
12.2 Geometry and Mass-Selective Characteristics of the AREX LIT 575
12.3 Hybrid Mass Spectrometers combined with the AREX LIT 579
    12.3.1 Duty Cycle Enhancements of the Orthogonal-Acceleration
        Mass Spectrometer 580
    12.3.2 Two-Dimensional Mass Spectrometry 582
12.4 Ion Fragmentation in the AREX LIT 583
    12.4.1 Collision-Induced Dissociation 584
    12.4.2 Electron-Capture Dissociation 587
12.5 Conclusions 588
References 589

12.1 INTRODUCTION

The advantages of a linear ion trap (LIT) compared to the conventional three-dimensional quadrupole ion trap (3D QIT) have been discussed in several reports [1–3], in which higher ion trap capacity and higher trapping efficiency of the LIT compared to those of 3D QIT are described. An LIT consists of multipole rod electrodes and end-cap electrodes, wherein radial resonant excitation in the quadrupole potential enables the mass-selective isolation, activation, and ejection of ions [2,3]. In particular, mass-selective ejection techniques are important practically because they enhance instrument sensitivities by over 10 times that of the conventional 3D QIT [3]. Linear ion traps (LITs) have been incorporated in tandem not only with stand-alone ion traps, but also in hybrid instruments with analyzers such as time-of-flight mass spectrometers (TOF-MS) [4,5], orbitrap mass spectrometers [6], and Fourier transform ion cyclotron resonance mass spectrometers (FT-ICR-MS) [7], and these have had a huge impact on mass spectrometry.*

---

* See also Volume 4, Chapter 11: Linear Ion Trap Mass Spectrometry with Mass-Selective Axial Ejection, by James W. Hager; Volume 4, Chapter 14: Electrically-induced Nonlinear Ion Traps, by Gregory J. Wells and August A. Specht.

There is no axial potential in the middle of the rod assembly in the original types of LIT, however, multipole rod electrodes with an axial potential were developed by the following groups. Gerlich developed an LIT with an additional cylindrical electrode around the rods by which an axial DC potential was formed [8]. This arrangement enabled the confinement of ions in a limited region inside the array of multipole rod electrodes. Thomson and coworkers have developed many variations of multipole rod electrodes with a unidirectional potential along the rod axis for a collision cell. These comprise the use of segmented rod sets [9], conical rods [10], and rods with additional electrodes (T-rods) [11], which are now called the LINAC™.*

In an ion guide with a unidirectional potential, ions pass through much faster than in the absence of an axial field. Since the ions are ejected from the collision cell within a few milliseconds, high-speed switching between different forms of MS/MS analysis (for example, multiple reaction monitoring (MRM) and selected-ion monitoring (SIM)) is possible. Recently another type of collision cell, which has a unidirectional potential, was also reported [12], where the axial potential was formed from the resistive coating on the surface of the rod electrodes.

In the previous studies noted above, the axial potentials influence the motion of the ions mass-independently; however, our group has developed an approach that enables mass-selective operation using an axial potential. When the axial potential in the middle of a set of multipole rod electrodes is approximated to a harmonic potential, it will enable mass-selective operations such as ion isolation, activation, and ejection. For this purpose, we have developed an 'axially resonant excitation' LIT, which we have termed the "AREX LIT" [13]. The AREX LIT has an axial harmonic potential that is produced by applying a DC voltage to vane-shaped electrodes inserted between the rod electrodes. This AREX LIT shows several characteristic properties arising from its capability for axial resonant excitation, which are beneficial for practical use. Recently, other configurations for an axial harmonic potential were also reported [14,15]. Dobson and Enke created an axial harmonic DC potential in the middle of the array of rods by applying DC voltages to both the entry and exit end-cap electrodes with an unique shape [14]. In this report, the trap operated mass-independently, where it was used for focusing ions in the axial direction. Since the trap confines the ions axially, they are introduced effectively with a narrow temporal distribution into a TOF-MS. Green et al. have developed an arrangement comprising many segmented sets of multipole rod electrodes to which different DC offset potentials are applied [15]. They demonstrated mass-selective ejection with dipolar excitation and parametric excitation in the axial harmonic potential. This chapter describes the AREX LIT, which has been developed in our laboratories as a mass-selective and ion fragmentation device, and discusses its applications.

---

* See also Volume 4, Chapter 10: Trapping and Processing Ions in Radio Frequency Ion Guides, by Bruce A. Thomson, Igor V. Chernushevich and Alexandre V. Loboda.

12.2   GEOMETRY AND MASS-SELECTIVE CHARACTERISTICS OF THE AREX LIT

Our AREX LIT, depicted in Figure 12.1, has 'incap' and 'endcap' lenses, four quadrupole rod electrodes, and four pairs of vane electrodes that are inserted between the rods. The rod length is 44 mm, the rod radius (R) is 9.17 mm, and the distance between the rod and the center axis (r_0) is 8 mm. Two phases of the RF trapping voltage (amplitude 200–600 V (zero-to-peak), frequency typically 770 kHz) are applied to the rod pairs, which confine the ions in the radial direction. The vane electrodes consist of a set of four 'front' vanes and another set of four 'rear' vanes, each with a thickness of 0.5 mm, and of shapes that were optimized by electric potential calculations [13]. The four vane electrodes comprising each set are at the same electric potential. The vane potential (D_0), which is an applied DC potential of the vane electrodes relative to the quadrupole-rods' DC offset, generates a potential that is approximately harmonic along the central axis of the quadrupole field. Since the vane electrodes are replaced easily without changing the multipole rod electrodes, various types of vane electrodes can be tested readily by experiment. A trapping efficiency of over 90% for reserpine ions with m/z 609 was obtained at a helium bath gas pressure of 2 mTorr [13], which is similar to that of the other LIT reports [1–3,5].

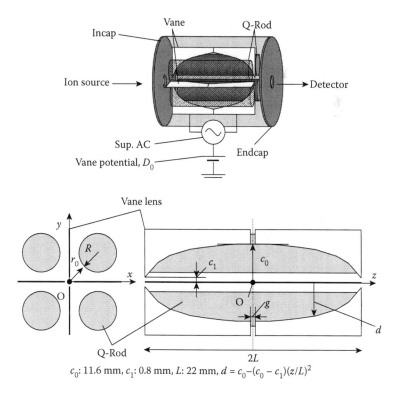

c_0: 11.6 mm, c_1: 0.8 mm, L: 22 mm, $d = c_0 - (c_0 - c_1)(z/L)^2$

FIGURE 12.1   Schematic diagram of the AREX LIT.

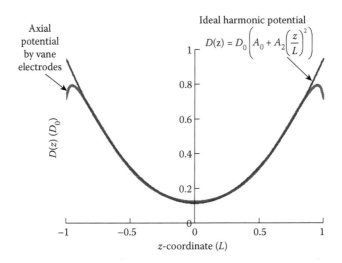

FIGURE 12.2 Axial DC potential along the quadrupole central z-axis of the AREX LIT, $x = y = 0$, $A_0 = 0.114$, $A_2 = 0.833$.

Figure 12.2 shows the shape of the axial DC potential along the center axis produced by the optimized vane electrodes. The axial potential is defined by the expression

$$D(z) = D_0 \left(A_0 + A_2 \left(\frac{z}{L} \right)^2 \right) \tag{12.1},$$

where A_0 and A_2 are free fitting parameters.

When the value of z is between $-0.8\,L$ and $0.8\,L$, the vane potential approximates sufficiently to a harmonic potential. When the value of z is less than $-0.8\,L$ or more than $0.8\,L$, it deviates from the harmonic near the ends of the rod. This deviation can be reduced by moving each vane closer to the ends, which causes ions to hit the vane electrodes during ejection. In order to retain a high ion ejection efficiency, it was determined that the minimum gap between the vane electrodes should be set at 1.6 mm to prevent the ions from hitting the vane electrodes. In the AREX LIT, the axial ion motion is given by

$$m \frac{d^2 z}{dt^2} = -ne \frac{\partial}{\partial z} D(z) = -2neA_2 D_0 \frac{z}{L^2} \tag{12.2},$$

where m is the mass of an ion, n is the charge number, e is the electronic charge, and f is the secular oscillation radial frequency in the axial direction. From Equation 12.2, the axial secular frequency is obtained as

$$f = \frac{1}{2\pi} \sqrt{\frac{2neA_2 D_0}{mL^2}} \tag{12.3}.$$

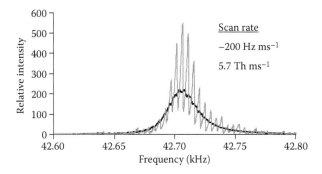

FIGURE 12.3 Frequency spectrum observed for reserpine ions (m/z 609). Bold line: average spectrum of 128 scans; thin line: single scan spectrum.

Since ions with specific m/z-value have specific secular frequency f in the axial direction, as stated in Equation 12.3, only ions that have a limited value of m/z can be excited kinetically by means of resonance oscillation.

To test the mass-selective ejection characteristics of the AREX LIT, a supplemental AC voltage (amplitude <20.0 V (zero-to-peak), frequency 5–100 kHz) was applied between the front and rear vanes. When a frequency matches the frequency of the secular motion corresponding to a given m/z-value, the ions are oscillated resonantly and ejected out of the trap in an axial direction. Scanning the frequency from 91.6 to 29.0 kHz corresponds to the mass-selective scanning from m/z 100 to m/z 1000 at $D_0 = 100$ V in our experimental setup.

Equation 12.3 can be differentiated to give

$$\frac{m}{dm} = -\frac{1}{2}\frac{f}{df} \tag{12.4}.$$

Therefore, the mass resolving power resulting from a frequency sweep is defined by $f_0/2\Delta f$, where f_0 is the center frequency of the peak and Δf is the full-width of the peak at half-maximum. Figure 12.3 shows a frequency sweep mass spectrum of a reserpine peak near m/z 609 at a decreasing frequency scan rate of −200 Hz ms^{-1}, which corresponds to an m/z-value scan rate of 5.7 Th ms^{-1}. The fine line is the spectrum obtained from a single scan, and the bold line is a mass spectral peak averaged over 128 scans. A calculated resonant frequency of 42.33 kHz was obtained from Equation 12.3, which corresponds well with the observed f_0 range of between 42.65 and 42.75 kHz shown in Figure 12.3. Several sub-peaks were observed in the single-scan mass spectrum. The time interval between sub-peaks was 23–24 μs, which corresponds to the period of the supplemental AC potential applied to the sets of vane electrodes. These sub-peaks appeared because ions are ejected during a specific phase range of the supplemental AC field independently of the phase of the main RF drive field. The mass resolving power calculated from the widths of the average peaks was determined for different scan rates, and the resulting graph is shown in Figure 12.4. Although, a lower mass scan rate achieves

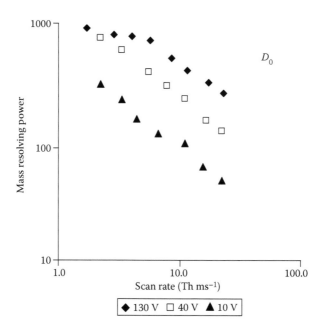

FIGURE 12.4 Plot of mass-resolving power *versus* scan rate of the AREX LIT for the ion *m/z* 609.3. (Reproduced from Hashimoto, Y., Hasegawa, H., Baba, T., and Waki, I., *J. Am. Soc. Mass Spectrom.* 2006, *17*, 685–690. With permission from Elsevier.)

a higher mass resolving power, the mass resolving power saturates below 800. Above a scan rate of 6 Th ms^{-1} with $D_0 = 130$ V, the mass resolution degrades in inverse proportion to the scan rate. Figure 12.4 shows also that the mass resolving power is proportional to the square root of D_0 at higher scan rates. Ejection efficiencies were in excess of 60% at $D_0 = 130$ V and greater than 80% at $D_0 = 20$ V [13,16]. These high ejection efficiencies are possible because the oscillation and ejection directions are the same, whereas the oscillation and ejection directions are perpendicular to one another in the other axial ejection method that uses a fringing field [2]. More than 50% of the ions are ejected forward in the direction of the detector, not backward, because a higher potential was applied to the incap lens than that to the end-cap. It is observed that the loss of ions is caused mainly by the collisions with the end-cap lens; this observation is supported by our simulation results.

Figure 12.5 shows the electrospray mass spectrum of reserpine ions. In addition to the reserpine molecular ion peak at *m/z* 609, fragment ion peaks at *m/z* 397, *m/z* 415, *m/z* 436, and *m/z* 448 were observed, which were produced using skimmer collision-induced dissociation (CID). When the scan period was 50 ms, the mass resolving power of each peak ranged from 350 to 500. These mass resolving power values were not as high as those reported for other radial ion excitation LITs. We suppose that this low resolving power comes from both electronic noise and from imperfections of the axial harmonic potential which deviates from the harmonic near the ends of the rod electrodes, as shown in Figure 12.2.

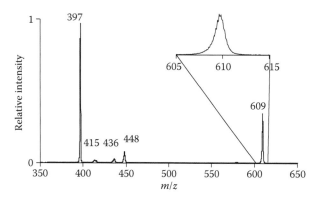

FIGURE 12.5 Electrospray mass spectrum of reserpine derived from a supplemental frequency sweep. Scan range: 57–42 kHz; scan speed: −0.3 kHz ms^{-1} (equivalent to 8.6 Th ms^{-1} at m/z 609.3); scan period: 50 ms; D_0: 130 V; V_{AC}: 15 V. (Reproduced from Hashimoto, Y., Hasegawa, H., Baba, T., and Waki, I., *J. Am. Soc. Mass Spectrom.* 2006, *17*, 685–690. With permission from Elsevier.)

12.3 HYBRID MASS SPECTROMETERS COMBINED WITH THE AREX LIT

It is well known that the sensitivity of a mass spectrometer is reduced because of low transmission and low duty cycle. In this context 'low duty cycle' means, for example, that when a target ion is transmitted selectively for detection by filtering techniques, the other ions are rejected without detection at the same time, thereby causing a reduction in sensitivity. When an ion trap with a mass-selective capability replaces the filtering techniques, sensitivity can be enhanced because the ion trap can eject mass-selectively desirable ions while holding the other ions inside the trap. However, a high transmission process is essential overall in order to obtain the benefit of this enhancement effect. Thus, an ion trap with not only high transmission but also good energy focusing of the ejected ions is required, because an ion distribution that is dispersed energetically reduces the transmission through the subsequent components of the ion beam path, such as a collision cell and a TOF-MS. According to our understanding, an ion trap that achieves simultaneously high transmission and energy focusing has not hitherto been reported because conventional resonant ion excitation causes a broad energy distribution due to the perturbations of the trapping RF field.

Figure 12.6 shows simulation results of the ejection energy distribution of ions from the 3D QIT and from our AREX LIT, which were calculated by SIMION version 7 (Techscience Ltd, Saitama, Japan). The 3D QIT has a broad energy distribution from 0 to 200 eV, which is too dispersed to obtain high transmission in the subsequent components of a hybrid instrument. On the other hand, the AREX LIT ejects mass-selectively ions with a much narrower energy distribution (less than 5 eV) than that of the conventional 3D QIT. We suppose that this difference arises because the mass-selective potential of the AREX LIT is formed by a DC field, whereas the potentials used to eject the ions from other types of ion traps are formed by an RF field. This energy distribution is narrow enough to attain high transmission in the subsequent

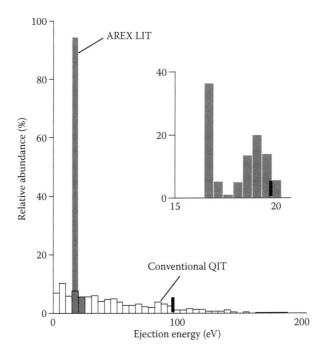

FIGURE 12.6 Simulated energy distribution of ions ejected from the AREX LIT and from a conventional QIT. The inset shows an expanded zone for the AREX LIT. The simulation parameters were: ion, m/z 500; cross-section, 200 Å^2; number of ions, 1000; bath gas pressure, 0.40 Pa (helium). AREX LIT dimensions: inscribed circle radius (r_0) 7.0 mm, rod radius (R) 8.2 mm; AREX LIT voltages: main RF drive amplitude applied to the quadrupole rods 1000 V (zero-to-peak), main drive frequency 1.0 MHz, D_0 20 V, supplemental AC amplitude applied between the front and rear vanes 3.0 V (zero-to-peak), supplemental AC frequency ca 18.5 kHz. Conventional QIT dimension: ring radius (r_0) 7.0 mm, half-distance between the end-cap electrodes (z_0) 5.0 mm; conventional QIT voltages: main RF drive amplitude applied to ring electrode ca 2000 V (zero-to-peak), main RF frequency 1.0 MHz, supplemental AC amplitude applied between the end-cap electrodes 4.0 V (zero-to-peak), supplemental AC frequency 400 kHz ($\beta = 0.80$). (Reproduced from Hashimoto, Y., Sugiyama, M., and Hasegawa, H., *J. Mass Spectrom., Soc. Jpn.* 2007, 55, 369–374. With permission from The Mass Spectrometry Society of Japan.)

components of the instrument. By utilizing both the energy focusing of the ejected ions and the high ejection efficiency described in the last section, we have demonstrated the sensitivity enhancement that is possible with tandem mass spectrometers that incorporate the AREX LIT.

12.3.1 Duty Cycle Enhancements of the Orthogonal-Acceleration Mass Spectrometer

An orthogonal-acceleration time-of-flight mass spectrometer (oa-TOF-MS) and a quadrupole time-of-flight mass spectrometer (QqTOF-MS), which incorporates an oa-TOF-MS, are now popular tools for proteome and metabolite analyses, as well as

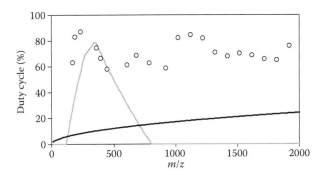

FIGURE 12.7 Comparison of the duty cycles of an oa-TOF-MS. Circles: oa-TOF-MS with AREX LIT; bold line: theoretical non-trap mode; thin line: theoretical pulsing mode.

other fields of application [17,18]. One of the most important performance characteristics of an oa-TOF-MS is its sensitivity, because poor sensitivity degrades the quality of the mass spectral acquisition. An inherently poor sensitivity requires, therefore, a long acquisition time in order to obtain a good S/N ratio in the mass spectrum. The sensitivity of an oa-TOF-MS depends strongly on the duty cycle, which in this context is defined by the ratio of the number of ions registered by the detector of the TOF-MS to the number of ions introduced to the TOF-MS's 'pusher.' Figure 12.7 shows examples of the duty cycles of various oa-TOF-MS systems. Since most of the ions are lost during the time delay between successive acceleration pulse periods of the pusher, and because they pass through the pusher without being accelerated, the duty cycle of a conventional oa-TOF-MS is typically less than 20%. The details concerning the duty cycle of an oa-TOF-MS are explained clearly in other reports [16,19]. It is evident, therefore, that improvement of this low duty cycle is one of the most important means for enhancing the sensitivity. Ion trapping combined with an oa-TOF-MS can enhance the duty cycle over a limited range of m/z-values by setting the optimum pulsing parameters. This technique is often called the 'pulsing' technique; for example, an efficiency in excess of 40% can be achieved over a limited m/z-range between m/z 200 and m/z 550, while virtually not compromising efficiencies below m/z 150 and above m/z 800, as shown by the thin line in Figure 12.7. With this method, however, the m/z-range with duty cycle enhancement is not sufficient for molecular-profiling applications such as the identification of peptide sequences, where ion peaks are distributed over a wide range of m/z-values.

In order to obtain a high duty cycle over a wide enough m/z-range for use in molecular profiling, we have developed an oa-TOF-MS as shown in Figure 12.8 [16], which consists of our AREX LIT, a collision cell with ion trapping capability, and an oa-TOF-MS incorporating an ion pulsing technique. Helium gas at a pressure of about 40 mTorr is introduced into the collision cell to collisionally focus the ions without dissociation. When the frequency of the supplemental AC potential applied to the vane electrodes is swept from higher to lower values, ions from lower m/z- to higher m/z-values are introduced to the oa-TOF-MS. Since the setting of the pulsing parameters is scanned synchronously with the m/z-scan of the AREX ejection process, high duty cycles of 60–90% are obtained over a wide m/z-range, for example

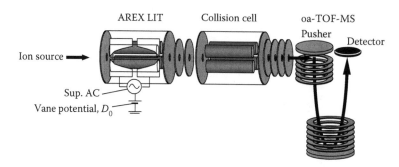

FIGURE 12.8   Schematic diagram of the AREX LIT/TOF-MS.

between m/z 174.1 and m/z 1922.0 as shown in Figure 12.7. This technique enhances the sensitivity by a factor of three to 10 times (depending upon the m/z-range) over a wide range of m/z-values compared to that achievable with a conventional oa-TOF-MS. An oa-TOF-MS, which incorporates this technique, is now installed in a commercial LIT/TOF mass spectrometer (the NanoFrontier™) available from Hitachi High-Technologies Corporation.

12.3.2 TWO-DIMENSIONAL MASS SPECTROMETRY

In conventional tandem mass spectrometry, only limited scan modes selected from product ion, neutral loss, precursor ion scans, and an MRM experiment can be conducted simultaneously, because many scan modes lead to a reduction in the duty cycle (that is, the sensitivity). Therefore, fragmentation data arising from minor ion peaks are often not obtained. To prevent such observations from being missed, ideally one should obtain all fragmentation data from all of the ions [20]. Several groups have demonstrated the concept of a two-dimensional mass spectrometer and its analog with a novel configuration, which consists of an ion trap acting as a source and having a mass-selective ion ejection capability, a collision cell, and a mass analyzer [21,22]. Such a device is expected to generate all the fragmentation data for all of the precursor ions over 100 times more efficiently than do conventional tandem mass spectrometers because an ion trap with mass-selective ion ejection introduces all the ions sequentially. On the one hand, the mass filtering technique rejects ions other than the filtered ions and, therefore, the duty cycle for a conventional tandem spectrometer with a mass filter, such as a QqTOF-MS, decreases as the precursor ion scan range increases. By contrast, the duty cycle of a two-dimensional mass spectrometer is not affected by enlarging the scanning m/z-range. A preliminary study of two-dimensional mass spectrometry has been performed using the AREX LIT as the source ion trap [23]. Although, the configuration of the instrument was the same as that shown in Figure 12.8, nitrogen gas was introduced into the collision cell to dissociate ions efficiently. Figure 12.9 show the two-dimensional mass spectra obtained by the instrument for different collision energies of 10 eV and 30 eV. After the AREX LIT had accumulated ions that were produced from two peptides once, it ejected them sequentially into the collision cell. The abscissa indicates the m/z-values that were calculated from the frequency of supplemental potential applied to the AREX LIT, which is labeled as the 'precursor

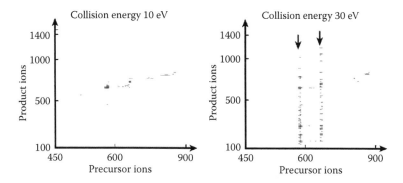

FIGURE 12.9 Two-dimensional (2D) mass spectra obtained at different collision energies. The collision energy was set to 10 eV (left), and 30 eV (right); the acquisition period for each spectrum was one second. (Reproduced from Hashimoto, Y., Sugiyama, M., and Hasegawa, H., *J. Mass Spectrom. Soc. Jpn.* 2007, *55*, 369–374. With permission from The Mass Spectrometry Society of Japan.)

ions.' The 'precursor ions' corresponds to the m/z-values of the ions before dissociation inside the collision cell. The ordinate indicates the m/z-values, which is labeled as the 'product ions,' calculated from the flight times of the ions in the oa-TOF-MS; these correspond to the m/z-values of product ions formed after dissociation inside the collision cell. In the mass spectrum at a low collision energy of 10 eV, the precursor ion m/z- and product ion m/z-values were the same, which suggests that under low energy conditions dissociation does not occur within the collision cell. Unlike the situation at low collision energy, there are many peaks in the mass spectrum obtained with at high collision energy of 30 eV, which suggests that fragmentation proceeds without any difficulty under these conditions. Each acquisition period for these 2D mass spectra was one second, which is much shorter than the 40 seconds given in another report where a 3D QIT was used as a source ion trap [21]. We assume that this difference arises both from the high ejection efficiency and the low energy dispersion of the AREX LIT.

Since, a 2D mass spectrum contains all the fragmentation data obtained from all the ions, various data such as product ion, neutral loss, and precursor ion scans can be extracted. As an example, Figure 12.10 shows the product ion mass spectra that were extracted from a 2D mass spectrum that was obtained with collision energy of 30 eV. These mass spectra show good sequence coverage of both peptides, which reveals a high enough sensitivity for the indicated identifications to be made.

In this section, examples have been shown where the high ion extraction efficiency and low energy dispersion of the AREX ejection greatly enhance the sensitivity of the oa-TOF-MS and 2D-MS, and it is expected that this capability can be applied to other hybrid mass spectrometers for enhancing their performance.

12.4   ION FRAGMENTATION IN THE AREX LIT

Thus far, usage of the AREX LIT as a mass-selective ion ejection device has been described in the preceding sections. LITs are, however, used not only as mass-selective

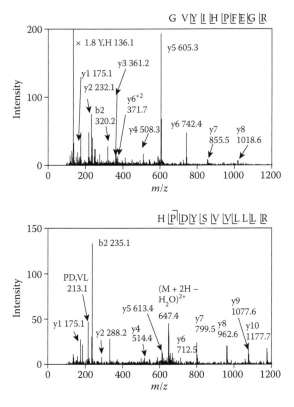

FIGURE 12.10   Product ion mass spectra resulting from the electrospray ionization of the indicated peptides, corresponding to the precursor ion marked by arrows on Figure 12.9. The collision energy was 30 eV, and the sequence information obtained for each of the peptides is shown above each spectrum. (Reproduced from Hashimoto, Y., Sugiyama, M., and Hasegawa, H., *J. Mass Spectrom. Soc. Jpn.* 2007, *55*, 369–374. With permission from The Mass Spectrometry Society of Japan.)

ion ejection devices but also as dissociation devices when incorporated into high-resolution hybrid mass spectrometers [4–7]. We have observed that the AREX LIT has unique properties as such a dissociation device [24].

12.4.1   COLLISION-INDUCED DISSOCIATION

The detection of low *m/z*-fragment ions, such as immnonium ions and iTRAQ™ reporter ions, is very useful for peptide analysis, but it is well known that low *m/z*-fragment ion detection inside a 3D QIT and a 'conventional' LIT is difficult [25]. Since the detectable *m/z*-range and the dissociation efficiency are in a mutual 'trade-off' situation, several CID methods have been developed recently for low *m/z*-fragment ion detection inside QITs [26,27]. Effective dissociation requires a high potential, where ions are excited kinetically and then activated by collisions with the

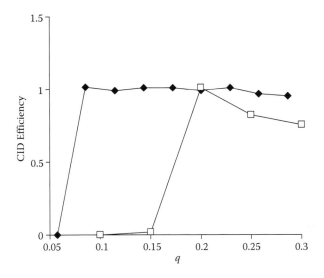

FIGURE 12.11   Plots of CID efficiencies *versus* q for the AREX LIT (◆) and a conventional LIT (□). Sample: reserpine (*m/z* 609.3); D_0: 40 V; CID period: 5 ms.

bath gas in the trap. The minimum *m/z*-value of fragment ions, M_{min}, which can be stabilized inside QITs, is

$$M_{min} = \frac{q}{q_{ej}} M_p \qquad (12.5),$$

where M_p is the *m/z*-value of the precursor ion, q is the stability parameter for the precursor ion, and q_{ej} is a constant value of 0.908. The oscillation potential of a QIT, which corresponds to the pseudopotential formed by the quadrupole field [28], is proportional to the square of the stability parameter q for the precursor ion. Figure 12.11 shows plots of the CID efficiencies *versus* the stability parameter, q, that were obtained experimentally from both a conventional LIT and our AREX LIT. These results show that the dissociation of the ions in a conventional LIT only proceeded when q was greater than 0.20, because the amplitude of the oscillating potential was insufficient to effect dissociation at a low value of q; the pseudopotential at q = 0.20 is calculated to be 18.1 V using the Dehmelt equation [28]. In order to obtain a high CID efficiency with a reasonable range of *m/z*-values for the product ions, a value of q between 0.20 and 0.25 is often chosen. For q = 0.23, Equation 12.5 becomes

$$M_{min} = 0.25 M_p \qquad (12.6).$$

We deduce from this equation that fragment ions whose values of *m/z* are less than about one quarter of that of the precursor ion are neither trapped nor detected. On the other hand, Figure 12.11 shows that effective dissociation occurred with low q-values, of less than 0.10, in the AREX-LIT because the oscillation potential formed

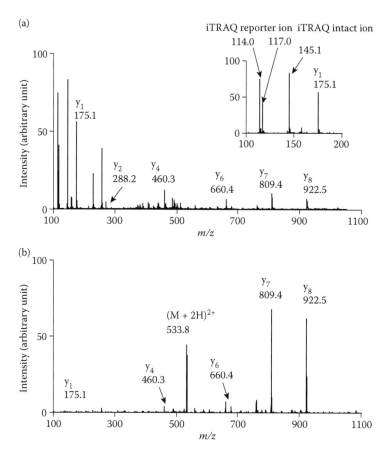

FIGURE 12.12 Product ion mass spectra (electrospray ionization) obtained with the AREX-LIT (a), and with a conventional LIT (b). Sample: iTRAQ-LCTVATLR; m/z 533.8; D_0: 40 V; CID period: 5 ms; $q = 0.10$ (AREX-LIT CID); $q = 0.23$ (conventional LIT CID). (Reproduced from Hashimoto, Y., Hasegawa, H., Sugiyama, M., Satake, H., Baba, T., and Waki, I., *J. Mass Spectrom. Soc. Jpn.* 2007, *55*, 339–342. With permission from The Mass Spectrometry Society of Japan.)

by the vane DC potential did not affect the trapped mass range. Collision-induced dissociation in the AREX LIT proceeded efficiently when the vane DC potential was above 17 V (data not shown). This threshold of the potential barrier is similar to the maximum value of the kinetic energy of ions oscillating in a pseudopotential well having a depth of 18.1 V calculated for the 3D QIT.

Figure 12.12 shows product ion mass spectra for a peptide labeled with an iTRAQ reagent [24]. In the AREX-LIT, low-m/z fragment ions, such as iTRAQ reporter ions (m/z 114.1 and m/z 117.1), and iTRAQ intact ions (m/z 145.1) were detected as strong peaks. This product ion mass spectrum (Figure 12.12a) is similar to the product ion mass spectra of iTRAQ-labeled samples obtained with a QqTOF-MS or tandem time-of-flight mass spectrometer (TOF/TOF), where low-m/z containment problem does not exist. On the other hand, Figure 12.12b shows no iTRAQ reporter ions (m/z 114.1 and m/z 117.1) because we chose $q = 0.23$ for effective CID in the conventional LIT experiment; the conventional LIT experiment was performed in the home-built LIT/TOF-MS described in reference [5]. Equation 12.6 indicates that no fragment peaks with m/z of less than 130 would be observed, which is consistent with Figure 12.12b. As described above, the AREX LIT is evidently effective for peptide identification and quantification when low-m/z product ions exist.

12.4.2 ELECTRON-CAPTURE DISSOCIATION

Electron capture dissociation (ECD) is a powerful sequence-analyzing technique for proteomics, which embraces a wide coverage of peptides and proteins with and without post-translational modifications [29,30]. Our group has performed ECD successfully in a conventional LIT by superimposing an axial magnetic field of 50–200 mTesla [31,32], where high-speed ECD within tens of milliseconds was achieved by introducing low energy electrons from the filament electron source outside the end-cap electrode. Post-translational modification analysis has been reported also using this high-speed ECD device [33]. Inside the AREX-LIT, ions can be oscillated mass-selectively in an axial direction for isolation and CID. The ion motion induced by axial resonant excitation is affected little by the axial magnetic field, because the directions of the excited velocity component and the magnetic field are the same. On the other hand, a mass shift, that was estimated to about 10 Th for m/z 1000 at $q = 0.10$, was observed when radial excitation was performed in a conventional LIT under an axial magnetic field of 200 mTesla [34]. To demonstrate sequential ECD/CID, which consists of an ECD step, followed by an isolation step and a CID step, we developed an arrangement in which a cylindrical permanent magnet was placed around the AREX LIT, producing an axial magnetic field of 150 mTesla [24]. Figure 12.13 shows the mass spectrum after ECD (a), and after sequential ECD/CID (a). The ECD target was triply charged ions of $[M + 3H]^{3+}$ that were produced from neurotensin by means of electrospray ionization. Dissociation sites identified by ECD and sequential ECD/CID are depicted in the upper right corner of each mass spectrum. Not only product ions but also charge-reduced radical ions $[M + 3H]^{2+ \cdot}$ were observed in the ECD mass spectrum (Figure 12.13a). After ECD, the charge-reduced radical ions were isolated by axial resonant excitation in order for them to act as the precursor ions of the following CID experiment, in which they were collisionally dissociated by axial resonant excitation, forming c and z product ions that are characteristic of ECD/CID fragmentation. This observation is in contrast to normal CID, in which b and y product ions are observed to predominate. The use of sequential ECD/CID improved the sequence coverage of the peptide compared with that of ECD-alone (note that the cleavages marked as bold lines in Figure 12.13b are dissociation sites observed with sequential ECD/CID).

In this Section, examples have been presented where axial oscillation inside the AREX LIT achieved low m/z product ion CID and sequential ECD/CID; it is expected that these unique fragmentation properties using the AREX LIT will be

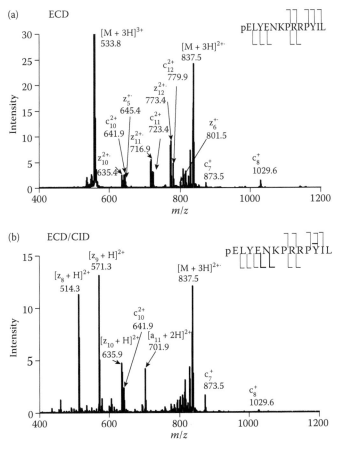

FIGURE 12.13 Electrospray mass spectra with ECD (a), and sequential ECD/CID (b). Sample, neurotensin, m/z 553.8; ECD period, 15 ms; CID period, 5 ms. (Reproduced from Hashimoto, Y., Hasegawa, H., Sugiyama, M., Satake, H., Baba, T., and Waki, I., *J. Mass Spectrom. Soc. Jpn.* 2007, *55*, 339–342. With permission from The Mass Spectrometry Society of Japan.)

effective for identifying complex molecular such as peptides with post-translational modifications.

12.5 CONCLUSIONS

In this chapter we have described the development of a novel LIT with axial resonant excitation, the AREX LIT. This new approach was accomplished using the axial harmonic DC potential formed by 'vane' electrodes positioned between the quadrupole rod electrodes of a conventional LIT arrangement. The resolving power of mass-selective ion ejection from the AREX LIT was between 350 and 500 at a typical scan rate, which was not as high as that obtained with other mass-selective ion ejection techniques. On the other hand, the AREX LIT ejects ions with a high efficiency, in excess of 80%, with a low energy distribution (less than 5 eV), as a consequence of which it enhances greatly the duty cycle when combined with an orthogonal acceleration TOF-MS and the transmission of a two-dimensional mass spectrometer. The AREX LIT has demonstrated also CID with low m/z-value product ions and sequential ECD/CID, both of which are known to present difficulties in conventional LITs.

REFERENCES

1. Douglas, D.J.; Frank, A.J.; Mao, D.M. Linear ion traps in mass spectrometry. *Mass Spectrom. Rev.* 2004, *23*, 1–29.
2. Hager, J.W. A new linear ion trap mass spectrometer. *Rapid Commun. Mass Spectrom.* 2002, *16*, 512–526.
3. Schwartz, J.C.; Senko, M.W.; Syka, J.E.P. A two-dimensional quadrupole ion trap mass spectrometer. *J. Am. Soc. Mass Spectrom.* 2002, *13*, 659–669.
4. Campbell, J.M.; Collings, B.A.; Douglas, D.J. A new linear ion trap time-of-flight system with tandem mass spectrometry capabilities. *Rapid Commun. Mass Spectrom.* 1998, *12*, 1463–1474.
5. Hashimoto, Y.; Hasegawa, H.; Waki, I. Dual linear ion trap/orthogonal acceleration time-of-flight mass spectrometer with improved precursor ion selectivity. *Rapid Commun. Mass Spectrom.* 2005, *19*, 1485–1491.
6. Makarov, A.; Denisov, E.; Kholomeev, A.; Baischun, W.; Lange, O.; Strupat K.; Horning, S. Performance evaluation of a hybrid linear ion trap/orbitrap mass spectrometer. *Anal. Chem.* 2006, *78*, 2113–2120.
7. Syka, J.E.P.; Marto, J.A.; Bai, D.L.; Horning, S.; Senko, M.W.; Schwartz, J.C.; Ueberheide, B.; Garcia, B.; Busby, S.; Muratore, T.; Shabanowitz, J.; Hunt, D.F. Novel linear quadrupole ion trap/FT mass spectrometer: performance characterization and use in the comparative analysis of Histone H3 post-translational modifications. *J. Proteome Res.* 2004, *3*, 621–626.
8. Gerlich, D. Inhomogeneous RF fields: a versatile tool for the study of processes with slow ions. *Adv. Chem. Phys.* 1992, *82*, 1–176.
9. Javahery, G.; Thomson, B. A Segmented radiofrequency-only quadrupole collision cell for measurements of ion collision cross section on a triple quadrupole mass spectrometer. *J. Am. Soc. Mass Spectrom.* 1997, *8*, 697–702.
10. Mansoori, B.A.; Dyer, E.W.; Lock, C.M.; Bateman, K.; Boyd, R.K.; Thomson, B. Analytical performance of a high-pressure RF-only quadrupole collision cell with an axial field applied using conical rods. *J. Am. Soc. Mass Spectrom.* 1998, *9*, 775–788.
11. Loboda, A.; Krutchinsky, A.; Loboda, O.; McNabb, J.; Spicer, V.; Ens,W.; Standing, K. Novel Linac II electrode geometry for creating an axial field in a multipole ion guide. *Eur. J. Mass Spectrom.* 2000, *6*, 531–536.
12. Russ, B.; Fjeldsted, J. Design and validation of high speed collision cell for triple quad and QqTOF instrumentation. *Proc. 54th ASMS Conf. Mass Spectrometry and Allied Topics,* Seattle, May 28–June 1, 2006, WP15 302.
13. Hashimoto, Y.; Hasegawa, H.; Baba, T.; Waki, I. Mass selective ejection by axial resonant excitation from a linear ion trap. *J. Am. Soc. Mass Spectrom.* 2006, *17*, 685–690.
14. Dobson, G.S.; Enke, C.G. Axial ion focusing in a miniature linear ion trap. *Anal. Chem.* 2007, *79*, 3779–3785.
15. Green, M.R.; Bateman, R.H.; Scott, G. Characterization of mass selective axial ejection from a linear ion trap with superimposed axial quadratic DC potential. *Proc. 55th ASMS Conf. Mass Spectrometry and Allied Topics,* Indianapolis, June 3–7, 2007, ThP051.
16. Hashimoto, Y.; Hasegawa, H.; Satake, H.; Baba, T.; Waki, I. Duty cycle enhancement of an orthogonal acceleration TOF mass spectrometer using an axially-resonant excitation linear ion trap. *J. Am. Soc. Mass Spectrom.* 2006, *17*, 1669–1674.

17. Morris, H.R.; Paxton, T.; Dell, A.; Langhorne, J.; Berg, M.; Bordoli, R.S.; Hoyes, J.; Bateman, R.H. High sensitivity collisionally-activated decomposition tandem mass spectrometry on a novel quadrupole/orthogonal-acceleration time-of-flight mass spectrometer. *Rapid Commun. Mass Spectrom.* 1996, *10*, 889–896.

18. Shevchenko, A.; Chernushevich, I.; Ens, W.; Standing, K.G.; Thomson, B.; Wilm, M.; Mann, M. Rapid *de novo* peptide sequencing by a combination of nanoelectrospray, isotopic labeling and a quadrupole/time-of-flight mass spectrometer. *Rapid Commun. Mass Spectrom.* 1997, *11*, 1015–1024.

19. Chernushevich, I.V. Duty cycle improvement for a quadrupole-time-of-flight mass spectrometer and its use for precursor ion scans. *Eur. J. Mass Spectrom.* 2000, *6*, 471–479.

20. Schwartz, J.C.; Wade, A.P.; Enke, C.G.; Cooks, R.G. Systematic delineation of scan modes in multidimensional mass spectrometry. *Anal. Chem.* 1990, *62*, 1809–1818.

21. Wang, H.; Kennedy, D.S.; Nugent, K.D.; Taylor, G.K.; Goodlett, D.R. A Qit-q-Tof mass spectrometer for two-dimensional tandem mass spectrometry. *Rapid Commun. Mass Spectrom.* 2007, *21*, 3223–3226.

22. Krutchinsky, A.N.; Cohen, H.; Chait, B. A novel high-capacity ion trap-quadrupole tandem mass spectrometer. *Int. J. Mass Spectrom.* 2007, *268*, 93–105.

23. Hashimoto, Y.; Sugiyama, M.; Hasegawa, H. Development of a two-dimensional mass spectrometer using an axially resonant excitation linear ion trap. *J. Mass Spectrom. Soc. Jpn.* 2007, *55(6)*, 369–374 (in Japanese).

24. Hashimoto, Y.; Hasegawa, H.; Sugiyama, M.; Satake, H.; Baba, T.; Waki, I. Tandem mass spectrometry using an axially resonant excitation linear ion trap. *J. Mass Spectrom. Soc. Jpn.* 2007, *55(5)*, 339–342.

25. Payne, A.H.; Glish, G.L. Thermally assisted infrared multiphoton photodissociation in a quadrupole ion trap. *Anal. Chem.* 2001, *73*, 3542–3548.

26. Cunningham, C.; Glish, G.L.; Burinsky, D.J. High amplitude short time excitation: a method to form and detect low mass product ions in a quadrupole ion trap mass spectrometer. *J. Am. Soc. Mass Spectrom.* 2006, *17*, 81–84.

27. Schwartz, J.C.; Syka, J.E.P.; Quarmby, S.T. Improving the fundamentals of MS^n on 2D linear ion traps: new ion activation and isolation techniques. *Proc. 53rd ASMS Conf. Mass Spectrometry and Allied Topics*, San Antonio, June 5–9, 2005, TODam 11:35.

28. Dehmelt, H.G. Radiofrequency spectroscopy of stored ions. *Adv. At. Mol. Phys.* 1967, *3*, 53–72.

29. Zubarev, R.A.; Keller, N.L.; McLafferty, F.W. Electron capture dissociation of multiply charged protein cations - a nonergodic process. *J. Am. Chem. Soc.* 1998, *120*, 3265–3266.

30. Zubarev, R.A. Reactions of polypeptide ions with electrons in the gas phase. *Mass Spectrom. Rev.* 2003, *22*, 57–77.

31. Baba, T.; Hashimoto, Y.; Hasegawa, H.; Hirabayashi, A.; Waki, I. Electron capture dissociation in a radio frequency ion trap. *Anal. Chem.* 2004, *76*, 4263–4266.

32. Satake, H.; Hasegawa, H.; Hirabayashi, A.; Hashimoto, Y.; Baba, T.; Masuda, K. Fast multiple electron capture dissociation in a linear radio frequency quadrupole ion trap. *Anal. Chem.* 2007, *79(22)*, 8755–8761.

33. Deguchi, K.; Ito, H.; Baba, T.; Hirabayashi, A.; Nakagawa, H.; Fumoto, M.; Hinou, H.; Nishimura, S. Structural analysis of O-glycopeptides employing negative- and positive-ion multi-stage mass spectra obtained by collision-induced and electron-capture dissociations in linear ion trap time-of-flight mass spectrometry. *Rapid Commun. Mass Spectrom.* 2007, *21*, 691–698.

34. Baba, T.; Satake, H.; Hashimoto, Y.; Hasegawa, H.; Hirabayashi, A.; Waki, I. Sequential electron capture- and collision induced dissociations in a linear radio-frequency quadrupole ion trap. *Proc. 53rd ASMS Conf. Mass Spectrometry and Allied Topics*, San Antonio, June 5–9, 2005, WP08 135.

Part V

3D-Quadrupole Ion Trap Mass Spectrometry

13 An Examination of the Physics of The High-Capacity Trap (HCT)

Desmond A. Kaplan, Ralf Hartmer,
Andreas Brekenfeld, Jochen Franzen,
and Michael Schubert

CONTENTS

13.1 Introduction 593
13.2 Background 594
    13.2.1 Ion Influences on Analytical Performance 594
    13.2.2 Geometry of High-Capacity Ion Traps 595
        13.2.2.1 Geometry Definitions 595
        13.2.2.2 Ion Capacity: A Geometric Exploration 595
    13.2.3 Ion Ejection 597
        13.2.3.1 Resonant Ion Ejection 597
        13.2.3.2 Ion Ejection in Non-Linear Fields 598
13.3 The Spherical High-Capacity Ion Trap: Experimental Evidence 600
13.4 Model for 'Ideal' Ion Ejection 600
13.5 Simulations 604
13.6 Experimental Evidence for A New Ion Ejection Model 606
    13.6.1 Methods for Studying the Ion Ejection Process 606
    13.6.2 Ejection Delay 607
    13.6.3 Influences on Figures of Merit 608
13.7 Targeted Applications 610
    13.7.1 Electron Transfer Dissociation (ETD) 610
    13.7.2 Increasing Operational Speed and Duty Cycle 614
13.8 Summary and Conclusion 615
References 616

13.1 INTRODUCTION

Quadrupole ion trap mass spectrometry has been a continually growing technology since its invention in 1953 and, over the past two decades following its commercialization, it has developed to be a work-horse instrumental technique in many analytical laboratories [1–4]. The electrodynamic fields which confine and control the ion motion in an ion trap have also been well characterized and described throughout the literature [2,5,6]. For more than a decade researchers have shown that the fields within a 'quadrupole' ion trap used for mass spectrometry have to contain contributions from non-linear fields (that is, higher-order fields) and, thus, are not purely quadrupolar [5,7]. A detailed description of the non-linear fields that are possible within quadrupole ion traps has been covered in volume I of this series [5]. There are many advantages, and in some instances challenges, associated with the use and introduction of non-linear fields into ion traps. The aim of this chapter is to describe the implementation of a combination of fields that leads to a higher number of ions being acceptable for and analyzed in the ion trap, or in other words the creation of a high-capacity ion trap (HCT).

13.2   BACKGROUND

13.2.1   Ion Influences on Analytical Performance

Before understanding the mechanisms for increasing *ion capacity* it is important to understand what is meant by this term, and the major factor that influences ion capacity. In addition to the trapping and excitation fields in the ion trap influencing ion trajectories, the electric field produced by the ions themselves has a significant impact upon their motion. The interaction of ions with other ions as they are being stored, excited, and/or ejected is one of the key factors that influence the ion capacity of the ion trap. The ion capacity defines the maximum number of ions that can be confined in the device and still achieve a set level of performance. There are three main different ion capacities to be defined: the 'spectral limit,' the 'isolation limit,' and the 'storage limit' [8]. The storage limit is perhaps the simplest to understand: this limit is the maximum number of ions that can be confined within the ion trap. Any further ions are either pushed out of the ion trap by Coulombic forces or cause already-trapped ions to be pushed out. This limit does not constrain the mass spectrometric applications of ion traps, because there are other limitations, as we will see below, restricting the number of ions to much lower levels. The isolation and spectral limits depend on the definition of spectral quality. The isolation limit is the number of ions that can be trapped and still allow a given isolation efficiency. For instance, the isolation limit might be defined as the number of ions that can be trapped and be isolated within a 1 Th window. The isolation limit is affected by ion/ion coupling, which prevents resonant ion selection and thereby achieving the desired isolation-width. Typically, the isolation limit is roughly 10-fold lower than the storage limit. Finally, the spectral limit is the number of ions that can be trapped while still achieving the desired resolution, scan speed, and mass accuracy. This limit is the strictest in that, as ions are being ejected, the ion/ion interaction can retard the resonant ejection process, cause ion coupling, and cause mass shifts. This limit can be 50 to a few hundred times smaller than the isolation limit. In mass spectrometry we are mainly interested in the quality of the final mass spectrum, and thus the spectral limit is the most important limit with respect to mass spectrometric operation of the ion trap. The significance of this statement must not be underestimated: the most important value is the number of ions that you can eject from the ion trap with at a certain resolution and speed. For example, it does not matter if the ion trap can store 1 million ions if it can deliver only 10,000 ions during ion ejection and still achieve peak widths of less than 0.5 Th at the desired scan speed. For the remainder of this chapter the term ion capacity will refer to this definition of spectral limit.

13.2.2 GEOMETRY OF HIGH-CAPACITY ION TRAPS

13.2.2.1 Geometry Definitions

There are two types of geometry of HCTs [9,10]. The first is a quadrupole mass analyzer with lenses at each end of the device; this is termed a '2D' ion trap because the electrodynamic fields confine the ions only in the x- and y-dimensions. The 2D ion trap has also been referred to synonymously as a linear ion trap (LIT).* Within the discussions in this chapter the term LIT will be used, since this term describes the fact that the ions trapped in these devices will form essentially a linear elongated cloud. As will be described below, the LIT is considered to be a high-capacity trap because of its volume. The second geometry of the HCT comes from the conventional geometry of the quadrupole ion trap, which is sometimes referred to as the '3D' ion trap. The term '3D' has been used because the electrodynamic trapping forces themselves are acting naturally on the ions in the x-, y-, and z-dimensions. The term '3D' does not, in fact, describe how the ions are really trapped; therefore, for the remainder of this chapter the term 'spherical' ion trap will be used because it describes the way in which the ions in a pure quadrupolar 3D trap find a perfect spherical confinement. The spherical *HCT* is achievable through the use of a unique combination of non-linear fields.

13.2.2.2 Ion Capacity: A Geometric Exploration

For the storage capacity limit, a simple evaluation of the geometry can lead to some assumptions of the capacity of the ion trap [1,10–13]. The maximum number of ions that can be stored in an ideal quadrupolar spherical ion trap is defined in Equation 13.1, where n_i is the number density, $\bar{D}_z$ is the pseudopotential well depth in the z-dimension, z_0 is the distance from the center of the trap to an end-cap electrode, e is the electronic charge, and ε_0 is the permittivity of free space

$$n_{i,\text{max}} = \frac{3\bar{D}_z \varepsilon_0}{ez_0^2} \tag{13.1}$$

An analogous equation for the LIT is shown in Equation 13.2, where $\bar{D}_{x,y}$ is the pseudopotential well depth in the (x,y) plane and r_0 is the distance from the central axis of the quadrupole rod set to the inner surface of a rod electrode

$$n_{i,\text{max}} = \frac{4\bar{D}_{x,y} \varepsilon_0}{er_0^2} \tag{13.2}$$

---

* See Volume 1, Chapter 11: Linear Ion Trap Mass Spectrometry with Mass-Selective Axial Ejection, by James W. Hager; Volume 1, Chapter 12, Axially-resonant Excitation Linear Ion Trap (AREX LIT), by Yuichiro Hashimoto; Volume 4, Chapter 14: Electrically-induced Nonlinear Ion Traps, by Gregory J. Wells and August A. Specht.

By simple inspection it is clear that both ion traps are capable of reaching similar maximum storage densities, assuming that all parts of the volumes of the two instruments could achieve such a condition. Due to the similarities of the equations for the maximum storage densities (Equations 13.1 and 13.2), Campbell *et al.* proposed the expression given in Equation 13.3 [14], which is basically a ratio of the volumes of the two devices

$$\frac{N_{\text{lin}}}{N_{\text{sph}}} = \frac{r_0^2 l}{z_0^3} \tag{13.3}$$

where N_{lin} and N_{sph} represent the number of ions in the LIT and in the spherical ion trap, respectively. The assumptions inherent in this conclusion are that the operating conditions of the ion traps are similar, namely that for the two devices the ions have similar secular frequencies and Mathieu q-values, the relative field errors due to space charge are similar, and that all parts of the ion traps can achieve the maximum storage density, as noted above. From Equation (13.3) it would seem that by simply increasing the length of the linear device it should be possible always to have more ions in a linear device. However, using the published dimension of one LIT [10] ($r_0 = 0.4$ cm, $l = 3.7$ cm) and a modern spherical ion trap ($r_0 = 1$ cm, $z_0 = 0.783$ cm) from the same vendor, Equation 13.3 results in a ratio that suggests there are 1.23 times more ions in the LIT.

It is not the purpose of this chapter to compare linear and spherical devices, but it is important to consider what is possible within the high-capacity devices available today. As mentioned above, the limits of the device must be defined with given expectations of performance. In a LIT, ions can be ejected radially or axially. For linear devices that utilize radial ejection, it is important to note that the longer the device the more likely that the ions secular frequencies will vary across the device due to tolerances in making a long LIT and to the resulting field perturbations. In axial-ejection LITs, the ions are shifted to the end of the LIT and are then excited resonantly and radially so as to be close to the rods. The radial motion is converted into axial motion utilizing disturbances of the quadrupolar field, thus causing the ions to be ejected axially. For axial ejection, the field homogeneity along the axis is not a major issue, but this ejection process is slow, so it is possible to have more ions but at the expense of scan speed.

Equations 13.1 and 13.2 do not take into account that, for mass spectrometric purposes, ion traps are operated under conditions for which the ion clouds are shrunk significantly by collisions with damping gas. Additionally, these two equations describe situations in which ions are under strong space-charge conditions, whereas for mass spectrometry the ion traps have to be run under weak space-charge conditions. As a result of these inherent weaknesses in this model, researchers have developed more practical equations which are shown in Equations 13.4 [12] and 13.5 [10],

$$\frac{N_{\text{lin}}}{N_{\text{sph}}} = \frac{3 l_c}{5 R_{\text{sph}}} \tag{13.4}$$

and

$$\frac{N_{lin}}{N_{sph}} = \frac{R_{lin}^2 l_c}{2R_{sph}^3}$$ (13.5),

where R_{sph} and R_{lin} represent the radii of the ion clouds in the devices spherical and linear devices, respectively, and l_c represents the length of the ion cloud in the linear device.

Estimating the real size of the ion cloud is difficult, and to do so accurately one should use photo-tomographic techniques. Cooks and coworkers showed that the radius of a cloud of ions with m/z 105 at a q_z of 0.3 in a spherical ion trap was ca 0.6 mm [15]. If we assume a similar shrinking factor for the radius of the ion cloud in the LIT, that is the ion cloud should shrink from a radius of 4 mm to a radius of 0.24 mm, and if we use the dimensions suggested above, application of Equation 13.5 results in a 'theoretical' ratio of maximum storage densities of four to five [1]. However, the aforementioned measurements were made when the ions were stored and 'cooled', not while the ions are being ejected. In fact, the limitation for ion capacity (spectral limit) is the condition of the ion cloud during ion ejection, in a process where the target ions are excited kinetically. Knowing what happens to the ions during the ejection process is an important part of understanding the ion capacity of an ion trap. The conclusion from this simple equation-based view of capacity is that the derived ion capacities between the two HCT geometries are too close to judge the performance just by the simple inspection of geometry factors. In reality, if the ion ejection process were better understood it may be possible to develop the model further.

13.2.3 ION EJECTION

13.2.3.1 Resonant Ion Ejection

Most ion trap mass spectrometers today employ a resonance technique to eject ions from the ion trap. In this method ions are excited resonantly using a dipole excitation field while the amplitude of the RF drive potential applied to the ion trap is increased linearly to eject ions out the ion trap at increasing values of m/z. On initial consideration one might think that the excited ions will have an equal probability of being ejected through holes in either end-cap electrode. As will be discussed later, however, there are methods to direct the ions so that they are ejected only through the holes in one 'exit' end-cap electrode. Most ion trap theories suggest that the ions should be excited and ejected rapidly so as to minimize ion/ion interactions during the ejection process. One method with which to decrease the ejection time is to increase the magnitude of the dipolar excitation field. This method is effective; however, it is possible that the ion will gain too much kinetic energy before reaching the proper secular frequency, and be ejected early under pre-resonant conditions. Such early ejection can result in a broadening of the peak within the mass spectrum. There is, therefore, an optimal magnitude of the dipolar excitation field for ejecting an ion with a given value of m/z. By decreasing the scan rate of the ion trap it is possible to improve the peak width, mainly because the ions spend a longer time in the excitation field. This method does reach limiting returns and has the negative side-effect that the scans take longer. The ideal ejection process would allow ions to be ejected unidirectionally through the exit end-cap electrode as quickly as possible in order to experience no space-charge effects, while spending as much time in the excitation field as possible so as to gain optimal resonant selectivity. The result is a mass spectrum that is acquired rapidly, with the peaks having a good resolution. The only question is how to achieve the ideal ejection process.

13.2.3.2   Ion Ejection in Non-Linear Fields [5]

The effects of non-linear fields in spherical ion traps have been well-described. Most notably, non-linear fields can be used to help improve ejection efficiency, ejection speed, and in some cases MS/MS efficiencies. The purpose of this Section is to review the most relevant of the effects and the impacts they have upon the performance of a spherical ion trap. Resonant ion ejection at non-linear resonance points on the stability diagram of the ion trap [5] meets several of the ejection criteria listed above. The non-LITs from Bruker Daltonik use a modified hyperbolic angle electrode geometry, and the fields in the device contain a combination of hexapole and octopole components. Resonant ion ejection in the Bruker ion traps is done at the hexapole non-linear resonance ($\beta_z = 2/3$). When the ions' secular frequency matches that of the dipole excitation voltage, the ions experience both the linear amplitude increase from the dipole field as well as the rapid hyperbolic growth from the non-linear resonance. The non-linear resonance causes a rapid uptake of energy and a resulting increase in the amplitude of the motion of the ion, so that the ions are ejected quickly from the ion trap. As shown in Figure 13.1 [5], ions are ejected in less than 400 μs for an auxiliary AC amplitude of 2 V_{0-p} applied between the end-cap electrodes to create the dipolar field. The scan rate employed in this simulation is at 33,000 Th s^{-1}, which illustrates further the speed that is achievable with excitation using non-linear resonances. The simulation illustrates also that unidirectional ejection can occur when an ion is excited resonantly using a dipolar excitation field (applied potential >0.7 V_{0-p}) in combination with the non-linear resonances arising from the hexapole and octopole fields present in the ion trap. In this simulation, with an applied potential of 4 V_{0-p} it is easy to see that the ions already have a beating oscillation before the ejection point. This is important, because it shows that the ions are being excited by the fields before they are completely in resonance. During earlier discussions the focus of modeling the ejection process from ion traps had always been directed to utilizing the higher-order fields to achieve an accelerated ion ejection of the targeted ions. Therefore the starting conditions were chosen so that the ion motion was in resonance with the higher-order fields exactly when the ion reached the resonance point with the dipolar excitation field. The excitation arising at non-linear resonances helps to eject the ions rapidly from the ion trap, as well as inducing unidirectional ejection. This ejection process had been implemented in all previous generations of Bruker non-LITs. By contrast, the spherical HCT takes the principles of this excitation process and makes additional use the higher-order fields to create a two-step excitation event, increasing the effective interaction time with the activation field and improving the space-charge limitations that are present during the ejection process.

The relative phase of the dipolar excitation and that of the main RF drive potential has an impact upon the performance of the spherical ion traps utilizing higher-order fields [5,16]. Since the resonance ejection process in the HCT*ultra*™ (Bruker

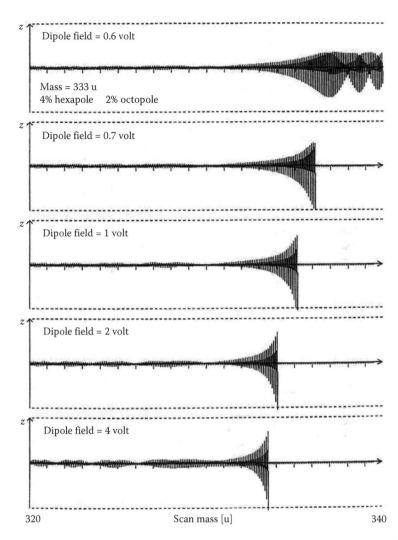

FIGURE 13.1   Simulations of ion ejection with combined superimposed hexapole and octopole fields for different amplitudes of the excitation RF applied to the end-cap electrodes. The
ejection is unidirectional for all dipolar excitation voltages. The voltage of 0.6 V_{0-p} does not
eject the ion (top), but 0.7 V_{0-p} already ejects the ion. The values 2 and 4 V_{0-p} (below) result
in sharp ejection, causing only very few beating oscillations before the point of ejection is
reached. (Reproduced from Franzen, J., Gabling, R-H., Schubert, M., and Wang, Y., *Practical
Aspects of Ion Trap Mass Spectrometry.* CRC Press, Boca Raton, 1, Chapter 3, 49–167. 1995.
Reprinted with permission.)

Daltonik, Bremen, Germany) is performed with an applied dipolar frequency that
is 1/3 of that of the main drive frequency ($\beta_z = 2/3$), the relative phase is a relevant
operational parameter. At every third cycle of the main RF drive potential, this and
the dipolar excitation potential realign. At a scanning speed of 26,000 Th s^{-1} this
synchrony occurs 10 times per mass/charge unit!

13.3 THE SPHERICAL HIGH-CAPACITY ION TRAP: EXPERIMENTAL EVIDENCE

In 2003, Bruker introduced a non-linear spherical ion trap with a high ion capacity, termed the High-Capacity Trap, HCT™ [9]. This geometry-optimized ion trap operates at 26,000 Th s^{-1}, which is double the scan rate of previous generation ion traps from Bruker. One figure of merit used to monitor the number of ions that the ion trap is able to analyze is to look at the relative mass shift as a function of the number of ions injected into the ion trap. A previous generation ion trap, the Esquire 3000plus, was compared with the spherical HCT. A noticeable effect of an inflated amount of ions in the trap during the ejection process, observable in the mass spectra, is the apparent shift of the measured mass positions, which is caused by the intensified ion/ion coupling, due mainly to Coulombic interaction, damping the motion of the ion to be ejected and, thereby, delaying the ejection process. Illustrated in Figure 13.2 is the plot of the measured m/z-values of the ions from protonated reserpine molecules of both the pure ^{12}C-isotopomer at the nominal m/z 609 and the isotopomer containing one ^{13}C-atom at m/z 610 as a function of the value of the parameter 'ICC actual' for a course of experiments with increasing accumulation time. The data were acquired with the spherical HCT and its predecessor, the Esquire 3000plus. The recorded 'ICC actual', which is the total ion signal intensity for each individual spectrum, is related directly to the overall number of ions that were detected in the mass spectrum.

As indicated by the vertical lines, for both instruments the data can be divided into two regimes, the first, at lower trap charging levels, reflects a slow increase of the observed mass shift with increased trap charging, and a second steep shift of the recorded mass position at higher charging levels. The crossing of both lines marks the ion charge density at which the mass spectra start to be distorted and, therefore, become useless for mass spectrometric analysis. For the Esquire 3000plus this limit is reached at an ICC actual of about 1.6×10^5 arbitrary units. The comparable ICC actual limit for the spherical HCT is at about 6.5×10^5 arbitrary units; thus, the HCT is able to analyze roughly four to five times more ions than its predecessor. Depicted in Figure 13.3b and c are two mass spectra taken from the data set for the Esquire 3000plus and from the HCT, respectively, each at an ICC actual of 1.8×10^5 arbitrary units. While the mass spectrum from the Esquire 3000plus is distorted highly with the ion peak of reserpine being shifted strongly, the reserpine peak is recorded with very good performance in the HCT mass spectrum. Even at an ICC actual of 7×10^5 arbitrary units the peak of reserpine appearing at a nominal value of 609 Th is well resolved in the HCT data set (Figure 13.3c). The ongoing discussion will explain the physics behind the HCT that has enabled this improvement in ion capacity.

13.4 MODEL FOR 'IDEAL' ION EJECTION

As described in Section 13.2.2.2, from a geometric standpoint it is possible to increase the storage capacity simply by increasing the geometric volume of a device. However, a larger geometric volume does not necessarily result in a higher spectral limit when it is the criterion for ejection of the ion during the mass spectral acquisition that determines the ion capacity. The ejection of ions from an ion trap must

(a)

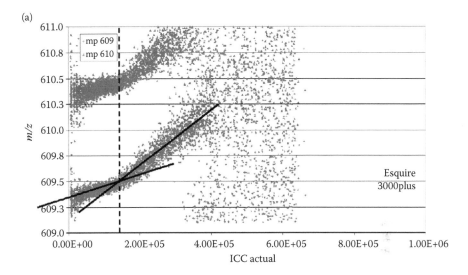

(b)

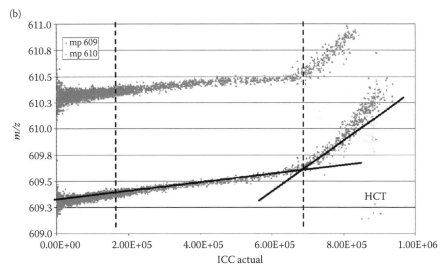

FIGURE 13.2   Plots showing the influence on the observed mass position by the ICC actual (related to the number of ions) for the Esquire 3000plus (a) and the spherical HCT™ (b). The plots show protonated reserpine molecules at m/z 609 and its ^{13}C isotopomer peak at m/z 610. In each diagram two straight lines mark regimes of normal mass shift and distortion of spectral performance. The crossing point is used to indicate the spectral capacity limit for each instrument.

be done so as to avoid space-charge effects, while keeping the ejection process as selective and rapid as possible. Thus, to achieve a higher capacity it is imperative to improve the ejection process.

One way to reduce ion/ion interaction and to improve the ejection process is the judicious use of non-linear fields and proper phase-correlation of the resonant

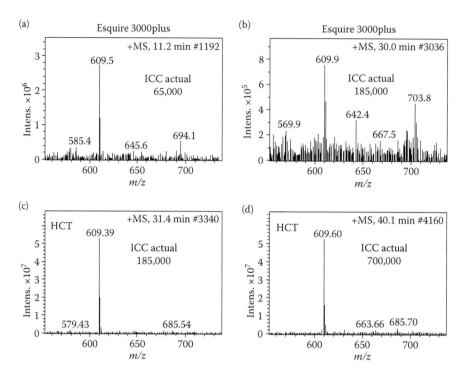

FIGURE 13.3   Individual mass spectra from the data set shown in Figure 13.2 for different ICC values for the Esquire 3000plus (a) and (b) and the spherical HCT (c) and (d). Plots (b) and (c) represent the same number of ions present in the trap during ejection. While the mass spectrum from the Esquire 3000plus is already distorted, the HCT gives well-resolved mass spectra even at much higher ion charge numbers (plot (d)).

excitation process with the RF drive voltage used to create the RF field for confining the ions, as put into practice in Bruker's spherical HCTs. A model for this approach is illustrated in Figure 13.4 as a simplified time-sequence of events during the acquisition of a mass spectrum.

According to this concept, the ions of interest are first excited by the dipolar resonance excitation (t_1) away from the other ions into a higher orbit as shown in Figure 13.4a and b. The ions of the target m/z-value, represented in grey in Figure 13.4, are excited resonantly to a higher orbit for a prolonged residence time outside the center of the ion trap and with high velocity while passing through the remaining ions in the center. As a result of this spatial separation and curtailed residence duration in the ion cloud, the interaction of these ions with those ions that are not resonantly excited, and remain in the center of the ion trap, is reduced significantly. Thus, only a narrow range of m/z-values populates the enlarged volume occupied by the cloud of excited ions, reducing further the effective ion density during ejection. The ions are kept at the excitation level for some time (typically a few hundred microseconds). As the ion orbit slowly continues to increase ($t_1 + t_d$) ions are ejected precisely utilizing the higher-order fields that are greater in strength near the end-cap electrodes,

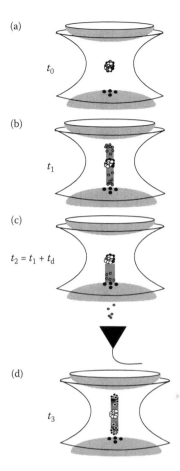

FIGURE 13.4   Non-linear ion ejection is a two-step process. First, (a), ions are cooled and located in the trap center. Second, (b), during the scan, the RF drive field is ramped and ions of a first mass (grey) come into resonance with the dipolar excitation RF field. These ions are driven to a higher orbit, while the ions of higher mass (black, white) remain cooled at the center. Third, the excited ions (grey, (c)) are picked up by the non-linear resonance and are ejected instantly. The cooled ions in the center stay unaffected. Then, the process repeats itself for higher masses (black, (d)). The ion capacity is improved because immediately before the non-linear ejection the ions populate a large volume (b) with reduced space-charge effects for these ions and a prolonged interaction time with the exciting RF field.

Figure 13.4c. This process will then begin again for ions of larger m/z-value, Figure 13.4d, represented in black, which were in the center of the trap in Figure 13.4a through c.

To achieve this process it is necessary to have two distinct resonant activation events. In this instance, the first excitation process is achieved by means of a simple dipolar resonance excitation, and the second is achieved by utilizing the progressive excitation at the non-linear resonance point, which is inherent to the Bruker Daltonik's HCT spherical ion trap geometry and ejection scheme.

13.5   SIMULATIONS

Simulations of the dipolar excitation process coupled with a non-linear resonance are used to gain an insight of the extent to which the ejection process can be accelerated by the influence of higher-order fields or, under different conditions, to discover the range of ejection delays that can be expected, in order to support the model depicted in the previous Section. During this simulation exercise we will restrict the discussion to motions in the z-direction only. While for a pure dipolar excitation field the driving force is to be assumed to be constant across the entire volume under investigation, the contribution of a hexapolar field to the driving force on an ion at a given location is proportional to the square of its distance from the center of the device. Having the origin of the z-axis in the center of the trap, the contribution of the hexapolar field would be represented by a force proportional to z^2 [5]. Moreover, as these higher-order fields are caused *via* geometric distortions of the main RF field, these higher-order fields are proportional to the RF drive field and, in particular, proportional to the time-dependence of the RF drive field [2,5,6]. The z^2-dependency of the force due to the hexapolar field has two important consequences. First of all the contribution of this field becomes very small near the center of the device. Therefore, for just slightly deformed geometries, the effects arising from higher-order fields are relevant only when ions are excited to reach larger displacements from the center. The other important consequence is that the sign, that is direction, (and value) of the driving force at a negative z-position is the same as for the force acting at the equivalent positive position at the same time. In other words, when the force is directed in the positive z-direction at a positive z-position (outward), at the same time the force at the negative z-position will be directed also in the positive z-direction (inward!). As a consequence of this behavior, an ion trapped by the enclosing RF field and oscillating with a secular frequency of 1/3 of the main RF drive field can pick up (or lose) energy from the main RF potential due the coupling by the hexapolar field contribution. This situation can be seen easily: assuming an outward-directed force at a given time, one half of a secular cycle later (the ion will be at the equivalent position on the other side of the ion trap) the main RF field has developed 3/2 cycles and has, therefore, inverted its sign. Therefore, the contribution of the higher-order field will be directed outward again. Consequently, the contribution of the higher-order field to the ion motion is not compensated every other half-cycle, but accumulates over time. The result of the assumed outward-directed force on the ion motion (either damping or escalating the ion motion) depends upon the phase of the ion motion at the given time. When the velocity of the ion is directed outward, the ion will pick up energy from the higher-order field for its motion, otherwise, when the velocity of the ion is directed to the center, the ion motion will be damped. Obviously the same holds true if, at a different time, the contribution of the higher-order field is directed toward the center. As seen, the result from the contribution of the higher-order field depends upon the phase-correlation between the ion motion and the phase of the higher-order field, which itself is coupled strictly to the main RF drive field. As mentioned before, when exciting ions that have been cooled down previously to the center of the trap, the ions will see only the dipolar field at the beginning of the ejection process, and the higher-order fields will be initially totally negligible. As there is just a weak (collisional) damping influence, the ion motion follows the excitation RF field with a phase shift of 90°. The motion of the ions when they reach a displacement from the center, where the contribution of the hexapolar field becomes relevant is, therefore, determined by the dipolar excitation RF field. Consequently, it is not surprising that the relative phase between the excitation voltage (determining the initial ion motion) and the main RF voltage (determining the phase of the higher-order fields) has a strong impact upon the final behavior of the ions.

For the simulation shown in Figure 13.5 we assumed a confining force proportional to the displacement z, directed toward the center and adjusted with respect to its strength to get the target ion to the desired oscillation frequency. In addition, we applied a dipolar excitation force (constant across the volume) at the aforementioned frequency and an additional hexapolar force, proportional to the square of z with a frequency of three-times the oscillation frequency. The important variable parameter for the simulations was the relative phase between the dipolar and hexapolar force (in addition to their relative amplitudes). The resulting equations of motion were integrated numerically to achieve the resulting trajectories, assuming that the ions started in the center with zero velocity under the initial conditions (which also could be altered). Two results from these simulations are shown in Figure 13.5, where the development of the displacement in z (expressed in arbitrary units) is plotted against the excitation time. In both cases, one can see the initial nearly linear increase of the oscillation amplitude. At an oscillation amplitude of about 15 the hexapolar

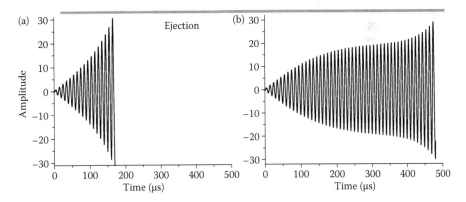

FIGURE 13.5 Simulations of the dipolar ion excitation and hexapolar ejection process, for two different settings of the relative phase between the excitation dipolar RF and the main RF drive field. In these simulations, ions are ejected as the amplitude of their motion reaches a value of 30 (arbitrary units), noted by the solid line. For the simulations shown in diagram (a), the dipolar RF excitation and the hexapolar ejection fields are in phase, and ions are ejected rapidly in a continuous process. For the diagram (b), the relative phase between the two fields is shifted by 180° and the hexapolar ejection is delayed by several hundred microseconds ('ejection delay'). During this time, the ions fill a large volume with reduced ion/ion interaction, before being ejected by the hexapolar resonance. (Note the micromotion associated with the RF drive field is not modeled in these simulations.)

field contribution exerts a significant influence. In the left example, Figure 13.5a, the phase of the ion motion is such that it can pick up energy immediately from the hexapolar field contribution, causing a rapid increase of the amplitude. As the contribution from the hexapolar field increases with the square of the displacement z, the increasing rise of the amplitude can be observed once the hexapolar field has become dominant. For the example in Figure 13.5b, the relative phase between the dipolar and the hexapolar contributions has been shifted by 180° relative to the conditions on the left side. In contrast to example (a), the simulation (b) shows that at an oscillation amplitude of about 15 the ion does not pick up energy from the hexapolar field but the motion is damped, so that the dipolar excitation cannot continue increasing the amplitude. The presence of the hexapolar field contribution causes a slight deformation of the effective potential. The ion motion, therefore, experiences a slight phase shift over time relative to the hexapolar frequency. After several cycles of delay, the ion can finally pick up energy as fast as in example (a). In these examples, an oscillation amplitude of about 30 represents the ejection of the ions. While the ion is ejected in example (a) after about 150 μs excitation, the ejection is delayed in simulation (b) by more than 300 μs. During this time, the ion is decoupled from the potential arising from the other ions in the center of the device and undergoes an extensive interaction with the exciting resonance field. Thus the essential criteria are met in order to achieve a high-capacity ion ejection and optimal resolving power. The situation illustrated in Figure 13.5b is described in the simulation model as comprising two distinct, decoupled excitation events, one due to the dipolar excitation and the other due to the progressive energy pick-up from the hexapolar field.

13.6   EXPERIMENTAL EVIDENCE FOR A NEW ION EJECTION MODEL

13.6.1   Methods for Studying the Ion Ejection Process

These experiments will explore the exact timing of the ejection of ions from the ion trap. In order to verify the model of a two-step delayed ejection event and to elucidate the differences to the previous generation ion traps, experiments were performed with the electrode geometries of a Bruker Esquire 3000plus and of a Bruker HCT*Ultra* ion trap. These experiments were conducted in the same instrumental system, with only changes to the electrode configuration between sets of experiments. The excitation RF voltage applied during activation and ejection of the ions was derived from the fundamental RF voltage for all experiments. The scan rate of 26,000 Th s^{-1} was used for both electrode configurations. A standard electrospray ionization (ESI) Tuning Mix (Agilent; Palo Alto, CA), which has ions covering the mass range from m/z 100 to 3000, was used in these experiments. The exact timing of the ejection process was analyzed by varying the phase between the fundamental RF voltage and the resonance excitation RF voltage. The relative mass position, mass resolving power, and ion abundance were monitored, at each phase. These data and the results from the simulations/calculations provide insight into how the ion capacity can be influenced as a result of the ejection process, and also how it is improved with proper experimental design.

13.6.2   EJECTION DELAY

An improvement of the ion capacity can be achieved during resonance ejection at a specific non-linear resonance with a strict control of the phase-correlation. As suggested by the simulations, the ions of interest are first excited away from the other ions into a higher orbit by the dipolar resonance excitation. After the ions are excited to this higher orbit (in position and momentum space), they are ejected later from the ion trap by the progressive energy pick-up from the non-linear resonance. Importantly, the non-linear resonance ejection occurs with high temporal precision. The separation of the initial excitation and the ejection stages can be probed by altering the relative phase between the dipolar excitation RF potential and the main RF drive potential. As shown in the simulations above, under certain phase relationships it is possible for ions to be ejected rapidly, while under other phase conditions ion ejection can be retarded for some time. As was illustrated in Figure 13.1, during the resonance ejection process ions are excited and de-excited continually until their secular frequency is close to the frequency of the dipolar excitation voltage. Consequently, there is always a delay from the time the ions first experience excitation to the time at which they are ejected, and this delay occurs for all ions. In these experiments, as the relative phase is changed, the delay in the ejection process varies from the 'normal' value (for which the instrument is calibrated) to a shorter or longer delay. Experimentally, this delay would be manifested as an apparent shift in the entire mass spectrum. A zoom-in of the m/z 1522 ion is shown in Figure 13.6 for mass spectra acquired with two different phase relationships. The mass shift is clear; however, because the

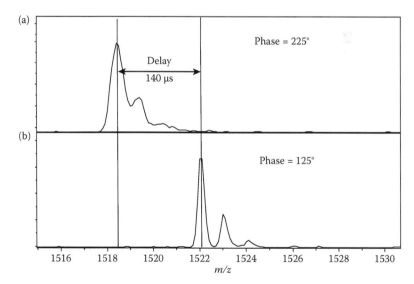

FIGURE 13.6   Mass spectra of an ESI Tune Mix showing the ion at m/z 1522 and how the ejection delay is determined by monitoring the relative mass shift. When the relative phase difference between the dipolar RF and the main RF fields is 225°, the ^{12}C isotopomer peak is at ca 1518.5 Th (a); for the relative phase difference of 125°, the ^{12}C isotopomer peak is at ca 1,522.1 Th (b). The instrument is scanning at 26,000 Th s^{-1}, therefore with a mass shift of 3.6 Th the ejection delay is 140 µs.

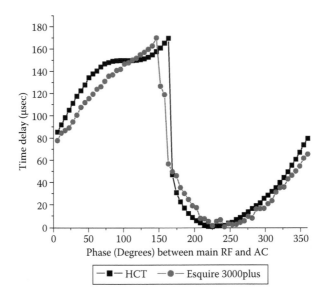

FIGURE 13.7 Ejection delay as function of phase difference between dipolar excitation RF field and the main RF drive field. Instruments are best operated at a point of zero derivative. For the Esquire 3000plus, one such point exists (at 225°). For the high-capacity ion trap (HCT), two points exist (at 120°, 225°). Operation at the additional point at 120° results in a large ejection delay correlating with improved ion capacity.

scan rate is known (26,000 Th s⁻¹) the relative ejection delay can be calculated. The ejection delay between these two data sets is about 140 μs.

Figure 13.7 shows the ejection delays as a function of phase between the main RF drive and the activation voltages observed for the m/z 1522 ion ejected from the HCT*Ultra* and from the Esquire 3000plus mass spectrometers. For both instruments, the fastest ejection (smallest delay) occurs at the same phase angle *ca* 225°. The maximum delay for both instruments was approximately 170 μs longer. This maximum delay corresponds to roughly 44 main RF cycles, so this delay is not just a small retardation as a result of Coulombic interaction; in fact the ions are likely to have a coherent motion during this time. This prolongation of the ejection process corresponds to the difference in the simulated ejection time points (Figure 13.5) for different relative phases. Furthermore, these measurements illustrate the need for strong phase-control during the excitation and ejection process. Without it mass spectra could be acquired at random phases, which would result in a variety of ejection delays (wider mass peaks) [16]. The excitation RF and the main RF potentials are derived electronically from the same fundamental frequency base in these instruments in order to ensure precise phase-control.

13.6.3 Influences on Figures of Merit

The real question that arises is which phase is a good phase to utilize? From a reliability and stability standpoint operating in a region that has minimal change in the ejection delay for the greatest change in phase is ideal (horizontal tangent). There is

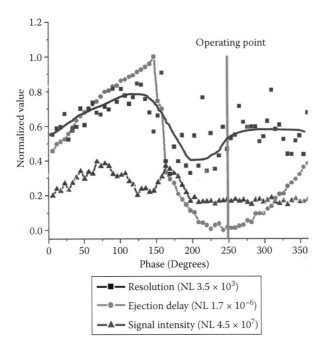

FIGURE 13.8 Ejection delay, mass resolution and ion abundance as functions of the phase between dipolar excitation RF and main RF, for Esquire 3000plus (operated at 26,000 Th s⁻¹).

only one option for operation of Esquire 3000plus, whereas for the HCT*Ultra* there are two options. Figures 13.8 and 13.9 show how the other figures of merit, abundance, and resolving power change as a function of relative phase for the Esquire 3000plus and for the HCT*Ultra*, respectively. From a phase angle of 50° to 140° the resolving power of the Esquire 3000plus is increasing, reaching a maximum at around 125°; however, neither the signal intensity nor the ejection delay reach steady plateaux in this phase region. Stable and robust operation for the Esquire 3000plus would be at about 240°. The Esquire 3000plus does show improvement in resolving power over a range of phases; however, the improvements occur on a slope and it should not be expected that this result would be reproducible. Whereas, because of the geometry improvements, in the HCT the resolving power and signal intensity both reach stable plateaux. It is important to note that the absolute performance of the Esquire 3000plus in these experiments is poorer than with the standard production instruments because it is operating at the faster scan rate of 26,000 Th s⁻¹ (the scan rate for the HCT). The general trends do not change at the standard scan rates for the Esquire 3000plus (13,000 Th s⁻¹) but the absolute resolving power and intensities become greater.

In the HCT*Ultra* both signal intensity and resolving power reach a maximum at 120°. Both of these parameters also reach plateaux for about 15°, showing further robustness at this operating point. The resolving powers at the operating points for the HCT and the Esquire 3000plus were 3500 and 1600, respectively. The resolving power of 3500 for the HCT occurs at an ICC value that is five-fold greater for the

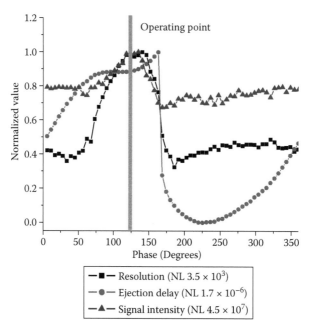

FIGURE 13.9 Ejection delay, mass resolution and ion abundance as functions of the phase between dipolar excitation RF and main RF, for HCT (operated at 26,000 Th s^{-1}). For the HCT, resolution, delay (correlated to the ion capacity) and ion abundance all maximize at the same operating point (120°), resulting in the highest scan speeds, highest ion capacity and highest sensitivity, all at the same settings. (NL: Normalization Level).

HCT than for the Esquire 3000plus. This result shows that this ejection delay leads to a higher spectral limit with improved resolving power.

These measurement results, therefore, support the model described above, together with the assumption that the ejection delay provoked by the specific interference with higher-order fields improves the ion capacity of the HCT. In this delayed ejection process, ions are excited initially to a higher orbit separating them from other ions in the trap. The excited ions still pass through the remaining ion cloud in a lower orbit but at a higher speed. Thus, the excited ions cannot couple with other ions and the final excitation is performed in a lower charge density environment, which is the key for achieving the higher capacity ion trap.

13.7   TARGETED APPLICATIONS

13.7.1   ELECTRON TRANSFER DISSOCIATION (ETD)

Ion traps have become a work-horse for the investigation of the molecular structure of biological molecular compounds. One typical application is the mass spectrometric analysis of peptides from a digestion of an entire protein; this proteomics approach is described as 'bottom-up.' Typically these peptides have lengths of 5–15 amino acids. During a typical analysis, each peptide is isolated and fragmented in the ion trap to generate the structural information of the amino acid sequence for each individual peptide. The most frequently used fragmentation technique in such tandem mass spectrometric approaches is collision-induced dissociation (CID) [17–19]. The CID-fragmentation of peptides results in sufficient sequence information, particularly for doubly-charged precursor peptide ions whose fragmentation mechanism has been studied extensively [20–22]. However, with increasing precursor charge state, collisional activation leads to less sequence information because, for multiply charged peptides, losses of small neutral molecules, such as water or ammonia, are the preferred cleavage channels. Electron Transfer Dissociation (ETD) is another relatively new fragmentation technique for the sequence-characterization, particularly of higher-charged peptides and even of proteins [23–27]. Chapter 15 in this volume addresses specifically different types of fragmentation techniques in ion traps.*

Of these techniques, ETD is particularly suitable for spherical ion traps and will, therefore, be discussed briefly in this chapter. Ion fragmentation using the ETD method can be performed easily in spherical ion traps because, by their three-dimensional trapping nature, these devices allow the mutual storage of the peptide cations and the subsequent filling of the reagent anions [23–25]. In a spherical ion trap, ion clouds of both polarities can be compressed easily toward the center of the trap just by increasing the RF drive voltage applied to the ring electrode. As a result of this increase the pseudopotential well becomes steeper and both ion clouds, that is the reagent anions as well as the peptide cations, are forced toward the center of the trap. As a result, these conditions are ideal for the reaction conditions due to the optimal overlap of both ion clouds.

The key feature of this new ETD fragmentation technique is the transfer of an electron from a reactant anion to a multiply-protonated peptide molecule, whereby the even-electron multiply-protonated peptide is transformed into an odd-electron intermediate species [26,28]. Due to the electronic configuration of the intermediate peptide radical cation, randomly distributed N–C alpha bond cleavages at each amino acid position of the peptide backbone are obtained. Fragmentation induced by ETD results typically in one cleavage per peptide, but each individual cleavage occurs throughout the entire amino acid sequence.† Therefore, for the sequence-characterization of an entire peptide, a high number of peptide cations have to be fragmented in the ion trap in order to generate the required pattern. The number of peptide cations necessary increases with the length of the amino acid chain. This requirement for higher numbers of ions makes the high ion capacity of a modern mass spectrometric device absolutely mandatory.

---

* Volume 4, Chapter 15: Fragmentation Techniques for Protein Ions Using Various Types of Ion Trap, by J. Franzen and K. P. Wanczek.

† See also Volume 5: Chapter 1, Ion/Ion Reactions in Electrodynamic Ion Traps, by Jian Liu and Scott A. McLuckey; Volume 5, Chapter 3, Methods for Multi-Stage Ion Processing Involving Ion/Ion Chemistry in a Quadrupole Linear Ion Trap, by Graeme C. McAlister and Joshua J. Coon; Volume 5, Chapter 4, Chemical Derivatization and Multistage Tandem Mass Spectrometry for Protein Structural Characterization, by Jennifer M. Froelich, Yali Lu and Gavin E. Reid; Volume 5, Chapter 5: Fourier transform ion cyclotron resonance mass spectrometry in the analysis of peptides and proteins, by Helen J. Cooper.

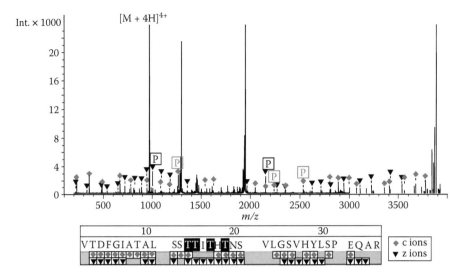

FIGURE 13.10   ESI ETD MS/MS data from a quadruply phosphorylated peptide from Protein Kinase C (tryptic digest); parent cation m/z 974.2 $[M + 4H]^{4+}$. This peptide has 12 different amino acids where phosphorylation is possible, and ETD assigned unambiguously the phosphate groups at the positions T13, T14, T16, and T18 marked by an outlined in black. All the identified c and z-type fragments are designated by diamonds and triangles, respectively. Sequence: VTDFG IATAL SSTTITHTNS VLGSV HYLSP EQAR.

Figure 13.10 shows the ETD MS/MS spectrum of a quadruply-phosphorylated peptide ion formed by ESI from a tryptic digest of Protein Kinase C. The amino acid chain analyzed consists of 12 different amino acids where phosphorylation is possible ($5 \times S$, $6 \times T$, and $1 \times Y$). In the ETD MS/MS data presented in Figure 13.10 phosphorylation is assigned unambiguously to positions T13, T14, T16, and T18 [29]. The clear identification of the multiple sites of phosphorylation demonstrates markedly the advantage of ETD. With ETD, labile, bounded post-translational modifications remain attached to the peptide backbone and can be localized unambiguously.

Another ETD mass spectrum from a larger peptide subjected to ESI, including the sequence length of 60 amino acids, is given in Figures 13.11 and 13.12 [29]. Nearly the entire amino acid sequence can be read out of the ETD MS/MS-data of the Galanin-like peptide, in Figure 13.12. Particularly for such a large number of fragments, it is essential that the ion trap should be capable of working with a high number of product ions simultaneously, and that these multiply-charged product ions can be ejected out of the ion trap in a high-resolution, seamless, scan mode with good sensitivity. Especially for such large peptide cations, and with the improved resolution in the enhanced mode as well as the wide scan range of up to m/z 3000, which is standard for the HCT, the analysis has demonstrated the advantages gained from employing higher charge-state ions.

For ETD and other types of ion/ion reactions, an excess of reagent anions is needed to perform efficient reactions. With the spherical HCT, a three- to four-fold excess of anions compared to the total number of precursor cations is typically sufficient.

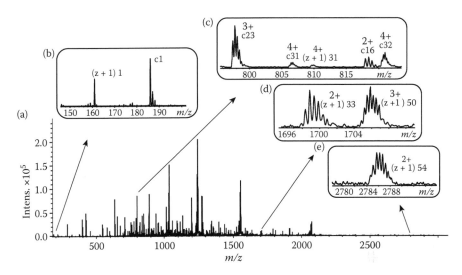

FIGURE 13.11 ESI ETD MS/MS of a Galanin-like peptide with a total length of 60 amino acids; parent ion m/z 776.0, $[M + 8H]^{8+}$; full mass range of the seamless mass spectrum (a) and selected enlarged subview into the ETD MS/MS spectrum (b), (c), (d) and (e). ETD-fragment ions up to the quadruply charged state are resolved isotopically.

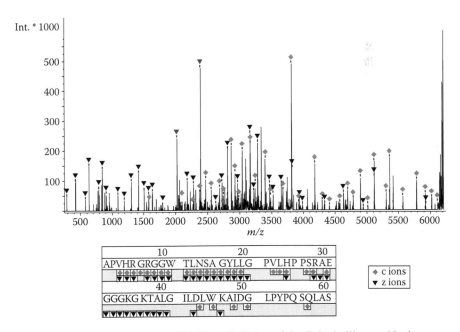

FIGURE 13.12 The deconvoluted ETD MS/MS data of the Galanin-like peptide that were shown in Figure 13.11 are illustrated. All the identified c- and z-type fragments are designated by diamonds and triangles, respectively. Sequence: APVHR GRGGW TLNSA GYLLG PVLHP PSRAE GGGKG KTALG ILDLW KAIDG LPYPQ SQLAS.

Under this condition the density of the anion cloud in a spherical ion trap is significantly less than the charge capacity limitation. Some groups, working with a linear ion storage device, have suggested that the number of anions needed for ETD should be as high as possible, to such an excess that the reagent anion reaches the space-charge limit [30,31]. This is not the case for the spherical HCT, which indicates that spherical ion traps are particularly suitable for utilizing ion/ion reactions.

13.7.2 Increasing Operational Speed and Duty Cycle [32]

Analytical demands, especially those involving complex biological mixtures, are pushing mass analyzers to the limit with respect to many analytical properties. To demonstrate that the enhancement of spherical ion traps is an ongoing progress, the following discussion illustrates the further evolution of the achievable speed of acquisition of MS/MS information for a typical application over the past few years.

Complete proteomes analyzed in 2D nanoLC-ESI-MS/MS runs may contain several thousands of proteins, which need a fast mass spectrometric analysis in order to identify as many peptides as possible. Furthermore, even peptides in low abundance have to be analyzed by MS/MS to a sufficient degree to avoid the situation that only the non-specific higher abundant proteins are identified. To achieve this goal, the number of MS/MS spectra that can be acquired per unit time in data-dependent 'Auto MS(N)' experiments is an important performance factor. Auto MS(N) experiments are where the instrument identifies the most important, valuable, and/or interesting ions in the MS spectrum and automatically performs tandem mass spectrometry-based on user-defined settings.

With the improvements of the HCT*Ultra*, about 200 MS/MS spectra per minute can be acquired for typical proteomics applications in the data-dependent Auto MS(N) mode. This rate represents a gain in the number of MS/MS spectra acquired per unit time by more than a factor of two over the past few years, and an even greater improvement compared to its predecessor of the Esquire3000 series. Analytically, the boost in performance is reflected by the number of unique identified proteins in an analysis of complex protein mixtures.

For the evaluation of the improvement of the overall speed of acquiring MS and MS/MS spectra within the HCT series, a soluble fraction of an *E. coli* lysate was used as a complex protein mixture and about 3 μg of protein were alkylated chemically and digested with trypsin following standard procedures. Chromatographically separated peptides were then analyzed under Auto MS(N) conditions, and the number of identified peptides and proteins used to judge the performance with respect to the duty cycle in Auto MS(N) experiments.

Figure 13.13 shows the increase of unique identified peptides as a function of the Mascot MS/MS score. Interestingly, the improvement gain becomes higher for higher MS/MS scores (database search performed with MASCOT 2.1.). Higher MS/MS scores result from higher quality MS/MS spectra and, as a result, we obtain a dramatic increase in the number of proteins identified by means of multiple peptides, which reflects the best identification certainty, as shown in Figure 13.13.

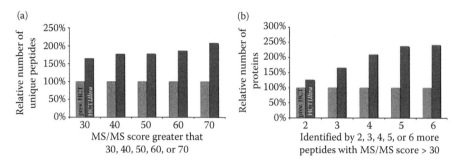

FIGURE 13.13 Number of the identified peptides (a) and proteins (b) from a complex protein mixture, analyzed with Bruker's HCT*Ultra* ™ (dark bar) and a previous HCT-instrument (light bar). A trypic digest of *E. coli* cell lysate sample was separated by a single 105 min gradient with a capillary liquid chromatograph (LC). The eluting peptides from this cell lysate were analyzed in the Auto MS(N)-mode by the two different instruments; all data have been acquired and processed with the HCT/esquire Compass 1.1 software. Database searches have been performed with MASCOT 2.1 and the extracted protein lists have been created by ProteinScape 1.3. With the increased duty cycle for Auto MS(N) of the HCT*Ultra*, the number of identified peptides for a given mascot score is increased more than two-fold, (a). Even the protein identification (represented by the given number of identified peptides) is significantly higher in this investigation, when the sample is analyzed with the HCT*Ultra*, (b).

13.8 SUMMARY AND CONCLUSION

A model for and experimental evidence of the physics behind the method for achieving a higher mass spectral capacity for the spherical ion trap have been presented. This method involves decreasing the ion density of the ions as they are being excited resonantly just before ejecting them from the ion trap, and decreasing the coupling to those ions that have not yet been excited. The simulations presented were in good agreement with this model of a delayed two-stage ion ejection. Experimentally, the ejection timing can be controlled by setting the relative phase between the dipolar excitation voltage and the main RF drive voltage.

The delayed ejection is advantageous because ions spend longer in the excitation field, resulting in a better resolving power and a reduced ion/ion interaction leading to a higher spectral ion capacity. For the spherical HCT there is an optimal phase and ejection delay where both the resolving power and ion abundance are at a maximum. This maximum has a plateau, which makes for stable robust operation. In these experiments, under the optimal operating conditions, the HCT showed a greater than five-fold increase in ion capacity over an Esquire 3000plus (for a given spectral limit), yielding a resolving power of 3500 for m/z 1522. These results show that with a precise design of the ion trap geometry, as well as the ejection methodology controlling the coupling to higher-order fields, one achieves a robust HCT without sacrificing scan speed, resolving power, or mass accuracy. Thus, in 2009 a new ion trap was developed that further pushes forward the limits of ion trap technology. This new spherical ion trap, called amaZon™, is capable of scanning at 52,000 Th sec⁻¹ while maintaining better than unit mass resolving power. This next generation ion trap achieves these scan speeds by taking advantage of the higher-order fields of the spherical HCT described in this chapter while further optimizing the collision gas environment in the ion trap. These most recent developments show that, while the ion trap has matured a lot over the past fifteen years, there is still room for growth and development into the future."

REFERENCES

1. Douglas, D.J.; Frank, A.J.; Mao, D.M. Linear ion traps in mass spectrometry. *Mass Spectrom. Rev.* 2005, *24*, 1–29.
2. March, R.E. Quadrupole ion trap mass spectrometry: theory, simulation, recent developments and applications. *Rapid Commun. Mass Spectrom.* 1998, *12*, 1543–1554.
3. March, R.E. Quadrupole ion trap mass spectrometry. In *Encyclopedia of Analytical Chemistry*, Ed. Meyers, R.A., 11848–11872. John Wiley & Sons Ltd.: Chichester, 2000.
4. Paul, W.; Steinwedel, H. Apparatus for Separating Charges Particles of Different Specific Charges. *U.S.Patent* 1960, 2,939,952.
5. Franzen, J.; Gabling, R-H.; Schubert, M.; Wang, Y. Nonlinear ion traps. In *Practical Aspects of Ion Trap Mass Spectrometry,* Eds. March, R E., Todd, J.F.J., Volume 1, Chapter 3, 49–167. Boca Raton, FL: CRC Press, 1995.
6. March, R.E.; Londry, F.A. Theory of quadrupole mass spectrometry. In *Practical Aspects of Ion Trap Mass Spectrometry*, Eds. March, R E., Todd, J.F.J., Volume 1, Chapter 2, 25–48. Boca Raton, FL: CRC Press, 1995.
7. Louris, J.N.; Schwartz, J.; Stafford, Jr.; G.C.; Syka, J.E.P.; Taylor, D. The Paul ion trap mass selective instability scan: trap geometry and resolution. *Proc. 40th ASMS Conference on Mass Spectrometry and Allied Topics*, Washington, DC, 1992, 1003–1004.
8. March, R E.; Todd, J.F.J. *Quadrupole Ion Trap Mass Spectrometry*. 2nd Edn. Hoboken, NJ: John Wiley & Sons, 2005.
9. Baessman, C.; Brekenfeld, A.; Zurek, G.; Schweiger-Hufnagel, U.; Lubeck, M.; Ledertheil, T.; Hartmer, R.; Schubert, M. A non-linear ion trap mass spectrometer with high ion storage capacity. *A workshop presentation at the 51st ASMS Conference on Mass Spectrometry and Allied Topics*, Montreal, PQ, 2003.
10. Schwartz, J.C.; Senko, M.W.; Syka, J.E.P. A two-dimensional quadrupole ion trap mass spectrometer. *J. Am. Soc. Mass Spectrom.* 2002, *13*, 659–669.
11. Prestage, J.D.; Janik, G.R.; Dick, G.J.; Maleki, L. Linear ion trap for 2nd-order doppler-shift reduction in frequency standard applications. *IEEE Trans Ultrason Ferroelectrics Freq Contr.* 1990, *37*, 535–542.
12. Prestage, J.D.; Dick, G.J.; Maleki, L. New ion trap for frequency standard applications. *J. App. Phys.* 1989, *66*, 1013–1017.
13. Collings, B.A.; Campbell, J.M.; Mao, D.M.; Douglas, D.J. A combined linear ion trap time-of-flight system with improved performance and MSⁿ capabilities. *Rapid Commun. Mass Spectrom.* 2001, *15*, 1777–1795.
14. Campbell, J.M.; Collings, B.A.; Douglas, D.J. A new linear ion trap time-of-flight system with tandem mass spectrometry capabilities. *Rapid Commun. Mass Spectrom.* 1998, *12*, 1463–1474.
15. Cleven, C.D.; Cooks, R.G.; Garrett, A.W.; Nogar, N.S.; Hemberger, P.H. Radial distributions and ejection times of molecular ions in an ion trap mass spectrometer: a laser tomography study of effects of ion density and molecular type. *J. Phys. Chem.* 1996, *100*, 40–46.

16. Londry, F.A.; March, R.E. Systematic factors affecting high mass-resolution and accurate mass assignment in a quadrupole ion trap. *Int. J. Mass Spectrom. Ion Processes* 1995, *144*, 87–103.

17. Wells, J.M.; McLuckey, S.A. Collision-induced dissociation (CID) of peptides and proteins. *Methods Enzymol.* 2005, *402*, 148–185.

18. Biemann, K.; Martin, S.A. Mass spectrometric determination of the amino acid sequence of peptides and proteins. *Mass Spectrom. Rev.* 1987, *6*, 1–76.

19. Aebersold, R.; Mann, M. Mass spectrometry-based proteomics. *Nature FIELD Publication* 2003, *422*, 198–207.

20. Wysocki, V.H.; Tsaprailis, G.; Smith, L.L.; Breci, L.A. Mobile and localized protons: a framework for understanding peptide dissociation. *J. Mass Spectrom.* 2000, *35*, 1399–1406.

21. Tsaprailis, G.; Nair, H.; Somogyi, A.; Wysocki, V.H.; Zhong, W.; Futrell, J.H.; Summerfield, S.G.; Gaskell, S.J. Influence of secondary structure on the fragmentation of protonated peptides. *J. Am. Chem. Soc.* 1999, *121*, 5142–5154.

22. Huang, Y.Y.; Triscari, J.M.; Pasa-Tolic, L.; Anderson, G.A.; Lipton, M.S.; Smith, R.D.; Wysocki, V.H. Dissociation behavior of doubly-charged tryptic peptides: correlation of gas-phase cleavage abundance with Ramachandran plots. *J. Am. Chem. Soc.* 2004, *126*, 3034–3035.

23. Brekenfeld, A.; Hartmer, R.; Ledertheil, T. Combination of ergodic and non-ergodic ion activation: new approach for PTM-peptide identification. *Proc. 53rd ASMS Conference on Mass Spectrometry and Allied Topics*, San Antonio, TX, 2005.

24. Hartmer, R.; Lubeck, M.; Baessmann, C.; Brekenfeld, A. Sequence characterization of intact proteins by consecutive ion/ion reactions of ETD and PTR. *Proc. 55th ASMS Conference on Mass Spectrometry and Allied Topics*, Indianapolis, IN, 2007.

25. Pitteri, S.J.; Chrisman, P.A.; Hogan, J.M.; McLuckey, S.A. Electron transfer ion/ion reactions in a three-dimensional quadrupole ion trap: reactions of doubly and triply protonated peptides with SO_2^-. *Anal. Chem.* 2005, *77*, 1831–1839.

26. Syka, J.E.P.; Coon, J.J.; Schroeder, M.J.; Shabanowitz, J.; Hunt, D.F. Peptide and protein sequence analysis by electron transfer dissociation mass spectrometry. *Proc. Natl. Acad. Sci. USA.* 2004, *101*, 9528–9533.

27. McLuckey, S.A.; Stephenson, Jr.; J.L. Ion/ion chemistry of high-mass multiply charged ions. *Mass Spectrom. Rev.* 1998, *17*, 369–407.

28. Pitteri, S.J.; McLuckey, S.A. Recent developments in the ion/ion chemistry of high-mass multiply charged ions. *Mass Spectrom. Rev.* 2005, *24*, 931–958.

29. Wilson, J.; Brekenfeld, A.; Baessman, C.; Hartmer, R. Electron transfer dissociation for sequence characterization of larger peptides and small proteins. *Proc. 54th ASMS Conference on Mass Spectrometry and Allied Topics*, Seattle, WA, 2006.

30. Coon, J.J.; Syka, J.E.P.; Schwartz, J.C.; Shabanowitz, J.; Hunt, D.F. Anion dependence in the partitioning between proton and electron transfer in ion/ion reactions. *Int. J. Mass Spectrom.* 2004, *236*, 33–42.

31. Good, D.M.; Wirtala, M.; McAlister, G.C.; Coon, J.J. Performance characteristics of electron transfer dissociation mass spectrometry. *Mol. Cell Proteomics* 2007, *6*, 1942–1951.

32. Lubeck, M.; Baessman, C.; Brekenfeld, A.; Schubert, M. A novel high-speed duty cycle of ion traps for a more detailed analysis of complex protein mixtures. *Proc. 53rd ASMS Conference on Mass Spectrometry and Allied Topics*, San Antonio, TX, 2005.

14 Electrically Induced Nonlinear Ion Traps

Gregory J. Wells and August A. Specht

CONTENTS

14.1   Introduction 620
    14.1.1   Why are Nonlinear Trapping Fields of Interest? 620
    14.1.2   Means of Forming Nonlinear Trapping Fields 621
14.2   Theory of Nonlinear Trapping Fields 621
    14.2.1   Multipole Field in a Three-Dimensional Ion Trap 621
    14.2.2   Dipole Component of the Trapping Field......................................624
    14.2.3   Hexapole and Octopole Components of the Trapping Field 627
14.3   Electrically Induced Nonlinear Fields in a Three-Dimensional
    Ion Trap..628
    14.3.1   Overview 628
    14.3.2   Triple Resonance Ion Ejection 629
    14.3.3   Ion Trajectory Simulations..630
        14.3.3.1   Frequency Spectrum with a Nonlinear
                Trapping Field..630
        14.3.3.2   Ion Ejection..630
14.4   Experimental Results..632
    14.4.1   Effects of Trapping Field Dipole..632
        14.4.1.1   Dipole Effects 632
        14.4.1.2   Supplementary Quadrupole Potential..........................635
        14.4.1.3   Triple Resonance 636
    14.4.2   Scan Performance 638
    14.4.3   Charge Capacity Measurements 640
    14.4.4   Summary..641
14.5   Electrically Induced Nonlinear Fields in a Two-Dimensional Ion Trap 643
    14.5.1   Overview 643
    14.5.2   Theory of Triple Resonance Ion Ejection Scanning 645
        14.5.2.1   Dipole Component of the Trapping Field 649
        14.5.2.2   Hexapole Component of the Trapping Field................652
    14.5.3   A Degeneracy Dilemma..653
    14.5.4   Implementation of Triple Resonance Ion Ejection Scanning........655
14.6   Ion Simulations 656
    14.6.1   Frequency Spectrum 656
    14.6.2   Degeneracy Effects 658

14.7   Experimental Results of Triple Resonance Scanning in a 2D Trap 660
        14.7.1   Effects with and Without Degeneracy 660
        14.7.2   Tuning 661
        14.7.3   Mass Spectral Performance 664
        14.7.4   Charge Capacity Measurements 665
14.8   Summary 667
References 668

14.1   INTRODUCTION

Ions are confined within an electrodynamic three-dimensional quadrupole field of rotational symmetry when their trajectories are bounded in the radial (r) and axial (z) directions. The ion motion in the trapping field is pseudo-harmonic. In a pure quadrupolar trapping field the ion motions in both the radial and axial directions are independent of each other. The equations of motion for a single ion in the trapping field can be resolved into a pure axial motion and a pure radial motion, which have identical mathematical forms described by the Mathieu equation [1]. The Mathieu equation for the axial motion contains two parameters, a_z and q_z, which characterize the solutions in the axial direction. Similar parameters, a_r and q_r, exist for the radial motions. These parameters define a two-dimensional region in (a_i, q_i) space for the coordinate ($i = r$ or z), in which the ion motions are bounded and therefore stable. For small values of q_i, the pseudo-harmonic motion of an ion can be characterized by the dominant fundamental frequency for motion in the (i) coordinate [1]. A point in (a_i, q_i) space defines the operating, or 'working,' point for the ion. The amplitude of the ion motion in the radial or axial direction can be increased by the application of a supplementary alternating field having a symmetry and a frequency that are in resonance with one of the frequencies of the ion motion. If the amplitude of the ion motion is increased sufficiently, the ion will be driven to the surface of an electrode. When a hole exists in the electrode to which the ion is directed, the ion will escape the trapping field altogether and exit the trap whereupon it can be detected. This manipulation of the ion motion forms the basis for mass scanning in ion trap mass spectrometers.

14.1.1   WHY ARE NONLINEAR TRAPPING FIELDS OF INTEREST?

Ideal quadrupole fields are difficult to achieve in practical trapping devices or mass filters. In this latter class of mass analyzers, designers have sought to minimize the presence of non-ideal fields by utilizing hyperbolic shaped electrodes. The early scientific literature abounds with references to the detrimental effects of non-ideal quadrupole fields, and these effects have been the subject of much theoretical and experimental research [2,3]. Non-ideal fields can be viewed as a quadrupole field upon which higher-order multipole fields are superimposed. The early work of von Busch and Paul [2], and later that of Dawson and Whetten [3], was directed to identifying the effects of these higher-order multipole fields and finding means to minimize their undesired effects. More recently, Yang and Franzen [4] have demonstrated how higher-order multipole fields can be used advantageously in three-dimensional ion trapping devices to improve mass spectral performance.

Multipole components have an effect both on the ion motion, as well as generating nonlinear resonances in the trapping field. The latter are operating points within the normal stability region in which the ions will increase their amplitude of oscillation without the need for any supplementary dipolar resonance field. A hexapole component added to the trapping field has been found to be particularly advantageous in that the associated nonlinear resonance can be used to eject ions for mass scanning [5–8]. The addition of an octopole component to the field is known to improve the coherence of ion ejection and to reduce the time distribution of ions of the same mass-to-charge ratio that are ejected by dipolar resonance ejection even when the operating points are not at the octopole nonlinear resonance [5–8]. Multipolar trapping fields are now used advantageously in all commercial three-dimensional ion traps of rotational symmetry. A new generation of linear two-dimensional ion traps is being developed commercially; however, these devices have yet to exploit fully all of the advantages of multipole trapping fields.

14.1.2 MEANS OF FORMING NONLINEAR TRAPPING FIELDS

Several different methods of forming additional multipole components are known. Thermo Finnigan ion traps are intentionally 'stretched' (that is, by defining a specific end-cap electrode shape and spacing to produce a pure quadrupolar field and then extending the value of z_o (by a set amount) to produce weak, even multipoles [9]. Wang *et al.* [10,11] have used electrodes for three-dimensional quadrupolar fields of rotational symmetry which are shaped to provide additional multipoles. This 'mechanical' approach utilizes end-cap electrodes or the ring electrode in which the asymptotic cone angle deviates from that of the ideal quadrupolar value. An alternative means described by Wells *et al.* [12,13] used an electrical approach which adds both higher and lower multipole components to the primary quadrupolar trapping field. This 'electrical' approach to multipole formation applies a dipole to the end-cap electrodes at the same frequency as the trapping field. The end-cap and ring electrodes have a geometry appropriate for the generation of an ideal quadrupole field—assuming they were infinite sheets. However, because the hyperbolic end-cap electrodes neither extend to infinity nor are free of orifices, additional higher-order multipoles are formed in the trapping field in addition to the lower-order dipole component. Because the additional multipoles are formed by electrical means, these fields can be optimized easily and can be turned on or off.

14.2 THEORY OF NONLINEAR TRAPPING FIELDS

14.2.1 MULTIPOLE FIELD IN A THREE-DIMENSIONAL ION TRAP

The canonical form of the electrodynamic trapping potential for a time-dependent field in a cylindrical coordinate system (r, z) is given by [14]

$$V_T(r,z,t) = \sum_{N=0}^{\infty} A_N \Phi_N(r,z) \Pi(t) + \sum_{N=0}^{\infty} B_N U_N(r,z) \tag{14.1},$$

where $\Pi(t) = \cos(\Omega t)$ expresses the temporal variations of the field with drive frequency Ω, $\Phi_N(r, z)$, and $U_N(r, z)$ represent the dynamic and static spatial variations of the field, and A_N, B_N are normalized constants. The spatial terms are related to the Legendre polynomials $P_N \cos(\theta)$ of order N in a field with rotational symmetry. The terms of the polynomial are expressed here as a function of the cylindrical coordinates (r, z) and the arbitrary distance s necessary to fix the boundary conditions [15]. The first six terms of the series are:

Dipole $P_1(\cos\theta) = z/s$ (14.2a);

Quadrupole $P_2(\cos\theta) = (2z^2 - r^2)/2s^2$ (14.2b);

Hexapole $P_3(\cos\theta) = (2z^3 - 3zr^2)/2s^3$ (14.2c);

Octopole $P_4(\cos\theta) = (8z^4 - 24z^2r^2 + 3r^4)/8s^4$ (14.2d);

Decapole $P_5(\cos\theta) = (8z^5 - 40z^3r^2 + 15zr^4)/8s^5$ (14.2e);

Dodecapole $P_6(\cos\theta) = (16z^6 - 120z^4r^2 + 90z^2r^4 - 5r^6)/16s^6$ (14.2f).

An ideal hyperbolic ion trap with electrodes extending to infinity would only have the quadrupole term Equation 14.2b, with $N = 2$ in Equation 14.1. Real ion traps with electrodes of finite extent will have additional terms with $N > 2$. Terms with order greater than 2 will cause a coupling between the radial and axial ion motion, and the superposition of multipolar fields will generate nonlinear resonances in the trapping field. When the operating point of the ion motion is at a nonlinear resonance, ions may absorb energy from the trapping field alone, resulting in an increase in the amplitude of the ion motion.

Wang has shown that nonlinear resonances in the trapping field are found when

$$n_r \omega_r + n_z \omega_z = \upsilon \Omega \tag{14.3},$$

where ω_r and ω_z are the radial and secular frequencies and υ, n_r, and n_z are integers [5]. In particular, the next highest multipole field after the quadrupole is the hexapole, where $N = 3$, Equation 14.2c. The use of this nonlinear resonance to eject ions axially from the ion trap has been discussed in detail in publications by Franzen, where the hexapole component has been increased by shaping the end-cap electrodes with a different angle [6–8].

The general form of the equations of motion can be obtained from the vector equation

$$m\frac{\partial^2 \vec{R}}{\partial t^2} + e\vec{\nabla}V_T = 0 \tag{14.4},$$

where the position vector is $\vec{R}(r,z)$, m is the ion mass and e is the charge on the ion. Referring to Equations 14.2 and 14.4, it can be seen that all components of force due to multipole fields greater than the dipole have a zero value at the center of the ion trap.

Equation 14.1 corresponds to an ideal quadrupole trapping field extending to infinity when $A_N = 0$ and $B_N = 0$ for all values of N except $N = 2$, and the periodic voltage $\Pi(t) = \cos(\Omega t)$ is applied to the ring electrode. The potential inside the ion trap with the end-cap electrodes grounded, $V_T(z,r,t) = V_{TFQ}$, is given from [16]

$$V_{TFQ} = \frac{V_{ac}}{(r_0^2 + 2z_0^2)}\left[r^2 - 2(z^2 - z_0^2)\right]\cos(\Omega t) + \frac{U}{(r_0^2 + 2z_0^2)}\left[r^2 - 2(z^2 - z_0^2)\right] \quad (14.5),$$

where V_{ac} is the zero-to-peak amplitude of the electrodynamic potential with frequency Ω, r_0 is the radius of the ring electrode, U is the DC potential applied to the ring electrode, and z_0 is the separation of the end-cap electrodes from the origin. The equations of the ion motion for this case allow the separation of the motion into the radial and axial components. The axial equation of motion for an ion is described by the differential equation

$$\frac{\partial^2 z}{\partial t^2} - \left(\frac{e}{m}\right)\left(\frac{4U}{r_0^2 + 2z_0^2}\right)z - \left(\frac{e}{m}\right)\left(\frac{4V_{ac}}{r_0^2 + 2z_0^2}\right)z\left[\cos(\Omega t)\right] = 0 \quad (14.6).$$

Defining the dimensionless parameters ζ, q_z, and a_z as [1]

$$\zeta = \frac{\Omega t}{2} \quad (14.7a)$$

$$q_z = +8eV_{ac}/[m(r_0^2 + 2z_0^2)\Omega^2] \quad (14.7b)$$

$$a_z = -16\,eU/[m(r_0^2 + 2z_0^2)\Omega^2] \quad (14.7c)$$

and substituting the ζ, q_z, and a_z from Equation 14.7 into Equation 14.6, the equation of motion for an ion can be rearranged to

$$\frac{d^2 z}{d\zeta^2} + \left[a_z - 2q_z\cos(2\zeta)\right]z = 0 \quad (14.8).$$

This second-order differential equation is the well-known Mathieu equation [1]. The stable solutions to the equation in the axial direction are characterized by the parameters q_z and a_z; the value of the parameters define the operating point of the ion within the stability region. The general solution to Equation 14.8 is

$$z(\zeta) = A\sum_{n=-\infty}^{+\infty}C_{2n}\cos(2n+\beta)\zeta + B\sum_{n=-\infty}^{+\infty}C_{2n}\sin(2n+\beta)\zeta \quad (14.9),$$

where A, B, and C_{2n} are normalization constants. The secular frequency of the ion motion in the z-direction, $\omega_{z,n}$, can be determined from the value of β_z [1] by

$$\omega_{z,n} = (n + \beta_z / 2)\Omega, \text{ or } \omega_{z,n} = (n - \beta_z / 2)\Omega, \text{ where } n = 0, 1, 2 \ldots \quad (14.10).$$

The value of β_z is a function of the working point in (a_z, q_z) space and can be computed from a well-known continuing fraction [1].

14.2.2 DIPOLE COMPONENT OF THE TRAPPING FIELD

To simplify the discussion, it is assumed that the ion trap is operated without the DC field, so that in Equation 14.5, $U = 0$. When the dipole component of the trapping field, where the spatial component is $V_{TFD} = V_{da} (z/z_0)\cos(\Omega t + \varphi)$ (Equation 14.2a), is added to the quadrupole trapping field, V_{TFQ} (Equation 14.5), the potential inside the ion trap $V_T (z, r, t)$ can be expressed as

$$V_T(z,r,t) = V_{TFQ} + V_{TFD} = \frac{V_{ac}}{(r_0^2 + 2z_0^2)}\left[r^2 - 2(z^2 - z_0^2)\right]\cos(\Omega t) = \frac{V_{da}z}{z_0}\cos(\Omega t + \varphi)$$

$$(14.11),$$

where V_{da} is the zero-to-peak amplitude of the trapping field dipole (TFD) applied between the end-cap electrodes. The TFD is phase-shifted by $+\varphi$ with respect to the quadrupole trapping field, V_{TFQ}. Restricting the phase to values of φ to $\varphi = N\pi$, where $N = 0, 1, 2, \ldots$; $V_d = V_{da}(-1)^N$, the instantaneous electric field acting on an ion in the axial direction due to the trapping potential $V_T (z, r, t)$ is

$$E_z = -\frac{\partial V}{\partial z} = -\left(-\frac{4V_{ac}}{(r_0^2 + 2z_0^2)}z + \frac{V_d}{z_0}\right)\cos(\Omega t) \quad (14.12),$$

and the equation of ion motion becomes, therefore,

$$m\frac{d^2z}{dt^2} = -e\left(-\frac{4V_{ac}}{(r_0^2 + 2z_0^2)} + \frac{V_d}{z_0}\right)\cos(\Omega t) \quad (14.13).$$

Substituting $\zeta = \frac{\Omega t}{2}$ we obtain

$$\frac{d^2z}{dt^2} = \frac{\Omega^2}{4}\frac{d^2z}{d\zeta^2} \quad (14.14),$$

so that by substitution of Equation 14.14 into Equation 14.13 and writing $2\zeta = \Omega t$, the basic equation of an ion motion in the axial direction becomes

$$\frac{d^2z}{d\zeta^2} = \frac{4e}{m\Omega^2}\left(\frac{4V_{ac}}{\left(r_0^2+2z_0^2\right)}z - \frac{V_d}{z_0}\right)\cos(2\zeta) \tag{14.15}.$$

On rearranging

$$\frac{d^2z}{d\zeta^2} - \left[2\left(\frac{8eV_{ac}}{m\left(r_0^2+2z_0^2\right)\Omega^2}z - \frac{2eV_d}{mz_0\Omega^2}\right)\right]\cos(2\zeta) = 0 \tag{14.16},$$

and defining

$$q_z = \frac{8eV_{ac}}{m\left(r_0^2+2z_0^2\right)\Omega^2} \tag{14.7b},$$

$$q_d = -\frac{2eV_d}{mz_0\Omega^2} \tag{14.17},$$

substitution of Equation 14.7b and 14.17 into Equation 14.16, gives an equation similar to the Mathieu equation

$$\frac{d^2z}{d\zeta^2} - 2\left(q_z z + q_d\right)\cos(2\zeta) = 0 \tag{14.18}.$$

With the definition and substitution $u = (q_z z + q_d)$ and $(d^2u/d\zeta^2) = q_z\,(d^2u/d\zeta^2)$ into Equation 14.18 we obtain the form of the Mathieu equation

$$\frac{d^2u}{d\zeta^2} - 2q_z u\cos(2\zeta) = 0 \tag{14.19}.$$

The axial displacement of the ion can found to be the sum of two terms

$$z = \frac{u - q_d}{q_z} = \frac{u}{q_z} - \frac{q_d}{q_z} \tag{14.20}.$$

The first term represents the normal time-dependent oscillatory solution, u (ζ), as in Equation 14.9, whilst the second term is an additive offset value which expresses the axial displacement of the ion motion due to the TFD as

$$-\frac{q_d}{q_z} = -\left(-\frac{2eV_d}{mz_0\Omega^2}\right)\left[\frac{m\left(r_0^2+2z_0^2\right)\Omega^2}{8eV_{ac}}\right] = \frac{\left(r_0^2+2z_0^2\right)V_d}{4z_0V_{ac}} \tag{14.21}.$$

During mass analysis ions are ejected sequentially at the same operating point, q_z, and as a consequence V_{ac}/m will be constant. In the special case in which $V_d = \eta V_{ac}$ Equation 14.21 becomes

$$-\frac{q_d}{q_z} = \frac{r_0^2+2z_0^2}{4z_0}\eta$$

and thus

$$z = \frac{u}{q_z} + \frac{r_0^2 + 2z_0^2}{4z_0}\eta \qquad (14.22).$$

When the TFD is phased properly and is present as a constant fraction η of the trapping field, it can be seen from Equation 14.22, that the ion motion is displaced uniformly in the axial direction by a constant amount. The magnitude and sign of the displacement is independent of the mass-to-charge ratio and the polarity of the ion charge. The displacement depends only on the percentage η of TFD and the geometry of the ion trap assembly. The direction of the displacement can be altered by changing the phase of the dipole from 0 to π.

The first-order effect of adding the dipole component to the trapping field is to displace the ions toward the end-cap electrode that has the TFD component in phase with the RF voltage applied to the ring electrode. Figure 14.1 shows a SIMION calculation of equipotential surfaces within an ion trap using the Saturn 2000 electrodes ($r_0 = 10$ mm and $z_0 = 7.07$ mm) to define the hyperbolic surfaces [17]. In this model the ring electrode is at a static potential of 1 volt, and the end-cap electrodes are at −0.075 and +0.075 volt, respectively; the entire electrode assembly is surrounded by a grounded cylinder (not shown). The potential on the end-cap electrodes creates a 15% TFD along the z-axis. The end-cap electrodes were displaced an additional 0.75 mm. along the z-axis, away from the center. This arrangement is often referred to as the 'stretched trap' geometry [9]. The equipotential lines plotted in Figure 14.1 show

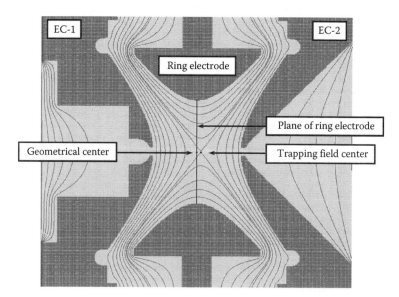

FIGURE 14.1 SIMION calculation of the equipotential surfaces in the Saturn 2000 electrodes, using static voltages, with the addition of a 15% trapping field dipole: ring electrode = +1 V, End-cap electrode 1 (EC-1) = +0.075 V, End-cap electrode 2 (EC-2) = −0.075 V. The potentials on the end-cap electrodes create a 15% TFD along the z-axis.

that the trapping field center is displaced toward the exit end-cap electrode (EC-2). Additionally, the axial pseudopotential well is lower toward the exit end-cap electrode relative to the entrance end-cap electrode. Therefore, ions will approach the exit end-cap electrode more closely than they do the entrance end-cap electrode (EC-1). The ions will be ejected preferentially through the hole in the exit end-cap electrode because their kinetic energy exceeds first the axial restoring force acting in the direction of the exit end-cap electrode. Unidirectional ion ejection with the TFD is observed experimentally when the ejection occurs by axial instability at $\beta_z = 1$, or when resonance ejection occurs using a supplementary dipole field at lower operating points.

14.2.3 HEXAPOLE AND OCTAPOLE COMPONENTS OF THE TRAPPING FIELD

The potentials at the grid points in a selected rectangular region along the z-axis were used in a multivariate regression fit to the first six terms of the expression for the multipole expansion of the potential given by in Equation 14.1. Table 14.1 lists the coefficients calculated for four different electrode conditions. The coefficients for the asymmetric trapping field, reported in the last column of Table 14.1, were calculated for a truncated, 'stretched' geometry (made by increasing the value of z_0, originally equal to 7.07 mm, by a further 0.75 mm) with a 15% TFD. It can be seen that the truncation of the hyperbolic surfaces introduced even-order multipole fields, corresponding to a predominately positive octapole contribution. The 'stretched' geometry has an octapole component that has changed sign compared to that in the ideal truncated version. Moreover, for the stretched and asymmetric trapping field geometries, a positive monopole term is obtained. In addition to the large dipole coefficient obtained when the TFD is present, a substantial negative hexapole coefficient of −0.04 is found also.

TABLE 14.1
A_N Coefficients for Equation (14.1)

N	Multipole Field	Ideal[a]	Truncated[b]	Stretched[c]	Asymmetric[d]
			Trapping Field Geometry		
0	Monopole	0	0	0.112	0.112
1	Dipole	0	0	0	−0.10
2	Quadrupole	−2	−2.003	−1.71	−1.75
3	Hexapole	0	0	0	−0.04
4	Octapole	0	0.004	−0.035	−0.03
5	Decapole	0	0	0	−0.008

[a] Ideal ion trap geometry with hyperbolic electrode surfaces extended to infinity, $r_0^2 = 2z_0^2$ with $r_0 = 10$ mm.

[b] Ion trap geometry with hyperbolic electrode surfaces truncated such that the length of each asymptote is $2.8r_0$, with $r_0 = 10$ mm.

[c] Ion trap geometry as for b, but stretched to $2z_0 = 15.64$ mm.

[d] Ion trap geometry as for c, with a 15% TFD.

14.3   ELECTRICALLY INDUCED NONLINEAR FIELDS IN A THREE-DIMENSIONAL ION TRAP

14.3.1   OVERVIEW

March and co-workers [18,19] used *dipolar resonant excitation* to eject unwanted ions from a quadrupole ion trap. In these studies, a supplementary alternating voltage was applied to the end-cap electrodes of the ion trap, out of phase, to produce an alternating dipole field in the axial direction. Resonant ejection occurs only for those ions having an axial frequency of motion equal to the frequency of the supplementary alternating field. The ions in resonance with the supplementary field increase the amplitude of their axial oscillation until the kinetic energy of the ions exceeds the restoring force of the RF trapping field, and ion ejection occurs in the axial direction.

Jefferts [20] used a quadrupole ion trap to distinguish ions of different mass-to-charge ratios that were formed by photodissociation inside of the trap. The trapping field frequency was swept and ions of successive mass-to-charge ratios were made unstable in the axial direction and were ejected sequentially from the trap and detected by an electron multiplier. Subsequently, Ensberg and Jefferts [21] used dipolar resonant excitation to eject ions out the trap to an external detector by applying an axial resonant field to the end-cap electrodes. The frequency of the applied field was swept and ions of successive mass-to-charge ratios were ejected from the trap. A variant of these methods is used in commercial ion trap mass spectrometers to eject ions by dipolar resonant excitation. The amplitude of the RF trapping field is increased linearly to move the operating point (q_z, a_z) of the ions until the fundamental secular frequency of ion motion comes into resonance with a supplementary alternating voltage on the end-cap electrodes and resonant ejection occurs.

Ion ejection by *quadrupolar resonant excitation* can be effected by the application of a supplementary alternating voltage applied in phase to the end-cap electrodes. *Parametric resonant excitation* by a supplementary quadrupole field causes ion amplitudes to increase in the axial direction when the ion frequency is one half of the supplementary quadrupole frequency [22]. Parametric resonant excitation has been investigated theoretically [22,23] and the resonant frequencies for quadrupolar excitation are

$$\omega_p = 2\omega_z / N \qquad (14.23),$$

where $N = 1, 2, 3, \ldots$, and ω_z is the axial dipole secular frequency defined in Equation 14.10.

The supplementary quadrupole field has a value of zero at the center of the ion trap. When a buffer gas such as helium is used to damp the ion trajectories to the center of the trap, parametric excitation is ineffectual due to the vanishing strength of the supplementary quadrupole field. It is necessary to displace the ions from the center of the supplementary quadrupole field to a location where the field has a non-zero value, in order to have a finite excitation force applied to the ions.

A weak resonant dipole field having a frequency of one half of the parametric frequency ($N = 1$, Equation 14.23), displaces ions from the center of the trap when the operating point of the ions is changed to bring the ion fundamental frequency of ion motion into resonance with the dipole field [22,24]. Since the parametric frequency is twice the dipole frequency the ions will absorb power from the supplementary quadrupole field. It is possible also at the unique operating point of $\beta_z = 2/3$ to use the first side-band frequency ($n = 1$, Equation 14.10) to displace ions from the trapping center. This mode of ion ejection, in which power is absorbed sequentially from the dipole and then from the quadrupole field, is adequate for ion ejection in a static trapping field, where the fundamental frequency of the ion motion is not changing due to the amplitude of the RF field. However, it is not optimal when the trapping field amplitude is changing, as is normally the case for mass scanning. In this case, the RF trapping field amplitude is increased to increase the fundamental secular frequency of the ion motion, bringing it into resonance first with the dipole field. The dipole field displaces the ion from the center of the trap, where the quadrupole field is zero. After the ion has been displaced from the center, it can then absorb power from the supplementary quadrupole field if it is in resonance with the parametric resonance. Therefore, it is necessary to fix the dipole resonant frequency at a value less than one half of the parametric resonance [24,25] so that as the fundamental frequency of the ion motion is increased, by increasing the trapping field RF amplitude, the ion motion will be in resonance sequentially with the dipole field, and then with the quadrupole field.

The geometry of commercial Varian ion traps electrodes are chosen to have higher-order fields, primarily an octopole component, obtained by increasing the separation of the end caps while maintaining the ideal hyperbolic surface. These surfaces have asymptotes at 35.26° with respect to the symmetric radial plane of the ion trap. Alternatively, the surfaces of the end-cap electrodes can be shaped with an angle of 35.96° [26–28] while maintaining the ideal separation between the end-cap electrodes. For either geometry the trapping field is symmetric with respect to the radial plane.

14.3.2 TRIPLE RESONANCE ION EJECTION

A new ion ejection method [12,13,29] is presented that uses a trapping field which is asymmetric with respect to the radial plane. The asymmetric trapping field is generated by adding an alternating voltage out-of-phase to each end-cap electrode at the same frequency as the ring electrode. This trapping field dipole component (TFD) causes the center of the trapping field to be not coincident with the geometric center of the ion trap electrode assembly. The first-order effect of adding the dipole component to the trapping field is to displace the ions toward the end-cap electrode that has the TFD component out-of-phase with the RF voltage applied to the ring electrode. A second-order effect is to superimpose a substantial hexapole field [30,31]. The resulting multipole trapping field has a nonlinear resonance at the operating point of $\beta_z = 2/3$ [30]. Because the ions are displaced already from the geometrical center of the trap by the asymmetric trapping field, the ions may now became subject to a hexapole resonance of finite value. Likewise, at this operating point of $\beta_z = 2/3$, a parametric resonance due to a supplementary quadrupole field will also have a non-zero value. Finally, the addition of a supplementary dipole field of the appropriate frequency will cause the ions to experience dipolar resonant excitation also. All three fields will have non-zero values at the operating point of $\beta_z = 2/3$ and, therefore, a triple resonance condition exists. An ion moved to this operating point will be in resonance with, and will absorb power from, all three fields simultaneously.

At the operating point of the triple resonance, power absorption by the ions is nonlinear. The amplitude of the axial ion motion increases nonlinearly and the ion is ejected rapidly from the trap. As ions suffer fewer collisions during rapid ejection their trajectories are less affected by collisions with the damping gas in the region of the resonance, and resolution is improved. Traditionally, damping gas is used during the trapping process to reduce the large initial spatial distribution of ions. During the ejection process, however, the damping gas will again increase the spatial distribution of ions causing a broadening of the spectral peak. Finally, the displacement of the trapping center toward the exit end-cap electrode causes the ions to be ejected exclusively through this electrode; thus doubling the number of ions detected.

14.3.3   Ion Trajectory Simulations

14.3.3.1   Frequency Spectrum with a Nonlinear Trapping Field

The axial trajectory of an ion of m/z 100 confined in a stretched ion trap without and with an asymmetric trapping field was computed using SIMION [17]. The TFD was 15%, the trap frequency was 1050 kHz, and the operating point of the ion in the stability diagram was $\beta_z = 0.51$. The maximum axial excursion of the ion was 6 mm from center of the field. The frequency spectrum of the ion trajectory obtained without the TFD (see Figure 14.2a) was obtained by Fourier analysis of 4000 data points of the ion trajectory. The frequency spectrum ranges from 0 to 2000 kHz, and the fundamental secular frequency of the ion motion $\omega_{z,0}$, is observed at approximately 270 kHz. The two side-bands, $(\Omega - \omega_{z,0})$ and $(\Omega + \omega_{z,0})$ are observed, as expected. The frequency spectrum of the ion trajectory with the TFD shown in Figure 14.2b possesses a greater number of frequencies because the TFD introduces a hexapole component in the trapping field. The drive frequency, Ω, at 1050 kHz and the second, third, and fourth harmonics, $2\omega_{z,0}$, $3\omega_{z,0}$, and $4\omega_{z,0}$, respectively, are observed also. The observation of these frequencies is to be expected from the superposition of odd-multipole components on the trapping field [6–8,10]. The side-bands $(\Omega - 3\omega_{z,0})$, $(\Omega - 2\omega_{z,0})$, and $(\Omega + 2\omega_{z,0})$ are observed also, though with lower intensity. In addition, the more intense side-bands of higher harmonics, $(2\Omega - \omega_{z,0})$, $(2\Omega - 2\omega_{z,0})$, $(2\Omega - 3\omega_{z,0})$, and $(2\Omega - 4\omega_{z,0})$, are observed. Franzen reported a remarkably similar frequency spectrum obtained from a modified-angle ion trap with non-ideal electrodes [32].

14.3.3.2   Ion Ejection

Two simulations of the ion trajectory for m/z 100 were computed to compare the dipolar resonant ejection and the triple resonant ejection modes using the frequencies given in Table 14.2. The model included collisions with the helium buffer gas having a Boltzmann velocity distribution at 373 K and a Langevin collision cross-section, with isotropic scattering in the center-of-mass that was transformed into the laboratory coordinates. All fields were generated using the stretched geometry for both ion ejection modes. Initially, the operating point was $\beta_z = 0.25$ and the ion

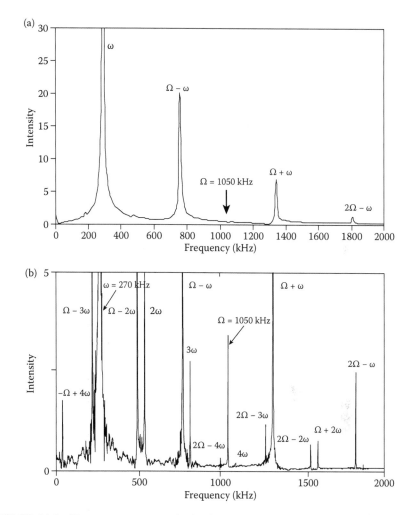

FIGURE 14.2 Frequency spectrum obtained by Fourier analysis of the calculated ion trajectory acquired using SIMION. The ion of m/z 100 was confined in a stretched 3D trap operating at main quadrupole trapping frequency of 1050 kHz and the voltage selected such that the ion was held at the operating point $\beta_z = 0.51$. (a) Frequency spectrum resulting from a standard quadrupole potential. (b) Frequency spectrum resulting from the addition of a 15% TFD. The addition of the TFD causes a hexapolar field component, which results in the observation of a number of side-bands and higher harmonics.

underwent approximately 70 collisions with the helium atoms during a 1 ms cooling period. The RF trapping voltage was increased rapidly during 0.1 ms to attain an operating point of the ion corresponding to 7 Th below either $\beta_z = 2/3$ or $\beta_z = 0.92$. Thereafter, the RF trapping voltage was increased at a rate of 0.18 Th ms^{-1}.

The ion trajectory was calculated in three dimensions, and the axial displacements and kinetic energies obtained with dipolar resonant excitation are shown in

TABLE 14.2

Frequencies of Ion Motion in the 3D Ion Trap (kHz), for the Triple Resonant Ejection and the Dipolar Resonant Ejection Modes Employed in the Simulation of Two Trajectories of the Ion m/z 100 (see text)

Frequencies	Triple Resonant Operating Point: $\beta_z = 2/3$	Dipolar Resonant Operating Point: $\beta_z = 0.92$
Ω	1050	1050
$\omega_{z,0}$	350	485
$(\Omega - \omega_{z,0})$	700	565
$\Delta\omega^a$	350	80
$2\omega_{z,0}$	700	970

[a] Beats between the side-band and the fundamental frequency components $\Delta\omega = (\Omega - 2\omega_{z,0})$.

Figure 14.3a and b. Figure 14.4a and b show the corresponding quantities obtained with triple resonant excitation. The triple resonant ejection mode shows the axial position displaced 1 mm away from the center of the ion trap as a result of the 15% TFD. The ion axial motion is limited to small excursions, ±1 mm, and ion kinetic energy is less than 10 eV up to 0.025 ms prior to ejection. In contrast, dipolar resonant excitation causes larger ion excursions and kinetic energies above 10 eV up to 0.25 ms prior to ejection. Collisions that occur during the period prior to ejection, in which the ion energy exceeds 10 eV, will cause dissociation of the ions. In this event, the resulting lower mass ions are usually at an operating point outside of the stability boundary and will be ejected in the axial direction within a few microseconds [23,32]. Generally, the width at the base of a mass peak is 0.12 ms for the triple resonant ejection mode and 0.18 ms for the dipolar resonant ejection mode. Ions ejected in the triple resonance mode from collisional dissociation during the 0.025 ms prior to the normal ejection will have only a small effect on the peak symmetry. However, in the dipolar resonant mode, ions ejected during the 0.25 ms prior to normal ejection will cause a shift in the mass centroid toward low mass.

14.4 EXPERIMENTAL RESULTS

14.4.1 Effects of Trapping Field Dipole

14.4.1.1 Dipole Effects

The mass spectra of hexachlorobenzene (HCB) formed by electron ionization (EI) in the interior of the trap, shown in Figure 14.5, were obtained by applying a supplementary dipole field between the end-cap electrodes having a frequency of 485 kHz. The mass spectra in Figure 14.5a and b were obtained with and without a 15% TFD, respectively. In both mass spectra the characteristic chlorine isotopic cluster is observed due to the molecular ion $C_6Cl_6^+$, with m/z 282, 284, 286, 288, 290, 292; the fragment ions $C_6Cl_5^+$ with m/z 247, 249, 251, 253, 255; $C_6Cl_4^+$ with

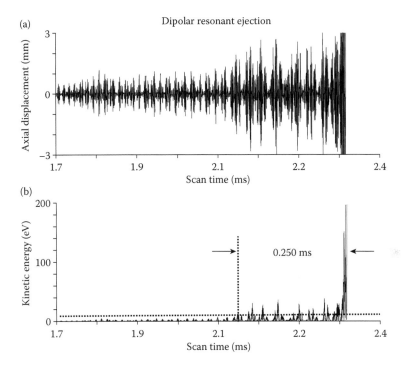

FIGURE 14.3 SIMION simulation of the last 0.7 ms of a quadrupolar voltage-ramped dipole-induced ejection at $\beta_z = 0.92$. (a) The z-axis displacement during the last 0.25 ms shows large excursions of the ion location. (b) The kinetic energy during the ejection process shows energies above 10 eV up to 0.25 ms prior to ejection. The horizontal dashed line represents a 10 eV kinetic energy.

m/z 212, 214, 216, 218; $C_6Cl_3^+$ with m/z 177, 179, 181; $C_6Cl_2^+$ with m/z 142, 144; C_6Cl^+ with m/z 107, 109. The relative ion abundances are very similar in both mass spectra. However, the absolute abundance of all ions is doubled when the TFD is added. Unidirectional ejection was observed also with the TFD, in the absence of the supplementary dipole field, when ions are ejected at the stability boundary at $\beta_z = 1$.

Figure 14.6a shows the effects of different percentages of TFD on the signal intensity of m/z 69 (obtained by EI of FC-43, perfluorotributylamine, a mass spectrometry calibration compound) as a function of the amplitude of supplementary dipole field at the side-band frequency of 700 kHz ($\Omega - \omega_{z,0}$). Ion ejection at $\beta_z = 2/3$ is the result of the double resonance of the side-band resonance from the supplementary dipole field, and the hexapole nonlinear resonance of the trapping field. The phase relationship between the supplementary dipole and trapping field was optimized to maximize the signal intensity. In the absence of any TFD, the threshold voltage required to eject ions at $\beta_z = 2/3$ was 5 V_{p-p}. Acceptable resolution was obtained when the amplitude was 15 V_{p-p}. As the TFD content in the trapping field was increased, the threshold for ion ejection was observed to decrease. The reduction of threshold voltage is a consequence of the increased strength of

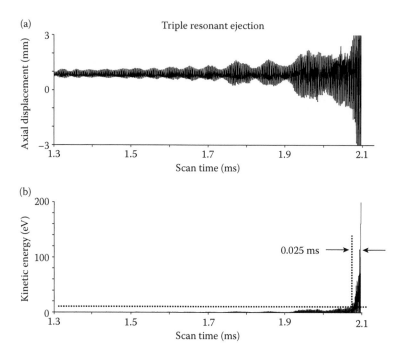

FIGURE 14.4 SIMION simulation of the last 0.7 ms of a quadrupolar voltage ramped triple resonance induced ejection at $\beta_z = 2/3$. (a) The z-axis displacement remains relatively small until only the last 0.025 ms prior to ejection. The average ion position is offset because of the trapping field dipole. (b) The kinetic energy during the ejection process shows energies less than 10 eV even up to 0.025 ms prior to ejection.

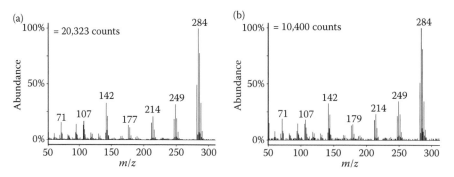

FIGURE 14.5 EI mass spectra of hexachlorobenzene obtained with a supplementary dipole amplitude of 4 V_{p-p} and frequency of 485 kHz. (a) TFD = 15%; base peak is 20,323 ion counts. (b) The TFD = 0; base peak is 10,400 ion counts.

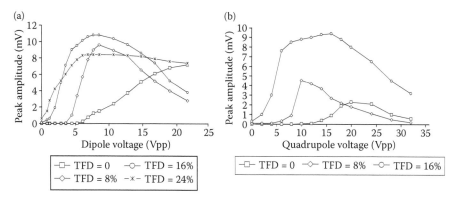

FIGURE 14.6 Ion signal intensity of m/z 69 (formed by EI of FC-43) ejected at $\beta_z = 2/3$ as a function of the amplitude of the 700 kHz supplementary field applied to the end-cap electrodes. Each curve is obtained with a different percentage of TFD. Ions are ejected by a double resonance with the hexapole resonance of the trapping field together with (a) the dipolar sideband resonance, and (b) the parametric resonance due to the supplementary quadrupole field.

the hexapole nonlinear resonance, as well as the increased displacement of the ions from the geometrical center. Similar results were obtained with m/z 414 (also from FC-43). Figure 14.6b shows similar results obtained using quadrupolar resonant excitation.

A greater influence from the odd-order fields is expected when increasing the percentage of the TFD. The advantage of forming the TFD electronically is that the circuit can be adjusted so that the percentage of TFD added can be varied. Figure 14.7 shows the effect of varying the TFD from 0 to 49% on the experimental EI mass spectra of CF_3^+ (m/z 69 from FC-43), with no supplementary voltages applied to the end-cap electrodes. The helium damping gas pressure was maintained at 2 mTorr. In Figure 14.7a there was no TFD, and the ions are ejected only at the stability boundary, $\beta_z = 1$. Figure 14.7b through d show that the effect of increasing the TFD component was to eject more of the ions at the nonlinear resonance at $\beta_z = 2/3$. The damping gas cools the ions to the center of the trapping field. However, since the center of the trapping field is displaced from the geometrical center of the trap electrodes, the hexapole nonlinear resonance has a non-zero strength and many of the ions are ejected at $\beta_z = 2/3$.

14.4.1.2 Supplementary Quadrupole Potential

Figure 14.8a shows the EI mass spectrum of FC-43 with 15% TFD and with a 2 V_{p-p} dipolar resonant excitation at a frequency of 700 kHz ($\Omega - \omega_{z,0}$). There is a double resonance with the supplementary dipole and the trapping field hexapole nonlinear resonance. However, because the amplitude of the supplementary voltage was near the ejection threshold (see Figure 14.6a) most of the ions were ejected at the stability boundary, at $\beta_z = 1$, rather than at the $\beta_z = 2/3$ nonlinear resonance. Figure 14.8b shows the same mass spectrum using a double resonance with a supplementary

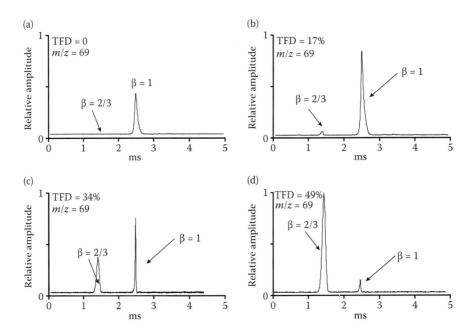

FIGURE 14.7 Mass spectrum of the CF_3^+ (m/z 69) ion formed during the EI of FC-43 for a series of TFD values with no supplementary voltage used. As the TFD% increases, the observed mass spectrum changes as ions move from being ejected at $\beta_z = 1$ to $\beta_z = 2/3$. (a) TFD = 0%; (b) TFD = 17%; (c) TFD = 34%; (d) TFD = 49%.

quadrupole potential having an amplitude of 10 V_{p-p} and frequency of 700 kHz ($2\omega_{z,0}$), and the trapping field hexapole nonlinear resonance. This voltage is far above the ejection threshold (see Figure 14.6b). The parametric resonance and the hexapole trapping field nonlinear resonance produce good resolution because the ions have been displaced from the center of the trap and all are in resonance simultaneously. The mass spectrum in Figure 14.8c resulted from combining the supplementary dipole and quadrupole excitation fields used in Figure 14.8a and b to affect a triple resonance.

14.4.1.3 Triple Resonance

Table 14.2 summarizes the operating frequencies for the dipolar resonant ejection and triple resonant ejection modes. The dipolar resonant ejection mode operating point is at $\beta_z = 0.92$, and the trapping frequency is 1050 kHz. The fundamental secular frequency, $\omega_{z,0} = 485$ kHz, is very close the first side-band frequency, $(\Omega - \omega_{z,0}) = 565$ kHz. The contributions, C_{2n} (Equation 14.9), to the ion motion in the axial direction for both these frequencies are almost identical because these coefficients are both approaching 1 when β_z is increased to 0.92. The amplitudes of both axial motions are quite large and the resulting beat frequency, $\Delta\omega = (\Omega - 2\omega_{z,0}) = 80$ kHz can be observed in the ion signal profile. The triple resonance ejection uses an operating point at $\beta_z = 2/3$ and the first side-band frequency is 700 kHz, which is far removed from the fundamental frequency of the secular motion of the ions, 350 kHz.

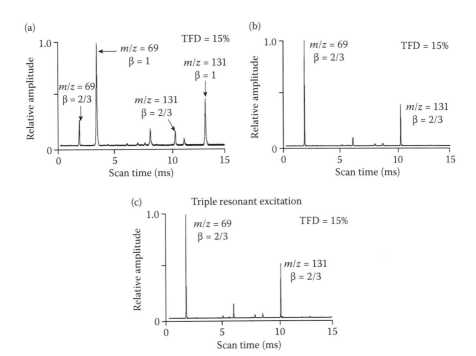

FIGURE 14.8 Mass Spectrum (EI) of FC-43 with 15% TFD. (a) 2 V_{p-p} dipolar resonance excitation is added. The voltage of this waveform alone is not enough to eject totally all of the ions at $\beta_z = 2/3$. (b) A 10 V quadrupole resonance excitation is added. The quadrupolar field in addition to the TFD induced hexapolar field is strong enough to produce a good mass spectrum at $\beta_z = 2/3$. (c) Mass spectrum resulting from combining the supplementary dipole and quadrupole from (a) and (b) above.

As a consequence, the beat frequency is 350 kHz and is easily removed from the detected ion signal by electronic filtering.

Metastable ions formed by EI, such as the molecular ion of n-butylbenzene (m/z 134) and the fragment ion $C_4F_9^+$ (m/z 219) from FC-43, have symmetrical peak shapes when they are mass-analyzed using the triple resonant ejection mode. In contrast, an asymmetric peak shape is observed with the dipolar resonant ejection mode, where the mass peak is skewed and tailing toward the low mass side of the mass spectrum due to dissociation prior to normal ejection.

Figure 14.9 compares the mass spectrum of the molecular ion (m/z 156) of undecane for dipolar resonant and triple resonant ion ejection. The molecular ions of normal hydrocarbons are often weak or absent in mass spectra obtained from an ion trap. The helium damping gas pressure was calculated to be approximately 5–6 mTorr, and the trap temperature was 150°C. The ion signal observed for the triple resonant ejection is considerably greater than that obtained with the dipolar resonant ejection mode. Only a factor of two was expected because of the unidirectional ejection occurring with the triple resonant ejection mode. As discussed earlier, collisions that occur during the long ejection time and the energy extremes experienced during the beating of the ion motion resulting from dipolar resonant excitation can

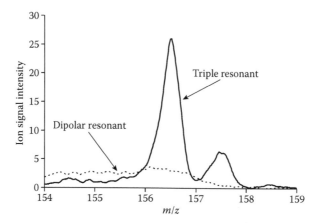

FIGURE 14.9   Mass spectrum of the molecular ion (*m/z* 156) of undecane, obtained by EI, acquired by both dipolar resonant and triple resonant ejection. The signal from the dipolar ejection is considerably smaller due to fragmentation of the ions during the ejection process.

cause dissociation early in the excitation process, and the resulting product ions will be ejected immediately by instability and detected. This loss of ions causes a reduction in the mass spectral peak intensity and a shift of the mass centroid toward low mass. The ion ejection time for triple resonance is significantly shorter, and few collisions of ions having sufficient kinetic energy to cause dissociation will occur. Therefore, the mass resolution is improved significantly for ions that have low thresholds for dissociation.

14.4.2   Scan Performance

Figure 14.10a and b compare the results of triple resonance and dipole scanning, respectively, at 5000 Th s^{-1} for singly-charged ions of a sulfide-bridged peptide formed by electrospray ionization and injected into the trap. The nonlinear ion ejection causes ions of the same mass-to-charge ratio to be ejected exponentially in time in a very narrow time packet, resulting in improved mass resolution (Figure 14.10a) compared to simple dipolar ion ejection (Figure 14.10b). Faster scan rates can be obtained while still maintaining good mass resolution because the ions are ejected from the trap in a very short time; for example Figure 14.11 shows a mass spectrum of the same peptide scanned at 15,000 Th s^{-1} utilizing the triple resonance-scan mode. The resolution observed at 15,000 Th s^{-1} utilizing the triple resonance-scan mode is equivalent to the resolution obtained at one third of the speed using the normal dipolar scanning method.

    High mass resolution can be obtained by reducing the scan rate when using the triple resonance-scan mode. Figure 14.12a shows a mass spectrum of the +3 charge state of insulin scanned at 250 Th s^{-1} with a resolution of *ca* 17,200. Figure 14.12b shows a spectrum of the +6 charge state of insulin at 250 Th s^{-1} with a resolution of *ca* 10,000. Figure 14.13a shows a mass spectrum of the +2 charge state of a

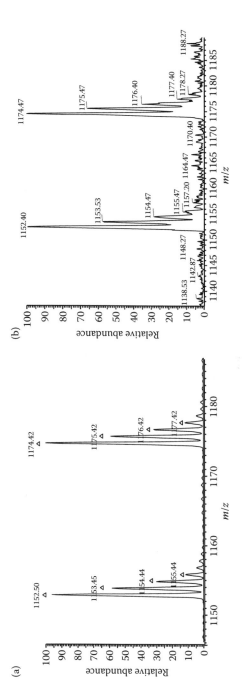

FIGURE 14.10 Mass spectrum of ions from a singly-charged sulfide-bridged peptide formed by external electrospray ionization and ejected from the trap at *ca* 5000 Th s^{-1}. (a) Mass spectrum acquired with the triple resonance-scan mode. (b) Mass spectrum acquired using the dipole ejection scan mode.

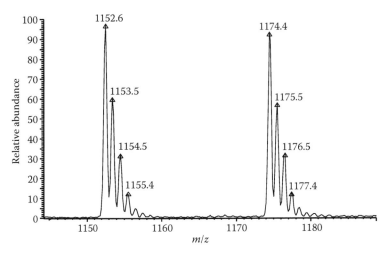

FIGURE 14.11 Mass spectrum of a singly-charged sulfide-bridged peptide formed by electrospray ionization and injected into the trap. The mass spectrum was acquired with the triple resonance-scan mode and ions were ejected from the trap at 15,000 Th s^{-1}.

sulfide-bridged peptide scanned at 5000 Th s^{-1} utilizing the triple resonance scanning mode. Figure 14.13b shows the same mass spectrum utilizing dipole scanning. The triple resonance-scan method induces ions to become ejected during a shorter time than the normal dipole method. As a result, the individual packets of charge relating to a specific species' isotopes are less spread out in space by collisions with helium and thus produce a higher spectral resolution. This significantly improved resolution translates into more reliable charge state determination under normal trap operating conditions.

14.4.3 CHARGE CAPACITY MEASUREMENTS

Tests were conducted in which identical ion fluxes were introduced into two ion traps having ring and end-cap electrodes of similar dimensions and curvature (that is the same values of r_0 and z_0 were used to define the shape of the electrodes). Each electrode system was optimized for the particular scan mode used. A figure of merit for the influence of space charge is the mass shift as a function of the number of ions in the ion trap. For the purposes of this experiment, with a fixed ion flux, the ion accumulation time was adjusted to increase or decrease the total number of ions in the trap. In these experiments, a solution of Ultramark was infused and electrosprayed into the mass spectrometer. Ultramark is a fluorinated polymer with repeating units every 100 Th between 922 Th and 2022 Th. For the purposes of measuring space charge, the ion at m/z 1022 was ejected and measured while the higher m/z-value species were still in the trap. Figure 14.14 compares dipole mass scanning and triple resonance mass scanning. Plots of mass shift *vs.* normalized ion accumulation time (*i.e.*, total charge) for each of the two modes of scanning are shown (the ion accumulation time was normalized relative to the base accumulation time required

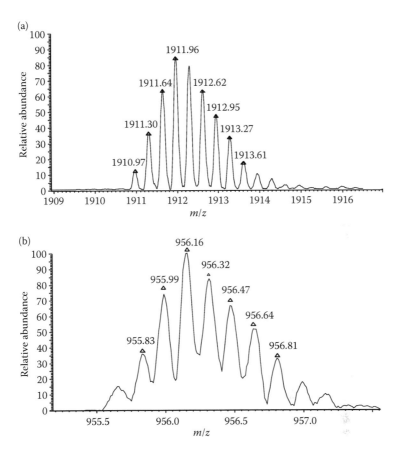

FIGURE 14.12 High resolution mass spectrum of insulin acquired at scan speed of 250 Th s^{-1} using the triple resonance-scan method. (a) The +3 charge state. (b) The +6 charge state.

to achieve a standard mass spectral signal-to-noise ratio for each system). It was observed that a 10 × increase in charge relative to that with dipole scanning was necessary with the triple resonance-scan mode in order to observe the same mass shift. Because both ion traps used identical electrodes, the ability of the triple resonance-scan mode to maintain a good peak shape and equivalent mass shift to the dipole scan mode with 10 times more charges must be related to the way in which ions of a given mass-to-charge ratio are positioned relative to each other as the ejection point is approached. Although the ions are ejected in a very short time, in the former they are dispersed along the axis of symmetry over a large distance; in this way the effects of space charge are reduced.

14.4.4 SUMMARY

The data presented indicate that the addition of a TFD component to the normal 'stretched' ion trap hyperbolic electrode geometry will generate a dipole and a

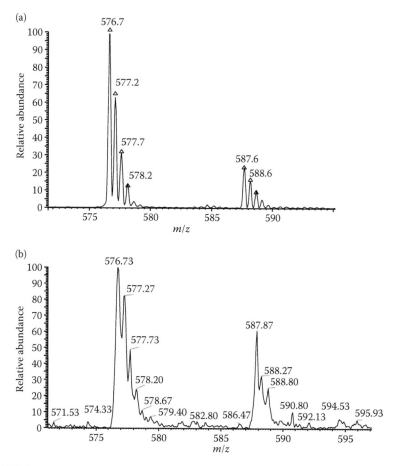

FIGURE 14.13 Mass spectrum of a doubly-charged sulfide-bridged peptide formed by electrospray ionization and ejected from the trap at *ca* 5000 Th s⁻¹. (a) Mass spectrum acquired with the triple resonance-scan mode. (b) Mass spectrum acquired using the dipole ejection scan mode.

significant hexapole component in the trapping field, in addition to the stretched induced octopole component. Furthermore, because of the asymmetric trapping field, the ions are not focused collisionally toward the geometric center of the electrode assembly, but rather toward a region between the electrodes displaced along the axis of the ion trap, away from the center, where the higher-order multipole fields, including the hexapole, have non-zero values. Therefore, when a supplementary quadrupole field is applied at a frequency such that the parametric oscillation corresponds to an operating point of $\beta_z = 2/3$, an ion located at this operating point will absorb power simultaneously both from the hexapole resonance component of the trapping field and the quadrupole parametric resonance. The addition of a weak dipole field corresponding to the first side-band resonance at an operating point of $\beta_z = 2/3$ will be in resonance simultaneously with the other two fields

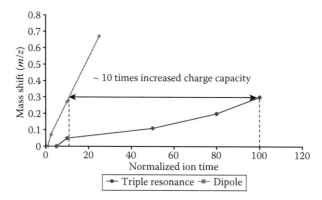

FIGURE 14.14 Plot of mass shift *vs.* normalized ion accumulation time for dipole and triple resonant scanning. The triple resonance mode was observed to be able to accommodate ten times the amount of charge compared to that with the dipole scanning mode before the test ion shifted 0.3 Th (see text for details).

also. This 'triple resonance' results in a nonlinear power absorption, which causes ion ejection.

In general, the triple resonance mass peak width is narrower and more symmetric than that obtained by dipolar resonant ejection. The mass peak width and symmetry for quadrupolar double resonance ejection is similar to that of the triple resonance and occurs without the need for a supplementary dipole field. The removal of the supplementary dipole field causes a slight reduction in resolution, predominately at higher levels of space charge.

14.5 ELECTRICALLY INDUCED NONLINEAR FIELDS IN A TWO-DIMENSIONAL ION TRAP

14.5.1 OVERVIEW

The analysis of the three-dimensional ion trap described previously can also be applied to a linear two-dimensional ion trap. A linear ion trap is formed from four elongated electrodes of hyperbolic cross-section arranged as shown in Figure 14.15a.* Electrodes 1 and 2 are connected electrically, as are electrodes 3 and 4. An alternating voltage $V(\Omega)$ with radiofrequency Ω is applied between rods sets 1, 2 and 3,4. The alternating electric field thus generated creates restoring forces on an ion, which are directed toward the center of the electrode structure. The quadrupolar restoring field is equivalent to a trapping field that traps the ions in the direction transverse to the central axis. When plates are located at the ends of the rod structures and have a

---

* See also Volume 4, Chapter 10: Trapping and Processing Ions in Radio Frequency Ion Guides, by Bruce A. Thomson, Igor V. Chernushevich and Alexandre V. Loboda; Volume 4, Chapter 11: Linear Ion Trap Mass Spectrometry with Mass-Selective Axial Ejection, by James W. Hager; Volume 4, Chapter 12: Axially-resonant Excitation Linear Ion Trap (AREX LIT), by Yuichiro Hashimoto.

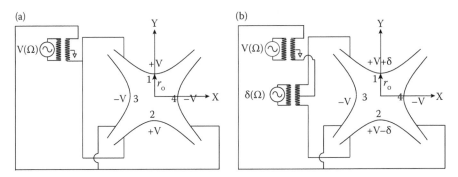

FIGURE 14.15 A linear ion trap is formed from four elongated electrodes of hyperbolic cross-section. Electrodes 1 and 2 are connected electrically, as are electrodes 3 and 4. (a) A radiofrequency alternating voltage $V(\Omega)$ with frequency Ω is applied between rods set 1, 2 and 3, 4. (b) In addition to the main radiofrequency voltage $V(\Omega)$, a supplementary alternating voltage $\delta(\Omega)$ is applied between electrodes 1 and 2, at the same frequency as the radiofrequency trapping field voltage applied between the sets of electrodes as shown.

DC voltage applied to them, a force will be applied to an ion that is directed along the axis of the rods. Thus, ions will be confined along the x-axis and y-axis directions due to the alternating voltage gradient, and along the z-axis by means of the DC potential applied to the end plates. When an additional alternating voltage $\delta(\Omega)$ is applied between electrodes 1 and 2 at the same frequency as the trapping field voltage applied between the sets of electrodes, as shown in Figure 14.15b, then two additional components are added to the trapping field so that it is no longer purely quadrupolar. The first of these is a dipolar component that has the effect of displacing the electrical center of the trapping field away (radially) from the geometric axis of symmetry along the z-axis of the electrodes; the second is a hexapolar component. It will be shown that the hexapolar component generates nonlinear resonances in the trapping field, and that these can be used to eject ions from the ion trap through a slot in one of the rods.

Ion ejection by dipolar resonant excitation can be effected by the application of a supplementary alternating voltage applied out-of-phase across a pair of opposing electrodes. Similarly, quadrupolar resonant excitation can be effected by the application of a supplementary alternating voltage applied in-phase across a pair of opposing electrodes. Parametric resonant excitation by a supplementary quadrupole field causes ion amplitudes to increase in the transverse direction when the ion frequency is one half of the supplementary quadrupole frequency [22]. Parametric resonant excitation in two dimensions is similar that discussed previously for the three-dimensional case. The supplementary quadrupole field has a value of zero at the center of the ion trap. When a buffer gas such as helium is used to damp the ion trajectories to the center of the trap, parametric excitation is ineffectual due to the vanishing strength of the supplementary quadrupole field. As in the three-dimensional case, it is necessary to displace the ions from the center of the supplementary quadrupole field to a location where the field has a non-zero value, in order to have a finite excitation force applied to the ions. A weak resonant dipole field having a frequency of one half of the parametric frequency, displaces ions from the center of the trap when the operating point of the ions is changed to bring the ion fundamental frequency into resonance with the dipole field [25].

In a manner similar to the geometry of the three-dimensional ion trap electrodes, the spacing between the pair of electrodes to which the supplementary voltages are applied can be modified to introduce an octopole component into the trapping field to enhance mass resolution [9,32].

A new ion ejection method [33], uses a trapping field that is asymmetric with respect to the transverse plane. The asymmetric trapping field is generated by adding an alternating voltage out-of-phase to each of an opposing pair of electrodes. This TFD component causes the center of the trapping field to be no longer coincident with the geometric center of the ion trap electrode assembly. A second-order effect is to superimpose a substantial hexapole field [30,31].

In the case of a linear ion trap, ions are confined within an electrodynamic quadrupolar field when their trajectories are bounded in the x- and y-directions. In a pure quadrupole trapping field, the x- and y-components of the ion motion are independent of each other, and the ion motion in the trapping field is pseudo-harmonic. The equations of motion for a single ion in the trapping field can be resolved, therefore, into a pure x-motion and a pure y-motion, which have identical mathematical forms described by the Mathieu equation [1]. The Mathieu equation for the y-axis motion depends on two parameters a_y and q_y, which characterize the solutions in the y-axis direction; similar parameters, a_x and q_x, exist for the x-axis motions. In order to trap ions it is required that their trajectories be stable simultaneously in both the x- and y-directions. It is known that non-ideal hyperbolic electrodes, or electrodes of circular shape that are used to approximate hyperbolic fields, generate nonlinear resonances within the field. It is known further that these nonlinear resonances degrade the performance of quadrupole mass filters [2,3]. It is not obvious that nonlinear resonances can be useful in linear ion traps. Additionally, the difference in symmetry between the three-dimensional system, with an axis of rotational symmetry, and that of the two-dimensional system will have a profound difference in how nonlinear resonances can be employed.

14.5.2 Theory of Triple Resonance Ion Ejection Scanning

The potential Φ in the space between electrodes disposed symmetrically about a central axis must, in general, satisfy Laplace's equation in cylindrical coordinates

$$\nabla^2\Phi\,(r,\theta,z) = \frac{1}{r}\frac{\partial}{\partial r}\left(r\frac{\partial\Phi}{\partial r}\right) + \frac{1}{r^2}\frac{\partial^2\Phi}{\partial\theta^2} + \frac{\partial^2\Phi}{\partial z^2} = 0 \qquad (14.24).$$

A general solution to Laplace's equation is given by [14]

$$\Phi(r,\theta) = \sum_{N=0}^{\infty}[(A_N'r^N + B_N'r^{-N})(C_N\cos(N\theta) + D_N\sin(N\theta)) + A_0\ln\left(\frac{r}{a}\right) \qquad (14.25).$$

If electrodes 1 and 2 are connected electrically to each other while electrodes 3 and 4 are connected electrically to each other, and further when arbitrary alternating and DC potentials are applied to electrodes 1 and 2 while the equal and opposite potentials are applied to electrodes 3 and 4 then the entire time dependant potential can be written by

$$V_t(r,\theta,t) = \sum_{n=0}^{\infty} \Phi(r,\theta)\left[a_n + b_n \cos\left[\frac{n\Omega}{2}(t - t_n) \right] \right] \qquad (14.26),$$

where Ω is the frequency of the alternating potential, n is the index of the expansion, and a_n and b_n are expansion coefficients. Limiting the harmonic content of the alternating potential to only the fundamental ($n = 2$) reduces the potential to the form

$$V_t(r,\theta,t) = \Phi(r,\theta)\left[U + V \cos\left[\Omega(t - t_2) \right] \right] \qquad (14.27),$$

where U is the DC voltage and V is the alternating voltage both applied equally (but in opposite polarity) to the electrode sets $1 + 2$ and electrodes $3 + 4$.

Since the potential must be finite at the origin, therefore

$$A'_N = 0 \quad \text{for} \quad N = 0$$

and

$$B'_N = 0 \quad \text{for} \quad N \geq 0.$$

Let

$$A'_N C_N = \left(\frac{1}{r_0} \right)^N A_n \quad \text{and} \quad A'_N D_N = \left(\frac{1}{r_0} \right)^N B_n,$$

therefore

$$\Phi(r,\theta) = \sum_{N=0}^{\infty} \left(\frac{r}{r_0} \right)^N \left[A_N \cos(N\theta) + B_N \sin(N\theta) \right] \qquad (14.28).$$

The general form of the electrodynamic potential for a time-dependent field in a cylindrical coordinate system (r, θ) is given by [14]

$$V_t(r,\theta,t) = \sum_{N=0}^{\infty} \left(\frac{r}{r_0} \right)^N \left[A_N \cos(N\theta) + B_N \sin(N\theta) \right]\left[U + V_{\cos}[\Omega(t - t_n)] \right] \qquad (14.29).$$

Since

$$r^N \cos(N\theta) = x^N - \binom{N}{2} x^{N-2} y^2 + \binom{N}{4} x^{N-4} y^4 - \binom{N}{6} x^{N-6} y^6 + \dots \quad (14.30a),$$

and

$$r^N \sin(N\theta) = \binom{N}{1} x^{N-1} y - \binom{N}{3} x^{N-3} y^3 + \binom{N}{5} x^{N-5} y^5 - \dots \quad (14.30b),$$

where the binomial coefficients are given by

$$\binom{N}{n} = \frac{N!}{(N-n)!n!},$$

substituting Equations 14.30a and b into Equation 14.28 and using the first three terms ($N = 1, 2, 3$) yields

$$\Phi(x,y) = \frac{A_1}{r_0} x + \frac{B_1}{r_0} y + \frac{A_2}{r_0^2} x^2 - \frac{B_2}{r_0^2} y^2 + \frac{A_3}{r_0^3}(x^3 - 3xy^2) + \frac{B_3}{r_0^3}(3x^2 y - y^3) \quad (14.31).$$

The coefficients can be determined from the electrode shapes. If the electrodes are hyperbolic sheets extending to infinity, oriented along the x- and y-axes, then their shapes are determined by

$$\frac{x^2}{r_0^2} - \frac{y^2}{r_0^2} = -1 \text{ for the electrodes along the } y\text{-axis} \quad (14.32a),$$

and

$$\frac{x^2}{r_0^2} - \frac{y^2}{r_0^2} = +1 \text{ for the electrodes along the } x\text{-axis} \quad (14.32b).$$

Using the electrodes as boundary conditions in Equation 14.31 gives

$$\Phi(x,y) = -\frac{1}{r_0^2}(x^2 - y^2) \quad (14.32c).$$

The general form of the quadrupole potential V_t is

$$V_t(x,y,t) = -\left[\frac{1}{r_0^2}(x^2 - y^2)\right][U + V \cos[\Omega(t - t_n)]] \quad (14.33).$$

The canonical form of the equations of motion for ions in an ideal quadrupole potential V_t field can be obtained from the vector equation

$$m\frac{\partial^2 \vec{R}}{\partial t^2} + e\vec{\nabla}V_t = 0 \quad (14.34),$$

where the position vector is $\vec{R}(x, y, z)$, m is the ion mass and e is the charge of the ion. The form of the potential allows the independent separation of the equations of the ion motion into the x-and y-components

$$\vec{E}_x = -\frac{\partial V_t}{\partial x} = +\frac{2x}{r_0^2}(U + V\cos[\Omega(t - t_n)]) \qquad (14.35a),$$

$$\vec{E}_y = -\frac{\partial V_t}{\partial y} = -\frac{2y}{r_0^2}(U + V\cos[\Omega(t - t_n)]) \qquad (14.35b),$$

and

$$\vec{E}_z = 0 \qquad (14.35c).$$

The canonical form of these equations when Equation 14.35 is substituted into Equation 14.34 is

$$\frac{d^2u}{d\zeta^2} + [a_u - 2q_u\cos(2\zeta)]u = 0 \qquad (14.36),$$

which is the well-known Mathieu equation (see also Equation 14.8), where the dimensionless parameters ζ, a_u, and q_u are

$$\zeta = \frac{\Omega t}{2} \qquad (14.37a),$$

$$\frac{d^2u}{dt^2} = \frac{\Omega^2}{4}\frac{d^2u}{d\zeta^2} \qquad (14.37b),$$

$$q_u = \psi_v 4eV/[mr_0^2\Omega^2] \qquad (14.37c),$$

and

$$a_u = \psi_v 8eU/[mr_0^2\Omega^2] \qquad (14.37d),$$

with $\Psi_x = +1$ for $u = x$, and $\Psi_y = -1$ for $u = y$.

This Mathieu equation is a second-order differential equation that has stable solutions characterized by the parameters a_u and q_u. The values of the parameters define the operating point of the ion within the stability region. By analogy with Equation 14.9, the general solution to Equation 14.36 is

$$u(\zeta) = A\sum_{n=-\infty}^{+\infty} C_{2n}\cos(2n + \beta_u)\zeta + B\sum_{n=-\infty}^{+\infty} C_{2n}\sin(2n + \beta_u)\zeta \qquad (14.38)$$

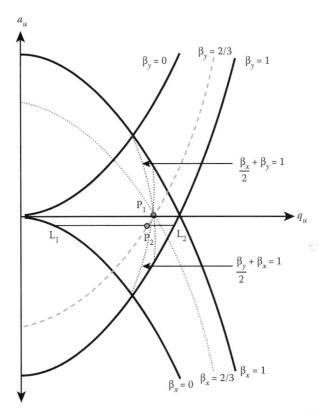

FIGURE 14.16 First Mathieu stability region for a two-dimensional ion trap. When $a_u = 0$ the $\beta_x = 2/3$ and $\beta_y = 2/3$ resonances are degenerate.

The secular frequency of the ion motion, ω_n can be determined from the value of β:

$$\omega_n = \left(n + \frac{\beta_u}{2}\right)\Omega \quad \text{and} \quad \omega_n = \left(n - \frac{\beta_u}{2}\right)\Omega \quad \text{where } n = 0, 1, 2, 3 \ldots \quad (14.39)$$

The value of β_u is a function of the operating point in (a_u, q_u) space and can be computed from a well-known continuing fraction [1].

The lower stability region of (a_u, q_u) space that is illustrated in Figure 14.16 shows the independent stable region for x- and y-motions. Ions must be stable simultaneously in both the x- and y-directions in order to be trapped, therefore, only operating points corresponding to (a_x, q_x) and (a_y, q_y) that lie in overlapping regions of stability can be used. These regions are bounded in the x-direction by $\beta_x = 0$ and $\beta_x = 1$, and in the y-direction by $\beta_y = 0$ and $\beta_y = 1$.

14.5.2.1 Dipole Component of the Trapping Field

If an additional alternating potential of amplitude $\delta(\Omega)$ is added to electrode 1 (see Figure 14.15b) in phase with the fundamental RF drive potential, and subtracted from electrode 2, then the coefficients in Equation 14.31 will change. Application of the boundary conditions to Equation 14.31 gives

$$\Phi(x,y) = \frac{\delta}{r_0}\left(\frac{1}{2\sqrt{2}}+1\right)y - \frac{V}{r_0^2}\left(x^2 - y^2\right) + \frac{\delta}{2\sqrt{2}r_0^3}\left(3x^2y - y^3\right) \qquad (14.40).$$

The general form of the new potential V_t, in which the DC potential (U) and the initial phase of the fundamental RF drive potential (t_n) are zero, is

$$V_t(x,y,t) = \left[\frac{\delta}{r_0}\left(\frac{1}{2\sqrt{2}}+1\right)y - \frac{V}{r_0^2}\left(x^2 - y^2\right) + \frac{\delta}{2\sqrt{2}r_0^3}\left(3x^2y - y^3\right)\right]\cos(\Omega\,t) \qquad (14.41).$$

Taking only the first two terms for now and substituting them in Equation 14.31 gives the instantaneous electric field acting on an ion in the axial direction due to the potential field V_t as

$$E_x = -\frac{\partial V_t}{\partial x} = +\frac{2x}{r_0^2}V\cos(\Omega\,t) \qquad (14.42a)$$

$$E_y = -\frac{\partial V_t}{\partial y} = -\frac{2y}{r_0^2}V\cos(\Omega t) - \frac{\delta}{r_0}\left(\frac{1}{2\sqrt{2}}+1\right)\cos(\Omega t) \qquad (14.42b),$$

and the equation of the ion motion in the y-direction becomes

$$m\frac{d^2y}{dt^2} = \left(\frac{-e2yV}{r_0^2} - \frac{e\delta}{r_0}\left(\frac{1}{2\sqrt{2}}+1\right)\right)\cos(\Omega\,t) \qquad (14.43).$$

Substituting $\zeta = (\Omega t/2)$ gives

$$\frac{d^2y}{dt^2} = \frac{\Omega^2}{4}\frac{d^2y}{d\zeta^2} \qquad (14.44).$$

By substitution of Equation 14.44 in Equation 14.43 and writing $2\zeta = \Omega t$, the basic equation of the ion motion in the y-direction is obtained

$$\frac{d^2y}{d\zeta^2} - 2\left(\frac{-4eV}{mr_0^2\Omega^2}y - \frac{2e\delta}{m\Omega^2r_0}\left(\frac{1}{2\sqrt{2}}+1\right)\right)\cos(2\zeta) = 0 \qquad (14.45).$$

Defining

$$q_y = \frac{-4eV}{m\Omega^2r_0^2} \qquad (14.46a)$$

and

$$q_{yD} = -\frac{2e\delta}{m\Omega^2 r_0}\left(\frac{1}{2\sqrt{2}}+1\right)$$

(14.46b),

substitution of Equations 14.46a and b into Equation 14.45, an equation similar to the Mathieu equation is obtained

$$\frac{d^2y}{d\zeta^2} - 2\left(q_y y + q_{yD}\right)\cos(2\zeta) = 0$$

(14.47).

With the definition and substitution

$$u = \left(q_y y + q_{yD}\right) \quad \text{and} \quad \frac{d^2u}{d\zeta^2} = q_y \frac{d^2y}{d\zeta^2}$$

into Equation 14.47, a form of the Mathieu equation is obtained

$$\frac{d^2u}{d\zeta^2} - 2q_y\, u\cos(2\zeta) = 0$$

(14.48).

The axial displacement of the ion can found to be the sum of two terms

$$y = \frac{u - q_{yD}}{q_y} = \frac{u}{q_y} - \frac{q_{yD}}{q_y}$$

(14.49),

in which the first term represents the normal time-dependent oscillatory solution, $u\,(\zeta)$, as in Equation 14.38, the second term is an additive offset value which expresses the displacement of the ion along the y-axis due to the dipole

$$-\frac{q_{yD}}{q_y} = \frac{-\delta r_0}{2V}\left(\frac{1}{2\sqrt{2}}+1\right)$$

(14.50).

During mass analysis, it is common to increase the AC voltage of the guiding field as a function of mass. In the special case in which $\delta = \eta V_{ac}$ Equation 14.50 becomes

$$-\frac{q_{yD}}{q_y} = -\left(\frac{1}{2\sqrt{2}}+1\right)\frac{r_0}{2}\eta$$

(14.51),

and thus

$$y = \frac{u}{q_y} - \left(\frac{1}{2\sqrt{2}}+1\right)\frac{r_0}{2}\eta$$

(14.52).

When the dipole is phased properly and is present as a constant fraction, η, of the trapping field, it can be seen from Equation 14.52, that the ion motion is displaced uniformly along the y-axis by a constant amount. The magnitude and sign of the displacement is independent of the mass-to-charge ratio and the polarity of the ion charge. The displacement depends only on the fraction (η) of the dipole contribution and the geometric dimensions of the electrode structure. The direction of the displacement can be altered by changing the phase of the dipole from 0 to π.

14.5.2.2 Hexapole Component of the Trapping Field

When the all three terms of the potential expressed in Equation 14.41 are included in Equation 14.34, the equations of motion will now become

$$m\frac{\partial^2 x}{\partial t^2} + e\left(-\frac{2x}{r_0^2} + 6\left(\frac{\delta}{V}\right)\left(\frac{xy}{2\sqrt{2}r_0^3}\right)\right)V\cos(\Omega t) = 0 \qquad (14.53a)$$

and

$$m\frac{\partial^2 x}{\partial t^2} + e\left(\frac{1}{r_0}\left(\frac{\delta}{V}\right)\left(\frac{1}{2\sqrt{2}}+1\right)+\frac{2y}{r_0^2}+\frac{3}{2\sqrt{2}r_0^3}\left(\left(\frac{\delta}{V}\right)(x^2-y^2)\right)\right)V\cos(\Omega t) = 0$$
$$(14.53b).$$

The three terms in brackets in Equation 14.53b correspond to the dipole, the quadrupole, and the hexapole contributions. Because each of Equations 14.53a and b contain terms that are not exclusively functions of the x- or y-coordinates, the motions in these respective directions are coupled. Rearranging Equation 14.53 and substituting Equation 14.37 gives

$$\frac{d^2 x}{d\zeta^2} - 2q_x x\cos(2\zeta)x = -\left(\frac{12e}{m\Omega^2 r_0^3 \sqrt{2}}\right)\left(\frac{\delta}{V}\right)(xy)\cos(2\zeta) \qquad (14.54a)$$

and

$$\frac{d^2 y}{d\zeta^2} - 2q_y \cos(2\zeta)y = -\frac{4e}{m\Omega^2 r_0^2}\left(r_0\left(\frac{\delta}{V}\right)\left(\frac{1}{2\sqrt{2}}+1\right)+\frac{3}{2r_0\sqrt{2}r_0}\left(\frac{\delta}{V}\right)(x^2-y^2)\right)$$
$$\cos(2\zeta) \qquad (14.54b),$$

which are now forms of the driven Mathieu equation, with the driving force on the right sides of the expressions.

The solutions to coupled nonlinear equations of the type of Equation 14.54 are known from the theory of nonlinear betatron oscillations in alternating gradient circular accelerators [34,35] and their mechanical analog [35,36]. The higher-order geometrical terms in Equation 14.54 produce singularities in the denominator of the solutions, indicating nonlinear resonances. The amplitude of oscillation of an ion at the operating point corresponding to a nonlinear resonance will be caused to increase without bounds in the direction of an electrode. The amplitude increase with time is not linear, as with simple dipole resonance ejection, but rather it increases at a rate depending on the order of the nonlinear resonance.

Nonlinear resonances will occur at the operating points having the relationship [5,10]

$$\beta_y n_y + n_x \beta_x = 2\nu \qquad (14.55a),$$

where $|n_y| + |n_x| = N$. Therefore, since $\omega = (\beta/2)\Omega$, and for $\nu = 1$

$$\Omega \frac{\beta_y}{2} K + (N - K)\Omega \frac{\beta_y}{2} = \Omega \qquad (14.55a),$$

or

$$\omega_y K + (N - K)\omega_x = \Omega \qquad (14.55c),$$

where $K = N, N - 2, N - 4, \ldots$. Thus, the third-order resonances ($N = 3$) generated in the field are

$$\beta_y = \frac{2}{3} \quad \text{for} \quad K = 3 \qquad (14.56a),$$

a pure resonance affecting only the y-coordinate, and

$$\frac{\beta_y}{2} + \beta_x = 1 \quad \text{for} \quad K = 1 \qquad (14.56b),$$

a coupled resonance affecting both the x- and y-coordinates (shown as dotted lines in Figure 14.16). Thus it is seen that the linear trapping field has a nonlinear resonance at $\beta_y = 2/3$, similar to that in the three-dimensional field. This nonlinear resonance can be used to eject ions in the direction of one of the electrodes. If an additional alternating potential $\delta(\Omega)$ of Figure 14.15b is applied between two opposing electrodes (1 and 2 of Figure 14.15b) at the frequency of ion oscillation in the trapping field, ions will be displaced in the direction of the electrodes.

14.5.3 A DEGENERACY DILEMMA

Equation 14.56 indicates that an ion at the operating point corresponding to $\beta_y = 2/3$ (Equation 14.56a) along the q_y-axis of the stability region (that is $a_y = 0$ when the

DC potential $U = 0$) will experience also a coupled resonance (Equation 14.56b) corresponding to $\beta_x = 2/3$, shown as point P_1 in Figure 14.16. Therefore, the two resonances are degenerate at this operating point, unlike the three-dimensional trap case (Figure 14.17) in which the $\beta_z = 2/3$ and the $\beta_z + \beta_r = 1$ resonance lines are separated along the q_z-axis. In the two-dimensional case, it is undesirable for an ion to be located at $\beta_x = \beta_y = 2/3$ because, at this operating point, an increase in amplitude in the y-direction will cause an increase in amplitude in the x-direction. A small DC potential added to the trapping field, and scanned such that the ratio of the DC to RF is held constant, will move the operating point at which ions are ejected when the RF/DC potential is scanned from the q_u axis (where $U = 0$) down to the point P_2 in Figure 14.16. At this operating point, ejection at $\beta_y = 2/3$ will occur at a lower value of q_y, thus needing a lower RF voltage to achieve ejection. The two nonlinear resonance lines are no longer degenerate at the new operating point and a pure $\beta_y = 2/3$ nonlinear resonance will be encountered before the coupled resonance. If a supplementary alternating potential is applied across opposing electrodes at a frequency corresponding to the operating point P_2 in Figure 14.16, then an increase in amplitude of the y-coordinate oscillations will occur without a concomitant increase in the x-coordinate oscillation, or *vice versa*.

Equations 14.37c and d indicate that if the ratio of V/m and U/m remain constant in time, then a_u and q_u will also remain constant in time. Mass scanning can be effected by causing ions of successive mass-to-charge ratios to pass through the same operating point linearly in time. Increasing the amplitude of the RF drive potential (V) and of the DC (U) linearly with time, such that their ratio V/U is constant, will result in ion ejection that is a linear function of m/z-value. The operating point (a_y, q_y) for ejection should correspond to $\beta_y = 2/3$. A supplementary resonance frequency corresponding to the fundamental secular value (ω) or one of the sidebands (for example ($\Omega - \omega$)) will result in an increase in the amplitude of the ion oscillation due both to the supplementary dipole resonance and to the nonlinear hexapole resonance of the trapping field, thereby effecting ion ejection through a slot in one of the electrodes.

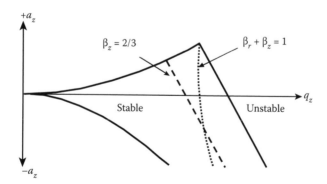

FIGURE 14.17 First Mathieu stability region for a three-dimensional ion trap showing two nonlinear resonance lines.

14.5.4 IMPLEMENTATION OF TRIPLE RESONANCE ION EJECTION SCANNING

A linear ion trap prototype was built at Varian and is shown in Figure 14.18. A set of four elongated hyperbolic electrodes is in the center of the assembly ('Center Electrodes' in Figure 14.18), and opposing electrodes are connected electrically and form an electrode pair. The fundamental RF drive voltage is applied between the electrode pairs to form a quadrupolar electric field. Two sets of four shorter hyperbolic electrodes ('End Electrodes' in Figure 14.18) are disposed at each end of the central set. These rod sets are coupled in the same manner to the fundamental RF drive voltage. Located at each end of the short electrodes is a plate with a circular aperture to which a DC voltage is applied to form a potential barrier along the z-axis (positive for positive ions and negative for negative ions). A voltage DC-1 is applied to the Entrance Aperture, and DC-2 is applied to the End Aperture. An additional DC voltage DC-3 is applied to all four rods in each set of End Electrodes. This arrangement of electrodes and plates forms the basic linear trap. An additional RF voltage (δ in Figure 14.15b) having the same frequency and phase as the fundamental RF potential is applied between all three pairs in the y-electrodes to form the dipole and hexapole components in the resultant electric field. A DC voltage is applied equally (but of opposite sign) to all six electrodes in the y- and x-directions to form a quadrupolar DC field and thus to shift the operating point from the q_y-axis to a line below (that is to shift from $a_y = 0$ to $a_y < 0$).

The triple resonance-scan function utilizes also two additional low voltage supplementary waveforms. The dipolar waveform is applied across the six y-axis electrodes and is chosen to be in resonance with either the fundamental (ω) or with one of the side-bands (that is ($\Omega - \omega$)) when the ion is at $\beta_y = 2/3$. A supplementary quadrupolar waveform is also applied in phase to all six y-axis electrodes. This waveform is applied at the quadrupolar resonance frequency and is equal to twice the secular frequency when the ion is at $\beta_y = 2/3$.

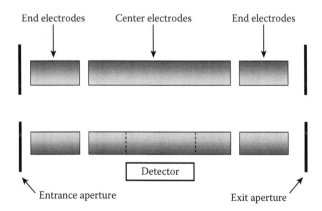

FIGURE 14.18 Basic two-dimensional linear ion trap assembly consisting of three sets of four elongated hyperbolic rods. DC potentials, applied to the entrance aperture and exit aperture end plates (as labeled) help to contain trapped ions in the axial z-dimension.

14.6   ION SIMULATIONS

14.6.1   FREQUENCY SPECTRUM

The trajectories of an ion of m/z 100 confined in a linear ion trap with an asymmetric trapping field were computed using the ion simulation program SIMION [17]. The trapping field dipole (TFD = δ/V) was 0%, $U = 0$ (DC), the trap frequency was 1050 kHz, and the operating point of the ion in the stability diagram was $\beta_y = 0.51$. Figure 14.19a and b show the Fourier transform of the component of ion motion in the x- and y-directions, respectively, obtained by Fourier analysis of 4000 data points of the ion trajectory, when there is no trapping field dipole applied to the electrodes. The frequency spectrum ranges from 0 to 2000 kHz, and the fundamental secular frequency of the ion motion ω, is observed at approximately 280 kHz. Only the fundamental ω and the side-band frequencies ($\Omega - \omega$) and ($\Omega + \omega$) are present in

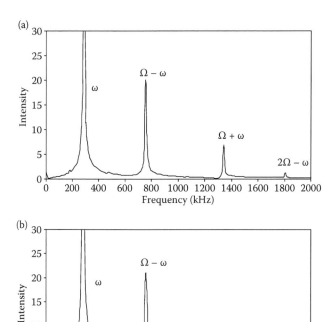

FIGURE 14.19 Frequency spectrum obtained by Fourier analysis of an ion trajectory calculated using SIMION. The ion of m/z 100 was confined in a linear trap operating at a main quadrupole trapping frequency of 1050 kHz and the amplitude of the voltage selected such that the ion was held at the operating point $\beta_y = 0.51$. (a) Frequency spectrum of the ion motion in the x-direction resulting from application of a standard quadrupole potential. (b) Frequency spectrum of the ion motion in the y-direction resulting from application of a standard quadrupole potential.

the ion motions. Figure 14.20a and b shows the Fourier transform of the component of ion motion in the x- and y-directions, respectively, when there is a 30% trapping field dipole applied to the electrodes. The TFD introduces a hexapole component in the trapping field and, therefore, in addition to the fundamental ω and the side-band frequencies $(\Omega - \omega)$ and $(\Omega + \omega)$, there are overtones in the ion motions present at 2ω, 3ω, and 4ω, as well as side-bands of higher harmonics. A nonlinear resonance occurs at an operating point if the harmonics of the ion's motional frequencies match side-band frequencies. The matching will occur for entire groups of harmonics and side-bands. It should be noted that the drive frequency Ω is observed in the

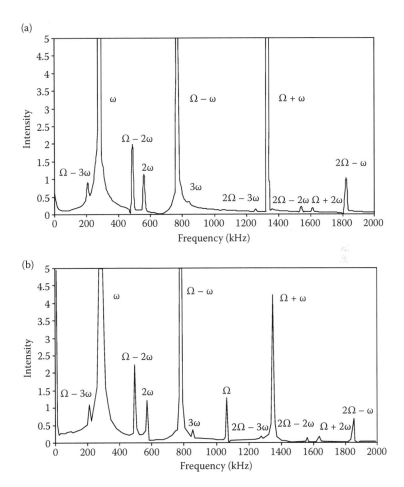

FIGURE 14.20 Frequency spectrum obtained by Fourier analysis of an ion trajectory calculated using SIMION. The ion of m/z 100 was confined in a linear trap operating at main quadrupole trapping frequency of 1050 kHz and the RF drive voltage selected such that the ion was held at the operating point $\beta_y = 0.51$. A TFD of 30% was applied to the electrodes as shown in Figure 14.15. The addition of the TFD causes a hexapolar field that results in a number of side bands and higher harmonics being observed. (a) Frequency spectrum of the ion motion in the x-direction. (b) Frequency spectrum of the ion motion in the y-direction.

y-direction motions, but not in the x-direction motions. This is consistent with an odd-order multipole in the field in the y-direction, but not the x-direction.

14.6.2 DEGENERACY EFFECTS

Figure 14.21 shows a simulation of ion motion corresponding to scanning through the operating point P_1 in Figure 14.16. The ion is being driven in the y-direction by both the supplementary resonant field (700 kHz corresponding to $a_y = 0$ and $q_y = 0.7846$, that is $\beta_y = 2/3$), as well as the pure and coupled nonlinear resonances. The displacement of the ion motion from the geometric center due to the trapping field dipole can be observed. The ion is being driven only in the x-direction by the coupled resonance. The result is an increase in the coordinates in both the x- and y-directions, with a significant displacement in the transverse direction at the time the ion approaches the electrode. Figure 14.22 shows a simulation of ion motion when a DC potential of 5 V is added to the electrode pair oriented in the y-direction, corresponding to the ion ejection point P_2 in Figure 14.16 ($a_y = 0.03$ and $q_y = 0.75$, that is on $\beta_y = 2/3$. Thus when the DC potential is added the $\beta_y = 2/3$ resonance condition is achieved at a lower q_y-value. No significant increase in the amplitude of the ion motion in the transverse direction is observed at this operating point.

Figure 14.23a shows a plot of the y-coordinate as a function of time in a linear quadrupole ion trap with 0% TFD, no collisions and 2 V of supplementary dipole voltage. The ions are excited at $\beta_y = 2/3$, but they are not ejected until the y-stability

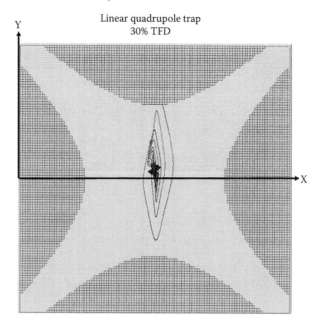

FIGURE 14.21 SIMION simulation of the ion motion corresponding to scanning through the operating point P1 in Figure 14.16. The ion is being driven in the y-direction by the supplementary dipole resonant field as well as by the pure and coupled nonlinear resonances. As the ion approaches the ejection point it gains energy in both the x- and y-dimensions.

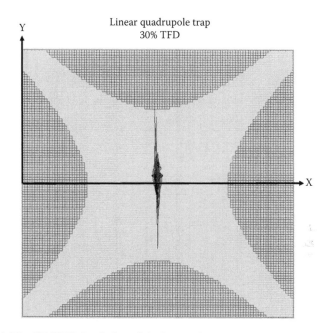

FIGURE 14.22 SIMION simulation of the ion motion corresponding to scanning through the operating point P_2 in Figure 14.16. The ion is being driven in the y-direction by the supplementary dipole resonant field as well by as the pure and coupled nonlinear resonances. An additional quadrupolar 5 V DC voltage is applied to the electrode in the y-direction. In this case, no significant increase in the transverse x-direction is observed at this operating point.

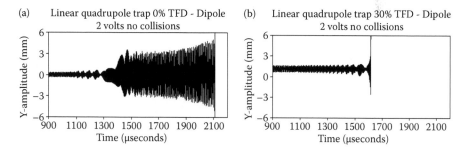

FIGURE 14.23 SIMION simulation of the last 1200 μs of a quadrupolar RF drive voltage ramp-induced ion ejection (57.7 volts ms^{-1}, peak-to-peak). (a) The y-coordinate as a function of time when the TFD = 0, no helium gas collisions, and 2 V_{p-p} of supplementary dipole voltage. The ions are excited at $\beta_y = 2/3$ but are not ejected. (b) The y-coordinate as a function of time with the TFD = 30% applied, 2 V_{p-p} of supplementary dipole, and no helium gas collisions. The presence of the hexapole non-linear resonance now causes the ion to be ejected at $\beta_y = 2/3$.

boundary ($\beta_y = 1$) is reached due to the absence of the nonlinear resonance and the extremely small supplementary voltage applied. Figure 14.23b shows the ejection of the ion when a 30% TFD is applied. The presence of the hexapole nonlinear resonance now causes the ion to be ejected at $\beta_y = 2/3$ utilizing the same 2-volt supplementary dipole voltage used in Figure 14.23a. Figure 14.24a shows the effect of

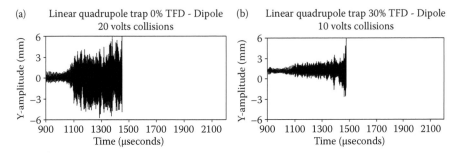

(a)   Linear quadrupole trap 0% TFD - Dipole   (b)   Linear quadrupole trap 30% TFD - Dipole
                    20 volts collisions   10 volts collisions

FIGURE 14.24   SIMION simulation of the last 1200 μs of a quadrupolar RF drive voltage ramp-induced ion ejection. (a) The y-coordinate as a function of time when the TFD = 0, helium gas collisions were included, and 20 V$_{p-p}$ of supplementary dipole voltage. (b) The y-coordinate as a function of time when the TFD = 30%, helium gas collisions were included, and 10 V$_{p-p}$ of supplementary dipole voltage. Ion ejection can now occur with only 10 V$_{p-p}$ of supplementary dipole, even in the presence of helium collisions.

collisions on the ejection process. A simulation with 0% TFD and 20 V of supplementary resonant dipole potential results in ion ejection at $\beta_y = 2/3$. Supplementary potentials below 10 V were insufficient to eject the ions at $\beta_y = 2/3$ and resulted in ion ejection at the $\beta_y = 1$ by instability, as in Figure 14.23a. When a TFD of 30% is added (Figure 14.24b), ion ejection occurs even at 10 V of supplementary dipole resonant potential with collisions due to the formation of a strong nonlinear resonance at $\beta_y = 2/3$.

14.7   EXPERIMENTAL RESULTS OF TRIPLE RESONANCE SCANNING IN A 2D TRAP

In order to examine the performance capabilities of the triple resonance mode of scanning an experimental two-dimensional ion trap was created. This device was constructed so that the shortest distance from the mechanical trap center to the nearest point on the x-electrodes was $x_0 = 6$ mm. Similarly, the distance from the trap center to the y-electrodes was $y_0 = 7$ mm. The y-axis, the dimension along which the ions were detected by a single detector, was stretched by an additional 1 mm to induce an octapolar field. Helium was leaked into the device at a rate of 0.5 mL/min, resulting in an estimated internal trap pressure of ca 1 mTorr. The device was driven at 781.25 kHz with the supplementary dipole frequency set to 260.4 kHz and the supplementary quadrupole frequency set to 520.8 kHz. The trapping field dipole was set to be 14% of the amplitude of the main RF drive voltage.

14.7.1   Effects With and Without Degeneracy

To demonstrate the importance of removing the degeneracy effect, an experiment was performed where the relative ion signal at m/z 242 (formed from an electrosprayed solution of tetrabutylammonium bromide) was monitored as a function of the quadrupolar DC voltage during ion ejection (Figure 14.25). For low DC values

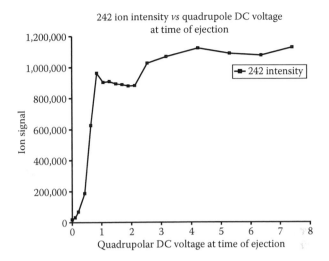

FIGURE 14.25 Ion intensity at *m/z* 242 (from an electrosprayed solution of tetrabutylammonium bromide) *vs.* the quadrupolar DC voltage applied at the time of ion ejection. At low DC voltage values the resonance coupling between the ion motions in the *x*- and *y*-directions causes very little ion signal to be observed at the detector.

virtually no intensity is observed in the mass spectrum. This observation is consistent with a coupling of the ion motion between the *y*- and *x*-dimensions, as shown in Figure 14.16. As the quadrupolar DC voltage is increased the β_y and β_x degeneracy begins to break, resulting in a higher signal intensity being observed as more ions are ejected in the direction of the *y*-axis. The pause in the rate of signal increase between 1 V and 2 V has been observed several times, and is likely to be the result of first removing the $\beta_x = 2/3$ resonance condition, followed by removing the resonance at $\beta_y/2 + \beta_x = 1$.

14.7.2 TUNING

The triple resonance-scan function uses two tuned supplementary waveforms in addition to the trapping field dipole. Both the relative phase angles as well as the amplitudes for the two externally applied waveforms must be selected carefully in order to achieve high performance. A unique 'score function,' which takes several desirable mass spectral features into account, has been created so as to provide a means of characterizing this complicated parameter space. The score function includes parameters such as the Full Width at Half Maximum (FWHM), the position of the next nearest isotope, and the mass shift as a function of space charge (Figure 14.26). In order to characterize the system, we have examined both the score function and peak amplitude at a number of dipole and quadrupole amplitudes and phase angles.

An example of the types of complicated patterns observed during the optimization process is shown in Figure 14.27. This figure shows a contour plot with the score function plotted at a number of different dipole and quadrupole amplitudes and phases. Within each smaller square in Figure 14.27 (and Figure 14.28) the dipole

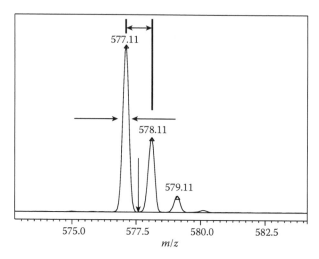

FIGURE 14.26 Example of the application of a unique spectral tuning score function. The score function includes parameters such as the FWHM, the distance to the next nearest isotope, and the mass shift between a low ion accumulation time and high ion accumulation time.

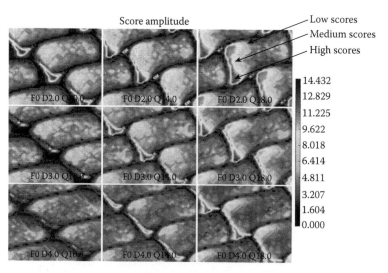

FIGURE 14.27 A contour plot with the score function plotted at a number of different dipole and quadrupole amplitudes and phases. Within each smaller square the dipole phase increases along the y-axis from 0° to 360° while the quadrupole phase increases along the x-axis from 0° to 360°. The larger squares represent steps of 2 V_{p-p} supplementary dipole voltage in the vertical direction and 8 V_{p-p} supplementary quadrupole voltage in the horizontal direction (see text for details).

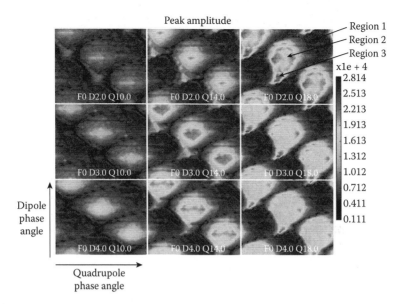

FIGURE 14.28 A contour plot with the peak height plotted at a number of different dipole and quadrupole amplitudes and phases. Within each smaller square the dipole phase increases along the y-axis from 0° to 360° while the quadrupole phase increases along the x-axis from 0° to 360°. The larger squares represent steps of 2 V_{p-p} for the supplementary dipole in the vertical direction and of 8 V_{p-p} for the supplementary quadrupole in the horizontal direction (see text for details).

phase increases along the y-axis from 0° to 360°, while the quadrupole phase increases along the x-axis from 0° to 360°. The larger squares within Figure 14.27 represent steps of 2 V in the supplementary dipole contribution in the vertical direction, and steps of 8 V in the supplementary quadrupole contribution in the horizontal direction. This type of experiment has been performed with a number of trap stretches and TFD values. In general, each of these experiments has yielded similar results, however the exact position of the high score regions shifts in complicated ways with TFD value and trap dimensions.

Figure 14.28 shows the relative peak heights obtained during the experiment described above. These phase maps show three types of observed peak heights. Region 1 indicates virtually no ion intensity and result from the ions being ejected in a direction opposite the detector. Region 2 shows locations where roughly 50% of the ions are observed at the detector, as is typical with standard ion trap operation. Finally, Region 3 in Figure 14.28 shows locations where virtually 100% of the ions are ejected toward the detector. These relative intensity statements have been confirmed by operating the ion trap with the TFD value having the opposite polarity. Reversing the TFD values is equivalent in effect to moving the detector to the opposite side of the trap. No intensity is observed when the TFD is reversed at phase-angle combinations which show 100% ion ejection when the TFD was in the correct direction. Key to operating the linear ion trap at peak performance is the fact that locations exist in four-dimensional space which demonstrate both 'high score values' and 100% unidirectional ejection.

These high performance regions are reproducible and may be understood in terms of a combination of both frequency domain and time domain influences of the fields on the ion orbits. The symmetries in the plots are tied to the harmonic relations of the applied waveforms. While the TFD does create an asymmetry in the electric field, which moves the cooled ion cloud a small distance away from the axis, at higher dipole and quadrupole voltages, however, the ejection direction is dominated by the relative phase angles of the supplementary waveforms.

As an example of the dramatic effect that changing the phase angle of either waveform can have on spectral performance, a series of mass spectra, which varied only in the relative supplementary quadrupole phase, were collected (see Figure 14.29a through e). In this case, changing from 180° to 140° quadrupole phase angle (relative to the RF phase positive zero crossing) caused the peak signal intensity to vary by more than a factor of six, and caused a dramatic change in mass spectral peak areas.

14.7.3 MASS SPECTRAL PERFORMANCE

Using the optimization routines described above, the triple resonance-scan function has been demonstrated to yield good results when operated on a 2D trap. Figure 14.30 shows a portion of a mass spectrum from an electrosprayed mixture of peptides obtained when scanning the 2D trap at 5000 Th s^{-1}. As described earlier, the triple resonance-scan function results in very rapid ion excitation and ejection. As a consequence, one would expect these operational conditions to be manifest in very high spectral resolution. To demonstrate this benefit, the mass spectrum in Figure 14.30 was acquired from a mixture of peptides, and shows a highly resolved +1 charge state in addition to a base line-resolved +3 charge state.

In the field of mass spectrometry, there has been a growing desire to increase the data acquisition speed in order to accommodate the elution peaks of shorter duration achieved with modern LC and GC techniques. Specifically, in the case of ion trap mass spectrometry, an increase in the spectral scanning speed can result in an increased duty cycle and an overall sensitivity improvement. Historically, however, for ion traps it has been well known that increasing the scan speed will result in lower spectral resolution. As a result, users have been forced to make trade offs between scan speed and spectral performance. In the case of the triple resonance-scan function, it has been observed that the ions are ejected very quickly and the method may lend itself to ultra-fast scanning.

Figure 14.31 shows a mass spectrum of a singly-charged electrosprayed ion of tetradecylammonium bromide recorded at 50,000 Th s^{-1}. This mass spectrum is baseline resolved, with a FWHM of 0.4 Th. Interestingly, at this scan speed, the spectrum begins to show the ultimate spectral resolution limit of the ion trap under these operating conditions. In the case of this work, at 50,000 Th s^{-1} with an applied RF frequency of 781.25 kHz and at $\beta_y = 2/3$, there are only approximately five cycles of the secular frequency between unit masses. As a result, when we observe mass spectral peaks of only 0.4 Th FWHM, most of the detected charge arrives in two to three individual

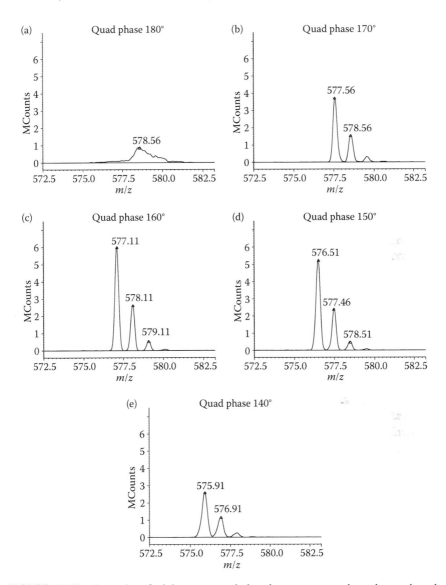

FIGURE 14.29 Examples of triple resonance-induced mass spectra where the quadrupole phase is adjusted to various phase angles. (a) Quadrupole Phase 180°, (b) Quadrupole Phase 170°, (c) Quadrupole Phase 160°, (d) Quadrupole Phase 150°, and (e) Quadrupole Phase 140°.

ion pulses. Indeed, by using a fast oscilloscope for data acquisition, the individual ion packets associated with each secular frequency cycle have been observed.

14.7.4 CHARGE CAPACITY MEASUREMENTS

One of the most important figures of merit of an ion trap mass spectrometer is it spectral charge capacity; that is, the number of charges which can be trapped

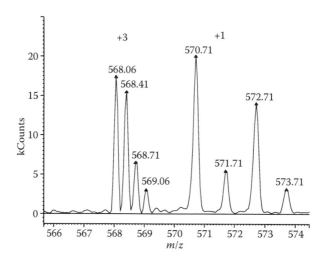

FIGURE 14.30 Electrospray mass spectrum of a mixture of peptides acquired at 5000 Th s⁻¹ scan speed using the triple resonance-scan function. This mass spectrum shows a base line-resolved triply-charged ion in addition to a highly-resolved singly-charged ion.

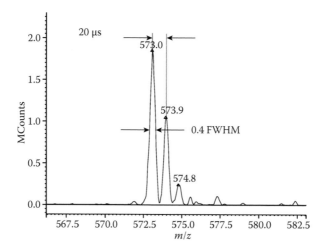

FIGURE 14.31 Mass spectrum of a sample acquired at 50,000 Th s⁻¹ scan speed with the triple resonance-scan function.

simultaneously without leading to spectral distortions. While there is no standard definition for measuring charge capacity, one approach is to infuse a complex sample and increase the ion accumulation time (which provides a measure of the number of ions in the trap) and then observe the relative mass shift for a low mass ion. For this study, two sets of trapping conditions were applied to the 2D trap. In one case, a 100 V DC potential difference was applied between the center section and end sections of the 2D trap. This potential difference resulted in a highly compressed ion cloud and, under these conditions, the 2D trap should emulate the performance of a

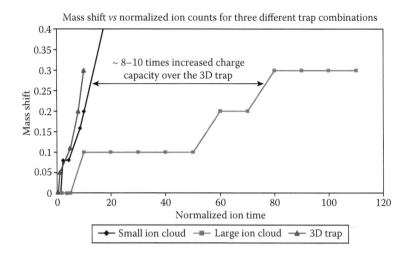

Mass shift *vs* normalized ion counts for three different trap combinations

FIGURE 14.32 Plot of mass shift *vs.* ion accumulation time for three scanning conditions: a 3D ion trap operated in the triple resonance mode, a linear trap operated in the triple resonance mode with a highly axially compressed ion cloud, and a linear trap operated in triple resonance mode with the ion cloud allowed to fill the entire linear trap volume (see text for details).

3D trap. A second set of conditions was also examined. In this case, a potential difference of less than 3 V was applied between the central and end rod sections. As a result, the ion cloud was allowed to expand over a wide region with the benefit of an expanded charge capacity.

Figure 14.32 shows the mass shift of an ion at *m/z* 1022 formed from electrospraying a solution of Ultramark under both the compressed and uncompressed ion cloud conditions as a function of the normalized ion signal. For reference, data obtained on a 3D trap operated with the triple resonance-scan function have been normalized for total ion counts and has been included also. In this experiment, it appears that the expanded ion cloud 2D trap has a charge capacity that is eight to 10 times greater than either the 3D trap or the small ion cloud system. The 3D trap in this case used the triple resonance-scan method which was shown previously (Figure 14.14) to have already a 10 times greater charge capacity than a standard, dipole-scanning, 3D trap. This result seems consistent with the work of Schwartz *et al.* [38], where moving from a 3D to 2D system with the simple dipole ejection method led to a 20 times increase in charge capacity. To demonstrate the value of having a very large spectral charge capacity, an illustrative mass spectrum (Figure 14.33) acquired by electrospraying Ultramark 1522 on the 2D trap in the triple resonance mode shows no evidence of space charging of the *m/z* 1508 peak despite the presence of a 100-fold greater intensity peak only 14 Th away.

14.8 SUMMARY

The work presented here indicates that the triple resonance-scan function can be applied equally to both a 3D and 2D trap geometry. The addition of the trapping

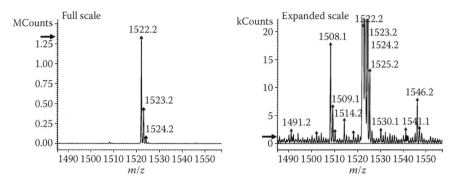

FIGURE 14.33 An example of a mass spectrum recorded with a linear trap in the triple resonance mode showing a large spectral dynamic range.

field dipole creates a significant hexapole component in the trapping field. In the case of a 3D trap, this hexapole field can be utilized easily to promote rapid ion ejection when combined with appropriately phased dipole and quadrupole waveforms at $\beta_y = 2/3$. For a 2D trap, however, the symmetrical nature of the device results in a coupling resonance condition between the x- and y-dimensions at $\beta_y = 2/3$ (at $a = 0$). As a result, it is essential to add a small quadrupole DC offset to the system to prevent almost total loss of signal due to the ions being displaced in both the x- and y-directions.

In practice, the triple resonance-scan function results in a complicated four-dimensional space optimization problem, however parameter combinations can be selected such that the instrument produces high resolution peaks with nearly 100% unidirectional ejection and a high resistance to space-charge effects. Fortunately, these parameter combinations have been found to be highly reproducible, and to depend only on the magnitude and sign of the trapping field dipole and the trap geometry.

Once optimized, the triple resonance scanning 2D ion trap has been shown to yield excellent mass spectral performance even at high scanning speed. In terms of charge capacity, it has been shown that a 2D trap has a larger charge capacity than a 3D ion trap when operated in an equivalent scan mode. Similarly, we have observed that, by careful selection of both the dipole and quadrupole phases and amplitudes, a 3D trap operated in the triple resonance-scan mode can yield a substantially higher charge capacity than an equivalent trap operated in a simple dipole mode. The work here suggests that by combining the high performance of the triple resonance-scan function with a 2D trap one can produce an ion trap with extraordinary trapping capacity and mass spectral performance.

REFERENCES

1. March, R.E.; Hughes, R.J. *Quadrupole Storage Mass Spectrometry*. New York: Wiley 1989, pp. 31–63; March, R.E.; Todd, J.F.J. *Quadrupole Ion Trap Mass Spectrometry*, 2nd Edn. Hoboken, NJ: John Wiley & Sons, Inc., 2005, pp. 34–72.

2. von Busch, F.; Paul, W. Über nichtlineare Resonanzen im electrischen Massfilter als Folge von Feldfehlern. *Z. Phys.* 1961, *164*, 588–594.

3. Whetten, N.R.; Dawson, P.H. Non-linear resonances in quadrupole mass spectrometers due to imperfect fields I. The quadrupole ion trap. *J. Vac. Sci. Technol.* 1969, *2*, 45–59.

4. Wang, Y.; Franzen, J. The non-linear ion trap. Part 3. Multipole components in three types of practical ion trap. *Int. J. Mass Spectrom. Ion Processes* 1994, *132*, 155–172.

5. Wang, Y. Non-linear resonance conditions and their relationships to higher multipole fields in ion traps and quadrupole mass filters. *Rapid Commum. Mass Spectrom.* 1993, *7*, 920–928.

6. Franzen, J. Simulation study of an ion cage with superimposed multipole fields. *Int. J. Mass Spectrom. Ion Processes* 1991, *106*, 63–78.

7. Franzen, J. The non-linear ion trap. Part 4. Mass selective instability scan with multipole superposition. *Int. J. Mass Spectrom. Ion Processes* 1993, *125*, 165–170.

8. Franzen, J. The non-linear ion trap. Part 5. Nature of non-linear resonances and resonant ion ejection. *Int. J. Mass Spectrom. Ion Processes* 1994, *130*, 15–40.

9. Louris, J.N.; Schwartz, J.; Stafford, G.C.; Syka, J.; Taylor, D. The Paul ion trap mass selective instability scan: Trap geometry and resolution. *Proc. 40th ASMS Conference on Mass Spectrometry and Allied Topics*, Washington, DC. May 31–June 5, 1992, p. 1003–1004.

10. Wang, Y.; Franzen, J.; Wanczek, K.P. The non-linear resonance ion trap. Part 2. A general theoretical analysis. *Int. J. Mass Spectrom. Ion Processes* 1993, *124*, 125–144.

11. Wang, Y.; Franzen, J. Ion excitation in a linear ion trap with a substantially quadrupole field having an added hexapole or higher order field. *Int. J. Mass Spectrom. Ion Processes* 1994, *132*, 155–172.

12. Wells, G.J.; Wang, M.; Maquette, E. Mass scanning method using an ion trap mass spectrometer. *U.S. Patent* 1998, 5,714,755.

13. Wells, G. A new ion ejection method employing an asymmetric trapping field. *Proc. 44th ASMS Conference on Mass Spectrometry and Allied Topics*, Portland, OR. May 12–18, 1996, p. 126.

14. Jackson, J.D. *Classical Electrodynamics.* New York: Wiley, 1967, pp. 27–93.

15. Beaty, E.C. Calculated electrostatic properties of ion traps. *Phys. Rev.* 1966, *A33*, 3645–3656.

16. Knight, R.D. The general form of the quadrupole ion trap potential. *Int. J. Mass Spectrom. Ion Processes* 1983, *51*, 127–131.

17. Beta version of SIMION 6, kindly provided by David Dahl, Idaho National Laboratory.

18. Fulford, J.E.; March, R.E. A new mode of operation for the three dimensional quadrupole ion storage (QUISTOR): The selective ion reactor. *Int. J. Mass Spectrom. Ion Phys.* 1978, *26*, 155–162.

19. Fulford, J.E.; Hoa, D.N.; Hughes, R.J.; March, R.E.; Bonner, R.F.; Wong, G.J. Radiofrequency mass selective excitation and resonant ejection of ions in a three-dimensional quadrupole ion trap. *J. Vac. Sci. Technol.* 1980, *17*, 829–835.

20. Jefferts, K.B. Rotational hfs spectra of H_2+ molecular ions. *Phys. Rev. Lett.* 1968, *20*, 39–41.

21. Ensberg, E.S.; Jefferts K.B. The visible photodissociation spectrum of ionized methane. *Astrophys. J.* 1975, *195*, L89–L91.

22. Langmuir, D.B.; Langmuir, R.V.; Shelton, H.; Wuerker, R.F. Containment device. *U.S. Patent* 1962, 3,065,640.

23. Alfred, R.L.; Londry, F.A.; March, R.E. Resonance excitation of ions stored in a quadrupole ion trap. Part IV. Theory of quadrupolar excitation. *Int. J. Mass Spectrom. Ion Processes* 1993, *125*, 171–185.

24. Franzen, J. Ejection of ions from ion traps by combined electrical dipole and quadrupole fields. *U.S. Patent* 1995, 5,468,957.

25. Kelley, P.E. Mass spectrometry method with two applied trapping fields having same spatial form. *U.S. Patent* 1995, 5,381,007.

27. Franzen, J.; Gabling, R.H.; Grasberg, G.H.; Weiss, G. Method and instrument for mass analyzing samples with a quistor. *U.S. Patent* 1990, 4,975,577.

28. Franzen, J.; Gabling, R.H.; Grasberg, G.H.; Weiss, G. Method for mass-spectroscopic examination of a gas mixture and mass spectrometer intended for carrying out this method. *U.S. Patent* 1991, 5,028, 777.

29. Franzen, J. Mass spectrometric high-frequency quadrupole cage with overlaid multipole fields. *U.S. Patent* 1992, 5,170,054.

30. Wang, M.; Marquette, E.G. Ion trap mass spectrometer method and apparatus for improved sensitivity. *U.S. Patent* 1994, 5,291,017.

31. Julian, R.K. *High-Performance Ion Trap Mass Spectrometry*. PhD. Thesis, Purdue University, 1993, 16–116.

32. Williams, J.D.; Reiser, H.P.; Kaiser, R.E.; Cooks, R.G. Resonance effects during ion injection into an ion trap mass spectrometer. *Int. J. Mass Spectrom. Ion Processes* 1991, *108*, 199–219.

33. Franzen, J.; Gabling, R-H.; Schubert, M.; Wang, Y. Nonlinear ion traps. In *Practical Aspects of Ion Trap Mass Spectrometry*, eds. R.E. March and J.F.J. Todd, Vol. 1, Chapter 3, pp. 49–167. Boca Raton: CRC Press, 1995.

34. Wells, G.J. Linear ion trap apparatus and method utilizing an asymmetrical trapping field. *U.S. Patent* 2006, 7,034,293.

35. Barbier, M.; Schoch, A. *CERN Technical Report,* 1958, *58-5,* 1–14.

36. Hagedorn, R. *CERN Technical Report, Part I & II,* 1957, *57-1,* 1–54.

37. Goldstein, H. *Classical Mechanics*, 215–272. Cambridge, MA: Addison-Wesley, 1965.

38. Schwartz, J.C.; Senko, M.W.; Syka, J.E.P. A two-dimensional quadrupole ion trap mass spectrometer. *J. Am. Soc. Mass Spectrom.* 2002, *13*, 659–669.

15 Fragmentation Techniques for Protein Ions Using Various Types of Ion Trap

Jochen Franzen and Karl Peter Wanczek

CONTENTS

15.1   Introduction 672
15.2   The Types of Ion Traps Used as Fragmentation Cells 675
    15.2.1   Kingdon Traps Need Ultrahigh Vacuum 675
    15.2.2   Magnetic Traps Need Ultrahigh Vacuum at Least in Some
            Measurement Phases 676
    15.2.3   Paul Traps are Operated Usually with Damping Gas 678
15.3   Fragmentation of Proteins 680
    15.3.1   Nomenclature of Protein Fragmentation 680
    15.3.2   The Basic Types of Protein Fragmentation 681
    15.3.3   What is Ergodic Fragmentation? 682
    15.3.4   Types of Electron-Induced Fragmentation Processes 684
            15.3.4.1   Electron Capture Dissociation (ECD) 684
            15.3.4.2   Electron Detachment Dissociation of
                      Negative Ions (EDD) 685
            15.3.4.3   Fragmentation by Ion/Ion Reactions 685
            15.3.4.4   Electron Transfer Dissociation (ETD) 686
            15.3.4.5   Inverse Electron Transfer Dissociation of
                      Negative Ions (IETD) 687
            15.3.4.6   Electron Transfer in Metastable Atom Induced
                      Dissociation (MAID) 687
15.4   The Various Fragmentation Techniques Possible in Different
    Types of Ion Trap 688
    15.4.1   FT-ICR Mass Spectrometers 688
            15.4.1.1   Ion Reactions and Structure 688
            15.4.1.2   Ergodic Fragmentation Processes 688
            15.4.1.3   Blackbody Infrared Radiative Dissociation (BIRD) 689

	15.4.1.4	Sustained Off-Resonance Irradiation (SORI)	
		Collisional Activation	690
	15.4.1.5	Surface-Induced Dissociation (SID)	690
	15.4.1.6	Electron-Induced Dissociation in ICR	691
15.4.2	Radiofrequency Paul Traps		691
	15.4.2.1	Ergodic Fragmentation by Gas Collisions	692
	15.4.2.2	Ergodic Fragmentation by Infrared Photon	
		Absorption	693
	15.4.2.3	Ergodic Fragmentation by Ion Bombardment	693
	15.4.2.4	Electron Capture Dissociation (ECD)	694
	15.4.2.5	Electron Transfer Dissociation (ETD)	695
	15.4.2.6	ETD in Combination with Deprotonation	696
15.4.3	Kingdon Ion Traps		698

15.5   Some Remarks on Automated Product Ion Spectrum Measurements in
       Tandem Mass Spectrometers 699
15.6   Conclusion 699
References 700

15.1   INTRODUCTION

The major problem in the mass spectrometric analysis of large molecules and biomolecules is the transfer of the sample particles, neutral molecules or ions, into the gas phase, because mass spectra can only be generated using particles in the gas phase. The best approach is to combine the vaporization and ionization processes. Several methods for this purpose have been developed: secondary ion mass spectrometry (SIMS), 1963, Benninghoven and Kirchner [1]; ^{252}Cf Plasma Desorption, 1976, Macfarlane and Torgerson [2]; fast atom bombardment (FAB), 1981, Morris *et al.* [3]; and field ionization (FI) and field desorption (FD), 1973, Beckey [4]. These methods made possible the analysis of biomolecules for a wider range of applications; they were, however, difficult to perform. FI was and is applied in the petroleum industry with great success and, because of many improvements of the FI technique [5], it has recently experienced a renaissance.

The explosive growth in the application of mass spectrometry to the analysis of large molecules, especially biomolecules, only became possible by the invention of two new ionization techniques; first, the spraying of analyte molecule solutions (electrospray ionization, ESI) and, second, the desorption of ions from layers of analytes on surfaces by means of laser irradiation without or with a matrix compound (matrix-assisted laser desorption/ionization, MALDI). For their development of methods for identification and structure analyses of biological macromolecules, John B. Fenn [6] and Koichi Tanaka [7] were each awarded one-quarter of the Nobel Prize in Chemistry in 2002. Parenthetically, the remaining half of the Nobel Prize in Chemistry was awarded to Kurt Wüthrich for the application of nuclear magnetic resonance spectrometry to the structure determination of biological macromolecules. The application of these techniques has been extended to the ionization of the most diverse kinds of organic and inorganic compounds, metal organic complexes, polymers, and small particles.

These new ionization techniques often generate protonated molecules. The ions that result from the protonation of biomolecules are heavier than the neutral molecule, and the mass increases with the number of protons added. For negative ions, the biomolecules are deprotonated and are correspondingly lighter. In all cases, however, there is (almost) no fragmentation of the biomolecules; this behavior is very welcome for molecular mass determination and the analysis of mixtures of biomolecules, for example, for the analysis of peptide mixtures obtained from larger biomolecules by enzymatic digestion. The importance of generating protonated molecules in large yields was realized much earlier. The first process, which employed a simple ion/molecule reaction technique, was chemical ionization (CI). Munson and Field [8] developed the CI technique for analytical applications, the process of CI itself having been discovered by Talroze in 1952 [9].

For the structural elucidation of biomolecules, however, we need insight into the molecule itself, which can be obtained by the study of fragment, or 'product', ions formed as a result of the dissociation of protonated molecules inside a so-called 'tandem mass spectrometer'. Here, the fragmentation of biomolecular ions begins to play a major role.

The key for today's biomolecular research is tandem mass spectrometry in its diverse embodiments. In tandem mass spectrometry, the so-called 'parent', or precursor, ions are first mass selected and isolated; these precursor ions are then fragmented, and the resulting product ions (originally referred to as 'daughter') ions are mass analyzed. Such an analysis can be done either in an instrument having two physically separated mass analyzers with a fragmentation cell in between ('tandem-in-space'), or in trap-type mass spectrometers able to select, fragment, and analyze ions in a single cell, step after step ('tandem-in-time').

In the early days of organic mass spectrometry, fragmentation processes could be studied with the aid of metastable ion decay [10]. At that time, electron impact ionization (EI) was predominant. In EI positive ions are generated by electron impact of the neutral molecules in the gas phase; electrons of ca 70 eV kinetic energy are utilized, because at that energy the ionization efficiency is optimal and the yields of the ions formed exhibit plateaux. Negative ions are generated by electron attachment. In mass spectrometers with a single analyzer, such as magnetic sector or quadrupole instruments, two kinds of ions are detected: molecular and fragment ions. Both are used for structure elucidation of molecules that can be transferred into the gas phase without thermal decomposition [11–13].

In these early investigations, only tandem-in-space mass spectrometry was used, investigating the fragmentation of 'metastable' ions that were decaying during their flight through field-free regions between the different parts of the mass spectrometer. Such studies were limited in early tandem-in-space mass spectrometers, where the magnetic sector was preceded by an electrostatic sector for enhanced energy resolution. A double-focusing instrument of reverse geometry, the 'ZAB-2F', in which the magnetic sector preceded the electric sector, was developed by the group of Beynon;

this configuration permitted precursor-ion mass selection, in terms of momentum/charge ratio, from the beam of ions exiting the magnetic field. Thus it was possible to study the kinetic energy released in the decay of a mass-analysed ion during its traverse of the field-free region between the magnetic and the electric sectors by mass-analyzed ion kinetic energy spectroscopy (MIKES) [10].*

Ion/molecule or ion/atom reactions at high collision energy have been employed for structural studies since early times of mass spectrometry [14]. These experiments are best performed in static mass spectrometers with at least two stages, where a high-energy ion beam is formed for mass analysis with magnetic or electric sectors. A collision cell can be placed in the field-free region, with the result that collision-induced dissociation (CID) at a high energy of several kiloelectronvolts takes place. Alternatively, triple-stage quadrupole instruments can be employed for this type of analysis, but at much lower collision energy, in which the second quadrupole is employed as a collision cell. CID at low-to-medium kinetic energies can be performed also in ion traps, and will be discussed later.

Ions may become metastable by obtaining an excess of internal energy during the ionization process, in most cases ionization by electron bombardment. Early research was directed to physico-chemical processes such as 'kinetic energy release' during the unimolecular dissociation of these excited species, and the elucidation of ion structures, as well as various other analytical applications, became a rapidly growing field. Today, this decay of metastable ions is still studied in special MALDI-TOF/TOF (time-of-flight) spectrometers, wherein kiloelectronvolt CID mass spectra can be obtained. For electrospray applications, however, a wide variety of fragmentation processes is used, and many of these are studied in different kinds of ion traps.

Fragmentation methods work best on multiply-charged ions, as produced by ESI. Singly-charged ions are much harder to fragment, and deliver a lower abundance of product ions because there is only one charge. With singly-charged protein ions, the product ion mass spectrum shows considerable gaps. Singly-charged ions delivered by MALDI are difficult to dissociate, and do not generate product ion mass spectra best-suited for analysis. It should be mentioned, however, that MALDI delivers metastable ions with good characteristics for generating high-quality product ion mass spectra in specialized mass spectrometers, and that there exists an in-source MALDI fragmentation method (in-source dissociation (ISD)) for generating mass spectra similar to those formed *via* electron capture dissociation (ECD). These kinds of fragmentation methods will not be elucidated here.

In the following discussion we concentrate on the fragmentation of protein ions, the most essential and diverse biomolecules in living bodies, within specialized ion traps. The fragmentation processes possible within the various types of ion traps depend very much on the characteristic features of the ion traps, and it is not possible for all fragmentation processes to be applied in all types of traps.

---

* See Volume 4, Chapter 1: An Appreciation and Historical Survey of Mass Spectrometry, by Raymond E. March and John F.J. Todd.

15.2 THE TYPES OF ION TRAPS USED AS FRAGMENTATION CELLS

Today, there are four well-known classes of devices with which to trap ions [15]:

- electrostatic Kingdon traps;
- magnetic Penning, or ion cyclotron resonance (ICR) traps;
- radiofrequency Paul traps; and
- electron beam traps.

The fourth type, the electron beam ion trap, until now, has been employed mainly in nuclear physics. The ions are trapped in a continuous or pulsed electron beam of high intensity and energy. Due to the high electron energy, multiply charged atomic ions can be produced in high yield. Recently, the fragmentation of organic molecular ions [16] produced in a pulsed electron beam trap has been demonstrated. The method has a great potential to develop into applications for bioanalysis.

15.2.1 KINGDON TRAPS NEED ULTRAHIGH VACUUM

From the early publication by Kingdon [17] in 1923 we know that ions can be stored for indefinitely long times in electrostatic fields, provided they have the correct kinetic energy to move in orbits around a central electrode inside an outer electrode box, preferably around a thin wire. The ions' kinetic energy should not be sufficient for them to reach the potential of the outer electrode box. When the device is operated under ultrahigh vacuum conditions the ions cannot lose kinetic energy by collisions with residual gas molecules and, therefore, they move forever around the central thin wire electrode; only in rare cases will the ions impinge on this wire during their introduction into the Kingdon trap. Once within the trap they will never impinge on the wire unless through a possible deflection in a close encounter with another ion, which may lead to changes in the directions of flight of both ions. When there is residual gas within the box, the ions will slowly and steadily lose kinetic energy by successive collisions with residual gas molecules, decrease the diameter of their orbiting motion and finally hit the central electrode.

There are modern forms of Kingdon traps that can be used as mass spectrometers. An excellent example is the Makarov Orbitrap™ [18,19], introduced by Thermo Fisher Scientific in 1999.* In an electrode arrangement of sophisticated form, the ions circle radially in orbits around a central lengthy electrode, and the axially orbiting ions oscillate harmonically in a specially formed electrostatic field between the central electrode and a completely encasing outer electrode housing. By choosing the correct shapes of all electrodes, the oscillating movement in the axial direction is completely independent of the orbiting movement in the radial direction. The image current of the ions resulting from the axial oscillations can be detected on suitable electrodes as a time domain signal which, when subjected to Fourier transformation into the

---

* See Volume 4, Chapter 3: Theory and Practice of the Orbitrap Mass Analyzer, by Alexander Makarov.

frequency domain, generates a mass spectrum. As with most Fourier transform mass spectrometers, the Orbitrap exhibits excellent mass resolution, comparable to ICR mass spectrometry.

The Orbitrap with its single central electrode is not the only way to decouple the axial and radial movement in Kingdon traps; investigations show a plurality of possible embodiments with more than one inner electrode. All these embodiments can be used to design mass spectrometers [20]. In some of these new types of Kingdon traps, the ions do not orbit at all around the inner electrodes; instead, they just oscillate in a plane between two or more inner electrodes. This planar disposition of trapped ions makes it easy to introduce ions into the trap and opens applications that include fragmentation processes of ions inside the Kingdon trap. In other types of Kingdon traps, the ions may perform even more complicated movements than just orbits around two, four, or more inner electrodes.

15.2.2   Magnetic Traps Need Ultrahigh Vacuum at Least in Some Measurement Phases

The first investigations on magnetic traps were performed early in the twentieth century. In 1921, Hull [21] investigated the motion of electrons between coaxial cylinders in a uniform magnetic field. In 1934, Penning and Addink [22–24] commenced investigations on the trapping of charged particles in structures placed in a magnetic field. In honor of Penning, traps employing magnetic and DC electric fields, are called Penning traps [25]. In these traps, the ions are trapped in two spatial dimensions by a high magnetic field; in the third dimension, an electrostatic trapping field has to be applied.

Penning traps were utilized in nuclear and atomic physics for precision measurement of atomic constants and for trapping exotic particles like antimatter. Originally they were built with three hyperbolic-shaped electrodes, two end-cap electrodes and a ring electrode. In the ideal case, these electrodes generate a quadrupolar field, which allows the cyclotron, magnetron, and trapping motions of the charged particles in the trap to be decoupled. This decoupling and the precise knowledge of any disturbances of the ion motion by field imperfections and space charge are of crucial importance for precision measurements of fundamental constants. Cylindrical traps with compensation electrodes were introduced in 1984 by Gabrielse [26]. Since of the nature of the trapping field, Penning traps can confine only ions of one polarity. In order to trap simultaneously particles of both polarities, more complicated designs are necessary. A cylindrical trap having at least five electrodes is necessary to generate the two potential extremes required to trap positive and negative particles at different sites in the trap. For interactions to occur, charged particles of one polarity have to be transferred to a site in the trap where they are sufficiently close to those of the other polarity. An advanced construction of a cylindrical trap with up to 32 ring electrodes was employed by the ATRAP (Antihydrogen Trap) collaboration in order to produce and to study antihydrogen [27]. In a more simple arrangement for ion/ion interaction studies, the potential configuration to trap ions of both polarities was reached with double trapping electrodes, the inner electrode constructed from a mesh, the outer of solid material [28]. Also trapping electrodes made of concentric rings can be utilized [29]. Another recent development for the study of ion/ion reactions uses a cylindrical five-electrode trap with electrodes of various diameters [30]. A different approach is the 'dynamic' trapping (combined mode) by application of a radiofrequency potential (pseudopotential, see next Section) to the trapping electrodes of a standard ICR cell [31].

In the 1960s, almost independently of the development of Penning traps in physics, a technique for the study of ion chemistry was developed in chemistry, namely ICR spectrometry [32,33]. The first instruments had a drift ICR cell with the axis perpendicular to the field of an electromagnet. A mass scan was performed by scanning the magnetic field. Three major inventions improved the technique: the introduction of a trapped ion ICR cell and pulsed instrument operation by McIver [34]; the application of the Fourier transform technique for mass analysis by Comisarow and Marshall [35]; and the introduction of superconducting magnets with high solenoidal fields by Allemann, Kellerhals, and Wanczek [36]. The best geometrical fit to the solenoid magnet was obtained with a cylindrical ICR cell [37]. This type of ICR cell has replaced almost completely the cells with plane electrodes, although the field is far from being quadrupolar. This technique was and is the method of first choice for the study of ion/molecule reactions at low pressures in the gas phase.

The ion motion in an ICR cell is, in principle, a superposition of the cyclotron, the magnetron, and the trapping motions. For the detection of ions, the cyclotron motion of the ions is excited and the image current induced by the phase-coherent motion of the ions in the ICR cell electrodes parallel to the magnetic field is measured as a function of time; the transient signal generated is then Fourier transformed. In this way, one obtains a frequency spectrum that is converted into a mass spectrum after internal or external calibration.* After detection, the ions are still trapped in the ICR cell and available for further experiments, such as ion/molecule reactions, interaction with photons, or fragmentation. The mass resolution is inversely proportional to the m/z (mass-to-charge) ratio of the ions, and proportional to magnetic field strength and the time for which the transient can be measured. Therefore, very high resolution can be obtained only at low pressure in the 10^{-10} mbar region, where the number of collisions is low, trapping efficiency is high, and trapping time long. With a 7 T magnetic field and a pressure of 10^{-11} mbar, a resolution of $>3 \times 10^8$ (FWHM) has been achieved at m/z 132 of xenon [38]. In analytical applications, which are predominant today, the ICR cell is kept normally at very low pressure to obtain high resolution. Ions are usually generated at elevated pressure to increase ion yield and sensitivity, therefore the ICR cell must be pumped differentially. For this differential pumping it is necessary to generate the ions outside the ICR cell or, better, outside the room temperature bore in front of the superconducting magnet using an external ion source [39]. The book by Marshall and Verdun [40] gives a good introduction into the basics of the technique. Commercially available ICR mass spectrometers are at present delivered with superconducting magnets up to 18 T field strength.

---

* See Volume 4, Chapter 7: Ion Accumulation Approaches for Increasing Sensitivity and Dynamic Range in the Analysis of Complex Samples, by Mikhail E. Belov, Yehia M. Ibrahim and Richard D. Smith; Volume 5, Chapter 5: Fourier transform ion cyclotron resonance mass spectrometry in the analysis of peptides and proteins, by Helen J. Cooper.

15.2.3   Paul Traps are Operated Usually with Damping Gas

When Wolfgang Paul and coworkers first published the development of the quadrupole mass spectrometer [41] in 1953 and the quadrupole ion trap [42] in 1958, a new era in mass spectrometry had begun. Mass spectrometry and ion trapping became possible without magnetic fields. W. Paul and H.G. Dehmelt each received one-quarter of the 1989 Nobel Prize in Physics for their developments of ion trapping techniques. Here, the Paul ion traps will be discussed only briefly; detailed descriptions of the technique and its applications are given in the first three volumes of this series [43].

The Paul trap comprises three hyperbolical-shaped electrodes. The application of appropriate voltages to this set of electrodes generates an inhomogeneous radiofrequency electric field for ion confinement; in the ideal case, a quadrupolar field is formed. The action of this field is best understood in terms of the so-called 'pseudopotential' theory. Pseudopotentials are not real potentials, but they act in the same manner. Pseudopotentials are mathematical fictions, derived from integration over the acceleration forces on ions within areas with gradients of radiofrequency fields. Unlike real potentials, pseudopotentials are not effective at every instant of time; instead, the radiofrequency field causes the ions to exhibit forced oscillations (the 'fundamental oscillations').

Integration of the acceleration forces over at least one radiofrequency period reveals a pseudoforce, driving the ions located in the region of the inhomogeneous field toward areas having lower field gradients. The corresponding pseudopotential forms a well inside the ion trap, with a minimum at the center of the trap and, in the absence of a 'damping', or 'buffer', gas, the ions can oscillate freely in this pseudopotential well.

These oscillations are called 'secular oscillations'; they are superimposed, at least in the outer trap regions, upon the faster 'fundamental oscillations' having the frequency of the storage radio frequency. Under a given set of conditions the secular oscillation frequencies of the ions are approximately inversely proportional to the value of m/z; approximation formulae for the determination of the m/z ratio are available to a very high degree of precision. Through knowledge of the secular frequency, ions can be excited resonantly by a corresponding high frequency voltage applied across the ion trap in the axial direction, either to eject the ions mass-selectively, or, more gently, for multiple collisions with the damping gas.

Most essential is the fact that pseudopotentials can trap ions of both polarities at the same time. This property of the pseudopotential makes possible the study of reactions between positive and negative ions, for example to effect electron transfer dissociation (ETD).

Paul traps are operated usually with a buffer gas to damp the free secular oscillations of the ions inside the pseudopotential well of the trap. The ions lose the kinetic energy of their oscillations, and gather in the center of the trap, forming a small cloud. The diameter of the ion cloud results from a balance between the repulsive Coulombic forces and the centripetal forces of the focusing pseudopotential well.

In a Paul trap, only a limited m/z-range can be stored simultaneously. This range is described by the Mathieu equation, which defines mainly its relationship to the amplitude and frequency of the applied radiofrequency potential and the dimensions of the trap. The equation gives sets of parameters that govern whether the ions inside the trap have stable trajectories of limited amplitude, or have unstable trajectories of rapidly increasing amplitude. There exists a lower m/z-threshold below which storage is no longer possible. Light ions below this well-defined threshold of m/z, are accelerated so strongly within one half-period of the radiofrequency voltage that they cross the trap and either impinge on electrodes, or are re-accelerated in phase with the applied RF field in such a way that they gather kinetic energy continuously until they leave the ion trap through holes in one of the electrodes. The lower mass threshold is directly proportional to the amplitude of the RF voltage. When a high RF voltage is needed for the fragmentation process, the lower mass threshold precludes the storage of low mass ions, and the product ion mass spectrum is attenuated so that it does not show these low m/z-value ions.

For high mass ions, the pseudopotential forces get weaker and weaker the higher the value of m/z. This behavior has the effect that the ion cloud in the trap is not a homogeneous mixture of light and heavy ions, instead, the light ions gather in the center of the cloud, and heavier ions collect in layers in the outer regions of the cloud. Depending upon the number of ions inside the trap and their Coulombic repulsion forces, there exists in practice an upper mass limit for the ions to be stored, because the highest m/z-value ions in the outermost shell are no longer held by the pseudopotential against the Coulombic forces. In contrast to the well-defined lower mass threshold, this upper limit smears out over a broad range of masses.

Within the three-dimensional Paul trap made from a ring electrode and two end-cap electrodes, the pseudopotential rises in all three spatial dimensions; the ions are completely enclosed and no further trapping field is needed to store the ions. After damping, the ion cloud gathered in the central well has a spherical form.

In contrast, two-dimensional Paul ion traps [44] (described also as 'linear ion traps') usually consist of pairs of lengthy rod electrodes, like in a quadrupole mass filter, connected in turn to the phases of the radiofrequency voltage. The linear trap design has descended from a race-track configuration Paul trap, first used by Drees and Paul [45]. A linear trap holds the ions only in the two spatial dimensions, perpendicular to the rod axis, by corresponding pseudopotentials; in the third dimension, additional electrostatic or RF fields must be applied to prevent the ions escaping in this direction. The cloud of ions takes the form of a thread along the axis of the system. The linear ion traps [46] can be employed as a mass spectrometer like a three-dimensional trap or to direct and transport ions [47].*

Such two-dimensional linear traps when filled with collision gas are used widely as fragmentation cells in various tandem mass spectrometers with an axial injection of accelerated ions. A quadrupole rod set constitutes one of the most frequently used types of fragmentation cell. There exist versions of these devices with DC

---

* See also Volume 4, Chapter 10: Trapping and Processing Ions in Radio Frequency Ion Guides, by Bruce A. Thomson, Igor V. Chernushevich and Alexandre V. Loboda; Volume 4, Chapter 11: Linear Ion Trap Mass Spectrometry with Mass-Selective Axial Ejection, by James W. Hager; Volume 4, Chapter 12: Axially-resonant Excitation Linear Ion Trap (AREX LIT), by Yuichiro Hashimoto; Volume 4, Chapter 14: Electrically-induced Nonlinear Ion Traps, by Gregory J. Wells and August A. Specht.

gradients along the axis, driving the ions in one direction. Potential gradients can be used to send the ions backwards and forwards, thereby increasing the number of collisions; they can be employed also to eject the product ions in the direction of the mass analyzer.

Besides these systems of multipole rod sets, there are other embodiments of two-dimensional ion traps, for example, made from a series of lengthy parallel ring electrodes, or from wires wound into helices; but here also the ions must be trapped in the third (axial) dimension by means of electrostatic fields or by applying, for example, an RF voltage between the ends of the rod-sets. Ions of both polarities can only be stored simultaneously by trapping with pseudopotentials in the axial direction. When the electrodes of the linear Paul trap are segmented, so that they can be operated at different DC potentials along the axis, it is possible to store positive and negative ions at the same time, but in different trap segments.

Both types of Paul traps, three-dimensional and two-dimensional, can be used as mass spectrometers on their own, using well-known mass-selective ejection processes for directing the ions toward detectors outside the traps. These mass spectrometers are relatively inexpensive and, therefore, are widespread; however, their mass precision and mass resolution are limited and sometimes do not fulfill requirements of modern biological mass spectrometry.

In three-dimensional Paul traps, the ions are detected with the aid of a secondary electron multiplier after mass-selective ejection. The ejection techniques are elaborate and allow rapid scanning of a mass range with great sensitivity, reproducibility, and medium to high mass resolution. One of the reasons for the breakthrough in ICR mass spectrometry to give high precision and ultrahigh resolution techniques was the application of Fourier transform techniques for mass analysis. Several groups have applied the Fourier transformation technique and image current detection for ion detection at low pressure inside the Paul trap. After detection the ions are still present in the trap. The secular frequencies of the ions under study are excited, and the image current is detected and Fourier transformed for mass analysis. Cooks and coworkers [48] described such a method for a three-dimensional RF ion trap. In their arrangement the detection of the secular frequencies was disturbed greatly by the presence of the high RF trapping voltage. Similar difficulties are described by Zerega *et al.* [49, 50].* The employment of a new type of RF ion trap, together with highly sophisticated electronics, by the group of Glasmachers is very promising [51,52].

15.3 FRAGMENTATION OF PROTEINS

15.3.1 Nomenclature of Protein Fragmentation

The most essential biological substances for which mass spectral fragmentation studies are carried out are proteins and peptides. (Poly)peptides are small proteins, an expression commonly applied to proteins digested by enzymes. The nomenclature

---

* Volume 4, Chapter 9: A Fourier Transform Operating Mode Applied to a Three-Dimensional Quadrupole Ion Trap, by Y. Zerega, J. Andre, M. Carette, A. Janulyte and C. Reynard.

for fragment ions formed from proteins and peptides follows the suggestions made by Roepstorff and Fohlman [53], and modified by Biemann and co-workers [54]; for more details the reader is referred to the text by Kinter and Sherman [55].

The weakest bonds within the chain of amino acids are the peptidic bonds between two successive amino acids, that is, the bond between the acid group of the N-terminal branch and the amino group of the C-terminal branch; a dissociation delivers the b and y ions. A charged N-terminal fragment is designated as a 'b ion', while a charged C-terminal fragment is called a 'y ion'. All ergodic fragmentations deliver b and y ions (see Section 15.3.3).

When the bond between the two carbon atoms of the amino acid is split, the corresponding C-terminal and N-terminal ions are designated as 'a ions' and 'x ions', respectively. These a and x fragment ions are produced, for example, by electron detachment dissociation (EDD). The dissociation between the amino group and the rest of the amino acid, as occurring predominantly in all electron-induced dissociations, delivers the 'c ions' and the 'z ions'.

15.3.2 THE BASIC TYPES OF PROTEIN FRAGMENTATION

The fragmentation process in a mass spectrometer is controlled by thermodynamic and kinetic factors, as well as by the time window of the instrument. There are two general categories of reactions that involve the participation of the gas phase that are important in mass spectrometric analyses: interactions in the gas phase, and interactions between the gas and solid phases.

Unimolecular reactions lead to dissociation of gas-phase ions after resonant or non-resonant deposition of internal energy (*e.g.*, metastable ions). In many cases, these reactions obey statistical laws and are, therefore, called 'ergodic' reactions. Bimolecular reactions are reactions between two particles, such as electrons, ions, and neutral molecules. Of importance are ion/molecule reactions, ion/ion reactions, and electron/ion reactions. FAB, SIMS, and surface-induced dissociation (SID) are examples for gas–solid interactions, which yield characteristic ions for mass spectrometric analysis.

Within the various types of ion traps we can perform different types of fragmentation processes upon the ions. Here, we restrict our description to the fragmentation of protein ions, for which we know of four basic types of fragmentation, each including a number of subtypes.

a. Unimolecular fragmentations are ergodic in most cases, and include energization of the dissociating species by (among others) CID, SID, infrared multi-photon dissociation (IRMPD), and blackbody infrared radiative dissociation (BIRD). Ergodic fragmentation splits the parent ions into b and y product ions, in most cases losing side chains like those of post-translational modifications (PTM).

b. Electron induced fragmentations include, for example, ECD and ETD. Electron induced fragmentation of positive protein ions dissociates the precursor ions predominantly, but not exclusively, into c and z product ions, with some additional a ions keeping the side chains intact. For the fragmentation of negative protein ions, EDD, and inverse electron transfer dissociation (IETD) deliver *a* and *x* product ions.

c. High-energy collision fragmentation, as well as hard collisions with gas molecules (HE-CID) and collisions with walls (high-energy surface induced decomposition) produce mainly *b* and *y* type product ions, but with considerable yields of 'internal' product ions, adding third-generation ('granddaughter') ions to the product ion mass spectrum.

d. Reactive fragmentation, most often involving bimolecular reactions between positive protein ions and negative reactant ions, generating product ions as products of the reaction.

The ergodic and the electron-induced fragmentation processes generate completely different product ion mass spectra, giving complementary information. For proteins and peptides, a most interesting combination of fragmentation processes is applying an ergodic and an electron-induced fragmentation process in parallel. Comparison of the two product ion mass spectra not only shows readily widely separated parts of the amino acid sequence, but above all it reveals kinds and locations of PTM.

Each type of protein fragmentation process requires the use of a variety of different methods and techniques, the details of which depend upon the type of ion trap used for the fragmentation process.

15.3.3 What is Ergodic Fragmentation?

Unimolecular dissociation processes have been described in terms of statistical models such as Rice–Ramsperger–Kassel and Marcus (RRKM) theory, Quasi-Equilibrium Theory (QET) and phase-space theories [56]. Since the large number of internal degrees of freedom of large molecules and the non-resonant nature of many ion activation techniques, the fragmentation process obeys the Boltzmann ergodic theorem of statistical mechanics.

Ergodic fragmentation is the dissociation of an analyte ion that occurs because the amount of internal energy exceeds that needed to break one of the bonds between the atoms of the ion. This process is not instantaneous, and it exhibits half-life times from nanoseconds to milliseconds depending on the amount of internal energy. There are several ways to 'pump' energy into an ion in order to make it fragment: multiple collisions with residual gas (CID); photon absorption, particularly by absorption of a multitude of infrared photons (IRMPD); bombardment with medium-energy electrons; and bombardment with ions, preferably involving multiple low-energy collisions with monatomic ions of opposite polarity.

Boltzmann originally formulated the so-called ergode hypothesis in terms of the statement: "In a closed system any combination of energetic states which can be realized by the amount of energy inside the closed system will at some time be established". The ergode hypothesis has long-since been proven mathematically, and has become a well-established theory in thermodynamics. It may take a long time for a certain combination of energetic states to become populated, and the half-life time is dependent upon the excess of available energy. The internal excitation of an ion leads therefore to a random distribution of energy into the internal degrees of freedom before fragmentation occurs.

An analyte ion that does not experience any exchange of energy with its surroundings is a closed system. According to the ergode theory, if it contains enough energy to decay, it will do so eventually because the combination of electronic and vibrational states that leads to the breaking of a bond will, at some time, become populated. The fragmentation process itself, however, destroys the closed system of the analyte ion; this process is irreversible.

Thus we see that an analyte ion with sufficient internal energy is no longer stable; it decays with a certain half-life. For unstable analyte ions that decay within a certain time window, the expression 'metastable' was coined several decades ago. Indeed, the term goes back to the early days of mass spectrometry, in which the decay of analyte ions in the straight path (field-free region) between the electric and magnetic deflecting fields was studied extensively: those ions that decayed within the time window needed to pass through this straight flight path were called 'metastable ions'. Most of these early studies were directed to the elucidation of the energetic processes involved in the decay, and measurements of the 'kinetic energy release', that is, the energy given to the fragment ion by the decay process, was one of the main goals. Today, we may use the expression metastable to describe those ions that decay within a pre-selected time window inside an ion trap, after sufficient energy has been pumped into the ion.

The probability of a given bond breaking depends upon its binding energy. Statistically, the weakest bonds will break first, and there is a high probability that only the weakest bonds will break at all. In proteins, the weakest bonds are most often the bonds associated with PTM and other side chains; ergodic fragmentation, therefore, involves initially losses of almost all the PTM, such as phosphorylations, glycosylations, etc. Within the amino acid backbone chain of the proteins, the weakest bonds are the so-called peptidic bonds between the amino acids; scission of these peptidic bonds results in formation of fragment ions of the b- or y-series. These peptidic bonds too are broken during ergodic fragmentation of proteins. The different peptidic bonds between different amino acids exhibit slightly different dissociation energies; therefore, some of the bonds will have a slightly higher probability of breaking than do others, resulting in different intensities of the fragment ions produced thereby. Non-peptidic bonds with their somewhat higher binding energy will be broken so rarely that, in practice, the resulting product ions cannot be observed within the product ion mass spectra.

In the literature, the question of whether unimolecular fragmentation of peptide radical cations is ergodic or not has been the subject of some controversy. Weinkauf *et al.* [57]. concluded from their photoionization experiments of small peptides that this process is non-ergodic, without energy dissipation. Laskin *et al.* [58] measured time- and energy-resolved fragmentation efficiency curves for an octapeptide radical cation; their experiments are in good agreement with statistical RRKM calculations, hence the observed processes are ergodic. McLafferty and co-workers [59] concluded that the unusual fragmentation during ECD results from a non-ergodic process, while Syrstad and Turecek [60], using high level *ab initio* calculations, demonstrated that ECD can be rationalized without assuming non-ergodic behavior.

15.3.4 Types of Electron-Induced Fragmentation Processes

The way in which to internally excite an analyte ion depends very much on the nature and operation of the ion trap. The different processes will be presented in Section 15.4. In the following account, the various fragmentation methods that will be discussed are:

- electron/ion reactions: ECD and EDD;
- ion/ion reactions: ETD, IETD; and
- ion/neutral reactions: 'metastable atom-induced dissociation' (MAID).

15.3.4.1 Electron Capture Dissociation (ECD)

'Dissociative recombination', that is dissociation of low mass molecular ions after electron capture, has been known for a long time and is well understood in physics [61]. Approximately 10 years ago McLafferty and co-workers [62] and, more recently, Zubarev and co-workers [63], successfully applied this kind of reaction to protein analysis. They called the method ECD. This method is a non-threshold, non-ergodic fragmentation process, induced by the capture of low-energy electrons by associated protons. Statistically, every amino acid may carry such an associated proton with about equal probability. This associated proton is neutralized by electron capture and lost as a neutral hydrogen atom. After rearrangement, the amino acid chain is cleaved (exception proline [64]). This cleavage does not occur at the peptide bond, but at a neighboring bond, leading mainly to the production of c and z fragment ions.

The most essential feature of this new type of fragmentation is that it breaks only the amino acid chain of the backbone: there is no cleavage of the side chains. All PTM stay intact and in their correct places in the new product ions formed. The fragmentation is fast: it takes place in less than 10^{-7} s. All fragmentation processes are stopped immediately when the supply of low-energy electrons ceases.

When the precursor ions are doubly-charged, that is, the molecule is doubly-protonated, one of the two fragments created remains as an ion. The fragmentation follows very simple rules, so that it is relatively simple to elucidate the structure of the molecule from the fragmentation pattern. It is often very simple to read off the sequence of peptides or proteins directly from the mass differences of the exceptionally large c product ion signals, in contrast to the evaluation of collisional product ion mass spectra. It is significantly easier to interpret these ECD product ion mass spectra than the corresponding CID product ion mass spectra. Since ECD product ions do not lose side chains, such as those formed by PTM, whereas CID product ion mass spectra regularly show losses of these side chains, the ECD product ion mass spectra contain information that is complementary to that found in CID product ion mass spectra. It is particularly useful to have both types of product ion mass spectra available for evaluation.

It is also possible to fragment triply- and higher multiply-charged precursor ions by means of ECD. With triply charged precursor ions, most ECD-induced fragmentations give two singly-charged product ions each containing one of the two terminal groups, that is, there will be one c-ion and one z-ion; however, there are cases where a triply-charged precursor ion decays into one doubly-charged product ion and one neutral particle.

It is particularly easy to carry out this type of fragmentation in ICR mass spectrometers [65], as the low-energy electrons in the range from 0.1 to 3 eV emitted from a thermionic cathode can simply be guided along the magnetic field lines to the stored cloud of analyte ions. Fragmentation by means of ECD can be implemented in RF ion traps only with some difficulty, as the strong RF fields do not allow the majority of the electrons to reach the cloud of analyte ions with low-energy. There are, nevertheless, a variety of approaches to ECD fragmentation in RF ion traps, but each of them requires more costly equipment and, until now, they have not offered sufficient sensitivity.

With electrons of kinetic energies between 3 and 10 eV hot electron capture dissociation (HECD) [66,67] is observed. Simultaneous electronic excitation of the precursor ions and electron capture alters the fragmentation mechanism.

15.3.4.2   Electron Detachment Dissociation of Negative Ions (EDD)

When multiply-charged negative protein ions are bombarded with electrons of moderate kinetic energies higher than *ca* 20 eV, one electron of the ion is ejected, and the backbone of the protein will break; this process is known as EDD. In contrast to ECD and ETD, the product ions formed by EDD belong to the *a*- and *x*-type of ions. This very interesting fragmentation process can be performed in Penning ion traps [68], however, it is much harder to get this reaction to work in 3D Paul ion traps, and requires magnetic focusing of the electrons [69]. A particularly significant application concerns structure elucidation of glycanes, because EDD breaks up glycane structures in a highly specific way, in contrast to CID.

15.3.4.3   Fragmentation by Ion/Ion Reactions

Ion/ion or ion/electron reactions are observed in interstellar space, in planetary atmospheres, and in plasmas [70]. Many of the ion/ion reactions yield neutral products, and are, therefore, difficult to study. This difficulty does not exist for reactions of multiply-charged ions with singly- or multiply-charged ions of opposite polarity. With ESI, multiply charged ions resulting from large molecules can be formed easily in the gas phase. While there are similarities between ion/molecule reactions and ion/ion reactions, ion/ion reactions are, however, in a category of their own. The dynamics and thermodynamics are quite different, and offer a great potential for the understanding of ion chemistry and structure.

Paul traps are able to store simultaneously ions of both charge polarities in the same region of the trap, in contrast to Penning traps. McLuckey and co-workers [71–73] have developed the study of ion/ion reactions in Paul traps into a method for analysis of high mass multiply-charged ions. Ion/ion reactions can be used as a tool to transform ions into a state best-suited for analysis, and they can lead directly to analytically useful information.*

---

* See Volume 5, Chapter 1: Ion/Ion Reactions in Electrodynamic Ion Traps, by Jian Liu and Scott A. McLuckey.

In the case of biopolymers, several types of single collision reactions can be distinguished:

- proton transfer, (proton transfer dissociation (PTD));
- multiple-proton transfer;
- electron transfer, (ETD, and IETD);
- metal ion transfer; and
- molecular modifications.

The names of methods that employ the dissociation reaction types listed are shown in brackets.

Several reaction mechanisms can be distinguished:

- direct hard sphere 'sticky' collision;
- proton- or electron-hopping at a crossing point of potential energy surfaces (at distances of ca 100 Å); and
- a Coulombically-bound orbit.

In the last-mentioned case, the orbit formation is the rate-determining step in many cases, and its radius is normally too large to allow chemistry to take place. Here only electron-induced fragmentation arising from ion–ion reactions will be discussed.

15.3.4.4   Electron Transfer Dissociation (ETD)

Recently, a method was published for the fragmentation of ions in RF ion traps that delivers the same kind of product ions as ECD, but by means of different reactions: ETD [74,75].* These reactions can be performed easily in radiofrequency Paul ion traps by introducing suitable negative ions in addition to the stored positive analyte ions. Other methods for this type of process have been described in two patents [76,77]. The product ions here, as in the case of ECD, also belong to the c- and z-series and are, therefore, very different from the product ions of the b- and y-series obtained by ergodic fragmentation. The product ions in the c- and z-series have significant advantages for determining the amino acid sequence from the mass-spectrometric data, not least because ETD product ion mass spectra can extend more easily down to smaller masses than those observed in CID product ion mass spectra.

The fragmentation of protein ions by electron transfer in RF ion traps is performed in a very simple manner by reactions between multiply-charged positive protein ions and suitable negative ions. These negative ions are most often radical anions, such as those of fluoranthene, fluorenone, anthracene, or other polyaromatic compounds. In radical anions, the chemical valences are not saturated, which permits the easy donation of electrons in order to reach an energetically favorable non-radical form. These radical anions are generated in negative chemical ionization

---

* See also Volume 5, Chapter 3: Methods for Multi-Stage Ion Processing Involving Ion/Ion Chemistry in a Quadrupole Linear Ion Trap, by Graeme C. McAlister and Joshua J. Coon.

(NCI) ion sources, using special conditions, most probably by simple electron capture or by electron transfer. In principle, the design of NCI ion sources is the same as for positive chemical ionization (CI ion sources) but, in operation, the former require the addition of a suitable gas such as methane in order to obtain large quantities of low-energy electrons. Negative chemical ionization sources are also referred to as electron attachment ion sources.

Quite often, the fragmentation process will not proceed to completion by itself: the fragments, partly ions, partly neutral particles, remain bound together by van der Waals forces or by hydrogen bridges. The ions generated are reduced in charge. Knowing the charge and mass, the ion may be excited resonantly to collide softly with the collision gas, separating the fragments and generating c-type and z-type ions. The expression 'c-type' and 'z-type' ions is used here because sometimes the bound fragment particles are not separated species, but are z-type ions of radical nature, sometimes designated as $z + 1$. Such is the case particularly in combinations with neutral fragments where there are no repulsive Coulomb forces to explode the bound fragments, as encountered after electron transfer to doubly-charged parent ions. A further hydrogen exchange reaction between the two particles can even produce a non-radical '$z + 2$' ion or a radical '$c - 1$' ion.

15.3.4.5   Inverse Electron Transfer Dissociation of Negative Ions (IETD)

Another most interesting dissociation type for negative ions is IETD. Instead of adding an electron, an electron is removed from the multiply-charged negative protein ion in a reaction with a positive ion having a high electron affinity. The positive ion may be a monatomic positive ion such as Xe^+ [76], but the reaction works successfully also with certain positive organic radical ions. The reaction runs efficiently in 3D Paul ion traps; a favored method uses the positive radical ions of fluoranthene, generated by the same ion source which produces, under different operating conditions, the negative radical ions of fluoranthene for ETD [75].

15.3.4.6   Electron Transfer in Metastable Atom Induced Dissociation (MAID)

It has been found that electron transfer can take place from highly excited neutral particles, for example by highly excited helium atoms from a FAB particle source [78,79]. This type of fragmentation is sometimes abbreviated to MAID ('metastable atom induced dissociation'). Here again, the product ion mass spectra are at least in some respect similar to those obtained from ECD, however, sometimes intermixed with CID-type product ions. The exact mechanism for this type of fragmentation still needs some research. It appears, however, that the source of the electron is irrelevant for the non-ergodic fragmentation process induced by neutralization of a proton by an electron.

In spite of the fact that ECD, ETD, and MAID fragmentation methods may commonly be referred to as 'electron-induced' fragmentation methods, there are some differences between ETD and MAID. Thus MAID can be used successfully to fragment singly-charged ions, for instance, formed by MALDI.

15.4   THE VARIOUS FRAGMENTATION TECHNIQUES POSSIBLE IN DIFFERENT TYPES OF ION TRAP

15.4.1   FT-ICR MASS SPECTROMETERS

15.4.1.1   Ion Reactions and Structure

Collisions between ions and molecules occurring in an ICR cell can lead to ion/molecule reactions or to CID, depending on the collision energy. Both techniques have been employed since the early days. The use of ion/molecule reactions for structure elucidation resembles the chemical methods for structure elucidation: the detailed study of reactivity, synthesis and degradation, in use before instrumental analysis became possible [80].

Mid-range infrared spectroscopy is a direct probe of ionic and molecular structures. Nowadays, infrared spectra of ions can be obtained with free electron lasers; for example Oomens *et al.* [81] have described charge state-resolved mid-IR spectra of a protein with 10 amino acids.* This approach is very promising and will induce an increase in the application of ion IR spectra and reactions for structural studies. However, the application of ion reactions and IR spectra for structural studies will not be discussed here.

15.4.1.2   Ergodic Fragmentation Processes

The yields of collisional processes and ion/molecule reactions increase with trapping time and pressure in the ICR cell of an ICR mass spectrometer. On the other hand, the mass resolution increases with the free time between collisions and, therefore, decreases with increasing pressure. Therefore, the task that has to be accomplished in high resolution ICR mass spectrometry of the CID products ions is to effect the dissociation step at medium pressure, $p > 10^{-7}$ mbar, while measuring the mass spectra at a lower pressure, $p < 10^{-8}$ mbar. The several methods that have been described for performing CID in Fourier transform ion cyclotron resonance (FT-ICR) instruments can be divided into two groups: (i) fragmentation inside the ICR cell; and (ii) fragmentation outside the ICR cell, followed by ion injection after fragmentation.

For fragmentation inside the ICR cell, one or several precursor ions to be fragmented first must be selected. All the other, non-required, ions are excited until they are quenched by collisions at the cell walls. This operation can be done mass-selectively by the stored waveform inverse Fourier transform [82] (SWIFT) technique. If necessary, the kinetic energy of the ions is increased by irradiation with the appropriate cyclotron frequencies. Then the pressure in the cell is raised by admitting a pulse of collision gas to permit CID, followed by mass analysis and detection of the CID products after pump down [83]. This technique is slow, and can result in ion losses. Rempel and Gross [84] described an RF-only-mode-event, similar to the operation of 'combined Paul traps' [15], to trap ions in the ICR cell at higher pressures. In this mode, either CID or ion/molecule reactions can be performed at a much

---

* See Volume 5, Chapter 7: Structure and Dynamics of Trapped Ions, by Joel H. Parks; Volume 5, Chapter 9: Spectroscopy of Trapped Ions, by Matthew W. Forbes, Francis O. Talbot and Rebecca A. Jockusch.

higher pressure without reducing the trapping efficiency of the ICR cell. For detection, the RF voltage is switched off and the pressure reduced.* Nowadays, sustained off-resonance irradiation (SORI) [85,86] is employed widely; this method results in sequential activation by multiple collisions at low energies, due to acceleration–deceleration cycles without quenching.

Many modern ICR instruments have been equipped with differentially pumped external ion sources [39]. This arrangement is favorable for ionization methods that have a high gas load, such as ESI and, at the same time, allows one to keep the pressure in the ICR cell low. Here the ICR cell is used only as a high-resolution mass spectrometer. A collision cell can be placed between the external ion source and the ICR cell to perform CID [87]. For further details, see the review by Laskin and Futrell [88].

Ergodic fragmentation can be achieved very conveniently in ICR spectrometers by IRMPD [89] or by BIRD, see below. For IRMPD, systems incorporating carbon dioxide lasers are available commercially for ICR mass spectrometers, in which the infrared radiation is admitted by means of a window built in the wall of the vacuum chamber. The infrared beam may be adjusted along the axis of the trap, passing lengthwise through the thread-like ensemble of precursor ions. The internal energy of the analyte ions is increased in a stepwise manner through the sequential absorption of some 10–20 infrared photons.

McLafferty and co-workers [90] showed that photodissociation of electrosprayed melittin and ubiquitin with UV radiation at 193 nm yields new c and z ions, in contrast to those observed with infrared irradiation. These product ions provide additional sequence information.

15.4.1.3 Blackbody Infrared Radiative Dissociation (BIRD)

BIRD [91], introduced for use with ICR in the 1990's as an approach to quantitative multiphoton dissociation of ions at thermal energies without the need for use of a laser, permitted the observation of thermal unimolecular dissociations at very low pressures. This method is especially well suited for the study of weakly-bound systems such as guest–host complexes. The experiments are conducted generally between 25 and 250°C, which corresponds to a binding energy of 20–60 kcal mol^{-1}. The internal energy of the ions studied must be as low as possible, therefore ESI ionization is preferred. Dunbar [92] realized that irradiation with a monochromatic IR laser is kinetically equivalent to irradiation by a blackbody source, which in the simplest case is the vacuum manifold, wherein the ICR cell is located. The BIRD process is, however, slow compared to methods which employ higher energies. Delay times ≥ 1 s are necessary to observe products in sufficient amounts. The BIRD technique is a useful approach with which to obtain structurally informative secondary fragmentation. McLafferty and coworkers [93] showed this for a 42 kDa protein, in which the fragmentation scheme compared very well with data obtained using IRMPD. The same group [94] reported similar results in experiments using BIRD

---

* See Volume 4, Chapter 8: Radio Frequency-Only-Mode Event and Trap Compensation in Penning Fourier Transform Mass Spectrometry, by Adam M. Brustkern, Don L. Rempel and Michael L. Gross.

(150°C, 60 s) and IRMPD on ions from even larger molecules. Interesting studies of the dissociation energetics of leucine enkephalin have been published by the Williams group [95], who used BIRD to determine the effective temperature of ions after SORI experiments [96]. The method is especially well suited for the study of weak interactions. For example, Kitova *et al.* [97] utilized BIRD with functional group replacement (BIRD/FGR) to study intermolecular hydrogen bonds within a desolvated non-covalent protein–ligand complex of a recombinant scFv protein of monoclonal antibody Se 155-4. The interactions were identified and their strengths were quantified.

15.4.1.4 Sustained Off-Resonance Irradiation (SORI) Collisional Activation

CID can be implemented with on-[98] and off-resonance [99,100] methods. With on-resonance CID, the cyclotron frequencies of the ions under study are excited until the desired kinetic energy is reached. The yield of CID is reduced, because the kinetic energy of the colliding ions reduces after turning off the excitation voltage. A technique to overcome this problem is repeated re-excitation, called multiple excitation collisional activation (MECA). This technique was first demonstrated for CID with ^{252}Cf plasma desorption [101]. McLafferty and coworkers [86] observed a strong increase of the CID product ion yield (up to 80%) from ubiquitin in the +11 charge state, in contrast to use of conventional CID (30%). The group compared MECA, CID, and SORI-CID, and concluded that SORI is the method of choice for large multiply-charged biomolecules because it gives high yield and selectivity, and is relatively simple to implement. Wu *et al.* [102] have employed SORI-CID to study variants of cytochrome *c* differing in only three of 104 amino acids. They obtained clearly different MS^4 spectra. Shin and Han [103] employed the oscillation of the ions' cyclotron radii during SORI irradiation to radially separate ions of slightly different *m/z* for photodissociation.

15.4.1.5 Surface-Induced Dissociation (SID)

Surface-induced dissociation was developed first by the group of Cooks [104]. With CID in the gas phase, a collision gas must be introduced into the ion trap. As we have seen above, this is problematic, especially in ICR. The energy deposited in a large ion does not produce in many cases an analytically useful amount of dissociation, even at collision energies in the kiloelectronvolt range. The energy deposited is distributed over a fairly broad range of degrees of freedom. These problems are reduced greatly when SID is employed. When the ions are directed to collide with a surface at a suitable energy and angle, a very fast dissociation occurs after single-step excitation. The physical and chemical properties of the surface play an important role, and organic self-assembled monolayers are especially well suited for this purpose. Coupled with FT-ICR MS, the lowest energy dissociation pathway can be reached and detailed information about the energetics and dynamics of peptide fragmentation can be obtained [105]. Most SID product ion mass spectra of peptides contain *b* and *y* product ions, and are often similar to SORI-CID mass spectra [106]. Even at collision energies of ≥50 eV, backbone fragmentation dominates the mass spectra.

An interesting development from the SID technique is soft landing, which has been introduced also by the group of Cooks [107], and has been reviewed recently by Laskin *et al.* [108]* It is not an analytical but a preparative tool, which employs techniques derived from the SID methodology. All kinds of projectile ions have been deposited on surfaces, small molecules, clusters, peptides, proteins, oligonucleotides, and viruses.

15.4.1.6   Electron-Induced Dissociation in ICR

The ICR cell is ideally suited for ECD, because the ions are or can be concentrated on the axis of the trap; indeed ECD was invented using FT-ICR instrumentation [62]. The electrons may be emitted by a hot cathode [109] mounted within the vacuum manifold in such a way that the electrons can move along the magnetic field lines to interact with the thread-like ensemble of parent ions. There are two favorable kinetic energy regimes for the successful capture of electrons: 0–3 eV and 10–13 eV.

As already mentioned above, ECD and IRMPD should be applied alternately to obtain product ion mass spectra with high information content. A favorable instrumental embodiment combines an axial infrared beam for IRMPD, with an electron-generating hot cathode that has a central hole for the infrared beam. For optimum ECD yield, the parent ions may be excited to small cyclotron orbits to match the diameter of the electron-emitting cathode ring. Alternatively, a central hot cathode may be used together with an infrared beam that penetrates the axial collection of parent ions at a small angle.

Excitation of the ions with an IR laser during ECD [110,111] promotes the fragmentation yield, because the fragmentation rates of long-lived fragment ion complexes are increased. O'Connor and coworkers [112] have investigated the kinetics of fragmentation of such complexes with the aid of ICR double resonance experiments (DR-ECD).

For multiply-charged negative parent ions, EDD may be applied. The ICR mass spectrometer is well suited for this interesting fragmentation mode. EDD works best in the kinetic energy regime of 20–30 eV.

15.4.2   Radiofrequency Paul Traps

There are various fragmentation methods capable of use with RF Paul ion traps. When the ion trap is employed as a mass spectrometer, the measurement of the mass spectrum of the product ions can be done with the RF ion trap itself. A range of scan methods is known for this operation, almost all of which are based on a rapid sequence of mass-selective ion ejections into an external detector. The product ions can, however, be measured also by coupling the ion trap to other types of mass analyzer. Well-known and commercially available examples are combinations of RF ion traps with ICR mass spectrometers, with Kingdon cells, and with time-of-flight (TOF) mass spectrometers.

---

* See also Volume 4, Chapter 2: Ion Traps for Miniature, Multiplexed and Soft Landing Technologies, by Scott A. Smith, Chris C. Mulligan, Qingyu Song, Robert J. Noll, R. Graham Cooks and Zheng Ouyang.

15.4.2.1   Ergodic Fragmentation by Gas Collisions

The conventional method of fragmenting analyte ions in RF ion traps is ergodic fragmentation, initiated by an excess of internal energy introduced into the analyte ions through numerous collisions with a collision gas in the RF ion trap. The collisions must occur with a certain minimum collision energy. The collision energy is conventionally created indirectly by a weak, resonant excitation of the secular ion oscillations of the precursor ions by means of a dipolar alternating voltage. The steady increase of the ion amplitude caused by the excitation transfers the ions into regions where the superimposed RF voltage causes fast and energetic oscillations giving rise to hard collisions. Thus resonantly excited ions experience a large number of collisions with the collision gas that prevent the ions from being ejected from the ion trap. The ions can accumulate energy through these collisions, resulting finally in ergodic decomposition of the ions and the creation of product ions. The collisions, however, do not only pump energy into the ions, they can also take away energy out of the ions, thus cooling the inner temperature of the ions. Heating the ions toward ergodic decay is always a complicated balance. The larger the ion, the more complicated is the heating, requiring higher and higher collision energies. On the other hand, ions with higher mass oscillate slower, thus reducing the average collision energy.

For many years, this means for carrying out CID was the only known method of fragmentation in RF ion traps. Collision-induced dissociation in RF ion traps has also disadvantages, however. For larger analyte ions, for example, it is necessary to use a very high RF voltage amplitude for storing the ions in order to create sufficiently strong collision conditions. A high RF voltage creates a deeper well of the pseudopotential, so that oscillations with equal amplitude have higher kinetic energies. As a consequence, the RF ion trap now has an increased low m/z threshold, so that ions having a value below this threshold can no longer be stored and are lost. According to a rule of thumb, the product ion mass spectrum has a low m/z-limit corresponding to an m/z-value that is around one third of that of the analyte precursor ion; the product ion mass spectrum no longer yields any information about the light product ions, as these ions have been lost.

There are several ways around this low-mass cut-off handicap, making use of the non-instantaneous nature of the ergodic decay. When the process of energizing the analyte precursor ions is enhanced and accelerated by a suitable means, the RF voltage may be decreased substantially and rapidly after the excitation process, now making it possible to retain the light product ions formed as a result of the decay. In this manner, it is possible to store even immonium ions, consisting of one amino acid each and created by secondary fragmentation of the initial product ions. Different special methods have recently become known to use this slow, metastable decay of the ions by the ergodic fragmentation process for storing light ions [113].

Multiply-charged heavy analyte ions often have a low m/z ratio of only around 600–1200 Th because of the large number of added protons; these analyte ions cannot be fragmented at all, as the RF voltage amplitude cannot be set high enough to generate sufficiently energetic collisions to excite the ions.

In quadrupolar linear Paul ion traps used as mass spectrometers, much the same methods may be applied, using two opposite rods for excitation of the ions. Linear multipole Paul ion traps are used often on their own as collision cells for ergodic fragmentation, most often with a quadrupole mass filter up front in order to select the precursor ions. In these fragmentation cells, the precursor ions are accelerated and injected axially into the device, which can now be operated with considerably higher collision gas pressure. The kinetic energy of the ions is chosen to be between 30 and 100 eV. In order to increase the yield with the aid of multiple ion passes through the trap, the ions are usually reflected at the exit electrode by application of an electrical potential to this electrode.

Some specialized types of linear Paul ion traps are designed so as be able to apply a DC voltage gradient along the axis [114,115]. Such specialized linear Paul ion traps may even assist in the ejection of the product ions toward the analyzing mass spectrometer.

15.4.2.2   Ergodic Fragmentation by Infrared Photon Absorption

March and co-workers [116–118] pioneered IRMPD in Paul traps in the years 1982 and 1983, and Hofstadler and Drader [119] have described an interesting improvement of IRMPD in RF ion traps. The infrared radiation of a carbon dioxide laser is introduced into a three-dimensional RF ion trap in a simple manner, through an evacuated hollow fiber containing an optically reflective internal coating. This use of IR radiation offers a further means for effecting ergodic fragmentation in RF ion traps; in 3D ion traps, where the ion cloud is small and spherical, the yield of fragment ions can be high with this method: Young *et al.* [120] achieved 100% dissociation of proton-bound dimers in 20 ms. This type of IRMPD fragmentation appears to be very favorable, as it can be carried out at low RF voltages, with the result that the low m/z product ions are then also stored. However, the electrodes of the ion traps must be very clean because the infrared photons release adsorbed substances from the inner walls of the trap, and these desorbed molecules may react in many ways with the analyte ions. Photodissociation efficiencies can be low, but these can be improved by pulsed addition of a collision gas [121].

Up to this point in time there have been no RF ion trap mass spectrometers marketed commercially that feature this method of fragmentation.

15.4.2.3   Ergodic Fragmentation by Ion Bombardment

Instead of exciting analyte ions to oscillations in order to make them collide energetically with collision gas atoms or molecules, ions of medium kinetic energy can be used to bombard the collisionally-cooled cloud of analyte ions within the ion trap, thereby causing them to become internally excited [122]. This procedure is, in principle, possible with ions of the same polarity as the analyte ions, but much easier, when ions of the opposite charge polarity are used. To avoid complex reactions of the bombardment ions with the analyte ions, monatomic ions may be used. For positive protein ions, F^-, Cl^-, Br^-, or, preferably, I^- ions are first choice. Complexes of the analyte ions with the Cl^- or Br^- ions can be recognized easily by their characteristic isotope pattern. On the other hand, mono-isotopic ions may have advantages in certain situations.

In 3D Paul ion traps, the bombarding ions are introduced in a manner that is similar to that used for the injection of the radical anions used for ETD (see below).

They may be generated even in the same ion source as the ETD radical anions, by operating the ion source for normal negative chemical ionization. Otherwise, a suitable ion source may be installed additionally within the vacuum system of the mass spectrometer. An electrospray ion source may even be used to deliver the bombarding ions. For example, by spraying a solution of CsI, negative iodide ions are formed, and these may be 'purified' in a mass filter on their way to the 3D Paul ion trap.

Inside the 3D Paul ion trap, the bombarding ions oscillate wildly before being cooled by the damping gas. Before these oscillations are fully damped, it is favorable to excite these ions by applying a supplementary radiofrequency voltage between the end-cap electrodes of the Paul ion trap, in resonance with the secular motion of the bombarding ions. By choice of the excitation voltage amplitude, the collisional kinetic energy of the bombarding ions can to some extent be adjusted. The duration, number of ions introduced, and kinetic energy have to be determined experimentally, in a manner analogous to the tuning of the optimum conditions for conventional collision-induced fragmentation.

In 2D Paul ion traps used as mass spectrometers, the same methods of exciting the ions may be applied. In 2D Paul ion traps operated as fragmentation cells, the bombarding ions may be injected easily into the cell along its axis, thereby accelerating them to suitable kinetic energies.

For negative protein ions, fragmentation may be initiated by bombardment with positive ions, for example, Li^+, Na^+, K^+, Rb^+, or, preferably, Cs^+ ions.

15.4.2.4   Electron Capture Dissociation (ECD)

The introduction of low-energy electrons into the spherical cloud of ions inside an RF Paul ion trap presents a special difficulty. The electrical potential around and within this cloud is known and can be used, in principle, to introduce electrons. The electrons can be emitted, for instance from hot cathodes, at such temporal phases of the RF drive voltage that they arrive at the cloud with the desired low kinetic energy. One method is to inject electrons during the zero-amplitude phase of the RF waveform and to apply a magnetic field along the instrument axis [63,123]. Baba *et al.* [123] applied a magnetic field of 50 mT along the axis of the linear trap, and Bushey *et al.* [124] also studied ECD in a linear ion trap located in the field of a permanent magnet. The authors compared the results with FT-ICR measurements which were carried out with the same sample, and found a great a similarity between the fragmentation patterns obtained with each method. Operation of a Paul trap in the presence of a magnetic field is called the 'combined mode' of operation, and it enlarges the stability region available for the ions. Analytical applications of the combined mode of operation of ion traps have not been studied in any detail, however, it is a very promising methodology.

Until now, experiments on electron capture in Paul traps without the use of additional magnetic fields have failed, and such experiments need to be developed further in order to compete with the capability of FT-ICR MS. In Paul traps, the process has a low yield, and is somewhat discouraging because the ETD process is so much easier to perform and gives a much better yield of product ions.

15.4.2.5   Electron Transfer Dissociation (ETD)

ETD, as described above in some detail, can be performed easily in 3D and 2D Paul ion traps. The original inventors of ETD, Syka, Hunt and co-workers [74,125], used a 2D trap to perform ETD. The 2D quadrupole ion trap is divided along its axis into three sections in which three independent adjustable DC potentials determine the axial potential. With equal axial potentials throughout the linear ion trap at the beginning of the experiment, it is possible to collect analyte ions, and to select and to isolate the precursor ions for the ensuing fragmentation; these ions can then be stored in one of the end sections by changing the local axial potential. Subsequently, the negative radical ions are introduced and, if necessary, selected, isolated, and stored in the other end section. The two clouds of ions are then intermixed by equalizing the potentials within the three sections. Fragmentation takes place during the mixing process; each reaction leads immediately to fragmentation because the spontaneous process is taking place in less than 10^{-7} s.

Surprisingly, this process of ETD works better in 3D Paul traps than in the 2D ion trap. Up to now, the reason for this disparity can only be assumed hypothetically. The key characteristic difference between these two types of traps is the probability of a product ion generated through ETD being exposed to further reactions with fresh radical anions.

Within a 3D trap, freshly-introduced anions oscillate wildly within the trap, they cross the inner spherical analyte cation cloud several times and will have an equal chance of interacting with any cation, whether it be a precursor molecular ion, or a product ion [126]. Once the process starts, only precursor analyte ions are present initially and they undergo reactions with the negative radical ions, forming first-generation product ions. With the supply of more and more negative radical ions, more and more product ions are generated, giving rise to secondary reactions of these ions with further negative radical ions. Some of the product ions of the first generation may thus be dissociated to yield second-generation product ions (originally called 'granddaughter' ions). All reaction probabilities for the different types of initial precursor or product ions are strongly proportional to the concentrations of these ions within the ion cloud.

For protein sequence analyses, the presence of second-generation ions, and the later presence of third-generation ions, reduces the quality of the product ion mass spectrum because the signals of 'internal' fragments, not containing one of the terminal amino acids, disturb somewhat the well-known identification programs that operate with protein data banks.

As is known from the kinetics of successive ('chain') reactions, there is a point in time where the product ions of the first generation will attain maximum concentration within the ion cloud, concurrent with low ion densities of precursor ions and product ions of the second and third generations. When the fragmentation process is stopped at this point in time, a very favorable product ion mass spectrum will be obtained. The fragmentation process can be stopped by sudden ejection of all the negative radical ions, for example by raising the radio frequency voltage amplitude to a value where these ions are no longer trapped, or by resonant ejection of these ions. The resonant ejection may be preferable because it makes possible the trapping of light product ions with values of m/z below that of the negative radical ions.

This process of successive reactions of the product ions differs significantly in 3D and 2D traps because, in the latter, in general, one cloud of positive ions is penetrated slowly by a second cloud of negative ions. Both clouds have the form of lengthy filament-like ensembles of ions; as noted previously, the two clouds are stored at the same time in different sections of the 2D trap such that the positive protein ions may be stored inside the 2D trap, and the negative ions supplied axially into the trap volume. In both cases, the two clouds start to penetrate each other with their 'heads' only, initializing the reactions that lead to fragmentation. But then fresh negative radical ions enter the reaction region in the cloud head, where now relatively high amounts of product ions may undergo further reactions. The precursor protein ions residing behind the head region can be fragmented only by negative radical ions that have not already reacted with ions in the head region. Although the reaction cross-sections are not very favorable, the nascent product ions react further, and further, and may even disappear completely by discharging. Only then will unused negative radical ions penetrate further along the axis and react with those protein ions situated in the regions behind the cloud head. Under very unfavorable conditions, in the end only the protein ions from the cloud tail will deliver fragment ions for the final product ion mass spectrum.

There are ways, however, in which the quality of the ETD product ion mass spectra of the 2D Paul ion trap may be improved. In a segmented linear ion trap, the radical anions can be excited resonantly during the mixing process and, instead of introducing the radical anions as a fine beam along the axis of the cell, they can be injected at a certain angle, or at a short distance away from the axis, or as a non-focused beam.

For the generation of the radical anions employed in ETD, use is made of electron attachment ion sources mounted in the vacuum system of the mass spectrometer.

15.4.2.6 ETD in Combination with Deprotonation

It is easy to interpret the product ion mass spectra when they are generated from precursor ions having between two and about four charges, because the product ions can be recognized as such from the differences in the m/z-values between adjacent peaks in their isotope patterns, even in RF ion trap mass spectrometers, and because the product ion mass spectra are not too complex. It is a different situation when highly charged precursor ions having, for instance, 10–30 charges, are subjected to this fragmentation procedure. The number of different types of product ion is extremely high and the great majority of ions are crowded in an m/z range of ca 600–1200 Th. The product ion mass spectrum is so complex that, in most cases, it is impossible to untangle it, even with the highest possible mass resolution. So many product ions, each with its isotopic pattern, are superimposed that even the best de-convolution algorithms are unable to cope with this mixture of signals.

Larger molecules, proteins in particular, yield multiply-charged ions in electrospray ion sources; as a rule of thumb, we can assume that every increase in mass by 800–1000 Da leads to a mean increase in charge of one proton. A protein with a mass of 10,000 Da has accepted, therefore, about 10–12 protons at the peak of the charge distribution but, in most cases, there is a broad distribution of ions with various numbers of charges, most of them being in the m/z ratio range from 600 to 1200 Th. Doubly- or triply-charged ions occur here with vanishingly small abundances and, therefore, cannot be used practically for generating the product ions. For these reasons, fragmentation studies encounter great difficulties with protein molecules in the range of molecular masses between 5 and 100 kDa, even though the highly charged analyte ions can be dissociated extremely well, for example, by electron transfer. The fragment ions created this way, in particular the heavy fragment ions, are themselves predominantly highly charged, and form the complex product ion mass spectra described above.

It has long been known that ions with multiple charges can be converted by continued deprotonation ('charge stripping') into ions having a single or a low number of charges. Charge stripping is effected very easily by continued proton transfer from multiply-protonated molecules to special kinds of negatively-charged ions, most particularly to non-radical anions, which are thereby neutralized. The reaction cross-section for such a proton transfer reaction is proportional to the square of the number of proton charges on the positively-charged ion; the deprotonation reaction, therefore, occurs very quickly for highly-charged ions, while the reaction rate is reduced sharply when the ions have lower charges. When the supply of negative reactant ions for deprotonation is stopped once singly-charged ions are formed, the measurements in the mass analyzer will yield mass spectra that can be evaluated easily, as these now contain essentially only the signals of singly-charged ions.

For the ETD of medium-sized proteins in ion traps, it has been demonstrated that this charge-reduction process can be applied also to the product ions created in this way. After highly charged ions formed from the large protein molecules have been stored, ETD is performed by injecting suitable radical anions; then, non-radical anions are injected for deprotonation until the charge states of the ions have been reduced to but a single charge. This process yields mass spectra of the ETD product ions that can be interpreted easily. An account of this technique has been reported by Hunt [127]. A further interesting method has been reported recently in which the highly charged multiply protonated molecules of a substance that are present with various levels of charge can be deprotonated simultaneously in an RF ion trap, and the process of deprotonation halted at a particular, defined, level of charge. In this way, all the multiply protonated molecules from higher levels of charge accumulate at this lower charge level in a partially deprotonated state. To do this, it is necessary to apply gentle resonant excitation of the secular oscillations, by means of a dipolar alternating voltage, at the m/z ratio corresponding to this defined charge level of the multiply protonated molecules. The ions that are then in forced oscillation at this charge level are no longer able to participate in further reactions with deprotonating reactant anions, as deprotonation requires a low relative velocity of the participating particles. This method is described by McLuckey et al. [128] Such a conversion of highly charged multiply protonated molecules of various charge levels to a specified level of charge brings, at the same time, an enhanced sensitivity, as the signals due to analyte ions in all the higher charge levels are now accumulated, with a relatively high yield, in the chosen level. Yields of more than 50% can be achieved. Furthermore, when highly-charged ions of several different substances are present, using this approach, it is possible to select only the specified analyte ions, because, if the reaction time is long enough, the ions of the other substances are not collected but undergo deprotonation until they are neutralized.

The electron transfer dissociation is then performed with this collection of precursor ions of the given charge state. The charge state should be selected so as to be not too high if possible, preferably in the range of three to five charges, because such ions result in product ion mass spectra that are easiest to interpret. For proteins with high molecular masses, the quadruply-charged precursor or product ions may already have a value of m/z which no longer lies within the scanning mass range for the Paul ion trap mass spectrometer, nevertheless, the alternating excitation voltage can be chosen so that these ions are collected effectively by means of deprotonation (this process is referred to as 'ion parking') [129].

A product ion mass spectrum containing singly-to quadruply-charged ions may be converted easily, for example by a computer program, into a mass spectrum consisting of the singly-charged ions only. This virtual product ion mass spectrum is ideal-for interpretation, and it extends up to values of m/z far beyond the practical scanning range of the mass spectrometer.

15.4.3   KINGDON ION TRAPS

In Kingdon ion traps with ions orbiting around a central electrode, any fragmentation will change the diameter of the orbits proportional to the change in the value of m/z, making most of the product ions impinge on either the inner or the outer electrodes. Only a fraction of the product ions in a mass range close to the original value of m/z are retained in orbit. In general, fragmentation here is not possible without losing large parts of the lower and higher portions of the mass range of the product ion mass spectrum. Therefore Kingdon traps are employed for mass analysis, after ion fragmentation has been performed in another device, such as in a linear ion trap.

Only in Kingdon traps in which the ions oscillate in a plane between pairs of inner electrodes, is fragmentation of the ions possible without detrimental losses of the product ions [20]. Ions decaying near to the maximum amplitude of the oscillation just remain oscillating fully, with the same amplitude; only those ions that decay in the center of the trap are lost. When ions decay during these oscillations, in total more than 60% of the product ions continue oscillating because most of the ions decay near the points of maximum oscillation amplitude, where they have a minimum speed and spend the maximum proportion of their time.

Fragmentation is, however, hard to initiate in Kingdon traps. Candidate methods are IRMPD and MAID, but more favorable is a method in which metastable ions are introduced into the central region, allowed to decay, and only then excited into their axial oscillations. In this way, for instance, a Kingdon ion trap could be used to generate product ion mass spectra of very high mass precision and resolution from ions generated by ionization through MALDI.

15.5 SOME REMARKS ON AUTOMATED PRODUCT ION SPECTRUM MEASUREMENTS IN TANDEM MASS SPECTROMETERS

Modern tandem mass spectrometers offer programs to measure automatically series of first generation and even second-generation product ion mass spectra, with sufficient speed to meet the requirements of substance delivery by separation methods such as chromatography or capillary electrophoresis. The analytes will be delivered by the separation methods in broad peaks, corresponding to increases and decreases in concentration, and lasting for about 5 to 30 s; peaks may also overlap. Depending upon the type of instrument used and the concentration of the substance, the measurement of a product ion mass spectrum takes about 0.2–2 s; in most cases about half a second.

The automatic product ion measurement programs have to perform a complicated series of tasks. First, they have to detect the appearance of a substance peak, investigate the most probable peak shape from the initial slope, determine the optimum time to measure the product ion mass spectrum while, at the same time, taking into account overlapping peaks of other substances. Second, a mass spectrum of the analyte is acquired and investigated to determine the optimum charge state for fragmentation. Third, fresh precursor ions are now isolated, fragmented according to a first pre-chosen method, and the product ion mass spectrum is acquired. Depending upon the fragmentation method, the quality of the product ion mass spectrum is investigated for its information content in respect to the analytical problem to be solved: a CID product ion mass spectrum may, for instance, show only one predominant ion species, generated by the loss of water. This mass spectrum has no value for the elucidation of the protein structure and, therefore, the program should decide to take a second generation product ion mass spectrum of this predominant first-generation ion species. When the fragmentation is being performed *via* ETD, the program has to inspect the product ion mass spectrum in order to determine the stopping point for the final dissociation because the neutralized proton still may be weakly bonded to the ion as described above. The program now should decide whether or not to repeat this measurement to help the dissociation to proceed to completion by some excitation of the resulting radical ions to undergo collisions.

After a first product ion mass spectrum that has good evaluation quality has been acquired, the program may contain instructions to take a second product ion mass spectrum using a different method of dissociation in order to generate complementary information about the protein structure.

15.6 CONCLUSION

Ion traps can be divided into four groups, according to their principles of operation: magnetic or Penning traps, electrodynamic or Paul traps, electrostatic or Kingdon traps, and electron beam traps. The first three kinds of traps are well established as tandem-in-time spectrometers for mass and structure analysis of all kinds of chemical compounds. Electron beam ion traps are just appearing in this field. Ion traps store ions, generated in or outside the trap. The stored ions can be selected, energized, fragmented, reacted, and detected in consecutive steps (tandem-in-time) without loss of the selected ions from the trap. The m/z-values of every stored ion species can be determined with high to very high resolution and precision. Since the invention of soft ionization techniques for large molecules, the ion trap techniques have developed very rapidly into an advanced and widely employed analytical method for biological material, especially peptides, proteins and related compounds. To perform this type of analysis, many different methods have been described. They have been discussed here. All of them use single-step or consecutive fragmentation processes of ions, which should contain molecular mass information, to obtain the sequence and structure of the analyte. Controlled fragmentation is the central task in this kind of analysis. Ion traps are especially well suited for this, because tandem-in-time operation is the inherent mode of operation. In this chapter, fragmentation methods have been discussed in general, and in their application in the different types of traps in the light of their characteristic features. The development of the technique is already highly advanced. However, there is a clear perspective for more specificity, sensitivity, mass resolution and precision, greater dynamic range, and faster operating speed. The methods of ion fragmentation by ion/molecule, ion/ion, ion/electron, and ion/photon interactions have still the potential for further elaboration. Infrared spectroscopy will be a very useful complementary method, if it could be employed routinely, as would be ESR or NMR spectroscopy of the gas-phase particles. Application of very low temperatures ($\ll 100$ K) is still to be explored. The trapping and analysis of neutral molecules at low temperatures can also increase the advantages and the range of application of the method.

REFERENCES

1. Benninghoven, A.; Kirchner, F. The energy distribution of atomized neutral and charged products. *Z. Naturforsch.* 1963, *18A*, 1008–1010.
2. MacFarlane, R.D.; Torgerson, D.F. Californium-252-plasma desorption time-of-flight MS. *Int. J. Mass Spectrom. Ion Phys.* 1976, *21*, 81–92.
3. Morris, H.R.; Panico, M.; Barber, M.; Bordoli, R.S.; Sedgwick, R.D. FAB: A new mass spectrometric method for peptide sequence analysis. *Biochem. Biophys. Res. Commun.* 1981, *101*, 623–631.
4. Beckey, H.D. *Principles of Field Ionization and Field Desorption Mass Spectrometry.* Pergamon Press, Oxford, 1973.
5. Schaub, T.M.; Linden, H.B.; Hendrickson, C.L.; Marshall, A.G. Continuous-flow sample introduction for field desorption/ionization mass spectrometry. *Rapid Commun. Mass Spectrom.* 2004, *18*, 1641–1644.
6. Fenn, J.B. Electrospray: Wings for molecular elephants, (Nobel Lecture). *Angew. Chem. Int. Edit.* 2003, *42*, 3871–3894.
7. Tanaka, K. Origin of macromolecular ionization by laser irradiation, (Nobel Lecture). *Angew. Chem. Int. Edit.* 2003, *42*, 3861–3870.
8. Munson, M.S.B.; Field, F.H. Reactions of gaseous ions XV: Methane +1% Ethane, and Methane +1% Propane. *J. Am. Chem. Soc.* 1965, *87*, 3294–3299.
9. Talroze, V.L.; Ljubimova, A.K. Secondary processes in the ion source of a mass spectrometer, (Reprint of the original paper 1952). *J. Mass Spectrom.* 1998, *33*, 502–504.
10. Cooks, R.G.; Beynon, J.H.; Caprioli, R.M.; Lester G.R. *Metastable Ions.* Elsevier, Amsterdam, 1973.

11. Kienitz, E.; Ed. *Massenspektrometrie* (Mass Spectrometry), Verlag Chemie, Weinheim, 1968.

12. McLafferty, F.W. *Interpretation of Mass Spectra*, 3rd Ed. University Science Books, Mill Valley, CA, 1980.

13. Beynon, J.H. *Mass Spectrometry and its Application to Organic Chemistry*, Elsevier, Amsterdam, 1960.

14. Cooks, R.G.; Ed. *Collision Spectroscopy*, Plenum Press, New York, 1978.

15. Major, F.G.; Gheorghe, V.N.; Werth, G. *Charged Particle Traps*, Springer, Berlin, 2005.

16. Kreller, M.; Zschornack, G.; Kentsch, U.; Heller R. Molecule fragmentation at the Dresden EBIS-A. *Rev. Sci. Instrum.* 2008, *79*, 02A702.

17. Kingdon, K.H. A method for the neutralization of electron space charge by positive ionization at very low gas pressures. *Phys. Rev.* 1923, *21*, 408–418.

18. Makarov, A.A. Electrostatic axially harmonic orbital trapping: A high-performance technique for mass spectrometry. *Anal. Chem.* 2000, *72*, 1156–1162.

19. Makarov, A.A. Mass spectrometer. *U.S. Patent* 1999, 5,886,346.

20. Köster, C. Massenspektrometer mit einer elektrostatischen Ionenfalle. *Patent Application DE* 2007, 10 2007 024 858.1; Köster, C. The concept of electrostatic non-orbital harmonic ion trapping. *Int. J. Mass Spectrom.* 2009, *287*, 114-118.

21. Hull, A.W. The effect of a uniform magnetic field on the motion of electrons between coaxial cylinders. *Phys. Rev.* 1921, *18*, 31–57.

22. Penning, F.M.; Addink, C.C.J. The starting potential of the glow discharge in the Neon-Argon mixtures between large parallel plates. Pt. I. Results. *Physica* 1934, *I*, 1007–1027.

23. Penning, F.M. The starting potential of the glow discharge in the Neon-Argon mixtures between large parallel plates. Pt. II. Discussion on the ionization and excitation by electrons and metastable ions. *Physica* 1934, *I*, 1028–1044.

24. Penning, F.M. Ein neues Manometer für niedrige Gasdrücke, insbesondere zwischen 10^{-3} und 10^{-5} mm. *Physica* 1937, *IV*, 71–75.

25. Dehmelt, H.G. Radio frequency spectroscopy of stored ions. *Adv. At. Mol. Phys.* Part I: Storage. 1967, *3*, 53–72 and Part II: Spectroscopy. 1969, *5*, 109–153.

26. Gabrielse, G.; Macintosh, C.F. Cylindrical Penning traps with orthogonalized anharmonicity compensation. *Int. J. Mass Spectrom. Ion Processes* 1984, *57*, 1–17.

27. Gabrielse, G.; Bowden, N.S.; Oxley, P.; Speck, A.; Storry, C.H.; Tan, J.N.; Wessels, M.; Grzonka, D.; Oelert, W.; Schepers, G.; Sefzick, T.; Walz, J.; Pittner, H.; Haensch, T.W.; Hessels, E.A. Driven production of cold antihydrogen and the first measured distribution of antihydrogen states, *Phys. Rev. Lett.* 2002, *89*, 233401.

28. Wang, Y.; Wanczek, K.P. A new ion cyclotron resonance cell for simultaneous trapping of positive and negative ions. *Rev. Sci. Instrum.* 1993, *64*, 883–889.

29. Malek, R.; Wanczek, K.P. FT-ICR spectrometry with simultaneous trapping of positive and negative ions. *Int. J. Mass Spectrom. Ion Processes* 1996, *157/158*, 199–214.

30. Kanawati, B.; Wanczek, K.P. Characterisation of a new open cylindrical ICR cell with unusual geometry. *Rev. Sci. Instrum.* 2007, *78*, 074102.

31. Gorshkov, M.V.; Guan, S.; Marshall, A.G. Dynamic ion trapping for Fourier transform ion cyclotron resonance mass spectrometry: Simultaneous positive- and negative-ion detection. *Rapid Comm. Mass Spectrom.* 1992, *6*, 166–172.

32. Graham, J.R.; Malone, D.P.; Wobshall, D.C. Ion cyclotron resonance in weakly ionized gases. *Bull. Am. Phys. Soc.* 1962, *7*, 69.

33. Hartmann, H.; Wanczek, K.P. Eds. *Ion Cyclotron Resonance Spectrometry*, Lecture Notes in Chemistry 7, Springer-Verlag, Berlin, 1978.

34. McIver, R.T., Jr. A trapped ion analyzer cell for ion cyclotron resonance spectroscopy. *Rev. Sci. Instrum.* 1970, *41*, 555–558.

35. Comisarow, M.B.; Marshall, A.G. Fourier transform ion cyclotron resonance spectroscopy. *Chem. Phys. Lett.* 1974, *25*, 282–283.

36. Allemann, M.; Kellerhals, H.P.; Wanczek, K.P. A new Fourier transform mass spectrometer with a superconducting magnet. *Chem. Phys. Lett.* 1980, *75*, 328–331.

37. Lee, S-H.; Wanczek, K.P.; Hartmann, H. A new cylindrical trapped ion ICR cell, *Adv. Mass Spectrom.* 1980, *8*, 1645–1649.

38. Knobeler, M. Fundamental studies of electrospray ionization and matrix-assisted laser desorption/ionization Fourier transform ion cyclotron resonance mass spectrometry, *Thesis*, University of Bremen, 1996.

39. Kofel, P.; Allemann, M.; Kellerhals, H.P.; Wanczek, K.P. External ion source for ion cyclotron resonance mass spectrometry. *Int. J. Mass Spectrom. Ion Processes* 1989, *87*, 237–247.

40. Marshall, A.G.; Verdun, F.R. *Fourier Transforms in NMR, Optical, and Mass Spectrometry*, Elsevier, Amsterdam, 1990.

41. Paul, W.; Steinwedel, H. Ein neues Massenspektrometer ohne Magnetfeld. *Z Naturforsch.* 1953, *A8,* 448–450.

42. Paul, W.; Osberghaus, O.; Fischer E. Ein Ionenkäfig (An Ion Cage). *Forschungsberichte Wirtschafts- und Verkehrministeriums, Nordrhein-Westfalen, Nr. 415 (Research Reports, Department of Commerce and Transport, North Rhine-Westphalia, No. 415)*, Westdeutscher Verlag, Köln und Opladen, 1958.

43. March, R.E.; Todd, J.F.J.; Eds. *Practical Aspects of Ion Trap Mass Spectrometry,* Volumes I, II, III, CRC Press, Boca Raton, FL, 1995.

44. Raizen, M.G.; Gilligan, H.M.; Bergquist, J.C.; Itano, W.M.; Wineland, D.J. Linear ion trap for high-accuracy spectroscopy of stored ions. *J. Mod. Optics* 1992, *39*, 233–242.

45. Drees, J.; Paul, W. Acceleration of electrons in a plasma betatron. *Z. Phys.* 1964, *180*, 340–361.

46. Schwartz, J.C.; Senko, M.W.; Syka, J.E.P. A two-dimensional quadrupole ion trap mass spectrometer. *J. Am. Soc. Mass Spectrom.* 2002, *13*, 659–669.

47. Huang, Y.; Guan, S.; Kim, H.S.; Marshall, A.G. Ion transport through a strong magnetic field gradient by radio-frequency only octupole ion guides. *Int. J. Mass Spectrom. Ion Processes* 1996, *152*, 121–133.

48. Soni, M.; Frankevich, V.; Nappi, M.; Santini, R.E.; Amy, J.W.; Cooks, R.G. Broad band Fourier transformation quadrupole ion trap mass spectrometry. *Anal. Chem.* 1996, *68*, 3314–3320.

49. Zerega, Y.; Perrier, P.; Carette, M. Performance enhancement of a Fourier transform ion trap mass spectrometer using indirect ion-motion frequency measurement. *Measurement Sci. Technol.* 2003, *14*, 323–328.

50. Janulyte, A.; Zerega, Y.; Carette, M. Harmful influences of confinement field non-linearities in mass identifying for a Fourier transform quadrupole ion trap mass spectrometer. *Int. J. Mass Spectrom.* 2007, *263*, 243–259.

51. Aliman, M.; Glasmachers, A. A novel electric ion resonance cell design with high signal-to-noise ratio and low distortion for Fourier transform mass spectrometry. *J. Am. Soc. Mass Spectrom.* 1999, *10*, 1000–1007.

52. Glasmachers. A.; Laue, A. Linear quadrupole ion trap for Fourier transform mass spectrometry. *Proc. 56th ASMS Conference on Mass Spectrometry and Allied Topics*, June 1–5, 2008, Denver, CO, WP 028.

53. Roepstorff, P.; Fohlman, J. Proposal for a common nomenclature for sequence ions in mass spectra of peptides. *Biomed. Mass Spectrom.* 1984, *11*, 601.

54. Johnson, R.S.; Martin, S.A.; Biemann, K. Collision-induced fragmentation of $(M + H)^+$ ions of peptides. Side chain specific sequence ions. *Int. J. Mass Spectrom. Ion Processes* 1988, *86*, 137–154.

55. For Example: Kinter, M.; Sherman, N.E. *Protein Sequencing and Identification Using Tandem Mass Spectrometry*, Wiley, New York, 2000.
56. A very good overview: *Principles of Mass Spectrometry Applied to Biomolecules*, Eds. Laskin, J.; Lifshitz, C. Wiley-Interscience, Hoboken, NJ, 2006.
57. Weinkauf, R.; Schlag, E.W.; Martinez, T.J.; Levine, R.D. Nonstationary electronic states and site-selective reactivity. *J. Phys. Chem.* 1997, A *101*, 7702–7710.
58. Laskin, J.; Futrell, J.H.; Chu, I.K. Is dissociation of peptide radical cations an ergodic process? *J. Am. Chem. Soc.* 2007, *129*, 9598–9599.
59. Breuker, K.; Oh, H.B.; Lin, C.; Carpenter, B.K.; McLafferty, F.W. Nonergodic and conformational control of the electron capture dissociation of protein cations. *Proc. Natl. Acad. Sci. U.S.A.* 2004, *101*, 14011–14016.
60. Syrstad, E.A.; Turecek, F.J. Toward a general mechanism of electron capture dissociation. *J. Am. Soc. Mass Spectrom.* 2005, *16*, 208–224.
61. McDaniel, E.W. *Collision Phenomena in Ionized Gases*, Wiley, New York, 1964.
62. Zubarev, R.A.; Kelleher, N.L.; McLafferty, F.W. ECD of multiply charged protein cations: A nonergodic process. *J. Am. Chem. Soc.* 1998, *120*, 3265–3266.
63. Silivra, O.A.; Kjeldsen, F.; Ivonin, I.A.; Zubarev, R.A. Electron capture dissociation of polypeptides in a three-dimensional quadrupole ion trap: Implementation and first results. *J. Am. Soc. Mass Spectrom.* 2005, *16*, 22–27.
64. Hayakawa, S.; Hashimoto, M.; Matsubara, H.; Turecek, F. Dissecting the proline effect: Dissociations of proline radicals formed by electron transfer to protonated pro-gly and gly-pro dipeptides in the gas phase. *J. Am. Chem. Soc.* 2007, *129*, 7936–7949.
65. Hakansson, K.; Emmet, M.R.; Hendrickson, C.L.; Marshall, A.G. High sensitivity ECD tandem FT ICR MS of microsprayed peptides. *Anal. Chem.* 2001, *73*, 3605–3610.
66. Kjeldsen, F.; Budnik, B.A.; Haselmann, K.F.; Jensen, F.; Zubarev, R.A. Dissociative capture of hot (3-13 eV) electrons by polypeptide polycations: An efficient process accompanied by secondary fragmentation. *Chem. Phys. Lett.* 2002, *356*, 201–206.
67. Kjeldsen, F.; Haselmann, K.F.; Sorensen, E.S.; Zubarev, R.A. Distinguishing of Ile/Leu amino acid residues in the PP3 protein by hot electron capture dissociation in Fourier transform ion cyclotron resonance mass spectrometry. *Anal.Chem.* 2003, *75*, 1267–1274.
68. Budnik, B.A.; Haselmann, K.F.; Zubarev, R.A. Electron detachment dissociation of peptide dianions: An electron-hole recombination phenomenon. *Chem. Phys. Lett.* 2001, *342*, 299–302.
69. Kjeldsen, F.; Silivra, O.A.; Ivonin, I.A.; Haselmann, K.F.; Gorshkov, M.; Zubarev, R.A. C_α-C backbone fragmentation dominates in electron detachment dissociation of gas-phase polypeptides polyanions. *Chem. Eur. J.* 2005, *11*, 1803–1812.
70. Adams, N.G.; Babcock, L.M.; Molek, C.D. Ion-ion recombinations. *Encyclopedia of Mass Spectrometry, vol. 1*, Armentrout, R.B., Ed. 2003, 555–561.
71. McLuckey, S.A.; Stephenson, J.L. Ion/ion chemistry of high-mass multiply charged ions. *Mass Spectrom. Rev.* 1998, *17*, 369–407.
72. Pitteri, S.J.; McLuckey, S.A. Recent developments in the ion-ion chemistry of high-mass multiply charged ions. *Mass Spectrom. Rev.* 2005, *24*, 931–958.
73. McLuckey, S.A. Gas phase bioion-ion reactions: The hows and whys of reagent selection. *Proc. 56th ASMS Conference on Mass Spectrometry and Allied Topics*, Denver, CO, June 1–5, 2008, MOB 8:30.
74. Syka, J.E.P; Coon, J.J.; Schoeder, M.J.; Shabanowitz, J.; Hunt, D.F. Peptide and proton sequence analysis by electron transfer dissociation mass spectrometry. *Proc. Natl. Acad. Sci. U.S.A.* 2004, *101*, 9528–9533.
75. Syka, J.E.P. Confining positive and negative ions with fast oscillating electric potentials. *US Patent* 2004, 7,026,630 B2.

76. Hartmer, R.; Brekenfeld, A. Ionenfragmentierung durch Elektronentransfer in Ionenfallen. *Patent DE* 2005, 10 2005 004 324.0.

77. Hunt, D.F.; Coon, J.J.; Syka, J.E.P.; Marto, J.A. Electron transfer dissociation for biopolymer sequence analysis. *U.S. Patent Application* 2005, 20050199804 A1.

78. Misharin, A.S.; Silivra, O.A.; Kjeldsen, F.; Zubarev, R.A. Dissociation of peptide ions by fast atom bombardment in a quadrupole ion trap. *Rapid Commun. Mass Spectrom.* 2005, *19*, 2163–2171.

79. Zubarev, R.A.; Misharin, A.; Silivra, O.; Kjeldsen, P. Ionenfragmentierung durch Beschuß mit Neutralteilchen *Patent DE* 2005, 10 2005 005 743 A1.

80. Russell, D.H.; Ed. *Gas Phase Inorganic Ion Chemistry*, Plenum Press, New York, 1989.

81. Oomens, J.; Polfer, N.; Moore, D.T.; van der Meer, L.; Marshall, A.G.; Eyler, J.R.; Meijer, G.; von Helden, G. Charge-state resolved mid infrared spectroscopy of gas phase-ions. *Phys. Chem. Chem. Phys.* 2005, *7*, 1345–1348.

82. Guan, S.; Marshall, A.G. Stored waveform inverse Fourier transform (SWIFT) ion excitation in trapped ion mass spectrometry. *Int. J. Mass Spectrom. Ion Processes* 1996, *156/157*, 5–37.

83. Cody, R.B.; Freiser, B.S. CID in an FT ICR mass spectrometer. *Int. J. Mass Spectrom. Ion Phys.* 1982, *41*, 199–204.

84. Rempel, D.L.; Gross, M.L. High-pressure trapping in Fourier transform mass spectrometry: A radio frequency-only-mode event. *J. Am. Soc. Mass Spectrom.* 1992, *3*, 590–594.

85. Gauthier, J.W.; Trautman, T.R.; Jacobson, D.B. Sustained off-resonance irradiation for collision-activated dissociation involving FT ICR MS. CID technique that emulates infrared multiphoton dissociation. *Anal. Chim. Acta.* 1991, *246*, 211–225.

86. Senko, M.W.; Spier, J-P.; McLafferty, F.W. Collisional activation of large multiply charged ions using FT MS. *Anal. Chem.* 1994, *66*, 2801–2808.

87. Wang, Y.; Shi, S.D.H.; Hendrickson, C.L.; Marshall, A.G. Mass-selective ion accumulation and fragmentation in a linear octopole ion trap external to an FT ICR mass spectrometer. *Int. J. Mass Spectrom.* 2000, *198*, 113–120.

88. Laskin, J.; Futrell, J.H. Activation of large ions in FT ICR mass spectrometry. *Mass Spectrom. Rev.* 2005, *24*, 135–167.

89. Little, D.P.; Senko, M.W.; O'Conner, P.B.; McLafferty, F.W. IRMPD of multiply charged ions for biomolecule sequencing. *Anal. Chem.* 1994, *66*, 2809–2815.

90. Guan, Z.; Kelleher, N.L.; O'Connor, P.B.; Aasrud, D.J.; Little, D.P.; McLafferty, F.W. 193 nm photodissociation of larger multiply-charged biomolecules. *Int. J. Mass Spectrom. Ion Processes* 1996, *157/158*, 357–364.

91. Review: Dunbar, R.C. BIRD (Blackbody infrared radiative dissociation): Evolution, principles, and applications. *Mass Spectrom. Rev.* 2004, *23*, 127–158.

92. Dunbar, R.C. Kinetics of low intensity infrared photodissociation. The thermal model and applications of the Tolman theorem. *J. Chem. Phys.* 1991, *95*, 2537–2548.

93. Ge, Y.; Horn, D.M.; McLafferty, F.W. Blackbody infrared radiative dissociation of larger (42 kDa) multiply charged proteins. *Int. J. Mass Spectrom.* 2001, *210/211*, 203–214.

94. Aaserud, D.J.; Guan. Z.; Little, D.P.; McLafferty, F.W. DNA sequencing with blackbody infrared radiative dissociation of electrosprayed ions. *Int. J. Mass Spectrom. Ion Processes* 1997, *167/168*, 705–712.

95. Schnier, P.D.; Price, W.D.; Strittmatter, E.F.; Williams, E.R. Dissociation energetics and mechanisms of leucine enkephalin $(M + H)^+$ and $(2M + X)^+$ ions (X = H, Li, Na, K, and Rb) measured by blackbody infrared radiative dissociation. *J. Am. Soc. Mass Spectrom.* 1997, *8*, 771–780.

96. Schnier, P.D.; Jurchen, J.C.; Williams, E.R. The effective temperature of peptide ions dissociated by sustained off-resonance irradiation collisional activation in Fourier transform mass spectrometry. *J. Phys. Chem.* 1999, *B 103*, 737–745.

97. Kitova, E.N.; Seo, M.; Roy, P-N.; Klassen, J.S. Elucidating the intermolecular inter-actions within desolvated protein-ligand complex. An experimental and computational study. *J. Am. Chem. Soc.* 2008, *130*, 1214–1226.

98. Cody, R.B.; Amster, I.J.; McLafferty, F.W. Peptide mixture sequencing by tandem Fourier transform mass spectrometry. *Proc. Natl. Acad. Sci. U.S.A.* 1985, *82*, 6367–6370.

99. Heck, A.J.R.; de Koning, L.J.; Pinkse, F.A.; Nibbering, N.M.M. Mass specific selection of ions in Fourier transform ion cyclotron resonance mass spectrometry. *Rapid Commun. Mass Spectrom.* 1991, *5*, 406–414.

100. Heck, A.J.R.; Derrick, P.J. Ultrahigh mass accuracy in isotope selective collision-induced dissociation using correlated sweep excitation and sustained off-resonance irradiation: An FT ICR MS case study of the $(M + 2H)^{2+}$ ion of bradykinin. *Anal. Chem.* 1997, *69*, 3603–3607.

101. Williams, R.E.; McLafferty, F.W. High resolution and tandem FT mass spectrometry with Californium-252 plasma desorption. *J. Am. Soc. Mass Spectrom.* 1990, *1*, 427–430.

102. Wu, Q.; van Orden, S.; Cheng, X.; Bakhtiar, R.; Smith, R.D. Characterization of cyto-chrome c variants with high-resolution FTICR mass spectrometry: Correlation of frag-mentation and structure. *Anal. Chem.* 1995, *67*, 2498–2509.

103. Shin, S.K.; Han, S.J. Application of sustained off-resonance irradiation: The beat fre-quency measurement and radial separation of mass-selected ions. *J. Am. Soc. Mass Spectrom.* 1997, *8*, 86–89.

104. Mabud, M.A.; Dekrey, M.J.; Cooks, R.G. SID of molecular ions. *Int. J. Mass Spectrom. Ion Processes* 1985, *67*, 285–294.

105. Bailey, T.H.; Laskin, J.; Futrell, J.H. Energetics of selective cleavage at acidic residues studied by time- and energy-resolved surface-induced dissociation in FT ICR MS. *Int. J. Mass Spectrom.* 2003, *222*, 313–327.

106. Laskin, J.; Futrell, J.H. Surface-induced dissociation of peptide ions: Kinetics and dynamics. *J. Am. Soc. Mass Spectrom.* 2003, *14*, 1340–1347.

107. Miller, S.A.; Luo, H.; Pachuta, S.J.; Cooks, R.G. Soft landing of polyatomic ions at fluorinated self-assembled monolayer surfaces. *Science* 1997, *275*, 1447–1450.

108. Laskin, J.; Wang, P.; Hadjar, O. Soft-landing of peptide ions onto self-assembled mono-layer surfaces: An overview. *Phys. Chem. Chem. Phys.* 2008, *10*, 1079–1090.

109. Tsybin, Y.O.; Hakansson, P.; Budnik, B.A.; Haselmann, K.F.; Kjeldsen, F.; Gorshkov, M.; Zubarev, R.A. Improved low energy injection systems for high rate electron capture dissociation in Fourier transform ion cyclotron resonance mass spectrometry. *Rapid Commun. Mass Spectrom.* 2001, *15*, 1849–1854.

110. Tsybin, J.O.; Witt, M.; Baykut, G.; Kjeldsen, F.; Hakansson, P. Combined infrared mul-tiphoton dissociation and electron capture dissociation with a hollow electron beam in Fourier transform ion cyclotron resonance mass spectrometry. *Rapid Commun. Mass Spectrom.* 2003, *17*, 1759–1768.

111. Hakansson, K.; Chalmers M.J.; Quinn, J.P.; McFarland, M.A.; Hendrickson, C.L.; Marshall, A.G. Combined electron capture and infrared multiphoton dissociation for multistage MS/MS in a Fourier transform ion cyclotron resonance mass spectrometer. *Anal. Chem.* 2003, *75*, 3256–3262.

112. Lin, C.; Cornoyer, J.J.; O'Connor, P.B. Probing the gas phase folding kinetics of peptide ions by IR activated DR-ECD. *J. Am. Soc. Mass Spectrom.* 2008, *19*, 780–789.

113. Brekenfeld, A. Measuring light fragments with ion traps. *GB Patent* 2007, 2,428,515.

114. Thomson, B.A.; Jolliffe, C.L. Quadrupole with axial DC field. *US Patent* 2000, 6,111,250.

115. Franzen, J. Method and device for the reflection of charged particles on surfaces. *US Patent* 2002, 5,572,035 A1.

116. Hughes, R.J.; March, R.E.; Young, A.B. Multiphoton dissociation of ions derived from 2-propanol in a QUISTOR with low power CW infrared laser radiation. *Int. J. Mass Spectrom. Ion Phys.* 1982, *42*, 255–263.

117. Hughes, R.J.; March, R.E.; Young, A.B. Multiphoton dissociation of ions derived from *iso*-propanol and deuterated analogues in a QUISTOR with low power CW infrared laser radiation. *Can. J. Chem.* 1983, *61*, 834–845.

118. Hughes, R.J.; March, R.E.; Young, A.B. Optimization of ion trapping characteristics for studies of ion photodissociation. *Can. J. Chem.* 1983, *61*, 824–833.

119. Hofstadler, S.A.; Drader, J.J. Systems and methods for introducing infrared multiphoton dissociation with hollow fiber waveguide. *Patent WO* 2002, 02/101 787 A1.

120. Young, A.B.; March, R.E.; Hughes R.J. Studies of infrared multiphoton dissociation rates of protonated aliphatic alcohol dimers. *Can. J. Chem.* 1985, *63*, 2324–2331.

121. Boue, S.M.; Stenphenson, J.L.; Yost, R.A. Pulsed helium introduction into a quadrupole ion trap for reduced collisional quenching during IR multiphoton dissociation of electrosprayed ions. *Rapid Commun. Mass Spectrom.* 2000, *14*, 1391–1397.

122. Brekenfeld, H.; Hartmer, R. Private communication.

123. Baba, T.; Hashimoto, Y.; Hasegawa, H.; Waki, I. Electron capture dissociation in a radio frequency ion trap. *Anal. Chem.* 2004, *76*, 4263–4266.

124. Bushey, J.M.; Takashi, B.; Baba, T.; Glish, G.L. Electron capture dissociation in a radio frequency ion trap versus a Fourier transform ion cyclotron resonance mass spectrometer. *Proc. 56th ASMS Conference on Mass Spectrometry and Allied Topics*, June 1–5, 2008, Denver, CO, WOC 8:50 am.

125. Coon, J.J.; Syka, J.E.P.; Schwartz, J.C.; Shabanowitz, J.; Hunt, D.F. Anion dependence in the partitioning between proton and electron transfer in ion/ion reactions. *Int. J. Mass Spectrom.* 2004, *236*, 33–42.

126. Franzen, J. Unpublished material.

127. Hunt, D. F. Comparative analysis of post-translationally modified proteins and peptides by mass spectrometry: New technology (electron transfer dissociation) and applications in the study of cell migration, the histone code and cancer vaccine development. *17th International Mass Spectrometry Conference,* Aug. 27–Sept. 1, 2006, Prague, Aug. 28, 9:00 am.

128. McLuckey, S.A.; Reid, G.E.; Wells, J.M. Method for selectively inhibiting reaction between ions. *U.S. Patent* 2006, 7,064,317 B2.

129. McLuckey, S.A.; Reid, G.E.; Wells, J.M. Ion parking during ion-ion reaction in electrodynamic ion trapss. *Anal. Chem.* 2002, *74*, 336–346.

16 Unraveling the Structural Details of the Glycoproteome by Ion Trap Mass Spectrometry

Vernon Reinhold, David J. Ashline, and Hailong Zhang

CONTENTS

16.1 Introduction 708
    16.1.1 Functional Relationships of Molecular Glycosylation 708
    16.1.2 Glycan Structural Complexity 709
    16.1.3 Synchrony of Glycan Structure and Operational
             Characteristics of the Ion Trap 710
    16.1.4 Basic Energetics 713
    16.1.5 Collision-Induced Dissociation of Glycans 713
16.2 Sample Preparation 717
    16.2.1 Release of N, and O-Linked Glycans 717
    16.2.2 Reduction and Methylation 718
    16.2.3 Mass Spectrometry 719
16.3 MS^n Disassembly Applications 719
    16.3.1 The Glycosylation Finger-Print 719
    16.3.2 Glycoprotein Standards 720
    16.3.3 Extracting the Details of Molecular Structure by MS^n 722
    16.3.4 Physiological Modulations 727
            16.3.3.1 Differentiation Profiles in Metastasis 727
            16.3.3.2 Differentiation Profiles in Adult Stem Cells 728
16.4 Software Tools: General Considerations 729
    16.4.1 Glycoscreen 729
    16.4.2 FragLib Tool Kit 731
    16.4.3 Composition Finder 731
    16.4.4 Application Examples 734
16.5 Conclusion 736
References 736

16.1 INTRODUCTION

Carbohydrates were the first biopolymer to be identified, by Lavoisier in 1750, and long before any knowledge of genes or proteins; in 1902, the second Nobel Prize to be awarded for Chemistry was won by Emil Fischer for his study of the basic components of carbohydrates. Now a century later, with sequence strategies for both a genome and proteome in place, a comprehensive approach for sequencing carbohydrate structures is taking shape slowly. Such components have provided the analytical challenge of the century; replete with stereo and structural isomers, a multiplicity of linkage positions, and all embedded in the intricacies of branching. This chapter outlines and summarizes progress made recently due mainly to our ability to ionize large polar molecules, thanks to the researches of John Fenn and Koichi Tanaka, who were awarded the 2002 Nobel Prize in Chemistry, and our understanding of the modes of operation of ion traps (ITs) developed by Wolfgang Paul, Norman Ramsey, and Hans Dehmelt, who were awarded the Nobel Prize in Physics in 1989. During the past two decades, with facile ionization, multiple steps of disassembly (that is multiple stages of mass selection, isolation, and dissociation, usually induced by collisions, or tandem mass spectrometry (MS^n)), and a series of software tools, an instrumental strategy is now evolving that can be identified clearly as a sequenator for the glycoproteome.

16.1.1 FUNCTIONAL RELATIONSHIPS OF MOLECULAR GLYCOSYLATION

One gene, many proteins, countless glycans. Genes provide the code for proteins, which, in turn, perform most of the functions of the genome. Protein expression relies on a genetically encoded template, but it is the post-translational modifications (PTM) that amplify dramatically functional diversity; in this respect, PTMs are often referred to as 'added informational space.' Because the glycome represents the main class of PTM, this flow of chemical information provides access to a vast field of unexplored function at minimum genetic cost; that is, while all living organisms have approximately the same number of genes, PTMs provide for an expression of diversity over and beyond a fixed template. Genes yield the roadmap, but PTMs permit one to steer around the potholes! Thus, it is not surprising to note that the much of recent research into biological function has been directed to the domain of oligosaccharide structure. Interestingly, in this chemically directed flow to functional activity (genome to the glycome), a template-driven system gets lost to a temporal, epigenetic series of events providing a structural meaning to biological plasticity (that is, the ability of organisms to change without a direct template message or epigeneticity) with an outcome of individual specificity. A corollary of these considerations is the fact that linking gene expression to biological function gets more diffuse and analytically challenging with each step [1,2]. Thus, in making the genetic relationship to glycoproteomic disease states, it may be more cost effective to consider this as a *reverse* genomic problem: that is start with a precise understanding of functional epitopes (glycans) and backtrack to genetic expression through exacting conduits, for example, glycopeptides to genes. We suggest that this might be a more effectual research endeavor.

The complexity of the glycoproteome presents challenges not seen in genomic and proteomic analyzes. The proliferation of reports attributing biological function to oligosaccharide epitopes continues as conjugates of proteins, lipids, and independently as proteoglycans. Such structures exhibit exacting domains, but yet intriguing overall heterogeneity. Initiation of function may be temporal by modulation of an existing structure or *de novo* synthesis in synchrony with development, differentiation, and transformation. Studies of these regulated and cell-specific activities have repeatedly generated excellent data documenting interactions in multiple tissues and environments. Because the specificity of these interactions is the basis of form and function, a knowledge of topology, linkage, stereo and structural isomers represents the chemical end-game to understanding. Topology in this context relates to monomer connectivity.

16.1.2 GLYCAN STRUCTURAL COMPLEXITY

Thus, to appreciate fully these specific biological roles, an improved accounting of carbohydrate structural detail needs greater scrutiny and more judicious reporting. Unfortunately, a comprehensive strategy for carbohydrate sequencing has been lacking. Over the past decades, the absence of an integrated approach to define fully a sequence has been given little attention, probably due to overwhelming and direct functional lineage from nucleic acid to protein polymers. Even selective strategies to assign components of glycan structure show little focus toward congruency. Thus, the reporting of a partial sequence, assumed motifs, arrays absent of linkage and branching information, and multiple structures enclosed in brackets are conclusions that are acceptable currently in journal reports. Such descriptions diminish the driving force for the development of improved and more comprehensive methodologies. In that regard, a glycan sequence should provide all the components of structure necessary for synthesis, including monomer identification, positions of inter-residue linkage (including anomericity), and topologies with linear and branching information. When considering molecular glycosylation, sites of conjugation also need to be identified. In numerous recent reports, alternative strategies for glycan structure have been considered, most frequently using interactive combinations of chromatography [3–5], lectins [6,7], enzymes, and their combinations [8,9]. We have specifically avoided considering these approaches because physical separations will always be challenged by a glycome's complexity. Moreover, the lack of exacting specificity offered with many biological tools (lectins), and their inability to define positions of antennal release (enzymes) coupled with the ultimate need for product characterization, introduce irresolvable complications. For a number of years we have proposed that gas-phase multi-stage tandem mass spectrometry (MS) of glycan ions, formed by stereospecific metal ion adduction, should provide the most comprehensive approach to a glycan's connectivity and subsequent structural understanding [10–12]. Probably the most important consideration for focusing upon quadrupole IT technology is its overall simplicity. Thus, a single instrument, the ion trap mass spectrometer (ITMS), is poised to provide high throughput (HTP) analysis and full automation.

16.1.3 Synchrony of Glycan Structure and Operational Characteristics of the Ion Trap

Mass spectrometry has proven to be an adaptable technique, and, based on past record, the prospects for improved sensitivity and specificity can, almost certainly, be assured. Correlation of glycan mass profiles with physiological change, or gene modulation, provides a powerful approach to glycan function but, in using such strategies, a representative glycome remains fundamental, as do all components that comprise each ion, for example, isomers. This specific structural problem is an often-overlooked aspect in molecular characterization because of its transparency to mass measurement, where 'mass transparency' is the inability to distinguish between isomers on the basis of mass/charge ratio value alone. It must be recognized that each connecting monomer may create a new topology (branching), institute different stereochemistry (equatorial or axial hydroxyl groups) that extends to linkage as α or β anomers, or establish a different structural isomer through its linkage position. Although not common, the occurrence of rings of different sizes, for example, pyrans and furans, in natural products adds an additional isomeric problem. As a consequence, a comprehensive structure elucidation can be a demanding undertaking. In this chapter, we summarize ways to approach this complexity by using the fragment details (or product ions) within a disassembly pathway to provide oligomer topology (connectivity), branching, and inter-residue linkage using a single ion trap (IT) instrument. As detailed below, a fundamental outcome is the molecular continuity compiled from discrete steps of disassembly that provides information for *in silico* reassembly. Of equal significance is the fact that each disassembly step is coupled with effective energy enhancements providing maximum structural detail to spectral end-products; that is, in each disassembly (or mass-selective) stage a product ion, formed by a fragmentation pathway of low activation energy, is selected and subjected to collision-induced dissociation (CID) to yield further product ions formed also by fragmentation pathways of low activation energy. These small oligomer product ions are compiled into a searchable library, (see Section 16.4. Software Tools).

For a number of years selective strategies and instrumental approaches have been coalescing gradually, leading to a better understanding of oligosaccharide structures. The earlier developments of collisional activation (CA) MS, introduced as CID by Jennings [13] and by Haddon and McLafferty [14], and tandem MS [15] were interesting adjunct techniques in MS, but the value to biopolymer analysis could not be appreciated until the advent of advances in high mass ionization. Field desorption, chemical ionization, and fast atom bombardment (FAB) were interim strategies that provided some direction for solid sample analysis and avoided the pyrolysis of heat vaporization and high energy of electron bombardment. Both electrospray ionization (ESI) and matrix-assisted laser desorption ionization (MALDI) produced mainly protonated and deprotonated molecules with very little fragmentation; thus, CID became a major strategy for the characterization of biopolymers. The first indications of the combined power of intact ionization with gas-phase disassembly for carbohydrates were reported in 1985 [16] using FAB ionization and two coupled analyzers (BE-EB). FAB ionization and the first analyzer (BE) provided precursor (or parent) ions free from matrix contaminants, and structural details of these precursors were observed after collision and product ion analysis in a second coupled analyzer (EB). Starting from complex mixtures, this instrumental approach, BE-CID-EB, contributed significantly to the glyco-sylphosphatidylinositol anchor structure obtained from human acetylcholinest-erase [17] and the analysis of rhizobial signaling molecules in nitrogen fixation [18]. These exciting structure-functional relationships heralded the prospects of gas-phase structural analysis in the absence of chromatography, complete with the sensitivity expected from electron multiplier detection. The developments, however, which have brought the most significant impact to carbohydrate analy-sis, were those leading to the dynamic stabilization of ions in three-dimensional radiofrequency quadrupole fields by Wolfgang Paul. These achievements led to the subsequent developments of the quadrupole mass filter and the quadrupole ion trap (QIT) [19,20]. Thus, with the ability to ionize samples at high mass and to observe fragments by CID in Paul traps, all the components were in place for a full and detailed investigation of carbohydrate structure. The first commercially avail-able IT instruments were sold in 1984, but it was another decade, when combined with ESI, that carbohydrate analysis moved beyond the early successes with triple quadrupole instruments. Although early reports demonstrated the advantages of ion trap mass spectrometry (ITMS), little attention was focused on defining the limits of carbohydrate structure. Since the commercial introduction of ESI-ITMS instrumentation (Finnigan's LCQ and Bruker-Franzen's ESQUIRE) in 1995, we have considered the application of this approach to have the greatest promise for carbohydrate sequencing.

The basic operating characteristics of the QIT instrument have features essential for carbohydrate sequencing; for example, tandem MS to the nth degree permits observation of nth-generation product ions at a collision energy lower than that which would be required to form such product ions directly from the *initial* precur-sor ion. Furthermore, by definition, product ions are of lower mass than the corre-sponding precursor ion and have fewer oscillators, thus at each successive stage of tandem MS, the new precursor ion having fewer oscillators than its predecessor will yield usually more abundant product ions. Around this technology we have assem-bled a set of chemical modifications to amplify our understanding of these charac-teristics; these modifications are summarized here. Foremost: (i) collision energies are enhanced effectively with disassembly providing greater structural detail upon progression to fewer oligomers; (ii) pathways of disassembly can be directed so that they define domains of structure, and provide antennal specificity; (iii) samples prepared by methylation and reduction can localize monomers in a disassembly pathway; (iv) incorporation of metal ions (for example, Na^+), enhances sensitivity, fragmentation, and adduct stereo-specificity, thereby providing fragment stereo-chemistry; (v) isomeric product ions exposed during disassembly that fail to define a single topology are indicators of structural isomers, which can be isolated and char-acterized; (vi) a pathway of disassembly releases preferentially labile residues pro-viding renormalized product ion mass spectra with more comparable bond stability for enhanced detection (see Section 16.3.3); (vii) signal averaging and static infu-sion (which obviates chromatographic interfacing, for example with ion mobility

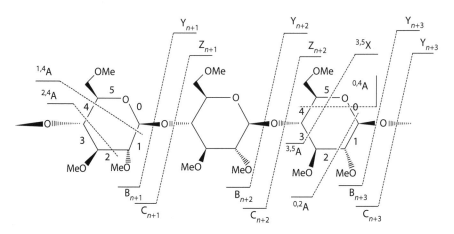

FIGURE 16.1   Scheme for major carbohydrate product ions following low energy collisions (CID) and resonance activation. Facile bond ruptures are mostly glycosidic bonds, B-/Y-type and C-/Z-type. Double cleavage cross-ring product ions are observed more commonly under higher energies or with glycans of smaller size where collision energy dissipation is constrained to fewer oscillators.

(IM),* liquid chromatography (LC) or gas chromatography (GC)† together provide an opportunity to capture product ions of low abundance; and (viii) creation of a searchable library of small oligomer product ions anchors the details of molecular disassembly. Thus, the IT generates a relational set of product ion compositions compiled in a library, thereby providing a foundation to structural understanding.

Each product ion composition can be defined simply as the sum of methylated monomers plus or minus any mass change due to bond rupture at, or near, a former linkage. We refer to these small mass identifiers as *scars*, for example, an open hydroxyl as in C- and Y-type ions, or a pyranene residue characteristic of B- and Z-type ions, (Figure 16.1). The ability to profile all ions in a sample glycome coupled with the facility (and need) to dig deep into each ion and to inquire about details of structure generates a wealth of data. These records need tools that can assimilate and integrate product ions for sequential meaning. Thus, bioinformatic tools are being built around this understanding, and such tools remain fundamental for any consideration of HTP analysis, (see Section 16.4. Software Tools).

---

* See Volume 4, Chapter 5: High-Field Asymmetric Waveform Ion Mobility Spectrometry, by Randall W. Purves; Volume 5, Chapter 8: Applications of Traveling Wave Ion Mobility Mass Spectrometry, by Konstantinos Thalassinos and James H. Scrivens; Volume 5, Chapter 13: The Study of Ion/Molecule Reactions at Ambient Pressure with Ion Mobility Spectrometry and Ion Mobility/Mass Spectrometry, by Gary A. Eiceman and John A. Stone.
† See Volume 5, Chapter 15: Technology Progress and Application in GC/MS and GC/MS/MS, by Mingda Wang and John E. George III.

16.1.4 Basic Energetics

Collisional activation of gas-phase ions has added significantly to the elucidation of molecular structure, but the constrained level of energy delivered in a single tandem mass spectral analysis (MS/MS) fails to capture the diversity of structure in large oligomers. The deposition of higher energies induce fragmentation to such a degree that molecular continuity is lost, while that of lower energies fails to detect double-bond ruptures needed for understanding linkage details. The IT, with its sequential approach to disassembly (MS^n), proportions effectively the highest energy to the smallest oligomers, allowing a cross-section of bonds to be probed in detail, with molecular continuity tracked by the precursor-product relationships in every step. This fundamental feature of the IT is based on the quasi-equilibrium theory (QET), which states that the availability of pathways for energy dissipation by fragmentation is inversely proportional to the number of oscillators. Consistent with that principle, we have explored structure-fragment relationships of small methylated oligomers and pursued the limits of informative fragmentation in linear (2D) and Paul (3D) ITs. Methylation fixes structural features and exposes points of rupture throughout disassembly, principles that add connectivity to the iterative steps within precursor-product relationships.

16.1.5 Collision-Induced Dissociation of Glycans

Developments leading to high mass ionization focused the greater need for improved techniques of oligomer depolymerization. Enzymatic proteolysis has largely filled this void for proteins, and peptide mass fingerprints are a sustaining technology for characterization. Comparable endoglycosidases are rare, but more challenging is the fact that monomers exist as multiple stereo and structural isomers. The mass transparency of linkage and branching (that is, inability to distinguish between isomers) makes a simple MS/MS fingerprint approach ineffective. In addition, presumptive amino acid sequences can be obtained from genomic information, whereas no such analogous sequence or linkage message is available for oligosaccharides. Over the decades, the literature is replete with chemical attempts (acid, base, reductive, oxidative) for partial oligosaccharide depolymerization with the hope of obtaining sequence overlap to define a single topological array. None has proven sustainable, and invariably polymer damage precludes any universally useful application. But the greater complications arise from the need to characterize a range of structural features, from the stereochemistry within monomers to the abundant isomeric connectivity within multimers.

Collisional activation has helped significantly to resolve this problem, but the constrained level of energy delivered with any one method fails to identify their structural diversity. High energy collisions, as provided by time-of-flight (TOF)-TOF analysis, or within modified ionization techniques, such as electron detachment dissociation (EDD), fragment samples to such a degree that molecular continuity is lost [21–23]. Uniquely, the IT, with its step-wise approach to disassembly, accesses the dissociative reaction pathway of lowest activation energy in each successive fragmentation stage thereby providing maximum structural insight to the smaller oligomers. This fundamental feature explains our results, and is clarified by the QET for predicting fragmentation of molecular ions in the excited state. Thus, the acquired internal energy (electronic, vibrational, and rotational) is distributed statistically to various oscillators, and bond breakage is assumed to occur due to vibrational excitation alone. The corollary of this assumption is that available pathways for energy dissipation are inversely proportional to the number of oscillators. Consistent with that principle, we explored and compiled initially a mass spectral library of small oligomer product ions where the linkage details were extensive, thereby providing an informative end-point to disassembly. This approach is consistent with the principle that, unless every intervening linkage is evaluated, a sequence cannot be considered to be comprehensive. Once sample methylation has replaced all available protons, only linkage disassembly (or CID) can render new hydroxylated products; thus, the observation of nascent hydroxyl formation, or scar, is indicative of a former linkage.

Usually, fragmentations can be rationalized on general principles of bond instability, but this is not the case with ion adduction. Such features introduce considerable disparity in product ions and their abundance but, importantly, the mass spectra are consistent and reproducible. A summarized scheme of possible product ions, which accounts for the most probable bond ruptures, is presented in Figure 16.1. This figure is a modification of that proposed originally [24], with changes in methyl derivatization, incrementing monomer numbers, and greater detail in cross-ring product ions. Dashed lines identify the points of bond rupture, which are designated with terminal letters of the alphabet.

Product ions that rupture on the C-1 side (monomer ring numbering) of the glycosidic oxygen are identified as B-, and Y-type ions, while those on the opposing side are identified as C- and Z-type ions. Collisional activation would be expected to produce all product ion types, with the resultant mass spectra being a function of ion source parameters. B- and Y-type ions are common product ions in positive ion mode, while negative ion mode usually exhibits a profile of C- and Z-type product ions. These product ions provide insight into molecular connectivity, but bring little understanding to the mass transparent (the inability to distinguish between isomers) features of linkage, branching, and stereochemistry. As mentioned above, ion abundance is a consequence of precursor bond lability and product ion stability, however, contributing most significantly to a comprehensive understanding of carbohydrate structure are the chemical properties of adduct ions, which bring both their own specificity and new opportunities for modifying product ion mass spectra. Usually a product ion mass spectrum is a consequence of indigenous precursor properties, while metal ion adduction provides an opportunity to probe unavailable features of structure, in particular, the stereochemical, linkage, and branching features that are frequently mass transparent. Thus, adduction (positive or negative) introduces a new qualitative measure with variables in *adduct* mass, positive or negative ion mode, and stereospecificity of binding. Such is not the case for all adducts and, as expected, protonation provides little linkage information with mainly the products of glycosidic cleavage. In contrast, adducted phosphate anions ($H_2PO_4^-$) show few glycosidic product ions and multiple monomer ring product ions [25]. These differences suggest altered energy distributions within the precursor upon adduction. Product ions with the greater adduct avidity are reflected in their abundance, alternative products of the same precursor are lost as neutrals. It is not uncommon, however, to see both products of a precursor ion in the mass spectrum, for example, the adducting species showing comparable affinity to different parts of the precursor (Na^+, Figure 16.2a). In this case, a hexose disaccharide product ion, m/z 445, serves as a precursor to a $^{3,5}A$ cross-ring cleavage product ion, m/z 329. The m/z 139 ion, composed of $[m = (445\ Da–329\ Da) + Na]^+$ is generally of low abundance or absent, presumably due to the weaker affinity for the metal ion.

Double cross-ring ruptures, or cleavages, (A- and X-type ions, as shown in Figure 16.1) brought about by CID are essential for determining linkage and, on many occasions, product ions can be rationalized as a series of retro-Diels-Alder (RDA) rearrangements. The formation of these multiple product ions is initiated by the unsaturation (1,2-pyrene) of the B-type glycosidic bond rupture (Figure 16.2a), which varies with linkage position in the precursor, (RO-, Figure 16.2b). Note that the $^{3,5}A$ (or $^{3,5}X$) product ions could not alone differentiate a linkage as both types (4-O, and 6-O) yield an identical mass, but uniquely the 6-linkage also exhibits

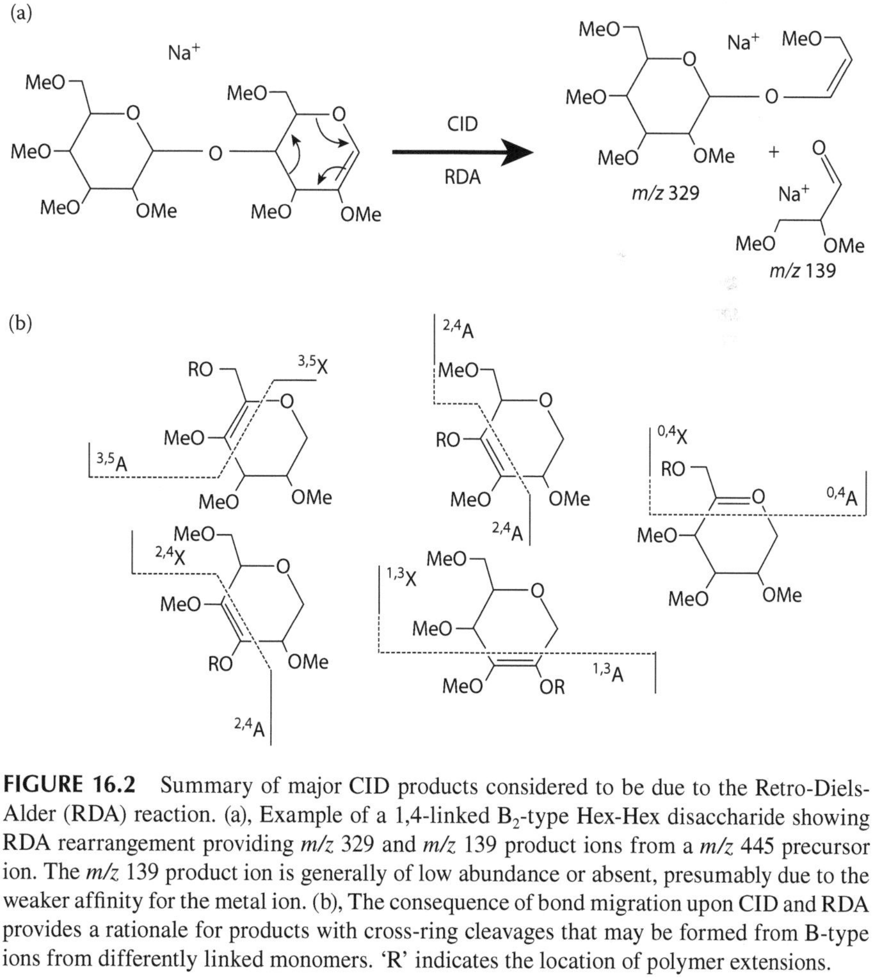

FIGURE 16.2 Summary of major CID products considered to be due to the Retro-Diels-Alder (RDA) reaction. (a), Example of a 1,4-linked B_2-type Hex-Hex disaccharide showing RDA rearrangement providing m/z 329 and m/z 139 product ions from a m/z 445 precursor ion. The m/z 139 product ion is generally of low abundance or absent, presumably due to the weaker affinity for the metal ion. (b), The consequence of bond migration upon CID and RDA provides a rationale for products with cross-ring cleavages that may be formed from B-type ions from differently linked monomers. 'R' indicates the location of polymer extensions.

a ^{0,4}A-type rupture (Figure 16.2b), and both product ions would support a 6-position linkage. An alternative product ion, D-type, that appears to have operational value has been reported [26], but it does not represent a different type of bond rupture (Figure 16.3A). In this case, a product ion was understood to originate from two independent bond ruptures leaving the trisaccharide ion, *m/z* 348, (Figure 16.3b). This fragmentation must be a feature of the facile $C_{1\beta}$ rupture and relative stability of the $C_{1\alpha}$ bond.

Ion abundance variations are important for differentiation of structures and, in this case, provided a strategy to define the blood group H, Le^a, Le^b, Le^x, and Le^y determinants with fucosylated analogues [26]. Such opportunities are exposed upon detailed studies of related structures and such outcomes, at this time, are not predictable. In a comparable manner, specific antennae within N-linked glycans can be defined (Figure 16.3c). In this example, the product ions *m/z* 737 (^{3,5}A) and *m/z* 709 (^{0,4}A) combine to identify the 6-linked antennae with a composition of three hexoses. The 3-linked antenna can also be specified with a combined rupture of ^{3,5}A and B$_3$-type

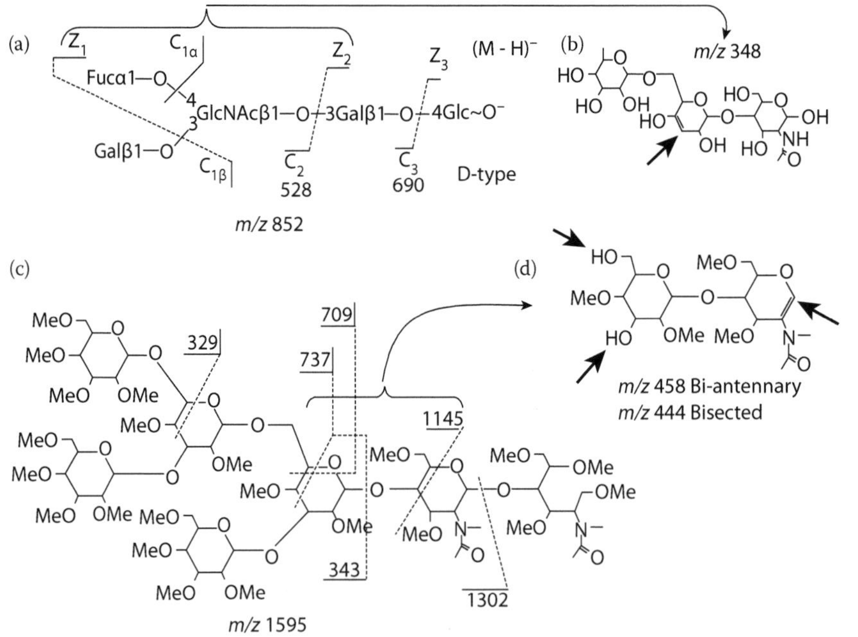

FIGURE 16.3 Summary of selected MS^n product ions that yield specific markers for N-linked glycan structures. (a), Fucosyl blood group glycans characterized with D-type product ions used as indicators of structural isomers (b). Not a new product ion type, but kinetically specific [18]. (c), Common product ions of N-linked glycans that provide markers for specifying antennae and their size. Identifiers of 6-linked antennae are the ^{3,5}A and ^{0,4}A-type product ions, and identifiers for a 3-linked antenna are a product ion pair (B-, and ^{3,5}A-Type) summed as a single ion. All product ions vary with monomer type and number. (d), Disassembly to N-linked core product ions Man-GlcNAc provide markers for antennae number, for example, two antennae, *m/z* 458, and three antenna *m/z* 444 (frequently referred to as bisected).

product ions, for example, m/z 343 (Figure 16.3c). Both of these identifying product ions are altered with antennal length (see Section 16.3.3, Figures 16.9 and 16.10). In addition, key product ions can identify and confirm the number of antennae extending from the central core mannose. As an example, a doubly branched core disaccharide would provide a product ion, m/z 458 (Figure 16.3d), while a triply branched glycan would provide the product ion m/z 444. The arrows in this figure indicate a former point of linkage and are referred to as scars. These product ion relationships are considered when setting pathways for disassembly, and are the basis for writing code to resolve glycan topology. For example, the $^{3,5}A$ product ions m/z 329, 737, and 1145 are related by 408 Da indicating a branched chain (Figure 16.3c) and such signature relationships are observed frequently. Structural associations can be confirmed by selecting alternative disassembly pathways that can reinforce preliminary assignments. Ion relationships or continuity are not always found in a single mass spectrum, but must be contiguous when including all pathways of disassembly. Importantly, when such relationships do not exist (anomalous ions), the likelihood of an isomeric structure is indicated. Resolving such complexities in the IT instrument is unparalleled, for such ions can be isolated and studied further in the absence of arduous chromatography. Isomeric structures are often missed because either they are unresolved chromatographically, or more abundant ions offer a satisfying singular solution. Low abundance ions could have decisive meaning for biomarker detection of an impending disease and carbohydrate epitopes are likely to play a major role in this regard. Such circumstances would warrant static infusion and signal averaging to obtain high quality data for spectral matching. These situations are largely incompatible with coupled LC-MS.

16.2 SAMPLE PREPARATION

16.2.1 RELEASE OF N, AND O-LINKED GLYCANS

N-linked glycans were released with PNGase F prepared with a glycoprotein denaturing buffer (5% sodium dodecyl sulfate (SDS), 0.4 M dithiothreitol (DTT)) and stored in 10% NP-40 (the commercially available detergent nonyl phenoxylpolyethoxylethanol). Sample glycoproteins were dissolved in a denaturing buffer 10 μL (10 × conc.), followed by 90 μL high performance liquid chromatography (HPLC) water (dilute to 1 × conc.). The mixture was vortexed and heated to 100°C for 20 min in a heating block. The mixture was centrifuged and cooled to room temperature. 10 μL of 10 × G7 buffer was added along with an equal volume of 10% NP-40 buffer, followed by 3 μL PNGase F and incubated overnight at 37°C. The sample was then run through a solid-phase extraction step (Sep-Pak C18 cartridge), which retains the peptides and proteins while passing the glycans in the flow-through. The volume was dried under vacuum in preparation for reduction and methylation. O-linked glycans are released by two methods using basic solutions with different overall goals. With stability concern for only the glycans, the classical conditions developed in 1966 by Carlson are still effective, using sodium hydroxide under reducing conditions with alkaline borohydride solution (1 M NaOH and 0.1 M NaBH$_4$) [27]. This *in situ* reduction prevents the released glycan, a hemiacetal, from degradation, a process identified as *peeling*. The solution was heated at 45°C for 12 h to release the O-linked glycans, followed by the addition of a 1 M solution of HCl to decompose the excess borohydride. The product alditols were purified as described above using a Sep-Pak cartridge. Small glycans were collected in the first column volumes, larger glycans were retained slightly and higher concentrations of acetonitrile helped to offset this retention. Typically, 25% (v/v) acetonitrile in water with 0.1% trifluoroacetic acid released all glycans in the column wash. Because proteins lack a specific consensus sequence for O-links, these general protocols have considerable merit for indentifying former positions and earlier studies had shown that β-elimination could provide this information [28]. Regrettably, the reductant degrades some peptides, and failed with larger proteins [29]. A milder base was considered, namely using ammonia which adds to the double-bond created [30,31]. Thus, the aminylated products allow proteolytic cleavage, and the identification of previous O-glycosylation sites. The label is chemically stable under the conditions of MALDI and ESI MS, and also under CID or post-source decay (PSD) conditions when peptide sequencing is considered. We have adapted these general protocols for greater specificity by using isotopically labeled d_6-dimethylamine (DMA), and carried out the release using microwave heating. This procedure shortened significantly the reaction conditions for quantitative release and diminished the peeling reaction with the shorter time (unpublished report). Dry samples were dissolved in 40% aqueous DMA in a 10-mL Pyrex sample tube, and a small stir bar was added to equalize heating. The microwave temperature was set to 70°C with a reaction time of one hour. Peptides are retained on C-18 tips and O-linked glycans, and peptide fractions are dried under a stream of nitrogen. Oligosaccharide samples can be labeled directly for HPLC separation and fluorescent detection, as well as being reduced and methylated for sequencing by MS^n. The peptides samples were eluted, dried, and dissolved in a matrix for MALDI sequencing and, in a recent innovation, electron transfer dissociation (ETD)* analysis provides O-linked glycan composition data [32,33]. By observing the glycopeptide sequence with the corresponding O-linked glycan intervals, a direct correlation can now be made to the specific site coupled with a detailed glycan sequence resolved by MS^n.

16.2.2 REDUCTION AND METHYLATION

The dried sample glycans were reduced by adding 200 μL of a sodium borohydride solution (10 mg mL^{-1} in 0.01 M NaOH) and were allowed to stand at room temperature overnight. The reaction was quenched by the addition of acetic acid with the sample in an ice bath, followed by addition of 4 mL ethanol, and evaporated to

---

* See Volume 4, Chapter 3: Theory and Practice of the Orbitrap™ Mass Analyzer, by Alexander Makarov; Volume 4, Chapter 10: Trapping and Processing Ions in Radio Frequency Ion Guides, by Bruce A. Thomson, Igor V. Chernushevich and Alexandre V. Loboda; Volume 4, Chapter 13: An Examination of the Physics of the High-Capacity Trap (HCT), by Desmond A. Kaplan, Ralf Hartmer, Andreas Brekenfeld, Jochen Franzen, Michael Schubert; Volume 4, Chapter 15: Fragmentation Techniques for Protein Ions Using Various Types of Ion Trap, by J. Franzen and K. P. Wanczek; Volume 5, Chapter 1: Ion/Ion Reactions in Electrodynamic Ion Traps, by Jian Liu and Scott A. McLuckey; Volume 5, Chapter 3: Methods for Multi-Stage Ion Processing Involving Ion/Ion Chemistry in a Quadrupole Linear Ion Trap, by Graeme C. McAlister and Joshua J. Coon.

dryness by vacuum centrifugation. Borates were removed by adding 3 mL 1% acetic acid in methanol and drying under vacuum; this procedure was repeated twice more. Subsequently, 3 mL of toluene were added and the sample was dried under nitrogen; this procedure was repeated three times. The sample was desalted by passage through a cation exchange resin (DOWEX 50W). For methylation of the glycan alditols, 1 mL of dimethyl sulfoxide (DMSO) was added to the dried sample, followed by freshly ground sodium hydroxide and vortexing for 20 s. Iodomethane was then added (100 µL) to the reaction mixture, followed by vortexing for one hour. The reaction was stopped by the addition of 2 mL of cold water. The methylated oligosaccharides were isolated by liquid–liquid extraction with dichloromethane, repeated at least three times. The combined organic layers were then washed multiple times with water. The dichloromethane was evaporated under nitrogen gas.

16.2.3 MASS SPECTROMETRY

Mass spectra were obtained from an LTQ (Thermo Fisher Scientific, Waltham, MA) equipped with a nanospray ion source. Samples were infused at flow rates ranging between 0.30 and 0.60 µL min^{-1}, and mass spectra were collected using Xcalibur 1.4 and 2.0 software (Thermo Fisher Scientific). In addition, some samples were infused using a Triversa Nanomate (Advion, Ithaca, NY). In selected cases, mass spectra were collected automatically as described. Ten scan events were programed with the first event, Scan event 1, being a profile analysis. Subsequent ion scans were selected automatically on the basis of the most abundant neutral loss or from a pre-programed neutral loss list. In situations where more than one product mass/charge ratio satisfied the neutral loss list, the more intense product ion was selected as the precursor for a subsequent scan. Signal averaging was accomplished by increasing the number of microscans for each scan and was generally between 5 and 50. Improved spectral quality was necessary at the higher orders of MS^n due to diminishing signal, particularly beyond MS^5 or MS^6. The minimum signal threshold for precursors was lowered to 10 counts for all levels of stages. Collision parameters were left at default values with normalized collision energy set to 35% or to a value leaving a minimal precursor ion peak intensity. The Mathieu trapping parameter, q_z, (see Chapter 1) was set to a value of 0.25 and the duration of resonance excitation was 30 ms. All ions were sodium adducts.

16.3 MS^n DISASSEMBLY APPLICATIONS

16.3.1 THE GLYCOSYLATION FINGER-PRINT

It is highly probable that a glycan MS profile, that is, a mass scan or MS^1 spectrum, complete with detailed structural analysis of each ion, would become the most exacting identifier of a sample or tissue. The absence of a restricting template to define structure, and the well-known modulation during growth, environmental changes, and metabolic factors suggests strongly there could be no better biomarker, that is, a chemical measure that is specific to the sample, tissue or disease state. By qualifying such a biomarker as 'better' means the biomarker identifies all subtle structural alterations whether metabolic, environmental, or genome induced. Unfortunately, established sequence protocols are seriously lacking, and a tradition of reporting has developed without standards with the result that analytical efforts can be summarized as a cottage industry.

16.3.2 GLYCOPROTEIN STANDARDS

Although all glycan samples render the inherent problems of linkage and branching, an equally important problem is that of structural isomers which are largely transparent to mass spectrometric characterization. This observation of isomers is not new, but its prevalence in common structures [10–12] suggests that this component of structure needs more serious consideration and analytical evaluation. Thus, for a comprehensive glycan analysis, it may be appropriate to argue that all possible isomers exist unless proven otherwise. Shared standard samples are an excellent way to compare and evaluate novel protocols, new instruments, and this challenging sequencing endeavor. Bovine ribonuclease B and egg ovalbumin (Sigma-Aldrich, St. Louis, MO) provide that option and are readily available. These glycoproteins offer reproducible glycan ion profiles and can be considered effective standards for this task. Following sample preparation of the bovine ribonuclease B and egg ovalbumin glycans, ESI-MS profiles of these product ions should look like those in Figures 16.4 and 16.5, respectively. The first profile, Figure 16.4, exhibits a single

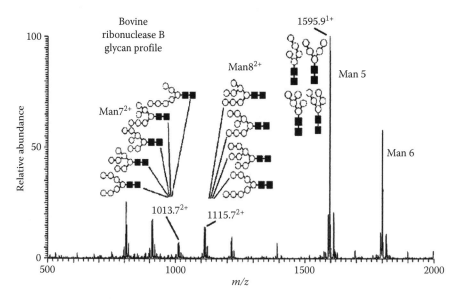

FIGURE 16.4 MS profile of N-linked glycans released from a 'standard' sample of bovine ribonuclease B. Products recognized as high-mannose type glycans were prepared for MS^n disassembly by reduction and methylation. The mass range shows two sets of ions, doubly- and singly-charged. Insert cartoons are structures determined by MS^n for the profiled ions $Man_5GlcNAc_2$, $Man_7GlcNAc_2$, and $Man_8GlcNAc_2$ with the detection of 4, 5, and 4 isomers, respectively. All structures were resolved in the gas-phase by ITMS with no chromatographic interfacing. Symbols represented by squares = GlcNAc; circles = Man, (see Figure 16.6).

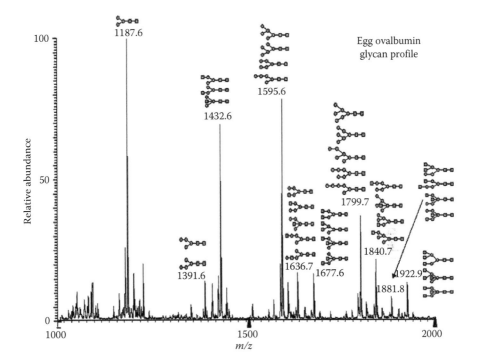

FIGURE 16.5 MS profile of N-linked glycans released from a 'standard' egg ovalbumin. Products recognized as complex N-linked glycans were prepared for MS^n analysis by reduction and methylation. Ten ion types were disassembled by MS^n and the isomers characterized are set as inserts adjacent to each precursor ion. Symbols represented by squares = GlcNAc; circles = Man, (see Figure 16.6).

set of doubly-charged glycomers between the mass range 800 Th and 1240 Th and a set of singly-charged glycomers extending beyond the scan range selected. The topologies determined (as inserts) are set as cartoon inserts for the ion compositions Man$_7$GlcNAcGlcNAcol, Man$_8$GlcNAcGlcNAcol, and Man$_5$GlcNAcGlcNAcol, (m/z 1013.7^{2+}, 1115.7^{2+}, and 1595.9^{1+}, respectively). For ovalbumin, Figure 16.5, 10 profiled ions are shown resolved. Topology in this context relates to monomer connectivity and is not meant to define linkage, although some linkage is implicit from biological insight, for example, N-linked core. Cartoon representation of structures has become common for obvious reasons and these abbreviations suggest some understanding of structure, but most caricatures are simple abbreviations. An earlier report has considered a more comprehensive representation with bond angle projected for linkage position, solid or dashed lines at the C-1 position to indicate anomericity, and even a wavy line when unknown [34]. Color-coding along with a graphic shape has been described also and standards have been proposed with many examples [35]. Unfortunately, these cartoons, although convenient, are not as successful for discussing ring product ions. The full structure presented in Figure 16.3c could be represented by one of four topologies (lower left) which were uncovered within the ion m/z 1595.9^{1+} (Figure 16.4).

16.3.3 EXTRACTING THE DETAILS OF MOLECULAR STRUCTURE BY MS^n

Although comprised of only a few basic units, for hexose, N-acetyl-hexosamine, deoxyhexose, etc., glycan analysis is a challenging endeavor due to the multiple linkage positions, branching, anomericity, and differences in monomer stereochemistry. Compounding this problem is the abundance of constitutional isomers found in biological samples (Figures 16.4 and 16.5). Constitutional isomers, commonly referred to as structural isomers, are glycans with identical composition arranged in different structural configurations. We have referred to these as topological isomers that simply relate to monomer connectivity. Isomer characterization is most often considered outside the realm of MS, and is usually reserved for separation through chromatographic methods. However, recent reports have demonstrated that sequential ion trap mass spectrometry (MS^n) provides both the sensitivity and attributes to resolve structures *de novo* from complex biological samples [10,11]. Upon fragmentation, methylated product ions (see Section 16.2) leave scars at the point of fracture. As an example, a glycosidic bond cleavage gives rise to a hydroxyl group (Y$^-$ ion) and a double-bond (B$^-$ ion). In the structural evaluation of bovine ribonuclease B glycans, it has been long established that Man$_7$GlcNAc$_2$ and Man$_8$GlcNAc$_2$ possess three isomers each and a single structure for Man$_5$GlcNAc$_2$ and Man$_6$GlcNAc$_2$ [36]. These conclusions have been supported recently using a combination of HPLC coupled with FT-ICR MS [37,38]. In a renewed effort to demonstrate the unique capabilities of ion trapping MS, 13 high-mannose isomers have been characterized structurally within the same three glycomers, six previously unreported (Figure 16.6). The usual set from Man7 and Man8 (three each) have been identified plus one unreported Man7 isomer (Figure 16.6). Additionally, incomplete α-glucosidase activity on the Man6 and Man7 glycoproteins appears to account for two additional isomeric structures. The preeminence of ITs for detail analysis was demonstrated further by resolving three undetected isomers within the Man5 glycomer summing to the six previously unreported structures in this glycoprotein (designated with stars, Figure 16.6) [10].

All reported structures represent a distribution of Golgi processing remnants that fall within the Man$_9$GlcNAc$_2$ footprint (Figure 16.6). The Golgi is a cellular organelle that synthesizes and degrades glycans structures. This degradation is called processing as it rebuilds the glycan (resynthesized) specific to new and different cellular demands. However, in the degradation it appears to be incomplete, leaving partially degraded products. Topologies were defined by ion compositions along a disassembly pathway, while linkage and branching details were aided by spectral identity in a small oligomer product ion library. Egg ovalbumin offers another readily available glycoprotein and the array of glycans released are replete with structural isomers. Ten ions were selected for detailed study and nine indicated multiple isomers (Figure 16.5). The detection of structural isomers is simply a process of observing aberrant ions along the pathway of disassembly that are inconsistent with a single structure. These ions may be evident at anytime during disassembly, and this is the basic reason why tandem MS fails to resolve such detail. An example of this approach is outlined for the single ion, *m/z* 1923.1 (Figure 16.5) and summarized with disassembly pathways, product ions, and concluding structures (Figure 16.7).

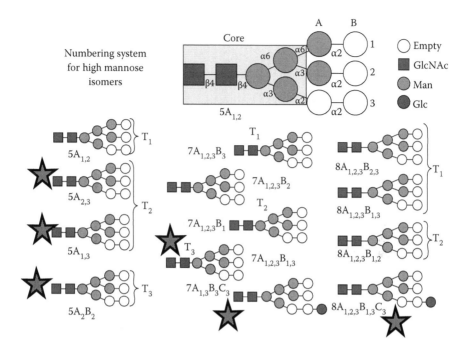

FIGURE 16.6 Numbering system that identifies all processing isomers set with cartoon representation for N-linked high mannose glycans. Top central insert represents the cartoon pattern and specific linkages along with a designated N-linked core region. Filled circles and squares represent the glycan $Man_5GlcNAc_2$. Set below are the 13 structural isomers detected for this glycoprotein (Figure 16.4) set over a template of empty circles representing the $Man_9GlcNAc_2$ precursor. Nomenclature for defining all Golgi-processed isomers (beyond core region) are identified as columns A and B with subscript rows 1, 2, and 3. Structures can be divided into topologies (monomer connectivity) and further subdivided into structural isomers (linkage). Identified $Man_5GlcNAc_2$ structures include three topologies and four structural isomers: T1 ($5A_{1,2}$); T2 ($5A_{2,3}$, $5A_{1,3}$); and T_3 ($5A_2B_2$). For Man_7 $GlcNAc_2$, three topologies were identified also with four structural isomers. For $Man_8GlcNAc_2$ two topologies have been identified that include three structural isomers. The stars represent isomeric structures not reported previously for this 'standard' glycoprotein.

At the bottom of Figure 16.7 is a listing of monomer product ions along with their cartoon notation. In the right hand side columns are the mass changes that occur as a result of a neutral loss, or as a charge-carrying metal ion adduct. Such product ions represent common features and signatures of structural components. As one example, observation of the bottom two product ions (*m/z* 458, 444, Man-GlcNAc) near the end of a pathway would signify a di- and a tri-antenna glycan, respectively. Thus, every product ion has not only a summed monomer mass, but a variable mass that relates to its connectivity in the oligomer. The top part of the table details four pathways that identified specifically the three topologies (**A, B, C**), and further identified two isomers with the same topology (**C**). By following the pathway, an assignment of topology is provided. It is important to note that the linkage detail that allows

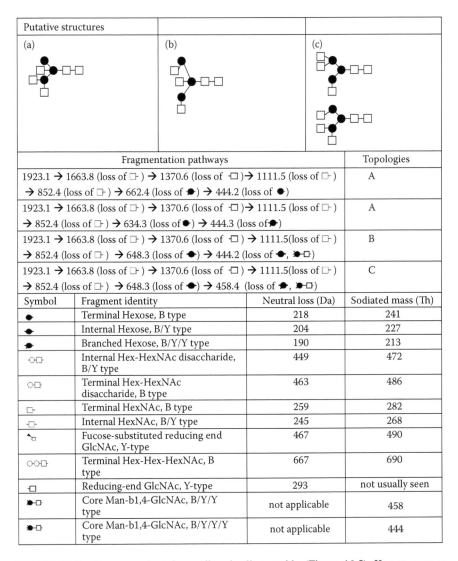

Putative structures		
(a)	(b)	(c)

Fragmentation pathways	Topologies
1923.1 → 1663.8 (loss of ☐⊦) → 1370.6 (loss of ⊡)→ 1111.5 (loss of ☐⊦) → 852.4 (loss of ☐⊦) → 662.4 (loss of ●) → 444.2 (loss of ●)	A
1923.1 → 1663.8 (loss of ☐⊦) → 1370.6 (loss of ⊡)→ 1111.5 (loss of ☐⊦) → 852.4 (loss of ☐⊦) → 634.3 (loss of ●) → 444.3 (loss of●)	A
1923.1 → 1663.8 (loss of ☐⊦) → 1370.6 (loss of ⊡) → 1111.5(loss of ☐⊦) → 852.4 (loss of ☐⊦) → 648.3 (loss of ●) → 444.2 (loss of ●, ●⊡)	B
1923.1 → 1663.8 (loss of ☐⊦) → 1370.6 (loss of ⊡) → 1111.5(loss of ☐⊦) → 852.4 (loss of ☐⊦) → 648.3 (loss of ●) → 458.4 (loss of ●, ●⊡)	C

Symbol	Fragment identity	Neutral loss (Da)	Sodiated mass (Th)
●	Terminal Hexose, B type	218	241
●	Internal Hexose, B/Y type	204	227
●	Branched Hexose, B/Y/Y type	190	213
○☐	Internal Hex-HexNAc disaccharide, B/Y type	449	472
○☐	Terminal Hex-HexNAc disaccharide, B type	463	486
☐⊦	Terminal HexNAc, B type	259	282
⊣☐⊦	Internal HexNAc, B/Y type	245	268
⬥☐	Fucose-substituted reducing end GlcNAc, Y-type	467	490
○○☐⊦	Terminal Hex-Hex-HexNAc, B type	667	690
⊣☐	Reducing-end GlcNAc, Y-type	293	not usually seen
●⊣☐	Core Man-b1,4-GlcNAc, B/Y/Y type	not applicable	458
●⊣☐	Core Man-b1,4-GlcNAc, B/Y/Y/Y type	not applicable	444

FIGURE 16.7 Summary chart for ovalbumin disassembly (Figure 16.5). Key to cartoon representation of product ions, pathways of disassembly, and products determined by MS^n for the precursor *m/z* 1923.1 (Figure 16.5). Top boxes include three topologies identified (A, B, and C) that included four isomers. Four pathways were followed that defined the isomers shown in the right hand 'Topologies' column. Bottom area depicts cartoon abbreviations, ion masses (Da) for neutral loss, and mass/charge ratio (Th) for sodium-adduct product ions.

resolution of structure **C** is a consequence of a library matching (see Section 16.4.2. FragLab Tool Kit).

Establishing the details and defining pathways as outlined in Figure 16.7 originates from MS^n spectra in a data-dependent manner. The following three figures explain how one isomer was determined of the four established previously for the

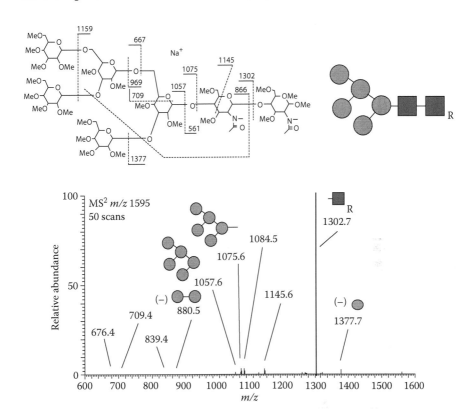

FIGURE 16.8 MS² of the canonical Man₅GlcNAc-GlcNAc<u>ol</u> glycan, *m/z* 1595.9, obtained from *C. elegans* (Figure 16.4). Facile rupture of terminal GlcNAc<u>ol</u> provides the base product ion, *m/z* 1302.7, to which the mass spectrum is renormalized, making other product ions unobservable. Cartoon inserts are proposed structures related to product ions.

Man₅GlcNAc-GlcNAc<u>ol</u> glycan (Figure 16.4) [10]. Isolation of this precursor and CID analysis indicated a facile loss of the terminal -GlcNAc<u>ol</u>, (*m/z* 1302), and a few minor product ions as indicated by the cartoon inserts (Figure 16.8). Activation of this base ion and MS³ analysis showed the major components of the canonical Man5 structure, and an improved evaluation of an (encircled) aberrant product ion, *m/z* 880. This ion, *m/z* 880, could be accounted for only by considering a neutral loss of a fully methylated disaccharide, a structural feature not present in the purported precursor. When one considers an N-linked core and a fully methylated disaccharide as components of the Man5 structure, the observation of *m/z* 880 suggests strongly that the aberrant precursor should have a topology different than shown in Figure 16.9. Isolation of this product ion, *m/z* 880, in an attempt to understand the remainder of the structure, together with subsequent MS⁴ analysis provided supporting evidence for this conjecture (Figure 16.10). The CID analysis indicates two major products from the *m/z* 880 precursor. The base ion, *m/z* 635, can be considered a B-type ion formed from the facile neutral loss of a terminal GlcNAc residue leaving the mannose trisaccharides of identical composition, but

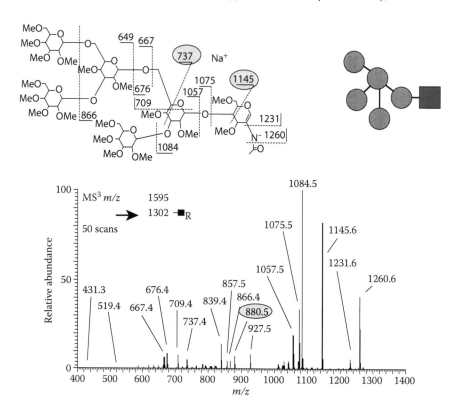

FIGURE 16.9 MS³ spectrum of *m/z* 1302.7 (neutral loss of reducing-end GlcNAcol) that was isolated and analyzed from Figure 16.8. Collision energies were held constant, but note the greater abundance of product ions from a precursor ion having multiple dissociation pathways of similar activation energy. Circled product ion, *m/z* 880, now visible represents neutral loss of a fully methylated disaccharide. A product ion of this type could not arise from the canonical structure and was expected to originate from an isomeric structure. Such ions provide an opportunity for the selective isolation of isomeric structures and for disassembly of the aberrant precursor ion. This point is demonstrated in Figure 16.10.

of differing isomeric structure. The key to their structures lies in the three circled product ions, *m/z* 505, 533, and 547. All three ions are common product ions representing, respectively, a ^{4,0}A-Type, a ^{3,5}A-Type, and the 3-linked antennal signature product ion (Figure 16.3), *m/z* 547. This ion, *m/z* 547, appears 204 Th higher indicating a fully methylated disaccharide linked to the central core mannose through a 3-linkage. This product ion appears to be a combination of a B, and a ^{3,5}A-Type product ion providing a single mass and comparable to the D-Type product ion mentioned previously (Figure 16.3). All three product ions would require further disassembly to ascertain the linkage between the mannose residues. This *m/z* 547 ion has been disassembled through the sequence *m/z* 547, 463, *via* MS⁶ and MS⁷ to isolate a Manα(1–2)Man(OH) C-Type product ion with every expectation of being a single isomer [10].

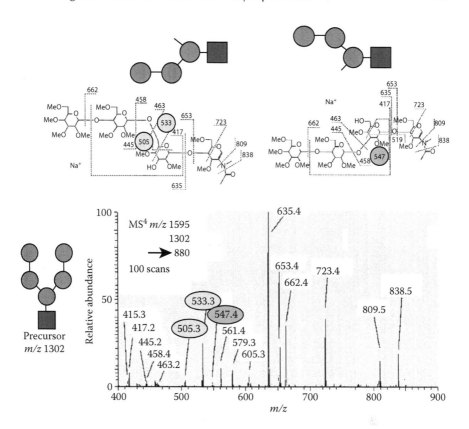

FIGURE 16.10 MS⁴ spectrum of the aberrant ion, m/z 880.5, isolated from Figure 16.9. Assuming N-linked Man$_5$GlcNAc$_2$ as the precursor and having lost a fully methylated disaccharide moiety, few structurally defining topologies remain for the Man$_3$GlcNAc$_2$ glycan. The mass spectrum appears to indicate two product ion structures, shown at the top of the figure, and the presence of both product ions is indicated by observation of the signature product ions, m/z 505.3, 533.3, and 547.4. These three ion species represent marker product ions for defining N-linked structures (Figure 16.3) along with the ion m/z 458 indicating a biantennary precursor. From these major identifiers, along with several others, it can be concluded that the aberrant isomer, m/z 1302, has the structure shown at the bottom left-hand-corner of the figure. Further disassembly (MS⁵) and spectral matching with library standards would define the linkage positions.

16.3.4 PHYSIOLOGICAL MODULATIONS

16.3.3.1 Differentiation Profiles in Metastasis

Detecting glycosylation changes during differentiation might be considered to be comparable to studies of biomarkers within metastatic tissues. There have been numerous reports of glycan profiling to characterize malignant cells and tissues. Most of these reports suggest and write structures (cartoons) for which there is little analytical support. Ancillary techniques for resolving the inherent complexities can advance structural understanding, but most protocols (chromatography, methylation analysis, enzymology, MS/MS), supplement rather than finalize an exact structure. Analytical transparency

Glycosylation tissue specificity---H6N3F1 *m/z* 1121.1

> ➤ Uncovered an anomalous set of N-linked glycans distinct to malignant non-metastatic and metastatic tumors; non-chromatographically.

>   ➤ MS^n for biomarker discovery and cancer therapeutics?

H6N3F1 *m/z* 1121.1

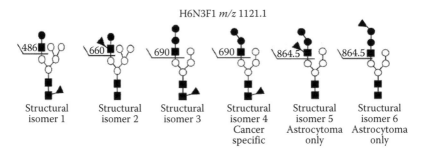

| Structural isomer 1 | Structural isomer 2 | Structural isomer 3 | Structural isomer 4 Cancer specific | Structural isomer 5 Astrocytoma only | Structural isomer 6 Astrocytoma only |

FIGURE 16.11 Glycan specificity shown within murine normal and metastatic tissues [12]. Control brain astrocytes, non-metastatic, and metastatic tumors cells were evaluated for specific glycan markers using the protocols described. Disassembly by MS^n of one selected ion, *m/z* 1121.1, from comparative tissues exemplifies that specificity. A facile cleavage at the GlcNAc linkage would differentiate some tissues using MS/MS, *m/z* 486, 660, 690, and 864, but further details would be unavailable. Chromatographic techniques and enzymology would be challenging considering the tiny amounts of sample available. Structures 1, 2, and 3 were found in all tissues, while structure 4 was specific for cancer tissues, and 5 and 6 were specific for astrocytomas.

to any one linkage exposes the possibility of a missed opportunity either to understand the role of linkages in tumor progression, or to uncover a unique cancer biomarker. We have reported on a murine cancer model that possesses all the untoward features of malignancy (spontaneous tumor generation, local invasion, intravasation, immune system survival, extravasation, secondary tumor formation, and metastasis) [11]. This specificity of glycosylation was found in multiple tissues and ions' compositions and data for the glycan Hex$_6$HexNAc$_3$Fuc, *m/z* 1121.1 are presented in Figure 16.11.

16.3.3.2   Differentiation Profiles in Adult Stem Cells

Despite the importance of stem cells both in cell biology and in clinical medicine, the molecular details underlying the characteristics that maintain pluripotency, or those that regulate differentiation, remain largely unknown. Thus, discovery of specific biomarkers remains an on-going challenge. Such identities are desired highly by biologists as essential tools for the study of function and by clinicians as targets for diagnostic and prognostic indicators of carcinogenicity and new avenues for metabolic intervention. Of the several methodologies that have been applied to characterize embryonic stem cells, the most widely used approach analyzes cell surface antigens by flow cytometry and assesses gene expression profiling by reverse transcription-polymerase chain reaction (RT-PCR) or microarrays. Many cell surface antigens used were detected first with antibodies prepared against preimplantation mouse embryos and/or against mouse or human embryonal carcinoma cells. Although the roles of those antigens in the maintenance of undifferentiated human embryonal carcinoma cells are not necessarily evident, they may represent useful markers for the identification of pluripotent stem cells.

Recent developments in genomic and proteomic technologies have provided tools to analyze gene expression profiles and to determine novel molecular markers that can serve as tools to detect, classify, and isolate subpopulations and to monitor their state of differentiation. A differential glycome analysis using the protocols outlined above was conducted on released N-linked glycans from the three lines of adult stem cells, SYM, ASYM, and p53SYM. Selected ions were disassembled sequentially from each MS profile and approximately 1000 mass spectra were collected. Software tools described below were used to assist in data analysis. The results indicated the overall profiles to be similar, but each profile contained ion species that exhibited important abundance differences; it was to these ion species that our attention was directed. When these ion species were selected for disassembly, they showed multiple isomers. One of these ion species was the m/z 1810.6 ion with an N-linked glycan composition of $Hex_4Gn_3Fuc_1$ (Figure 16.12). Two of these isomers were shared by all three samples (S_1 and S_2, right columns, Figure 16.12); one was shared by the SYM and ASYM cells, (S_3); one is shared by ASYM and p53SYM cells, (S_4); and one isomer was found to be ASYM-specific, that is, undetected in either SYM states, (S_5). Disassembly pathways were employed to resolve the isomeric structures and, although most of the mass spectra showed similarities between samples at every MS^n level, critical isomeric structural differences were detected. Here, we use the mass spectra and the structures identified within the ion m/z 1810.6 as an example. The ion m/z 1810.6, corresponding to an N-linked glycan of composition $H_4N_3F_1$, was isolated from each of the profiles and examined. These preliminary results (from a small sampling) indicated that the isomeric differences within adult stem cell samples can be unraveled easily, but require a deeper level of disassembly, that is MS^n, where $n = 3$–5.

16.4 SOFTWARE TOOLS: GENERAL CONSIDERATIONS

Sequencing carbohydrates by ITMS requires extensive training, and even the experts find the application very labor-intensive. The manual interpretation of multiple mass spectra remains an intimidating barrier especially when pathway decisions are decided in a data-dependent manner. Software tools should be able to overcome much of this workload, and this Section summarizes some of those considerations.

16.4.1 GLYCOSCREEN

Glycoscreen is 'in-house' nomenclature for a screening tool that has been designed to filter out background contaminants and consider only those ions possessing carbohydrate compositions. Common carbohydrates samples and their modifications are considered as input data. The output defines and accepts only carbohydrate specific isotopic distributions to rebuild MS profiles, thereby facilitating data analysis. The general features include fitting profile data to carbohydrate-specific ions and assigning compositions to each profiled ion. The software supports both MALDI and ESI profiles. The code was developed in C# using Visual Studio 2005 in conjunction with Microsoft .Net 2.0 and SQL Server 2005 Express Edition, which is required for installing and running the tools properly.

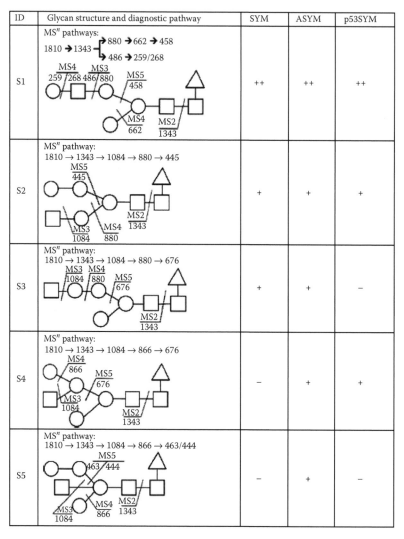

FIGURE 16.12 Differential glycome analysis (using the protocols described) for the detection of specific markers of adult stem cells. Released N-linked glycans from three self-renewal states, SYM, ASYM, and p53SYM, were profiled and several ions disassembled. The overall profiles appeared to be similar, but each with important abundance differences which focused attention to the inherent ions in each profile. Shown in this figure are the data for ion, m/z 1810, which provided a composition Hex_4HexNA_3Fuc. Isomers were detected as described (Figure 16.10) to reveal the five topologies shown (S1–S5). Two of these isomers were shared by all three samples (S1 and S2, right columns, Figure 16.12); one was shared by the SYM and ASYM cells, (S3); one is shared by ASYM and p53SYM cells, (S4); and one isomer was found to be ASYM specific, for example, undetected in either SYM states, (S5). Many of the known marker ions were observed to ascertain topologies while spectral library matching supported identification of the structural details. A small portion of the total glycome has undergone analysis, but, as with the cancer tissues, clear differences were readily detected.

16.4.2 FragLib Tool Kit

As considered above, MS^n spectra of carbohydrate samples can be utilized as an exacting metabolic fingerprint. Structural details, including monomer identification, branching, linkage, and stereochemical features, can be determined by a combination of glycan disassembly and spectral matching of the products. The library has been generated from commercial samples (small oligomers) and those prepared from well-characterized glycoproteins. The library uses the generic XML format to organize the information. The mass spectral data can be exported in various standard formats, which are fully compatible with the existing popular mass spectral searching engines including the NIST MS search tool. These search tools are available at http://chemdata.nist.gov/mass-spc/Srch_v1.7/index.html. A series of associated software tools has been designed to manage raw spectral files by capturing the essential mass spectral information and converting it into one abbreviated NIST MS tool. A recent update has indicated 91,292 mass spectra requiring only 36.7 MB; this compilation is easily searchable and is mobile.

An application of this tool is detailed for characterizing carbohydrate product ions in the presence of isomeric structures (Figure 16.13) [39]. In the top left corner (A) is an unknown MS^4 product ion mass spectrum (m/z 660) isolated along the pathway m/z 1157→882→660. This latter mass spectrum, Figure 16.13a, was submitted for a reference match in the library. The mass spectra of the two top hits are shown in Figure 16.13b and c, which are labeled Fuc-α(1-2)-Gal-β(**1-4**)-GlcNAc- and Fuc-α(1-2)-Gal-β(**1-3**)-GlcNAc-, respectively. The only difference between these two structures is on the Gal-GlcNAc inter-residue linkage. Comparisons between the unknown and the two reference mass spectra suggest that the unknown could be a mixture of the two isomeric structures. To assess that possibility, simulated mass spectra were constructed by mixing the reference mass spectra with incremental changes in each component. The spectral similarities (R scores) between the unknown and each simulated mass spectrum were calculated subsequently. In Figure 16.13d is shown a plot of R score *vs* the mixture composition: maximum spectral similarity was achieved at the ratio of 13% Fuc-α(1-2)-Gal-β(**1-3**)-GlcNAc- to 87% Fuc-α(1-2)-Gal-β(**1-4**)-GlcNAc-, which suggests the ratio of the two structures. In Figure 16.13e is shown a side-by-side comparison of the mass spectrum of the unknown with the simulated mass spectrum of the mixture (13%:87%); clearly, the two mass spectra exhibit great similarity. This comparison supports the probability that both isomeric compounds are present and suggests their relative ratios in the unknown sample.

16.4.3 Composition Finder

From determined mass/charge ratios, this web-based application generates glycan compositions from MS and MS^n data. Currently, the neutral N-linked and O-linked glycans are supported. The tool returns all possible compositions for a single precursor or MS^n product ions within a user-specified mass error window. For MS^n data, the software considers the constraints of the precursor ion and returns only substructure products. For determining topology, both the observed ion and neutral loss compositions are reported.

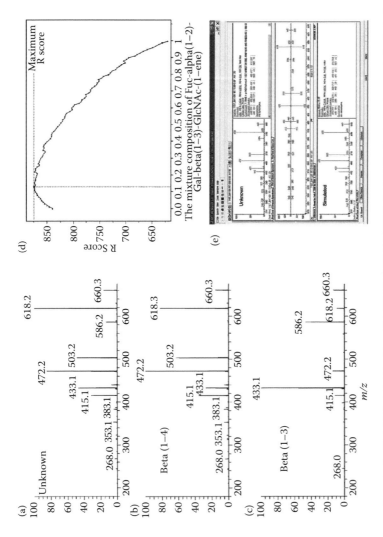

FIGURE 16.13 Library characterization of glycan samples using FragLab Tool kit [31]. Application summarizes carbohydrate fragment identification among the presence of isomeric structures. (a), Mass spectrum obtained using MS⁴ submitted for spectral matching detected (b) and (c) for the structures Fuc-α(1-2)-Gal-β(**1-4**)-GlcNAc- and Fuc-α(1-2)-Gal-β(**1-3**)-GlcNAc-. Comparisons suggest the unknown to be a mixture of the two reference isomers. A simulation, (d), is a plot of R score *vs* the mixture with a ratio of 13% Fuc-α(1-2)-Gal-β(**1-3**)-GlcNAc- to 87% Fuc-α(1-2)-Gal-β(**1-4**)-GlcNAc-. (e) is a spectral comparison of the unknown with the simulated spectrum.

UNH glycomics center glycoinformatics server

[Introduction | query by mass | query in batch | query in breadth | query in depth | reference]

Computation result report:

Released glycan type: N-linked glycan
Glycan reducing terminal type: reduced
Query mass type: monoisotopic
Error tolerance: +/− 0.5

		Observed peak (Na$^+$)		Neutral loss	
Depth	Mass	Composition	Mass	Composition	
MS^1	1800	H6N1R1	n/a	n/a	
MS^2	1506.73	*H6N1-(Pos1DB)1	293.27	*R1-(Pos2/3/4/6OH)1	
MS^3	1084.55	**H4N1-(Pos2/3/4/6OH)1-(Pos1DB)1	422.18	*H2-(Pos1DB)1	
MS^4	839.45	**H4-(Pos2/3/4/6OH)1-(Pos1DB)1	245.1	**N1-(Pos2/3/4/6OH)1-(Pos1DB)1	
MS^5	621.27	***H3-(Pos2/3/4/6OH)2-(Pos1DB)1	218.18	*H1-(Pos1DB)1	
MS^6	431.18	**H2-(Pos2/3/4/6OH)1-(Pos1DB)1	190.09	***H1-(Pos2/3/4/6OH)2-(Pos1DB)1	
MS^7	259.09	*H1-(Pos1OH)1	172.09	no match	

FIGURE 16.14 A screenshot of Composition Finder result on the MS^n pathway of m/z 1800.00→1506.7→1084.5→839.4→621.3→431.2→259.1 from a chicken ovalbumin N-linked glycan sample (Figure 16.5). Abbreviations used in linear notation of ion compositions; Pos1 = position 1 on the ring; DB = double bond; OH = open hydroxyl (likely to occur at different ring positions, except 5 for pyrans).

Designed to handle single MS^n pathways, the tool assists in the determination of glycan topology and, depending on the results, one is able to derive parent ion topology using logic reasoning. As an example, Figure 16.14 is a screenshot for the MS^n pathway of m/z 1800.0→1506.7→1084.6→839.5→621.3→431.2→259.1 from an ovalbumin N-linked glycan sample (Figure 16.5). The sample was prepared as usual and sodium cluster ions were formed. Composition Finder was capable of determining compositions both of neutral losses and of remaining product ions for each MS^n disassembly step. The syntax of the composition coding is: [Type of monomer] [Number of monomer]–[Type of scar*] [Number of scar*]. For instance, the composition of the ion m/z 1506.73 is denoted as H6N1–(Pos1DB)1, which means the glycan product ion has six hexoses, one HexNAc, with one single scar at Position 1 (reducing end) and the scar is a double-bond. By tracking the changes of compositions and using logical reasoning, the putative glycan topology can be derived from this pathway. The tool was developed in Python (http://www.python.org/) and PHP (http://php.net/), the backend relational database management system is MySQL (http://www.mysql.com/) and can be accessed *via* a web interface on the client side and the computation is performed on the server side. Thus, the requirement on client PCs is minimal: a modern web browser is adequate for using the tool.

16.4.4 APPLICATION EXAMPLES

The N-linked glycans obtained from chicken ovalbumin were selected to demonstrate the use the tools jointly. The sample was prepared as usual and MS profiles were run (Figure 16.5). From the MS profile, 21 distinct glycan compositions were identified and compiled. The output data detailed the *m/z*-value, relative intensity, ion type, distribution fitting score, and relative mass error. Each putative composition was selected as a precursor ion for further MS^n disassembly. In this example, the ion *m/z* 1800 (H6N2) was selected. As noted in the previous Section, the product ion mass spectra along the pathway *m/z* 1800.0→1506.7→1084.5→839.5→621.3→431.2 →259.1 were obtained and submitted to Composition Finder in order to determine the corresponding product ions' compositions. The result was returned as is illustrated in Figure 16.14. By tracking the change of compositions along the given pathway and using logical reasoning, the putative glycan topology can be derived. The reasoning process and the proposed topology are summarized in Figure 16.15. The 'cloud' represents the carbohydrate portion whose structural details remain unresolved. As the MS^n disassembly proceeds, the topology hidden in the 'cloud' reveals itself. Starting with the intact glycan, composition H6N1R1, (R denotes the reducing end HexNAc), we are able to differentiate it from the non-reducing end HexNAc (N) because the sample has been reduced. In MS^2, the reducing end HexNAc is detected as a neutral loss. The composition within the cloud shrinks to H6N1; in MS^3, two hexoses are located on the non-reducing end. As a result, the cloud contains H4N1 with two scars; in MS^4, the HexNAc of the N-glycan core is located; in MS^5, another

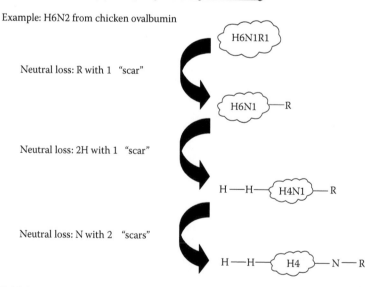

FIGURE 16.15   Change of compositions along the given pathway provides logical reasoning from which the putative glycan topology can be derived. The process and proposed topology are summarized. The 'cloud' represents the carbohydrate portion unresolved which diminishes upon further disassembly.

terminal hexose is detected; and in MS^6, an additional neutral loss of hexose is identified. The number of scars within the cloud drops from three to two. Therefore, three scars must go with this hexose; this conclusion means the hexose should form a branch with the previously detected two arms. Inside the cloud, only two hexoses remain. Finally, in MS^7, the observation of one terminal hexose only suggests that the two hexoses within the cloud should be branched, not linear. Thus, the putative topology of one structure within the ion m/z 1800 has been established.

As mentioned, ions that cannot fit a proposed topology indicate the presence of isomers. Such ions are selected for MS^n disassembly in order to ascertain the structures of all components of the sample. Newly obtained MS^n spectra are compiled using the FragLab Tool Kit and are submitted to the library for identification (Figure 16.16). Any top hits returned are examined to confirm with the topologies

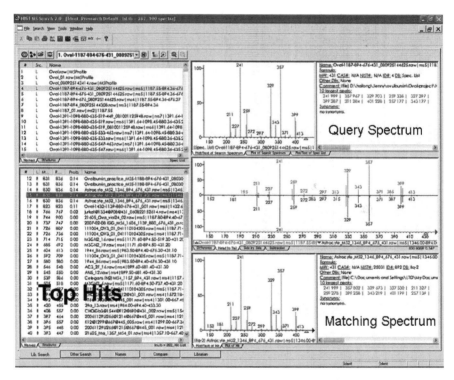

FIGURE 16.16 A screenshot result using a library research tool that compares and contrasts mass spectra. In this case a spectrum from a murine brain astrocyte cell sample contrasted with a glycan spectrum from egg ovalbumin. *Top left*: a list of spectra newly obtained from an egg ovalbumin sample. One spectrum is selected as the query spectrum to search against the FragLib collection. *Bottom left*: the top hits of the spectral library searching result. The query spectrum matches various reference spectra very well. One matching spectrum obtained from a murine brain astrocyte sample previously is highlighted from the hits list. *Top right*: the visualization and the MS^n pathway detail of the query spectrum. *Bottom right*: the visualization and the MS^n pathway detail of the selected matching spectrum. *Middle right*: The Head-to-Tail spectral comparison of the query spectrum and the matching spectrum. The two mass spectra are almost identical.

concluded from logical reasoning. The library search routine searches by mass and pathway providing an opportunity to evaluate each step where selected matches can be closely contrasted (middle right panel, Figure 16.16). Figure 16.16 is a screenshot of one representative library searching result where a similar structure was detected in a murine brain astrocyte and an egg ovalbumin sample.

16.5   CONCLUSION

For resolving molecular complexity in the field of MS, the analytical *coup de grace* has focused persistently on higher resolution and/or chromatographic interfacing. This chapter demonstrates that such attributes have little influence when considering the utmost challenges of a carbohydrate polymer. Here, the brilliant technology of the IT reigns supreme for having a solution long before it could connect to the problem, reminiscent of Pirandello's 'Six Characters in Search of an Author.' This carbohydrate application exemplifies the clever development of IT technology by providing component resolution exceeding the best attempts with chromatography, coupled with structural understanding of linkage, branching, stereo and structural isomers. Who would have dreamed that successive cleavage of moiety after moiety would be possible, and all in the gas-phase?

REFERENCES

1. Turnbull, J.E.; Field, R.A. Emerging glycomics technologies. *Nat. Chem. Biol.* 2007, *3*, 74–77.
2. Service, R.F. Proteomics. Will biomarkers take off at last? *Science* 2008, *321*(5897), 1760.
3. Mechref, Y.; Novotny, M.V. Glycomic analysis by capillary electrophoresis-mass spectrometry. *Mass Spectrom Rev*. Editorial, 2008, DOI: 10.1002/mas.20209.
4. Novotny, M.V.; Soini, H.A.; Mechref, Y. Biochemical individuality reflected in chromatographic, electrophoretic and mass-spectrometric profiles. *J. Chromatogr. B Analyt. Technol. Biomed. Life Sci.* 2008, *866*, 26–47.
5. Volpi, N.; Maccari, F.; Linhardt, R.J. Capillary electrophoresis of complex natural polysaccharides. *Electrophoresis* 2008, *29*, 3095–3106.
6. Gemeiner, P.; Mislovicova, D.; Tkac, J.; Svitel, J.; Patoprsty, V.; Hrabarova, E.; Kogan, G.; Kozar, T. Lectinomics II. A highway to biomedical/clinical diagnostics. *Biotechnol Adv.* 2009, *27*, 1–15.
7. Hirabayashi, J. Concept, strategy and realization of lectin-based glycan profiling. *J. Biochem.* 2008, *144*, 139–147.
8. Park, S.; Lee, M.R.; Shin, I. Chemical tools for functional studies of glycans. *Chem. Soc. Rev.* 2008, *37*, 1579–1591.
9. Domann, P.J.; Pardos-Pardos, A.C.; Fernandes, D.L.; Spencer, D.I.; Radcliffe, C.M.; Royle, L.; Dwek, R.A.; Rudd, P.M. Separation-based glycoprofiling approaches using fluorescent labels. *Proteomics* 2007, 7 Suppl. 1, 70–76.
10. Prien, J.M.; Ashline, D.J.; Reinhold, V.N.; Lapadula, A.J.; Zhang, H. The high mannose glycans from bovine ribonuclease B; Isomer characterization by ion trap MS. *J. Am. Soc. Mass Spectrom.* 2009, *20*, 539–556.
11. Ashline, D.J.; Lapadula, A.J.; Liu, Y.H.; Lin, M.; Grace, M.; Pramanik, B.; Reinhold, V.N. Carbohydrate structural isomers analyzed by sequential mass spectrometry. *Anal. Chem.* 2007, *79*, 3830–3842.

12. Prien, J.M.; Huysentruyt, L.C.; Ashline, D.J.; Lapadula, A.J.; Seyfried, T.N.; Reinhold, V.N. Differentiating N-linked glycan structural isomers in metastatic and non-metastatic tumor cells using sequential mass spectrometry. *Glycobiology* 2008, *18*, 353–366.

13. Jennings, K.R. Collision-induced decompositions of aromatic molecular ions. *Int. J. Mass Spectrom. Ion Phys.* 1968, *1*, 227–235.

14. Haddon, W.F.; McLafferty, F.W. Metastable ion characteristics. VII. Collision-induced metastables. *J. Am. Chem. Soc.* 1968, *90*, 4745–4746.

15. Futrell, J.; Miller, C. Tandem mass spectrometer for study of ion-molecule reactions. *Rev. Sci. Instrum.* 1966, *37*, 1521–1532.

16. Carr, S.A.; Reinhold, V.N.; Green, B.N.; Hass, J.R. Enhancement of structural information in FAB ionized carbohydrate samples by neutral gas collision. *Biomed. Mass Spectrom.* 1985, *12*, 288–295.

17. Roberts, W.L.; Santikarn, S.; Reinhold, V.N.; Rosenberry, T.L. Structural characterization of the glycoinositol phospholipid membrane anchor of human erythrocyte acetylcholinesterase by fast atom bombardment mass spectrometry. *J. Biol. Chem.* 1988, *263*, 18776–18784.

18. Spaink, H.P.; Sheeley, D.M.; van Brussel, A.A.; Glushka, J.; York, W.S.; Tak, T.; Geiger, O.; Kennedy, E.P.; Reinhold, V.N.; Lugtenberg, B.J. A novel highly unsaturated fatty acid moiety of lipo oligosaccharides signals determines host specificity of Rhizobium. *Nature* 1991, *354*, 125–130.

19. Paul, W.; Steinwedel, H. A new mass spectrometer without a magnetic field. *Z. Naturforsch.* 1953, *8A*, 448–450.

20. Paul, W. Electromagnetic traps for charged and neutral particles (Nobel Lecture) *Angew. Chem., Int. Ed. Engl.* 1990, *29*, 739–748.

21. Mechref, Y.; Novotny, M.V.; Krishnan, C. Structural characterization of oligosaccharides using MALDI-TOF/TOF tandem mass spectrometry. *Anal. Chem.* 2003, *75*, 4895–4903.

22. Novotny, M.V.; Mechref, Y. New hyphenated methodologies in high-sensitivity glycoprotein analysis. *J. Sep. Sci.* 2005, *28*, 1956–1968.

23. Harvey, D.J.; Royle, L.; Radcliffe, C.M.; Rudd, P.M.; Dwek, R.A. Structural and quantitative analysis of N-linked glycans by matrix-assisted laser desorption ionization and negative ion nanospray mass spectrometry. *Anal. Biochem.* 2008, *376*, 44–60.

24. Domon, B.; Costello, C.E. A systematic nomenclature for carbohydrate fragmentations in FAB-MS/MS spectra of glycoconjugates. *Glycoconjugate J.* 1988, *5*, 397–409.

25. Harvey, D.J.; Baruah, K.; Scanlan, C.N. Application of negative ion MS/MS to the identification of N-glycans released from carcinoembryonic antigen cell adhesion molecule 1 (CEACAM1). *J. Mass Spectrom.* 2008, *44*, 50–60.

26. Chai, W.; Pistarev, V.; Lawson, A.M. Negative-ion electrospray mass spectrometry of neutral underivatized oligosaccharides. *Anal. Chem.* 2001, *73*, 651–657.

27. Carlson, D.M. Oligosaccharides isolated from pig submaxillary mucin. *J. Biol. Chem.* 1966, *241*, 2984–2986.

28. Rademaker, G.; Haverkamp, J.; Thomas-Oates, J.E. Determination of glycosylation sites in O-linked glycopeptides: a sensitive mass spectrometric protocol. *Org. Mass Spectrom.* 1993, *28*, 1536–1541.

29. Rademaker, G.J. Mass spectrometry: a modern approach to solving biological structural problems, Ph.D. Thesis, Utrecht University, 1996.

30. Rademaker, G.J.; Pergantis, S.A.; Blok-Tip, L.; Langridge, J.I.; Kleen, A.; Thomas-Oates, J. E. Mass spectrometric determination of the sites of O-glycan attachment with low picomolar sensitivity. *Anal. Biochem.* 1998, *257*, 149–160.

31. Hanisch, F-G.; Jovanovic, M.; Peter-Katalinic, J. Glycoprotein identification and localization of O-glycosylation sites by mass spectrometric analysis of deglycosylated/alkylaminylated peptide fragments. *Anal. Biochem.* 2001, *290*, 47–59.

32. Wu, S-L.; Hhmer, A.F.R.; Hao, Z.; Karger, B.L. On-line LC–MS approach combining collision-induced dissociation (CID), electron-transfer dissociation (ETD), and CID of an isolated charge-reduced species for the trace-level characterization of proteins with post-translational modifications. *J. Proteome Res.* 2007, *6*, 4230–4244.

33. Perdivara, I.; Petrovich, R.; Allinquant, B.; Deterding, L.J.; Tomer, K.B. Elucidation of O-glycosylation structures of the beta-amyloid precursor protein by liquid chromatography-mass spectrometry using electron transfer dissociation and collision induced dissociation. *J. Proteome Res.* 2009, *8*, 631–642.

34. Butler, M.; Quelhas, D.; Quelhas, D.; Critchley, A.J.; Carchon, H.; Hebestreit, H.F.; Hibbert, R.G.; Vilarinho, L.; Teles, E.; Matthijs, G.; Schollen, E.; Argibay, P.; Harvey, D.J.; Dwek, R.A.; Jaeken, J.; Rudd, P.M. Detailed glycan analysis of serum glycoproteins of patients with congenital disorders of glycosylation indicates the specific defective glycan processing step and provides an insight into pathogenesis. *Glycobiology* 2003, *13*, 601–622.

35. Nomenclature for representation of glycan structure; consortium for functional glycomics. http://www.functionalglycomics.org/static/consortium/Nomenclature.shtml (accessed December 10, 2009).

36. Fu, D.; Chen, L.; O'Neill, R.A. A detailed structural characterization of ribonuclease B oligosaccharides by 1H NMR spectroscopy and mass spectrometry. *Carbohydr. Res.* 1994, *261*, 173–186.

37. Costello, C.E.; Contado-Miller, J.M.; Cipollo, J.F. A glycomics platform for the analysis of permethylated oligosaccharide alditols. *J. Am. Soc. Mass Spectrom.* 2007, *18*, 1799–1812.

38. Zhao, C.; Xie, B.; Chan, S.Y.; Costello, C.E.; O'Connor, P.B. Collisionally activated dissociation and electron capture dissociation provide complementary structural information for branched permethylated oligosaccharides. *J. Am. Soc. Mass Spectrom.* 2008, *19*, 138–150.

39. Zhang, H.; Singh, S.; Reinhold, V.N. Congruent strategies for carbohydrate sequencing. 2. FragLib: an MS^n spectral library. *Anal Chem.* 2005, *7*, 6263–6270.

17 Collisional Cooling in the Quadrupole Ion Trap Mass Spectrometer (QITMS)

Philip M. Remes and Gary L. Glish

CONTENTS

17.1 Introduction 739
    17.1.1 Helium Bath Gas and the Quadrupole Ion Trap Mass
           Spectrometer (QITMS)... 739
    17.1.2 Why Is Collisional Cooling Important?..................................... 741
    17.1.3 Internal Energy in MS/MS 741
17.2 Theory... 745
17.3 Collisions 746
    17.3.1 Collisional Cooling Rate Constants 748
    17.3.2 Probing Collisional Cooling Experimentally.............................. 751
    17.3.3 Collisional Cooling Rate Constants as a Function of Ion Size 758
17.4 Conclusions.. 762
Acknowledgments.. 763
References 763

17.1 INTRODUCTION

17.1.1 HELIUM BATH GAS AND THE QUADRUPOLE ION TRAP MASS SPECTROMETER (QITMS)

The addition of high pressures of helium (*ca* 1×10^{-3} Torr) to quadrupole ion traps is a practice that is unique to this type of mass analyzer. The fact that a mass spectrometer can operate at all in this pressure regime was surprising to the mass spectrometry (MS) community. Most mass spectrometers operate at background gas pressures at least three and up to eight orders of magnitude lower ($1 \times 10^{-6} - 1 \times 10^{-11}$ Torr) in order to minimize the scattering effects caused by collisions of the ions with background gases. The effects of collisions have been observed since the early days in quadrupole ion trap mass spectrometer (QITMS) development. Increased background gas pressure was first seen to stabilize arrays of charged iron and aluminum

739

particles in a QITMS [1], and was later shown to have a profound effect on both the number of trapped ions and their storage lifetimes [2–5]. The effect of added neutral gas pressure on the peak width and height of N_2^+ ions analyzed with the mass-selective storage technique was investigated, and the peak height was found to maximize at around 1×10^{-4} Torr, while peak width remained largely constant [6]. Simulation work showed that charge-exchange collisions caused ions to concentrate to the center of the trap, decreasing kinetic energy and increasing extraction efficiency [7–9]. Nonetheless, most QITMS systems were in fact operated in the 10^{-5} Torr range for many years. In the early 1980s, when the QITMS was being developed as a gas chromatography detector by Finnigan MAT (now Thermo Fisher Scientific), experiments to determine the effect of the helium carrier gas on mass spectra obtained with the mass-selective instability scan mode yielded the surprising discovery that increasing background pressure into the milliTorr range actually enhanced simultaneously the sensitivity *and* the resolution [10,11]. Both effects have been rationalized on the basis of kinetic damping of the ion motion to the center of the ion trap. This damping increases trapping efficiency and causes the ions to experience a more ideal and homogeneous quadrupolar field, away from the higher-order field perturbations caused by the exit and entrance holes in the end-cap electrodes and electrode truncation. Kinetic damping proved later to be indispensable for achieving high dissociation efficiencies with infrared multiphoton dissociation (IRMPD), see below, collision-induced dissociation (CID) in mass spectrometry/mass spectrometry (MS/MS) [12], and for trapping ions formed externally to the ion trapping volume *via* techniques such as electrospray ionization (ESI) [13] and matrix-assisted laser induced desorption ionization (MALDI) [14–16]. Specific mention must be made to several previous simulation studies of trapping externally formed ions in a QITMS [17–20]; however, these works were mostly not focused on the role of kinetic damping through collisions but instead elaborated on the optimum RF-phase relation and timing for external injection of ions.*

As well as the damping of the ion kinetic energy, a second process caused by a high helium gas pressure is the dissipation of ion *internal* energy.† Calculations give the frequency of helium/ion collisions to be of the order of 20 per millisecond [21], with the typical ion residence time in the QITMS being several tens to hundreds of milliseconds. In fact, during normal 'storage' conditions in a QITMS, one can assume generally that ions have been quasi-thermalized, such that the distribution of ion internal energies can be described with Boltzmann's equation at the temperature of the background gas [22–24]. These hundreds to thousands of helium/ion collisions that occur before an ion is mass analyzed can have appreciable effects on MS and MS/MS spectra. The goal of this chapter is to give the reader a solid foundation for understanding collisional cooling: what it is, why it is important to consider, and how it has been measured theoretically and experimentally.

---

* See also Volume 4, Chapter 18: 'Pressure Tailoring' for Improved Ion Trap Performance, by Dodge L. Baluya and Richard A. Yost.

† See also Volume 4, Chapter 21: Photochemical Studies of Metal Dication Complexes in an Ion Trap, by Guohua Wu, Hamish Stewart and Anthony J. Stace.

17.1.2 WHY IS COLLISIONAL COOLING IMPORTANT?

Although MS is hailed widely for its ability to identify analytes and to obtain structural connectivity information about them, the nature of MS is such that noticeably different mass spectra for the same analytes can be obtained depending upon the instrument parameters. This situation arises because MS, and in particular tandem MS, is fundamentally a chemical reaction kinetics experiment which asks: "What happens to an ion formed from analyte A when excited to internal energy E in the presence of species B and C for time t?" Although the intended result is, usually, to observe an intact A ion or product ions resulting from the unimolecular dissociation of A, the intensity and identity of these product ions may vary, and reactions with species B and C are possible. The result depends on the internal energy of the ion and the time available for these reactions, as can be understood using Rice–Ramsperger–Kassel–Marcus (RRKM) theory [25–27]. The experimental time is a variable which can be measured easily and controlled in a QITMS, but the internal energy of a species is harder to quantify and control. However, when the internal energy and the factors affecting it are known, theory can make reasonable estimates of results, conditions can be optimized for a particular experiment, and the information contained in a mass spectrum is more readily interpretable. Understanding the role that collisional cooling plays in determining ion internal energy is important in this context.

17.1.3 INTERNAL ENERGY IN MS/MS

The data shown in Figure 17.1 illustrate how internal energy can have a dramatic effect on the product ion pathways and intensities in MS/MS. The product ion mass spectrum of the protonated peptide YGGFL after dissociation by conventional CID in a quadrupole ion trap is shown in Figure 17.1a. In a slow heating method such as conventional CID, generally a few product ion species of relatively high mass-to-charge ratio are formed through the dissociation pathways with the smallest activation energies [28]. Figure 17.1b is the product ion mass spectrum of protonated YGGFL using high amplitude short time excitation (HASTE) [29]. The method of

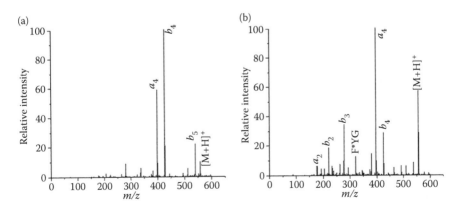

FIGURE 17.1 ESI spectra showing (a) conventional resonance excitation CID mass spectrum of protonated YGGFL; (b) HASTE CID mass spectrum of protonated YGGFL.

HASTE CID is performed with a supplementary voltage a factor of 10–50 greater than that for conventional CID but with activation times of 1–2 ms *versus* standard times of 20–40 ms. In addition, the q_z working point is lowered following the HASTE CID process to allow lower mass-to-charge product ions to be trapped and detected. This technique activates the parent ion to higher internal energies in a shorter period of time, for example 2 ms *versus* 40 ms used for conventional CID. Higher energy product ion formation pathways may be accessed with HASTE CID, and product ions are often formed with enough internal energy to dissociate further. The internal temperature gain of HASTE relative to conventional CID for several protonated peptides has been estimated to be *ca* 100 K [30].

As will be discussed below, collisional activation and collisional cooling are, fundamentally, two outcomes of the same energy transfer process. When the ions have low internal energy and undergo high relative kinetic energy collisions, the most probable outcome is a gain of internal energy. Likewise, when ions have high internal energy and undergo low relative kinetic energy collisions, a loss of internal energy is more probable. Some of the observations from Figure 17.1 can be rationalized on the basis of these facts. For example, when a product ion is formed by CID in a QITMS, immediately this ion is no longer of the correct value of *m/z* for it to be subjected to resonant excitation and it loses kinetic energy rapidly with each collision with the background gas. Collisional cooling effects diminution of both the kinetic and the internal energies such that the likelihood of further dissociation of this product ion is reduced [31]. In this manner, conventional CID product ion mass spectra are, in many cases, dominated by first generation product ions. Energy is added slowly, such that the excess of internal energy over the dissociation threshold is low. When an ion's internal energy is increased slowly, the reaction channels accessed initially are those of lowest activation energy; furthermore, as most ion dissociation reactions are endothermic, the reverse activation energy is less than the forward activation energy such that the product ions formed have little, if any, internal energy. Product ions, in turn, have little excess internal energy when they are formed, and this energy is dissipated rapidly. With HASTE CID the same rationale can be applied, except internal energy is added in larger increments per collision during the activation process, because the collision energy in the center-of-mass frame is greater due to the enhanced ion kinetic energy. Product ions are thereby formed more rapidly with HASTE and with a greater excess of internal energy. The product ions in this case are more likely to dissociate before being cooled below their dissociation threshold, and second and third generation products are more common. Although these are general trends, as the size of the parent ion increases the excess of internal energy in the product ions is greater and multiple generations of product ions may be observed, even with slow heating methods. This observation is due to the increased time required to randomize the internal energy in a large ion, or the 'kinetic shift', leading to a greater excess of internal energy above the dissociation threshold [32].

One of the clearest examples of the influence of collisional cooling in MS is observed in laser dissociation experiments such as IRMPD [33–39]. The technique of IRMPD is an attractive option for MS/MS in a QITMS because it decouples the ion activation event from the ion trapping parameters and thus offers a method for observing low-*m/z* product ions that are not formed and/or retained in conventional CID. During IRMPD, the kinetic energy distribution of the ions stays relatively constant; the ions are not given any supplementary kinetic excitation. As the ion internal energy is raised slowly at first through photon absorption, the most probable outcome of the low relative kinetic energy ion/helium interactions is collisional cooling. At the typical operating pressure of 1×10^{-3} Torr of helium, the rate of internal energy loss *via* collisions can exceed the rate of internal energy gain *via* IR photon absorption, and little or no dissociation is observed for peptide ions at laser powers up to at least 50 W [40]. Several solutions to this problem are possible. The simplest is to decrease the operating pressure in the ion trap, thereby decreasing the rate of internal energy loss through collisions. Mass spectral performance, especially sensitivity (from reduced trapping of injected ions and decreased efficiency of ion detection), deteriorates at lower trapping pressures, and this approach is of limited utility outside of fundamental studies of IRMPD. Another solution is to increase the equilibrium internal energy of the ions by heating the helium gas by 100–200°C. The average kinetic energy of ion and neutral is increased at higher temperature, given approximately by Equation 17.1, where k is Boltzmann's constant and T is the kinetic temperature of the bath gas,

$$ E = \frac{3}{2} kT \tag{17.1} $$

The internal energy of the ions is raised with a heated bath gas, so that a further increase in internal energy up to the dissociation threshold constitutes a smaller excursion from equilibrium than if the ion were heated from room temperature to dissociation. The rate of collisional cooling is, therefore, lower in the heated bath gas case, and ion dissociation proceeds more readily than from room temperature. This technique has been termed thermally assisted IRMPD (TA-IRMPD) [40], and can be carried out experimentally with the addition of a bake-out heating bulb in the vacuum chamber. The utility of TA-IRMPD can be seen from the data in Figure 17.2. At room temperature and normal helium pressure, no dissociation is observed (Figure 17.2a). Dissociation is observed when the helium pressure is decreased (Figure 17.2b), but the sensitivity is increased by an order of magnitude when the normal helium pressure is used and the system is heated to 160°C (Figure 17.2c).

The effects of collisional cooling can be overcome for IRMPD by increasing the photon fluence; obviously using a higher power laser will help, but also focusing the laser beam can increase dissociation efficiency. Other wavelengths of light can also be used to dissociate ions. Ultraviolet (UV) light has been used to dissociate peptide ions [41–43]. Photodissociation at more energetic wavelengths does not suffer from collisional cooling because the dissociation threshold can be surpassed with one photon. UV dissociation is not a slow heating dissociation method, and the fragmentation patterns observed are different than those for IRMPD and CID, which have been shown to form more or less the same product ions [36].

Another example of the influence of collisional cooling takes place during atmospheric pressure matrix-assisted laser desorption ionization (AP-MALDI). Ions are believed to be formed *via* MALDI with an excess of internal energy that is proportional to the difference in gas-phase basicity of the analyte ions *versus* the matrix ions [44]. This relationship has been demonstrated using conventional MALDI with a Fourier transform ion cyclotron resonance (FT-ICR) mass spectrometer [44].

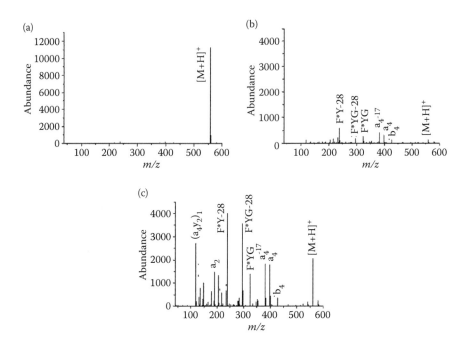

FIGURE 17.2 Mass spectra resulting from the IRMPD of protonated leucine enkephalin under three sets of experimental conditions: (a) normal helium pressure and temperature; (b) reduced helium pressure, room temperature; (c) normal helium pressure, increased temperature. (Reproduced from Payne, A.H. and Glish, G.L., *Anal. Chem.* 2001, *73*, 3542–3548. With permission from the American Chemical Society.)

The experiment used MS/MS to probe the relative internal energy of ions formed with different matrices, and the results could be rationalized on the basis that ions with larger internal energy will fragment with less excitation energy than will ions with less internal energy. However, such a comparison of ions formed using different matrices for AP-MALDI in a QITMS appears to have been made precipitously [45], overlooking the fact that any ion internal energy differences imposed by the matrix have probably been erased by thousands of collisions from the time the ions are formed in the source, at atmosphere, until final mass analysis, in the QITMS. One of the main advantages of AP-MALDI over conventional MALDI is this thermalizing effect, which decreases the amount of analyte fragmentation.

An interesting result of collisional cooling in the QITMS is that differences in ion conformations can be probed, using the knowledge that internal energy equilibration has taken place. It has been shown using surface-induced dissociation (SID), where there is no collisional cooling, that ESI-generated ions have less internal energy than ions that have been generated by liquid secondary ion mass spectrometry (LSIMS) [46]. This disparity in internal energy has been demonstrated by comparing the dissociation efficiency as a function of collision energy. However, in analogous experiments in the QITMS different results are observed. Due to the collisional cooling and quasi-thermalization in the QITMS, it might be assumed that there should be no differences in the fragmentation efficiency of LSIMS- and ESI-generated ions as a function of CID excitation voltage. What has been observed, however, is that LSIMS-generated ions require a little more energy to dissociate than do ESI-generated ions; just the opposite of the expected trend based on internal energy from ionization. This disparity in internal energy has been attributed to different ion conformations with different local minima on the potential energy surface [47].

17.2 THEORY

The dissociation of an ion can be thought of as having an activation step and a dissociation step [48]. In the activation step, shown in Equation 17.2, the internal energy of the parent ion ABC^+ is increased by collisions with the neutral bath gas N; in the dissociation step (Equation 17.3), one or more bonds in ABC^+ are broken, yielding product ions AB^+ and neutral species C [25–26].

$$ABC^+ + N \longrightarrow ABC^{*+} \tag{17.2};$$

$$ABC^{*+} \longrightarrow AB^+ + C \tag{17.3}.$$

The rate of the dissociation step can be understood using RRKM theory. This theory is based on the central premise that the internal energy in a molecule or ion is distributed statistically throughout all its bonds. The RRKM equation, Equation 17.4, takes into account the probability of occupying the various vibrational states at a given internal energy to give a reaction rate constant as a function of internal energy,

$$k(E) = \frac{\sigma N^{\neq}(E - E_0)}{h \cdot \rho(E)} \tag{17.4}.$$

In Equation 17.4, σ is the degeneracy factor, E is the internal energy, E_0 is the critical energy for dissociation, h is Planck's constant, and $N^{\neq}(E-E_0)$ and $\rho(E)$ are, respectively, the summation of individual states for the transition state and the density of states of the parent ion, both calculated by the direct count method [26]. In other words, $N^{\neq}(E-E_0)$ is the number of possible arrangements of internal energy with enough energy to break the bond, and $\rho(E)$ is all the other possible internal energy arrangements.

While computational programs, for example Gaussian 03, can be used to find transition state structures and their vibrational frequencies [49], such calculations are computationally expensive, especially for larger structures. However, the transition state frequencies can be estimated from the parent ion frequencies by deleting a vibration assumed to be the reaction coordinate and scaling the first few frequencies (which have the most effect on $N^{\neq}$) by a factor S [26]. Values of $S > 1$ give a transition state with entropy $\Delta S^{\neq} < 0$, while values of $S < 1$ give a transition state with $\Delta S^{\neq} > 0$. This treatment of the transition state is acceptable because RRKM theory is sensitive only to $\Delta S^{\neq}$, not to the absolute transition state frequencies [23]. When E_0 and S are known for (a) particular dissociation(s), then the calculation of a product ion mass spectrum with a certain internal energy distribution can be performed.

17.3   COLLISIONS

The collision of an ion with a neutral gas atom or molecule is one of the most important means for energy transfer in MS, with the most common application of this
process being CID. When an ion in a QITMS collides with a neutral species such as
a helium atom, kinetic and internal energy may be inter-converted. Since energy is
conserved, Equation 17.5 can be written, where KE_0, KE_F, IE_0, and IE_F are, respectively, the initial kinetic energy, final kinetic energy, initial internal energy, and final
internal energy of the ion (ION) and the colliding helium atom (He). Helium, being
monatomic, has no internal energy unless electronically excited in the collision.

$$KE_0^{ION} + IE_0^{ION} + KE_0^{He} = KE_F^{ION} + IE_F^{ION} + KE_F^{He} \qquad (17.5).$$

The change in internal energy of the ion is the quantity of interest, and so
Equation 17.5 may be rearranged to give Equation 17.6, where Δ denotes the final
value minus the initial value for each type of energy. Consideration of Equation 17.6
shows that an ion may be excited in a collision through a conversion (reduction) of
the kinetic energy of either the ion or of the helium atom and, conversely, an ion may
be cooled by transferring its internal energy into an increase in kinetic energy of the
ion or of the helium atom.

$$\Delta IE^{ION} = -\Delta KE^{ION} - \Delta KE^{He} \qquad (17.6).$$

Whether an ion gains or loses internal energy in a collision depends upon the probability of energy transfer between states, and this dependence is a problem that has
been considered theoretically for a QITMS [22,50,51]. These studies use modeling
to simulate the time evolution of the internal energy of an ion ensemble. The change
in ion internal energy after a collision can be calculated *via* a Monte Carlo method,
using the set of probabilities for energy transfer from the current energy state to all the
other energy states. In a derivation by Plass and Cooks, the energy transfer probability
density is denoted $P(E, K, \Delta E)$, where E is the internal energy, K is the relative kinetic
energy of the ion and a helium atom, and ΔE is the change in internal energy from the
collision [22]. The terms E, K, and η, the conversion efficiency of kinetic to internal
energy for positive internal energy steps, are the main adjustable parameters in this
simulation, where determining the energetic outcome of each collision (ΔE) is the
goal. Factors such as the bath gas mass, temperature, and ion/neutral collision cross-
section are used only to determine the ion/neutral collision rate and, thus, the number
of times that the Monte Carlo program will be called for a given time period.

The basic shape of the $P(E, K, \Delta E)$ function is designed to have the highest value
at $\Delta E = 0$ (elastic collision), and to decrease exponentially on either side, which is
the so-called 'weak collider model' [50]. The $P(E, K, \Delta E)$ curves for *n*-butylbenzene
under different E and K conditions are shown in Figure 17.3. In the plot shown in
Figure 17.3a, E is 0.187 eV, the average internal energy for *n*-butylbenzene at room
temperature, and K is set to various values between 0.037 eV, the average gas kinetic
energy at room temperature, and 0.930 eV. At $K = 0.037$ eV, $P(E, K, \Delta E)$ is roughly

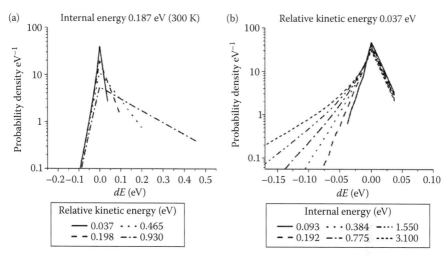

FIGURE 17.3 (a) *n*-Butylbenzene molecular ion probability density curves for collisions at constant room temperature, ion internal energy, and increasing collision energies; (b) probability density curves for constant room temperature, collision energy, and increasing ion internal energy.

symmetric, so that the probability of an energy up-step or energy down-step is the same, that is the ion has reached thermal equilibrium with the helium bath gas. As the collision energy increases, there is a greater probability of an energy up-step, and collisional activation is the dominant process. In the plot shown in Figure 17.3b, a constant $K = 0.037$ eV is set, corresponding to the average kinetic energy of a room temperature gas, and the ion internal energy is increased. In this case, as the ion internal energy is increased the probability of an energy down-step increases, and collisional cooling is the dominant process. Each $P(E, K, \Delta E)$ curve in Figure 17.3 is normalized such that Equation 17.7 holds for given values of E and K,

$$\int_{-E}^{K} P(E,K,\Delta E)d(\Delta E) = 1 \qquad (17.7).$$

The limits of the integration reflect the situation that the most energy that could conceivably be lost in a collision is $-E$, the internal energy in the ion, while the maximum internal energy that could conceivably be gained in a collision is K. To simulate the time-evolution of ion internal energy for collisional cooling, first the total number of collisions taking place can be calculated with a collision theory such as the Langevin model, given in Equation 17.8, where υ is collision frequency (s^{-1}), q_{ion} is ion charge, N_{neut} is gas number density, α_{neut} is the atom/molecule polarizability, μ is the reduced mass of ion and neutral, and ε_0 is the permittivity of free space,

$$\upsilon = 2.210 \cdot \pi \cdot q_{ion} \cdot N_{neut} \cdot \sqrt{\frac{\alpha_{neut}}{(4 \cdot \pi \cdot \varepsilon_0) \cdot \mu}} \qquad (17.8).$$

An initial ion internal energy E, relative kinetic energy K, and up-step energy transfer efficiency η are specified, which, along with the ion vibrational frequencies, determine the probability density distribution. For a given set of conditions, the energy step ΔE for a collision is selected from its normalized distribution (that is, from $P(E, K, \Delta E)$) by selecting a random number from the range 0 to 1. The new ΔE is added to the current internal energy E, which modifies $P(E, K, \Delta E)$ for the next time step in the simulation. The Monte Carlo process is repeated for the number of collisions taking place during the experimental time period. The entire simulation can be repeated many times, and the ensemble of ion internal energies at any point can be used for further calculations.

An interesting application of these calculations was presented in Plass and Cooks' work, where the effect of collisional excitation on QITMS peak shapes was simulated [22]. The large kinetic energies obtained during ejection from the QITMS have been known to cause both collisional dissociation of fragile ions and shifts in observed mass-to-charge ratio, which has been termed 'chemical mass shift' [52–56]. The mechanism at work has been described as arising from a prolonged ejection time in some QITMS geometries that is sufficient for ions to dissociate during the ejection process. Since the ion being ejected is at the low-mass cut-off of the ion trap, product ions of lower mass-to-charge ratio formed in this way will not have stable trajectories and will be ejected slightly before the main group of ions. Plass and Cooks used the above Monte Carlo method embedded in ITSIM [57] software to simulate the ejection and internal energy evolution of n-butylbenzene, and the RRKM theory was used to calculate dissociation rates for n-butylbenzene, which has well-known energetics [58]. The simulated mass spectrum using an up-step transfer efficiency of $\eta = 0.40$ was remarkably similar to the experimental mass spectrum [22].

17.3.1 COLLISIONAL COOLING RATE CONSTANTS

The rate constant for collisional cooling of a vibrationally excited ion can be extracted from a plot of average internal energy *versus* time by fitting the points to an exponential decay equation of the form in Equation 17.9. In this equation, E is the average internal energy, which at time (t) zero is represented by the constants $(A + C)$, B is the collisional cooling rate constant, and C is the internal energy at very long times,

$$E(t) = A \cdot e^{-Bt} + C \tag{17.9}$$

Throughout this discussion, one must be careful to differentiate between the rate of collisional cooling and the collisional cooling rate constant. The rate changes with time: for example, when an excited ion is allowed to cool, initially internal energy is lost quickly and at later times internal energy is lost more slowly as equilibrium is approached. The rate constant, in contrast, describes how fast internal energy decays past a given point, as in Equation 17.10 (derived from Equation 17.9, where B is the rate constant. Equation 17.10 shows that every $1/B$ seconds, the internal energy will fall by a factor of 2.72:

$$\left(\frac{E\left(\frac{1}{B}\right)}{E_{initial}} \right) = \exp(1) = 2.72 \qquad (17.10).$$

The most general mechanisms of energy transfer in a QITMS have been described before [50,59] and are outlined in the following equations, where M^+ is the singly charged precursor ion, M^{+*} is the activated precursor ion, and He is a helium bath gas atom.

$$M^+ + He \xrightarrow{k_{CA}} M^{+*} + He \qquad (17.11),$$

$$M^+ + h\upsilon \xrightarrow{k_{IR}} M^{+*} \qquad (17.12),$$

$$M^{+*} + He \xrightarrow{k_{ccool}} M^+ + He \qquad (17.13),$$

$$M^{+*} \xrightarrow{k_{radiative}} M^+ + h\upsilon \qquad (17.14),$$

$$M^{+*} \xrightarrow{k_{dissociation}} products \qquad (17.15).$$

The total rate of change of the concentration of M^{+*} is the summation

$$\frac{d[M^{+*}]}{dt} = k_{CA}[M^+][He] + k_{IR}[M^+][h\upsilon]$$

$$-k_{ccool}[M^{+*}][He] - k_{radiative}[M^{+*}] - k_{dissociation}[M^{+*}] \qquad (17.16),$$

where the rate constants, k_{CA}, k_{IR}, k_{ccool}, $k_{radiative}$, and $k_{dissociation}$ refer, respectively, to the rates of collisional activation, infrared activation, collisional cooling, radiative emission, and unimolecular dissociation. A simple scenario suitable for modeling is the one where an ion has received an increase in internal energy from a laser photon and is allowed to cool under normal QITMS storage conditions. The assumptions made are that the rate of collisional activation, infrared activation, radiative emission, and dissociation are small, and that the dominant energy transfer process is collisional cooling, such that Equation 17.16 becomes reduced to Equation 17.17,

$$\frac{d[M^{+*}]}{dt} = -k_{ccool}[M^{+*}][He] \tag{17.17}$$

At constant pressure of helium, the pseudo-first order rate constant, k'_{ccool}, for collisional cooling of the vibrationally excited ion is given in Equation 17.18, and has units of s^{-1}. This term is synonymous with 'collisional cooling rate constant'

$$k'_{ccool} = k_{ccool}[He] \tag{17.18}$$

Integration of Equation 17.17 between zero time and time t gives an exponential decay process of the form of Equation 17.19

$$[M^{+*}]_t = [M^{+*}]_0 \cdot e^{-k'_{ccool} \cdot t} \tag{17.19}$$

Assuming $[M^{+*}]$ is proportional to internal energy, Equation 17.19 is interchangeable with Equation 17.9, with the caveat that $[M^{+*}]$ can decay to zero, while the internal energy will decay until the ion reaches thermal equilibrium with the temperature of the helium bath gas, that is, not to zero.

Goeringer and McLuckey used the above equations to calculate rate constants for collisional cooling in a QITMS for a series of alanine–glycine peptide ions $(AG)_n$ $(n = 8–32)$ [50]. Two conditions at 1 mTorr of helium pressure were investigated, one where the ions started with the average internal energy at 450 K, and the other where the ions had the average internal energy at 300 K plus 4 eV, which simulates the absorption of a 308 nm photon. For the ions with the average internal energy corresponding to 450 K, the value of k'_{ccool} ranged from 530 s^{-1} for $(AG)_8$, to 290 s^{-1} for $(AG)_{32}$ [50]. The conclusion can be drawn from these data that smaller ions are cooled collisionally faster than are larger ions. This conclusion could make sense in terms of consideration of the number of degrees of freedom. Generally, more activation energy is necessary to dissociate larger ions than smaller ions, because the activation energy is randomized throughout all the vibrational modes such that each mode receives, on average, less activation energy than for a smaller ion. In the same manner, if roughly the same amount of internal energy is carried away per collision with a helium atom, followed by internal energy randomization, the average energy per vibrational mode will be depleted more slowly in a large ion than in a small ion. Indeed, when the calculated rate constants are plotted as a function of the number of degrees of freedom, the rate constants do not decrease proportionally, but rather the rate constants of high mass ions deviate to the high side of a linear decrease. Two assumptions were made in this modeling work [50] which, if modified, could alter the results: that the efficiency of energy transfer, η, is the same for all sizes of ion, and that each encounter with a helium atom constitutes only one exchange of internal energy. As will be described in Section 17.3.3, experiments have been performed that suggest these assumptions are not entirely valid, and more work needs to be done to understand the relationship between the collisional cooling rate constant and ion size.

17.3.2 Probing Collisional Cooling Experimentally

Collisional cooling can be probed experimentally by raising the internal energy of an ion and monitoring the rate of internal energy loss. Such experiments have been carried out in a selected-ion flow tube (SIFT) mass spectrometer for the collisional cooling of N_2^+ with several noble gases [60]. Laser-induced fluorescence (LIF) was used to monitor the population of vibrational states $v = 0-4$ for N_2^+ ions excited during electron ionization. A specific vibronic transition could be probed with an excimer-pumped dye laser, and fluorescence yield measured.

Another way of determining internal energy loss over time is to measure a chemical property dependent upon internal energy, such as the ion dissociation rate (Equation 17.2). The so-called 'two-pulse' photodissociation experiment falls under this category, and has been used to investigate rate constants for radiative relaxation [61] and collisional relaxation [59,62]. Although there are several variations of the two-pulse method, the general scheme consists of four steps: (1) excite a selected population of ions by a first pulse of photons; (2) allow the ions to dissipate internal energy for a period of time; (3) excite the ions with a second pulse of photons; and (4) measure the ion fragmentation efficiency. Fragmentation efficiency, FE, is defined in Equation 17.20, where $\sum F$ is the sum of the product ion intensities and P is the precursor ion intensity, as determined from the product ion mass spectrum,

$$FE = \frac{\sum F}{\sum F + P} \qquad (17.20).$$

Steps 2–4 are repeated for a suitable range of cooling times. When using fragmentation efficiency as a measure of internal energy, several considerations are important. A change in internal energy does not always result in a proportional change in fragmentation efficiency, as a glance at a RRKM plot of dissociation rate *versus* internal energy will show [25–27]. Generally speaking, the dissociation rate does not change as much at high internal energy, and changes much faster at lower internal energies. Using ion fragmentation as a probe of internal energy places constraints on the range of internal energies that may be observed. An ion may have significantly more internal energy than when at thermal equilibrium with the bath gas, but have less internal energy than that necessary to undergo dissociation. Under this condition the observed rate constant for collisional cooling will be artificially high [63]. When the ion cooling is exponential, then Equation 17.21 gives the relationship between the observed rate constant, k_{obs}, and the actual rate constant, k_{actual}, where E_{limit} is the internal energy limit below which changes cannot be seen and $E_{initial}$ is the initial internal energy

$$k_{actual} = -\ln\left(\frac{E_{limit}}{E_{initial}}\right) k_{obs} \qquad (17.21).$$

The first technical embodiment of the two-pulse method is illustrated in Figure 17.4. This scheme holds the output power of the laser constant, and controls the activation rate through the duration of the laser pulses [64]. The experimental timing diagram is shown in Figure 17.4a. The first laser pulse time is held constant and after each successive cooling time the ions are irradiated with a second laser pulse of variable duration. Figure 17.4b is a depiction of average ion internal energy during the course of the experiment. At the limit of the cooling time between laser pulses, the ions have re-equilibrated with the bath gas and there are no further changes in fragment ion abundance with increasing cooling time. The resultant data are of the form given in Figure 17.4c, which shows plots of fragmentation efficiency *versus* second laser pulse duration for a range of cooling times. Data at a given second laser pulse duration are extracted to create a plot of fragmentation efficiency *versus* cooling time, from which the cooling rate constant can be obtained. One advantage of this method is that the ions can be excited to just under the dissociation threshold by tuning the first laser pulse width. A reproducible starting point for observing the effects of cooling is established and the theoretical complications of dissociation on the cooling rate constant are not encountered until the second laser pulse, which may simplify rationalization of the results. Additionally, the solid curve in Figure 17.4c acts as a standard by which the quality of the experiment may be judged. This curve represents a set of data taken with no first laser pulse. As cooling time between pulses is increased, eventually the once-excited ions are cooled to equilibrium, and the two-pulse curves should approach the solid curve.

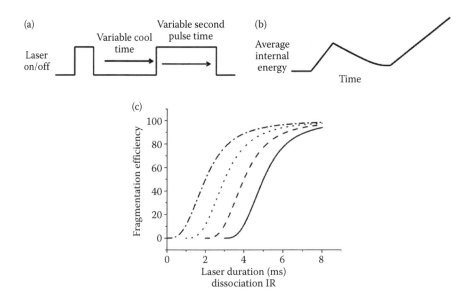

FIGURE 17.4   (a) Experimental timing diagram for the constant laser power, variable cool time, and variable second laser pulse duration method; (b) idealization of average ion internal energy over the course of the experiment; (c) plots of fragmentation efficiency *vs* second laser pulse duration. The dash-dot-dash curve is for a short cooling time, the two dotted and dashed curves are for longer cooling times, and the solid curve is for no first laser pulse.

A variation on the above method for determining cooling rates is to use two laser pulses of constant time width, separated by a variable cooling time. The timing of the experiment is the same, except the second pulse width is fixed. This method reduces greatly the time needed to complete the experiment, so that many repetitions can be taken and averaged within a short length of time. A 'cooling curve' for the protonated molecule ion of the peptide FLLVPLG, taken using this second two-pulse method is shown in Figure 17.5, where the squares are data points and the line is a fit to an exponential decay. The collisional cooling rate constant is extracted from this exponential decay fit. A reproducible initial ion excitation level can be obtained by setting the laser power such that a given fragmentation efficiency is obtained at 0 ms cool time. This initial condition allows for more reproducible results in day-to-day experiments, and can be used also to estimate the initial internal energy level of the ions. As will be described in more detail below, the observed rate constants are to some extent dependent upon the initial ion internal energy level. This second method has been more successful than the first method for characterizing ion cooling at faster time scales (<5 ms).

The first two-pulse cooling measurements were performed by Dunbar and co-workers [65] in FT-ICR mass spectrometers, where the ultra-high vacuum conditions dictate that collisional cooling will be negligible and energy relaxation takes place principally through radiative emission (Equation 17.14). Under these conditions, the two-pulse method yields decay rates corresponding to the rate constant for energy loss by radiative emission. This value has been determined for several ions, and is generally of the order of 1 s^{-1} [61,63,64,66,69]. When the two-pulse experiment is performed in a QITMS, the situation is reversed, and the rate constants are two to three orders of magnitude larger [59,62]. Collisional cooling is the dominant internal energy loss pathway in the QITMS and, under most conditions, the

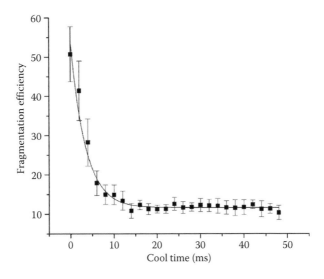

FIGURE 17.5 Plot of fragmentation efficiency as a function of cooling time in the two-pulse photodissociation experiment for the peptide FLLVPLG.

radiative emission process is small enough to ignore. These details were elucidated in a study of collisional cooling rate constants carried out under different temperature and pressure conditions for the $[M + H]^+$ ion of the peptide leucine enkephalin [59]. The plot in Figure 17.6 a summarizes the results for varying temperature while holding the helium pressure constant at 2.8×10^{-4} Torr. The collisional cooling rate constant decreases linearly over the range of temperatures studied (293–438 K). The results for varying pressure at a constant temperature of 438 K are shown in Figure 17.6b. The data shown in 17.6b are evidence that at this temperature the rate of collisional cooling at lower pressures is small enough that radiative emission is a significant contributor to internal energy dissipation. The cooling rate does not change with further reductions in pressure below 2.8×10^{-4} Torr indicating that, for leucine enkephalin, the rate constant for internal energy loss by radiative emission is *ca* 30 s^{-1}.

In contrast, the collisional cooling rate constant was 400 s^{-1} for protonated leucine enkephalin at 1 mTorr when activated for 5 ms with a 50 W CO$_2$ laser [59]. This result agrees with the master equation modeling described above [50], where extrapolation of Goeringer and McLuckey's results for an ion with the number of degrees of freedom of YGGFL yields $k'_{ccool} = 600$ s^{-1} for a 300 K activation.

These experiments were extended to sub-ambient temperatures using a QITMS designed for performing IRMPD spectroscopy [62]. The rate constant for collisional cooling of the molecular ion of *n*-butylbenzene was determined over a range of temperatures from 296 K down to 36 K by the constant laser pulse width method. The ions were excited to 50% fragmentation efficiency with zero cool time, giving a dissociation pattern (*m/z* 91/92 ratio) which corresponds to an average internal energy of 2.14 eV [70]. Calculating average internal energy as a function of internal temperature for *n*-butylbenzene using Equation 17.22 gives *ca* 960 K for this average

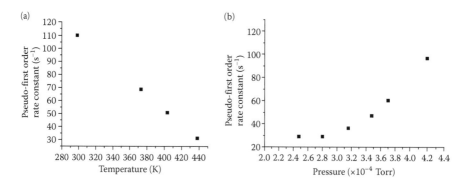

FIGURE 17.6 (a) Plot of pseudo-first-order cooling rate constant for the ESI-generated $[M + H]^+$ ion of leucine enkephalin as a function of temperature at a helium pressure of 2.8×10^{-4} Torr; (b) Plot of pseudo-first-order cooling rate constant for the ESI-generated $[M + H]^+$ ion of leucine enkephalin as a function of helium bath gas pressure at 165°C. (Reproduced from Black, D.M., Payne, A.H., and Glish, G.L., *J. Am. Soc. Mass Spectrom.* 2006, *17*, 932–938. With permission from the American Society for Mass Spectrometry.)

internal energy, where T is internal temperature, k is Boltzmann's constant, and $\rho(E)$ is the density of states

$$\langle E \rangle = \frac{\int\limits_{0}^{\infty} \rho(E) \cdot E \cdot e^{-E/kT} dE}{\int\limits_{0}^{\infty} \rho(E) e^{-E/kT} dE} \tag{17.22}.$$

The results are summarized in the plot in Figure 17.7a, where each curve corresponds to data taken at a different temperature. The far left curve (squares) is for 36 K, the far right curve left-facing triangles) is for 296 K, and the other curves are at intermediate temperatures (for details, see caption). Figure 17.7b shows a plot of the pseudo-first-order rate constants as a function of temperature for a helium pressure of 5.8×10^{-4} Torr. As in Figure 17.6 at higher temperatures, the rate constant increases as the temperature is decreased. The rates appear to be non-linear, especially at lower temperatures, which agrees with measurements of the resonance excitation voltage necessary to reach a given internal energy level as a function of temperature. Although the rate constants are smaller at low pressure, no clear plateau was observed to signify that the collisional cooling rate was less than the rate of radiative emission. The rate constant for radiative emission for n-butylbenzene had previously been found to be 0.97 s^{-1} [70]. The rate constants for collisional cooling of n-butylbenzene at 296 K and at helium pressures of 0.7 and 1 mTorr, respectively, were 420 and 590 s^{-1}, respectively.

The significance of these results is that they give an indication of the time period within which an excited ion will lose its internal energy in a QITMS under different conditions. According to Equation 17.10, at 296 K and typical operating pressures of 0.7–1.1 mTorr of helium, the number of product ions detected for n-butylbenzene

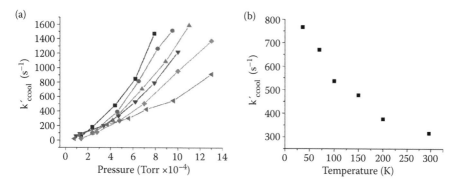

FIGURE 17.7 (a) Plots of pseudo-first-order rate constants *vs* helium pressure determined at 36 K (squares), 70 K (circles), 100 K (up-triangle), 150 K (down-triangle), 200 K (diamond), and 296 K (left-facing triangle); (b) plots of pseudo-first-order rate constants *vs* temperature at a helium pressure of 5.8×10^{-4} Torr. (Reproduced from Remes, P.M. and Glish, G.L., *Int. J. Mass. Spectrom.* 2007, *265*, 176–181. With permission from Elsevier.)

and leucine enkephalin ions is halved every 1–3 ms. The times needed for complete relaxation were observed to be on the order of 5–20 ms. As mentioned above, one must be cautious in relating fragmentation efficiency decay to internal energy decay, so simulations were performed to decipher this relationship.

The Monte Carlo method described above was used to obtain collisional cooling rate constants, derived from internal energies and fragmentation efficiencies, in order that some conclusions can be made about the validity of measurements performed with the two-pulse experimental method. Some initial calculations were performed to choose a value of η, the up-step energy transfer efficiency. Previous work suggested a value of 0.40 [22]. To test this, 2000 ions were allowed to undergo 600 collisions with helium atoms at a pressure of 0.7 mTorr, a relative kinetic energy of $K = 0.037$ eV and an initial internal energy $E = 0.187$ eV. The collision rate at this pressure of helium using Equation 17.8 is 1.357×10^4 s^{-1}. The resulting distributions of internal energies for three values of η, 0.30, 0.35, and 0.40, are shown in Figure 17.8 as segmented lines, and the solid line represents a Boltzmann distribution of internal energies at 296 K. The data for $\eta = 0.35$ appeared to match most closely the Boltzmann distribution, so this was the value chosen for further simulations. Next, the relative kinetic energy that results in an internal energy distribution giving 50% fragmentation efficiency was calculated. Equation 17.5 was used to determine the rates of dissociation of n-butylbenzene by its two major pathways, the rearrangement ion at 92 Th, and simple cleavage ion at 91 Th.

These ions have critical energies for their formation by dissociation of 0.99 eV and 1.61 eV, respectively, and the parent ion vibrational frequencies are also known [58]. Using the room temperature internal energy distribution shown in Figure 17.8 as a

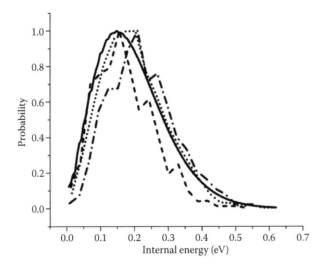

FIGURE 17.8 Internal energy distributions for n-butylbenzene ions, obtained for different values of η, up-step energy transfer coefficient, using initial $E = 0.187$ eV, $K = 0.037$ eV, and 600 collisions with helium atoms. A Boltzmann distribution of internal energies at 296 K is shown as the solid line, the dashed, dotted, and dash-dotted lines correspond to $\eta = 0.30$, 0.35, 0.40, respectively.

starting point, an internal energy distribution yielding 52% fragmentation efficiency was reached after 100 collisions (7.3 ms) at a relative kinetic energy of $K = 0.11$ eV. This collisional excitation at $K = 0.11$ eV simulates the photoexcitation by the laser. The internal energy distribution after 100 collisions represents the situation after two laser pulses and zero cooling time. One laser pulse was taken, therefore, as being equivalent to 50 collisions at $K = 0.11$ eV. This 50-collision distribution was then allowed to cool for increasing amounts of time (numbers of collisions) at $K = 0.037$ eV, from 0 to 300 collisions at 10-collision intervals. After each interval, the ions were subjected to 50 collisions at $K = 0.11$ eV, representing the second laser pulse in the two-pulse experiment. The resulting internal energy distributions illustrating this process are shown in Figure 17.9. The dashed lines with markers represent internal energies after one 'laser pulse' of 50 collisions at $K = 0.11$ eV, and followed by a variable length of cooling time. The solid lines with markers represent internal energies after the second laser pulse. The data points shown as squares are for zero cooling time between pulses, the circles are for 40 collisions (2.95 ms) between pulses, and the triangles are for 80 collisions (5.90 ms) between pulses. The fragmentation efficiency was measured for each of the final distributions (solid lines with markers).

The results of this simulation are shown in Figure 17.10. The squares are for the average internal energy after the first laser pulse and the given amount of cooling time, normalized to 1.043 eV, and the circles are for the fragmentation efficiency that would result from the corresponding ion internal energy distribution undergoing the second laser pulse, normalized to 52%. The best fits to an exponential decay are shown as lines, which give collisional cooling rate constants of 248 s^{-1} for the internal energy and 231 s^{-1} for the fragmentation efficiency. The calculated fragmentation

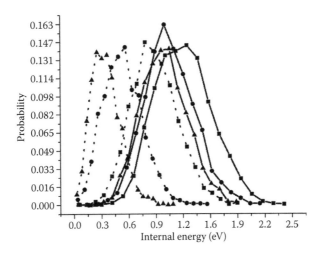

FIGURE 17.9 Internal energy distributions at different points in the two-pulse photodissociation simulation. The dashed lines are internal energies after one simulated 'laser pulse' of 50 collisions at $K = 0.11$ eV and various cooling times. The solid lines with markers are internal energies after the second simulated laser pulse of 50 collisions at $K = 0.11$ eV: squares are for 0 collisions between pulses, circles are for 40 collisions, and triangles are for 80 collisions.

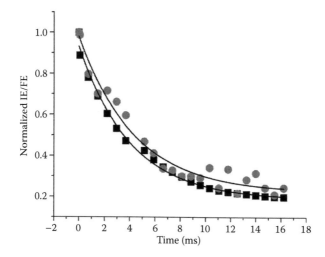

FIGURE 17.10 Results of a simulation of the two-pulse photodissociation experiment showing decay of internal energy and fragmentation efficiency of n-butylbenzene ions: the squares are for internal energy, and the circles are for fragmentation efficiency. The solid lines are the best fit to an exponential decay.

efficiency rate constant is in fair agreement with the experimental two-pulse value, $420 \ s^{-1}$, obtained under these same conditions.

The 'graininess' of the fragmentation efficiency plot (shown as circles) at long cooling times is due to the coarse energy-binning of the internal energy histograms and the limited number of ions used, which was set to 2000 to speed up calculations. The conclusion from these calculations is that, under the influence of collisions with helium atoms, the fragmentation efficiency decays with a similar rate constant to the decay of internal energy for n-butylbenzene. These rate constants, subject as they are to the calculated ion/neutral collision frequency of $13.57 \ ms^{-1}$, are comparable to the experimentally measured values.

17.3.3 COLLISIONAL COOLING RATE CONSTANTS AS A FUNCTION OF ION SIZE

The term 'ion size' is one which may encompass a host of physical properties, among them molecular weight, geometric size, and degrees of freedom. The assumption made in this Section, unless mention is made to the contrary, is that all of these quantities have greater magnitudes when one ion has a larger 'size' than another. The dependence of the collisional cooling process on these properties has not been explored thoroughly as yet. There is a rationale that could support an inherent increase or a decrease in collisional cooling rate constant with ion size. Convincing evidence for a decrease in collision cooling rate constant with increase in the number of degrees of freedom of the ion was shown with a stochastic model for internal energy transfer, as described above [50]. Likewise, the basic physics of inelastic collisions shows that the greater the mass of an ion the more slowly it will cool. The maximum kinetic energy is transferred when two species collide together and form a complex. Consider an ion of mass m_1 and velocity v_1 which collides with a resting neutral particle of mass m_2, forming a complex having velocity v_2. Conservation of momentum leads to Equation 17.23, which, on rearrangement, gives Equation 17.24:

$$m_1 v_1 = (m_1 + m_2) v_2 \qquad (17.23);$$

$$v_2 = \frac{m_1}{(m_1 + m_2)} v_1 \qquad (17.24).$$

The ratio of final kinetic energy KE_f, to initial kinetic energy KE_i is given in Equation 17.25, which is used to give Equation 17.26, the fraction of kinetic energy available for transfer in a collision

$$\frac{KE_f}{KE_i} = \frac{0.5(m_1 + m_2)\left(\frac{m_1}{(m_1 + m_2)} v_1\right)^2}{0.5 m_1 v_1^2} = \frac{m_1}{(m_1 + m_2)} \qquad (17.25)$$

$$\frac{KE_i - KE_f}{KE_i} = \frac{\left(1 - \frac{m_1}{(m_1 + m_2)}\right) \cdot KE_i}{KE_i} = \frac{m_2}{(m_1 + m_2)} \qquad (17.26)$$

This relationship shows that a heavier ion can transfer a smaller fraction of its kinetic energy to internal energy per collision, and should cool more slowly; and vice versa for a heavier neutral species such that the ion cools more rapidly.

Experiments have been performed, however, that suggest an energy transfer efficiency that changes and actually increases with degrees of freedom [32,71]. The underlying principle is a proposed increasing compressibility of the larger molecules [32]. These experimental data are for ions having 45–180 degrees of freedom without an immediate indication of what happens for even larger ions, while the stochastic simulations above were done on ions having 411–1635 degrees of freedom. Trajectory simulations predict that for large ions, several collisions per encounter with a neutral gas atom are possible and even likely [32]. Improvements to the Monte Carlo method could evidently be made with a better understanding of these collisional phenomena. It is worth noting that the Monte Carlo calculations support a limited up-step energy transfer efficiency of $\eta \leq 0.50$. If necessary, however, efficiencies higher than this could be simulated with multiple collisions [50].

An experimental attempt to observe the dependence of ion size on collisional cooling rate constant was made recently [72]. Ions formed by protonation of the peptides GGGG (247 Da), leucine enkephalin (YGGFL, 556 Da), angiotensin III anti-peptide (GVYVHPV, 771 Da), and des-Arg[1] Bradykinin (PPGFSPFR, 905 Da) were investigated using the constant laser pulse width, two-pulse photodissociation method in a Finnigan ITMS™ at 0.7 mTorr. A method of relating the CID voltages to ion internal temperature was applied to these peptides to standardize the initial conditions, deriving an internal temperature from the fragmentation efficiency with zero cool time between laser pulses. This method draws upon the first-order relationship between CID voltage and ion internal temperature in an ideal QITMS, which was described previously, and is given in Equation 17.27, where T_{eff} is the effective temperature, T is the neutral gas temperature, M is the neutral gas mass, m is the ion mass, k is Boltzmann's constant, Γ is a geometric factor of 0.82, e is the fundamental electronic charge, E is the electric field from the auxiliary AC voltage, and $\xi(T)$ is the temperature-dependent reduced collision frequency [72].

$$T_{eff} = T + \frac{M}{3km} \cdot \frac{\Gamma^2 e^2 E^2}{2m\xi(T)^2} \tag{17.27}$$

The plot in Figure 17.11a shows Equation 17.27 applied to a number of peptides at a range of CID excitation levels. It has been noted that the temperature scale is not absolute, and calibration may be necessary [73]; temperature scale calibration was carried out for various ions [74]. A comparison of internal energy as a function of CID voltage using Equation 17.27 *versus* the above temperature scale calibration was undertaken, showing fair agreement, especially at higher temperatures (data not shown). Equation 17.27, while admittedly providing only a rough estimate of ion temperature, allows a standardization of the two-pulse measurements. CID was performed on the above peptide ions at a range of excitation amplitudes, such that the fragmentation efficiency range from 0 to 100% was covered. The plot of fragmentation efficiency *versus* CID excitation amplitude, shown in Figure 17.11b, can be converted to a graph of fragmentation efficiency *versus* ion internal temperature *via* Equation 17.27. The two-pulse photodissociation method

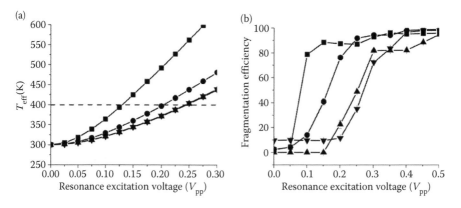

FIGURE 17.11 (a) Plots of ion internal temperature (K) *vs* resonance excitation voltage calculated using Equation 17.23; (b) Plots of fragmentation efficiency *vs* resonance excitation voltage for various singly protonated peptide molecules formed by electrospray ionization. The squares are for the peptide GGGG, circles for YGGFL, up-triangles for GVYVHPV, and down-triangles for FLLVPLG. The dashed, horizontal line in (a) is at $T_{eff} = 400$ K, the initial condition for the two-pulse relaxation measurements.

for measurement of the relaxation rate constant was applied also to the same ions, where the laser power from a 50 W CO_2 laser was adjusted to give a fragmentation efficiency corresponding to a given internal temperature at zero cooling time between laser pulses.

The results of the experiment are shown Figure 17.12. In general for each peptide ion, the pseudo-first order rate constant for collisional cooling becomes smaller as the initial excitation level is increased, that is, more time is needed to cool hot ions below their dissociation thresholds. At a given initial excitation level, the smaller ions require more time to cool than do the larger ones. Part of this effect could be explained by the kinetic shift causing higher dissociation thresholds for the larger ions, which have more degrees of freedom. The observation window of internal energies would be reduced for this reason, and the cooling time for the large ions would be made smaller. The results for the singly-and doubly-charged peptide PPGFSPFR are not the same: the doubly-charged ion takes much longer to cool than does the singly-charged ion. This result was largely unexpected, because both ions have almost the same number of vibrational degrees of freedom and mass. One could argue that the doubly-charged ion has a more extended structure [75] due to charge repulsion, undergoes more collisions and, therefore, cools faster than does the singly charged ion. Yet the data show just the opposite trend. As above, the result can be rationalized on the premise that there is a much lower dissociation threshold for the doubly-charged ion (due to charge repulsion), which allows observation of the

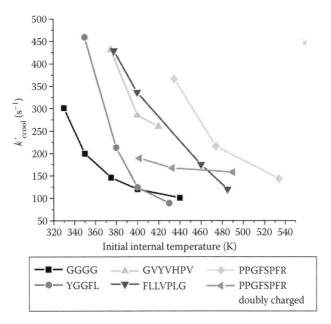

FIGURE 17.12 Plots of pseudo-first-order collisional cooling rate constants for a set of protonated peptides as a function of ion internal temperature at 0 ms cooling time. The squares are for GGGG, circles for YGGFL, up-triangles for GVYVHPV, down-triangles for FLLVPLG, diamonds for singly charged PPGFSPFR, and left-triangles for doubly charged PPGFSPFR.

cooling process at lower internal energies. Part of the differences in collisional cooling rate constants for ions of different size could be due to other factors, however. The increased collision rate for large molecules with their larger collision cross-section could play a role. More significantly, the efficiency of collisional de-activation could be inherently greater for larger molecules, as was proposed for collisional activation [32]. For large ionized molecules, the multiple collisions per ion/neutral encounter and increased compressibility that were demonstrated for internal energy uptake could very well have the inverse effect under de-activating conditions: more internal energy could be carried away per ion/neutral encounter as the size of ions increases.

The above data are a first indication that perhaps larger ions are cooled more efficiently by collisions than are smaller ions. The current experimental/theoretical methods must be modified to reduce the uncertainties arising from the disparate ranges of internal energy that may be observed as number of degrees of freedom changes. More experiments are needed to understand how the collisional cooling process changes with the properties of ion size, that is, collision cross-section, mass, and degrees of freedom. From a practical standpoint, however, larger ions appear to require less cooling time (because of the increased cross-sections and the possibility of multiple ion/neutral collisions per encounter) to stop dissociating than small ions, with the time needed for zero fragmentation of most ions being around 5–20 ms at 0.7 mTorr of helium, becoming even less at higher pressures.

17.4 CONCLUSIONS

The use of a 'high' pressure (*ca* 1 mTorr) bath gas in a mass analyzer is unique to the QITMS. It is well known that the bath gas improves the analytical performance of the instrument, increasing both resolution and sensitivity; this contrasts with the situation for most analytical instruments, where resolution and sensitivity are inversely related, that is increasing resolution decreases sensitivity and *vice versa*. What is often ignored is that when there is kinetic energy cooling there is also internal energy cooling. The internal energy cooling is most often manifested by a reduced extent of dissociation.

In conventional CID in the QITMS, the effect of the reduced extent of dissociation is that fewer dissociation pathways are accessible, not a reduced amount of dissociation of the parent ion. Unlike CID experiments in other types of mass spectrometers, the precursor ion subjected to resonant excitation in the QITMS is accelerated between collisions and thus collisional cooling is not an issue in this stage of the experiment. However, once the precursor ion dissociates, the product ions are not in resonance with the CID excitation voltage and thus they are cooled collisionally, with respect to both kinetic and internal energy. The effect of this process is that product ions are less likely to undergo further dissociation. Thus, QITMS CID product ion mass spectra often have less chemical information than that provided by other types of mass spectrometers because there is less consecutive dissociation. This limitation can be overcome by performing MS^n experiments.

Infrared multiphoton dissociation is an alternative dissociation method to CID and can be very efficient in a QITMS. The effect of collisional cooling is more readily apparent in these experiments because the parent ion is not excited kinetically, only internally. In the extreme case, no dissociation occurs at all because the rate of collisional cooling of the internal energy is faster than the rate of internal energy gain by photon absorption. Rate constants of collisional cooling have been calculated theoretically and measured experimentally. The results from theory and experiment are in reasonable agreement with one another, with collisional cooling rate constants in the hundreds to thousands per second. Thus, to achieve dissociation the internal energy of the precursor has to be raised to a level such that the dissociation rate is greater than the cooling rate under the given experimental conditions.

ACKNOWLEDGMENTS

The authors would like to thank Maria Demireva for obtaining some of the unpublished data presented in this chapter, and Doug Goeringer for help with the Monte Carlo simulations. This work was supported in part by a grant from the National Science Foundation—CHE-0431825.

REFERENCES

1. Wuerker, R.F.; Shelton, H.; Langmuir, R.V. Electrodynamic containment of charged particles. *J. Appl. Phys.* 1959, *30*, 342–349.
2. Major, F.G.; Dehmelt, H.G. Exchange-collision technique for the RF spectroscopy of stored ions. *Phys. Rev.* 1968, *170*, 91–107.
3. Dawson, P.H.; Whetten, N.R. The three-dimensional quadrupole ion trap. *Naturwiss.* 1969, *3*, 109–112.
4. Dehmelt, H.G. Radiofrequency spectroscopy of stored ions. I. Storage. In *Advances in Atomic and Molecular Physics*, eds. D.R. Bates and I. Estermann, Academic Press, New York, 1967, Vol. 3, pp. 53–72.
5. Blatt, R.; Schmeling, U.; Werth, G. Sensitivity of ion traps for spectroscopic applications. *Appl. Phys.* 1979, *20*, 295–298.
6. Dawson, P.H.; Lambert, C. High-pressure characteristics of the quadrupole ion trap. *J. Vac. Sci. Technol.* 1975, *12*, 941–942.
7. Bonner, R.F.; March, R.E.; Durup, J. Effect of charge exchange reactions on the motion of ions in three-dimensional quadrupole electric fields. *Int. J. Mass Spectrom. Ion Phys.* 1976, *22*, 17–34.
8. Bonner, R.F.; March, R.E. Effects of charge-exchange collisions on motion of ions in 3-dimensional quadrupole electric-fields. Part II. Program improvements and fundamental results. *Int. J. Mass Spectrom. Ion Phys.* 1977, *25*, 411–431.
9. Doran, M.C.; Fulford, J.E.; Hughes, R.J.; Morita, Y.; March, R.E.; Bonner, R.F. Effects of charge-exchange reactions on the motion of ions in 3-dimensional quadrupole electric-fields. 3. A 2-ion model. *Int. J. Mass Spectrom. Ion Processes* 1980, *33*, 139–158.
10. Stafford, G. Ion trap mass spectrometry: A personal perspective. *J. Am. Soc. Mass Spectrom.* 2002, *13*, 589–596.
11. Stafford, Jr., G.C.; Kelley, P.E.; Syka, J.E.P.; Reynolds, W.E.; Todd, J.F.J. Recent improvements in and analytical applications of advanced ion trap technology. *Int. J. Mass Spectrom. Ion Processes* 1984, *60*, 85–98.
12. Louris, J.N.; Cooks, R.G.; Syka, J.E.P.; Kelley, P.E.; Stafford, G.C.; Todd, J.F.J. Instrumentation, applications, and energy deposition in quadrupole ion trap mass spectrometry. *Anal. Chem.* 1987, *59*, 1677–1685.

13. Van Berkel, G.J.; Glish, G.L.; McLuckey, S.A. Electrospray ionization combined with ion trap mass spectrometry. *Anal. Chem.* 1990, *62*, 1284–1295.

14. Chambers, D.M.; Goeringer, D.E.; McLuckey, S.A.; Glish, G.L. Matrix-assisted laser desorption of biological molecules in the quadrupole ion trap mass spectrometer. *Anal. Chem.* 1993, *65*, 14–20.

15. Lennon, J.D.I.; Glish, G.L. A MALDI probe for mass spectrometers. *Anal. Chem.* 1997, *69*, 2525–2529.

16. Qin, J.; Chait, B.T. Matrix-assisted laser desorption ion trap mass spectrometry: Efficient isolation and effective fragmentation of peptide ions. *Anal. Chem.* 1996, *68*, 2108–2112.

17. Quarmby, S.T.; Yost, R.A. Fundamental studies of ion injection and trapping of electro-sprayed ions on a quadrupole ion trap. *Int. J. Mass Spectrom.* 1999, *190/191*, 81–102.

18. O, C.S.; Schuessler, H.A. Confinement of ions created externally in a radio-frequency ion trap. *J. Appl. Phys.* 1981, *52*, 1157–1166.

19. Kishore, M.N.; Ghosh, P.K. Trapping of ions injected from an external source into a three-dimensional R.F. quadrupole field. *Int. J. Mass Spectrom. Ion Processes* 1979, *29*, 345–350.

20. Todd, J.F.J.; Freer, D.A.; Mather, R.E. The quadrupole ion store (QUISTOR). XII. The trapping of ions injected from an external source – a description in terms of phase-space dynamics. *Int. J. Mass Spectrom. Ion Phys.* 1980, *36*, 371–386.

21. March, R.E. An introduction to quadrupole ion trap mass spectrometry. *J. Mass Spectrom.* 1997, *32*, 351–369.

22. Plass, W.R.; Cooks, R.G. A Model for energy transfer in inelastic molecular collisions applicable at steady state or non-steady state and for an arbitrary distribution of collision energies. *J Am. Soc. Mass Spectrom.* 2003, *14*, 1348–1359.

23. Laskin, J.; Byrd, M.; Futrell, J. Internal energy distributions resulting from sustained off-resonance excitation in FTMS. I. Fragmentation of the bromobenzene radical cation. *Int. J. Mass Spectrom.* 2000, *195/196*, 285–302.

24. Asano, K.G.; Goeringer, D.E.; McLuckey, S.A. Thermal dissociation in the quadrupole ion trap: Ions derived from leucine enkephalin. *Int. J. Mass Spectrom.* 1999, *185/186/187*, 207–219.

25. Baer, T.; Hase, W.L. *Unimolecular Reaction Dynamics,* Oxford University Press, Oxford, 1996.

26. Baer, T.; Mayer, P.M. Statistical Rice-Ramsperger-Kassel-Marcus quasi-equilibrium theory calculations in mass spectrometry. *J. Am. Soc. Mass Spectrom.* 1997, *8*, 103–115.

27. McLafferty, F.W.; Turecek, F. *Interpretation of Mass Spectra,* University Science Books, Mill Valley, CA, 1993.

28. McLuckey, S.A. Principles of collisional activation in analytical mass spectrometry. *J. Am. Soc. Mass Spectrom.* 1992, *3*, 599–614.

29. Cunningham, Jr., C.; Glish, G.L.; Burinsky, D.J. High amplitude short time excitation: A method to form and detect low mass product ions in a quadrupole ion trap mass spectrometer. *J. Am. Soc. Mass Spectrom.* 2006, *17*, 81–84.

30. Remes, P.M.; Cunningham, Jr., C.; Glish, G.L. Theoretical estimation of peptide internal temperature during collision induced dissociation in a quadrupole ion trap. *Proc. 53rd ASMS Conference on Mass Spectrometry and Allied Topics,* San Antonio, TX, June 2005. Abstract TP12.

31. Asano, K.G.; Goeringer, D.E.; Butcher, D.J.; McLuckey, S.A. Bath gas temperature and the appearance of ion trap tandem mass spectra of high-mass ions. *Int. J. Mass Spectrom.* 1999, *190/191*, 281–293.

32. Marzluff, E.M.; Beauchamp, J.L. Collisional activation studies of large molecules. In *Large Ions: Their Vaporization, Detection and Structural Analysis,* eds. T. Baer, C-Y. Ng, and I. Powis, John Wiley and Sons, Chichester, UK, 1996, 115–143.

33. Grant, E.R.; Schulz, P.A.; Sudbo, A.S.; Shen, Y.R.; Lee, Y.T. Is multiphoton dissociation of molecules a statistical thermal process? *Phys. Rev. Lett.* 1978, *40*, 115–118.

34. Black, J.G.; Yablonovitch, E.; Bloembergen, N. Collisionless multiphoton dissociation of SF_6: A statistical thermodynamic process. *Phys. Rev. Lett.* 1977, *38*, 1131–1134.

35. Gabryelski, W.; Li, L. Photoinduced dissociation of electrospray-generated ions in an ion trap/time-of-flight mass spectrometer using a pulsed CO_2 laser. *Rapid Commun. Mass Spectrom.* 2002, *16*, 1805–1811.

36. Goolsby, B.J.; Brodbelt, J.S. Tandem infrared multiphoton dissociation and collisionally activated dissociation techniques in a quadrupole ion trap. *Anal. Chem.* 2001, *73*, 1270–1276.

37. Little, D.P.; Speir, J.P.; Senko, M.W.; O'Connor, P.B.; McLafferty, F.W. Infrared multiphoton dissociation of large multiply charged ions for biomolecule sequencing. *Anal. Chem.* 1994, *66*, 2809–2815.

38. Hughes, R.J.; March, R.E.; Young, A.B. Multiphoton dissociation of ions derived from 2-propanol in a quistor with low-power CW infrared laser radiation. *Int. J. Mass Spectrom. Ion Phys.* 1982, *42*, 255–263.

39. Young, A.B.; March, R.E.; Hughes, R.J. Studies of infrared multiphoton dissociation rates of protonated aliphatic alcohol dimers. *Can. J. Chem.* 1985, *63*, 2324–2331.

40. Payne, A.H.; Glish, G.L. Thermally assisted infrared multiphoton photodissociation in a quadrupole ion trap. *Anal. Chem.* 2001, *73*, 3542–3548.

41. Cui, W.; Thompson, M.S.; Reilly, J.P. Pathways of peptide ion fragmentation induced by vacuum ultraviolet light. *J. Am. Soc. Mass Spectrom.* 2005, *16*, 1384–1398.

42. Thompson, M.S.; Cui, W.; Reilly, J.P. Factors that impact vacuum ultraviolet photofragmentation of peptide ions. *J. Am. Soc. Mass Spectrom.* 2007, *18*, 1439–1452.

43. Thompson, M.S.; Cui, W.; Reilly, J.P. Fragmentation of singly-charged peptides by photodissociation at $\lambda = 157$ nm. *Angew. Chem. Int. Edn.* 2004, *43*, 4791–4794.

44. Stevenson, E.; Breuker, K.; Zenobi, R. Internal energies of analyte ions generated from different matrix-assisted laser desorption/ionization matrices. *J. Mass Spectrom.* 2000, *35*, 1035–1041.

45. Konn, D.O.; Murrell, J.; Despeyroux, D.; Gaskell, S.J. Comparison of the effects of ionization mechanism, analyte concentration, and ion "cool-times" on the internal energies of peptide ions produced by electrospray and atmospheric pressure matrix-assisted laser desorption ionization. *J. Am. Soc. Mass Spectrom.* 2005, *16*, 743–751.

46. Jones, J.L.; Dongré, A.R.; Somogyi, A.; Wysocki, V.H. Sequence dependence of peptide fragmentation efficiency curves determined by electrospray ionization/surface-induced dissociation mass spectrometry. *J. Am. Chem. Soc.* 1994, *116*, 8368–8369.

47. Danell, A.S.; Glish, G.L. Evidence for ionization-related conformational differences of peptide ions in a quadrupole ion trap. *J. Am. Soc. Mass Spectrom.* 2001, *12*, 1331–1338.

48. Douglas, D.J., Mechanism of the collision-induced dissociation of polyatomic ions studied by triple quadrupole mass-spectrometry. *J. Phys. Chem.* 1982, *86*, 185–191.

49. Foresman, J.B.; Frisch, A. *Exploring Chemistry with Electronic Structure Methods,* Gaussian, Pittsburgh, PA, 1993.

50. Goeringer, D.E.; McLuckey, S.A. Relaxation of internally excited high-mass ions simulated under typical quadrupole ion trap storage conditions. *Int. J. Mass Spectrom.* 1998, *177*, 163–174.

51. Laskin, J.; Futrell, J.H. On the efficiency of energy transfer in collisional activation of small peptides. *J. Chem. Phys.* 2002, *116*, 4302–4310.

52. Li, H.; Plass, W.E.; Patterson, G.E.; Cooks, R.G. Chemical mass shifts in resonance ejection experiments in the quadrupole ion trap. *J. Mass Spectrom.* 2002, *37*, 1051–1058.

53. Wells, J.M.; Plass, W.R.; Patterson, G.E.; Ouyang, Z.; Badman, E.R.; Cooks, R.G. Chemical mass shifts in ion trap mass spectrometry: Experiments and simulations. *Anal. Chem.* 1999, *71*, 3405–3415.

54. Wells, J.M.; Plass, W.R.; Cooks, R.G. Control of chemical mass shifts in the quadrupole ion trap through selection of resonance ejection working point and RF scan direction. *Anal. Chem.* 2000, *72*, 2677–2683.

55. Cleven, C.D.; Cox, K.A.; Cooks, R.G.; Bier, M.E. Mass shifts due to ion/ion interactions in a quadrupole ion-trap mass spectrometer. *Rapid Commun. Mass Spectrom.* 1994, *8*, 451–454.

56. Cox, K.A.; Cleven, C.D.; Cooks, R.G. Mass shifts and local space charge effects observed in the quadrupole ion trap at higher resolution. *Int. J. Mass Spectrom. Ion Processes* 1995, *144*, 47–65.

57. Forbes, M.; Sharifi, M.; Croley, T.; Lausevic, Z.; March, R.E. Simulation of ion trajectories in a quadrupole ion trap: A comparison of three simulation programs. *J. Mass Spectrom.* 1999, *34*, 1219–1239.

58. Baer, T.; Dutuit, O.; Mestdagh, H.; Rolando, C. Dissociation dynamics of normal-butylbenzene ions – the competitive production of m/z 91-fragment and 92-fragment ions. *J. Phys. Chem.* 1988, *92*, 5674–5679.

59. Black, D.M.; Payne, A.H.; Glish, G.L. Determination of cooling rates in a quadrupole ion trap. *J. Am. Soc. Mass Spectrom.* 2006, *17*, 932–938.

60. Kato, S.; Bierbaum, V.M.; Leone, S.R. Laser fluorescence and mass spectrometric measurements of vibrational relaxation of N_2^+ (v) with He, Ne, Ar, Kr, and Xe*. *Int. J. Mass Spectrom. Ion Processes* 1995, *149/150*, 469–486.

61. Dunbar, R.C. Infrared radiative cooling of isolated polyatomic molecules. *J. Chem. Phys.* 1989, *90*, 7369–7375.

62. Remes, P.M.; Glish, G.L. Collisional cooling rates in a quadrupole ion trap at sub-ambient temperatures. *Int. J. Mass. Spectrom.* 2007, *265*, 176–181.

63. Dunbar, R.C. Infrared radiative cooling of gas-phase ions. *Mass Spec. Rev.* 1992, *11*, 309–339.

64. Faulk, J.F.; Dunbar, R.C. Ion cyclotron resonance ion trap measurements of energy relaxation in gas-phase ions: Three techniques compared for thiophenol ion. *J. Phys. Chem.* 1989, *93*, 7785–7789.

65. Dunbar, R.C.; Chen, J.H. Two-pulse photodissociation measurement of infrared fluorescence relaxation in bromobenzene ion. *J. Phys. Chem.* 1984, *88*, 1401–1404.

66. Asamoto, B.; Dunbar, R.C. Observation of the infrared radiative relaxation of iodobenzene ions using two-light-pulse photodissociation. *J. Phys. Chem.* 1987, *91*, 2804–2807.

67. Dunbar, R.C.; Chen, J.H.; So, H.Y.; Asamoto, B. Infrared fluorescence relaxation of photoexcited gas-phase ions by chopped-laser two-photon dissociation. *J. Chem. Phys.* 1987, *86*, 2081–2086

68. Dunbar, R.C. Time-resolved photodissociation study of relaxation processes in gas-phase styrene ion. *J. Chem. Phys.* 1989, *91*, 6080–6085.

69. Dunbar, R.C.; Zaniewski, R.C. Infrared multiphoton dissociation of styrene ions by low-power continuous CO_2 laser irradiation. *J. Chem. Phys.* 1992, *96*, 5069–5075.

70. Uechi, G.T.; Dunbar, R.C. Thermometric study of CO_2 laser heating and radiative cooling of n-butylbenzene ions. *J. Chem. Phys.* 1993, *98*, 7888–7897.

71. Marzluff, E.M.; Campbell, S.; Rodgers, M.T.; Beauchamp, J.L. Collisional activation of large molecules is an efficient process. *J. Am. Chem. Soc.* 1994, *116*, 6947–6948.

72. Demireva, M.P.; Remes, P.M.; Glish, G.L. Internal energy relaxation of different mass peptide ions in quadrupole ion trap mass spectrometry. *Proc. 55th ASMS Conference on Mass Spectrometry and Allied Topics,* Indianapolis, IN, June 3–7, 2007. Abstract MPF098.

73. Goeringer, D.E.; McLuckey, S.A. Evolution of ion internal energy during collisional excitation in the paul ion trap: A stochastic approach. *J. Chem. Phys.* 1996, *104*, 2214–2221.

74. Gabelica, V.; Karas, M.; De Pauw, E. Calibration of ion effective temperatures achieved by resonant activation in a quadrupole ion trap. *Anal. Chem.* 2003, *75*, 5152–5159.

75. Valentine, S.J.; Counterman, A.E.; Clemmer, D.E. A database of 660 peptide ion cross sections: Use of intrinsic size parameters for bona fide predictions of cross sections. *J. Am. Soc. Mass Spectrom.* 1999, *10*, 1188–1211.

18 'Pressure Tailoring' for Improved Ion Trap Performance

Dodge L. Baluya and Richard A. Yost

CONTENTS

18.1　Overview..769
18.2　Role of Buffer Gas in Ion Trap Operation 770
　　　18.2.1　Original Discovery..770
　　　18.2.2　Ion Injection and Trapping Efficiency: Role of Buffer Gas 770
　　　18.2.3　Collision-Induced Dissociation: Role of Buffer Gas 771
　　　18.2.4　Reagent Gas: Use of Pulsed Valves..772
18.3　The Concept of Pressure Tailoring...773
18.4　Experimental Section 776
　　　18.4.1　Argon/Nitrogen Charge-Exchange 780
　　　18.4.2　Pulsed Helium Buffer Gas Effects..780
　　　18.4.3　Pressure Effects on Fragile Ions 782
18.5　Results and Discussion 782
　　　18.5.1　Optimization of the Gas Pulse Profile 782
　　　18.5.2　Multi-Pulse Experiments 785
　　　18.5.3　Pressure Effects on Fragile Ions 788
18.6　Summary and Conclusions 789
References..790

18.1　OVERVIEW

The presence of a buffer gas in quadrupole ion traps (QITs) cools injected ions by reducing the ions' kinetic energy, thereby enhancing trapping efficiency, and serves also as the collision gas for collision-induced dissociation (CID) in tandem mass spectrometry. However, during other events of the analytical scan, having buffer gas inside the QIT can be disadvantageous. During the mass isolation and ejection events, fragile ions can collide with buffer gas atoms/molecules with enough kinetic energy to cause fragmentation and to degrade mass resolution. Currently, commercially available QITs maintain a constant buffer gas pressure of *ca* 1 mTorr, a compromise between efficient cooling and limiting possible fragmentation of ions.

Previous research has shown significant improvement of mass spectrometric (MS) analysis with the QIT when pulsed buffer gas introduction techniques have been employed [1–4], but all these studies have employed single gas pulses from a single pulsed valve. This pulsed technique, however, is not used commercially due to the added costs and complexity of operation. Furthermore, precise control of the buffer gas pressure cannot be achieved by a single valve alone. This chapter explores the potential of using multiple pulsed valves to tailor the buffer gas pressure during each segment of the MS scan function.

18.2 ROLE OF BUFFER GAS IN ION TRAP OPERATION

18.2.1 ORIGINAL DISCOVERY

In the 1950s when Paul and Steinwedel invented the quadrupole ion trap and the quadrupole mass filter [5], but the use of buffer gas was not even considered in their design. There were several investigations into the influence of the presence of trace gases on the behavior of trapped ions in the ensuing two decades. It was only in 1984, when the quadrupole ion trap was initially presented as a commercial detector for GC, that the use of buffer gas at significant pressures was introduced as an aid to improving mass spectral performance.* The presence of 1 mTorr of helium in the trap was not by design, but rather the result of GC carrier gas introduced (along with the compounds eluting from the GC column) directly into the trap with limited vacuum pumping (a 60 L s⁻¹ turbo pump) to minimize the size and price of the instrument. The ability of this lightweight gas (helium or hydrogen) to enhance both mass resolution and sensitivity was an unexpected observation [6,7]. At that time, the reasons behind the improvement had not been fully explained, but it was suspected that the buffer gas had a major effect on ion injection, trapping, and ejection [8].

18.2.2 ION INJECTION AND TRAPPING EFFICIENCY: ROLE OF BUFFER GAS

Buffer gas plays a key role in trapping ions, especially those which are created externally and injected into the trap. Before the importance of using buffer gas was established, ions were created within the cavity of the ion trap. This ion-generation technique did not allow for the flexibility of coupling ionization sources such as electrospray ionization (ESI) or matrix-assisted laser desorption ionization (MALDI). On the other hand, creating ions externally away from the ion trap prevents or reduces ion/molecule reactions inside the trap. With neutral (sample) molecules being introduced outside (rather than within) the ion trap volume, the population of neutrals inside the trap can be minimized, thereby reducing ion/molecule reactions during trapping and mass analysis steps. Several groups have looked into the fundamentals of trapping externally created ions; the general finding in those studies was that reduction of the kinetic energy of injected ions was necessary for trapping [9–12]. This kinetic energy, which is needed in order for the ions to penetrate the trap RF

---

* See Volume 4, Chapter 17: Collisional Cooling in the Quadrupole Ion Trap Mass Spectrometer (QITMS), by Philip M. Remes and Gary L. Glish.

field, is also sufficient enough to allow them to escape. A number of collisions with a low molecular weight gas can reduce the kinetic energy. Within a short period after ion injection, the displacement of the ions collapses to the trap center. The center is where field imperfections are at a minimum, thus diminishing ion losses due to nonlinear resonances [13].

On the other hand, having a pressure of buffer gas that is too high can be detrimental to ion trap performance. The real problem is the loss of resolution due to collisional broadening of the peak (in time) during the mass-selective instability ejection. A tight packet of ions (of a particular m/z-value) being ejected toward the detector can experience collisions, delaying the ejection time and resulting in peak broadening [14].

Another reason for peak broadening is ion fragility, which has been reported previously to be a cause of mass shifts [15–18]. As the name suggests, fragile ions tend to dissociate easily into fragment ions. Ion fragility becomes more problematic in ion trap mass spectrometry during resonant ejection, when the fragile ions approach the border of the stability diagram. The ions gain kinetic energy, experience more energetic collisions with the buffer gas and, if fragile, can dissociate into fragment ions prior to ejection. The working points of fragment ions are already beyond the boundary of the stability diagram and thus fragment ions will be ejected earlier than the fragile parent ion. The resulting effect on the mass spectrum is 'peak fronting' or mass shift to lower apparent mass, which is intensified with the increase in the resonance voltage amplitude [18]. Ion fragility can lead also to ion losses during mass isolation for tandem mass spectrometry (MS/MS).

Studies using gases other than helium as a buffer gas have been conducted. The addition of small amounts of heavy target gases (neon, argon, krypton, or xenon) to the helium buffer gas improves trapping efficiency of cesium iodide cluster ions of high m/z ratio as well as improving performance in CID [19]. Other researchers have attempted to replace helium entirely with air or argon as a buffer gas. The use of heavier gases was found to improve sensitivity and CID efficiency at the expense of mass resolution [20].

18.2.3 Collision-Induced Dissociation: Role of Buffer Gas

Another aspect of QIT mass spectrometry (MS) that highlights the importance of buffer gas is tandem MS, wherein structures of ions can be elucidated further by fragmentation using CID [21]. The technique involves isolation of mass-selected ions after ionization, and resonant excitation of those ions *via* the application of a supplementary sinusoidal potential between the end-cap electrodes. This potential, which is called the resonant excitation or 'tickle' voltage, has a frequency that is tuned to the selected ion's fundamental secular frequency. The amplitude of the tickle voltage is small enough just to move the ions away from the trap center but not to be ejected. In this way the ions acquire kinetic energy from the RF drive potential and collide with neutral buffer gas atoms. The resulting collisions, if energetic enough, produce product ions, which are trapped and then analyzed in the detection step. Effective fragmentation requires both the use of a buffer gas and the application of the tickle voltage. Since it is straightforward to optimize the tickle voltage, there have been a few attempts to optimize the buffer gas pressure for higher CID efficiency. Furthermore, the optimum pressure for CID is limited to a few mTorr because higher pressures during other events in the scan function can compromise mass resolution and sensitivity [16].

Another disadvantage of higher buffer gas pressures for CID is lower internal energy deposition, which becomes a problem for CID experiments of larger m/z-value ions of biomolecules [14]. Higher m/z-value ions typically require greater energy deposition to induce fragmentation, which is difficult to achieve at higher pressures. Increased buffer gas pressure lowers the distance of travel before a collision can occur, thereby lowering the number of energetic collisions; the increased frequency of collisions with buffer gas atoms can also cause de-excitation of previously excited ions.

While it is practical to use the existing buffer gas as the CID gas, some studies have shown advantages of using gases heavier than helium for more effective fragmentation [2]. Typically, CID is carried out at q_z-values between 0.2 and 0.6 for high fragmentation efficiency [19,21,22]. Fragmentation of higher mass ions, such as peptide ions, is often performed at lower q_z-values in order to trap the parent ions and observe lower m/z-value product ions, but at the expense of lower excitation energy. A study showed that peptide ion CID fragmentation efficiency was increased when heavy gases, such as argon and xenon, were introduced *via* a pulsed valve during CID [2]. It was noted also that the use of a pulsed valve for CID gas delivery was important to avoid the negative influence of heavy gases acting as the buffer gas during other periods in the ion trap scan function.

18.2.4 REAGENT GAS: USE OF PULSED VALVES

Applications using reagent gases introduced directly into the ion trap for targeted ion/molecule reactions have been reported; these applications take advantage of the QIT's ability to isolate analyte ions and to serve as a reaction chamber for gas-phase reactions [23,24]. Pulsed introduction of reagent gas into a QIT was reported as a method for enhancing control over reactions occurring inside the trap [4]. Unwanted ions can be removed by using mass-isolation techniques, but elimination of neutral molecules required a different approach. Neutrals can be eliminated only by the action of the high vacuum pump. In an investigation to locate the positions of carbon–carbon double bonds in alkenes, ionized alkenes were isolated prior to injecting neutral reagents. In this way, ion/molecule reactions were more controlled and, therefore, reduced the product ion complexity [25]. Another application is the differentiation of enantiomers within the QIT by mass-selecting chiral reactant ions and generating Diels–Alder reaction products [26]. Although it was reported to be inconclusive as regards enhanced chiral selectivity, the importance of having a reproducible means of injecting neutral reagents into the trap was demonstrated.

For all of the different roles of gases in the ion trap, a common theme is that different events in the scan function require different pressures, or even different gases. The continuous introduction of gases, particularly buffer gas, is a compromise between the requirements of trapping and detection. Moving to a more dynamic approach would permit optimization of the pressures for the individual events in the scan function. The tailoring of buffer gas pressure is designed to maximize the efficiency of each event in the analytical scan.

18.3 THE CONCEPT OF PRESSURE TAILORING

A typical scan function for obtaining a full scan mass spectrum with the ion trap mass spectrometer, also known as a timing diagram, is shown in Figure 18.1 [27]. The scan function is a representation of the temporal variation of a sequence of events that takes place in order to obtain a mass spectrum. To perform a MS analysis, the sample has to be ionized and the analyte ions must be guided into the trap, cooled, and then ejected for detection. This series of steps involves the critical timing of various potentials applied to the lenses, QIT electrodes, and electron multiplier. The scan function is embedded in the software that controls the QIT instrument; some values can be changed by the user. The steps are classified arbitrarily into four major categories: pre-injection, injection, post-injection, and detection events. Each step has a particular time period, which may differ from scan to scan, depending on the experimental settings, for example, automatic gain control (AGC) mode, scanned mass range, etc.

The pre-injection event initializes the electrode potentials to ensure the same starting values on each scan; in particular, the gate lens, as well as the multiplier, is turned off to prevent ions from being injected into the trap and being detected erroneously. This step is important when the user needs to implement certain functions (for example, triggering a pulsed valve) prior to ionization, thus, any commands related to that procedure can be inserted in this event.

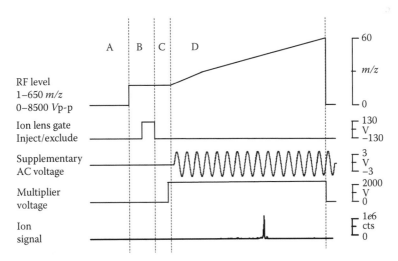

FIGURE 18.1 A typical scan function for the QIT, which is categorized into four divisions: (A), pre-injection; (B), injection; (C), post-injection; and (D), detection event. Also shown are the potentials associated with each step. The traces are not to scale. (Adapted from Yates, N.A., *Ph.D. Dissertation*, University of Florida, 1994.)

The injection event involves turning on the filament, for instance, and the gate lens to produce and to inject ions into the trap. The duration of this portion of the scan function depends upon whether the user is using AGC or fixed ionization times. AGC is used to prevent the ion trap from overfilling with ions, in order to avoid space charging [28]. The AGC method adjusts the injection time by pre-sampling with a fixed injection time and fast scan out, comparing the resulting current with a preset target value, and adjusting the injection time based on the difference.

The post-injection event involves the cooling phase, which comes immediately after the end of injection. This event allows time for the kinetically energetic ions to cool down by colliding with buffer gas inside the trap, increasing the chances of trapping. The multiplier is also warmed up by placing a voltage on it and allowing it to stabilize for a short time.

During the detection event the trapped ions are ejected sequentially, in order of increasing m/z-value from the trap. The trapped ions begin with a low value of q_z, and then are moved to the q_z-value for ejection and detection by ramping up the RF amplitude. In most cases, QITs use resonant ejection, in which an auxiliary waveform is applied between the end-cap electrodes in order to eject ions at a q_z-value below the stability limit at $q_z = 0.908$. Using resonant ejection during detection, sometimes called 'axial modulation', gives better mass resolution and sensitivity [29].

Conventionally, the buffer gas pressure is constant throughout the scan function. This pressure is set at a level that is a compromise between signal intensity and mass resolution. The concept of 'pressure tailoring' is defined as implementing different buffer gas pressures that are optimized for each event in the scan function. The method of pressure tailoring is shown in Figure 18.2, where the simplified scan function includes four events: pre-injection, injection, post-injection, and detection events. The major timed components are also listed. Another timed parameter, buffer gas pressure, has been added to the scan function in this figure. This proposed parameter can be treated like a potential that can be turned 'on' and 'off'. It can be anticipated that the time constant for the buffer gas pressure will be longer than for an electric potential, as the rates of pressure changes will be affected by 'impedances', such as limited conductance into and out of the ion trap, limited pumping speed, and the response time for opening and closing the pulsed valve.

When ions are created externally and then injected into the trap, they must be injected with enough kinetic energy to penetrate the RF field within the QIT. Once inside the trap, however, the ions need to be slowed down in order for them to be trapped successfully. During ion injection, therefore, the presence of buffer gas inside the QIT is critical to reducing effectively the ions' axial and radial motion by collisional cooling; thus the buffer gas must be turned 'on' for this event, as shown in Figure 18.2. At the post-injection event, buffer gas is turned 'off' and a delay time is added to pump away the buffer gas before detection. In the detection event where the ions are ejected sequentially from the QIT, the presence of buffer gas is not needed and can even be detrimental to the analysis because, as noted above, collisions with gas molecules can promote fragmentation of fragile ions during mass-selective ejection, affecting the mass resolution and mass assignment [15]. Thus, the buffer gas supply is kept turned off until the next scan. Clearly, there are potential advantages to having buffer gas present during some portions of the scan function and absent during others.

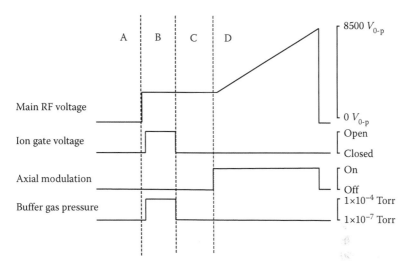

FIGURE 18.2 A simplified QIT scan function for a full scan MS using the mass-selective instability technique with the proposed buffer gas pressure tailoring scheme. It can be divided into four general events: (A), pre-injection; (B), injection; (C), post-injection; and (D), detection. This scan function includes the concept that the buffer gas pressure could be added as another parameter varied during the scan function, rather than kept constant, as is done typically. Buffer gas pressure scales are approximate values based on the minimum and maximum level readings for the QIT. In this scan function, buffer gas pressure is increased only during ion injection.

A typical scan function for MS/MS with CID has the same events as the full MS scan function, plus additional events for mass isolation and CID after ion injection and cooling. Thus, the buffer gas pressure parameter should be adjusted accordingly (Figure 18.3). The post-injection event (C) for the MS/MS CID scan function has three steps in addition to cooling the injected ions, (C_1): isolation (C_2) and excitation (C_3) of precursor ions for fragmentation, and then cooling the fragment ions (C_4) before scanning them out for detection. For CID and cooling (C_3 and C_4), the buffer gas pressure is raised then lowered prior to detection.

Previous investigations in our laboratory have examined the effects of pulsed introduction of buffer gas on ion storage and detection efficiencies in a QIT [3]. These studies determined strategic points within the scan function where the presence of buffer gas is important, and monitored the effects on ion signal intensity as the presence and pressure of buffer gas were varied. These previous experiments have all been performed on ion trap systems using internal ionization. They demonstrated that a higher ion signal can be obtained by using the pulsed introduction of buffer gas as compared to operation at a constant pressure. However, with a single pulsed valve, control of the pressure and pulse duration is limited to the supply pressure applied to the pulsed valve and the valve-open time. There are several limitations to this approach, as discussed below.

Firstly, the amount of gas that can be delivered will be limited by the supply flow. The valve has only two states, open and closed, wherein only two levels of buffer gas

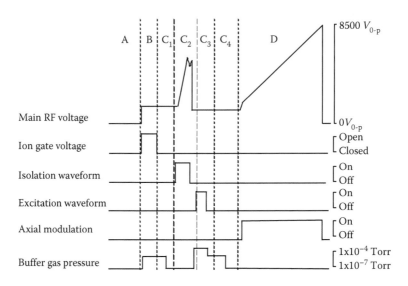

FIGURE 18.3 A simplified QIT scan function for MS/MS scan with CID. Similar to the full MS scan, it is divided into four general events but with some changes: (A), pre-injection; (B), injection; (C), CID version of post-injection (cooling C_1, isolation C_2, excitation C_3, and cooling C_4); and (D), detection. This scan function shows the proposed variation of buffer gas pressure during different portions of the scan function. Buffer gas pressure scales are approximate values based on the minimum and maximum level readings for the QIT.

flow can be achieved. Secondly, the amount of gas delivered is not proportional to the valve-open time, because of the pressure build-up behind the valve during the times when the valve is closed. As a result, the longer the valve is opened, the higher the apex of the gas pulse instead of having a longer gas pulse of the same height [30].

The gas pressure profile from the pulsed valve has similarities to a sinusoidal peak, and in concept could be combined with other peaks to produce another gas profile. This method is similar to combining multiple sinusoidal waveforms to form a square wave. Figure 18.4 illustrates this concept, using two pulsed gas profiles from different sources of varying supply pressures, which are then combined to produce a new profile. For example, Valve 1 outputs a single gas pulse (Figure 18.4a) while Valve 2 is opened twice to produce a gas profile that has two smaller peaks, one before and one after the first peak with a particular interval between (Figure 18.4b). Summing both valves' profiles will produce a new profile (Figure 18.4c) that is only attainable by combination of two valves. The dotted lines in Figure 19.4c show the outline of the intended gas profile. The various combinations can provide the means to tailor the buffer gas pressure with more flexibility than employing just a single valve.

18.4 EXPERIMENTAL SECTION

The quadrupole ion trap used for this study was a research-grade QIT with a custom-built differentially pumped vacuum chamber, as shown in Figure 18.5, and controlled with Finnigan GCQ electronics. For clarity, Figure 18.5a shows only one valve

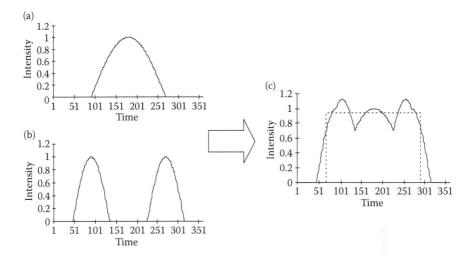

FIGURE 18.4 Simulated profile of using two pulsed valves with varying gas pulse patterns to generate a 'square' gas pulse profile. (a), ion intensity profile obtained from a single gas pulse initiated at *ca* 100 time units; (b), ion intensity profile obtained from two similar gas pulses initiated at *ca* 50 and *ca* 225 time units; (c) ion intensity profile obtained from new gas pulse profile formed by combining the timed gas pulses from (a) and (b) in a manner similar to the construction of a square wave from sine waves. An overlay of the desired square pulse profile is shown (dashed line).

installed. A second valve was installed in the same manner, as shown in Figure 18.5c. The two pulsed valves used were three-way high performance Series 9 from General Valve (Fairfield, NJ) with an orifice of 0.060 in. The pulsed valve has three connections: a 'common input', C_{in}, (Figure 18.5a, inset) to which the gas supply is connected, and two outputs between which the valve switches during open and closed states.

During the course of the overall research project, two different configurations for coupling the valves and the various gas lines were employed, and these have been termed 'Mode I' and 'Mode II'. Thus there were two different connections to the common (C_{in}) input employed during the course of the research. The common input of the first pulsed Valve (marked '1' in Figure 18.5c) was connected to the same supply as the continuous helium buffer gas supply, as shown in Figure 18.5c. Shut-off valves leading to the continuous and to the first pulsed valve lines were installed in order to select the supplies without venting. For example, during instrument calibration and tuning, a continuous supply of helium is needed. This supply passes through a 1/16 in. i.d. stainless steel tubing and is inserted into the vacuum chamber through a 1/8 in. bored-through Swagelok (Solon, OH) O-seal connector. This supply input flow was regulated using a pressure regulator valve (Porter Instruments Company, Hartfield, PA) connected to a capillary restrictor. The common input of the second pulsed valve (Valve 2 in Figure 18.5c) was connected to a helium gas cylinder with 1/8 in. i.d. stainless steel tubing *via* a Granville-Philips (Boulder, CO) Series 203 variable leak valve ('GP'); the helium supply pressure was controlled with a Matheson (Montgomeryville, PA) two-stage pressure regulator on the cylinder.

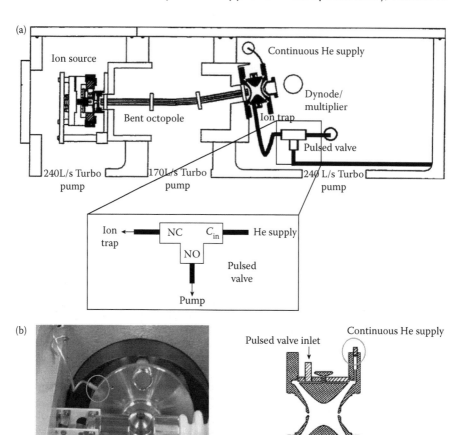

FIGURE 18.5 (a), Diagram of a three-way pulsed valve placement inside the vacuum chamber of the QIT. The NC output was connected to the QIT with Teflon tubing (2.5 in. × ¼ in. i.d.) that passed through a hole drilled in one of the nonconductive spacers between the trap electrodes. The normally open (NO) output was connected to a mechanical pump *via* a variable leak valve. The common (C_{in}) input was connected to a helium gas supply *via* another variable leak valve. The head pressure on the helium supply was controlled by a pressure regulator; (b), Picture (and cross sectional view) of the trap showing the continuous supply line connected to the side of the end-cap electrode (in circle); (c), Schematic diagram of the gas line connections and valve placement for two pulsed valves, labeled 1 and 2. (Mode II). Also shown are the pressure regulators, and the shut-off valves, together with the capillary restrictor, and the variable leak valves GP and GL.

The normally closed (NC) outputs for the pulsed valves were connected also to the ion trap with different configurations. As shown in Figure 18.5b, the continuous buffer gas supply is connected typically through the outer edge of an end-cap electrode where there is a 1/16 in. diameter hole leading to one side of the hyperbolic surface. In Mode I, Valve 1 was connected through another hole in the end-cap electrode but

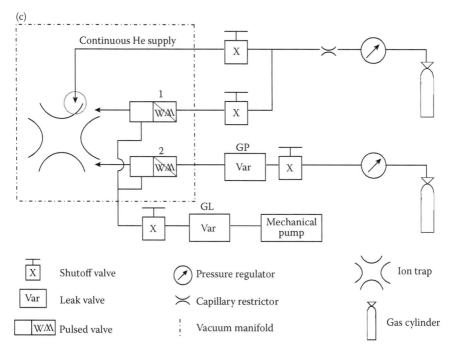

FIGURE 18.5 (Continued)

of the same design as for the hole for the continuous supply, and using a 2.5 in. × 1/8 in. i.d. Teflon tube with a stainless steel adapter. The NC output for Valve 2 was connected to the ion trap using a 2.5 in. × 1/4 in. i.d. Teflon tube, which was anchored by a PEEK adapter (1/8 in. i.d.) and mounted in a hole bored into a Delrin end-cap spacer. In Mode II, this method of coupling Valve 1 to the ion trap was implemented, as shown in Figure 19.5c. This change in connection was used in later experiments to increase the conductance into the trap for the gas pulse.

Unless specified otherwise, for both Modes I and II the normally open (NO) outputs of both Valve 1 and Valve 2 were connected together and coupled to an Alcatel mechanical pump with 1/8 in. i.d. stainless steel tubing passing through a welded 1/4 in. bored-through connector on the flange on the vacuum chamber; the flow rate to the pump was regulated by a Granville-Philips variable leak valve ('GL'). In this manner, the amount of buffer gas could be controlled by varying the setting on the variable leak valve.

The Custom Tune software (Thermo, San Jose, CA) was used to allow software control of the transistor–transistor logic (TTL) signal generated by the GCQ processor. This TTL signal was sent to the trigger input of an SRS Model DS345 (Sunnyvale, CA) function generator, in which a customized waveform allowed for setting the length of the pulse width (open time) and timing of the solenoid valve. The signal from the function generator was then amplified to the appropriate DC voltage (24 VDC) using a custom-built circuit to drive the solenoid valves. The minimum pulse widths for the first and second pulsed valves to open were 1.9 ms and 1.8 ms, respectively, which were within the specification limits listed by the manufacturer.

18.4.1 Argon/Nitrogen Charge-Exchange

The initial goal in implementing the pressure tailoring system was to characterize the pulsed gas profile that can be attained with the solenoid valves. Ideally, monitoring the pressure profile of the pulsed valve would involve monitoring the pressure inside the ion trap directly with a pressure gauge. To the knowledge of the authors, however, there is no pressure gauge that will allow measurement in this pressure range on the millisecond timescale. Thus, a technique involving the study of ion/molecule reactions occurring within the trap as a function of the pressure of added gas was used, as employed previously in this laboratory [1,26]. For these experiments, argon ions were injected into the trap using Valve 1 and allowed to undergo charge-exchange with nitrogen molecules pulsed into the trap. Using the Mode I configuration, the source of nitrogen used was that present as a residual impurity in the helium buffer gas pulsed into the trap. Helium itself was not used as the pulsed gas to be monitored because the QIT will not scan down to a sufficiently low value of m/z; furthermore, because helium has such a high ionization energy that it would not be ionized by charge-exchange with any other species of singly charged ion.

Changes had to be made in the default GCQ scan function to accommodate the triggering of the pulse valves. The valve was triggered by a pulse from the MS and an appropriate delay time (t_{delay}) was employed before ion injection step, as shown in Figure 18.6a. The ion-detect time is defined as the amount of time that argon ions are allowed to react with the pulsed neutral N_2 molecules (from the start of the injection time of the Ar^+ ions until the N_2^+ ion at m/z 28 is scanned out). Considering the mass scan range from m/z 10 to 28 (3.3 ms @ 0.180 ms amu^{-1}), injection time (1 ms), cooling time (3 ms), multiplier warm-up time (2 ms) and ramp time (1 ms), the ion-detect time was calculated to be 10.3 ms. To help the reader understand better, the ion-detect event could be treated in a manner similar to a sampling window that is used when measuring waveform signals, and the ion-detect time corresponds to the sampling resolution. The concept is that the gas pulse has the same temporal profile for each repetition; because it is periodic, by shifting t_{delay} sequentially across the gas profile, one can sample the pulsed gas profile by monitoring the product of the charge-exchange reaction, in this case, the N_2^+ ion at m/z 28 (Figure 18.6b). The series of data were then extracted from the files and plotted in Excel for the reconstruction of the gas profile. Post scan times were lengthened to 1000 ms to ensure no carry over of pulsed gas between scans.

18.4.2 Pulsed Helium Buffer Gas Effects

In this series of experiments, using Mode II, evaluation of the ability of the pulsed valve system to create different pressure profiles for the use of helium as buffer gas was carried out. The perfluorotributylamine, (PFTBA), fragment ion, CF_3^+ at m/z 69, was selected due to its stability and its ready formation by electron ionization (EI). The effects of the pulsed helium buffer gas on the intensity of injected CF_3^+ ions was monitored; without helium, the ions are not trapped efficiently. Another objective was to create new gas profiles from the synchronized operation of the two pulsed valves. Individual valve gas profiles, as well as combined gas profiles, were monitored in a manner similar to the Ar^+/N_2 charge-exchange experimental method.

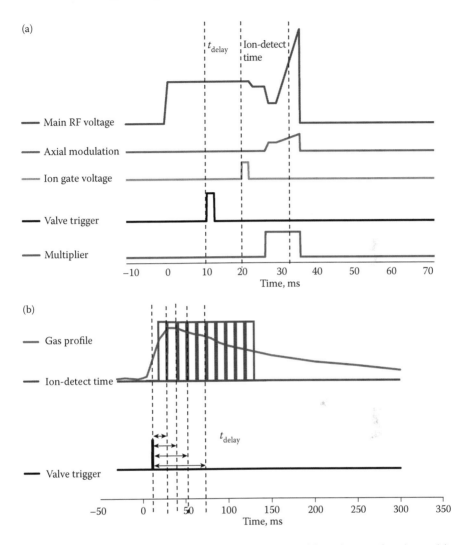

FIGURE 18.6 Diagrams showing how the t_{delay} is inserted into the scan function and its operation. (a), A scan function that shows the timing of opening the valve for ion/molecule reactions between Ar^+ ions and N_2 introduced *via* a pulsed valve. Using 24 VDC to drive the pulsed valve, the minimum pulse time to open the valve was 1.9 ms. As shown on the diagram, t_{delay} was set at 10 ms; (b), N_2^+ signal intensity as a function of pressure. The continuous line can be interpreted as a neutral gas profile with a series of ion-detect times overlayed on the profile. The ion-detect times (in the pulse waveform line) shows how the gas profile is sampled across its peak by shifting t_{delay} sequentially, relative to the valve trigger.

For these experiments, the two pulsed valves were used with two different helium gas supplies, see Figure 18.5c. The supply of gas to Valve 1 was from a helium gas tank controlled by a pressure regulator (Porter Instruments Company, Hartfield, PA) followed by a 6 in. long fused silica capillary (0.050 mm i.d.). The supply of gas to Valve 2 was from another helium tank *via* a Granville Philips variable leak valve (GP). The NC outputs of both Valves 1 and 2 were connected to the ion trap *via* 2.5 in. × 1/4 in. i.d. Teflon tubing through holes drilled in the end-cap electrode spacers with PEEK adapters. The NO outputs of both Valves 1 and 2 were connected together inside the vacuum chamber using 1/8 in. copper tubing and a Tee union, and then connected to a mechanical pump outside the vacuum chamber *via* another Granville Philips variable leak valve (GL).

Each valve had its own triggering circuit, comprising an SRS function generator that was coupled with a custom-built amplifier. This arrangement was used so that each valve could be operated independently, in terms of timing and pulse width. The general settings for the function generators were as follows: mode = arbitrary/burst mode with point value format, amplitude = 10 V peak to peak, frequency = 10 kHz, trigger mode = positive in (triggers on the rising edge). With these settings, pulse widths in increments of 0.1 ms with an amplitude of 5 V_{0-p} could be created (data point values 1000 is equivalent to 5 V).

18.4.3 Pressure Effects on Fragile Ions

The benefit of using buffer gas pressure tailoring is most evident in the analysis of fragile ions. The parent molecular ion of *n*-butylbenzene (*m/z* 134) has been reported to produce fragile ions [15,18]. The compound was purchased from Sigma Aldrich (St. Louis, MO) and was used to monitor the behavior of fragile ions in terms of both intensity and mass shifts when using pulsed helium buffer gas. Again using the Mode II configuration, static pressure effects were investigated by increasing the head pressure of the continuous buffer gas supply. For the pulsed valve experiment, the Valve 1 was used with a helium supply (2.5 psi head pressure, from the continuous buffer gas supply), and the peak signal intensities and mass shifts were monitored.

18.5 RESULTS AND DISCUSSION

An important aspect of using the pulsed valve system was to evaluate the valve performance in terms of its ability to produce a gas peak. A single scan time for the ion trap MS from the pre-ionization to post-scan events is around 150 ms for a scan range of *m/z* 50–650 for this QIT system. The events where gas is required (ion injection and cooling) have durations typically from 1–25 ms. On the other hand, the duration of the events where gas is not needed (ion ejection and detection) is approximately 100 ms. From these times, the ideal gas pulse width is seen to be approximately 10–30 ms FWHM (matching the maximum ion injection time) in order to be used practically in MS experiments. For example, in detecting a typical chromatographic peak with a peak width of 1 s, having a scan duration of 150 ms would give six data points across the peak. Increasing the scan duration would lower the number of data points and thus result in a loss of resolution for the chromatographic peak.

18.5.1 Optimization of the Gas Pulse Profile

The Ar/N_2 charge-exchange reaction was used for the experiments to characterize the output of a single pulsed valve. The initial configuration of the pulsed valve

(Valve 1) was with the NO output capped with a stainless steel plug, thus converting the three-way into a two-way pulsed valve. Using the Mode I configuration, the NC output was connected to the trap *via* the same entrance used for the continuous buffer gas flow, that is, through one of the end-cap electrodes. The continuous helium gas supply line leading to the ion trap was split with a Tee union and connected with shut-off valves, which enabled it to be connected to the pulsed valve and to the ion trap at the same time. With the shut-off valve configuration, the continuous supply and pulsed supply could be used independently, while sharing the same source. Initially, a nitrogen gas cylinder was connected as the source, but even with the lowest setting on the pressure regulator (0.5 psi), no ion signal either for N_2^+ or for the injected Ar^+ ions was obtained, leading to the conclusion that nitrogen, at least at the pressures used, was not a suitable buffer gas. As noted previously, nitrogen and other more massive buffer gases are not commonly used with QITs. To overcome, as noted in Section 18.4.1, this problem, the residual nitrogen impurity ($< 0.01\%$) in the helium gas supply was monitored instead. The advantage of using residual nitrogen in the helium buffer gas was that it provided both a low-mass buffer gas, helium, plus the target charge-exchange gas, nitrogen. This arrangement may also have allowed a lower partial pressure of nitrogen in the trap than could have been achieved with the admission of pure nitrogen, even at the lowest head pressure setting.

Mass spectra for the residual N_2^+ ions formed by charge-exchange were obtained and the peak width (FWHM) of the N_2^+ ion signal (and thus the peak width of the gas pulse) as affected by the head pressure of the pulsed valve is shown in Figure 18.7a. As the figure shows, control of the head pressure only did not reduce significantly the gas pulse profile width at half maximum, with the average FWHM peak widths > 100 ms. The absolute ion intensities for the 5, 2.5, and 1.5 psi peaks at 100% peak height were 3.1×10^6, 2.5×10^6, and 3.4×10^6 counts, respectively; the variations in ion intensity with pressure were generally reproducible, and may reflect the effect of the varying pressure of helium buffer gas introduced along with the nitrogen.

The next experimental arrangement examined was designed to reduce the peak width by unplugging the NO output of Valve 1 and letting the gas supplied to the valve during the closed valve states be discarded into the vacuum chamber (the right hand chamber in Figure 18.5). This setup was designed to reduce the pressure behind the poppet of the pulsed valve. However, the continuous leak of helium from the NO output raised the overall trap chamber pressure to $> 1 \times 10^{-5}$ Torr. Therefore, the gas supply connection was changed to a helium gas supply with a variable leak valve in front of the fused silica restriction, with the head pressure set to 2.5 psi, for finer control of the supply pressure.

Data for this experimental setup are shown in Figure 18.7b. Having the NO output uncapped did not allow the pulse valve to build-up enough backing pressure before the valve was triggered, thus explaining the slower rise times of the gas pulse (70 ms to reach 100% ion intensity, compared to 10 ms observed with the previous setup). Another reason for the slower rise time was the impedance brought about by the small diameter orifice used by the continuous supply to put buffer gas inside the trap (Figure 18.5b). It was also noted that the signal intensity for the pulsed N_2^+ was only 1.5 times higher than the baseline signal of N_2^+ (arising from the nitrogen contained within the gas flowing from the NO output into the chamber). The setting on the

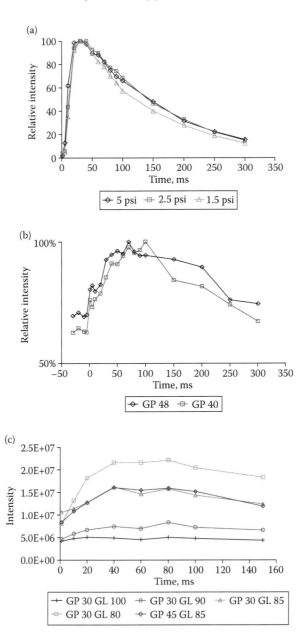

FIGURE 18.7 Preliminary data for gas pulse optimization, showing intensity of N_2^+ (m/z 28) formed by charge-exchange with Ar^+ as a function of delay time. The pressure for the common (C_{in}) inlet was varied by controlling the head pressure or setting of the GP leak valve. (a), NO closed and varying head pressure; (b), NO open and varying GP supply (head pressure 2.5 psi); (c), NO and C_{in} with variable leak gauges: GP = setting for the supply restriction, GL = setting for the leak restriction (head pressure 10 psi).

variable leak valve could not be increased too much as it would raise the overall trap chamber pressure too high, which would pose a problem for the turbo pumps. It was concluded from these initial experiments that, as a compromise, a restriction between the NO output and a vacuum source would offer better control of the gas pressure behind the valve.

The following changes were made, therefore, to the valve setup: (1) connecting the gas supply to the C_{in} inlet of Valve 2 (in order to keep the previous coupling for performance comparison without breaking vacuum to reconnect Valve 1) *via* a variable leak valve, with no capillary restrictor in between them; (2) connecting the NO valve output of Valve 2 *via* another leak valve to a mechanical pump instead of dumping into the vacuum chamber; and (3) increasing the conductance of the connecting tubing between the valve and the ion trap. The previous tubing (2.5 in. × 1/8 in. i.d.) that was connected to a 1/16 in. diameter hole through the end-cap electrode (see Figure 18.5b for the location) was replaced by Teflon tubing that was 2.5 in. long with an internal diameter of 1/4 in. As shown in Figure 18.5b, a hole was drilled into the one of the end-cap electrode spacers and a PEEK tube with screw threads was attached to the hole in order to accommodate the larger diameter tube. Using a two times larger diameter tube should increase the conductance of the pulsed gas inlet by a factor of four and thereby decrease the rise times. Figure 18.7c shows the data with the preliminary settings of this setup. The parameter GP is the setting for the inflow leak valve and GL is the setting for the outflow leak valve. Notice that with less input supply flow and more output leak flow (GP 30, GL 100) the pulse profile signal did not rise much above the baseline signal (not shown). On the other hand, decreasing the outflow restriction improves the pulse profile rise time. After evaluating several combinations for both settings, it was concluded that the smallest pulse width achieved was *ca* 100 ms FWHM, with GP setting of 40 and GL setting of 200.

Until the gas pulse profile can be reduced to around 10 ms, which is the average cooling time and excitation time for CID, the pulsed valve system can be used only for limited research applications. The ion trap scan function can be adjusted to accommodate time intervals suited for the current gas pulse width, but scan times will be 500 ms longer, which is not practical for applications such as LC/MS or GC/MS experiments where faster scan rates are needed to obtain better chromatographic data. Nevertheless, the current setup permits proof-of-principle demonstration of pressure tailoring.

18.5.2 MULTI-PULSE EXPERIMENTS

Experiments were performed using the Mode II configuration to investigate the reproducibility of the gas pulses when they were within 100 ms of each other and to determine if the gas pulses from a valve could be added. The experimental setup was similar to that of the single pulse experiment; multiple pulses from the same valve were used in the scan function. The signal intensity of the PFTBA fragment ion at *m/z* 69 was monitored as a function of delay time in order to show how the former depends upon the amount of pulsed gas. As shown in Figure 18.8, the function generator was configured to produce two pulses, with the second pulse 70 ms after

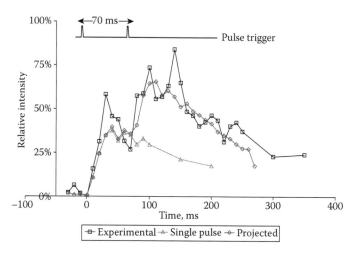

FIGURE 18.8 Gas profile of the multi-pulse experiment, measured by monitoring the CF_3^+ intensity ion with helium buffer gas pulsed into the ion trap. Experimental data for a single pulse were taken and used to create the projected two-pulse data and plotted with the experimental two-pulse data. The pulsed valve was triggered twice with the second pulse 70 ms after the first. The experimental data for the two-pulse delivery of buffer gas followed a trend that is similar to the projected profile, indicating reproducibility of the pulsed gas profile. Shown above the figure is the pulse trigger timing that was sent to the pulse valve.

the first one. Experimental data from a single pulse experiment were used to create a projected two-pulse profile and plotted to compare with the experimental data for two pulses. It can be concluded from the agreement between these two plots that two pulses for a single valve provide the pressure one would expect. The next step in the development was to implement control of two independent pulsed valves.

The results from the two-valve experiment are shown in Figure 18.9. The two pulsed valves (Valves 1 and 2) were triggered either simultaneously or with 100 ms delay between them. The CF_3^+ ion intensity profile was measured in separate experiments for each valve; those profiles were summed to generate the projected profile to compare the data when both valves were triggered. The trigger pulse from the MS was connected to two function generators. The supply head pressure for Valve 1 was set to 2.5 psi; the settings for Valve 2 were GP = 40 and GL = 200, with a head pressure of 10 psi. The delay time for opening each pulsed valve was adjusted with its function generator. Each data point in the profile was an average of 30 scans. The post scan time was reduced from 1000 ms to 500 ms to speed up data collection.

For the data in Figure 18.9a, both valves were triggered at time 0 ms. The ion intensity profiles for each valve are plotted as well as the projected profile (sum of the CF_3^+ ion intensities of using two valves) for comparison. The intensity for the experimental data was *ca* 25% lower than the projected profile. This is readily explained by considering the data on the effect of buffer gas pressure on ion intensity presented by Stafford *et al.* [7] At lower buffer gas pressures, the signal intensity increases approximately linearly with increasing buffer gas pressure. However, at higher pressures, the increase in the signal intensity drops and eventually the signal reaches a maximum.

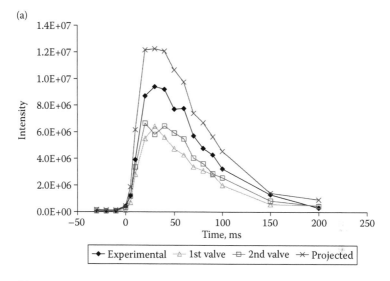

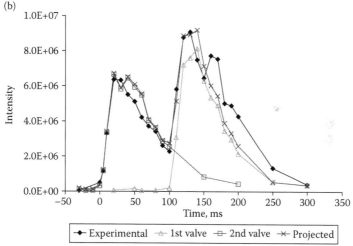

FIGURE 18.9 Gas profile of two pulsed valves operated synchronously as measured by monitoring the CF_3^+ ion intensity as function of delay time. The supply head pressure for Valve 1 was set to 2.5 psi; the supply settings for Valve 2 were set to GP = 40 and GL = 200 with a head pressure of 10 psi. (a), Both valves were opened at time 0; (b), the first valve had a delay time of 100 ms in relation to the second valve.

In Figure 18.9b, a delay of 100 ms was set for the first valve relative to the timing of the second valve. Again, individual profiles for each valve as well as the projected profile were plotted with the experimental profile. The experimental profile has good agreement with the projected profile. With the delay of 100 ms, most of the gas from the second valve was already pumped away and thus, the pressure contributed by both valves remains in the linear range for ion intensity [7].

18.5.3 PRESSURE EFFECTS ON FRAGILE IONS

For this study, again using the Mode II configuration, the $M^{+\bullet}$ ion of n-butylbenzene (m/z 134) was chosen as a model compound since it has been studied previously and classified as a fragile ion [15]. The QIT was calibrated using PFTBA under normal buffer gas conditions. The analyte was put into a 1/8 in. i.d. $\times$ 3 in. long glass tube and the vapor headspace was admitted into the ion source *via* a Granville-Philips Series 203 variable leak valve to yield ion gauge pressures of 5×10^{-6} to 1×10^{-5} Torr. All the data points in the graphs were averaged from 100 scans.

The 'zoom scan' mode [31] was used to acquire the mass spectra. Zoom scan is defined as a method of acquiring mass spectral data with slower mass scan rates, thus having more data points to define the mass peaks, thereby resulting in higher mass resolution. The normal mass scan rate of the GCQ is 0.180 ms amu^{-1}, whereas in the zoom scan mode the scan rate is slowed down 10-fold to 1.8 ms amu^{-1}. The zoom scan method is not an inherent function in the GCQ tune software, but the Custom Tune software has advanced feature controls to change the mass scan rates. Note that the mass assignments in the spectra needed to be calibrated after the data were taken, as the default mass calibration was performed at the normal mass scan rate.

Figure 18.10a shows the effects of variation of static helium buffer gas pressure on signal intensity and shift of mass assignment of the $M^{+\bullet}$ ion formed from n-butylbenzene. The signal intensity increases as more buffer gas is available in the QIT for higher trapping efficiency. However it evident also that at higher pressures of buffer gas, the mass assignment shifts downwards as collisions with buffer gas lead to increased dissociation of the fragile molecular ion and thus earlier ejection of fragment ions. The mass spectra of n-butylbenzene are shown in Figure 18.10b and C: the mass spectrum taken with higher head pressure (10 psi) has the centroid mass assignment shifted slightly toward the lower m/z value, when compared to the mass spectrum for 2.5 psi.

Figure 18.11 shows the effect of pulsed buffer gas during the detection event (see Figure 18.2) in the scan function. The cooling time is defined as the length of time after ion injection and before the RF amplitude is ramped for detection. Post scan time is defined as the length of time after the end of the RF amplitude ramp before the start of the next scan function. In these experiments, the first pulsed valve was used and operated with a pulse width of 1.9 ms. The helium head pressure behind the valve was set to 2.5 psi. Each data point was an average of 30 scans. The ion injection time was fixed to 1 ms.

The conditions for the data shown in Figure 18.11a were set to have buffer gas present during both the ion injection and the detection event in the scan function. The cooling time was set to 1 ms and the post scan time was set to 1000 ms to ensure that the pulsed gas for each scan did not carry over to the next scan. It was observed that the m/z 134 ion intensity profile followed the pulse gas profile; in contrast, the mass shift decreased at longer delay times as the amount of buffer gas during the detection event decreased. In Figure 18.11b, the cooling time was set to 1000 ms, which allowed the buffer gas that was introduced during injection to be pumped away before starting the detection event. The results showed that there was minimal mass shift of the n-butylbenzene ion, even at the highest buffer gas level (indicated by the signal intensity profile).

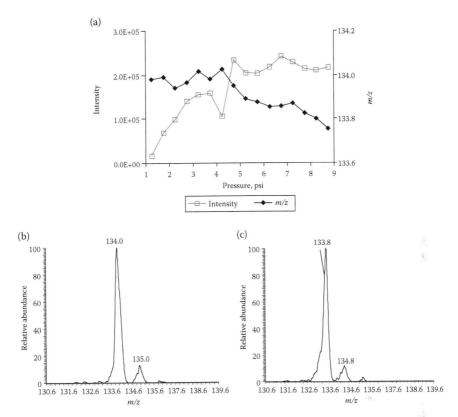

FIGURE 18.10 Static buffer gas experimental results. (a) Effects of static buffer gas pressure on signal intensity and mass assignment of the $M^{+\bullet}$ ion of n-butylbenzene (m/z 134). The mass placement shifted downwards as the static pressure was increased. The open square trace indicates the intensity and the closed diamond trace is for the mass-to-charge. Lower traces: (b), zoom (slow) scan mass spectra of n-butylbenzene at 2.5 psi; and (c), the corresponding mass spectra of n-butylbenzene at 10 psi buffer gas head pressure. Centroid mass assignments were used.

These findings demonstrate clearly that the buffer gas is needed during ion injection in order to trap the ions efficiently, while the presence of buffer gas during ion detection can affect analysis of fragile ions and cause mass shifts. Implementing a pulsed valve system allows the gas pressure to be tailored for different events in the scan function. Unfortunately, the current pulsed valve system does not provide a sufficiently narrow buffer gas pulse to tailor the pressure without adding delays in the scan function.

18.6 SUMMARY AND CONCLUSIONS

The potential of tailoring the buffer gas pressure with the use of multiple pulsed valves has been demonstrated. Buffer gas, when introduced at the right point in the scan function, can enhance the trapping of injected ions, as well as increase the fragmentation efficiency for CID and MS/MS. With tailored buffer gas pressure, the

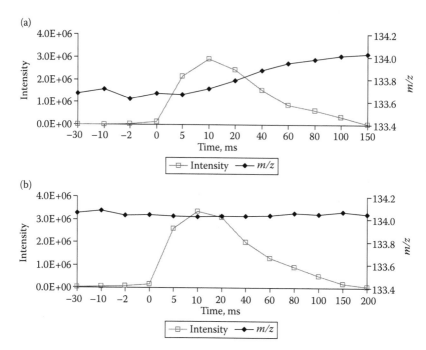

FIGURE 18.11 Experiments using pulsed buffer gas introduction and its effect on analysis of fragile ions. The open square trace indicates the intensity (left-hand ordinate), and the closed diamond trace is for the mass-to-charge ratio (right-hand ordinate). The intensity and mass assignment of the M$^{+\bullet}$ ion (*m/z* 134) of *n*-butylbenzene is plotted *vs* delay time: (a), cooling time was 1 ms and post scan time was 1000 ms; (b), cooling time was 1000 ms and post scan time was 1 ms.

possibility of using increased pressure during trapping and CID events without the detrimental effects of buffer gas during detection can be achieved. The effects are highlighted in the analyses of fragile ions.

The gas pulse was profiled by a technique using the QIT as a pressure gauge. Even though the pulse can be reproduced and tailored, the narrowest gas peak width achieved was 100 ms FWHM, five times the ideal peak width of 20 ms. Thus, applications for the current setup are limited. The scan function can be lengthened to accommodate the wide buffer gas pulse, but scan rates will be decreased below practical speeds for GC/MS and LC/MS. Nevertheless, improved ion trap performance with pulsed buffer gas tailoring may justify the added complexity in QIT operation.

REFERENCES

1. Coopersmith, B.I.; Yost, R.A. Internal pulsed valve sample introduction on a quadrupole ion-trap mass-spectrometer. *J. Am. Soc. Mass Spectrom.* 1995, *6*, 976–980.
2. Doroshenko, V.M.; Cotter, R.J. Pulsed gas introduction for increasing peptide CID efficiency in a MALDI quadrupole ion trap mass spectrometer. *Anal. Chem.* 1996, *68*, 463–472.
3. Williams, T.L.; Stephenson, J.L.; Yost, R.A. The effects of pulsed introduction of buffer gas on ion storage and detection efficiencies in a quadrupole ion trap. *J. Am. Soc. Mass Spectrom.* 1997, *8*, 532–538.

4. Emary, W.B.; Kaiser, R.E.; Kenttamaa, H.I.; Cooks, R.G. Pulsed gas introduction into quadrupole ion traps. *J. Am. Soc. Mass Spectrom.* 1990, *1*, 308–311.

5. Paul, W.; Steinwedel, H. Apparatus for separating charged particles of different specific charges. *U.S. Patent* 1960, 2,939,952.

6. Stafford, G.C.; Kelley, P.E.; Syka, J.E.P.; Reynolds, W.E.; Todd, J.F.J. Recent improvements in and analytical applications of advanced ion trap technology. *Int. J. Mass Spectrom. Ion Processes* 1984, *60*, 85–98.

7. Stafford, G.C.; Kelley, P. E.; Stephens, D.R. Method of mass analyzing a sample by use of a quadrupole ion trap. *U.S. Patent* 1985, 4,540,884.

8. Stafford, G.C. Ion trap mass spectrometry: a personal perspective. *J. Am. Soc. Mass Spectrom.* 2002, *13*, 589–596.

9. Louris, J.N.; Amy, J.W.; Ridley, T.Y.; Cooks, R.G. Injection of ions into a quadrupole ion trap mass-spectrometer. *Int. J. Mass Spectrom. Ion Processes* 1989, *88*, 97–111.

10. Fischer, E. Die dreidimensionale Stabilisierung von Ladungstragern in einem Vierpolfeld. *Z. Phys.* 1959, *156*, 1–26.

11. Todd, J.F.J.; Waldren, R.M.; Freer, D.A.; Turner, R.B. The quadrupole ion store (QUISTOR). Part X. Space charge and ion stability. B. On the theoretical distribution and density of stored charge in RF quadrupole fields. *Int. J. Mass Spectrom. Ion Phys.* 1980, *35*, 107–150.

12. Quarmby, S.T. Fundamental studies of ion injection and trapping of electrosprayed ions on a quadrupole ion trap mass spectrometer. *Ph.D. Dissertation,* University of Florida, 1997.

13. Plass, W.R.; Li, H.Y.; Cooks, R.G. Theory, simulation and measurement of chemical mass shifts in RF quadrupole ion traps. *Int. J. Mass Spectrom.* 2003, *228*, 237–267.

14. Brodbelt, J.S. Effects of collisional cooling on ion detection. In *Practical Aspects of Ion Trap Mass Spectrometry*, eds. R.E. March and J.F.J. Todd, Vol. 1, Chapter 5, pp. 209–220. CRC Press, Boca Raton, FL, 1995.

15. Murphy, J.P.; Yost, R.A. Origin of mass shifts in the quadrupole ion trap: dissociation of fragile ions observed with a hybrid ion trap/mass filter instrument. *Rapid Commun. Mass Spectrom.* 2000, *14*, 270–273.

16. Murphy, J.P. Fundamental studies of the quadrupole ion trap mass spectrometer: compound dependent mass shifts and space charge. *Ph.D. Dissertation,* University of Florida, 2002.

17. McClellan, J.E.; Murphy, J.P.; Mulholland, J.J.; Yost, R.A. Effects of fragile ions on mass resolution and on isolation for tandem mass spectrometry in the quadrupole ion trap mass spectrometer. *Anal. Chem.* 2002, *74*, 402–412.

18. Li, H.Y.; Plass, W.R.; Patterson, G.E.; Cooks, R.G. Chemical mass shifts in resonance ejection experiments in the quadrupole ion trap. *J. Mass Spectrom.* 2002, *37*, 1051–1058.

19. Morand, K.L.; Cox, K.A.; Cooks, R.G. Efficient trapping and collision-induced dissociation of high-mass cluster ions using mixed target gases in the quadrupole ion trap. *Rapid Commun. Mass Spectrom.* 1992, *6*, 520–523.

20. Danell, R.M.; Danell, A.S.; Glish, G.L.; Vachet, R.W. The use of static pressures of heavy gases within a quadrupole ion trap. *J. Am. Soc. Mass Spectrom.* 2003, *14*, 1099–1109.

21. Louris, J.N.; Cooks, R.G.; Syka, J.E.P.; Kelley, P.E.; Stafford, G.C.; Todd, J.F.J. Instrumentation, applications, and energy deposition in quadrupole ion-trap tandem mass-spectrometry. *Anal. Chem.* 1987, *59*, 1677–1685.

22. Todd, J.F.J. Introduction to practical aspects of ion trap mass spectrometry. In *Practical Aspects of Ion Trap Mass Spectrometry*, eds. R.E. March and J.F.J. Todd, Vol. 1, Chapter 1, pp. 3–24. CRC Press, Boca Raton, FL, 1995.

23. Bauerle, G.F.; Brodbelt, J.S. Evaluation of steric and substituent effects in phenols by competitive reactions of dimethyl ether ions in a quadrupole ion-trap. *J. Am. Soc. Mass Spectrom.* 1995, *6*, 627–633.

24. Vedel, F.; Vedel, M.; Brodbelt, J.S. Ion/molecule reactions. In *Practical Aspects of Ion Trap Mass Spectrometry*, eds. R.E. March and J.F.J. Todd, Vol. 1, Chapter 8, pp. 345–399. CRC Press, Boca Raton, FL, 1995.

25. Einhorn, J.; Kenttamaa, H.I.; Cooks, R.G. Information on the location of carbon-carbon double-bonds in C6-C23 linear alkenes from carbon addition-reactions in a quadrupole ion trap equipped with a pulsed sample-inlet system. *J. Am. Soc. Mass Spectrom.* 1991, 2, 305–313.

26. Guckenberger, G.B. Chemical ionization and ion molecule reactions in the quadrupole ion trap mass spectrometer with ion injection. *Ph.D. Dissertation,* University of Florida, 1999.

27. Yates, N.A. Methods for gas chromatography/tandem mass spectrometry on the quadrupole ion trap. *Ph.D. Dissertation,* University of Florida, 1994.

28. Huston, C.K. Manipulation of ion trap parameters to maximize compound-specific information in gas-chromatographic mass-spectrometric analyses. *J. Chromatog.* 1992, 606, 203–209.

29. Kaiser, R.E.; Cooks, R.G.; Stafford, G.C.; Syka, J.E.P.; Hemberger, P.H. Operation of a quadrupole ion trap mass-spectrometer to achieve high mass charge ratios. *Int. J. Mass Spectrom. Ion Processes* 1991, 106, 79–115.

30. Papanastasiou, D.; Ding, L.; Raptakis, E.; Brookhouse, I.; Cunningham, J.; Robinson, M. Dynamic pressure measurements during pulsed gas injection in a quadrupole ion trap. *Vacuum* 2006, 81, 446–452.

31. Schwartz, J.C.; Syka, J.E.P.; Jardine, I. High-resolution on a quadrupole ion trap mass-spectrometer. *J. Am. Soc. Mass Spectrom.* 1991, 2, 198–204.

19 A Quadrupole Ion Trap/Time-of-Flight Mass Spectrometer Combined with a Vacuum Matrix-Assisted Laser Desorption Ionization Source

Dimitris Papanastasiou, Omar Belgacem,
Helen Montgomery, Mikhail Sudakov,
and Emmanuel Raptakis

19.1   Introduction 793
19.2   Instrumentation 794
          19.2.1   Measuring Pressure Transients in a QIT 796
          19.2.2   Ion Injection and Ejection Scheme 798
          19.2.3   Suppressing Metastable Ion Decay 802
19.3   Theory of Equilibrated Ion Clouds 803
19.4   Applications 812
          19.4.1   Glycobiology 812
          19.4.2   Lipidomics 813
          19.4.3   Other Applications 819
19.5   Conclusions 819
References 820

19.1   INTRODUCTION

The development of soft laser desorption [1] and the advancement of the matrix-assisted laser desorption ionization (MALDI) [2] source as an integral part of modern mass spectrometers [3] has been realized using time-of-flight (TOF) mass analyzers, because TOF is compatible with the pulsed nature of lasers and capable of making high-mass measurements. High-energy collision-induced dissociation (CID) in tandem TOF MS [4–7] has proved to be a complementary technique to low-energy CID by introducing unique fragmentation pathways of macromolecules; however, despite the high mass resolving power and high mass-accuracy of MALDI/TOF instruments, their performance is counterbalanced by the low efficiency of high-energy CID, and the degradation of resolution in the second stage of TOF mass analysis. In contrast, the ability of the quadrupole ion trap (QIT) to isolate sequentially and to dissociate precursor ions with high efficiency, enabling tandem-in-time mass spectrometry to be performed in a single unit, offers an alternative solution to when combined with a MALDI source. The mass-resolving power of the QIT is independent of the number of cycles executed throughout an experiment. Nevertheless, wide mass range injection from an external pulsed source, efficient trapping, and suppression of the metastable character of thermally labile molecular ions amongst others, impose decisive technical challenges.

The advantages of combining a vacuum MALDI source with a hybrid QIT/TOF MS are multifold. Vacuum MALDI does not suffer from low efficiency in transporting ions from the atmosphere into the first pumping stage of the instrument and should, therefore, appear as a highly sensitive ion source. The complication of the mass spectra and the reduction in ion intensities of the analyte molecules as a result of adduct formation with matrix ions is not observed at pressures below 1 Torr [8]. In parallel, the QIT/TOF MS has the potential of combining the diverse functions that can be executed with stored ions, and the high resolution, mass-accuracy, and speed of a TOF mass analyzer. The performance characteristics of a hybrid vacuum MALDI QIT/TOF MS are illustrated in the light of a number of applications, following a discussion of the obstacles encountered and a few practical solutions that emerged in developing the instrument.*

19.2   INSTRUMENTATION

The instrument described in the present study is a modified vacuum MALDI QIT/TOF manufactured by KRATOS Analytical (Manchester, UK), also known as the MALDI AXIMA QIT™. A description of the original system is available in the literature [9]. Here, the focus is primarily on (a) the modifications made to the QIT to control variable pressure conditions; (b) ion injection efficiency; (c) the restrictions imposed by the metastable character of ions generated by MALDI, and, finally; (d) ion ejection efficiency and the properties of the TOF mass analyzer when fed from a QIT. A schematic diagram of the vacuum MALDI QIT/TOF MS is shown in Figure 19.1. The ion optical system for injecting ions in the QIT comprises a 'pyramid mirror' situated between two sets of axially symmetric electrodes, to which are applied appropriate potentials, forming two successive Einzel lenses. The pyramid mirror reflects the laser beam (N_2, 337 nm, 5 ns pulse) in the direction of the ion optical axis and also allows for optical access to the MALDI target by a CCD camera. The pressure in the ion source as measured by a hot cathode gauge is ca 10^{-4} mTorr.

---

* See also Volume 4, Chapter 12: Axially-resonant Excitation Linear Ion Trap (AREX LIT), by Yuichiro Hashimoto.

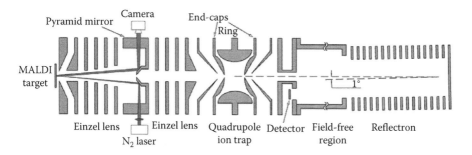

FIGURE 19.1   Schematic diagram of the vacuum MALDI QIT/TOF MS.

The ions are focused through the electrodes using high-voltages (e.g., 5 keV). A weak electric field is established on top of the target plate to push gently the ions through the dense MALDI plume.

The QIT has a three-dimensional (3D) unstretched geometry with a 10 mm inscribed radius of the ring electrode. Ions are decelerated prior to entering the QIT to kinetic energies of <50 eV. The RF voltage is switched off initially and the ions are further decelerated in the QIT by a 'reflectron' field established by a weak DC voltage-pulse applied to the exit end-cap electrode. The range of mass-to-charge (m/z) ratios stored is controlled by varying the time delay introduced between the laser pulse and switching-on the RF trapping field. The amplitude of the RF voltage and the corresponding low-mass cut-off (LMCO) are adjusted accordingly. Available electronics can generate the RF voltage at full amplitude (for example, $V_{RF} = 5$ kV zero-to-peak, $f = 500$ kHz) within a few tens of nanoseconds.

Isolation of ions in the QIT is based on the Filtered Noise Field (FNF) algorithm [10]. The waveform is applied to each of the two end-cap electrodes with reversed polarity, that is, in dipolar mode. A Fourier transformation of the FNF produces typically a histogram with frequencies ranging from 1 to 250 kHz with a variable step, usually set at 0.5 kHz. The notch appears at 70 kHz where the ions at approximately $q_z = 0.39$ (see Section 19.3 for the definition of q_z) satisfy the relationship $(m/z)/V_{RF} = 1$, that is, the amplitude of the RF voltage supplied to the ring and the mass-to-charge ratio of ions oscillating with a secular frequency of 70 kHz have the same numerical value (m/z measured in Th and V_{RF} in V). Prior to dissociation, the secular frequency of precursor ions is reduced to 35 kHz by decreasing the amplitude of the RF-voltage to half, $(m/z)/V_{RF} = 2$; thus, lowering the LMCO. Resonant excitation for inducing fragmentation is established by a low-voltage sinusoidal waveform at frequencies lower (*ca* 1%) than the secular frequency of the precursor ion. The waveform is applied to the end-cap electrodes in three steps of increasing amplitude for 60 ms. Ions are activated *via* collisions during a pressure transient established by pulsing argon.

The ejection of precursor and fragment ions for TOF mass analysis is achieved by two simultaneous high-voltage pulses applied to the end-cap electrodes, while the amplitude of the RF voltage is gradually reduced and forced to ground potential through a high-voltage switch. The extraction voltage-pulses establish a 16 kV difference across the 14 mm separation distance between the end-cap electrodes.

The flight tube is maintained at −10 kV (10 kV) for the analysis of positive (negative) ions and the total flight path from the QIT to the detector is 1800 mm. The TOF analyzer incorporates a gridless two-stage reflectron, tilted by *ca* 1.0°. The ions hit a dual micro-channel plate (MCP) at the end of their travel and the signal is processed by a 1 GHz transient recorder.

19.2.1 MEASURING PRESSURE TRANSIENTS IN A QIT

The diverse functions executed during the course of an experiment in a QIT may require different pressures for enhancing the overall performance. For example, the ability of the QIT to cool kinetically and to store ions is enhanced at *ca* 1 mTorr [11], while dissociation of ions appears enhanced both at reduced [12] and elevated pressures [13]. A similar situation arises in a QIT/TOF mass spectrometer where vacuum conditions are necessary for ejecting ions into the TOF mass analyzer to avoid scattering, thereby causing loss of mass-resolving power and sensitivity. The technique of pulsing gas into a trapping device to satisfy the demand for localized time-dependent pressures was implemented originally in ion cyclotron resonance (ICR) instruments [14], and extended later to QIT mass spectrometers [15], high-energy collision cells [6], and ion sources as a means for deactivating metastable ions [16]. Up to this point, despite the wide use of pulsed introduction of gases in mass spectrometry, there was no information available on the actual pressure profiles.*

A method for the determination of the dynamic pressure variations during pulsed introduction of gases in a QIT was reported recently [17]. The measurements were carried out on the MALDI AXIMA QIT, specially modified to reflect the actual variations in pressure during mass spectrometric analysis. The principle of the measurement relies on the operational aspects of a hot cathode gauge [18]. In brief, the QIT is enclosed in a shroud, which is differentially pumped by a 70 Ls^{-1} turbomolecular pump. The shroud accommodates pipes for connecting a leak valve to control the background pressure, together with two modified solenoid valves (Parker-Hannifin, Warwick, UK) to pulse gas during injection of ions in the QIT and during resonance excitation for the CID experiments. Figure 19.2 shows a cross-section of the ion-trap gauge and electron/ion trajectories calculated using SIMION [19]. Electrons emitted from the filament are accelerated to *ca* 80 eV and focused through the end-cap electrode. Collisions between electrons and the pulsed gas particles generate ions. A weak DC quadrupolar field is established by floating the end-cap electrodes at 25 V positive with respect to the ring electrode, which is maintained at ground potential. The time-dependent ion current collected on the ring and measured by an oscilloscope, $i_p(t)$, is proportional to the gas density and pressure, $p(t)$. The electron current, $i_e(t)$, is measured at the 'stop' electrode, which is connected to and situated behind the far end-cap electrode, and provides a measure of the number of electrons available for ionization. The sensitivity of the ion-trap gauge, s, is determined at steady-state conditions by admitting gas to increase gradually the pressure relative

---

* See also Volume 4, Chapter 17: Collisional Cooling in the Quadrupole Ion Trap Mass Spectrometer (QITMS), by Philip M. Remes and Gary L. Glish; Volume 4, Chapter 18: 'Pressure Tailoring' for Improved Ion Trap Performance, by Dodge L. Baluya and Richard A. Yost.

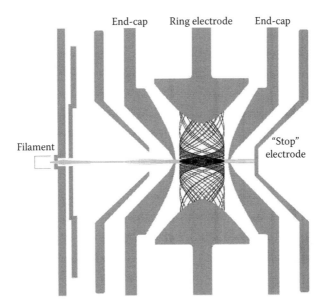

FIGURE 19.2 Trajectories in SIMION of electrons emitted from the filament and ions formed inside the ion-trap gauge.

to the background levels, p_0. If the ion and electron current transients are known, the pressure profiles can then be calculated using the expression [17]

$$p(t) = p_0 + \frac{i_p(t) - i_{p_0}}{s\, i_e(t)} \qquad (19.1).$$

The two pressure profiles in Figure 19.3 correspond to helium pulses obtained with the original and with modified instruments, respectively, using the same pulsed valve. Based on the pressure profile obtained from the original design, the gas lines and the pumping system were upgraded in order to increase the range of pressures accessible, to shorten the rise time of the pressure transient, and to minimize the residence time of the gas in the QIT. Consequently, the gas load to the surrounding compartments of the instrument was reduced considerably, the time frame for injecting/cooling ions and performing CID was shortened, and the sensitivity of the instrument improved. The improved sensitivity is due, in part, to the increased pressures, which confine those ions with initially unstable trajectories, by reducing their limits of excursion before they exceed the inner dimensions of the device, and presumably, in part, to the reduction in the amount of gas leaking from the introduction end-cap electrode and the degree of scattering the ions experience upon their approach. Moreover, the ability to repeat the pulses several times and to admit different gases throughout an experiment eradicates the compromise imposed by operating the QIT at a fixed pressure necessary to satisfy all possible functions during the course of an experiment. Finally, gas pulses (30 ms full-width, 10th-maximum) can serve as fast

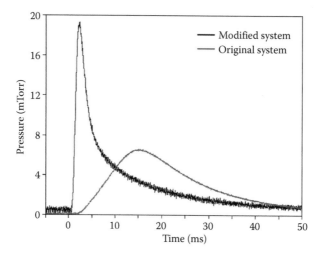

FIGURE 19.3 Pressure transients for helium measured before and after the modification of the gas introduction line and pumping system.

intermediate steps in between the different functions executed in the QIT during the course of an experiment, without adding appreciably to the overall analysis time. For example, the sensitivity of the instrument appears to be enhanced when the fragment ions generated with a broad distribution in their kinetic energies and positions in the QIT are cooled collisionally prior to ejection for mass analysis.

19.2.2 Ion Injection and Ejection Scheme

The efficiency of injection of ions into a QIT and ejection for TOF mass analysis determines to a great extent the sensitivity of the instrument. The process of injecting ions generated by MALDI into a QIT through an electrostatic lens in vacuum [20–25] has several distinguishable features. Ions generated by MALDI are characterized by a common velocity distribution [26] and the kinetic energy scales linearly with m/z-value. Collimation of such a non-monoenergetic ion beam through the end-cap electrode of a QIT is problematic since the angular divergence increases with m/z and strong electric fields are required; these are necessary also for minimizing the arrival time difference between packets of ions of different m/z ratios in order to inject a wider mass range. Storing a wide mass range is complicated due to the deflection the ions experience by the oscillating trapping field as they enter the QIT. Specially designed waveforms [22,24] have been developed to minimize such deflection, or even to eliminate the effect, as in the 'rapid RF start-up technique' implemented in the MALDI AXIMA QIT [9]. Ideally, elevated pressures are necessary during the injection process to enhance the trapping efficiency. The process of ejecting trapped ions for TOF mass analysis, and the mass resolving power that can be obtained, is limited by the variable initial conditions of the thermalized ions in the QIT, and the difficulty of damping the high-frequency, high-amplitude RF voltage instantaneously at the onset of the extraction voltage pulses applied to the end-cap electrodes. The design characteristics of the instrument address many of the complications involved in developing the hybrid MALDI QIT/TOF MS, for example the ability to satisfy the demand for variable pressure conditions in the QIT. The method of ion injection is another critical feature and has been characterized using simulations. Estimation of the ion ejection efficiency and the properties of the QIT as a source for the TOF mass analyzer have been investigated experimentally.

Simulation of ion trajectories through the electrostatic lens and the QIT was performed using AXSIM [27]. Snapshots of the injection process are shown in Figure 19.4 for a wide range of m/z-values extending from 1700 Th to 11,500 Th at (a) 2 µs; (b) 9 µs; (c) 20 µs; (d) 30 µs; and (e) 200 µs, respectively. The injection process is initiated by pulsing helium gas into the shroud enclosing the QIT. The laser pulse is synchronized to the pulsed valve based on the pressure profiles determined experimentally. Following the laser shot, ions are accelerated initially by a weak electric field (a), and acquire a maximum kinetic energy of a few kiloelectronvolts as they traverse the pyramid mirror. The ion beam elongates progressively before the lighter and faster ions are decelerated at the entrance of the QIT (b). The size of the beam contracts and the trapping volume is filled slowly with ions (c). The ions first entering the QIT experience a "reflectron" field established by a weak DC voltage-pulse applied to the exit end-cap electrode, while the heavier ones are fed continuously through the introduction end-cap electrode hole. The time interval that the ions are allowed to fill the trapping volume is defined by the onset of the RF trapping field relative to the laser pulse; in this example 23.5 µs (d). The flight time for ions of m/z in excess of 10,000 Th exceeds this time interval and the trapping efficiency drops considerably due to the strong deflection experienced by the RF fringe fields at the entrance hole on the end-cap electrode. Ions that survive the first 200 µs in the QIT can be stored efficiently (e).

The trapping efficiency across the mass range examined is shown in Figure 19.4f. The amplitude of the RF voltage is set at 4 kV and the drive frequency at 500 kHz, which correspond to a LMCO of 1722 Th. The timing of the electrical pulses (application of the RF trapping field and duration of the weak DC voltage-pulse applied to the exit end-cap electrode) is adjusted accordingly to store this lowest mass in the QIT. AXSIM investigations employing the hard-sphere collision model [28] indicate that the effect of gas on the trapping efficiency is not critical during the first 200 µs of the trapping motion. The estimated losses can be attributed solely to the wide initial ion kinetic energy spread and the corresponding angular divergence of the incoming ion beam. Minimum losses are also observed for ions present in the QIT prior to the application of the RF trapping field, despite their rather wide kinetic energy and spatial spreads. The variability of the trapping efficiency is due to a statistical error related to the number of ions used in the simulation. Averages values of the trapping efficiency exceeding 80% are considered typical.

The efficiency of the ion ejection process and that of transmission through the TOF mass analyzer determine the range of m/z-values of ions arriving at the detector; optimum efficacies for both processes are essential for exploiting in full the property of the TOF analyzer to accommodate an unlimited mass range, in this case the range of m/z-values stored in the QIT. Unfavorable to this exploitation is the size of the ion cloud stored in the QIT that increases as the pseudopotential well grows

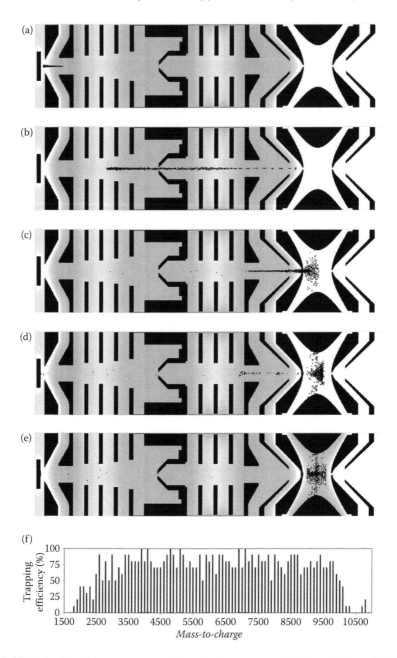

FIGURE 19.4   Simulation snapshots of the injection process at (a) 2 μs; (b) 9 μs; (c) 20 μs; (d) 30 μs; and (e) 200 μs relative to the laser shot. The trapping efficiency of injected ions is also shown (f).

shallower for the heavier ions (see Section 19.3). Transmission can be restricted, therefore, by the diameter of the hole on the exit end-cap electrode, which must remain sufficiently small to prevent penetration of the field external to the QIT electrodes that would cause distortion of the quadrupolar field. In addition, the two-stage gridless reflectron acts as a divergent lens unless the ions remain tightly on-axis. The latter effect becomes important as the initial conditions of the ions for the TOF experiment are characterized by variable spatial and velocity distributions across the entire mass range stored in the QIT.

The (a,q) stability diagram and the operating line of the QIT running at 500 kHz and $V_{RF} = 4$ kV (zero-to-peak) are shown in Figure 19.5a. Experiments with

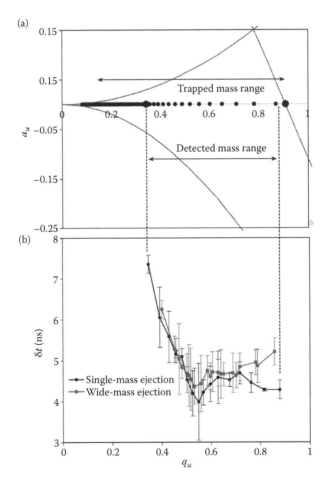

FIGURE 19.5 (a) Stability diagram and operating line of the QIT running at 500 kHz and $V_{RF} = 4$ kV. Arrows indicate the simulated mass range stored in the QIT, and the mass range detected experimentally. The LMCO (1722 Th) and the highest m/z-value detected experimentally (5000 Th) are highlighted on the operating line; (b) variations in the arrival time spread of the ions (FWHM) measured by the MCP as a function of the stability parameter q_u (equivalent to q_z for motion along the z-axis of the ion trap).

polyethyleneglycol (PEG) samples to identify the range of masses transmitted through the TOF analyzer are compared to the trapping efficiency estimated using simulations. As indicated by the arrows in the Figure 19.5a, the detected mass-to-charge ratio range corresponds to 40% of the range stored in the QIT that extends from the LCMO of 1722 Th to ca 10,000 Th, which is equivalent to $q_u = q_z \approx 0.15$ (see Section 19.3 for the definition of q_z).

Variations in the arrival time spread of the ions as measured by the MCP are shown in Figure 19.5b and indicate the mass-dependent initial conditions when a QIT serves as a source for a TOF mass analyzer. The two sets of results illustrated correspond to trapping and ejecting the entire mass-to-charge ratio range, and trapping, isolating, cooling, and ejecting a single mass for TOF analysis. The minimum arrival time spread, measured at full-width half-maximum (FWHM), in both cases is 3.5 ns and corresponds to $q_u \approx 0.55$ (m/z ca 2840 Th).

19.2.3 SUPPRESSING METASTABLE ION DECAY

Metastable ions in mass spectrometry are internally excited species, sufficiently stable to leave the ionization region, but with half-lives shorter than their TOF to the detector [29,30]. TOF instruments incorporating a reflectron expose the metastable character of ions generated by MALDI and, despite the low mass accuracy of the fragment ions, serve as a tool for molecular structure investigations [31]. The general notion is that the appearance of fragment ions in the mass spectrum is problematic unless it can be controlled. The intensity of such unwanted fragmentation in MALDI TOF MS, which becomes problematic especially in the analysis of complex mixtures, can be reduced significantly by lowering the laser fluence, thus extending the half-life of metastable ions; however, such ions still lack the long-term stability necessary for trapping devices. Collisional activation of metastable ions injected into an ion trap can accelerate the kinetics of unimolecular dissociation and enhance the problem. Efforts to deactivate the ions generated by vacuum MALDI were focused initially on controlling the working pressure of the ion trap analyzer [32,33]. Finally, efficient deactivation and stabilization was demonstrated by operating the ion source at elevated pressures (ca 1 Torr) and with the simultaneous use of multipole ion guides [34–37].

Deactivation and stabilization of ions generated by vacuum MALDI was demonstrated recently in a QIT by pulsing gas to achieve pressures exceeding the threshold of 10 mTorr for ca 3 ms [38]. The key element for controlling unimolecular dissociation of internally excited molecules relies on the fast transition to an environment where the injected ions undergo thermalizing collisions with the buffer gas present in the QIT. Experiments showed that the suppression of the metastable character of the ions is no longer possible unless the process of collisional cooling, the damping of the initial high-kinetic energy of the ions, is completed within a few hundreds of microseconds. For the preservation of intact molecular ions it is critical, not only to exceed the threshold of 10 mTorr in ca 1 ms or less, but to synchronize the timing of ion injection to the peak of the pressure transient. The effect of the two pressure transients shown in Figure 19.3 on the quality of the mass spectra is presented in Figure 19.6. Six standard peptides were used to demonstrate the

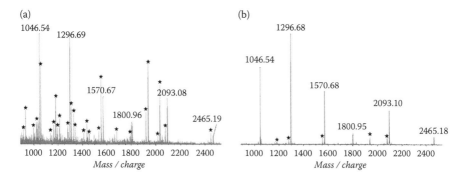

FIGURE 19.6 (a) Mass spectrum showing the extensive dissociation of six peptides injected in the QIT obtained with the original system. The peptides precursor ions are: *m/z* 1046.54 (Angiotensin II), *m/z* 1296.68 (Angiotensin I), *m/z* 1570.68 (Glu-fibrinopeptide), *m/z* 1800.95 (N-acetyl renin), *m/z* 2093.10 (ACTH fragment 1-17), and *m/z* 2465.18 (ACTH fragment 18-39); (b) Mass spectrum demonstrating deactivation and stabilization of the injected peptides obtained with the modified system. Both samples were mixed with CHCA. Fragment ions are indicated by asterisks.

suppression of ion dissociation induced upon injection into the QIT; their *m/z* ratios expressed in Th are: 1046.54 (Angiotensin II), 1296.68 (Angiotensin I), 1,570.68 (Glu-fibrinopeptide), 1800.95 (N-acetyl renin), 2093.10 (ACTH fragment 1–17), and 2465.18 (ACTH fragment 18–39). The matrix used was α-cyano-4-hydroxycinnamic acid (CHCA). The mass spectrum in Figure 19.6a was obtained with the original MALDI AXIMA QIT™, and is indicative of the extensive dissociation of all the peptides present in the sample. The most abundant fragment ions are marked with an asterisk. Figure 19.6b shows that the pressure transient established following the modifications to the instrument has reduced dissociation of the ions nearly to baseline levels, even when using CHCA as a matrix and which is known to enhance the metastable character of the ions.

19.3 THEORY OF EQUILIBRATED ION CLOUDS

The behavior of ions in equilibrium with the buffer gas particles is important for understanding the fundamental aspects of the operation of RF ion traps. Accurate theoretical investigations concerned with the time-dependent properties of an ensemble of ions stored in quadrupole fields exist [39,40]. A simpler approach to the problem is described herein, which was presented recently in a more concise form [38]. The analysis is based on the general properties of ion motion in RF fields and basic physical principles. The Boltzmann distribution function of non-interacting particles in thermal equilibrium at temperature T interacting with a potential field $\Phi(\mathbf{r})$ is [41]

$$f(\mathbf{r},\mathbf{v}) = C \cdot \exp\left[-\beta\left(\frac{m\mathbf{v}^2}{2} + \Phi(\mathbf{r})\right)\right] \qquad (19.2),$$

where $\beta = (kT)^{-1}$, k is the Boltzmann constant and $\mathbf{v},\mathbf{r}$ are velocity and position of the particles, respectively. The time-dependent potential in the QIT when the RF voltage is applied to the ring electrode is

$$\Phi(\mathbf{r},t) = V_{RF} \cos \Omega t \cdot \left(0.5 + 0.5 \frac{x^2 + y^2 - 2z^2}{r_0^2} \right) \tag{19.3},$$

where, V_{RF} is the zero-to-peak amplitude of the RF voltage (DC voltage is zero), Ω is the frequency of the RF potential, and r_0 is the inscribed radius of the ring electrode.

Since the equilibrium of the ions in a buffer gas is non-stationary, so must be the distribution function. The 'effective potential', or 'pseudopotential', theory [42] provides considerable simplification and the averaged motion of a particle is described in terms of a pseudopotential function

$$\Phi_{eff}(\mathbf{r}) = \int_T \frac{e^2}{m} \left| \nabla \Phi(\mathbf{r},t) \right|^2 dt \tag{19.4},$$

where integration takes place over one complete cycle. In the case of a quadrupole field the effective potential function using Equation 19.4 becomes

$$\Phi_{eff} = \bar{D}_z \left(\frac{z^2}{z_0^2} + 0.5 \frac{x^2 + y^2}{r_0^2} \right) \tag{19.5},$$

where $\bar{D}_z = q_z V_{RF}/8$ is the full depth of the pseudopotential well in the z-direction, $q_z = 4eV_{RF}/m\Omega^2 r_0^2$ is the so-called 'stability parameter', and $z_0 = r_0/\sqrt{2}$ is half the separation distance between the inner surfaces of the end-cap electrodes of the trap in the z-direction. In a pure quadrupole field of a 3D ion trap the stability parameters in all three directions are related, $q_x = q_y = -0.5q_z$. For all the following calculations we use a single parameter q, defined as the stability parameter in the z-direction, $q = q_z$. The density distribution of the "average" positions and velocities of the ions in equilibrium with the buffer gas at temperature T is obtained by substituting Equation 19.5 in Equation 19.2

$$n(\mathbf{r}) = N \left(\frac{e\bar{D}_z}{\pi kT z_0^2} \right)^{1/2} \cdot \exp \left[-\frac{e\bar{D}_z}{kT} \cdot \frac{z^2}{z_0^2} \right] \cdot \frac{0.5e\bar{D}_z}{\pi kT r_0^2} \cdot \exp \left[-\frac{0.5e\bar{D}_z}{kT} \cdot \frac{x^2 + y^2}{r_0^2} \right] \tag{19.6}.$$

Here, N is the total number of ions. Equation 19.6 states that the distribution density is normal in each direction with standard deviations $\sigma_z = 0.5r_0 \sqrt{kT/e\bar{D}_z}$ and $\sigma_r = r_0 \sqrt{kT/e\bar{D}_z}$ in the axial and radial directions, respectively. An estimation of the upper m/z ratio of an ion cloud, which occupies the full radius of the QIT, can be made by considering that more than 99% of the ions appear within three standard deviations from the center

$$r_c = 3\sigma_r = 3r_0 \sqrt{\frac{kT}{e\bar{D}_z}} \tag{19.7}.$$

where r_c is the radius of the ion cloud, whose diameter $d_c = 2r_c$. The diameter of an ion cloud, d_c, increases with mass, m, as the depth of the pseudopotential well grows shallower, $\bar{D}_z \propto (1/m)$. The threshold where the diameter of the ion cloud exceeds that of the ejection hole on the end-cap electrode is a critical parameter for ejecting ions efficiently in the TOF mass analyzer. The condition for the upper mass ejected without losses through the end-cap electrode hole (diameter d_h) is obtained by comparing the two diameters, $d_c \leq d_h$. Using Equation 19.7 we obtain that, $6r_0\sqrt{kT/e\bar{D}_z} \leq d_h$ and the expressions for the value of the q parameter and the corresponding mass are, respectively,

$$q \geq 288 \left(\frac{r_0}{d_h}\right)^2 \frac{kT}{eV_{RF}}, \quad \text{or} \quad m \leq \frac{eV}{72\Omega^2 r_0^2} \left(\frac{d_h}{r_0}\right)^2 \frac{eV_{RF}}{kT} \tag{19.8}.$$

For simplifying the numerical calculations, the last expression 19.8 can be re-arranged as follows:

$$m_{[Th]} \leq 1.313 \frac{V_{RF\,[V]}^2}{f_{[MHz]}^2\, r_{0\,[mm]}^2} \left(\frac{d_h}{r_0}\right)^2 \frac{300}{T_{[K]}} \tag{19.9},$$

where $f = \Omega/2\pi$. The numerical factor in this last expression is obtained by substituting for the unit charge, $e = 1.60217 \times 10^{-19}$ C, $\pi = 3.14159$, and converting SI units to those used in the brackets as subscripts. For the ion trap under consideration ($f = 0.5$ MHz, $r_0 = 10$ mm) and by setting the zero-to-peak amplitude of the RF voltage at $V_{RF} = 4000$ V, at $T = 300$ K, Equation 19.9 predicts that ions below 8400 Th will be transmitted through the 1 mm diameter hole without losses. The estimated transmission efficiency for the ca 10,000 Th ion, the heaviest mass trapped according to the simulation results, is $d_h/d_c \approx 0.92$. The upper m/z-value determined experimentally is ca 5000 Th. This discrepancy between theory and experiment is attributed partly to the increased size of the ion cloud as a result of reducing the amplitude of the RF voltage before ion ejection, and partly to the mass-dependent transmission of the ions through the reflectron TOF mass analyzer. The upper m/z-value observed experimentally can also be limited by space-charge interactions, which increase the size of the ion cloud for the heavier m/z-values when the lower ones are abundant.

Further analysis of Equation 19.8 shows the importance of increasing the amplitude of the RF voltage for enhancing transmission. Unfortunately, there is a trade-off between V_{RF} and the LMCO. When V_{RF} is increased to improve ejection efficiency, the value of the LMCO can be maintained by increasing the frequency according to $f \propto \sqrt{V_{RF}}$. Equation 19.8 can be combined with the stability conditions to estimate the mass range transmitted through the end-cap electrode hole with 100% efficiency. The lowest mass is determined by the LMCO at $q = 0.908$, or $m = 4eV_{RF}/0.908\Omega^2 r_0^2$. This last expression for the lowest mass trapped efficiently can be combined with Equation 19.8 to give the ratio of highest/lowest masses (relative mass range), which can be transmitted with 100% efficiency from the ion trap into TOF:

$$\frac{m_{high}}{m_{low}} = \frac{0.908}{288} \left(\frac{d_h}{r_0}\right)^2 \frac{eV_{RF}}{kT}, \quad \text{or} \quad \frac{m_{high}}{m_{low}} = 0.122 \left(\frac{d_h}{r_0}\right)^2 V_{RF\,[V]} \frac{300}{T_{[K]}} \tag{19.10}.$$

For the conditions examined above, the relative mass/charge range calculated theoretically is nearly five times the LMCO, while the experimental value is greater by a factor of three. Assuming a fixed frequency and amplitude of the RF voltage, Equation 19.10 shows the possibility of increasing the mass range without affecting the LMCO by using a colder buffer gas. Reducing the temperature is an attractive solution since lower temperatures can also reduce the spread in ion positions and velocities prior to ejection for TOF mass analysis, and consequently improve the mass resolving power; however, the use of cold buffer gases is associated with many technological challenges and has not been implemented so far in commercial instruments.

Another important consequence of the Boltzmann distribution function, Equation 19.2, relates to the single value of the temperature used to describe both the velocity distribution of the ions in equilibrium and that of the buffer gas. It must be noted that while spatial micro-oscillations due to the periodic RF potential are comparatively small (the ratio of the amplitude of micro-oscillations with respect to the amplitude of the 'slow' secular oscillation in the z-direction is of the order of $q_z/2$ for $q_z < 0.4$), the amplitude of velocity of micro-oscillations is of the same order as the amplitude of velocity of the 'secular' motion.

Derivation of the equations of ion motion in the time-dependent quadrupole field defined in Equation 19.3 is as follows. The equation of ion motion in the x-direction is

$$m\frac{d^2x}{dt^2} = -e\frac{\partial}{\partial x}\Phi(\mathbf{r},t), \quad \text{or} \quad m\frac{d^2x}{dt^2} = -\left(\frac{eV_{\text{RF}}}{r_0^2}\cos\Omega t\right)\cdot x \qquad (19.11).$$

Assuming that the solution to the equation of motion can be expressed as a sum of the 'slow' secular motion, $X_s(t)$, and the 'fast' micro-oscillations, $h_x(t)\cos\Omega t$, we may write

$$x(t) = X_s(t) + h_x(t)\cos\Omega t \qquad (19.12).$$

The 'fast' motion has a characteristic frequency equal to the RF frequency, while the frequency of the 'slow' secular motion is much smaller. Taking the second derivative with respect to time we obtain

$$\frac{d^2x(t)}{dt^2} = \frac{d^2X_s(t)}{dt^2} - h_x(t)\Omega^2\cos\Omega t. \qquad (19.13).$$

The assumption made here is that $h_x(t)$ is approximately independent of time. Substituting Equation 19.13 into Equation 19.11 the equation of motion in the x-direction becomes

$$m\frac{d^2X_s(t)}{dt^2} - mh_x(t)\Omega^2\cos\Omega t = -e\frac{V_{\text{RF}}}{r_o^2}\cos\Omega t \cdot \left[X_s(t) + h_x(t)\cos\Omega t\right],$$

or

$$\frac{d^2 X_s(t)}{dt^2} - h_x(t)\Omega^2 \cos\Omega t = -\frac{eV_{RF}}{mr_0^2} \cdot [X_s(t)\cos\Omega t + h_x(t)\cos^2\Omega t] \quad (19.14).$$

Equation 19.14 contains terms with frequencies significantly different than the frequency of oscillation. The functions $X_s(t)$ and $h_x(t)$ are 'slow' with frequencies equal to the frequency of the secular motion, while terms proportional to $\cos\Omega t$ have a frequency of oscillation close to the RF frequency, and terms proportional to $\cos^2\Omega t$ have twice that frequency. With the use of the following trigonometric relationship, $\cos^2\alpha = 0.5 + 0.5\cos 2\alpha$ Equation 19.14 can be rearranged to

$$\frac{d^2 X_s(t)}{dt^2} + \frac{eV_{RF}}{mr_0^2} \cdot h_x(t)[0.5 + 0.5\cos 2\Omega t] - h_x(t)\Omega^2 \cos\Omega t + \frac{eV_{RF}}{mr_0^2} \cdot X_s(t)\cos\Omega t = 0$$

$$(19.15).$$

The 'slow' terms of Equation 19.15 have no $\cos\Omega t$ factor and should satisfy the equation

$$\frac{d^2 X_s(t)}{dt^2} + 0.5\frac{eV_{RF}}{mr_0^2} \cdot h_x(t) = 0 \quad (19.16a).$$

Similarly, the 'fast' oscillating terms are proportional to $\cos\Omega t$, and can be written as

$$-h_x(t)\Omega^2 \cos\Omega t + \frac{eV_{RF}}{mr_0^2} \cdot X_s(t)\cos\Omega t = 0,$$

or

$$h_x(t) = \frac{eV_{RF}}{m\Omega^2 r_0^2} X_s(t) \quad (19.16b).$$

The terms proportional to $\cos 2\Omega t$ function are oscillating with frequency twice as that of the RF potential. In the following calculations these terms are neglected, assuming that the fast oscillation does not contribute significantly to the ion motion. Finally, with the use of Equation 19.16b, the equation of the 'slow' motion, Equation 19.16a, becomes

$$\frac{d^2 X_s(t)}{dt^2} + \frac{e^2 V_{RF}^2}{2m^2 r_0^4} \cdot X_s(t) = 0 \quad (19.17).$$

Substituting for the stability parameter q, the equations of motion are

$$x(t) = X_s(t) \cdot \left[1 + 0.25q\cos\Omega t\right] \tag{19.18a},$$

$$m\frac{d^2X_s(t)}{dt^2} + e\frac{qV_{RF}}{8r_0^2} \cdot X_s(t) = 0 \tag{19.18b}.$$

The last expression of the 'slow' secular motion is an equation of harmonic vibrations with constant frequency $\omega_{s,x} = \sqrt{qeV_{RF}/8mr_0^2}$. Such type of ion motion is possible only in a parabolic potential, as that defined in Equation 19.5.

So far, we have derived the equation for ion motion in the x-direction only. For the y-direction the equation of secular motion will be the same as that in Equation 19.18b. For the axial motion, z-direction, and using the above formalism, the result is

$$z(t) = Z_s(t) \cdot \left[1 - 0.5q\cos\Omega t\right] \tag{19.18c},$$

$$m\frac{d^2Z_s(t)}{dt^2} + e\frac{qV_{RF}}{2r_0^2} \cdot Z_s(t) = 0 \tag{19.18d}.$$

The pseudopotential function, defined in Equation 19.5, can be derived directly from the equations of motion, Equation 19.18b and Equation 19.18d.

For the computation of the average kinetic energy of the ions we use Equation 19.18a and Equation 19.18c. The velocity of the ion motion in the RF field is calculated as follows. Starting with Equation 19.18a for the ion motion in the x-direction we must take into account the fact that the variation of the function $X_s(t)$ is much slower than the rate of change of the factor $\cos\Omega t$ when taking derivatives. It follows then that the velocity of the particle is

$$v_x(t) = \frac{dx(t)}{dt} \approx \frac{dX_s(t)}{dt} - \frac{q}{4}X_s(t)\sin\Omega t \tag{19.19}.$$

Introducing the velocity of the 'slow' motion $V_{s,x} = dX_s/dt$ Equation 19.19 can be expressed as

$$v_x(t) \approx V_{s,x}(t) - \frac{q}{4}X_s(t)\Omega\sin\Omega t \tag{19.20}.$$

The kinetic energy in the x-direction, K_x, is

$$K_x = \frac{1}{2}mv_x^2 = \frac{1}{2}mV_{s,x}^2 - \frac{q\Omega}{4}V_{s,x}X_s\sin\Omega t + \frac{q^2\Omega^2}{32}X_s^2\sin^2\Omega t \tag{19.21}.$$

The time-dependent average value of the kinetic energy of the ions in equilibrium can be obtained by considering the statistical average from the left and right parts of Equation 19.21

$$\langle K_x \rangle = \frac{1}{2}m\langle V_{s,x}^2 \rangle - m\frac{q\Omega}{4}\langle V_{s,x}X_s \rangle \sin\Omega t + m\frac{q^2\Omega^2}{32}\langle X_s^2 \rangle \sin^2\Omega t \qquad (19.22).$$

For the computation of the average value of $\langle V_{s,x}^2 \rangle$ we need to use the Maxwell velocity distribution for particles in equilibrium at temperature T. In the X direction the probability distribution for particles with velocities from $V_{s,x}$ to $V_{s,x} + dV_{s,x}$ is

$$dP(V_{s,x}, dV_{s,x}) = f_1(V_{s,x}) \cdot dV_{s,x},$$

where

$$f_1(V_x) = \sqrt{\frac{m}{2\pi kT}}\exp\left(-\frac{mV_x^2}{2kT}\right) \qquad (19.23).$$

Computation of the average value of $V_{s,x}^2$ is carried out with the use of Equation 19.23 as follows

$$\langle V_{s,x}^2 \rangle = \int_{-\infty}^{+\infty} V_{s,x}^2 f_1(V_{s,x}) dV_x = \int_{-\infty}^{+\infty} V_{s,x}^2 \sqrt{\frac{m}{2\pi kT}}\exp\left(-\frac{mV_x^2}{2kT}\right) dV_{s,x}$$

$$= \int_{-\infty}^{+\infty}\left(\xi\sqrt{\frac{2kT}{m}}\right)^2\sqrt{\frac{m}{2\pi kT}}\exp(-\xi^2)\sqrt{\frac{2kT}{m}} \cdot d\xi = \frac{2kT}{m}\cdot\frac{1}{\sqrt{\pi}}\cdot\int_{-\infty}^{+\infty}\xi^2\exp(-\xi^2)d\xi = \frac{kT}{m}$$

$$(19.24).$$

Here, we have introduced the variable $\xi = V_x\sqrt{m/2kT}$ and used the relationship [43] $\int_{-\infty}^{+\infty}\xi^2\exp(-\xi^2)d\xi = \sqrt{\pi}/2$. Finally, for the first term of Equation 19.22 we obtain

$$\frac{1}{2}m\langle V_{s,x}^2 \rangle = \frac{kT}{2} \qquad (19.25).$$

The result is a simple consequence of the fundamental statistical law which states that at steady-state conditions and at temperature T the average value of energy per degree of freedom is equal to $0.5kT$.

By taking into account that the position and velocity are independent statistical variables, hence $\langle V_{s,x}X_s \rangle = \langle V_{s,x} \rangle \cdot \langle X_s \rangle$ and the average value $\langle V_{s,x} \rangle$ is zero, we obtain that

$$\langle V \rangle = \int_{-\infty}^{+\infty} V_{s,x} f_1(V_x) dV_{s,x} = \int_{-\infty}^{+\infty} V_{s,x}\sqrt{\frac{m}{2\pi kT}}\exp\left(-\frac{mV_{s,x}^2}{2kT}\right) dV_{s,x} = 0 \qquad (19.26)$$

The function under the integral is odd and since integration takes place over a symmetrical (infinite) range the value of the integral is zero.

For the computation of the value of $\langle X_s^2 \rangle$ for a single particle we need to use the distribution function, Equation 19.6:

$$\langle X_s^2 \rangle = \frac{1}{N} \int x^2 \cdot n(\mathbf{r}) dx \, dy \, dz$$

$$= \left(\frac{e\bar{D}_z}{\pi kTz_0^2} \right)^{1/2} \cdot \frac{0.5e\bar{D}_z}{\pi kTr_0^2} \int_{-\infty}^{+\infty} \exp\left[-\frac{e\bar{D}_z}{kT} \cdot \frac{z^2}{z_0^2} \right] dz$$

$$\cdot \int_{-\infty}^{+\infty} \exp\left[-\frac{0.5e\bar{D}_z}{kT} \cdot \frac{y^2}{r_0^2} \right] dy \cdot \int_{-\infty}^{+\infty} x^2 \exp\left[-\frac{0.5e\bar{D}_z}{kT} \cdot \frac{x^2}{r_0^2} \right] dx \qquad (19.27).$$

Working in the z-direction

$$\int_{-\infty}^{+\infty} \exp\left[-\frac{e\bar{D}_z}{kT} \cdot \frac{z^2}{z_0^2} \right] dz = z_0 \sqrt{\frac{kT}{e\bar{D}_z}} \int_{-\infty}^{+\infty} \exp(-\xi^2) d\xi = z_0 \sqrt{\frac{kT}{e\bar{D}_z}} \sqrt{\pi} \; . \qquad (19.28).$$

Here, we have introduced and used the relationship [43] $\int_{-\infty}^{+\infty} \exp(-\xi^2) d\xi = \sqrt{\pi}$.
Working in the y-direction

$$\int_{-\infty}^{+\infty} \exp\left[-\frac{0.5e\bar{D}_z}{kT} \cdot \frac{y^2}{r_0^2} \right] dy = r_0 \sqrt{\frac{kT}{0.5e\bar{D}_z}} \int_{-\infty}^{+\infty} \exp(-\xi^2) d\xi = r_0 \sqrt{\frac{kT}{0.5e\bar{D}_z}} \sqrt{\pi} \qquad (19.29),$$

in which the new parameter introduced above $\xi = y/r_0 \sqrt{0.5e\bar{D}_z/kT}$.
With the use of Equations 19.28 and 19.29, the integral Equation 19.27 becomes

$$\langle X_s^2 \rangle = \sqrt{\frac{0.5e\bar{D}_z}{\pi kTr_0^2}} \cdot \int_{-\infty}^{+\infty} x^2 \exp\left[-\frac{0.5e\bar{D}_z}{kT} \cdot \frac{x^2}{r_0^2} \right] dx = \xi = \frac{x}{r_0} \sqrt{\frac{0.5e\bar{D}_z}{kT}}$$

$$= r_0^2 \frac{kT}{0.5e\bar{D}_z} \frac{1}{\sqrt{\pi}} \int_{-\infty}^{+\infty} \xi^2 \exp\left(-\xi^2 \right) d\xi = r_0^2 \frac{kT}{e\bar{D}_z} \qquad (19.30).$$

The last term in Equation 19.22 without the $\sin^2\Omega t$ factor writes as follows:

$$m\frac{q^2\Omega^2}{32}\langle X_s^2 \rangle = m\frac{q^2\Omega^2}{32} r_0^2 \frac{kT}{e\bar{D}_z} = m\frac{4eV_{RF}}{m\Omega^2 r_0^2} \frac{q\Omega^2}{32} r_0^2 \frac{kT}{e\bar{D}_z} = \frac{qeV_{RF}}{8} \cdot \frac{kT}{e\bar{D}_z} = kT \quad (19.31).$$

Finally, we obtain the equation for the average kinetic energy in the x-direction

$$\langle K_x \rangle = \frac{kT}{2} + kT \sin^2 \Omega t, \quad \text{or} \quad \langle K_x \rangle = \frac{kT}{2} + \frac{kT}{2} \cdot \left[1 - \cos\left(2\Omega t \right) \right]. \quad (19.32).$$

Computation of the average kinetic energy in y- and z-directions can be carried out in a similar fashion and the result is equivalent to that of Equation 19.32. Consequently, the total kinetic energy of the ions is

$$\langle K \rangle = \langle K_x \rangle + \langle K_y \rangle + \langle K_z \rangle = \frac{3kT}{2} + \frac{3kT}{2} \left[1 - \cos(2\Omega t) \right] \quad (19.33).$$

It follows that the average value of the total kinetic energy of the ions in equilibrium with the buffer gas at temperature T oscillates, completing two cycles every period of the RF trapping field and has a maximum value of $9kT/2$, which is three times greater than the average thermal energy $3kT/2$.

Let us consider the consequences of this result in terms of the present investigation, where externally generated MALDI ions dissociate upon injection in the ion trap. Ions generated by MALDI emerge through a laser-induced explosion of the matrix carrying the analyte. The temperature of the exploding material is high and, as a result, the analyte ions become sufficiently excited internally to dissociate. This metastable character of the ions generated by MALDI can be suppressed inside the ion trap, where their excess vibrational energy can be transferred to the cold buffer gas. In order for this vibrational-to-translational energy transfer to be effective, the kinetic energy of the ions must be sufficiently low. Initially, the kinetic energy of the ions injected in the QIT is rather high and the buffer gas molecules cannot absorb part their internal energy efficiently. The rate of kinetic energy reduction depends on the ion collision cross-section, pressure and the mass of the buffer gas particles. According to our simulations and experimental results, the process of translational cooling for preserving intact molecular ions, is completed within less than 1 ms at 15 mTorr He. At this stage the average kinetic energy of the ions is defined by Equation 19.33. At room temperature of 300 K the value for $3kT/2 = 0.039$ eV. The ion kinetic energy oscillates as a result of the RF field, but even at its maximum value it is only three times higher and, therefore, remains well below 1 eV.

At such low kinetic energies Langevin collisions dominate [44] and the collision cross-section becomes inversely proportional to the velocity [45], in contrast to hard-sphere collisions where the cross-section is constant across a wide energy range. Activation/deactivation processes in ion traps are phenomena that occur simultaneously, even during resonance excitation conditions [46,47]; however, it has been suggested that internal heating of ions *via* collisions persists until they are kinetically cooled [48]. Theoretical studies suggest that the deactivation rate in a QIT is favored over activation for kinetically cooled ions [49], and the effect was demonstrated recently by experiment [50]. Consequently, the faster the transition to such thermalized collisions, the greater the possibility of stabilizing metastable ions formed externally to the QIT.

19.4 APPLICATIONS

The MALDI QIT/TOF instrument is employed in a broad range of biological and inorganic analytical applications. Its main advantages are the high sensitivity and the ability to isolate precursor ions with isotopic resolution over a wide mass range, as well as the power to maintain high mass resolving power at relatively high masses (*ca* 6000 Th), independent of the number of cycles during MS^n experiments. Given these advantages, the MALDI QIT/TOF appears to be suited ideally for high-end proteomics studies as well as less conventional studies with polymers and lipids.

19.4.1 GLYCOBIOLOGY

An important application in proteomics is the detection and characterization of post-translational modifications (PTM) [51]. Glycosylation is one of the major PTM that has significant effects on protein folding and ultimately on protein activity. The MALDI-QIT/TOF has been used extensively in carbohydrates research [52–58], where glycoproteins were isolated and digested subsequently with various endoproteinases in order to characterize O-linked and/or N-linked oligosaccharides. The analysis of oligosaccharide moieties attached to the peptide backbone can help in the identification of the number of sugar residues, the anomericity of the glycosidic bond (α-*versus* β-linkage), the position of the glycosidic bond as well as some branching information. Sequential fragmentation information as illustrated in Figure 19.7 can be obtained from glycopeptides and/or isolated oligosaccharides entities.*

The use of MALDI-MS^n leads not only to the N-linked oligosaccharide's structure but also to its position within the amino-acid sequence. In such experiments, several particularities are observed usually. Firstly, sialic acid (SA) residues attached at the end of the N-glycans are very labile and tend to be lost during the analysis process, although there is evidence that such losses may be minimized by judicious choice of the matrix [59]. Such metastable ion decay is present in almost all TOF

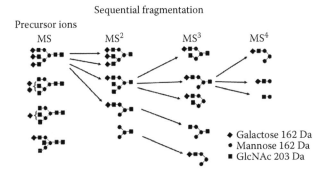

Sequential fragmentation

Precursor ions

MS   MS^2 MS^3 MS^4

♦ Galactose 162 Da
● Mannose 162 Da
■ GlcNAc 203 Da

FIGURE 19.7 Schematic representation of the fragmentation pathway of a glycan in an MS^4 experiment in the MALDI QIT/TOF MS.

---

* See also Volume 4, Chapter 16: Unraveling the Structural Details of the Glycoproteome by Ion Trap Mass Spectrometry, by Vernon Reinhold, David J. Ashline, and Hailong Zhang.

mass spectrometers incorporating a reflectron, where the flight time is sufficiently long to observe extensive loss of SA. Secondly, the fragmentation of glycopeptides leads generally to two sets of concomitant fragmentation pathways. The first one, concerned with the glycan structure, is characterized by the presence of fragments in the higher mass range of the MS^2 spectra. A second fragmentation pathway, related to the amino acid sequence, can be observed generally at the lower end of the MS^2 mass range.

The human transferrin glycopeptide (GP2) was characterized using MS^2 and MS^3 experiments illustrating the phenomenon described above. The MS^2 spectrum of a human transferrin tryptic glycopeptide (m/z 4136) is shown in Figure 19.8. The ion subjected to MS^2 analysis was $[M + H\text{-}2SA]^+$. The identification of the N-glycosylation site showing an enhancement of the peptide fragmentation processes of an ion carrying only one sugar residue (GlcNAc) was possible using the mass spectrum in Figure 19.9. In this case, the peptide selected for the MS^3 experiment (m/z 2701) has already lost a large part of the glycans moiety during the first cycle of the CID experiment. The mass spectrum also exhibits cleavage of the amino acid backbone *via* a series of ions without the remaining GlcNAc residue. The observation of the two portions of the y-ion series separated by a gap of 318 Da (1793.29–1475.19 Da), equivalent to an aspartic acid moiety attached to a GlcNAc sugar residue, indicates the exact location of the N-linked glycan within the peptide sequence. Figure 19.10 represents an example of how MS^3 can be used in order to obtain peptide sequence information starting with the MS^2 experiment of a glycopeptide. In this latter case, the selection of ions (from the MS^2 spectra) that do not carry any sugar moieties will lead to a fragmentation containing exclusively amino acid sequence information.

The fragmentation of a specific diagnostic ion using MS^4 experiments can provide valuable information regarding to the type of branching present in the oligosaccharide as is illustrated in Figure 19.11, where successive steps of isolation/dissociation of a high mannose glycan $(Man)_9(GlcNAc)_2$ are performed. The fragmentation of a diagnostic ion (m/z 923.4) was obtained using MS^4 experiments and provided valuable information regarding to the type of branching (see the inset) present in the high mannose sugar [60].

19.4.2 LIPIDOMICS

The lipids represent probably one of the most important classes of compounds in life sciences [61]. Due to their physico-chemical properties and structure, they exhibit a broad range of vital biological functions, ranging from energy storage to essential players in human metabolism [62,63]. In recent years, the study of lipids using mass spectrometry has increased considerably [64]. As a direct consequence, the use of MALDI mass spectrometry for the study of lipids [65] has gained popularity among the lipid research community. MALDI is a versatile tool for the analysis of lipids due to its fast and reliable method for the identification of lipids, as well as its tolerance to contaminants. It can be seen as a method that could, in the future, replace the need for HPLC separation, which is typically used in lipids' research [66]. In that respect, MALDI QIT/TOF MS offers a unique tool that allows the researcher to use the advantages of the MALDI technique, and the ability to fragment

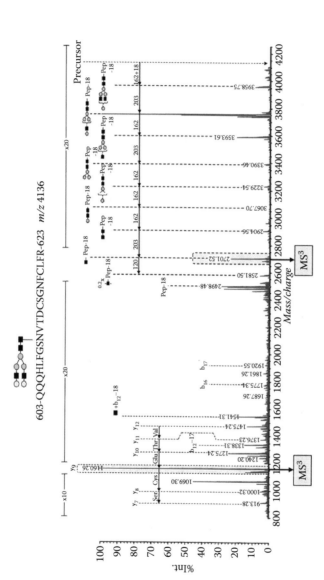

FIGURE 19.8 MS² spectrum of the human transferrin glycopeptide [M + H-2SA] + (m/z 4136) that has already lost two sialic acid moieties. The spectrum exhibits the fragmentation of the glycan part attached to the peptide.

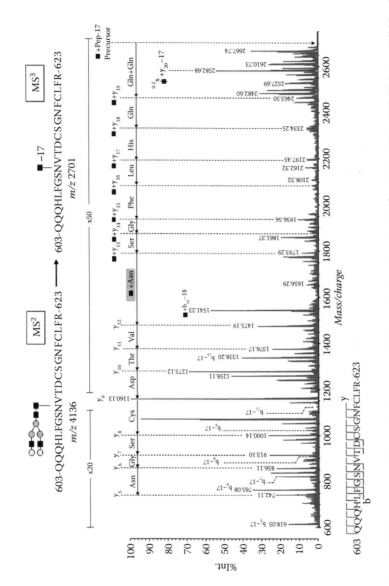

FIGURE 19.9 Positive MS³ spectrum of the product ion at m/z 2701 (originating from the MS² analysis of the 4139 Th ion), exhibiting fragmentation of the peptide backbone. The detection of specific fragments allowed us to identify the glycosylation site.

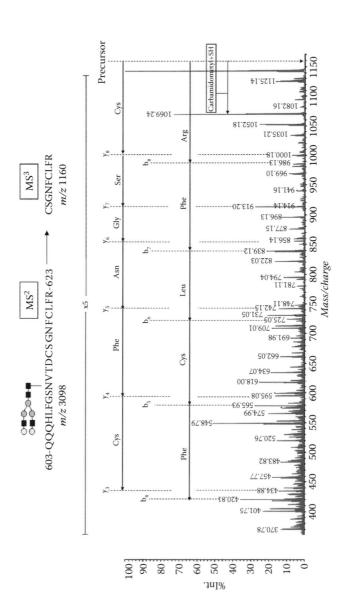

FIGURE 19.10 Positive ion MS³ spectrum of a product ion (*m/z* 1160), exhibiting fragmentation of the peptide backbone. The peptide sequence is completely assigned.

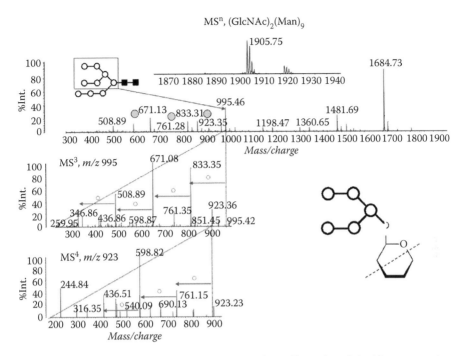

FIGURE 19.11 Fragmentation of a high mannose glycan. Detection of significant cross-ring cleavage ion at m/z 923.4. This ion is significant diagnostically as it contains only the antenna from the six-branch of the core mannose and thus giving information related to the branching. Product ions originating from cross-ring cleavages are indicated with shaded circles. Open circles are representing successive losses of mannose from a selected precursor.

selectively and successively mono-isotopes of interest. Mono-isotopic selection of lipids is crucial owing to the natural occurrence of C–C bond unsaturation in the fatty acid chain. Such unsaturation leads to ions in the mass spectrum separated by 2 Th, which are usually difficult to isolate and to fragment separately using standard MALDI–TOF instrumentation.

Figure 19.12 represents a typical MALDI mass spectrum of a crude mixture of lipids derived from human plasma. The matrix used for this purpose is 2,4,6-tri-hydroxy-acetophenone (THAP) spiked with di-ammonium hydrogen citrate (DAHC). The use of THAP matrix [67] in MALDI is well suited for the analysis of various classes of lipids. Plasma lipid extracts in MALDI exhibit mostly phospholipids and triacylglycerols (triglycerides or TAG). Depending on the experimental conditions (matrix used, solvent, cations, etc.), it is possible to ionize selectively certain classes of lipids individually from a mixture. In this case, the chosen conditions allowed for the detection of mainly the phosphatidylcholine and TAG species. The advantages of the MALDI QIT/TOF are the high tolerance to the required salts added during the analysis and the high sensitivity of the technique. Under these conditions it is still possible to select mono-isotopically each precursor ion in order to obtain the class and the fatty acid composition of the lipid. A positive MS^2 spectrum of a TAG (16:0, 18:2, 18:1) is shown in Figure 19.13. The sodium adduct [M + Na]+ (m/z 879.74) was detected and

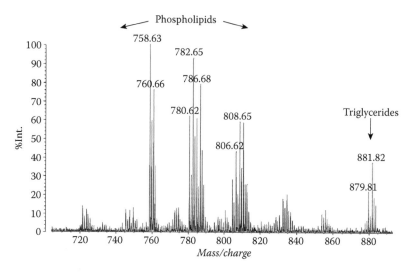

FIGURE 19.12 Positive ion mass spectrum of lipids derived from human plasma. The matrix employed was THAP.

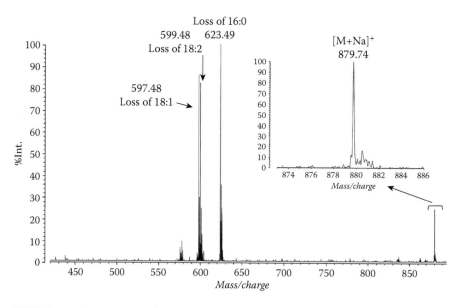

FIGURE 19.13 Positive MS² spectrum of a plasma TAG (16:0, 18:2, 18:1). The losses of three different fatty acid residues are observed.

selected for MS² with isotopic resolution in order to perform the CID experiment. The product ion mass spectrum exhibits clearly the loss of the three different fatty acid chains (loss of RCOOH) where RCOOH can be either a 16:0, an 18:1 or an 18:2 fatty acid. Such a method can be used as a fast screen for different mixtures of lipids, originating not only from plasma but also from various other sources. Experiments employing MS³ could also be used to detect and to characterize unequivocally isobaric species present at the MS1 stage.

19.4.3 Other Applications

The technique of QIT/TOF MS has been used mainly in the field of glycobiology, and especially carbohydrate research. This area of investigation is probably one of the most exciting and challenging parts of proteomics studies because it gives insights into the functions and interactions of proteins, in addition to peptide sequence information. More classical proteomics studies [68–74] involving the determination of the amino acid structures as well as *de novo* sequencing have found great advantages in using this technique. Other application areas, such as the analysis of polymers, remain rare and only few attempts [75] have been carried out. In general, polymer analyses are done using lithium, sodium, and potassium as additives in order to generate different adducts and to enhance the efficiency of fragmentation in MS² experiments. In some cases, [73] lithium adducts appear to be highly suitable for such enhancement, similar to the results that have been observed for the analysis of lipids [67]. The analysis of polymers, especially co-polymers, using MS³ is an area that will show growth in the future.

Product ion mass spectra of small molecules such as flavonoids have been investigated [76] and the low-energy CID spectra generated in the QIT have been compared to high-energy CID spectra obtained with a tandem TOF MS. Other small molecules' studies involve the investigation of antioxidants used in Chinese medicine [77].

Decoupling the ion source from the mass analyzer introduces distinctive features and has unique advantages for the investigation of molecules desorbed from uneven surfaces, for example, proteins desorbed directly from membranes [78]. The combination of product ion mass spectra of high quality obtainable directly from a tissue surface with the ability to perform MS^n experiments of high sensitivity establishes the MALDI QIT/TOF instrument as a promising tool for mass spectrometric imaging experiments [79,80]. Finally, this technique has been coupled successfully to thin layer chromatography, allowing the analysis of glycosphingolipids directly from the TLC plate [81].

19.5 CONCLUSIONS

The combination of a vacuum MALDI ion source and a QIT coupled to a reflectron TOF mass analyzer presents a powerful configuration, provided that key technical challenges associated with the design, construction, and performance of this type of an instrument are resolved. In this work, the metastable character of ions generated by vacuum MALDI and the kinetics of unimolecular dissociation are restrained using fast pressure transients established by pulsing gas in the QIT to enhance vibrational-to-translational energy transfer between the analyte species and the buffer gas particles. Intact ions of macromolecules can be stored efficiently for further manipulation. Principal aspects of this particular instrument configuration related to the injection and ejection efficiency of ions in and from of the QIT are discussed in detail, highlighting advantages and limitations of the technique, and also indicating areas of interest for further research and instrument development.

A simple mathematical approach to describe the average kinetic energy of the ions trapped in RF fields in equilibrium with the buffer gas is derived within the framework of the 'pseudopotential theory'. It is found that the kinetic energy of the ions oscillates completing two cycles in every RF period. The lowest value of the kinetic energy is equal to $3kT/2$, the energy of the particles in the absence of an external field. The maximum kinetic energy is three times higher the moment the RF potential passes through zero, in other words, when the rate of change of the RF potential is the greatest. Oscillations of the ions kinetic energy are due to the micromotion of the particles in the RF field.

Finally, the main advantages of the QIT to isolate sequentially precursor ions with unit resolution for MS^n experiments, as well as the ability to record mass spectral lines with mass resolving power independent of the number of cycles executed during the course of an experiment, are illustrated in the applications section.

REFERENCES

1. Tanaka, K.; Waki, H.; Ido Y.; Akita, S.; Yoshida, Y.; Yoshida, T.; Matsuo, T. Protein and polymer analyses up to m/z 100,000 by laser ionization time-of-flight mass spectrometry. *Rapid Commun. Mass Spectrom.* 1998, 2, 151–153.
2. Karas, M.; Hillenkamp, F. Laser desorption ionization of proteins with molecular masses exceeding 10,000 daltons. *Anal. Chem.* 1988, 60, 2299–2301.
3. O'Connor, P.B.; Hillenkamp, F. MALDI Mass Spectrometry Instrumentation. In Hillenkamp, F.; Peter-Catalinic, J. (Eds.) *MALDI MS, A Practical Guide to Instrumentation Methods and Applications*. Wiley, 2007, Chapter 2, pp. 29–75.
4. Medzihradszky, K.; Campbell, J.; Baldwin, M.; Juhasz, P.; Vestal, M.; Burlingame, A.L. The characteristics of peptide collision-induced dissociation using a high-performance MALDI-TOF/TOF tandem mass spectrometer. *Anal. Chem.* 2000, 72, 552–558.
5. Suckau, D.; Resemann, A.; Schuerenberg, M.; Hufnagel, P.; Franzen, J.; Holle, A. A novel MALDI LIFT-TOF/TOF mass spectrometer for proteomics. *Anal. Bioanal. Chem.* 2003, 376, 952–965.
6. Cotter, R.J.; Gardner, B.D.; Iltchenko, S.; English, R.D. Tandem time-of-flight mass spectrometry with a curved field reflectron. *Anal. Chem.* 2004, 76, 1976–1981.
7. Belgacem, O.; Bowdler, A.; Brookhouse, I.; Brancia, F.E.; Raptakis, E. Dissociation of biomolecules using a ultraviolet matrix-assisted laser desorption/ionisation time-of-flight/curved field reflectron tandem mass spectrometer equipped with a differential-pumped collision cell. *Rapid Commun. Mass Spectrom.* 2006, 20, 1653–1660.
8. Loboda, A,V.; Chernushevich, I.V. Investigation of the mechanism of matrix adduct formation in MALDI at elevated pressure. *Int. J. Mass Spectrom.* 2005, 240, 101–105.
9. Ding, L.; Kawatoh, E.; Tanaka, K.; Smith, A.J.; Kumashiro, S. High efficiency MALDI-QIT-TOF mass spectrometer. *Proc. SPIE–Int. Soc. Opt. Eng.* 1999, 3777, 144–154.
10. Hoekman, D.J.; Kelley, P.E. Method for generating filtered noise signal and broadband signal having reduced dynamic range for use in mass spectrometry. *US Patent* 1997, 5,703,358.
11. March, R.E.; Todd, J.F.J. *Quadrupole Ion Trap Mass Spectrometry*. 2nd Edition, Wiley, Hoboken, NJ, 2005, pp. 113–117.
12. Brodbelt, J.S. Effects of Collisional Cooling on Ion Detection. In March, R.E.; Todd, J.F.J. (Eds.) *Practical Aspects of Ion Trap Mass Spectrometry*. Vol. 1, CRC Press, Boca Raton, FL, 1995, Chapter 5, pp. 209–220.

13. Doroshenko, V.M.; Cotter, R.J. Pulsed gas introduction for increasing peptide CID efficiency in a MALDI/quadrupole ion trap mass spectrometer. *Anal. Chem.* 1996, *68*, 463–472.

14. Carlin, T.J.; Freiser, B.S. Pulsed valve addition of collision and reagent gases in Fourier transform mass spectrometry. *Anal. Chem.* 1983, *55*, 571–574.

15. Emary, W.B.; Kaiser, R.E.; Kenttamaa, H.I.; Cooks, R.G. Pulsed gas introduction into quadrupole ion traps. *J. Am. Soc. Mass Spectrom.* 1990, *1*, 308–311.

16. O'Connor, P.B.; Costello, C.E. A high pressure matrix-assisted laser desorption/ionization Fourier transform mass spectrometry ion source for thermal stabilization of labile biomolecules. *Rapid Commun. Mass Spectrom.* 2001, *15*, 1862–1868.

17. Papanastasiou, D.; Ding, L.; Raptakis, E.; Brookhouse, I.; Cunningham, J.; Robinson, M. Dynamic pressure measurements during pulsed gas injection in a quadrupole ion trap. *Vacuum* 2006, *81*, 446–452.

18. Fitch, R.K.; Chambers, A. Measurement of Pressure. In Chambers, A.; Fitch, R.K.; Halliday, B.S. (Eds.) *Basic Vacuum Technology*. 2nd Edition, Institute of Physics Publishing, Bristol and Philadelphia, 1998, Chapter 4, pp. 86–114.

19. Dahl, D., SIMION 7, INEEL, 2000.

20. Jonscher, K.; Currie, G.; McCormarck, A.L.; Yates, J.R. Matrix-assisted laser desorption of peptides and proteins on a quadrupole ion trap mass spectrometer. *Rapid Commun. Mass Spectrom.* 1993, *7*, 20–26.

21. Schwartz, J.C.; Bier, M.E. Matrix-assisted laser desorption of peptides and proteins using a quadrupole ion trap mass spectrometer. *Rapid Commun. Mass Spectrom.* 1993, *7*, 27–32.

22. Doroshenko, V.M.; Cotter, R.J. A new method of trapping ions produced by matrix-assisted laser desorption ionization in a quadrupole ion trap. *Rapid Commun. Mass Spectrom.* 1993, *7*, 822–827.

23. Qin, J.; Steenvoorden, R.J.J.M.; Chait, B.T. A practical ion trap mass spectrometer for the analysis of peptides by matrix-assisted laser desorption/ionization. *Anal. Chem.* 1996, *68*, 1784–1791.

24. Qin, J.; Chait, B.T. Matrix-assisted laser desorption ion trap mass spectrometry: efficient trapping and ejection of ions. *Anal. Chem.* 1996, *68*, 2102–2107.

25. He, L.; Liu, Y.; Zhu, Y.; Lubman, D.M. Detection of oligonucleotides by external injection into an ion trap storage/reflectron time-of-flight device. *Rapid Commun. Mass Spectrom.* 1997, *11*, 1440–1448.

26. Beavis, R.C.; Chait, B.T. Velocity distributions of intact high mass polypeptide molecule ions produced by matrix assisted laser desorption. *Chem. Phys. Lett.* 1991, *181*, 479–484.

27. Sudakov, M. AXSIM Computer Simulation Program, *7th International Conference of Charged Particle Optics*, Cambridge, UK, 24–28 July, 2006.

28. He, L.; Lubman, D.M. Simulation of external ion injection, cooling and extraction processes with SIMION 6.0 for the ion trap/reflectron time-of-flight mass spectrometer. *Rapid Commun. Mass Spectrom.* 1997, *11*, 1467–1477.

29. Beynon, J.H.; *Mass Spectrometry and its Applications to Organic Chemistry*. Elsevier, Amsterdam, 1960, Chapter 7, pp. 238–290.

30. Cooks, R.G.; Beynon, J.H.; Caprioli, R.M.; Lester, G.R. *Metastable Ions*. Elsevier, Amsterdam, 1973.

31. Spengler, B. Post-source decay analysis in matrix-assisted laser desorption/ionization mass spectrometry of biomolecules. *J. Mass Spectrom.* 1997, *32*, 1019–1036.

32. Li, Y.; Tang, K.; Little, D.P.; Koster, H.; Hunter, R.; McIver, Jr., R.T. High-resolution MALDI Fourier transform mass spectrometry of oligonucleotides. *Anal. Chem.* 1996, *68*, 2090–2096.

33. Li, Y.; Hunter, R.; McIver, Jr., R.T. Ultrahigh-resolution Fourier transform mass spectrometry of biomolecules above m/z 5,000. *Int. J. Mass Spectrom. Ion Processes.* 1996, *157/158*, 175–188.

34. Krutchinsky, A.N.; Loboda, A.V.; Spicer, V.L.; Dworschak, R.; Ens, W.; Standing, K.G. Orthogonal injection of matrix-assisted laser desorption/ionization ions into a time-of-flight spectrometer through a collisional damping interface. *Rapid Commun. Mass Spectrom.* 1998, *12*, 508–518.

35. Verentchikov, A.; Smirnov, I.; Vestal, M. *Proc. 47th ASMS Conference on Mass Spectrometry and Allied Topics*, Dallas, TX, 13–17 June, 1999, ThPC 061.

36. Loboda, A.V.; Krutchinsky, A.N.; Bromirski, M.; Ens, W.; Standing, K.G. A tandem quadrupole/time-of-flight mass spectrometer with a matrix-assisted laser desorption/ionization source: design and performance. *Rapid Commun. Mass Spectrom.* 2000, *14*, 1047–1057.

37. Krutchinsky, A.N.; Kalkum, M.; Chait, B.T. Automatic identification of proteins with a MALDI-quadrupole ion trap mass spectrometer. *Anal. Chem.* 2001, *73*, 5066–5077.

38. Papanastasiou, D.; Belgacem, O.; Sudakov, M.; Raptakis, E. Ion thermalization using pressure transients in a quadrupole ion trap coupled to a vacuum matrix-assisted laser desorption ionization source and a reflectron time-of-flight mass analyzer. *Rev. Sci. Instrum.* 2008, *79*, 055103.

39. Alekseev, V.A.; Krylova, D.D. Quasi-energy and space-velocity distributions in an oscillating parabolic potential. *Phys. Scripta T.* 1998, *57*, 32–35.

40. Goeringer, D.; Vienland, L. Moment theory of ion motion in traps and similar devices: III. Two-temperature treatment of quadrupole ion traps. *J.Phys. B: At. Mol. Opt. Phys.* 2005, *38*, 4027–4044.

41. Chapman, S.; Cowling, T.G. *The Mathematical Theory of Non-Uniform Gases*, 3rd Edition, Cambridge University Press, New York, 1970, Chapter 4, pp. 67–85.

42. Landau, L.D.; Lifshitz, E.M. *Mechanics*, Pergamon, Oxford, 1960.

43. Abramovitz, M.; Stigan, I.A. (Eds), *Handbook of Mathematical Functions with Formulas, Graphs, and Mathematical Tables*, Dover, New York, 1964.

44. Julian, R.K.; Nappi, M.; Weil, R.; Cooks, R.G. Multiparticle simulation of ion motion in the ion trap mass spectrometer: resonant and direct current pulse excitation. *J. Am. Soc. Mass Spectrom.* 1995, *6*, 57–70.

45. Gioumousis, G.; Stevenson, D.P. Reactions of gaseous molecule ions with gaseous molecules. *J. Chem. Phys.* 1958, *29*, 294–299.

46. Goeringer, D.E.; McLuckey, S.A. Evolution of ion internal energy during collisional excitation in the Paul ion trap: a stochastic approach. *J. Chem. Phys.* 1996, *104*, 2214–2221.

47. Goeringer, D.E.; McLuckey, S.A. Kinetics of collision-induced dissociation in the Paul trap: a first-order model. *Rapid Commun. Mass Spectrom.* 1996, *10*, 328–334.

48. McLuckey, S.A.; Van Berkel, G.J.; Goeringer, D.E.; Glish, G.L. Ion trap mass spectrometry of externally generated ions. *Anal. Chem.* 1994, *66*, 689A–696A.

49. Goeringer, D.E.; McLuckey, S.A. Relaxation of internally excited high-mass ions simulated under typical quadrupole ion trap storage conditions. *Int. J. Mass Spectrom.* 1988, *177*, 163–174.

50. Black, D.M.; Payne, A.H.; Glish, G.L. Determination of cooling rates in a quadrupole ion trap. *J. Am. Soc. Mass Spectrom.* 2006, *17*, 932–938.

51. Larsen, M.R.; Trelle, M.B.; Thingholm, T.E.; Jensen, O.N. Analysis of posttranslational modifications of proteins by tandem mass spectrometry. *Biotechniques.* 2006, *40*, 790–797.

52. Kameyama, A.; Kikuchi, N.; Nakaya, S.; Ito, H.; Sato, T.; Shikanai, T.; Takahashi, Y.; Takahashi, K.; Narimatsu, H. A strategy for identification of oligosaccharide structures using observational multistage mass spectral library. *Anal. Chem.* 2005, *77*, 4719–4725.

53. Ito, H.; Kameyama, A.; Sato, T.; Sukegawa, M.; Ishida, H.; Narimatsu, H. Strategy for the fine characterization of glycosyltransferase specificity using isotopomer assembly. *Nature Methods.* 2007, *4*, 577–582.

54. Suzuki, Y.; Miyasaki, M.; Ito, E.; Suzuki, M.; Yamashita, T.; Taira, H.; Suzuki, A. Structural characterization of N-glycans of cauxin by MALDI-TOF mass spectrometry and nano LC-ESI-Mass Spectrometry. *Biosci. Biotechnol. Biochem.* 2007, *71*, 811–816.

55. Zehl, M.; Pittenauer, E.; Jirovetz, L.; Bandhari, P.; Singh, B.; Kaul, V.K.; Rizzi A.; Allmaier, G. Multistage and tandem mass spectrometry of glycosylated triterpenoid saponins isolated from *Bacopa monnieri*: comparison of the information content provided by different techniques. *Anal. Chem.* 2007, *79*, 8214–8221.

56. Takemori, N.; Komori, N.; Matsumoto, H. Highly sensitive multistage mass spectrometry enables small-scale analysis of protein glycosylation from two-dimensional polyacrylamide gels. *Electrophoresis* 2006, *27*, 1394–1406.

57. Suzuki, Y.; Suzuki, M.; Nakahara, Y.; Ito, Y.; Ito, E.; Goto, N.; Miseki, K.; Iida, I.; Suzuki, A. Structural characterization of glycopeptides by N-terminal protein ladder sequencing. *Anal. Chem.* 2006, *78*, 2239–2243.

58. Ojima, N.; Masuda, K.; Tanaka, K.; Nishimura, O. Analysis of neutral oligosaccharides for structural characterization by matrix-assisted laser desorption/ionization quadrupole ion trap time-of-flight mass spectrometry. *J. Mass Spectrom.* 2005, *40*, 380–388.

59. Fukuyama, Y.; Nakaya, S.; Yamazaki, Y.; Tanaka, K. Ionic liquid matrixes optimized for MALDI-MS of sulfated/sialylated/neutral oligosaccharides and glycopeptides. *Anal. Chem.* 2008, *80*, 2171–2179.

60. Harvey, D.J.; Martin, R.L.; Jackson, K.A.; Sutton C.W. Fragmentation of N-linked glycans with a matrix-assisted laser desorption/ionization ion trap time-of-flight mass spectrometer. *Rapid Commun. Mass Spectrom.* 2004, *18*, 2997–3007.

61. Nicolaou, A.; Kokotos, G. *Bioactive Lipids.* Oily Press, Bridgewater, 2004.

62. Maxfield, F.R.; Tabas, I. Role of cholesterol and lipid organization in disease. *Nature* 2005, *438*, 612–621.

63. Zeisel, S.H.; Blusztajn, J.K. Choline and human nutrition. *Annu. Rev. Nutr.* 1994, *14*, 269–296.

64. Murphy, R.C.; Fiedler, J.; Hevko, J. Analysis of nonvolatile lipids by mass spectrometry. *Chem. Rev.* 2001, *101*, 479–526.

65. Schiller, J.; Süß, R.; Arnhold, J.; Fuchs, B.; Leßig, J.; Müller, M.; Petković, M.; Spalteholz, H.; Zschörnig, O.; Arnold, K. Matrix-assisted laser desorption and ionization time-of-flight (MALDI-TOF) mass spectrometry in lipid and phospholipid research. *Progress in Lipid Research* 2007, *43*, 449–488.

66. Schiller, J.; Süß, R.; Fuchs, B.; Muller, M.; Zschornig, O.; Arnold, K.; MALDI-TOF MS in lipidomics. *Front. Biosci.* 2007, *12*, 2568–2579.

67. Stubiger, G.; Belgacem, O. Analysis of lipids using 2,4,6-trihydroxyacetophenone as a matrix for MALDI mass spectrometry. *Anal. Chem.* 2007, *79*, 3206–3213.

68. Kreunin, P.; Yoo, C.; Urquidi, V.; Lubman, D.M.; Goodison, S. Proteomic profiling identifies breast tumor metastasis associated factors in an isogenic model. *Proteomics* 2007, *7*, 299–312.

69. Takemori, N.; Komori, N.; Thompson, J.N.; Yamamoto, M.; Matsumoto, H. Novel eye-specific calmodulin methylation characterized by protein mapping in Drosophila melanogaster. *Proteomics* 2007, *7*, 2651–2658.

70. Suzuki, T.; Ito, M.; Ezure, T.; Shikata, M.; Ando, E.; Utsumi, T.; Tsunasawa, S.; Nishimura, O. Protein prenylation in an insect cell-free protein synthesis system and identification of products by mass spectrometry. *Proteomics* 2007, *7*, 1942–1950.

71. Watanabe, S.; Yamada, M.; Ohtsu, I.; Makino, K. Ketoglutaric semialdehyde dehydrogenase isozymes involved in metabolic pathways of D-glucarate, D-galactarate, and hydroxy-L-proline. *J. Biol. Chem.* 2007, *282*, 6685–6695.

824 Practical Aspects of Trapped Ion Mass Spectrometry, Volume IV

72. Takahashi, H.; Kumagai, T.; Kitani, K.; Mori, M.; Matoba, Y.; Sugiyama, M. Cloning and characterization of a streptomyces single module-type non-ribosomal peptide synthetase catalyzing a blue pigment synthesis. *J. Biol. Chem.* 2007, *282*, 9073–9081.

73. Komori, N.; Takemori, N.; Kim, H.K.; Singh, A.; Hwang, S.; Foreman, R.D.; Chung, K.; Chung, J.M.; Matsumoto, H. Proteomics study of neuropathic and nonneuropathic dorsal root ganglia: altered protein regulation following segmental spinal nerve ligation injury. *Physiol. Genomics.* 2007, *29*, 215–230.

74. Yoo, C.; Patwa, T.H.; Kreunin, P.; Miller, F.R.; Huber, C.G.; Nesvizhskii, A.I.; Lubman, D.M. Comprehensive analysis of proteins of pH fractionated samples using monolithic LC/MS/MS, intact MW measurement and MALDI-QIT-TOF MS. *J. Mass Spectrom.* 2007, *42*, 312–334.

75. Okuno, S.; Kiuchi, M.; Arakawa, R. Structural characterization of polyethers using MALDI-QIT-TOF mass spectrometry. *Eur. J. Mass Spectrom.* 2006, *12*, 181–187.

76. March, R.E.; Li, H.; Belgacem O.; Papanastasiou, D. High-energy and low-energy collision-induced dissociation of protonated flavonoids generated by MALDI and by electrospray ionization. *Int. J. Mass Spectrom.* 2007, *262*, 51–66.

77. Cai, Y.; Xing, J.; Sun, M.; Zhan, Z.; Corke, H. Phenolic antioxidants (hydrolyzable tannins, flavonols, and anthocyanins) identified by LC-ESI-MS and MALDI-QIT-TOF MS from *Rosa chinensis* flowers. *J. Agric. Food Chem.* 2005, *53*, 9940–9948.

78. Nakanishi, T.; Ohtsu, I.; Furuta, M.; Ando, E.; Nishimura, O. Direct MS/MS analysis of proteins blotted on membranes by a matrix-assisted laser desorption/ionization-quadrupole ion trap-time-of-flight tandem mass spectrometer. *J. Proteome Res.* 2005, *4*, 743–747.

79. Shimma, S.; Sugiura, Y.; Hayasaka, Y.; Zaima, N.; Matsumoto, M. Mass imaging and identification of biomolecules with MALDI-QIT-TOF-based system. *Anal. Chem.* 2008, *80*, 878–885.

80. Aoki, Y.; Toyama, A.; Shimada, T.; Sugita, T.; Aoki, C.; Umino, Y.; Suzuki, A.; Aoki, D.; Daigo, Y.; Nakamura, Y.; Sato, T. A novel method for analyzing formalin-fixed paraffin embedded (FFPE) tissue sections by mass spectrometry imaging. *Proc. Jpn. Acad. Ser. B* 2007, *83*, 205–214.

81. Nakamura, K.; Suzuki, Y.; Goto-Inoue, N.; Yoshida-Noro, C.; Suzuki, A. structural characterization of neutral glycosphingolipids by thin-layer chromatography coupled to matrix-assisted laser desorption/ionization quadrupole ion trap time-of-flight MS/MS. *Anal. Chem.* 2006, *78*, 5736–5743.

Part VI

Photochemistry of Trapped Ions

20 Photodissociation in Ion Traps

Jennifer S. Brodbelt

CONTENTS

20.1 Introduction.. 827
20.2 IRMPD Applications 831
    20.2.1 Small Molecules/Drugs.. 831
    20.2.2 Peptides and Proteins 832
    20.2.3 Oligosaccharides 836
    20.2.4 Nucleic Acids.. 837
20.3 UVPD Applications.. 838
20.4 Conclusions... 841
Acknowledgments.. 842
References.. 843

20.1 INTRODUCTION

The search for alternatives to collision-induced dissociation (CID) for the structural characterization of ions has led to the development of photodissociation (PD), an activation method in which a laser is used to energize ions. The accumulation of internal energy *via* absorption of one or multiple photons offers an efficient, tunable strategy that couples well with ion trap mass spectrometry. Both continuous wave (cw) and pulsed lasers, with wavelengths ranging from the infrared to ultraviolet (UV), have been used for PD. The energization processes promoted by collisional activation and photoactivation are fundamentally quite different, although the desired outcome is the same: deposition of sufficient energy to cause bond cleavages that lead to interpretable, diagnostic product ions for elucidation of ion structures. Collision-induced dissociation has remained the most widespread ion activation method in ion trap mass spectrometers, in large part because of its rich history and relatively well-understood mechanism that entails conversion of translational energy of mass-selected precursor ions into internal energy through collisions with an inert target gas, typically helium [1]. CID is accomplished by the application of a supplemental RF voltage, matched to the ion's secular frequency, to the end-cap electrodes [1]. In a conventional quadrupole ion trap, the ion trap is operated nominally at *ca* 1 mTorr of helium pressure which enhances the sensitivity and resolution of the ion trap due to collisional cooling of ions to the center of the trap, thus focusing the ions for storage and ejection [2].* This pressure is also favorable for CID experiments where activating collisions over a 20–100 ms period are desirable. Since the very nature of CID is collision-based and because CID depends on the excitation of an ion's kinetic energy in order to encourage more energetic collisions with the target gas, there exists the opportunity for some ion losses due to scattering events leading to ion trajectories that are either unstable or exhibit excursions beyond the confines of the ion trap, especially when exploiting conditions that maximize energy deposition. Moreover, the m/z-range of ions stored in the ion trap is determined by the magnitude of the RF voltage applied to the trapping electrodes, and it is this same RF voltage that defines the energy transfer during CID based on the energetics of the ion/target collisions. Collision-based energy deposition is maximized at the expense of trapping the lower m/z-range. Typical CID conditions in an ion trap result in truncation of the lower quarter to third of the m/z-range, thus resulting in loss of some potentially diagnostic product ions. CID is also operated generally in a resonant mode, meaning that the frequency of the supplemental RF voltage is tuned to match the frequency of motion of a selected precursor ion. As a consequence, any product ions produced will not be activated further because their secular frequencies will overlap neither with the frequency of the selected precursor ion nor with that of the applied supplementary RF voltage. MS^n strategies may be employed to energize and dissociate any uninformative or dead-end product ions.

Photodissociation is an alternative ion activation method that entails exposing an ion population to photons of a selected wavelength until sufficient internal energy is accumulated to cause dissociation. An ion may accumulate energy through either the absorption of one or several high energy UV or visible photons (ca 2–10 eV photon^{-1}) or dozens or hundreds of low energy IR photons (ca 0.1 eV photon^{-1}). The activation period may range from nanoseconds to hundreds of milliseconds depending on the photon flux of the laser, the competition between ion activation and collisional cooling, and the energy deposition per photon. Photoactivation offers several inherent advantages compared to CID. Unlike CID, ion activation *via* photon absorption is independent of the RF trapping voltage used to store ions. A low RF trapping voltage can be applied during photoactivation, thus allowing storage of very low-m/z product ions that might be lost during CID. In addition, photoactivation is a non-resonant process, meaning that the selected precursor ions and the resulting product ions may absorb photons, leading to secondary dissociation and the potential for a greater array of diagnostic product ions while also alleviating the need for discrete multi-step MS^n experiments. Also, PD requires neither the translational excitation of ions nor collisions with target gas molecules, alleviating the possibility of unstable ion trajectories and/or ion scattering effects.†

---

* See also Volume 4, Chapter 17: Collisional Cooling in the Quadrupole Ion Trap Mass Spectrometer (QITMS), by Philip M. Remes and Gary L. Glish; Volume 4, Chapter 18: 'Pressure Tailoring' for Improved Ion Trap Performance, by Dodge L. Baluya and Richard A. Yost.

† See also Volume 4, Chapter 21: Photochemical Studies of Metal Dication Complexes in an Ion Trap, by Guohua Wu, Hamish Stewart and Anthony J. Stace; Volume 5, Chapter 7: Structure and Dynamics of Trapped Ions, by Joel H. Parks; Volume 5, Chapter 9: Spectroscopy of Trapped Ions, by Matthew W. Forbes, Francis O. Talbot and Rebecca A. Jockusch.

To date, there have been more studies that have utilized IR lasers for PD [3–28] than UV or visible lasers, although the latter are gaining in popularity due to the significantly higher energy per photon [29–43]. The framework for infrared multiphoton photodissociation (IRMPD) in quadrupole ion traps was established in the 1980s in several fundamental studies undertaken by March and co-workers using a quadrupole ion store (QUISTOR)-type mass spectrometer [44–48]. The use of IRMPD in a commercial QIT instrument was first reported in 1994, in which a multi-pass optical arrangement was used to enhance PD efficiencies [3]. Introduction of laser light into an ion trap may be accomplished *via* a hole in the ring electrode or end-cap electrode of a three-dimensional quadrupole ion trap or by axial admittance of the laser beam through a two-dimensional linear ion trap, see Figure 20.1. The addition of a several mm-diameter hole in the ring electrode does not seriously perturb the performance of an ion trap, other than a change in capacitance that must be matched to the RF drive circuit. The laser must be triggered or gated in a reproducible way, typically *via* a TTL pulse from the control software, such that the ions are irradiated during the selected segment of the scan function. More specific details about implementation of PD are available in past reports for a variety of commercial ion trap mass spectrometers [4–8].

Economical cw CO_2 lasers are the most common choice for infrared multiphoton dissociation. The laser beam is introduced typically into the vacuum chamber *via* a ZnSe window. Absorption of each 10.6 μm photon, a wavelength that is absorbed by a wide range of organic molecules, deposits approximately 0.12 eV. The irradiation

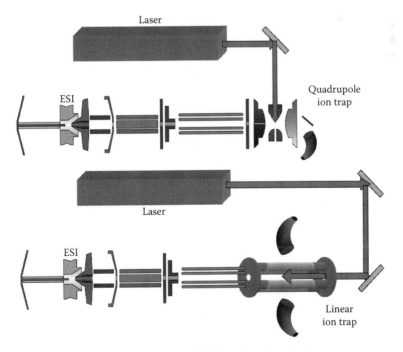

FIGURE 20.1   Schematics of ion traps modified for photodissociation.

period may range from 2 ms to several seconds depending on the laser power, beam diameter, the IR absorptivity of the ions, and the trap pressure which influences the extent of collisional cooling that counteracts ion energization. To offset this latter problem, which has limited particularly the success of IRMPD in quadrupole ion traps for ions that have high critical energies, several methods of enhancing energy deposition during IRMPD have been explored. The buffer gas used to optimize ion trap performance can be heated [9], or the laser can be focused *via* a hollow fiber waveguide to increase the rate of photoabsorption relative to the rate of collisional cooling [6]. Pulsing the buffer gas is another effective option in which ions are accumulated at higher pressures but irradiated at lower pressures [10,11], in addition to strategies that combine CID with IRMPD (collision-assisted IRMPD) to afford greater energy deposition [4,11]. The intrinsic photoabsorptivities can be increased artificially also by attachment of strong IR chromophores to the analytes of interest [12,13]. For instance, it is now well known that the $O = P-OH$ and $P-O-C$ bond stretches that occur in phosphate moieties absorb strongly at 10.6 μm. Phosphorylated peptides [14,15] and oligonucleotides [5] naturally possess these chromophores that make them absorb highly and result in high PD efficiencies. On the other hand, IRMPD can be facilitated by the attachment of strongly IR-absorbing functionalities, such as *via* coordination of phosphorylated bipyridines to flavonoid diglycosides [12], or *via* derivatization of oligosaccharides with boronic acids [13]. For example, flavonoids, phenolic phytochemicals with molecular weights ranging from 300 to 700 Da, exhibit very poor IRMPD efficiencies, and thus IRMPD is not a viable alternative to CID for differentiation of isomeric flavonoids. To overcome this problem, a strategy was developed to increase the IR absorptivities of the flavonoids by non-covalent addition of strong IR chromophores [12]. Chromogenic phosphonate groups are incorporated into bipyridyl chelating ligands that are coordinated to the flavonoids *via* metal complexation prior to ESI-MS analysis [12]. The resulting complexes, [metal^{2+} (flavonoid–H) (4,4′-bis(diethylmethylphosphonate)-2,2′-bipyridine]$^{+}$ have high IR absorptivities that facilitate IRMPD. As illustrated in Figure 20.2, IR irradiation of the neodiosmin complex that contains 2,2′-bipyridine causes no dissociation. In contrast, the neodiosmin complex that contains the phosphorylated chelating ligand undergoes very efficient IRMPD, resulting in a diagnostic fragmentation pattern that allows differentiation of neodiosmin from other isomeric flavonoids.

The use of Nd:YAG, excimer, and optical parametric oscillator (OPO):YAG lasers for UV and visible PD in ion trap mass spectrometers has also been reported [29–43]. The energies of UV photons are 20–100 fold greater than IR photons, thus requiring that far fewer photons be absorbed to cause ion dissociation. Some of the commonly reported UV wavelengths include 266 nm (the fourth harmonic of a Nd:YAG laser, 4.66 eV), [42] 355 nm (the third harmonic of a Nd:YAG laser, 3.5 eV) [31–33], 157 nm (F_2 excimer, 7.9 eV per photon) [30,34,37,41] and 193 nm (ArF excimer, 6.4 eV). An OPO-Nd:YAG laser offers a tunable range from 205 nm to 2550 nm (6.0–0.49 eV) [29,38,40,43]. For implementation of UV or visible photodissociation (VisPD), an optical window with appropriate transmissive properties is used to introduce the laser beam into the ion trap.

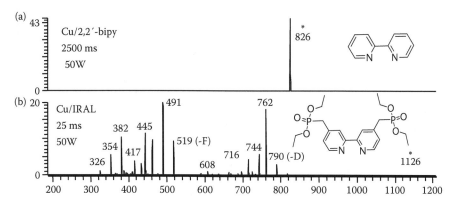

FIGURE 20.2 IRMPD of (a) [Cu^{2+}(neodiosmin − H) 2,2-bipy]$^+$, 50 W for 2500 ms, and (b) [Cu^{2+} (neodiosmin − H) IRAL-2]$^+$, 50 W for 25 ms; IRAL means infrared-absorbing ligand. −F indicates loss of the flavonoid; −D indicates loss of the disaccharide moiety. (Reproduced from Pikulski, M., Wilson, J.J., Aguilar, A., and Brodbelt, J.S., *Anal Chem.* 2006, *78*, 8512–8517. With permission from American Chemical Society.)

20.2 IRMPD APPLICATIONS

20.2.1 SMALL MOLECULES/DRUGS

The use of IRMPD has proven to be a versatile ion activation method for a number of applications, including characterization of small organic molecules such as natural products and drugs to larger biological molecules [5,7,8,12,25,27]. Some of the first applications of IRMPD focused on the analysis of drugs [16–18,20,24], ranging from tetracyclines to erythromycins to penicillins, among others. Examples illustrating comparisons of CID and IRMPD mass spectra for two drugs, erythromycin (a macrolide antibiotic) and paromomycin (an aminoglycoside antibiotic), are shown in Figures 20.3 and 20.4, respectively. The CID mass spectrum of protonated erythromycin (*m/z* of the precursor ion is 734) is dominated by the loss of the cladinose sugar, resulting in the product ion of *m/z* 576. Multiple losses of water are also prevalent. The IRMPD mass spectrum also reveals the loss of the cladinose sugar, but the most prominent product ion (*m/z* 158) represents protonated desosamine, the other major functionality of erythromycin. The CID and IRMPD mass spectra obtained for protonated paromomycin both display the losses of one or more amino sugar groups which may occur in conjunction with additional water losses (Figure 20.4). The terminal amino sugar also appears as a significant product ion (*m/z* 163) in the IRMPD spectrum, thus demonstrating that IRMPD promotes cleavage of each glycosidic bond and allowing the entire aminoglycoside sequence to be mapped. As illustrated in Figures 20.3 and 20.4, IRMPD results typically in the same types of product ions produced by CID, with two notable differences. First, because the photoactivation process is independent of the RF trapping voltage, a much lower RF trapping voltage can be applied during the IRMPD period and thus the ion trapping range extends to lower *m/z*-values. This operational mode in IRMPD experiments allows the storage and

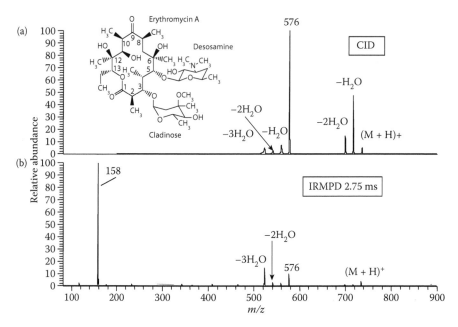

FIGURE 20.3 (a) CID and (b) IRMPD mass spectra of protonated erythromycin acquired in a quadrupole ion trap mass spectrometer equipped with a cw CO_2 laser. (Data collected by Carol Parr, University of Texas at Austin.)

detection of diagnostic low-m/z product ions that would not be observed in conventional CID experiments. Second, due to the non-resonant nature of photoactivation, primary product ions have the opportunity to absorb photons and dissociate into secondary product ions. This phenomenon explains why PD mass spectra often display a greater array of product ions than do CID mass spectra; in addition, the less informative "dead-end" product ions, such as those resulting from dehydration, are converted into more informative ions upon PD. The secondary absorption and dissociation promoted by the non-resonant nature of photoactivation also reduces the need for multistep MS^n strategies.

The energy deposited during photoactivation is influenced by the power of the laser (total photon flux), the duration of the photoirradiation period or number of laser pulses, the pressure within the ion trap, the overlap of the laser beam with the ion cloud, the wavelength of the laser (energy per photon), and the photoabsorptivity of the ions of interest. The first five parameters are generally user-controlled variables that can be adjusted to alter the energy deposition, thus affording some degree of control over the resulting dissociation patterns of ions, similar to the variation of collision energy in CID. The last parameter (photoabsorptivity) is an intrinsic property of the analyte ions.

20.2.2 Peptides and Proteins

Infrared multiphoton dissociation has also been used successfully for the structural characterization of peptides, nucleic acids, and oligosaccharides [5,7,8,13–15,21,

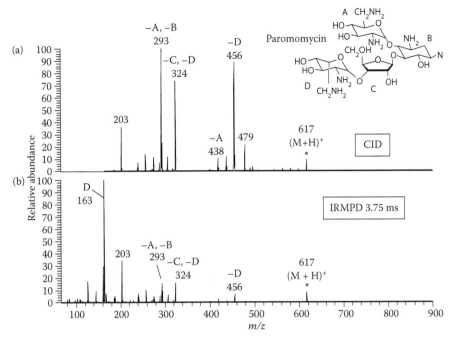

FIGURE 20.4 (a) CID and (b) IRMPD mass spectra of protonated paromomycin acquired in a quadrupole ion trap mass spectrometer equipped with a cw CO_2 laser. (Data collected by Carol Parr, University of Texas at Austin.)

25,27]. For example, IRMPD has been used to sequence peptides and, in some cases, more extensive sequence coverage is obtained by IRMPD in comparison to CID because of the extension of the lower m/z-range afforded by IRMPD. This extended trapping range allows detection of the terminal b and y ions that help identify the N- and C-terminal residues of peptides, as well as immonium ions that are useful for direct identification of specific amino acids and modified residues [14]. An example is illustrated in Figure 20.5 for the analysis of tyrosine kinase receptor JAK2, in which the y_1, y_2, b_1, and b_2 ions are not detected upon conventional CID. Moreover the CID mass spectrum is dominated by a product ion attributed to a dehydration/solvent adduction process, and the abundance of this particular ion is reduced significantly in the corresponding IRMPD spectrum.

Derivatization strategies can also be coupled with IRMPD to facilitate peptide sequencing [7]. For example, an N-terminal sulfonation reaction lowers the critical energies of the resulting derivatized peptides, making them dissociate more readily, as well as fixing a negatively charged site to the N-terminus [7]. The presence of a negative charge site at the N-terminus means that the b fragments formed upon dissociation of singly charged peptides will exist as neutral species, not charged ions. The resulting product ion mass spectra exhibit an easily tracked series of y ions that are conducive to interpretation by *de novo* sequencing algorithms. A comparison of the fragmentation patterns generated by CID and IRMPD for N-terminal sulfonated fibrinopeptide A (ADSGEGDFLAEGGGVR) is shown in Figure 20.6. Both product

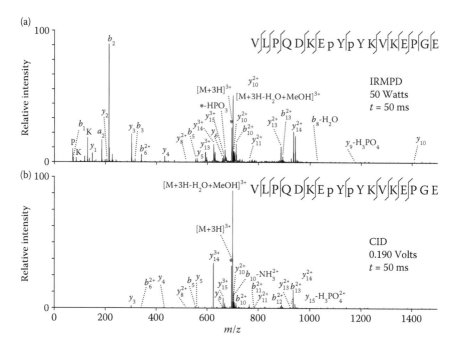

FIGURE 20.5 (a) IRMPD and (b) CID mass spectra of triply protonated tyrosine kinase receptor JAK 2. The precursor ions are indicated by asterisks. (Reproduced from Crowe, M., and Brodbelt, J.S., *J. Am. Soc. Mass Spectrom.* 2004, *15*, 1581–1592. With permission from Elsevier.)

ion mass spectra display continuous series of y ions, but the y_1 to y_5 ions are not detected in the CID mass spectrum because they fall below the lower m/z limit.

The technique of IRMPD can be exploited also for the characterization of cross-linked proteins based on using IR-chromogenic reagents [8]. For example, a novel cross-linker was designed as a homobifunctional reagent with two terminal NHS-esters that react with primary amines and a spanner arm-length that potentially bridges up to *ca* 28 Å between α-carbons of two lysine residues in proteins. The cross-linker also contains a key phosphate chromophore that absorbs at 10.6 μm [8]. Upon reaction of the cross-linker with amines, stable amide bonds are formed that covalently link lysines that are constrained spatially within the protein. Subsequent tryptic digestion of the protein results in formation of intermolecularly cross-linked peptides in which the two constituent peptides are labeled α and β, respectively. Cleavage along the backbone of the α (or β) peptide yields products ions such as $y_{n\alpha}$ or $b_{n\alpha}$ (or $y_{n\beta}$ or $b_{n\beta}$). The IRMPD mass spectrum for one cross-linked peptide obtained upon tryptic digestion of ubiquitin is shown in Figure 20.7. The IRCX-cross-linked peptide containing $T^{55}LSDYNIQK^{63}ESTLHLVLR^{72}$ (α peptide) and $M^1QIFVK^6TLTGK^{11}$ (β peptide) dissociated efficiently upon IR irradiation, yielding solely *b* and *y* product ions. Cleavage occurred at 13 of the 14 amide bonds C-terminal to the cross-linked lysines, providing sequence tags for each of the two

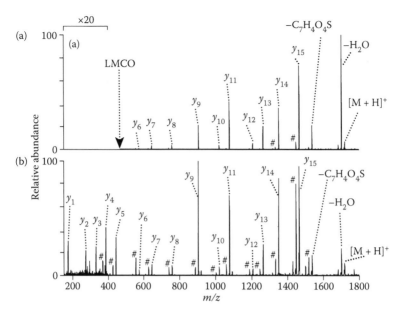

FIGURE 20.6 ESI-MS/MS mass spectra of N-terminal sulfonated fibrinopeptide A (ADSGEGDFLAEGGGVR) (a) CID and (b) IRMPD. Magnification scales apply to each pair of mass spectra for a given peptide over the mass range indicated. Satellite peaks due to loss of H_2O are labeled with a # symbol. (Reproduced from Wilson, J.J. and Brodbelt, J.S., *Anal Chem.* 2006, *78*, 6855–6862. With permission from American Chemical Society.)

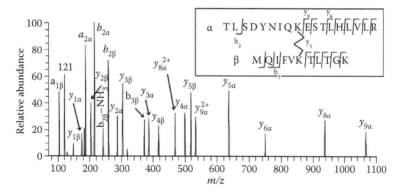

FIGURE 20.7 IRMPD mass spectrum (30 ms, 50 W, $q = 0.1$) of a tryptic peptide from ubiquitin crosslinked by IRCX acquired during data-dependent LC-IRMPD-MS experiment: $[T^{55}LSDYNIQK^{63}ESTLHLVLR^{72} \otimes IRCX \otimes M^1QIFVK^6TLTGK^{11} + 5H]^{5+}$ (*m/z* 746.1) where the first peptide listed is the α peptide and the second peptide is referred to as the β peptide. (Reproduced from Gardner, M., Vasicek, L., Shabbir, S., Anslyn, E., and Brodbelt, J.S., *Anal Chem.* 2008, *80*, 4807–4819. With permission from American Chemical Society.)

constituent peptides. As demonstrated in this example, IRMPD provides an effective ion activation method for characterization of the cross-linked products.

20.2.3 OLIGOSACCHARIDES

The use of IRMPD has proven to be successful for the structural characterization of isomeric oligosaccharides based on a derivatization procedure that uses a boronic acid ligand (infrared absorbing boronic acid, IRABA) that contains an IR-chromogenic phosphonate group [13]. The IRABA ligand reacts with the oligosaccharide by formation of two covalent bonds between the boronic acid and the diol functionalities of the reducing sugar in conjunction with two water losses [13]. The IRMPD mass spectra for the resulting IRABA-derivatized lacto-N-fucopentaoses (LNFP) oligosaccharides are illustrated in Figure 20.8. Two general fragmentation pathways are observed. The dominant pathway (highlighted with bold labels) corresponds to sequential cleavages from the non-reducing end of the oligosaccharides, and a secondary pathway entails cleavages that begin at the reducing end of the oligosaccharides.

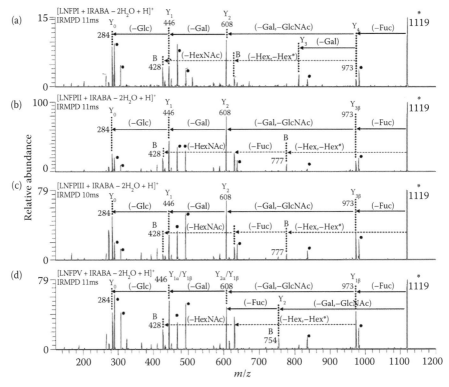

FIGURE 20.8 IRMPD mass spectra of LNFP oligosaccharides after derivatization with IRABA: (a) [LNFPI + IRABA-2H$_2$O + H]$^+$, (b) [LNFPII + IRABA − 2H$_2$O + H]$^+$, (c) [LNFPIII + IRABA − 2H$_2$O + H]$^+$, (d) [LNFPV + IRABA-2H$_2$O + H]$^+$. Product ions that stem from loss of a phosphonate moiety are indicated with a '•'. (Reproduced from Pikulski, M., Hargrove, A., Shabbir, S., Anslyn, E., and Brodbelt, J.S, J. *Am. Soc. Mass Spectrom*. 2007, *18*, 2094–2106. With permission from Elsevier.)

The products that arise from the first fragmentation pathway occur solely from the non-reducing end of the oligosaccharide, and the resulting sequential losses correspond to masses of individual residues that allow one to sequence directly the oligosaccharide. The boronic acid derivatization protocol in conjunction with IRMPD is a versatile strategy for identification of isomeric oligosaccharides.

20.2.4 NUCLEIC ACIDS

Infrared multiphoton dissociation is an extremely efficient ion activation method for nucleic acids [5,21,25]. The phosphate backbone promotes high IR absorptivity at 10.6 μm, thus affording efficient energy deposition with irradiation times as low as a few milliseconds for IRMPD. An example of a direct comparison of fragmentation trends for variable energy deposition from CID and IRMPD for the duplex d(GCGGGAATTGGGCG)/(CGCCCAATTCCCGGC) (5-charge state) is shown in Figure 20.9 [21]. The abundances of precursor ions, base loss ions, and both lower m/z and higher m/z sequence ions (i.e., $(a–B)$ and w ions) are mapped as a function of collision energy (CID voltage) for CID (Figure 20.9a) and irradiation time for IRMPD (Figure 20.9b). In the energy-variable CID experiment, the percentage of base loss ions increases greatly for the CID data as a function of CID voltage, but the percentage of sequence ions increases only slightly. In contrast, a large percentage of base loss ions are formed upon IRMPD during the first 1.0–1.5 ms, but these ions dissociate into diagnostic sequence ions with longer irradiation times. The total

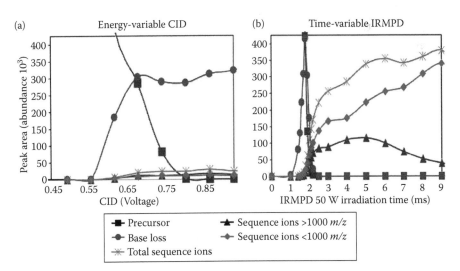

FIGURE 20.9 Energy-variable dissociation comparison for the $[M - 5H]^{5-}$ ion of duplex d(GCGGGAATTGGGCG)/(CGCCCAATTCCCGGC) obtained by (a) CID (30 ms activation period and $q_z = 0.25$) and (b) IRMPD (50 W laser power and $q_z = 0.1$). Product ion types are grouped by ion type including: base loss, sequence ions with $m/z > 1000$, sequence ions with $m/z < 1000$, and total sequence ions as summarized in the legend. (Reproduced from Wilson, J.J. and Brodbelt, J.S., *Anal Chem.* 2007, *79*, 2067–2077. With permission from American Chemical Society.)

abundance of sequence ions exhibits little change beyond 5 ms of irradiation time, but there is a notable shift from higher m/z ions to lower m/z ions that is consistent with secondary dissociation.

20.3 UVPD APPLICATIONS

Although the energy deposited per photon is low for IRMPD, most molecules exhibit some absorptivity in the IR range, making IRMPD a reasonably universal ion activation method. Absorption of UV or visible photons is a more selective process, and thus ultraviolet photodissociation (UVPD) requires either more careful matching of the wavelength to the existing chromophores of the analyte molecules or derivatization of the analyte molecules by attachment of suitable chromophores. Nonetheless, there has been increasing interest in the development and application of UVPD and visible PD in ion traps with recent studies aimed at the analysis of nucleic acids, peptides, and oligosaccharides [29–43].

For example, a quadrupole ion trap instrument equipped with an OPO-Nd:YAG laser [29,38,40,43] has been used to characterize oligonucleotides and peptides by UVPD. UV photoactivation of single- and double-strand oligonucleotide anions in the range of 250–285 nm causes electron detachment from the selected precursor ions, and subsequent CID of the resulting charge-reduced radical anions leads to the formation of w, a, d, and z ions that are complementary to $(a–B)$ and w ions encountered typically by CID or IRMPD experiments [39]. UVPD of protonated peptides in the range of 220–280 nm causes hydrogen-atom losses and side-chain cleavages, resulting in radical product ions that are not observed typically upon CID [29,40,43].

Photons of 193 nm (6.4 eV) produced by an ArF excimer laser are sufficiently energetic to cause ion dissociation after absorption of a single photon, and the activation process does not cause losses of key post-translational modifications that occur commonly upon CID of peptides. An example is shown in Figure 20.10 for the protonated phosphopeptide GRTGRRNpSIHDIL analyzed in a linear ion trap mass spectrometer. Collision-induced dissociation produces predominantly b ions that occur in conjunction with the loss of H_3PO_4, while UVPD (a single pulse at 193 nm) results in the formation of some b ions similar to CID, but also produces a series of a-type product ions, which allow the phosphorylation site to be pinpointed. The a-type ions reveal that phosphate modification resides on the serine at position 8 based on the a_8, a_9, a_{10}, and a_{11} ions, all which retain the phosphate modification, and the a_5, a_6, and a_7 ions, none of which retain the phosphate modification. A sequence map of GRTGRRNpSIHDIL is shown in (Figure 20.10c). The b ions produced by CID do not retain the phosphate group, making it difficult to identify the site of phosphorylation. By contrast, UVPD at 193 nm creates a richer array of diagnostic product ions, including the series of a-type ions and the contiguous eight-residue sequence tag that retains the phosphate modification that are observed in Figure 20.10b.

The application of UVPD using photons of 157 nm (7.9 eV) produced from an F_2 excimer laser has been reported also for the structural characterization of protonated peptides and oligosaccharides in a linear ion trap mass spectrometer [30,34,37,41]. Singly charged peptides produced mainly x- and a-type ions, unlike the y- and b-type

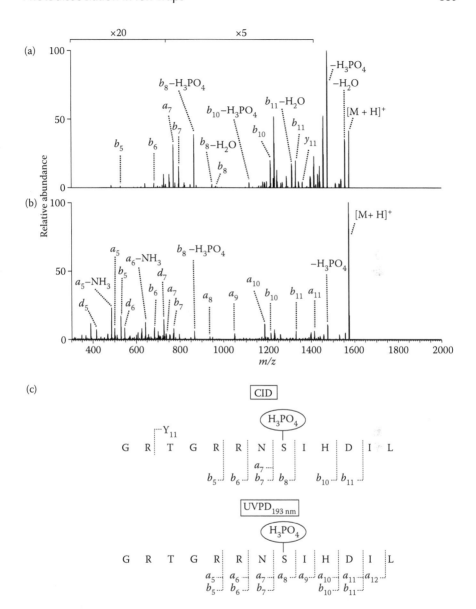

FIGURE 20.10 ESI-MS/MS spectra of a phosphorylated peptide (GRTGRRNpSIHDIL) by (a) CID and by (b) UVPD at 193 nm using 1 pulse at 8 mJ/pulse. The magnification scale applies to both mass spectra over the indicated mass range. The sequence coverage map is shown in (c). (Data collected by Jeff Wilson, University of Texas at Austin.)

ions observed commonly upon CID. The UVPD mass spectra of sodium-cationized glycans display a rather complicated array of product ions due to cross-ring cleavages, yielding A and X type product ions that revealed specific isomeric linkages in the oligosaccharides [37]. From a practical standpoint, implementation of UVPD at 157 nm is more challenging because molecular oxygen absorbs efficiently in this range, thus necessitating the use of nitrogen purging or a high vacuum interface to remove oxygen from the beam path of the laser.

Molecules that do not absorb strongly at convenient UV wavelengths can be derivatized by attachment of suitable chromophores prior to UVPD [31]. For example, non-absorbing peptides were converted to highly UV-absorbing peptides *via* N-terminal or C-terminal derivatization using chromophore-containing reagents [31], such as the sulfonation reagent Alexa Fluor 350 (AF350). The resulting peptides absorb strongly at 355 nm (3.5 eV per photon), the wavelength of a pulsed frequency-tripled Nd:YAG laser. For doubly protonated peptides, UVPD leads to greatly simplified fragmentation patterns in which series of either *y* or *b* ions are detected depending on the position of the chromophore at the N- or C-terminus, instead of the complex array of both *y* and *b* ions observed normally upon CID. When the UV-chromophore is attached to the N-terminus, only the *y* ions are observed because the chromophore-containing *b* ions undergo extensive secondary dissociation and, ultimately, are annihilated upon exposure to multiple laser pulses. For example, the CID mass spectrum of protonated FSWGAEGQR is rather complex and cluttered due to a wide variety of *b* and *y* ions, as well as secondary losses of water and internal fragments (Figure 20.11a). The N-terminal AF350-derivatized peptide yields only a series of *y* ions upon CID, as shown in Figure 20.11b. The UVPD mass spectrum is the most impressive because the AF350-modified peptide yields the entire series of *y* ions down to the y_1 ion and the uninformative losses of H_2O and SO_3 are decreased significantly. It is interesting also to note that UVPD in an ion trap exhibits no dependence on the helium pressure, unlike IRMPD in which loss of internal energy *via* collisional deactivation is a substantial problem. The lack of pressure dependence is attributed to the fast time scale of UV energy deposition: the 10-nanosecond period of a Q-switched YAG laser pulse means energy deposition occurs in 10 nanoseconds with photons that each deposit 3.5 eV. In contrast, infrared photons deposit approximately 0.1 eV per photon, requiring absorption of dozens or even hundreds of photons for sufficient energy deposition to promote dissociation and thus allowing greater opportunity for competitive collisional deactivation.

UV activation can be used also to initiate site-specific reactions of modified biological molecules in an ion trap [42]. For example, UVPD at 266 nm (4.66 eV per photon produced by a Nd:YAG laser) of proteins containing iodo-tyrosine residues resulted in radical-directed dissociation that promoted selective cleavage at tyrosine and histidine residues [42]. Cytochrome c was iodinated at its tyrosine side-chains prior to introduction into a linear ion trap mass spectrometer *via* ESI, then subjected to UVPD using a pulsed Nd:YAG laser. Photoactivation of the iodo-tyrosine-containing proteins promoted the loss of I·, leading to radical product ions that were amenable to subsequent CID.

Reductive amination of oligosaccharides using fluorophore reagents, such as 6-aminoquinoline (6-AQ), 2-amino-9(10H)-acridone (AMAC), and 7-amino-methylcoumarin (AMC), converts non-absorbing oligosaccharides into ones that undergo efficient UVPD at 355 nm [33]. As an example, the resulting UVPD mass spectra for three sodium-cationized fluorophore-labeled lacto-N-difucohexaoses (LNDFH) are shown in Figure 20.12 [33]. The UVPD mass spectra for the AMC-derivatized LNFDH oligosaccharides exhibit substantial differences in their

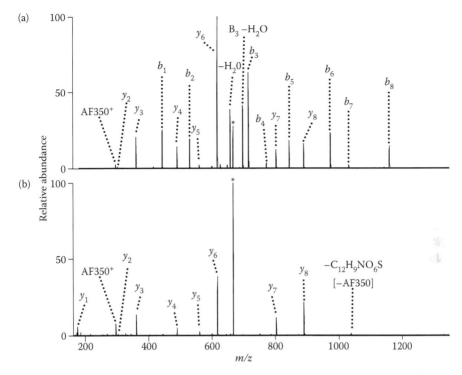

FIGURE 20.11 ESI-MS/MS spectra of doubly protonated N-terminally modified AF350-FSWGAEGQR peptide by (a) CID (0.50 V) and (b) UVPD (15 pulses at 10 Hz). UVPD shows a complete series of y ions with minimal complexity due to elimination of the redundant b ions. The peak with the $-C_{12}H_9NO_6S$ label corresponds to the loss of the charged AF350 moiety from the N-terminus of the peptide, thus producing the unmodified, singly charged peptide species. An asterisk (*) is used to signify the precursor ion. (Reproduced from Wilson, J.J. and Brodbelt, J.S., *Anal. Chem.* 2007, *79*, 7883–7892. With permission from American Chemical Society.)

fragmentation patterns that allow differentiation of the isomers. The dominant product ions incorporate the non-reducing end of the oligosaccharide, giving A- and C-type ions as opposed to the Y-type ions traditionally created by CID. Cross-ring cleavages of the reducing sugar containing the fluorophore for the LNDFH isomers produce two diagnostic ions, $^{0,2}A_5$ for LNFDH-Ia and LNFDH-Ib and $^{0,1}A_4/Y_{3\alpha''}$ for LNFDH-II. A second series of product ions (labeled ∇, $C_{3\alpha}$ ions) allows the LNDFH–Ia and –Ib isomers to be distinguished readily from LNDFH–II.

20.4 CONCLUSIONS

The advantages offered by PD, such as the independence of the photoactivation process on the trapping voltage, which allows storage of ions having a broader *m/z*-range, have made it a versatile alternative to CID in ion trap mass spectrometers. A growing number of studies have demonstrated the utility of PD for characterization of

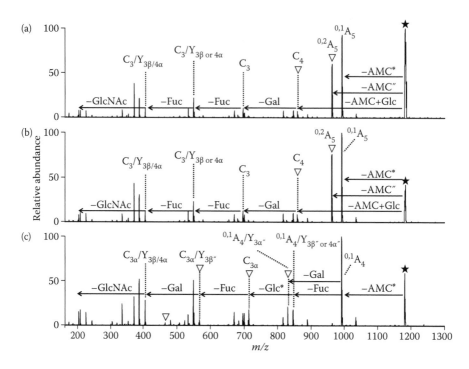

FIGURE 20.12 ESI-MS/MS spectra of (a) [LNDFH-Ia + Na + AMC]$^+$; (b) [LNDFH-Ib + Na + AMC]$^+$ and (c) [LNDFH-II + Na + AMC]$^+$ by UVPD with 20 pulses at 60 mJ/pulse for all three mass spectra. The triangle symbol (∇) represents unique fragments useful in isomeric differentiation. Loss of the fluorophore reagent by cross-ring cleavage of the reducing sugar is labeled –AMC* and –AMC″, resulting in formation of ^{0,1}A and ^{0,2}A type product ions, respectively. Loss of the fluorophore reagent and reducing sugar glucose is labeled –AMC + Glc. A star symbol ($\star$) is used to signify the precursor ion. (Reproduced from Wilson, J. and Brodbelt, J.S., *Anal. Chem.* 2008, *80*, 5186–5196. With permission from American Chemical Society.)

an impressive array of molecules, from small organic molecules such as drugs to larger biological molecules including oligosaccharides, peptides, and nucleic acids. Although many of the original applications entailed the use of IR lasers, attention has shifted recently to the use of UV lasers, which allow deposition of more energy per photon. As newer low-cost bench-top lasers are introduced, it is anticipated that the development and application of numerous PD strategies will become even more commonplace.

ACKNOWLEDGMENTS

Support from the Welch Foundation (F1155) and the National Science Foundation (CHE-0718320) is gratefully acknowledged.

REFERENCES

1. Louris, J.N.; Cooks, R.G.; Syka, J.E.P.; Kelley, P.E.; Stafford, G.C.; Todd, J.F.J. Instrumentation, applications, and energy deposition in quadrupole ion-trap tandem mass spectrometry. *Anal. Chem.* 1987, *59*, 1677–1685.

2. Stafford, G.C.; Kelley, P.E.; Syka, J.E.P.; Reynolds, W.E.; Todd, J.F.J. Recent improvements in and analytical applications of advanced ion trap technology. *Int. J. Mass Spectrom. Ion Processes* 1984, *60*, 85–98.

3. Stephenson, J.L., Jr.; Booth, M.M.; Shalosky, J.A.; Eyler, J.R.; Yost, R.A. Infrared multiple photon dissociation in the quadrupole ion trap via a multipass optical arrangement. *J. Am. Soc. Mass Spectrom.* 1994, *5*, 886–893.

4. Hashimoto, Y.; Hasegawa, H.; Yoshinari, K.; Waki, I. Collision-activated infrared multiphoton dissociation in a quadrupole ion trap mass spectrometer. *Anal. Chem.* 2003, *75*, 420–425.

5. Keller, K.M.; Brodbelt, J.S. Collisionally activated dissociation and infrared multiphoton dissociation of oligonucleotides in a quadrupole ion trap. *Anal. Biochem.* 2004, *326*, 200–210.

6. Drader, J.J.; Hannis, J.C.; Hoftstadler, S.A. Infrared multiphoton dissociation with a hollow fiber waveguide. *Anal. Chem.* 2003, *75*, 3669–3674.

7. Wilson, J.J.; Brodbelt, J.S. Infrared multiphoton dissociation for enhanced de novo sequence interpretation of N-terminal sulfonated peptides in a quadrupole ion trap. *Anal. Chem.* 2006, *78*, 6855–6862.

8. Gardner, M.; Vasicek, L.; Shabbir, S.; Anslyn, E.; Brodbelt, J.S. Chromogenic crosslinker for the characterization of protein structure by infrared multiphoton dissociation mass spectrometry. *Anal. Chem.* 2008, *80*, 4807–4819.

9. Payne, A.H.; Glish, G.L. Thermally assisted infrared multiphoton photodissociation in a quadrupole ion trap. *Anal. Chem.* 2001, *73*, 3542–3548.

10. Boue, S.M.; Stephenson, J.L.; Yost, R.A. Pulsed helium introduction in a quadrupole ion trap for reduced collisional quenching during infrared multiphoton dissociation of electrosprayed ions. *Rapid Commun. Mass Spectrom.* 2000, *14*, 1391–1397.

11. Hashimoto, Y.; Hasegawa, H.; Waki, I. High sensitivity and broad dynamic range infrared multiphoton dissociation for a quadrupole ion trap. *Rapid Commun. Mass Spectrom.* 2004, *18*, 2255–2259.

12. Pikulski, M.; Wilson, J.J.; Aguilar, A.; Brodbelt, J.S. Amplification of infrared multiphoton dissociation efficiency in a quadruple ion trap by using IR-active ligands. *Anal. Chem.* 2006, *78*, 8512–8517.

13. Pikulski, M.; Hargrove, A.; Shabbir, S.; Anslyn, E.; Brodbelt, J.S. Sequencing and characterization of oligosaccharides using infrared multiphoton dissociation and boronic acid derivatization in a quadrupole ion trap. *J. Am. Soc. Mass Spectrom.* 2007, *18*, 2094–2106.

14. Crowe, M.; Brodbelt, J.S. Infrared multiphoton dissociation (IRMPD) and collisionally activated dissociation of peptides in a quadrupole ion trap with selective IRMPD of phosphopeptides. *J. Am. Soc. Mass Spectrom.* 2004, *15*, 1581–1592.

15. Crowe, M.C.; Brodbelt, J.S. Differentiation of phosphorylated and unphosphorylated peptides by high performance liquid chromatography/electrospray ionization/infrared multiphoton dissociation in a quadrupole ion trap. *Anal. Chem.* 2005, *77*, 5726–5734.

16. Goolsby, B.J.; Brodbelt, J.S. Characterization of β-lactams by photodissociation and collision activated dissociation in a quadrupole ion trap. *J. Mass Spectrom.* 1998, *33*, 705–712.

17. Goolsby, B.J.; Brodbelt, J.S. Analysis of protonated and alkali metal cationized aminoglycoside antibiotics by collisional activated dissociation and photodissociation in a quadrupole ion trap. *J. Mass Spectrom.* 2000, *35*, 1011–1024.

18. Vartanian, V.H.; Goolsby, B.; Brodbelt, J.S. Identification of tetracycline antibiotics by electrospray ionization in a quadrupole ion trap. *J. Am. Soc. Mass Spectrom.* 1998, *9*, 1089–1098.

19. Crowe, M.C.; Goolsby B.J.; Hergenrother, P.; Brodbelt, J.S. Characterization of erythromycin analogs by collisional activated dissociation and infrared multiphoton photodissociation in a quadrupole ion trap. *J. Am. Soc. Mass Spectrom.* 2002, *13*, 630–649.

20. Shen, J.; Brodbelt, J.S. Characterization of ionophore/metal complexes by infrared multiphoton photodissociation and collisionally activated dissociation in a quadrupole ion trap mass spectrometer. *Analyst* 2000, *125*, 641–650.

21. Wilson, J.J.; Brodbelt, J.S. Infrared multiphoton dissociation of duplex DNA/drug complexes in a quadrupole ion trap. *Anal. Chem.* 2007, *79*, 2067–2077.

22. Colorado, A.; Shen, J.X.; Vartanian, V.H.; Brodbelt, J.S. Use of infrared multiphoton dissociation with SWIFT for electrospray ionization and laser desorption applications in a quadrupole ion trap mass spectrometer. *Anal. Chem.* 1996, *68*, 4033–4043.

23. Gabelica, V.; Rosu, F.; De Pauw, E.; Lamaire, J.; Gillet, J-C.; Poully, J-C.; Lecomte, F.; Gregoire, G.; Schermann, J-P.; Desfrancois, C. Infrared signature of DNA G-quadruplexes in the gas phase. *J. Am. Chem. Soc.* 2008, *130*, 1810–1811.

24. Goolsby, B.J.; Brodbelt, J.S. Tandem infrared multiphoton dissociation and collisionally activated dissociation techniques in a quadrupole ion trap. *Anal. Chem.* 2001, *73*, 1270–1276.

25. Mazzitelli, C.L.; Brodbelt, J.S. Probing ligand binding to duplex DNA using $KMnO_4$ reactions and electrospray ionization tandem mass spectrometry. *Anal. Chem.* 2007, *79*, 4636–4647.

26. Remes, P.M.; Glish, G.L. Collisional cooling in a quadrupole ion trap at sub-ambient temperatures. *Int. J. Mass Spectrom.* 2007, *265*, 176–181.

27. Stephenson, J.L.; Booth, M.M.; Boue, S.M.; Eyler, J.R.; Yost, R.A. Analysis of biomolecules using electrospray ionization-ion trap mass spectrometry and laser photodissociation. In *Biochemical and Biotechnological Applications of Electrospray Ionization Mass Spectrometry*; Snyder, A.P.; Ed. *ACS Symposium Series*: American Chemical Society Washington, DC. 1996, 601–644.

28. Brodbelt, J.S.; Wilson, J.J. Infrared multiphoton dissociation in quadrupole ion traps. *Mass Spectrom. Rev.* 2008, *28*, 390–424.

29. Tabarin, T.; Antoine, R.; Broyer, M.; Dugourd, P. Specific photodissociation of peptides with multi-stage mass spectrometry. *Rapid Commun. Mass Spectrom.* 2005, *19*, 2883–2892.

30. Thompson, M.S.; Cui, W.; Reilly, J.P. Fragmentation of singly charged peptide ions by photodissociation at 157 nm. *Angew. Chem. Int. Ed.* 2004, *43*, 4791–4794.

31. Wilson, J.J.; Brodbelt, J.S. MS/MS simplification by 355 nm ultraviolet photodissociation of chromophore-derivatized peptides in a quadrupole ion trap. *Anal. Chem.* 2007, *79*, 7883–7892.

32. Wilson, J.; Kirkovits; G.J.; Sessler, J.L.; Brodbelt, J.S. Non-covalent complexation and energy transfer for peptide photodissociation in the gas phase. *J. Am. Soc. Mass Spectrom.* 2008, *19*, 257–260.

33. Wilson, J.; Brodbelt, J.S. Ultraviolet photodissociation at 355 nm of fluorescently-labeled oligosaccharides. *Anal. Chem.* 2008, *80*, 5186–5196.

34. Zhang, L.; Cui, W.; Thompson, M.S.; Reilly, J.P. Structures of α-type ions formed in the 157 nm photodissociation of singly-charged peptide ions. *J. Am. Soc. Mass Spectrom.* 2006, *17*, 1315–1321.

35. Zhang, L.; Reilly, J.P. Use of 157-nm photodissociation to probe structures of y- and b-type ions produced in collision-induced dissociation of peptide ions. *J. Am. Soc. Mass Spectrom.* 2008, *19*, 695–702.

36. Cui, W.; Thompson, M.S.; Reilly, J.P. Pathways of peptide ion fragmentation induced by vacuum ultraviolet light. *J. Am. Soc. Mass Spectrom.* 2005, *16*, 1384–1398.

37. Devakumar, A.; Thompson, M.S.; Reilly, J.P. Fragmentation of oligosaccharide with 157 nm vacuum ultraviolet light. *Rapid Comm. Mass Spectrom.* 2005, *192*, 2313–2320.

38. Gabelica, V.; Rosu, F.; Tabarin, T.; Kinet, C.; Antoine, R.; Broyer, M.; De Pauw, E.; Dugourd, P. Base-dependent electron photodetachment from negatively charged DNA strands upon 260-nm laser irradiation. *J. Am. Chem. Soc.* 2007, *129*, 4706–4713.

39. Gabelica, V.; Tabarin, T.; Antoine, R.; Rosu, F.; Compagnon, I.; Broyer, M.; De Pauw, E.; Dugourd, A. Electron photodetachment dissociation of DNA polyanions in a quadrupole ion trap mass spectrometer. *Anal. Chem.* 2006, *78*, 6564–6572.

40. Joly, L.; Antoine, R.; Broyer, M.; Dugourd, P.; Lemoine, J. Specific UV photodissociation of tyrosyl-containing peptides in multistage mass spectrometry. *J. Mass Spectrom*, 2007, *42*, 818–824.

41. Kim, T.Y.; Thompson, M.S.; Reilly, J.P. Peptide photodissociation at 157 nm in a linear ion trap mass spectrometer. *Rapid. Commun. Mass Spectrom.* 2005, *19*, 1657–1665.

42. Ly, T.; Julian, R.R. Residue-specific radical-directed dissociation of whole proteins in the gas phase. *J. Am. Chem. Soc.* 2008, *130*, 351–358.

43. Lemoine, J.; Tabarin, T.; Antoine, R.; Broyer, M.; Dugourd, P. UV photodissociation of phospho-seryl-containing peptides: laser stabilization of the phosphor-seryl bond with multistage mass spectrometry. *Rapid Commun. Mass Spectrom.* 2006, *20*, 507–511.

44. Hughes, R.J.; March, R.E.; Young, A.B. Multiphoton dissociation of ions derived from 2-propanol in a QUISTOR with low power CW infrared laser radiation. *Int. J. Mass Spectrom. Ion Phys.* 1982, *42*, 255–263.

45. Hughes, R.J.; March, R.E.; Young, A.B. Multiphoton dissociation of ions derived from *iso*-propanol and deuterated analogues in a QUISTOR with low power CW infrared laser radiation. *Can. J. Chem.* 1983, *61*, 834–845.

46. Hughes, R.J.; March, R.E.; Young, A.B. Optimization of ion trapping characteristics for studies of ion photodissociation. *Can. J. Chem.* 1983, *61*, 824–833.

47. March, R.E. Multiphoton induced dissociation of protonated dimers of 2-propanol. NATO Advanced Study Institute, September 1982, Vimeiro, Portugal. Ionic processes in the gas phase. M.A.A. Ferreira (ed.) Series C: Mathematical and Physical Sciences, 1983, *118*, 359–360.

48. Young, A.B.; March, R.E.; Hughes, R.J. Studies of infrared multiphoton dissociation rates of protonated aliphatic alcohol dimers. *Can. J. Chem.* 1985, *63*, 2324–2331.

21 Photochemical Studies of Metal Dication Complexes in an Ion Trap

Guohua Wu, Hamish Stewart, and Anthony J. Stace

CONTENTS

21.1 Introduction 847
21.2 Description of the Apparatus...848
    21.2.1 Pick-Up Process and Ionization 849
    21.2.2 Ion Selection and Transmission 850
    21.2.3 Ion Trapping, Storage, and Isolation 850
    21.2.4 Total Ion Current...851
    21.2.5 Buffer Gas Influence on Trapping Efficiency and
        Photodissociation 851
    21.2.6 The Effects of Residual Water on the Behavior of Dication
        Complexes..853
    21.2.7 Cooling the Ion Trap with Liquid Nitrogen 854
21.3 Photodissociation of Trapped Ions 855
    21.3.1 IR Photodissociation of Trapped Ions..856
    21.3.2 UV Photodissociation of Trapped Ions 858
21.4 Conclusion 860
Acknowledgments..860
References...860

21.1 INTRODUCTION

Experiments designed to study the structure and chemistry of gas-phase multiply-charged metal/molecule complexes must surmount necessarily several experimental challenges [1], most of which arise because of the strong propensity for such complexes to undergo charge-separation reactions where fragmentation is accompanied by one or more units of charge moving away from the central metal cation. Similar problems do not arise in the study of singly-charged complexes and, as a consequence, considerable progress has been made using such techniques as collision-induced dissociation (CID) [2,3] and photodissociation to investigate the properties of these ions [4,5].

A number of ion trap experiments (ICR and quadrupole) have used infrared multi-photon dissociation (IRMPD) to provide important information on the structures of ions, particularly when complemented with theoretical techniques to calculate vibrational frequencies. Early experiments used cw line-tunable CO_2 lasers to promote IRMPD [6–10], and the advantage of using an ion trap is that stored ions sit within the radiation field and are able to accumulate large numbers (>30) of low energy photons [11]. Shin and Beauchamp [7], for example, obtained IRMPD spectra for the $Mn(CO)_4CF_3^-$ ion generated from different precursors, and used the spectral similarity as evidence of a common structure. More recently, Free Electron Lasers (FEL) have been used in conjunction with ion traps to generate IR spectra covering a much wider wavelength range than is accessible with a CO_2 laser [12,13].

To date, most spectroscopic experiments on metal cations have been restricted mainly to singly-charged species, which contrasts with the fact that the +2 charge state is observed more commonly for a significant fraction of metals in the periodic table, as part of their condensed phase chemistry and biochemistry. Since the second ionization energy (IE) of most metal atoms exceeds the first IE of most ligands, charge transfer often prevents the observation of stable multiply-charged complexes in the gas phase. Many small dications are, in fact, metastable with respect to dissociation into two singly-charged ions, and it is this metastability that makes them particularly vulnerable to partial charge transfer when they are stored for the purposes of either a spectroscopic or kinetic experiment.

Electrospray ionization (ESI) has been shown by Kebarle and co-workers to be very effective at producing high yields of solvated, multiply-charged ions [14,15]. The mechanism of ESI has been discussed widely [16–19], and although the technique is very versatile in terms of the range of ligands that can be used as solvents, there are many simple molecules, *e.g.*, NH_3, which cannot be attached to metal dications. However, when used in conjunction with an ion trap, ESI has been shown to be very effective at generating metal dication complexes for both kinetic or spectroscopic experiments [20,21]. Recent experiments by Walker *et al.* have demonstrated also the potential of laser vaporization as a mechanism for generating selected doubly charged metal complexes [22].

As an alternative to ESI as a means of generating multiply-charged metal/ligand complexes, our group has adopted the pick-up technique [23–26], which we have shown to be capable of producing complexes from a wide variety of metals and ligands. Examples of the types of di- and tri-cation complexes that can be generated range from very simple complexes with inert gas, *e.g.*, $[Cu(Ar)_n]^{2+}$ for n in the range 1–6 [25], through to $[Al(CH_3CN)_4]^{3+}$ [27], which has an established solid-state chemistry. In addition, the pick-up technique is capable of generating complexes in charge states that are otherwise difficult to access for example, those of silver(II) [28] and gold(II) [29]. To produce multiply-charged ions in preparation for chemical and spectroscopic experiments, the pick-up technique has been coupled to a quadrupole mass filter and an ion trap. The procedures outlined here demonstrate how this combination of techniques is capable of producing trapped ions in sufficient number density and to photodissociate them using infrared and UV/visible wavelengths in order to obtain mass spectra.

21.2   DESCRIPTION OF THE APPARATUS

The apparatus shown schematically in Figure 21.1 has been used successfully to study both the IRMPD and UV spectroscopy of dication complexes; however, some

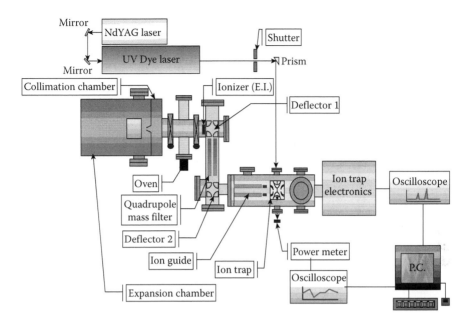

FIGURE 21.1   Schematic diagram of the apparatus, showing the pick-up region (oven), the quadrupole mass filter, and the ion trap, together with ancillary equipment.

aspects of the equipment are not yet refined fully and are the subject of continuing development. The instrument comprises of a series of differentially pumped vacuum chambers, covering the following processes: supersonic expansion, beam collimation, metal vaporization and pick-up, quadrupole mass selection, and ion trapping.

21.2.1   Pick-Up Process and Ionization

All experiments to date have demonstrated quite clearly that argon as a carrier gas is central to the success of the pick-up technique. Typically, a mixture of the gas with solvent vapor undergoes supersonic expansion through a continuous nozzle to give a beam of mixed clusters, that is then collimated before passing through to an oven chamber containing metal vapor produced from a Knudsen effusion cell. The pick-up process involves a collision with a metal atom followed by the evaporation of argon atoms to stabilize the neutral metal-containing complexes. This general approach appears to be suitable for all metals capable of generating sufficient vapor pressure at temperatures below 2000°C. Notable advantages of the pick-up technique over, for example, ESI are: (i) it can be used to study a range of metal oxidation states in association with molecules that are not readily accessible in the condensed phase, for example CO_2 has proved to be an extremely effective ligand [30]; and (ii) the pick-up technique can be used also to prepare complexes in oxidation states that are difficult to achieve in the condensed phase, for example those of gold(II) [29,31]. A disadvantage is that the technique is restricted currently to ligands that are volatile; however, some success has been obtained with a dual pick-up system where involatile ligands are held in a heated reservoir inside the vacuum chamber that houses the nozzle [32].

A molecular dynamics simulation of the pick-up processes has shown that clusters undergo a solid/liquid phase transition and that the colliding species then moves from the surface through the liquid to the center of the cluster. Stability is then regained by the evaporation of argon atoms [33].

Experiments have shown [26,30] that the optimum partial pressure of metal required for the production of mixed solvent/metal clusters lies in the range 10^{-1} and 10^{-2} mbar. At higher pressures, the cluster beam intensity drops due to scattering while at lower pressures, insufficient numbers of metal atoms are picked up. Experience has shown that adequate pumping at the base and throat of the Knudsen cell is essential if the latter is to function at high temperatures and moderately high background pressures (ca 10^{-5} mbar when the cluster beam is transmitted).

The collimated neutral cluster beam enters a molecular beam ionizer (ABB Extrel, Pittsburgh, PA, USA), which is positioned approximately 100 cm downstream from the expansion nozzle. Optimum ionization efficiency is achieved using four filament coils connected in a series/parallel combination and mounted on a quadrupole deflector. The latter ensures that un-ionized material does not contaminate the quadrupole rods. For high ion signal intensity, the filament emission current is set to 5 A and the electron energy to 100 eV. Under these conditions, a quasi-plasma is produced in the ionization region and ion space charge appears to dominate over electron space charge, with the former playing a major role in determining extraction efficiency and final ion signal intensity. With the correct ion source conditions, it is observed frequently that singly- and doubly-charged complexes have similar ion signal intensities.

21.2.2 Ion Selection and Transmission

Ions with a pre-set kinetic energy are extracted from the ionization region and pass through a quadrupole deflector into a quadrupole mass filter within which ions of a specific mass-to-charge ratio are isolated for further study. Mass discrimination on the part of the quadrupole mass filter is minimized by operating at low resolution. The mass-selected ions are turned through 90° by a second quadrupole deflector before passing through a quadrupole ion guide and focusing lens and into a three-dimensional radio-frequency quadrupole ion trap (Finnigan ITMS system, San Jose, CA, USA). After accumulation, the trapped ions and any reaction products are ejected toward an electron multiplier located just beyond the exit end-cap electrode. At each stage of acceleration and focusing, the energy and dimensions of the ion beam can be optimized for acceptance by tuning the beam through the ion trap and directly on to the electron multiplier. For this purpose, the ion trap is held at ground potential. Mass spectra and selected ion intensities can also be monitored on a detector (not shown) located at 180° from the ion guide.

21.2.3 Ion Trapping, Storage, and Isolation

The ion trap is operated in the total storage mode, *i.e.*, with zero DC voltage on the two end-cap electrodes and an RF voltage oscillating at 1.1 MHz is applied to the ring electrode. To help trap incoming ions, high purity helium is admitted directly to the trap *via* a precision leak valve and, depending on the ion signal intensity, ions are accumulated continuously for times of between 200 ms and 1000 ms. A set of deflection plates, located immediately before the ion trap is used to 'stop and start' the continuous ion beam, such that ions only enter the ion trap during the 'acquisition' sequence of an experiment. Ion complexes of interest are trapped preferentially either by DC isolation or by ramping the RF voltage to an appropriate value. This isolation step removes unwanted fragment ions formed by CID or/and ion/molecule reactions between injected ions and background molecules; the most reactive being O_2 and water. The latter proved to be a particular problem within the quadrupole mass filter and the ion trap chambers, and steps taken to minimize the effects of residual water have included cryopumping, baking, and vapor traps on all gas lines. The quadrupole mass filter and ion trap vacuum chambers are baked continuously and pumped routinely to a background pressure below 3×10^{-8} mbar (uncorrected).

21.2.4   Total Ion Current

Experiments have shown that good signal-to-noise ratios can be achieved in these experiments when approximately 2×10^4 ions enter the trap with a trapping efficiency that is assumed to be *ca* 1% (see below) during each accumulation cycle. Since the injected ion beam is continuous, the total number of ions, N, reaching the ion trap during the accumulation stage needs to be

$$N = I \times t \times 10^{-19}/(1.602 \times 10^{-19} \times G) \qquad (21.1)$$

where I is the ion current in pA, as measured at the second detector with the ring electrode on the ion trap at ground potential, t is the accumulation time in seconds, and G is the multiplier gain. A minimum initial current of approximately 370 pA is required when the accumulation time is 1 second and G is 10^5. Using the methods outlined above, this initial condition is not difficult to achieve for many of the metal dication complexes of interest. Moreover, the accumulation time can be modified (within limits) to suit the initial signal intensities of selected ions.

21.2.5   Buffer Gas Influence on Trapping
Efficiency and Photodissociation

Ion traps are operated typically at moderately high pressures (*ca* 10^{-3} mbar) using helium as a buffer gas, which serves to improve performance (sensitivity and mass resolution) by damping the motions of ions and directing their movement to the center of the ion trap. Difficulties in accurate pressure measurement can arise because helium is bled directly into the ion trap and the subsequent gas pressure is monitored on an ion gauge situated some distance from the ion trap. A more accurate measurement of helium pressure in an ion trap can be achieved by calculating the flow conductance and pumping speed, and treating the various orifices (for example the end-cap ion entrance and exit holes) as apertures for the purposes of determining

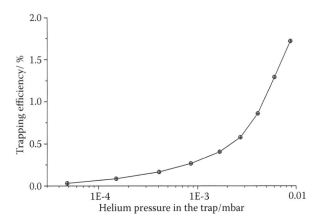

FIGURE 21.2 The effect of helium pressure on the trapping efficiency of the dication complex $[^{64}Zn(pyridine)_4]^{2+}$. Ions were injected into the ion trap with a laboratory-frame kinetic energy of 22 eV.

molecular flow. The flows and pump speed can then be used to scale remote pressure measurements using the equation [34]

$$P_1 = (1 + S_n \times C_p/(S_n + C_p)C) \times P_{ion}/\beta \qquad (21.2),$$

where P_1 is the estimated pressure in the ion trap, C is the molecular flow conductance of an aperture in L s^{-1}, S_n is the nominal pumping speed, C_p is the conductance of the tubing between the pump and the vacuum chamber, P_{ion} is the pressure reading on the ionization gauge, and β is the relative sensitivity of the gauge to helium, which is 0.25.

Figure 21.2 shows the relation between helium pressure, P_1, and the trapping efficiency of $[Zn(pyridine)_4]^{2+}$ complexes. The presence of helium focuses ions to the center of the ion trap both radially and axially by quenching their kinetic energy through collisions [35]. The overall shape of the plot matches previous observations by Cooks and co-workers for singly-charged ions [36]. As can be seen from Figure 21.2, high pressures of helium increase the trapping efficiency; however, with reference to laser-based experiments (see below), it should be noted that helium is also responsible for collisional deactivation during experiments that involve IRMPD.* To separate energy quenching from photoexcitation, Yost and co-workers [37] incorporated a pulsed helium inlet arrangement into their ion trap scan function. The results showed improvements in both the trapping efficiency of ions injected from an ESI source and in the yield of photofragments.† Experiments on IRMPD performed in

---

* See Volume 4, Chapter 17: Collisional Cooling in the Quadrupole Ion Trap Mass Spectrometer (QITMS), by Philip M. Remes and Gary L. Glish.
† See also Volume 4, Chapter 18: 'Pressure Tailoring' for Improved Ion Trap Performance, by Dodge L. Baluya and Richard A. Yost.

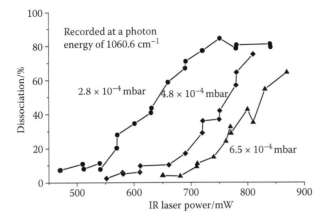

FIGURE 21.3 Plot showing the effects of helium pressure and laser power on photofragment yield during an IRMPD experiment on $[^{64}Zn(pyridine)_4]^{2+}$. The ions have been excited with a CO_2 laser operating at a photon energy of 1060.6 cm^{-1} and the ion trap was at room temperature.

the apparatus described above show that collisional deactivation begins to dominate events once the helium pressure is higher than 6.5×10^{-4} mbar. However, as Figure 21.3 shows, the degree to which collisional deactivation influences ion dissociation depends also quite critically on laser power. Most of the IRMPD experiments discussed here have been undertaken at a nominal trap pressure of 1.5×10^{-4} mbar. The ability to maximize ion number density at the center of the trap is particularly significant in photodissociation experiments because the laser beam typically enters and exits the ion trap *via* holes in the ring electrode. In contrast to IRMPD experiments, photodissociation at UV wavelengths benefits from ions being accumulated at the center of the ion trap without experiencing the additional problem of collisional relaxation; the reason being that photoexcitation and fragmentation operate on much faster timescales than do collision frequencies.

21.2.6 THE EFFECTS OF RESIDUAL WATER ON THE BEHAVIOR OF DICATION COMPLEXES

Figure 21.4a shows a mass spectrum recorded following injection of the complex $[^{64}Zn(pyridine)_2]^{2+}$ into an ion trap and where it has been stored for 1000 ms. In addition to a very weak precursor ion at m/z 111, the mass spectrum has additional ions at m/z 120 and 129, which are assigned as $[^{64}Zn(H_2O)(pyridine)_2]^{2+}$ and $[^{64}Zn(H_2O)_2(pyridine)_2]^{2+}$, respectively. These products arise from precursor ions that have picked up one and two water molecules during their residence time in the ion trap. At room temperature, the coordination does not increase beyond four molecules, but the intensity of $[^{64}Zn(H_2O)_2(pyridine)_2]^{2+}$ continues to rise as a function of trapping time, which suggests that four coordinate Zn^{2+} may be a stable structure under these conditions. Similarly, when $[^{64}Zn(pyridine)_3]^{2+}$ is injected into the ion trap [34], this ion picks up only a single water molecule. When $[^{24}Mg(pyridine)_4]^{2+}$ is

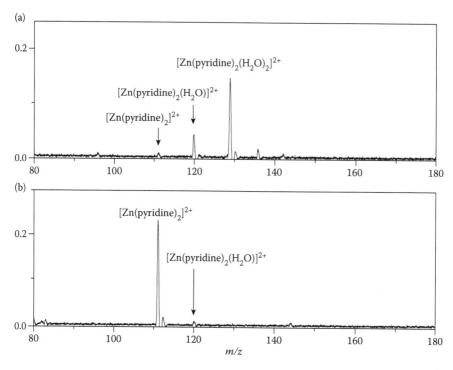

FIGURE 21.4 Mass spectra recorded following the injection and storage of [^{64}Zn(pyridine)$_4$]$^{2+}$ in an ion trap: (a) at room temperature; (b) after cooling the end-cap electrodes with liquid nitrogen.

injected into the ion trap, the result is very different from that seen for Zn^{2+} complexes [34]. In addition to the precursor ion, there is a doubly-charged CID product ion, [^{24}Mg(pyridine)$_3$]$^{2+}$, and these ions are accompanied also by a range of charge-transfer fragments, namely: MgOH(pyridine)$_3^+$, MgOH(pyridine)$_2^+$, and protonated pyridine (C$_5$H$_5$NH$^+$). It would appear that [Mg(pyridine)$_4$]$^{2+}$ is much more susceptible than is [Zn(pyridine)$_4$]$^{2+}$ to attack by water, and that the barrier to hydrolysis for Mg^{2+} is much lower than that for Zn^{2+}. It is interesting to note that in the bulk phase, Zn^{2+} is seen as a stronger acid (lower pK$_h$) than is Mg^{2+}; however, the outcome of interactions of the type described above may be driven, in part, by ligand preference, with Zn^{2+} favoring coordination with nitrogen-donating ligands [38], whereas Mg^{2+} prefers oxygen as a donor atom [30].

21.2.7 COOLING THE ION TRAP WITH LIQUID NITROGEN

The generation of ions by electron impact means that they emerge from the ion source with high levels of internal excitation. Compounding this problem is the fact that ligand dication binding energies are frequently very large (ca 200–300 kJ mol^{-1}) and, therefore, evaporation is not an effective mechanism for the dissipation of residual energy. As a result, spectral features recorded in the absence of any cooling can be very broad, in some cases covering many thousands of wavenumbers [39,40]. Moreover, a broad internal energy distribution may lead also to the occurrence of structural isomers. The only advantage to be gained from the presence of internal energy is that single photon dissociation can be achieved with a broader range of wavelengths than is possible when the ions are cold [40]. In order to improve the resolution of photodissociation spectra, we have experimented with cooling the ion trap with liquid nitrogen. There are a number of advantages to be gained from achieving cold ions.* These are (i) the ion internal energies will be low, which will improve spectral resolution; (ii) ions will have low kinetic energies and so will reside close to the bottom of the potential well that is defined by the trapping voltages; (iii) ions with low binding energies might be trapped more successfully when the damping collisions with helium are at low energies; and (iv) cooling can help to reduce the amount of residual water vapor present in the ion trap. With respect to (ii) above, it should be noted that because the bottom of the potential well coincides with the center of the ion trap, there is optimum overlap with the passage of a laser beam through the apparatus. Furthermore, when ions are concentrated at the center of the ion trap, there should be an improvement in mass resolution upon ion ejection; however, space charge effects may limit ultimately spectral resolution when ions are irradiated in the center of the ion trap.

To cool the ion trap, a small copper reservoir was constructed such that it could be push-fitted on to one of the end-cap electrodes, and then thermal contact between the two end-cap electrodes was maintained *via* three copper connecting rods. An external reservoir was filled with liquid nitrogen that then circulated through into the trap reservoir *via* two stainless steel bellows. Collisions with the cold end-cap electrodes cooled the helium that, in turn, cooled the trapped ions. The internal temperature of the latter is estimated to be *ca* 100 K, which compares favorably with the > 400 K internal temperature of ions emerging from the ion source. The latter temperature has been estimated from the observation that many of the ion complexes exhibit unimolecular fragmentation [41] and, therefore, have internal energies that are at least equivalent to the high binding energies of the individual ligands.

The consequences of cooling can be seen most graphically in Figure 21.4 where $[Zn(pyridine)_2]^{2+}$ has been injected into the ion trap. A mass spectrum recorded in the absence of liquid nitrogen in the reservoir is shown in Figure 21.4a, where, as discussed above, the precursor ion is almost completely eliminated by reaction with residual water. Figure 21.4b shows that, following cooling, any background water is effectively removed by cryopumping and it becomes possible to trap $[Zn(pyridine)_2]^{2+}$ ions. Similar results have been obtained for other co-ordinately unsaturated ions, such as $[Mn(pyridine)_3]^{2+}$ [42].

21.3   PHOTODISSOCIATION OF TRAPPED IONS

For an ion trap operating over a timescale of hundreds of milliseconds and at a finite gas pressure, dissociation following photoexcitation is in competition with several

---

* See Volume 5, Chapter 10: Sympathetically-Cooled Single Ion Mass Spectrometry, by Peter Frøhlich Staanum, Klaus Højbjerre and Michael Drewsen.

significant relaxation mechanisms. In the presence of laser radiation and at a gas pressure where collisions are important, the time evolution of the population $A(E, t)$ of a molecular or ionic energy level at an energy E, can be described in terms of the following master equation [11]:

$$\frac{\partial A(E,t)}{\partial t} = \omega \int \left[q(E,E')A(E') - q(E',E)A(E) \right] dE' - k(E)A(E) - A(E)/\tau_R$$

$$- \left[\varphi_A(E) + \varphi_S(E) \right] A(E) + \varphi_A(E - h\nu) \, A(E - h\nu) \qquad (21.3),$$

$$+ \varphi_S(E + h\nu) A(E + h\nu) + A(E + h\nu)/\tau_R$$

where ω is a collision frequency, $q(E,E')$ is the probability of collisional energy transfer from level E' to E, $k(E)$ is the rate constant for unimolecular decay of those ions with energy $\geq \varepsilon_0$, the critical energy of reaction, $\varphi_A(E)$ is the rate coefficient for the absorption of radiation, $\varphi_S(E)$ is the rate coefficient for stimulated emission, and τ_R is the radiative lifetime.

The details of how Monte Carlo methods can be applied to Equation 21.3 in order to model IRMPD in an ion trap have been presented earlier [11]. The same approach has also been used more recently to examine the significance of kinetic shifts when IRMPD in an ion trap is used to study processes that have high activation energies [43]. Depending on the wavelength of the radiation being used, different aspects of Equation 21.3 come into play. Figure 21.3 shows that, at IR wavelengths, collisions are the dominant relaxation mechanism, whilse at UV/visible wavelengths (not shown), radiative decay becomes significant and collisions are too infrequent to compete. Equation 21.3 can be used also to underpin the benefits of cooling. For many dication complexes, the binding energies of ligands, $[M(L_n)]^{2+}$, beyond $n = 4$ drop quite significantly when compared with smaller ions, and it can prove almost impossible to trap such species at room temperature. The model shows that for a trapping time of 1 s the flux of blackbody radiation at 300 K is sufficient to promote the dissociation of $[M(L_n)]^{2+}$ complexes over that time scale; however, once the ion trap temperature drops below ca 100 K blackbody effects are no longer significant [42].

21.3.1   IR PHOTODISSOCIATION OF TRAPPED IONS

Many experiments on IR photodissociation in an ion trap have used radiation from a line-tunable CO_2 laser [6–9]. In the experiments we have reported the laser has operated in a chopped cw mode with a pulse duration timed to match the residence time of ions in the ion trap [43,44]. The laser beam (<5 mm in diameter) is focused by a ZnSe lens and passes radially through the center of the ion trap with an entrance and exit that are defined by two 1 mm-diameter holes in the ring electrode. All photodissociation experiments have been undertaken with the ion trap operating in a mass-selective axial instability mode that is initiated 2 ms after the laser pulse. The ion signals were amplified and recorded on a digital oscilloscope (LeCroy LT374M,

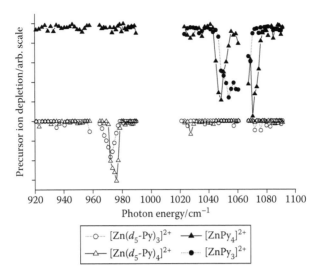

FIGURE 21.5 Photofragmentation spectra recorded following the IRMPD of [Zn(pyridine)$_{3,4}$]$^{2+}$ and [Zn(d_5-pyridine)$_{3,4}$]$^{2+}$ complexes with a CO_2 laser. Each spectrum is presented in the form of a depletion in precursor ion signal and covers the complete tuning range of the CO_2 laser. Spectra recorded from complexes where pyridine is the ligand are represented by closed circles and triangles, and spectra where d_5-pyridine is the ligand are represented by open circles and triangles.

New York, NY, USA), which is triggered, together with the CO_2 laser, by TTL signals from a delay/pulse generator (Stanford Research DG535, Sunnyvale, CA, USA). The delay generator receives a TTL pulse from the scan acquisition printed circuit board (PCB) on the ITMS system, which in turn is controlled within a single ITMS scan function. Typical IRMPD timing sequences for the ion deflector, the laser, and the ion trap duty cycle have been give in an earlier publication [34], and photofragmentation mass spectra represent typically averages taken of 200 ion trap scan functions (termed microscans within the ITMS software).

This approach has been used to record the first gas phase IR spectra of a series of dication complexes [44], and typical results comparing [Zn(pyridine)$_{3,4}$]$^{2+}$ and [Zn(d_5-pyridine)$_{3,4}$]$^{2+}$ are shown in Figure 21.5. The spectra were recorded with a line-tunable CO_2 laser (PL4, Edinburgh Instruments, Edinburgh, UK) and cover the available tuning range (920–1100 cm^{-1}) with a resolution of ca 1.5 cm^{-1}. There is a gap in the photon energy range at ca 1000 cm^{-1} where the laser switches between two ro-vibrational branches. In all cases the infrared-active modes are on the pyridine ring and, as can be seen, the results are quite sensitive both to the size of the complex and to isotopic substitution. The complexes involving pyridine absorb predominantly in the range 1040–1080 cm^{-1}, but these features appear to shift to ca 970 cm^{-1} when d_5-pyridine is used as a ligand. In a further refinement of the experiment, a short section of the IR spectrum has been repeated with the ion trap having been cooled with liquid nitrogen. The result is shown in Figure 21.6, where it can be seen that cooling narrows the absorption peak. Interestingly, a side effect of cooling is that the IRMPD of cold [Zn(pyridine)$_4$]$^{2+}$ complexes requires more photons to obtain

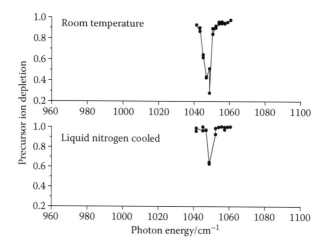

FIGURE 21.6 Comparison between an infrared absorption feature recorded for [Zn(pyridine)$_4$]$^{2+}$ at room temperature and the same feature when recorded with the ion trap end-cap electrodes cooled with liquid nitrogen.

the same photodissociation yield as an experiment performed at room temperature. Both of the results shown in Figure 21.6 were obtained using the same laser power (400 mW) but, as can be seen, the photofragment intensity is lower for the cold ions [42]. Depending on their initial internal energy, the trapped ions will need to absorb 30–40 IR photons in order to dissociate [11,43]. As the cold ions begin this process with less residual internal energy than those ions that are held in the trap at room temperature, fewer cold ions will absorb sufficient photons to dissociate during their time in the trap.

21.3.2   UV PHOTODISSOCIATION OF TRAPPED IONS

To obtain UV/visible photodissociation mass and optical spectra using a pulsed laser, such as an Nd:YAG laser, in conjunction with an ion trap requires a very different approach to that discussed above. The timing sequences are different and need to be gated more precisely than is required for IRMPD. The events in one cycle can be divided into three stages: (i) accumulation and trapping; (ii) photodissociation; and (iii) detection. During stage (i), a deflector in front of the ion trap is grounded for approximately 300 ms and, during that time, an optical shutter at the laser exit remains closed. This procedure ensures that the complex ion of interest can be injected continuously into the ion trap without being photodissociated. During stage (ii), an 800 ms-long DC pulse of 50 V, is applied on the deflector to exclude further ions from the ion trap. The optical shutter is opened for 700 ms, during which time mass-selected ions held in the ion trap are subjected to UV photoexcitation by the unfocused output of a pulsed dye laser (Sirah, Lasers, Kaarst, Germany). The latter is pumped by either the second or third harmonic of a Nd:YAG laser (Spectra-Physics Lasers, Didcot, Oxfordshire, UK) operating at 10 Hz and with a 7 ns pulse length. During stage (iii), the

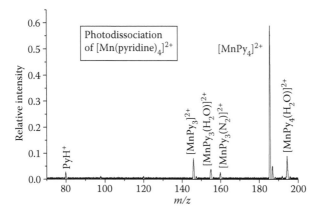

FIGURE 21.7 Ion trap mass spectra recorded following the UV photodissociation of [Mn(pyridine)$_4$]$^{2+}$ at 39,742 cm^{-1}. The most significant photofragments are labeled.

precursor and all fragment ions are ejected from the trap by ramping the RF voltage amplitude. This step provides mass analysis and the signal can be averaged by repeating stages (i)–(iii). A typical duty cycle is approximately 1200 ms in duration, and each cycle is repeated 200 times to yield an averaged photofragment mass spectrum. All mass channels related to photofragmentation are monitored continuously as the wavelength of the laser is scanned, and the experimental data are normalized usually with respect to precursor ion intensity and laser power. The approach outlined above gives the photodissociation spectra a multiplex advantage in that all photofragments are captured and mass assigned in a single sweep, and a final mass spectrometric stage is not required. The system as a whole appears to be more stable when the laser and the ion trap duty cycle are allowed to operate independently. The optical shutter then controls the number of laser pulses entering the ion trap but, within individual ion trap cycles, this value may vary slightly because the units are independent. However, the final signal is averaged over a sufficiently large number of laser shots that any small fluctuation has no significant effect on photofragment yield.

Figure 21.7 shows a photofragmentation mass spectrum recorded when [Mn(pyridine)$_4$]$^{2+}$ has been injected into the ion trap and photoexcited at a photon energy of 39,742 cm^{-1} [42]. Despite the ion trap being cold, this ion and its photofragments appear to be more susceptible to the attachment of water than their Zn^{2+} counterparts. In addition, the [pyridine + H]$^+$ ion appears in the photodissociation mass spectrum, suggesting that a charge-transfer transition is being excited. Figure 21.8 shows the short preliminary section of a UV photofragment spectrum recorded for [Mn(pyridine)$_4$]$^{2+}$, where it can be seen that there is considerable (as yet unassigned) structure, the appearance of which we attribute to the ions being cold and the experiment is, therefore, capable of resolving transitions between discrete energy states. A more complete UV spectrum has been recorded recently for [Zn(pyridine)$_4$]$^{2+}$ where it has been possible to assign structure to transitions between individual electronic states of the dication [45].

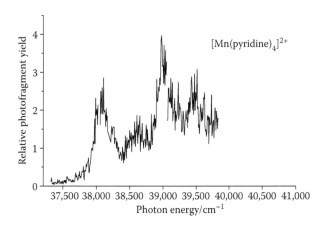

FIGURE 21.8 Short section of a UV photofragmentation spectrum recorded for cold $[Mn(pyridine)_4]^{2+}$ ions. The absorption features are, as yet, unassigned.

21.4 CONCLUSION

We have described here an apparatus designed to study the photo-induced fragmentation of mass-selected metal dication complexes. An initial evaluation of the experiment demonstrates that the pick-up technique is capable of yielding sufficient ion signal intensity for quantitative experiments on the photophysics of a range of metal dications, such as Cu^{2+}, Mg^{2+}, and Zn^{2+} complexed with pyridine.

We show that IRMPD can be applied to systems with high binding energies (*ca* 300 kJ mol^{-1}) and, with comparatively modest laser powers, photofragmentation can be promoted in dication complexes held within an ion trap. Cooling the ions appears to be an effective way of solving several problems associated with background water vapor and with high internal energy ions as a consequence of the method of preparation. Spectra recorded at UV/visible wavelengths appear to benefit from improved resolution as the result of ion cooling.

ACKNOWLEDGMENTS

The authors thank EPSRC for financial support for the construction of the ion trap apparatus and the mechanical workshop in the Department of Chemistry at Nottingham University for construction of the liquid nitrogen reservoir on the ion trap.

REFERENCES

1. Stace A.J. Metal ion solvation in the gas phase: the quest for higher oxidation states. *J. Phys. Chem.* 2002, *106*, 7993–8005.
2. Leary, J.A.; Armentrout, P.B. Gas phase metal ion chemistry: from fundamentals to biochemical interactions. *Int. J. Mass. Spectrom.* 2001, *204*, ix.
3. Rodgers, M.T.; Armentrout P.B. Noncovalent metal-ligand bond energies as studied by threshold collision-induced dissociation. *Mass Spectrom. Rev.* 2000, *19*, 215–247.
4. Dunbar, R.C. Photodissociation of trapped ions. *Int. J. Mass. Spectrom.* 2000, *200*, 571–589.

5. Duncan, M.A. Frontiers in the spectroscopy of mass-selected molecular ions. *Int. J. Mass Spectrom.* 2000, *200*, 545–569.

6. Wight, C.A.; Beauchamp, J.L. Infrared-spectra of gas-phase ions and their use in eluci-dating reaction-mechanisms – identification of C_7H_7-structural isomers by multi-photon electron detachment using a low-power infrared-laser. *J. Am. Chem. Soc.* 1981, *103*, 6499–6501.

7. Shin, S.K.; Beauchamp, J.L. Identification of $Mn(CO)_4CF_3^-$, $Mn(CO)_5CF_3^-$ structural isomers by IR multiphoton dissociation, collision-induced dissociation, and specific ligand displacement-reactions – studies of the trifluoromethyl migratory decarbo-nylation reaction in the gas-phase. *J. Am. Chem. Soc.* 1990, *112*, 2057–2066.

8. Peiris, D.M.; Cheeseman, M.A.; Ramanathan, R.; Eyler, J.R. Infrared multiple-photon dissociation spectra of gaseous-ions. *J. Phys. Chem.* 1993, *97*, 7839–7843.

9. Peiris, D.M.; Riveros, J.M.; Eyler, J.R. Infrared multiple photon dissociation spectra of methanol-attached anions and proton-bound dimer cations. *Int. J. Mass Spectrom. Ion Processes* 1996, *159*, 169–183.

10 Young, A.B.; March, R.E.; Hughes, R.J. Studies of infrared multiphoton disso-ciation rates of protonated aliphatic alcohol dimmers. *Can. J. Chem.* 1985, *63*, 2324–2331.

11. Stace, A.J. Infrared photophysics in an ion trap. *J. Chem. Phys.* 1998, *109*, 7214–7223.

12. Oomens, J.; van Roij, A.J.A.; Meijer, G.; von Helden, G. Gas-phase infrared photodis-sociation spectroscopy of cationic polyaromatic hydrocarbons. *Astrophys. J.* 2000, *542*, 404–410.

13. van Heijnsbergen, D.; von Helden, G.; Meijer, G.; Maitre, P.; Duncan, M.A. Infrared spectra of gas-phase V^+-(benzene) and V^+-(benzene)$_2$ complexes. *J. Am. Chem. Soc.* 2002, *124*, 1562–1563.

14. Blades, A.T.; Jayaweera, P.; Ikonomou, M.G.; Kebarle, P. Studies of alkaline-earth and transition-metal M^{++} gas-phase ion chemistry. *J. Chem. Phys.* 1990, *92*, 5900–5906.

15. Blades, A.T.; Jayaweera, P.; Ikonomou, M.G.; Kebarle, P. Ion-molecule clusters involv-ing doubly charged metal-ions (M^{2+}). *Int. J. Mass Spectrom. Ion Processes* 1990, *102*, 251–267.

16. Iribarne, J.V.; Thomson, B.A. Evaporation of small ions from charged droplets. *J. Chem. Phys.* 1976, *64*, 2287–2294.

17. Kebarle, P.; Tang L. From ions in solution to ions in the gas-phase – the mechanism of electrospray mass-spectrometry. *Anal. Chem.* 1993, *65*, 972A–A986.

18. Smith, J.N.; Flagan, R.C.; Beauchamp, J.L. Droplet evaporation and discharge dynam-ics in electrospray ionization. *J. Phys. Chem. A* 2002, *106*, 9957–9967.

19. Fenn, J.B. Mass spectrometric implications of high-pressure ion sources. *Int. J. Mass Spectrom.* 2000, *200*, 459–478.

20. Thompson, C.J.; Husband, J.; Aquirre, F.; Metz, R.B. Photodissociation dynamics of hydrated Ni^{2+} clusters: $Ni^{2+}(H_2O)_n$ ($n = 4$–7). *J. Phys. Chem. A* 2000, *104*, 8155–8159.

21. Combariza, M.Y.; Vachet, R.W. Effect of coordination geometry on the gas-phase reac-tivity of four-coordinate divalent metal ion complexes. *J. Phys. Chem. A* 2004, *108*, 1757–1763.

22. Walker, N.R.; Grieves, G.A.; Jaeger, J.B.; Walters, R.S.; Duncan, M.A. Generation of "unstable" doubly charged metal ion complexes in a laser vaporization cluster source. *Int. J. Mass Spectrom.* 2003, *228*, 285–295.

23. Woodward, C.A.; Dobson, M.P.; Stace, A.J. Chemistry of Mg^+ and Mg^{2+} in association with methanol clusters. *J. Phys. Chem. A* 1997, *101*, 2279–2287.

24. Stace, A.J.; Walker, N.R.; Firth, S. $[Cu(H_2O)_n]^{2+}$ clusters: the first evidence of aqueous Cu(II) in the gas phase. *J. Am. Chem. Soc.* 1997, *119*, 10239–10240.

25. Walker, N.R.; Wright, R.R.; Barran, P.E.; Cox, H.; Stace, A.J. Unexpected stability of [CuAr]²⁺, [AgAr]²⁺, [AuAr]²⁺, and their larger clusters. *J. Chem. Phys.* 2001, *114*, 5562–5567.

26. Wright, R.R.; Walker, N.R.; Firth, S.; Stace, A.J. Coordination and chemistry of stable Cu(II) complexes in the gas phase. *J. Phys. Chem. A* 2001, *105*, 54–64.

27. Puškar, L.; Tomlins, K.; Duncombe, B.J.; Cox, H.; Stace, A.J. What is required to stabilize Al³⁺? A gas-phase perspective. *J. Am. Chem. Soc.* 2005, *127*, 7559–7569.

28. Walker, N.R.; Wright, R.R.; Stace, A.J. Stable Ag(II) coordination complexes in the gas phase. *J. Am. Chem. Soc.* 1999, *121*, 4837–4844.

29. Walker, N.R.; Wright, R.R.; Barran, P.E.; Stace, A.J. Stable gold(II) complexes in the gas phase. *Organometallics* 1999, *18*, 3569–3571.

30. Walker, N.R.; Dobson, M.P.; Wright, R.R.; Barran, P.E.; Murrell, J.N.; Stace, A.J. A gas phase study of the co-ordination of Mg²⁺ with oxygen- and nitrogen-containing ligands. *J. Am. Chem. Soc.* 2000, *122*, 11138–11145.

31. Walker, N.R.; Wright, R.R.; Barran, P.E.; Murrell, J.N.; Stace, A.J. Comparisons in the behavior of stable copper(II), silver(II), and gold(II) complexes in the gas phase: are there implications for condensed-phase chemistry? *J. Am. Chem. Soc.* 2001, *123*, 4223–4227.

32. Duncome, B.J.; Duale, K.; Stace, A.J. unpublished results.

33. Del Mistro, G.; Stace, A.J. Cluster-molecule collisions - a molecular-dynamics analysis of a pick-up experiment. *Chem. Phys. Lett.* 1992, *196*, 67–72.

34. Wu, G.; Chapman, D.; Stace, A.J. Trapping and recording the collision- and photo-induced fragmentation patterns of multiply charged metal complexes in the gas phase. *Int. J. Mass Spectrom.* 2007, *262*, 211–219.

35. Doran, M.C.; Fulford, J.E.; Hughes, R.J.; Morita, Y.; March, R.E.; Bonner, R.F. Effects of charge-exchange reactions on the motion of ions in three-dimensional quadrupole electric fields. Part III. A two-ion model. *Int. J. Mass Spectrom. Ion Phys.* 1980, *33*, 139–158.

36. Louris, J.N.; Amy, J.W.; Ridley, T.Y.; Cooks, R.G. Injection of ions into a quadrupole ion trap mass-spectrometer. *Int. J. Mass Spectrom.* 1989, *88*, 97–111.

37. Boue, S.M.; Stephenson, J.L.; Yost, R.A. Pulsed helium introduction into a quadrupole ion trap for reduced collisional quenching during infrared multiphoton dissociation of electrosprayed ions. *Rapid Commun. Mass Spectrom.* 2000, *14*, 1391–1397.

38. Dumcombe, B.J.; Puškar, L.; Wu, B.; Stace, A.J. Gas-phase experiments on the chemistry and coordination of Zn(II) by aprotic solvent molecules. *Can. J. Chem.* 2005, *83*, 1994–2004.

39. Thompson, C.J.; Faherty, K.P.; Stringer, K.L.; Metz, R.B. Electronic spectroscopy and photodissociation dynamics of Co²⁺-methanol clusters: Co²⁺(CH₃OH)ₙ (n = 4–7). *Phys. Chem. Chem. Phys.* 2005, *7*, 814–818.

40. Guan, J.; Puškar, L.; Esplugas, R.; Cox, H.; Stace A.J. Ligand field photofragmentation spectroscopy of [Ag(L)ₙ]²⁺ complexes in the gas phase: Experiment and theory. *J. Chem. Phys.* 2007, *127*, Article No. 064311.

41. Duncombe, B.J.; Duale, K.; Buchanan-Smith, A.; Stace, A.J. The solvation of Cu²⁺ in gas-phase clusters of water and ammonia. *J. Phys. Chem. A.* 2007, *111*, 5158–5165.

42. Wu, G.; Stewart, H.; Stace, A.J. unpublished results.

43. Wu, G.; Stace, A.J. IRMPD study of the furan cation in an ion trap: Evidence of the extreme effect a competitive shift can have on reaction pathway. *Chem. Phys. Lett.* 2005, *412*, 1–4.

44. Wu, G.; Guan, J.; Aitken, G.D.C.; Cox, H.; Stace, A.J. Infrared multiphoton spectra from metal dication complexes in the gas phase. *J. Chem. Phys.* 2006, *124*, Article No. 201103.

45. Wu, G.; Norris, C.; Stewart, H.; Cox, H.; Stace, A.J. State-resolved UV photofragmentation spectrum of the metal dication complex [Zn(pyridine)₄]²⁺. *Chem. Commun.* 2008, 4153–4155.

Author Index*

Aberth, W., 45
Addink, C.C.J., 676
Aguilar, A., 831
Ahern, A.J., 37
Alimpiev, S.S., 253
Allemann, M., 677
Al-Omair, A.S., 47
Alpert, D., 74
Amy, J.W., 225, 226
Anders, L.R., 76, 77
Anderson, G.A., 409, 412
Andre, J., 75, 105, 448, 469, 496, 499, 500,
    503–506, 680
Anslyn, E., 835, 836
Appell, J., 65
Armitage, M.A., 103
Armstrong, J.T., 77
Ashcroft, A.E., 39
Ashline, D.J., 106, 265, 707, 812
Aston, F.W., 13–16, 31, 35
Aubry, C., 55, 56, 58
Audoin, C., 106
Austin, D.E., 111, 148, 177, 194, 202, 204, 373,
    390, 392, 394
Austin, W.E., 99
Axelsson, B.O., 365
Baba, T., 416, 417, 578, 579, 586, 588, 694
Badman, E.R., 174, 217, 218, 220
Bahadur, K., 38
Bahr, R.E., 124
Bainbrdge, K.T., 18, 32
Baker, F.A., 53, 54
Bakker, J.M.B., 89
Baldeschwieler, J.D., 75, 76
Ball, G.W., 75, 95
Baluya, D.L., 102, 740, 769, 796, 828, 852
Banner, A.E., 99
Barber, M., 32, 33, 45
Barnett, D.A., 327, 329, 331, 332, 334–337, 349,
    354, 356–359, 361
Bartky, W., 19
Bayard, R.T., 74
Bazakutska, V.A., 120
Beauchamp, J.L., 76–78, 848
Beckey, H-D., 39, 672
Behre, E., 124
Behrisch, R., 253

Belford, M., 335–337, 343, 349
Belgacem, O., 286, 415, 448, 793
Belov, M.E., 355, 401, 409, 410, 413, 418, 419,
    422–424, 677
Benilan, M-N., 106
Benner, W.H., 253
Bennett, W.H., 94
Benninghoven, A., 44, 45, 672
Berkling aus Leipzig, K., 96
Berton, A., 107, 285
Beynon, J.H., 9–12, 15, 17, 22, 24, 26, 28, 30, 33,
    34, 54, 673
Bhansali, S., 205
Biemann, K., 30, 681
Bier, M.E., 109, 111, 376
Black, D.M., 754
Blain, M.G., 182, 194, 388
Blake, R.S., 92
Blauth, E.W., 67, 68, 87, 93, 95
Bleakney,W., 29
Blears, J., 22
Blümel, R., 124–126
Bonner, R.F., 106
Born, M., 8, 9
Borsdorf, H., 311
Bourne, A.J., 54
Bouzid, S., 513–516
Bowers, M.T., 77
Boyd, R.K., 33, 92
Boyd, R.L., 94
Bradbury, N.E., 420
Brancia, F.L., 107, 148, 273, 285, 291, 293, 295,
    296, 301, 302
Brekenfeld, A., 95, 374, 593, 718
Brenton, A.G., 63
Brincourt, G., 496, 499, 500
Brockman, A.H., 363
Brodbelt, J.S., 43, 827, 831, 834–837, 841, 842
Brooks, P.R., 116
Browning, E.B., 6, 7
Brownnutt, M., 199
Brustkern, A.M., 18, 75, 402, 433,
    463–465, 689
Buchanan, M.V., 79, 80
Budzikiewicz, H., 14, 30
Bumgarner, J., 195
Buryakov, I.A., 315–317

---

* The names listed here refer only to authors whose names appear in the text and/or in the captions.

Busch, K.L., 31
Bushey, J.M., 694
Cady, W.G., 115
Cameron, E.A., 86
Campbell, J.M., 410, 596
Capellen, J.M., 48
Caprioli, R.M., 47
Caravatti, P., 408
Cardenas, M., 195
Carette, M., 75, 105, 448, 469, 490, 496, 499,
         500, 503, 504–506, 509, 511–516, 680
Carlson, D.M., 717
Castro, M.E., 453, 455
Cha, B.C., 410
Champarnaud, E., 364
Chang, J.P., 194
Chappell, W.J., 180, 206, 207, 223
Chaudhary, A., 195, 203, 205
Chernushevich, I.V., 525, 574, 643, 679, 718
Chipuk, J.E., 43
Christofilos, N.C., 96
Chupka, W.A., 37, 38
Church, D.A., 117, 377
Cleven, C.D., 166
Clow, R.P., 77
Clowers, B.H., 421, 422, 424
Collings, C.A., 410
Comisarow, M.B., 78, 434, 677
Compton, K.T., 115
Condon, E.U., 22, 23, 56
Contreras, J.A., 178
Conzemius, R.J., 48
Cooks, R.G., 14, 31, 61, 107, 123, 166, 169, 171,
         174, 180, 188, 189, 206, 207, 209,
         211–218, 220–227, 229, 259, 375, 471,
         597, 680, 690, 691, 746, 748, 852
Coon, J.J., 108, 285, 611, 686, 718
Cooper, H.J., 75, 300, 416, 611, 677
Cornish, T.J., 91
Costa, A.B., 223
Cotter, R.J., 91
Courant, E.D., 96
Covey, T.R., 52, 352
Craig, R.D., 26, 27, 36
Creaser, C.S., 420
Crowe, M., 834
Cruz, D., 192, 194
Curcuruto, O., 50
Dahan, M., 253
Daly, N.R., 38
Damm, C.C., 84
Danby, C.J., 54
Danielson III, W.F., 422, 424
Davies. A.C., 115
Dawson, J.H., 91, 92
Dawson, P.H., 95, 102, 492, 500, 620
Day, S., 314, 340

Dehmelt, H.G., 125, 374, 585, 678
Delfosse, J., 29
Demirev, P., 47
Dempster, A.J., 16, 18, 19, 35
Derrick, P.J., 60
Deuschle. T., 200
Devienne, F.M., 47
Dewar, J., 11
Dillon, A.F., 45
Ding, L., 107, 148, 273, 275, 280, 288, 289, 291,
         293, 295, 296, 301, 302, 327, 329, 357
Djerassi, C., 30
Dobson, G.S., 574
Dole, M., 51
Douglas, D.J., 352, 410, 526
Drader, J.J., 693
Drees, J., 679
Drewsen, M., 855
Dunbar, R.C., 689, 753
Duncan, J.S., 221, 222
Dunoyer, L., 8
Dunyach, J.J., 335–337, 349
Durup, J., 63, 65
Earnshaw, S., 252
Edelefsen, N.E., 72
Eggers, Jr., D.F., 86
Eiceman, G.A., 311, 333, 339, 420, 536, 712
Elliott, R.M., 32, 34
Ellis, H.W., 313
Ells, B., 327, 329, 332, 354, 357
Enke, C.G., 574
Ensberg, E.S., 628
Errock, G.A., 26
Evans, Jr., C.A., 51
Eyring, H., 28
Fehsenfeld, F.C., 65
Feigl, P., 48
Fenn, J.B., 49–51, 672
Fico, C.D., 221, 222
Fico, M., 194, 205–207, 222–226
Field, F.H., 40, 42, 43, 673
Fies, W.J., 471
Finnigan, R.E., 68
Fischer, E., 101, 183
Fizeau, H.L., 80
Fletcher, J., 125
Fohlman, J., 681
Forbes, M.W., 688, 828
Fournier, P., 65
Franzen, J., 36, 95, 113, 265, 374, 416, 492, 507,
         508, 593, 599, 611, 620, 622, 630,
         671, 718
Friedland, S.S., 86
Friedman, L., 28
Froelich, J.M., 611
Fujiwara, M., 455
Futrell, J.H., 75, 77, 689

Gabling, R-H., 599
Gabrielse, G., 406, 676
Gabryelski, W., 361
Gaede, W., 15
Gao, L., 227, 229
Gardner, M., 835
Garrick-Bethell, I., 125, 126
George III, J.E., 712
Gerlach, W., 9
Gerlich, D., 526, 574
Ghaderi, S., 435
Gibson, J.R., 152
Giles, R., 291, 293, 295, 296
Gill, P., 199
Gillig, J.J., 119, 120
Glasmachers, A., 680
Glish, G.L., 102, 739, 754, 755, 770, 796, 828, 852
Goeringer, D.E., 183, 750, 754
Gomer, R., 38
Good, M.L., 96
Goode, G.C., 77
Gooden, J.K., 445
Goodwin, M.P., 221, 222
Gorshkov, M.P., 315
Gorshkov, M.V., 409
Goudsmit, S.A., 81, 82
Gray, A.L., 37
Green, M.R., 574
Griep-Raming, J., 211, 212
Grigsby, R.D., 14
Gross, M.L., 18, 75, 402, 433, 435, 444–447, 463–465, 688, 689
Guan, S., 80
Guevrement, R., 314, 323–325, 327, 329, 332, 333, 340, 342, 343, 353, 354, 356, 357, 359, 361
Guilhaus, M., 91, 92
Guymon, A.J., 194, 211, 212
Haddon, W.F., 710
Hadjarab, F., 405, 456
Hager, J.W., 110, 545, 547, 548, 552–554, 557, 564, 573, 595, 643, 679
Hallegatte, R., 513–516
Hamdam, M., 50
Han, S.J., 690
Hansen, B.J., 390, 394
Hanson, C.D., 453, 455
Hargrove, A., 836
Harkewicz, R., 412
Harrington, D.B., 87, 88
Harrington, P.B., 312
Harris, F.M., 54
Harrison, A.G., 43, 54, 62
Harrison, W.W., 37
Hartmer, R., 95, 374, 593, 718
Hasegawa, H., 416, 417, 578–580, 583, 584, 586, 588

Hashimoto, Y., 110, 300, 415–417, 546, 573, 578–580, 583, 584, 586, 588, 595, 643, 679, 794
Hasted, J.B., 53, 54
Hatsis, P., 362, 363
Hawkins, A.R., 202, 204, 390, 392, 394
Hays, E.E., 82
Hazama, M., 287
Heil, O., 53
Heinrich, L., 96
Hendricks, C.D., 51
Hensinger, W.K., 198
Herb, R.G., 116
Herschbach, D.R., 116
Herzog, R.F.K., 19, 44, 100
Hiller, J., 107
Hipple, J.A., 22, 23, 56, 73, 74, 78
Hirabayashi, A., 417
Hoaglund, C.S., 420
Hofstadler, S.A., 693
Højbjerre, K., 855
Holliman, C.L., 446, 447
Holme, A.E., 99
Holmes, J.L., 55, 56, 58
Homer, J.B., 88
Honig, R.E., 44, 45
Hoover, Jr., H., 20
Hooverman, R.H., 116
Horn, L.A., 166
Horning, E.C., 42
Horton, F., 115
Hu, Q., 120
Hüber, G., 200
Hughes, A., 89
Hughes, R.J., 58, 95
Hull, A.W., 114, 115, 676
Hunt, D.F., 43, 695, 697
Hunter, R.L., 449–452
Ibrahim, Y.M., 355, 401, 417–419, 422, 677
Inghram, M.G., 37, 38
Inoue, M., 455
Iribarne, J.V., 51
Iwamoto, K., 287
Jackson, C.L., 403
Jackson, G.S., 458–460
Jacoby, C.B., 446, 447
James, A.T., 26
James, J.A., 36
Janulyte, A., 75, 105, 448, 469, 503–506, 509, 511, 513–516, 580
Jefferts, K.B., 628
Jeffries, J.B., 438
Jennings, K.R., 32, 33, 710
Jinno, M., 287
Jockusch, R.A., 688, 828
Johnson, E.G., 25
Jones, J.L., 385

Jordan, E.B., 18, 32
Jornton-Karlsson, M., 365
Judy, J.W., 193
Kang, H., 407, 461, 462
Kapitza, P.L., 526
Kaplan, D.A., 95, 374, 593, 718
Kapron, J.T., 362
Karachevtsev, G.V., 60
Karas, M., 49
Katzenstein, H.S., 86
Kebarle, P., 848
Kellerhals, H.P., 677
Kelley, P.E., 103
Kennedy, B.C., 75
Kerley, E.L., 453, 455
King, A.T., 339
King, R., 362
Kingdon, K.H., 67, 113, 114, 254, 675
Kinstle, T.H., 75
Kinter, M., 681
Kirchner, F., 672
Kistemaker, P.G., 48
Kitova, E.N., 690
Klaassen, T., 358
Knight, R.D., 117–120, 161, 253, 254, 256
Kodera, K., 287
Kofel, P., 408
Koppius, O.G., 81
Korsunskii, M.I., 120
Kothari, S., 225, 226
Koyanagi, G.K., 60, 61
Krauss, M., 28
Krueger, F.R., 47
Kulkarni, P.S., 435
Kumashiro, S., 107, 275, 280, 288, 289, 291, 293
Lammert, S.A., 111, 148, 177, 178, 373, 377,
        381–385, 471
Langevin, P., 53
Langmuir, D.B., 106
Langmuir, I., 15, 113, 115
Laskin, J., 683, 689, 691
Later, D.W., 178
Laude, D.A., 405, 454–456
Laughlin, B.C., 212–216
Lavoisier, A., 708
Lawrence, E.O., 70–72
Lawson, G., 99, 100
LeBlanc, J.C.Y., 557
Leck, J.H., 99
Ledford, Jr., E.B., 435
Lee, E.D., 178, 202, 204, 385, 392
Lee, M.L., 178, 202, 204, 385, 392
Levin, D.S., 333
Li, A.A., 355, 364
Li, H.Y., 188, 189
Liebl, J., 44
Lindemann, F.A., 13

Liu, J., 108, 265, 611, 685, 718
Livingston, M.S., 72, 96
Liyu, A.V., 413, 424
Loboda, A.V., 525, 530, 574, 643, 679, 718
Londry, F.A., 110, 552–554
Long, F.A., 28
Lourette, N.M., 407, 461, 462
Louris, J.N., 106
Low, Y.L., 197, 198
Lu, Y., 611
Lyon, P.A., 44
Lyubimova, A.K., 40
Maas, J.D., 202, 204, 223, 224, 392
Macfarlane, R.D., 47, 672
Madsen, M.J., 198
Mahan, E.A., 362
Makarov, A.A., 67, 112, 119, 121, 123, 190, 251,
        254, 471, 508, 675, 718
Mamyrin, B.A., 90, 91
Mann, A.K., 24
Mann, M., 52
March, R.E., 3, 54, 58, 61, 62, 95, 97, 101, 103,
        104, 109, 152, 157, 163–167, 184, 225,
        374, 375, 440, 628, 674, 693, 829
Marshall, A.G., 78, 80, 403, 434, 449, 450, 459,
        460, 677
Martin, A.J.P., 26
Mason, E.A., 311, 314
Mason, R.S., 33
Masselon, C., 412
Mather, R.E., 102
Mathieu, E., 148
Mattauch, J., 19
Matyjaszczyk, M.S., 314, 340
May, M.A., 253
Mayer, P.M., 55, 56, 58
McAlister, G.C., 265, 611, 686, 718
McDaniel, E.W., 108, 311, 314
McIlraith, A.H., 117
McIver, Jr., R.T., 42, 77, 78, 408, 409, 451,
        452, 677
McLafferty, F.W., 28, 29, 31, 34, 683, 684, 689,
        690, 710
McLaren, I.H., 87
McLuckey, S.A., 31, 108, 265, 535, 611, 685, 697,
        718, 750, 754
McMahon, A.W., 61
McMahon, T.B., 78
McSheehy, S., 359
Mehl, J.T., 362
Melton, C.E., 32
Melzner, F., 253
Mercer, R.S., 62
Mester, Z., 359, 361
Meyerson, S., 20, 30
Mie, A., 364, 365
Miller, I.W., 390, 394

Miller, R.A., 339
Misharin, A.S., 209, 212–216, 224
Mitchell, D.W., 406
Moller, S.P., 253
Monroe, C., 198
Montgomery, H., 286, 415, 448, 793
Morgan, R.P., 9–12, 15, 17
Morgan, T.G., 54
Morris, H.R., 44, 672
Mosca, G., 115
Moxom, J., 197, 198
Muller, E.W., 38
Muller, J., 389
Mulligan, C.C., 107, 169, 259, 375, 691
Munson, M.S.B., 40, 673
Murray, J.A., 178
Myung, S., 421
Naito, Y., 454, 455
Nappi, M., 166
Nazarenko, L.M., 253
Nazarov, E.G., 339
Ney, E.P., 24
Nguema, T., 499, 500
Nibbering, N.M.M., 60
Nielsen, R.A., 420
Niemann, H.B., 75
Nier, A.O., 18, 21, 24, 25, 35, 74
Noll, R.J., 107, 169, 259, 375, 691
O'Connor, P.B., 691
O'Donnell, R.M., 311, 312
Oksman, P.A., 253
Oliphant, J.L., 178
Oliphant, J.R., 385
Oliphant, M.L.E., 45
Olmschenk, S., 198
Olson, A.R., 19
Oomens, J., 688
Ottinger, C., 60
Ouyang. Z., 107, 169, 171, 180, 206, 207,
    212–216, 227, 229, 259, 375, 691
Pace, D., 115
Pai, C.S., 197, 198
Palmer, G.H., 38
Papanastasiou, D., 286, 415, 448, 793
Parks, J.H., 688, 828
Parr, C., 832, 833
Paša-Tolić, L., 407, 412, 461, 462
Patterson, H., 38
Pau, S., 196–198, 201, 388
Paul, W., 96, 101, 124, 170, 620, 678, 679,
    711, 770
Payan, J.C., 496
Payne, A.H., 744, 754
Peik, E., 125
Peng, Y., 390, 393, 394
Penning, F.M., 676
Perrier, P., 496, 499, 500

Perrin, J., 8
Perry, R.H., 119
Petritis, B.O., 424
Pikulski, M., 831, 836
Pipich, C.W., 314, 340
Plass, W.R., 178, 188, 189, 377, 381–384, 746, 748
Popov, B., 35
Porter, C.J., 62
Post, R.F., 96
Posthumus, M.A., 48
Pringle, S.D., 421
Prior, D.C., 413, 418, 419, 422, 424
Prókai, L., 39
Purves, R.W., 309, 314, 317, 323–325, 327, 329,
    332, 333, 335–337, 340, 342, 343,
    349, 353, 354, 356, 357, 359, 361, 420,
    536, 712
Quayle, A., 21, 24, 27
Rabchuk, J.A., 198
Ramendik, G., 37
Ramsey, J.M., 197, 198, 201
Raptakis, E., 286, 415, 448, 793
Raveane, L., 107, 285
Ray, A., 365
Rayleigh, Lord, 50
Redhead, P.A., 94
Reed, R.I., 30
Reichle, R., 200
Reid, G.E., 611
Reilly, P.T.A., 197, 198
Reimann, C.T., 365
Reiner, E.I., 62
Reinhold, V., 106, 265, 707, 812
Remes, P.M., 102, 739, 755, 770, 796, 828, 852
Rempel, D.L., 18, 75, 402, 433, 444–447,
    463–465, 688, 689
Reynard, C., 75, 105, 448, 469, 503–506, 680
Richards, J.A., 107
Richardson, O.W., 37
Ridgeway, M., 346
Rinehart, Jr., K.L., 75
Roberts, T.R., 24, 25, 74
Robinson, E.W., 353, 407, 413, 461, 462
Roboz, J., 30, 34
Rock, S.M., 22, 26, 27
Rockwood, A.L., 14, 202, 204, 385, 390, 392, 394
Roepstorff, P., 681
Rojansky, V., 89
Romer, A., 358
Rosenstock, H.M., 28, 29, 32
Russell, D.H., 117, 453, 455
Sakurai, T., 253
Sanzone, G., 88
Satake, H., 416, 417, 586, 588
Satoh, T., 89, 253
Saunders, J.H., 30
Saxton, J.J., 116

Schaub, T.M., 80
Schmidt-Kaler, F., 200
Schnitzler, W., 200
Schubert. M., 95, 374, 593, 599, 718
Schuy, K.D., 36
Schwab, K., 198
Schwartz, J.C., 109, 183, 667
Scrivens, J.H., 312, 420, 712
Sekioka, T., 119
Sekiya, S., 287
Shabbir, S., 835, 836
Sharp, T.E., 78
Shen, Y.F., 412
Sheretov, E.P., 102, 107
Sherman, M.G., 451, 452
Sherman, N.E., 681
Shin, S.K., 690, 848
Short, R.T., 195, 205
Shvartsburg, A.A., 332, 343, 345–347, 353, 355
Sichtermann, W., 45
Singer, K., 200
Sloan, D.H., 72
Smalley, R.E., 408
Smith, L.G., 82, 83, 88
Smith, R.D., 343, 345, 355, 401, 406, 407, 409,
    412, 413, 418, 419, 422, 424, 461,
    462, 677
Smith, S.A., 107, 169, 223, 227, 229, 259,
    375, 691
Smyth, H.D., 19, 31
Smythe, W.R., 93
Snyder, H.S., 96
Soddy, F., 13
Sommer, H., 73, 74
Song, Q., 107, 169, 225–227, 229, 375, 691
Specht, A.A., 95, 546, 573, 595, 679
Staanum, P.F., 855
Stace, A.J., 740, 828, 847
Stafford, Jr., G.C., 103, 175, 225, 226, 786
Steiner, U., 377
Steinwedel, H., 96, 101, 170, 770
Stephens, D.R., 103
Stephens, W.E., 84, 86
Stern, O., 9
Stewart, H., 740, 828, 847
Stewart, H.R., 19
Stick, D., 196, 198
Stone, J.A., 311, 420, 536, 712
Stoney, G.J., 10
Su, C-S., 253
Sudakov, M., 280, 286, 288, 289, 291, 293, 415,
    448, 512, 793
Sugiyama, M., 580, 583, 584, 586, 588
Sultan, J., 361
Sun, X., 312
Syka, J.E.P., 109, 111, 376, 471, 695
Syrstad, E.A., 683

Tabert, A.M., 209–212, 221, 222
Takáts, Z., 212–216
Talbot, F.O., 688, 828
Tal'roze, V.L., 40, 60, 673
Tanaka, K., 49, 287, 672
Tang, K., 345, 355, 358, 421
Taylor, D., 225, 226
Taylor, S., 152
Terenin, A.N., 35
Thalassinos, K., 312, 420, 712
Thibault, P., 358
Thomas, H.A., 73, 74
Thompson, C.V., 178, 377, 381–384
Thompson, R.C., 199
Thomson, B.A., 51, 525, 535, 574, 643, 679, 718
Thomson, J.J., 10–14, 35, 40, 44, 56, 62
Thrall, B., 412
Tipler, P.A., 115
Todd, J.F.J., 3, 42, 47, 95, 97, 99–102, 104, 109,
    152, 157, 163–167, 178, 202, 204, 374,
    375, 440, 498, 674
Tolic, N., 412
Tolley, H.D., 392
Tolley, S.E., 178, 202, 204, 385, 390, 392, 394
Tolmachev, A.V., 407, 418, 419, 458, 461, 462,
    530, 560
Torgerson, D.F., 672
Traldi, P., 107, 285
Turecek, F.J., 683
Udseth, H.R., 409
van Amerom, F.H.W., 195, 205, 388
Van Dyck, Jr., R.S., 457
Vartanian, V.H., 405, 454–456
Vasicek, L., 835
Vastola, F.F., 48
Vedel, F., 9, 436
Verbeck, G.F., 389
Verbueken, A.H., 37, 48
Verdun, F.R., 80, 677
Verentchikov, A.N., 253
Vestal, M.I., 50, 52
Viehbock, F.P., 44
Viehland, L.A., 334
Viggiano, A.A., 311
Vilkov, A., 212–216
Volný, M., 227, 229
von Busch, F., 620
von Zahn, U., 99
Waite, R.W., 385
Waki, I., 416, 578, 579, 586, 588
Waldren, R.M., 102
Walker, N.R., 848
Wanczek, K-P., 75, 113, 117, 265, 390, 416, 471,
    611, 671, 677, 718
Wang, M., 202, 204, 385, 392, 712
Wang, Y., 390, 391, 471, 492, 507, 599,
    620, 621

Washburn, H.W., 20
Weil, C., 166
Weinkauf, R., 683
Welling, M., 410
Wells, G.J., 95, 546, 573, 595, 621, 679
Wells, J.M., 174
Werth, G., 436
Whetten, N.R., 102, 492, 500, 620
White, D., 88
White, F.A., 32
White, F.M., 459, 460
White, R.L., 435
Whitten, W.B., 197, 198, 201
Wiley, W.C., 87, 88
Wilkins, C.L., 435
Williams, A.E., 26, 28, 30
Williams, D.H., 30
Williams, E.R., 353, 364, 690
Williams, J.L., 36
Wilpers, G., 199
Wilson, C.T.R., 11
Wilson, H.W., 38
Wilson, J.J., 831, 835, 837, 839, 841, 842
Wise, M.B., 178, 377, 381–384
Wobschall, D., 74, 76

Wolff, M.M., 84, 86
Wollnik, H., 90, 106, 253
Woodgate, S.S., 75
Wu, G., 740, 828, 847
Wu, G.X., 186
Wu, J.T., 363
Wu, Q., 690
Wu, S., 407, 461, 462
Wuerker, R.F., 106
Wyttenbach, T., 421
Xu, W., 180, 206
Yamaguchi, S., 253
Yamashita, M., 49
Yang, L., 253
Yates, N.A., 773
Yost, R.A., 102, 740, 769, 796, 828, 852
Young, A.B., 62, 693
Yu, M., 205, 206
Yu, X., 534, 535
Zeleny, J., 50
Zerega, Y., 75, 105, 448, 469, 490, 496, 499, 500,
        503–506, 509, 511–516, 680
Zhang, H., 106, 265, 707, 812
Zhao, R., 412
Zubarev, R.A., 684

Subject Index

(*a,q*) stability diagram, 801
α-Cyano-4-hydroxycinnamic acid, CHCA, 803
(*a_i, q_i*) space, 620
[Glu1]-fibrinopeptide, 532, 542
[Glu1]-fibrinopeptide, doubly-protonated, MS^2
    and MS^3 spectra of, 533
$[H_2O]_n H^+$, 323, 326
1,2-Dichlorobenzene, 237
1,3-Butadiene, 443, 444
1,3-Dichlorobenzene, 187, 210–212, 240, 243
1,4-Difluorobenzene, 222, 244
$^{111}Cd^+$ ion, Doppler-cooled, 199
12–90G(DF) mass spectrometer, 34
$^{12}CH_2^+$, 24
$^{134}Xe^+$, 188, 189
$^{14}N^+$, 24
15% TFD, *see* Trapping field dipole
$^{16}O_2^+$, 83
18-Crown-6, 421
2,4,6-Tri-hydroxy-acetophenone, THAP, 817, 818
2.5D Technique, 191
^{20}Na, decay of, 47
^{252}Cf Plasma Desorption, 672, 690
^{252}Cf spontaneous fission, half-life of, 47
2-Chlorobutane, 333
3,4-Dichlorobenzene, 238
^{32}S, accurate mass of, 83
$^{32}S^+$, 83
$^{35}Cl^-$, 359
$^{37}Cl^-$, 359
$^{40}Ca^+$, ion crystals, Doppler-cooled, 201
4-Vinylcyclohexyl radical cation, isomerization
    of, 444
^{63}Ni foil, 42

A

a, 108, 125, 275, 383
a ion, 681, 682, 685
A_N coefficients, 627
Ab initio calculations, 683
AB Sciex, 109, 110
Absolute abundance, 633
AC dipolar excitation of axial motion in the
    Orbitrap, simulation of, 190
Accelerating potential, electron, 35
    potentials, ratio of, 24
Acceleration region kinetics, 60
    spectrometers, 93

Accelerator, linear resonance, 72
    multi-stage linear, 70
Accurate mass, 24
    detector, Orbitrap analyzer, 263
    mass determination, 30, 82, 83
Acetaminophen, 239, 243, 362
Acetic acid, 213
Acetic acids, trihalogenated, 331
Acetone, 39, 49, 538
    vapor, 242
Acetophenone, 210, 212, 237, 238, 243, 244
Achilles' heel, 258
Acid/base chemistry, 41
Acquisition time, 461, 581
Acrolein, 241
Acrylonitrile, 241
Activated reaction intermediate, 14
Activation energy, 710, 713, 726, 741, 742, 750
    reverse, 59
Activation rate, 752
    step, 745
    time, 742
    voltages, 608
Active hydrogen atoms, 43
Actual CF, 325
Additional multipoles, 621
Adduct formation, 794
    ions, 714
    stereo-specificity, 711
    formation, 443, 444
Adhesion layer, phosphorus-doped nickel, 204
Adiabatic approximation, 508
Adult stem cell, 728–730
Advion, Ithaca, NY, USA
AEI, *see* Associated Electrical Industries Ltd
Affinities, chloride, 66
    fluoride, 66
    hydride, 65
    proton, 65
AGC, *see* Automatic gain control
AGC-IFT mode, 419
Agilent, Palo Alto, CA, USA, 606
Air, 771
Al^+, 118
Alachlor, 239
Alcohols, primary, 43
    secondary, 43
    tertiary, 43
AlGaAs layer, 197

Algorithm, inverse transform, 423
  multiplexing, 423
Alkali metal ions, 694
All ions MS/MS, 267
Allene, 29
Alternating electric field, 643
  gradient circular accelerators, 653
  voltage gradient, 644
Alumina, 203
Aluminum, 739
AmaZon spherical ion trap, 616
Ambient pressure, 712, 713
Americal Petroleum Institute Project 44 Tables of
    Mass Spectral Data, 22
American Physical Society, 72, 84
American Society for Mass Spectrometry,
    103, 128
American Society for Mass Spectrometry,
    Distinguished Contribution
    Award, 112
Amines, 538
Amines, trace amounts of in atmospheric air, 315
Amino acid, 681, 684
  backbone, 813
  protein backbone, 683, 684
  protein backbone, fragmentation, 690
  sequence, 686, 812, 813
  sequence characterization, 611
Amino acids, 45, 48, 213
Ammonia, 43, 536, 538
Ammonium acetate, 323
Amphetamine, 362
Amplification factor (Multiplying factor), 70, 72
Amplitude evolution of secular frequency peaks,
    509, 511
Amplitude evolutions, 289
Amplitude-scanning devices, 289
AMT (accurate mass tag) approach, 413
A_n terms, 392
Analog-to-digital converter (ADC), 416
Analyte ion, 683
Analytical instruments, multiplexed, 207
Analytical scan, 769, 773
Analyzer region, *see* Ion separation region
Anchor points, 194
Anchor structure, 711
Angular focusing, second-order, 25
Angular momentum, conservation of, 115
Angular velocity, 81
Anharmonicities, 405
Aniline, 222, 244
Annular radius of curvature, 376
Anomer, 710
Anomericity, 722
Antennal specificity, 711, 715
Anthracene, 686
Antibodies, 264, 728

Antihydrogen, 676
Antimatter, 676
Antioxidant, 819
Antonczak, M., 268
APCI, +/−, 227
APCI, *see* Chemical ionization, atmospheric
    pressure
Apex isolation, RF/DC, 216
Apex Precision Power; Cirrus Logic, Inc., Austin,
    TX, USA, 494
APGDI, *see* Atmospheric pressure glow
    discharge ionization
API III LC/MS/MS system, 526
API, *see* American Petroleum Institute Project
    44, or *see* Chemical ionization,
    atmospheric pressure
APIE, *see* Atmospheric pressure ion evaporation
Apodization function, 479, 480, 483
  Blackman, 487
Apodization, spectral, 465
Application examples, 734
Applications, 812, 819
Applied Biosystems, Foster City, CA, USA, 348
Applied Biosystems/MDS Analytical
    Technologies, 539, 541
Applied dipole, 621
Appreciation and historical survey of mass
    spectrometry, 674
Aqua regia, 204
a_r, 161, 474, 620
Ar⁺, 44, 74
Ar⁺ ion beams, 45
AREX LIT as source ion trap in 2D mass
    spectrometry, 582
AREX LIT with axial magnetic field, 587
AREX LIT, axial DC potential along, 576
  axial ion motion, expression for, 576
  collision-induced dissociation in, 584
  frequency spectrum for reserpine ions, 577
  geometry of, 575
  high ejection efficiency of, 583
  hybrid mass spectrometers combined
      with, 579
  ion fragmentation in, 583
  low energy dispersion of, 583
  mass-selective characteristics of, 575
  mass-selective DC potential of, 579
  schematic diagram of, 575
  secular frequency, expression for, 576
AREX LIT, *see* Axially resonant excitation
    linear on trap
AREX LIT/TOF-MS, schematic diagram of, 582
AREX, *see* Axially resonant excitation, 574
AREX-LIT, lack of low-mass cut-off due to use
    of vane DC potential, 586
Arg(8)-vasotocin, 411
Argine ions, ESI-generated, 207

Arginine, 213, 239, 243
trimer of, 244
Arginine clusters, protonated, 223
monomer, protonated, 243
protonated, 206, 215
Argon, 771, 772, 780, 782, 784, 795, 849, 850
Argon atoms, metastable, 43
ARK, *see* Acceleration region kinetics
Aromatic molecular ions, 32
Array of detectors, 172
Array of ionization sources, 172
Array, ion trap, monolithic, 198
micro ion trap, 171
micro-CIT, 194, 197, 198
serial, 219
Arrayed geometries, 191
Arrays of smaller ion traps, 375
Arrays, CITs, 209
CITs of different sizes, parallel, 216
CITs of the same size, parallel, 210
CITs, serial, 219
micro-scale CITs, micromachined, 209
RITs, 175, 221
RITs, parallel, 221
Arrival time, 798, 801, 802
Arsenic-containing compounds, 359
Arsine, 239
Aspirin, 239, 243
Associated Electrical Industries Ltd, AEI, 21, 27
Associated proton, 684
ASTM-E14, Committee, 75
Aston, F.W., 13, 15, 16, 18, 31, 35, 127
Aston's third mass spectrograph, 16
Astrocytoma, 728
Asymmetical toroidal ion trap, 383, 384
operating parameters, 387
Asymmetric trapping field, 627, 629, 630, 642, 645, 656, 664
Asymmetric waveform, 317, 344
Asymmetric waveforms, representations of, 345
Asymptote, 627, 629
Asymptotes, changing the angle of, 382
common, 148, 151, 158
Asymptotic cone angle, 621
Atmospheric emissions, 517
Atmospheric interface, 213
Atmospheric pressure glow discharge ionization, APGDI, 500
Atmospheric pressure interfaces, 172
Atmospheric pressure ion evaporation, APIE, 51
Atmospheric pressure ionization, API, *see* Chemical ionization, atmospheric pressure
Atmospheric pressure ionization mass spectrometry, 525
Atmospheric pressure ionization sources, 364

Atmospheric pressure matrix-assisted laser desorption ionization, AP-MALDI, *see* MALDI, high-pressure
Atmospheric pressure, ion trapping at, 327
Atom, origin of the word, 8
Atom/molecule polarizability, 747
Atomic clocks, 9, 31
Atomic constants, precision measurement, 676
Atomic number, 13
Atomic physics, applications in, 117
Atmospheric pressure (chemical) ionization, AP(C)I, *see* Chemical ionization, atmospheric pressure
ATRAP, antihydrogen trap, 676
Atrazine, 239
a_u, 149, 374
Auger electron spectroscopy, 64, 65
Automatic compound-specific identification, 218
Automatic gain control, AGC, 105, 374, 412, 773, 774
AutoMS(N) experiments, 614, 615
Auxiliary electrodes, 530
Auxiliary excitation signals, 547
Auxiliary RF voltage, 105
Auxiliary waveform, 760, 774
Auxilliary wiggles, 479
Average internal energy, 748, 750, 752, 754
a_x, 156, 304
Axial and radial energies, interconversion of, 558
Axial barrier, 418
Axial component of electric field, 551
Axial confinement, 417
Axial DC potential, expression for, 576
shape of, 576
Axial displacement, 631–636, 664
Axial ejection, mass-selective, 545, 679, 691
Axial electric field exerience by an ion, total, 551
Axial electric field, net positive, 551
Axial field homogeneity, 596
Axial forces, q-dependent, 548
Axial frequencies, 167
Axial harmonic motion, 121
Axial injection, 679
Axial instability, 627
Axial ion ejection, 628, 630
volume for, 553
Axial ion trajectory, 630
Axial ion-motion spectrum, 513
Axial magnetic field, 587
Axial modulation, 304, 774–776, 781
Axial modulation resonance ejection, 380
Axial motion, 808
equation for, 508
Axial oscillation, 254, 402, 403, 628
Axial potential well, 411
Axial resonant ejection of ions, 118, 577
Axial space-charge field, 404

Axial stability, 103
Axialisation, quadrupolar, 438
Axially resonant excitation linear ion trap,
       AREX LIT, 110, 300, 415, 546,
       573–595, 643, 794
Axis, central, approximately harmonic potential
       along, 575
    unidirectional potential along, 574
AXSIM software, 799
a_y, 304
a_y, 156
a_z, 276
a_z, 103, 104, 161, 474, 620
Azimuthal dipolar electric potential, 408
Azimuthal electric fields, alternating, 405
Azimuthal quadrupolarity, 459
Azimuthal symmetry, 406
Azobenzene radical anions, 535

B

b ion, 681–683, 686, 690
$B_5H_9^+$, 83
B16 cells, mouse, 411
Background gas, 31, 421, 742
Background gas pressure, 739
Background gas temperature, 740, 751
Background gas, collisions with, 252
Background noise, 301, 491
Background pressure, 434
Background species, isobaric, 357
Background, chemical, 358
Bacterial lipopolysaccharides (LPS), 364
BAD, *see* Boundary-activated dissociation
Bainbridge, K.T., 35
Balance CF, 325
Balance condition, dependence upon RF
       configuration, 555
Balance conditions, 326
Balance points, 325
Balschun, W., 268
Barber-Elliott scan, 32, 33
Basic blue, 242
Basic violet, 242
Basicities, relative, 41
Bath gas, 176, 188, 575, 585, 596, 678, 694, 743,
       745, 746, 751, 752; *see also* Buffer gas;
       Carrier gas; Damping gas
Bath gas heated, 743
Bath gas, collisions with, 188
    helium, 443
    interactions with ions, 314
    rôle of, 330
Batteries, 8
Battery-powered micro ion trap, 230
Bayard-Alpert ionization gauge, 74
BDCA⁻, *see* Bromodichloroacetate anion, BDCA⁻

BDCAA, *see* Bromodichloroacetic acid, BDCAA
Be⁺, 118
Beading, 14, 31
Beam bender, quadrupole, 417
Bear Instruments, USA, 377
Beat frequency, 637
Beating oscillation, 598
Beckman mass spectrometer, 69, 94
Bendix Aviation Corporation, USA, 87
Bendix time-of-flight mass spectrometer, 87
Bennett mass spectrometer, 69, 94
Bent octopole, 778
Benzene, 239, 396, 452
Benzene in BTX mixtures, 242
Benzene ions, 446, 447
Benzene molecular ion, 452
Benzene radical cations, 445
Benzene vapor, 242
Benzene, toluene, and xylene (BTX)
       mixtures, 238
Benzo[a]pyrene, 58, 59
BEQQ mass spectrometer, 62
Berzelius, J.J., 8
Bessel function, 462
β, 108, 441
$β_2$-microglobulin conformers, 352
β-distonic ion, 443, 444
$β_r$, 103, 104, 474
$β_u$, 156
$β_x$, 225
$β_z$, 103, 104, 289, 474, 509
Beynon, John H., 26, 127
Bimolecular reaction, 681, 682
Binary pseudo-random sequence, 423
Binding energy, *see* Nucleus, binding energy of,
Binomial coefficients, 647
Bioinformatic tools, 712
Biological analysis, 525
Biological compounds, 45, 47, 106
Biological macromolecule, 672
Biological mixtures, complex, 614
Biological plasticity, 708
Biologically active species, preservation of the
       conformations of, 228
Biomarker detection, 350, 717, 719, 727, 728
Biomarker discovery, 425
Biomolecules, 672
Biopolymer, 686, 708
Biotransformation products, 568
Biotransformation profiling, 264
Biotransformations, 567
BIRD, *see* Blackbody infrared radiative
       dissociation
Bisinusoidal waveform, 344, 346, 347
Black hole, 492
Blackbody infrared radiative dissociation, BIRD,
       681, 689, 890

Blackbody radiation, 856
Blackett, P.M.S., 9
Blackman function, 483, 487
Blade holder, 200
Blanc's Law, 331, 332
Blood group, 716
Blood plasma, human, 413, 424
Blumel, R., 123, 128
Boltzmann constant, 313, 420, 481, 743, 760, 804
Boltzmann distribution, 756
Boltzmann distribution function, 803, 806
Boltzmann ergodic theorem of statistical mechanics, 682
Boltzmann velocity distribution, 630
Boltzmann's equation, 740
Bombesine, doubly-charged, 224
Bond cleavages, simple, 57
Bottom-up technique, 264, 610
Boundary conditions, 650
Boundary ejection, 187, 279
Boundary-activated dissociation, BAD, 107, 285
Bounded trajectory, 374
Bovine ribonuclease B, 720, 722
Bovine serum albumin, BSA, 287, 412, 413, 537
Bovine ubiquitin +8 charge state, 352, 353
Bovine ubiquitin, multiply-charged states of, 333
Box-car detector, 102
Boyd mass spectrometer, 69, 94
Bradbury-Nielsen gate, 420
Bradykinin, 241, 295, 352, 542
Bradykinin, doubly-charged, 354
Bradykinin, doubly-protonated, 290, 292, 295, 296, 297
Bradykinin, sodium-cationized, 295
Branched ion guide, see Ion guide, branched
Branching, 708, 710, 713, 714, 720, 722, 731, 736
Broadband RF signal ('chirp'), 79
Broadening, space-charge-induced, 423
Bromate, 323, 358, 359
Bromobenzene, 210, 212, 243
Bromobenzene molecular ion, 455
Bromochloroacetate anion, 326
Bromocriptine, protonated, 558
Bromodichloroacetate anion, BDCA$^-$, 331, 361
Bromodichloroacetic acid, BDCAA, 331, 361
Brønsted acid, 43
Brookhaven National Laboratory, USA, 81
Brubaker lens, 229
Bruker, 382
Bruker Daltonik, Bremen, Germany, 598
Bruker-Franzen ESQUIRE, 711
BSA, see Bovine serum albumin
BTX mixtures, benzene in, 242
BTX mixtures, see Benzene, toluene, and xylene mixtures

Buffer gas, 102, 105, 106, 109, 116, 266, 310, 319, 628, 630, 644, 678, 769–771, 778–780, 782, 783, 786, 788–790, 795, 803, 804, 806, 811, 820, 830, 852
Buffer gas atom, 771
Buffer gas collisions, effects of, 184
Buffer gas pressure, 188, 291, 769–776, 786, 788
Buffer gas, cold, 806, 811
    pulsed introduction, 769, 770, 772–778, 782
    see also Bath gas
    see also Damping gas
    single pulse, 770
Bulk gases, 331
Bulk micromachining, 192
Butane, 19, 57
Butane, iso, see Isobutane

C

c ion, 681, 684, 686, 687, 689
C. elegans, 725
C^+, 118
c-1 ion, 687
$C_{12}H_{20}^{3+}$, 58, 59
$C_2H_2^+$, 58
$C_2H_5^+$, 41
$C_3H_3^+$, 58
$C_3H_4^+$, 29
$C_3H_5^+$, 58
$C_3H_6^+$, 29
$C_3H_7^+$, 58
$C_4H_9^+$, 58
$C_5H_7^+$, 58
$C_6F_6^+$, 460
$C_6H_{10}^+$, 58
$C_7H_7^+$, 61
$C_7H_7^{2+}$, 14
$C_7H_8^+$, 61
CAD, see Computer-aided design
Caffeine, 239, 242, 243, 245
Caffeine, protonated, 229
Calibrant ions, 412
Calibrant peptide, 541
Calibration, stability of, 257
Calutron ion source, 117
Canal-rays, 44
Canalstrahlen, 8, 44; see also Kanalstrahlen
Cancer biomarker, 728
Cancer therapeutics, 728
Cancer tissue, 730
Cancer-specific structural isomer, 728
Candy mint, methyl salicylate from, 240
Cameron, A.E., 84
Canonical form, 648
Cantilevered electrodes, 198, 200
Cantilevers, 199

Capacitance, 180, 458
   parasitic, 180
   toroidal ion trap, 380
   total trap, 213
CAPI, *see* Continuous atmospheric pressure inlet
Capillary electrophoresis, 699
Capillary electrophoresis electrospray mass
      spectrometry, CE-ES-MS, 364
Capillary electrophoresis-FAIMS-MS,
      CE-FAIMS-MS, 364
Capillary restrictor, 777–779, 781, 785
Carbohydrate, 708
Carbohydrate epitopes, 717
Carbohydrate polymer, 736
Carbohydrate product ion nomenclature, 712
Carbohydrate sequencing, 711, 729, 732
Carbohydrate specific isotopic distribution, 729
Carbohydrate structure, 711, 714
Carbohydrates research, 812, 819
Carbon, 11, 26
Carbon dioxide, 849
Carbonic anhydrase, 287
Carrier gas, 319, 330
Carrier gas, nitrogen, 333
Cartesian coordinate system, 480
Catalyst surface fabrication, 228
Cathode ring, electron-emitting, 691
Cation exchange resin, 719
CCD camera, 794
CDBAA, *see* Chlorodibromoacetic acid
CEC, *see* Consolidated Engineering Corporation
Cell anharmonicity, 406
Cell surface antigen, 728
Cell, cubic, 404
   hyperbolic, 404
   infinity, 405
   multi-electrode geometries, 404
   orthorhombic, 404
Center electrodes, 655
Center section, 666, 667
Center-of-mass frame, 742
Center-of-mass system, 630
Central electrode, 122, 123, 380, 255, 267, 675
Centrifugal force, 122, 123
Centrifugal separation, 13
Centripetal force, 123, 678
Ceramic disks, 202
   laser-cut, 177
Ceramic particles, glass-encapsulated, 203
Ceramic plates, parallel, 392
Ceramic substrates, 389
Ceramics, laser-machined, 202
Cerebrospinal fluid, artificial, 411
CERN, Switzerland, 73
Cesium iodide cluster ion, 771
Cesium ion, 335
CF, actual, 325

CF, balance, 325
CF, *see* Compensation field
CF/DF plots, 335, 336
CH_4^+, 41
CH_5^+, 40, 41
CH_5^+ ion, dependence of relative intensity on
      methane pressure, 40
Chaos Theory, 123
Chaos, sea of, 125
Chaotic behavior, 125
Characteristic curves, 162
Characteristic radius, 120, 254
Characteristic values, *see* Characteristic curves
Charge capacity, 418, 549, 665–668
Charge capacity measurements, 640, 665
Charge containment, 545
Charge density, 181, 296, 416
Charge density control, 565
Charge evaporation model, 51
Charge exchange, 780, 782, 783
Charge exchange reaction, 210
Charge permutation, 12
Charge reduction through proton transfer, 537
Charge repulsion, 761
Charge separation, 59
Charge state determination, 640
Charge state separation techniques, 525
Charge stripping, 697
Charge transfer, 848, 854
Charge transfer reactions, ion/molecule, 536
Charge transfer, double-, 63
   partial, 848
Charge, elementary, *see* Elementary
      (electronic) charge
Charged particles, 740
Charge-exchange collisions, 740
Charge-induced peak broadening, 549
Charge-localization, 28
Charge-reduced ions, 538
Charge-reduced radical ions, isolation of by axial
      resonant excitation, 587
Charge-separation reaction, 847
Charge-stripping, 538
Charge-to-mass, ratio of, 10
Charles Stark Draper Laboratory, Cambridge,
      MA. USA, 315
$CHCl_3$, 226
Chemical background, 358
Chemical derivatization, 611
Chemical end-game, 709
Chemical ionization, CI, 40, 41, 243, 262, 673,
      500, 687, 710
   atmospheric pressure, APCI, 42, 177, 226,
      229, 245, 262, 266
   desorption, DCI, 42
   direct, *see* Chemical ionization, desorption
   high pressure, 42

low pressure, 42
   negative, NCI, 686, 687, 694
   reagent gases for, 43
Chemical mass shift, 748
Chemical noise, 536, 537
Chemical structure, 340
Chemical vapor deposition, CVD, 194
Chemical warfare agents, detection of, 351
Chemical-mechanical polishing, CMP,
      194, 196
Cherwell, Baron, 14
Cherwell, Viscount, 14
Chicken ovalbumin, 733, 734
Chinese medicine, 819
Chip production, protein, 228
Chip-based trap, 199
Chiral reactant ions, 772
Chiral reference compounds, 364
Chirp, *see* Broadband RF signal
Chlorate, 323, 358, 359
Chloride affinities, 66
Chlorine, 16
   atomic weight of, 16
   isotopes of, 16
Chlorodibromoacetic acid, CDBAA, 332
Chromatographic peak, 782
Chromatography, 228, 699, 709, 711, 717, 722,
      727, 728, 736
Chromatography, vapor-phase, *see* Gas
      Chromatography
Chromium, evpaporated, 203
Chromogenic phosphonate groups, 830
Chronotron, 82
Churchill, Winston, 14
CI, *see* Chemical ionization
CID efficiency, 771, 772
CID, *see* collision-induced dissociation
Circuitry, stability of electronic, 183
CIT analyzers, array of, 213
CIT array, 217, 243
CIT array mass spectrometer, 212, 216
CIT array, different sized traps, 244
   four-channel, 211
   multiple ion monitoring with, 218
   multiplexed, 211
   serial, 244
CIT arrays, micro-scale, 196
CIT geometry, parameters needed to
      descibe, 187
CIT mass analyzers, improved, parallel array
      of, 214
CIT, half-size, 175
   potential distribution inside, 187
   see cylindrical ion trap
   serial array, ion transfer within, 220
CITs of different sizes, parallel arrays of, 216
CITs of the same size, parallel arrays of, 210

CITs, micro-, 192
   micro-scale, micromachined arrays of, 209
   parallel and serial coupling of, 209
   serial arrays of, 219
Claritin, 243
Clay pipes, diffusion through the stems of, 13
Closed cylindrical trap, 461
Closed trap, 454
Cloud, 734, 735
Cloud-focusing, benefits of, 447
Cluster ion, 364, 691
Cluster series, methanol, 66
   water, 66
Clustering behavior, 340
Clustering, enthalpies of, 65
Clusters, 335, 315, 328
Clusters, solvent, 358
CMP, *see* Chemical-mechanical polishing
CO^+, 32
CO_2, 331
Coaxial ion traps, 393, 395
Coalescence of closely-separated m/z-values, 258
Cocaine, 241–243, 245
Cocaine, protonated, 229
Cockroach neuropeptides, 80
Codeine, 331, 362
Coherent ion motion, 471
Coherent ion packets, 256
Cold trap, 15
Collection, non-destructive, 228
Collings, B.A., 569
Collision cell, 264, 574, 674, 689, 692
   gas, 62
Collision cross-section, 74, 352, 762
   average, expression for, 312
   elastic, 188
   measurement of, 529
Collision, elastic, 746
Collision energies, different, 558
Collision energy, 583, 674, 688, 711, 712, 719,
      726, 747, 837
   average, 692
   inappropriate choices of, 559
Collision frequency, 549, 747, 760, 853, 856
   normalized reduced, 441, 442
   reduced (ξ_n), 439
Collision gas, 679, 690, 692, 693, 769
Collision gas, pressure, 693
   pulse, 688, 693
Collision, inelastic, 758
Collision multi-particle, stochastic, 185
Collision phenomena, 31
Collision rate, 762
Collisional activation, 710, 713, 714, 742, 747,
      762, 802, 827
Collisional activation by resonance excitation,
      requirement for, 556

Collisional cooling, 102, 116, 393, 414, 464, 491, 497, 526, 528, 529, 541, 739–744, 746, 747, 750, 751, 753, 758, 761–763, 770, 774–776, 788, 790, 796, 798, 802, 827, 830
Collisional cooling rate constant, 748–755, 757–759, 761–763
Collisional cooling, absence of, 259
  rate of, 748, 749, 753, 754
Collisional damping, 438, 605
Collisional deactivation, 762, 840, 853
Collisional dissociation, 632
Collisional excitation, 32, 106, 748, 757
Collisional focusing, 438, 441, 445, 526, 529, 541, 642
Collisional fragmentation, 528
Collisional migration, 498, 596
Collisional migration of ions, 116
Collisional peak broadening, 771
Collisional processes, 688
Collisional relaxation, 751
Collisional stabilzation, 444
Collision-induced dissociation, CID, 14, 30, 32, 62, 83, 106, 265–267, 274, 377, 416, 448, 482, 534, 611, 674, 681, 682, 684, 686–688, 690, 692, 694, 699, 708, 710, 712–715, 718, 725, 741–746, 759, 760, 762, 769, 771, 772, 775, 776, 785, 789, 790, 794, 797, 813, 818, 819, 827, 828, 830–835, 837–839, 841, 847, 851, 854
  high energy, 794, 819
  low energy, 794, 819
  multiple, 556
  skimmer, 578
Collisions, 252, 491, 630–632, 637, 638, 640, 659, 660, 675, 677, 708, 739, 746–748, 755, 757
Collisions of buffer gas, effects of, 184
Collisions with background neutral species, 472
Collisions with neutral bath gas, 188
Collisions, absence of, 57
  compound-dependent effect on ejection delay, 188
  inelastic, 185
  ion/neutral, see Ion/neutral collisions
  ion-ion, 404
  number of, 680
  time between, effect of pressure on, 258
Collisonal cooling, 105, 106, 443, 550
Combination of fields, implementation of, 594
Combined DMS with IMS, 350
Combined mode, 694
Combined Paul traps, 688
Combustion, 81
Commercial FAIMS systems for use with mass spectrometry, 347

Commercial ion trap mass spectrometers, 628
Commercial work, first application of mass spectrometry to, 20
Compensated trapped-ion cell, open, 456
Compensation, 447
Compensation electrodes, 404, 407, 454, 676
Compensation field, CF, 324, 325
Compensation potentials, 199
Compensation voltage, CV, 319, 320, 322, 328, 338, 353, 457
Compensation, trap, 448
Compensation, trap, electrical, 446
Complementary frequencies, 168
Complex, 847
Complex samples, analysis of, 401
Complex samples, analysis of, 355
Complex, doubly-charged, 848, 850
  metal dication, 847, 848, 850, 853
  metal dication, $^{24}Mg^{2+}$, 853, 854
  metal dication, $^{64}Zn^{2+}$, 853–855, 857
  metal dication, Cu^{2+}, 831
  metal dication, Mn^{2+}, 855, 859, 860
  metal/molecule, 847
  multiply-charged, 847, 848
  singly-charged, 850
Components, higher-order, 162
Composition Finder, 731, 733, 734
Compound identification, basic procedures for, 27
Compound-specific identification, automatic, 218
Computational program, 745
Computer-aided design, CAD, 205
COMSOL 3.0a software, 185
COMSOL field solver, 3D, 190
COMSOL, Inc., USA, 185
Concentric cylinder electrodes, see Electrodes, concentric cylinder
Concentric metal rings, 389
Conductance, 774, 779, 785
Cone of reflection, 551, 552
  change of RF balance conditions, 554
  effect on by changing fringing-fields, 553
Confined ions, number of, dependence on deceleration potential, 506
Confinement stage, 482
Confinement time, 105, 479
Confinement, zones of poor, 492
Conformation, changes in, 328
Conformations, protein, gas-phase, 343
Connecting rod, copper, 855
Conservation of momentum, 759
Consolidated Engineering Corporation, CEC, 20–22, 27
Constant neutral-loss, 33
Contaminants, 568
Continued deprotonation, 697
Continuing fraction, 624, 649

Continuous atmospheric pressure inlet, CAPI, 241, 243–245
Contour plot, 661–663, 668
Control grid, 87
Cooks, R.G., 63
Cooling period, 631
Cooling time, 751–754, 757, 760, 761, 790
Cooling, collisional, 105, 106
Coordinate system for toroidal ion trap, 379
Corkscrew trajectory, 90
Corona discharge, 311, 315–317, 323, 324
Corpuscles, 10
Co-trapped ions, 549
Coulomb repulsion, 330, 344
Coulombic energy of repulsion, 59
Coulombic forces, 594, 678, 687
Coulombic interaction, 503, 600, 608
Coulombic interactions between isotopic ions, 297
Coulombic repulsion, 458, 491
Coulombically-bound orbit, 686
Coupled analyzers, BE-EB, 710
Coupled nonlinear equations, 652, 654
Coupled nonlinear resonances, 658, 659
Coupling between axial and radial directions, 507
Coupling between x- and y-coordinates, 560
Coupling of components of radial motion by magnetic induction, 438
Coupling of radial and axial degrees of freedom, 546, 560
Coupling peak, amplitude of, 509
Coupling peaks, higher amplitude, 513
Coupling resonance condition, 668
Cr/Au layer, 196
Cracking pattern, 20
Craig, R.D., 24
Criteria-dependent scans, 567
Criterion, Shannon, 479, 490
Criterion, Nyquist, 479
Critical energy for dissociation, 563, 745
Critical mass (m_c), 439
Cross-ring cleavage, 712, 714, 715
Cross-ring cleavage ion, 817
Cross-ring product ion, 712, 714, 715
Cross-section, hyperbolic, 643, 644
    Langevin, 53, 186
Cross-talk, 208, 210, 214, 243
Cross-talk contamination, 212
Cross-talk peak, 222
    ionic, 208, 214
    level of, dependence on injection time, 216
    neutral, 208, 211, 214
Cryopanels, 409
Crystal formation, 125
Crystallized ions, cylindrical sheet of, 125
Cs^+ ions, fast, 46

CsI, 542, 694
C-terminal branch, 681
C-terminal residue, 832, 840
C-trap, 122, 261–263, 266
    additional analytical capabilities provided by, 264
    fragmentation in, 264
Cube electrode, 349
Cubic cell, 404
    ion motion in, 403
Cubic ion cell, *see* Cubic trap
Cubic trap, 437, 445, 455, 458, 460
    screened elctrostatic, 404
Cubic trapped ion analyzer, schematic of, 435
Curtain gas, 330, 354
Curvature of inner and outer ring elctrode surfaces, differences between, 383
Curved electrodes, 340
Cut-off, high mass-to-charge, 442
    low mass, 443
CV spectra with flat cylindrical electrodes for hypothetical ion types, 322
CV spectra with flat plate electrodes for hypothetical ion types, 320
CV spectra, effect of dispersion voltage upon, 323
CV spectrum, 320–322, 324, 332, 326, 360
CV windows, 343
CV, optimum, 336
CV, *see* Compensation voltage
CV/DV plots, 335
CV/DV values, 324
CVD, *see* Chemical vapor deposition
CV-value, 330, 336, 352, 357, 359, 362
CV-values, polarity of, 321
Cyanogen chloride, 241
Cyclohexane, 19
Cyclopropane radical cation, 444
Cyclotron, 70, 71, 96
Cyclotron equation, 434
Cyclotron frequency, 72, 73, 81, 405, 448–450, 452, 457, 688, 690
    normalized, 442
    normalized unperturbed (w), 439
Cyclotron frequency of the critical mass, normalized unperturbed (w_{nc}), 439
Cyclotron frequency shift, 404, 452
Cyclotron mode, 448
Cyclotron motion, 402, 435, 676, 677
    damping of, 437
Cyclotron orbit, 691
Cyclotron orbit size, 454, 458
Cyclotron radius, 457
Cyclotron resonance absorption, 77
Cyclotron resonance cell, 76
Cyclotron resonance, ion, *see* Ion cyclotron resonance, ICR

Cyclotron shift, dependence on ion mode amplitudes in the trap, 466
Cyclotron-mode amplitude, 436, 462
Cyclotron-orbit size, 453
Cylindical electrodes, non-homogeneous electric field between, 323, 324
Cylindical FAIMS device, ion focusing in, 325
Cylindrical anode, 114
Cylindrical coordinate system, 621
Cylindrical coordinates, 378, 380, 622, 645
Cylindrical electrode, 341
Cylindrical FAIMS configuration, 327
Cylindrical ICR cell, 677
   compensated, 407
Cylindrical ion trap arrays, 209
Cylindrical ion trap, CIT, 96, 106, 170, 173, 174, 179, 182, 184, 203, 204, 237–239
   miniature, 186
Cylindrical trap, 676
Cylindrical tube, 106
CYP 1A2 marker assays, 362
Cytochrome c, 243, 287, 352, 408, 409, 465, 690, 840
Czemper, F., 268

D

D^+, 16
$D^+ – H_2^+$ doublet, 74
D_0, 575
D_2O, 354
d_6-Dimethylamine, DMA, 718
Dalton, John, 8
Damascene processing, 194
Damping gas, 498, 630, 635, 637, 644; *see also* Buffer gas; Bath gas
Damping gas pressure, 635, 637
Damping, weak, collisional, 605
Daniell cell, electromotive force of, 50
DAPI, *see* Discontinuous atmospheric pressure interface
Dark state, 123
Darwin, Charles, 67, 127
Data acquisition rate, 664
Data-dependent experiments, 614
Data-dependent ion accumulation, 411
Data-dependent ion ejection, 411
Data-dependent scanning, 413
Davis, S., 268
DAWI, *see* Digital asymmetric waveform isolation
DC barrier, 525, 530
DC electric field gradients, 310
DC offset potentials, 574
DC offset voltage, *see* Compensation voltage
DC potential harmonicity, 458
DC pulse tomography, 471

DC voltage, 644–646, 654, 655, 660, 661, 666, 668
DC voltage gradient, 693
D-camphor, 222, 244
DC-blade, 200
DCI, *see* Chemical ionization, desorption
De novo sequencing, 819
Dead space, 230
Dead time, 475
Decapole component of potential, 162
Decapole field, 173
Decapole term, 622
Decarboxylated anions, 331, 332
Decelerating spectrometer, 94
Deceleration potential, 506
Deceleration potential configurations, 502
Decomposition, 312
De-convolution algorithm, 52, 696
Decoupled motions, 676
Deep reactive ion etching, DRIE, 196, 199
DEET, *see* N,N-diethyl-*m*-toluamide
De-excitation, 191
Deflection, side-kick, 408
Deflector electrode, 339
Degeneracy dilemma, 653
Degeneracy effects, 658, 660
Degeneracy factor, 745
Degrees of freedom, 28, 59, 300, 682, 683, 750, 754, 758, 759, 761, 762, 809
Dehmelt approximation, 162
Dehmelt equation, 585
Dehmelt, H.G., 9, 708
Delay time, 784, 785, 787, 788, 790
Delayed two-stage ejection, 615
Delrin end-cap spacer, 779
δ_{sc}, 403
Demetriades, N., 268
Dempster, A.J., 13, 15, 16, 18,
Denisov, E., 268
Density of states, 745
Density, separation of isotopes by, 13
Deoxyhexose, 722
Deprit perturbation series, 462
Deprotonated molecules, 673, 710
Deprotonation, 695
DESI, *see* Desorption electrospray ionization
Desolvation gas, 330
Desolvation region, 330
Desorption electrospray ionization, DESI, 177, 241, 243
Desorption model, 51
Desorption process, 43
Detect plates, 435
Detection, 256
Detection events, 403, 773–776, 782, 788–790
Detection limit, *see* Limit of detection, LOD

Detection, biomarker, 350
  chemical warfare agents, 351
  electrometric, 350
  non-destructive, 101
  non-destructive, FT, 471
  peptide, 350
  principle of in ESTs, 253
  toxic chemicals, 351
  trace explosive vapors, 351
Detector, box-car, 102
Detectors, array of, 172
Deuterated internal standards, 360
Deuterium incorporation, 353
Deuterium oxide, 43
DF, see Dispersion field
DFT, see Discrete Fourier transform
Diagnostic ion, 813, 817
Di-ammonium hydrogen citrate, DAHC, 817
Diamond microcrystals, 201
Diastereomeric complexes, 364
Dibutylamine, 241
Dichloromethane, 203, 391
Diels–Alder reaction, 772
Differential glycome analysis, 729, 730
Differential mobility spectrometry, DMS,
    316, 339
Differentially pumped atmospheric pressure inlet,
    DPAPI, 239, 243
Differentiation profile, 727, 728
Diffusion, 13
Diffusion of ions, 313, 321, 330, 338, 344
Diffusion, packet size increase, 258
Digital asymmetric waveform, 284
Digital asymmetric waveform isolation, DAWI,
    298, 299
Digital fequency scan, 285
Digital ion trap, DIT, 69, 96, 107, 148, 273, 275,
    283, 288
Digital ion trap mass spectrometer, 107, 148, 285
Digital linear ion trap, 302, 304
Digital signal generator, DSG, 285
Digital trapping waveform, 285
Digital waveform, 275
Dimensionless parameters, 623, 648
Dimethyl methylphosphonate, DMMP, 201, 202,
    237, 239, 240, 242
Dimethyl sulfoxide, 719
Diocotron motion, 402
Dioxins, 517
Dipolar alternating voltage, 692
Dipolar ejection potential, 482
Dipolar electric field, uniform, 501
Dipolar electric potential, azimuthal, 408
Dipolar excitation, 203, 283, 287, 288, 290, 291,
    298, 410, 414, 574
Dipolar excitation process, simulation of, 604
Dipolar excitation waveform, 109

Dipolar excitation, optimum magnitude of, 597
  relative phase of, 598
  supplementary, 108
Dipolar ion excitation, simulation of, 605
Dipolar potential, 477
Dipolar resonance ejection, 621, 628, 630, 632,
    637–639, 642–644
Dipolar resonance excitation, 549, 550, 602
Dipolar resonant excitation, 628, 631, 635–637,
    655, 660
Dipolar sideband resonance, 635
Dipole amplitude, 661–663, 668
Dipole component of potential, 162
Dipole component of the trapping field, 624, 629,
    642, 644, 649, 652
Dipole effect, 632
Dipole excitation field, 597
Dipole field, 629, 645, 652
Dipole linearity, deviation from, 459
Dipole mass scanning, 640
Dipole phase, 662, 663, 668
Dipole term, 622, 623, 655
Dipole voltage, 635
Dipole-induced ejection, 633
Dipoles, electric, reversible locking of, 352
Dipping, silver ink, 206
Direct digital synthesizing technology, 285
Direct insertion probe, 30
Direction-focusing, 17
Dirichlet boundary condition, 462
Disaccharide, 715, 717, 725–727, 735
Disassembly application, 719, 728–731, 733–735
Disassembly pathway, 710, 711, 716, 717, 721,
    722, 724, 726–729, 734
Disassembly stage, 710, 733
Discharge tubes, 8
Discharge, DC low pressure, 35
Discontinuous atmospheric pressure interface,
    DAPI, 176, 177, 239, 242, 243
Discovery rates, false, 413
Discrete Fourier transform, DFT, 479
Discrimination, mechanical, 495
Discrimination, TOF, 496
Dispersion of trajectory phase, 485
Dispersion voltage, DV, 317, 318, 320, 322,
    323, 346
Disperson field, DF, 324, 325
Disperson of trajectory amplitude, 485
Displacement direction, 626, 630
Dissociation pathway, 741, 742, 762
Dissociation step, 745
Dissociation threshold, 638, 742, 743, 752, 761
Dissociation, boundary-activated, 107
  critical energy for, 563
  electron transfer, 108
Dissociative reaction pathway, 713, 726
Dissociative recombination, 684

Distillation, 13
Distribution density, 804
Distribution function, 809
DIT circuitry, schematic diagram of, 275
DIT prototype, 284, 286
DIT, electrodes, different geometries for, 287
DIT, *see* Digital ion trap
DIT, stability diagram, first region for, 280
Dithiothreitol, DTT, 717
DKT, *see* Kingdon trap, dynamic
DMF, *see* N,N-dimethylformamide
DMMP, *see* Dimethyl methylphosphonate
DMS, *see* Differential mobility spectrometry
DMS/IMS², 350
Dodecapolar component, negative, 187
Dodecapolar field, 190
Dodecapole component of potential, 162
Dodecapole term, 622
Dodo, 67, 93, 101, 127
Dome electrode, 341
Doped polysilicon, 196
Doppler-cooled ¹¹¹Cd⁺ ion, 199
Doppler-cooled ion crystals of ⁴⁰Ca⁺, 201
Double resonance, 77, 633, 635, 643
Double trap, 112, 395
Double-charge transfer, 63
Double-field ion source, 87
Double-focusing, 24
Double-focusing instrument of reverse geometry,
    ZAB-2F, 673
Double-resonance experiment, 437
Doublet, resolution of, 24
Doublets, 22
Doubly-charged glycomers, 721
Doubly-charged ion, 761
Doubly-charged precursor ion, 684, 685, 687
Doubly-protonated molecule, 684
Dow Chemical Company, USA, 28
DPAPI, *see* Differentially pumped atmospheric
    pressure inlet
$\overline{D_r}$, 181, 528
DREAMS, *see* Dynamic range expansion applied
    to mass spectrometry
DRIE, *see* Deep reactive ion etching
Drift cell, 312, 437
Drift gas, 312
Drift ICR cell, 677, 689, 691
Drift time, 420
Drift tube, length of, 311
Drift-cell ICR analyzer, 75
Drift-tube ion mobility spectrometry, DT-IMS,
    310, 319–321
Drift-tube IMS experiment, 311
Drive frequency, 622, 630, 656
Drugs, 45
Drugs of abuse, 265
Drugs, anti-cancer, 362

DSG, *see* Digital signal generator
DSM Somos, New Castle, DE, USA, 206
DT-IMS, *see* Drift-tube ion mobility
    spectrometry
Dual quadrupole ion trap mass spectometer, 498
Dual source interface, 540
Dual source, Ionspray and oMALDI,
    simultaneous operation of, 542
Dual-cell design, 80, 408
Dual-LIT/oa-TOF, 416
Dump pulse, 118
Duty cycle, 108, 173, 228, 277, 284, 305, 409,
    420. 421, 425, 538, 581, 664
Duty cycle enhancement, 580, 581, 589
Duty cycle penalty, 557, 566
Duty cycle, improvement in oaTOF systems, 530
    increasing, 614
    low, 579
    reduction of, 582
DV values, 323
DV, *see* dispersion voltage
$\overline{D_{x,y}}$, 557
Dye laser, 751
Dynamic Kingdon trap, *see* Kingdon trap,
    dynamic
Dynamic mass spectrometers, *see* Mass
    spectrometers, dynamic
Dynamic range, 203, 267, 401
Dynamic range expansion applied to mass
    spectrometry, DREAMS, 411, 413, 414
Dynamic range, increasing, 355
Dynamics, non-linear, 125
$\overline{D_z}$, 181, 595

E

e, Elementary charge 14
E. coli (*Escherichia coli*), lipid extract from, 225,
    226, 245
E. coli lysate, tryptic digest of, 614, 615
E/N, 337, 346
    definition of, 313
    importance of, 334
E/N-values, 331, 351
Earnshaw's theorem, 252, 526
Earth, age of, 37
ECD fragments, 304
ECD in a conventional LIT with axial magnetic
    field, 587
ECD, *see* Electron capture dissociation
ECD/CID, sequential, 587, 588
EDD, *see* Electron detachment dissociation
Edge effects, 393
Edison's experiments, 9
EDM, *see* Electrical discharge machining
EE, *see* Ions, even-electron, EE
Effective gap width, 322

Effective potential, 125, 181
Effective potential well, *see* Pseudopotential well
Effective potential, deformation by hexapolar
     contribution, 606
Effective temperature, 760
Efficiency of ion transfer, 408
Efficiency, extraction, 182
    trapping, 182
Egg ovalbumin, 720–722, 724, 735, 736
Eggers, Jr., D.F., 84
EHI, *see* Ionization, electrohydrodynamic
EI, *see* Electron ionization
Eigenfunction of the matrix, 279
Einzel lens, 794, 795
Ejection delay, 605, 607
    effect of relative phase difference, 608–610
Ejection efficiency, 548, 578
Ejection energy distribution, 579
Ejection process, delay in, 600
Ejection q-value, *see* q-value, ejection
Ejection scan, resonant, 391
Ejection threshold, 635, 636
    boundary, 187
    delayed two-stage, 615
    resonant dipolar, 492
    resonant ion, 187
Elastic collision, 746
Electodes, end-cap, apertures in, 507
Electrodes, vane, *see* Vane electrodes
Electrodes, flat plate, absence of ion
    focusing, 326
Electric field between two concentric cylinders,
    equation for, 323
Electric field produced by ions, 594
Electric field strength, 330
Electric field, axial component of, 551
    difference type, 121
    periodic inhomogeneous, 97
    retarding, 448
    subtractive type, 121
Electric fields, higher order, 95
    linear, 95
    nonlinear, 95
    optimization and modification of, 389
    toroidal, 391
Electric perturbation, effect on motion
    modulations, 512
Electric potential, 149
Electrical breakdown, 332
Electrical center, 644
Electrical charging, 45
Electrical discharge machining, EDM, 388
Electrical trap compensation, 446
Electrically compensated trap to eighth
    order, 461
Electrode designs, FAIMS, other, 343
Electrode displacement, 507

Electrode imperfections, 326
Electrode set, temperature-controlled, 360
Electrode surfaces, 158
Electrode surfaces, equations for, QIT, 158
Electrode truncation, 390, 507, 740
Electrode, central, 122, 380
    cylindrical, 341
    cylindrical with hemispherical terminus, 327
    deflector, 339
    dome, 341
    external, 294
    extraction, 288
    FAIMS, inner, radius of, 326
    ring, 156
    spindle, 111, 380
Electrode, virtual, 123
Electrodes of circular shape, 645
Electrodes, auxiliary, 530
    cantilevered, 198
    compensation, 404, 407
    compensation, optimization of trap
        with, 454
    compensation, pair of, 456
    concentric cylinder, 321, 322, 324
    'D'-shaped, 70
    endcap, *see* End-cap electrode(s)
    flat parallel plate, *see* Parallel plate
        electrodes, flat
    hyperbolic, 148
    incap, 110
    non-equipotential, 389, 396
    planar, 338, 339
    polarization, 339
    shimming, 454
    vane, 110
    vane-shaped, 574
    curved, 340
Electrodynamic field of rotational symmetry, 620
Electrodynamic ion funnel, 408, 411, 418, 421
Electrodynamic ion traps, 108, 265, 611,
    685, 718
Electrodynamic potential, V_{ac}, zero-to-peak
    amplitude, 623
Electrodynamic potential, Ω, frequency, 623
Electrodynamic trapping potential, 621
Electroless copper deposition, 206
Electroless plating, 204
Electrolytic cells, 8
Electrometer, quadrant, 18
Electrometric detection, 350
Electron, discovery of, 10, 14
Electron affinity, 64, 687
Electron attachment, 673
Electron attachment, ion source, 687
Electron beam trap, 53, 54, 675, 699
Electron bombardment, 710
Electron capture, 687

Electron capture dissociation, ECD, 274, 299, 300–302, 416, 448, 587, 674, 681, 683–687, 691, 694

Electron detachment dissociation of negative ions, 685

Electron detachment dissociation, EDD, 681, 682, 684, 691, 713

Electron energy, ionizing, 19

Electron impact ionization, 35, 673, 854; *see also* Electron ionization

Electron injection, 694

Electron injection trajectories, 301

Electron ionization, EI, 22, 35, 40, 52, 65, 237, 238, 243, 436, 632, 634–638, 751, 780

Electron multiplier, 628, 773, 774, 778, 780, 781

Electron multiplier detection, 711

Electron multiplier with magnetic field, 88

Electron multiplier, fast, 472

Electron paramagnetic resonance spectrometer (EPR), 76

Electron removal, 687

Electron space charge, 850

Electron spin resonance, 699

Electron trajectory, 797

Electron transfer, 687

Electron transfer dissociation, ETD, 108, 252, 262, 264, 266, 274, 535, 610, 611, 678, 681, 684–687, 693–699, 718

Electron, name of, 10

Electron/ion reaction, 681, 684

Electron-hopping, 686

Electronic charge, 760

Electronic filtering, 637

Electron-induced dissociations, 681, 682, 684

Electrophoresis, 228

Electroplating, copper, 206

Electrospray interface, 541

Electrospray ion source, 49

Electrospray ionization mass spectrometry, 315

Electrospray ionization, ESI, 11, 50, 175, 177, 215, 226, 241, 243, 245, 292, 295, 302, 311, 316, 328, 352, 407, 436, 500, 525, 541, 550, 578, 584, 588, 638–640, 642, 660, 661, 664, 666, 667, 672, 674, 684, 685, 689, 696, 710, 711, 718, 720, 729, 740, 741, 744, 770, 830, 835, 839–842, 848, 849, 852; *see also* Ionspray

Electrospray ionization, Tuning Mix, 606

Electrospray needle, 342

Electrospray source, 81

Electrospray technique, 49

Electrostatic analyzer, ESA, 25, 26, 32, 33

Electrostatic field, 81

Electrostatic field, 675

Electrostatic Kingdon trap, *see* Kingdon trap, electrostatic

Electrostatic sector, 673, 674

Electrostatic trap, EST, 252
   closed, 252, 253
   decay of ion packets in, 257
   open, 252, 253
   space-charge effects in, 258
   tandem mass spectrometry inside, 259

Electrostatic traps, common properties of, 257
   deflection, 252, 253
   orbital, 252, 253
   reflection, 252, 253
   taxonomy of, 252, 253

Elementary (electronic) charge, e, 14

Elementary experiments, 105, 473, 475

Elements, 'isotopic', name of, 13

Elongated trapped-ion trap, 449, 451

Elution peaks, 664

Embryonal carcinoma cell, human, 728, 729
   mouse, 728

Embryonic stem cell, 728

Emission current, electron, 35

Enantiomer amplification, 228

Enantiomer differentiation, 772

End aperture, 655

End electrodes, 655

End plates, 644, 655

End section, 666, 667

End-cap electrode, 621–623, 626, 628, 629, 632, 635, 640, 740, 771, 774, 778, 782, 783, 785, 795–799, 801, 805, 827, 829, 850, 851, 854, 858; *see also* End-cap electrodes

End-cap electrode, entrance, 627
   exit, 627
   holes, 196, 288

End-cap electrodes, 100, 110, 111, 114, 156, 276, 676, 679, 694; *see also* End-cap electrode
   $2z_0$, separation of, 623, 640
   hyperbolic, 621
   hyperboloidal, 106
   image currents induced between, 471

End-cap electrodes, mesh, 175

End-cap electrodes, stretching axial separation of, 382, 393

Endcap lens, 575

End-cap spacing, 188

End-caps, shimming, 393

Endogenous interferences, 362

Energetics, 713

Energy absorption, 435

Energy distribution, simulated, of ejected ions, 580

Energy focusing, 579

Energy of repulsion, Coulombic, 59

Energy randomization, 300

Energy resolution, 86, 673

Energy transfer, 742, 746, 748, 811, 819

Energy transfer efficiency, 748, 759
Energy transfer probability density, 746
Energy transfer process, 750
Energy, internal, 64
Energy, rapid uptake of, 598
Energy, translational, 64
Energy-loss techniques, 352
Energy-resolved fragmentation efficiency
    curve, 683
Energy-resolved metastable ion peaks, 57
Enhanced detection, 711
Enrico Fermi Award (1986), 96
Enthalpies of clustering, 65
Enthalpy change, 66
Entrance aperture, 655
Entropy, 745
Environmental monitoring, 230
Enzymatic digestion, 673
Enzymatic proteolysis, 713
Enzymes, 709
Enzymology, 727, 728
Ephedrine (−), 239
EPSI, *see* Ionization, electrospray
Equation setting, 501
Equation, hyper-logarithmic; *see also* Equation,
    quadro-logarithmic
    quadro-logarithmic, 120, 254
    quadro-logarithmic; *see also* Equation,
    hyper-logarithmic
    second-order differential, 148
Equations of motion, 622, 623, 645–648, 652
Equilibrium constant, 66
Equilibrium, non-stationary, 252
    stable stationary, 252
Ergode theory, 683, 692
Ergodic fragmentation, 681–683, 686, 688, 689,
    692, 693
Ergodic reaction, 681
Erythromycin, 831, 832
ESA, *see* Electrostatic analyzer
Escherichia coli, see E. coli
ESI ETD MS/MS data, 612, 613
ESI sources, 213
ESI, *see* Electrospray ionization
ESI, +/−, 227
ESI-FAIMS-MS, 335, 336, 352, 357
ESI-FAIMS-TOFMS, 361
ESI-generated ions, 740, 741, 744, 745, 754
Esquire 3000plus mass spectrometer, 600, 602,
    606, 608, 609
Esso Research and Engineering Company,
    USA, 40
EST, *see* Electrostatic trap
η, single trapping parameter in DKTs, 125
ETD capabilities, 265
ETD, *see* electron transfer dissociation
Ethene, 443, 444

Ethyl acetate, 538
Ethyl parathion, 241
Ethylene oxide, 241
Ethylmethylamine, non-covalent complexes
    of, 421
Even-electron ions, EE, *see* Ions, even-electron
Even-electron multiply-protonated peptide, 611
Even-order multipole fields, 627
Even-order non-linear fields, 381
Event, Radio-frequency-only-mode, 433, 436, 437
Events, high-pressure, 436
Exactive mass spectrometer, 266
Excedrin, 239, 243
Excitation by injection, 122, 260
Excitation by off-axis injection, 256
Excitation energy, 744
Excitation energy, internal, 33
Excitation energy, vertical, 64
Excitation event, 403
    two-step, 598
Excitation events, decoupled, distinct, 606
Excitation kinetic, resonance oscillation by, 577
Excitation of ions, 256
Excitation of ions to higher orbit, 607
Excitation of ions, resonant quadrupolar, 512
Excitation period, 404
Excitation plates, 453
Excitation voltage, 276, 745, 762
Excitation waveform, 776
Excitation, de-, 191
    dipolar, 283, 287, 288, 290, 291, 298
    fixed-frequency resonance, 292
    IR multiphoton, 299
    parametric, 410
    progressive, at non-linear resonance
        point, 603
    quadrupolar, 283
    re-, 191
    resonant quadrupolar, 409
    resonant, 106
    resonant dipolar, 409
    supplementary dipolar, 108
    supplementaty quadrupolar, 514–516
Excite plates, 435
Exit and entrance holes, 740
Explosions, 81
Explosive vapors, trace, detection of, 351
Explosives, nitro-organic, 333
External accumulation of ions, 401, 402
External accumulation, selective, 411
External electrode, 294
External ion injection, 404
External ion source, 677
External RF storage, 259
Externally generated ions, trapping efficiency
    for, 545
    quadrupole mass filtering of, 409

Extraction efficiency, 182, 583
Extraction electrode, 288
Extraction region, 547
Extraction voltage-pulse, 795, 798
Extravasation, 728

F

FAB, *see* Fast atom bombardment
Fabrication errors, 183
Fabrication of miniature ion traps, 191
Fabrication tolerances, 178
Fabrication, monolithic, 191
Facilitating substance, 48
FAE, *see* Field-adjusting electrode
FAIMS (High-field asymmetric waveform ion
        mobility spectrometry), 31, 309, 311,
        420, 536, 712
FAIMS, bioanalytical applications, 362
    cube design, 348, 349
    cylindrical, device, 328
    development of, 315
    dome design, 348
    electric field-driven, 351
    environmental applications, 360
    illustration of ion separation in, 318
    ion focusing in, 321
    ion separation in, 316
    ion trapping in, 327
    microchip-sized, 350
    origins of, 315
    resolving power, concept of, 321
    separation, sFAIMS, 328
    side-to-side, 329
    tandem arrangement, 328
    trapping, 328
    universal model, 332
FAIMS and mass spectrometry, complementary
        natures of, 364
FAIMS apparatus, quadrupole, 343
FAIMS as a gas-phase ion filter, 351
FAIMS coupled to QqTOFMS instrument, 357
FAIMS device, reduced pressure, 343
    segmented, 343
FAIMS devices, mathematical models for planar
        and cylindrical, 330
FAIMS electrode designs, other, 343
FAIMS hardware, 337
FAIMS ion trapping device, cross-sectional view
        of, 327
FAIMS ion trapping, half-life of, 328
FAIMS-mass spectrometry, FAIMS-MS, 315, 341
FAIMS resolution, concept of, 321
FAIMS system, schematic of basic components
        of, 317
FAIMS systems, tandem, 342
False discovery rates, 413

Family history, 127
Faraday collector, 20, 22
Faraday cup, 94
Faraday cup detector, 192
Faraday plate detector, 311
Fast atom bombardment ionization, FAB, 44–46,
        672, 681, 687, 710
Fast atom bombardment, flowing, 47
Fast Fourier transform, FFT, 256
Fast injection, 260, 261
Fast neutral atoms, 45
Fatty acid chain, 817
Fatty acid residue, 818
FD, *see* Field desorption
FC-43, *see* Perfluorotributylamine
Fe^+, 118
Feedback loop, 184
FEL, *see* Free electron laser
Fenn, J., 49–51, 708
FFR, *see* Field-free region
FI, *see* Field ionization
Fiber inlet mass spectrometry, FIMS, 238
Fibrinopeptide, 835
Field components, 173
    dodecapole, 389
    higher-order, 187
    individual variation of, 389
    octopole, 389
    quadrupole, 389
Field desorption, FD, 38, 39, 43, 672, 710
Field desorption mass spectra, transience of, 46
Field gradients, DC electric field, 310
Field imperfections, 676, 771
Field interpolation, 185
Field ion microscopy, 38
Field ion spectrometer, FIS, 315
Field ionization kinetics, 60
Field ionization, FI, 38, 39, 672
Field optimization, 383
Field penetration, 187
Field shape, 449
Field simulation programs, 381
Field, dodecapolar, 190
    fringing, 190
    hexapole component of, 598
    internal electrical, 173
    nonlinear, 258
    octopolar, 190
    octopole component of, 598
Field-adjusting electrode, FAE, 286,
        288–290, 298
Field-corrected ion cell, 455
Field-corrected trap, 453, 454
Field-free region, FFR, 12, 33, 673, 674, 683
Field-induced deviations in mobility, effect of
        different gases, 331
Field-penetration, 77

Fields, combination of, 594
  higher-order, 105, 149, 594
  higher-order postive, 287, 292
  net, virtual, 328
  non-homogeneous, 338
  non-homogeneous, modes of operation for, 327
  nonlinear, 594
  ralative magnitudes of higher-order
       components, 179
  time-dependent, 257
Figures of merit, influences on, 608
Filament, 774
Filament design, 'boat', 37, 38
  three-, 37, 38
Filament temperature, 37
Filaments, rhenium, 37
  tungsten, 35, 37, 38
Filars, 225
Filars, independently controlled DC potentials
       applied to, 225
Filtered noise field, FNF, 298
Filtered Noise Field, algorithm, 795
FIMS, *see* Fiber inlet mass spectrometry
Finite excitation force, 628, 644
Finnigan Corporation, USA, 103, 500
Finnigan GCQ electronics, 776, 779, 780, 788
Finnigan ITMS system, 759, 850
Finnigan LCQ; *see also* Thermo Fisher
       Scientific, 711
Finnigan MAT, 381
First analyzer, BE, 711
First side-band frequency, 629
FIS, *see* Field ion spectrometer
Fischer, E., 708
Flat-bottom distribution, 417
Flavonoid, 819, 830, 831
Flavonoid diglycoside, 830
Flight time, 813
Flight time for an ion, 477
Flight time information, 476
Flip chip bonding, 196
Flow cytometry, 728
Flow injection analysis, 362
Flowing afterglow, 65
Flowing FAB, 47
Fluoranthene, 686, 687
Fluorenone, 686
Fluorescence yield, 751
Fluorescent detection, 718
Fluoroanthene anion, 265
Fluoride affinities, 66
FNF, *see* Filtered noise field
Focusing errors, 62
Focusing, direction-, 19
  time, 89
  time-lag, 86, 87
  velocity-, 19

Force acting upon an ion, 154, 160
Force, centrifugal, 122, 123
  centripetal, 123, 678
  dipolar excitation, 605
  hexapolar, dependence upon z, 605
Ford, H., 8
Formaldehyde, 241
Forward geometry, *see* Geometry, forward
Forward scanning, 291, 297
Four-cell system, 77
Four-channel CIT array, 211
Four-dimensional space, 664, 668
Fourier analysis, 630, 631, 656, 657
Fourier transform, FT, 252, 435, 472, 471, 513,
       656, 657, 795
Fourier transform ion cyclotron resonance mass
       spectrometer, FT-ICR-MS, 31, 53, 70,
       78–80, 109, 113, 117, 122, 184, 253,
       256–258, 267, 268, 300, 311, 401, 416,
       446, 471, 677, 688, 690, 691, 694, 722,
       743, 753, 688
Fourier transform operating mode, 75, 78, 105,
       448, 469, 680
Fourier transform technique, 677, 680
Fourier transform, discrete, *see* Discrete Fourier
       transform
  fast, FFT, 256
Fourier transformation, 675–677
Four-region cell, 77
Fragile ions, 748, 769, 771, 774, 782,
       788–90
FragLab Tool Kit, 724, 731, 732, 735
Fragment ion, *see* Product ion
Fragment ion abundance, 752
Fragment stereochemistry, 711
Fragmentation, 328
Fragmentation cell, 673, 675, 693, 694
Fragmentation mechanisms, 24
Fragmentation pathway, 794, 812, 813, 817,
       836, 837
Fragmentation pattern, 830, 833, 841
Fragmentation process, 679
Fragmentation studies, 697
Fragmentation techniques, 671, 674, 688
Fragmentation techniques for protein ions, 113,
       265, 416, 611, 718
Fragmentation, extent of, 40
  proteins, 680
Fragmentation, time-delayed, *see* Time-delayed
       fragmentation
Franck-Condon Principle, 22, 41
Franck-Condon transition, 28
Free electron laser, 688, 848
Free energy, Gibbs, 66
Freeze-drying effect, 51
Freqency of radial motion, 254, 255
Frequencies of ion motion, 620, 632

Frequencies, axial, 167
  complementary, 168
Frequencies, radial, 167
  secular, *see* Secular frequencies
Frequency broadening, 510
Frequency components, 284
Frequency domain, 676, 677
Frequency domain influence, 664
Frequency line, shape of, 479
Frequency of axial oscillations, 254
Frequency of rotation, 254, 255
Frequency resolution, 486
Frequency shift, 453, 454, 458, 491
Frequency shifts, cyclotron, 404
  cyclotron, space-charge-induced, 412
  negative, 406
Frequency spectrum, 630, 631, 656, 657
Frequency sweep, 74
Frequency, cyclotron, *see* Cyclotron frequency
  diocotron, 403; *see also* Diocotron motion
  ion cloud rotation, 403
  magnetron, 403; *see also* Magnetron mode
  secular, *see* Secular frequency
Frequency, unperturbed, 403
  winding, 126
Fringe induction, 437
Fringing field modification, dynamic, 298
Fringing field, 291, 546, 578
  exit, 546, 547
Fringing fields, 190
  electric, 99
Fringing regions, reduction of 2D quadrupolar
  potential in, 551
Froehlich, U., 269
FT operating mode, typical wave train for, 493
  requirements of, 493
FT, *see* Fourier transform
FT-ICR, *see* Fourier transform ion cyclotron
  resonance mass spectrometry
FT-ICR detection, external accumulation of ions
  for, 402
FT-ICR experiments, 119
FT-ICR experiments, broadband, 406
FT-ICR instrument, 408
FT-ICR mass spectrometers, 573
FT-ICR mass spectrometry, 402
FT-ICR mass spectrometry of peptides and
  proteins, 300, 416, 611
FT-ICR, method of operation, 75
FTMS, *see* Fourier transform ion cyclotron
  resonance mass spectrometer
Fucosyl blood group, 716
Function generator, 779, 782, 785, 786
Function, boundary, 279
  hyperbolic, 278
  sinusoidal, 278
Fundamental frequency components, 632

Fundamental frequency for ion motion, 620, 629
Fundamental oscillation, 678
Fundamental RF drive potential, 649, 650, 655
Fundamental secular frequency, 628, 629, 636,
  645, 656; *see also* Secular frequency
Furan, 710
Fused silica restriction, 783

G

GaAs substrate, 197, 198
Gadolinium, 38
Galanin-like peptide, 612, 613
Gallium-arsenide, 196, 197
Gap width, 316, 346
  effective, 322
Gas chromatography columns, efflluents from, 81
Gas chromatography detector, 740
Gas chromatography, GC, 26, 274, 664, 712,
  770, 785
Gas chromatography/mass spectrometry, GC/
  MS, 103, 712, 785, 790
Gas chromatography/tandem mass spectrometry,
  GC/MS/MS, 712
Gas collison cells, 62
Gas introduction line, 796, 798
Gas mixture, binary, effect on CV spectra, 332
Gas mixtures, 336
  effect on FAIMS efficiency, 331
Gas modifier, 333
Gas number density, 310, 311, 330, 334, 747
Gas peak, 782
Gas pressure profile, 776, 777, 780–783,
  785–788, 790
Gas pulse, 776, 777, 779, 780, 783, 786, 790
Gas pulse duration, 776, 782, 786, 790
Gas pulse optimization, 784
Gas pulses, 785–787
Gas pulsing, 796, 797, 799, 802, 819
Gas, curtain, 330
  drift, *see* Drift gas
Gases, bulk, 331
  trace, 333
Gas-phase basicity, 743
Gas-phase disassembly, 710
Gas-phase protein structures, 352
Gas-phase reaction, 772
Gas-phase structural analysis, 711
Gas-solid interactions, 681
Gated trapping, 409
Gated-trapping technique, 408
Gating time, width of, 313
Gauss' Law, 449, 457
Gaussian dispersions, 485
Gaussian distribution, 166, 480–483, 503, 510
Gaussian distribution, zero-centered, 483
Gaussian peaks, 57–59, 322

GC carrier gas, 770
GC column, 770
GC, *see* Gas Chromatography
GC/MS, 361
GC/MS system, portable, 177
GC/MS, *see* Gas chromatography/mass
      spectrometry
Gd^+, 38
GDEI, *see* Glow discharge electron impact
      ionization
GdO^+, 38
Gene expression profiling, 728, 729
Genealogical approach, 127
General Electric Company, USA, 102, 113
Genes, 708
Geometric axis of symmetry, 644
Geometric effects, 183
Geometric size, 758
Geometrical center, 626, 629, 635, 642–645, 658
Geometrical evolution of ion traps, 171
Geometries, arrayed, 191
    non-hyperbolic, 406
Geometry constant, G_T, 438
Geometry defintions, 595
Geometry, circular, 373
    forward, 34
    inverse, *see* Geometry reversed
    Nier-Johnson, 55
    non-stretched, 286, 473
    non-stretched DIT, 290
    optimization of, 171
    reversed, 34, 45, 46, 55
    RIT, optimized, 175
    stretched, 206, 287
    un-stretched, 286, 473
Germanium, 45
Germanium coating, 202
Germanium layer, resistive, 389
Germanium, semiconductor, 391
Germmanium-silicon alloy, 45
Gettering action, 117
Ghost peaks, 374
Gibbs free energy, 66
Glass-encapsulated ceramic particles, 203
Glow discharge electron impact ionization,
      GDEI, 240
GLP, *see* Good Laboratory Practice
Glucuronide metabolite, 362, 363
Glu-fibrinopeptide B, 290, 291, 293
Glutamine, 213–215, 243
Glutathione-trapped reactive metabolites, 568
Glycan, 708, 709, 713, 723–727
Glycan analysis, 722
Glycan compositions, 731
Glycan mass profiles, 710, 720, 727
Glycan mass scan, 719
Glycan sequence, 718

Glycan specificity, 728
Glycan structural complexity, 709, 722–727
Glycan structure, 710, 722–727, 813, 817
Glycerol, 49
Glycobiology, 812, 819
Glycolic acid, 361
Glycome, 712
Glycopeptide, 812, 813
Glycoprotein, 717, 723, 812
Glycoprotein standards, 720
Glycoproteome, 707, 709, 812
Glycoproteome, structural details, 106, 265, 707
Glycoproteomic disease states, 708
Glycoscreen, 729
Glycosides, 48
Glycosidic cleavage, 714, 722
Glycosphingolipid analysis, 819
Glycosylation, 265, 683, 812, 813, 815
Glycosylation change, 727
Glycosylation finger print, 719
Glycosylphosphatidylinositol, 711
Glyerol solvent matrix, 46
Gold, 35, 203, 204
Gold electroplating, 223
Gold sputtering, 206, 223
Gold wires, 197
Gold(II), 848, 849
Gold/tungsten alloy, 202, 203
Goldstein, E, 8
Golgi organism, 722
Golgi processing remnants, 722
Good Laboratory Practice, GLP, 358
Gramicidin S, 52, 328, 329, 335
Graviational field of double stars, 117
Grayson, M.A., 128
Green, B., 26
Griep-Raming, J., 268
Grounded cylinder, 626
G_T, *see* Geometry constant
Guard electrodes, 114
Guardion-7 GC/MS system, 385, 386
Guest-host complex, 689

H

H^-, 14, 63, 64
$H\bullet$, 63, 64
H/D exchange, 352–354
H^+, 14, 32, 39, 63, 64
H_2, 14
H_2^+, 14, 16, 31, 32, 39
H_3^+, 31, 32
H_3O^+, 43
Hadamard de-modulation, 253
Hadamard transform, 252
Half life, decay of excited ion, 683
Half-life for ion trapping in FAIMS, 328

Halide ions, 693, 694

Halo ion trap, 69, 111, 112, 177, 202, 204, 230, 388, 389, 391, 392

Haloacetate anions, 332

Haloacetic acids, 314

Hamiltonian, electric potential in, 462

Hamlet, 7

Hamming function, 483, 487

Hand-held miniature mass spectrometers, *see* Mass spectrometers, hand-held miniature

Hard sphere 'sticky' collision, 686

Hardman, M., 268

Hard-sphere collision model, 799, 811

Harmonic vibration, 808

Harmonics, 630, 631

HASTE, *see* High amplitude short time excitation

HCD, *see* Higher-energy collisional dissociation

HCD collision cell, 266, 267

HCT, *see* High-capacity trap,

HCT, spherical, 602

HCT*ultra* mass spectrometer, 598, 600, 606, 608, 609, 614, 615

HD Technologies, Manchester, UK, 268

Headspace vapor, 238, 240

Heat vaporization, 710

Heated ionization sources, 334

Helical path, 81, 82, 89

Helices, wire-wound, 680

Helium, 11, 38, 105, 109, 114, 116, 188, 291, 331, 385, 498, 548, 575, 628, 630, 631, 635, 637, 640, 644, 659, 660, 687, 739, 740, 743, 746, 750, 754, 762, 770–772, 777–783, 786, 788, 797–799, 851, 852, 855

Helium bath gas, 443, 739, 740, 746, 747, 749, 750

Helium buffer gas, 840

Helium pressure, 219, 414, 420, 743, 744, 750, 754, 756, 762

Helium pulse, 444

Helium, target gas, 827

Helium/ion collision frequency, 740

Helium/ion collisions, 740, 755, 758

Helmholtz coils, 76

Hemenway, E., 268

Hemoglobin, pig, 357

Hengelbrock, O., 268

Heroin, 229, 242, 245

Herschbach, D.R., 9

Hexachlorobenzene, HCB, 632, 634

Hexafluorobenzene molecular cation, 460

Hexapolar ejection process, simulation of, 605

Hexapolar field, consequences of, 604

    contribution of, 604

    coupling by, 604

    direction of force on an ion, 604

Hexapole component, 162, 621, 627, 629–631, 637, 642, 644, 652, 657, 668

Hexapole nonlinear resonance, 635, 636, 654, 659

Hexapole resonance, 629, 633, 635, 636

Hexapole term, 622, 645, 655, 657

Hexapole, RF-only, 411

Hexose, 722

Hg^+, 70

High amplitude short time excitation, HASTE, 741, 742

High energy collisional fragmentation, HE-CID, 682

High mass detection, 285

High mass ionization, 710, 713

High mass-to-charge cut-off, 442

High performance liquid chromatography, HPLC, 717, 718, 722, 813

High resolution, 373

High transmission and energy focusing, simultaneous, 579

High vacua, 15

High-capacity trap, HCT, 95, 108, 374, 593

    geometry of, 595

    physics of, 593

High-density/low-energy plasma, 196

High-energy wall collisions, HE-SID, 682

Higher order fields, negative, 294

    positive, 287, 292

Higher-energy collisional dissociation, HCD, 262, 265, 266

Higher-order differential ion mobility separations, HODIMS, 347

Higher-order field components, 187

Higher-order field perturbation, 740

Higher-order field, contribution not compensated for, 604

Higher-order fields, 594

    influence of, 604

Higher-order geometrical terms, 653

Higher-order multipole fields, 620, 629, 642

High-field asymmetric waveform ion mobility spectrometry, *see* FAIMS

High-field Orbitrap, 267

Highly-charged ions, 117

High-pressure events, 436

High-pressure RF 2D quadrupole, trapping and moving ions in, 528

High-resolution analysis, 374

High-resolution frequency spectra, 498

High-to-low field mobility, ratio (K_h/K), 331, 340, 352

Historical survey of mass spectrometry, 375

Hitachi High-Technologies Corporation, Tokyo, Japan, 109, 110, 582

Hitachi Perkin-Elmer, 34

HODIMS, *see* Higher-order differential ion mobility separations

Hoffman, A., 268
Hole in electrode, 679
Hoover, Herbert Clark, 31st President of the
     United States, 20
Hoover, Herbert Clark, Jr., 20
Horning, S., 268
Horse heart myoglobin, 285, 286
Hot cathode gauge, 796
Hot cathode, electron-generating, 691, 694
Hot electron capture dissociation, HECD, 685
Hour-glass IFT, 421
HPLC *see* High performance liquid
     chromatography
Huels, W., 268
Hughes, J., 268
Human acetylcholinesterase, 711
Human metabolism, 813
Human plasma lipids, 817, 818
Human transferrin glycopeptide, GP2,
     813, 814
Hurst, J., 268
Hydraulic compression, 203
Hydride affinities, 65
Hydride extractor, 43
Hydride-transfer reaction, 41
Hydrocarbon analysis, mass spectrometric
     analysis of, 19
Hydrocarbon mixtures, 27
Hydrocarbons, 20, 39, 637
Hydrofluoric acid, 194, 197
Hydrogen, 12, 26, 114, 770
Hydrogen bridge, 687
Hydrogen cyanide, 241
Hydrogen nuclei, 14
Hydrogen/deuterium exchange, gas-phase, 43
Hydroxide extractor, 43
Hyperbolae, rectangular, 151
Hyperbolic angle electrode geometry, 598
Hyperbolic cell, 404
Hyperbolic electrodes, 148
Hyperbolic electrodes, four elongated, 655
Hyperbolic function, 278
Hyperbolic trap, 406, 457
Hyperbolic, non-, geometries, 406
Hyperbolic-shaped electrodes, 676, 678
Hyperboloid mass analyzers, 275
Hyperboloid, single-sheet, 100
     two-sheet, 100
Hyperboloidal geometry, 156
Hyperthermal energies, 228
Hypodermic needle, 49

I

IC fabrication methods, 195
IC, *see* Integrated circuit
ICC (Ion charge control), 601

ICPMS, *see* Inductively coupled plasma mass
     spectrometry
ICR, *see* Ion cyclotron resonance
ICR analyser, drift-cell, 75
ICR analyzer, 70
ICR cell, 74
     compensated-open cylindrical, 407
     infinity, 407
     open, 407
ICR double resonance, 691
ICR spectrometers, trapped-cell, 55
ICR trap, 402, 435, 847
     orthogonalized, 408
ICR trap configurations, assorted, 450
ICR, drift-cell, 75
ICR-9 mass spectrometer, 77
Ideal end-cap electrodes' separation, 629
Ideal hyperbolic ion trap, 622, 641
Ideal hyperbolic surface, 629, 647
Ideal ion trap geometry, 327
Ideal quadrupole field, 621, 623, 645, 647
IE, *see* Ionization energy,
IETD, *see* Inverse electron transfer dissociation
IFT, *see* Ion funnel trap
IFT charge capacity, 423
IFT coupled to oa-TOFMS, 417
IFT, hour-glass, 421
     pseudopotential in, 417
     trapping efficiency of, 422
IKES, *see* Ion kinetic energy spectroscopy
Im(A), 440
Image charge, 402; *see also* Image current
Image current, 78, 122, 253, 256, 435, 471, 675,
     677, 680
Image current detection, 258, 454
Image current, Fourier transfomation of, 122
Image signal, 105
Image signal of confined ions, 477
Image signal of motions of simultaneously
     trapped ions, 472
Image signal, amplitude of, 483
     electron ionization, 488
     expression for, 479
     temporal, 479
Imaging gas, 38
Immune system survival, 728
Impedance, 774, 783
Impedance mis-matching, 180, 195
Impending disease, 717
Imperfections mechanical, effect on ion motion
     nonlinearities, 507
Imprisoned positive ions, 113
IMS, *see* Ion mobility spectrometry
IMS drift-tube, 311
IMS separation timescale, 423
IMS-TOFMS, dynamically multiplexed, 424, 425
in silico reassembly, 710

Incap electrodes, 110
Incap lens, 575
Increased temperature, 744
Inductively coupled plasma mass spectrometry, ICPMS, 37
Inelastic collision, 758
Inert gas, pulsing of, 408
Infinite sheets, 621
Infinity cell, 405
Infrared absorbing boronic acid, IRABA, 836
Infrared absorbing ligand, IRAL, 831
Infrared absorptivity, 830, 837
Infrared activation, rate of, 749
Infrared beam, 691, 693
    axial, 691
Infrared chromophore, 830, 834, 836
Infrared irradiation, 689, 693, 830, 834
Infrared laser, 691
Infrared laser, monochromatic, 689
Infrared multiphoton dissociation, IRMPD, 101, 299, 448, 681, 682, 689–691, 693, 698, 740, 742–744, 754, 762, 829–838, 847, 848, 852, 853, 856–858
    collision-assisted, 830
Infrared photodissociation, 856
Infrared photon, 828
Infrared photon absorption, 743
Infrared spectroscopy, 699
Infrared spectrum, 857
Infrared spectrum of ions, 688
Inhomogeneous electric field, periodic, 97
Initial conditions, confinement, 480
    dispersion of, 485
Injected ion trajectory, shape of, 503
Injection system, 30
Injection time, effect on cross-talk, 216
Injection, excitation by, 122
    off-axis, 256
    tangential, 122
Ink, 243
Inlet system, 30
Inorganic compounds, 673
Inscribed circle, radius of (r_0), 151
In-source dissociation, ISD, 674
Instability zones, 105
    regions of, 98
    trajectory, 164
Insulin, 47, 638, 641
Insulin, +3 charge state, 641
Insulin, +6 charge state, 638, 641
Insulin-β-chain, 298, 299
Intact proteins, analysis of, 264
Integrated circuit, IC, 178, 191, 192, 201
Integrated System for Ion Simulation, ISIS, 501
Interferences, reduction/removal of, 362
Intermolecular hydrogen bonds, 690
Internal calibration, 264

Internal energy, 64, 674, 682, 683, 714, 740–751, 753, 756–758, 761, 762
Internal energy distribution, 563, 745, 757, 772, 855
Internal energy distribution, time evolution of, 564
Internal energy randomization, 750
Internal energy transfer, 758
Internal energy, collisional removal of, 437
Internal excitation, 682
Internal heating, 811
Internal ionization, 775
Internal standards, deuterated, 360
Inter-residue linkage, 709, 710
Interstellar space, 685
In-trap ion processing, 559
Intravasation, 728
Inverse electron transfer dissociation, IETD, 682, 684, 686
    negative ion, 687
Inverse transform algorithm, 423
Inversion problem, 406
Iodate, 323, 358, 359
Iodine voltameter, 73
Iodomethane, 719
Iomefloxacin, 239
Ion abundance, effect of relative phase difference, 609, 610
Ion acceleration, 93
Ion acceleration spectrometers, radiofrequency, 94
Ion accumulation, 219, 355, 401
Ion accumulation approaches, 677
Ion accumulation time, 600, 640, 643, 662, 666
Ion activation, 574, 742, 795
Ion activation method, 828, 836–838
Ion adduction, 714
Ion axial displacement, 625, 626
Ion axial motion, 632, 633
Ion axial oscillations, 675
Ion beam, 799
Ion bombardment, 682, 693
Ion capacity, 375, 595
    HCT, improved, 610
    limitation of during ion ejection, 597
    meaning of the term, 594
Ion charge, 747
    control, ICC, 601
Ion chemistry, 677
    gas-phase, 21, 53
Ion cloud, 664, 666, 667, 678, 679, 685, 693, 695, 696, 799, 803–805, 832
Ion cloud confined on nonlinear resonance line, measured frequencies of, 510
Ion cloud diameter, 805
Ion cloud excitations, 471
Ion cloud manipulation, 445

Ion cloud rotation frequency, 403
Ion cloud, axially compressed, 667
    charge density of, 416
    collisionally-cooled, 693
    expanded, 667
    form of a thread, 679, 689, 696
    shrinkage of, 596
    spherical, 693, 694
Ion clouds, initial conditions of, 475
Ion clouds, two, opposite polarity, 695, 696
Ion collision cross-section, 811
Ion confinement, 474
Ion conformation, 744, 745
Ion conformations, gas-phase, 312
Ion cooling, 294, 635, 664, 740, 746, 751, 769,
        770, 774–776, 780, 785, 788, 790,
        802, 811
Ion coupling, 594
Ion creation, 474
Ion creation cell, 472
Ion cyclotron resonance, ICR, 42, 69, 77, 847
    instrument, 796
    mass spectrometry, 75, 676, 677, 680, 690
    trap, 675, 677, 685, 688, 689, 691
    instruments based upon, 69
Ion damping, 679
Ion decay, metastable, 57
Ion density, 503, 504
Ion density during ejection, effective, 602
Ion density, analytical limits to, 374
Ion density, dependence on m/z-value, 505
Ion detection, 256, 620
Ion detection efficiency, 775
Ion diffusion, 529
Ion discrimination, 411
Ion dissociation, 632, 637, 638, 673, 708
Ion dissociation efficiency, 743
Ion dissociation rate, 751
Ion dissociation reaction, 742
Ion distribution at end of injection, 503
Ion distribution, injection, at the end of, 503
Ion ejection, 475, 574, 595, 621, 625, 627–630,
        633, 634, 637, 638, 641, 643–645, 653,
        654, 659–661, 664, 668, 678, 748, 769,
        770, 774, 798, 802, 805
Ion ejection away from detector, 663
Ion ejection delay, beneficial effects of, 610
Ion ejection direction, 664
Ion ejection efficiency, 576, 805, 819
Ion ejection in lon-linear fields, 598
Ion ejection time, 638, 771
Ion ejection to and away from detector, 663
Ion ejection to detector, 663
Ion ejection, accelerated by higher-order fields, 598
    axial, 596
    data-dependent, 411
    direction of, 597

    early, 597
    high-capacity, achievement of, 606
    mass-selective, *see* Mass-selective
        ion ejection
    model for ideal process of, 600
    non-linear two-step process, model for, 603
    radial, 596
    regions of, 551
    resonant, 288
Ion energization, 681
Ion ensemble, 746, 803
Ion excursion, 632, 633
Ion extraction efficiency, 326, 740
Ion flow, driven by diffusion, 541
    driven by electric fields, 541
    driven by gas flow, 541
    reversing, 531
Ion flux, 640
Ion flux in transmission-only quadrupole
        experiment, 566
Ion focusing, 337
Ion focusing in cylindrical FAIMS device, 325
Ion focusing in FAIMS, 321
Ion focusing, absence with flat plate
        elctrodes, 326
    atmospheric pressure, 364
Ion fragility, 771, 774, 782
Ion fragmentation, 638, 673, 677, 688, 691, 711,
        713, 714, 722, 744, 751, 769, 771,
        774, 775
Ion fragmentation device, 574
Ion fragmentation efficiency, 740, 744, 751–754,
        756–758, 760, 761, 772, 789
Ion fragmentation pathway, 710
Ion fragmentation, yield, 691, 693
Ion fundamental frequency, 628
Ion fundamental secular frequency, *see* Secular
        frequency, fundamental
Ion funnel, 227, 355, 527
Ion funnel trap, IFT, 401, 418
    schematic diagram of, 418, 422
Ion funnel, electrodynamic, 408, 411, 418, 421
Ion gate, 420
Ion gate voltage, 775, 776, 781
Ion gauge pressure, 788
Ion gauge, Orbitron, 116
Ion guide branched, five ion sources, 539
    three ion sources, 539
Ion guide, axial potential of, 530
    hexapole, 526, 527
    octopole, 526, 527
    quadrupole, 527
    RF-only, 227
Ion guides as collision cells, 526
Ion guides between two quadrupole mass
        filters, 526
Ion guides, branched, 539

Ion guides, higher-order multipoles as, 526
    multipoles and ring guides, 526
    quadrupole, multiplexed input/output, 540
    radiofrequency, 414, 525, 526, 530, 574, 643,
        679, 718
Ion injection, 474, 740, 769–771, 773–776, 780,
        782, 788, 789, 798, 799
Ion injection efficiency, 819
Ion injection process, 799, 800
Ion injection, external, 404
    fast, 259
    slow, 259
Ion intensity profile, 786
Ion internal temperature, 760
Ion introduction, 294
Ion isolation, 216, 574, 708, 769, 771, 775, 802
    multiple stage, 173
Ion kinetic energy, 12, 631–634, 675, 679, 740,
        742, 746, 769–771, 774, 808, 809,
        811, 820
Ion kinetic energy spectroscopy, IKES,
        33, 34
Ion lens gate, 773
Ion lifetimes, 115
Ion loss, 771
Ion loss, prevention of, 527
Ion losses, 319, 328, 578
Ion micromotion, unintended oscillatory, 200
Ion migration, 678
Ion mirror (reflection EST), 253
Ion mirror, *see* Reflectron ion mirror
Ion mobility, 536, 711
Ion mobility at high electric fields, 313
Ion mobility constant, 311
Ion mobility drift tubes, 525
Ion mobility equation, 311
Ion mobility spectometry/mass spectrometry,
        311, 420, 536
Ion mobility spectrometry, differential, *see*
        Differential mobility spectrometry
    high-field asymmetric waveform,
        see FAIMS
    IMS, 31, 310, 311, 401, 420, 712
    traveling wave, 311, 420, 536
Ion mobility spectrum, 311
Ion mobility, changes in, 317, 340
    dependence upon electric field, power series
        for, 313
    duty cycle of, 421
    hypothetical dependence of different ion
        types upon electric field, 314
Ion mobility/mass spectrometry, 712
Ion motion, 620, 649, 650, 656, 658, 659, 661
Ion motion equation, 806–808
Ion motion frequency, 828
Ion motion in a rectangular wave quadrupole
        field, 276

Ion motion, coherent, 471
    damping, 604
    escalating, 604
    nonlinearities impaired by, 493
    perturbation of, 491
    phase of, 604
    phase-correlation with field, 604
    types of, 68
Ion number, 640, 679
Ion optical axis, 794
Ion orbit, 664, 675
Ion packet rotation, radius of, squeezing, 261
Ion packets, 420, 421
    coherent, 256
    decay of in an EST, 257
    short, 257
Ion parking, 698
Ion photoabsorptivity, 832
Ion population, 425, 530
Ion probability density, 747
Ion processing, in-trap, 559
Ion reaction, 219
Ion rearrangement, 57
Ion reflection, regions of, 551
Ion residence time, 740
Ion scattering, 797
Ion scattering effects, 828
Ion secular frequency, *see* Secular frequency
Ion selected-CV spectra, IS-CV, 322,
        332, 353
Ion separation in FAIMS, 316
Ion separation region, 316
Ion separation, multiply-charged, 560
    role of temperature in, 335
Ion signal intensity, 635, 664
Ion simulations, 656
Ion size, 751, 758, 762
Ion soft-landing, 245
    serial RITs, 227
Ion source, 312, 779
    calutron, 117
    double-field, 87
    electrospray, *see* Electrospray ion source
    external, 472
    Nier-type, 53
    space-charge trapping in, 53
    tight, 42
Ion space charge, 850
Ion storage capacity, 374
Ion storage volumes, 393
Ion storate ring, 377
Ion structure, 827, 847
Ion structure elucidation, 674, 688
Ion target value, 774
Ion thermodynamic properties, 64
Ion trajectory, 620, 628, 631, 632, 645, 656, 657,
        748, 797

Ion trajectory in RF-only quadrupole,
    calculated, 529
Ion trajectory simulation program, *see* ITSIM
Ion trajectory, simulation of, 630, 656, 799
Ion transfer, bi-directional, 534
Ion transfer process, efficiency of, 534
Ion transfer within serial CIT array, 220
Ion transmission, 99, 579
Ion trap, 9, 683, 684, 690, 707, 708, 710, 712, 713,
    717, 718, 743, 769–75, 778–780, 782,
    783, 785, 786, 790, 827, 830, 838, 840,
    847–849, 853–859
Ion trap array resolution, 180
Ion trap array, monolithic, 198
Ion trap arrays, cylindrical, 209
    multiplexed, 207
Ion trap assembly, 626, 645
Ion trap capacity, 555, 573
Ion trap center, 628–630, 632, 636, 642, 644,
    645, 658, 660, 740, 771, 827, 851–853,
    855, 856
Ion trap chamber pressure, 785
Ion trap dimensions, 663, 679
Ion trap electrode truncation, 510
Ion trap electrodes, 778
Ion trap equipotential surfaces, 626
Ion trap geometry, 668
Ion trap mass spectrometer, 620, 665
Ion trap mass spectrometry, 106, 265, 664
Ion trap mass spectrometry, ITMS, 707, 709, 711,
    713, 718, 720, 722, 729, 736, 771, 812,
    827, 829, 830, 841
Ion trap mass spectrum, 859
Ion trap operation, 770
Ion trap operation characteristics, 710, 722
Ion trap performance, 740, 769–771, 790, 796
    improved, 102
    imulations of, 183
Ion trap scan function, 852, 857
Ion Trap Simulation Program, *see* ITSIM
Ion trap volume, 770
Ion trap, 3D, 804
    cylindrical, *see* Cylindrical ion trap,
    digital, *see* Digital ion trap,
    external, accumulation of ions in, 408
    filling of, 555
    halo, *see* Halo ion trap
    high-capacity, *see* High-capacity ion trap
    ICR, 402
    Kingdon, *see* Kingdon trap
    Knight electrostatic, *see* Knight electrostatic
        ion trap
    linear, *see* Linear ion trap
    Low-temperature co-fired ceramic,
        LTCC, 205
    micro array, 171
    miniature, ideal attributes of, 230

modified for photodissociation, 829
    quadrupole, *see* Quadrupole ion trap
    spherical, 595
    toroidal, *see* Toroidal ion trap
    various types of, 113, 265, 416, 611
Ion trap/oa-TOFMS instrument, dual, 414
Ion trap/oa-TOFMS, hybrid, 414
Ion trap/TOF technology, 414
Ion trapping, 770, 802
Ion trapping at atmsopheric pressure, 327
Ion trapping capacity, 170
Ion trapping devices, 620, 643
Ion trapping efficiency, 677, 689, 740, 769–771,
    788, 789, 798–800, 802, 852
Ion trapping in FAIMS, 327
Ion trapping parameters, 742
Ion trapping, definition of, 7, 310
    inefficiency of from continuous beam, 408
    selective, 328
    space-charge, 53
Ion traps, 671, 674
Ion traps with circular geometries, 373
Ion traps, circular geometry, 111, 148, 177
    coaxial, 393, 395
    electrodynamic, 108, 265, 611
    geometrical evolution of, 171
    micro, 230
    miniature, 107, 169, 186, 259, 375, 691
    miniature, fabrication of, 191
    multiple, 170
    multiplexed, 107, 169, 173, 259, 375, 691
    nonlinear, electrically induced, 95, 546,
        573, 595
    stereolithographic apparatus,
        SLA, 206, 207
    smaller, arrays of, 375
    soft landing, *see* Soft landing
    types of, 675, 681, 682
Ion, 'aged', 54
    damping motion of, 600
    flight time of, 55
    force acting upon, 154
    sequence, 837, 838
Ion/atom reaction, 674
Ion/electron reactions, 525
Ion/ion chemistry, 108, 265, 611, 686, 718
Ion/ion coupling, 594
Ion/ion interaction during ejection process,
        mimimization of, 597
Ion/ion interaction studies, 676, 677, 681
Ion/ion interactions, 594, 601
    reduced, 615
Ion/ion reaction, 265, 274, 684, 685, 718
    dynamics, 685
    thermochemistry, 685
Ion/ion reactions, 108, 208, 265, 525, 611
Ion/ion reactions with bi-directional flow, 534

Ion/ion reactions, suitability of spherical ion
    traps for, 614
Ion/mobility spectrometry, 311, 420, 536
Ion/molecule collisions, 15
Ion/molecule reaction, 12, 19, 40, 53, 54, 70, 75,
    77, 208, 214, 216, 219, 237, 374, 421,
    437, 525, 673, 674, 677, 681, 685, 688,
    712, 770, 772, 780, 781, 852
Ion/molecule reaction cell, 264
Ion/molecule reactions at ambient pressure, 311
Ion/molecule reactions of trapped ions, 536
Ion/molecule reactions, unwanted, 15
Ion/neutral collision cross-section, 746
Ion/neutral collision rate, 746
Ion/neutral collisions, 7, 739
Ion/neutral interactions, 408
Ion/neutral reaction, 684
Ion/photon reactions, 525
Ionanalyitics Coporation, Ottawa, ON, Canada,
    315, 347
Ion-cloud distribution, precise tailoring of, 466
Ion-detect time, 781
Ion-gas interaction, 313
Ion-induced dipole, 53
Ion-spray, 52; *see also* Ionization, electrospray
Ion-ion collisions, 404
Ionization, 708
Ionization efficiency, 19, 38, 63, 673, 848
Ionization efficiency curve, 40
Ionization energy, IE, 37, 63, 65
    second, 65, 848
Ionization process, 672
    'hard', 40
    'soft', 41
Ionization sources, array of, 172
    heated, 334
Ionization, bombardment, 43
    chemical, *see* Chemical ionization
    desorption electrospray, *see* Desorption
        electrospray ionization
    electrohydrodynamic, EHI, 51
    electron impact, *see* Electron impact
        ionization
    electron, *see* Electron ionization
    electrospray, *see* Electrospray ionization
    external, 210, 219
    field, 38
    glow discharge, *see* Glow discharge electon
        ionization
    internal, 219
    laser, 47
    methods of, 34
    Penning, *see* Penning ionization
    processes of, time-dependence of, 57
    selective, 43
    spark source, 35, 36
    spray, 49–53

    surface, 37
    thermal, 37
    thermospray, TSP, 52
Ion-lifetime studies, 472
Ion-loss mechanisms, 321
Ion-motion spectrum, axial, 513
Ion-neutral interactions, 335
Ions at atmospheric pressure, 310
Ions created externally, 740, 770, 774
Ions, charge-reduced, 538
    coherent ensemble of, 406
    collisional migration of, 116
    confined, image signal of, 477
    continuous drift of, 77
    diffusion of, 313
    doubly-charged, 27, 59
    even-electron, EE, 27, 29, 41
    excited, dissociation of, 61
    external accumulation of, 401
    highly-charged, 117
    imprisoned, 113
    maximum number of, 594
    metastable, 55
    near monoenergetic beam of, 17
    odd-electron, OE, 27, 29
    polyatomic, collisionally activated, 13
    polyatomic, metastable, 13
    prefiltering, 350
    primary, 44, 47
    relative stability of, 27
    secondary, 44, 45
    spatial separation of, 602
    trapping and processing, 574
    triply-charged, 59
Ion-separation, gas-phase, 310
Ionspray, 539, 549; *see also* Electrospray
    ionization
Ion-trap gauge, 796, 797
Ion-trapping kinetics, 328
IR, *see* Infrared
IRABA, *see* infrared absorbing boronic acid
IRAL, *see* Infrared absorbing ligand
Iridium, 35
IRMPD, *see* Infrared multiphoton dissociation
Iron, 739
Irradiation, 688
IS-CV, *see* Ion selected-CV (IS-CV) spectra
ISIS, *see* Integrated System for Ion Simulation
Islands, trapping, 125
    stable, 126
Isobaric background species, 357
Isobaric interference, 360
Isobaric ions, 362
iso-β_x line, 225
Isobutane, 32, 43
Isolation efficiency, 594
Isolation limit, 594

Isolation waveform, 776
Isoleucine, 362
Isomer differentiation, 830, 841, 842
Isomeric product ions, 711, 726, 731
Isomers, 710, 714, 717, 720, 726, 729, 732
Isotope abundances, 201
Isotope envelope, 298
Isotope patterns, 695
Isotope ratio analysis, 30
Isotope ratios, 24, 27
Isotope separation, 13
Isotope, name of, 13
Isotopes, enrichment of, 359
    separation of, 359
    stable, 19
    abundances, 16, 18, 22, 26
Isotopic cluster, 632
Isotopic resolution, 258
Isotopic substitution, 857
Isotopomers, 600
Isotropic release of translational energy, 59
Isotropic scattering, 630
Iterative simulations, 173
ITMS, Finnigan, 189
iTRAQ reporter ions, 584, 586
ITSIM (Ion trajectory simulation program),
    167, 501
ITSIM 5.0, features of, 185
ITSIM 6.0, 190
ITSIM software, 172, 184, 186, 190, 383, 748
ITSIM, development timeline of, 185

J

Jardine, I., 268
JEOL, Tokyo, Japan, time-of-flight mass
    spectrometer, 350
Jet, free, terminal velocity in, 50
Jung, G., 268

K

K, ion mobility constant, 311
K_0, reduced ion mobility, expression for 311, 312
Kanalstrahlen, 8, 44, *see also* Canalstrahlen
Katz, C., 268
Kelley, P.E., 103
Kellman, M., 269
Kent Uiversity, UK, 27
K_h/K, high-to-low field mobility, ratio of, 331
Kholomeev, A., 268
Kiloelectronvolt CID, 674
Kinetic damping, 740
Kinetic energy distribution, 742, 798
Kinetic energy release, 61, 674, 683
Kinetic energy, constant, time-of-flight mass
    spectrometer, 84

Kinetic energy, conversion of, 64
Kinetic excitation and colllisional activation, 584
Kinetic shift, 742, 761, 856
Kinetic theory of ion transport, 355
Kinetics, field ionization, 60
Kingdon equation, 124, 125
Kingdon trap, 67, 112, 113, 117, 122, 125, 127
    applications in atomic physics and
        FT-ICR, 117
    dynamic, DKT, 113, 118, 122, 123, 126
    dynamic, physical principles of, 124
    electrostatic, 675, 676
    plurality of possible embodiments, 676, 691,
        698, 699
    static, SKT, 113, 114, 116, 123
Knight electrostatic ion trap, 118, 119
Knight trap, design of, 120
Knight trap, potential distribution in, 118
Knudsen effusion cell, 849, 850
KRATOS Analytical, UK, 21, 794
Krypton, 771

L

Laboratory coordinates, 630
LabVIEW 6.0.2i, 210
LabVIEW software, 480, 494
Lactones, 28
LAMMA 1000 instrument, 49
Lammert, S.A., 376
Lange, O., 268
Langevin collision, 811
Langevin collision cross-section, 53, 186, 630
Langevin model, 747
Langmuir, I., 113
LANL, *see* Los Alamos National
    Laboratory, USA
Laplace condition, 148–150, 158
Laplace equation, 150, 161, 379, 389, 407, 462,
    501, 507, 645
Laplace solver, 185
Large hadron collider, LHC, 73
Large molecules, 351
Larmor rotation, 439
Laser, 827, 829
Laser beam, 794, 829, 830, 832, 840, 855, 856
Laser cooling, 7
Laser desorption, 793
Laser fluence, 802
Laser ionization, 47
Laser photon, 749
Laser photon flux, 828, 832
Laser polymerization, 205
Laser power, 743, 752, 753, 761, 837, 838, 853,
    858, 859
Laser pulse, 752, 757, 760, 761, 795, 799
Laser pulse duration, 752

Laser pulse width, 752, 754, 759
Laser radiation, 856
Laser source, pulsed, 256
Laser vaporization, 848
Laser, carbon dioxide, 689, 693, 754, 761, 829,
        832, 833, 848, 853, 856, 857
    cw, 827, 832
    excimer, 830, 838
    infrared, 829, 842
    pulsed, 827, 858
    pulsed dye, 858
    TEA-CO$_2$, 48
    ultraviolet, 829, 842
    visible, 829
    YAG, 830, 838, 840
Laser-induced explosion, 811
Laser-induced fluorescence, LIF, 751
Laser-machined ceramics, 202
Lavoisier, A, 8
Lawther, R., 268
LC/MS/MS, *see* Liquid chromatography/tandem
        mass spectrometry
LC-APCI-FAIMS-MS, 362
L-carvone, 222, 244
LC-ESI-FT-ICR, 411
LC-MALDI, 266
LC-MS/MS, 362
LCQ Duo electronics, 212
LDR, *see* Linear dynamic range
Lead isotopes, 13
Leak detectors, 74, 93
Leak inlet, 237, 243, 244
Leak valve, 796
LeBlanc, J.C.Y., 569
Lectin, 709
Legendre polynomial, 162, 186, 378, 406, 502,
        508, 622
Lens, endcap, 575
Lens, incap, 575
Leucine, 362
Leucine and isoleucine isomers, differentiation
        between, 535
Leucine enkephalin, 364, 690, 744, 754,
        756, 759
Level of performance, conditions for, 594
Leybold-Heraeus, Germany, 75, 94
Library, 710, 712, 714, 722, 724, 727, 730–732,
        735, 736
Library screening, 207
Library searching, 556
Lifetimes of excited precursor ions, 60
Light bulb, gas-filled incandescent, 113
Light, speed of, 71
Limit of detection, LOD, 20, 45, 237–244, 257,
        361, 413
Limit of excursion, 797
Limit, isolation, 594

    isolation space-charge, 374
    mass analysis storage, 374
    maximum storage, 374
    spectral, 594
    storage, 594
LINAC collision cell, 529
    tilted rod, 530
Linear accelerator, 93
Linear dynamic range, LDR, 238, 241, 242
Linear ion trap mass spectrometry, 573, 595, 643,
        644, 679
Linear ion trap prototype, 655, 658
Linear ion trap, 2D, 92
    AB Sciex, 110
Linear ion trap, axially-resonant excitation, *see*
        Axially resonant excitation ion linear
        ion trap, AREX LIT
Linear ion trap, digital, 302, 304
    Hitachi axially resonant excitation,
        AREX, 110
    LIT, 69, 108, 150, 171, 259, 263, 410, 545,
        573, 595, 596, 621, 643–653, 659, 660,
        664, 666, 668, 679, 693, 694, 713, 718,
        829, 838, 840
    multiple fills of, 558
    quadrupole, 108, 265, 611
    standard, first stability diagram for, 305
    ThermoFisher Scientific, 109
Linear ion traps, LIT and RIT, 375
Linear multipole Paul ion trap, 692
Linear quadrupole ion trap, 410
Linear time-of-flight mass spectrometers, 83
Linear trap array, unit cell of, 199
Linear trapping field, 653
Line-of-sight, 199, 211
Linkage disassembly, 714, 715
Linkage position, 708, 709, 714, 722, 731, 736
Linked scan, 33
Liouville's theorem, 528
Lipid solution, 226
Lipid study, 812, 813, 818, 819
Lipidomics, 813
Liquid chromatography, LC, 52, 664, 712,
        717, 785
Liquid chromatography/mass spectrometry,
        LC/MS, 717, 785, 790
Liquid chromatography/tandem mass
        spectrometry, LC/MS/MS, 525
Liquid nitrogen, 15, 854, 855, 857
Liquid secondary ion mass spectrometry, LSIMS,
        744, 745
Lissajous curve, 165
LIT, multisegment comprising four
        PCBs, 200
LIT, *see* Linear ion trap
Lithography, 192
Livingston, M. S., 72

Llewellyn, P.M., 75
LCMO, *see*, Low-mass cut-off
Local invasion, 728
LOD, *see* Limit of Detection
Logical reasoning, 733, 734, 736
Lorentzian peak shape, 491
Los Alamos National Laboratory, Los Alamos, NM, USA, 185, 190, 382
Low temperature, 699
Low temperature probe, LTP, 177
Low-energy electrons, 684, 687, 694
Lowest energy dissociation pathway, 690
Low-field seeker, 124
Low-mass cut-off, LMCO, 164, 305, 443, 556, 679, 692, 748, 795, 799, 801, 802, 805, 806
  expression for, 585
Low-pressure chemical vapor deposition, LPCVD, 192, 195
Low-temperature co-fired ceramic, LTCC, 203, 204
LPCVD, *see* Low-pressure chemical vapor deposition
LSIMS, Mass spectrometry, liquid secondary ion
LTCC CIT ring electrode, fabrication process for, 205
LTCC, *see* Low-temperature co-fired ceramic
LTP, *see* Low temperature probe
LTQ and Orbitrap mass spectrometers, parallel operation of, 263
LTQ Orbitrap hybrid mass spectrometer, 263
LTQ Orbitrap mass spectrometer, 262
LTQ Orbitrap, applications of, 264
LTQ Orbitrap, extensions of, 265
Lycopodium spores, charged, 126
Lysine, 242
Lysine residue, 834
Lysozyme, 49

M

m/z, *see* Mass/charge ratio
m/z-range, 182
m/z-range, detectable, and dissociation efficiency, trade-off between, 584
Macroions, 50
Magnet, Alnico, 35
Magnetic field generated by a solenoid, 81
Magnetic field lines, 685
Magnetic focusing of electrons, 685
Magnetic Penning trap, 675
Magnetic period mass spectrometer, 82
Magnetic sector, 673, 674
Magnetic time-of-flight mass spectrometer, 81
Magnetic traps, 676

Magnetron drift, 402
Magnetron excitation probe, 445
  duration of, 446
Magnetron mode amplitude, 436, 462
Magnetron mode, 448
  excitation of, 445
Magnetron motion, 402, 436, 676, 677
Magnetron orbit, 403
Makarov, Alexander, 128
Makarova, Anna, 269
MALDI, *see* Matrix-assisted laser desorption ionization
MALDI AXIMA QIT, 794, 796, 798, 799, 803, 812, 813, 817, 819
MALDI, high-pressure, 539, 541, 743, 744
MALDI, high vacuum, *see* Matrix-assisted laser desorption ionization, intermediate vacuum, vMALDI
MALDI, in-field, 446, 447
MALDI/TOF instrument, 794, 798, 802, 811, 813, 817
MALDI-TOF/TOF, spectrometer, 674
Malek, R., 269
Malignant cell, 727
Malignant metastatic tumor, 728
Malignant non-metastatic tumor, 728
Mannose, 813, 817
March, R.E., 518
Mars-96 program, 275
Marshall, A.G., 79, 128
MASCOT 2.1 software, 614, 615
Mascot MS/MS score, 614
Mask, 192
Mass accuracy, 267, 436, 594, 794
Mass accuracy histograms, 407
Mass accuracy, higher, 453
  reduced, 404
Mass analysis, 625, 637, 651, 673, 680, 688, 739, 762, 770, 773, 793, 798, 819
Mass analyzer, 620, 680
Mass assignment, 774, 788–790
Mass assignment, changes in, 549
Mass calibration equation, 435
Mass centroid, 632
  shift, 638
Mass correction, channel-to-channel, procedure used for, 225
Mass discrimination, minimizing, 447
Mass filter, mass-selective stability mode of operation, 176
Mass measurement, accuracy of, 412, 413
  high precision, 457
Mass peak, 632, 643
Mass placement shift, 789
Mass precision, 680
Mass range, 170, 267
  extended, 285

Mass resolution, 183, 290, 313, 413, 488, 514, 630, 636, 638, 640, 643, 645, 666, 668, 676, 677, 680, 688, 740, 762, 769–772, 774, 788, 855; *see also* Resolution
  effect of relative phase difference, 609, 610
  increase, 492
  maximum, 291
  optimum, 490
  ultrahigh, 680, 688
Mass resolving power, 267, 436, 454, 458, 464, 577, 578, 794, 798, 806, 812, 820; *see also* Resolving power
  theoretical limit, 461
Mass scan, 677, 775, 776
Mass scan rate, 788, 790
Mass scanning, 620, 621
Mass scanning, triple resonance, 640, 645, 655
Mass selection, 219, 708
Mass shift, chemical, 187, 189
Mass shift, 179, 184, 186, 188, 189, 258, 374, 381, 452, 594, 600, 640, 641, 643, 661, 662, 666, 667, 748, 771, 782, 788, 789
  apparent, 188
  chemical, 187, 189
Mass spectal distortion, 600
Mass spectra, 854
Mass spectral library, 714, 727
Mass spectral noise, reduction of, 364
Mass spectral peak, influence of ICC upon, 601
Mass spectral performance, 620, 632–642, 660–668, 743, 770
Mass spectrograph, 15, 16,
Mass spectrometer operations, combinations of, parallel/serial, 209
Mass spectrometer, 673, 679, 681, 739, 793
  Bendix time-of-flight, 87
  CIT array, 212
  digital ion trap, 273
  dual quadrupole ion trap, 498, 499
  Esquire 3000plus, 600
  Exactive, 266
  four-channel parallet RIT, 221, 222
  hand-held miniature, 176
  HCT*ultra*, 598, 600, 606, 608, 609, 614, 615
  ICR, 65, 72
  linear ion trap, 545
  LTQ-Orbitrap, 262
  magnetic period, 82
  miniature portable, 170
  monopole, 96, 99
  Orbitrap, *see* Orbitrap mass analyzer
  orthogonal acceleration time-of-flight, *see* Time-of-flight mass spectrometer, orthogonal acceleration
  QUISTOR/quadrupole, 103
  sector, BE, 55
  selected-ion flow tube, SIFT, 53, 751

  spiratron, 89
  tandem-in-space, 673
  tandem-in-time, 673
  thermal emission, 38
  TSQ-70 triple quadrupole, 377
Mass spectrometers with ion focusing, 15
Mass spectrometers, double-focusing, 19
Mass spectrometers, dynamic, 67
  classification of types of, 68
  definition of, 67
  development of, 67
  hand-held miniature, 176, 230
  static, 68
  static, definition of, 67
  types of, 67
  time-of-flight, *see* Time-of-flight mass spectrometer
Mass spectrometers, sector, study of gas-phase ion chemistry, 53
  static, definition of, 67
  time-of-flight, 80, 83, *see also* Time-of-flight, TOF, mass spectrometers
Mass spectrometric analysis, 672, 769
Mass spectrometric imaging, 819
Mass spectrometry, biomolecular structure studies, 710–719
  fragmentation methods, 672–674
  structure interpretation software, 729–736
Mass spectrometry calibration compound, 633
Mass spectrometry comes of age, 21
Mass spectrometry community, 739
Mass spectrometry, analytical applications of, 21
  analytical, development of, 21
  appreciation and historical survey of, 3, 375
  dynamic, 21
  Fourier transform ion cyclotron resonance, 78
  genesis of, 9
  inductively coupled, *see* Inductively coupled plasma mass spectrometry
  ion cyclotron resonance, 75
  liquid secondary ion, LSIMS, 46
  neutralization-reionization, 62
  organic, 30
  plasma desorption, californium-252, 44, 47
  secondary ion, SIMS, 44, 541, 672, 681
  secondary neutral, SNMS, 44
  time-of-flight, 48
  two-dimensional, 582
Mass spectrum, CID, 832–835
  effects of ion fragility, 771, 788, 789
  high resolution, 638, 641, 664
  IRMPD, 832–836
  scan function, 773
  scanning of, 18
  shift in, effect of relative phase, 607
  simulated, 748

stability of ions, 802, 803
UVPD, 840
Mass transparency, 710, 714
Mass/charge ratio, 11, 13, 14
Mass-analysed ion, 674
Mass-analyzed ion kinetic energy spectroscopy,
    MIKES, 34, 674
Mass-independence of axial potential, 574
Mass-selected ion, 771
Mass-selective axial ejection, efficiency of, 548
    MSAE, 110, 545, 549, 552, 554, 569, 573,
        595, 643
    resolution of, 553
    sensitivity of, 553
Mass-selective axial instability mode, 856
Mass-selective detection, 101
Mass-selective ejection, 101, 103, 104, 109, 252
Mass-selective instability, 165
Mass-selective instability ejection, 740, 771,
    774, 775
Mass-selective instability scan, 175, 274, 275
Mass-selective ion activation, 573
Mass-selective ion ejection, 573, 680
Mass-selective ion ejection, axial, 546
    radial, 546
Mass-selective ion isolation, 573
Mass-selective modes, 173
Mass-selective modes, non-, 173
Mass-selective operation using axial
    potential, 574
Mass-selective stage (disassembly pathway), 710
Mass-selective storage, 101, 102
Mass-selectivity stability mode, 176
Mass transparency, 710
Master equation, 856
Matching mass spectrum, 735
Mathematical models for planar and cylindrical
    FAIMS devices, 330
Mathieu equation, 98, 100, 148, 149, 155, 160,
        162, 278, 281, 374, 375, 438, 439, 620,
        623, 625, 645, 648, 651, 652, 654, 678
Mathieu stability parameters, 103, 180, 275
Mathieu trapping parameter, 719
Matrix clusters, 286
Matrix contaminants, 711
Matrix ions, 743, 794
Matrix transform method, 278
Matrix, eigenfunction of, 279
Matrix-assisted laser desorption ionization,
        MALDI, 39, 44, 48, 81, 90, 262, 285,
        293, 414, 436, 463, 500, 530, 672, 674,
        687, 698, 710, 718, 729, 740, 744, 770,
        793, 802, 819
    intermediate vacuum, vMALDI, 286, 415, 448,
        793–795, 802, 812, 813, 818, 819
    orthogonal injection, oMALDI, 541, 542
Matrix-shimmed ion trap, 458, 460

Mattauch-Herzog geometry, 19, 51
Maximum ion axial excursion, 630
Maximum number of ions, 594
Maximum storage limit, 374
MBE, see Molecular beam epitaxy
MBSA, see Molecular beam solid analysis
m_c, Critical mass, 439
McKnight, B., 268
McLafferty rearrangement, 29
McLeod gauge, 114
MCP, see Microchannel plate,
MDS Sciex API 3000 triple quadrupole mass
    spectrometer, 334
Mean-free-path, 19
Mechanical discrimination, 495
Mechanisms, fragmentation, 24
Medical diagnostics, in vivo, 230
Medical implants, surface passification for,
    bio-compatible, 228
Medium-energy electron bombardment, 682
Melittin, 542, 689
Melting point, 193
Membrane, 819
Membrane inlet mass spectrometry, MIMS, 237,
    242, 243
MEMS, see Micro-electromechanical systems
Mercury, 11
Mercury diffusion pump, 15
Mercury vapor, 114, 116
Metabolic fingerprint, 731
Metabolic networks, 265
Metabolism studies, 567
Metabolites, predicted, 568
Metal complexation, 830
Metal dication complexes, 740
Metal ion, 715, 724
Metal ion adduction, 709, 714, 723, 724
Metal ion incorporation, 711
Metal ion transfer, 686
Metal organic complex, 673
Metal-coated polymers, 190
Metallization, 196, 206, 223
Metalobolomics, 207
Metal-oxide-semiconductor field-effect
    transistor, MOFSET, 284, 285
Metastable, 13
Metastable atom-induced dissociation, MAID,
    684, 687, 698
Metastable ion, 55, 637, 674, 681, 683, 796, 802,
    811, 819
Metastable ion analysis, 30
Metastable ion decay, 32, 57, 673, 683, 692,
    802, 812
Metastable ion intensities, pressure dependence
    upon, 32
Metastable ion peaks, 22
Metastable ion peaks, energy-resolved, 57

Metastable ion suppression, 802, 803, 811
Metastable ion transition peaks, 27
Metastable ions, lifetime of, 471
Metastable peak height, variation with
    pressure, 23
Metastable reaction products, 437
Metastable, description of the adjective, 57
Metastasis, 727
Metastatic tissue, 728
Methamphetamine, 242, 245
Methamphetamine, protonated, 229, 241
Methane, 40, 43, 687
Methanol, 39
Methanol cluster series, 66
Methanol-water solvent mixture, 52, 213, 323
Methotrexate, 337, 338
Methyl salicylate, 237–240
Methylated oligosaccharides, 719
Methylation analysis, 727
Methylene chloride, 333
Metrological parameters, 482
Metropolitan Vickers Electrical Company Ltd,
    UK, 21, 22, 26, 38
Micro ion trap array, 171
Micro ion traps, future of, 230
Micro-channel plate, MCP, 117, 796, 801, 802
Micro-CIT array, 194, 195, 197
Micro-CITs, 192
Micro-electromechanical systems, MEMS, 178,
    191, 192, 201, 339
Microfabricated traps, 180
Microfabrication, 171
Microfabrication methods, 388, 389
Microfabrication technology, 112
Micromachined arrays of micro-scale CITs, 209
Micromaching, 191
Micromachining process, 350
Micromachining, bulk, 192
    surface, 192
Micromotion, 125
Micro-oscillation, 806
Microprobe mass analysis, 48
Micro-scale CIT arrays, 196
Microscopy, field ion, *see* Field ion microscropy
Microsoft.Net 2.0, 729
Microsystins, 361
Microwave heating, 718
Mid-range infrared spectroscopy, 688
Migration of ions, collisional, 116
MIKES spectrum, 57, 58
MIKES, *see* Mass-analyzed ion kinetic energy
    spectroscopy
MIMS, *see* Membrane inlet mass spectrometry
Mini 5 mass spectrometer, 237
Mini 7 mass spectrometer, 237, 238
Mini 8 mass spectrometer, 239
Mini 10 mass spectrometer, 177, 242, 243

Mini 11 mass spectrometer, 177, 243
Miniature cylindrical ion traps, 186
Minimum energy to ionize a hydrogen atom, 12
Mining Safety Appliance Company (MSA),
    Pittsburgh, USA, 315, 340, 347
Minoxidil, d_{10}-, protonated, 360
Minoxidil, protonated, 360
Minus 10% rule, 187
Misalignment tolerance, 194
Mixture composition, 731
Mobility, reduced, *see* Reduced mobility,
Mode 1, 327
Mode 2, 327
Mode I, 777–779, 783
Mode II, 777–780, 782, 785, 788
Mode N1, 327
Mode N2, 327, 351
Mode P1, 327
Mode P2, 327, 351
Modeling ion behavior and ion losses in
    FAIMS, 328
Modeling study, 746
Modeling, supercomputer, 406
Modes of operation in non-homogeneous electric
    fields, 327
Modes of oscillation, three, 435
Modes, compensenated and uncompensated,
    differences in performance, 463
Modified-angle ion trap, 630
Moehring, T., 269
MOFSET, *see* Metal-oxide-semiconductor
    field-effect transistor
Molecular beam epitaxy, MBE, 196, 197
Molecular beam ionizer, 850
Molecular beam solid analysis, MBSA, 47
Molecular connectivity, 714
Molecular disassembly, 712
Molecular dynamics simulation, 850
Molecular flow, 852
Molecular glycosylation, 708, 709
Molecular ion, 632, 637, 638, 673, 674
Molecular ions, decomposition of, 60
Molecular mass determination, 673
Molecular modification, 686
Molecular profiling applications, 581
Molecular rays, *see* Rays, molecular
Molecular structure, 722
Molecular weight, 758
Momentum transfer theory, 332
Momentum, constant, time-of-flight mass
    spectrometer, principle of, 85
Momentum-to-charge separation, 55
Mono-frequency perturbation, 512
Mono-isotopic ion selection, 817, 818
Monolithic 2D trap, chip-based, 199
Monolithic fabrication, 191
Monolithic ion trap array, 198

Monomer identification, 731
Monomer ring numbering, 714
Monomer ring product ions, 714
Monomer stereochemistry, 722
Monopole component of potential, 162, 627
Monopole mass spectrometer, *see* Mass
    spectrometer, monopole
Monte Carlo method, 746, 748, 756, 759, 856
Morgan, R.P., 127
Morphine, 331, 362, 363
Mothball, 240
Motion modulations, electric perturbation,
    induced by, 512
Motion of trapped ions, 254
Motion, axial, 181
    axial oscillations, 402
    four basic types of, 402
    radial, 181, 254
    random thermal, 404
    rotational, 254
    rotational around central axis, 402
Mouse B16 cells, 411
MRM, *see* Multiple reaction monitoring
MS, *see* Mass spectrometry
MS/MS efficiency, 598
MS/MS, *see* Tandem mass spectrometry
MS2 mass spectrometer, 22
MS^2, *see* Tandem mass spectrometry
MS^3 analysis (MS/MS/MS), 531, 560, 568, 725,
    726, 729
MS3 mass spectrometer, 22
MS30 mass spectrometer, 26
MS^4 analysis, 725, 727, 729, 731, 732, 734
MS^4 spectra, 690
MS^5 analysis, 719, 727, 729
MS5 mass spectrometer, 38
MS^6 analysis, 719, 726, 735
MS^7 analysis, 725, 733–735
MS7 mass spectrometer, 36, 51
MS8 mass spectrometer, 26, 27
MS9 mass spectrometer, 26, 27, 32
MS902 mass spectrometer, 27, 45
MSA, *see* Mining Safety Appliance Company
    (MSA), Pittsburgh, USA
MSAE, *see* Mass-selective axial ejection
MS^n, *see* Tandem mass spectrometry, multiple
    stages of
Muenster, H., 269
Muliple ionic species selection mode, 227
Multichannel devices, 365
Multi-particle simulations, stochastic
    collision, 185
Multiple charging, 52
Multiple excitation collisional activation,
    MECA, 690
Multiple fills, 264
Multiple ion monitoring, 218

Multiple ion sources, branched ion guides
    for, 539
Multiple reaction monitoring, MRM, 555, 566,
    567, 574
Multiple sinusoidal waveforms, 776, 777
Multiple steps of disassembly, 708
Multiple-proton transfer, 686
Multiplet peak spacing, 53
Multiplexed arrays, 221
Multiplexed ion trap arrays, 207
Multiplexed mass spectrometers, possible
    configurations of, 208
Multiplexing algorithm, 423
Multiplexing of ion traps, 173
Multipole expansion, electrical potential, 186
Multiply-charged atomic ion, 675
Multiply-charged ion, 674
Multiply-charged ion separation, 560
Multiply-charged positive protein ion, 686
Multiply-charged precursor ion, 684
Multiplying factor, *see* Amplification factor
Multipolar trapping field, 621, 623, 629
Multipole components, 621
    higher order, relationship with plate
        spacing, 393
Multipole expansion coefficients, 187
Multipole expansion for potential in ion trap with
    cylindrical symmetry, 378
Multipole fields, higher-order, 560
Multipole ion guide, 802
Multipole potential, diminishing, 549
Multipole rod arrays, 549
Multipole rod set, 680
Multipoles, higher-order, 390, 393, 394
Multi-pulse experiment, 785–787, 789, 790
Multi-stage ion processing, 108, 265, 611
    686, 718
MULTUM mass spectrometer, 253
MULTUM-2 mass spectrometer, 253
Murine brain astrocyte cell, 735, 736
Murine cancer model, 728
Mutual storage of cations and anions, 611
Mycotoxin zearalenone, 361
Myoglobin, 243, 408, 409, 541
Myoglobin, multiply-charged ions, 542
MySQL system, 733

N

N,N-diethyl-*m*-toluamide, DEET, 239
N,N-dimethylformamide, DMF, 46
N_2, 63, 64, 331
N_2^+, 32, 61, 64, 74
N_2^{2+}, 64
N_2O, 331
N_2O, doubly-charged state of, 61
N_2O^+, unimolecular decomposition of, 61

N-acetyl-hexosamine, 722
Nanoelectrospray, *see* Nanospray ionization
nano-ESI, 177
Nanoform 15120 polymer, 206
NanoFrontier mass spectrometer, 582
nanoLC-MS, 358
Nanospray ion source, 719
Nanospray ionization, nESI, 243, 358, 362, 436
Naphthalene, aqueous, 243
Naphthalene, 238, 240
Nascent hydroxyl formation, 714
National Bureau of Standards, Washington, DC, USA, 73
National Centre for Atmospheric Science, Leeds, UK, 128
National High Magnetic Field Laboratory, Tallahassee, USA, 128
National Insitute for Standards and Technology (NIST), Boulder, CO, USA, 128, 387
National Instruments Corporation, Austin, TX, USA, 480
National Research Council of Canada (NRC), Ottawa, ON, Canada, 315
n-Butane, 23, 32; *see also* Butane
n-Butylbenzene, 61, 240, 380, 382, 384, 637, 746–748, 754, 756, 758, 782, 788–790
n-butylbenzene ion, 188
N-C alpha bond cleavages, randomly distributed, 611
NCI, *see* Chemical ionization, negative
Negative chemical ionization, *see* Chemical ionization, negative
Negative higher-order fields, 294
Negative ion mode, 714
Negative ions, 686, 695, 696
    multiply-charged, 691
Negative protein ions, 682
NeH_2, 13
Neon, 11, 13, 15, 38, 114, 771
Neon, four lines appearing from, 16
Neon, isotopic consitution of, 16
nESI, *see* Nanospray ionization
Net fields, virtual, 328
Neurotensin, 588
Neutral atom beams, bombardment with fast, 45
Neutral loss list, 719
Neutral loss scan, 395, 555
    constant, 567
Neutralization-reionization mass spectrometry, NRMS, 62, 63
Newton's equations, 186
Next nearest isotope position, 661, 662
N-glycan, 812, 813
$n_{i,max}$, 595
Nier, A.O., 26
Nier-Johnson geometry, 19

NIST MS search tool, 731
NIST, *see* National Institute of Standards and Technology, Boulder, CO, USA
Nitrate, 323, 358, 359
Nitric oxide, 43
Nitroaromatic compounds, 190
Nitrobenzene, 14, 26, 238, 239, 266, 323, 354, 547, 548, 751, 780, 782–784, 840
Nitrogen collision gas, 582
Nitrogen fixation, 711
Nitrogen gas pulse, 464
Nitro-organic explosives, 333
NL (normalization level), 609, 610
N_{lin}, 596
N-linked glycan, 716, 720, 723, 727–729, 731, 733, 734
N-linked glycan release, 717, 730
NMR, *see* Nuclear magnetic resonance spectrometry
NO, 63
NO^+, 32, 61
Nobel Prize, Chemistry (1902), 708
    Chemistry (1908), 13
    Chemistry (1921), 13
    Chemistry (1922), 13
    Chemistry (1932), 113
    Chemistry (2002), 48, 49, 672, 708
    Physics (1904), 50
    Physics (1906), 13
    Physics (1943), 9
    Physics (1944), 9
    Physics (1989), 9, 678, 708
Noble gases, 751
Noise, 257
Noise induced by operating mode, 484
Noise induced by shape of spectral line, 482
Noise, background, 301, 491
    chemical, 536, 537
    electronic, 257, 578
    mass spectral, reduction of, 364
    imperfections of the axial harmonic potential, 578
    internal, 257
Nolting, D., 269
Nomenclature for Golgi-processed isomers, 723
n-Nonane, 217, 218, 244
Non-covalent complexes, 421
Non-destructive collection, 228
Non-destructive FT detection, 471
Non-equipotential electrodes, planar, 396
Non-ergodic behavior, 683
Non-ergodic fragmentation process, 684, 687
Non-homogeneous fields, 338
Non-ideal electrodes, 630
Non-ideal field, 620
Non-ideal quadrupole fields, 620

Nonlinear betatron oscillations, 652
Non-linear dynamics, 125
Nonlinear field, 258, 595
Nonlinear fields, even-order, 381
  ion ejection in, 598
  phase correlation with, 601
Nonlinear ion ejection, 638
Nonlinear ion trap, 619
Nonlinear ion traps, electrically-induced, 95,
  546, 573, 595, 679
Nonlinear power absorption, 630, 643
Nonlinear resonance, 101, 621, 622, 629, 635,
  644, 645, 653, 654, 657–660, 771
Nonlinear resonance line, 654
Nonlinear resonance point, 189
Nonlinear resonance, ion ejection at, 598
Nonlinear resonances in the trapping field, 621
Nonlinear resonances, coupled, 658, 659
Nonlinear trapping field, 620, 628, 630, 643
  theory, 621
Nonlinearities, ion motion, induced by
  mechanical imperfections, 507
  unexpected, 492
Nonlinearity, electrically-induced, 619,
  628, 643
Non-mass-selective modes, 173
Non-peptidic bonds, 683
Non-stretched DIT geometry, 290
Non-stretched geometry, 286
Nonyl phenoxylpolyethoxylethanol, 717
Normalization constants, 624
Normalized ion count, 667
Notch, 795
Notched waveforms, 176
N-oxide metabolite, 362
NRC, see National Research Council of Canada
  (NRC), Ottawa, ON, Canada,
NRMS, see Neutralization-reionization mass
  spectrometry
N_{sph}, 596
N-terminal branch, 681
N-terminal residue, 832, 840, 841
Nuclear and atomic physics, 672, 675, 676
Nuclear magnetic resonance, NMR,
  spectrometry, 73, 76, 672, 699
Nuclear resonance absorption, 74
Nucleic acid, 832, 842
Nucleotides, 48
Nucleus, 13
Nucleus, binding energy of, 16
Nuclide Corporation, USA, 34
Nuclides, 82
Nuclidic masses, 16
Number density of ions, expression for, 595
Number of ions detected, maximum, 497
  minimum, 497
Numbering system, 723

Numerical simulations of FAIMS devices, 330
Nyquist criterion, 479

O

O_2, 63, 331
Oak Ridge National Laboratory, 376
oaTOF analyzer, duty cycle losses in, 532
oaTOF mass spectrometer combined with AREX
  LIT, 581
oaTOF mass spectrometer, duty cycle
  enhancement of, 580
oaTOF mass spectrometer, sensitivity of, 581
Octapeptide radical cation, 683
Octopolar component, positive, 187
Octopolar field, 173, 190
Octopole component, 621, 627, 642, 645
Octopole component of potential, 162
Octopole nonlinear resonance, 621
Octopole term, 622, 629
Odd-electron intermediate species, 611
Odd-order fields, 635
OE, see Ions, odd-electron
Off-axis injection, excitation by, 256
Off-central-axis rotation, 376
Oligomer depolymerization, 713
Oligomer product ion library, 722
Oligomer topology, 710, 713
Oligonucleotide, 691, 832
Oligopeptides, 48
Oligosaccharide, 48, 812, 813, 832, 836, 837,
  840–842
Oligosaccharide structure, 708, 710
O-linked glycan release, 717, 718, 731
oMALDI, see Matrix-assisted laser desorption
  ionization, orthogonal injection
ω, axial frequency in the Orbitrap 254, 403
ω, cyclotron frequency, normalized
  unperturbed, 439
ω_-, 436
ω_+, 436
ω_D diocotron frequency, 403
ω_{ICR}, detected cyclotron frequency, 403
ω_M, magnetron frequency, 403
w_{nc}, cyclotron frequency of the critical mass,
  normalized unperturbed, 439
ω_φ, rotational motion frequency in the
  Orbitrap, 254
ω_r, 410
ω_r, radial motion frequency in the Orbitrap, 254
ω_R, ion cloud rotation frequency in the Penning
  trap, 403
Omegatron, 69, 73, 74, 93
ω_u, generalized secular frequency along the
  u-coordinate of the Paul trap, 156
ω_z, axial secular frequency in the Paul trap,
  479, 509

One-sheeted hyperboloid, 438
On-resonance CID, 690
Open cells, 456
Open cylindrical trap, 454, 455
Open cylindrical trap, compensated, 458
Open trap, 454
Open trapped-ion cell, compensated, 456
Operating line, 225, 801
Operating mode sequence, 473
Operating point, 620, 621, 623, 625, 627–632,
  636, 642, 645, 649, 653–659
Operational amplifier, linear power, 494
Operational speed, increasing, 614
Orbitap, electrostatic potential distribution
  in, 121
Orbitrap mass spectrometer, 31, 53, 67, 69, 109,
  112, 113, 117–119, 127, 190, 251, 471,
  573, 675, 676
Orbitrap mass analyzer, 718
Orbitrap mass analyzer as a stand-alone mass
  spectrometer, 266
Orbitrap mass analyzer with external RF
  storage, 259
Orbitrap mass analyzer, analytical parameters
  of, 257
  operation of, 254
  theory and practice of, 67, 112, 190, 251, 471
  combination with linear ion trap, 123
  common properties of, 257
  equation for potential within, 120
  equations for the electrode surfaces of, 120
  very high resolution, 122
Orbitron ion gauge, 116
Orbitron vacuum pump, 116
Organic compounds, industrial, 22
Organic electronics, thin film, 228
Orthogonal injection matrix-assisted laser
  ionization, oMALDI, *see* Matrix-
  assisted laser desorption ionization,
  orthogonal injection
Orthogonal-acceleration technique, 90
Orthogonal-acceleration time-of-flight
  mass spectrometer, oaTOF, *see*
  Time-of-flight mass spectrometer,
  orthogonal-acceleration
Orthorhombic cell, 404
Oscillations, axial, 254
Oscillations, harmonic, 253
Oscillator, 713, 714
Oscilloscope, 796
OSHA D air (Occupational Safety and Health
  Administration), 239
Outer electrode box, 675
Outer electrodes, split (OE-1, OE-2), 255
Out-of-phase ion trajectories, 517
Owlstone Nanotech Lonestar, 351
Owlstone Nanotech, Cambridge, UK, 347, 350

Oxoanions, 322, 359
Oxygen, 11, 26, 839, 840, 851, 854

P

PA, *see* Proton affinity
Pacific Northwest National Laboratories,
  Richland, WA, USA, 316
Packing fraction, 83
PAHs, *see* Polycyclic aromatic hydrocarbons
Pair of rod electrodes, 643–645
Palladium, 35
Parabola mass spectra, beading in, 14, 31
Parabola method, Thomson, 35
Parabolas, brightening in, 14
Parabolic curves, 11
Parabolic potential, 808
Parallel array of CIT mass analyzers,
  performance of, 215
Parallel array of improved CIT mass
  analyzers, 214
Parallel arrays of CITs of different sizes, 216
Parallel arrays of CITs of the same size, 210
Parallel arrays of RITs, 221
Parallel operation of LTQ and Orbitrap mass
  spectrometers, 263
Parallel plate electrodes, 435
Parallel plate electrodes, flat, 317, 321
Parallel RIT mass spectrometer,
  four-channel, 221
Parameter space, characterization, 661
Parametric frequency, 629
Parametric resonance, 629, 635, 636, 642
Parametric resonant excitation, 410, 574,
  628, 644
Parent ion, *see* Precursor ion
Parent ion, thread-like ensemble, 691
Paromomycin, 831, 833
Particle-in-a-box, PIB, 404, 405, 449
Paschen curve, 230, 351
Paschen's Law, 346
Path stability, 96
Paul ion trap, linear multipole, 692
Paul mode, 437
Paul trap, 117, 125, 126, 181, 186, 389, 393, 413,
  414, 421, 437, 448, 466, 678–680, 694,
  698, 699, 711
Paul trap in a magnetic field, stability diagram
  for, 438
Paul trap, shrinking the size of, 170
Paul trap, three-dimensional, 679–681, 685, 687,
  693, 694
Paul trap, two-dimensional, 679, 680, 694–696
Paul traps, combined, 688
Paul, W., 9, 72, 273, 708
Paul-type ion trap, 390
Pb^+, 118

PCB, *see* Printed circuit board
PCB materials, off-gassing of, 202
PCI, *see* Chemical ionization
PD, *see* Photodissociation
PD efficiency, enhancement of, 829
Peak amplitude, 663, 668
Peak amplitude fluctuation, 486
Peak amplitude near nonlinear resonance line, 509
Peak broadening, 59, 179, 374, 597
Peak broadening, charge-induced, 549
Peak coalescence, 186
Peak height, 740
Peak mis-assignment, 224
Peak multiplicity, 52
Peak performance, 663, 664
Peak resolution, computed, 489
Peak shape, 748
    dish-topped, 57–59
    flat-topped, 57, 58
    Gaussian, 57, 58
Peak shapes, various, 58
Peak spacing, multiplet, *see* Multiplet peak
    spacing
Peak splitting, 458, 461
    nonlinearity of the confinement field, arising
        from, 498
    resonance effects caused by, 511
    space-charged induced, 405
Peak symmetry, 632, 637, 641
Peak width, 313, 549, 597, 740
    computed, 489
PECVD, *see* Plasma-enhanced chemical vapor
    deposition,
Peeling, 718
PEG, *see* Poly(ethylene glycol)
Penicillins, 831
Penning Fourier transform mass spectometry, 18,
    75, 402, 689
Penning mode, 442, 448
Penning ionization, 43
Penning trap, 117, 126, 402, 403, 406, 433, 434,
    437, 466, 471, 685, 699
    compensated hyperbolic, 457
    hyperbolic, 457
    open-endcap, 406
Peptide, 45, 312, 314, 357, 358, 568, 584,
    666, 680–682, 690, 691, 717, 718,
    741–743, 750, 753, 754, 756, 759–761,
    802, 812–814, 832–835, 839, 842
Peptide backbone, 815, 816
Peptide backbone, direct cleavage of, 299
Peptide detection, 350
Peptide fragmentation, dynamics, 690
Peptide ion, 300, 772
Peptide ions, doubly-charged precursor, 611
Peptide mass fingerprints, 713
Peptide mixture, 673

Peptide pairs, 411
Peptide radical cation, 683
Peptide sequence, 816, 819
Peptide sequences, identification of, 581
Peptide sequencing, 718
Peptide standard mixture, 802, 803
Peptide, ETD mass spectrum of, 535
Peptide, iTRAQ reagent labeled with, 586
Peptide, sulfide-bridged, 638–640, 642, 664
Peptides and proteins, mass spectrometry of, 300,
    416, 611
Peptides, 2D mass spectral analysis of, 582
Peptides, *de novo* sequencing of, 264
Peptides, phosphorylated, 567
Peptides, tryptic, 355, 413
Peptidic bond, 681, 683
Percentages of TFD, 633
Perchlorate ion, 314
Perchlorate ions in water, 360
Perfluorotributylamine, PFTBA; FC-43, 183, 188,
    189, 237, 240, 243, 633, 635–637, 780,
    785, 788
Period of RF confinement, 478
Periodic inhomogeneous electric field, 97
Periodic table, 13, 848
Periodic temporal signal, Fourier transformation
    of, 486
Permittivity of free space, 747
Perturbation method, 508
Perturbation of ion motion, 491
Perturbation voltage, 512
Perturbation, electric, effect on motion
    modulations, 512
Perturbations, electrical, 491
Perturbations, mechanical, 492
Pesch, R., 268
Petroleum fractions, 80
Petroleum industry, 672
PFTBA, *see* Perfluorotributylamine
PGD2, prostanoid ions, 362
PGE2, prostanoid ions, 362
Phase angle, 661–665
Phase map, 663
Phase relationship, 633
Phase space trajectory plots, 503, 504
Phase space, stable islands in, 126
Phase stability, 93
Phase, relative, for dipolar excitation, 598
    relative, of excitation voltage, 605
Phase-correlation, strict control of, 607
Phase-locking, 406
Phase-modulation, 406
    avoidance of, 514
    supplementary, 513
Phase-shift, 624
Phase-shifted harmonic, 344
Phase-space volume, 528

Phosgene, 241
Phosphatidylglycerol, 226
Phosphoethanolamine, 226
Phosphorus-doped nickel adhesion layer, 204
Phosphorylated peptide, 832
Phosphorylation, 264, 265, 683
Phosphotyrosine, 568
Photoactivation, 832, 840, 841
Photochemical studies, 740, 847
Photodissociation efficiency, 693
Photodissociation reactions, 526
Photodissociation, PD, 628, 690, 743, 832, 841, 847, 853
Photodissociation, single photon, 855
Photoexcitation, 757
Photofragment, 852
Photofragment yield, 853, 859
Photofragmentation spectrum, 857
Photographic recording, 25
Photoionization, PI, 41, 65
Photolithographic patterning, 193
Photolithography, 192, 194, 199, 202
Photon, 750
Photon absorption, 682, 743
Photon bombardment, 44
Photon fluence, 743
Photons, low energy, 848
Photopolymerization, 206
Photoresist, 192, 203, 204
Photo-tomography, 597
PHP language, 733
Phthalic acid, *ortho-*, *meta-*, and *para-* isomers of, 331
Physics of the high-capacity trap, HCT, 718
Physiological modification, 727
PI, *see* Photoionization,
PIB, *see* Particle-in-a-box
Pick-up technique, 848, 849
Pig hemoglobin, 357
Pirandello, L., 736
Planar electrodes, 338, 339, 394
Planar quadrupole ion trap, 393, 394
Planck's constant, 745
Planetary atmosphere, 685
Plasma, 362, 685
Plasma desorption mass spectrometry, californium-252, 47
Plasma lipid extract, 817
Plasma, high-energy/low-density, 196
Plasma-enhanced chemical vapor deposition, PECVD, 193
Plass, W., 382
Plate spacing, relationship with higher-order multipole components, 393
Plates, detect, 435
    excitation, 453
    receiver, 79, 435

    transmitter, 79, 435
    trap, 453
Plating, electroless *see* Electroless plating
Platinum, 35
Pluripotency, 728
Pluripotent stem cell, 729
p-Nitrotoluene, 237
Poisson distribution, 480
Poisson software, 190, 382
Poisson/Superfish, 185
Polarity of waveform, 326
Polarity switching, 267
Polarization electrodes, 339
Poly(ethylene glycol), PEG, 48, 52, 364, 802
Polycarbonate window, 221
Polychlorinated dibenzofurans, 517
Polychlorinated dibenzo-*p*-dioxins, 517
Polycyclic aromatic hydrocarbons (PAHs), 218
Polyimide, copper-coated, 200
Polymer study, 812, 819
Polymers, 39, 673
Polypropylene glycol ammonium adduct, 550
Polypropylene glycols, mixture of, 561
Polysilicon, doped, 196
Polyurethane pad, 194
Population, time evolution, 856
Positive chemical ionization, PCI, *see* Chemical ionization
Positive ion mode, 714
Positive ions, emission of from hot surfaces, 37
Positive rays, *see* Rays positive
Post-injection, 773–776
Post-ionization, 48
Post-source decay, PSD, 718
Post-translational modification, PTM, 284, 411, 588, 681, 683, 684, 708, 812, 838
Post-translation modification, unambiguous localization of, 612
Potential barrier, 655
Potential configurations, deceleration, 502
Potential distribution, 118
Potential distribution inside CIT, 187
Potential energy surface, 745
Potential energy surface, crossing point, 686
Potential functions, radial, 54
Potential gradient, 680
Potential surface for a quadrupole ion trap, pure quadrupole, 166
Potential well, axial, 411
Potential well, 255, 692
Potential, decapole component of, 162
    dipole component of, 162
    dodecapole component of, 162
    effective, 125, 181; *see also* Pseudopotential
    hexapole component of, 162
    higher-order terms of, 161
    multipole expansion for, 378

octopole component of, 162
pseudo-, *see* Pseudopotential
quadrupolar, 277
radial, logarithmic dependence of, 115
Potentials, compensation, 199
Potential distribution, quadro-logarithmic, 254
Potholes, 708
Power, 180
Power absorption by materials of fabrication, 200
Power dissipation, contribution from different
        components, 181
    general formula for, 180
Power mass spectrum, experimental, 490
    simulated, 496
Power series expansion, 459
Power series for dependence of mobility on
        electric field, 313
Power spectrum, evolution of, simulation results,
        513, 515, 516
Precessional frequency, nuclear resonance, 73
Precise mass measurement, 74
Precursor ion (Parent ion), 77, 417, 528, 583, 673,
        674, 684, 685, 688, 691–693, 695, 696,
        711, 715, 719, 721, 724–727, 731, 733,
        734, 742, 745, 749, 751, 756, 762, 763,
        771, 772, 775, 795, 803, 817, 820, 828,
        831, 841, 842
Precursor ion isolation, 298
    high mass resolution for, 298
Precursor ion scan, 33, 555, 567
Precursor ion, activated, 749
    collisionally robust, 556
    first-generation, 560
    fragmentation of, 55
    second-generation, 560
Precursor ions, different time slices of, 563
    highly excited, 563
    lifetimes of excited, 60
    lower energy, 563
Precursor-ion selection, 453
Precursor-product relationship, 713
Pre-filter, RF-only, 99
Preimplantation mouse embryo, 728
Pre-injection, 773–776
Pre-Raphaelites, 63
Pre-resonant conditions, 597
Pre-scan, 566
Pressure effects, 788
Pressure gauge, 780, 790
Pressure profile, 796, 797, 799, 780
Pressure regulator valve, 777–779, 781, 783
Pressure tailoring, 102, 740, 769, 773–776, 780,
        782, 785, 789, 790, 796
Pressure transient, 796–798, 802, 803, 819
Pressure, background, 434
Pressure, FAIMS device, reduced, 343
Pressure, residual, time between collisions, 258

Pressure-dependence, second-order, 83
Pressure-focusing, 441
Primary amine, 834
Printed circuit board, PCB, 196, 201, 223
Printed circuit board processes, 200
Prior, M.H., 117
Probability density distribution, 748
Product ion (Fragment ion), 33, 528, 534, 583,
        632, 637, 638, 673, 683, 692, 695, 710,
        711, 714–717, 719, 720, 722–727, 731,
        734, 741, 742, 745, 748, 751, 762, 771,
        772, 788, 795, 798, 803, 815–817, 832,
        841, 854
Product ion analysis, 219
Product ion ejection, 680
Product ion mass spectrum, 674, 679, 682–684,
        686, 687, 690–692, 696, 711, 714, 715,
        717, 725, 727, 731, 740–742, 745, 751,
        762, 813, 817–819, 833
    composite, 556
    ECD, 684
    virtual, 698
Product ion scan, summed, 227
Product ion spectrum measurements,
        automated, 698
Product ion, carbohydrate, 731
    first-generation, 532
Product ions, lack of trapping or detection of, 585
    mass-selective isolation of, 531
    reversing flow of, 531
    time-resolution of, 563
Products, metastable reaction, 437
Projectile ions, 691
Propane, 19, 29
Propene radical cation, 444
n-Propylamine, non-covalent complexes of, 421
Propylene, 29
Protein, 680–682, 688, 708, 717, 718, 819, 832, 840
Protein activity, 812
Protein analysis, 348, 684, 695
Protein chip production, 228
Protein confomers, separation of, 529
Protein conformations, gas-phase, 343
Protein data bank, 695
Protein digestion, enzymes, 680
Protein expression, 708
Protein folding, 352, 812
Protein fragmentation, nomenclature, 680, 681
Protein interactions, 819
Protein ions, 671, 686, 687, 691, 693
    fragmentation, 113, 265, 416, 611, 674, 681
    negative, 685, 694
Protein Kinase C, tryptic digest of, 612
Protein mass spectra, 176
Protein mixtures, 287
Protein structural characterization, 611
Protein structures, gas-phase, 352

Protein–ligand complex, 690
Proteins, 228, 258, 264, 312, 314, 353, 407, 568
Proteins, large, 49
Proteins, tryptic digest of, 411
ProteinScape 1.3 platform, 615
Proteolytic cleavage, 718
Proteolytic digests, 423
Proteome, *Shewanella oneidensis*, 419
Proteomic analyses, 560
Proteomic studies, top-down, 411
Proteomics studies, 207, 358, 567, 587, 610, 812, 819
Proton abstractors, 538
Proton affinity, PA, 65, 66, 536
Proton transfer, 686
Proton transfer dissociation, PTD, 686
Proton transfer reactions, 536, 537
Proton, name of, 14
Protonated molecule, 673, 710
Protonation, enthalpy change for, 66
Proton-bound dimer, 214, 693
Proton-hopping, 686
Protons, 70
Proton-transfer reaction, 41, 65
Proton-transfer reaction instrument, 92
Proust's Law of Definite Proportions, 8
Pseudo-first order rate constant, 750, 754, 761
Pseudoforce, 678, 679
Pseudo-harmonic ion motion, 620
Pseudopotential, 125, 181, 411, 417, 418, 677, 678
Pseudopotential field, 530
Pseudopotential function, 804, 808
Pseudopotential model, 526
Pseudopotential theory, 804, 820
Pseudopotential well depth, 181, 195, 206
Pseudopotential well depth in the z-direction, 804, 805
Pseudopotential well depth, radial, 557, 558
Pseudopotential well in 2D RF quadrupole, expression for, 528
Pseudopotential well in $2n$-multipole, expression for, 527
Pseudopotential well, 7, 279, 282, 283, 409, 410, 442, 526, 527, 595, 611, 678, 799
Pseudopotential, quadrupole field formed by, 585
Pseudo-tandem mass analyzer, 395
PTR, *see* Proton-transfer reaction
Pulse trigger, 786
Pulsed instrument operation, 677
Pulsed laser source, 256
Pulsed valve, 770, 772–778, 780–783, 786–789, 797, 799
Pulsed valve, three-way, 778, 779, 782, 783
Pulsing technique, 581
Pump, Gaede rotary mercury piston, 11
Pumping speed, 774
Pure gases various, used for transport through FAIMS, 331

Pure ion axial motion, 620
Pure ion radial motion, 620
Pure quadrupolar trapping field, 620
Pure quadrupole field, 804
Pusher region, 91
Pye, W.G. and Co., UK, 16
Pyramid mirror, 794, 795, 799
Pyran, 710
Pyrex wafers, 339
Pyridine, 239, 852–855, 857–859
Pyridine, protonated, 444
Pyridine-d_5, 857
Pyridine radical anion, 443
Pyrolysis, 710
Python system, 733

Q

q, 125, 182, 275, 585
Q, 440
q_{eject}-value, 189, 552, 553
QIT, *see* Quadrupole ion trap
QIT using FT techniques for non-destructive detection, 471
QIT with planar geometry, 201
QIT, potential inside, 474
QITMS, *see* Quadrupole ion trap mass spectrometer
QIT, structure of, 156
QIT/TOF MS, 794, 819
q_M, Mathieu parameter, 528
QMF, *see* Quadrupole mass filter
QMF, structure of, 152
Q-q-Q/LIT, 555, 556, 567, 568
QqTOF mass spectrometer, 586
q_r, 161, 181, 410, 474, 620
QStar Elite quadrupole-TOF mass spectrometer, 539
QTRAP instrument, 555, 559
q_u, 149, 374, 801, 802
Quadrant electrometer, *see* Electrometer, quadrant
Quadratic potential well, 378
Quadro-logarithmic potential distribution, 120, 254
Quadrupole ion trap mass spectrometer, 102
Quadrupolar (parametric) resonance excitation, 549, 550
Quadrupolar axialization, 438
Quadrupolar component of DC trapping potential, 438
Quadrupolar devices, 149
Quadrupolar electric field, 655, 676, 678, 740
Quadrupolar excitation, 283, 409, 410 supplementary, 514–517
Quadrupolar potential, 148, 153, 158, 277
Quadrupolar potential distribution, ideal, 97

Quadrupolar resonant excitation, 628, 635, 637, 644
Quadrupolar restoring field, 643
Quadrupolar trapping field, 620
Quadrupolar voltage-ramped, 633–635
Quadrupolar well, 457
Quadrupolar, definition of the term, 95, 149
Quadrupolarity, azimuthal, 459
Quadrupole amplitude, 661–663, 668
Quadrupole beam bender, 417
Quadrupole collision cell, 547
Quadrupole deflector, 850
Quadrupole devices, radiofrequency, 92
    theory of, 148
Quadrupole excitation field, 636
Quadrupole FAIMS apparatus, 343
Quadrupole field, 620, 629, 711, 803, 804, 806
Quadrupole instruments, 673
Quadrupole ion guide, 850
Quadrupole ion store, QUISTOR, 100, 101, 440, 829
Quadrupole ion trap mass spectrometer, QITMS, 102, 739–742, 744, 746, 748–750, 753, 754, 760, 762, 770, 771, 773, 796, 828, 832, 833; *see also* Quadrupole ion trap, QIT
Quadrupole ion trap theory, and alternative approach to, 161
Quadrupole ion trap, QIT, 43, 173, 177, 184, 273, 276, 410, 595, 628, 637–641, 645, 678, 679, 696, 709, 711, 713, 718, 722. 739, 741, 769, 770, 772–776, 778–780, 782, 783, 788, 790, 793–799, 801–805, 811, 819, 820, 827, 829, 833, 838, 847, 848, 850, 851; *see also* Quadrupole ion trap mass spectrometer, QITMS
    electrode surfaces, 158
    Fourier transform mode of operating, 105
    mass spectrometry with, 101
    new FT operating mode for, 472
    planar, 393
    stability diagram for, 104
    structure of, 156
    tandem mass spectrometry with, 106
    three-dimensional, 75, 92, 100, 105, 156, 448, 469, 573
    unstretched, 276
Quadrupole ion trap/TOF mass spectrometer, 286, 415, 448, 793
Quadrupole linear ion trap, 686, 692
Quadrupole mass filter, QMF, 53, 55, 67–69, 92, 96, 100, 108, 149–151, 156, 173, 273, 498, 620, 645, 679, 693, 711, 770, 848–851
    comparison of FAIMS with, 320
    electrode arrangement of, 97

mass-selective stability mode of operation, 176
    resolution of, 99
    round rods, 152
    standard transmission type, 555
    structure of, 152
Quadrupole mass spectrometer, 678
Quadrupole phase, 662–665, 668
Quadrupole rod set, 645, 679
Quadrupole term, 622, 652
Quadrupole trapping region, linear fields for, 381
Quadrupole traps, traditional, 126
Quadrupole ion trap systems, 530
Quadrupole/time-of-flight systems, 530
Quantum computing, 199, 230
Quantum mechanics, 126
Quantum processing, 199
Quasi-continuous beam, 416
Quasi-Equilibrium Theory, QET, 682, 713
Quasi-plasma, 850
Quenching, 79
Query mass spectrum, 735
QUISTOR mode, 498
QUISTOR/quadrupole mass spectrometer, 103
q-value, 548, 596
q-value, ejection (q_{eject}) , 189, 552, 553
q_x, 156, 304, 804
q_y, 156, 304, 804
q_z, 103, 104, 161, 181, 274, 276, 374, 474, 620, 719, 742, 772, 774, 795, 804

R

R, annular radius of curvature for toroidal trap, 376
r_0, 151, 156, 181, 274, 374, 473, 527, 804
Racetrack, 377
Racetrack configuration ion trap, 679
Radial amplitude excitation, mass-selectivity of, 551
Radial and axial degrees of freedom, coupling of, 560
Radial and axial ion motion coupling, 622
Radial diffusion, 408
Radial electric field, anharmonicity in, 461
Radial excursions, 560
Radial frequencies, 167, 261
Radial motion, 254
Radial potential functions, 54
Radial secular frequency measurement, 495
Radial separation, 411
Radial space-charge field, 530
Radial stability, 103
Radial stratification, 411
Radially diffused ion cloud, 445
Radially dispersed ion clouds, collisional focusing of, 445

Radiationless transitions, 28
Radiative emission, rate of, 749, 753, 754
Radiative relaxation, 751
Radical anions, 693, 694
Radical localization, 28
Radioactive foils, 311, 316
Radiofrequency alternating voltage, 644, 646
Radiofrequency devices, 92
Radiofrequency field, 678
Radiofrequency ion acceleration
        spectrometers, 94
Radiofrequency ion guide, *see* Ion guides,
        radiofrequency
Radiofrequency Paul trap, 675, 678–680, 685,
        686, 691, 692, 694, 696, 697
Radiofrequency potential, 679
Radiofrequency quadrupolar potentials,
        instruments employing, 95
Radiofrequency quadrupole devices, historical
        development of, 95
Radiofrequency quadrupole devices, theory of, 148
Radiofrequency-ion mobility spectrometry,
        RF-IMS, 316
Radio frequency-only-mode event, 18, 75, 402,
        433, 436, 437
Radius of curvature, annular, R, 376
Rain-drop, 50
Ramsey, N., 9, 708
Random distribution of energy, 683
Random number, 748
Raphus cucullatus, 67
Rare earths, 35
Rate of increase of radial amplitude,
        exponential, 549
Rate of increase of radial amplitude, linear, 549
Rate-determining step, 686
Rayleigh, Third Baron, 50, 51
Rays, canal, 11
    cathode, 10
    molecular, 8
    positive, 11, 13
RCA Laboratories, USA, 45
Re(A), 440
Reaction chamber, 772
Reaction channel, 742
Reaction coordinate, 745
Reaction cross-section, 697
Reaction kinetics, 741
Reaction probability, 695
Reaction rate constant, 745
Reaction region, 77
Reaction time, 741
Reactive fragmentation, 682
Reagent gas, 772
Rearrangement ions, 28, 29
Rearrangement process, 27, 28
Receive rods, 453, 454

Receiver plates, 79, 435
Recoding, electrical, 36
Recoiling fragments, 47
Recombinant scFv protein, monoclonal antibody
        Se 155–4, 690
Recording, photographic, 36
Rectangular asymmetric waveform, 317
Rectangular hyperbolae, 151
Rectangular voltage, 279
Rectangular wave, 277, 318
Rectangular wave cycles, 281
Rectangular wave quadrupole field, ion motion
        in, 276
Rectangular waveform, 107, 148, 273, 275,
        344, 346
Rectilinear ion trap arrays, 221
Rectilinear ion trap, RIT, 171, 175, 179, 190, 228,
        240, 242, 375
Redhead mass spectrometer, 69, 93, 94
Reduce mobility, equation for, 312
Reduced helium pressure, 744
Reduced mass, 313, 747
Reduced mobility, 311
Reduced pressure FAIMS device, 343
Re-excitation, 191, 464
Reflecton technology, 91
Reflector, two-section, 90
Reflectron field, 795
Reflectron ion mirror, 81, 414, 796, 799, 801,
        805, 813
Reflectron, *see* Time-of-flight mass spectrometer,
        reflectron
Regions, non-field-free, 55
Reionization, 63
Relative ion abundance, 633
Relative kinetic energy, 742, 746–748, 755, 757
Relaxation rate constant, 761
Reliability, choice of phase, 608
Re-phasing of ions, 123
Reproducibility, 680
Reserpine, 333, 548
Reserpine ions, 575, 578
Reserpine, protonated, 553, 600
Reservoir, copper, 855
Residence time, 183, 320, 529, 797
Residence time of ions outside trap center, 602
Residual gas analyzer, RGA, 68, 74, 75, 93
Residual gas molecules, 675, 682
Resistively coated rods, 529
Resistivity, 193
Resolution limit, 510
Resolution; *see also* Mass resolution
    energy, 86
    enhanced, 471
    frequency, 486
    guaranteed, 27
    high, 24

high mass, 19
  increasing, 490
  ion trap array, 180
  space, 86
  spectral, 594
Resolving power, 609; *see also* Mass
      resolving power
  higher, 80
  mass, 267
  maximum for the Orbitrap, 268
  optimal, achievement of, 606
Resonance activation, 712
Resonance condition, coupling, 668
  triple, 630
Resonance coupling, 661
Resonance ejection, 183, 189, 289, 627
  axial modulation, 380
Resonance excitation, 678, 687, 692, 697, 719
  auxiliary, 546
  dipolar, 549, 550
  fixed-frequency, 292
  signal, 550
  quadrupolar (parametric), 549, 550
Resonance ion ejection, triple, 629, 632, 634,
      636–643, 645, 655, 660, 661, 664–668
Resonance mass scanning, triple, 640, 645, 655
Resonance phenomena arising from
      coupling, 507
Resonance voltage amplitude, 771
Resonance, nonlinear, *see* Nonlinear resonance
Resonant activation events, two distinct, 603
Resonant dipole field, 628
Resonant ejection, 288, 597, 695, 771, 774
Resonant ejection scan, 391
Resonant energy absorption, 74
Resonant excitation, 106, 253, 741, 742, 760, 762,
      771, 795, 796, 811
Resonant ion selection, prevention of, 594
Resonant quadrupolar excitation of ions, 512
Restoring force, 628, 643
Retarding electric field, 448
Retro-Diels-Alder, RDA, rearrangement, 715
Reverse activation energy, 59, 742
Reverse genomic problem, 708
Reverse scan direction, 189
Reverse scanning, 291, 294, 297
Reverse transcription-polymerase chain reaction,
      RT-PCR, 728
Reversed geometry, *see* Geometry, reversed
Reversing ion flow, 531
RF amplitude, 774, 775, 788
RF barriers, 525
RF drive circuit, 829
RF drive frequency, 657
RF drive potential, 771
RF field, 770, 771, 774
RF frequency, 806, 807

RF fringe field, 799
RF fringing-fields, 549
RF ion guides, charge capacity of, 530
  high-pressure applications of, 526
  low-pressure applications of, 526
  trapping and processing of ions in, 525
RF ion trap, 803
RF ion trap, new type, 680
RF level, 773, 775
RF multipole ion guides, 539
RF power modulation, 444
RF storage device, radial ejection from, 261
RF trapping field, 628, 629, 795, 799, 808,
      811, 820
RF trapping voltage, 631, 659
RF voltage, 626, 657, 660, 679, 689, 694, 775,
      776, 781, 795, 798, 804–806
  auxiliary, 105
  rapid curtailment, 692
RF/DC (apex) isolation, 216
RF-blade, 200
RF-IMS, *see* Radiofrequency-ion mobility
      spectrometry
RF-only ion guide, 227
RF-only mode, 99
RF-only multipoles, 266
RF-only-mode-event, 437, 446, 688, 689
RF-only-mode-event, potential analytical
      applications of, 447
RF-phase relation, 740
RGA, *see* Residual gas analyzer
Rhizobial signaling molecule, 711
Rice–Ramsperger–Kassel–Marcus theory, 682,
      741, 745, 748, 751; *see also* RRKM
Rickard, Andrew R., 128
Ring electrode, 100, 106, 111, 156, 473, 621, 623,
      626, 629, 640, 676, 679, 795, 796, 804,
      829, 850, 856
  gold-plated ceramic, 202, 203
  r_0, radius, 623, 640
Ring electrodes, concentric, 201
  lengthy parallel series, 680
Ring guide, 527
Ring product ion, 721
Rings, concentric metal, 389
RIT, *see* Rectilinear ion trap
RIT analyzers, four-channel multiplexed, 224
RIT array, 12-channel, 223
  polymer, 244
RIT arrays, 175, 205
RIT geometry, optimized, 175
RIT, 4-channel, multiplexed, 245
  four-electrode, 190
  interelectrode-stretch in, 175
RITs, advantages of, 221
  miniature, constructed using SLA, 222
  parallel arrays of, 221

reduced scale, 205
   serial and ion soft-landing, 227
   serially coupled, 245
   stretched, 223
RK4, *see* Runge-Kutta fourth-order method
RMU-7 mass spectrometer, 34
Roadmap, 708
Roberts, T.R., 26
Rock, S.M., 28
Rod electrodes, 679
Rod electrodes with resistive coating, 574
Rod electrodes, circular, 152
Rod electrodes, hyperbolic, 152
Rod electrodes, segmented, 574
Rod offset voltage, scanning of, 528
Rod set axis, 644
Rods, circular cross-sections, 546
   hyperbolic, 546
   receive, 453, 454
   resistively coated, 529
   segmented, 529
   tilted, 529
Room temperature., 743, 744, 746,
      747, 858
Rotation frequency, ion cloud, 403
Rotational frequencies, 261
Rotational motion, 254
Rotational motion around a central axis, 402
Rotational symmetry, 622
Royal Institution of Great Britain, UK, 128
RRKM calculation, statistical, 683
RRKM kinetics, 184
Runge-Kutta fourth-order method, RK4, 185,
      186, 482, 483, 508, 513
Rutherford, E., 13, 14
Rydberg electron attachment, 472
Rydberg states, 472

S

S/N, *see* Signal-to-noise ratio
Sample molecules, 770
Saturn 2000 electrodes, 626
Scaling consideration, 178
Scaling laws, 191
Scan direction, 295
Scan function, 294, 770, 772–776, 781, 785,
      788–790
Scan line, 104
Scan mode, 640
Scan performance, 638
Scan rate, decrease of, 597
Scan rate, 549, 577, 598–600, 609, 638–642,
      664, 666; *see also* Scan speed
Scan speed, 267, 291, 293, 594, 664, 668; *see also*
      Scan rate
Scan time, 782

Scan, neutral loss, 555
   precursor ion, 555
Scanning electron microscopy, SEM, 194
Scanning, forward, 291
   reverse, 291
   voltage amplitude-, 492
Scans, criteria-dependent, 567, 568
   dependent, 567, 568
   neutral-loss, 395
   survey, 567
Scar, 712, 714, 717, 722, 733–735
Scattering effects, 739
Scattering gas, 32
Scattering, backward, 59
   forward, 59
Schroeder, E., 269
Schwartz, J.C., 268
Scientific Instrument Services, Ringoes, NJ,
      USA, 230, 391
Scigelova, M., 269
Scintillation detector, 14
Score function, 661, 662, 664
Screened trap, 453
Searchable library, 710, 712, 730
Second analyzer, EB, 711
Second ionization energies, 65
Second, T., 268
Secondary electron multiplier, 680
Secondary electrons, emission of, 12
Secondary ion mass spectrometry, SIMS, *see*
      Mass spectrometry, secondary ion
Secondary tumor formation, 728
Secular axial frequency information, 478
Secular axial motion, 479
   approximate solution of equation for, 508
Secular frequencies, axial, difference between, 489
   variation of across device, 596
Secular frequency, 7, 156, 165, 167, 281–283,
      292, 297, 410, 509, 604, 624, 649, 655,
      664, 680, 692, 795, 827, 828
Secular frequency dispersion, 553
Secular frequency measurement, axial, 475
Secular frequency peaks, amplitude of evolution
      of, 509
Secular frequency shift, 294
Secular frequency spectrum, axial, 476, 477, 478
Secular frequency, axial, 472, 477, 479
   changes in with radial displacment, 560
   dependence on maximum amplitude of ion
      motion, 507
   fundamental, 771
   radial, measurement of, 495
Secular ion motion, axial, 471
   radial, 471
Secular motion, 279, 281, 283, 806, 808
   higher harmonic components of, 283
   oscillatory, 410

Secular oscillation, 678
Segmented electrodes, 680
Segmented FAIMS device, 343
Segmented quadrupole collision cell as drift or
    mobility cell, 529
Segmented rods, 529
Seismographic survey, 27
Selected-ion flow tube, SIFT, mass spectrometer,
    see Mass spectrometer, selected ion
    flow tube
Selected-ion monitoring, SIM, 574
Selective external accumulation, 411
Selective ion trapping, 328
Selectivity, 207
Selectra, 339
Self-bunching, *see* Synchronization
SEM, *see* scanning electron microscopy
Semiconductor processes, 192
Semiconductor substrate, 196
Senko, M.W., 268
Sensitivity, 207, 313, 401, 664, 680, 710, 711, 722,
    740, 743, 762, 770–772, 774
  increasing, 355
  reduction in, 579
Sensor preparation, surface modification for, 228
Separation FAIMS (sFAIMS), *see* FAIMS,
    separation
Sequenator for the glycoproteome, 708
Sequence strategies, 708
Sequencing algorithm, de novo, 833
Serial array, RITs, 227
    CITs, 219, 220
Serial CIT array, ion transfer within, 220
Serial RITs and ion soft-landing, 227
Serne, Natasha, 128
SF_5^+, 476, 486–488, 505, 506, 511, 514
SF_6, 331, 488, 497
sFAIMS, *see* FAIMS, separation
Shannon criterion, 479, 490
Shaped electrodes, 621
Shell Research Ltd, 21, 22
Shewanella oneidensis proteome, 419
Shimadzu Corporation, Japan, 21, 48
Shimadzu quadupole ion trap, 286
Shimming electrodes, 454
Shock-waves, 81
Shutoff valve, 779, 783
Shutter grid, 312, 313
Si, 196
Si_3N_4, 195
Sialic acid, SA, residue, 812, 813
SID, *see* Surface-induced dissociation
Side-bands, 630, 631, 633, 636, 642, 656, 657, 684
Side-chain losses, 536
Side-kick deflection, 408
SIFT, *see* Mass spectrometer, selected ion
    flow tube,

Signal averaging, 711, 717, 719
Signal intensity, 774, 775, 786, 788–790
Signal processing, 475
Signal pulse, rectangular shape, expression
    for, 478
Signal sampling, 478
Signal-to-noise ratio, *S/N*, 183, 214, 223, 405,
    409, 411, 419, 423, 475, 482, 514,
    581, 641
  estimated from simulated spectra, 485
SILAC, *see* Stable isotope labeling using amino
    acids in cell culture,
Silica-on-silicon processes, 199
Silicon, 194
Silicon wafer, 192, 195
Silver, 45
Silver epoxy, 196
Silver ink dipping, 206
Silver ink spraying, 206
Silver spray-paint, 223
Silver target, 45
Silver(II), 848
Silver/germanium bond pads, 197
SIM, *see* Selected ion monitoring
SIMION, 501
SIMION 6.0, 184, 453, 630, 631, 656, 657
SIMION 7.0, 185, 300, 391, 392, 579, 796, 797
SIMION 8.0, 230
SIMION calculation, 455, 626
SIMION equipotential plots, 118, 119
SIMION modeling, 120
SIMION simulation, 633, 634, 658–660
Simple harmonic motion, axial (*z*-mode),
    436, 448
SIMS, *see* Mass spectrometry, secondary ion
SIMS, dynamic, 44
SIMS, static, 44, 46
Simulated mass range, 801
Simulated product ion mass spectrum, 732
Simulated profile, 777
Simulation, 480, 799, 802, 805, 811
Simulation and experimental results, comparison
    of, 505
Simulation process, schematic representation
    of, 481
Simulation snapshot, 800
Simulation studies, 740, 746, 748, 755, 757, 758
Simulations, 183, 184
  iterative, 173
  numerical, of FAIMS devices, 330
Simultaneous ion trapping, 678
Simultaneous storage, positive and negative
    ions, 685
SiN, 194
Single ion mass spectrometry, 31
Single topological array, 713
Single-charge mass spectrum, 697

Single-step excitation monolayers, 690
Singly-charged glycomers, 721
Singly-charged ion, 761
Singly-charged protein ion, 674
Sinusoidal function, 278
Sinusoidal RF trap, 283
Sinusoidal waveform, 275, 344
Sinusoidal waveform, clipped, 350
SiO_2, 194–196, 199
SiO_2 insulators, 196
Sionex Corporation, Cambridge, MA, USA, 315, 339, 347, 350
Six Characters in Search of an Author, 736
Skimmer voltage, 531
Skin depth, 193
SKT, *see* Kingdon trap, static,
SL, *see* Soft-landing
SL methods, 228
SLA, Stereolithography apparatus, 179
    using, construction miniature RITs, 222
SLA fabricated, 243, 244
SLA metallization processes, 206
SLA prototyping methods, 190
SLA rapid prototyping, 224
Sloan, D.H., 72
Slot, 644, 654
Slow injection, 259, 260
Small molecules, loss of, 611
SNMS, *see* Mass spectrometry,
        secondary neutral
Soddy, F., 13
Sodium adduct, 719
Sodium dodecyl sulfate, SDS, 717
Soft landing, SL, 227, 691
Soft-landing instrument, 176
Soft-landing technologies, 107, 169, 259, 375, 691
Soft-landing, ion, serial RITs, 227
Software tools, 729
Soil samples, 20
Solenoid valve, 779, 780, 796
Solenoid, magnetic field generated by, 81
Solid organic materials, vapor from, 30
Solid phase micro extraction, SPME, 238, 387
Solid-phase extraction, 717
Solvation, 312
Solvent clusters, 358
Sorbent tube inlet, 241
SORI-CID, *see* Sustained off-resonance
        irradiation-collision-induced
        dissociation
Source, election ionization, 35
    electron bombardment, 18
Space charge, 295, 374, 402, 491, 640, 641, 643, 661, 667, 668, 676, 774
    negligible, 420
    relative field errors due to, 596
Space charge in miniature traps, 375

Space resolution, 86
Space, (a_u, q_u), 149
Space-charge conditions, 596
Space-charge effect, 77, 203, 214, 257, 296, 297, 322, 330, 549
Space-charge effects in an EST, 258
Space-charge effects, absence of, 410
    avoidance of, 601
Space-charge effects, reduction of, 328
Space-charge field, axial, 404
Space-charge induced peak splitting, 405
Space-charge interaction, 805
Space-charge interactions, 258
Space-charge ion trapping, 53
Space-charge limit, 296, 396
Space-charge limit, isolation, 374
Space-charge minimization, 566
Space-charge propulsion, 529
Space-charge repulsion, 7, 319, 321, 338, 411, 423
Space-charge-induce broadening, 423
Space-charge-induced cyclotron frequency shifts, 412
Spacing, end-cap, increasing, 188
Spark source mass spectrometer, 35, 36
Spatial distribution of ions, 630
Spatial frequency, 179
Specificity, 710
Spectra sidebands, 404
Spectral apodization, 465
Spectral capacity, higher, 615
Spectral dynamic range, 668
Spectral limit, 594, 597
    higher, 610
Spectral performance, 664, 668
Spectral quality, definition of, 594
Spectral resolution, 594
Spectral similarity, R score, 731
Spectroscopy, translational energy-loss, 63
Spherical confinement, 595
Spherical coordinates, 186, 378
Spherical harmonic contribution, fourth-, sixth-, and eighth-order, 463
Spherical ion trap, 595, 615
Spindle electrode, central, 111
Spindle-like trapping volume, 119
Spiral motion, 255
Spiral trajectory, 255
Spiratron, *see* Time-of-flight mass spectrometers, spiratron
SPME, *see* Solid phase micro extraction
Spontaneous tumor generation, 728
Spray ionization, 49–53
Spraying, silver ink, 206
Sputtering, 44, 194
SQL Server 2005 Express Edition, 729
Square wave, 277, 776, 777

Square wave quadrupole field, 282
ξ$_n$, *see* Collision frequency, reduced
Srega, J., 268
Stability boundary, 632, 633, 635
Stability diagram, 98, 103, 163, 165, 279, 280,
      282, 305, 375, 438, 439, 598, 771
Stability diagram border, 771
Stability diagram, quadrupole ion trap, 104
Stability limit, 774
Stability of electronic circuitry, 183
Stability of voltages, 257
Stability parameter, 103, 108, 274, 439, 804, 808
Stability region, 149, 163, 621, 623, 649, 653,
      654, 656
Stability, choice of phase, 608
      axial, 103
      ion trajectory, regions of, 162
      path, 96
      phase, 93
      radial, 103
      regions of, 98
      trajectory, 93, 164
Stable isotope labeling using amino acids in cell
      culture (SILAC), 264
Stable region, 279
Stable trajectories, 98, 374, 474, 679
Stacked-ring assembly, 414, 421
Stafford, Jr., G.S., 103
Standard deviation, 804
Standard Oil Company, USA, 20
Standard quadrupole potential, 631
Static buffer gas pressure, 782, 788, 789
Static infusion, 711, 717
Static Kingdon trap, *see* Kingdon trap, static,
Static mass spectrometers, *see* Mass
      spectrometers, static
Statistical laws, 681
Steady state, 14
Steel spheres, charged, 126
Stephens, D., 103
Stephens, W.E., 84
Stereo isomers, 708, 709, 713, 714, 731, 736
Stereolithography, 175
Stereolithography apparatus, *see* SLA
Steroids, 30
Stochastic model, 758
Stop electrode, 797
Storage lifetime, 740
Storage limit, 594
Storage limit, maximum, 374
Storage volume, ion, 393
Storage volume, toroidal, 393
Stored wave inverse Fourier transform, SWIFT,
      216, 227, 228, 237, 298, 299, 688
Stray charge buildup, 200
Stretched ion trap, 621, 630, 631, 663
Stretched RITs, 223

Stretched trap geometry, 206, 287, 626, 627, 630,
      641, 663
Stretching axial separation of end-cap
      electrodes, 382
Strube, S., 269
Structural elucidation, 556
Structural isomers, 708, 709, 711, 713, 722,
      723, 736
Structure analysis, 672, 673
Structure determination, 672, 673
Structure elucidation, 710
Structure elucidation of glycanes, 685
Strupat, K., 268
Strutt, J. W., 50
Sub-peaks, phase-dependent appearance of, 577
Substance P, 183, 301–303, 542
Succession modes, 475
Sugar residue, 812, 813
Sulfation, 265
Sulfur dioxide, 83, 241
Summary chart, 724
Superconducting magnets, 434, 677
Superimposed RF voltage, 692
Superposition of odd-multipole components, 630
Supplemental AC voltage, 577
Supplemental RF voltage, 827, 828
Supplemental RF voltage, frequency, 828
Supplementary AC voltage, 760, 773
Supplementary alternating field, 620, 644
Supplementary alternating voltage, 628,
      644, 654
Supplementary dipolar excitation, 108
Supplementary dipolar resonance field, 621, 627,
      632–637, 643, 658–660
Supplementary dipole voltage, 635, 636, 644,
      655, 658, 660–663
Supplementary phase-modulation, 513
Supplementary quadrupolar excitation, 514–516
Supplementary quadrupole field, 629, 635, 642
Supplementary quadrupole frequency, 628
Supplementary quadrupole potential, 635, 654,
      655, 662, 663
Supplementary radiofrequency voltage, 694
Supplementary sinusoidal potential, 742, 771
Supplementary waveforms, 664
Surface imperfections, 183
Surface micromachining, 192
Surface roughness, 179, 180, 191, 203
Surface-induced dissociation, SID, 681, 690,
      691, 744
Surface ionization, *see* Ionization, surface
Survey scans, 567
Sustained off-resonance irradiation, SORI,
      689, 690
Sustained off-resonance irradiation-
      collision- nduced dissociation
      (SORI- CID), 448

SWIFT, *see* Stored wave inverse Fourier
transform
Switched-potential, DC, 472, 475
Switching circuits, 302, 304
Switching voltage, 284
Syka, J.E.P., 268
Symmetry, azimuthal, 406
Synchrometer, 83
Synchronization (self-bunching), 258
Synchronous scanning, 581
Synchrotron, 83
Synthetic polymer anaylsis, 447
Syrotron, 76

T

TAGA 6000 mass spectrometer, 526
Tanaka, K., 48, 708
Tandem FAIMS arrangement, 328, 342
Tandem IMS, 355
Tandem mass analyzer, pseudo-, 395
Tandem mass spectrometry, 30, 62, 106, 170, 191,
216, 223, 224, 244, 256, 263, 264, 274,
299, 373, 453, 531, 559, 586, 598, 673,
679, 699, 708, 710, 711, 713, 722, 727,
728, 740–742, 744, 769, 771, 775, 776,
789, 813, 814, 832, 841, 842
automatic, data-dependent, 614
disassembly applications, 719–729
multiple stages of (MS^n), 559, 563, 611, 708,
709, 711, 713, 715, 718–720, 722, 724,
728, 729, 731, 733, 734, 762, 812–816,
819, 820, 828, 833
RF-only-mode-event, role of, 448
Tandem mass spectrometry inside an EST, 259
Tandem TOF/TOF mass spectrometer, 586
Tandem-in-space mass spectrometry, 106, 395
Tandem-in-time mass spectrometry, 106, 794
Tangential injection, 122
Tank circuit, 76
Targeted analysis, 567
Taurocholic acid, 337, 338, 556, 557
Taurocholic acid, anion of, 336
Taxol, 362
Taylor, L., 268
TBAA, *see* Tribromooroactic acid
TCAA, *see* Trichloroactic acid
TDF, *see* Time-delayed fragmentation
Teflon, 242
Teflon tubing, 778, 779, 782, 785
Tellurium, 35
Teloy, E., 124
TELS, *see* Translational energy-loss
spectroscopy
Temperature, 754
role in ion separation, 335
Temperature-controlled electrode set, 360

Temporal distribution, narrow, 574
Temporal precision, high, 607
Temporal signal, 477
Temporal variations of the field, 622
Tenax, 242
Tennessee Eastman Corporation, USA, 84
Terbutaline enantiomers, 364
Terbutaline enantiomers, separation of, 365
Terpinene, alpha, 238
Tetrabutylammonium bromide, 660, 661
Tetradecylammonium bromide, 664
Texas A & M University, USA, 117
T_{excite}, 404
tFAIMS, *see* FAIMS, trapping
TFD value of opposite polarity, 663
Theory and practice of the orbitrap mass
analyzer, 675
Thermal conductivity, 193
Thermal decomposition, 30
Thermal desorption, 517
Thermal equilibrium, 747
Thermal ionization, *see* Ionization, thermal
Thermal motion, random, 404
Thermal stability, 45
Thermally sensitive compounds, 47
Thermally-assisted IRMPD, TA-IRMPD, 743
Thermionic cathode, 685
Thermo Electron, 316, 348
Thermo Finnigan ion traps, 621
ThermoFisher Scientific, LCQ, 719
Thermo Masslab Ltd, Manchester, UK, 268
Thermochemical data, 214
Thermodynamic properties, ion, 64
ThermoFisher Scientific, Bremen, Germany, 269
ThermoFisher Scientific, Germany and USA,
109, 110, 112, 128, 269, 675
ThermoFisher Scientific, San Jose, USA, 268,
316, 413
Thermospray ionization, TSP, *see* Ionization,
thermospray
Thin film organic electronics, 228
Thin layer chromatography, TLC, 819
Thomson, J.J., 9, 13–15, 19, 31, 45, 50, 53, 127
thomson (Th), unit of, 14
Thornton Research Centre, UK, 21
Three-dimensional ion trapping device, 620, 628,
632, 643, 645, 654, 667, 668
Three-dimensional ion traps of rotational
symmetry, 621
Three-dimensional Paul trap, 679–681, 685, 687,
693, 694
Three-dimensional quadrupole field, 620
Three-dimensional quadrupole ion trap, 681,
695, 696
Three-dimensional radiofrequency quadrupole
field, 711
Three-stage instrument, 32

Throughput, 207
Tickle voltage, 512, 771
Tickle voltage amplitude, 771
Tickling, *see* Resonant excitation
Tilted rods, 529
Time delay, 795
Time dispersion, methods of, 68
Time domain influence, 664
Time domain signal, 675
Time evolution, 746, 747
Time focusing, 89
Time packet, 638
Time window, 681, 683
Time, inter-collision, 688
Timed parameter, 774, 775
Time-delayed fragmentation, TDF, 562, 565
Time-dependent field, 621
Time-dependent fields, 257
Time-lag focusing, 86, 87
Time-of-flight, TOF, 184, 793, 802, 819
Time-of-flight analyzer, 300
Time-of-flight analyzer with electrospray
        ionization, 91
Time-of-flight detection, 253
Time-of-flight mass analyzer, 311
Time-of-flight mass spectrometer, TOF, 48, 80,
        83, 573, 691
    constant momentum, 85
    cyclotron principle, using, 81
    helical path device, 81
    linear, 69, 83
    linear, basic principle of, 83
    linear, time-lag focusing, 86
    magnetic, 69, 81
    orthogonal-acceleration, oaTOF, 91, 92, 414
    reflectron, 81, 90, 796, 799, 801, 805, 813
    spiratron, 69, 81, 88, 89, 253
Time-of-flight mass spectrometry, TOFMS,
        793, 794
Time-of-flight, constancy of in magnetic field, 81
Time-of-flight, TOF/TOF, analysis, 713
Time-of-flight analysis, ion trapping prior to, 401
Time-resolved fragmentation efficiency curve, 683
Time-to-digital converter, TDC, 416
TiN, 193
Titanium, gettering action of, 117
Todd, J.F.J., 518
TOF, *see* Time-of-flight mass spectrometers
TOF digitization, 495
TOF discrimination, 476, 496
TOF discrimination, need for suppression of, 496
TOF expression for one ion, 477
TOF histogram, 105
TOF histogram, expression for, 477
TOF histograms, 472, 475, 482, 486
TOF histograms, experimental, 476
TOF histograms, numerical processing of, 479

TOF mass analysis, 794–796, 799, 802, 805,
        806, 813
TOF mass spectrometer, comparison of DT-IMS
        with, 320
Toluene, 217, 218, 237–239, 242, 244, 386, 387
Toluene solutions, headspace over, 238
Toluene vapor headspace, 240
Top hits, 735
Topatron, 93–95
Top-down proteomic studies, 411
Topler pump, 9
Topology, 709–711, 717, 721, 723, 725, 731,
        733–735
Toriodal trapping volume, 395
Torion Technologies Inc, American Fork, UT,
        USA, 385
Toroidal electric fields, 391
Toroidal ion trap, 69, 111, 148, 177, 178, 184,
        202, 376
    asymmetical, operating parameters for, 387
    asymmetrical, 383
    capacitance of, 380
    conception and theory of, 376
    coordinate system for, 379
    development and performance of, 380
    miniature, major components of, 386
    reduced size, 385
Toroidal ion traps, additional parameters for, 378
Toroidal storage volume, 393
Toroidal volume, 383
Total ion ejection, 472
Total pressure curve, 99
Total storage mode, 850
townsend (Td), definition of, 313
Toxic chemicals, detection, 351
Trace Atmospheric Gas Analyzer, TAGA
        6000, 526
Trace gases, 333
Trace impurities, 40
Trajector stability, 93
Trajectories of ions, axial components of, 103
    radial components of, 103
Trajectories, bound, 474
    stable, 98, 152, 474
    unstable, 98
Trajectory instability, 164
Trajectory of ion, power-spectral Fourier analysis
        of, 167
Trajectory simulations, 759
Trajectory stability, 164
Trajectory, bounded, 374
    corkscrew, 90
    stable, 374
Transition matrix, 440
Transition state, 745
Transition state frequency, 745
Transition state structure, 745

Translational energy, 64
Translational energy, release of, 59
Translational energy-loss spectroscopy, TELS, 63
Transmission efficiency, 805
Transmission RF-only quadrupole, 546
Transmitter plates, 79, 435
Transverse components, 93
Trap capacitance, total, 213
Trap compensation, 433, 448, 465, 689
Trap depth, 54
Trap dimensions, varying, mass analysis by, 217
Trap exit, expression for radial position of, 495
Trap optimization using compensation
        electrodes, 454
Trap plate, 35, 453
Trap, Achilles' heel of, 258
    closed, 454
    closed cylindrical, 461
    conventional, 452
    double, 395
    electrically compensated to eighth order, 461
    elongated trapped-ion, 449, 451
    field-corrected, 453, 454
    internal volume of, 546
    Kingdon, see Kingdon trap
    Knight electrostatic, see Knight
        electrostatic trap
    matrix-shimmed ion, 458, 460
    one-ring, 201
    open, 454
    open cylindrical, 454, 455
    Paul see Paul trap
    Penning, see Penning trap
    screened, 453
    screened electrostatic, 452
    three-ring, 201
Trapped ion analyzer cell, 78
Trapped ion cell, cubic, 79
    pulsed modes of operation of, 77
Trapped ion ICR cell, 677
Trapped ion number, 740
Trapped ion, cooled, 855
    definition of, 7, 310
Trapped ions, motion of, 254
    number density, 848, 853
    photodissociation, 848, 853
    ions, spectroscopy of, 688
    ions, structure and dynamics, 688
    ions, trajectory of, 7
Trapped-cell ICR spectrometers, 77
Trapped-ion cell analyzer, 75
Trapping and processing ions in RF ion guides, 574
Trapping capacity, 182, 546
    high, 177
Trapping devices, 802
Trapping efficiency, 182, 545, 546, 548, 557,
        573, 575

Trapping efficiency of IFT, 422
Trapping electrodes, 677
Trapping FAIMS, tFAIMS, 328, 330
Trapping field, 402, 405, 620, 633, 643, 644, 649,
        652, 653, 668
Trapping field center, 626, 627, 629, 635, 644,
        645, 660
Trapping field dipole, TFD, 624–626, 629–637,
        641, 645, 656–661, 663, 664, 668
Trapping field frequency, 621, 628
Trapping field, DC, anharmonicities of, 405
Trapping island, primary, 125
Trapping islands, 125
Trapping motion, 676, 677
Trapping of charged particles, 676
Trapping parameter, single, 125
Trapping parameters, 149
Trapping potential, 54
Trapping pressure, 688
Trapping region, 54
Trapping rings, 456
Trapping time, 677, 688
Trapping voltage, 403
Trapping voltage waveform, 108
Trapping volume, 799
Trapping volume, spindle-like, 119
Trapping well, 437
Trapping, gated-, 408
Traps, microfrabricated, 180
    Magnetic, 676, 677
    miniature, spage-charge in, 375
Trap-type mass spectrometers, 673
Traveling wave ion mobility mass
        spectrometry, 712
Triacylglycerol, TAG, 817, 818
Tribromoacetic acid, TBAA, 332
Tributylamine, 241, 244
Tributylamine, protonated, 223
Trichloroactic acid, TCAA, 332
Trichloroethylene, 394, 395
Triple quadrupole mass spectrometer, 99, 109,
        526, 674, 711
    inherent sensitivity while scanning, 567
    observation of product ions with, 562
Triple quadrupole systems, low pressure collision
        cells in, 528
Triple resonance condition, 630
Triple resonance ion ejection, 629, 632, 634,
        636–643, 645, 655, 660, 661, 664–668
Triple resonance mass scanning, 640, 645, 655
Triple-frequency focus, 462
Triple-quadrupole systems, 530
Triply-charged precursor ion, 684, 685
Trisaccharide ion, 716, 725
Triterpenoids, 30
Triversa Nanomate, 719
Truncated electrodes, 627

Trypsin, 407
Tryptic digestion, 834
Tryptic peptide, doubly-protonated, 563, 565
Tryptic peptides, 355
TSP, *see* Ionization, thermospray
TSQ Quantum triple quadrupole mass
    spectrometer, 316, 348
TSQ-70 triple quadrupole mass spectrometer, 377
Tucker, E.B., 20
Tumor progression, 728
Tungsten, 194
Tuning, 661
Turnaround time, 414
Two ion traps, 640
Two-dimensional linear ion trap, 643–645, 649,
    654–657, 660
Two-dimensional mass spectrometry, 582
Two-dimensional Paul ion trap, 679, 680, 694–696
Two-pulse photodissociation experiment,
    751–753, 755, 757–760
Two-pulse profile, 786
Two-sheeted hyperboloid, 438
Two-step excitation event, 598
TXB2 anomers, dyanamically interchanging, 362
Type A ions, 314
Type B ions, 314
Type C behavior, 351
Type C ions, 314
Tyrosine kinase receptor, 833, 834

U

Ubiquitin, 689, 690, 834, 835
Ubiquitin ion conformers, 355
UFMP, *see* Ultrafine metal powder
Ultra high vacuum, 437
Ultra-fast scanning, 664
Ultrafine metal powder (UFMP), 48, 49
Ultrahigh vacuum conditions, 675
Ultramark, 243, 640, 667
Ultraviolet light, 743
Ultraviolet radiation, 689
Undecane, 637, 638
Under-sampling, 490
Unidirectional ion ejection, 627, 630, 633, 664, 668
Unimolecular dissociation, 57, 674, 683, 689, 741,
    802, 819
Unimolecular dissociation, rate of, 749
Unimolecular fragmentation, 28, 855
Unimolecular ion dissociation, 33
Unimolecular reaction, 681, 682
Unintended oscillatory ion micromotion, 200
Unit charge, e, 805; *see also* Elementary charge
United States Department of Energy, 380
University de Paris-Sud, France, 63
University of Bonn, Germany, 96
University of California Berkeley, USA, 70

University of California Radiation Laboratory,
    USA, 96
University of New South Wales, Australia, 91
University of Pennsylvania, USA, 84
University of Purdue, West Lafayette, IN, USA,
    172, 382, 383
Unstable trajectories, *see* Trajectories,
    unstable
Unstable trajectory, 679, 797
Unstretched geometry, 473
Unstretched quadrupole ion trap, 276
Uranium hexafluoride enrichment, 21
Urine, 362
US National Academy of Sciences, USA, 72
UV chromophore, 832
UV photodissociation, 832, 841, 842, 858
UV photon, 830
UV spectroscopy, 848, 853
UV/vis photodissociation, 858
UVPD, *see* Photodissociation,

V

vMALDI QIT/TOF MS, 794, 795
Vacuum pump, orbitron, 116
Vacuum spark method, Dempster, 37
Vacuum, high quality, 472
Valence electrons, even number of, 29
Valve trigger, 781
van der Waals forces, 687
Vandermey, J., 569
Vane electrodes, 110, 111, 574, 576, 588
Vane oscillating potential, frequency scanning
    of, 577
Vane potential, 575
Vanes, front, 575
Vanes, rear, 575
Vaporization process, 672
Variable leak valve, 777–779, 781, 782, 784,
    785, 788
Varian Associates, USA, 75, 77
Varian Inc., Walnut Creek, CA, USA, 377
Varian ion trap, commercial, 629
    commercial, geometry of, 629
Varian prototype linear ion trap, 655
Vasopressin, 463, 464
V_{da}, zero-to-peak amplitude of trapping field
    dipole, TFD, 624–626
VEGA program, 275
Velocitron, 86
Velocity distribution, 798
Velocity focusing, 15
Velocity focusing, first-order, 25
Velocity selection, use of radiofrequency electric
    fields for, 93
Versatility of analysis, 207
VG Micromass, UK, 34, 56

Vias, 202, 203
Vibrational frequency, 745, 748, 756
Vibrational states, 745, 748
Vibrational motion, 57
Vinyl acetate, 239
Vinyl methyl ether radical cation, 443, 444
Virtual electrodes, 123
Virtual net fields, 328, 329
Virus, 691
Visible photon, 828, 838
Visual Studio 2005, 729
Vitamin B12, 49
Vitamins, 45
vMALDI, *see* Matrix-assisted laser desorption
    ionization, intermediate vacuum
Voltage gradient, 312
Voltage, trapping waveform, 108
Voltages, stability of, 257
Volume of accessible points in $(\text{In}(A), \text{Re}(A))$
    space, 441
Volume, toroidal, 383

W

Waste incineration plants, 517
Water, 43
Water, 851, 853, 854
Water cluster ions, 324
Water cluster series, 66
Water disinfection, chlorinated and brominated
    by products from, 360
Water, vapor, 855, 859
Waters Micromass ZQ, 350
Waters Q-TOF mass spectrometers, 348
Wave, rectangular, 277
    sinusoidal, 344
    square, 277
Waveform generator, 310
Waveform, asymmetric, 317
    asymmetric, rectangular, 317
    bisinusoidal, 344, 346, 347
    digital, 275
    digital asymmetric, 284
    dipolar excitation, 109
    polarity of, 326
    rectangular, 273, 344, 346
    sinusoidal, 275
Waveforms, asymmetric representations of, 345
    modes P1 and N2 for, 327
    modes P2 and N1 for, 327
    notched, 176
    rectangular RF, 275
    types of, 344
Wave-train generator, 494
Weak collider model, 746
Weak interactions, 690
Weakest bonds, 683

Wesleyan University, CT, USA, 124, 128
Westinghouse, USA, 21, 74
Wieghaus, A., 268
Wien filter, 63
WIG cell, 118
WIG, *see* Wire-ion guide
Wiggles, auxilliary, 479, 487
Willemite screen, 12
Winding frequency, 126
Wineland, D. J., 128
Wintergreen, *see* Methyl salicylate,
Wire-ion guide, WIG, 117
Wise, M.B., 376
Working point, 620, 742, 771
Wüthrich, K., 672

X

x ion, 681, 682, 685
Xcalibur 1.4 and 2.0 software, 719
Xcalibur control software, 212
Xe^+, 490, 500
Xenon, 677, 687, 771, 772
Xenon atoms, fast, 46
Xenon ions, 196, 202
XO_3^-, 322
y ion, 681–683, 686, 690
Yamashita, M., 50
York University, Ontario, Canada, 529
Yoshida, Y., 48

Z

z ion, 681, 684, 686, 687, 689
Z coordinate, 439
$z+1$ ion, 687
$z+2$ ion, 687
z_0, 156, 181, 274, 374, 473, 804
ZAB mass spectrometer, 34
ZAB, meaning of, 62
ZAB-1F, 45
ZAB-2F mass spectrometer, 34, 57
ZAB-2FQ mass spectrometer, 62
Zabrouskov, V., 268
z-axis, 801
Zeller, M., 269
Zeolitic materials, selective, 517
zeptomol (zmol), definition of, 408
Zero filling, 479
Zero padding, 479
Ziberna, T., 268
z-mode, *see* Simple harmonic motion, axial
z-mode amplitude, 436, 462
    re-excitation of, 464
ZnSe window, 829
Zones of poor confinement, 492
Zoom scan mode, 788, 789

Printed and bound by CPI Group (UK) Ltd, Croydon, CR0 4YY

18/10/2024

01776243-0020